**insul-** island: insula
**inter-** among, between: interphase
**intra-** inside: intracellular
**irid-** rainbow, colored circle: iris
**is-** equal: isotope
**jugul-** neck, throat: jugular vein
**junct-** yoke, join: junctional fiber
**juxta-** near, beside: juxtaglomerular nephron
**kerat-** horn: keratin
**kyph-** hump, humpbacked: kyphosis
**labi-** lip: labia
**labr-** lip: glenoidal labrum
**labyrinth-** maze: bony labyrinth
**lacrima-** tears: lacrimal gland
**lact-** milk: prolactin
**lacun-** pool: lacuna
**lamell-** thin plate: lamella
**lanug-** down: lanugo
**laten-** hidden: latent
**-lemm** rind, peel: neurolemma
**leuk-** white: leukocyte
**lingu-** tongue: lingual tonsil
**lip-** fat: lipids
**-logy** study of: physiology
**lord-** bent back: lordosis
**lun-** moon: semilunar valve
**lute-** yellow: macula lutea
**ly-** loose, dissolve: lysosome
**-lyt** dissolvable: electrolyte
**macr-** large: macrophage
**macula** spot, stain: macula lutea
**mal-** bad, abnormal: malnutrition
**malle-** hammer: malleus
**mamm-** breast: mammary gland
**man-** hand: manubrium
**mandib-** jaw: mandible
**mast-** breast: mastitis
**maxill-** jawbone: maxillary bone
**meat-** a passage: auditory meatus
**medi-** middle: adrenal medulla
**melan-** black: melanin
**men-** month: menstrual
**mening-** membrane: meninges
**ment-** mind: mental foramen
**mes-** middle: mesoderm
**meta-** after, beyond, accompanying: metabolism
**meter** measure: calorimeter
**micr-** small: microfilament
**mict-** to pass urine: micturition
**mid-** middle: midbrain
**milli-** one-thousandth: millimeter
**mit-** thread: mitosis
**mon-** one: monozygotic
**mons-** an eminence: mons pubis
**morul-** mulberry: morula
**mot-** move: motor neuron
**multi-** many: multipolar neuron
**mut-** change: mutation
**my-** muscle: myofibril
**nas-** nose: nasal
**nat-** to be born: prenatal

**ne-** new, young: neonatal period
**nephr-** kidney: nephron
**neur-** nerve: neuron
**neuter** neither one nor the other: neutral
**nod-** knot: nodule
**nucle-** kernel: nucleus
**nutri-** nourish: nutrient
**obes-** fat: obesity
**occipit-** back of the head: occipital lobe
**oculi-** eye: orbicularis oculi
**odont-** tooth: odontoid process
**-oid** form: odontoid process
**olfact-** to smell: olfactory
**olig-** few, small: oligodendrocyte
**oo-** egg: oogenesis
**or-** mouth: oral cavity
**orb-** circle: orbital
**-osis** abnormal condition: leukocytosis
**oss-** bone: osseous tissue
**-ous** having qualities of: cancerous
**ov-** egg: synovial fluid
**palpebra** eyelid: levator palpebrae superioris
**papill-** nipple: papillary muscle
**para-** beside: parathyroid glands
**pariet-** wall: parietal membrane
**path-** disease, sickness: pathogen
**pell-** skin: pellagra
**pelvi-** basin: pelvic cavity
**peri-** around: pericardial membrane
**pept-** digest: peptic ulcer
**phag-** eat: phagocytosis
**phen-** show, be seen: phenotype
**phleb-** vein: phlebitis
**phot-** light: photoreceptor
**phren-** mind, midriff: phrenic nerve
**pia-** pious, gentle: pia mater
**pino-** drink: pinocytosis
**pleur-** rib, side: pleural membrane
**plex-** strike: choroid plexus
**plic-** fold: plicae circularis
**poie-** make, produce: hematopoiesis
**poly-** many: polyunsaturated
**popl-** ham (knee): popliteal artery
**por-** passage, channel: pore
**post-** after: postnatal period
**pre-** before: prepatellar bursa
**prime** first: primordial follicle
**pro-** before: prophase
**prox-** nearest: proximal
**pseud-** false: pseudostratified epithelium
**pter-** wing: pterygoid process
**puber-** adult: puberty
**pulmo-** lung: pulmonary artery
**pyl-** door, orifice: pyloric sphincter
**quadr-** four: quadriceps femoris
**ramus** branch: gray ramus
**rect-** straight: rectum
**ren-** kidney: renal cortex
**ret-** net: rete testis

**reticul-** network: sarcoplasmic reticulum
**rhin-** nose: rhinitis
**sacchar-** sugar: monosaccharide
**saltator** dancer: saltatory conduction
**sarc-** flesh: sarcoplasm
**scler-** hard: sclera
**seb-** grease: sebaceous gland
**sect-** cut: section
**sella-** saddle-shaped: sella turcica
**semi-** half: semitendinosus
**sen-** old: senescence
**sens-** feeling: sensory neuron
**sin-** hollow: sinus
**-some** body: ribosome
**sorpt-** to soak up: absorption
**squam-** scale: squamous epithelium
**sta-** halt, make stand: hemostasis
**strat-** spread out: substrate
**stria-** groove: striated muscle
**sub-** under: substrate
**sulc-** furrow: sulcus
**super-** above: superior
**supra-** above: supraspinatus muscle
**sutur-** sewing: suture
**syn-** together: synthesis
**syndesm-** binding together: syndesmosis
**systol-** contraction: systole
**tachy-** rapid: tachycardia
**tal-** ankle: talus
**tetan-** stiff: tetanic
**thalam-** chamber: thalamus
**theo-** sheath: theca externa
**therm-** heat: thermoreceptor
**thromb-** clot: thrombocyte
**thyr-** shield: thyroid gland
**toc-** childbirth: oxytocin
**tom-** cut: anatomy
**ton-** stretch: isotonic
**trans-** across: transverse
**tri-** three: tricuspid valve
**trigon-** triangle: trigone
**trop-** turn, react: adrenocorticotropic
**troph-** nurture: hypertrophy
**tuber-** swelling: tuberculosis
**tympan-** drum: tympanic membrane
**umbil-** navel: umbilical cord
**un-** one: unipolar neuron
**ur-** urine: ketonuria
**uter-** womb: uterine tube
**vag-** to wander: vagus nerve
**valent** having power: electrovalent
**vas-** vessel: vasa recta
**vesic-** bladder: vesicle
**vill-** hair: villus
**visc-** internal organs: viscera
**vitre-** glass: vitreous humor
**voluntar-** of free will: voluntary muscle
**xiph-** sword: xiphoid process
**zon-** belt: zona pellucida
**zym-** ferment: enzyme

# Hole's human

eighth edition

## Anatomy & Physiology

**David Shier**
Washtenaw Community College

**Jackie Butler**
Grayson County College

**Ricki Lewis**
The University at Albany

WCB
McGraw-Hill

Boston  Burr Ridge, IL  Dubuque, IA  Madison, WI  New York  San Francisco  St. Louis
Bangkok  Bogotá  Caracas  Lisbon  London  Madrid
Mexico City  Milan  New Delhi  Seoul  Singapore  Sydney  Taipei  Toronto

# WCB/McGraw-Hill

*A Division of The **McGraw·Hill** Companies*

HOLE'S HUMAN ANATOMY & PHYSIOLOGY

 This book is printed on acid-free paper.

2 3 4 5 6 7 8 9 0 QPH/QPH 9 3 2 1 9

ISBN 0–697–34193–3

Vice president and editorial director: *Kevin T. Kane*
Executive editor: *Colin H. Wheatley*
Sponsoring editor: *Kristine Tibbetts*
Developmental editor: *Connie Balius-Haakinson*
Marketing manager: *Heather Wagner*
Senior project manager: *Jayne Klein*
Production supervisor: *Mary E. Haas*
Designer: *K. Wayne Harms*
Photo research coordinator: *John C. Leland*
Art editor: *Jodi K. Banowetz*
Compositor: *Shepherd, Inc.*
Typeface: *10/12 Melior*
Printer: *Quebecor Printing Book Group/Hawkins, TN*

Interior/cover design: *Chris Reese*
Cover image: *Rob Lang/FPG International LLC*
Illustrators: *Todd Buck, Chris Creek, Peg Gerrity, Illustrious Inc., Elizabeth Morales, Thomas Waldrop, Observatory Group Inc.*

The credits section for this book begins on page 1019 and is considered an extension of the copyright page.

**Library of Congress Cataloging-in Publication Data**

Shier, David.
    Hole's human anatomy and physiology.—8th ed. / David Shier, Jackie Butler, Ricki Lewis.
      p.    cm.
    Rev. ed. of: Hole's human anatomy & physiology. 7th ed. ©1996.
    Includes index.
    ISBN 0–697–34193–3
    1. Human physiology.   2. Human anatomy.   I. Butler, Jackie.
  II. Lewis, Ricki.   III. Shier, David.   Hole's human anatomy & physiology.   IV. Title.
    [DNLM: 1. Anatomy.   2. Physiology.   QS 4 S555h 1999]
  QP34.5H63     1999
  612—dc21
  DNLM/DLC
  for Library of Congress                98–5318
                               CIP

http://www.mhhe.com

# New Version 2.0!

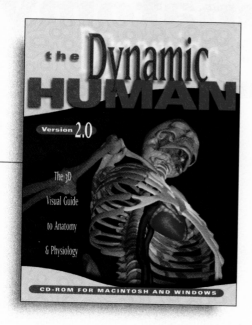

*by WCB/McGraw-Hill and Engineering Animation, Inc.*

1998 • CD-ROMs (two CD set) for Macintosh and Windows • ISBN 0-697-38935-9

This powerful new version of *The Dynamic Human* CD-ROM builds upon the best in anatomy and physiology technology. **Version 2.0** features added detail and depth— expanding upon a variety of topics and incorporating the World Wide Web.

## Features

- The World Wide Web provides the ideal opportunity to implement and greatly extend the powers of this updated CD-ROM. The all-new *Dynamic Human Version 2.0* web site features an immense amount of supplemental information, including:
  - approximately 20 to 30 readings per body system
  - extensive links to additional informational sites on the World Wide Web
  - supplemental visuals, including animations, video, and vivid illustrations
  - correlation guides for many WCB/McGraw-Hill texts
  - on-line self-testing
- An updated design and interface make navigation much simpler with easy-to-use pop-up menus.
- Each body system can be studied through four standard content areas: anatomy, explorations, clinical concepts, and histology.
- A self-quizzing section is available for each body system.
- Powerful 3D images and even more animations enhance the thorough content and draw your students into the wonder of anatomy and physiology.

## Contents

Human Body • Skeletal System • Muscular System • Nervous System • Endocrine System • Cardiovascular System • Lymphatic System • Digestive System • Respiratory System • Urinary System • Reproductive System

## System Requirements

| IBM/PC or compatible | Macintosh |
| --- | --- |
| 486/66 or better (Pentium recommended) | Power PC |
| Windows 95 or newer | System 7.1 or newer |
| 16 MB RAM or better | 16 MB RAM or better |
| 640 x 480 x 256 color monitor | 640 x 480 x 256 color monitor |
| CD-ROM drive | CD-ROM drive |
| (transfer rate of 300 kbs or better) | (transfer rate of 300 kbs or better) |
| SoundBlaster compatible audio card | Mouse |
| Mouse | |

This distinctive icon appears in appropriate figure legends of the new edition of *Hole's Human Anatomy and Physiology* and correlates information to specific modules found on the new *Dynamic Human CD-ROM, Version 2.0.*

# Look
## at the great anatomy and physiology study tools
## WCB/McGraw-Hill has to offer!

## WCB Life Science Animations Videotapes

This series of six videotapes contains animations of complex physiological processes. The animations make challenging concepts easier to understand.

Tape 1 *Chemistry, The Cell, and Energetics*
**ISBN 0-697-25068-7**

Tape 2 *Cell Division, Heredity, Genetics, Reproduction, and Development*
**ISBN 0-697-25069-5**

Tape 3 *Animal Biology #1*
**ISBN 0-697-25070-9**

Tape 4 *Animal Biology #2*
**ISBN 0-697-25071-7**

Tape 5 *Plant Biology, Evolution, and Ecology*
**ISBN 0-697-26600-1**

Tape 6 *Physiological Concepts of Life Science*
**ISBN 0-697-21512-1**

## New!
## Life Science Animations Videotape 3D

**ISBN 0-07-290652-9**

Featuring 42 animations of key biological processes, this tape contains realistic three-dimensional animations and is fully narrated. *Life Science Animations Videotape 3D* is perfect for your lecture or can be placed on reserve in your library for student reference.

## New!
## The Web Resource Center

**http://www.mhhe.com/biosci/abio/**

### A powerful tool

*For instructors*—Our Web Resource Center is a wonderful resource for new ideas. Here you'll find links to hot topics relating to anatomy and physiology, teaching aids, information about our products, a means to communicate directly with WCB/McGraw-Hill, and much more. The Human Anatomy and Physiology Society (HAPS) web site is just a click away for the latest information about conventions and other professional issues.

*For students*—Do you need additional study questions to help prepare for tests? Are you interested in doing extra research? Check out our Resource Center. Best of all, ***it's free!***

# QuickStudy

Computerized Study Guide for *Hole's Human Anatomy and Physiology,* 8th edition

*by Louis Giacinti*
*Milwaukee Area Technical College*

Focus study time only in the areas where you need the most help.

Available on diskette:

IBM ISBN: 0-697-34212-3
Macintosh ISBN: 0-697-34213-1

## How does it work?

- Take the computerized test for each chapter.
- The program presents you with feedback for each answer.
- A page-referenced study plan is then created for all incorrect answers.
- Look up the answers to the questions you missed.
- This easy-to-use *QuickStudy* is not a duplicate of the printed study guide that accompanies your text. It is full of new study aids.

- **Learning Objectives** outline the major points of each text chapter.
- **Key Terms** in each chapter are listed and defined.
- **Review** examines important concepts and facts in detail.
- **Take Quiz** allows testing on any combination of—or all—chapters.

*Four different review methods are available.*

*The program provides instant feedback on quiz results.*

# Look at the great anatomy and physiology study tools WCB/McGraw-Hill has to offer!

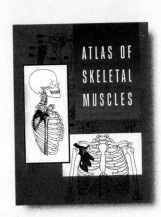

## Atlas of Skeletal Muscles, 2e

*by Judith Stone and Robert Stone*
**ISBN 0-697-13790-2**

The precise drawings of this inexpensive atlas show the origin, insertion, action, and innervation of all human skeletal muscles. The illustrations will help you locate muscles and understand their actions

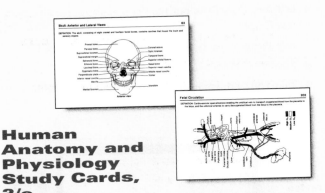

## Human Anatomy and Physiology Study Cards, 3/e

*by Kent Van De Graaff et al.*
**ISBN 0-697-26447-5**

Make studying a breeze with this boxed set of 300, two-sided study cards. Each card provides a complete description of terms, clearly labeled drawings, pronunciation guides, and clinical information on diseases. The *Study Cards* offer a quick and effective way for you to review human anatomy and physiology.

## Student Study Guide for Hole's Human Anatomy and Physiology

*by Nancy A. Sickles Corbett*
**ISBN 0-697-34206-9**

For each chapter in this text, a corresponding study guide chapter contains an overview, chapter objectives, a focus question, a mastery test, study activities, and answer keys.

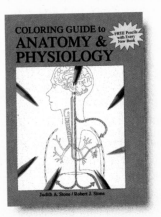

## Coloring Guide to Anatomy and Physiology

*by Judith Stone and Robert Stone*
**ISBN 0-697-17109-4**

The *Coloring Guide* provides a thorough review of anatomical and physiological concepts, and emphasizes learning through the process of color association. By labeling and coloring each drawing, you will easily learn key anatomical and physiological structures and functions.

**To order any of these products, contact your bookstore manager or call our Customer Service Department at 800-338-3987.**

# BRIEF CONTENTS

CONTENTS   VIII
LIST OF CLINICAL APPLICATIONS   XXI
PREFACE   XXIII
CORRELATION GUIDES   XXIX

## UNIT ONE
### LEVELS OF ORGANIZATION   1
1  INTRODUCTION TO HUMAN ANATOMY AND PHYSIOLOGY   1
2  CHEMICAL BASIS OF LIFE   36
3  CELLS   59
4  CELLULAR METABOLISM   101
5  TISSUES   135

## UNIT TWO
### SUPPORT AND MOVEMENT   160
6  SKIN AND THE INTEGUMENTARY SYSTEM   160
7  SKELETAL SYSTEM   184
8  JOINTS OF THE SKELETAL SYSTEM   256
9  MUSCULAR SYSTEM   281

## UNIT THREE
### INTEGRATION AND COORDINATION   343
10  NERVOUS SYSTEM I:
    Basic Structure and Function   343
11  NERVOUS SYSTEM II:
    Divisions of the Nervous System   374
12  SOMATIC AND SPECIAL SENSES   431
13  ENDOCRINE SYSTEM   479

## UNIT FOUR
### TRANSPORT   520
14  BLOOD   520
15  CARDIOVASCULAR SYSTEM   553
16  LYMPHATIC SYSTEM AND IMMUNITY   620

## UNIT FIVE
### ABSORPTION AND EXCRETION   655
17  DIGESTIVE SYSTEM   655
18  NUTRITION AND METABOLISM   703
19  RESPIRATORY SYSTEM   738
20  URINARY SYSTEM   778
21  WATER, ELECTROLYTE, AND ACID-BASE BALANCE   813

## UNIT SIX
### THE HUMAN LIFE CYCLE   835
22  REPRODUCTIVE SYSTEMS   835
23  HUMAN GROWTH AND DEVELOPMENT   891
24  GENETICS   928

APPENDICES   986
GLOSSARY   993
CREDITS   1019
INDEX   1023

# CONTENTS

LIST OF CLINICAL APPLICATIONS XXI
PREFACE XXIII
CORRELATION GUIDES XXIX

## UNIT ONE — LEVELS OF ORGANIZATION

### CHAPTER 1

### INTRODUCTION TO HUMAN ANATOMY AND PHYSIOLOGY

CHAPTER OBJECTIVES   1
KEY TERMS   1
UNDERSTANDING WORDS   1

ANATOMY AND PHYSIOLOGY   3
CHARACTERISTICS OF LIFE   4
MAINTENANCE OF LIFE   5
   Requirements of Organisms   5
   Homeostasis   5

LEVELS OF ORGANIZATION   8
ORGANIZATION OF THE HUMAN BODY   10
   Body Cavities   10

CLINICAL APPLICATION 1.1
   ULTRASONOGRAPHY AND MAGNETIC RESONANCE IMAGING: A TALE OF TWO PATIENTS   10

   Thoracic and Abdominopelvic Membranes   13
   Organ Systems   14

ANATOMICAL TERMINOLOGY   18
   Relative Position   18
   Body Sections   19
   Body Regions   22

SOME MEDICAL AND APPLIED SCIENCES   24
   CHAPTER SUMMARY   25
   CRITICAL THINKING QUESTIONS   26
   REVIEW EXERCISES   27

REFERENCE PLATES
   The Human Organism   28

### CHAPTER 2

### CHEMICAL BASIS OF LIFE

CHAPTER OBJECTIVES   36
KEY TERMS   36
UNDERSTANDING WORDS   36

STRUCTURE OF MATTER   37
   Elements and Atoms   37
   Atomic Structure   37
   Isotopes   38
   Bonding of Atoms   39

CLINICAL APPLICATION 2.1
   RADIOACTIVE ISOTOPES REVEAL PHYSIOLOGY   40

   Molecules and Compounds   42

CLINICAL APPLICATION 2.2
   IONIZING RADIATION: A LEGACY OF THE COLD WAR   43

   Chemical Reactions   44
   Acids, Bases, and Salts   44
   Acid and Base Concentrations   45

CHEMICAL CONSTITUENTS OF CELLS   46
   Inorganic Substances   46
   Organic Substances   47

CLINICAL APPLICATION 2.3
   CT SCANNING AND PET IMAGING   54
   CHAPTER SUMMARY   57
   CRITICAL THINKING QUESTIONS   58
   REVIEW EXERCISES   58

# CHAPTER 3

## CELLS

CHAPTER OBJECTIVES 59
KEY TERMS 59
UNDERSTANDING WORDS 59

### A COMPOSITE CELL 60

*Cell Membrane 63*

#### CLINICAL APPLICATION 3.1
FAULTY ION CHANNELS CAUSE DISEASE 66

#### CLINICAL APPLICATION 3.2
THE BLOOD-BRAIN BARRIER 68

*Cytoplasm 68*

#### CLINICAL APPLICATION 3.3
DISEASE AT THE ORGANELLE LEVEL 74

*Cell Nucleus 77*

### MOVEMENTS INTO AND OUT OF THE CELL 78

*Diffusion 78*
*Facilitated Diffusion 80*
*Osmosis 80*
*Filtration 81*
*Active Transport 83*
*Endocytosis 83*
*Exocytosis 85*
*Transcytosis 85*

### THE CELL CYCLE 88

*Interphase 88*
*Mitosis 88*
*Cytoplasmic Division 92*
*Cell Differentiation 92*

### CONTROL OF CELL REPRODUCTION 93

#### CLINICAL APPLICATION 3.4
CLONING 95

*Health Consequences of Loss of Cell Reproduction Control 96*
CHAPTER SUMMARY 98
CRITICAL THINKING QUESTIONS 99
REVIEW EXERCISES 99

# CHAPTER 4

## CELLULAR METABOLISM

CHAPTER OBJECTIVES 101
KEY TERMS 101
UNDERSTANDING WORDS 101

### METABOLIC PROCESSES 102

*Anabolism 102*
*Catabolism 102*

### CONTROL OF METABOLIC REACTIONS 104

*Enzyme Action 104*
*Cofactors and Coenzymes 105*
*Factors that Alter Enzymes 105*

### ENERGY FOR METABOLIC REACTIONS 106

*Release of Chemical Energy 106*

### AN OVERVIEW OF CELLULAR RESPIRATION 106

*Glycolysis 106*
*Aerobic Respiration 106*
*ATP Molecules 106*

### A CLOSER LOOK AT CELLULAR RESPIRATION 108

*Carbohydrate Pathways 108*

#### CLINICAL APPLICATION 4.1
OVERRIDING A BLOCK IN GLYCOLYSIS 111

*Lipid Pathways 113*
*Protein Pathways 114*
*Regulation of Metabolic Pathways 115*

### NUCLEIC ACIDS AND PROTEIN SYNTHESIS 117

*Genetic Information 117*
*Genetic Code 119*

#### CLINICAL APPLICATION 4.2
DNA MAKES HISTORY 120

*RNA Molecules 120*
*Protein Synthesis 122*
*DNA Replication 126*

### CHANGES IN GENETIC INFORMATION 126

*Nature of Mutations 126*

#### CLINICAL APPLICATION 4.3
GENE AMPLIFICATION REVOLUTIONIZES BIOMEDICINE 128

*Effects of Mutations 129*

### THE HUMAN GENOME PROJECT 130

#### CLINICAL APPLICATION 4.4
PHENYLKETONURIA 131
CHAPTER SUMMARY 132
CRITICAL THINKING QUESTIONS 134
REVIEW EXERCISES 134

# CHAPTER 5

## TISSUES

CHAPTER OBJECTIVES 135
KEY TERMS 135
UNDERSTANDING WORDS 135

### EPITHELIAL TISSUES 136

*General Characteristics 136*
*Simple Squamous Epithelium 137*
*Simple Cuboidal Epithelium 137*

Simple Columnar Epithelium   137
Pseudostratified Columnar Epithelium   138
Stratified Squamous Epithelium   139
Stratified Cuboidal Epithelium   139
Stratified Columnar Epithelium   140
Transitional Epithelium   140
Glandular Epithelium   142

**CONNECTIVE TISSUES   144**

General Characteristics   144
Major Cell Types   144
Connective Tissue Fibers   145

### CLINICAL   APPLICATION   5.1
DISORDERS OF ORDERLY COLLAGEN   146

Loose Connective Tissue   147
Adipose Tissue   148
Reticular Connective Tissue   148
Dense Connective Tissue   149

Elastic Connective Tissue   149
Cartilage   150
Bone   151
Blood   152

**MUSCLE TISSUES   153**

General Characteristics   153
Skeletal Muscle Tissue   153
Smooth Muscle Tissue   153
Cardiac Muscle Tissue   154

**NERVOUS TISSUES   154**

### CLINICAL   APPLICATION   5.2
TISSUE ENGINEERING   155

CHAPTER SUMMARY   157
CRITICAL THINKING QUESTIONS   158
REVIEW EXERCISES   159

## UNIT TWO   SUPPORT AND MOVEMENT

### CHAPTER 6

# SKIN AND THE INTEGUMENTARY SYSTEM

CHAPTER OBJECTIVES   160
KEY TERMS   160
UNDERSTANDING WORDS   160

**TYPES OF MEMBRANES   161**

Skin and Its Tissues   161
Epidermis   163

### CLINICAL   APPLICATION   6.1
SKIN CANCER   166

Dermis   166
Subcutaneous Layer   168

**ACCESSORY ORGANS OF THE SKIN   168**

Hair Follicles   168
Nails   169

### CLINICAL   APPLICATION   6.2
HAIR LOSS   170

Skin Glands   170

### CLINICAL   APPLICATION   6.3
ACNE   172

**REGULATION OF BODY TEMPERATURE   173**

Heat Production and Loss   173
Problems in Temperature Regulation   174

### CLINICAL   APPLICATION   6.4
ELEVATED BODY TEMPERATURE   175

**SKIN COLOR   175**

Genetic Factors   175
Environmental Factors   176
Physiological Factors   176

**HEALING OF WOUNDS AND BURNS   177**

Cuts   177
Burns   177

**COMMON SKIN DISORDERS   179**

CHAPTER SUMMARY   181
CRITICAL THINKING QUESTIONS   182
REVIEW EXERCISES   183

### CHAPTER 7

# SKELETAL SYSTEM

CHAPTER OBJECTIVES   184
KEY TERMS   184
UNDERSTANDING WORDS   184

**BONE STRUCTURE   185**

Bone Classification   185
Parts of a Long Bone   185
Microscopic Structure   186

**BONE DEVELOPMENT AND GROWTH   188**

Intramembranous Bones   188
Endochondral Bones   189
Growth of an Endochondral Bone   190
Homeostasis of Bone Tissue   192
Factors Affecting Bone Development, Growth,
    and Repair   192

**BONE FUNCTION   193**

Support and Protection   193

---
**CLINICAL   APPLICATION   7.1**
FRACTURES   194

---

Body Movement   196
Blood Cell Formation   197
Inorganic Salt Storage   197

---
**CLINICAL   APPLICATION   7.2**
OSTEOPOROSIS   198

---

**SKELETAL ORGANIZATION   199**

Number of Bones   199
Divisions of the Skeleton   199

**SKULL   201**

Cranium   201
Facial Skeleton   208
Infantile Skull   213

**VERTEBRAL COLUMN   214**

A Typical Vertebra   215
Cervical Vertebrae   216
Thoracic Vertebrae   217
Lumbar Vertebrae   218
Sacrum   218
Coccyx   218

---
**CLINICAL   APPLICATION   7.3**
DISORDERS OF THE VERTEBRAL COLUMN   220

---

**THORACIC CAGE   220**

Ribs   220
Sternum   220

**PECTORAL GIRDLE   222**

Clavicles   222
Scapulae   222

**UPPER LIMB   223**

Humerus   223
Radius   225
Ulna   225
Hand   226

**PELVIC GIRDLE   227**

Coxal Bones   227
Greater and Lesser Pelves   228
Differences between Male and Female Pelves   231

**LOWER LIMB   231**

Femur   231
Patella   231
Tibia   232
Fibula   233
Foot   233

**CLINICAL TERMS RELATED TO THE SKELETAL SYSTEM   235**

CHAPTER SUMMARY   237
CRITICAL THINKING QUESTIONS   239
REVIEW EXERCISES   240

---
**REFERENCE   PLATES**
Human Skull   241

---

**C H A P T E R   8**

# JOINTS OF THE SKELETAL SYSTEM

CHAPTER OBJECTIVES   256
KEY TERMS   256
UNDERSTANDING WORDS   256

**CLASSIFICATION OF JOINTS   257**

Fibrous Joints   257
Cartilaginous Joints   258
Synovial Joints   259

**GENERAL STRUCTURE OF A SYNOVIAL JOINT   260**
**TYPES OF SYNOVIAL JOINTS   262**
**TYPES OF JOINT MOVEMENTS   262**
**EXAMPLES OF SYNOVIAL JOINTS   268**

Shoulder Joint   268
Elbow Joint   269
Hip Joint   271

---
**CLINICAL   APPLICATION   8.1**
REPLACING JOINTS   273

---

Knee Joint   273

**CLINICAL TERMS RELATED TO JOINTS   275**

---
**CLINICAL   APPLICATION   8.2**
JOINT DISORDERS   276

---

CHAPTER SUMMARY   278
CRITICAL THINKING QUESTIONS   279
REVIEW EXERCISES   279

---

**C H A P T E R   9**

# MUSCULAR SYSTEM

CHAPTER OBJECTIVES   281
KEY TERMS   281
UNDERSTANDING WORDS   281

**STRUCTURE OF A SKELETAL MUSCLE   282**

Connective Tissue Coverings   282
Skeletal Muscle Fibers   284
Neuromuscular Junction   286
Motor Units   286

**SKELETAL MUSCLE CONTRACTION   287**

Role of Myosin and Actin   287
Stimulus for Contraction   289

Energy Sources for Contraction   291
Oxygen Supply and Cellular Respiration   292
Oxygen Debt   292
Muscle Fatigue   293
Fast and Slow Muscles   293
Heat Production   294

**MUSCULAR RESPONSES   294**

Threshold Stimulus   294
All-or-None Response   294
Recording a Muscle Contraction   294
Staircase Effect   295
Summation   295
Recruitment of Motor Units   295
Sustained Contractions   295
Types of Contractions   296

**CLINICAL   APPLICATION   9.1**
USE AND DISUSE OF SKELETAL MUSCLES   297

**SMOOTH MUSCLES   297**

Smooth Muscle Fibers   297
Smooth Muscle Contraction   298

**CARDIAC MUSCLE   298**

**SKELETAL MUSCLE ACTIONS   300**

Origin and Insertion   300
Interaction of Skeletal Muscles   300

**MAJOR SKELETAL MUSCLES   301**

Muscles of Facial Expression   302
Muscles of Mastication   303

**CLINICAL   APPLICATION   9.2**
TMJ SYNDROME   304

Muscles that Move the Head and Vertebral Column   305
Muscles that Move the Pectoral Girdle   308
Muscles that Move the Arm   311
Muscles that Move the Forearm   312
Muscles that Move the Hand   313
Muscles of the Abdominal Wall   316
Muscles of the Pelvic Outlet   319
Muscles that Move the Thigh   320
Muscles that Move the Leg   322
Muscles that Move the Foot   326

**CLINICAL TERMS RELATED TO THE MUSCULAR
SYSTEM   331**

CHAPTER SUMMARY   333

CRITICAL THINKING QUESTIONS   335

REVIEW EXERCISES   335

**REFERENCE   PLATES**
Surface Anatomy   338

# UNIT THREE   INTEGRATION AND COORDINATION

## CHAPTER 10
## NERVOUS SYSTEM I: BASIC STRUCTURE AND FUNCTION

CHAPTER OBJECTIVES   343

KEY TERMS   343

UNDERSTANDING WORDS   343

**GENERAL FUNCTIONS OF THE NERVOUS
SYSTEM   344**

Neuron Structure   345

**CLINICAL   APPLICATION   10.1**
MIGRAINE   346

**CLINICAL   APPLICATION   10.2**
MULTIPLE SCLEROSIS   349

**CLASSIFICATION OF NEURONS AND
NEUROGLIA   349**

Classification of Neurons   350
Classification of Neuroglia   351
Regeneration of Nerve Fibers   353

**CELL MEMBRANE POTENTIAL   354**

Distribution of Ions   354
Resting Potential   355

Local Potential Changes   355
Action Potentials   357
Refractory Period   359
All-or-None Response   359
Impulse Conduction   359

**CLINICAL   APPLICATION   10.3**
FACTORS AFFECTING IMPULSE CONDUCTION   360

**THE SYNAPSE   360**

Synaptic Potentials   360
Neurotransmitters   361

**CLINICAL   APPLICATION   10.4**
MYASTHENIA GRAVIS   364

Neuropeptides   365

**IMPULSE PROCESSING   365**

Neuronal Pools   365
Facilitation   365
Convergence   365
Divergence   366

**CLINICAL   APPLICATION   10.5**
OPIATES IN THE HUMAN BODY   366

**CLINICAL   APPLICATION   10.6**
DRUG ADDICTION   367

CHAPTER SUMMARY   371

CRITICAL THINKING QUESTIONS   372

REVIEW EXERCISES   372

CHAPTER SUMMARY 427
CRITICAL THINKING QUESTIONS 429
REVIEW EXERCISES 429

# CHAPTER 11

# NERVOUS SYSTEM II: DIVISIONS OF THE NERVOUS SYSTEM

# CHAPTER 12

# SOMATIC AND SPECIAL SENSES

CHAPTER OBJECTVES 374
KEY TERMS 374
UNDERSTANDING WORDS 374

**MENINGES 376**
**SPINAL CORD 377**
  Structure of the Spinal Cord 377
  Functions of the Spinal Cord 378

**CLINICAL APPLICATION 11.1**
  SPINAL CORD INJURIES 382

**REFLEX ARCS 383**
  Reflex Behavior 383

**CLINICAL APPLICATION 11.2**
  USES OF REFLEXES 386

**BRAIN 387**
  Brain Development 387
  Structure of the Cerebrum 387
  Functions of the Cerebrum 390
  Basal Ganglia 394
  Ventricles and Cerebrospinal Fluid 394

**CLINICAL APPLICATION 11.3**
  CEREBRAL INJURIES AND ABNORMALITIES 395

**CLINICAL APPLICATION 11.4**
  PARKINSON'S DISEASE 396

  Diencephalon 398

**CLINICAL APPLICATION 11.5**
  CEREBROSPINAL FLUID PRESSURE 400

  Brain Stem 401
  Cerebellum 403

**CLINICAL APPLICATION 11.6**
  BRAIN WAVES 406

**PERIPHERAL NERVOUS SYSTEM 407**
  Structure of Peripheral Nerves 407
  Nerve Fiber Classification 407
  Cranial Nerves 408
  Spinal Nerves 412

**AUTONOMIC NERVOUS SYSTEM 418**
  General Characteristics 418

**CLINICAL APPLICATION 11.7**
  SPINAL NERVE INJURIES 419

  Autonomic Nerve Fibers 419
  Autonomic Neurotransmitters 423
  Actions of Autonomic Neurotransmitters 424
  Control of Autonomic Activity 425

**CLINICAL TERMS RELATED TO THE NERVOUS SYSTEM 426**

CHAPTER OBJECTVES 431
KEY TERMS 431
UNDERSTANDING WORDS 431

**RECEPTORS AND SENSATIONS 432**
  Receptor Types 432
  Sensory Impulses 433
  Sensations 433
  Sensory Adaptation 433

**SOMATIC SENSES 434**
  Touch and Pressure Senses 434
  Temperature Senses 434
  Sense of Pain 435

**CLINICAL APPLICATION 12.1**
  CANCER PAIN AND CHRONIC PAIN 438

  Stretch Receptors 438

**SPECIAL SENSES 439**

**CLINICAL APPLICATION 12.2**
  MIXED-UP SENSES—SYNESTHESIA 440

**SENSE OF SMELL 440**
  Olfactory Receptors 440
  Olfactory Organs 440
  Olfactory Nerve Pathways 442
  Olfactory Stimulation 442

**SENSE OF TASTE 442**
  Taste Receptors 442
  Taste Sensations 444
  Taste Nerve Pathways 445

**CLINICAL APPLICATION 12.3**
  SMELL AND TASTE DISORDERS 445

**SENSE OF HEARING 446**
  External Ear 446
  Middle Ear 446
  Auditory Tube 448
  Inner Ear 448
  Auditory Nerve Pathways 450

**SENSE OF EQUILIBRIUM 451**
  Static Equilibrium 453

**CLINICAL APPLICATION 12.4**
  HEARING LOSS 454

  Dynamic Equilibrium 455

**SENSE OF SIGHT 457**
  Visual Accessory Organs 457
  Structure of the Eye 461
  Light Refraction 468

**CONTENTS**

*Visual Receptors   469*

### CLINICAL APPLICATION 12.5
REFRACTION DISORDERS   470

*Visual Pigments   470*
*Stereoscopic Vision   473*
*Visual Nerve Pathways   473*

## CLINICAL TERMS RELATED TO THE SENSES   475

CHAPTER SUMMARY   475
CRITICAL THINKING QUESTIONS   477
REVIEW EXERCISES   478

# ENDOCRINE SYSTEM
# 13

CHAPTER OBJECTIVES   479
KEY TERMS   479
UNDERSTANDING WORDS   479

## GENERAL CHARACTERISTICS OF THE ENDOCRINE SYSTEM   480
## HORMONE ACTION   480

*Chemistry of Hormones   480*
*Actions of Hormones   482*

### CLINICAL APPLICATION 13.1
STEROIDS AND ATHLETES—AN UNHEALTHY COMBINATION   485

*Prostaglandins   487*

## CONTROL OF HORMONAL SECRETIONS   487

*Control Sources   488*
*Negative Feedback Systems   488*

## PITUITARY GLAND   488

*Anterior Pituitary Hormones   490*

### CLINICAL APPLICATION 13.2
GROWTH HORMONE UPS AND DOWNS   493

*Posterior Pituitary Hormones   494*

## THYROID GLAND   495

*Structure of the Gland   495*
*Thyroid Hormones   495*

## PARATHYROID GLANDS   499

*Structure of the Glands   499*
*Parathyroid Hormone   499*

## ADRENAL GLANDS   501

*Structure of the Glands   501*
*Hormones of the Adrenal Medulla   502*
*Hormones of the Adrenal Cortex   503*

### CLINICAL APPLICATION 13.3
DISORDERS OF THE ADRENAL CORTEX   506

## PANCREAS   507

*Structure of the Gland   507*
*Hormones of the Islets of Langerhans   507*

### CLINICAL APPLICATION 13.4
DIABETES MELLITUS   509

## OTHER ENDOCRINE GLANDS   510

*Pineal Gland   510*
*Thymus Gland   510*
*Reproductive Glands   510*
*Digestive Glands   510*
*Other Hormone-Producing Organs   510*

### CLINICAL APPLICATION 13.5
MISREPRESENTING MELATONIN   511

## STRESS AND ITS EFFECTS   511

*Types of Stress   511*
*Responses to Stress   512*

## CLINICAL TERMS RELATED TO THE ENDOCRINE SYSTEM   512

CHAPTER SUMMARY   516
CRITICAL THINKING QUESTIONS   518
REVIEW EXERCISES   518

# UNIT FOUR TRANSPORT

# BLOOD
# 14

CHAPTER OBJECTIVES   520
KEY TERMS   520
UNDERSTANDING WORDS   520

## BLOOD AND BLOOD CELLS   521

*Blood Volume and Composition   521*
*The Origin of Blood Cells   522*

*Characteristics of Red Blood Cells   522*
*Red Blood Cell Counts   524*
*Destruction of Red Blood Cells   524*
*Red Blood Cell Production and Its Control   526*
*Dietary Factors Affecting Red Blood Cell Production   526*
*Types of White Blood Cells   527*

### CLINICAL APPLICATION 14.1
KING GEORGE III AND PORPHYRIA   529

*White Blood Cells Fight Infection   531*
*White Blood Cell Counts   532*
*Blood Platelets   532*

## BLOOD PLASMA   532

*Plasma Proteins   532*

### CLINICAL APPLICATION 14.2
LEUKEMIA   533

Gases and Nutrients   535
Nonprotein Nitrogenous Substances   536
Plasma Electrolytes   537

**HEMOSTASIS   537**

Blood Vessel Spasm   537
Platelet Plug Formation   537
Blood Coagulation   537
Prevention of Coagulation   542

**BLOOD GROUPS AND TRANSFUSIONS   542**

Antigens and Antibodies   542

**CLINICAL   APPLICATION   14.3**
THE RETURN OF THE MEDICINAL LEECH   543

**CLINICAL   APPLICATION   14.4**
COAGULATION DISORDERS   544

ABO Blood Group   544
Rh Blood Group   546

**CLINICAL TERMS RELATED TO THE BLOOD   547**

**CLINICAL   APPLICATION   14.5**
REPLACING BLOOD   548

CHAPTER SUMMARY   549
CRITICAL THINKING QUESTIONS   551
REVIEW EXERCISES   551

# CARDIOVASCULAR SYSTEM

C H A P T E R 15

CHAPTER OBJECTIVES   553
KEY TERMS   553
UNDERSTANDING WORDS   553

**STRUCTURE OF THE HEART   554**

Size and Location of the Heart   554
Coverings of the Heart   555
Wall of the Heart   556
Heart Chambers and Valves   557
Skeleton of the Heart   560
Path of Blood through the Heart   561
Blood Supply to the Heart   562

**HEART ACTIONS   565**

Cardiac Cycle   566
Heart Sounds   568
Cardiac Muscle Fibers   568
Cardiac Conduction System   568

**CLINICAL   APPLICATION   15.1**
HEART TRANSPLANTS   569

Electrocardiogram   571
Regulation of the Cardiac Cycle   571

**BLOOD VESSELS   573**

Arteries and Arterioles   573

**CLINICAL   APPLICATION   15.2**
ARRHYTHMIAS   574

Capillaries   577
Exchanges in the Capillaries   579
Venules and Veins   580

**CLINICAL   APPLICATION   15.3**
BLOOD VESSEL DISORDERS   582

**BLOOD PRESSURE   582**

Arterial Blood Pressure   582
Factors that Influence Arterial Blood Pressure   584
Control of Blood Pressure   585

**CLINICAL   APPLICATION   15.4**
MEASUREMENT OF ARTERIAL BLOOD PRESSURE   586

**CLINICAL   APPLICATION   15.5**
SPACE MEDICINE   588

**CLINICAL   APPLICATION   15.6**
HYPERTENSION   590

Venous Blood Flow   590
Central Venous Pressure   592

**PATHS OF CIRCULATION   592**

Pulmonary Circuit   592

**CLINICAL   APPLICATION   15.7**
EXERCISE AND THE CARDIOVASCULAR SYSTEM   593

Systemic Circuit   594

**ARTERIAL SYSTEM   595**

Principal Branches of the Aorta   595
Arteries to the Neck, Head, and Brain   597
Arteries to the Shoulder and Upper Limb   600
Arteries to the Thoracic and Abdominal Walls   601
Arteries to the Pelvis and Lower Limb   602

**VENOUS SYSTEM   604**

Characteristics of Venous Pathways   604
Veins from the Head, Neck, and Brain   604
Veins from the Upper Limb and Shoulder   606
Veins from the Abdominal and Thoracic Walls   606
Veins from the Abdominal Viscera   607
Veins from the Lower Limb and Pelvis   608

**CLINICAL TERMS RELATED TO THE CARDIOVASCULAR SYSTEM   609**

**CLINICAL   APPLICATION   15.8**
MOLECULAR CAUSES OF CARDIOVASCULAR DISEASE   612

**CLINICAL   APPLICATION   15.9**
CORONARY ARTERY DISEASE   614

CHAPTER SUMMARY   616
CRITICAL THINKING QUESTIONS   618
REVIEW EXERCISES   618

# CHAPTER 16

# LYMPHATIC SYSTEM AND IMMUNITY

CHAPTER OBJECTIVES   620
KEY TERMS   620
UNDERSTANDING WORDS   620

**LYMPHATIC PATHWAYS   621**

*Lymphatic Capillaries   621*
*Lymphatic Vessels   621*
*Lymphatic Trunks and Collecting Ducts   621*

**TISSUE FLUID AND LYMPH   625**

*Tissue Fluid Formation   625*
*Lymph Formation   625*
*Lymph Function   625*

**LYMPH MOVEMENT   626**

*Lymph Flow   626*
*Obstruction of Lymph Movement   626*

**LYMPH NODES   626**

*Structure of a Lymph Node   626*
*Locations of Lymph Nodes   627*
*Functions of Lymph Nodes   628*

**THYMUS AND SPLEEN   628**

*Thymus   628*
*Spleen   629*

**BODY DEFENSES AGAINST INFECTION   630**
**NONSPECIFIC DEFENSES   630**

*Species Resistance   630*
*Mechanical Barriers   631*
*Chemical Barriers   631*
*Fever   631*
*Inflammation   631*
*Phagocytosis   632*

**SPECIFIC DEFENSES (IMMUNITY)   632**

*Antigens   632*
*Lymphocyte Origins   632*
*T Cells and Cell-Mediated Immunity   633*
*B Cells and Antibody-Mediated Immunity   634*

CLINICAL   APPLICATION   16.1
IMMUNOTHERAPY   638

*Immune Responses   640*
*Types of Immunity   642*
*Allergic Reactions   643*
*Transplantation and Tissue Rejection   644*
*Autoimmunity   646*

**CLINICAL TERMS RELATED TO THE LYMPHATIC SYSTEM AND IMMUNITY   647**

CLINICAL   APPLICATION   16.2
IMMUNITY BREAKDOWN: AIDS   648

CHAPTER SUMMARY   651
CRITICAL THINKING QUESTIONS   653
REVIEW EXERCISES   654

# UNIT FIVE ABSORPTION AND EXCRETION

# CHAPTER 17

# DIGESTIVE SYSTEM

CHAPTER OBJECTIVES   655
KEY TERMS   655
UNDERSTANDING WORDS   655

**GENERAL CHARACTERISTICS OF THE ALIMENTARY CANAL   656**

*Structure of the Wall   656*
*Movements of the Tube   657*
*Innervation of the Tube   657*

**MOUTH   658**

*Cheeks and Lips   658*
*Tongue   658*
*Palate   660*
*Teeth   661*

CLINICAL   APPLICATION   17.1
DENTAL CARIES   664

**SALIVARY GLANDS   665**

*Salivary Secretions   665*
*Major Salivary Glands   665*

**PHARYNX AND ESOPHAGUS   666**

*Structure of the Pharynx   666*
*Swallowing Mechanism   667*
*Esophagus   667*

**STOMACH   670**

*Parts of the Stomach   670*
*Gastric Secretions   670*
*Regulation of Gastric Secretions   672*
*Gastric Absorption   673*

CLINICAL   APPLICATION   17.2
OH, MY ACHING STOMACH!   674

*Mixing and Emptying Actions   674*

**PANCREAS   676**

*Structure of the Pancreas   676*
*Pancreatic Juice   676*
*Regulation of Pancreatic Secretion   677*

**LIVER   678**

*Liver Structure   678*
*Liver Functions   680*
*Composition of Bile   681*
*Gallbladder   681*

## CLINICAL APPLICATION 17.3

HEPATITIS 682

*Regulation of Bile Release* 683

## CLINICAL APPLICATION 17.4

GALLBLADDER DISEASE 684

*Functions of Bile Salts* 684

## SMALL INTESTINE 685

*Parts of the Small Intestine* 685
*Structure of the Small Intestinal Wall* 686
*Secretions of the Small Intestine* 688
*Regulation of Small Intestinal Secretions* 688
*Absorption in the Small Intestine* 689
*Movements of the Small Intestine* 691

## LARGE INTESTINE 692

*Parts of the Large Intestine* 692
*Structure of the Large Intestinal Wall* 693
*Functions of the Large Intestine* 694
*Movements of the Large Intestine* 695
*Feces* 695

## CLINICAL APPLICATION 17.5

DISORDERS OF THE LARGE INTESTINE 696

## CLINICAL TERMS RELATED TO THE DIGESTIVE SYSTEM 697

CHAPTER SUMMARY 699

CRITICAL THINKING QUESTIONS 701

REVIEW EXERCISES 702

## CHAPTER 18

### NUTRITION AND METABOLISM

CHAPTER OBJECTIVES 703

KEY TERMS 703

UNDERSTANDING WORDS 703

## WHY WE EAT 704

*Carbohydrates* 705
*Carbohydrate Sources* 705
*Carbohydrate Utilization* 705
*Carbohydrate Requirements* 706

## LIPIDS 706

*Lipid Sources* 706
*Lipid Utilization* 706
*Lipid Requirements* 707

## PROTEINS 707

*Protein Sources* 708
*Nitrogen Balance* 709
*Protein Requirements* 709

## ENERGY EXPENDITURES 710

*Energy Values of Foods* 710
*Energy Requirements* 710
*Energy Balance* 711
*Desirable Weight* 711

## CLINICAL APPLICATION 18.1

OBESITY 712

## VITAMINS 713

*Fat-Soluble Vitamins* 714

## CLINICAL APPLICATION 18.2

DO VITAMINS PROTECT AGAINST HEART DISEASE AND CANCER? 715

*Water-Soluble Vitamins* 717

## MINERALS 722

*Characteristics of Minerals* 722
*Major Minerals* 722
*Trace Elements* 724

## HEALTHY EATING—THE FOOD PYRAMID AND READING LABELS 728

*Malnutrition* 729
*Starvation* 730

## CLINICAL APPLICATION 18.3

NUTRITION AND THE ATHLETE 732

## CLINICAL TERMS RELATED TO NUTRITION AND METABOLISM 732

CHAPTER SUMMARY 733

CRITICAL THINKING QUESTIONS 736

REVIEW EXERCISES 736

## CHAPTER 19

### RESPIRATORY SYSTEM

CHAPTER OBJECTIVES 738

KEY TERMS 738

UNDERSTANDING WORDS 738

## WHY WE BREATHE 739

## ORGANS OF THE RESPIRATORY SYSTEM 740

*Nose* 740
*Nasal Cavity* 740
*Sinuses* 741
*Pharynx* 741

## CLINICAL APPLICATION 19.1

THE EFFECTS OF CIGARETTE SMOKING ON THE RESPIRATORY SYSTEM 742

*Larynx* 742
*Trachea* 744
*Bronchial Tree* 746
*Lungs* 750

## BREATHING MECHANISM 751

*Inspiration* 751

## CLINICAL APPLICATION 19.2

LUNG IRRITANTS 753

*Expiration* 756
*Respiratory Volumes and Capacities* 757
*Alveolar Ventilation* 758
*Nonrespiratory Air Movements* 759

CONTROL OF BREATHING   760

Respiratory Center   760

CLINICAL   APPLICATION   19.3
RESPIRATORY DISORDERS THAT DECREASE VENTILATION   761

Factors Affecting Breathing   762

ALVEOLAR GAS EXCHANGES   764

Alveoli   764

CLINICAL   APPLICATION   19.4
EXERCISE AND BREATHING   765

Respiratory Membrane   765
Diffusion through the Respiratory Membrane   765

GAS TRANSPORT   767

Oxygen Transport   767

CLINICAL   APPLICATION   19.5
DISORDERS IMPAIRING GAS EXCHANGE   768

Carbon Monoxide   769
Carbon Dioxide Transport   769

CLINICAL TERMS RELATED TO THE RESPIRATORY
SYSTEM   774

CHAPTER SUMMARY   774

CRITICAL THINKING QUESTIONS   776

REVIEW EXERCISES   776

C  H  A  P  T  E  R   **20**

# URINARY SYSTEM

CHAPTER OBJECTIVES   778

KEY TERMS   778

UNDERSTANDING WORDS   778

KIDNEYS   779

Location of the Kidneys   779
Kidney Structure   780
Functions of the Kidneys   780
Renal Blood Vessels   782

CLINICAL   APPLICATION   20.1
CHRONIC KIDNEY FAILURE   783

Nephrons   785

CLINICAL   APPLICATION   20.2
GLOMERULONEPHRITIS   787

URINE FORMATION   789

Glomerular Filtration   789
Filtration Pressure   791
Filtration Rate   792
Regulation of Filtration Rate   793
Tubular Reabsorption   794

CLINICAL   APPLICATION   20.3
THE NEPHROTIC SYNDROME   795

Regulation of Urine Concentration and Volume   796
Urea and Uric Acid Excretion   799
Tubular Secretion   800

Urine Composition   801

CLINICAL   APPLICATION   20.4
RENAL CLEARANCE   802

ELIMINATION OF URINE   802

Ureters   802
Urinary Bladder   803
Micturition   805

CLINICAL   APPLICATION   20.5
URINALYSIS: CLUES TO HEALTH   806

Urethra   806

CLINICAL TERMS RELATED TO THE URINARY
SYSTEM   809

CHAPTER SUMMARY   809

CRITICAL THINKING QUESTIONS   811

REVIEW EXERCISES   811

C  H  A  P  T  E  R   **21**

# WATER, ELECTROLYTE, AND ACID-BASE BALANCE

CHAPTER OBJECTIVES   813

KEY TERMS   813

UNDERSTANDING WORDS   813

DISTRIBUTION OF BODY FLUIDS   814

Fluid Compartments   814
Body Fluid Composition   815
Movement of Fluid between Compartments   815

WATER BALANCE   816

Water Intake   816
Regulation of Water Intake   816
Water Output   817
Regulation of Water Output   818

ELECTROLYTE BALANCE   819

Electrolyte Intake   819
Regulation of Electrolyte Intake   819
Electrolyte Output   819
Regulation of Electrolyte Output   819

CLINICAL   APPLICATION   21.1
WATER BALANCE DISORDERS   820

CLINICAL   APPLICATION   21.2
SODIUM AND POTASSIUM IMBALANCES   823

ACID-BASE BALANCE   823

Sources of Hydrogen Ions   823
Strengths of Acids and Bases   824
Regulation of Hydrogen Ion Concentration   824

CLINICAL   APPLICATION   21.3
ACID-BASE IMBALANCES   829

CLINICAL TERMS RELATED TO WATER
AND ELECTROLYTE BALANCE   828

CHAPTER SUMMARY   832

CRITICAL THINKING QUESTIONS   833

REVIEW EXERCISES   833

## CHAPTER 22

# REPRODUCTIVE SYSTEMS

CHAPTER OBJECTIVES   835

KEY TERMS   835

UNDERSTANDING WORDS   835

**ORGANS OF THE MALE REPRODUCTIVE SYSTEM   836**

**TESTES   836**

Descent of the Testes   836
Structure of the Testes   836
Formation of Sperm Cells   838
Structure of a Sperm Cell   841

**MALE INTERNAL ACCESSORY ORGANS   842**

Epididymis   842
Vas Deferens   842
Seminal Vesicle   842
Prostate Gland   843

### CLINICAL APPLICATION 22.1
PROSTATE ENLARGEMENT   844

Bulbourethral Glands   843
Semen   843

**MALE EXTERNAL REPRODUCTIVE ORGANS   845**

Scrotum   845
Penis   845
Erection, Orgasm, and Ejaculation   845

### CLINICAL APPLICATION 22.2
MALE INFERTILITY   846

**HORMONAL CONTROL OF MALE REPRODUCTIVE FUNCTIONS   848**

Hypothalamic and Pituitary Hormones   848
Male Sex Hormones   850
Actions of Testosterone   850
Regulation of Male Sex Hormones   850

**ORGANS OF THE FEMALE REPRODUCTIVE SYSTEM   850**

**OVARIES   850**

Ovary Attachments   852
Ovary Descent   852
Ovary Structure   852
Primordial Follicles   852
Oogenesis   852
Follicle Maturation   853
Ovulation   855

**FEMALE INTERNAL ACCESSORY ORGANS   855**

Uterine Tubes   855
Uterus   856
Vagina   858

**FEMALE EXTERNAL REPRODUCTIVE ORGANS   859**

Labia Majora   859
Labia Minora   859
Clitoris   859
Vestibule   859
Erection, Lubrication, and Orgasm   859

**HORMONAL CONTROL OF FEMALE REPRODUCTIVE FUNCTIONS   860**

Female Sex Hormones   860
Female Reproductive Cycle   861
Menopause   863

**PREGNANCY   864**

Transport of Sex Cells   864
Fertilization   865
Early Embryonic Development   866
Hormonal Changes during Pregnancy   868
Other Changes during Pregnancy   869
Birth   869

**MAMMARY GLANDS   871**

Location of the Glands   871
Structure of the Glands   871
Development of the Breasts   871

### CLINICAL APPLICATION 22.3
FEMALE INFERTILITY   872

### CLINICAL APPLICATION 22.4
BREAST CANCER UPDATE   874

Milk Production and Secretion   876

**BIRTH CONTROL   877**

Coitus Interruptus   877

### CLINICAL APPLICATION 22.5
HUMAN MILK—THE PERFECT FOOD FOR HUMAN BABIES   878

Rhythm Method   878
Mechanical Barriers   879
Chemical Barriers   880
Oral Contraceptives   880
Injectable Contraception   880
Contraceptive Implants   881
Intrauterine Devices   881
Surgical Methods   881

**SEXUALLY TRANSMITTED DISEASES   882**

**CLINICAL TERMS RELATED TO THE REPRODUCTIVE SYSTEMS   883**

CHAPTER SUMMARY   885

CRITICAL THINKING QUESTIONS   889

REVIEW EXERCISES   889

## C H A P T E R 23

# HUMAN GROWTH AND DEVELOPMENT

CHAPTER OBJECTIVES   891
KEY TERMS   891
UNDERSTANDING WORDS   891

**PRENATAL PERIOD   892**
  *Period of Cleavage   892*

### CLINICAL APPLICATION 23.1
  PREIMPLANTATION GENETIC DIAGNOSIS   894

  *Embryonic Stage   896*

### CLINICAL APPLICATION 23.2
  ASSISTED REPRODUCTIVE TECHNOLOGIES   898

  *Fetal Stage   907*

### CLINICAL APPLICATION 23.3
  SOME CAUSES OF BIRTH DEFECTS   908

  *Fetal Blood and Circulation   912*

**POSTNATAL PERIOD   913**
  *Neonatal Period   914*

### CLINICAL APPLICATION 23.4
  JOINED FOR LIFE   916

  *Infancy   918*
  *Childhood   918*
  *Adolescence   919*
  *Adulthood   919*
  *Senescence   921*

**AGING   922**
  *Passive Aging   922*
  *Active Aging   923*
  *The Human Life Span   923*

### CLINICAL APPLICATION 23.5
  OLD BEFORE THEIR TIME   924

**CLINICAL TERMS RELATED TO GROWTH AND DEVELOPMENT   924**

  CHAPTER SUMMARY   925
  CRITICAL THINKING QUESTIONS   927
  REVIEW EXERCISES   927

## C H A P T E R 24

# GENETICS

CHAPTER OBJECTIVES   928
KEY TERMS   928
UNDERSTANDING WORDS   928

**THE EMERGING ROLE OF GENETICS IN MEDICINE   929**
**GENES AND CHROMOSOMES   930**
**CHROMOSOMES AND SEX CELL PRODUCTION   931**
  *Meiosis   931*

### CLINICAL APPLICATION 24.1
  IT'S ALL IN THE GENES   932

  *Meiosis Leads to Genetic Variability   936*
  *Chromosomes and Genes Come in Pairs   936*

### CLINICAL APPLICATION 24.2
  MAPS, MARKERS, AND MEDICINE   938

**MODES OF INHERITANCE   940**
  *Dominant and Recessive Inheritance   940*
  *Different Dominance Relationships   941*

**GENE EXPRESSION   942**
  *Penetrance and Expressivity   942*
  *Pleiotropy   943*
  *Genetic Heterogeneity   943*

**COMPLEX TRAITS   943**

### CLINICAL APPLICATION 24.3
  ADOPTION AND TWIN STUDIES   945

**MATTERS OF SEX   945**
  *Sex Determination   945*
  *Sex Chromosomes and Sex-Linked Traits   945*

**GENDER EFFECTS ON PHENOTYPE   947**
  *Sex-Limited Traits   947*
  *Sex-Influenced Traits   947*

**CHROMOSOME DISORDERS   948**
  *Polyploidy   948*
  *Aneuploidy   949*

### CLINICAL APPLICATION 24.4
  FRAGILE X SYNDROME AND EXPANDING GENES   950

**MEDICAL GENETICS   951**
  *Diagnosis   951*

### CLINICAL APPLICATION 24.5
  DOWN SYNDROME   952

  *Genetic Counseling   954*
  *Gene Therapy   955*

  CHAPTER SUMMARY   960
  CRITICAL THINKING QUESTIONS   962
  REVIEW EXERCISES   962

## R E F E R E N C E   P L A T E S
  *Human Cadavers   963*

**APPENDIX A: PERIODIC TABLE OF ELEMENTS   986**

**APPENDIX B: UNITS OF MEASUREMENT AND THEIR EQUIVALENTS   987**

**APPENDIX C: LABORATORY TESTS OF CLINICAL IMPORTANCE   989**

**GLOSSARY   993**

**CREDITS   1019**

**INDEX   1023**

# CLINICAL APPLICATIONS

**CHAPTER 1**

1.1: Ultrasonography and Magnetic Resonance Imaging: A Tale of Two Patients   10

**CHAPTER 2**

2.1: Radioactive Isotopes Reveal Physiology   40
2.2: Ionizing Radiation: A Legacy of the Cold War   43
2.3: CT Scanning and PET Imaging   54

**CHAPTER 3**

3.1: Faulty Ion Channels Cause Disease   66
3.2: The Blood-Brain Barrier   68
3.3: Disease at the Organelle Level   74
3.4: Cloning   95

**CHAPTER 4**

4.1: Overriding a Block in Glycolysis   111
4.2: DNA Makes History   120
4.3: Gene Amplification Revolutionizes Biomedicine   128
4.4: Phenylketonuria   131

**CHAPTER 5**

5.1: Disorders of Orderly Collagen   146
5.2: Tissue Engineering   155

**CHAPTER 6**

6.1: Skin Cancer   166
6.2: Hair Loss   170
6.3: Acne   172
6.4: Elevated Body Temperature   175

**CHAPTER 7**

7.1: Fractures   194
7.2: Osteoporosis   198
7.3: Disorders of the Vertebral Column   220

**CHAPTER 8**

8.1: Replacing Joints   273
8.2: Joint Disorders   276

**CHAPTER 9**

9.1: Use and Disuse of Skeletal Muscles   297
9.2: TMJ Syndrome   304

**CHAPTER 10**

10.1: Migraine   346
10.2: Multiple Sclerosis   349
10.3: Factors Affecting Impulse Conduction   360
10.4: Myasthenia Gravis   364
10.5: Opiates in the Human Body   366
10.6: Drug Addiction   367

**CHAPTER 11**

11.1: Spinal Cord Injuries   382
11.2: Uses of Reflexes   386
11.3: Cerebral Injuries and Abnormalities   395
11.4: Parkinson's Disease   396
11.5: Cerebrospinal Fluid Pressure   400
11.6: Brain Waves   406
11.7: Spinal Nerve Injuries   419

**CHAPTER 12**

12.1: Cancer Pain and Chronic Pain   438
12.2: Mixed-up Senses—Synesthesia   440
12.3: Smell and Taste Disorders   454
12.4: Hearing Loss   454
12.5: Refraction Disorders   470

**CHAPTER 13**

13.1: Steroids and Athletes—An Unhealthy Combination   485
13.2: Growth Hormone Ups and Downs   493
13.3: Disorders of the Adrenal Cortex   506
13.4: Diabetes Mellitus   509
13.5: Misrepresenting Melatonin   511

**CHAPTER 14**

14.1: King George III and Porphyria   529
14.2: Leukemia   533
14.3: The Return of the Medicinal Leech   543
14.4: Coagulation Disorders   544
14.5: Replacing Blood   548

**CHAPTER 15**

15.1: Heart Transplants   569
15.2: Arrhythmias   574
15.3: Blood Vessel Disorders   582
15.4: Measurement of Arterial Blood Pressure   586
15.5: Space Medicine   588
15.6: Hypertension   590
15.7: Exercise and the Cardiovascular System   593
15.8: Molecular Causes of Cardiovascular Disease   612
15.9: Coronary Artery Disease   614

**CHAPTER 16**

16.1: Immunotherapy   638
16.2: Immunity Breakdown: AIDS   648

**CHAPTER 17**

17.1: Dental Caries   664
17.2: Oh, My Aching Stomach!   674
17.3: Hepatitis   682
17.4: Gallbladder Disease   684
17.5: Disorders of the Large Intestine   696

**CHAPTER 18**

18.1: Obesity   712
18.2: Do Vitamins Protect Against Heart Disease and Cancer?   715
18.3: Nutrition and the Athlete   732

**CHAPTER 19**

19.1: The Effects of Cigarette Smoking on the Respiratory System   742
19.2: Lung Irritants   753
19.3: Respiratory Disorders that Decrease Ventilation   761
19.4: Exercise and Breathing   765
19.5: Disorders Impairing Gas Exchange   768

## Chapter 20

20.1: Chronic Kidney Failure   783
20.2: Glomerulonephritis   787
20.3: The Nephrotic Syndrome   795
20.4: Renal Clearance   802
20.5: Urinalysis: Clues to Health   806

## Chapter 21

21.1: Water Balance Disorders   820
21.2: Sodium and Potassium Imbalances   823
21.3: Acid-Base Imbalances   829

## Chapter 22

22.1: Prostate Enlargement   844
22.2: Male Infertility   846
22.3: Female Infertility   872
22.4: Breast Cancer Update   874
22.5: Human Milk—The Perfect Food for Human
   Babies   878

## Chapter 23

23.1: Preimplantation Genetic Diagnosis   894
23.2: Assisted Reproductive Technologies   898
23.3: Some Causes of Birth Defects   908
23.4: Joined for Life   916
23.5: Old before Their Time   924

## Chapter 24

25.1: It's All in the Genes   932
25.2: Maps, Markers, and Medicine   938
25.3: Adoption and Twin Studies   945
25.4: Fragile X Syndrome and Expanding Genes   950
25.5: Down Syndrome   952

**CLINICAL APPLICATIONS**

# PREFACE

Revising a textbook is a little like caring for a car. Sometimes a complete overhaul is needed; other times, a simple tune-up will do. For this eighth edition of *Hole's Human Anatomy and Physiology,* we felt that a tune-up and polish were in order. The last edition was changed substantially. This time around, we considered what adopters liked and what they didn't like, while updating the factual material and greatly improving both the verbal and visual clarity of the presentation. The result, we hope, is an even more exciting introduction into the workings of the human body.

## FAVORITE FEATURES

**Introductory Vignettes**　True-to-life vignettes begin each chapter with a glimpse of relevancy. Despite the popularity of many of the vignettes from the last edition, we have replaced several—too many exciting new events in medical science forced us to! Some vignettes explore molecular and cellular explanations for whole-body phenomena: how a inflamed pancreas reflects a faulty gene (chapter 2); why some people can't get AIDS (chapter 3); a teen who dies when a bacterial toxin in spoiled spaghetti halts mitochondrial function (chapter 4). Others explore illness: a woman who lost her skin (chapter 6), a girl who couldn't smile (chapter 9), and a description of gout—in *Tyrannosaurus rex* (chapter 8)! Several vignettes preview emerging technologies: the Visible Human Project on the Internet (chapter 5); new skull base surgery (chapter 7); cord blood banking (chapter 14), and a living parent-to-child organ transplant (chapter 16).

**Clinical Applications**　These practical sidebars apply chapter concepts by describing new technology, and sometimes bring in a historical perspective. Many clinical applications address common illnesses, such as migraines, neurological disorders, prostate enlargement, and breast cancer. Often Clinical Applications introduce molecular information, such as "Disease at the Organelle Level" (chapter 3) and "Molecular Causes of Cardiovascular Disease" (chapter 15). The new addition to chapter 3, "Faulty Ion Channels Cause Disease," weds the latest discoveries in cell biology to diverse disorders.

Some of the Clinical Applications new to this edition reflect recent headlines. For example, "Space Medicine" (chapter 15) explores the effects of prolonged microgravity on physiology in an astronaut. Clinical Application 3.4 discusses cloning, 13.5 investigates the hype in "Misrepresenting Melatonin," and 23.4 presents the Hensel twins, little girls who are joined from the neck down. New Clinical Applications on illnesses discuss acne (chapter 6), Parkinson's disease (chapter 11), and lung irritants (chapter 19).

**InnerConnections**　These graphics, which appear at the ends of selected chapters, review and connect material at a glance. They conceptually link the highlighted organ system to every other, which reinforces the dynamic interplays between groups of organs. InnerConnections can be used as springboards for class discussion, ideas for further study or term papers, review of chapter concepts, and reinforcement of the "big picture" in learning and applying the concepts of anatomy and physiology.

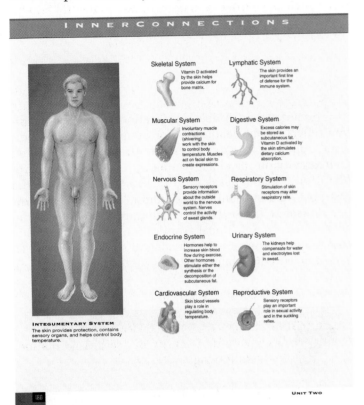

**INTEGUMENTARY SYSTEM**
The skin provides protection, contains sensory organs, and helps control body temperature.

**Skeletal System**
Vitamin D activated by the skin helps provide calcium for bone matrix.

**Lymphatic System**
The skin provides an important first line of defense for the immune system.

**Muscular System**
Involuntary muscle contractions (shivering) work with the skin to control body temperature. Muscles act on facial skin to create expressions.

**Digestive System**
Excess calories may be stored as subcutaneous fat. Vitamin D activated by the skin stimulates dietary calcium absorption.

**Nervous System**
Sensory receptors provide information about the outside world to the nervous system. Nerves control the activity of sweat glands.

**Respiratory System**
Stimulation of skin receptors may alter respiratory rate.

**Endocrine System**
Hormones help to increase skin blood flow during exercise. Other hormones stimulate either the synthesis or the decomposition of subcutaneous fat.

**Urinary System**
The kidneys help compensate for water and electrolytes lost in sweat.

**Cardiovascular System**
Skin blood vessels play a role in regulating body temperature.

**Reproductive System**
Sensory receptors play an important role in sexual activity and in the suckling reflex.

UNIT TWO

The blood cell micrographs in chapter 14 are taken at similar magnifications for ease of comparison, and use truer colors, for authenticity. Chapter 16's topics are reorganized to improve flow, and the discussion of AIDS is completely updated. Chapter 19 presents a clarified explanation of tidal volume, and improved figures of microscopic and macroscopic lung anatomy. Chapter 21 includes a new figure on the distribution of body water, and a clearer explanation of acid-base balance.

Many changes and updates to the text come straight from the headlines, enabling students to see how the material in the classroom relates directly to real life. For example, discussion of the blood-brain barrier in Clinical Application 3.2 has new information on Gulf War syndrome. Chapter 3 also includes transcytosis in the discussion on membrane transport, which became better understood as researchers learned more about how HIV enters human cells. Chapter 10's vignette on Alzheimer's disease is expanded to cover the several genetic causes of this common illness, and Clinical Application 10.6 describes the biochemistry of addiction, including nicotine addiction. Chapter 14 discusses a new way to coat red blood cells to fashion a "universal" blood type, and chapter 17 mentions a new breath test to detect *Helicobacter pylori,* the bacterium that causes ulcers. Also in chapter 17, coverage of hepatitis includes newly discovered types of this illness. The coverage of obesity in chapter 18 discusses diet drugs, including those that have failed.

**New Tables**   Several tables have been added to review chapter concepts or to add detail. These include Sleep Disorders (11.6), Clotting Factors (14.11), Types of Cytokines (16.4), Transplant Types (16.9), and Five Leading Causes of Death (23.6).

## CONTENT, UPDATING, AND EMPHASIS CHANGES

Clarity is a major goal of a textbook, and we have made efforts to improve the readability and information content of certain difficult areas. Chapter 3 now features an updated discussion of membrane transport, and a clearer visual presentation of the pH scale. Cellular respiration, DNA synthesis and protein synthesis—by many accounts the most challenging subjects—are presented more clearly in chapter 4. The micrographs of tissues in chapter 5 are in sharper focus, and therefore easier to understand. Chapter 9 has a new, circular figure for the mechanism of muscle contraction, as well as a new accompanying discussion.

   In chapter 11, a new discussion and figure describe the crossover point for ascending and descending nerve tracts, and the figures in this chapter are more consistent.

## LEARNING TOOLS

A variety of tools make the study of human anatomy and physiology easier and more enjoyable. These aids help with mastering the basic concepts necessary to progress to more difficult material.

### Understanding Words

This feature, found at the chapter beginnings and on the endsheets, helps build vocabulary. "Understanding Words" includes root words, stems, prefixes, and suffixes that reveal word meanings and origins. Each root and an example word that uses the root are defined. Knowing the roots from these lists helps in remembering scientific word meanings and in understanding newly encountered terms.

### Chapter Objectives

Chapter objectives indicate what the reader should be able to do after mastering the information within the narrative. The review exercises at the end of each chapter are also phrased as detailed objectives, and these may be valuable to read before beginning a chapter. The chapter objectives and review exercises are guides to important sections of the narrative.

## Introductory Vignettes

Stories that open each chapter vividly introduce the topic. These vignettes are all interesting real-life events, often taken from the headlines or from reports in medical or scientific journals.

## Key Terms

Key terms and their phonetic pronunciations at the beginning of each chapter help build science vocabulary. The words in the lists are used in the chapter, and may be found in subsequent chapters. The glossary explains phonetic pronunciation.

## Review Questions Within the Narrative

Review questions at the ends of major sections within each chapter reinforce main points. The ability to answer these questions without checking back indicates mastery of the material.

## Illustrations, Photographs, and Tables

A textbook on human anatomy and physiology is more than a collection of written descriptions; it is also a visual experience. The many illustrations, photographs, and tables in each chapter are located near their related textual discussion. The art is designed and placed to help the reader visualize structures and processes, to clarify complex ideas, to represent how structures relate to each other, to summarize sections of the narrative, or to present pertinent data. Sets of special reference plates illustrate the structures and locations of the major internal organs of the body, depict the structural detail of the human skull, and help locate major features of the body surface and visualize organs exposed by the dissection of a cadaver.

Frequent use of icons in the art orient the reader and establish a sense of scale. Color is consistent from chapter to chapter; if a lymphatic pathway is green in one chapter, it is green elsewhere, also. Careful selection of new micrographs and rendering of new art better correlates the two.

## Boxed Information

Short paragraphs in colored boxes appear throughout each chapter. Several are new to this edition. Some of these short boxes apply chapter ideas to clinical situations. Others discuss changes in organ structure and function that occur with aging. Many small boxes introduce new medical technology or discuss interesting medical experiments.

## Clinical Applications

Throughout the chapters, longer boxed sections add flavor, depth, and perspective to major concepts by providing information on related pathology, offering historical insights, or exploring technological applications of knowledge of anatomy and physiology.

## InnerConnections

These multipurpose illustrations, found at the ends of selected chapters, conceptually link the highlighted body system to every other system, reinforcing the dynamic interactions of groups of organs. These graphic representations review chapter concepts, make connections, and stress the "big picture" in learning and applying the concepts and facts of anatomy and physiology.

## Clinical Terms

Lists of related terms often used in clinical situations end many chapters. These terms, along with their phonetic pronunciations and brief definitions, expand understanding of medical terminology.

## Critical Thinking Questions

These questions at the end of each chapter apply main concepts to clinical or research situations, taking the student beyond memorization to utilization of knowledge.

## Review Exercises

The review exercises at the end of each chapter check understanding of the major ideas in the narrative, in the sequence in which they are presented.

## Appendices, Glossary, and Index

The appendices contain a variety of useful information. They include the following:

A. Periodic Table of the Elements

B. Units of Measurement and Their Equivalents

C. Laboratory Tests of Clinical Importance

## Multimedia Correlations

This eighth edition introduces the *Dynamic Human, Version 2.0, The 3-D Visual Guide to Anatomy and Physiology CD-ROM,* which interactively illustrates the complex relationships between anatomical structures and their functions in the human body. This program covers each body system, demonstrating clinical concepts, histology, and physiology. The *Dynamic Human* (dancing man) icon 𝓍 appears in appropriate figure legends to alert the reader to the corresponding information. A list of correlating figures to specific sections of *The Dynamic Human, Version 2.0,* follows this preface on p. xxix.

A set of six videotapes contains nearly 100 animations of physiological processes integral to the study of human anatomy and physiology. Entitled *WCB's Life Science Animation (LSA) Videotape Series,* these videotapes cover such topics as cell division, genetics, and reproduction. A new LSA 3D Videotape with forty-two key biological processes is included in these correlations. A videotape icon ▭ appears in appropriate figure legends to alert the reader to these animations. A list of the figures that relate to the animations follows this preface on p. xxxiii.

## World Wide Web

*Hole's Human Anatomy & Physiology* is featured in the Applied Biology Resource Center at http://www.mhhe.com/biosci/abio/. Instructors will find links to hot topics relating to anatomy and physiology, teaching aids, product information, and a means to communicate with WCB/McGraw-Hill. Students will find additional study questions to help with test preparation. Visit our resource center regularly to get the latest updates.

## SUPPLEMENTARY MATERIALS

The following supplementary materials are designed to help the instructor plan class work and presentations and to aid students in learning.

1. *Laboratory Manual for Hole's Human Anatomy and Physiology* (0-697-34217-4) by Terry R. Martin is designed to accompany the eighth edition of *Hole's Human Anatomy and Physiology.* The lab manual has been thoroughly revised.

2. *Instructor's Manual and Test Item File* (0-697-34210-7) by Michael F. Peters, Baker College of Port Huron, contains lecture suggestions and guidelines, application questions, and critical thinking issues for each chapter objective. It contains additional questions and issues to initiate student discussion. The Instructor's Manual also contains test items for each chapter to evaluate student understanding of the subject matter.

    A *Visuals Testbank* is contained at the end of the Instructor's Manual. These transparency masters feature line art from the text with labels deleted for student quizzing or practice.

    The *Answer Key for Chapter Review Exercises*, prepared by Connie Vinton-Schoepske, is also included within the Instructor's Manual.

3. *Student Study Guide* (0-697-34206-9) by Nancy A. Sickles Corbett contains chapter overviews, chapter objectives, focus questions, mastery tests, study activities, and answer keys corresponding to the chapters of the text.

4. *A computerized test* generator is offered free upon request to qualified adopters of this textbook. A complete test item file is available on computer diskette for Windows (0-697-34215-8) and Macintosh (0-697-34216-6) computers.

5. *Transparencies* (0-697-34209-3) include a set of 300 acetate sheets that complement classroom lectures or can be used for short quizzes.

6. Visit *Hole's Human Anatomy and Physiology* at www.mhhe.com/biosci/abio/. The web site has additional quizzing questions, links to hot topics, and other aids.

7. *Instructor's Manual for the Laboratory Manual* (0-697-34211-5) by Terry R. Martin provides a chart to correlate the laboratory manual with chapters of the textbook. For each exercise there is a list of required materials, the approximate time for completion, topics for discussion, and answers to the questions in the lab report.

8. *Histology Color Slides* (0-697-28931-1) include a set of seventy micrographs of tissues, organs, and other body features described in the textbook.

9. *Visual Resource Library* (0-697-42203-8) is a CD-ROM containing all of the line art in this textbook. The CD-ROM contains an easy-to-use program to view images quickly. Images may be imported into PowerPoint to customize multimedia presentations.

10. *Intelitool Supplement to accompany the Laboratory Manual* (0-697-33368-X) by Terry R. Martin contains four Intelitool laboratory exercises on muscle physiology, reflex physiology, electrocardiography, and spirometry.

Other learning aids available from WCB/McGraw-Hill include:

11. *The Dynamic Human CD-ROM, Version 2.0* (0-697-38935-9) illustrates the important relationship between anatomical structures and their functions in the human body. Realistic computer visualization and three-dimensional visualization are the premier features of this learning tool.

12. *The Dynamic Human Videodisc* (0-697-37994-9) contains more than twenty-five animations, 130 histological micrographs, clinical footage, and line art from *Hole's Human Anatomy and Physiology.*

13. *WCB's Life Science Animations (LSA)* contains fifty-three animations on five VHS videocassettes; Chemistry, The Cell, and Energetics (0-697-25068-7); Cell Division, Heredity, Genetics, Reproduction, and Development (0-697-25069-5); Animal Biology No. 1 (0-697-25070-9); Animal Biology No. 2 (0-697-25071-7); and Plant Biology, Evolution, and Ecology (0-697-26600-1). Another available videotape is Physiological Concepts of Life Science (0-697-21512-1). A new 3D videotape (0-07-290652-9) is also available with forty-two key biological processes all narrated and animated in vibrant color with dynamic three-dimensional graphics.

14. *Anatomy and Physiology Videodisc* (0-697-27716-X) is a four-sided videodisc containing more than thirty animations of physiological processes, as well as line art and micrographs.

15. *Virtual Physiology Lab CD-ROM* (0-697-37994-9) has ten simulations of animal-based experiments common in the physiology component of a laboratory course; allows students to repeat experiments for improved mastery.

16. *Laboratory Atlas of Anatomy and Physiology,* second edition, (0-697-39480-8) by Douglas Eder et al. is a full-color atlas containing histology, human skeletal anatomy, human muscular anatomical dissections, and reference tables.

17. *Case Histories in Human Physiology,* third edition, by Donna Van Wynsberghe and Gregory Cooley is a web-based workbook that stimulates analytical thinking through case studies and problem solving; includes an instructor's answer key. (www.mhhe.com/biosci/ap/vanwyn/)

18. *The McGraw-Hill Learning Architecture* is a browser-based product that is a solution for delivering educational content over networked environments. The Learning Architecture connects students with each other as well as their instructor in an integrated environment. In addition to providing support and collaboration tools for users, such as built-in messaging and discussion lists, the Learning Architecture also manages all students on the server as well as the course material assigned to them. The benefits of this system to both instructors and students are tremendous (0–07–450946–2).

19. *WCB Anatomy and Physiology Videotape Series* consists of four videotapes, free to qualified adopters, including Blood Cell Counting, Identification and Grouping (0-697-11629-8); Introduction to the Human Cadaver and Prosection (0-697-11177-6); Introduction to Cat Dissection: Cat Musculature, (0-697-11630-1); and Internal Organs and Circulatory System of the Cat (0-697-13922-0).

20. *Human Anatomy and Physiology Study Cards,* third edition (0-697-26447-5), by Kent Van De Graaff, Ward Rhees, and Christopher Creek, is a boxed set of 300 illustrated cards ($3 \times 5$ in.), each of which concisely summarizes a concept of structure or function, defines a term, and provides a concise table of related information.

21. *Explorations in Human Biology CD-ROM* (0-697-37907-8 Macintosh and 0-697-37906-X Windows) by George Johnson consists of sixteen interactive animations of human biology. *Explorations in Cell*

*Biology and Genetics CD-ROM* (0-697-37908-6) by George Johnson contains seventeen animations that afford an engrossing way for students to delve into these often-challenging topics.

22. *Life Science Living Lexicon CD-ROM* (0-697-37993-0) by William Marchuk provides interactive vocabulary-building exercises. It includes the meanings of word roots, prefixes, and suffixes with illustrations and audio pronunciations.

23. *Survey of Infectious and Parasitic Diseases* (0-697-27535-3) by Kent M. Van De Graaff is a black-and-white booklet that presents the essential information on 100 of the most common and clinically significant diseases. A one-page presentation that includes pronunciation, derivation, definition, life cycle, description, signs and symptoms, laboratory diagnoses, and prevention/treatment is devoted to each of these diseases.

24. *Coloring Guide to Anatomy and Physiology* (0-697-17109-4) by Robert and Judith Stone emphasizes learning through the process of color association. The Coloring Guide provides a thorough review of anatomical and physiological concepts.

25. *Atlas of the Skeletal Muscles,* second edition (0-697-13790-8) by Robert and Judith Stone is a guide to the structure and function of human skeletal muscles. The illustrations help students locate muscles and understand their actions.

## ACKNOWLEDGMENTS

Any textbook is the result of hard work by a large team. Although we directed the revision, many "behind-the-scenes" people at WCB/McGraw-Hill were indispensable to the project. We would like to thank our editorial team of Michael Lange, Colin Wheatley, Kris Tibbetts, and Connie Haakinson; our production team, which included Jayne Klein, Jodi Banowetz, Wayne Harms, and John Leland; the copyeditor, Jane Matthews; and most of all, John Hole, for giving us the opportunity and freedom to continue his classic work. We also thank our wonderfully patient families for their support.

*David Shier*
*Jackie Butler*
*Ricki Lewis*

## REVIEWERS

We would like to acknowledge the valuable contributions of the reviewers for the eighth edition who read either portions or all of the manuscript as it was being prepared, and who provided detailed criticisms and ideas for improving the narrative and the illustrations. They include the following:

Janice Asel
*Mitchell Community College*
Beth M. Atkin
*Washington State Community College*
Gordon Atkins
*Andrews University*
Stephanie Sajdak Baiyasi
*Delta College*
Anna Bartosh
*Howard County Junior College*
William R. Belzer
*Clarion University of Pennsylvania—Venango Campus*
Edwin Bessler
*Franciscan University of Steubenville*
E. Beth Bonner
*Delgado Community College*
Ray D. Burkett
*Shelby State Community College*
Rebecca M. Burt
*Southeast Community College—Beatrice Campus*
Jennifer Carr Burtwistle
*Northeast Community College*
Michael S. Capp
*Carlow College*
Holly Carmichael
*Wilson Technical Community College*
Melvin C. Chambliss
*Michigan State University's Veterinary Technology Program*
William H. Chrouser
*Warner Southern College*
Lu Anne Clark
*Lansing Community College*
Barbara J. Cohen
*Delaware County Community College*
Mary Catharine Cox
*Wingate University*
Allen R. Crooker, Jr.
*Hartwick College*
Lin Doyle
*Northwest College*
Duane A. Dreyer
*Durham Technical Community College*
Peter I. Ekechukwu
*Horry-Georgetown Technical College*
Barbara F. Ensley
*Haywood Community College*
Gary Estep
*Lubbock Christian University*
Louis A. Giacinti
*Milwaukee Area Technical College*

William A. Gibson
*University of New Orleans*
Susan K. Gilmore
*University of Pittsburgh at Bradford*
*Jamestown Community College*
David E. Harris
*Lewiston-Auburn College,*
*University of Southern Maine*
George E. Heath
*University of Maryland Eastern Shore*
Drusilla Beal Jolly
*Forsyth Technical Community College*
Beverly W. Juett
*Midway College*
Kamal I. Kamal
*Valencia Community College, West Campus*
Gary M. Kiebzak
*Miller Orthopaedic Clinic*
*Charlotte, NC*
Glenn E. Kietzmann
*Wayne State College*
Alan Knowles
*Pensacola Christian College*
Kristin Krause
*Saint Thomas Aquinas College*
Gopal Krishna
*Moberly Area Community College*
Barbara Lax
*Community College of Allegheny County*
Nancy Longlet
*Concordia College*
Lisa Lupini
*Baker College of Flint*
Bradford D. Martin
*La Sierra University*
William J. Mathena
*Kaskaskia College*
Julie A. Medlin
*Northwestern Michigan College*
Jim Miller
*College of the Southwest*
Eli C. Minkoff
*Bates College*
Robert Moldenhauer
*Washtenaw Community College*
James (Jym) C. Moon
*Western Iowa Technical Community College*
David Mork
*Saint Cloud State University*
C. Aubrey Morris
*Pensacola Junior College*
Tony E. Morris
*Fairmont State College*
Steve C. Nunez
*Sauk Valley Community College*

Nicole J. Okazaki
*Southeastern Louisiana University*
Charles M. Page
*El Camino College*
Mark A. Paulissen
*McNeese State University*
Mary S. Rea
*Sage Junior College*
Donald Rodd
*University of Evansville*
Connie E. Rye
*Bevill State Community College*
David A. Sandmire
*University of New England*
Soma Sanyal
*Penn State–Altoona*
Marilyn Shopper
*Johnson County Community College*
Richard Sims
*Jones County Junior College*
Katherine Smalley
*Emporia State University*
Denise L. Smith
*Skidmore College*
Michael E. Smith
*Valdosta State University*
Paul M. Spannbauer
*Hudson Valley Community College*
Marian Spozio
*Jefferson Community College*
Sarah Anne Staples
*Andrew College*
John R. Steele
*Ivy Tech State College*
Dennis M. Sullivan
*Cedarville College*
P. Alleice Summers
*Dyersburg State Community College*
Patricia J. Thomas
*Delgado Community College*
William R. Tobin
*West Valley Central School*
Don Varnado
*Southern Ohio College—Northern Kentucky Campus*
Dianne L. Vermillion
*School of Nursing—University of Rochester*
Garry M. Wallace
*Northwest College*
Norma J. Weekly
*Wilkes Community College*
Christine A. Wilson
*Community College of Allegheny County—Boyce Campus*
Barbara Wineinger
*Vincennes University Jasper*
Clarence C. Wolfe
*Northern Virginia Community College Annandale Campus*

| Figure Number | Title | Figure Number | Title |
|---|---|---|---|
| 1.9 | Human Body/Explorations/Visible Human/Female/Thorax | 7.2 | Skeletal/Explorations/Cross Section of a Long Bone |
| | Human Body/Explorations/Visible Human/Male/Thorax | 7.3 | Skeletal/Histology/Compact Bone |
| 1.10 | Human Body/Explorations/Visible Human/Female/Abdomen | | Skeletal/Histology/Spongy Bone |
| | | 7.5 | Skeletal/Histology/Compact Bone |
| | Human Body/Explorations/Visible Human/Male/Abdomen | 7B | Skeletal/Clinical Applications/Fractured Femur |
| 1.12 | Muscular/Anatomy | 7.17 | Skeletal/Anatomy/Gross Anatomy |
| | Skeletal/Anatomy | 7.19 | Skeletal/Anatomy/3D Viewer: Cranial Anatomy |
| 1.13 | Nervous/Anatomy | | Skeletal/Anatomy/Gross Anatomy/Axial Skeleton/Skull |
| | Endocrine/Anatomy | 7.20 | Skeletal/Anatomy/Gross Anatomy/Axial Skeleton/Skull |
| 1.14 | Cardiovascular/Anatomy | 7.21 | Skeletal/Anatomy/3D Viewer: Cranial Anatomy |
| | Immune & Lymphatic/Anatomy | | Skeletal/Anatomy/Gross Anatomy/Axial Skeleton/Skull |
| 1.15 | Digestive/Anatomy | 7.22 | Skeletal/Anatomy/3D Viewer: Cranial Anatomy |
| | Respiratory/Anatomy | | Skeletal/Anatomy/Gross Anatomy/Axial Skeleton/Skull |
| | Urinary/Anatomy | 7.23 | Skeletal/Anatomy/Gross Anatomy/Axial Skeleton/Skull |
| 1.16 | Reproductive/Anatomy | 7.24 | Skeletal/Anatomy/Gross Anatomy/Axial Skeleton/Skull |
| 1.18 | Human Body/Explorations/Anatomical Orientation/Planes | 7.26 | Skeletal/Anatomy/Gross Anatomy/Axial Skeleton/Skull |
| 1.19 | Human Body/Explorations/Anatomical Orientation/Planes | 7.29 | Skeletal/Anatomy/Gross Anatomy/Axial Skeleton/Skull |
| 2D | Human Body/Clinical Applications/CT | 7.30 | Skeletal/Anatomy/Gross Anatomy/Axial Skeleton/Skull |
| 2E | Human Body/Clinical Applications/PET | 7.31 | Skeletal/Anatomy/Gross Anatomy/Axial Skeleton/Skull |
| 3.1 | Human Body/Anatomy/Cell Size | 7.32 | Human Body/Clinical Applications/X-Ray |
| 3.2 | Human Body/Anatomy/Cell Shape | 7.34 | Skeletal/Anatomy/Gross Anatomy/Axial Skeleton/Vertebral Column |
| 3.3 | Human Body/Anatomy/Cell Components | 7.35 | Skeletal/Anatomy/Gross Anatomy/Axial Skeleton/Vertebral Column |
| 3.7 | Human Body/Anatomy/Cell Components | | |
| 3.10 | Human Body/Anatomy/Cell Components | 7.38 | Skeletal/Anatomy/Gross Anatomy/Axial Skeleton/Vertebral Column |
| 3.11 | Human Body/Anatomy/Cell Components | 7.39 | Skeletal/Anatomy/Gross Anatomy/Axial Skeleton/Vertebral Column |
| 3.13 | Human Body/Anatomy/Cell Components | | |
| 3.14 | Human Body/Anatomy/Cell Components | 7.40 | Skeletal/Anatomy/3D Viewer: Thoracic Anatomy |
| 3.15 | Human Body/Anatomy/Cell Components | 7.41 | Skeletal/Anatomy/Gross Anatomy/Axial Skeleton/Thoracic Cage |
| 3.16 | Human Body/Anatomy/Cell Components | 7.42 | Skeletal/Anatomy/Gross Anatomy/Appendicular Skeleton/Pectoral Girdle |
| 3.17 | Human Body/Anatomy/Cell Components | | |
| 5.1 | Human Body/Histology/Simple Squamous Epithelium | 7.43 | Skeletal/Anatomy/Gross Anatomy/Appendicular Skeleton/Pectoral Girdle |
| 5.3 | Human Body/Histology/Simple Columnar Epithelium | 7.44 | Skeletal/Anatomy/Gross Anatomy/Appendicular Skeleton/Upper Limb |
| 5.5 | Human Body/Histology/Pseudostratified Ciliated Columnar Epithelium | 7.45 | Skeletal/Anatomy/Gross Anatomy/Appendicular Skeleton/Upper Limb |
| 5.6 | Human Body/Histology/Stratified Squamous Epithelium | 7.46 | Skeletal/Anatomy/Gross Anatomy/Appendicular Skeleton/Upper Limb |
| 5.9 | Human Body/Histology/Transitional Epithelium | 7.47 | Skeletal/Anatomy/Gross Anatomy/Appendicular Skeleton/Upper Limb |
| 5.23 | Human Body/Histology/Hyaline Cartilage | | |
| 5.24 | Human Body/Histology/Elastic Cartilage | 7.49 | Skeletal/Anatomy/Gross Anatomy/Appendicular Skeleton/Pelvic Girdle |
| 5.25 | Human Body/Histology/Fibrocartilage | | |
| 5.26 | Skeletal/Histology/Compact Bone | | |
| 5.28 | Muscular/Histology/Skeletal Muscle | | |
| 5.29 | Muscular/Histology/Smooth Muscle | | |
| 5.30 | Muscular/Histology/Cardiac Muscle | | |
| 5.31 | Nervous/Histology/Spinal Neuron | | |
| 7.1 | Skeletal/Explorations/Cross Section of a Long Bone | | |

| Figure Number | Title | Figure Number | Title |
|---|---|---|---|
| 7.50 | Skeletal/Anatomy/Gross Anatomy/Appendicular Skeleton/Pelvic Girdle | 9.32 | Muscular/Anatomy/Body Regions/Abdomen and Back |
| 7.52 | Skeletal/Anatomy/Gross Anatomy/Appendicular Skeleton/Lower Limb | 9.33 | Muscular/Anatomy/Body Regions/Pelvic Outlet Male and Female |
| 7.53 | Skeletal/Anatomy/Gross Anatomy/Appendicular Skeleton/Lower Limb | 9.34 | Muscular Anatomy/Body Regions/Thigh |
| 7.54 | Skeletal/Anatomy/Gross Anatomy/Appendicular Skeleton/Lower Limb | 9.35 | Muscular Anatomy/Body Regions/Thigh |
| 7.55 | Skeletal/Anatomy/Gross Anatomy/Appendicular Skeleton/Lower Limb | 9.36 | Muscular Anatomy/Body Regions/Thigh |
| 7.56 | Skeletal/Anatomy/Gross Anatomy/Appendicular Skeleton/Lower Limb | 9.37 | Human Body/Explorations/Visible Human/Female/Thigh |
| | | | Human Body/Explorations/Visible Human/Male/Thigh |
| 8.1 | Skeletal/Explorations/Fibrous Joints | 9.38 | Muscular/Anatomy/Body Regions/Leg |
| 8.2 | Skeletal/Explorations/Fibrous Joints | 9.39 | Muscular/Anatomy/Body Regions/Leg |
| 8.3 | Skeletal/Explorations/Fibrous Joints | 9.40 | Muscular/Anatomy/Body Regions/Leg |
| 8.4 | Skeletal/Explorations/Fibrous Joints | 9.41 | Human Body/Explorations/Visible Human/Male/Leg |
| 8.5 | Skeletal/Explorations/Cartilaginous Joints | 10.1 | Nervous/Histology/Spinal Neuron |
| 8.6 | Skeletal/Histology/Fibrocartilage | 11.2 | Nervous/Anatomy/Spinal Cord Anatomy |
| 8.7 | Skeletal/Explorations/Synovial Joints/Generic Synovial Joint | 11.3 | Nervous/Anatomy/Spinal Cord Anatomy |
| 8.9 | Skeletal/Explorations/Synovial Joints/Types of Synovial Joints | 11.4 | Nervous/Anatomy/Spinal Cord Anatomy |
| | | | Nervous/Histology/Spinal Cord Anatomy |
| 8.10 | Skeletal/Explorations/Synovial Joints/Synovial Joint Motion | 11.6 | Nervous/Explorations/Motor and Sensory Pathways |
| 8.11 | Skeletal/Explorations/Synovial Joints/Synovial Joint Motion | 11.7 | Nervous/Explorations/Motor and Sensory Pathways |
| 8.20 | Skeletal/Clinical Applications/MRI of the Knee | 11.8 | Nervous/Explorations/Reflex Arc |
| 8A | Skeletal/Clinical Applications/Joint Disorders (Arthritis) | 11.13 | Nervous/Anatomy/Gross Anatomy of the Brain |
| | | 11.14 | Nervous/Anatomy/Gross Anatomy of the Brain |
| 9.2 | Muscular/Anatomy/Skeletal Muscle | 11.23 | Nervous/Anatomy/Gross Anatomy of the Brain |
| 9.4 | Muscular/Anatomy/Skeletal Muscle | 12.6 | Nervous/Explorations/Olfaction |
| 9.7 | Muscular/Explorations/Neuromuscular Junction | 12.8 | Nervous/Explorations/Taste |
| 9.10 | Muscular/Explorations/Sliding Filament Theory | 12.9 | Nervous/Histology/Vallate Papillae |
| 9.17 | Muscular/Explorations/Isometric vs. Isotonic Contraction | 12.10 | Nervous/Explorations/Zones of Taste |
| | | 12.11 | Nervous/Explorations/Hearing |
| 9.18 | Muscular/Histology/Cardiac Muscle | 12.16 | Nervous/Histology/Organ of Corti |
| 9.19 | Muscular/Explorations/Muscle Action Around Joints | 12.20 | Nervous/Explorations/Static Equilibrium |
| | | 12.23 | Nervous/Explorations/Dynamic Equilibrium |
| 9.22 | Muscular/Anatomy/Body Regions/Head and Neck | 12.24 | Nervous/Explorations/Vision |
| 9.23 | Muscular/Anatomy/Body Regions/Abdomen and Back | 12.27 | Nervous/Histology/Eye |
| 9.24 | Muscular/Anatomy/Body Regions/Pectoral Girdle and Upper Arm | 12.35 | Nervous/Histology/Retina |
| 9.25 | Muscular/Anatomy/Body Regions/Abdomen and Back | 12A | Nervous/Clinical Applications/Nearsighted vs. Farsighted |
| | Muscular/Anatomy/Body Regions/Pectoral Girdle and Upper Arm | 12B | Nervous/Clinical Applications/Nearsighted vs. Farsighted |
| 9.26 | Muscular/Anatomy/Body Regions/Pectoral Girdle and Upper Arm | 12.43 | Nervous/Explorations/Vision |
| 9.27 | Muscular/Anatomy/Body Regions/Pectoral Girdle and Upper Arm | 13.2 | Endocrine/Anatomy/Gross Anatomy |
| | | 13.4 | Endocrine/Explorations/Endocrine Function |
| 9.28 | Muscular/Anatomy/Body Regions/Pectoral Girdle and Upper Arm | 13.6 | Endocrine/Explorations/Endocrine Function |
| | | 13.9 | Endocrine/Anatomy/Gross Anatomy/Hypothalamus and Pituitary Gland |
| 9.29 | Muscular/Anatomy/Body Regions/Forearm | 13.12 | Endocrine/Histology/Pituitary Gland |
| 9.30 | Muscular/Anatomy/Body Regions/Forearm | 13.13 | Endocrine/Explorations/Endocrine Function |

| Figure Number | Title | Figure Number | Title |
|---|---|---|---|
| 13.14 | Endocrine/Explorations/Hypothalamic-Pituitary-Thyroid Axis | 17.1 | Digestive/Anatomy/Gross Anatomy |
| | | | Digestive/Anatomy/3D Viewer: Digestive Anatomy |
| 13.16 | Endocrine/Anatomy/Gross Anatomy/Thyroid Gland | 17.3 | Digestive/Anatomy/Gross Anatomy |
| 13.17 | Endocrine/Histology/Thyroid Gland | 17.5 | Digestive/Explorations/Oral Cavity |
| 13.22 | Endocrine/Anatomy/Gross Anatomy/Thyroid Gland | 17.7 | Digestive/Anatomy/Gross Anatomy |
| | | 17.10 | Digestive/Histology/Tooth |
| 13.26 | Endocrine/Anatomy/Gross Anatomy/Adrenal Gland | 17.11 | Digestive/Anatomy/Gross Anatomy/Salivary Gland |
| 13.27 | Endocrine/Histology/Adrenal Cortex | 17.12 | Digestive/Histology/Submandibular Gland |
| | Endocrine/Histology/Adrenal Medulla | 17.15 | Digestive/Anatomy/Gross Anatomy/Esophagus |
| 13.32 | Endocrine/Anatomy/Gross Anatomy/Pancreas | 17.16 | Digestive/Histology/Esophagus |
| 14.15 | Immune & Lymphatic/Explorations/Non-Specific Immunity/Phagocytosis | 17.17 | Digestive/Anatomy/Gross Anatomy/Stomach |
| | | 17.20 | Digestive/Histology/Fundic Stomach |
| 14.20 | Cardiovascular/Explorations/Generic Vasculature/Medium-Sized Artery | 17.21 | Digestive/Explorations/Digestion |
| | | 17.24 | Digestive/Anatomy/Gross Anatomy/Pancreas |
| | Cardiovascular/Histology/Vasculature | 17.27 | Human Body/Explorations/Visible Human/Female/Abdomen |
| 15.3 | Cardiovascular/Anatomy/3D Viewer: Thoracic Anatomy | | Human Body/Explorations/Visible Human/Male/Abdomen |
| | Cardiovascular/Anatomy/Gross Anatomy of the Heart | 17.28 | Digestive/Anatomy/Gross Anatomy/Liver |
| 15.5 | Cardiovascular/Anatomy/Gross Anatomy of the Heart | 17.29 | Digestive/Histology/Liver |
| 15.6 | Cardiovascular/Anatomy/Gross Anatomy of the Heart | 17.31 | Digestive/Clinical Applications/Gallstones |
| | | 17.32 | Digestive/Explorations/Digestion |
| 15.10 | Cardiovascular/Explorations/Heart Dynamics/Blood Flow | 17.33 | Digestive/Anatomy/Gross Anatomy/Small Intestine |
| 15.12 | Cardiovascular/Anatomy/Gross Anatomy of the Heart | 17.38 | Digestive/Histology/Duodenal Villi |
| 15.17 | Cardiovascular/Explorations/Heart Dynamics/Cardiac Cycle | 17.40 | Digestive/Anatomy/Gross Anatomy/Small Intestine |
| 15.18 | Cardiovascular/Explorations/Heart Dynamics/Heart Sounds | 17.41 | Digestive/Explorations/Digestion |
| | | 17.42 | Digestive/Explorations/Digestion |
| 15.19 | Cardiovascular/Explorations/Heart Dynamics/Conduction Cycle | 17.43 | Digestive/Explorations/Digestion |
| | | 17.44 | Digestive/Explorations/Digestion |
| 15.23 | Cardiovascular/Explorations/Heart Dynamics/Electrocardiogram | 17.45 | Digestive/Anatomy/Gross Anatomy/Large Intestine |
| 15.25 | Cardiovascular/Explorations/Generic Vasculature | 17C | Digestive/Explorations/Barium Radiograph |
| | Cardiovascular/Histology/Vasculature | 19.1 | Respiratory/Anatomy/Gross Anatomy |
| 15.26 | Cardiovascular/Explorations/Generic Vasculature | 19.2 | Respiratory/Anatomy/Gross Anatomy |
| | Cardiovascular/Histology/Vasculature | 19.5 | Respiratory/Anatomy/Gross Anatomy |
| 15.29 | Cardiovascular/Explorations/Generic Vasculature | 19.6 | Respiratory/Anatomy/Gross Anatomy |
| | Cardiovascular/Histology/Vasculature | 19.8 | Respiratory/Anatomy/Gross Anatomy |
| 15.32 | Cardiovascular/Histology/Vasculature | 19.10 | Respiratory/Histology/Trachea |
| 15.33 | Cardiovascular/Explorations/Generic Vasculature | 19.12 | Respiratory/Anatomy/Gross Anatomy |
| 15.59 | Cardiovascular/Explorations/Generic Portal System | 19.15 | Respiratory/Histology/Alveoli |
| | | 19.16 | Respiratory/Explorations/Gas Exchange |
| 16.4 | Immune & Lymphatic/Anatomy/Gross Anatomy | 19.18 | Respiratory/Histology/Bronchiole |
| 16.8 | Immune & Lymphatic/Explorations/Lymph Formation and Movement | 19.19 | Respiratory/Anatomy/3D Viewer: Thoracic Anatomy |
| 16.9 | Immune & Lymphatic/Histology/Lymph Node | 19.22 | Respiratory/Explorations/Mechanics of Breathing |
| 16.12 | Immune & Lymphatic/Anatomy/Gross Anatomy | 19.23 | Respiratory/Explorations/Mechanics of Breathing |
| 16.13 | Immune & Lymphatic/Histology/Thymus | 19.24 | Respiratory/Explorations/Mechanics of Breathing |
| 16.14 | Immune & Lymphatic/Anatomy/Gross Anatomy | 19.26 | Respiratory/Clinical Applications/Spirometry |
| 16.15 | Immune & Lymphatic/Histology/Spleen | 19.34 | Respiratory/Explorations/Gas Exchange |
| | | 20.1 | Urinary/Anatomy/Gross Anatomy |

## DYNAMIC HUMAN, VERSION 2.0
## CORRELATION GUIDE 🏃

| Figure Number | Title | Figure Number | Title |
|---|---|---|---|
| 20.4 | Urinary/Anatomy/Kidney Anatomy | 22.15 | Reproductive/Anatomy/3D Viewer: Female Anatomy |
| 20.6 | Urinary/Anatomy/Kidney Anatomy | | Reproductive/Anatomy/Female Reproductive Anatomy |
| 20.9 | Urinary/Anatomy/Nephron Anatomy | | |
| 20.10 | Urinary/Histology/Renal Cortex | 22.17 | Reproductive/Explorations/Female: Oogenesis |
| | Urinary Histology/Renal Medulla | 22.18 | Reproductive/Histology/Ovary |
| 20.15 | Urinary/Explorations/Urine Formation | 22.19 | Reproductive/Histology/Ovarian Follicle |
| 20.29 | Urinary/Clinical Applications/Ultrasound of Bladder | 22.20 | Reproductive/Histology/Ovary |
| | | | Reproductive/Histology/Ovarian Follicle |
| 20.30 | Urinary/Histology/Bladder | 22.22 | Reproductive/Explorations/Female: Ovarian Cycle |
| 22.1 | Reproductive/Anatomy/3D Viewer: Male Anatomy | 22.24 | Reproductive/Histology/Uterine Tube |
| | Reproductive/Anatomy/Male Reproductive Anatomy | 22.29 | Reproductive/Explorations/Female: Ovarian Cycle |
| 22.4 | Reproductive/Histology/Testis | | Reproductive/Explorations/Female: Menstrual Cycle |
| 22.5 | Reproductive/Explorations/Male: Spermatogenesis | | |
| | Reproductive/Histology/Testis | 22.43 | Reproductive/Clinical Applications/Male: Vasectomy |
| 22.11 | Reproductive/Anatomy/Male Reproductive Anatomy | 22C | Reproductive/Clinical Applications/Female: Breast Cancer |

# LIFE SCIENCE ANIMATIONS
# CORRELATION GUIDE 📼

| Figure Number | Tape | Concept | Title |
|---|---|---|---|
| 2.3 | Tape 1 | Concept 1 | Formation of an Ionic Bond |
| 3.12 | Tape 1 | Concept 4 | Cellular Secretion |
|  | Tape 6 | Concept 4 | Cellular Secretion |
| 3.23 | Tape 6 | Concept 2 | Osmosis |
| 3.27 | Tape 6 | Concept 3 | Active Transport |
| 3.31 | Tape 1 | Concept 3 | Endocytosis |
| 3.32 | Tape 1 | Concept 4 | Cellular Secretion |
|  | Tape 6 | Concept 4 | Cellular Secretion |
| 3.35 | Tape 2 | Concept 12 | Mitosis |
| 4.5 | Tape 6 | Concept 1 | Lock and Key Model of Enzyme Action |
| 4.6 | Tape 1 | Concept 5 | Glycolysis |
| 4.7 | Tape 1 | Concept 11 | ATP as an Energy Carrier |
| 4.8 | Tape 1 | Concept 11 | ATP as an Energy Carrier |
| 4.10 | Tape 1 | Concept 5 | Glycolysis |
| 4.11 | Tape 1 | Concept 6 | Oxidative Respiration (including the Krebs Cycle) |
| 4.12 | Tape 1 | Concept 6 | Oxidative Respiration (including the Krebs Cycle) |
|  | Tape 1 | Concept 7 | The Electron Transport Chain and the Production of ATP |
|  | Tape 6 | Concept 5 | The Electron Transport Chain and Oxidative Phosphorylation |
| 4.24 | Tape 2 | Concept 16 | Transcription of a Gene |
| 4.25 | Tape 2 | Concept 17 | Protein Synthesis |
| 4.26 | Tape 2 | Concept 16 | Transcription of a Gene |
|  | Tape 2 | Concept 17 | Protein Synthesis |
| 4.27 | Tape 2 | Concept 17 | Protein Synthesis |
| 4.28 | Tape 2 | Concept 15 | DNA Replication |
| 9.2 | Tape 3 | Concept 29 | Levels of Muscle Structure |
| 9.9 | Tape 3 | Concept 31 | Regulation of Muscle Contraction |
| 9.10 | Tape 3 | Concept 30 | Sliding Filament Theory of Muscle Contraction |
| 10.4 | Tape 3 | Concept 22 | Formation of Myelin Sheath |
| 10.14 | Tape 6 | Concept 6 | Conduction of Nerve Impulses |
| 10.15 | Tape 6 | Concept 6 | Conduction of Nerve Impulses |
|  | Tape 3 | Concept 23 | Saltatory Nerve Conduction |
| 10.17 | Tape 6 | Concept 8 | Synaptic Transmission |
| 10.19 | Tape 3 | Concept 24 | Signal Integration |
| 11.8 | Tape 3 | Concept 25 | Reflex Arcs |
| 12.15 | Tape 3 | Concept 27 | Organ of Corti |
| 12.20 | Tape 3 | Concept 26 | Organ of Static Equilibrium |
| 13.4 | Tape 6 | Concept 10 | Action of Steroid Hormone on Target Cells |
| 13.6 | Tape 3 | Concept 28 | Peptide Hormone Action (cAMP) |
|  | Tape 6 | Concept 12 | Cyclic AMP Action |
| 14.21 | Tape 4 | Concept 40 | A, B, O Blood Types |
| 15.1 | Tape 4 | Concept 37 | Blood Circulation |
| 15.17 | Tape 4 | Concept 32 | The Cardiac Cycle and the Production of Sounds |
| 15.18 | Tape 4 | Concept 32 | The Cardiac Cycle and the Production of Sounds |
| 15.23 | Tape 4 | Concept 38 | Production of the Electrocardiogram |
| 16.18 | Tape 4 | Concept 41 | B-Cell Immune Response |
| 16.19 | Tape 4 | Concept 41 | B-Cell Immune Response |
| 16.20 | Tape 4 | Concept 42 | Structure and Function of Antibodies |
| 16.22 | Tape 4 | Concept 41 | B-Cell Immune Response |
| 17.4 | Tape 4 | Concept 33 | Peristalsis |
| 17.41 | Tape 4 | Concept 34 | Digestion of Carbohydrates |
| 17.42 | Tape 4 | Concept 35 | Digestion of Proteins |
| 17.43 | Tape 4 | Concept 36 | Digestion of Lipids |
| 22.5 | Tape 2 | Concept 19 | Spermatogenesis |
| 22.17 | Tape 2 | Concept 20 | Oogenesis |

| Figure Number | Tape | Title |
| --- | --- | --- |
| 2.2 | Module 1 | Atomic Structure and Covalent and Ionic Bonding |
| 2.3 | Module 1 | Atomic Structure and Covalent and Ionic Bonding |
| 2.4 | Module 1 | Atomic Structure and Covalent and Ionic Bonding |
| 3.12 | Module 3 | Cellular Secretion |
| 3.21 | Module 4 | Diffusion |
| 3.23 | Module 5 | Osmosis |
| 3.32 | Module 3 | Cellular Secretion |
| 3.36 | Module 10 | Mitosis |
| 4.5 | Module 7 | Enzyme Action |
| 4.12 | Module 9 | Electron Transport Chain |
| 4.19 | Module 13 | Structure of DNA |
| 4.22 | Module 13 | Structure of DNA |
| 4.24 | Module 18 | Transcription |
| 4.25 | Module 19 | Translation |
| 4.26 | Module 18 | Transcription |
|  | Module 19 | Translation |
| 4.27 | Module 19 | Translation |
| 4.29 | Module 16 | DNA Mutations |
| 9.4 | Module 40 | Muscle Contraction |
| 9.10 | Module 40 | Muscle Contraction |
| 10.14 | Module 39 | Action Potential |
| 13.4 | Module 41 | Hormone Action |
| 13.6 | Module 41 | Hormone Action |
| 16.18 | Module 35 | Clonal Selection |
| 16.19 | Module 34 | How T Lymphocytes Work |
| 16.21 | Module 33 | Complement System |
| 17.41 | Module 36 | Digestion Overview |
| 19.34 | Module 37 | Gas Exchange |
| 20.15 | Module 38 | Kidney Function |
| 20.20 | Module 38 | Kidney Function |
| 20.22 | Module 38 | Kidney Function |
| 22.5 | Module 10 | Mitosis |
|  | Module 11 | Meiosis |
| 22.17 | Module 11 | Meiosis |

# INTRODUCTION TO HUMAN ANATOMY AND PHYSIOLOGY

## CHAPTER 1

## CHAPTER OBJECTIVES

AFTER YOU HAVE STUDIED THIS CHAPTER, YOU SHOULD BE ABLE TO:

1. Define *anatomy* and *physiology*, and explain how they are related.

2. List and describe the major characteristics of life.

3. List and describe the major requirements of organisms.

4. Define *homeostasis* and explain its importance to survival.

5. Describe a homeostatic mechanism.

6. Explain the levels of organization of the human body.

7. Describe the locations of the major body cavities.

8. List the organs located in each major body cavity.

9. Name the membranes associated with the thoracic and abdominopelvic cavities.

10. Name the major organ systems and list the organs associated with each.

11. Describe the general functions of each organ system.

12. Properly use the terms that describe relative positions, body sections, and body regions.

## KEY TERMS

THE ACCENT MARKS USED IN THE PRONUNCIATION GUIDES ARE DERIVED FROM A SIMPLIFIED SYSTEM OF PHONETICS STANDARD IN MEDICAL USAGE. THE SINGLE ACCENT (′) DENOTES THE MAJOR STRESS, MEANING THAT EMPHASIS IS PLACED ON THE MOST HEAVILY PRONOUNCED SYLLABLE IN THE WORD. THE DOUBLE ACCENT (″) INDICATES SECONDARY STRESS. A SYLLABLE MARKED WITH A DOUBLE ACCENT RECEIVES LESS EMPHASIS THAN THE SYLLABLE THAT CARRIES THE MAIN STRESS, BUT MORE EMPHASIS THAN NEIGHBORING UNSTRESSED SYLLABLES.

**anatomy** (ah-nat′o-me)
**appendicular** (ap″en-dik′u-lar)
**axial** (ak′se-al)
**cardiovascular** (kahr″de-o-vas′ku-lur)
**digestion** (di-jest′yun)
**excretion** (ek-skre′shun)
**homeostasis** (ho″me-ō-sta′sis)
**metabolism** (mě-tab′o-lizm)
**negative feedback** (neg′ah-tiv fēd′bak)
**organelle** (or″gan-el′)
**organism** (or′gah-nizm)
**pericardial** (per″ĭ-kar′de-al)
**peritoneal** (per″ĭ-to-ne′al)
**physiology** (fiz″e-ol′o-je)
**pleural** (ploo′ral)
**reproduction** (re″pro-duk′shun)
**respiration** (res″pĭ-ra′shun)
**thoracic** (tho-ras′ik)
**visceral** (vis′er-al)

## UNDERSTANDING WORDS

**append-,** to hang something: *append*icular—pertaining to the upper limbs and lower limbs.

**cardi-,** heart: peri*cardi*um—membrane that surrounds the heart.

**cerebr-,** brain: *cerebr*um—largest portion of the brain.

**cran-,** helmet: *cran*ial—pertaining to the portion of the skull that surrounds the brain.

**dors-,** back: *dors*al—position toward the back of the body.

**homeo-,** same: *homeo*stasis—maintenance of a stable internal environment.

**-logy,** the study of: physio*logy*—study of body functions.

**meta-,** change: *meta*bolism—chemical changes that occur within the body.

**nas-,** nose: *nas*al—pertaining to the nose.

**orb-,** circle: *orb*ital—pertaining to the portion of skull that encircles an eye.

**pariet-,** wall: *pariet*al membrane—membrane that lines the wall of a cavity.

**pelv-,** basin: *pelv*ic cavity—basin-shaped cavity enclosed by the pelvic bones.

**peri-,** around: *peri*cardial membrane—membrane that surrounds the heart.

**pleur-,** rib: *pleur*al membrane—membrane that encloses the lungs within the rib cage.

**-stasis,** standing still: homeo*stasis*—maintenance of a stable internal environment.

**super-,** above: *super*ior—referring to a body part that is located above another.

**-tomy,** cutting: ana*tomy*—study of structure, which often involves cutting or removing body parts.

J udith R. had not been wearing a seat belt when the accident occurred because she had to drive only a short distance. She hadn't anticipated the intoxicated driver in the oncoming lane who swerved right in front of her. Thrown several feet, she now lay near her wrecked car as emergency medical technicians immobilized her neck and spine. Terrified, Judith tried to assess her condition. She didn't think she was bleeding, and nothing hurt terribly, but she felt a dull ache in the upper right part of her abdomen.

Minutes later, in the emergency room, a nurse gave Judith a quick exam, checking her blood pressure, pulse and breathing rate, and other vital signs and asking questions. These vital signs reflect underlying metabolic activities necessary for life, and they are important in any medical decision. Because Judith's vital signs were stable, and she was alert, knew who and where she was, and didn't seem to have any obvious life-threatening injuries, transfer to a trauma center was not necessary. However, Judith continued to report abdominal pain. The attending physician ordered abdominal X rays, knowing that about a third of patients with abdominal injuries show no outward sign of a problem. As part of standard procedure, Judith received oxygen and intravenous fluids, and a technician took several tubes of blood for testing.

A young physician approached and smiled at Judith as assistants snipped off her clothing. The doctor carefully looked and listened, and gently poked and probed. She was looking for cuts; red areas called hematomas where blood vessels had broken; and treadmarks on the skin. Had Judith been wearing her seat belt, the doctor would have checked for characteristic "seat belt contusions," crushed bones or burst hollow organs caused by the twisting constrictions that can occur at the moment of impact when a person wears a seat belt. Finally, the doctor measured the girth of Judith's abdomen. If her abdomen swelled later on, this could indicate a complication, such as infection or internal bleeding.

On the basis of a hematoma in Judith's upper right abdomen and the continued pain coming from this area, the emergency room physician ordered a computed tomography (CT) scan. The scan revealed a lacerated liver. Judith underwent emergency surgery to remove the small torn portion of this vital organ.

When Judith awoke from surgery, a different physician was scanning her chart, looking up frequently. The doctor was studying her medical history for any notation of a disorder that might impede healing. Judith's history of slow blood clotting, he noted, might slow her recovery from surgery. Next, the physician looked and listened. A bluish discoloration of Judith's side might indicate bleeding from her pancreas, kidney, small intestine, or aorta (the artery leading from the heart). A bluish hue near the navel would also be a bad sign, indicating bleeding from the liver or spleen. Her umbilical area was somewhat discolored.

The difference between life and death may depend on a health care professional's understanding of the human body.

The doctor gently tapped Judith's abdomen and carefully listened to sounds from her digestive tract. A drumlike resonance could mean that a hollow organ had burst, whereas a dull sound might indicate internal bleeding. Judith's abdomen produced dull sounds throughout. Plus, her abdomen had swollen, the pain intensifying when the doctor gently pushed on the area. With Judith's heart rate increasing and blood pressure falling, bleeding from the damaged liver was a definite possibility.

Blood tests confirmed the doctor's suspicions. Because blood is a complex mixture of biochemicals, it serves as a barometer of health. Injury or illness disrupts the body's maintenance of specific levels of various biochemicals. This maintenance is called homeostasis. Judith's blood tests revealed that her body had not yet recovered from the accident. Levels of clotting factors produced by her liver were falling, and blood was oozing from her incision, a sign of impaired clotting. Judith's blood glucose level remained elevated, as it had been in the emergency room. Her body was still reacting to the injury.

Based on Judith's blood tests, heart rate, blood pressure, reports of pain, and the physical exam, the doctor sent her back to the operating room. Sure enough, the part of her liver where the injured portion had been removed was still bleeding. When the doctors placed packing material at the wound site, the oozing gradually stopped. Judith returned to the recovery room and, as her condition stabilized, to her room. This time, all went well, and a few days later she was able to go home. The next time she drove, Judith wore her seat belt!

Imagine yourself as one of the health care professionals who helped identify Judith R.'s injury and get her on the road back to health. How would you know what to look, listen, and feel for? How would you place the signs and symptoms into a bigger picture that would suggest the appropriate diagnosis? Nurses, doctors, technicians, and other integral members of health care teams must have a working knowledge of the many intricacies of the human body. How can they begin to understand its astounding complexity? The study of human anatomy and physiology is a daunting, but fascinating and ultimately life-saving, challenge.

Our understanding of the human body has a long and interesting history (fig. 1.1). It began with our earliest ancestors, who must have been as curious about how their bodies worked as we are today. At first their interests most likely concerned injuries and illnesses, because healthy bodies demand little attention from their owners. Although they did not have emergency rooms to turn to, primitive people certainly suffered from occasional aches and pains, injured themselves, bled, broke bones, developed diseases, and contracted infections. At first, healers relied heavily on superstitions and notions about magic. However, as they tried to help the sick, these early medical workers began to discover useful ways of examining and treating the human body. They observed the effects of injuries, noticed how wounds healed, and examined dead bodies to determine the causes of death. They also found that certain herbs and potions could sometimes be used to treat coughs, headaches, and other common problems. These long-ago physicians began to wonder how these substances, the forerunners of modern drugs, affected body functions in general.

People began asking more questions and seeking answers, setting the stage for the development of modern medical science. Techniques for making accurate observations and performing careful experiments evolved and knowledge of the human body expanded rapidly.

This new knowledge of the structure and function of the human body required a new, specialized language. Early medical providers devised many terms to name body parts, describe their locations, and explain their functions. These terms, most of which originated from Greek and Latin, formed the basis for the language of anatomy and physiology. (A list of some of the modern medical and applied sciences appears on page 24.)

1 What factors probably stimulated an early interest in the human body?

2 What kinds of activities helped promote the development of modern medical science?

**FIGURE 1.1**

The study of the human body has a long history, as this illustration from the second book of *De Humani Corporis Fabrica* by Andreas Vesalius, issued in 1543, indicates. Note the similarity to the anatomical position (described later in this chapter).

## ANATOMY AND PHYSIOLOGY

As you read this book, you will begin to learn how the human body maintains life by studying two major areas of medical science, **anatomy** and **physiology.** Anatomy deals with the **structures** (morphology) of body parts—what are their forms, and how are they arranged? Physiology considers the **functions** of these body parts—what do they do, and how do they do it? Although anatomists tend to rely more on examination of the body and physiologists more on experimentation, together their efforts have provided us with a solid foundation upon which to build an understanding of how our bodies work as living organisms.

It is difficult to separate the topics of anatomy and physiology because anatomical structures make possible their functions. Parts form a well-organized unit—the **human organism**—and each part plays a role in the operation of the unit as a whole. This functional role depends upon the way the part is constructed. For example, the arrangement of parts in the human hand with its long, jointed fingers makes grasping possible. The heart's powerful muscular walls are structured to contract and propel

**FIGURE 1.2**
The structures of body parts make possible their functions: (*a*) the hand is adapted for grasping, (*b*) the heart for pumping blood, and (*c*) the mouth for receiving food.

blood out of the chambers and into blood vessels, and valves associated with these vessels and chambers ensure that the blood will move in the proper direction. The shape of the mouth enables it to receive food; teeth are shaped so that they break solid foods into smaller pieces; and the muscular tongue and cheeks are constructed to help mix food particles with saliva and prepare them for swallowing (fig. 1.2).

Anatomy and physiology are ongoing as well as ancient fields. Research frequently expands our understanding of physiology, particularly at the molecular level, and unusual, new anatomical findings are also reported. Recently, researchers discovered a previously unknown muscle between two bones in the head, providing physiologists with a new opportunity to understand body function.

1. What are the differences between an anatomist and a physiologist?

2. Why is it difficult to separate the topics of anatomy and physiology?

3. List several examples that illustrate how the structure of a body part makes possible its function.

4. How are anatomy and physiology both old and new fields?

## CHARACTERISTICS OF LIFE

A scene such as Judith R.'s accident and injury underscores the delicate balance that must be maintained in order to sustain life. In those seconds at the limits of life—the birth of a baby, a trauma scene, or the precise instant of death following a long illness—we often think about just what combination of qualities constitutes this state that we call life. Indeed, although this text addresses the human body, the most fundamental characteristics of

life are shared by all **organisms.** As living organisms, we can move and respond to our surroundings. We start out as small individuals and then grow, eventually to possibly reproduce. We gain energy by taking in or ingesting food, by breaking it down or digesting it, and by absorbing and assimilating it. The absorbed substances circulate throughout the internal environment of our bodies. We can then use them to grow and repair, or extract energy by the process of respiration. Finally, we excrete wastes from the body. Taken together, these physical and chemical events or reactions constitute **metabolism.** Table 1.1 summarizes these characteristics of life.

At the accident scene and throughout Judith R.'s hospitalization, health care workers repeatedly monitored her **vital signs**—observable body functions that reflect metabolic activities essential for life. Vital signs indicate that a person is alive. Assessment of vital signs includes measuring body temperature and blood pressure and monitoring rates and types of pulse and breathing movements. Absence of vital signs signifies death. A person who has died displays no spontaneous muscular movements (including those of the breathing muscles and beating heart), does not respond to stimuli (even the most painful that can be ethically applied), exhibits no reflexes (such as the knee-jerk reflex and pupillary reflexes of the eye), and generates no brain waves (demonstrated by a flat electroencephalogram, which reflects a lack of brain activity).

1. What are the characteristics of life?

2. What physical and chemical events constitute metabolism?

## TABLE 1.1 CHARACTERISTICS OF LIFE

| Process | Examples | Process | Examples |
|---|---|---|---|
| Movement | Change in position of the body or of a body part; motion of an internal organ | Digestion | Breakdown of food substances into simpler forms that can be absorbed and used |
| Responsiveness | Reaction to a change taking place inside or outside the body | Absorption | Passage of substances through membranes and into body fluids |
| Growth | Increase in body size without change in shape | Circulation | Movement of substances from place to place in body fluids |
| Reproduction | Production of new organisms and new cells | Assimilation | Changing of absorbed substances into chemically different forms |
| Respiration | Obtaining oxygen, using oxygen in releasing energy from foods, and removing carbon dioxide | Excretion | Removal of wastes produced by metabolic reactions |

## MAINTENANCE OF LIFE

With the exception of an organism's reproductive system, which perpetuates the species, all body structures and functions work to maintain life.

### Requirements of Organisms

Life depends upon the following environmental factors:

1. **Water** is the most abundant substance in the body. It is required for a variety of metabolic processes, and it provides the environment in which most of them take place. Water also transports substances within organisms and is important in regulating body temperature.

2. **Food** refers to substances that provide organisms with necessary chemicals (nutrients) in addition to water. Nutrients supply energy and raw materials for building new living matter.

3. **Oxygen** is a gas that makes up about one-fifth of the air. It is used in the process of releasing energy from nutrients. The energy, in turn, is used to drive metabolic processes.

4. **Heat** is a form of energy. It is a product of metabolic reactions, and it partly controls the rate at which these reactions occur. Generally, the more heat, the more rapidly chemical reactions take place. *Temperature* is a measure of the amount of heat present.

5. **Pressure** is an application of force on an object or substance. For example, the force acting on the outside of a land organism due to the weight of air above it is called *atmospheric pressure.* In humans, this pressure plays an important role in breathing. Similarly, organisms living under water are subjected to *hydrostatic pressure*—a pressure exerted by a liquid—due to the weight of water above them. In complex organisms, such as humans, heart action produces blood pressure (another form of hydrostatic pressure), which keeps blood flowing through blood vessels.

Although organisms require water, food, oxygen, heat, and pressure, these factors alone are not enough to ensure survival. Both the quantities and the qualities of such factors are also important. Table 1.2 summarizes the major requirements of organisms.

### Homeostasis

Some organisms exist as single **cells,** the smallest living units. Consider the amoeba, the simple, one-celled organism found in lakes and ponds (fig. 1.3). Despite its simple structure compared to a human, an amoeba has very specific requirements that must be met if it is to survive. As long as the outside world—its **environment**—supports its requirements, it flourishes. As environmental factors such as temperature, water composition, and food availability become unsatisfactory, its survival may be threatened. Although the amoeba has a limited ability to move from one place to another, environmental changes are likely to affect the whole pond, and with no place else to go, the amoeba dies.

In contrast to the amoeba, we humans are composed of about 70 trillion cells that form their own environment inside our bodies. This **internal environment** protects our cells (and us!) from changes in the outside world that would kill isolated cells such as the amoeba. Our cells interact to keep this internal environment relatively constant, despite an ever-changing outside environment. This maintenance of a stable internal environment is

TABLE 1.2

## REQUIREMENTS OF ORGANISMS

| Factor | Characteristic | Use | Factor | Characteristic | Use |
|--------|----------------|-----|--------|----------------|-----|
| Water | A chemical substance | For metabolic processes, as a medium for metabolic reactions, to transport substances, and to regulate body temperature | Oxygen | A chemical substance | To help release energy from food substances |
| | | | Heat | A form of energy | To help regulate the rates of metabolic reactions |
| Food | Various chemical substances | To supply energy and raw materials for the production of necessary substances and for the regulation of vital reactions | Pressure | A force | Atmospheric pressure for breathing; hydrostatic pressure to help circulate blood |

**FIGURE 1.3**

The amoeba is an organism consisting of a single cell (50× micrograph enlarged to 100×).

called **homeostasis,** and it is so important that most of our metabolic energy is spent on it. Many of the tests performed on Judith R. during her hospitalization assessed her body's return to homeostasis.

To better understand this idea of maintaining a stable internal environment, imagine a room equipped with a furnace and an air conditioner. Suppose the room temperature is to remain near 20°C (68°F), so the thermostat is adjusted to a **set point** of 20°C. Because a thermostat is sensitive to temperature changes, it will signal the furnace to start and the air conditioner to stop whenever the room temperature drops below the set point. If the temperature rises above the set point, the thermostat will cause the furnace to stop and the air conditioner to start.

As a result, a relatively constant temperature will be maintained in the room (fig. 1.4).

A similar **homeostatic mechanism** regulates body temperature in humans (fig. 1.5). The "thermostat" is a temperature-sensitive region in a control center of the brain called the hypothalamus. In healthy persons, the set point of this body thermostat is at or near 37°C (98.6°F).

If a person is exposed to a cold environment and the body temperature begins to drop, the hypothalamus senses this change and triggers heat-conserving and heat-generating activities. For example, blood vessels in the skin constrict so that blood flow there is reduced and heat is retained in deeper tissues. At the same time, small groups of muscle cells may be stimulated to contract involuntarily, an action called shivering. Such muscular contractions produce heat, which helps warm the body.

If a person becomes overheated, the hypothalamus triggers a series of changes that promote loss of body heat. For example, sweat glands in the skin secrete watery perspiration. As the water evaporates from the surface, heat is carried away and the skin is cooled. At the same time, blood vessels in the skin dilate. This allows the blood that carries heat from deeper tissues to reach the surface where more heat is lost to the outside. The hypothalamus also stimulates increased heart rate, causing a greater volume of blood to move into the surface vessels, and it stimulates an increase in breathing rate, allowing more heat-carrying air to be expelled from the lungs. Body temperature regulation is discussed in more detail in chapter 6.

Another homeostatic mechanism regulates the blood pressure in the blood vessels (arteries) leading away from the heart. In this instance, pressure-sensitive areas (sensory receptors) within the walls of these vessels

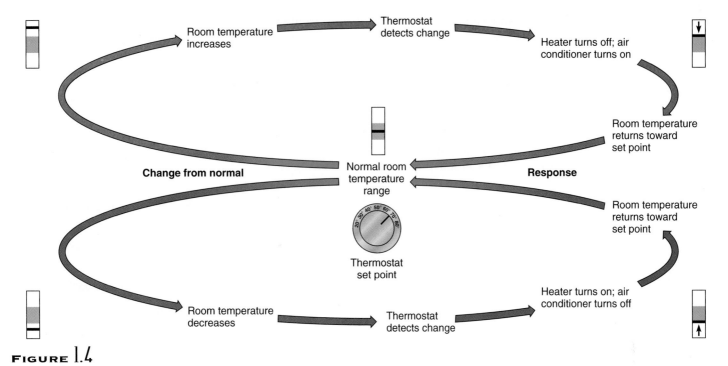

**FIGURE** 1.4

A thermostat that can signal an air conditioner and a furnace to turn on or off maintains a relatively stable room temperature. This system is an example of a homeostatic mechanism.

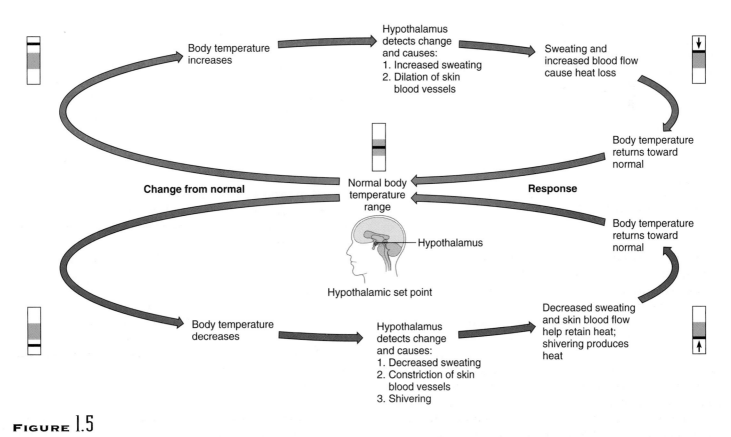

**FIGURE** 1.5

The homeostatic mechanism that regulates body temperature is another example of homeostasis.

sense changes in blood pressure and signal a pressure control center in the brain. If the blood pressure is above the pressure set point, the brain signals the heart, causing its chambers to contract less rapidly and with less force. Because of decreased heart action, less blood enters the blood vessels, and the pressure inside the vessels decreases. If the blood pressure is dropping below the set point, the brain center signals the heart to contract more rapidly and with greater force so that the pressure in the vessels increases. Chapter 15 discusses blood pressure regulation in more detail.

A homeostatic mechanism also regulates the concentration of the sugar glucose in blood. In this case, cells within an organ called the pancreas determine the set point. If, for example, the concentration of blood glucose increases following a meal, the pancreas detects this change and releases a chemical (insulin) into the blood. Insulin allows glucose to move from the blood into various body cells and to be stored in the liver and muscles. As this occurs, the concentration of blood glucose decreases, and as it reaches the normal set point, the pancreas decreases its release of insulin. If, on the other hand, the blood glucose concentration becomes abnormally low, the pancreas detects this change and releases a different chemical (glucagon) that causes stored glucose to be released into the blood. Chapter 13 discusses regulation of the blood glucose concentration in more detail.

In each of these examples, homeostasis is maintained by a self-regulating control mechanism that can receive signals (or feedback) about changes away from the normal set point and can cause reactions that return conditions to normal. Since the changes away from the normal state stimulate responses in the opposite direction, the responses are called *negative,* and the homeostatic control mechanism is said to act by a process of **negative feedback** (see fig. 4.15 and chapter 13).

---

Sometimes changes occur that stimulate still other similar changes. Such a process that causes movement away from the normal state is called a *positive feedback mechanism.*

Although most feedback mechanisms in the body are negative, a positive system operates for a short time when a blood clot forms, because the chemicals present in a clot promote still more clotting (see chapter 14). Another illustration of positive feedback is milk production. If a baby suckles with greater force or duration, the mother's mammary glands respond by making more and more milk. These examples are unusual. Because positive feedback mechanisms usually produce unstable conditions, most examples are associated with diseases and may lead to death.

---

Homeostatic mechanisms maintain a relatively constant internal environment, yet physiological values may vary slightly in a person from time to time or from one person to the next. Therefore, both normal values for an individual and the idea of a **normal range** for the general population are clinically important. The normal range icons in figures 1.4 and 1.5 are intended to reinforce this concept. Numerous examples of homeostasis are presented throughout this book, and normal ranges for a number of physiological variables are listed in Appendix C.

1. What requirements of organisms are provided from the external environment?

2. What is the relationship between oxygen use and heat production?

3. Why is homeostasis so important to survival?

4. Describe three homeostatic mechanisms.

---

## LEVELS OF ORGANIZATION

Early investigators, limited in their ability to observe small parts, focused their attention on larger body structures. Studies of small parts had to await invention of magnifying lenses and microscopes, which came into use about 400 years ago. These tools revealed that larger body structures were made up of smaller parts, which, in turn, were composed of even smaller ones.

Today, scientists recognize that all materials, including those that comprise the human body, are composed of chemicals. Chemicals consist of tiny, invisible particles called **atoms,** which are commonly bound together to form larger particles called **molecules;** small molecules may combine to form larger molecules called **macromolecules.**

Within the human organism, the basic unit of structure and function is a cell. Although individual cells vary in size and shape, all share certain characteristics. Human cells contain structures called **organelles** that carry on specific activities. These organelles are composed of aggregates of large molecules, including proteins, carbohydrates, lipids, and nucleic acids. All cells in a human contain a complete set of genetic instructions, yet use only a subset of them, allowing cells to develop specialized functions. All cells share the same characteristics of life, and must meet requirements to continue living.

Cells are organized into layers or masses that have common functions. Such a group of cells forms a **tissue.** Groups of different tissues form **organs**—complex structures with specialized functions—and groups of organs that function closely together comprise **organ systems.** Organ systems make up an **organism.**

A body part can be described at different levels. The heart, for example, contains muscle, fat, and nervous tissue. These tissues, in turn, are constructed of cells, which contain organelles. All of the structures of life are,

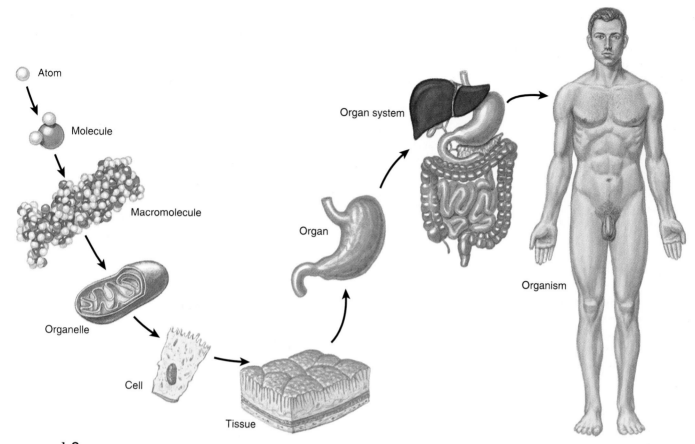

**FIGURE 1.6**

A human body is composed of parts within parts, which increase in complexity from the level of the atom to the whole organism.

| TABLE 1.3 | LEVELS OF ORGANIZATION | |
|-----------|-----------|-----------|
| **Level** | **Example** | **Illustration** |
| Atom | Hydrogen atom, lithium atom | Figure 2.1 |
| Molecule | Water molecule, glucose molecule | Figure 2.9 |
| Macromolecule | Protein molecule, DNA molecule | Figure 2.18 |
| Organelle | Mitochondrion, Golgi apparatus, nucleus | Figure 3.12 |
| Cell | Muscle cell, nerve cell | Figure 3.2 |
| Tissue | Simple squamous epithelium, loose connective tissue | Figure 5.1 |
| Organ | Skin, femur, heart, kidney | Figure 7.2 |
| Organ system | Integumentary system, skeletal system, digestive system | Figure 7.17 |
| Organism | Human | Figure 23.25 |

ultimately, composed of chemicals (fig. 1.6). Clinical Application 1.1 describes two technologies used to visualize differences among tissues.

Chapters 2–6 discuss these levels of organization in more detail. Chapter 2 describes the atomic and molecular levels; chapter 3 deals with organelles and cellular structures and functions; chapter 4 explores cellular metabolism; chapter 5 describes tissues; and chapter 6 presents membranes as examples of organs, and the skin and its accessory organs as an example of an organ system. Beginning with chapter 7, the structures and functions of each of the organ systems are described in detail. Table 1.3 lists the levels of organization and some corresponding illustrations in this textbook.

**1** How does the human body illustrate levels of organization?

**2** What is an organism?

**3** How do body parts that occupy different levels of organization vary in complexity?

# Ultrasonography and Magnetic Resonance Imaging: A Tale of Two Patients

The two patients enter the hospital medical scanning unit hoping for opposite outcomes. Vanessa Q., who has suffered several early pregnancy losses, hopes that an ultrasound exam will reveal a living embryo in her still-flat abdomen. Michael P., a 16-year-old who has been suffering from terrible headaches, is to undergo a magnetic resonance imaging (MRI) scan to assure his physician (and himself!) that the cause of the headache is not a brain tumor.

Both ultrasound and magnetic resonance imaging scans are noninvasive procedures that provide images of soft internal structures. Ultrasonography uses high-frequency sound waves that are beyond the range of human hearing. A technician gently presses a device called a transducer, which emits sound waves, against the skin and moves it slowly over the surface of the area being examined, which in this case is Vanessa's abdomen (fig. 1A).

Prior to the exam, Vanessa drank several glasses of water. Her filled bladder will intensify the contrast between her uterus (and its content) and nearby organs because as the sound waves from the transducer travel into the body, some of the waves reflect back to the transducer when they reach a border between structures of slightly different densities. Other sound waves continue into deeper tissues, and some of them are re-flected back by still other interfaces. As the reflected sound waves reach the transducer, they are converted into electrical impulses that are amplified and used to create a sectional image of the body's internal structure on a viewing screen. This image is known as a sonogram (fig. 1B).

Glancing at the screen, Vanessa yelps in joy. The image looks only like a fuzzy lima bean with a pulsating blip in the middle, but she knows it is the image of an embryo—and its heart is beating!

Vanessa's ultrasound exam takes only a few minutes, whereas Michael's MRI scan takes an hour. First he receives an injection of a dye that provides contrast so that a

**FIGURE 1A**

Ultrasonography uses reflected sound waves to visualize internal body structures.

## ORGANIZATION OF THE HUMAN BODY

The human organism is a complex structure composed of many parts. The major features of the human body include various cavities, various types of membranes, and a group of organ systems.

## Body Cavities

The human organism can be divided into an **axial portion,** which includes the head, neck, and trunk, and an **appendicular portion,** which includes the upper and lower limbs. Within the axial portion are two major cavities—a **dorsal cavity** and a larger **ventral cavity.** The organs within such a cavity are called **viscera.** The dorsal

**FIGURE** 1B
This image resulting from an ultrasonographic procedure reveals the presence of a fetus in the uterus.

**FIGURE** 1C
Falsely colored MRI image of a human head and brain (sagittal section).

radiologist examining the scan can distinguish certain brain structures. Then, a nurse wheels the narrow bed on which Michael lies into a chamber surrounded by a powerful magnet and a special radio antenna. The chamber, which looks like a metal doughnut, is the MRI instrument. As Michael settles back and closes his eyes, a technician activates the device.

The magnet generates a magnetic field that alters the alignment and spin of certain types of atoms within Michael's brain. At the same time, a second rotating magnetic field causes particular types of

atoms (such as the hydrogen atoms in body fluids and organic compounds) to release weak radio waves with characteristic frequencies. The nearby antenna receives and amplifies the radio waves, which are then processed by a computer. Within a few minutes, the computer generates a sectional image based on the locations and concentrations of the atoms being studied (fig. 1C). The device continues to produce

data, painting portraits of Michael's brain in the transverse, coronal, and sagittal sections.

Michael and his parents nervously wait two days for the expert eyes of a radiologist to interpret the MRI scan. Happily, the scan shows normal brain structure. Whatever is causing Michael's headaches, it is not a possibly life-threatening brain tumor. ■

cavity can be subdivided into two parts—the **cranial cavity,** which houses the brain, and the **vertebral canal** (spinal cavity), which contains the spinal cord and is surrounded by sections of the backbone (vertebrae). The ventral cavity consists of a **thoracic cavity** and an **abdominopelvic cavity.** Figure 1.7 shows these major body cavities.

The thoracic cavity is separated from the lower abdominopelvic cavity by a broad, thin muscle called the **diaphragm.** When it is at rest, this muscle curves upward into the thorax like a dome. When it contracts during inhalation, it presses down upon the abdominal viscera. The wall of the thoracic cavity is composed of skin, skeletal muscles, and bones. Within the thoracic

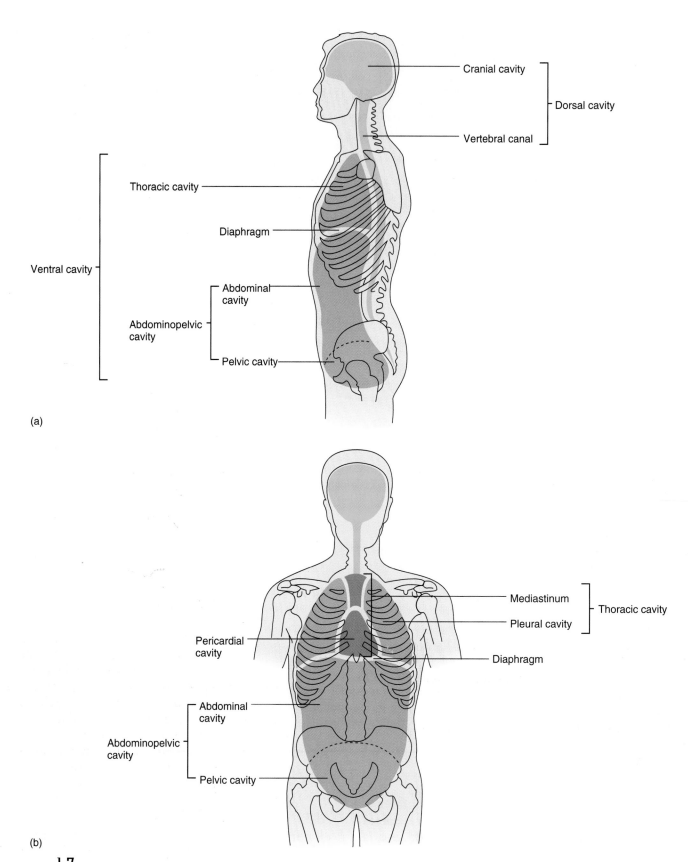

Cranial cavity

Dorsal cavity

Vertebral canal

Thoracic cavity

Diaphragm

Ventral cavity

Abdominal cavity

Abdominopelvic cavity

Pelvic cavity

(a)

Mediastinum

Thoracic cavity

Pleural cavity

Pericardial cavity

Diaphragm

Abdominal cavity

Abdominopelvic cavity

Pelvic cavity

(b)

**FIGURE 1.7**

Major body cavities. (*a*) Lateral view. (*b*) Coronal view.

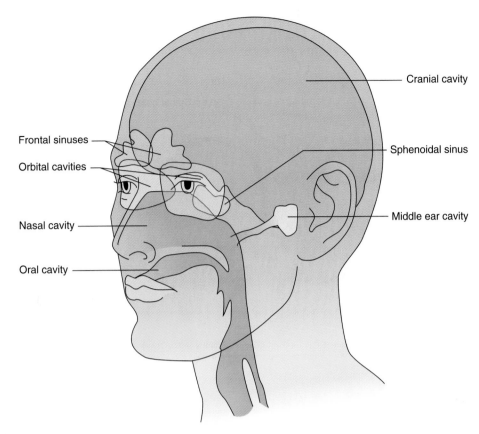

**FIGURE** 1.8
The cavities within the head include the cranial, oral, nasal, orbital, and middle ear cavities, as well as several sinuses.

cavity are the lungs and a region between the lungs, called the **mediastinum.** The mediastinum separates the thorax into two compartments that contain the right and left lungs. The remaining thoracic viscera—heart, esophagus, trachea, and thymus gland—are within the mediastinum.

The abdominopelvic cavity, which includes an upper abdominal portion and a lower pelvic portion, extends from the diaphragm to the floor of the pelvis. Its wall primarily consists of skin, skeletal muscles, and bones. The viscera within the **abdominal cavity** include the stomach, liver, spleen, gallbladder, and the small and large intestines.

The **pelvic cavity** is the portion of the abdominopelvic cavity enclosed by the pelvic bones. It contains the terminal end of the large intestine, the urinary bladder, and the internal reproductive organs.

Smaller cavities within the head include the following (fig. 1.8):

1. *Oral cavity,* containing the teeth and tongue.

2. *Nasal cavity,* located within the nose and divided into right and left portions by a nasal septum. Several air-filled sinuses are connected to the nasal cavity. These include the sphenoidal and frontal sinuses (see chapter 7).

3. *Orbital cavities,* containing the eyes and associated skeletal muscles and nerves.

4. *Middle ear cavities,* containing the middle ear bones.

## Thoracic and Abdominopelvic Membranes

The walls of the right and left thoracic compartments, which contain the lungs, are lined with a membrane called the *parietal pleura.* The lungs themselves are covered by a similar membrane called the *visceral pleura.* (*Note: Parietal* refers to the membrane attached to the wall of a cavity, and *visceral* refers to one that is associated with an internal organ, such as a lung.)

The parietal and visceral **pleural membranes** are separated by a thin film of watery fluid (serous fluid) that they secrete. Although there is normally no actual space between these membranes, the potential space between them is called the *pleural cavity.*

The heart, which is located in the broadest portion of the mediastinum, is surrounded by **pericardial membranes.** A thin *visceral pericardium* (epicardium) covers the heart's surface and is separated from the *parietal pericardium* by a small amount of serous fluid. The potential space between these membranes is called the *pericardial cavity.* The parietal pericardium is covered by a

**FIGURE** 1.9 ✗

A transverse section through the thorax reveals the serous membranes associated with the heart and lungs (superior view).

Image labels: Vertebra, Azygos v., Aorta, Esophagus, Right lung, Right atrium of heart, Right ventricle of heart, Visceral pleura, Pleural cavity, Parietal pleura, Sternum, Spinal cord, Mediastinum, Left lung, Rib, Left ventricle of heart, Visceral pericardium, Pericardial cavity, Parietal pericardium, Fibrous pericardium, Anterior, Plane of section

much thicker third layer, the *fibrous pericardium.* Figure 1.9 shows the membranes associated with the heart and lungs.

In the abdominopelvic cavity, the membranes are called **peritoneal membranes.** A *parietal peritoneum* lines the wall, and a *visceral peritoneum* covers each organ in the abdominal cavity. The potential space between these membranes is called the *peritoneal cavity* (fig. 1.10).

1. What does *visceral* mean?

2. Which organs occupy the dorsal cavity? The ventral cavity?

3. Name the cavities of the head.

4. Describe the membranes associated with the thoracic and abdominopelvic cavities.

5. Distinguish between the parietal and visceral peritoneum.

## Organ Systems

The human organism consists of several organ systems. Each system includes a set of interrelated organs that work together to provide specialized functions. The maintenance of homeostasis depends on these organ systems working together in a coordinated manner. A figure called **"InnerConnections"** featured at the end of certain chapters ties together the ways in which organ systems interact. As you read about each organ system, you may want to consult the illustrations of the human torso provided in reference plates 1–7 and locate some of the features listed in the descriptions.

### Body Covering

The organs of the **integumentary system** (fig. 1.11) include the skin and accessory organs such as the hair, nails, sweat glands, and sebaceous glands. These parts protect underlying tissues, help regulate body temperature, house a variety of sensory receptors, and synthesize certain products. The integumentary system is discussed in chapter 6.

### Support and Movement

The organs of the skeletal and muscular systems support and move body parts. The **skeletal system** (fig. 1.12) consists of the bones as well as the ligaments and cartilages that bind bones together at joints. These parts provide frameworks and protective shields for softer tissues,

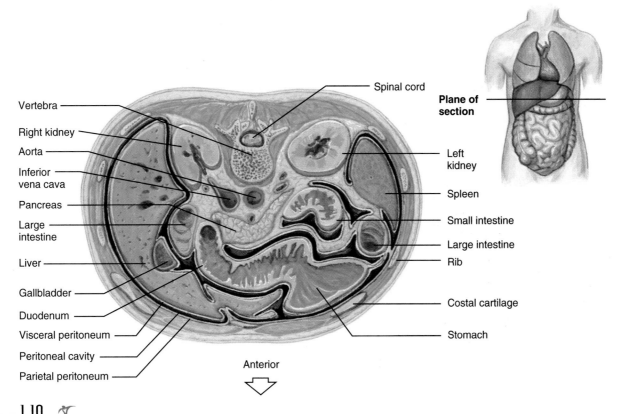

**Spinal cord**

Vertebra

Right kidney

Aorta

Inferior vena cava

Pancreas

Large intestine

Liver

Gallbladder

Duodenum

Visceral peritoneum

Peritoneal cavity

Parietal peritoneum

**Plane of section**

Left kidney

Spleen

Small intestine

Large intestine

Rib

Costal cartilage

Stomach

Anterior

**FIGURE** 1.10

A transverse section through the abdomen (superior view). Note that the large intestine is labeled twice.

Integumentary system

Skeletal system

Muscular system

**FIGURE** 1.11

The integumentary system forms the body covering.

**FIGURE** 1.12

Organ systems associated with support and movement.

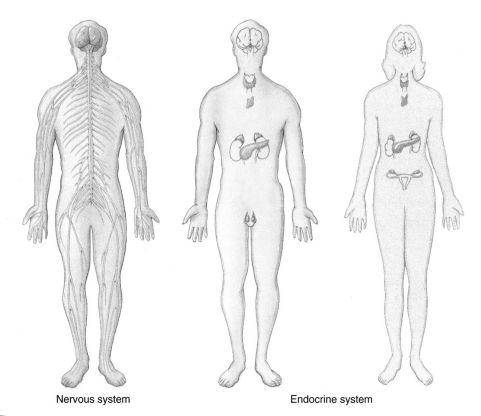

Nervous system        Endocrine system

**FIGURE** 1.13

Organ systems associated with integration and coordination.

serve as attachments for muscles, and act together with muscles when body parts move. Tissues within bones also produce blood cells and store inorganic salts.

The muscles are the organs of the **muscular system** (fig. 1.12). By contracting and pulling their ends closer together, they provide the forces that cause body movements. They also help maintain posture and are the main source of body heat. The skeletal and muscular systems are discussed in chapters 7, 8, and 9.

## Integration and Coordination

For the body to act as a unit, its parts must be integrated and coordinated. The nervous and endocrine systems control and adjust various organ functions from time to time to maintain homeostasis.

The **nervous system** (fig. 1.13) consists of the brain, spinal cord, nerves, and sense organs. Nerve cells within these organs use electrochemical signals called *nerve impulses* (action potentials) to communicate with one another and with muscles and glands. Each impulse produces a relatively short-term effect on its target. Some nerve cells act as specialized sensory receptors that can detect changes occurring inside and outside the body. Other nerve cells receive the impulses transmitted from these sensory units and interpret and act on the information. Still other nerve cells carry impulses from the brain or spinal cord to muscles or glands, stim-

ulating them to contract or to secrete products. Chapters 10 and 11 discuss the nervous system, and chapter 12 discusses sense organs.

The **endocrine system** (fig. 1.13) includes all the glands that secrete chemical messengers, called *hormones*. Hormones, in turn, travel away from the glands in body fluids such as blood or tissue fluid. Usually a particular hormone affects only a particular group of cells, called its *target tissue.* The effect of a hormone is to alter the metabolism of the target tissue. Compared to nerve impulses, hormonal effects occur over a relatively long time period.

Organs of the endocrine system include the pituitary, thyroid, parathyroid, and adrenal glands, as well as the pancreas, ovaries, testes, pineal gland, and thymus gland. These are discussed further in chapter 13.

## Transport

Two organ systems transport substances throughout the internal environment. The **cardiovascular system** (fig. 1.14) includes the heart, arteries, capillaries, veins, and blood. The heart is a muscular pump that helps force blood through the blood vessels. Blood transports gases, nutrients, hormones, and wastes. It carries oxygen from the lungs and nutrients from the digestive organs to all body cells, where these substances are used in metabolic processes. Blood also transports hormones from en-

Cardiovascular system

Lymphatic system

**FIGURE** 1.14 ✗

Organ systems associated with transport.

docrine glands to their target tissues, and carries wastes from body cells to the excretory organs, where the wastes are removed from the blood and released to the outside. Blood and the cardiovascular system are discussed in chapters 14 and 15.

The **lymphatic system** (fig. 1.14) is sometimes considered part of the cardiovascular system. It is also involved with transport and is composed of the lymphatic vessels, lymph fluid, lymph nodes, thymus gland, and spleen. This system transports some of the fluid from the spaces within tissues (tissue fluid) back to the bloodstream and carries certain fatty substances away from the digestive organs. Cells of the lymphatic system are called lymphocytes, and they defend the body against infections by removing disease-causing microorganisms and viruses from the tissue fluid. The lymphatic system is discussed in chapter 16.

## Absorption and Excretion

Organs in several systems absorb nutrients and oxygen and excrete wastes. The organs of the **digestive system** (fig. 1.15), for example, receive foods from the outside. Then they break down food molecules into simpler forms that can pass through cell membranes and thus be absorbed into the internal environment. Materials that are not absorbed are eliminated by being transported outside. Certain digestive organs (chapter 17)

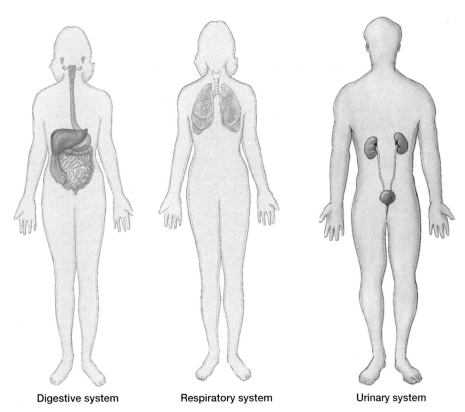

Digestive system

Respiratory system

Urinary system

**FIGURE** 1.15 ✗

Organ systems associated with absorption and excretion.

also produce hormones and thus function as parts of the endocrine system.

The digestive system includes the mouth, tongue, teeth, salivary glands, pharynx, esophagus, stomach, liver, gallbladder, pancreas, small intestine, and large intestine. Chapter 18 discusses nutrition and metabolism, considering the fate of foods in the body.

The organs of the **respiratory system** (fig. 1.15) take air in and out and exchange gases between the blood and the air. More specifically, oxygen passes from air within the lungs into the blood, and carbon dioxide leaves the blood and enters the air. The nasal cavity, pharynx, larynx, trachea, bronchi, and lungs are parts of this system, which is discussed in chapter 19.

The **urinary system** (fig. 1.15) consists of the kidneys, ureters, urinary bladder, and urethra. The kidneys remove wastes from blood and assist in maintaining the body's water and electrolyte balance. The product of these activities is urine. Other portions of the urinary system store urine and transport it outside the body. Chapter 20 discusses the urinary system.

Sometimes the urinary system is called the *excretory system.* However, excretion, or waste removal, is also a function of the respiratory system, and to a lesser extent the digestive and integumentary systems.

## Reproduction

Reproduction is the process of producing offspring (progeny). Cells reproduce when they divide and give rise to new cells. The **reproductive system** (fig. 1.16) of an organism, however, produces whole new organisms like itself (see chapter 22).

The male reproductive system includes the scrotum, testes, epididymides, vasa deferentia, seminal vesicles, prostate gland, bulbourethral glands, urethra, and penis. These structures produce and maintain the male sex cells, or sperm cells (spermatozoa). The male reproductive system also transfers these cells from their site of origin into the female reproductive tract.

The female reproductive system consists of the ovaries, uterine tubes, uterus, vagina, clitoris, and vulva. These organs produce and maintain the female sex cells (egg cells or ova), receive the male sex cells (sperm cells), and transport the female sex cells within the female reproductive system. The female reproductive system also supports development of embryos and functions in the birth process.

Table 1.4 summarizes the organ systems, the major organs that comprise them, and their major functions in the order you will read about them in this book. Figure 1.17 illustrates the organ systems in humans.

**1** Name the major organ systems and list the organs of each system.

**2** Describe the general functions of each organ system.

Male reproductive system          Female reproductive system

**FIGURE 1.16**
Organ systems associated with reproduction.

## ANATOMICAL TERMINOLOGY

To communicate effectively with one another, investigators over the ages have developed a set of terms with precise meanings. Some of these terms concern the relative positions of body parts, others refer to imaginary planes along which cuts may be made, and still others describe body regions. When such terms are used, it is assumed that the body is in the **anatomical position;** that is, it is standing erect, the face is forward, and the upper limbs are at the sides, with the palms forward.

### Relative Position

Terms of relative position are used to describe the location of one body part with respect to another. They include the following:

1. **Superior** means a part is above another part, or closer to the head. (The thoracic cavity is superior to the abdominopelvic cavity.)

2. **Inferior** means a part is below another part, or toward the feet. (The neck is inferior to the head.)

3. **Anterior** (or ventral) means toward the front. (The eyes are anterior to the brain.)

| TABLE 1.4 | ORGAN SYSTEMS |

| Organ System | Major Organs | Major Functions |
|---|---|---|
| Integumentary | Skin, hair, nails, sweat glands, sebaceous glands | Protect tissues, regulate body temperature, support sensory receptors |
| Skeletal | Bones, ligaments, cartilages | Provide framework, protect soft tissues, provide attachments for muscles, produce blood cells, store inorganic salts |
| Muscular | Muscles | Cause movements, maintain posture, produce body heat |
| Nervous | Brain, spinal cord, nerves, sense organs | Detect changes, receive and interpret sensory information, stimulate muscles and glands |
| Endocrine | Glands that secrete hormones (pituitary gland, thyroid gland, parathyroid glands, adrenal glands, pancreas, ovaries, testes, pineal gland, and thymus gland) | Control metabolic activities of body structures |
| Cardiovascular | Heart, arteries, capillaries, veins, | Move blood through blood vessels and transport substances throughout body |
| Lymphatic | Lymphatic vessels, lymph nodes, thymus, spleen | Return tissue fluid to the blood, carry certain absorbed food molecules, defend the body against infection |
| Digestive | Mouth, tongue, teeth, salivary glands, pharynx, esophagus, stomach, liver, gallbladder, pancreas, small and large intestines | Receive, break down, and absorb food; eliminate unabsorbed material |
| Respiratory | Nasal cavity, pharynx. larynx, trachea, bronchi, lungs | Intake and output of air, exchange of gases between air and blood |
| Urinary | Kidneys, ureters, urinary bladder, urethra | Remove wastes from blood, maintain water and electrolyte balance, store and transport urine |
| Reproductive | Male: scrotum, testes, epididymides, vasa deferentia, seminal vesicles, prostate gland, bulbourethral glands, urethra, penis | Produce and maintain sperm cells, transfer sperm cells into female reproductive tract |
| | Female: ovaries, uterine tubes, uterus, vagina, clitoris, vulva | Produce and maintain egg cells, receive sperm cells, support development of an embryo and function in birth process |

4. **Posterior** (or dorsal) is the opposite of anterior; it means toward the back. (The pharynx is posterior to the oral cavity.)

5. **Medial** relates to an imaginary midline dividing the body into equal right and left halves. A part is medial if it is closer to this line than another part. (The nose is medial to the eyes.)

6. **Lateral** means toward the side with respect to the imaginary midline. (The ears are lateral to the eyes.) **Ipsilateral** pertains to the same side (the spleen and the descending colon are ipsilateral), whereas **contralateral** refers to the opposite side (the spleen and the gallbladder are contralateral).

7. **Proximal** is used to describe a part that is closer to the trunk of the body or closer to another specified point of reference than another part. (The elbow is proximal to the wrist.)

8. **Distal** is the opposite of proximal. It means a particular body part is farther from the trunk or farther from another specified point of reference than another part. (The fingers are distal to the wrist.)

9. **Superficial** means situated near the surface. (The epidermis is the superficial layer of the skin.) **Peripheral** also means outward or near the surface. It is used to describe the location of certain blood vessels and nerves. (The nerves that branch from the brain and spinal cord are peripheral nerves.)

10. **Deep** is used to describe parts that are more internal. (The dermis is the deep layer of the skin.)

## Body Sections

To observe the relative locations and arrangements of internal parts, it is necessary to cut or section the body along various planes (figs. 1.18 and 1.19). The following terms are used to describe such planes and sections:

1. **Sagittal** refers to a lengthwise cut that divides the body into right and left portions. If a sagittal section passes along the midline and divides the body into equal parts, it is called median (midsagittal).

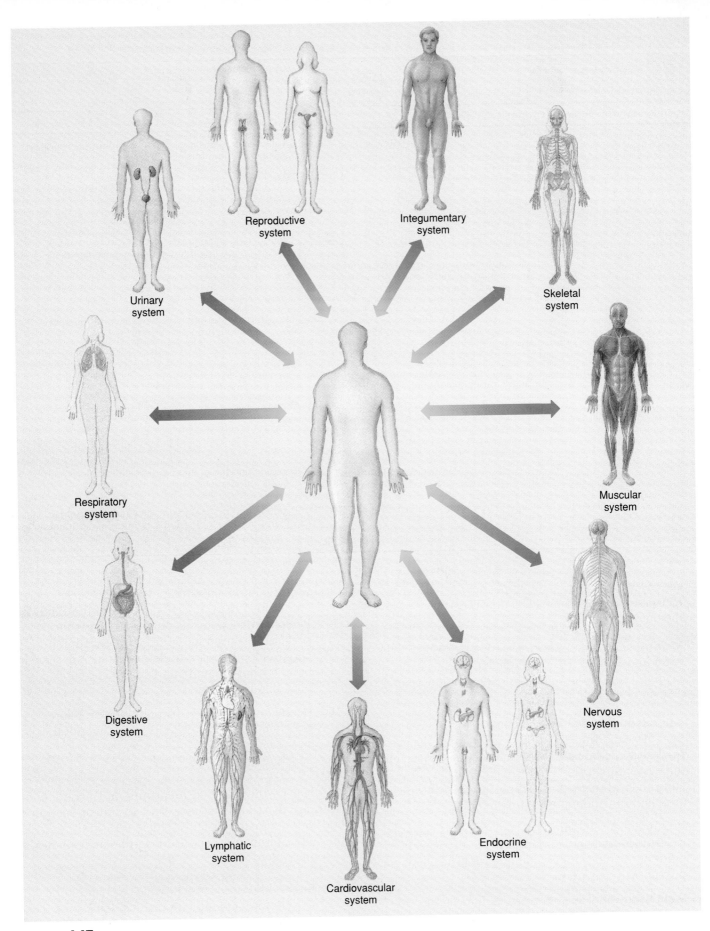

Reproductive system

Integumentary system

Urinary system

Skeletal system

Respiratory system

Muscular system

Digestive system

Nervous system

Lymphatic system

Endocrine system

Cardiovascular system

**FIGURE** 1.17

The organ systems in humans.

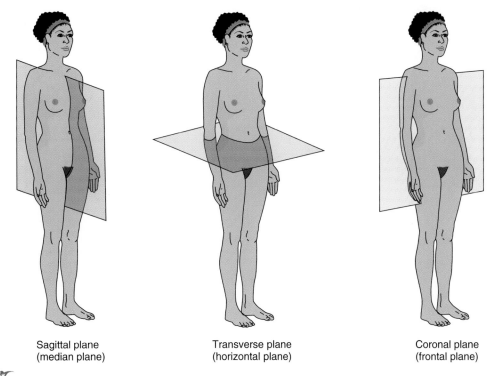

Sagittal plane
(median plane)

Transverse plane
(horizontal plane)

Coronal plane
(frontal plane)

**FIGURE** 1.18 𝒳

To observe internal parts, the body may be sectioned along various planes.

(a)

(b)

(c)

**FIGURE** 1.19 𝒳

A human brain sectioned along (a) the sagittal plane, (b) the transverse plane, and (c) the coronal plane.

**FIGURE 1.20**

Cylindrical parts may be cut in (*a*) cross section, (*b*) oblique section, or (*c*) longitudinal section.

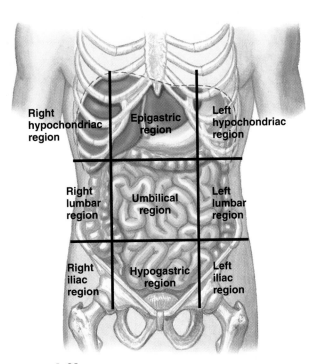

**FIGURE 1.21**

The abdominal area is subdivided into nine regions.

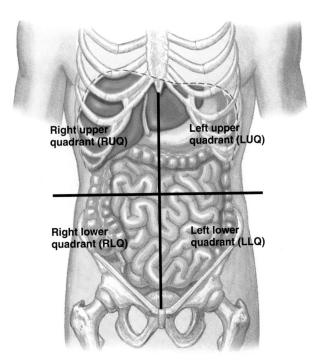

**FIGURE 1.22**

The abdominal area may be subdivided into four quadrants.

2. **Transverse** (or horizontal) refers to a cut that divides the body into superior and inferior portions.

3. **Coronal** (or frontal) refers to a section that divides the body into anterior and posterior portions.

Sometimes a cylindrical organ such as a blood vessel is sectioned. In this case, a cut across the structure is called a *cross section*, an angular cut is called an *oblique section,* and a lengthwise cut is called a *longitudinal section* (fig. 1.20).

## Body Regions

A number of terms designate body regions. The abdominal area, for example, is subdivided into the following regions, as shown in figure 1.21:

1. **Epigastric region**   The upper middle portion.

2. **Left** and **right hypochondriac regions**   On each side of the epigastric region.

3. **Umbilical region**   The central portion.

4. **Left** and **right lumbar regions**   On each side of the umbilical region.

5. **Hypogastric region**   The lower middle portion.

6. **Left** and **right iliac (or inguinal) regions**   On each side of the hypogastric region.

The abdominal area also may be subdivided into the following four quadrants, as illustrated in figure 1.22:

1. **Right upper quadrant** (RUQ).

2. **Right lower quadrant** (RLQ).

3. **Left upper quadrant** (LUQ).

4. **Left lower quadrant** (LLQ).

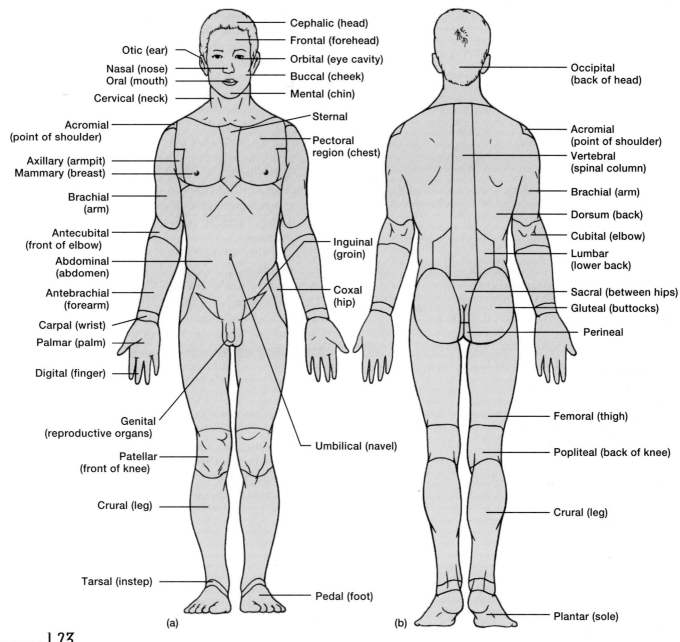

**FIGURE 1.23**

Some terms used to describe body regions. (*a*) Anterior regions. (*b*) Posterior regions.

The following terms are commonly used when referring to various body regions. Figure 1.23 illustrates some of these regions.

**abdominal** (ab-dom'ĭ-nal)   region between the thorax and pelvis.
**acromial** (ah-kro'me-al)   point of the shoulder.
**antebrachial** (an″te-bra′ke-al)   forearm.
**antecubital** (an″te-ku′bĭ-tal)   space in front of the elbow.
**axillary** (ak'sĭ-ler″e)   armpit.
**brachial** (bra′ke-al)   arm.

**buccal** (buk'al)   cheek.
**carpal** (kar'pal)   wrist.
**celiac** (se'le-ak)   abdomen.
**cephalic** (sĕ-fal'ik)   head.
**cervical** (ser'vĭ-kal)   neck.
**costal** (kos'tal)   ribs.
**coxal** (kok'sal)   hip.
**crural** (krōōr'al)   leg.
**cubital** (ku'bĭ-tal)   elbow.
**digital** (dij'ĭ-tal)   finger.
**dorsum** (dor'sum)   back.

**femoral** (fem′or-al) thigh.

**frontal** (frun′tal) forehead.

**genital** (jen′i-tal) reproductive organs.

**gluteal** (gloo′te-al) buttocks.

**inguinal** (ing′gwĭ-nal) depressed area of the abdominal wall near the thigh (groin).

**lumbar** (lum′bar) region of the lower back between the ribs and the pelvis (loin).

**mammary** (mam′er-e) breast.

**mental** (men′tal) chin.

**nasal** (na′zal) nose.

**occipital** (ok-sip′ĭ-tal) lower posterior region of the head.

**oral** (o′ral) mouth.

**orbital** (or′bi-tal) eye cavity.

**otic** (o′tik) ear.

**palmar** (pahl′mar) palm of the hand.

**patellar** (pah-tel′ar) front of the knee.

**pectoral** (pek′tor-al) chest.

**pedal** (ped′al) foot.

**pelvic** (pel′vik) pelvis.

**perineal** (per″ĭ-ne′al) region between the anus and the external reproductive organs (perineum).

**plantar** (plan′tar) sole of the foot.

**popliteal** (pop″lĭ-te′al) area behind the knee.

**sacral** (sa′kral) posterior region between the hipbones.

**sternal** (ster′nal) middle of the thorax, anteriorly.

**tarsal** (tahr′sal) instep of the foot.

**umbilical** (um-bil′ĭ-kal) navel.

**vertebral** (ver′te-bral) spinal column.

**1** Describe the anatomical position.

**2** Using the appropriate terms, describe the relative positions of several body parts.

**3** Describe three types of body sections.

**4** Describe the nine regions of the abdomen.

**5** Explain how the names of the abdominal regions describe their locations.

## SOME MEDICAL AND APPLIED SCIENCES

**cardiology** (kar″de-ol′o-je) Branch of medical science dealing with the heart and heart diseases.

**cytology** (si-tol′o-je) Study of the structure, function, and diseases of cells.

**dermatology** (der″mah-tol′o-je) Study of skin and its diseases.

**endocrinology** (en″do-krĭ-nol′o-je) Study of hormones, hormone-secreting glands, and the diseases involving them.

**epidemiology** (ep″ĭ-de″me-ol′o-je) Study of the factors determining the distribution and frequency of the occurrence of health-related conditions within a defined human population.

**gastroenterology** (gas″tro-en″ter-ol′o-je) Study of the stomach and intestines, as well as their diseases.

**geriatrics** (jer′e-at′riks) Branch of medicine dealing with older individuals and their medical problems.

**gerontology** (jer″on-tol′o-je) Study of the process of aging and the various problems of older individuals.

**gynecology** (gi″nĕ-kol′o-je) Study of the female reproductive system and its diseases.

**hematology** (hem″ah-tol′o-je) Study of blood and blood diseases.

**histology** (his-tol′o-je) Study of the structure and function of tissues.

**immunology** (im″u-nol′o-je) Study of the body's resistance to disease.

**neonatology** (ne″o-na-tol′o-je) Study of newborn infants and the treatment of their disorders.

**nephrology** (nĕ-frol′o-je) Study of the structure, function, and diseases of the kidneys.

**neurology** (nu-rol′o-je) Study of the nervous system in health and disease.

**obstetrics** (ob-stet′riks) Branch of medicine dealing with pregnancy and childbirth.

**oncology** (ong-kol′o-je) Study of cancers.

**ophthalmology** (of″thal-mol′o-je) Study of the eye and eye diseases.

**orthopedics** (or″tho-pe′diks) Branch of medicine dealing with the muscular and skeletal systems, and their problems.

**otolaryngology** (o″to-lar″in-gol′o-je) Study of the ear, throat, larynx, and their diseases.

**pathology** (pah-thol′o-je) Study of structural and functional changes within the body that disease causes.

**pediatrics** (pe″de-at′riks) Branch of medicine dealing with children and their diseases.

**pharmacology** (fahr″mah-kol′o-je) Study of drugs and their uses in the treatment of diseases.

**podiatry** (po-di′ah-tre) Study of the care and treatment of the feet.

**psychiatry** (si-ki′ah-tre) Branch of medicine dealing with the mind and its disorders.

**radiology** (ra″de-ol′o-je) Study of X rays and radioactive substances, as well as their uses in diagnosing and treating diseases.

**toxicology** (tok″sĭ-kol′o-je) Study of poisonous substances and their effects upon body parts.

**urology** (u-rol′o-je) Branch of medicine dealing with the urinary and male reproductive systems, and their diseases.

## Introduction (page 3)

1. When a patient arrives at a hospital with an unknown injury, medical staff must rapidly apply their knowledge of human anatomy and physiology to correctly diagnose the problem.
2. Early interest in the human body probably developed as people became concerned about injuries and illnesses.
3. Early doctors began to learn how certain herbs and potions affected body functions.
4. The idea that humans could understand forces that caused natural events led to the development of modern science.
5. A set of terms originating from Greek and Latin formed the basis for the language of anatomy and physiology.

## Anatomy and Physiology (page 3)

1. Anatomy deals with the form and organization of body parts.
2. Physiology deals with the functions of these parts.
3. The function of a part depends upon the way it is constructed.

## Characteristics of Life (page 4)

Characteristics of life are traits all organisms share.

1. These characteristics include:
   a. Movement—changing body position or moving internal parts.
   b. Responsiveness—sensing and reacting to internal or external changes.
   c. Growth—increasing in size without changing in shape.
   d. Reproduction—producing offspring.
   e. Respiration—obtaining oxygen, using oxygen to release energy from foods, and removing gaseous wastes.
   f. Digestion—breaking down food substances into forms that can be absorbed.
   g. Absorption—moving substances through membranes and into body fluids.
   h. Circulation—moving substances through the body in body fluids.
   i. Assimilation—changing substances into chemically different forms.
   j. Excretion—removing body wastes.
2. These activities constitute metabolism.

## Maintenance of Life (page 5)

The structures and functions of body parts maintain the life of the organism.

1. Requirements of organisms
   a. Water is used in many metabolic processes, provides the environment for metabolic reactions, and transports substances.
   b. Food supplies energy, raw materials for building substances, and chemicals necessary in vital reactions.
   c. Oxygen is used in releasing energy from food materials; this energy drives metabolic reactions.
   d. Heat is a product of metabolic reactions and helps control rates of these reactions.
   e. Pressure is an application of force; in humans, atmospheric and hydrostatic pressures help breathing and blood movements, respectively.

2. Homeostasis
   a. If an organism is to survive, the conditions within its body fluids must remain relatively stable.
   b. The tendency to maintain a stable internal environment is called homeostasis.
   c. Homeostatic mechanisms regulate body temperature, blood pressure, and blood glucose concentration.
   d. Homeostatic mechanisms employ negative feedback.

## Levels of Organization (page 8)

The body is composed of parts that can be considered at different levels of organization.

1. Matter is composed of atoms.
2. Atoms join to form molecules.
3. Organelles consist of aggregates of large molecules.
4. Cells, which are composed of organelles, are the basic units of structure and function of the body.
5. Cells are organized into layers or masses called tissues.
6. Tissues are organized into organs.
7. Organs form organ systems.
8. Organ systems constitute the organism.
9. These parts vary in complexity progressively from one level to the next.

## Organization of the Human Body (page 10)

1. Body cavities
   a. The axial portion of the body contains the dorsal and ventral cavities.
      (1) The dorsal cavity includes the cranial cavity and vertebral canal.
      (2) The ventral cavity includes the thoracic and abdominopelvic cavities, which are separated by the diaphragm.
   b. The organs within a body cavity are called viscera.
   c. Other body cavities include the oral, nasal, orbital, and middle ear cavities.

2. Thoracic and abdominopelvic membranes

   Parietal membranes line cavities; visceral membranes cover organs.
   a. Thoracic membranes
      (1) Pleural membranes line the thoracic cavity and cover the lungs.
      (2) Pericardial membranes surround the heart and cover its surface.
      (3) The pleural and pericardial cavities are potential spaces between these membranes.
   b. Abdominopelvic membranes
      (1) Peritoneal membranes line the abdominopelvic cavity and cover the organs inside.
      (2) The peritoneal cavity is a potential space between these membranes.

3. Organ systems

   The human organism consists of several organ systems. Each system includes interrelated organs.
   a. Integumentary system
      (1) The integumentary system covers the body.

(2) It includes the skin, hair, nails, sweat glands, and sebaceous glands.

(3) It protects underlying tissues, regulates body temperature, houses sensory receptors, and synthesizes substances.

b. Skeletal system

(1) The skeletal system is composed of bones and the ligaments and cartilages that bind bones together.

(2) It provides framework, protective shields, and attachments for muscles; it also produces blood cells and stores inorganic salts.

c. Muscular system

(1) The muscular system includes the muscles of the body.

(2) It moves body parts, maintains posture, and produces body heat.

d. Nervous system

(1) The nervous system consists of the brain, spinal cord, nerves, and sense organs.

(2) It receives impulses from sensory parts, interprets these impulses, and acts on them, stimulating muscles or glands to respond.

e. Endocrine system

(1) The endocrine system consists of glands that secrete hormones.

(2) Hormones help regulate metabolism by stimulating target tissues.

(3) It includes the pituitary gland, thyroid gland, parathyroid glands, adrenal glands, pancreas, ovaries, testes, pineal gland, and thymus gland.

f. Digestive system

(1) The digestive system receives foods, breaks down nutrients into forms that can pass through cell membranes, and eliminates materials that are not absorbed.

(2) Some digestive organs produce hormones.

(3) The digestive system includes the mouth, tongue, teeth, salivary glands, pharynx, esophagus, stomach, liver, gallbladder, pancreas, small intestine, and large intestine.

g. Respiratory system

(1) The respiratory system provides for intake and output of air, and for exchange of gases between the blood and the air.

(2) It includes the nasal cavity, pharynx, larynx, trachea, bronchi, and lungs.

h. Cardiovascular system

(1) The cardiovascular system includes the heart, which pumps blood, and the blood vessels, which carry blood to and from body parts.

(2) Blood transports oxygen, nutrients, hormones, and wastes.

i. Lymphatic system

(1) The lymphatic system is composed of lymphatic vessels, lymph nodes, thymus, and spleen.

(2) It transports lymph from tissue spaces to the bloodstream and carries certain fatty substances away from the digestive organs. Lymphocytes defend the body against disease-causing agents.

j. Urinary system

(1) The urinary system includes the kidneys, ureters, urinary bladder, and urethra.

(2) It filters wastes from the blood and helps maintain fluid and electrolyte balance.

k. Reproductive systems

(1) The reproductive system enables an organism to produce progeny.

(2) The male reproductive system includes the scrotum, testes, epididymides, vasa deferentia, seminal vesicles, prostate gland, bulbourethral glands, urethra, and penis, which produce, maintain, and transport male sex cells.

(3) The female reproductive system includes the ovaries, uterine tubes, uterus, vagina, clitoris, and vulva, which produce, maintain, and transport female sex cells.

## Anatomical Terminology (page 18)

Terms with precise meanings are used to help investigators effectively communicate with one another.

1. Relative position
   These terms describe the location of one part with respect to another part.

2. Body sections
   Body sections are planes along which the body may be cut to observe the relative locations and arrangements of internal parts.

3. Body regions
   Special terms designate various body regions.

## CRITICAL THINKING QUESTIONS

1. In many states, death is defined as "irreversible cessation of total brain function." How is death defined in your state? How is this definition related to the characteristics of life?

2. In health, body parts interact to maintain homeostasis. Illness may threaten homeostasis, requiring treatments. What treatments might be used to help control a patient's (a) body temperature, (b) blood oxygen concentration, and (c) water content?

3. Suppose two individuals have benign (noncancerous) tumors that produce symptoms because they occupy space and crowd adjacent organs. If one of these persons

has a tumor in her ventral cavity and the other has a tumor in his dorsal cavity, which patient would be likely to develop symptoms first? Why?

4. If a patient complained of a stomachache and pointed to the umbilical region as the site of the discomfort, which organs located in this region might be the source of the pain?

5. How could the basic requirements of a human be provided for a patient who is unconscious?

6. Assuming that the same information could be obtained by either method, what would be the advantage of using ultrasonography rather than X rays to visualize a fetus in the uterus?

## Part A

1. Briefly describe the early development of knowledge about the human body.
2. Distinguish between the activities of anatomists and physiologists.
3. How does a biological structure's form determine its function? Give an example.
4. List and describe ten characteristics of life.
5. Define *metabolism*.
6. List and describe five requirements of organisms.
7. Explain how the idea of homeostasis relates to the five requirements you listed in item 6.
8. Distinguish between heat and temperature.
9. What are two types of pressures that may act upon organisms?
10. How are body temperature, blood pressure, and blood glucose concentration controlled?
11. Describe how homeostatic mechanisms act by negative feedback?
12. How does the human body illustrate the levels of anatomical organization?
13. Distinguish between the axial and appendicular portions of the body.
14. Distinguish between the dorsal and ventral body cavities, and name the smaller cavities within each.
15. What are the viscera?
16. Where is the mediastinum?
17. Describe the locations of the oral, nasal, orbital, and middle ear cavities.
18. How does a parietal membrane differ from a visceral membrane?
19. Name the major organ systems, and describe the general functions of each.
20. List the major organs that comprise each organ system.
21. In what body region did Judith R.'s injury occur?

## Part B

1. Name the body cavity housing each of the following organs:

   a. stomach      f. rectum
   b. heart      g. spinal cord
   c. brain      h. esophagus
   d. liver      i. spleen
   e. trachea      j. urinary bladder

2. Write complete sentences using each of the following terms correctly:

   a. superior      h. contralateral
   b. inferior      i. proximal
   c. anterior      j. distal
   d. posterior      k. superficial
   e. medial      l. peripheral
   f. lateral      m. deep
   g. ipsilateral

3. Prepare a sketch of a human body, and use lines to indicate each of the following sections:

   a. sagittal      b. transverse      c. coronal

4. Prepare a sketch of the abdominal area and indicate the location of each of the following regions:

   a. epigastric      c. hypogastric      e. lumbar
   b. umbilical      d. hypochondriac      f. iliac

5. Prepare a sketch of the abdominal area and indicate the location of each of the following regions:

   a. right upper quadrant      c. left upper quadrant
   b. right lower quadrant      d. left lower quadrant

6. Provide the common name for the region described by the following terms:

   a. acromial      j. gluteal      s. perineal
   b. antebrachial      k. inguinal      t. plantar
   c. axillary      l. mental      u. popliteal
   d. buccal      m. occipital      v. sacral
   e. celiac      n. orbital      w. sternal
   f. coxal      o. otic      x. tarsal
   g. crural      p. palmar      y. umbilical
   h. femoral      q. pectoral      z. vertebral
   i. genital      r. pedal

 **For More Information Related to This Chapter, Please Visit** *Hole's Human Anatomy and Physiology* **Home Page**
http://www.mhhe.com/biosci/abio/

# THE HUMAN ORGANISM

■ *The following series of illustrations show the major organs of the human torso. The first plate illustrates the anterior surface and reveals the superficial muscles on one side. Each subsequent plate exposes deeper organs, including those in the thoracic, abdominal, and pelvic cavities.*

*Chapters 6–22 of this textbook describe the organ systems of the human organism in detail. As you read them, you may want to refer to these plates to help visualize the locations of organs and the three-dimensional relationships among them. You may also want to study the photographs of human cadavers in the reference plates that follow chapter 24. These photographs illustrate many of the larger organs of the human body.*

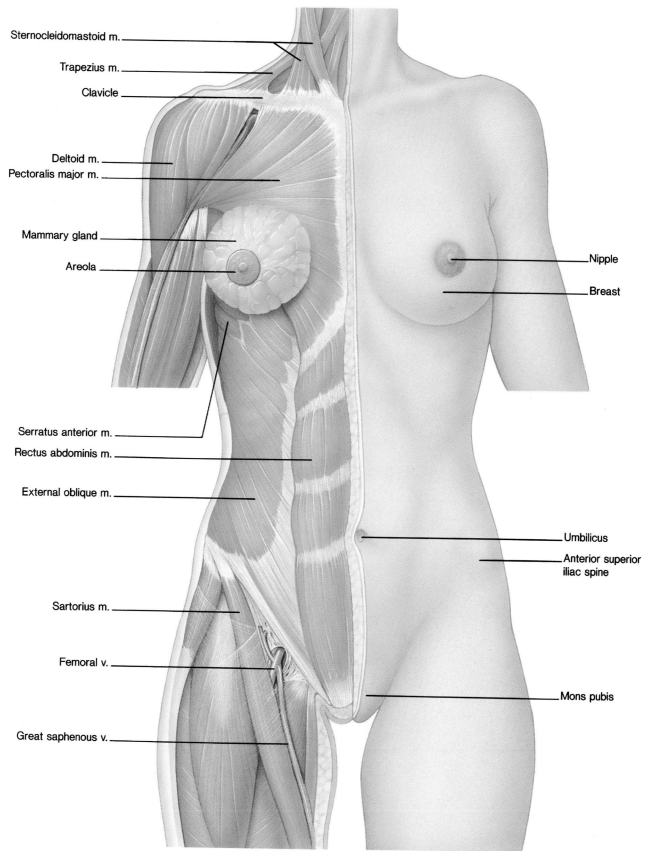

Sternocleidomastoid m.

Trapezius m.

Clavicle

Deltoid m.

Pectoralis major m.

Mammary gland

Areola

Nipple

Breast

Serratus anterior m.

Rectus abdominis m.

External oblique m.

Umbilicus

Anterior superior
iliac spine

Sartorius m.

Femoral v.

Mons pubis

Great saphenous v.

**PLATE One**

Human female torso, showing the anterior surface on one side and the superficial muscles exposed on the other side. (*m.* stands for *muscle; v.* stands for *vein.*)

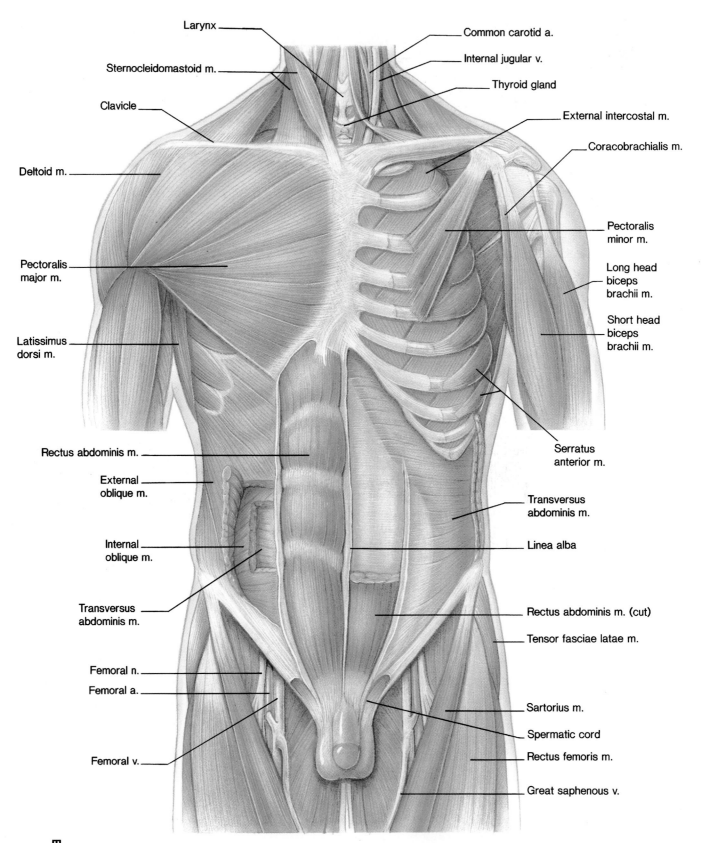

Larynx

Common carotid a.

Internal jugular v.

Sternocleidomastoid m.

Thyroid gland

Clavicle

External intercostal m.

Coracobrachialis m.

Deltoid m.

Pectoralis
minor m.

Long head
biceps
brachii m.

Pectoralis
major m.

Short head
biceps
brachii m.

Latissimus
dorsi m.

Serratus
anterior m.

Rectus abdominis m.

External
oblique m.

Transversus
abdominis m.

Internal
oblique m.

Linea alba

Transversus
abdominis m.

Rectus abdominis m. (cut)

Tensor fasciae latae m.

Femoral n.

Femoral a.

Sartorius m.

Spermatic cord

Rectus femoris m.

Femoral v.

Great saphenous v.

PLATE TWO

Human male torso, with the deeper muscle layers exposed. (*n.* stands for *nerve; a.* stands for *artery.*)

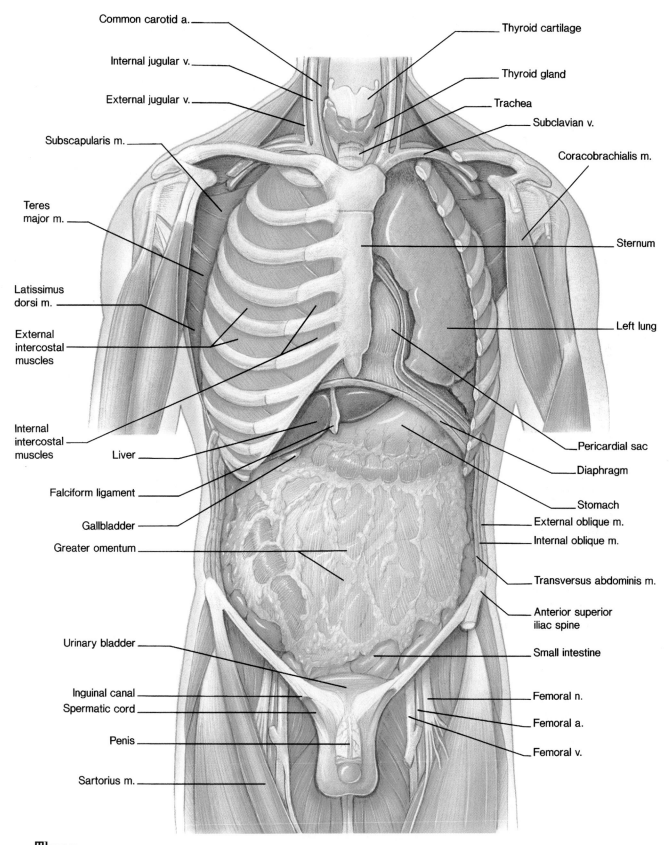

Common carotid a.

Internal jugular v.

External jugular v.

Subscapularis m.

Teres
major m.

Latissimus
dorsi m.

External
intercostal
muscles

Internal
intercostal
muscles

Liver

Falciform ligament

Gallbladder

Greater omentum

Urinary bladder

Inguinal canal

Spermatic cord

Penis

Sartorius m.

Thyroid cartilage

Thyroid gland

Trachea

Subclavian v.

Coracobrachialis m.

Sternum

Left lung

Pericardial sac

Diaphragm

Stomach

External oblique m.

Internal oblique m.

Transversus abdominis m.

Anterior superior
iliac spine

Small intestine

Femoral n.

Femoral a.

Femoral v.

## PLATE Three
Human male torso, with the deep muscles removed and the abdominal viscera exposed.

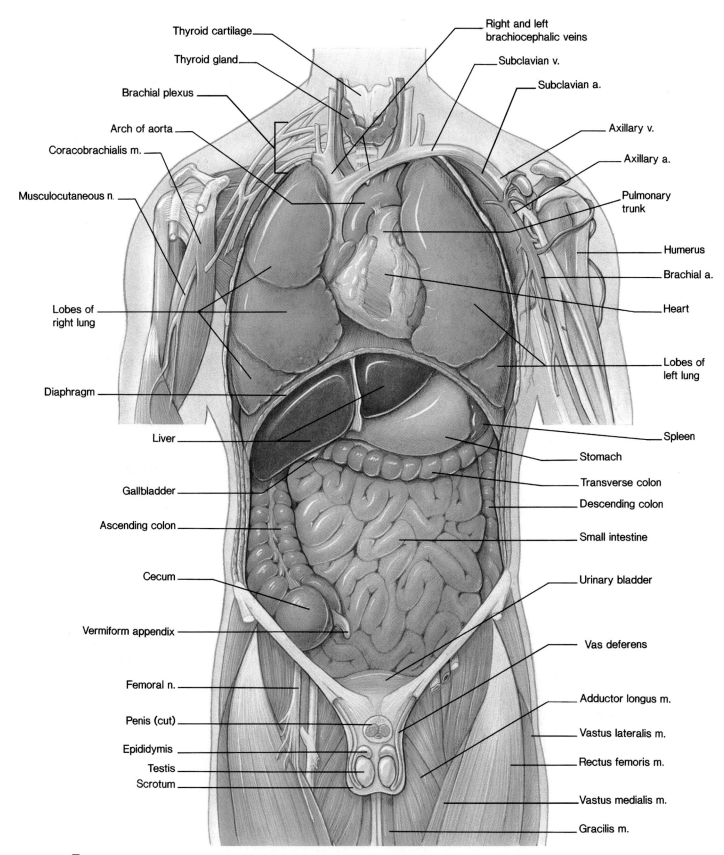

Thyroid cartilage

Thyroid gland

Brachial plexus

Arch of aorta

Coracobrachialis m.

Musculocutaneous n.

Lobes of
right lung

Diaphragm

Liver

Gallbladder

Ascending colon

Cecum

Vermiform appendix

Femoral n.

Penis (cut)

Epididymis

Testis

Scrotum

Right and left
brachiocephalic veins

Subclavian v.

Subclavian a.

Axillary v.

Axillary a.

Pulmonary
trunk

Humerus

Brachial a.

Heart

Lobes of
left lung

Spleen

Stomach

Transverse colon

Descending colon

Small intestine

Urinary bladder

Vas deferens

Adductor longus m.

Vastus lateralis m.

Rectus femoris m.

Vastus medialis m.

Gracilis m.

**PLATE Four**

Human male torso, with the thoracic and abdominal viscera exposed.

Right common carotid a.

Right subclavian a.

Brachiocephalic a.

Superior vena cava

Right atrium

Right ventricle

Lobes of liver

Gallbladder

Cystic duct

Duodenum

Ascending colon

Mesentery

Ileum (cut)

Cecum

Vermiform appendix

Common iliac a.

Ovary

Uterine tube

Femoral a.

Femoral v.

Adductor longus m.

Gracilis m.

Vastus medialis m.

Larynx

Trachea

Left subclavian a.

Arch of aorta

Pulmonary a.

Pulmonary trunk

Pulmonary v.

Left atrium

Lung

Left ventricle

Diaphragm

Spleen

Stomach

Transverse colon

Jejunum (cut)

Descending colon

Ureter

Sigmoid colon

Rectum

Uterus

Tensor fasciae latae m.

Round ligament of uterus

Urinary bladder

Great saphenous v.

Rectus femoris m.

Vastus lateralis m.

Sartorius m.

PLATE Five

Human female torso, with the lungs, heart, and small intestine sectioned and the liver reflected (lifted back).

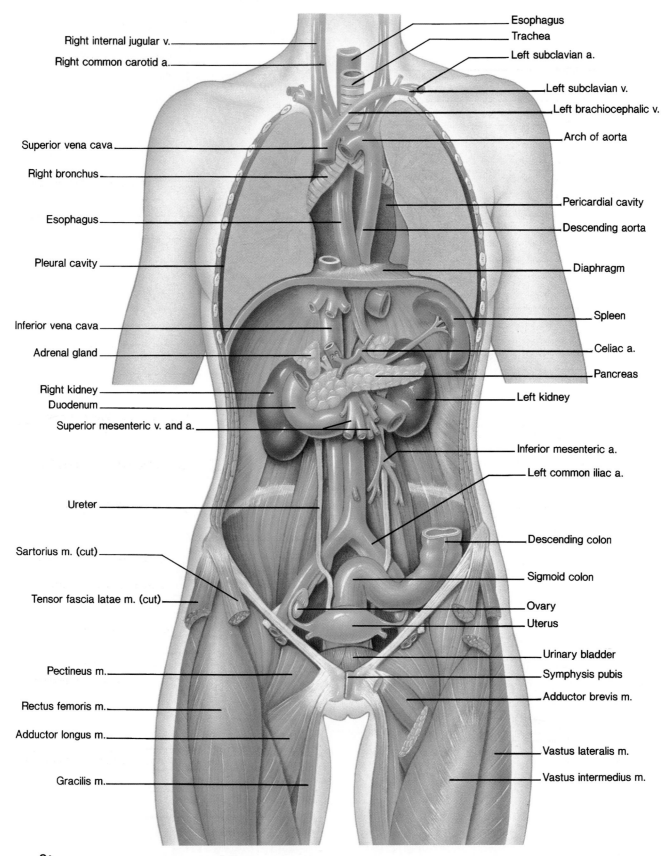

Right internal jugular v.

Right common carotid a.

Esophagus

Trachea

Left subclavian a.

Left subclavian v.

Left brachiocephalic v.

Arch of aorta

Superior vena cava

Right bronchus

Pericardial cavity

Descending aorta

Esophagus

Pleural cavity

Diaphragm

Spleen

Inferior vena cava

Adrenal gland

Celiac a.

Pancreas

Right kidney

Duodenum

Left kidney

Superior mesenteric v. and a.

Inferior mesenteric a.

Left common iliac a.

Ureter

Descending colon

Sartorius m. (cut)

Sigmoid colon

Tensor fascia latae m. (cut)

Ovary

Uterus

Urinary bladder

Pectineus m.

Symphysis pubis

Rectus femoris m.

Adductor brevis m.

Adductor longus m.

Vastus lateralis m.

Gracilis m.

Vastus intermedius m.

## PLATE SIX

Human female torso, with the heart, stomach, liver, and parts of the intestine and lungs removed.

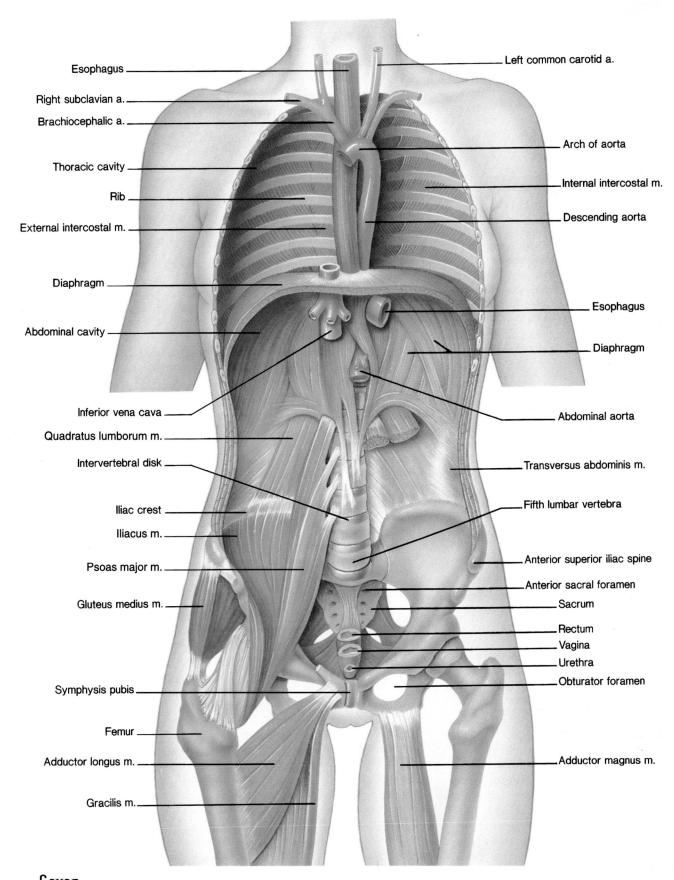

Esophagus

Right subclavian a.

Brachiocephalic a.

Thoracic cavity

Rib

External intercostal m.

Diaphragm

Abdominal cavity

Inferior vena cava

Quadratus lumborum m.

Intervertebral disk

Iliac crest

Iliacus m.

Psoas major m.

Gluteus medius m.

Symphysis pubis

Femur

Adductor longus m.

Gracilis m.

Left common carotid a.

Arch of aorta

Internal intercostal m.

Descending aorta

Esophagus

Diaphragm

Abdominal aorta

Transversus abdominis m.

Fifth lumbar vertebra

Anterior superior iliac spine

Anterior sacral foramen

Sacrum

Rectum

Vagina

Urethra

Obturator foramen

Adductor magnus m.

**PLATE** Seven

Human female torso, with the thoracic, abdominal, and pelvic viscera removed.

# CHEMICAL BASIS OF LIFE

## UNDERSTANDING WORDS

**bio-,** life: *bio*chemistry—branch of science dealing with the chemistry of life forms.

**di-,** two: *di*saccharide—compound whose molecules are composed of two saccharide units bound together.

**glyc-,** sweet: *glyc*ogen—complex carbohydrate composed of sugar molecules bound together.

**iso-,** equal: *iso*tope—atom that has the same atomic number as another atom but a different atomic weight.

**lip-,** fat: *lip*ids—group of organic compounds that includes fats.

**-lyt,** dissolvable: electro*lyte*—substance that dissolves in water and releases ions.

**mono-,** one: *mono*saccharide—compound whose molecule consists of a single saccharide unit.

**nucle-,** kernel: *nucle*us—central core of an atom.

**poly-,** many: *poly*unsaturated—molecule that has many double bonds between its carbon atoms.

**sacchar-,** sugar: mono*sacchar*ide—sugar molecule composed of a single saccharide unit.

**syn-,** together: *syn*thesis—process by which substances are united to form a new type of substance.

**-valent,** having power: co*valent* bond—chemical bond produced when two atoms share electrons.

## CHAPTER OBJECTIVES

AFTER YOU HAVE STUDIED THIS CHAPTER, YOU SHOULD BE ABLE TO:

1. Explain how the study of living material depends on the study of chemistry.

2. Describe the relationships between matter, atoms, and molecules.

3. Discuss how atomic structure determines how atoms interact.

4. Explain how molecular and structural formulas are used to symbolize the composition of compounds.

5. Describe three types of chemical reactions.

6. Define pH.

7. List the major groups of inorganic substances that are common in cells.

8. Describe the general functions of various types of organic substances in cells.

## KEY TERMS

**atom** (at′om)
**carbohydrate** (kar″bo-hi′drāt)
**compound** (kom′-pownd)
**decomposition** (de″kom-po-zish′un)
**electrolyte** (e-lek′tro-līt)
**element** (el′e-ment)
**inorganic** (in″or-gan′ik)

**ion** (i′on)
**isotope** (i′so-top)
**lipid** (lip′id)
**molecule** (mol′ě-kūl)
**nucleic acid** (nu-kle′ik as′id)
**organic** (or-gan′ik)
**protein** (pro′te-in)
**synthesis** (sin′thě-sis)

The reunion of the extended Slone family in Kentucky in the spring of 1994 was an unusual event. Not only did ninety relatives gather, but medical researchers also attended, sampling blood from everyone. The reason—the family is very rare in that many members suffer from hereditary pancreatitis, locally known as Slone's disease. In this painful and untreatable condition, the pancreas digests itself. This organ produces digestive enzymes, and hormones that regulate the blood glucose level. The researchers were looking for biochemical instructions, in the form of genes, that might explain how the disease arises. This information may also help the many thousands of people who suffer from nonhereditary pancreatitis.

Kevin Slone, who organized the reunion, knew well the ravages of his family's illness. In 1989, as a teenager, he was hospitalized for severe abdominal pain. When he was again hospitalized five years later, three-quarters of his pancreas had become scar tissue. Because many relatives also complained of frequent and severe abdominal pain, Kevin's father, Bobby, began assembling a family tree. Using a computer, he traced more than 700 relatives through nine generations. Although he didn't realize it, Bobby Slone was conducting sophisticated and invaluable genetic research.

David Whitcomb and Garth Ehrlich, geneticists at the University of Pittsburgh, had become interested in hereditary pancreatitis, and put the word out that they were looking for a large family in which to hunt for a causative gene. A colleague at a new pancreatitis clinic at the University of Kentucky put them in touch with the Slones and their enormous family tree. Soon after the blood sampling at the family reunion, the researchers identified the biochemical cause of hereditary pancreatitis.

Affected family members have a genetic glitch that blocks normal control of the manufacture of trypsin, a digestive enzyme that breaks down protein. When the powerful enzyme accumulates, it digests the pancreas. A disorder felt painfully at the whole-body level is caused by a problem at the biochemical level. Researchers are using the information provided by the Slone family to develop a diagnostic test that can identify the errant gene before symptoms begin. The genetic information may also someday lead to a treatment.

---

Chemistry considers the composition of substances and how they change. Although it is possible to study anatomy without much reference to chemistry, it is essential for understanding physiology, because body functions depend on cellular functions that in turn result from chemical changes.

As interest in the chemistry of living organisms grew, and knowledge of the subject expanded, a field of life science called biological chemistry, or **biochemistry,** emerged. Biochemistry has been important not only in helping explain physiological processes but also in developing many new drugs and methods for treating diseases.

**1**  Why is a knowledge of chemistry essential to understanding physiology?

**2**  What is biochemistry?

---

## STRUCTURE OF MATTER

**Matter** is anything that has weight and takes up space. This includes all the solids, liquids, and gases in our surroundings as well as in our bodies. All matter consists of particles that are organized in specific ways. Table 2.1 lists some particles of matter and their characteristics.

### Elements and Atoms

All matter is composed of fundamental substances called **elements.** As of early 1998, 112 such elements are known, although naturally occurring matter on earth includes only 92 of them. Among these elements are such common materials as iron, copper, silver, gold, aluminum, carbon, hydrogen, and oxygen. Some elements exist in a pure form, but these and other elements are more commonly parts of chemical combinations called **compounds.**

Elements required by the body in large amounts—such as carbon, hydrogen, oxygen, nitrogen, sulfur, and phosphorus—are termed **bulk elements.** These elements make up more than 95% (by weight) of the human body (table 2.2). Elements required in small amounts are called **trace elements.** Many trace elements are important parts of enzymes, which are proteins that regulate the rates of chemical reactions in living organisms. Some elements that are toxic in large amounts, such as arsenic, may actually be vital in very small amounts, and these are called **ultratrace elements.**

Elements are composed of particles called **atoms,** which are the smallest complete units of the elements. The atoms that make up each element are chemically identical to one another, but they differ from the atoms that make up other elements. Atoms vary in size, weight, and the way they interact with one another. Some atoms, for instance, can combine either with atoms like themselves or with other kinds of atoms.

### Atomic Structure

An atom consists of a central portion called the **nucleus** and one or more **electrons** that constantly move around the nucleus. The nucleus contains one or more relatively large particles, **protons** and usually **neutrons,** whose

## TABLE 2.1  SOME PARTICLES OF MATTER

| Name | Characteristic | Name | Characteristic |
|------|----------------|------|----------------|
| Atom | Smallest particle of an element that has the properties of that element | Neutron ($n^0$) | Particle with about the same weight as a proton; uncharged and thus electrically neutral; found within a nucleus |
| Electron ($e^-$) | Extremely small particle with almost no weight; carries a negative electrical charge and is in constant motion around an atomic nucleus | Ion | Particle that is electrically charged because it has gained or lost one or more electrons |
| Proton ($p^+$) | Relatively large atomic particle; carries a positive electrical charge and is found within a nucleus | Molecule | Particle formed by the chemical union of two or more atoms |

## TABLE 2.2  MAJOR ELEMENTS IN THE HUMAN BODY (BY WEIGHT)

| Major Elements | Symbol | Approximate Percentage of the Human Body |
|----------------|--------|------------------------------------------|
| Oxygen | O | 65.0 |
| Carbon | C | 18.5 |
| Hydrogen | H | 9.5 |
| Nitrogen | N | 3.2 |
| Calcium | Ca | 1.5 |
| Phosphorus | P | 1.0 — 99.9% |
| Potassium | K | 0.4 |
| Sulfur | S | 0.3 |
| Chlorine | Cl | 0.2 |
| Sodium | Na | 0.2 |
| Magnesium | Mg | 0.1 |

**Trace Elements**

| | | |
|---|---|---|
| Cobalt | Co | |
| Copper | Cu | |
| Fluorine | F | |
| Iodine | I | less than 0.1% |
| Iron | Fe | |
| Manganese | Mn | |
| Zinc | Zn | |

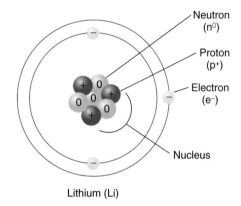

Lithium (Li)

### FIGURE 2.1

This simplified representation of an atom of lithium includes three electrons in motion around a nucleus that contains three protons and four neutrons. Circles depict electron shells.

weights are about equal, but which are otherwise quite different (fig. 2.1).

Electrons, which are so small that they have almost no weight, carry a single, negative electrical charge ($e^-$). Each proton carries a single, positive electrical charge ($p^+$). Neutrons are uncharged and thus are electrically neutral ($n^0$).

Because the nucleus contains protons, this part of an atom is always positively charged. However, the number of electrons outside the nucleus equals the number of protons, so a complete atom is said to have no net charge and is electrically neutral.

The atoms of different elements contain different numbers of protons. The number of protons in the atoms of a particular element is called its **atomic number.** Hydrogen, for example, whose atoms contain one proton, has atomic number 1; carbon, whose atoms have six protons, has atomic number 6.

The weight of an atom of an element is primarily due to the protons and neutrons in its nucleus, because the electrons have so little weight. For this reason, the number of protons plus the number of neutrons in each of an element's atoms essentially equals the **atomic weight** of that element. Thus, the atomic weight of hydrogen which has only one proton and no neutrons is approximately 1. The atomic weight of carbon with six protons and six neutrons is approximately 12 (table 2.3).

## Isotopes

All the atoms of a particular element have the same atomic number because they have the same number of protons and electrons. However, the atoms of an element

TABLE 2.3

## ATOMIC STRUCTURE OF ELEMENTS 1 THROUGH 12 (SEE ALSO APPENDIX A)

| Element | Symbol | Atomic Number | Atomic Weight | Protons | Neutrons | Electrons in Shells First | Second | Third |
|---------|--------|---------------|---------------|---------|----------|-------|--------|-------|
| Hydrogen | H | 1 | 1 | 1 | 0 | 1 | | |
| Helium | He | 2 | 4 | 2 | 2 | 2 (inert) | | |
| Lithium | Li | 3 | 7 | 3 | 4 | 2 | 1 | |
| Beryllium | Be | 4 | 9 | 4 | 5 | 2 | 2 | |
| Boron | B | 5 | 11 | 5 | 6 | 2 | 3 | |
| Carbon | C | 6 | 12 | 6 | 6 | 2 | 4 | |
| Nitrogen | N | 7 | 14 | 7 | 7 | 2 | 5 | |
| Oxygen | O | 8 | 16 | 8 | 8 | 2 | 6 | |
| Fluorine | F | 9 | 19 | 9 | 10 | 2 | 7 | |
| Neon | Ne | 10 | 20 | 10 | 10 | 2 | 8 (inert) | |
| Sodium | Na | 11 | 23 | 11 | 12 | 2 | 8 | 1 |
| Magnesium | Mg | 12 | 24 | 12 | 12 | 2 | 8 | 2 |

vary in the number of neutrons in their nuclei; thus, they vary in atomic weight. For example, all oxygen atoms have eight protons in their nuclei. Some, however, have eight neutrons (atomic weight 16), others have nine neutrons (atomic weight 17), and still others have ten neutrons (atomic weight 18). Atoms that have the same atomic numbers but different atomic weights are called **isotopes** of an element. Because a sample of an element is likely to include more than one isotope, the atomic weight of the element is often presented as the average weight of the isotopes present.

The ways atoms interact with one another are due largely to their numbers of electrons. Because the number of electrons in an atom equals its number of protons, all the isotopes of a particular element have the same number of electrons and chemically react in the same manner. For example, any of the isotopes of oxygen can have the same function in the metabolic reactions of an organism.

Isotopes of an element may be stable, or may have unstable atomic nuclei that decompose, releasing energy or pieces of themselves until they reach a stable form. Such unstable isotopes are called *radioactive,* and the energy or atomic fragments they emit are called *atomic radiation.* Elements that have radioactive isotopes include oxygen, iodine, iron, phosphorus, and cobalt. Some radioactive isotopes are used to detect and treat disease (Clinical Application 2.1).

Atomic radiation includes three common forms called alpha ($\alpha$), beta ($\beta$), and gamma ($\gamma$). Each kind of radioactive isotope produces one or more of these forms of radiation. Alpha radiation consists of particles from atomic nuclei, each of which includes two protons and two neutrons, that move relatively slowly and cannot easily penetrate matter. Beta radiation consists of much smaller particles (electrons) that travel faster and more deeply penetrate matter. Gamma radiation is similar to X-radiation and is the most penetrating of these forms.

1. What is the relationship between matter and elements?

2. Which elements are most common in the human body?

3. How are electrons, protons, and neutrons positioned within an atom?

4. What is an isotope?

5. What is atomic radiation?

## Bonding of Atoms

Atoms may combine with other atoms by forming **bonds.** When atoms form such bonds, they gain or lose electrons, or share electrons.

The electrons of an atom are found in one or more regions of space called *shells* around the nucleus. The maximum number of electrons that each of the first three inner shells can hold for elements of atomic number 18 and under is as follows:

| | |
|---|---|
| First shell (closest to the nucleus) | 2 electrons |
| Second shell | 8 electrons |
| Third shell | 8 electrons |

More complex atoms may have as many as eighteen electrons in the third shell.

Simplified diagrams such as those in figure 2.2 are used to show electron configuration in atoms. Notice that the single electron of a hydrogen atom is located in the first shell, the two electrons of a helium atom fill its first shell, and the three electrons of a lithium atom occur with two in the first shell and one in the second shell.

Atoms such as helium, whose outermost electron shells are filled, have stable structures and are chemically inactive or inert (they cannot form chemical bonds). Atoms with incompletely filled outer shells, such as those of hydrogen or lithium, tend to gain, lose, or share electrons in ways that empty or fill their outer shells. In this way they achieve stable structures.

# Radioactive Isotopes Reveal Physiology

Vicki L. arrived early at the nuclear medicine department of the health center. As she sat in an isolated cubicle, a doctor in full sterile dress approached with a small metal canister marked with numerous warnings. The doctor carefully unscrewed the top, inserted a straw, and watched as the young woman sipped the fluid within. It tasted like stale water, but was actually a solution containing a radioactive isotope, iodine-131.

Vicki's thyroid gland had been removed three months earlier, and this test was to determine whether any active thyroid tissue remained. The thyroid is the only part of the body to metabolize iodine, so if Vicki's body retained any of the radioactive drink, it would mean that some of her cancerous thyroid gland remained. By using a radioactive isotope, her physicians could detect iodine uptake using a scanning device called a scintillation counter (fig. 2A). Figure 2B illustrates iodine-131 uptake in a complete thyroid gland.

The next day, Vicki returned for the scan, which showed that a small amount of thyroid tissue was indeed left and was functioning. This meant another treatment would be necessary. Vicki would drink more of the radioactive iodine, enough to destroy the remaining tissue. This time she drank the solution while in an isolation room, which was lined with paper to keep her from contaminating the floor, walls, and furniture. The same physician administered the radioactive iodine.

**FIGURE 2A**

Physicians use scintillation counters such as this to detect radioactive isotopes.

Hydrogen (H)    Helium (He)    Lithium (Li)

**FIGURE 2.2**

The single electron of a hydrogen atom is located in its first shell. The two electrons of a helium atom fill its first shell. The three electrons of a lithium atom occur with two in the first shell and one in the second shell.

An atom of sodium, for example, has eleven electrons: two in the first shell, eight in the second shell, and one in the third shell. This atom tends to lose the single electron from its outer shell, which leaves the second shell filled and the form stable (fig. 2.3a).

A chlorine atom has seventeen electrons, with two in the first shell, eight in the second shell, and seven in the third shell. An atom of this type tends to accept a single electron, thus filling its outer shell and achieving stability.

Because each sodium atom tends to lose a single electron and each chlorine atom tends to accept a single electron, sodium and chlorine atoms react together. During this reaction, a sodium atom loses an electron and is left with eleven protons ($11^+$) in its nucleus and only ten

Vicki's physician had this job because his own thyroid had been removed many years earlier, and therefore the radiation couldn't harm him.

After two days in isolation, Vicki went home with a list of odd instructions. She was to stay away from her children and pets, wash her clothing separately, use disposable utensils and plates, and flush the toilet three times each time she used it.

These precautions would minimize her contaminating her family—mom was radioactive!

Iodine-131 is a medically useful radioactive isotope because it has a short *half-life,* a measurement of the time it takes for half of an amount of an isotope to decay to a nonradioactive form. The half-life of iodine-131 is 8.1 days. With the amount of radiation in Vicki's body dissipating by half every 8.1 days, after three months there

would be hardly any left. Doctors hoped that the remaining unhealthy thyroid cells would leave her body along with the radioactive iodine.

Isotopes of other elements have different half-lives. The half-life of iron-59 is 45.1 days; that of phosphorus-32 is 14.3 days; that of cobalt-60 is 5.26 years; and that of radium-226 is 1,620 years.

A form of thallium-201 with a half-life of 73.5 hours is commonly used to detect disorders in the blood vessels supplying the heart muscle or to locate regions of damaged heart tissue after a heart attack. Gallium-67, with a half-life of 78 hours, is used to detect and monitor the progress of certain cancers and inflammatory illnesses. These medical procedures inject the isotope into the blood and follow its path using detectors that record images on paper or film.

Radioactive isotopes are also used to assess kidney function, estimate the concentrations of hormones in body fluids, measure blood volume, and study changes in bone density. Cobalt-60 is a radioactive isotope used to treat some cancers. The cobalt emits radiation that damages cancer cells more readily than it does healthy cells. ■

Larynx

Thyroid gland

Trachea

(a)    (b)

**FIGURE 2B**

(a) A scan of the thyroid gland twenty-four hours after the patient receives radioactive iodine. Note how closely the scan in (a) resembles the shape of the thyroid gland as depicted in (b).

electrons ($10^-$). As a result, the atom develops a net electrical charge of $1^+$ and is symbolized $Na^+$. At the same time, a chlorine atom gains an electron, which leaves it with seventeen protons ($17^+$) in its nucleus and eighteen electrons ($18^-$). Thus, it develops a net electrical charge of $1^-$ and is symbolized $Cl^-$.

Atoms that have become electrically charged by gaining or losing electrons are called **ions.** Oppositely charged ions attract and may form a type of chemical bond called an **ionic bond** (electrovalent bond). Sodium ions ($Na^+$) and chloride ions ($Cl^-$) uniting in this manner form the compound sodium chloride (NaCl), or table salt (fig. 2.3*b*).

Similarly, a hydrogen atom may lose its single electron and become a hydrogen ion ($H^+$). Such an ion can

bond with a chloride ion ($Cl^-$) to form hydrogen chloride (HCl, hydrochloric acid).

Atoms may also bond by sharing electrons rather than by gaining or losing them. A hydrogen atom, for example, has one electron in its first shell, but requires two electrons to achieve a stable structure. It may fill this shell by combining with another hydrogen atom in such a way that the two atoms share a pair of electrons. As figure 2.4 shows, the two electrons then encircle the nuclei of both atoms, and each atom becomes stable. In this case the chemical bond between the atoms is called a **covalent bond.** One pair of electrons shared is a *single covalent bond;* two pairs of electrons shared is a *double covalent bond.*

Another type of chemical bond, called a *hydrogen bond,* is described later in this chapter. Clinical Application 2.2 examines how radiation that moves electrons can affect human health.

## Molecules and Compounds

When two or more atoms bond, they form a new kind of particle called a **molecule.** A **molecular formula** is used to depict the numbers and kinds of atoms in a molecule. Such a formula consists of the symbols of the elements in the molecule with numbers to indicate how many atoms of each element are present. For example, the molecular formula for water is $H_2O$, which means there are two atoms of hydrogen and one atom of oxygen in each molecule. The molecular formula for the sugar glucose is $C_6H_{12}O_6$, which means there are six atoms of carbon, twelve atoms of hydrogen, and six atoms of oxygen in a glucose molecule.

If atoms of the same element combine, they produce molecules of that element. Gases of hydrogen ($H_2$), oxygen ($O_2$), and nitrogen ($N_2$) consist of such molecules. If atoms of different elements combine, molecules of substances called **compounds** form. Two atoms of hydrogen, for example, can combine with one atom of oxygen to produce a molecule of the compound water ($H_2O$), as figure 2.5 shows. Table sugar, baking soda, natural gas, beverage alcohol, and most medical drugs are compounds.

A molecule of a compound always contains definite kinds and numbers of atoms. A molecule of water ($H_2O$), for instance, always contains two hydrogen atoms and one oxygen atom. If two hydrogen atoms combine with two oxygen atoms, the compound formed is not water, but hydrogen peroxide ($H_2O_2$).

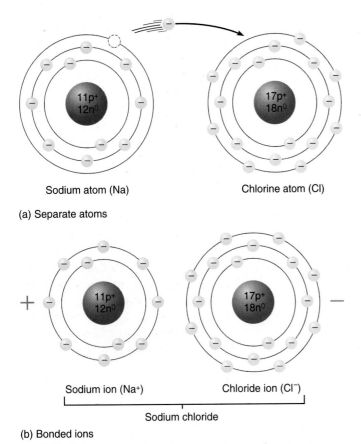

Sodium atom (Na)          Chlorine atom (Cl)

(a) Separate atoms

Sodium ion (Na⁺)          Chloride ion (Cl⁻)

Sodium chloride

(b) Bonded ions

**FIGURE 2.3**

(a) If a sodium atom loses an electron to a chlorine atom, the sodium atom becomes a sodium ion and the chlorine atom becomes a chloride ion. (b) These oppositely charged particles attract electrically and join by an ionic bond.

**1** What is an ion?

**2** Describe two ways that atoms may combine with other atoms.

**3** Distinguish between a molecule and a compound.

Usually atoms of each element form a specific number of chemical bonds. Hydrogen atoms form single bonds, oxygen atoms form two bonds, nitrogen atoms form three bonds, and carbon atoms form four bonds. Symbols and lines can be used to represent the bonding capacities of these atoms, as follows:

$$-\text{H} \qquad\qquad -\text{O}-$$

$$-\text{N}- \qquad\qquad -\overset{|}{\underset{|}{\text{C}}}-$$

Hydrogen atom    +    Hydrogen atom          Hydrogen molecule

**FIGURE 2.4**

A hydrogen molecule forms when two hydrogen atoms share a pair of electrons and join by a covalent bond.

# Ionizing Radiation: A Legacy of the Cold War

Alpha, beta, and gamma radiation are called ionizing radiation because their energy adds or removes electrons from atoms (fig. 2C). Electrons dislodged by ionizing radiation can affect nearby atoms, disrupting physiology at the chemical level in a variety of ways—causing cancer, clouding the lens of the eye, and interfering with normal growth and development.

In the United States, most people are exposed to very low levels of ionizing radiation, mostly from background radiation, which originates from natural environmental sources (Table 2A). This is not true, however, for people who live near sites of atomic weapons manufacture. Epidemiologists are now studying recently uncovered medical records that document illnesses linked to long-term exposure to ionizing radiation in a 1,200-square kilometer area in former East Germany. It is a frightening tale.

Today, the lake near Oberrothenback, Germany, appears inviting, but looks are deceiving. The lake contains enough toxins to kill thousands of people, its water polluted with heavy metals, low-level radioactive chemical waste, and 22,500 tons of arsenic. Radon, a radioactive byproduct of uranium, permeates the soil. High death rates among farm animals and pets have been traced to their drinking from the polluted lake. Cancer rates and respiratory disorders among the human residents nearby are far above normal. This isn't surprising, given the region's toxic history.

The lake in Oberrothenback once served as a dump for a factory that produced "yellow cake," a term for processed uranium ore, which was used to build atomic bombs for the former Soviet Union. In the early 1950s, nearly half a million workers labored here and in surrounding areas in factories and mines. Records released in 1989, after the reunification of Germany, reveal that workers were given perks, such as alcoholic beverages and better wages, to work in the more dangerous areas. The workers paid a heavy price: tens of thousands died of lung ailments.

Today, these health records may answer a long-standing question: What are the effects of exposure to long-term, low-level ionizing radiation? Until now, the risks of such exposure have been extrapolated from health statistics amassed for the victims, survivors, and descendants of the atomic blasts in Japan in the Second World War. But a single exposure, such as a bomb blast, may not have the same effect on the human body as extended exposure, such as the uranium workers experienced. The cold war may be over, but a lethal legacy of its weapons remains. ■

**FIGURE 2C**

(*a*) Ionizing radiation may dislodge an electron from an electrically neutral hydrogen atom. (*b*) Without its electron, the hydrogen atom becomes a positively charged hydrogen ion ($H^+$).

(a) Hydrogen atom (H)

(b) Hydrogen ion ($H^+$)

Ionizing radiation

Dislodged electron

| TABLE 2A | SOURCES OF IONIZING RADIATION | |
|---|---|---|
| Background (Natural environmental) | Cosmic rays from space | |
| | Radioactive elements in earth's crust | |
| | Rocks and clay in building materials | |
| | Radioactive elements naturally in the body (phosphorus-40, carbon-14) | |
| Medical and dental | X rays | |
| | Radioactive substances | |
| Other | Atomic and nuclear weapons | |
| | Mining and processing radioactive minerals | |
| | Radioactive fuels in nuclear power plants | |
| | Radioactive elements in consumer products (luminescent dials, smoke detectors, color TV components) | |

**FIGURE 2.5**

Under certain conditions, hydrogen molecules can combine with oxygen molecules to form water molecules.

| $H-H$ | $O=O$ | H⧵O⧸H | $O=C=O$ |
|:---:|:---:|:---:|:---:|
| $H_2$ | $O_2$ | $H_2O$ | $CO_2$ |

**FIGURE 2.6**

Structural formulas of molecules of hydrogen, oxygen, water, and carbon dioxide. Note the double covalent bonds.

Representations such as these show how atoms bond and arrange in various molecules. Single lines represent single bonds and double lines represent double bonds. Illustrations of this type are called **structural formulas** (fig. 2.6).

## Chemical Reactions

Chemical reactions form or break bonds between atoms, ions, or molecules. Those being changed by the chemical reaction are called **reactants.** Those formed at the reaction's conclusion are called **products.** When two or more atoms, ions, or molecules bond to form a more complex structure, as when hydrogen and oxygen atoms bond to form molecules of water, the reaction is called **synthesis.** Such a reaction can be symbolized this way:

$$A + B \rightarrow AB$$

If the bonds of a reactant molecule break to form simpler molecules, atoms, or ions, the reaction is called **decomposition.** For example, molecules of water can decompose to yield the products hydrogen and oxygen. Decomposition is symbolized as follows:

$$AB \rightarrow A + B$$

Synthetic reactions, which build larger molecules from smaller ones, are particularly important in growth of body parts and repair of worn or damaged tissues. De-

composition reactions occur when food substances are digested and they release energy.

A third type of chemical reaction is an **exchange reaction** (replacement reaction). In this reaction, parts of two different kinds of molecules trade positions. The reaction is symbolized as follows:

$$AB + CD \rightarrow AD + CB$$

An example of an exchange reaction is an acid reacting with a base, producing water and a salt. This type of reaction is discussed in the following section.

Many chemical reactions are reversible. This means the product or products can change back to the reactant or reactants. A **reversible reaction** is symbolized using a double arrow, as follows:

$$A + B \rightleftarrows AB$$

Whether a reversible reaction proceeds in one direction or another depends on such factors as the relative proportions of reactant (or reactants) and product (or products) as well as the amount of energy available. **Catalysts** are molecules that influence the rates of chemical reactions, but are not consumed in the reaction.

## Acids, Bases, and Salts

Some compounds release ions (ionize) when they dissolve in water or react with water molecules. Sodium chloride (NaCl), for example, ionizes into sodium ions ($Na^+$) and chloride ions ($Cl^-$) when it dissolves (fig. 2.7). This reaction is represented as:

$$NaCl \rightarrow Na^+ + Cl^-$$

Because the resulting solution contains electrically charged particles (ions), it will conduct an electric current. Substances that release ions in water are, therefore, called **electrolytes.** Electrolytes that release hydrogen

**FIGURE 2.7**

Crystals of table salt (NaCl) dissolve in water, releasing sodium ions (Na⁺) and chloride ions (Cl⁻).

| TABLE 2.4 | TYPES OF ELECTROLYTES | |
|---|---|---|
| | **Characteristic** | **Examples** |
| Acid | Substance that releases hydrogen ions (H⁺). | Carbonic acid, hydrochloric acid, acetic acid, phosphoric acid |
| Base | Substance that releases ions that can combine with hydrogen ions. | Sodium hydroxide, potassium hydroxide, magnesium hydroxide, sodium bicarbonate |
| Salt | Substance formed by the reaction between and acid and a base. | Sodium chloride, aluminum chloride, magnesium sulfate |

ions ($H^+$) in water are called **acids.** For example, in water the compound hydrochloric acid (HCl) releases hydrogen ions ($H^+$) and chloride ions ($Cl^-$):

$$HCl \rightarrow H^+ + Cl^-$$

Electrolytes that release ions that combine with hydrogen ions are called **bases.** The compound sodium hydroxide (NaOH) in water releases hydroxyl ions ($OH^-$). The hydroxyl ions, in turn, can combine with hydrogen ions to form water. Thus, sodium hydroxide is a base:

$$NaOH \rightarrow Na^+ + OH^-$$

(*Note:* Some ions, such as $OH^-$, contain two or more atoms. However, such a group usually behaves like a single atom and remains unchanged during a chemical reaction.)

Acids and bases can react to form water and electrolytes called **salts.** For example, hydrochloric acid and sodium hydroxide react to form water and sodium chloride:

$$HCl + NaOH \rightarrow H_2O + NaCl$$

Table 2.4 summarizes the three types of electrolytes.

## Acid and Base Concentrations

Concentrations of acids and bases affect the chemical reactions that constitute many life processes, such as those controlling breathing rate. Thus, the concentrations of these substances in body fluids are of special importance.

Hydrogen ion concentration can be measured in grams of ions per liter of solution. However, because hydrogen ion concentration can cover such a wide range (gastric juice is 0.01 grams $H^+$/liter; household ammonia has 0.00000000001 grams $H^+$/liter), a shorthand system called the **pH scale** has been developed. This system

tracks the number of decimal places in a hydrogen ion concentration without having to write them out. For example, a solution with a hydrogen ion concentration of 0.1 grams per liter has a pH value of 1.0; a concentration of 0.01 g $H^+$/L has pH 2.0; 0.001 g $H^+$/L has pH 3.0, and so forth. Between each whole number on the pH scale, which extends from pH 0 to pH 14.0, there is a tenfold difference in hydrogen ion concentration. Note that as hydrogen ion concentration increases, pH value decreases.

In pure water, which ionizes only slightly, the hydrogen ion concentration is 0.0000001 g/L, and the pH is 7.0. Because water ionizes to release equal numbers of acidic hydrogen ions and basic hydroxyl ions, it is *neutral.*

$$H_2O \rightarrow H^+ + OH^-$$

The concentrations of hydrogen ions and hydroxyl ions are always in balance, such that if one increases, the other decreases, and vice versa. Solutions with more hydrogen ions than hydroxyl ions are *acidic.* That is, acidic solutions have pH values less than 7.0 (fig. 2.8). Solutions with fewer hydrogen ions than hydroxyl ions are *basic* (alkaline); that is, they have pH values greater than 7.0.

Table 2.5 summarizes the relationship between hydrogen ion concentration and pH. Chapter 21 discusses the regulation of hydrogen ion concentrations in the internal environment.

Many fluids in the human body function within a narrow pH range. Illness results when pH changes. The normal pH of blood, for example, is 7.35 to 7.45. Blood pH of 7.5 to 7.8, called **alkalosis,** makes one feel agitated and dizzy. This can be caused by breathing rapidly at high altitudes, taking too many antacids, high fever, anxiety, or mild to moderate vomiting that rids the body of stomach acid. **Acidosis,** in which blood pH falls to 7.0 to 7.3, makes one feel disoriented and fatigued, and breathing may become difficult. This condition can result from

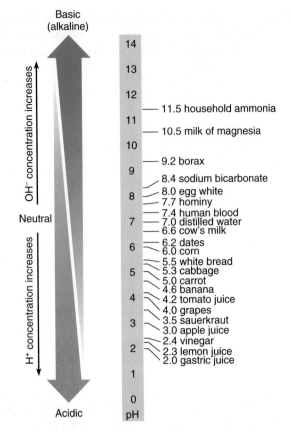

**FIGURE 2.8**

As the concentration of hydrogen ions ($H^+$) increases, a solution becomes more acidic, and the pH value decreases. As the concentration of hydrogen ion acceptors (such as hydroxyl or bicarbonate ions) increases, a solution becomes more basic, and the pH value increases. Note the pH of some common substances.

| TABLE 2.5 HYDROGEN ION CONCENTRATIONS AND PH | | |
|---|---|---|
| **Grams of H⁺ per Liter** | **pH** | |
| 0.00000000000001 | 14 | ↑ |
| 0.0000000000001 | 13 | |
| 0.000000000001 | 12 | |
| 0.00000000001 | 11 | Increasingly basic |
| 0.0000000001 | 10 | |
| 0.000000001 | 9 | |
| 0.00000001 | 8 | |
| 0.0000001 | 7 | Neutral—neither acidic |
| 0.000001 | 6 | nor basic |
| 0.00001 | 5 | |
| 0.0001 | 4 | |
| 0.001 | 3 | Increasingly acidic |
| 0.01 | 2 | |
| 0.1 | 1 | ↓ |
| 1.0 | 0 | |

severe vomiting that empties the alkaline small intestinal contents, diabetes, brain damage, impaired breathing, and lung and kidney disease.

**1** What is a molecular formula? A structural formula?

**2** Describe three kinds of chemical reactions.

**3** Compare the characteristics of an acid with those of a base.

**4** What is pH?

## CHEMICAL CONSTITUENTS OF CELLS

The chemicals that enter into metabolic reactions or are produced by them can be divided into two large groups. Generally, those that contain carbon and hydrogen atoms are called **organic;** the rest are called **inorganic.**

Inorganic substances usually dissolve in water or react with water to release ions; thus, they are *electrolytes.* Many organic compounds also dissolve in water, although as a group they are more likely to dissolve in organic liquids such as ether or alcohol. Organic compounds that dissolve in water usually do not release ions, and are therefore called *nonelectrolytes.*

### Inorganic Substances

Common inorganic substances in cells include water, oxygen, carbon dioxide, and inorganic salts.

#### Water

Water ($H_2O$) is the most abundant compound in living material and accounts for about two-thirds of the weight of an adult human. It is the major component of blood and other body fluids, including those within cells.

When substances dissolve in water, relatively large pieces of the substances break into smaller ones, and eventually, molecules or ions result. These particles, which may be ions, are much more likely to react with one another than were the original large pieces. Consequently, most metabolic reactions occur in water.

Water also plays an important role in transporting chemicals within the body. Blood, which is more than 90% water, carries many vital substances, such as oxygen, sugars, salts, and vitamins, from organs of the digestive and respiratory systems to cells. Water also carries waste materials, such as carbon dioxide and urea, from these cells to the lungs and kidneys, respectively, which remove them from the blood and release them outside the body.

In addition, water can absorb and transport heat. Blood carries heat released from muscle cells during exercise from deeper parts of the body to the surface. At the same time, water released by skin cells in the form of perspiration can carry heat away by evaporation.

## Oxygen

Molecules of oxygen gas ($O_2$) enter the internal environment through the respiratory organs and are transported throughout the body by the blood, especially by red blood cells. Within cells, organelles use oxygen to release energy from nutrient molecules. The released energy is used to drive the cell's metabolic activities. A continuing supply of oxygen is necessary for cell survival and, ultimately, for the survival of the organism.

NO (nitric oxide) and CO (carbon monoxide) are two small chemicals with bad reputations. NO is found in smog, cigarettes, and acid rain. CO is a colorless, odorless, lethal gas that is notorious for causing death when it leaks from home heating systems or exhaust pipes in closed garages. However, NO and CO are important in physiology as biological messenger molecules. NO is involved in digestion, memory, immunity, respiration, and circulation. CO functions in the spleen, which recycles old red blood cells, and in the parts of the brain that control memory, smell, and vital functions.

## Carbon Dioxide

Carbon dioxide ($CO_2$) is a simple, carbon-containing inorganic compound. It is produced as a waste product when energy is released during certain metabolic processes. As it moves from cells into surrounding body fluids and blood, most of the carbon dioxide reacts with water to form a weak acid (carbonic acid, $H_2CO_3$). This acid ionizes, releasing hydrogen ions ($H^+$) and bicarbonate ions ($HCO_3^-$), which blood carries to the respiratory organs. There, the chemical reactions reverse, and carbon dioxide gas is produced, eventually to be exhaled.

## Inorganic Salts

Inorganic salts are abundant in body fluids. They are the sources of many necessary ions, including ions of sodium ($Na^+$), chloride ($Cl^-$), potassium ($K^+$), calcium ($Ca^{+2}$), magnesium ($Mg^{+2}$), phosphate ($PO_4^{-3}$), carbonate ($CO_3^{-2}$), bicarbonate ($HCO_3^-$), and sulfate ($SO_4^{-2}$). These ions play important roles in metabolic processes, helping to maintain proper water concentrations in body fluids, pH, blood clotting, bone development, energy transfer within cells, and muscle and nerve functions.

These electrolytes are regularly gained and lost by the body, but must be present in certain concentrations, both inside and outside cells, to maintain homeostasis.

Such a condition is called **electrolyte balance.** Disrupted electrolyte balance occurs in certain diseases, and modern medical treatment places considerable emphasis on restoring it.

Table 2.6 summarizes the functions of some of the inorganic components of cells.

**1** What are the general differences between an organic molecule and an inorganic molecule?

**2** What is the difference between an electrolyte and a nonelectrolyte?

**3** Define electrolyte balance.

## Organic Substances

Important groups of organic substances in cells include carbohydrates, lipids, proteins, and nucleic acids.

### Carbohydrates

**Carbohydrates** provide much of the energy that cells require. They also supply materials to build certain cell structures, and they often are stored as reserve energy supplies.

Carbohydrates are water-soluble molecules that contain atoms of carbon, hydrogen, and oxygen. These molecules usually have twice as many hydrogen as oxygen atoms, the same ratio of hydrogen to oxygen as in water molecules ($H_2O$). This ratio is easy to see in the molecular formulas of the carbohydrates glucose ($C_6H_{12}O_6$) and sucrose ($C_{12}H_{22}O_{11}$).

Carbohydrates are classified by size. Simple carbohydrates, or **sugars,** include the **monosaccharides** (single sugars) and **disaccharides** (double sugars). A monosaccharide may include from three to seven carbon atoms, occurring in a straight chain or a ring (fig. 2.9). Monosaccharides include glucose (dextrose), fructose, and galactose. Disaccharides consist of two 6-carbon units. Sucrose (table sugar) and lactose (milk sugar) are disaccharides (fig. 2.10a and b).

Complex carbohydrates, also called **polysaccharides,** are built of simple carbohydrates (fig. 2.10c). Plant starch is an example. Starch molecules consist of highly branched chains, each containing many glucose units.

Animals, including humans, synthesize a polysaccharide similar to starch called *glycogen.* Its molecules also consist of branched chains of sugar units; each branch consists of a dozen or fewer glucose units.

### Lipids

**Lipids** are a group of organic chemicals that are insoluble in water but soluble in organic solvents, such as ether and chloroform. Lipids include a number of compounds, such as fats, phospholipids, and steroids, that have vital functions in cells and are important constituents of cell membranes (see chapter 3). The most common lipids are the *fats,* which are primarily used to supply energy for

TABLE 2.6

## INORGANIC SUBSTANCES COMMON IN CELLS

| Substance | Symbol or Formula | Functions |
|---|---|---|
| **I. Inorganic Molecules** | | |
| Water | $H_2O$ | Major component of body fluids (chapter 21); medium in which most biochemical reactions occur; transports various chemical substances (chapter 14); helps regulate body temperature (chapter 6) |
| Oxygen | $O_2$ | Used in release of energy from glucose molecules (chapter 4) |
| Carbon dioxide | $CO_2$ | Waste product that results from metabolism (chapter 4); reacts with water to form carbonic acid (chapter 19) |
| **II. Inorganic Ions** | | |
| Bicarbonate ions | $HCO_3^-$ | Help maintain acid-base balance (chapter 21) |
| Calcium ions | $Ca^{+2}$ | Necessary for bone development (chapter 7); muscle contraction (chapter 9) and blood clotting (chapter 14) |
| Carbonate ions | $CO_3^{-2}$ | Component of bone tissue (chapter 7) |
| Chloride ions | $Cl^-$ | Help maintain water balance (chapter 21) |
| Hydrogen ions | $H^+$ | pH of the internal environment (chapters 14, 19, 21) |
| Magnesium ions | $Mg^{+2}$ | Component of bone tissue (chapter 7); required for certain metabolic processes (chapter 18) |
| Phosphate ions | $PO_4^{-3}$ | Required for synthesis of ATP, nucleic acids, and other vital substances (chapter 4); component of bone tissue (chapter 7); helps maintain polarization of cell membranes (chapter 10) |
| Potassium ions | $K^+$ | Required for polarization of cell membranes (chapter 10) |
| Sodium ions | $Na^+$ | Required for polarization of cell membranes (chapter 10); helps maintain water balance (chapter 21) |
| Sulfate ions | $SO_4^{-2}$ | Help maintain polarization of cell membranes (chapter 10) and acid-base balance (chapter 21) |

**FIGURE 2.9**

(a) Molecules of the monosaccharide glucose ($C_6H_{12}O_6$) may have a straight chain of carbon atoms. (b) More commonly, glucose molecules form a ring structure. (c) This shape symbolizes the ring structure of a glucose molecule.

cellular activities. Fat molecules can supply more energy gram for gram than can carbohydrate molecules. This is why eating a fatty diet leads to weight gain.

Like carbohydrates, fat molecules are composed of carbon, hydrogen, and oxygen atoms. However, they contain a much smaller proportion of oxygen than do carbohydrates. The formula for the fat *tristearin,* $C_{57}H_{110}O_6$, illustrates these characteristic proportions.

The building blocks of fat molecules are **fatty acids** and **glycerol.** Although the glycerol portion of every fat molecule is the same, there are many kinds of fatty acids

and, therefore, many kinds of fats. All fatty acid molecules contain a carboxyl group (—COOH) at the end of a chain of carbon atoms. Fatty acids differ in the lengths of their carbon atom chains, although such chains usually contain an even number of carbon atoms. The fatty acid chains also may vary in the ways the carbon atoms join. In some cases, the carbon atoms are all linked by single carbon-carbon bonds. This type of fatty acid is *saturated;* that is, each carbon atom binds as many hydrogen atoms as possible and is thus saturated with hydrogen atoms. Other fatty acid chains do not bind their maximum num-

(a) Monosaccharide    (b) Disaccharide

(c) Polysaccharide

**FIGURE 2.10**

(a) A monosaccharide molecule consisting of one 6-carbon building block. (b) A disaccharide molecule consisting of two of these building blocks. (c) A polysaccharide molecule consisting of many building blocks, which may form branches.

(a) Saturated fatty acid

(b) Unsaturated fatty acid

**FIGURE 2.11**

(a) A molecule of saturated fatty acid and (b) a molecule of unsaturated fatty acid. Double bonds are shown in red. Note that they cause a "kink" in the shape of the molecule.

ber of hydrogen atoms. Therefore, they have one or more double bonds between carbon atoms. Fatty acids with one double bond are called *monounsaturated,* and those with two or more double bonds are *polyunsaturated* (fig 2.11).

Fatty acids and glycerol are united so that each glycerol molecule combines with three fatty acid molecules. The result is a single fat molecule or *triglyceride* (fig. 2.12). Fat molecules that contain only saturated fatty acids are called **saturated fats,** and those that include unsaturated fatty acids are called **unsaturated fats.** Each kind of fat molecule has distinct properties.

A *phospholipid* molecule is similar to a fat molecule in that it contains a glycerol portion and fatty acid chains. The phospholipid, however, has only two fatty acid chains, and in place of the third one has a portion containing a phosphate group. This phosphate-containing portion is soluble in water (hydrophilic) and forms the "head" of the molecule, whereas the fatty acid portion is insoluble in water (hydrophobic) and forms a "tail." Figure 2.13 illustrates the molecular structure of cephalin, a phospholipid in blood. Other phospholipids are important in cellular structures.

A diet rich in saturated fat increases a person's risk of developing atherosclerosis, a serious disease that causes obstruction of certain blood vessels. It is healthful to substitute unsaturated, particularly monounsaturated, fats for dietary saturated fats.

Saturated fats are more abundant in fatty foods that are solids at room temperature, such as butter, lard, and most other animal fats. Unsaturated fats are plentiful in fatty foods that are liquids at room temperature, such as soft margarine and seed oils, including corn oil and soybean oil. Coconut and palm oils, however, are exceptions—they are relatively high in saturated fat.

*Steroid* molecules are complex structures that include connected rings of carbon atoms (fig. 2.14). Among the more important steroids are cholesterol, which is in all body cells and is used to synthesize other steroids; sex hormones, such as estrogen, progesterone, and testosterone; and several hormones from the adrenal glands. Chapters 13, 14, 20, 21, and 22 discuss these

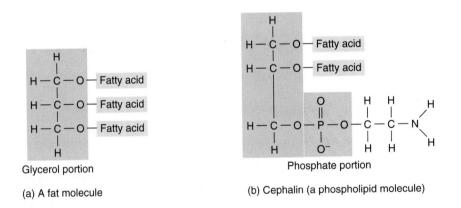

**FIGURE 2.12**

A triglyceride molecule (fat) consists of a glycerol and three fatty acid "tails."

(a) A fat molecule

(b) Cephalin (a phospholipid molecule)

**FIGURE 2.13**

(a) A fat molecule (triglyceride) contains a glycerol and three fatty acids. (b) In a phospholipid molecule, a phosphate-containing group replaces one fatty acid.

(a) Structure of a steroid

(b) Cholesterol

**FIGURE 2.14**

(a) The general structure of a steroid. (b) The structural formula for cholesterol, a steroid widely distributed in the body.

TABLE 2.7

## IMPORTANT GROUPS OF LIPIDS

| Group | Basic Molecular Structure | Characteristics |
|---|---|---|
| Triglycerides | Three fatty acid molecules bound to a glycerol molecule | Most common lipid in the body; stored in fat tissue as an energy supply. Fat tissue also provides insulation beneath the skin. |
| Phospholipids | Two fatty acid molecules and a phosphate group bound to a glycerol molecule (may also include a nitrogen-containing molecule attached to the phosphate group) | Used as structural components in cell membranes; large amounts are in the liver and parts of the nervous system. |
| Steroids | Four connected rings of carbon atoms | Widely distributed in the body with a variety of functions; includes cholesterol, sex hormones, and certain hormones of the adrenal glands. |

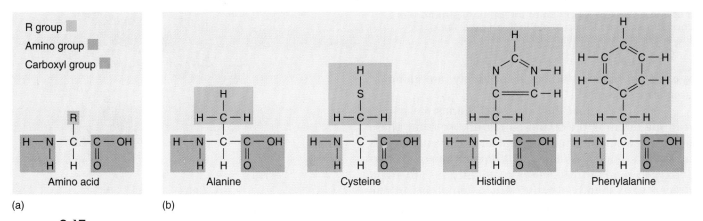

(a)  (b)

**FIGURE 2.15**

(a) The general structure of an amino acid. Note the amino group (pink) and carboxyl group (blue) that are common to all amino acid molecules. (b) Some representative amino acids and their structural formulas. Each amino acid molecule has a particular shape. Also note the different R groups.

steroids. Table 2.7 summarizes the molecular structures and characteristics of lipids.

## Proteins

**Proteins** have a great variety of functions. Some proteins serve as structural materials, energy sources, or chemical messengers (hormones). Other proteins combine with carbohydrates (glycoproteins) and function as receptors on cell surfaces that bind to particular kinds of molecules. Yet others act as weapons (antibodies) against substances that are foreign to the body. Many proteins play vital roles in metabolic processes as **enzymes.** Enzymes are molecules that act as catalysts in living systems. That is, they speed specific chemical reactions without being consumed in the process. (Enzymes are discussed in chapter 4.)

Like carbohydrates and lipids, proteins consist of atoms of carbon, hydrogen, and oxygen. In addition, proteins always contain nitrogen atoms, and sometimes contain sulfur atoms as well. The building blocks of proteins are smaller molecules called **amino acids.**

Each amino acid molecule has an amino group ($-NH_2$) at one end and a carboxyl group ($-COOH$) at the other end. Between these end groups is a single carbon atom (central carbon), and it, in turn, is bound to a hy-

drogen atom and to another group of atoms called a *side chain* or R group. The composition of the side chain distinguishes one type of amino acid from another (fig. 2.15). About twenty different kinds of amino acids comprise proteins in organisms.

Proteins have complex shapes, yet the way they are put together is surprisingly simple. Each type of protein molecule includes specific numbers of different amino acids connected by peptide bonds (special covalent bonds) in a particular linear sequence (primary structure), as figure 2.16 shows. These chains vary in length from less than 100 to more than 5,000 amino acids. The amino acid chain usually twists to form a coil (secondary structure) (fig. 2.17), and this, in turn, may fold into a three-dimensional form or **conformation** (tertiary structure) (fig. 2.18).

A protein's conformation determines its function. The amino acid sequence and interactions between the amino acids of a protein determine the conformation.

Different kinds of protein molecules have different conformations that make their particular functions possible. Some protein molecules are long and fibrous, such as the keratin protein that forms hair, and the threads of fibrin protein that knit a blood clot. Many proteins are globular. Myoglobin and hemoglobin, which transport

(a)

(b)

**FIGURE** 2.16

Primary structure of a protein. (a) A peptide bond between two amino acids. (b) A portion of the primary structure or amino acid sequence of a protein. Each different color represents a different amino acid.

oxygen in muscle and blood, respectively, are globular, as are many enzymes.

Weak attractions called **hydrogen bonds** are important in maintaining the conformation of a protein. These bonds form when electrons are not shared evenly in certain covalent bonds. As the result of such unequal sharing, oxygen tends to become very slightly negative and hydrogen very slightly positive. Hydrogen atoms in one part of a protein may then attract oxygen atoms elsewhere in the molecule (see fig. 2.17). These forces contribute to the molecule's overall shape.

Various treatments can cause the secondary and tertiary structures of a protein's conformation to fall apart, or *denature.* Because the primary structure (amino acid sequence) remains, sometimes the protein can regain its shape when normal conditions return. High temperature, radiation, pH changes, and certain chemicals (such as urea) can denature proteins.

A familiar example of irreversible protein denaturation is the response of the protein albumin to heat (for example, cooking an egg white). A permanent wave that curls hair also results from protein denaturation. Chemicals first break apart the tertiary structure formed when sulfur-containing amino acids attract each other within keratin molecules. This relaxes the hair. When the chemicals are washed out and the hair set, the sulfur bonds reform, but in different places, changing the appearance of the hair.

## Nucleic Acids

**Nucleic acids** constitute genes, the instructions that control a cell's activities, and play important roles in protein synthesis. These molecules are very large and complex. They contain atoms of carbon, hydrogen, oxygen, nitrogen, and phosphorus, which form building blocks called **nucleotides.** Each nucleotide consists of a 5-carbon sugar (ribose or deoxyribose), a phosphate group, and one of several organic bases (fig. 2.19). Such nucleotides, linked in a chain, form a **polynucleotide** (fig. 2.20).

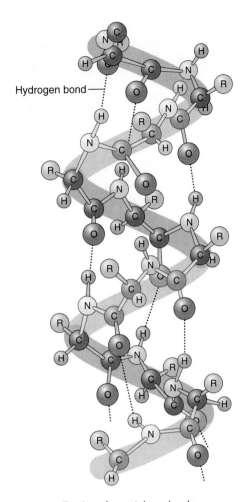

Hydrogen bond

Portion of a protein molecule

**FIGURE** 2.17

Secondary structure of a protein. The amino acid chain of a protein molecule sometimes twists, forming a coil held together by hydrogen bonds.

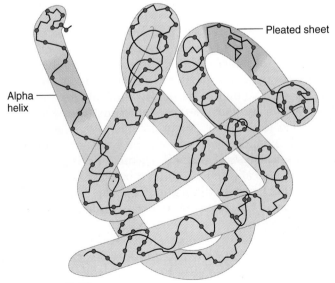

Pleated sheet

Alpha helix

**FIGURE** 2.18

Tertiary structure of a protein. The coiled amino acid chain of a protein molecule folds into a unique three-dimensional conformation.

There are two major types of nucleic acids. One type is composed of molecules whose nucleotides contain ribose sugar; it is called **RNA** (ribonucleic acid) and it is a single polynucleotide chain. The nucleotides of the second type contain deoxyribose sugar; nucleic acid of this type is called **DNA** (deoxyribonucleic acid) and it is a double

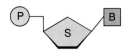

**FIGURE** 2.19

A nucleotide consists of a 5-carbon sugar (S), a phosphate group (P), and an organic base (B).

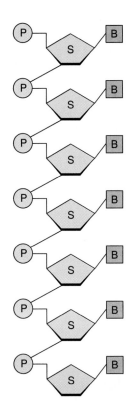

**FIGURE** 2.20

A schematic representation of a polynucleotide chain. A nucleic acid molecule consists of one (RNA) or two (DNA) polynucleotide chains.

polynucleotide chain. Figure 2.21 compares the structure of ribose and deoxyribose, which differ by one oxygen atom. DNA and RNA also differ in the types of bases they contain.

DNA molecules store information in a type of molecular code. Cells use this information to construct specific protein molecules, which have a wide variety of functions. RNA molecules help to synthesize proteins.

DNA molecules have a unique ability to make copies of, or replicate, themselves. They replicate prior to cell reproduction, and each newly formed cell receives an exact copy of the original cell's DNA molecules. Chapter 4 discusses the storage of information in nucleic acid molecules, use of the information in the manufacture of protein molecules, and the function of these proteins in controlling metabolic reactions.

Table 2.8 summarizes the four groups of organic compounds. Figure 2.22 shows three-dimensional (space-filling) models of some important molecules, illustrating their shapes. Clinical Application 2.3 describes two techniques used to view human anatomy and physiology.

**1** Compare the chemical composition of carbohydrates, lipids, proteins, and nucleic acids.

**2** How does an enzyme affect a chemical reaction?

**3** What is likely to happen to a protein molecule that is exposed to intense heat or radiation?

**4** What are the functions of DNA and RNA?

Ribose          Deoxyribose

**FIGURE** 2.21

The molecules of ribose and deoxyribose differ by a single oxygen atom.

| TABLE 2.8 | ORGANIC COMPOUNDS IN CELLS | | | | |
| --- | --- | --- | --- | --- | --- |
| **Compound** | **Elements Present** | **Building Blocks** | **Functions** | | **Examples** |
| Carbohydrates | C,H,O | Simple sugar | Provide energy, cell structure | | Glucose, starch |
| Lipids | C,H,O (often P) | Glycerol, fatty acids, phosphate groups | Provide energy, cell structure | | Triglycerides, phospholipids, steroids |
| Proteins | C,H,O,N (often S) | Amino acids | Provide cell structure, enzymes, energy | | Albumins, hemoglobin |
| Nucleic acids | C,H,O,N,P | Nucleotides | Store information for the synthesis of proteins, control cell activities | | RNA, DNA |

# CT Scanning and PET Imaging

Physicians use two techniques—computerized tomography (CT) scanning and positron emission tomography (PET imaging)—to paint portraits of anatomy and physiology, respectively.

In CT scanning, an X-ray emitting device is positioned around the region of the body being examined. At the same time, an X-ray detector is moved in the opposite direction on the other side of the body. As these parts move, an X-ray beam passes through the body from hundreds of different angles. Because tissues and organs of varying composition absorb X rays differently, the intensity of X rays reaching the detector varies from position to position. A computer records the measurements made by the X-ray detector and combines them mathematically. This creates on a viewing screen a sectional image of the internal body parts (fig. 2D).

Ordinary X-ray techniques produce two-dimensional images known as radiographs or X rays or films. A CT scan provides three-dimensional information. The CT scan can also clearly differentiate between soft tissues of slightly different densities, such as the liver and kidneys, which cannot be seen in a conventional X-ray image. Thus, a CT scan can often spot abnormal tissue, such as a tumor. For example, a CT scan can tell whether a sinus headache that does not respond to antibiotic therapy is caused by a drug-resistant infection or a tumor.

PET imaging uses radioactive isotopes that naturally emit positrons, which are atypical positively charged electrons, to detect biochemical activity in a specific body part. Useful isotopes in PET imaging include carbon-11, nitrogen-13, oxygen-15, and fluorine-18. When one of these isotopes releases a positron, it interacts with a nearby negatively charged electron. The two particles destroy each other, an event called annihilation. At the moment of destruction, two gamma rays appear and move away from each other in opposite directions. Special equipment detects the gamma radiation.

To produce a PET image of biochemically active tissue, a person is injected with a metabolically active compound that includes a bound positron-emitting isotope. To study the brain, for example, a person is injected with glucose-containing fluorine-18. After the brain takes up the isotope-tagged compound, the person rests her head within a circular array of radiation detectors. A device records each time two gamma rays are emitted simultaneously and travel in opposite directions (the result of annihilation). A computer collects and combines the data and generates a cross-sectional image. The image indicates the location and relative concentration of the radioac-

(a)

(b)

**FIGURE** 2D

CT scans of (a) the head and (b) the abdomen.

tive isotope in different regions of the brain, and can be used to study those parts metabolizing glucose.

PET images reveal the parts of the brain that are affected in such disorders as Huntington disease, Parkinson's disease, epilepsy, and Alzheimer's disease, and they are used to study blood flow in vessels supplying the brain and heart. The technology is invaluable for detecting the physiological bases of poorly understood behavioral disorders, such as obsessive-compulsive disorder. In this condition, a person repeatedly performs a certain behavior, such as washing hands, showering, locking doors, or checking to see that the stove is turned off. PET images of people with this disorder reveal intense activity in two parts of the brain that are quiet in the brains of unaffected individuals. Knowing the site of altered brain activity can help researchers develop more directed drug therapy.

In addition to highlighting biochemical activities behind illness, PET scans allow biologists to track normal brain physiology. Figure 2E shows that different patterns of brain activity are associated with learning and with reviewing something already learned. ■

**FIGURE 2E**

These PET images demonstrate brain changes that accompany learning. The top and bottom views show different parts of the same brain. The "naive" brain on the left has been given a list of nouns, and asked to visualize each word. In the middle column, the person has practiced the task, so he can picture the nouns with less brain activity. In the third column, the person receives a new list of nouns. Learning centers in the brain show increased activity.

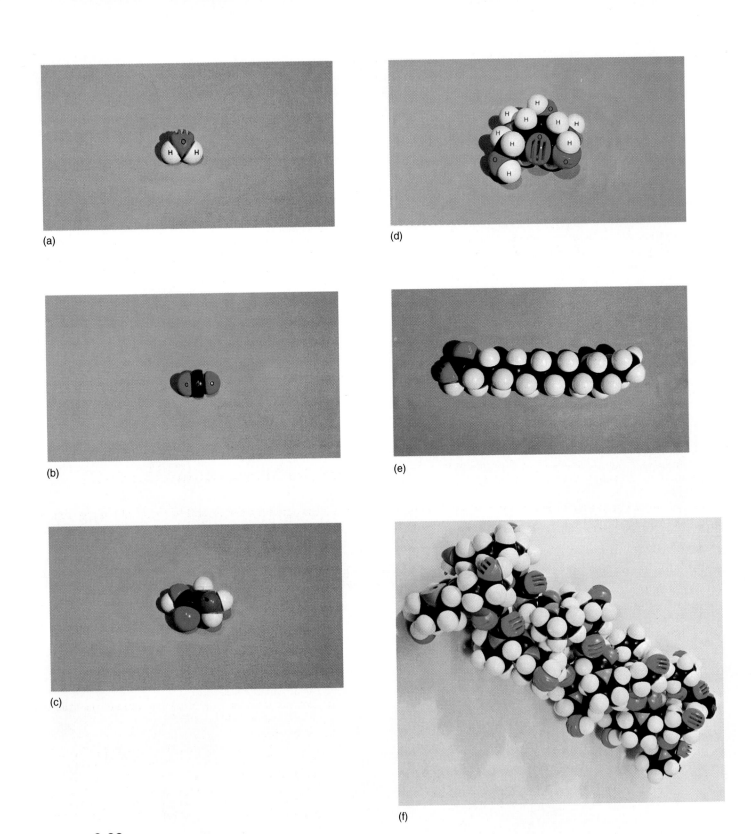

FIGURE 2.22

These three-dimensional (space-filling) models show the relative sizes of several important molecules: (a) water, (b) carbon dioxide, (c) glycine (an amino acid), (d) glucose (a monosaccharide), (e) a fatty acid, and (f) collagen (a protein). White = hydrogen, red = oxygen, blue = nitrogen, black = carbon.

## Introduction (page 37)

Chemistry deals with the composition of substances and changes in their composition. Biochemistry is the chemistry of living things.

## Structure of Matter (page 37)

Matter is anything that has weight and takes up space.

1. Elements and atoms
   a. Naturally occurring matter on earth is composed of ninety-two elements.
   b. Elements occur most frequently in chemical combinations called compounds.
   c. Elements are composed of atoms.
   d. Atoms of different elements vary in size, weight, and ways of interacting.
2. Atomic structure
   a. An atom consists of electrons surrounding a nucleus, which contains protons and neutrons. The exception is hydrogen, which contains only a proton in its nucleus.
   b. Electrons are negatively charged, protons positively charged, and neutrons uncharged.
   c. A complete atom is electrically neutral.
   d. The atomic number of an element is equal to the number of protons in each atom; the atomic weight is equal to the number of protons plus the number of neutrons in each atom.
3. Isotopes
   a. Isotopes are atoms with the same atomic number but different atomic weights (due to differing numbers of neutrons).
   b. All the isotopes of an element react chemically in the same manner.
   c. Some isotopes are radioactive and release atomic radiation.
4. Bonding of atoms
   a. When atoms combine, they gain, lose, or share electrons.
   b. Electrons are arranged in shells around a nucleus.
   c. Atoms with completely filled outer shells are inactive, whereas atoms with incompletely filled outer shells tend to gain, lose, or share electrons and thus achieve stable structures.
   d. Atoms that lose electrons become positively charged; atoms that gain electrons become negatively charged.
   e. Ions with opposite charges attract and join by ionic bonds; atoms that share electrons join by covalent bonds.
5. Molecules and compounds
   a. Two or more atoms of the same element that join form a molecule of that element. Atoms of different elements that unite form a molecule of a compound.
   b. Molecules contain definite kinds and numbers of atoms.
   c. A molecular formula represents the numbers and kinds of atoms in a molecule.
   d. A structural formula represents the arrangement of atoms within a molecule.
6. Chemical reactions
   a. In a chemical reaction, bonds between atoms, ions, or molecules break or form.
   b. Three kinds of chemical reactions are synthesis, in which larger molecules form from smaller particles; decomposition, in which smaller particles form from larger molecules; and exchange reactions, in which parts of two different molecules trade positions.
   c. Many reactions are reversible. The direction of a reaction depends upon the proportion of reactants and products, the energy available, and the presence or absence of catalysts.
7. Acids, bases, and salts
   a. Compounds that ionize when they dissolve in water are electrolytes.
   b. Electrolytes that release hydrogen ions are acids, and those that release hydroxyl or other ions that react with hydrogen ions are bases.
   c. Acids and bases react together to form water and electrolytes called salts.
8. Acid and base concentrations
   a. The concentration of hydrogen ions ($H^+$) and hydroxyl ions ($OH^-$) in a solution can be represented by pH.
   b. A solution with equal numbers of $H^+$ and $OH^-$ is neutral and has a pH of 7.0; a solution with more $H^+$ than $OH^-$ is acidic (pH less than 7.0); a solution with fewer $H^+$ than $OH^-$ is basic (pH greater than 7.0).
   c. There is a tenfold difference in hydrogen ion concentration between each whole number in the pH scale.

## Chemical Constituents of Cells (page 46)

Molecules containing carbon and hydrogen atoms are organic and are usually nonelectrolytes; other molecules are inorganic and are usually electrolytes.

1. Inorganic substances
   a. Water is the most abundant compound in cells. Many chemical reactions take place in water. Water transports chemicals and heat, and helps release excess body heat.
   b. Oxygen releases energy needed for metabolic activities from glucose and other molecules.
   c. Carbon dioxide is produced when energy is released during metabolic processes.
   d. Inorganic salts provide ions needed in a variety of metabolic processes.
   e. Electrolytes must be present in certain concentrations inside and outside of cells.
2. Organic substances
   a. Carbohydrates provide much of the energy required by cells; their building blocks are simple sugar molecules.
   b. Lipids, such as fats, phospholipids, and steroids, supply energy and are used to build cell parts; their building blocks are molecules of glycerol and fatty acids.
   c. Proteins serve as structural materials, energy sources, hormones, cell surface receptors, antibodies, and enzymes.
      (1) Enzymes initiate or speed chemical reactions without being consumed themselves.
      (2) The building blocks of proteins are amino acids.
      (3) Proteins vary in the numbers and kinds of amino acids they contain, the sequences in which these amino acids are arranged, and their three-dimensional structures.

(4) Protein molecules may be denatured by exposure to excessive heat, radiation, electricity, or certain chemicals.
  d. Nucleic acids constitute genes, the instructions that control cell activities, and play important roles in protein synthesis.
    (1) The two major kinds are RNA and DNA.
    (2) Nucleic acid molecules are composed of building blocks called nucleotides.

(3) DNA molecules store information that is used by cell parts to construct specific kinds of protein molecules.
(4) RNA molecules help synthesize proteins.
(5) DNA molecules are replicated and an exact copy of the original cell's DNA is passed to each of the newly formed cells, resulting from cell reproduction.

## CRITICAL THINKING QUESTIONS

1. What acidic and alkaline substances do you encounter in your everyday life? What foods do you eat regularly that are acidic? What alkaline foods do you eat?

2. Using the information on pages 48–49 to distinguish between saturated and unsaturated fats, try to list all of the sources of saturated and unsaturated fats you have eaten during the past twenty-four hours.

3. How would you reassure a patient who is about to undergo CT scanning for evaluation of a tumor, and who fears becoming a radiation hazard to family members?

4. Various forms of ionizing radiation, such as that released from X-ray tubes and radioactive substances, are commonly used in the treatment of cancer, yet such exposure can cause adverse effects, including the development of cancers. How would you explain the value of radiation therapy to a cancer patient in light of this seeming contradiction?

5. How would you explain the importance of amino acids and proteins in a diet to a person who is following a diet composed primarily of carbohydrates?

6. What clinical laboratory tests with which you are acquainted involve a knowledge of chemistry?

7. Explain why the symptoms of many inherited diseases result from abnormal protein function.

## REVIEW EXERCISES

1. Distinguish between chemistry and biochemistry.
2. Define *matter*.
3. Explain the relationship between elements and atoms.
4. Define *compound*.
5. List the four most abundant elements in the human body.
6. Describe the major parts of an atom.
7. Distinguish between protons and neutrons.
8. Explain why a complete atom is electrically neutral.
9. Distinguish between atomic number and atomic weight.
10. Define *isotope*.
11. Define *atomic radiation*.
12. Describe how electrons are arranged within atoms.
13. Explain why some atoms are chemically inert.
14. Distinguish between an ionic bond and a covalent bond.
15. Distinguish between a single covalent bond and a double covalent bond.
16. Explain the relationship between molecules and compounds.
17. Distinguish between a molecular formula and a structural formula.
18. Describe three major types of chemical reactions.
19. Define *reversible reaction*.
20. Define *catalyst*.
21. Define *acid, base, salt,* and *electrolyte*.
22. Explain what *pH* measures.
23. Distinguish between organic and inorganic substances.
24. Describe the roles played by water and by oxygen in the human body.
25. List several ions that cells require, and describe their general functions.
26. Define *electrolyte balance*.
27. Describe the general characteristics of carbohydrates.
28. Distinguish between simple and complex carbohydrates.
29. Describe the general characteristics of lipids.
30. Distinguish between saturated and unsaturated fats.
31. Describe the general characteristics of proteins.
32. Describe the function of an enzyme.
33. Explain how protein molecules may become denatured.
34. Describe the general characteristics of nucleic acids.
35. Explain the general functions of nucleic acids.

**For More Information Related to This Chapter, Please Visit** *Hole's Human Anatomy and Physiology* **Home Page**
**http://www.mhhe.com/biosci/abio/**

# CELLS

## CHAPTER OBJECTIVES

**AFTER YOU HAVE STUDIED THIS CHAPTER, YOU SHOULD BE ABLE TO:**

1. Explain how cells differ from one another.

2. Describe the general characteristics of a composite cell.

3. Explain how the components of a cell's membrane provide its functions.

4. Describe each kind of cytoplasmic organelle and explain its function.

5. Describe the cell nucleus and its parts.

6. Explain how substances move into and out of cells.

7. Describe the cell cycle.

8. Explain how a cell reproduces.

9. Describe several controls of cell reproduction.

## KEY TERMS

**active transport** (ak′tiv trans′port)
**centrosome** (sen′tro-sōm)
**chromosome** (kro′mo-sōm)
**cytoplasm** (si′to-plazm)
**cytoskeleton** (si′to-skel-i-tun)
**differentiation** (dif″er-en″she-a′shun)
**diffusion** (dĭ-fu′zhun)
**endocytosis** (en″do-si-to′sis)
**endoplasmic reticulum** (en′do-plaz′mik re-tik′u-lum)
**equilibrium** (e″kwi-lib′re-um)
**exocytosis** (ex-o-si-to′sis)
**extracellular** (eks″trah-sel′u-lar)
**facilitated diffusion** (fah-sil″i-tat′ed dĭ-fu′zhun)
**filtration** (fil-tra′shun)

**Golgi apparatus** (gol′je ap″ah-ra′tus)
**intercellular** (in″ter-sel′u-lar)
**intracellular** (in″trah-sel′u-lar)
**lysosome** (li′so-sōm)
**micrometer** (mi′kro-me″ter)
**mitochondrion** (mi″to-kon′dre-on); plural, **mitochondria** (mi″to-kon′dre-ah)
**mitosis** (mi-to′sis)
**nucleolus** (nu-kle′o-lus)
**nucleus** (nu′kle-us)
**osmosis** (oz-mo′sis)
**permeable** (per′me-ah-bl)
**phagocytosis** (fag″o-si-to′sis)
**pinocytosis** (pi″no-si-to′sis)
**ribosome** (ri′bo-sōm)
**transcytosis** (tranz-si-to′-sis)
**vesicle** (ves′i-k′l)

## UNDERSTANDING WORDS

**cyt-,** cell: *cyt*oplasm—fluid that occupies the space between the cell membrane and nuclear envelope.

**endo-,** within: *endo*plasmic reticulum—complex of membranous structures within the cytoplasm.

**hyper-,** above: *hyper*tonic—solution that has a greater osmotic pressure than the cytosol.

**hypo-,** below: *hypo*tonic—solution that has a lesser osmotic pressure than the cytosol.

**inter-,** between: *inter*phase—stage that occurs between mitotic divisions of a cell.

**iso-,** equal: *iso*tonic—solution that has an osmotic pressure equal to the cytosol.

**lys-,** to break up: *lys*osome—organelle that contains enzymes capable of breaking up molecules of protein, carbohydrate, and nucleic acid.

**mit-,** thread: *mit*osis—part of cell division during which threadlike chromosomes appear within a cell.

**phag-,** to eat: *phag*ocytosis—process by which a cell takes in solid particles.

**pino-,** to drink: *pino*cytosis—process by which a cell takes in tiny droplets of liquid.

**pro-,** before: *pro*phase—first stage of mitosis.

**som,** body: ribo*som*e—tiny, spherical organelle that contains protein and RNA.

**vesic-,** bladder: *vesic*le—small organelle that contains various substances.

Certain people seem incapable of contracting HIV infection. For example, a women received a blood transfusion in 1980 that was later found to be contaminated with HIV, the virus that causes AIDS. But the woman never became infected. Some intravenous drug users share needles with people who later develop AIDS, but never themselves become infected. Similarly, one man who has had hundreds of male sex partners since 1978, several of whom have died of AIDS, has stayed healthy.

We usually think of avoiding AIDS by avoiding activities that spread the virus and this is without doubt the best course. But what protects these people, all of whom have been exposed to HIV? A lucky few individuals cannot contract AIDS because of an abnormality in their cells, which, paradoxically, protects them.

When HIV enters a human body, it approaches certain white blood cells, called CD4 lymphocytes, that control the immune system. The virus binds first to receptors called CD4—the receptors are proteins that extend from the cell surface. Once bound, HIV moves down the CD4 receptor and binds another receptor, called CCR5. Only then can the virus enter the cell, and start the chain reaction of viral reproduction that ultimately topples immunity.

Thanks to heredity, one percent of Caucasians in the United States, and much lesser numbers of Asians, African Americans, and Native Americans, have cell surfaces that lack the crucial CCR5 HIV docking sites. These lucky few individuals cannot get AIDS, because HIV cannot enter their cells. Another 20% of the Caucasian population (less for others) have half the normal number of CCR5 receptors. These people can become infected, but remain healthy longer than is usual.

Researchers are now applying this knowledge of how AIDS begins at the cellular level to develop vaccines and new treatments. Understanding how HIV interacts with cells, the units of life, has revealed what might finally prove to be HIV's Achilles heel—a protein portal called CCR5.

---

An adult human body consists of about seventy-five trillion **cells**, the basic units of an organism. All cells have much in common, yet those in different tissues vary in a number of ways.

Cells vary considerably in size. We measure cell sizes in units called *micrometers*. A micrometer equals one thousandth of a millimeter and is symbolized μm. A human egg cell is about 140 μm in diameter and is just barely visible to an unaided eye. This is large when compared to a red blood cell, which is about 7.5 μm in diameter, or the most common types of white blood cells, which vary from 10 to 12 μm in diameter. On the other hand, smooth muscle cells can be between 20 and 500 μm long (fig. 3.1).

Cells also vary in shape, and typically their shapes make possible their functions (fig. 3.2). For instance, nerve cells often have long, threadlike extensions many centimeters long that transmit nerve impulses from one part of the body to another. Epithelial cells that line the inside of the mouth are thin, flattened, and tightly packed, somewhat like floor tiles. They form a barrier that shields underlying tissue. Muscle cells, which contract and pull structures closer together, are slender and rodlike, with their ends attached to the parts they move. Muscle cells are filled with contractile proteins. An adipose cell is little more than a blob of fat; a B lymphocyte is an antibody factory. The human body is a conglomeration of many types of cells.

## A COMPOSITE CELL

It is not possible to describe a typical cell, because cells vary so greatly in size, shape, content, and function. We can, however, consider a hypothetical composite cell that includes many known cell structures (fig. 3.3).

A cell consists of three major parts—the **nucleus,** the **cytoplasm,** and the **cell membrane.** The nucleus is innermost and is enclosed by a thin membrane called the nuclear envelope. The cytoplasm is a mass of fluid that surrounds the nucleus and is itself encircled by the even thinner cell membrane (also called a plasma membrane). Within the cytoplasm are specialized structures called **cytoplasmic organelles** that perform specific functions. The nucleus directs the overall activities of the cell by functioning as the hereditary headquarters, housing the genetic material.

A cell with a nucleus, such as those of the human body, is termed *eukaryotic,* meaning "true nucleus." In contrast are the *prokaryotic* ("before nucleus") cells of bacteria. Although bacterial cells lack a nucleus and other membrane-bound organelles and are thus simpler than eukaryotic cells, the bacteria are nevertheless quite a successful life form—they are literally everywhere. A third type of cell, termed *archaeal,* lacks a nucleus but has many features like those of eukaryotic cells.

1 Give two examples to illustrate how the shape of a cell makes possible its function.

2 Name the major parts of a cell.

3 What are the general functions of the cytoplasm and nucleus?

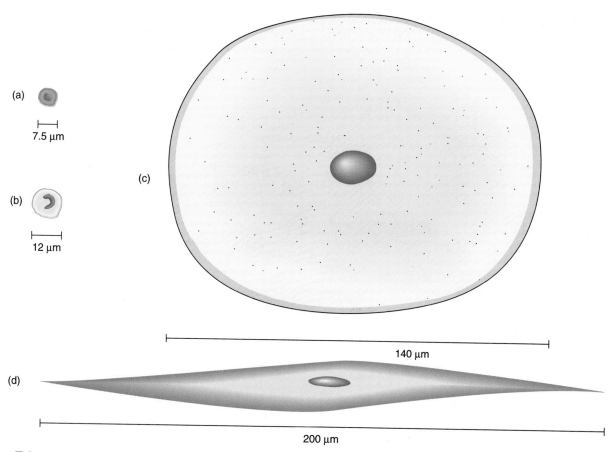

**FIGURE** 3.1 🏃

Cells vary considerably in size. This illustration shows the relative sizes of four types of cells. (*a*) Red blood cell, 7.5 μm in diameter; (*b*) white blood cell, 10–12 μm in diameter; (*c*) human egg cell, 140 μm in diameter; (*d*) smooth muscle cell, 20–500 μm in length.

**FIGURE** 3.2 🏃

Cells vary in shape and function. (*a*) A nerve cell transmits impulses from one body part to another. (*b*) Epithelial cells protect underlying cells. (*c*) Muscle cells pull structures closer.

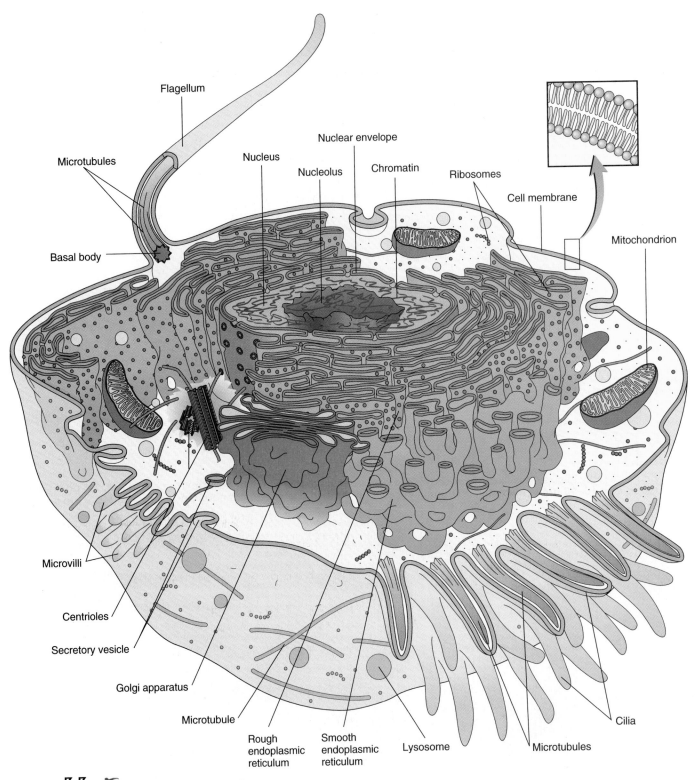

Flagellum

Microtubules

Basal body

Nuclear envelope

Nucleus

Nucleolus

Chromatin

Ribosomes

Cell membrane

Mitochondrion

Microvilli

Centrioles

Secretory vesicle

Golgi apparatus

Microtubule

Rough
endoplasmic
reticulum

Smooth
endoplasmic
reticulum

Lysosome

Microtubules

Cilia

**FIGURE 3.3**

A composite cell. Organelles are not drawn to scale.

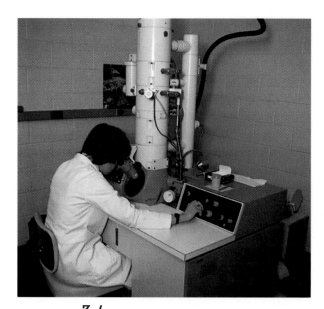

**FIGURE 3.4**

A transmission electron microscope.

(a)

Blood vessel wall

Red blood cells

(b)

## Cell Membrane

The cell membrane is the outermost limit of a cell, but it is more than a simple boundary surrounding the cellular contents. It is an actively functioning part of the living material, and many important metabolic reactions take place on its surfaces.

### General Characteristics

The cell membrane is extremely thin—visible only with the aid of an electron microscope (fig. 3.4)—but it is flexible and somewhat elastic. It typically has complex surface features with many outpouchings and infoldings that increase surface area. The membrane quickly seals minute breaks, but if it is extensively damaged, cell contents escape, and the cell dies.

Red blood cells

(c)

**FIGURE 3.5**

Human red blood cells as viewed using (a) a light microscope (1,200×), (b) a transmission electron microscope (2500×), and (c) a scanning electron microscope (1,900×). The cells in (a) are stained to make them easier to see.

The maximum effective magnification possible using a light microscope is about 1,200×. A transmission electron microscope (TEM) provides an effective magnification of nearly 1,000,000×, whereas a scanning electron microscope (SEM), can provide about 50,000×. Photographs of microscopic objects (micrographs) produced using the light microscope and the transmission electron microscope are typically two-dimensional, but those obtained with the scanning electron microscope have a three-dimensional quality (fig. 3.5).

"Heads" of phospholipid

"Tails" of phospholipid

Cell membrane

(a)

Cell membrane

(b)

**FIGURE 3.6**

(a) A transmission electron micrograph of a cell membrane (250,000× micrograph enlarged to 600,000×); (b) the framework of the membrane consists of a double layer of phospholipid molecules.

In addition to maintaining the integrity of the cell, the membrane controls the entrance and exit of substances, allowing some in while excluding others. A membrane that functions in this manner is *selectively permeable.* The cell membrane is crucial because it is a conduit between the cell and the extracellular fluids in the body's internal environment. It even allows the cell to receive and respond to incoming messages, a process called **signal transduction.**

## Membrane Structure

Chemically, the cell membrane is mainly composed of lipids and proteins, with a small quantity of carbohydrate. Its basic framework consists of a double layer (bilayer) of phospholipid molecules (see chapter 2 and fig. 2.13*b*) that self-assemble so that their water-soluble (hydrophilic) "heads," containing phosphate groups, form the surfaces of the membrane, and their water-insoluble (hydrophobic) "tails," consisting of fatty acid chains, make up the interior of the membrane (see figs. 3.3 and 3.6). The lipid molecules can move sideways within the plane of the membrane, and collectively they form a thin, but stable fluid film.

Because the interior of the cell membrane consists largely of the fatty acid portions of the phospholipid molecules, it is oily. Molecules that are soluble in lipids, such as oxygen, carbon dioxide, and steroid hormones, can pass through this layer easily; however, the layer is impermeable to water soluble molecules, such as amino

acids, sugars, proteins, nucleic acids, as well as various ions. Many cholesterol molecules embedded in the interior of the membrane also help make the membrane impermeable to water-soluble substances. In addition, the relatively rigid structure of the cholesterol molecules helps stabilize the membrane.

A cell membrane includes only a few types of lipid molecules, but many kinds of proteins (fig. 3.7). These proteins provide the special functions of the membrane, and they can be classified according to their shapes. One group of proteins, for example, consists of tightly coiled, rodlike molecules embedded in the phospholipid bilayer. These fibrous proteins may completely span the membrane; that is, they may extend outward from its surface on one side, while their opposite ends communicate with the cell's interior. Such proteins often function as *receptors* that are specialized to combine with specific kinds of molecules, such as hormones (see chapter 13).

Another group of cell membrane proteins are more compact and globular. Some of these proteins, called *integral proteins,* are embedded in the interior of the phospholipid bilayer. Typically, they span the membrane and provide mechanisms by which small molecules and ions can cross the otherwise impermeable phospholipid bilayer. For example, some of these integral proteins form "pores" in the membrane that allow water molecules to pass through. Other integral proteins are highly selective, and form channels that allow only particular ions to

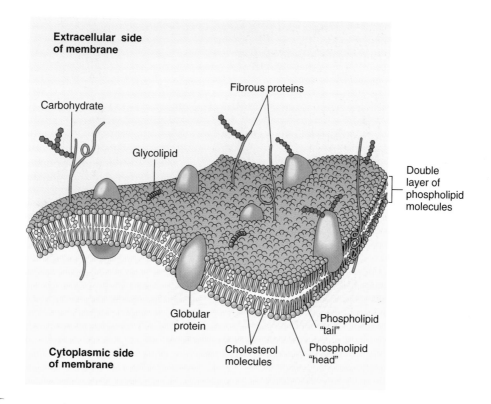

**Extracellular side of membrane**

Carbohydrate

Fibrous proteins

Glycolipid

Double layer of phospholipid molecules

Globular protein

**Cytoplasmic side of membrane**

Cholesterol molecules

Phospholipid "tail"

Phospholipid "head"

**FIGURE 3.7**

The cell membrane is composed primarily of phospholipids (and some cholesterol), with proteins scattered throughout the lipid bilayer and associated with its surfaces.

enter. In muscle and nerve cells, for example, selective channels control the movements of sodium and potassium ions, which are important in muscle contraction and nerve impulse conduction (see chapters 9 and 10). Clinical Application 3.1 discusses how abnormal ion channels can cause disease.

Yet other globular proteins, called *peripheral proteins*, associate with the surface of the cell membrane. These proteins function as enzymes (see chapter 4), and many are part of signal transduction. Carbohydrate groups associated with peripheral proteins form *glycoproteins* that help cells to recognize and bind to each other. This is important as cells aggregate to form tissues. Cell surface glycoproteins also mark the cells of an individual as "self." The immune system can distinguish between "self" cell surfaces, and "nonself" cell surfaces which may indicate a potential threat, such as the presence of infectious bacteria.

### Intercellular Junctions

Some cells, such as blood cells, are separated from each other in fluid-filled spaces (intercellular spaces). Many other cell types, however, are tightly packed, with structures called **intercellular junctions** connecting their cell membranes.

In one type of intercellular junction, called a *tight junction,* the membranes of adjacent cells converge and fuse. The area of fusion surrounds the cell like a belt, and the junction closes the space between the cells. Cells that form sheetlike layers, such as those that line the inside of the digestive tract, often are joined by tight junctions. The linings of tiny blood vessels in the brain are extremely tight (Clinical Application 3.2).

Another type of intercellular junction, called a *desmosome,* rivets or "spot welds" adjacent skin cells, so they form a reinforced structural unit. The membranes of certain other cells, such as those in heart muscle and muscle of the digestive tract, are interconnected by tubular channels called *gap junctions.* These channels link the cytoplasm of adjacent cells and allow ions, nutrients (such as sugars, amino acids, and nucleotides), and other small molecules to move between them (fig. 3.8). Table 3.1 summarizes these intercellular junctions.

### Cell Adhesion Molecules

Often cells must interact dynamically and transiently, rather than form permanent attachments. Proteins called **cell adhesion molecules,** or **CAMs** for short, guide cells on the move. Consider what happens when a white blood cell must travel in the bloodstream to the site of an

# Faulty Ion Channels Cause Disease

What do collapsing horses, irregular heartbeats in teenagers, and cystic fibrosis have in common? All result from abnormal ion channels in cell membranes.

Ion channels are tunnels through the lipid bilayer of a biological membrane that consist of protein (see fig. 10.10). These passageways permit electrical signals to pass in and out of membranes in the form of ions. An ion channel functions as a gate, opening or closing to a specific ion in response to certain conditions. Ten million ions can pass through an ion channel in one second. Events that can trigger an ion channel to open or close include a change in voltage across the membrane, binding of a molecule (ligand) to the cell membrane, or receiving biochemical messages from within the cell.

Abundant ion channels include those specific for calcium ($Ca^{+2}$), chloride ($Cl^-$), sodium ($Na^+$) or potassium ($K^+$). A cell may have a few thousand ion channels specific for each ion.

Many drugs act by affecting ion channels (table 3A). The distribution of specific ion channels on particular cell types explains the symptoms of illnesses that result from abnormal channels. Following are descriptions of three illnesses caused by malfunctioning ion channels.

## HYPERKALEMIC PERIODIC PARALYSIS AND SODIUM CHANNELS

The quarter horse was originally bred to run the quarter mile in the 1600s. Four particularly fast stallions were used to establish much of the current population of nearly 3 million animals. Unfortunately, one of the original stallions had an inherited condition called hyperkalemic periodic paralysis. The horse was indeed a champion, but the disease brought on symptoms undesirable in a racehorse—attacks of weakness and paralysis that caused sudden collapse.

Hyperkalemic periodic paralysis results from abnormal sodium channels in the cell membranes of muscle cells. But the trigger for the temporary paralysis is another ion: potassium. When the blood potassium level rises, as it may following intense exercise, it slightly alters the muscle

| TABLE 3A | DRUGS THAT AFFECT ION CHANNELS | |
|---|---|---|
| **Target** | **Indication** | |
| Calcium channels | Antihypertensives Anti-angina (chest pain) | |
| Sodium channels | Anti-arrhythmias, diuretics Local anesthetics | |
| Chloride channels | Anticonvulsants Muscle relaxants | |
| Potassium | Antihypertensives, anti-diabetics (non-insulin-dependent) | |

| TABLE 3.1 | TYPES OF INTERCELLULAR JUNCTIONS | |
|---|---|---|
| **Type** | **Function** | **Location** |
| Tight junctions | Close space between cells by fusing cell membranes | Cells that line inside of the small intestine |
| Desmosomes | Bind cells by forming "spot welds" between cell membranes | Cells of the outer skin layer |
| Gap junctions | Form tubular channels between cells that allow substances to be exchanged | Muscle cells of the heart and digestive tract |

cell membrane's electrical potential. Normally, this slight change would have no effect. In affected horses, however, the change causes sodium channels to open too widely, and admit too much sodium into the cell. The influx of sodium renders the muscle cell unable to respond to nervous stimulation for a short time—but long enough for the racehorse to fall.

Humans can inherit this condition too. In one affected family, several members collapsed after eating bananas! Bananas are very high in potassium, which triggered the symptoms of hyperkalemic periodic paralysis.

## LONG-QT SYNDROME AND POTASSIUM CHANNELS

A Norwegian family had four children, all born deaf. Three of the children died at ages 4, 5, and 9; the fourth so far has been lucky. All of the children inherited from their unaffected "carrier" parents a condition called "long-QT syndrome associated with deafness." They have abnormal potassium channels in the heart muscle and in the inner ear. In the heart, the malfunctioning channels cause fatal arrhythmia. In the inner ear, the abnormal channels alter the concentration of potassium ions in a fluid, impairing hearing.

The inherited form of long-QT syndrome in the Norwegian family is extremely rare, but other forms of the condition are more common, causing 50,000 sudden deaths each year, often in apparently healthy children and young adults. Several cases were attributed to an interaction between the antihistamine Seldane (terfenadine) and either an antibiotic (erythromycin) or an antifungal drug (ketoconazole), before Seldane was removed from the market in 1997.

Diagnosing long-QT syndrome early is essential because the first symptom may be fatal. It is usually diagnosed following a sudden death of a relative, or detected on a routine examination of the heart's electrical activity (an electrocardiogram, see fig. 15.23). Drugs, pacemakers, and surgery to remove certain nerves can treat the condition and possibly prevent sudden death.

## CYSTIC FIBROSIS AND CHLORIDE CHANNELS

A seventeenth-century English saying, "A child that is salty to taste will die shortly after birth," described the consequence of abnormal chloride channels in the inherited illness cystic fibrosis (CF). The disorder affects 1 in 2,500 Caucasians, 1 in 14,000 blacks, and 1 in 90,000 Asians, and is inherited from two unaffected parents who are carriers. The major symptoms of impaired breathing, respiratory infections, and a clogged pancreas result from extremely thick mucus secretions. Affected individuals undergo twice-daily exercise sessions to shake free the sticky mucus, and take supplemental digestive enzymes to aid pancreatic function. Strong antibiotics are used to combat their frequent lung infections.

In 1989, researchers identified the microscopic defect that causes CF as abnormal chloride channels in cells lining the lung passageways and ducts in the pancreas. The primary defect in the chloride channels also causes sodium channels to malfunction. The result is salt trapped inside affected cells, which draws moisture in, thickening the surrounding mucus. Several experimental gene therapies attempt to correct affected cells' instructions for building chloride channel proteins. ■

injury, where it is required to fight infection. Imagine that such a cell must reach a woody splinter embedded in a person's palm (fig. 3.9).

Once near the splinter, the white blood cell must slow down in the turbulence of the bloodstream. A CAM called *selectin* does this by coating the white blood cell and providing traction. The white blood cell slows to a roll, and binds to carbohydrates on the inner capillary surface. Clotting blood, bacteria, and decaying tissue at the injury site release biochemicals (chemoattractants) that attract the white blood cell. Finally, a CAM called *integrin* contacts an adhesion receptor protein protruding into the capillary space near the splinter and pushes up through the capillary cell membrane, grabbing the passing slowed white blood cell and directing it between the tilelike cells of the capillary wall. White blood cells collecting at an injury site produce inflammation and, with the dying bacteria, form pus.

Brooke Blanton was born lacking the CAMs that enable white blood cells to adhere to blood vessel walls. As a result, her sores do not heal, never forming pus because white blood cells never reach injury sites. Brooke's earliest symptoms were teething sores that did not heal. Today Brooke must be very careful to avoid injury or infection because her white blood cells, although plentiful and healthy, simply zip past her wounds.

# The Blood-Brain Barrier

Perhaps nowhere else in the body are cells attached as firmly and closely as they are in the 400-mile network of capillaries in the brain. The walls of these microscopic blood vessels are but a single cell thick. They form sheets that fold into minute tubules. A century ago, bacteriologist Paul Ehrlich showed the existence of the blood-brain barrier by injecting a dye intravenously. The brain failed to take up the dye, indicating that its blood vessels did not allow the molecules to leave and enter the brain's nervous tissue.

Studies in 1969 using the electron microscope revealed that in the brain, capillary cell membranes overlap to form a barrier of tight junctions. Unlike the cells forming capillary walls elsewhere in the body, which are pocked with vesicles and windowlike portals called clefts, the cells comprising this blood-brain barrier have few vesicles, and no clefts. Certain star-shaped brain cells called astrocytes contribute to this barrier as well.

The impenetrable barrier that the capillaries in the brain form shields delicate brain tissue from toxins in the bloodstream and from biochemical fluctuations that could be overwhelming if the brain had to continually respond to them. It also allows selective drug delivery to the periphery—for example some antihistamines do not cause drowsiness because they cannot breach the blood-brain barrier. But all this protection has a limitation—the brain cannot take up many therapeutic drugs that must penetrate to be effective. By studying the types of molecules embedded in the membranes of the cells forming the barrier, researchers are developing clever ways to sneak drugs into the brain. They can tag drugs to substances that can cross the barrier, design drugs to fit natural receptors in the barrier, or inject substances that temporarily relax the tight junctions forming the barrier. Drugs that can cross the blood-brain barrier could be used to treat Alzheimer's disease, Parkinson's disease, brain tumors, and AIDS-related brain infections.

A malfunctioning blood-brain barrier can threaten health. During the Persian Gulf War in 1991, response of the barrier to stress in soldiers caused illness. Many troops were given a drug to protect against the effects of nerve gas on peripheral nerves—those outside the brain and spinal cord. The drug, based on its chemistry, was not expected to cross the blood-brain barrier. However, 213 Israeli soldiers treated with the drug developed brain-based symptoms, including nervousness, insomnia, headaches, drowsiness, and inability to pay attention and to do simple calculations. Further reports from soldiers, and experiments on mice, revealed that under stressful conditions, the blood-brain barrier can temporarily loosen, admitting a drug that it would normally keep out. The barrier, then, is not a fixed boundary, but rather a dynamic structure that can alter in response to a changing environment. ∎

## Cytoplasm

When viewed through a light microscope, cytoplasm usually appears as clear jelly with specks scattered throughout. However, a transmission electron microscope (see fig. 3.4), which produces much greater magnification and ability to distinguish fine detail (resolution), reveals that cytoplasm contains networks of membranes and organelles suspended in a clear liquid called **cytosol.** Cytoplasm also contains abundant protein rods and tubules that form a supportive framework called the **cytoskeleton.**

The activities of a cell occur largely in its cytoplasm, where nutrient molecules are received, processed, and used in various metabolic reactions. Within the cytoplasm, the following organelles have specific functions:

1. **Endoplasmic reticulum.** The endoplasmic reticulum (ER) is a complex organelle composed of membrane-bound flattened sacs, elongated canals, and fluid-filled vesicles. These membranous parts are interconnected, and they communicate with the cell membrane, the nuclear envelope, and certain cytoplasmic organelles. ER is widely distributed through the cytoplasm, providing a tubular transport system for molecules throughout the cell.

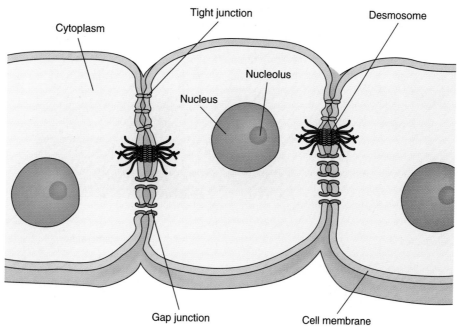

Tight junction

Desmosome

Nucleolus

Nucleus

Gap junction

Cell membrane

**FIGURE 3.8**

Some cells are joined by intercellular junctions, such as tight junctions that fuse neighboring membranes, desmosomes that serve as "spot welds," or gap junctions that allow small molecules to move between the cytoplasm of adjacent cells.

**Attachment (rolling)**

White blood cell

Selectins

**Adhesion**

Integrins

IL-8

Carbohydrates on capillary wall

Blood vessel lining cell

Adhesion receptor proteins

**Exit**

Splinter

**FIGURE 3.9**

Cell adhesion molecules (CAMs) direct white blood cells to injury sites, such as this splinter. Selectin proteins latch onto a rolling white blood cell and bind carbohydrates on the inner blood vessel wall at the same time, slowing the cell from moving at 2,500 micrometers per second to a more leisurely 50 micrometers per second. Then integrin proteins anchor the white blood cell to the blood vessel wall. Finally, the white blood cell squeezes between lining cells at the injury site and exits the bloodstream.

The endoplasmic reticulum also participates in the synthesis of protein and lipid molecules, some of which may be assembled into new membranes. Commonly, the outer membranous surface of the ER is studded with many tiny, spherical organelles called *ribosomes* that give the ER a textured appearance when viewed with an electron microscope. Such endoplasmic reticulum is termed *rough* ER. Endoplasmic reticulum that lacks ribosomes is called *smooth* ER (fig. 3.10).

The ribosomes of rough ER are sites of protein synthesis. The proteins may then move through the canals of the endoplasmic reticulum to the Golgi apparatus for further processing. Smooth ER, on the other hand, contains enzymes important in lipid synthesis.

2. **Ribosomes.** Besides being found on the endoplasmic reticulum, some ribosomes are scattered freely throughout the cytoplasm. All ribosomes are composed of protein and RNA, and provide a structural support and enzymes required to link amino acids to form proteins (see chapter 4).

3. **Golgi apparatus.** The Golgi apparatus is composed of a stack of half a dozen or so flattened, membranous sacs called *cisternae*. This organelle refines, packages, and delivers proteins synthesized by the ribosomes associated with the ER (fig. 3.11).

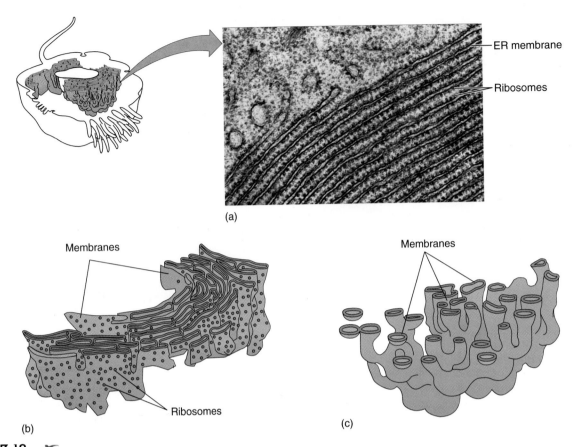

ER membrane

Ribosomes

(a)

Membranes

Membranes

Ribosomes

(b)

(c)

**FIGURE 3.10**

(a) A transmission electron micrograph of rough endoplasmic reticulum (ER) (28,000×). (b) Rough ER is dotted with ribosomes, whereas (c) smooth ER lacks ribosomes.

Endoplasmic reticulum

Golgi apparatus

Secretory vesicle

(a)

Rough endoplasmic reticulum

Secretory vesicle

Golgi apparatus

(b)

**FIGURE 3.11**

(a) A transmission electron micrograph of a Golgi apparatus (48,000×). (b) The Golgi apparatus consists of membranous sacs that continually receive vesicles from the endoplasmic reticulum and produce vesicles for secretion.

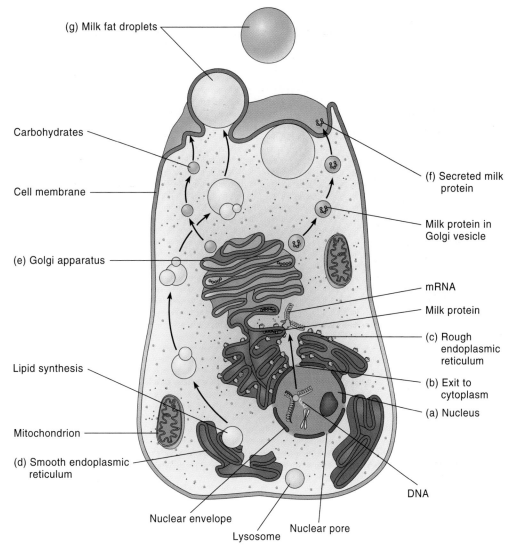

(g) Milk fat droplets

Carbohydrates

Cell membrane

(e) Golgi apparatus

Lipid synthesis

Mitochondrion

(d) Smooth endoplasmic reticulum

Nuclear envelope

Lysosome

Nuclear pore

DNA

(a) Nucleus

(b) Exit to cytoplasm

(c) Rough endoplasmic reticulum

Milk protein

mRNA

Milk protein in Golgi vesicle

(f) Secreted milk protein

**FIGURE 3.12**

Milk secretion illustrates how organelles interact to synthesize, transport, store, and export biochemicals. Secretion begins in the nucleus (*a*), where messenger RNA molecules bearing genetic instructions for production of milk proteins exit through nuclear pores to the cytoplasm (*b*). Most proteins are synthesized on membranes of the rough endoplasmic reticulum (ER) (*c*), using amino acids in the cytoplasm. Lipids are synthesized in the smooth ER (*d*), and sugars are synthesized, assembled, and stored in the Golgi apparatus (*e*). An active mammary gland cell releases milk proteins from vesicles that bud off of the Golgi apparatus (*f*). Fat droplets pick up a layer of lipid from the cell membrane as they exit the cell (*g*). When the baby suckles, he or she receives a chemically complex secretion—milk.

Proteins arrive at the Golgi apparatus enclosed in tiny vesicles composed of membrane from the endoplasmic reticulum. These sacs fuse to the membrane at the beginning or innermost end of the Golgi apparatus, which is specialized to receive proteins. Previously, these protein molecules were combined with sugar molecules as glycoproteins.

As the glycoproteins pass from layer to layer through the Golgi stacks, they are modified chemically. For example, sugar molecules may be added or removed from them. When the altered

glycoproteins reach the outermost layer, they are packaged in bits of Golgi apparatus membrane that bud off and form transport vesicles. Such a vesicle may then move to the cell membrane, where it fuses and releases its contents to the outside of the cell as a secretion. Other vesicles may transport glycoproteins to organelles within the cell (fig. 3.12). Movement of substances within cells by way of vesicles is called *vesicle trafficking.*

In some cells, including certain liver cells and white blood cells (lymphocytes), glycoprotein molecules are secreted as rapidly as they are

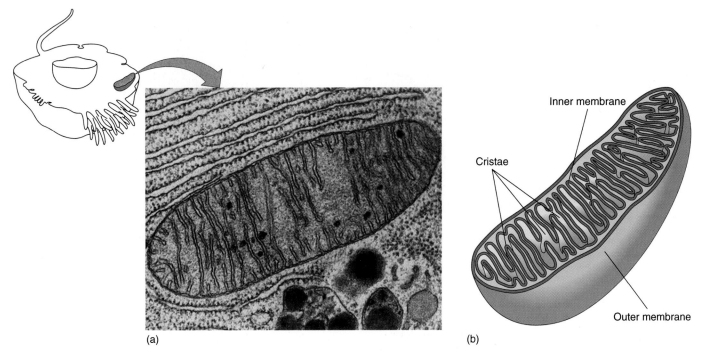

**FIGURE 3.13**

(a) A transmission electron micrograph of a mitochondrion (40,000×). (b) Cristae partition this saclike organelle.

synthesized. However, in certain other cells, such as those that manufacture protein hormones, vesicles containing newly synthesized molecules remain in the cytoplasm and release the stored substances only when the cells are stimulated. (Chapter 13 discusses hormone secretion.)

Secretory vesicles that originate in the ER not only release substances outside the cell, but also provide new cell membrane. This is especially important during cell growth.

◾ 1  What is a selectively permeable membrane?

◾ 2  Describe the chemical structure of a cell membrane.

◾ 3  What are the different types of intercellular junctions?

◾ 4  What are some of the events of cell adhesion?

◾ 5  What are the functions of the endoplasmic reticulum?

◾ 6  Describe how the Golgi apparatus functions.

4. **Mitochondria.** Mitochondria are elongated, fluid-filled sacs 2–5 μm long. They often move about slowly in the cytoplasm and can reproduce by dividing. A mitochondrion contains a small amount of DNA that encodes information for making a few kinds of proteins and specialized RNA. However, most proteins used in mitochondrial functions are encoded in the DNA of the nucleus. These proteins are synthesized elsewhere in the cell and then enter the mitochondria.

The membrane of a mitochondrion has two layers—an outer membrane and an inner membrane. The inner membrane is folded extensively to form shelflike partitions called *cristae.* Small, stalked particles that contain enzymes are connected to the cristae. These enzymes and others dissolved in the fluid within the mitochondrion control many of the chemical reactions that release energy from glucose and other organic nutrients. The mitochondria transform this newly released energy into a chemical form, the molecule **adenosine triphosphate** (ATP), that cells can readily use (fig. 3.13 and chapter 4). For this reason mitochondria are sometimes called the "powerhouses" of cells.

A typical cell has about 1,700 mitochondria, but cells with very high energy requirements, such as muscle, have many thousands of mitochondria. This is why a common symptom of illnesses affecting mitochondria is muscle weakness. Symptoms of these "mitochondrial myopathies" include exercise intolerance and weak and flaccid muscles.

Mitochondria are particularly fascinating to biologists because they provide glimpses into the past. Mitochondria are passed to offspring from mothers only, because these organelles are excluded from the part of a sperm that enters an egg cell. Evolutionary biologists study the DNA sequences of genes in mitochondria to trace human origins, back to a long-ago group of ancestors metaphorically called "mitochondrial Eve."

Mitochondria may provide clues to a past far more remote than the beginnings of humankind. According to the widely accepted endosymbiont theory, mitochondria are the remnants of once free-living bacteria-like cells that were swallowed by primitive eukaryotic cells. These bacterial passengers remain in our cells today, where they participate in energy reactions.

Lysosomes

**FIGURE 3.14**

In this falsely colored transmission electron micrograph, lysosomes appear as membranous sacs (14,100×).

5. **Lysosomes.** Lysosomes are the "garbage disposals" of the cell, whose function is to dismantle debris. They are sometimes difficult to identify because their shapes vary so greatly. However, they commonly appear as tiny, membranous sacs (fig. 3.14). These sacs contain powerful enzymes that break down proteins, carbohydrates, and nucleic acids, including foreign particles composed of these substances. Certain white blood cells, for example, engulf bacteria that are then digested by the lysosomal enzymes. This is one way that white blood cells help stop bacterial infections.

Lysosomes also destroy worn cellular parts. In fact, lysosomes in certain scavenger cells may engulf and digest entire body cells that have been injured. How the lysosomal membrane is able to withstand being digested itself is not well understood, but this organelle sequesters enzymes that can function only under very acidic conditions, preventing them from destroying the cellular contents around them. Human lysosomes contain forty or so different types of enzymes. An abnormality in just one type of lysosomal enzyme can be devastating to health (Clinical Application 3.3).

6. **Peroxisomes.** Peroxisomes are membranous sacs that resemble lysosomes in size and shape. Although present in all human cells, peroxisomes are most abundant in the liver and kidneys. Peroxisomes contain enzymes, called peroxidases, that catalyze metabolic reactions that release hydrogen peroxide ($H_2O_2$) as a by-product. Peroxisomes also contain an enzyme called catalase, which decomposes hydrogen peroxide, which is toxic to cells.

The outer membrane of a peroxisome contains some forty types of enzymes, which catalyze a variety of biochemical reactions, including:

- synthesis of bile acids, which are used in fat digestion

- breakdown of lipids called very long chain fatty acids

- degradation of rare biochemicals

- detoxification of alcohol

Abnormal peroxisomal enzymes can drastically affect health.

7. **Centrosome.** A centrosome (central body) is a structure located in the cytoplasm near the Golgi apparatus and nucleus. It is nonmembranous and consists of two hollow cylinders called *centrioles* built of tubelike proteins called microtubules. The centrioles usually lie at right angles to each other and function in cell reproduction. During this process, the centrioles move away from one another to either side of the nucleus, where they form "spindle fibers" that distribute *chromosomes,* which carry DNA information, to the newly forming cells (fig. 3.15).

Centrioles also form parts of hairlike cellular projections called cilia and flagella.

8. **Cilia** and **flagella.** Cilia and flagella are motile processes that extend from the surfaces of certain

# Disease at the Organelle Level

German physiologist Rudolph Virchow first hypothesized cellular pathology—disease at the cellular level—in the 1850s. Today, new treatments for many disorders are a direct result of understanding a disease process at the cellular level. Here, we examine how three abnormalities—in mitochondria, in peroxisomes, and in lysosomes—cause whole-body symptoms.

## MELAS AND MITOCHONDRIA

Sharon had always been small for her age, easily fatigued, slightly developmentally delayed, and had difficulty with schoolwork. She also had seizures. At age 11, she suffered what seemed to be a stroke. An astute physician who observed Sharon's mother, Lillian, suspected that the girl's symptoms were all related, and the result of abnormal mitochondria, the organelles that house the biochemical reactions that extract energy from nutrients.

The doctor noticed that Lillian was uncoordinated and had numb hands. When she asked if Lillian ever had migraines, she said that she suffered from them nearly daily, as did her two sisters and one brother. Lillian and her siblings also had diabetes mellitus and muscle weakness. Based on this information, the doctor ordered several blood tests for mother and daughter, which revealed that both had elevated levels of biochemicals (pyruvic acid and lactic acid) that indicated that they were unable to extract the maximal energy from nutrients. Muscle biopsies then showed the source of the problem—abnormal mitochondria. Accumulation of these mitochondria in smooth muscle cells in blood vessel walls in the brain caused Sharon's stroke, and was probably also causing her seizures.

All of the affected family members were diagnosed with a disorder called MELAS, which stands for the major symptoms—mitochondrial encephalomyopathy, lactic acidosis, and strokelike episodes. Their mitochondria cannot synthesize some of the proteins required to carry out the energy reactions. The responsible gene is part of the DNA in mitochondria, and Lillian's mother transmitted it to all of her children. But because mitochondria are inherited only from the mother, Sharon's uncle will not pass MELAS to his children.

## ADRENOLEUKODYSTROPHY (ALD) AND PEROXISOMES

For young Lorenzo Odone, the first sign of adrenoleukodystrophy was disruptive behavior in school. When he became lethargic, weak, and dizzy, his teachers and parents

(a)                                                                 (b)

— Centriole (cross section)

— Centriole (longitudinal section)

**FIGURE 3.15**

(a) A transmission electron micrograph of the two centrioles in a centrosome (142,000×). (b) Note that the centrioles lie at right angles to one another.

realized that his problem was not just temper tantrums. His skin darkened, blood sugar levels plummeted, heart rhythm altered, and the levels of electrolytes in his body fluids became imbalanced. He lost control over his limbs as his nervous system continued to deteriorate. Lorenzo's parents took him to many doctors. Finally one of them tested the child's blood for an enzyme normally manufactured in peroxisomes.

Lorenzo's peroxisomes lacked the second most abundant protein in the outer membrane of this organelle. Normally, the missing protein transports an enzyme into the peroxisome. The enzyme controls breakdown of a type of very long chain fatty acid. Without the enzyme, the fatty acid builds up in cells in the brain and spinal cord, eventually stripping these cells of their fatty sheaths, made of a substance called myelin. Without the myelin sheaths, the nerve cells cannot transmit messages fast enough. Death comes in a few years.

For Lorenzo and many other sufferers of ALD, eating a type of triglyceride from rapeseed oil slows the buildup of the very long chain fatty acids for a few years, stalling symptoms. But the treatment eventually impairs blood clotting and other vital functions, and fails to halt the progression of the illness.

## TAY-SACHS DISEASE AND LYSOSOMES

Michael was a pleasant, happy infant who seemed to be developing normally until about six months of age. Able to roll over and sit for a few seconds, he suddenly seemed to lose those abilities. Soon, he no longer turned and smiled at his mother's voice, and he did not seem as interested in his mobile. Concerned about Michael's reversals in development, his anxious parents took him to the doctor. It took exams by several specialists to diagnose Michael's Tay-Sachs disease, because, thanks to screening programs in the population groups known to have this inherited illness, fewer than ten new cases appear each year. Michael's parents were not among those ethnic groups and previously had no idea that they both were carriers of the gene that causes this very rare illness.

A neurologist clinched her suspicion of Tay-Sachs by looking into Michael's eyes, where she saw the telltale "cherry red spot" indicating the illness. A look at his cells provided further clues—the lysosomes, tiny enzyme-filled sacs, were swollen to huge proportions. Michael's lysosomes lacked one of the 40 types of lysosomal enzymes, resulting in a "lysosomal storage disease" that built up fatty material on his nerve cells. His nervous system would continue to fail, and he would be paralyzed and unable to see or hear by the time he died, before the age of four years.

The cellular and molecular signs of Tay-Sachs disease—the swollen lysosomes and missing enzyme—had been present long before Michael began to lag developmentally. The next time his parents expected a child, they had her tested before birth for the enzyme deficiency. They learned, happily, that she would be a carrier like themselves, but not ill. ■

---

cells. They are structurally similar and differ mainly in their length and the number present. Both consist of a constant number of microtubules organized in a distinct cylindrical pattern.

Cilia are abundant on the free surfaces of some epithelial cells. Each cilium is a tiny, hairlike structure about 10 μm long, which attaches just beneath the cell membrane to a modified centriole called a *basal body.*

Cilia occur in precise patterns. They have a "to-and-fro" type of movement that is coordinated so that rows of cilia beat one after the other, producing a wave that sweeps across the ciliated surface. For example, this action propels mucus over the surface of tissues that form the lining of the respiratory tract (fig. 3.16). Chemicals in cigarette smoke destroy cilia, causing respiratory ailments.

A cell usually has only one flagellum, which is much longer than a cilium. A flagellum begins its characteristic undulating, wavelike motion at its base. The tail of a sperm cell, for example, is a flagellum that causes the sperm's swimming movements (fig. 3.17 and chapter 22).

9. **Vesicles.** Vesicles (vacuoles) are membranous sacs that vary in size and contents. They may form when a portion of the cell membrane folds inward and pinches off. As a result, a tiny, bubblelike vesicle, containing some liquid or solid material that was formerly outside the cell, enters the cytoplasm. The Golgi apparatus and ER also form vesicles. Fleets of vesicles transport many substances into and out of cells.

10. **Microfilaments** and **microtubules.** Two types of thin, threadlike structures in the cytoplasm are microfilaments and microtubules. Microfilaments are tiny rods of the protein actin that typically occur in meshworks or bundles. They cause various kinds of cellular movements. In muscle cells, for example, microfilaments constitute *myofibrils,*

**FIGURE 3.16**

(a) Cilia, such as these (arrow), are common on the surfaces of certain cells that form the inner lining of the respiratory tract (10,000×). (b) Cilia have a power stroke and a recovery stroke that create a "to-and-fro" movement that sweeps fluids across the tissue surface.

**FIGURE 3.17**

Light micrograph of human sperm cells (1,000×). Flagella form the tails of these cells.

**FIGURE 3.18**

A transmission electron micrograph of microfilaments and microtubules within the cytoplasm (35,000×).

which cause these cells to shorten or contract. In other cells, microfilaments associated with the inner surface of the cell membrane aid cell motility (fig. 3.18).

Microtubules are long, slender tubes with diameters two or three times greater than those of microfilaments. They are composed of the globular protein tubulin. Microtubules are usually somewhat rigid, and form the cytoskeleton, which helps maintain the shape of the cell (fig. 3.19). In cilia and flagella, microtubules interact to provide movement (see figs. 3.16 and 3.17).

Cell membrane

Mitochondrion

Rough endoplasmic reticulum

Nucleus

Ribosomes

Microtubule

Microfilament

(a)

(b)

**FIGURE 3.19**

(a) Microtubules help maintain the shape of a cell by forming an internal "scaffolding," or cytoskeleton, beneath the cell membrane and within the cytoplasm. (b) A falsely colored electron micrograph of cells showing the cytoskeleton (250× micrograph enlarged to 750×).

Microtubules also move organelles and structures within the cell. For instance, microtubules are assembled from tubulin subunits in the cytoplasm during cellular reproduction, and help distribute chromosomes to the newly forming cells, a process described in more detail later in this chapter. A cell's microtubules provide conduits for organelles, like the tracks of a roller coaster.

11. **Other structures.** In addition to organelles, cytoplasm contains lifeless chemicals called *inclusions*. These usually are in a cell temporarily. Inclusions include stored nutrients such as glycogen and lipids, and pigments such as melanin in the skin.

**1** Why are mitochondria called the "powerhouses" of cells?

**2** How do lysosomes function?

**3** Describe the functions of microfilaments and microtubules.

**4** Distinguish between organelles and inclusions.

## Cell Nucleus

A nucleus is a relatively large, usually spherical structure that directs the activities of the cell. It is enclosed in a double-layered **nuclear envelope,** which consists of an inner and an outer lipid bilayer membrane. These two membranes have a narrow space between them, but are joined at places that surround relatively large openings called **nuclear pores.** These pores are not mere perforations, but channels consisting of more than 100 different types of proteins. Nuclear pores allow certain dissolved substances to move between the nucleus and the cytoplasm (fig. 3.20), most notably molecules of messenger RNA that carry genetic information.

The nucleus contains a fluid (nucleoplasm) in which other structures float. These structures include the following:

1. **Nucleolus.** A nucleolus ("little nucleus") is a small, dense body largely composed of RNA and protein. It has no surrounding membrane and is formed in specialized regions of certain chromosomes. It is the site of ribosome production. Once ribosomes form, they migrate through the nuclear pores to the cytoplasm. A cell may have more than one nucleolus. The nuclei of cells that synthesize large amounts of protein, such as those of glands, may contain especially large nucleoli.

2. **Chromatin.** Chromatin consists of loosely coiled fibers in the nuclear fluid. When cell reproduction begins, these fibers become more tightly coiled to form rodlike chromosomes. Chromatin fibers are composed of continuous DNA molecules wrapped around clusters of eight molecules of proteins

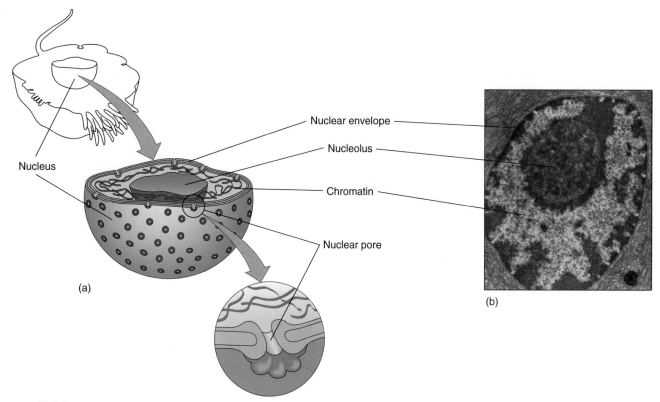

**FIGURE** 3.20

(*a*) The pores in the nuclear envelope allow certain substances to pass between the nucleus and the cytoplasm. (*b*) A transmission electron micrograph of a cell nucleus (8,000×). It contains a nucleolus and masses of chromatin.

called histones, giving the appearance of beads on a string. The DNA molecules contain the information for synthesis of proteins that promote cellular life processes.

Table 3.2 summarizes the structures and functions of organelles.

1. How are the nuclear contents separated from the cytoplasm?

2. What is the function of the nucleolus?

3. What is chromatin?

Cells die in different ways. *Apoptosis* is one form of cell death in which the cell manufactures an enzyme that cuts up DNA not protected by histones. This is an active process because a new substance is made. Apoptosis is important in shaping the embryo, in maintaining organ form during growth, and in developing the immune system and the brain.

*Necrosis* is a type of cell death that is a passive response to severe injury. Typically proteins lose their characteristic shapes, and the cell membrane deteriorates as the cell swells and bursts. Unlike apoptosis, necrosis causes great inflammation.

## MOVEMENTS INTO AND OUT OF THE CELL

The cell membrane is a barrier that controls which substances enter and leave the cell. Oxygen and nutrient molecules enter through this membrane, whereas carbon dioxide and other wastes leave through it. These movements involve *physical* (or passive) processes such as diffusion, facilitated diffusion, osmosis, and filtration, and *physiological* (or active) mechanisms such as active transport, endocytosis, and exocytosis. The mechanisms by which substances cross the cell membrane are important for understanding many aspects of physiology.

### Diffusion

**Diffusion** (also called simple diffusion) is the process by which molecules or ions spontaneously move in a liquid solution or in air. Diffusion is caused by random movement of molecules. Under natural conditions, molecules and ions are in constant motion. Each particle travels in a separate path along a straight line until it collides and bounces off some other particle. Then it moves in another direction, only to collide again and change direction once more. Diffusion mixes molecules.

Diffusing particles tend to move from regions where they are in higher concentration toward regions where they are in lower concentration. This process is

## TABLE 3.2 — STRUCTURES AND FUNCTIONS OF ORGANELLES

| Organelle | Structure | Function |
|---|---|---|
| Cell membrane | Membrane mainly composed of protein and lipid molecules | Maintains integrity of the cell, controls the passage of materials into and out of the cell, and provides for signal transduction |
| Endoplasmic reticulum | Complex of connected, membrane-bound sacs, canals, and vesicles | Transports materials within the cell, provides attachment for ribosomes, and synthesizes lipids |
| Ribosomes | Particles composed of protein and RNA molecules | Synthesize proteins |
| Golgi apparatus | Group of flattened, membranous sacs | Packages and modifies protein molecules for transport and secretion |
| Mitochondria | Membranous sacs with inner partitions | Release energy from food molecules and transform energy into usable form |
| Lysosomes | Membranous sacs | Contain enzymes capable of digesting worn cellular parts or substances that enter cells |
| Peroxisomes | Membranous vesicles | Contain enzymes called peroxidases, important in the breakdown of many organic molecules |
| Centrosome | Nonmembranous structure composed of two rodlike centrioles | Helps distribute chromosomes to new cells during cell reproduction and initiates formation of cilia |
| Cilia | Motile projections attached to basal bodies beneath the cell membrane | Propel fluids over cellular surface |
| Flagella | Motile projections attached to basal bodies beneath the cell membrane | Enable sperm cells to move |
| Vesicles | Membranous sacs | Contain substances that recently entered the cell and store and transport newly synthesized molecules |
| Microfilaments and microtubules | Thin rods and tubules | Support cytoplasm and help move substances and organelles within the cytoplasm |
| Nuclear envelope | Porous double membrane that separates the nuclear contents from the cytoplasm | Maintains the integrity of the nucleus and controls the passage of materials between the nucleus and cytoplasm |
| Nucleolus | Dense, nonmembranous body composed of protein and RNA molecules | Site of ribosome formation |
| Chromatin | Fibers composed of protein and DNA molecules | Contains cellular information for synthesizing proteins |

called *net diffusion,* and the difference in concentrations of a substance that causes it is called a *concentration gradient.*

Consider sugar (a solute) put into a glass of water (a solvent), as illustrated in figure 3.21. The sugar at first remains in high concentration at the bottom of the glass. As the sugar molecules move about, they may collide with each other or miss each other completely. Since they are less likely to collide with each other in areas where there are fewer sugar molecules, net diffusion gradually moves the sugar from areas of high concentration to areas of lower concentration (*down* the concentration gradient), and eventually the sugar molecules evenly distribute in the water. This state of uniform distribution of molecules is called equilibrium. Although molecules continue to move randomly (diffuse) after equilibrium is reached, no further *net* diffusion occurs, and concentrations no longer change.

Net diffusion of substances into or out of cells can occur if (1) the cell membrane is permeable to that substance and (2) if a concentration gradient for that substance exists, such that it is at a higher concentration either outside or inside of the cell. As body cells use oxygen, for example, the intracellular concentration of oxygen drops, whereas the blood maintains a high extracellular concentration of oxygen. Thus there is normally a concentration gradient for net diffusion of oxygen into cells.

The issue of permeability is somewhat more complex because of the selective nature of the cell membrane. Lipid-soluble substances, such as oxygen, carbon dioxide, steroids, and general anesthetics, freely cross the cell membrane by simple diffusion. Although not lipid-soluble, water molecules are small enough that they also move directly through the phospholipid bilayer by simple diffusion. Small solutes that are not lipid-soluble,

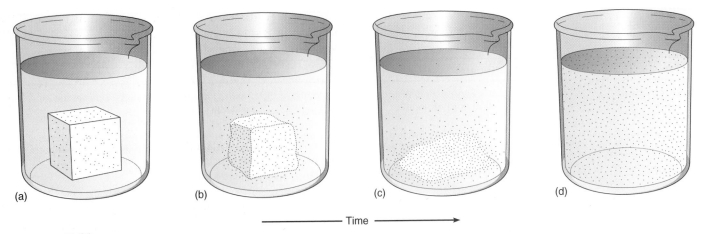

**FIGURE 3.21** [image]

An example of diffusion (*a, b,* and *c*). A sugar cube placed in water slowly disappears as the sugar molecules diffuse from regions where they are more concentrated toward regions where they are less concentrated. (*d*) Eventually, the sugar molecules distribute evenly throughout the water.

such as ions of sodium, potassium and chloride, may diffuse through protein channels in the membrane, described earlier. Water molecules may also diffuse through similar channels, called pores. Because this type of movement uses membrane proteins as "helpers," it is considered to be a form of another type of diffusion, called **facilitated diffusion.** Facilitated diffusion is very important not only for ions, but for larger water-soluble molecules, such as glucose and amino acids. It is described in more detail later.

A number of factors influence the diffusion rate, but those most important in the body are distance, the concentration gradient, and temperature. In general, diffusion is more rapid over shorter distances, larger concentration gradients, and at higher temperatures. Homeostasis maintains all three of these factors at optimum levels.

## Facilitated Diffusion

Most sugars and amino acids are insoluble in lipids, and they are too large to pass through membrane pores. Facilitated diffusion includes not only protein channels, but also certain proteins that function as "carriers" to bring such molecules across the cell membrane. In the facilitated diffusion of glucose, for example, glucose combines with a protein carrier molecule at the surface of the membrane. This union of glucose and carrier molecule changes the shape of the carrier that moves glucose to the other side of the membrane. The glucose portion is released, and the carrier molecule can return to its original shape to pick up another glucose molecule. The hormone *insulin,* discussed in chapter 13, promotes facilitated diffusion of glucose through the membranes of certain cells.

Facilitated diffusion is similar to simple diffusion in that it can only move molecules from regions of higher concentration toward regions of lower concentration. However, unlike simple diffusion, the number of carrier

**FIGURE 3.22**

Some substances move into or out of cells by facilitated diffusion, transported by carrier molecules from a region of higher concentration to one of lower concentration.

molecules in the cell membrane limits the rate of facilitated diffusion (fig. 3.22).

## Osmosis

**Osmosis** is a special case of diffusion. It occurs whenever there is net diffusion of water molecules from a region of higher water concentration to a region of lower water concentration across a selectively permeable membrane,

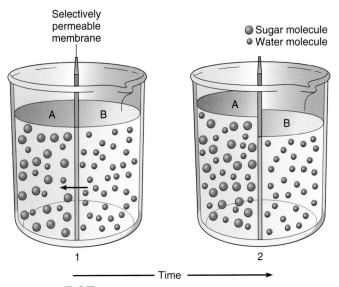

Selectively permeable membrane

● Sugar molecule
● Water molecule

A  B

A  B

1

2

Time

**FIGURE** 3.23

Osmosis. (*1*) A selectively permeable membrane separates the container into two compartments. At first, compartment A contains water and sugar molecules, whereas compartment B contains only water. As a result of molecular motions, water diffuses by osmosis from compartment B into compartment A. Sugar molecules remain in compartment A because they are too large to pass through the pores of the membrane. (*2*) Also, because more water is entering compartment A than is leaving it, water accumulates in this compartment. The level of liquid rises on this side.

such as a cell membrane. In the following example, assume that the selectively permeable membrane is permeable to water molecules (the solvent), but impermeable to glucose molecules (the solute).

In solutions, a higher concentration of solute (glucose in this case) means a lower concentration of water; a lower concentration of solute means a higher concentration of water. This is because the solute molecules take up space that water molecules would otherwise occupy.

Just like molecules of other substances, molecules of water will diffuse from areas of higher concentration to areas of lower concentration. In figure 3.23, the presence of glucose in compartment A means that the water concentration there is less than the concentration of pure water in compartment B. Therefore, water diffuses from compartment B across the selectively permeable membrane and into compartment A. In other words, water moves from compartment B into compartment A by osmosis. Glucose, on the other hand, cannot diffuse out of compartment A because the selectively permeable membrane is impermeable to it.

Note in figure 3.23 that as osmosis occurs, the level of water on side A rises. This ability of osmosis to generate enough pressure to lift a volume of water is called *osmotic pressure.*

The greater the concentration of nonpermeable solute particles (glucose in this case) in a solution, the

*lower* the water concentration of that solution and the *greater* the osmotic pressure. Water always tends to diffuse toward solutions of greater osmotic pressure.

Since cell membranes are generally permeable to water, water equilibrates by osmosis throughout the body, and the concentration of water and solutes everywhere in the intracellular and extracellular fluids is essentially the same. Therefore, the osmotic pressure of the intracellular and extracellular fluids is the same. Any solution that has the same osmotic pressure as body fluids is called **isotonic.**

Solutions that have a higher osmotic pressure than body fluids are called **hypertonic.** If cells are put into a hypertonic solution, there will be a net movement of water by osmosis out of the cells into the surrounding solution, and the cells shrink. Conversely, cells put into a **hypotonic** solution, which has a lower osmotic pressure than body fluids, tend to gain water by osmosis and swell. Although cell membranes are somewhat elastic, the cells may swell so much that they burst. Figure 3.24 illustrates the effects of the three types of solutions on red blood cells.

It is important to control the concentration of solute in solutions that are infused into body tissues or blood. Otherwise, osmosis may cause cells to swell or shrink, impairing their function. For instance, if red blood cells are placed in distilled water (which is hypotonic to them), water will diffuse into the cells and they will burst (hemolyze). On the other hand, if red blood cells are exposed to 0.9% NaCl solution (normal saline), the cells will remain unchanged because this solution is isotonic to human cells. Similarly, a 5% solution of glucose is isotonic to human cells. (The lower percentage is needed with NaCl to produce an isotonic solution, in part because NaCl ionizes in solution more completely and produces a greater number of solute particles than does glucose.)

## Filtration

Molecules move through membranes by diffusion or osmosis because of their random movements. In other instances, molecules are forced through membranes by the process of **filtration.**

Filtration is commonly used to separate solids from water. One method is to pour a mixture of solids and water onto filter paper in a funnel (fig. 3.25). The paper serves as a porous membrane through which the small water molecules can pass, leaving the larger solid particles behind. Hydrostatic pressure, which is created by the weight of water due to gravity, forces the water molecules through to the other side. Making coffee uses this method.

In the body, tissue fluid forms when water and dissolved substances are forced out through the thin, porous walls of blood capillaries, but larger particles

(a) Cell in hypertonic solution

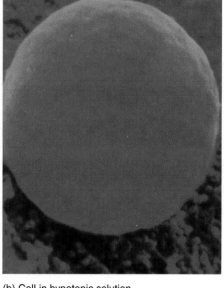
(b) Cell in hypotonic solution

(c) Cell in isotonic solution

**FIGURE** 3.24

(*a*) If red blood cells are placed in a hypertonic solution, more water leaves than enters, and the cells shrink (8,200×). (*b*) In a hypotonic solution, more water enters than leaves, and the cells swell, become spherical, and may burst (8,200×). (*c*) In an isotonic solution, water enters and leaves the cells in equal amounts, and their sizes and shapes remain unchanged (8,200).

**FIGURE** 3.25

In this example of filtration, gravity provides the force that pulls water through filter paper, while tiny openings in the paper retain the solids. This process is similar to the drip method of preparing coffee.

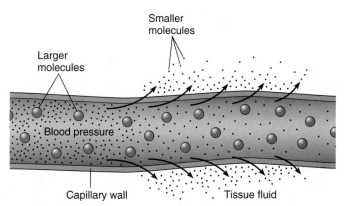

**FIGURE** 3.26

In this example of filtration, blood pressure forces smaller molecules through tiny openings in the capillary wall. The larger molecules remain inside.

such as blood protein molecules are left inside (fig. 3.26). The force for this movement comes from blood pressure, generated largely by heart action, which is greater within the vessel than outside it. (Although heart action is an active body process, filtration is still considered passive, because it can occur due to the pressure caused by gravity alone.) Filtration is discussed further in chapters 15 and 20.

**1** What kinds of substances most readily diffuse through a cell membrane?

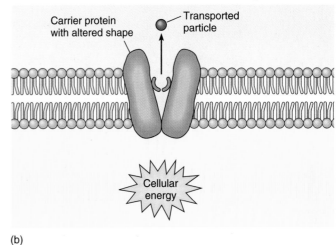

(a)

(b)

**FIGURE 3.27**

(a) During active transport, a molecule or ion combines with a carrier protein, whose shape is altered as a result. (b) This process, which requires energy, transports the particle through the cell membrane from an area of low concentration to an area of high concentration. Different substances move into or out of cells by this process.

2    Explain the differences between diffusion, facilitated diffusion, and osmosis.

3    Distinguish among hypertonic, hypotonic, and isotonic solutions.

4    Explain how filtration occurs in the body.

## Active Transport

When molecules or ions pass through cell membranes by diffusion, facilitated diffusion, or osmosis, their net movement is from regions of higher concentration to regions of lower concentration. Sometimes, however, the net movement of particles passing through membranes is in the opposite direction, from a region of lower concentration to one of higher concentration.

Sodium ions, for example, can diffuse slowly through cell membranes. Yet, the concentration of these ions typically remains many times greater on the outside of cells (in the extracellular fluid) than on the inside of cells (in the intracellular fluid). This is because sodium ions are continually moved through the cell membrane from the regions of lower concentration (inside) to the regions of higher concentration (outside). Movement of this type is called **active transport.** It requires energy that derives from cellular metabolism. In fact, up to 40% of a cell's energy supply may be used for active transport of particles through its membranes.

Active transport is similar to facilitated diffusion in that it uses carrier molecules within cell membranes. As figure 3.27 shows, these carrier molecules are proteins that have binding sites that combine with the specific particles being transported. Such a union triggers release of cellular energy, and this energy alters the shape of the carrier protein. As a result, the "passenger" molecules move through the membrane. Once on the other side, the

transported particles are released, and the carrier molecules can accept other passenger molecules at their binding sites. Because they transport substances from regions of low concentration to regions of higher concentration, these carrier proteins are sometimes called "pumps." A sodium/potassium pump, for example, transports sodium ions out of cells and potassium ions into cells.

Particles that are moved across cell membranes by active transport include sugars, amino acids, and sodium, potassium, calcium, and hydrogen ions. Some of these substances are actively transported into cells, and others are transported out. Movements of this type are important to cell survival, particularly maintenance of homeostasis. Some of these movements are described in subsequent chapters as they apply to specific organ systems.

## Endocytosis

Two processes use cellular energy to move substances into or out of a cell without actually crossing the cell membrane. In **endocytosis,** molecules or other particles that are too large to enter a cell by diffusion or active transport are conveyed within a vesicle that forms from a section of the cell membrane. In **exocytosis,** the reverse process secretes a substance stored in a vesicle from the cell.

The three forms of endocytosis are pinocytosis, phagocytosis, and receptor-mediated endocytosis. In **pinocytosis,** cells take in tiny droplets of liquid from their surroundings (fig. 3.28). When this happens, a small portion of cell membrane indents (invaginates). The open end of the tubelike part thus formed seals off and produces a small vesicle about 0.1 μm in diameter. This tiny sac detaches from the surface and moves into the cytoplasm.

For a time, the vesicular membrane, which was part of the cell membrane, separates its contents from the rest of the cell; however, the membrane eventually breaks

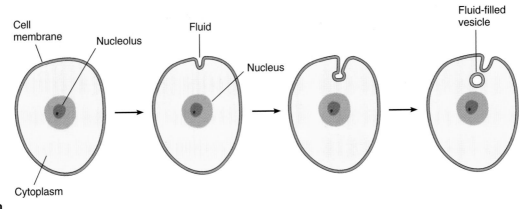

**FIGURE** 3.28

A cell may take in a tiny droplet of fluid from its surroundings by pinocytosis.

**FIGURE** 3.29

A cell may take in a solid particle from its surroundings by phagocytosis.

down and the liquid inside becomes part of the cytoplasm. In this way, a cell is able to take in water and the particles dissolved in it, such as proteins, that otherwise might be too large to enter.

**Phagocytosis** is similar to pinocytosis, but the cell takes in solids rather than liquid. Certain kinds of cells, including some white blood cells, are called *phagocytes* because they can take in solid particles such as bacterial cells and cellular debris. When a phagocyte first encounters such a particle, the particle attaches to the phagocyte's cell membrane. This stimulates a portion of the membrane to project outward, surround the particle, and slowly draw it inside the cell. The part of the membrane surrounding the solid detaches from the cell's surface, forming a vesicle containing the particle (fig. 3.29). Such a vesicle may be several micrometers in diameter.

Usually, a lysosome soon combines with such a newly formed vesicle, and lysosomal digestive enzymes decompose the vesicular contents (fig. 3.30). The products of this decomposition may then diffuse out of the lysosome and into the cytoplasm, where they may be used as raw materials in metabolic processes. Any remaining residue may be expelled from the cell by exocytosis. In this way, phagocytic cells can dispose of foreign objects, such as dust particles, remove damaged cells or

cell parts that are no longer functional, or destroy bacteria that might otherwise cause infections. Phagocytosis is an important line of defense against invasion by disease-causing microorganisms.

Pinocytosis and phagocytosis engulf anything. In contrast is the more discriminating **receptor-mediated endocytosis,** which moves very specific kinds of particles into the cell. In this mechanism, protein molecules extend through the cell membrane and are exposed on its outer surface. These proteins are receptors to which specific substances (ligands) from the fluid surroundings of the cell can bind. Molecules that can bind to the receptor sites selectively enter the cell; other kinds of molecules are left outside (fig. 3.31).

Entry of cholesterol molecules into cells illustrates receptor-mediated endocytosis. Cholesterol molecules synthesized in liver cells are packaged into large spherical particles called *low-density lipoproteins* (LDL). An LDL particle has a coating that contains a binding protein called *apoprotein-B.* The membranes of various body cells have receptors for apoprotein-B. When the liver releases LDL particles into the blood, cells with apoprotein-B receptors can recognize the LDL particles and bind them. Formation of such a receptor-ligand combination stimulates the cell membrane to indent and form a vesicle

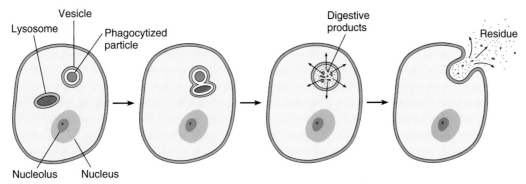

**FIGURE** 3.30

When a lysosome combines with a vesicle that contains a phagocytized particle, its digestive enzymes may destroy the particle. The products of this intracellular digestion diffuse into the cytoplasm. Any residue may be expelled from the cell by exocytosis.

around the LDL particle. The vesicle carries the LDL particle to a lysosome, where enzymes digest it and release the cholesterol molecules for cellular use.

Receptor-mediated endocytosis is particularly important because it allows cells with the appropriate receptors to remove and process specific kinds of substances from their surroundings, even when these substances are present in very low concentrations. In short, this mechanism provides specificity.

As a toddler, Stormie Jones already had a blood serum cholesterol level six times normal. Before she died at age 10, she had suffered several heart attacks and had undergone two cardiac bypass surgeries, several heart valve replacements, and finally a heart-liver transplant. The transplant lowered her blood cholesterol to a near-normal level, but she died from the multiple traumas suffered over her short lifetime.

Stormie had the severe form of familial hypercholesterolemia (FH), meaning simply too much cholesterol in the blood. Her liver cells lacked LDL receptors. Blocked from entering cells, cholesterol accumulated in her bloodstream, forming the plaques that caused her heart disease.

Stormie Jones was one in a million. Far more common are the one in 500 people who have the milder form of FH, in which liver cells have half the normal number of LDL receptors. These individuals are prone to suffer heart attacks in early adulthood. However, they can delay symptom onset by taking precautions to avoid cholesterol buildup, such as exercising, eating a low-fat diet, and not smoking. (These precautions may also benefit individuals not suffering from FH.)

## Exocytosis

Exocytosis is essentially the reverse of endocytosis. In exocytosis, substances made within the cell are packaged into a vesicle, which then fuses with the cell membrane, thereby releasing its contents outside the cell. Cells secrete some proteins by this process. Nerve cells use exocytosis to release the neurotransmitter chemicals that signal other nerve cells, muscle cells, or glands (fig. 3.32).

## Transcytosis

Endocytosis brings a substance into a cell, and exocytosis transports a substance out of a cell. Another process, **transcytosis,** combines endocytosis and exocytosis (fig. 3.33). Transcytosis is the selective and rapid transport of a substance or particle from one end of a cell to the other. It enables substances to cross barriers formed by tightly connected cells.

HIV, the virus that causes AIDS, may initially infect a human body by using transcytosis to cross lining (epithelial) cells in the anus and female reproductive tract. Experiments using tissues growing in laboratory culture show that HIV enters white blood cells in mucous secretions, and the secretions then carry the infected cells to an epithelial barrier. Near these lining cells, viruses rapidly exit the infected white blood cells, and are quickly enveloped by the lining cell membranes in receptor-mediated endocytosis. HIV particles are ferried, in vesicles, through the lining cell, without actually infecting (taking over) the cell, to exit from the cell membrane at the other end of the cell—in as little as 30 minutes! After transcytosis, the HIV particles infect white blood cells beyond the epithelial barrier.

Table 3.3 summarizes the types of movement into and out of the cell.

1. What type of mechanism maintains unequal concentrations of ions on opposite sides of a cell membrane?

2. How are facilitated diffusion and active transport similar? How are they different?

3. What is the difference between pinocytosis and phagocytosis?

4. Describe receptor-mediated endocytosis.

5. What does transcytosis accomplish?

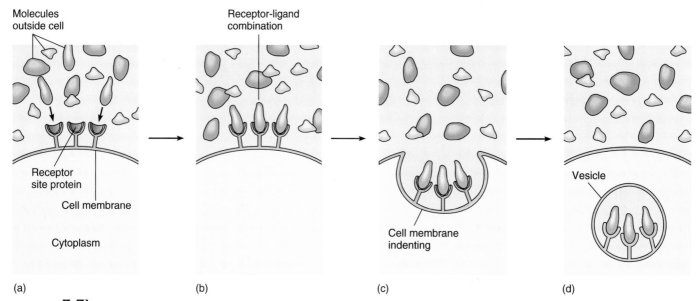

**FIGURE** 3.31

Receptor-mediated endocytosis. (*a*) A specific substance binds to a receptor site protein. (*b* and *c*) The combination of the substance with the receptor site protein stimulates the cell membrane to indent. (*d*) The resulting vesicle transports the substance into the cytoplasm.

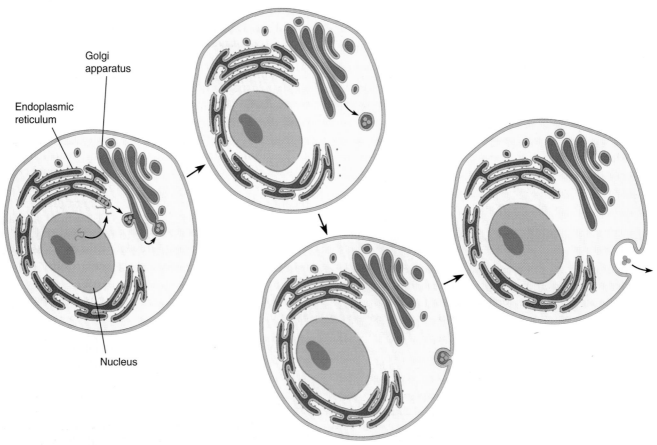

**FIGURE** 3.32

Exocytosis releases particles, such as newly synthesized proteins, from cells.

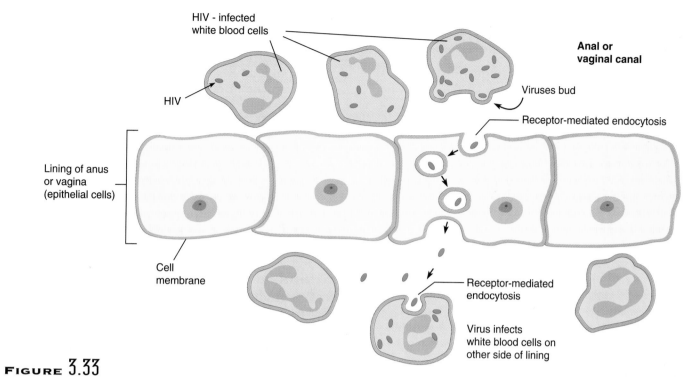

HIV - infected
white blood cells

Anal or
vaginal canal

HIV

Viruses bud

Receptor-mediated endocytosis

Lining of anus
or vagina
(epithelial cells)

Cell
membrane

Receptor-mediated
endocytosis

Virus infects
white blood cells on
other side of lining

**FIGURE 3.33**

Transcytosis transports HIV across the lining of the anus or vagina.

| TABLE **3.3** | MOVEMENTS INTO AND OUT OF THE CELL | | | |
|---|---|---|---|---|
| **Process** | **Characteristics** | **Source of Energy** | **Example** | |
| *I. Passive (Physical) Processes* | | | | |
| **A.** Simple diffusion | Molecules or ions move from regions of higher concentration toward regions of lower concentration. | Molecular motion | Exchange of oxygen and carbon dioxide in the lungs | |
| **B.** Facilitated diffusion | Molecules move across the membrane through channels or by carrier molecules from a region of higher concentration to one of lower concentration. | Molecular motion | Movement of glucose through a cell membrane | |
| **C.** Osmosis | Water molecules move from regions of higher concentration toward regions of lower concentration through a selectively permeable membrane. | Molecular motion | Distilled water entering a cell | |
| **D.** Filtration | Smaller molecules are forced through porous membranes from regions of higher pressure to regions of lower pressure. | Hydrostatic pressure | Molecules leaving blood capillaries | |
| *II. Active (Physiological) Processes* | | | | |
| **A.** Active transport | Carrier molecules transport molecules or ions through membranes from regions of lower concentration toward regions of higher concentration. | Cellular energy | Movement of various ions and amino acids through membranes | |
| **B.** Endocytosis | | | | |
| 1. Pinocytosis | Membrane engulfs droplets of liquid from surroundings. | Cellular energy | Membrane-forming vesicles containing large particles dissolved in water | |
| 2. Phagocytosis | Membrane engulfs solid particles from surroundings. | Cellular energy | White blood cell membrane engulfing bacterial cell | |
| 3. Receptor-mediated endocytosis | Membrane engulfs selected molecules combined with receptor proteins. | Cellular energy | Cell removing cholesterol-containing LDL particles from its surroundings | |
| **C.** Exocytosis | Vesicles fuse with membrane and release contents outside of the cell. | Cellular energy | Protein secretion, neurotransmitter release | |
| **D.** Transcytosis | Combines receptor-mediated endocytosis and exocytosis to ferry particles through a cell | Cellular energy | HIV crossing a cell layer | |

# THE CELL CYCLE

The series of changes that a cell undergoes from the time it forms until it reproduces is called the **cell cycle** (fig. 3.34). Superficially, this cycle seems rather simple—a newly formed cell grows for a time and then divides in half to form two new cells, which in turn may grow and divide. Yet the specific events of the cycle are quite complex. For ease of study, the cell cycle can be considered to consist of distinct stages, which include interphase, mitosis, cytoplasmic division, and differentiation.

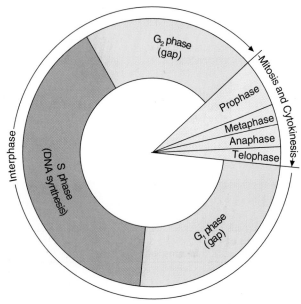

**FIGURE** 3.34

The cell cycle.

## Interphase

Before a cell actively divides, it must amass important biochemicals and duplicate much of its contents, so that two cells can form from one. This period of preparedness is called **interphase.** Once thought to be a time of rest, interphase is actually a time of great synthetic activity. During interphase, the cell maintains its routine "housekeeping" functions yet also takes on the tremendous task of replicating its genetic material. It also duplicates membranes, ribosomes, lysosomes, peroxisomes, and mitochondria. Interphase is divided into three phases. The S phase, when DNA replicates, is bracketed by two gap or growth periods, called $G_1$ and $G_2$, when other structures are duplicated (fig. 3.35).

## Mitosis

In cell reproduction, a cell divides to form two new, complete cells (daughter cells). This process includes two separate steps: division of the nuclear contents, which is called mitosis (karyokinesis), and division of the cytoplasm (cytokinesis). (Chapters 22 and 24 describe another type of cell division that includes a process called *meiosis,* which produces sex cells.)

Division of the nucleus by mitosis is, of necessity, very precise, because the nucleus contains information (in the form of DNA molecules) that "tells" the cell how to carry on life processes. Each new cell must have a complete copy of this information in order to survive. Although the chromosomes have already been copied in interphase, it is in mitosis that the copies are evenly distributed between the two new cells.

Mitosis is often described in stages, but the process is actually continuous without marked changes between one step and the next (fig. 3.36). The idea of stages is useful,

(a)

(b)

**FIGURE** 3.35

(*a*) Interphase lasts until a cell begins to undergo mitosis. (*b*) A micrograph of a cell in interphase (250× micrograph enlarged to 1000×). Although present, centrioles and chromatin fibers are not clearly visible at this magnification.

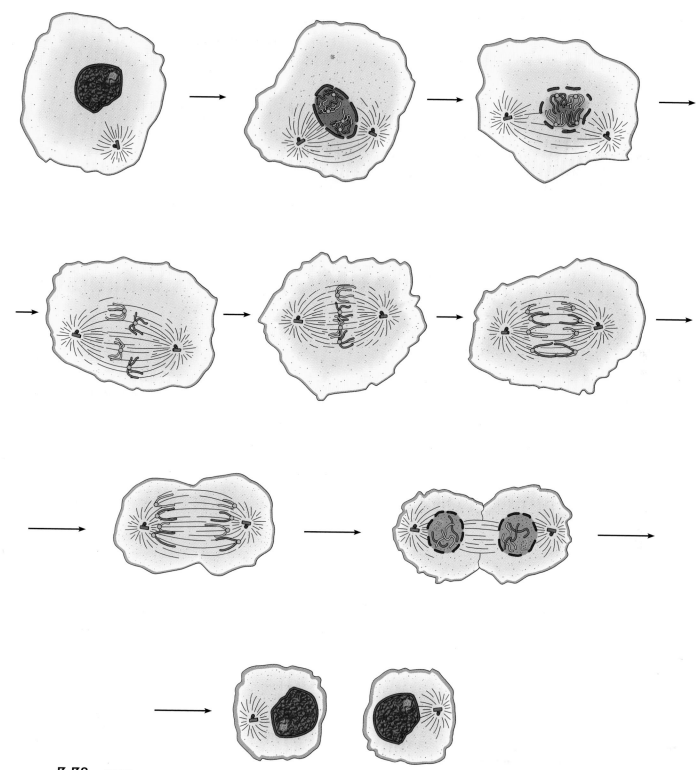

FIGURE 3.36 📼

Mitosis is a continuous process during which the replicated genetic material is divided into two equal portions. After reading about mitosis, identify the phases of the process and the cell parts shown in this diagram.

(a)

(b)

**FIGURE 3.37**

(a) In prophase, chromosomes form from chromatin in the nucleus, and the centrioles move to opposite sides of the cell. (b) A micrograph of a cell in prophase (250× micrograph enlarged to 1,000×).

(a)

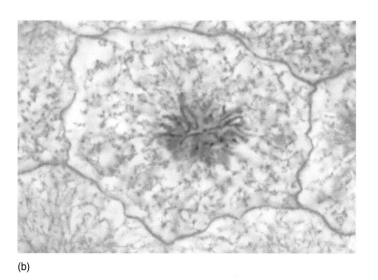

(b)

**FIGURE 3.38**

(a) Later in prophase, the nuclear envelope and nucleolus disappear. (b) A micrograph of a cell in late prophase (polar view) (280× micrograph enlarged to 1,000×).

however, to indicate the sequence in which major events occur. The stages of mitosis are:

1. **Prophase.** One of the first indications that a cell is going to reproduce is the appearance of *chromosomes*. These structures form as fibers of chromatin condense into tightly coiled rods. During interphase, the DNA molecules replicate (duplicate), so that each chromosome is composed of two identical structures, called chromatids, that are temporarily attached by a region on each called a *centromere*.

    The centrioles of the centrosome replicate just before the onset of mitosis, and during prophase, the two newly formed pairs of centrioles move to opposite sides of the cell. Soon the nuclear envelope and the nucleolus disperse and are no longer visible. Microtubules are assembled from tubulin proteins in the cytoplasm, and these structures associate with the centrioles and chromosomes (figs. 3.37 and 3.38). A spindle-shaped array of microtubules (spindle fibers) forms between the centrioles as they move apart.

2. **Metaphase.** Spindle fibers attach to the centromeres of the chromosomes so that a fiber accompanying one chromatid attaches to one centromere, and a fiber accompanying the other chromatid attaches to its centromere (fig. 3.39). The

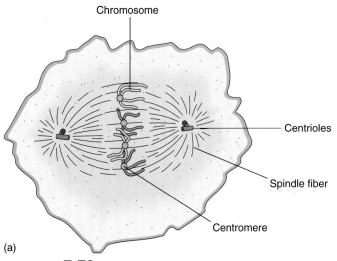

Chromosome

Centrioles

Spindle fiber

Centromere

(a)

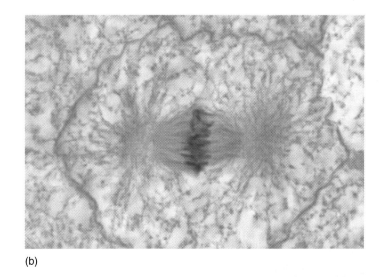

(b)

**FIGURE** 3.39

(a) In metaphase, chromosomes line up midway between the centrioles. (b) A micrograph of a cell in metaphase (280× micrograph enlarged to 1,000×). Although present, centrioles are not clearly visible at this magnification.

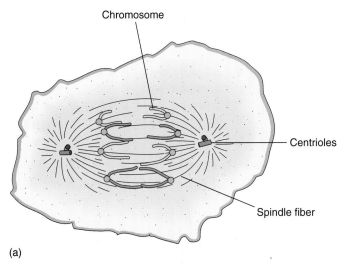

Chromosome

Centrioles

Spindle fiber

(a)

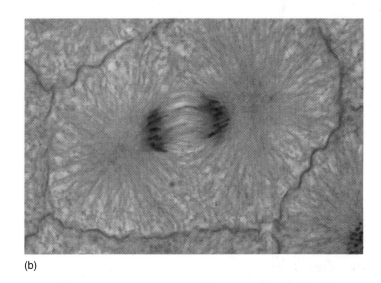

(b)

**FIGURE** 3.40

(a) In anaphase, centromeres divide, and the spindle fibers that have become attached to them pull the chromatids, now called chromosomes, toward the centrioles. (b) A micrograph of a cell in anaphase (280× micrograph enlarged to 1,100×).

chromosomes move along the spindle fibers and align about midway between the centrioles, as a result of microtubule activity.

3. **Anaphase.** Soon the centromeres of the chromatids separate, and these identical chromatids are now considered individual chromosomes. The separated chromosomes move in opposite directions, and once again the movement results from microtubule activity. The spindle fibers shorten and pull their attached chromosomes toward the centrioles at opposite sides of the cell (fig. 3.40).

4. **Telophase.** The final stage of mitosis begins when the chromosomes complete their migration toward the centrioles. It is much like prophase, but in reverse. As the chromosomes approach the centrioles, they begin to elongate and unwind from rodlike structures to threadlike structures. A nuclear envelope forms around each chromosome set, and nucleoli appear within the newly formed nuclei. Finally, the microtubules disassemble into free tubulin molecules (fig. 3.41).

Table 3.4 summarizes the phases of mitosis.

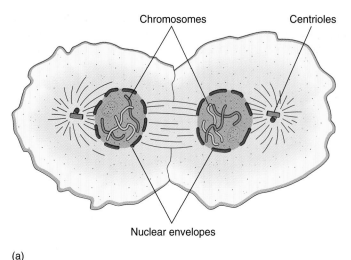

Chromosomes        Centrioles

Nuclear envelopes

(a)

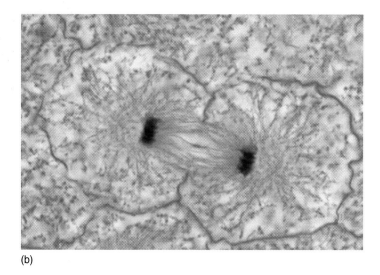

(b)

## FIGURE 3.41

(*a*) In telophase, chromosomes elongate to become chromatin threads, and the cytoplasm begins to distribute between the two newly forming cells. The replicate chromatids have separated to form the chromosomes of the two new cells.
(*b*) A micrograph of a cell in telophase (280× micrograph enlarged to 1,100×).

### TABLE 3.4 — MAJOR EVENTS IN MITOSIS AND CYTOKINESIS

| Stage | Major Events |
|---|---|
| Prophase | Chromatin condenses into chromosomes; centrioles move to opposite sides of cytoplasm; nuclear membrane and nucleolus disperse; microtubules appear and associate with centrioles and chromatids of chromosomes. |
| Metaphase | Spindle fibers from the centrioles attach to the centromeres of each chromosome; chromosomes align midway between the centrioles. |
| Anaphase | Centromeres separate, and chromatids of the chromosomes separate; spindle fibers shorten and pull these new individual chromosomes toward centrioles. |
| Telophase | Chromosomes elongate and form chromatin threads; nuclear membranes appear around each chromosome set; nucleoli appear; microtubules disappear. |

## Cytoplasmic Division

Cytoplasmic division (cytokinesis) begins during anaphase when the cell membrane starts to constrict around the middle. It continues to do so through telophase. The musclelike contraction of a ring of actin microfilaments pinches off two cells from one. The microfilaments assemble in the cytoplasm and attach to the inner surface of the cell membrane. The contractile ring forms at right angles to the microtubules that pulled the chromosomes to opposite ends of the cell during mitosis. As the ring pinches, it separates the two newly formed nuclei and apportions about half of the organelles into each of the new cells.

The newly formed cells may differ slightly in size and number of organelles and inclusions, but they have identical chromosomes and thus contain identical DNA information. Except for size, they are copies of the original cell (fig. 3.42).

## Cell Differentiation

Because all body cells (except egg and sperm) contain the same DNA information, they might be expected to look and function alike; obviously, they do not.

A human begins as a single cell—an egg cell that has been fertilized by a sperm cell. This cell divides to form two cells; they, in turn, divide into four cells; the four become eight; and so forth. Then, in the third to eighth weeks, the cells specialize, developing distinctive structures and beginning to function in different ways. Some become skin cells, others become bone cells, and still others become nerve cells (fig. 3.43). By the time of birth, a human has more than 200 types of cells. The process by which cells develop different structures and specialized functions is called **differentiation.**

Cell specialization reflects genetic control. Special proteins activate some genes and repress others, controlling the collection of biochemicals in the cell, and therefore its characteristics. In a nerve cell, the genes controlling neurotransmitter synthesis would be activated; in a bone cell, these genes would be silenced because it does not use neurotransmitters, but genes encoding the protein collagen, a major component of bone, would be very active. A differentiated cell can be compared to a library. It contains a complete collection

(a)

(b)

(c)

**FIGURE 3.42**

Following mitosis, the cytoplasm of a cell divides in two, as seen in these scanning electron micrographs (a 3,750×; b 3,750×; c 3,190×).

From *Scanning Electron Microscopy in Biology*, by R. G. Kessel and C. Y. Shih. © 1976 Springer-Verlag.

of information, but only some of that information is accessed. Different cell types interact in ways that aid survival of the individual.

**1** Why is precise division of nuclear materials during mitosis important?

**2** Describe the events that occur during mitosis.

**3** Name the process by which some cells become muscle cells and others become nerve cells.

**4** How does DNA control differentiation?

## CONTROL OF CELL REPRODUCTION

How often a cell reproduces is strictly controlled and varies with cell type. Skin cells, blood-forming cells, and cells that line the intestine, for example, reproduce often and continually. In contrast, cells of the liver reproduce a specific number of times, and then cease. If injury or surgery removes some liver cells, the remaining cells may be stimulated to reproduce again, regenerating the organ. Still other cells, such as nerve cells, apparently lose their ability to reproduce as they differentiate; therefore, damage to nerve cells usually permanently impairs nerve function. Clinical Application 3.4 considers cloning, a technique that places the nucleus of a differentiated cell into a fertilized egg cell lacking a nucleus, and regenerates a new individual from the altered cell.

Most types of human cells reproduce up to about 50 times when grown in the laboratory. Adherence to this limit can be startling. A connective tissue cell from a human fetus reproduces 35 to 63 times, the average being about 50 times. However, a similar cell from an adult reproduces only 14 to 29 times, as if the cells "know" how many times they have reproduced.

A physical basis for this mitotic clock is the DNA at the tips of chromosomes (telomeres), where the same six-nucleotide sequence repeats many hundreds of times. With each mitosis, up to 1,200 nucleotides are lost. When the chromosome tips wear down to a certain point, this somehow signals the cell to cease reproducing.

Other external and internal factors influence when a cell reproduces. Within cells, waxing and waning levels of proteins called kinases and cyclins control cell reproduction. Another internal influence is simply cell size, specifically the ratio between the surface area the cell membrane provides and the cell volume. The larger the cell, the more nutrients it requires to maintain the activities of life. However, a cell's surface area limits the amount of nutrients that can enter. Because volume increases faster than does surface area, a cell can grow too large to efficiently obtain nutrients. A cell can solve this growth problem by reproducing. The resulting cells are smaller than the original cell, and thus have a more favorable surface area-to-volume relationship.

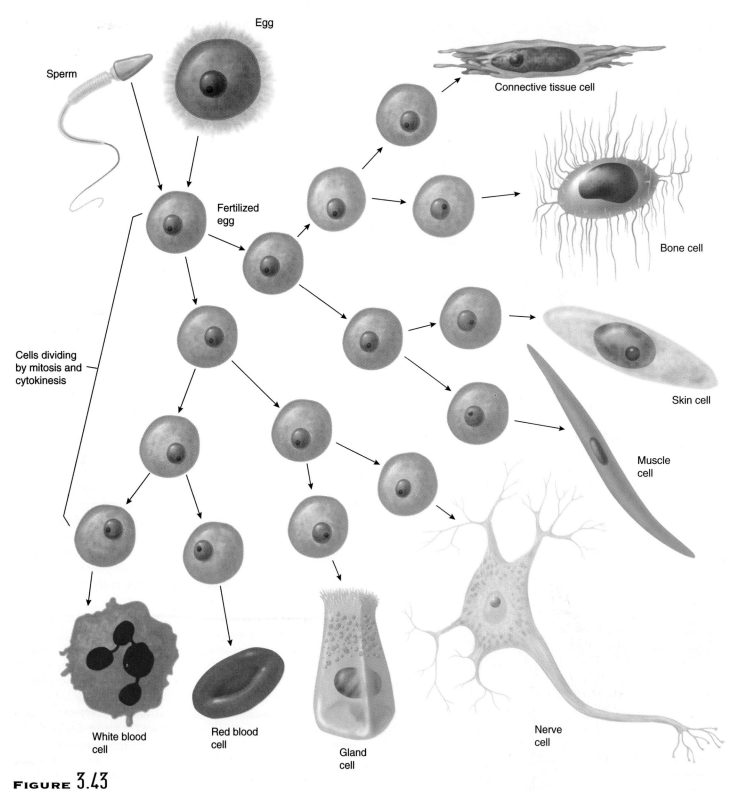

**FIGURE 3.43**

The trillions of cells in an adult human derive by mitosis from the original fertilized egg cell. As different genes turn on or off in different cells, the characteristics of specific cell types emerge. (Relative cell sizes are not to scale.)

# Cloning

The human body is built of more than 200 types of specialized, or differentiated, cell types. Once a cell activates certain subsets of the total genetic package present in all cells, there usually isn't any turning back. A nerve cell remains a nerve cell; an adipose cell stays an adipose cell. In contrast to a differentiated cell, a fertilized egg cell retains the potential to become any cell type—a little like a college student before declaring a major. Such a cell is called "pluripotent." Something about the cytoplasm of the cell keeps it in a state where it can follow any cell "fate."

What would happen if a nucleus from a differentiated cell was placed in a fertilized egg cell whose nucleus had been removed? Would the special egg cytoplasm literally turn back the developmental clock, returning the differentiated cell's nucleus to a pluripotent state, and possibly enabling it to specialize in a different way? In 1997, Scottish researchers did just that, in sheep. They took a cell from an adult sheep's breast and transferred its nucleus to a fertilized egg cell whose nucleus had been removed. The altered cell divided, and divided again, and when it was a ball of cells, it was implanted into an unrelated ewe. Development continued. On February 7, 1997, the result—a now-famous sheep named Dolly—graced the cover of *Nature* magazine, and immediately triggered worldwide controversy (fig. 3A).

Dolly is a clone of the sheep that donated the breast cell, which means that, except for the small amount of mitochondrial DNA, the two animals are genetically identical. Creating Dolly was quite a feat—it took 277 tries. Despite the difficulty of the procedure, and the fact that it had not been performed on humans, public reaction in the months following Dolly's debut was largely negative, as talk centered on cloning humans. Politicians called for "no-clone zones," editorials envisioned scenarios of farming replicas to harvest their organs for spare parts, and films and cartoons flourished (fig. 3B).

Lost in the fear of cloning was the fact that restoring developmental potential to a specialized cell can have medical applications. Cloning would enable rapid mass production of cows genetically altered to produce proteins that are of use to humans as drugs—such cows, for example, already make the human version of tissue plasminogen activator, used as an anticlotting agent to save lives following heart attacks and strokes. Pigs genetically altered to have cell surfaces compatible with humans are being considered as organ donors for humans—cloning would scale-up that technology. Someday, the ability to take a human cell back in time and perhaps give it a new specialization might lead to a way to regenerate injured tissue, or grow replacement parts.

Cloning humans raises a philosophical issue. To what extent do genes determine who we are? That is, how identical are individuals who have the same genes? That question has already been answered by naturally occurring human clones—identical twins. Extensive analysis of identical twins who were separated at birth demonstrates that although they are physically alike, and share many health characteristics and even some peculiar quirks, they also differ in many ways. The environment exerts powerful effects on who we are, contributing to characteristics such as personality traits that are harder to define biochemically. ■

**FIGURE 3A**
Dolly, a most unusual ewe.

**FIGURE 3B**
A view of cloning.

External controls of cell division include hormones and growth factors. Hormones are biochemicals manufactured in a gland and transported in the bloodstream to a site where they exert an effect. Hormones signal mitosis in the lining of a woman's uterus each month, building up the tissue to nurture a possible pregnancy. Similarly, a pregnant woman's hormones stimulate mitosis in her breasts when their function as milk-producing glands will soon be required.

Growth factors are biochemicals that are like hormones in function but act closer to their sites of synthesis. Epidermal growth factor, for example, stimulates growth of new skin beneath the scab on a skinned knee. Salivary glands also produce this growth factor. This is why an animal's licking a wound may speed healing.

> Growth factors are used as drugs. Epidermal growth factor (EGF), for example, can hasten healing of a wounded or transplanted cornea, a one-cell-thick layer covering the eye. Normally these cells do not reproduce. However, a damaged cornea treated with EGF undergoes mitosis, restoring a complete cell layer. EGF is also used to help the body accept skin grafts and to stimulate healing of skin ulcers that occur as a complication of diabetes.

Space availability is another external factor that influences the timing and rate of cell reproduction. Healthy cells do not reproduce if they are surrounded by other cells, a phenomenon called contact (density dependent) inhibition.

**1** How do cells vary in their rates of reproduction?

**2** What factors control the number of times and the rate at which cells reproduce?

## Health Consequences of Loss of Cell Reproduction Control

Control of cell reproduction is absolutely crucial to health. With too infrequent mitosis, an embryo could not develop, a child could not grow, and wounds would not heal. Too frequent mitosis produces an abnormal growth, or neoplasm, which may form a disorganized mass called a **tumor.**

Tumors are of two types. A *benign* tumor remains in place like a lump, eventually interfering with the function of healthy tissue. A *malignant,* or cancerous, tumor looks quite different—it is invasive, extending into surrounding tissue, resembling a crab with outreaching claws, which is where the name "cancer" comes from. Cancer cells, if not stopped, eventually reach the circula-

| TABLE 3.5 | CHARACTERISTICS OF CANCER CELLS |
|---|---|

Loss of cell cycle control

Heritability (a cancer cell reproduces to form more cancer cells)

Transplantability (a cancer cell implanted into another individual will cause cancer to develop)

Dedifferentiation (loss of specialized characteristics)

Loss of contact inhibition

Ability to induce local blood vessel formation (angiogenesis)

Invasiveness

Ability to metastasize (spread)

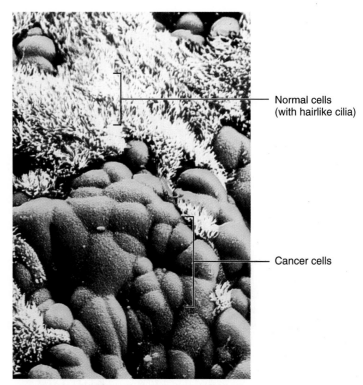

— Normal cells (with hairlike cilia)

— Cancer cells

**FIGURE 3.44**

A cancer cell is rounder and less specialized than surrounding healthy cells. It secretes biochemicals that cut through nearby tissue, and others that stimulate formation of blood vessels that nurture the tumor's growth (2,200×).

tion and spread, or metastasize, to other sites. Table 3.5 lists characteristics of cancer cells, and figure 3.44 illustrates how cancer cells infiltrate healthy tissue.

Cancer is a collection of disorders distinguished by their site of origin and the affected cell type. Many cancers are treatable with surgery, radiation, or chemicals or immune system substances used as drugs. Experimental gene therapy fights cancer by giving tumor cells surface molecules that attract an attack by the immune system.

At least two types of genes cause cancer. **Oncogenes** activate other genes that increase cell reproduction rate. **Tumor suppressor genes** normally hold mitosis in check. When tumor suppressor genes are removed or otherwise inactivated, this lifts control of the cell cycle, and uncontrolled reproduction leading to cancer results (fig. 3.45). Environmental factors, such as exposure to toxic chemicals or radiation, may induce cancer by altering (mutating) oncogenes and tumor suppressor genes. Cancer may also be the consequence of a failure of normal cell death, resulting in overgrowth.

Normal anatomy and physiology—in other words, health—ultimately depend upon both the quality and quantity of the cells that comprise the human body. Understanding the cellular bases of diseases suggests new diagnosis and treatment methods.

**1** How can too infrequent cell reproduction affect health?

**2** How can too frequent cell reproduction affect health?

**3** What is the difference between a benign and a cancerous tumor?

**4** What are two ways that genes cause cancer?

**5** How can factors in the environment cause cancer?

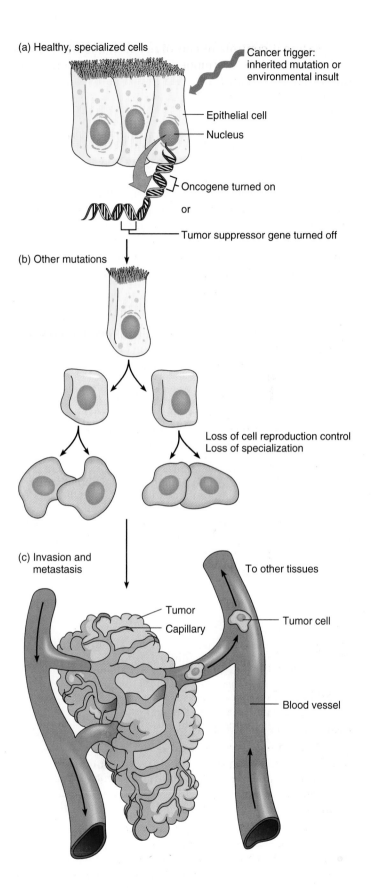

(a) Healthy, specialized cells

Cancer trigger: inherited mutation or environmental insult

Epithelial cell
Nucleus

Oncogene turned on

or

Tumor suppressor gene turned off

(b) Other mutations

Loss of cell reproduction control
Loss of specialization

(c) Invasion and metastasis

To other tissues

Tumor
Capillary

Tumor cell

Blood vessel

**FIGURE 3.45**

(a) In a healthy cell, oncogenes are not expressed and tumor suppressor genes are expressed. As a result, cell reproduction rate is under control. Cancer begins in a single cell when an oncogene is turned on or a tumor suppressor gene is turned off. This initial step may result from an inherited mutation or from exposure to radiation, viruses, or chemicals that cause cancer. (b) Malignancy often results from a series of genetic alterations (mutations). An affected cell reproduces more often than the cell type it descends from and eventually loses its specialized characteristics. (c) Cancers grow and spread by inducing formation of blood vessels to nourish them, and then breaking away from their original location. The renegade cells often alter their genetic makeup and surface characteristics as they travel. This changeable nature is why many treatments eventually cease to work, or a supposedly vanquished cancer shows up someplace in the body other than where it originated.

## Introduction (page 60)

Cells vary considerably in size, shape, and function. The shapes of cells are important in determining their functions.

## A Composite Cell (page 60)

1. A cell includes a nucleus, cytoplasm, and a cell membrane.
2. Cytoplasmic organelles perform specific vital functions, but the nucleus controls the overall activities of the cell.
3. Cell membrane
   a. The cell membrane forms the outermost limit of the living material.
   b. It acts as a selectively permeable passageway that controls the movements of substances between the cell and its surroundings, and thus is the site of signal transduction.
   c. It includes protein, lipid, and carbohydrate molecules.
   d. The cell membrane framework mainly consists of a double layer of phospholipid molecules.
   e. Molecules that are soluble in lipids pass through the membrane easily, but water-soluble molecules do not.
   f. Cholesterol molecules help stabilize the membrane.
   g. Proteins provide the special functions of the membrane.
      (1) Rodlike proteins form receptors on cell surfaces.
      (2) Globular proteins form channels for passage of various ions and molecules.
   h. Specialized intercellular junctions (tight junctions, desmosomes, and gap junctions) connect cells.
   i. Cell adhesion molecules oversee some cell interactions and movements.
4. Cytoplasm
   a. Cytoplasm contains networks of membranes and organelles suspended in fluid.
   b. Endoplasmic reticulum is composed of connected membranous sacs, canals, and vesicles that provide a tubular communication system and an attachment for ribosomes; it also functions in the synthesis of proteins and lipids.
   c. Ribosomes are particles of protein and RNA that function in protein synthesis.
   d. The Golgi apparatus is a stack of flattened, membranous sacs that package glycoproteins for secretion.
   e. Mitochondria are membranous sacs that contain enzymes that catalyze the reactions that release energy from nutrient molecules and transform it into a usable form.
   f. Lysosomes are membranous sacs containing digestive enzymes that destroy debris and worn-out organelles.
   g. Peroxisomes are membranous, enzyme-containing vesicles.
   h. The centrosome is a nonmembranous structure consisting of two centrioles that aids in the distribution of chromosomes during cell reproduction.
   i. Cilia and flagella are motile processes that extend from some cell surfaces.
      (1) Cilia are numerous tiny, hairlike structures that wave, moving fluids across cell surfaces.
      (2) Flagella are longer processes, such as the tail of a sperm cell.

   j. Vesicles are membranous sacs containing substances that recently entered the cell.
   k. Microfilaments and microtubules are threadlike processes that aid cellular movements and support and stabilize the cytoplasm.
   l. Cytoplasm may contain nonliving cellular products, such as nutrients and pigments, called inclusions.
5. Cell nucleus
   a. The nucleus is enclosed in a double-layered nuclear envelope that has nuclear pores that control movement of substances between the nucleus and cytoplasm.
   b. A nucleolus is a dense body of protein and RNA that is the site of ribosome synthesis.
   c. Chromatin is composed of loosely coiled fibers of protein and DNA that condense into chromosomes during cell reproduction.

## Movements Into and Out of the Cell (page 78)

Movement of substances into and out of the cell may involve physical or physiological processes.

1. Diffusion
   a. Diffusion is the random movement of molecules in air or liquid solution.
   b. Diffusion is movement of molecules or ions from regions of higher concentration toward regions of lower concentration (down a concentration gradient).
   c. It is responsible for exchanges of oxygen and carbon dioxide within the body.
   d. The most important factors determining the rate of diffusion in the body include distance, the concentration gradient, and temperature.
2. Facilitated diffusion
   a. Facilitated diffusion uses protein channels or carrier molecules in the cell membrane.
   b. This process moves substances such as ions, sugars, and amino acids from regions of higher concentration to regions of lower concentration.
3. Osmosis
   a. Osmosis is a special case of diffusion in which water molecules diffuse from regions of higher water concentration to lower water concentration through a selectively permeable membrane.
   b. Osmotic pressure increases as the number of particles dissolved in a solution increases.
   c. Cells lose water when placed in hypertonic solutions and gain water when placed in hypotonic solutions.
   d. A solution is isotonic when it contains the same concentration of dissolved particles as the cell contents.
4. Filtration
   a. In filtration, molecules move from regions of higher hydrostatic pressure toward regions of lower hydrostatic pressure.
   b. Blood pressure filters water and dissolved substances through porous capillary walls.
5. Active transport
   a. Active transport moves molecules or ions from regions of lower concentration to regions of higher concentration.
   b. It requires cellular energy and carrier molecules in the cell membrane.

6. Endocytosis
   a. In pinocytosis, a cell membrane engulfs tiny droplets of liquid.
   b. In phagocytosis, a cell membrane engulfs solid particles.
   c. In receptor-mediated endocytosis, receptor proteins combine with specific molecules in the cell surroundings. The membrane engulfs the combinations.
7. Exocytosis
   a. Exocytosis is the reverse of endocytosis.
   b. In exocytosis, vesicles containing secretions fuse with the cell membrane, releasing the substances to the outside.
8. Transcytosis
   a. Transcytosis combines endocytosis and exocytosis.
   b. In transcytosis, a substance or particle crosses a cell.
   c. Transcytosis is specific and rapid.

## The Cell Cycle (page 88)

1. The cell cycle includes interphase, mitosis, cytoplasmic division, and differentiation.
2. Interphase
   a. Interphase is the stage when a cell grows, DNA replicates, and new organelles form.
   b. It terminates when the cell begins mitosis.
3. Mitosis
   a. Mitosis is the division and distribution of DNA to new cells during cell reproduction.
   b. The stages of mitosis include prophase, metaphase, anaphase, and telophase.
4. The cytoplasm divides into two portions following mitosis.
5. Cell differentiation is the specialization of cell structures and functions.

## Control of Cell Reproduction (page 93)

1. Cellular reproductive capacities vary greatly.
2. Chromosome tips that shorten with each mitosis provide a mitotic clock, usually limiting the number of reproductions to 50.
3. Cell reproduction is limited, and controlled by both internal and external factors.
4. As a cell grows, its surface area increases to a lesser degree than its volume, and eventually the area becomes inadequate for the requirements of the living material within the cell. When a cell reproduces, the new cells have more favorable surface area-volume relationships.
5. Cancer is the consequence of a loss of cell cycle control.

## CRITICAL THINKING QUESTIONS

1. Which process—diffusion, osmosis, or filtration—accounts for the following situations?
   a. Injection of a drug that is hypertonic to the tissues stimulates pain.
   b. A person with extremely low blood pressure stops producing urine.
   c. The concentration of urea in the dialyzing fluid of an artificial kidney is kept low.
2. Which characteristic of cell membranes may explain why fat-soluble substances such as chloroform and ether rapidly affect cells?
3. A person exposed to excessive X rays may lose white blood cells and become more susceptible to infection. How are these effects related?
4. Exposure to tobacco smoke causes cilia to cease moving and degenerate. Why might this explain why tobacco smokers have an increased incidence of respiratory infections?
5. How would you explain the function of phagocytic cells to a patient with a bacterial infection?
6. How is knowledge of cellular reproduction important to an understanding of:
   a. growth
   b. wound healing
   c. cancer
7. Why are enlarged lysosomes a sign of a serious illness?

## REVIEW EXERCISES

1. Use specific examples to illustrate how cells vary in size.
2. Describe how the shapes of nerve, epithelial, and muscle cells are well suited to their functions.
3. Name the major components of a cell, and describe how they interact.
4. Discuss the structure and functions of a cell membrane.
5. How do cilia, flagella, and cell adhesion molecules move cells?
6. Distinguish between organelles and inclusions.
7. Define *selectively permeable.*
8. Describe the chemical structure of a membrane.
9. Explain how the structure of a cell membrane determines which types of substances it is permeable to.
10. Explain the function of membrane proteins.
11. Describe three kinds of intercellular junctions.
12. Describe the structures and functions of each of the following:
    a. endoplasmic reticulum
    b. ribosome
    c. Golgi apparatus
    d. mitochondrion
    e. lysosome
    f. peroxisome
    g. cilium
    h. flagellum
    i. centrosome
    j. vesicle
    k. microfilament
    l. microtubule

13. Describe the structure of the nucleus and the functions of its contents.
14. Distinguish between diffusion and facilitated diffusion.
15. Name three factors that increase the rate of diffusion.
16. Explain how diffusion aids gas exchange within the body.
17. Define *osmosis.*
18. Define *osmotic pressure.*
19. Explain how the number of solute particles in a solution affects its osmotic pressure.
20. Distinguish among solutions that are hypertonic, hypotonic, and isotonic.
21. Define *filtration.*
22. Explain how filtration moves substances through capillary walls.
23. Explain why active transport is called a physiological process, whereas diffusion is called a physical process.
24. Explain the function of carrier molecules in active transport.
25. Distinguish between pinocytosis and phagocytosis.
26. Describe *receptor-mediated endocytosis.* How might it be used to deliver drugs across the blood-brain barrier?
27. Explain how transcytosis includes endocytosis and exocytosis.
28. List the phases in the cell cycle. Why is interphase *not* a time of cellular rest?
29. Name the two processes included in cell reproduction.
30. Describe the major events of mitosis.
31. Explain how the cytoplasm is divided during cellular reproduction.
32. Explain what happens during interphase.
33. Define *differentiation.*
34. Explain how differentiation may reflect repression of DNA information.
35. How does loss of genetic control cause cancer?

**For More Information Related to This Chapter, Please Visit** *Hole's Human Anatomy and Physiology* **Home Page**

**http://www.mhhe.com/biosci/abio/**

# CELLULAR METABOLISM

## CHAPTER 4

## CHAPTER OBJECTIVES

### AFTER YOU HAVE STUDIED THIS CHAPTER, YOU SHOULD BE ABLE TO:

1. Distinguish between anabolism and catabolism.

2. Explain how enzymes control metabolic processes.

3. Explain how the reactions of cellular respiration release chemical energy.

4. Describe how energy is made available for cellular activities.

5. Describe the general metabolic pathways of carbohydrate, lipid, and protein metabolism.

6. Explain how metabolic pathways are regulated.

7. Describe how DNA molecules store genetic information.

8. Explain how protein synthesis relies on genetic information.

9. Describe how DNA molecules are replicated.

10. Explain how genetic information can be altered and how such a change may affect an organism.

## KEY TERMS

aerobic respiration (a-er-ō′bik res″pi-ra′shun)
anabolism (an″ah-bol′lizm)
anaerobic respiration (an″a-er-ō′bik res″pi-ra′shun)
apoptosis (ap″o-to′sis)
catabolism (kat″ah-bol′liz-m)
coenzyme (ko-en′zīm)
deamination (de-am″ĭ-na′shun)
dehydration synthesis (de″hi-dra′shun sin′the-sis)
DNA
energy (en′er-je)
enzyme (en′zīm)

gene (jēn)
genetic code (je-net′ik kōd)
glycolysis (gli-kol′ĭ-sis)
hydrolysis (hi-drol′ĭ-sis)
metabolism (me-tab′o-liz-m)
mutation (mu-ta′shun)
oxidation (ok″si-da′shun)
reduction (re-duk′shun)
replication (re″pli-ka′shun)
RNA
substrate (sub′strāt)
transcription (tranz-krip′-shun)
translation (tranz-lay′shun)

## UNDERSTANDING WORDS

**aer-,** air: *aer*obic respiration—respiratory process that requires oxygen.

**an-,** without: *an*aerobic respiration—respiratory process that does not require oxygen.

**ana-,** up: *ana*bolism—cellular processes in which smaller molecules are used to build up larger ones.

**cata-,** down: *cata*bolism—cellular processes in which larger molecules are broken down into smaller ones.

**co-,** with: *co*enzyme—substance that unites with a protein and completes the structure of an enzyme molecule.

**de-,** undoing: *de*amination—process by which nitrogen-containing portions of amino acid molecules are removed.

**mut-,** change: *mut*ation—change in the genetic information of a cell.

**-strat,** spread out: sub*strat*e—substance upon which an enzyme acts.

**sub-,** under: *sub*strate—substance upon which an enzyme acts.

**-zym,** causing to ferment: en*zym*e—protein that initiates or speeds up a chemical reaction without itself being consumed.

A physician and his 17-year-old son ate left-over spaghetti with homemade pesto sauce for several days, each time after it had been unrefrigerated for an hour or two. On the fourth day, the food had a peculiar odor, but the father heated it in a pan anyway. About a half hour after lunch, father and son developed severe abdominal pain. The father recovered, but the son began to behave strangely, becoming listless, then very sleepy. A yellow pallor indicated that his liver was malfunctioning.

Because of the rapid onset of abdominal pain after eating reheated food, food poisoning was likely. Indeed, the boy's body fluids and the pan used to reheat the spaghetti contained *Bacillus cereus,* a type of bacterium that pro-duced a toxin that can cause abdominal pain. In the boy, the toxin took a deadly turn to the liver.

To learn how the bacterial toxin harms the liver, researchers applied toxin from the boy to rat liver cells growing in culture. This experiment revealed that the toxin targets mitochondria, the organelles that house the biochemical reactions that extract energy from nutrients. Specifically, the toxin destroyed the mitochondria's ability to break down fats. Ironically, liver cells have many mitochondria to power the energy-requiring reactions that break down toxins. With his liver mitochondria severely impaired, the boy's liver literally ran out of energy and shut down. He died four days after the spaghetti meal.

## METABOLIC PROCESSES

In every human cell, even in the most sedentary individual, thousands of chemical reactions essential to life occur every second. A special type of protein called an **enzyme** controls the pace of each reaction. The sum total of chemical reactions within the cell constitutes **metabolism.**

Many metabolic reactions occur one after the other, with the products of one reaction serving as starting materials of another, forming intricate pathways and cycles that may intersect by sharing intermediate compounds. As a result, metabolism in its entirety may seem enormously complex. However, individual pathways of metabolism are fascinating to study because they reveal how cells function—in essence, how chemistry becomes biology. This chapter explores how metabolic pathways supply a cell with energy, and how other biochemical processes enable a cell to produce proteins—including the enzymes that make all of metabolism possible.

Metabolic reactions and pathways are of two types. In **anabolism,** larger molecules are constructed from smaller ones, requiring input of energy. In **catabolism,** larger molecules are broken down into smaller ones, releasing energy.

### Anabolism

Anabolism provides all substances required for cellular growth and repair. For example, an anabolic process called *dehydration synthesis* joins many simple sugar molecules (monosaccharides) to form larger molecules of glycogen. When a runner consumes pasta the night before a race, digestion breaks down the complex carbohydrates to monosaccharides, which can be absorbed. The bloodstream then carries these energy-rich molecules to body cells. Here, dehydration synthesis joins the monosaccharides to form glycogen, which stores energy that the runner may not need until later, as the finish line nears. When monosaccharide units join, an —OH (hydroxyl group) from one monosaccharide molecule and an —H (hydrogen atom) from an —OH group of another are removed. As the —H and —OH react to produce a water molecule, the monosaccharides are joined by a shared oxygen atom, as figure 4.1 shows (read from left to right). As the process repeats, the molecular chain extends.

Similarly, glycerol and fatty acid molecules join by dehydration synthesis in fat (adipose) tissue cells to form fat molecules. In this case, three hydrogen atoms are removed from a glycerol molecule, and an —OH group is removed from each of three fatty acid molecules, as figure 4.2 shows (read from left to right). The result is three water molecules and a single fat molecule, whose glycerol and fatty acid portions are bound by shared oxygen atoms.

In cells, dehydration synthesis also builds protein molecules by joining amino acid molecules. When two amino acid molecules are united, an —OH from one and an —H from the —NH₂ group of another are removed. A water molecule forms, and the amino acid molecules join by a bond between a carbon atom and a nitrogen atom (fig. 4.3; read from left to right). This type of bond, called a *peptide bond,* holds the amino acids together. Two such bound amino acids form a *dipeptide,* and many joined in a chain form a *polypeptide.* Generally, a polypeptide consisting of 100 or more amino acid molecules is called a *protein,* although the boundary between polypeptides and proteins is not precisely defined.

### Catabolism

Physiological processes that break down larger molecules into smaller ones constitute catabolism. An example of catabolism is **hydrolysis,** which can decompose carbohydrates, lipids, and proteins. A water molecule is used to split these substances into two parts.

The hydrolysis of a disaccharide, for instance, results in two monosaccharide molecules (see fig. 4.1; read from right to left). In this case, the bond between the simple sugars breaks, and the water molecule supplies a hydrogen atom to one sugar molecule and a hydroxyl group

**FIGURE 4.1**

Two monosaccharides may join by dehydration synthesis to form a disaccharide. In the reverse reaction, the disaccharide is hydrolyzed to two monosaccharides.

**FIGURE 4.2**

A glycerol molecule and three fatty acid molecules may join by dehydration synthesis to form a fat molecule (triglyceride). In the reverse reaction, fat is hydrolyzed to three fatty acids and glycerol.

**FIGURE 4.3**

When two amino acid molecules unite by dehydration synthesis, a peptide bond forms between a carbon atom and a nitrogen atom. In the reverse reaction, a dipeptide is hydrolyzed to two amino acids.

to the other. Thus, hydrolysis is the reverse of dehydration synthesis:

**Dehydration**
**synthesis**
**Monosaccharide + Monosaccharide ⇄ Disaccharide + Water**
**Hydrolysis**

Hydrolysis breaks down carbohydrates into monosaccharides; fats into glycerol and fatty acids (see fig. 4.2; read from right to left); proteins into amino acids

(see fig. 4.3; read from right to left); and nucleic acids into nucleotides.

Hydrolysis does not occur automatically, even though in the body water molecules are readily available to provide the —H and —OH necessary for hydrolysis. For example, water-soluble substances such as the disaccharide sucrose (table sugar) *dissolve* in a glass of water, but do not undergo hydrolysis. Hydrolysis (like dehydration synthesis) requires the help of specific enzymes, which are discussed in the next section.

The reactions of metabolism are often reversible. However, the enzyme that speeds, or catalyzes, the anabolic reaction is often different from that which catalyzes the catabolic reaction.

Both catabolism and anabolism must be carefully controlled so that the breakdown or energy-releasing reactions occur at rates that are adjusted to the requirements of the building up or energy-utilizing reactions. Any disturbance in this balance is likely to damage or kill cells.

1 What are the general functions of anabolism and catabolism?

2 What substance does the anabolism of monosaccharides form? Of glycerol and fatty acids? Of amino acids?

3 Distinguish between dehydration synthesis and hydrolysis.

## CONTROL OF METABOLIC REACTIONS

Different kinds of cells may conduct specialized metabolic processes, but all cells perform certain basic reactions, such as the buildup and breakdown of carbohydrates, lipids, proteins, and nucleic acids. These reactions include hundreds of very specific chemical changes that must occur in a particular sequence. Enzymes control the rates of these metabolic reactions.

### Enzyme Action

Like other chemical reactions, metabolic reactions require energy (activation energy) before they proceed. This is why heat is used to increase the rates of chemical reactions in laboratories. Heat energy increases the rate at which molecules move and the frequency of molecular collisions. These collisions increase the likelihood of interactions among the electrons of the molecules that can form new chemical bonds. The temperature conditions in cells are usually too mild to adequately promote the reactions of life. Enzymes make these reactions possible.

Certain microorganisms, colorfully called "extremophiles," live in conditions of extremely high or low heat, salinity, or pH. Their enzymes have evolved to function under these conditions, and are useful in industrial processes that are too harsh to use other enzymes.

Enzymes are almost invariably globular proteins (see chapter 2) that promote specific chemical reactions within cells by lowering the activation energy required to start these reactions. Enzymes can speed metabolic reactions by a factor of a million or more.

Enzymes are required in very small quantities, because as they work they are not consumed and can, therefore, function repeatedly. Also, each enzyme has *specificity,* acting only on a particular kind of substance, which is called its **substrate.** For example, the substrate of an enzyme called catalase (found in the peroxisomes of liver and kidney cells) is hydrogen peroxide, a toxic by-product of certain metabolic reactions. This enzyme's only function is to decompose hydrogen peroxide into water and oxygen, helping prevent accumulation of hydrogen peroxide that might damage cells.

Cellular metabolism includes hundreds of different chemical reactions, each controlled by a specific kind of enzyme. Often sequences of enzyme-controlled reactions, called **metabolic pathways,** lead to synthesis or breakdown of particular biochemicals (fig. 4.4). Thus, hundreds of different kinds of enzymes are present in every cell, and each enzyme must be able to "recognize" its specific substrate. This ability to identify a substrate depends upon the shape of an enzyme molecule. That is, each enzyme's polypeptide chain twists and coils into a unique three-dimensional form, or conformation, that fits the special shape of its substrate molecule.

The action of the enzyme catalase is obvious when using hydrogen peroxide to cleanse a wound. Injured cells release catalase, and when hydrogen peroxide contacts them, bubbles of oxygen are set free. The resulting foam removes debris from inaccessible parts of the wound.

During an enzyme-catalyzed reaction, regions of the enzyme molecule called **active sites** temporarily combine with portions of the substrate, forming an enzyme-substrate complex. This interaction strains chemical bonds in the substrate in a way that makes a particular chemical reaction more likely to occur. When it does, the enzyme is released in its original form, able to bind another substrate molecule (fig. 4.5).

Enzyme catalysis can be summarized as follows:

| Substrate molecule | + | Enzyme molecule | → | Enzyme-substrate complex | → | Product (changed substrate) | + | Enzyme molecule (unchanged) |
|---|---|---|---|---|---|---|---|---|

The speed of an enzyme-catalyzed reaction depends partly on the number of enzyme and substrate molecules in the cell. The reaction occurs more rapidly if the concentration of the enzyme or the concentration of the substrate increases. Also, the efficiency of different kinds of enzymes varies greatly. Thus, some enzymes can process only a few substrate molecules per second, whereas others can handle thousands or nearly a million substrate molecules per second.

**FIGURE 4.4**

A metabolic pathway consists of a series of enzyme-controlled reactions leading to a product.

(a) Substrate molecule

(b) Enzyme molecule

(c) Enzyme-substrate complex

(d) Product molecules

(e) Unaltered enzyme molecule

**FIGURE 4.5** ▭

(a) The shape of a substrate molecule fits (b) the shape of the enzyme's active site. (c) When the substrate molecule temporarily combines with the enzyme, a chemical reaction occurs. The result is (d) product molecules and (e) an unaltered enzyme. Many enzymatic reactions are reversible.

Enzyme names are often derived from the names of their substrates, with the suffix *-ase* added. For example, a lipid-splitting enzyme is called *lipase,* a protein-splitting enzyme is *protease,* and a starch-(amylum) splitting enzyme is *amylase.* Similarly, *sucrase* is an enzyme that splits the sugar sucrose, *maltase* splits the sugar maltose, and *lactase* splits the sugar lactose.

## Cofactors and Coenzymes

Often an enzyme is inactive until it combines with a nonprotein component that either helps the active site attain its appropriate shape or helps bind the enzyme to its substrate. Such a substance, called a **cofactor,** may be an ion of an element, such as copper, iron, or zinc, or it may be a small organic molecule, called a **coenzyme.** Coenzymes are often composed of vitamin molecules or incorporate vitamin molecules into their structures.

**Vitamins** are essential organic substances that human cells cannot synthesize (or cannot synthesize in sufficient quantities) and therefore must come from the diet. Since vitamins provide coenzymes that can, like enzymes, function again and again, cells require very small quantities of vitamins. Chapter 18 discusses vitamins further.

## Factors that Alter Enzymes

Almost all enzymes are proteins, and like other proteins, they can be denatured by exposure to excessive heat, radiation, electricity, certain chemicals, or fluids with extreme pH values. For example, many enzymes become inactive at 45°C, and nearly all of them are denatured at 55°C. Some poisons are chemicals that denature enzymes. Cyanide, for instance, can interfere with respiratory enzymes and damage cells by halting their energy-obtaining processes.

> The antibiotic drug penicillin interferes with enzymes that enable certain bacteria to construct cell walls. As a result, the bacteria die. In this manner, penicillin protects against certain bacterial infections. The drug does not harm human cells because these do not have cell walls.

1 What is an enzyme?

2 How can an enzyme control the rate of a metabolic reaction?

3 How does an enzyme "recognize" its substrate?

4 What is the role of a cofactor?

5 What factors can denature enzymes?

# ENERGY FOR METABOLIC REACTIONS

**Energy** is the capacity to change matter or move something; that is, it is the ability to do work. Therefore, we recognize energy by what it can do. Common forms of energy include heat, light, sound, electrical energy, mechanical energy, and chemical energy.

Although energy cannot be created or destroyed, it can be converted from one form to another. An ordinary incandescent light bulb changes electrical energy to heat and light, and an automobile engine converts the chemical energy in gasoline to heat and mechanical energy.

Changes occur in the human body as a characteristic of life—whenever this happens energy is being transferred. Thus, all metabolic reactions involve energy in some form.

## Release of Chemical Energy

Most metabolic processes depend on chemical energy. This form of energy is held in the chemical bonds that link atoms into molecules and is released when these bonds break. Burning releases the chemical energy held within the bonds of many substances as heat and light. Similarly, cells "burn" glucose molecules in a process called **oxidation.** The energy released by oxidation of glucose is used to promote cellular metabolism. There are, however, some important differences between the oxidation of substances inside cells and the burning of substances outside them.

Burning usually requires a relatively large amount of energy to begin, and most of the energy released escapes as heat or light. In cells, enzymes initiate oxidation by decreasing the amount of energy (activation energy) required to start the process. Also, by transferring energy to special energy-carrying molecules, cells are able to capture in the form of chemical energy almost half of the energy released. The rest escapes as heat, which helps maintain body temperature.

**Cellular respiration** is the process that releases energy from molecules such as glucose and transfers it to other molecules. The chemical reactions of cellular respiration must occur in a particular sequence, each one controlled by a different enzyme. Some of these enzymes are in the cell's cytosol, whereas others are in the mitochondria.

**1** What is energy?

**2** How does cellular oxidation differ from burning?

**3** Define cellular respiration.

# AN OVERVIEW OF CELLULAR RESPIRATION

Cellular respiration occurs in three distinct, yet interconnected, series of reactions: **glycolysis,** the **citric acid cycle,** and **oxidative phosphorylation.** The products of these reactions include heat, $CO_2$, and water, as well as energy in a form the cell can use. Some of the energy is lost as heat, but almost half is used to produce **ATP** (adenosine triphosphate), an energy-rich molecule that cells can use to power the reactions of life.

## Glycolysis

Glycolysis, or the "breaking of glucose," is a series of ten enzyme-catalyzed reactions that break down the 6-carbon glucose molecules into two 3-carbon pyruvic acid molecules. This phase of cellular respiration occurs in the cytosol, and is termed anaerobic because it does not require oxygen. It is sometimes called *anaerobic respiration.*

Energy from ATP is required to activate the reactions of glycolysis. However, more energy is released than is used. The excess energy is used to synthesize more ATP. Thus, there is a net gain of ATP from glycolysis (fig. 4.6).

## Aerobic Respiration

Following glycolysis, oxygen must be available for the process of energy extraction to continue. For this reason, the remaining reactions of cellular respiration are called *aerobic respiration.* They include two series of reactions that occur within the mitochondria—the citric acid cycle (Krebs cycle) and oxidative phosphorylation. These reactions release considerably more energy for synthesis of ATP molecules.

By the end of the aerobic reactions, glucose is broken down to carbon dioxide molecules and hydrogen atoms. The carbon dioxide diffuses out of the cell as a waste, and the hydrogen atoms combine with oxygen molecules to form water. Thus, the final products of glucose oxidation are carbon dioxide, water, heat, and ATP.

## ATP Molecules

For each glucose molecule that is decomposed completely, up to thirty-eight molecules of ATP can be produced. All but two are formed during the aerobic phase.

Each ATP molecule consists of three main parts—an adenine, a ribose, and three phosphates in a chain (fig. 4.7). (Note the similarity between the general structure of an ATP molecule and the structure of a nucleotide within a nucleic acid molecule described in chapter 2; see fig. 2.19.)

Almost half of the energy released during cellular respiration is used to synthesize ATP from **ADP** (adenosine diphosphate), which has only two phosphates. ATP formation attaches a phosphate to ADP, a process known as **phosphorylation.**

The third phosphate of ATP is attached by a high-energy bond, and the chemical energy stored in that bond may be quickly transferred to another molecule in a metabolic process. When such an energy transfer occurs, the terminal, high-energy phosphate bond of the ATP

**FIGURE 4.6**

Glycolysis (anaerobic respiration) occurs in the cytosol and does not require oxygen, whereas aerobic respiration occurs in the mitochondria only in the presence of oxygen.

**FIGURE 4.7**

An ATP molecule consists of an adenine, a ribose, and three phosphates. The wavy lines connecting the last two phosphates represent high-energy chemical bonds. When broken, these bonds provide energy for metabolic processes.

**FIGURE 4.8**

ATP provides energy for anabolic reactions and is regenerated by catabolic reactions.

molecule breaks, releasing its energy. Energy released from ATP molecules is used whenever cellular work is performed, such as when muscle cells contract, when membranes carry on active transport, or when cells synthesize and release secretions.

An ATP molecule that loses its terminal phosphate when it is used becomes an ADP molecule. However, the ADP molecule can be converted back into an ATP by capturing some energy and reattaching a phosphate. Thus, as figure 4.8 shows, ATP and ADP molecules shuttle back and forth between the energy-releasing reactions of cellular respiration and the energy-utilizing reactions of the cell.

ATP is not the only kind of energy-carrying molecule within a cell, but it is the primary one. Without ATP, cells die quickly.

1. What is anaerobic respiration? Aerobic respiration?

2. What happens to the energy that cellular respiration releases?

3. What are the final products of cellular respiration?

4. What is the function of ATP molecules?

## A CLOSER LOOK AT CELLULAR RESPIRATION

Anabolic and catabolic pathways in cells usually consist of a number of different steps that must occur in a particular sequence, such as the reactions of aerobic respiration. The enzymes controlling these reactions must act in a specific order. Such precision of activity suggests that the enzymes are physically positioned in the exact sequence as that of the reactions they control. Indeed, the enzymes responsible for some of the reactions of aerobic respiration are located in tiny, stalked particles on the membranes (cristae) within the mitochondria (see chapter 3).

Metabolic pathways are usually interconnected so that certain molecules may enter more than one pathway.

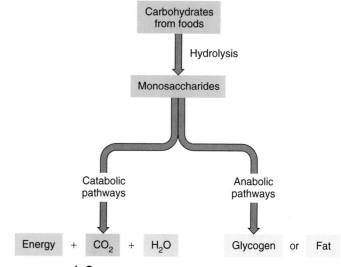

**FIGURE 4.9**

Hydrolysis breaks down carbohydrates from foods into monosaccharides. The resulting molecules may enter catabolic pathways and be used as energy sources, or they may enter anabolic pathways and be converted to glycogen or fat.

For example, carbohydrate molecules from foods may enter catabolic pathways and be used to supply energy, or they may enter anabolic pathways and be stored as glycogen, fat, or other molecules (fig. 4.9).

### Carbohydrate Pathways

The average human diet largely consists of carbohydrates, which are digested into monosaccharides such as glucose. Carbohydrates are primarily used as cellular energy sources, and therefore they usually enter the catabolic pathways of cellular respiration. First, in glycolysis, glucose is broken down to pyruvic acid in the cytosol, as previously discussed. Next, the pyruvic acid is transported into the mitochondrion and is broken down to yield a 2-carbon acetyl group in a transition reaction that is not part of glycolysis. The acetyl group then combines with a molecule of coenzyme A to form a

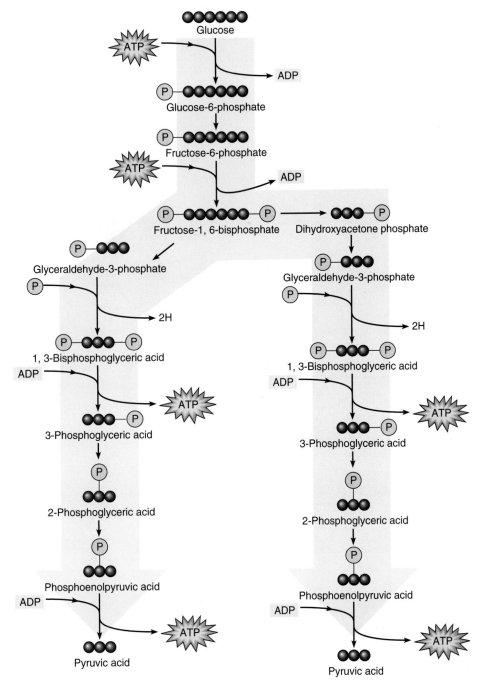

**FIGURE 4.10**

Chemical reactions of glycolysis. There is a net production of 2 ATP molecules from each glucose molecule. The four hydrogen atoms released provide high-energy electrons that may be used to generate ATP in the electron transport chain, described later. (*Note:* By convention, carbon atoms are sometimes numbered from right to left.)

substance called *acetyl coenzyme A.* It, in turn, forms several intermediate products through the chemical reactions of the citric acid cycle (Krebs cycle). These reactions release energy, and some of it is used to synthesize ATP. The rest of the energy is lost as heat. The products of this oxidation process are carbon dioxide and water. The next sections take a closer look at these events, which provide much of the energy in the human body.

## Glycolysis

Figure 4.10 illustrates the chemical reactions of glycolysis. In the early steps of this metabolic pathway, the original glucose molecule is altered by the addition of phosphate groups (*phosphorylation*) and by the rearrangement of its atoms. ATP supplies the phosphate groups and the energy to drive these reactions. The result is a molecule of fructose bound to two phosphate groups

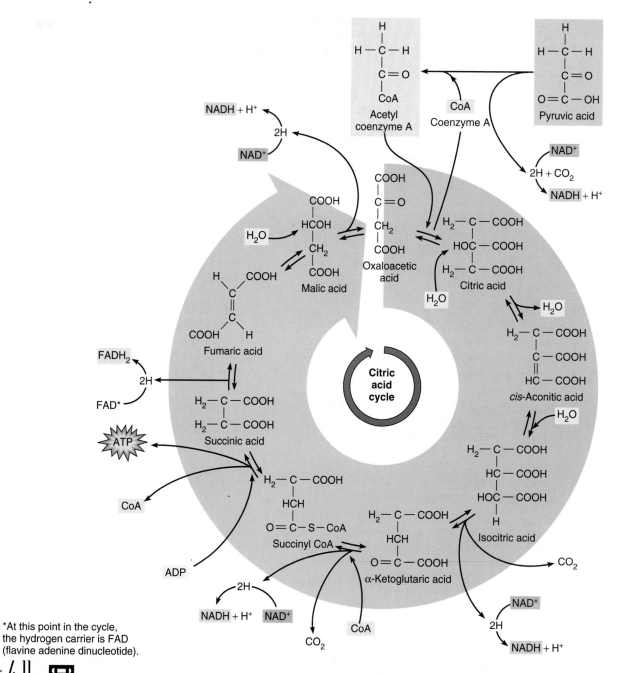

**FIGURE 4.11**

Chemical reactions of the citric acid cycle. NADH molecules carrying hydrogens are highlighted.

(fructose-1,6-bisphosphate). This molecule is split through two separate reactions into two 3-carbon molecules (glyceraldehyde-3-phosphate). Since each of these is converted to pyruvic acid, the following reactions, 1 through 5, must be counted twice to account for a single glucose molecule.

1. An inorganic phosphate group is added to glyceraldehyde-3-phosphate to form 1,3-bisphosphoglyceric acid, releasing two hydrogen atoms, to be used in ATP synthesis, described later.

2. 1,3-bisphosphoglyceric acid is changed to 3-phosphoglyceric acid. As this occurs, some energy in the form of a high-energy phosphate is transferred from the 1,3-bisphosphoglyceric acid to an ADP molecule, phosphorylating the ADP to ATP.

3. A slight alteration of 3-phosphoglyceric acid forms 2-phosphoglyceric acid.

4. A change in 2-phosphoglyceric acid converts it into phosphoenolpyruvic acid.

# Overriding a Block in Glycolysis

Michael P. was noticeably weak from his birth. He didn't move much, had poor muscle tone and difficulty breathing, and grew exhausted merely from the effort of feeding. At the age of two and a half months, he suffered his first seizure, staring and jerking his limbs for several frightening minutes. Despite medication, his seizures continued, occurring more frequently.

The doctors were puzzled because the results of most of Michael's many medical tests were normal—with one notable exception. His cerebrospinal fluid (the fluid that bathes the brain and spinal cord) was unusually low in glucose and lactic acid. These deficiencies told the physicians that Michael's cells were not performing glycolysis or anaerobic respiration.

Hypothesizing that a profound lack of ATP was causing the symptoms, medical researchers decided to intervene beyond the block in the boy's metabolic pathway, taking a detour to energy production. When Michael was seven and a half months old, he began a diet rich in certain fatty acids. Within four days, he appeared to be healthy for the very first time! The diet had resumed aerobic respiration at the point of acetyl coenzyme A formation by supplying an alternative to glucose. Other children with similar symptoms have since enjoyed spectacular recoveries similar to Michael's thanks to the dietary intervention, but doctors do not yet know the long-term effects of the therapy. This medical success story, however, illustrates the importance of the energy pathways—and how valuable our understanding of them can be. ∎

---

5. Finally, a high-energy phosphate is transferred from the phosphoenolpyruvic acid to an ADP molecule, phosphorylating it to ATP. A molecule of pyruvic acid remains.

Overall, one molecule of glucose is ultimately broken down to two molecules of pyruvic acid. Also, a total of four hydrogen atoms are released (step *a*), and four ATP molecules form (two in step *b* and two in step *e*). However, because two molecules of ATP are used early in glycolysis, there is a net gain of only two ATP molecules during this phase of cellular respiration.

In the presence of oxygen, each pyruvic acid molecule is oxidized to an acetyl group, which then combines with a molecule of coenzyme A (obtained from the vitamin pantothenic acid) to form acetyl coenzyme A. As this occurs, two more hydrogen atoms are released for each molecule of acetyl coenzyme A formed. The acetyl coenzyme A is then broken down by means of the citric acid cycle, which figure 4.11 illustrates.

Because obtaining energy for cellular metabolism is vital, disruptions in glycolysis or the reactions that follow it can devastate health. Clinical Application 4.1 tells how medical sleuths traced a boy's unusual combination of symptoms to a block in glycolysis.

## Citric Acid Cycle

An acetyl coenzyme A molecule enters the citric acid cycle by combining with a molecule of oxaloacetic acid to form citric acid. As citric acid is produced, coenzyme A is released and thus can be used again to form acetyl coenzyme A from pyruvic acid. The citric acid is then changed by a series of reactions back into oxaloacetic acid, and the cycle may repeat.

Steps in the citric acid cycle release carbon dioxide and hydrogen atoms. More specifically, for each glucose molecule metabolized in the presence of oxygen, two molecules of acetyl coenzyme A enter the citric acid cycle. The cycle releases four carbon dioxide molecules and sixteen hydrogen atoms. At the same time, two more molecules of ATP form.

The released carbon dioxide dissolves in the cytoplasm and leaves the cell, eventually entering the bloodstream. Most of the hydrogen atoms released from the citric acid cycle, and those released during glycolysis and during the formation of acetyl coenzyme A, supply electrons used to produce ATP.

## ATP Synthesis

Note that in figures 4.10 and 4.11 various metabolic reactions release hydrogen atoms. The electrons of these hydrogen atoms contain much of the energy associated with the chemical bonds of the original glucose molecule. To keep this energy in a usable form, these hydrogen atoms, with their high energy electrons, are passed in pairs to *hydrogen carriers*. One of these carriers is $NAD^+$ (nicotinamide adenine dinucleotide). When $NAD^+$ accepts a pair of hydrogen atoms, the two electrons and

**FIGURE** 4.12  ⬛

A summary of ATP synthesis by oxidative phosphorylation.

one hydrogen nucleus bind to NAD$^+$ to form NADH, and the remaining hydrogen nucleus (a hydrogen ion) is released as follows:

$$NAD^+ + 2H \rightarrow NADH + H^+$$

NAD$^+$ is a coenzyme obtained from a vitamin (niacin), and when it combines with the energized electrons it is said to be *reduced*. **Reduction** results from the addition of electrons, often as part of hydrogen atoms. Another electron acceptor, FAD (flavine adenine dinucleotide), acts in a similar manner, combining with two electrons and two hydrogen nuclei to form FADH$_2$ (fig. 4.11). In their reduced states, the hydrogen carriers NADH and FADH$_2$ now hold most of the energy once held in the bonds of the original glucose molecule.

Figure 4.12 shows that NADH can release the electrons and hydrogen nucleus. Since this reaction removes electrons, the resulting NAD$^+$ is said to be *oxidized*. **Oxidation** results from the removal of electrons, often as part of hydrogen atoms; it is the opposite of reduction. The two electrons this reaction releases pass to a series of electron carriers. The regenerated NAD$^+$ can once again accept electrons, and is recycled.

The molecules that act as electron carriers comprise an **electron transport chain.** As electrons are passed from one carrier to another, the carriers are alternately reduced and oxidized as they accept or release electrons. The transported electrons gradually lose energy as they proceed down the chain.

Among the members of the electron transport chain are several proteins, including a set of iron-containing molecules called **cytochromes.** The chain is located in the inner membranes of the mitochondria (see chapter 3). The folds of the inner mitochondrial membrane provide surface area on which the energy reactions take place. In a muscle cell, the inner mitochondrial membrane, if stretched out, may be as much as forty-five times as long as the cell membrane!

The final cytochrome of the electron transport chain (cytochrome oxidase) gives up a pair of electrons and causes two hydrogen ions (formed at the beginning of the sequence) to combine with an atom of oxygen. This process produces a water molecule:

$$2e^- + 2H^+ + \tfrac{1}{2} O_2 \rightarrow H_2O$$

Thus, oxygen is the final electron acceptor. In the absence of oxygen, electrons cannot pass through the electron transport chain, NAD$^+$ cannot be regenerated, and aerobic respiration halts.

Cyanide (CN$^-$) kills by binding to the electron acceptor in the electron transport chain that passes electrons to oxygen. Although there are antidotes, death is usually nearly instantaneous. Cyanide is used as a pesticide, for extracting silver and gold from ores, in electroplating, and in the textile and rubber industries.

Another respiratory poison, 2,4-dinitrophenol (DNP), "uncouples" passage of electrons down the electron transport chain from ATP synthesis. As a result, electrons move, but no ATP forms. When a person breathes DNP, the metabolic rate races as electrons are passed with no energy payoff. On a whole-body level, temperature soars, sweat pours, nausea and vomiting are severe, and the person collapses and dies. DNP is used in dye manufacture, wood preservation, and as an insecticide.

Note in figure 4.12 that as electrons pass through the electron transport chain, energy is released. Some of this energy is used by a mechanism involving the enzyme complex *ATP synthase* to combine phosphate and ADP by a high-energy bond (phosphorylation), forming ATP. Also note in figures 4.10 and 4.11 that twelve pairs of hydrogen atoms are released during the complete breakdown of one glucose molecule—two pairs from glycolysis, two pairs from the conversion of pyruvic acid to acetyl coenzyme A (one pair from each of two pyruvic acid molecules), and eight pairs from the citric acid cycle (four pairs for each of two acetyl coenzyme A molecules).

High-energy electrons from ten pairs of these hydrogen atoms produce thirty ATP molecules in the electron transport chain. Two pairs enter the chain differently and form four ATP molecules. Because this process of forming ATP involves both the oxidation of hydrogen atoms and the bonding of phosphate to ADP, it is called oxidative phosphorylation. Also, there is a net gain of two ATP molecules during glycolysis, and two ATP molecules form by direct enzyme action in two turns of the citric acid cycle. Thus, a maximum of thirty-eight ATP molecules form for each glucose molecule metabolized.

In the absence of oxygen, the electron transport chain does not work, and NADH cannot lose a hydrogen to become NAD⁺. Without NAD⁺, pyruvic acid molecules produced by glycolysis undergo a reaction that produces **lactic acid.** Human muscle cells that are working so strenuously that their production of pyruvic acid exceeds the oxygen supply begin to produce lactic acid. In this condition of "oxygen debt," the muscle cells are forced to depend solely on the less-efficient anaerobic pathway, which provides fewer ATPs per glucose molecule than does aerobic respiration. The accumulation of lactic acid contributes to the feeling of muscle fatigue and cramps. Walking after cramping at the end of a race can make a runner feel better by hastening the depletion of lactic acid. When oxygen is again available in sufficient quantity, lactic acid is converted back to pyruvic acid in the liver. From this pyruvic acid, the body can extract more energy.

Lactic acid formation occurs in an interesting variety of circumstances. Coaches measure lactic acid levels in swimmers' and sprinters' blood to assess their physical condition. Lactic acid accumulates to triple the normal levels in the bloodstreams of children who vigorously cry when they are being prepared for surgery, but not in children who are calm and not crying. This suggests that lactic acid formation accompanies stress. Cancer cells deep within a tumor, far from oxygen-delivering blood, also produce lactic acid by metabolizing glucose. The lactic acid forms a "halo" around the cancer cells, which inactivates immune system biochemicals that would otherwise fight the cancer.

## Carbohydrate Storage

Excess glucose in cells may enter anabolic carbohydrate pathways and be linked into storage forms such as glycogen. Most cells can produce glycogen, but liver and muscle cells store the greatest amounts. Following a meal, when blood glucose concentration is relatively high,

liver cells obtain glucose from the blood and synthesize glycogen. Between meals, when blood glucose concentration is lower, the reaction is reversed, and glucose is released into the blood. This mechanism ensures that cells throughout the body will have a continual supply of glucose to support respiration.

Glucose can also react to form fat molecules, which are later deposited in adipose tissues. This happens when a person takes in more carbohydrates than can be stored as glycogen or are required for normal activities. The body has an almost unlimited capacity to perform this type of anabolism, so overeating, even if mostly carbohydrates, can result in becoming obese (overweight).

## Lipid Pathways

Foods contain lipids in the form of phospholipids, cholesterol, or, most commonly, fats (triglycerides). A triglyceride consists of a glycerol portion and three fatty acids.

The liver controls lipid metabolism, removing lipids from the circulation and altering their molecular structures. For example, the liver can shorten or lengthen the carbon chains of fatty acid molecules or introduce double bonds into these chains, thus converting fatty acids from one form to another.

Lipids provide a variety of physiological functions; however, fats are mainly used to supply energy. Gram for gram, fats contain more than twice as much chemical energy as carbohydrates or proteins. This is why people trying to lose weight are advised to minimize fats in their diets.

Before a triglyceride molecule can release energy, it must undergo hydrolysis. As figure 4.13 shows, some of the resulting fatty acid portions can then react to form molecules of acetyl coenzyme A by a series of reactions called **beta oxidation,** which occurs in the mitochondria.

In the first phase of beta oxidation, fatty acids are activated. This change requires energy from ATP and a special group of enzymes called thiokinases. Each of the enzymes in this group can act upon a fatty acid with a particular carbon chain length.

Once fatty acid molecules have been activated, other enzymes called **fatty acid oxidases** that are located within mitochondria break them down. This phase of the reactions removes segments of fatty acid chains (containing two carbon atoms each). The liver converts some of these segments into acetyl coenzyme A molecules. Excess acetyl CoA is converted into compounds called **ketone bodies,** such as acetone, which later may be changed back to acetyl coenzyme A. In either case, the resulting acetyl coenzyme A can be oxidized by the citric acid cycle. The glycerol portions of the triglyceride molecules can also enter metabolic pathways leading to the citric acid cycle, or they can be used to synthesize glucose.

It is also possible for glycerol and fatty acid molecules resulting from the hydrolysis of fats to be changed back into fat molecules by anabolic processes and to be stored in fat tissue. Additional fat molecules can be synthesized from excess glucose or amino acids.

When ketone bodies form faster than they can be decomposed, some of them are eliminated through the lungs and kidneys. Consequently, the ketone acetone may cause the breath and urine to develop a fruity odor. This sometimes happens when a person fasts, forcing body cells to metabolize a large amount of fat, and in persons suffering from diabetes mellitus who may also develop a serious imbalance in pH called acidosis, which results from an over-accumulation of acetone and other acidic ketone bodies.

1   What is a metabolic pathway?

2   How do cells use carbohydrates?

3   What must happen to fat molecules before they can supply energy?

## Protein Pathways

When dietary proteins are digested, the resulting amino acids are absorbed and transported by the blood to cells. Many of these amino acids are used to form new protein molecules, as specified by DNA.

Protein molecules may also supply energy. To do this, they must first be broken down into amino acids. The amino acids then undergo **deamination,** a process that occurs in the liver that removes the nitrogen-containing portions (—NH$_2$ groups) from the amino acids. These —NH$_2$ groups are converted later into a waste called **urea.**

The liver produces urea from amino groups formed by deamination of amino acids. The blood carries urea to the kidneys, where it is excreted in urine. Certain kidney disorders impair the ability to remove urea from the blood, raising the blood urea concentration. A blood test called blood urea nitrogen (BUN) determines the blood urea concentration, and is often used to evaluate kidney function.

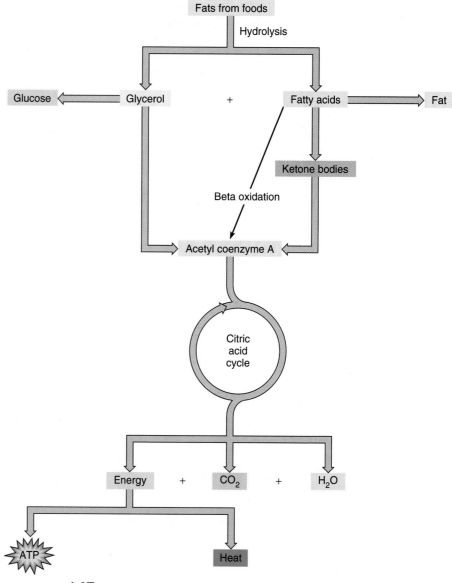

### FIGURE 4.13

We digest fats from foods into glycerol and fatty acids, which may enter catabolic pathways and be used as energy sources.

The remaining deaminated portions of the amino acids are broken down by one of several pathways. Some of these pathways lead to formation of acetyl coenzyme A, and others more directly lead to steps of the citric acid cycle. As energy is released from the cycle, some of it is captured in molecules of ATP (fig. 4.14). If energy is not required immediately, still other metabolic pathways may change the deaminated portions of the amino acids into glucose or fat molecules.

Glucose can be changed back into some amino acids if certain nitrogen-containing molecules are available. However, about eight necessary amino acids cannot be synthesized in adequate amounts in human cells and so must be provided in the diet. For this reason, these are called **essential amino acids.** They are discussed in chapter 18.

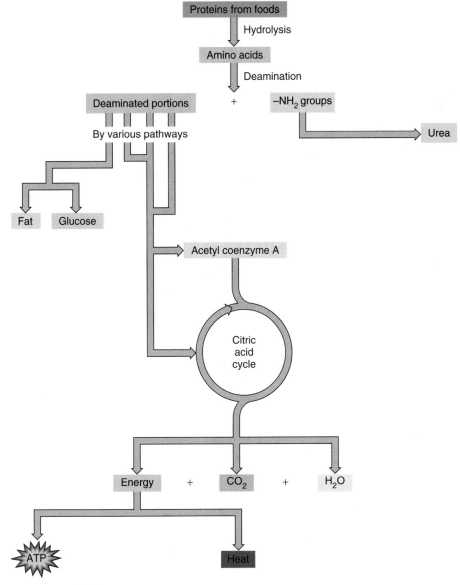

## FIGURE 4.14

We digest proteins from foods into amino acids, but before these smaller molecules can be used as energy sources, they must be deaminated.

## Regulation of Metabolic Pathways

The rate at which a metabolic pathway functions is often determined by a regulatory enzyme responsible for one of its steps. This regulatory enzyme is present in limited quantity. Consequently, it can become saturated with substrate molecules whenever the substrate concentration increases above a certain amount. Once the enzyme is saturated, increasing the amount of substrate will no longer affect the reaction rate. In this way, a single enzyme in a pathway can control the whole pathway.

As a rule, such a **rate-limiting enzyme** is the first enzyme in a series. This position is important because some intermediate substance of the pathway might accumulate if an enzyme occupying another location in the sequence were rate limiting.

Often the product of a metabolic pathway inhibits the rate-limiting regulatory enzyme. This type of control is called negative feedback. Accumulating product inhibits the pathway, and synthesis of the product falls. When the amount of product decreases, the inhibition lifts and more product is synthesized. This keeps the rate of production relatively stable (fig. 4.15).

Figure 4.16 summarizes catabolism of proteins, carbohydrates, and fats.

**1** How do cells utilize proteins?

**2** What is a rate-limiting enzyme?

**3** How can negative feedback control a metabolic pathway?

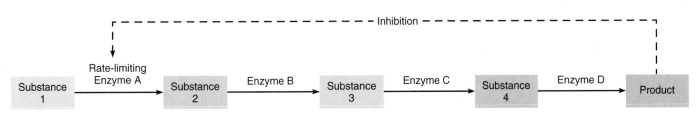

## FIGURE 4.15

A negative feedback mechanism may control a rate-limiting enzyme in a metabolic pathway. The product of the pathway inhibits the enzyme.

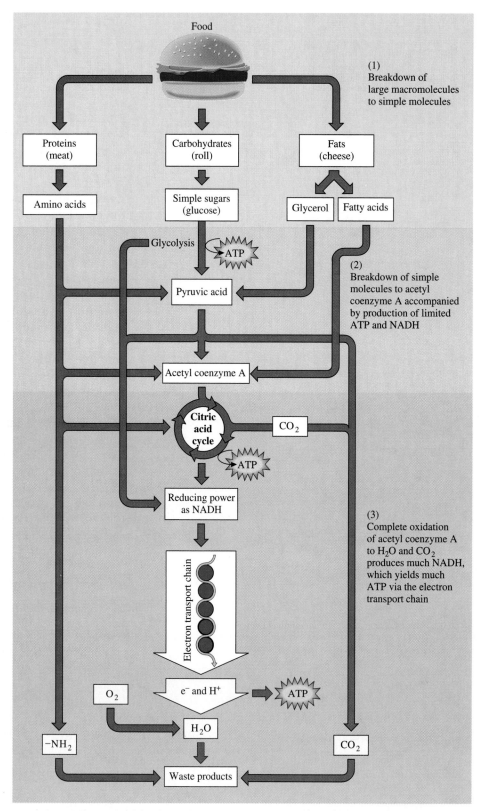

**FIGURE** 4.16

A summary of the breakdown (catabolism) of proteins, carbohydrates, and fats.

**FIGURE 4.17**

Each nucleotide of a nucleic acid consists of a 5-carbon sugar (S), a phosphate group (P), and an organic, nitrogenous base (B).

NH₂

Adenine

Thymine

NH₂

Cytosine

Guanine

**FIGURE 4.18**

Each nucleotide of a DNA molecule contains one of the four nitrogenous bases: adenine, thymine, cytosine, or guanine.

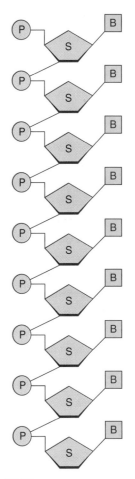

**FIGURE 4.19**

A single strand of DNA consists of a chain of nucleotides connected by a sugar-phosphate backbone.

## NUCLEIC ACIDS AND PROTEIN SYNTHESIS

Because enzymes control the metabolic pathways that enable cells to survive, cells must have information for producing these specialized proteins. Many other proteins are important in physiology as well, such as blood proteins, the proteins that form muscle and connective tissue, and the antibodies that protect against infection. The information that instructs a cell to synthesize a particular protein is held in the sequence of building blocks of **deoxyribonucleic acid** (DNA), the genetic material. As we will see later in this chapter, the correspondence between a unit of DNA information and a particular amino acid constitutes the **genetic code.**

### Genetic Information

Children resemble their parents because of inherited traits, but what actually passes from parents to a child is *genetic information,* in the form of DNA molecules from the parents' sex cells. As an offspring develops, mitosis passes the information on to new cells. Genetic information "tells" cells how to construct a great variety of protein molecules, each with a specific function. The portion of a DNA molecule that contains the genetic information for making a particular protein is called a **gene.**

Recall from chapter 2 that nucleotides are the building blocks of nucleic acids. A nucleotide consists of a 5-carbon sugar (ribose or deoxyribose), a phosphate group, and one of several organic, nitrogen-containing (nitrogenous) bases (fig. 4.17). In a DNA nucleotide, the base may be one of four types: adenine, thymine, cytosine, or guanine (fig. 4.18). DNA nucleotides form long strands (polynucleotide chains) by alternately joining their sugar and phosphate portions, which provides a "backbone" structure (fig. 4.19).

A DNA molecule consists of two polynucleotide chains. The nitrogenous bases project from the sugar-phosphate backbone of one strand and bind by hydrogen

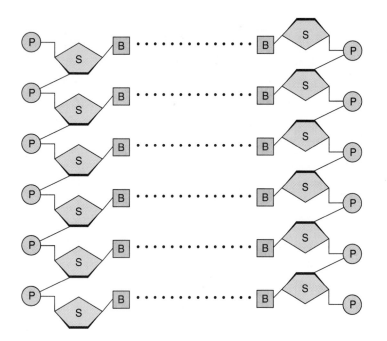

**FIGURE** 4.20

DNA consists of two polynucleotide chains. The nitrogenous bases of one strand are held to the nitrogenous bases of the second strand by hydrogen bonds (dotted lines). Note that the sugars point in opposite directions—that is, the strands are antiparallel.

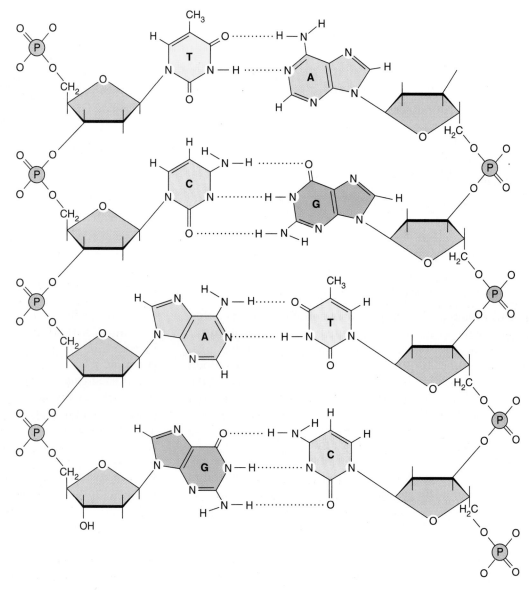

**FIGURE** 4.21

The nucleotides of a double-stranded DNA molecule pair so that an adenine (A) of one strand hydrogen bonds to a thymine (T) of the other strand, and a guanine (G) of one strand hydrogen bonds to a cytosine (C) of the other. The dotted lines represent hydrogen bonds.

(a)                                    (b)

**FIGURE 4.22**

(a) The molecular ladder of a double-stranded DNA molecule twists into a double helix. (b) A model of a portion of a DNA molecule.

bonds to the nitrogenous bases of the second strand (fig. 4.20). The resulting structure is somewhat like a ladder, in which the uprights represent the sugar and phosphate backbones of the two strands and the rungs represent the paired nitrogenous bases. Notice that the sugars forming the two backbones point in opposite directions. For this reason, the two strands are called *antiparallel.*

A DNA molecule is sleek and symmetrical because the bases pair in only two combinations, maintaining a constant width of the overall structure. Adenine (A), a two-ring structure, binds to thymine (T), a single-ring structure. Guanine (G), a two-ring structure, binds to cytosine (C), a single-ring structure (fig. 4.21). These pairs—A with T and G with C—are called **complementary base pairs.** Because of this phenomenon, the sequence of one DNA strand can always be derived from the other by following the "base pairing rules." For example, if the sequence of one strand of the DNA molecule is G, A, C, T, then the complementary strand's sequence is C, T, G, A.

The sequence of bases in only one of the DNA strands encodes the information for a protein. The double-stranded DNA molecule twists to form a double helix, resembling a spiral staircase (fig. 4.22). An individual DNA molecule may be several million base pairs long. Investigators can use DNA sequences to identify individuals (Clinical Application 4.2).

### Genetic Code

Genetic information specifies how to correctly position the amino acids in a polypeptide chain. Each of the twenty different amino acids is represented in a DNA molecule by a triplet code, consisting of sequences of three nucleotides. That is, the sequence C, G, T in a DNA strand represents one kind of amino acid; the sequence G, C, A represents another kind; and T, T, A still another kind (table 4.1). Other sequences encode instructions for beginning or ending the synthesis of a protein molecule.

Thus, the sequence of nucleotides in a DNA molecule dictates the sequence of amino acids of a particular protein molecule, and indicates how to start or stop the protein's synthesis. This method of storing information for protein synthesis is the genetic code. Because DNA molecules are located in the nucleus and protein synthesis

# DNA Makes History

In July 1918, the last tsar of Russia, Nicholas II, and his family, the Romanovs, met gruesome deaths at the hands of Bolsheviks in a town in the Ural Mountains of central Russia. Captors led the tsar, tsarina, four daughters and one son, plus the family physician and three servants, to a cellar and shot them, bayoneting those who did not die quickly. The executioners stripped the bodies and loaded them onto a truck, which would take them to a mine shaft where they would be left. But the truck broke down, and the bodies were instead placed in a shallow grave, then damaged with sulfuric acid so that they could not be identified.

In July 1991, two Russian amateur historians found the grave, and based on its location, alerted the government that the long-sought bodies of the Romanov family might have been found. An official forensic examination soon determined that the skeletons were from nine individuals. The sizes of the skeletons indicated that three were children. The porcelain, platinum, and gold in the teeth of some of the skeletons suggested that they were royalty. The facial bones were so decomposed from the acid that conventional forensic tests were not possible. But one very valuable type of evidence remained—DNA. Forensic scientists extracted DNA from bone cells and mass-produced it for study using a technique called the polymerase chain reaction (PCR) described in Clinical Application 4.3.

By identifying DNA sequences specific to the Y chromosome, which is found only in males, the DNA detectives could tell which of the skeletons were from males. Then they delved into the DNA in mitochondria. Because these organelles pass from mother to offspring, identifying a mitochondrial DNA pattern in a woman and children would establish her as their mother. This was indeed so for one of the women (with impressive dental work) and the children.

But a mother, her children, and some companions does not a royal family make. The researchers had to connect the skeletons to the royal family. Again they turned to DNA. Genetic material from one of the male skeletons shared certain rare DNA sequences with DNA from living descendants of the Romanovs. This man also had aristocratic dental work and shared DNA sequences with the children! The mystery of the fate of the Romanovs was apparently solved, thanks to the help of DNA. ■

---

occurs in the cytoplasm, the genetic information must somehow get from the nucleus into the cytoplasm. RNA molecules accomplish this.

> The genetic code is said to be universal, because all species on earth use the same DNA base triplets to specify the same amino acids. Researchers deciphered the code in the 1960s. When the media mentions an individual's genetic code, or that scientists are currently breaking the code, what they really are referring to is the sequence of DNA bases comprising a certain gene—not the genetic code (the correspondence between DNA triplet and amino acid) itself.

## RNA Molecules

**RNA** (ribonucleic acid) molecules differ from DNA molecules in several ways. RNA molecules are single-stranded, and their nucleotides contain ribose rather than deoxyribose sugar. Like DNA, RNA nucleotides each contain one of four organic bases, but whereas adenine, cytosine, and guanine nucleotides occur in both DNA and RNA, thymine nucleotides are found only in DNA. In place of thymine nucleotides, RNA molecules contain uracil (U) nucleotides (fig. 4.23).

The first step in the delivery of information from the nucleus to the cytoplasm is the synthesis of a type of RNA called **messenger RNA** (mRNA). In messenger RNA synthesis, RNA nucleotides form complementary base pairs with a section of a strand of DNA that encodes a particular protein. However, just as the words in a sentence must be read in the correct order to make sense, the base sequence of a strand of DNA must be "read" in the correct direction. Furthermore, only one of the two antiparallel strands of DNA contains the genetic message. An enzyme called RNA polymerase determines the correct DNA strand and the right direction for RNA synthesis.

In mRNA synthesis, RNA polymerase binds to a promoter, which is a DNA base sequence that begins a gene.

## TABLE 4.1 — SOME NUCLEOTIDE SEQUENCES OF THE GENETIC CODE

| Amino Acids | DNA Sequence | Complementary RNA Sequence |
|---|---|---|
| Alanine | CGT | GCA |
| Arginine | GCA | CGU |
| Asparagine | TTA | AAU |
| Aspartic acid | CTA | GAU |
| Cysteine | ACA | UGU |
| Glutamic acid | CTT | GAA |
| Glutamine | GTT | CAA |
| Glycine | CCG | GGC |
| Histidine | GTA | CAU |
| Isoleucine | TAG | AUC |
| Leucine | GAA | CUU |
| Lysine | TTT | AAA |
| Methionine | TAC | AUG |
| Phenylalanine | AAA | UUU |
| Proline | GGA | CCU |
| Serine | AGG | UCC |
| Threonine | TGC | ACG |
| Tryptophan | ACC | UGG |
| Tyrosine | ATA | UAU |
| Valine | CAA | GUU |
| **Instructions** | | |
| Start protein synthesis | TAC | AUG |
| Stop protein synthesis | ATT | UAA |

**FIGURE 4.23**

Each nucleotide of an RNA molecule contains one of four nitrogenous bases: adenine, uracil, cytosine, or guanine.

As a result of RNA polymerase binding, a section of the double-stranded DNA molecule unwinds and pulls apart, exposing a portion of the gene. RNA polymerase then moves along the strand, exposing other portions of the gene. At the same time, a molecule of mRNA forms as RNA nucleotides complementary to those along the DNA strand are strung together. For example, if the sequence of DNA bases is A, C, A, A, T, G, C, G, T, A, the complementary bases in the developing mRNA molecule will be U, G, U, U, A, C, G, C, A, U, as figure 4.24 shows. (The other strand of DNA is not used in this process, but it is important in DNA replication, discussed later in the chapter.)

RNA polymerase continues to move along the DNA strand, exposing portions of the gene, until it reaches a special DNA base sequence (termination signal) that signals the end of the gene. At this point, the RNA polymerase releases the newly formed mRNA molecule and leaves the DNA. The DNA then rewinds and assumes its previous double-helix structure. This process of copying DNA information into the structure of an mRNA molecule is called **transcription.**

Messenger RNA molecules can be hundreds or even thousands of nucleotides long. They exit the nucleus through the nuclear pores, and enter the cytoplasm (fig. 4.25). There they associate with ribosomes and act as

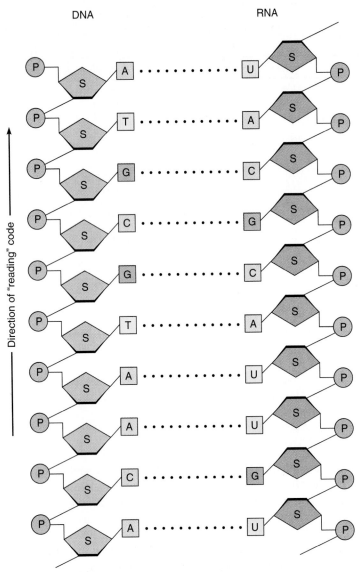

DNA          RNA

Direction of "reading" code

**FIGURE 4.24**

When an RNA molecule is transcribed from a strand of DNA, complementary nucleotides bond as in a double-stranded molecule of DNA, with one exception: RNA contains uracil nucleotides (U) in place of thymine nucleotides (T).

Until the early 1980s, all enzymes were thought to be proteins. Then, researchers found that a bit of RNA that they thought was contaminating a reaction in which RNA molecules are shortened actually contributed the enzymatic activity. The RNA enzymes were named "ribozymes." Because certain RNA molecules can carry information as well as function as enzymes—two biologically important properties—they may have been a bridge between chemicals and the earliest cell-like assemblages on earth long ago.

## Protein Synthesis

Synthesizing a protein molecule requires that the correct amino acid building blocks be present in the cytoplasm. Furthermore, these amino acids must align in the proper sequence along a strand of messenger RNA. A second kind of RNA molecule, synthesized in the nucleus and called **transfer RNA** (tRNA), does this. A transfer RNA molecule consists of only seventy to eighty nucleotides and has a complex three-dimensional shape, but it is the two ends of the tRNA molecule that are most important.

At one end, each transfer RNA molecule has a specific binding site for a particular amino acid. There is at least one type of transfer RNA molecule for each of the twenty amino acids. Before the transfer RNA can pick up its amino acid, the amino acid must be activated. Special enzymes catalyze this step. ATP provides the energy to form a bond between the amino acid and its transfer RNA molecule.

The other end of each transfer RNA molecule includes a region, called the **anticodon,** that contains three nucleotides in a particular sequence unique to that type of transfer RNA. These nucleotides bond only to a specific complementary mRNA codon. In this way, the appropriate transfer RNA carries its amino acid to the correct place in the sequence, as prescribed by the mRNA.

Although there are only 20 amino acids to be coded for, four bases can combine in triplets 64 different ways, so there are 64 different codons possible and all of them occur in mRNA. Because 3 of these codons do not have a corresponding transfer RNA, when they occur they provide a "stop" signal, indicating the end of protein synthesis, much like the period at the end of this sentence. A total of 61 different transfer RNAs are specific for the remaining 61 codons, which means that more than one type of tRNA can correspond to the same amino acid type. Because a given amino acid can be specified by more than one codon, the genetic code is "redundant." However, each type of tRNA can bind only its one particular amino acid, so the instructions are precise and the corresponding codon will only code for that amino acid.

The binding of tRNA and mRNA occurs in close association with a ribosome. A ribosome is a tiny parti-

patterns, or templates, for synthesizing proteins. Protein synthesis is called **translation** (fig. 4.26).

Because an amino acid corresponds to a sequence of three nucleotides in a DNA molecule, the same amino acid is represented in the transcribed messenger RNA by the complementary set of three nucleotides. Such a triplet of nucleotides in a messenger RNA molecule is called a **codon** (see table 4.1). Note that sixty-four possible DNA base triplets encode twenty different amino acids. This means that more than one codon can specify the same amino acid, a point we will return to soon. Table 4.2 compares DNA and RNA molecules.

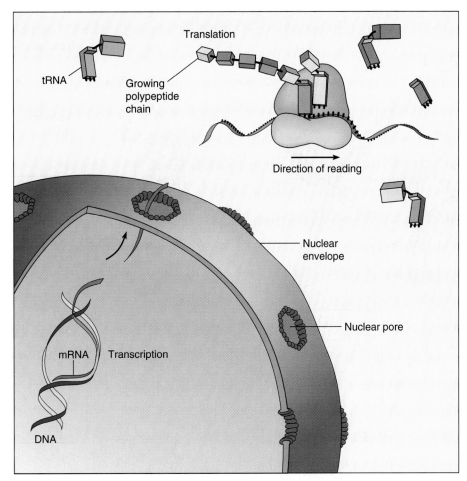

**FIGURE 4.25**

An mRNA molecule transcribed from a section of DNA exits the nucleus and enters the cytoplasm, where it associates with a ribosome to begin translation.

cle of two unequal-sized subunits composed of **ribosomal RNA** (rRNA) and protein. The smaller subunit of a ribosome binds to a molecule of messenger RNA near the codon at the beginning of the messenger strand. This action allows a transfer RNA molecule with the complementary anticodon to bring the amino acid it carries into position and temporarily join to the ribosome. A second transfer RNA molecule, complementary to the second codon on mRNA, then binds (with its activated amino acid) to an adjacent site on the ribosome. The first transfer RNA molecule then releases its amino acid, providing the energy for a peptide bond to form between the two amino acids (fig. 4.27). This process repeats again and again as the ribosome moves along the messenger RNA, adding amino acids one at a time to the developing protein molecule. The enzymatic activity necessary for bonding of the amino acids comes from ribosomal proteins and some RNA molecules (ribozymes) in the larger subunit of the ribosome. This subunit also holds the growing chain of amino acids.

A molecule of messenger RNA usually associates with several ribosomes at the same time. Thus, several copies of that protein, each in a different stage of development, may be present at any given moment.

As the protein molecule develops, proteins called *chaperones* fold it into its unique shape, and when the process is completed, the protein is released as a separate functional molecule. The transfer RNA molecules, ribosomes, mRNA, and the enzymes can function repeatedly in protein synthesis.

ATP molecules provide the energy for protein synthesis. Because a protein may consist of many hundreds of amino acids, and the energy from three ATP molecules is required to link each amino acid to the growing chain, a large fraction of a cell's energy supply supports protein synthesis.

The quantity of a particular protein that a cell synthesizes is generally proportional to the quantity of the corresponding messenger RNA present. The rate at which messenger RNA is synthesized in the nucleus (transcription) and the rate at which enzymes (ribonucleases) destroy the messenger RNA in the cytoplasm therefore controls protein synthesis. Proteins called *transcription factors* activate certain genes, thereby controlling which proteins a cell produces, and how

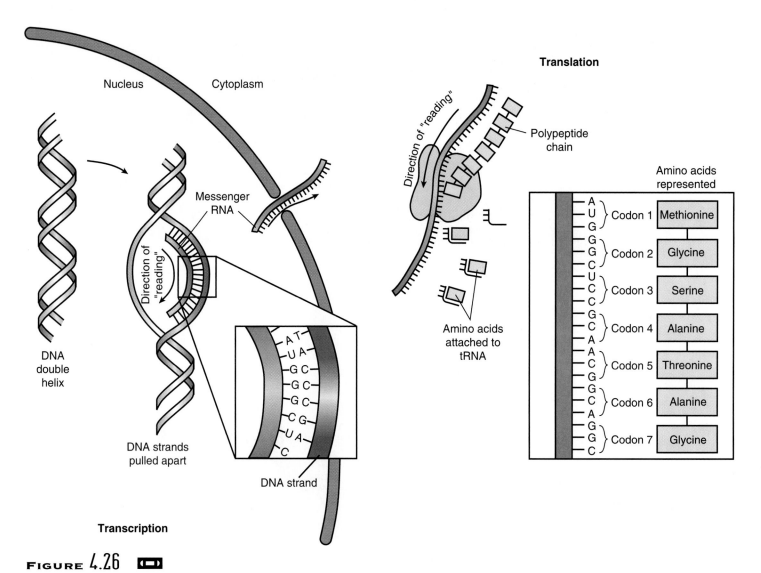

Nucleus    Cytoplasm

Direction of "reading"

Polypeptide chain

Amino acids represented

Messenger RNA

Direction of "reading"

| | AUG ⟩ Codon 1 | Methionine |
| | GGC ⟩ Codon 2 | Glycine |
| | UCC ⟩ Codon 3 | Serine |
| | GCA ⟩ Codon 4 | Alanine |
| | ACG ⟩ Codon 5 | Threonine |
| | GCA ⟩ Codon 6 | Alanine |
| | GGC ⟩ Codon 7 | Glycine |

Amino acids attached to tRNA

DNA double helix

AT
UA
GC
GC
GC
CG
UA
C

DNA strands pulled apart

DNA strand

Transcription

**FIGURE 4.26**

DNA information is transcribed into mRNA, which in turn is translated into a sequence of amino acids.

| TABLE 4.2 | A COMPARISON OF DNA AND RNA MOLECULES | |
| --- | --- | --- |
| | **DNA** | **RNA** |
| *Main location* | Part of chromosomes, in nucleus | Cytoplasm |
| *5-carbon sugar* | Deoxyribose | Ribose |
| *Basic molecular structure* | Double-stranded | Single-stranded |
| *Organic bases included* | Adenine, thymine, cytosine, guanine | Adenine, uracil, cytosine, guanine |
| *Major functions* | Contains genetic code for protein synthesis, replicates prior to mitosis | Messenger RNA carries transcribed DNA information to cytoplasm and acts as template for synthesis of protein molecules; transfer RNA carries amino acids to messenger RNA; ribosomal RNA provides structure and enzyme activity for ribosomes (ribozymes) |

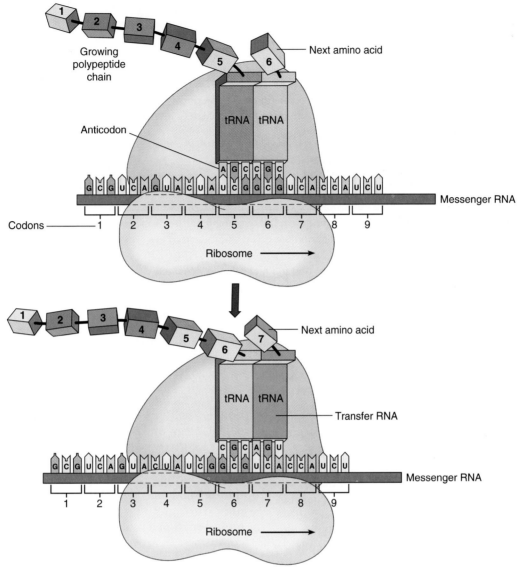

FIGURE 4.27

Molecules of transfer RNA (tRNA) attach to and carry specific amino acids, aligning them in the sequence determined by the codons of mRNA. These amino acids, connected by peptide bonds, form the polypeptide chain of a protein molecule.

many copies form. A connective tissue cell might have many messenger RNAs representing genes that encode the protein collagen; a muscle cell would have abundant messenger RNAs encoding muscle proteins. Extracellular signals such as hormones and growth factors activate transcription factors. Table 4.3 summarizes protein synthesis.

1 What is the function of DNA?

2 How is information carried from the nucleus to the cytoplasm?

3 How are protein molecules synthesized?

Some antibiotic drugs fight infection by interfering with bacterial protein synthesis, RNA transcription, or DNA replication. Rifampin is a drug that blocks bacterial transcription by binding to RNA polymerase, preventing the gene's message from being transmitted. Streptomycin is an antibiotic that binds a bacterium's ribosomal subunits, braking protein synthesis to a halt. Quinolone blocks an enzyme that unwinds bacterial DNA, preventing both transcription and DNA replication. Humans have different ribosomal subunits and transcription and replication enzymes than bacteria, so the drugs do not affect these processes in us.

| TABLE | 4.3 | PROTEIN SYNTHESIS |
|-------|-----|-------------------|

**Transcription (within the Nucleus)**

1. RNA polymerase binds to the base sequence of a gene.
2. This enzyme unwinds a portion of the DNA molecule, exposing part of the gene.
3. RNA polymerase moves along one strand of the exposed gene and catalyzes synthesis of an mRNA molecule, whose nucleotides are complementary to those of the strand of the gene.
4. When RNA polymerase reaches the end of the gene, the newly formed mRNA molecule is released.
5. The DNA molecule rewinds and closes the double helix.
6. The mRNA molecule passes through a pore in the nuclear envelope and enters the cytoplasm.

**Translation (within the Cytoplasm)**

1. A ribosome binds to the mRNA molecule near the codon at the beginning of the messenger strand.
2. A tRNA molecule that has the complementary anticodon brings its amino acid to the ribosome.
3. A second tRNA brings the next amino acid to the ribosome.
4. A peptide bond forms between the two amino acids, and the first tRNA is released.
5. This process is repeated for each codon in the mRNA sequence as the ribosome moves along its length, forming a chain of amino acids.
6. As the chain of amino acids grows, it folds, with the help of chaperone proteins, into the unique conformation of a functional protein molecule.
7. The completed protein molecule is released. The mRNA molecule, ribosome, and tRNA molecules are recycled.

## DNA Replication

When a cell reproduces, each newly formed cell must have a copy of the original cell's genetic information (DNA) so it will be able to synthesize the proteins necessary to build cellular parts and carry on metabolism. DNA **replication** occurs during interphase of the cell cycle.

As replication begins, hydrogen bonds break between the complementary base pairs of the double strands comprising the DNA molecule. Then the double-stranded structure unwinds and pulls apart, exposing unpaired nucleotide bases. New nucleotides pair with the exposed bases, forming hydrogen bonds. An enzyme, DNA polymerase, catalyzes this base pairing. Enzymes then knit together the new sugar-phosphate backbone. In this way, a new strand of complementary nucleotides extends along each of the old (original) strands. Two complete DNA molecules result, each with one new and one original strand (fig. 4.28). During mitosis, the two DNA molecules that form the two chromatids of each of the chromosomes separate so that one of these DNA molecules passes to each of the new cells. Clinical Application 4.3 discusses the polymerase chain reaction (PCR), a method for mass-producing, or amplifying, genes that has revolutionized biomedical science.

## CHANGES IN GENETIC INFORMATION

The amount of genetic information held within a set of human chromosomes is enormous, equal to twenty sets of *Encyclopaedia Britannica.* Because each of the trillions of cells in an adult body results from mitosis (except for egg and sperm), genetic information had to be replicated many times and with a high degree of accuracy. DNA can peruse itself for errors and correct them, a process termed DNA repair. Still, occasionally a replication mistake occurs or DNA is damaged, altering the genetic information. Such a change in DNA is called a **mutation.** Some mutations can cause devastating medical conditions; many others may be harmless variants.

## Nature of Mutations

Mutations can originate in a number of ways. In one common mechanism during DNA replication, a base may pair incorrectly with the newly forming strand, or extra bases may be added. Or, sections of DNA strands may be deleted, moved to other regions of the molecule, or even attached to other chromosomes. In any case, the consequences are similar—genetic information is changed. If a protein is constructed from this information, its molecular structure may be faulty and the function changed or absent. For example, the muscle weakness of Duchenne muscular dystrophy may result from a mutation in the gene encoding the protein dystrophin. The mutation may be a missing or changed nucleotide base or absence of the entire dystrophin gene. In each case, lack of dystrophin, which normally supports muscle cell membranes during contraction, causes the cells to collapse. The muscles weaken and atrophy. Figure 4.29 shows how the change of one base may cause another inherited illness, sickle cell disease.

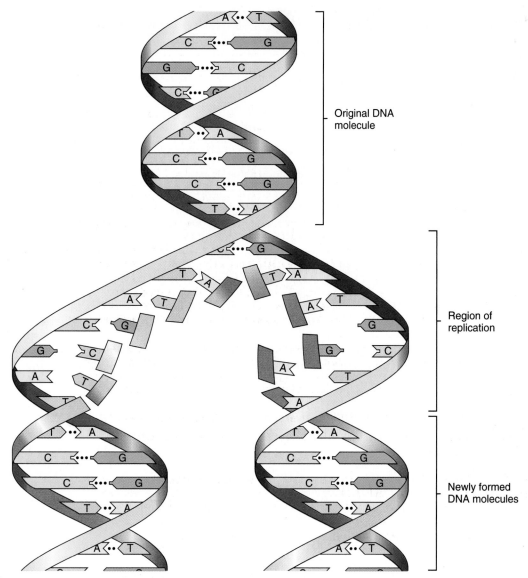

**FIGURE** 4.28

When a DNA molecule replicates, its original strands separate locally. A new strand of complementary nucleotides forms along each original strand.

Original DNA molecule

Region of replication

Newly formed DNA molecules

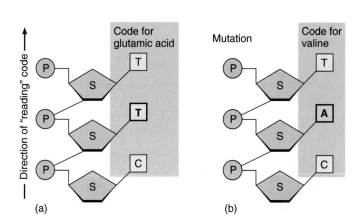

Direction of "reading" code →

Code for glutamic acid

Mutation

Code for valine

(a)

(b)

**FIGURE** 4.29

(a) The DNA code for the amino acid glutamic acid is CTT. (b) If something happens to change the first thymine in this section of the molecule to adenine, the DNA code changes to CAT, which specifies the amino acid valine. The resulting mutation, when it occurs in the DNA-encoding parts of the protein hemoglobin, causes sickle cell disease.

# Gene Amplification Revolutionizes Biomedicine

The polymerase chain reaction (PCR) is a procedure that borrows a cell's mechanism for DNA replication, allowing researchers to make many copies of a gene of interest. Starting materials are

- two types of short DNA pieces known to bracket the gene of interest, called primers
- a large supply of DNA bases
- the enzymes that replicate DNA

A simple test procedure rapidly builds up copies of the gene. Here's how it works.

In the first step of PCR, heat is used to separate the two strands of the target DNA—such as bacterial DNA in a body fluid sample from a person who has symptoms of an infection. Next, the temperature is lowered and the two short DNA primers are added. The primers bind by complementary base pairing to the separated target strands. In the third step, DNA polymerase and bases are added. The DNA polymerase adds bases to the primers and builds a sequence complementary to the target sequence. The newly synthesized strands then act as templates in the next round of replication, which is immediately initiated by raising the temperature. All of this is done in an automated device called a thermal cycler that controls the key temperature changes.

The pieces of DNA accumulate geometrically. The number of amplified pieces of DNA equals $2^n$, where n equals the number of temperature cycles. After just twenty cycles, one million copies of the original sequence are in the test tube. Table 4A lists some diverse applications of PCR.

PCR's greatest strength is that it works on crude samples of rare and minute sequences, such as a bit of brain tissue on the bumper of a car, which in one criminal case lead to identification of a missing person. PCR's greatest weakness, ironically, is its exquisite sensitivity. A blood sample submitted for diagnosis of an infection contaminated by leftover DNA from a previous run, or a stray eyelash dropped from the person running the reaction, can yield a false positive result. The technique is also limited in that a user must know the sequence to be amplified. ■

| TABLE 4A | PCR APPLICATIONS |
|---|---|

**PCR Has Been Used to Amplify:**

Genetic material from HIV in a human blood sample when infection has been so recent that antibodies are not yet detectable.

A bit of DNA in a preserved quagga (a relative of the zebra) and a marsupial wolf, which are recently extinct animals.

DNA in sperm cells found in the body of a rape victim so that specific sequences could be compared to those of a suspect in the crime.

Genes from microorganisms that cannot be grown or maintained in culture for study.

Mitochondrial DNA from various modern human populations. Comparisons of mitochondrial DNA sequences indicate that *Homo sapiens* originated in Africa, supporting fossil evidence.

DNA from the brain of a 7,000-year-old human mummy, which indicated that native Americans were not the only people to dwell in North America long ago.

DNA sequences unique to moose in hamburger meat, proving that illegal moose poaching had occurred.

DNA sequences in maggots in a decomposing human corpse, enabling forensic scientists to determine the time of death.

DNA in deteriorated road kills and carcasses washed ashore, to identify locally threatened species.

DNA in products illegally made from endangered species, such as powdered rhinoceros horn, sold as an aphrodisiac.

DNA sequences in animals that are unique to the bacteria that cause Lyme disease, providing clues to how the disease is transmitted.

DNA from genetically altered microbes that are released in field tests, to follow their dispersion.

DNA from a cell of an 8-celled human preembryo, to diagnose cystic fibrosis.

Y chromosome specific DNA from a human egg fertilized in the laboratory to determine the sex.

A papilloma virus DNA sequence present in, and possibly causing, an eye cancer.

Fortunately, cells detect damage in their DNA molecules, and use **repair enzymes** to clip out defective nucleotides in a single DNA strand and fill the resulting gap with nucleotides complementary to those on the remaining strand of DNA. This restores the original structure of the double-stranded DNA molecule.

If DNA is not repaired, illness may result. A class of disorders affects DNA repair. One such condition is xeroderma pigmentosum (XP).

When other youngsters burst out of their homes on a sunny day to frolic outdoors, a child who has XP must cover up as completely as possible, wearing pants and long sleeves even in midsummer, and must apply sunscreen on every bit of exposed skin. Moderate sun exposure easily leads to skin sores or cancer. Even with all the precautions, the child's skin is a sea of freckles. Special camps and programs for children with XP allow them to play outdoors at night, when they are safe.

## Effects of Mutations

The nature of the genetic code provides some protection against mutation. Sixty-one codons specify the twenty types of amino acids, and therefore some amino acids correspond to more than one codon type. Usually, two or three codons specifying the same amino acid differ only in the third base of the codon. A mutation that changes the third codon base can encode the same amino acid. For example, the DNA triplets GGA and GGG each specify the amino acid proline. If a mutation changes the third position of GGA to a G, the amino acid for that position in the encoded protein does not change—it is still proline.

If a mutation alters a base in the second position, the substituted amino acid is very often similar in overall shape to the normal one, and the protein is not changed significantly enough to affect its function. This mutation, too, would go unnoticed. (An important exception is the mutation shown in fig. 4.29.) Yet another protection against mutation is that a person has two copies of each chromosome, and therefore of each gene. If one copy is mutated, the other may provide enough of the gene's normal function to maintain health. (This is more complicated for the sex chromosomes, X and Y, discussed in chapter 24.) Finally, it also makes a difference whether a mutation occurs in a body cell of an adult or in a cell that is part of a developing embryo. In an adult, a mutant cell might not be noticed because many normally functioning cells surround it. In the embryo, however, the abnormal cell might give rise to many cells forming the developing body. In fact, all the cells of a person's body could be defective if the mutation were present in the fertilized egg.

Mutations may occur spontaneously if a chemical quirk causes a base in an original DNA strand to be in an unstable form just as replication occurs there. Certain

| TABLE 4.4 COMMONLY ENCOUNTERED MUTAGENS | |
| --- | --- |
| **Mutagen** | **Source** |
| Aflatoxin B | Fungi growing on peanuts and other foods |
| 2-amino 5-nitrophenol | |
| 2,4-diaminoanisole | |
| 2,5-diaminoanisole | Hair dye components |
| 2,4-diaminotoluene | |
| p-phenylenediamine | |
| Caffeine | Cola, tea, coffee |
| Furylfuramide | Food additive |
| Nitrosamines | Pesticides, herbicides, cigarette smoke |
| Proflavine | Antiseptic in veterinary medicine |
| Sodium nitrite | Smoked meats |
| Tris (2,3-dibromopropyl phosphate) | Flame retardant in children's sleepwear |

chemical substances, called **mutagens,** cause mutations. Researchers often use mutagens to intentionally alter gene function, in order to learn how a gene normally acts. Table 4.4 lists some mutagens.

Because a gene consists of a sequence of hundreds of building blocks, mutation can alter a gene in many ways—just like a typographical error can occur on this page in many ways. Different mutations in the same gene can produce different severities of symptoms. The most common mutation in the gene that causes cystic fibrosis, for example, causes severe lung infection and obstruction and digestive difficulties. Other mutations cause frequent bronchitis and pneumonia. This second group is less severely affected on a daily basis, and generally live longer than people with the more common mutation.

Ultraviolet radiation in sunlight is a familiar mutagen. It can cause an extra chemical bond to form between thymines that are adjacent on a DNA strand. This bond forms a kink, which can cause an incorrect base to be inserted during DNA replication. If sun damage is not extensive, repair enzymes remove the extra bonds, and replication proceeds. If damage is great, the cell undergoes apoptosis (cell death) and dies. We experience this as a peeling sunburn. If a sun-damaged cell cannot be repaired or does not die, it often turns cancerous. This is why many years of sunburns can cause skin cancer.

A type of disorder called an "inborn error of metabolism" results from inheriting mutations that alter an enzyme. Such an enzyme block in a biochemical pathway has two general effects: the substance that the

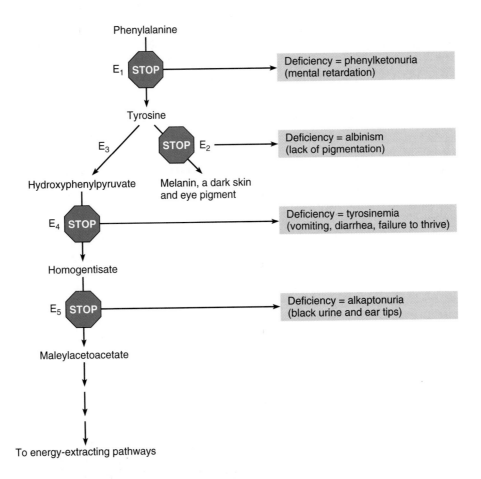

Phenylalanine

E₁ STOP → Deficiency = phenylketonuria (mental retardation)

Tyrosine

E₃    STOP E₂ → Deficiency = albinism (lack of pigmentation)

Hydroxyphenylpyruvate    Melanin, a dark skin and eye pigment

E₄ STOP → Deficiency = tyrosinemia (vomiting, diarrhea, failure to thrive)

Homogentisate

E₅ STOP → Deficiency = alkaptonuria (black urine and ear tips)

Maleylacetoacetate

To energy-extracting pathways

**FIGURE 4.30**

Four inborn errors of metabolism result from blocks affecting four enzymes in the pathway for the breakdown of phenylalanine, an amino acid. PKU results from a block of the first enzyme of the pathway ($E_1$). A block at $E_2$ leads to buildup of the amino acid tyrosine, and lack of its breakdown product, the pigment melanin, causes the pink eyes, white hair, and white skin of albinism. A block at $E_4$ can be deadly in infancy, and at $E_5$ leads to alkaptonuria, which causes severe joint pain and blackish deposits in the palate, ears, and eyes.

enzyme normally acts on builds up, and the substance resulting from the enzyme's normal action becomes scarce. It is similar to blocking a garden hose: water pressure builds up behind the block, but no water comes out after it.

The biochemical excesses and deficiencies that an inborn error of metabolism triggers can drastically affect health. The specific symptoms depend upon which pathways and substances are affected. Figure 4.30 shows how blocks of different enzymes in one biochemical pathway lead to different sets of symptoms. Clinical Application 4.4 describes phenylketonuria (PKU), one of these inborn errors about which we know a great deal.

1 How are DNA molecules replicated?

2 What is a mutation?

3 How do mutations occur?

4 What kinds of mutations are of greatest concern?

5 How does the nature of the genetic code protect against mutation, to an extent?

## THE HUMAN GENOME PROJECT

Researchers throughout the world are currently systematically locating, identifying, and deciphering the functions of each of the 70,000 genes that encode the biochemical blueprints of the human body. The project is expected to be completed by the year 2003. With each discovery, biologists can begin to explain how symptoms on a whole-body level result from abnormalities at the cellular and molecular levels. Although this genetic approach to understanding disease may apply directly to rare, inherited illnesses, the underlying mechanisms revealed will also explain normal physiology.

Unraveling the molecular underpinnings of disease lies at the forefront of biomedical research and is already entering and affecting the practice of medicine. The new

# Phenylketonuria

In Oslo, Norway, in 1934, an observant mother of two mentally retarded children noticed that their soiled diapers had an odd, musty odor. She mentioned this to Ivar Folling, a relative who was a physician and biochemist. Folling was intrigued. Analyzing the children's urine, he found large amounts of the amino acid phenylalanine, which is usually present only in trace amounts because an enzyme catalyzes a chemical reaction that breaks it down. The children lacked this enzyme because they had inherited an inborn error of metabolism called phenylketonuria, or PKU. The buildup of phenylalanine causes mental retardation.

Researchers wondered if a diet very low in phenylalanine might prevent the mental retardation. The diet would include the other nineteen types of amino acids so that normal growth, which requires protein, could occur.

The diet would theoretically alter the body's chemistry in a way that would counteract the overabundance of phenylalanine that the faulty genes caused.

In 1963, theory became reality when researchers devised a dietary treatment for this otherwise devastating illness (fig. 4A). The diet is very restrictive and difficult to follow, but it does prevent mental retardation. However, treated children may still have learning disabilities. We still do not know how long people with PKU should adhere to the diet, but it may be for their entire lives. ■

### FIGURE 4A

These three siblings have each inherited PKU. The older two siblings—the girl in the wheelchair and the boy on the right—are mentally retarded, because they were born before a diet that prevents symptoms became available. The child in the middle, although she also has inherited PKU, is of normal intelligence because she was lucky enough to have been born after the diet was invented.

field of molecular medicine is also changing the face of the traditional studies of anatomy and physiology. Wherever possible, this book provides molecular explanations of how the human body works.

**1** What is the human genome project?

**2** What will be the practical results of the human genome project?

# Introduction (page 102)

A cell continuously carries on metabolic processes.

# Metabolic Processes (page 102)

1. Anabolism
   a. Anabolism builds large molecules from smaller molecules.
   b. In dehydration synthesis, hydrogen atoms and hydroxyl groups are removed, water forms, and smaller molecules bind by sharing atoms.
   c. Complex carbohydrates are synthesized from monosaccharides, fats are synthesized from glycerol and fatty acids, and proteins are synthesized from amino acids.
2. Catabolism
   a. Catabolism breaks down larger molecules into smaller ones.
   b. In hydrolysis, a water molecule supplies a hydrogen atom to one portion of a molecule and a hydroxyl group to a second portion; the bond between these two portions breaks.
   c. Complex carbohydrates are decomposed into monosaccharides, fats are decomposed into glycerol and fatty acids, and proteins are decomposed into amino acids.

# Control of Metabolic Reactions (page 104)

Enzymes control metabolic reactions.

1. Enzyme action
   a. Metabolic reactions require energy to start.
   b. Enzymes are proteins that increase the rate of specific metabolic reactions.
   c. An enzyme acts upon a molecule by temporarily combining with it and distorting its chemical structure.
   d. The shape of an enzyme molecule fits the shape of its substrate molecule.
   e. When an enzyme combines with its substrate, the substrate changes, enabling it to react, forming a product. The enzyme is released in its original form.
   f. The rate of enzyme-controlled reactions depends upon the numbers of enzyme and substrate molecules and the efficiency of the enzyme.
   g. Enzymes are usually named according to their substrates, with -*ase* at the end.
2. Cofactors and coenzymes
   a. Cofactors are additions to some enzymes that are necessary for their function.
   b. A cofactor may be an ion or a small organic molecule called a coenzyme.
   c. Vitamins, which are the sources of coenzymes, usually cannot be synthesized by human cells in adequate amounts.
3. Factors that alter enzymes
   a. Enzymes are proteins and can be denatured.
   b. Factors that may denature enzymes include excessive heat, radiation, electricity, certain chemicals, and extreme pH values.

# Energy for Metabolic Reactions (page 106)

Energy is a capacity to produce change or to do work. Common forms of energy include heat, light, sound, electrical energy, mechanical energy, and chemical energy. Whenever changes take place, energy is being transferred from one part to another.

1. Release of chemical energy
   a. Most metabolic processes utilize chemical energy that is released when molecular bonds are broken.
   b. The energy glucose releases during cellular respiration is used to promote metabolism.
   c. Enzymes in the cytoplasm and mitochondria control cellular respiration.

# An Overview of Cellular Respiration (page 106)

1. Glycolysis
   a. The first phase of glucose catabolism occurs in the cytosol and does not require oxygen.
   b. Some of the energy released in glycolysis is transferred to molecules of ATP.
2. Aerobic respiration
   a. The second phase of glucose catabolism occurs within the mitochondria and requires oxygen.
   b. Considerably more energy is transferred to ATP molecules during aerobic respiration than during glycolysis.
   c. The products of glucose catabolism are heat, carbon dioxide, water, and energy.
3. ATP molecules
   a. Thirty-eight molecules of ATP can be produced for each glucose molecule that is catabolized.
   b. Energy is captured in the bond of the terminal phosphate of each ATP molecule.
   c. Captured energy is released when the terminal phosphate bond of an ATP molecule is broken.
   d. An ATP molecule that loses its terminal phosphate becomes an ADP molecule.
   e. An ADP molecule can be converted to an ATP molecule by capturing energy and a phosphate.

# A Closer Look at Cellular Respiration (page 108)

Metabolic processes usually have a number of steps that must occur in a specific sequence. A sequence of enzyme-controlled reactions is called a metabolic pathway. Typically, metabolic pathways are interconnected.

1. Carbohydrate pathways
   a. Carbohydrates may enter catabolic pathways and be used as energy sources.

b. Glycolysis breaks down a 6-carbon glucose molecule into two 3-carbon pyruvic acid molecules, resulting in a small net gain of ATP.

c. The citric acid cycle is a complex series of reactions that decompose molecules, release carbon dioxide, release hydrogen atoms that have high-energy electrons, and form ATP molecules.

d. Hydrogen atoms from the citric acid cycle become hydrogen ions, which, in turn, combine with oxygen to form water molecules.

e. Electrons from hydrogen atoms enter an electron transport chain. Energy released from the chain is used to form ATP.

f. Each glucose molecule metabolized yields a maximum of thirty-eight ATP molecules.

g. Excess carbohydrates may enter anabolic pathways and be polymerized into and stored as glycogen, or converted to fat.

2. Lipid pathways
   a. Most dietary fats are triglycerides.
   b. Before fats can be used as an energy source they must be broken down into glycerol and fatty acids.
   c. Beta oxidation decomposes fatty acids.
      (1) Beta oxidation activates fatty acids and breaks them down into segments of two carbon atoms each.
      (2) Fatty acid segments are converted into acetyl coenzyme A, which can be oxidized in the citric acid cycle.
   d. Fats can be synthesized from glycerol and fatty acids and from excess glucose or amino acids.

3. Protein pathways
   a. Proteins are used as building materials for cellular components, as enzymes, or as energy sources.
   b. Before proteins can be used as energy sources, they must be broken down into amino acids, and the amino acids deaminated.
   c. The deaminated portions of amino acids can be broken down into carbon dioxide and water, or used to produce glucose or fat.
   d. Human cells cannot synthesize eight essential amino acids in adequate amounts. These amino acids must come from foods.

4. Regulation of metabolic pathways
   a. A rate-limiting enzyme may regulate a metabolic pathway.
   b. A negative feedback mechanism in which the product of a pathway inhibits the regulatory enzyme may control the regulatory enzyme.
   c. The rate of product formation usually remains stable.

# Nucleic Acids and Protein Synthesis (page 117)

DNA molecules contain information that tells a cell how to synthesize proteins, including enzymes.

1. Genetic information
   a. DNA information specifies inherited traits.
   b. A gene is a portion of a DNA molecule that contains the genetic information for making one kind of protein.
   c. The nucleotides of a DNA strand are in a particular sequence.
   d. The nucleotides pair with those of the second strand in a complementary fashion.

2. Genetic code
   a. The sequence of nucleotides in a DNA molecule represents the sequence of amino acids in a protein molecule.
   b. RNA molecules transfer genetic information from the nucleus to the cytoplasm.

3. RNA molecules
   a. RNA molecules are usually single-stranded, contain ribose instead of deoxyribose, and contain uracil nucleotides in place of thymine nucleotides.
   b. Messenger RNA molecules, which are synthesized in the nucleus, contain a nucleotide sequence that is complementary to that of an exposed strand of DNA.
   c. Messenger RNA molecules move into the cytoplasm, associate with ribosomes, and are templates for the synthesis of protein molecules.

4. Protein synthesis
   a. Molecules of transfer RNA position amino acids along a strand of messenger RNA.
   b. A ribosome binds to a messenger RNA molecule and allows a transfer RNA molecule to recognize its correct position on the messenger RNA.
   c. The ribosome contains enzymes and ribozymes required for the synthesis of the protein and holds the protein until it is completed.
   d. As the protein forms, it folds into a unique shape.
   e. ATP provides the energy for protein synthesis.

5. DNA replication
   a. Each new cell requires a copy of the original cell's genetic information.
   b. DNA molecules are replicated during interphase of the cell cycle.
   c. Each new DNA molecule contains one old strand and one new strand.

# Changes in Genetic Information (page 126)

A DNA molecule contains a great amount of information. A change in the genetic information is a mutation.

1. Nature of mutations
   a. Mutations include several kinds of changes in DNA.
   b. A protein synthesized from an altered DNA sequence may function abnormally or not at all.
   c. Repair enzymes can correct some forms of DNA damage.

2. Effects of mutations
   a. The genetic code protects against some mutations.
   b. A mutation in a sex cell or fertilized egg or early embryo may have a more severe effect than a mutation in an adult because a greater proportion of the individual's cells are affected.

# The Human Genome Project (page 130)

Researchers from many nations are cooperating to identify and study all human genes.

1. Understanding the function of individual genes will help us fight inherited disorders.

2. Understanding gene function can explain normal anatomy and physiology.

## CRITICAL THINKING QUESTIONS

1. Because enzymes are proteins, they can denature. How does this explain the fact that changes in the pH of body fluids during illness may threaten life?

2. Some weight-reducing diets drastically limit intake of carbohydrates but allow many foods rich in fat and protein. What changes would such a diet cause in cellular metabolism? What changes would appear in the urine?

3. Why are vitamins that function as coenzymes in cells required in extremely low concentrations?

4. What changes in concentrations of oxygen and carbon dioxide would you expect to find in the blood of a person who is forced to exercise on a treadmill beyond his or her normal capacity? How might these changes affect the pH of the person's blood?

5. How do the antibiotic actions of penicillin and streptomycin differ?

6. A student is accustomed to running 3 miles each afternoon at a slow, leisurely pace. One day, she runs a mile as fast as she can. Afterwards she is winded, with pains in her chest and leg muscles. She thought she was in great shape! What has she experienced, in terms of energy metabolism?

7. In fructose intolerance, a missing enzyme makes a person unable to utilize fructose, a simple sugar abundant in fruit. Infants with the condition have very low mental and motor function. Older children are very lethargic and mildly mentally retarded. By adulthood, the nervous system deteriorates, eventually causing mental illness and death.

   Molecules that are derived from fructose are intermediates in the first few reactions of glycolysis. The enzyme missing in people with fructose intolerance would normally catalyze these reactions. Considering this information about the whole-body and biochemical effects of fructose intolerance, suggest what might be happening on a cellular level to these people.

8. Write the sequence of the complementary strand of DNA to the sequence A, G, C, G, A, T, T, G, C, A, T, G, C. What is the sequence of mRNA that would be transcribed from the given sequence?

9. Explain why exposure to ultraviolet light in tanning booths may be dangerous.

## REVIEW EXERCISES

1. Define *anabolism* and *catabolism.*
2. Distinguish between dehydration synthesis and hydrolysis.
3. Define *peptide bond?*
4. Define *enzyme.*
5. How does an enzyme interact with its substrate?
6. List three factors that increase the rates of enzyme-controlled reactions.
7. How are enzymes usually named?
8. Define *cofactor.*
9. Explain why humans require vitamins in their diets.
10. Explain how an enzyme may be denatured.
11. Define *energy.*
12. Explain how the oxidation of molecules inside cells differs from the burning of substances outside cells.
13. Define *cellular respiration.*
14. Distinguish between anaerobic and aerobic respiration.
15. Explain the importance of ATP to cellular processes.
16. Describe the relationship between ATP and ADP molecules.
17. Define *metabolic pathway?*
18. Describe the starting material and products of *glycolysis.*
19. State the products of the citric acid cycle.
20. How are carbohydrates stored?
21. Define *beta oxidation.*
22. Explain how fats may provide energy.
23. Define *deamination* and explain its importance.
24. Explain why some amino acids are called essential.
25. Explain how one enzyme can regulate a metabolic pathway.
26. Describe how a negative feedback mechanism can help control a metabolic pathway.
27. Explain the chemical basis of genetic information.
28. Describe the chemical makeup of a gene.
29. Describe the general structure and components of a DNA molecule.
30. Distinguish between the functions of messenger RNA and transfer RNA.
31. Distinguish between transcription and translation.
32. Explain two functions of ribosomes in protein synthesis.
33. Distinguish between a codon and an anticodon.
34. Explain how a DNA molecule is replicated.
35. Define *mutation,* and explain how mutations may originate.
36. Define *repair enzyme.*
37. Explain how a mutation may affect an organism's cells—or not affect them.

 **For More Information Related to This Chapter, Please Visit** *Hole's Human Anatomy and Physiology* **Home Page**
http://www.mhhe.com/biosci/abio/

# TISSUES

## CHAPTER OBJECTIVES

### AFTER YOU HAVE STUDIED THIS CHAPTER, YOU SHOULD BE ABLE TO:

1. Describe the general characteristics and functions of epithelial tissue.

2. Name the types of epithelium and identify an organ in which each is found.

3. Explain how glands are classified.

4. Describe the general characteristics of connective tissue.

5. Describe the major cell types and fibers of connective tissue.

6. List the types of connective tissue within the body.

7. Describe the major functions of each type of connective tissue.

8. Distinguish among the three types of muscle tissue.

9. Describe the general characteristics and functions of nervous tissue.

## KEY TERMS

**adipose tissue** (ad′ĭ-pōs tish′u)
**cartilage** (kar′ti-lij)
**chondrocyte** (kon′dro-sīt)
**connective tissue** (kŏ-nek′tiv tish′u)
**epithelial tissue** (ep″ĭ-the′le-al tish′u)
**fibroblast** (fi′bro-blast)

**fibrous tissue** (fi′brus tish′u)
**macrophage** (mak′ro-fāj)
**muscle tissue** (mus′el tish′u)
**nervous tissue** (ner′vus tish′u)
**neuroglia** (nu-rog′le-ah)
**neuron** (nu′ron)
**osteocyte** (os′te-o-sīt)
**osteon** (os′te-on)

## UNDERSTANDING WORDS

**adip-**, fat: *adip*ose tissue—tissue that stores fat.

**chondr-**, cartilage: *chondr*ocyte—cartilage cell.

**-cyt**, cell: osteo*cyt*e—bone cell.

**epi-**, upon, after, in addition: *epi*thelial tissue—tissue that covers all free body surfaces.

**-glia**, glue: neuro*glia*—cells that bind nervous tissue together.

**hist-**, web, tissue: *hist*ology—study of composition and function of tissues.

**hyal-**, resemblance to glass: *hyal*ine cartilage—flexible tissue containing chondrocytes.

**inter-**, among, between: *inter*calated disk—band of gap junctions located between the ends of adjacent cardiac muscle cells.

**macr-**, large: *macr*ophage—large phagocytic cell.

**neur-**, nerve: *neur*on—nerve cell.

**os-**, bone: *os*seous tissue—bone tissue.

**phag-**, to eat: *phag*ocyte—cell that engulfs and destroys foreign particles.

**pseud-**, false: *pseud*ostratified epithelium—tissue whose cells appear to be arranged in layers, but are not.

**squam-**, scale: *squam*ous epithelium—tissue whose cells appear flattened or scalelike.

**strat-**, layer: *strat*ified epithelium—tissue whose cells occur in layers.

**stria-**, groove: *stria*ted muscle—tissue whose cells are characterized by alternating light and dark cross-markings.

tudents preparing to enter health fields must memorize many details of human anatomy and physiology. Traditionally students have studied tissues and organs by using atlases of drawings and by dissecting cadavers. The National Library of Medicine offers a new way to study the human body. The idea for the Visible Human Project was born in 1986, to complement the vast printed medical literature. Eight years later, a digital image library consisting of 1,871 one-millimeter-thick sections of a newly deceased man appeared on the Internet. In addition to digital images/photographs of thick sections, magnetic resonance images, computerized tomography scans, and X rays were taken of the whole body.

In life, the Visible Man was Joseph Paul Jernigan, executed by lethal injection in 1993 at age 39 for killing an elderly man. After the execution, Jernigan's body was flown to the University of Colorado Medical Sciences Center where, within eight hours of death, the imaging began. Researchers saw it as a rare opportunity to obtain a healthy body soon after death. Medical ethicists, concerned with the rights of death row inmates, however, contend that when Jernigan willed his body to science, he may not have wanted every nuance of his anatomy to appear on computer screens everywhere—more likely he envisioned a fate on a dissection table. But Jernigan's lawyer maintains that his client willed his body to science to atone for his crime, and relatives of his victim agree.

The Visible Woman is a 59-year-old who died of a heart attack. Her sections were taken at 0.33-millimeter intervals, yielding more than 5,000 images of her anatomy. This distance is equal for all three axes in space, which makes it easier to derive three-dimensional reconstructions.

The Visible Human Project is described as a "unique interactive anatomical digital atlas." But much work lies ahead, to classify, identify, measure, and label all structures. This will entail tracing many structures through several slices, to reconstruct tissues in three dimensions. According to the National Library of Medicine, "The larger, long-term goal of the Visible Human Project [is] to transparently link the print library of functional-physiological knowledge with the image library of structural-anatomical knowledge into one unified resource of health information."

## TABLE 5.1 TISSUES

| Type | Function | Location | Distinguishing Characteristics |
|------|----------|----------|-------------------------------|
| Epithelial | Protection, secretion, absorption, excretion | Cover body surface, cover and line internal organs, compose glands | Lack blood vessels, readily reproduce, cells are tightly packed |
| Connective | Bind, support, protect, fill spaces, store fat, produce blood cells | Widely distributed throughout the body | Mostly have good blood supply, cells are farther apart than cells of epithelia with matrix in between |
| Muscle | Movement | Attached to bones, in the walls of hollow internal organs, heart | Contractile |
| Nervous | Transmit impulses for coordination, regulation, integration, and sensory reception | Brain, spinal cord, nerves | Cells connect to each other and other body parts |

In all complex organisms, cells are organized into layers or groups called **tissues.** Although the cells of different tissues vary in size, shape, arrangement, and function, those within a tissue are quite similar.

Usually tissue cells are separated by nonliving, intercellular materials that the cells produce. These intercellular materials vary in composition and amount from one tissue to another and may be solid, semisolid, or liquid. For example, a solid (mineral) separates bone tissue cells, whereas a liquid (plasma) separates blood tissue cells.

The tissues of the human body include four major types: *epithelial, connective, muscle,* and *nervous.* These tissues are combined into organs that have specialized functions. Table 5.1 compares the four major tissue types.

This chapter examines in detail epithelial and connective tissues. Subsequent chapters discuss other kinds of tissues.

**1** What is a tissue?

**2** List the four major types of tissue.

## EPITHELIAL TISSUES

### General Characteristics

**Epithelial tissues** are widespread throughout the body. Since epithelium covers organs, forms the inner lining of

Simple — ⎡
squamous ⎣
epithelium

Surface
of tissue

Nucleus

Basement —
membrane

Connective —
tissue

**FIGURE** 5.1 𝒳

Simple squamous epithelium consists of a single layer of tightly packed flattened cells (250× micrograph enlarged to 700×).

body cavities, and lines hollow organs, it always has a free surface—one that is exposed to the outside or to an open space internally. The underside of this tissue is anchored to connective tissue by a thin, nonliving layer called the **basement membrane.**

> One of the ways that cancer cells spread is by secreting a substance that dissolves basement membranes. This enables cancer cells to invade adjacent tissue layers. Cancer cells also produce fewer adhesion proteins, or none at all, which enables them to invade surrounding tissue.

As a rule, epithelial tissues lack blood vessels. However, nutrients diffuse to epithelium from underlying connective tissues, which have abundant blood vessels.

Because epithelial cells readily reproduce, injuries heal rapidly as new cells replace lost or damaged ones. Skin cells and the cells that line the stomach and intestines are continually being damaged and replaced.

Epithelial cells are tightly packed, with little intercellular material. In many places, desmosomes attach one to another (see chapter 3). Consequently, these cells form effective protective barriers in such structures as the outer layer of the skin and the inner lining of the mouth. Other epithelial functions include secretion, absorption, and excretion.

Epithelial tissues are classified according to the shape and number of layers of cells. Epithelial tissues that are composed of single layers of cells are *simple;* those with two or more layers of cells are *stratified;* those with thin, flattened cells are *squamous;* those with cube-like cells are *cuboidal;* and those with elongated cells are *columnar.* In the following descriptions, note that the free surfaces of epithelial cells are modified to reflect their specialized functions.

## Simple Squamous Epithelium

**Simple squamous epithelium** consists of a single layer of thin, flattened cells. These cells fit tightly together, somewhat like floor tiles, and their nuclei are usually broad and thin (fig. 5.1).

Substances pass rather easily through simple squamous epithelium, which is common at sites of diffusion and filtration. For instance, simple squamous epithelium lines the air sacs (alveoli) of the lungs where oxygen and carbon dioxide are exchanged. It also forms the walls of capillaries, lines the insides of blood and lymph vessels, and covers the membranes that line body cavities. However, because it is so thin and delicate, simple squamous epithelium is easily damaged.

## Simple Cuboidal Epithelium

**Simple cuboidal epithelium** consists of a single layer of cube-shaped cells. These cells usually have centrally located, spherical nuclei (fig. 5.2).

Simple cuboidal epithelium covers the ovaries and lines the kidney tubules and ducts of certain glands, such as the salivary glands, pancreas, and liver. In the kidneys, it functions in secretion and absorption; in glands, it secretes glandular products.

## Simple Columnar Epithelium

The cells of **simple columnar epithelium** are elongated; that is, they are longer than they are wide. This tissue is composed of a single layer of cells whose nuclei are usually located at about the same level, near the basement membrane (fig. 5.3).

Simple columnar epithelium lines the uterus, stomach, and intestines. Because its cells are elongated, this tissue is thick, which suits it to protect underlying tissues. Simple columnar epithelium also secretes digestive fluids and absorbs nutrients from digested foods.

Basement membrane
Surface of tissue
Simple cuboidal epithelium
Nucleus
Cell membrane
Connective tissue

**FIGURE** 5.2
Simple cuboidal epithelium consists of a single layer of tightly packed cube-shaped cells (100× micrograph enlarged to 320×).

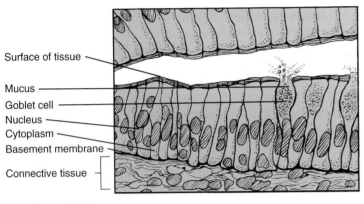

Surface of tissue
Mucus
Goblet cell
Nucleus
Cytoplasm
Basement membrane
Connective tissue

**FIGURE** 5.3
Simple columnar epithelium consists of a single layer of elongated cells (400×).

Simple columnar cells, specialized for absorption, often have many minute, cylindrical processes extending from their surfaces. These processes, called *microvilli,* are from 0.5 to 1.0 μm long. They increase the surface area of the cell membrane where it is exposed to substances being absorbed (fig. 5.4).

Typically, specialized, flask-shaped glandular cells are scattered among the columnar cells of simple columnar epithelium. These cells, called *goblet cells,* secrete a protective fluid called *mucus* onto the free surface of the tissue (see fig. 5.3).

## Pseudostratified Columnar Epithelium

The cells of **pseudostratified columnar epithelium** appear stratified or layered, but they are not. A layered effect occurs because the cell nuclei are at two or more levels in the row of aligned cells. However, the cells, which vary in shape, all reach the basement membrane, even though some of them may not contact the free surface.

**FIGURE** 5.4
A scanning electron micrograph of microvilli, which occur on exposed surfaces of some columnar epithelial cells (33,000×).

Surface of tissue
Cilia
Nucleus
Goblet cell
Basement membrane
Connective tissue

**FIGURE 5.5** 🏃

Pseudostratified columnar epithelium appears stratified because nuclei are at different levels (100× micrograph enlarged to 320×).

Surface of tissue
Squamous cells
Layer of reproducing cells
Basement membrane
Connective tissue

**FIGURE 5.6** 🏃

Stratified squamous epithelium consists of many layers of cells (70×).

The cells of pseudostratified columnar epithelium often are fringed with *cilia,* which are 7–10 µm in length. The cilia extend from the free surfaces of the cells, and they move constantly. Goblet cells scattered throughout this tissue secrete mucus, which the cilia sweep away (fig. 5.5).

Pseudostratified columnar epithelium lines the passages of the respiratory and reproductive systems. In the respiratory passages, the mucus-covered linings are sticky and trap dust and microorganisms that enter with the air. The cilia move the mucus and its captured particles upward and out of the airways. In the female reproductive tubes, cilia aid in moving egg cells through the oviducts to the uterus.

## Stratified Squamous Epithelium

Stratified epithelium is named for the shape of the cells forming the outermost layers. **Stratified squamous epithelium** consists of many layers of cells, making this tissue relatively thick. Cells nearest the free surface are flattened the most, whereas those in the deeper layers,

where cellular reproduction occurs, are cuboidal or columnar. As the newer cells grow, older ones are pushed farther and farther outward, where they flatten (fig. 5.6).

The outermost layer of the skin (epidermis) is stratified squamous epithelium. As the older cells are pushed outward, they accumulate a protein called *keratin,* then harden and die. This "keratinization" produces a covering of dry, tough, protective material that prevents water and other substances from escaping from underlying tissues and blocks chemicals and microorganisms from entering.

Stratified squamous epithelium also lines the oral cavity, throat, vagina, and anal canal. In these parts, the tissue is not keratinized; it stays soft and moist, and the cells on its free surfaces remain alive.

## Stratified Cuboidal Epithelium

**Stratified cuboidal epithelium** consists of two or three layers of cuboidal cells that form the lining of a lumen (fig. 5.7). The layering of the cells provides more protection than the single layer affords.

Basement membrane

Stratified cuboidal epithelium

Lumen

Connective tissue

Nucleus

**FIGURE 5.7**

Stratified cuboidal epithelium consists of two to three layers of cube-shaped cells surrounding a lumen (100× micrograph enlarged to 320×).

Lumen

Stratified columnar epithelium

Nucleus

Basment membrane

Underlying connective tissue

**FIGURE 5.8**

Stratified columnar epithelium consists of a superficial layer of columnar cells overlying several layers of cuboidal cells (250× micrograph enlarged to 1,000×).

Stratified cuboidal epithelium lines the larger ducts of the mammary glands, sweat glands, salivary glands, and pancreas. It also forms the lining of developing ovarian follicles and seminiferous tubules, which are parts of the female and male reproductive systems, respectively.

## Stratified Columnar Epithelium

**Stratified columnar epithelium** consists of several layers of cells (fig. 5.8). The superficial cells are elongated, whereas the basal layers consist of cube-shaped cells. Stratified columnar epithelium is in the vas deferens, part of the male urethra, and in parts of the pharynx.

## Transitional Epithelium

**Transitional epithelium** (uroepithelium) is specialized to change in response to increased tension. It forms the inner lining of the urinary bladder and lines the ureters and part of the urethra. When the wall of one of these organs contracts, the tissue consists of several layers of

cuboidal cells; however, when the organ is distended, the tissue stretches, and the physical relationships among the cells change. While distended, the tissue appears to contain only a few layers of cells (fig. 5.9). In addition to providing an expandable lining, transitional epithelium forms a barrier that helps prevent the contents of the urinary tract from diffusing back into the internal environment. Table 5.2 summarizes the characteristics of the different types of epithelial tissues.

A cancer originating in epithelium is called a *carcinoma,* and it is estimated that up to 90% of all human cancers are of this type. Most carcinomas begin on surfaces that contact the external environment, such as skin, linings of the airways in the respiratory tract, or linings of the stomach or intestines in the digestive tract. This observation suggests that the more common cancer-causing agents may not penetrate tissues very deeply.

Unstretched transitional epithelium

Basement membrane

Underlying connective tissue

(a)

Stretched transitional epithelium

Basement membrane

Underlying connective tissue

(b)

**FIGURE 5.9**

Transitional epithelium is (*a*) unstretched and consists of many layers when the organ wall contracts (250× micrograph enlarged to 600×). (*b*) The tissue stretches and appears thinner when the organ is distended (250× micrograph enlarged to 600×).

**TABLE 5.2 EPITHELIAL TISSUES**

| Type | Description | Function | Location |
|------|-------------|----------|----------|
| Simple squamous epithelium | Single layer, flattened cells | Filtration, diffusion, osmosis, covers surface | Air sacs of lungs, walls of capillaries, linings of blood and lymph vessels |
| Simple cuboidal epithelium | Single layer, cube-shaped cells | Secretion, absorption | Surface of ovaries, linings of kidney tubules, and linings of ducts of certain glands |
| Simple columnar epithelium | Single layer, elongated cells | Protection, secretion, absorption | Linings of uterus, stomach, and intestines |
| Pseudostratified columnar epithelium | Single layer, elongated cells | Protection, secretion, movement of mucus and cells | Linings of respiratory passages and reproductive tract |
| Stratified squamous epithelium | Many layers, top cells flattened | Protection | Outer layer of skin, linings of oral cavity, throat, vagina, and anal canal |
| Stratified cuboidal epithelium | 2–3 layers, cube-shaped cells | Protection | Linings of larger ducts of sweat glands, salivary glands, and the pancreas |
| Stratified columnar epithelium | Top layer of elongated cells, lower layers of cube-shaped cells | Protection, secretion | Vas deferens, part of the male urethra, and parts of the pharynx |
| Transitional epithelium | Many layers cube-shaped and elongated cells | Distensibility, protection | Inner lining of urinary bladder and linings of ureters and part of urethra |

Tissue surface

Duct

Secretory portion

Simple tubular

Simple
branched tubular

Simple
coiled tubular

Simple
branched alveolar

Compound tubular

Compound alveolar

**FIGURE** 5.10
Structural types of exocrine glands.

1. List the general characteristics of epithelial tissue.
2. Explain how epithelial tissues are classified.
3. Describe the structure of each type of epithelium.
4. Describe the special functions of each type of epithelium.

## Glandular Epithelium

**Glandular epithelium** is composed of cells that are specialized to produce and secrete substances into ducts or into body fluids. Such cells are usually found within columnar or cuboidal epithelium, and one or more of these cells constitutes a *gland*. Glands that secrete their products into ducts that open onto some internal or external surface are called **exocrine glands.** Glands that secrete their products into tissue fluid or blood are called **endocrine glands.** (Endocrine glands are discussed in chapter 13.)

An exocrine gland may consist of a single epithelial cell (unicellular gland), such as a mucus-secreting goblet cell, or it may be composed of many cells (multicellular gland). In turn, the multicellular forms can be structurally subdivided into two groups—simple and compound glands.

A *simple gland* communicates with the surface by means of an unbranched duct, and a *compound gland* has a branched duct. These two types of glands can be further classified according to the shapes of their secretory portions. Glands that consist of epithelial-lined tubes are called *tubular glands;* those whose terminal portions form saclike dilations are called *alveolar glands* (acinar glands). Branching and coiling of the secretory portions may occur as well. Figure 5.10 illustrates several types of exocrine glands classified by structure. Table 5.3 summarizes the types of exocrine glands, lists their characteristics, and provides an example of each type.

Exocrine glands are also classified according to the ways these glands secrete their products. Glands that release fluid products by exocytosis are called **merocrine glands.** Glands that lose small portions of their glandular cell bodies during secretion are called **apocrine glands.** Glands that release entire cells are called **holocrine glands.** After release, the cells containing accumulated secretory products disintegrate to release their secretions (figs. 5.11 and 5.12). Table 5.4 summarizes these glands and their secretions.

Most exocrine secretory cells are merocrine, and they can be further subdivided as either *serous cells* or *mucous cells.* The secretion of serous cells is typically

TABLE 5.3

## TYPES OF EXOCRINE GLANDS

| Type | Characteristics | Example |
|------|-----------------|---------|
| Unicellular glands | A single secretory cell | Mucus-secreting goblet cell (see fig. 5.3) |
| Multicellular glands | Glands that consist of many cells | |
|   Simple glands | Glands that communicate with surface by means of unbranched ducts | |
|    **1.** Simple tubular gland | Straight tubelike gland that opens directly onto surface | Intestinal glands of small intestine (see fig. 17.3) |
|    **2.** Simple coiled tubular gland | Long, coiled, tubelike gland; long duct | Eccrine (sweat) glands of skin (see fig. 6.11) |
|    **3.** Simple branched tubular gland | Branched, tubelike gland; duct short or absent | Mucous glands in small intestine (see fig. 17.3) |
|    **4.** Simple branched alveolar gland | Secretory portions of gland expand into saclike compartments along duct | Sebaceous gland of skin (see fig. 5.12) |
|   Compound glands | Glands that communicate with surface by means of branched ducts | |
|    **1.** Compound tubular gland | Secretory portions are coiled tubules, usually branched | Bulbourethral glands of male (see fig. 22.1) |
|    **2.** Compound alveolar gland | Secretory portions are irregularly branched tubules with numerous saclike outgrowths | Salivary glands (see fig. 17.12) |

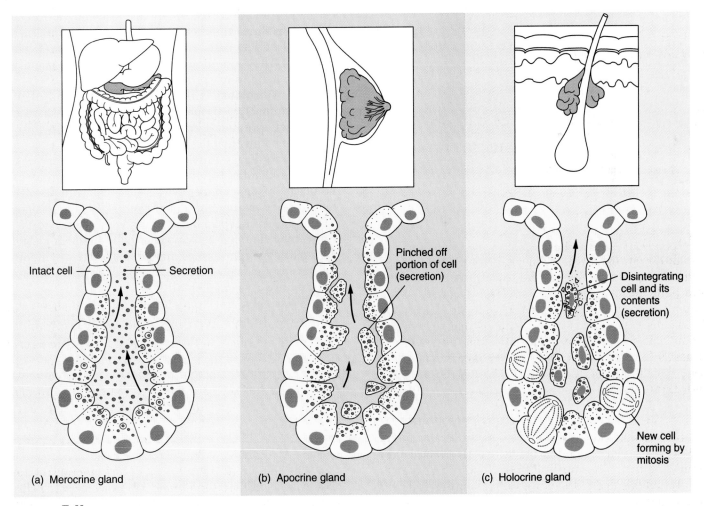

(a) Merocrine gland     (b) Apocrine gland     (c) Holocrine gland

## FIGURE 5.11

(a) Merocrine glands release secretions without losing cytoplasm. (b) Apocrine glands lose small portions of their cell bodies during secretion. (c) Holocrine glands release entire cells filled with secretory products.

watery, has a high concentration of enzymes, and is called *serous fluid.* Such cells are common in the linings of the body cavities. Mucous cells secrete a thicker fluid called *mucus.* This substance is rich in the glycoprotein *mucin* and is abundantly secreted from the inner linings of the digestive and respiratory systems.

**1** Distinguish between exocrine and endocrine glands.

**2** Explain how exocrine glands are classified.

**3** Distinguish between a serous cell and a mucous cell.

Sebaceous gland

Surface of skin

Hair follicle (hair shaft removed)

**FIGURE** 5.12

The sebaceous gland associated with a hair follicle is a simple branched alveolar gland that secretes entire cells (30×).

# CONNECTIVE TISSUES

## General Characteristics

**Connective tissues** occur throughout the body and are the most abundant type of tissue by weight. They bind structures, provide support and protection, serve as frameworks, fill spaces, store fat, produce blood cells, protect against infections, and help repair tissue damage.

Connective tissue cells are not adjacent to each other as epithelial cells are, and they have an abundance of intercellular material, or **matrix,** between them. This matrix consists of fibers and a ground substance whose consistency varies from fluid to semisolid to solid.

Connective tissue cells can usually reproduce. These tissues have varying degrees of vascularity, but in most cases, they have good blood supplies and are well nourished. Some connective tissues, such as bone and cartilage, are quite rigid. Loose connective tissue (areolar), adipose tissue, and dense connective tissue are more flexible.

## Major Cell Types

Connective tissues contain a variety of cell types. Some of them are called *fixed cells* because they are usually present in stable numbers. These include fibroblasts and mast cells. Other cells, such as macrophages, are *wandering cells.* They temporarily appear in tissues, usually in response to an injury or infection.

The **fibroblast** is the most common kind of fixed cell in connective tissues. It is a large, star-shaped cell. Fibroblasts produce fibers by secreting protein into the matrix of connective tissues (fig. 5.13).

**Macrophages** (histiocytes) originate as white blood cells (see chapter 14) and are almost as numerous as fibroblasts in some connective tissues. They are usually attached to fibers, but can detach and actively move about. Macrophages are specialized to carry on phagocytosis. Because they function as scavenger cells that can clear foreign particles from tissues, macrophages are an important defense against infection (fig. 5.14). They also play a role in immunity (see chapter 16).

**Mast cells** are large and are widely distributed in connective tissues, where they are usually located near

| TABLE 5.4 | TYPES OF GLANDULAR SECRETIONS | | |
|---|---|---|---|
| **Type** | **Description of Secretion** | | **Example** |
| Merocrine glands | A fluid product released through the cell membrane by exocytosis | | Salivary glands, pancreatic glands, sweat glands of the skin |
| Apocrine glands | Cellular product and portions of the free ends of glandular cells pinch off during secretion | | Mammary glands, ceruminous glands lining the external ear canal |
| Holocrine glands | Entire cells laden with secretory products disintegrate | | Sebaceous glands of the skin |

**FIGURE 5.13**

A scanning electron micrograph of a fibroblast (4,000×).

Cell membrane

Nucleus

Cytoplasmic granule

**FIGURE 5.15**

A transmission electron micrograph of a mast cell (5,000×).

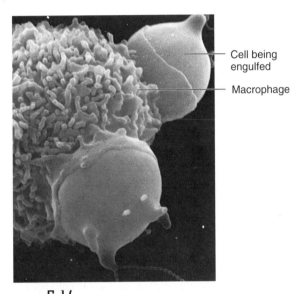

Cell being engulfed

Macrophage

**FIGURE 5.14**

Macrophages are scavenger cells common in connective tissues. In this scanning electron micrograph, a macrophage engulfs two cells (4,500×).

Elastic fibers

Collagen fibers

**FIGURE 5.16**

Falsely colored scanning electron micrograph of collagenous fibers (here yellow) and elastic fibers (here blue) (4,100×).

blood vessels (fig. 5.15). They release *heparin,* a compound that prevents blood clotting. Mast cells also release *histamine,* a substance that promotes some of the reactions associated with inflammation and allergies, such as asthma and hay fever (see chapter 16).

## Connective Tissue Fibers

Fibroblasts produce three types of connective tissue fibers: collagenous fibers, elastic fibers, and reticular fibers.

**Collagenous fibers** are thick threads of the protein collagen, which is the major structural protein of the body. Collagenous fibers are grouped in long, parallel bundles, and they are flexible but only slightly elastic (fig. 5.16). More importantly, they have great tensile strength—that is, they can resist considerable pulling

force. Thus, collagenous fibers are important components of body parts that hold structures together, such as ligaments (which connect bones to bones) and tendons (which connect muscles to bone).

Tissue containing abundant collagenous fibers is called *dense connective tissue.* Such tissue appears white, and for this reason collagenous fibers are sometimes called *white fibers. Loose connective tissue,* on the other hand, has sparse collagenous fibers. Clinical Application 5.1, figure 5.17, and table 5.5 concern disorders that result from abnormal collagen.

"Clinically proven, anti-aging collagen cream retexturizes skin, making it more resilient, and looking younger!" proclaims the advertisement that seems too good to be true. It is. Although any moisture on dry skin may temporarily improve its appearance, collagen molecules are far too large to actually penetrate the skin. The only thing a collagen cream is sure to affect is your wallet.

# CLINICAL APPLICATION

# Disorders of Orderly Collagen

Much of the human body consists of the protein collagen. It accounts for more than 60% of the protein in bone and cartilage, and provides 50–90% of the dry weight of skin, ligaments, tendons, and the dentine of teeth. Collagen is in the eyes, blood vessel linings, basement membranes, and connective tissue. It is not surprising, then, that defects in collagen cause a variety of medical problems.

Collagen abnormalities are devastating because this protein has an extremely precise structure that is easily disrupted, even by slight alterations that might exert little noticeable effect in other proteins. Collagen is sculpted from a precursor molecule called procollagen. Three procollagen chains coil and entwine to form a very regular triple helix.

Triple helices form as the procollagen is synthesized, but once secreted from the cell, the helices are trimmed. The collagen fibrils continue to associate outside the cell, building the networks that hold the body together. Collagen is rapidly synthesized and assembled into its rigid architecture. Many types of mutations can disrupt the protein's structure, including missing procollagen chains, kinks in the triple helix, failure to cut mature collagen, and defects in aggregation outside the cell.

Knowing which specific mutations cause disorders offers a way to diagnose the condition before symptoms arise. This can be helpful if early treatment can follow. A woman who has a high risk of developing hereditary osteoporosis, for example, might take calcium supplements or begin estrogen replacement (hormone) therapy before symptoms appear.

Aortic aneurysm is a more serious connective tissue disorder that can be presymptomatically diagnosed if the underlying mutation is detected. In aortic aneurysm, a weakened aorta (the largest blood vessel in the body, which emerges from the heart) bursts. Knowing that the mutant gene has not been inherited can ease worries—and knowing that it has been inherited can warn affected individuals to have frequent ultrasound exams so that aortic weakening can be detected early enough to correct with surgery. ■

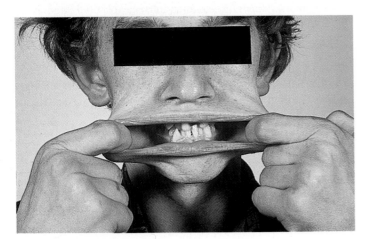

**FIGURE 5.17**
Abnormal collagen causes the stretchy skin of Ehlers-Danlos syndrome type I.

sume their original lengths and shapes when the force acting upon them is removed. Elastic fibers are common in body parts that are normally subjected to stretching, such as the vocal cords and air passages of the respiratory system. Elastic fibers are sometimes called yellow fibers, because tissues amply supplied with them appear yellowish (see fig. 5.16).

> Surgeons use elastin in foam, powder, or sheet form to prevent scar tissue adhesions from forming at the sites of tissue removal. Elastin is produced in bacteria that are genetically altered to contain human genes that instruct them to manufacture the human protein. This is cheaper than synthesizing elastin chemically, and safer than obtaining it from cadavers.

**Elastic fibers** are composed of bundles of microfibrils embedded in a protein called *elastin*. These fibers branch, forming complex networks in various tissues. They are weaker than collagenous fibers but very elastic. That is, they are easily stretched or deformed and will re-

**Reticular fibers** are very thin collagenous fibers. They are highly branched and form delicate supporting networks in a variety of tissues. Table 5.6 summarizes the components of connective tissue.

TABLE 5.5

## COLLAGEN DISORDERS

| Disorder | Molecular Defect | Symptoms |
|----------|------------------|----------|
| Chondrodysplasia | Collagen chains are too wide and asymmetric | Stunted growth; deformed joints |
| Dystrophic epidermolysis bullosa | Breakdown of collagen fibrils that attach skin layers to each other | Stretchy, easily scarred skin; lax joints |
| Hereditary osteoarthritis | Substituted amino acid in collagen chain alters shape | Painful joints |
| Osteogenesis imperfecta type I | Too few collagen triple helices | Easily broken bones; deafness; blue sclera (whites of the eyes) |
| Stickler syndrome | Short collagen chains | Joint pain; degeneration of retina and fluid around it |

TABLE 5.6

## COMPONENTS OF CONNECTIVE TISSUE

| Component | Characteristic | Function |
|-----------|----------------|----------|
| Fibroblasts | Widely distributed, large, star-shaped cells | Secrete proteins that become fibers |
| Macrophages | Motile cells sometimes attached to fibers | Clear foreign particles from tissues by phagocytosis |
| Mast cells | Large cells, usually located near blood vessels | Release substances that may help prevent blood clotting and promote inflammation |
| Collagenous fibers (white fibers) | Thick, threadlike fibers of collagen with great tensile strength | Hold structures together |
| Elastic fibers (yellow fibers) | Bundles of microfibrils embedded in elastin | Provide elastic quality to parts that stretch |
| Reticular fibers | Thin fibers of collagen | Form supportive networks within tissues |

Ground substance

Fibroblast

Elastic fiber

Collagenous fiber

**FIGURE 5.18**

Loose connective tissue, or areolar tissue, contains numerous fibroblasts that produce collagenous and elastic fibers (250× micrograph enlarged to 1,000×).

1. What are the general characteristics of connective tissue?

2. What are the major types of fixed cells in connective tissue?

3. What is the primary function of fibroblasts?

4. What are the characteristics of collagen and elastin?

## Loose Connective Tissue

**Loose connective tissue,** or **areolar tissue,** forms delicate, thin membranes throughout the body. The cells of this tissue, mainly fibroblasts, are located some distance apart and are separated by a gel-like ground substance that contains many collagenous and elastic fibers that fibroblasts secrete (fig. 5.18).

**FIGURE 5.19**

Adipose tissue cells contain large fat droplets that push the nuclei close to the cell membranes (75× micrograph enlarged to 300×).

Cell membrane
Nucleus
Fat droplet

Collagenous fibers
White blood cell
Fibroblast

**FIGURE 5.20**

Reticular connective tissue is a network of thin collagenous fibers, which contains numerous fibroblasts and white blood cells (250× micrograph enlarged to 1,000×).

Loose connective tissue binds the skin to the underlying organs and fills spaces between muscles. It lies beneath most layers of epithelium, where its many blood vessels nourish nearby epithelial cells.

## Adipose Tissue

**Adipose tissue,** or fat, is a specialized form of connective tissue. Certain cells within connective tissue (adipocytes) store fat in droplets within their cytoplasm. At first these cells resemble fibroblasts, but as they accumulate fat, they enlarge, and their nuclei are pushed to one side (fig. 5.19). When adipocytes become so abundant that they crowd out other cell types, they form adipose tissue. This tissue lies beneath the skin, in spaces between muscles, around the kidneys, behind the eyeballs, in certain abdominal membranes, on the surface of the heart, and around certain joints.

Adipose tissue cushions joints and some organs, such as the kidneys. It also insulates beneath the skin, and it stores energy in fat molecules.

A person is born with a certain number of fat cells. Because excess food calories are likely to be converted to fat and stored, the amount of adipose tissue in the body reflects diet. During a period of fasting, adipose cells may lose their fat droplets, shrink, and become more like fibroblasts again.

Infants and young children have a continuous layer of adipose tissue just beneath the skin, which gives their bodies a rounded appearance. In adults, this subcutaneous fat thins in some regions and remains thick in others. For example, in males adipose tissue usually thickens in the upper back, arms, lower back, and buttocks; in females it is more likely to develop in the breasts, buttocks, and thighs.

## Reticular Connective Tissue

**Reticular connective tissue** is composed of thin, collagenous fibers in a three-dimensional network. It supports the walls of certain internal organs, such as the liver, spleen, and lymphatic organs (fig. 5.20).

Fibroblasts

Collagenous fibers

**FIGURE 5.21**

Regular dense connective tissue consists largely of tightly packed collagenous fibers (250× micrograph enlarged to 1,000×).

Collagenous fibers

Elastic fibers

Fibroblast

**FIGURE 5.22**

Elastic connective tissue contains many elastic fibers with collagenous fibers between them (170× micrograph enlarged to 680×).

## Dense Connective Tissue

**Dense connective tissue** consists of many closely packed, thick, collagenous fibers, a fine network of elastic fibers, and a few cells, most of which are fibroblasts. Subclasses of this tissue are regular or irregular, according to how organized the fiber patterns are.

Collagenous fibers of regular dense connective tissue are very strong, enabling the tissue to withstand pulling forces (fig. 5.21). It often binds body parts together, as parts of tendons and ligaments. The blood supply to regular dense connective tissue is poor, slowing tissue repair. This is why a sprain, which damages tissues surrounding a joint, may take considerable time to heal.

Fibers of irregular dense connective tissue are thicker, interwoven, and more randomly organized. This allows the tissue to sustain tension exerted from many different directions. Irregular dense connective tissue is found in the dermis, the inner skin layer.

## Elastic Connective Tissue

**Elastic connective tissue** mainly consists of yellow, elastic fibers in parallel strands or in branching networks. Between these fibers are collagenous fibers and fibroblasts. This tissue is found in the attachments between vertebrae of the spinal column (ligamenta flava). It is also in the layers within the walls of certain hollow internal organs, including the larger arteries, some portions of the heart, and the larger airways, where it imparts an elastic quality (fig. 5.22).

**1** Differentiate between loose connective tissue and dense connective tissue.

**2** What are the functions of adipose tissue?

**3** Distinguish between reticular and elastic connective tissues.

Chondrocyte
in lacuna

Nucleus

Intercellular
material

**FIGURE 5.23**

Hyaline cartilage cells (chondrocytes) are located in lacunae, which are in turn surrounded by intercellular material containing very fine collagenous fibers (160× micrograph enlarged to 640×).

Elastic fibers

Intercellular
material

Nucleus

Chondrocyte

Lacuna

**FIGURE 5.24**

Elastic cartilage contains many elastic fibers in its intercellular material (400× micrograph enlarged to 1,600×).

## Cartilage

**Cartilage** is a rigid connective tissue. It provides support, frameworks, attachments, protects underlying tissues, and forms structural models for many developing bones.

Cartilage matrix is abundant and is largely composed of collagenous fibers embedded in a gel-like ground substance. This ground substance is rich in a protein-polysaccharide complex (chondromucoprotein) and contains a large amount of water. Cartilage cells, or **chondrocytes,** occupy small chambers called *lacunae,* and thus are completely within the matrix.

A cartilaginous structure is enclosed in a covering of connective tissue called *perichondrium.* Although cartilage tissue lacks a direct blood supply, blood vessels are in the surrounding perichondrium. Cartilage cells near the perichondrium obtain nutrients from these vessels by diffusion, which is aided by the water in the matrix. This lack of a direct blood supply is why torn cartilage heals slowly, and why chondrocytes do not reproduce frequently.

The three types of cartilage are distinguished by their different types of intercellular material. Hyaline cartilage has very fine collagenous fibers in its matrix, elastic cartilage contains a dense network of elastic fibers, and fibrocartilage has many large collagenous fibers.

**Hyaline cartilage** (fig. 5.23), the most common type, looks somewhat like white glass. It is found on the ends of bones in many joints, in the soft part of the nose, and in the supporting rings of the respiratory passages. Parts of an embryo's skeleton begin as hyaline cartilage "models" that bone gradually replaces. Hyaline cartilage is also important in the growth of most bones and in repair of bone fractures (see chapter 7).

**Elastic cartilage** (fig. 5.24) is more flexible than hyaline cartilage because its matrix contains many elastic fibers. It provides the framework for the external ears and parts of the larynx.

**Fibrocartilage** (fig. 5.25), a very tough tissue, contains many collagenous fibers. It is a shock absorber for structures that are subjected to pressure. For example, fibrocartilage forms pads (intervertebral disks) between the individual bones (vertebrae) of the spinal column. It also cushions bones in the knees and in the pelvic girdle.

Chondrocyte in lacuna

Nucleus

Collagenous fiber

Intercellular material

**FIGURE** 5.25

Fibrocartilage contains many large collagenous fibers in its intercellular material (450× micrograph enlarged to 1,800×).

Canaliculi

Osteocyte in lacuna

Osteonic canal

Lamellae

Nucleus

Osteocyte

Cell process in canaliculus

**FIGURE** 5.26

Bone matrix is deposited in concentric layers (lamellae) around osteonic canals (160×).

## Bone

**Bone** (osseous tissue) is the most rigid connective tissue. Its hardness is largely due to mineral salts, such as calcium phosphate and calcium carbonate, in its matrix. This intercellular material also contains a great amount of collagen, whose fibers flexibly reinforce the mineral components of bone.

Bone internally supports body structures. It protects vital structures in the cranial and thoracic cavities, and is an attachment for muscles. Bone also contains red marrow, which forms blood cells, and it stores and releases inorganic salts.

Bone matrix is deposited by bone cells (osteocytes) in thin layers called *lamellae,* which form concentric patterns around tiny longitudinal tubes called *osteonic (Haversian) canals.* Osteocytes are located in lacunae that are rather evenly spaced between the lamellae. Consequently, osteocytes also form concentric circles (fig. 5.26).

In a bone, the osteocytes and layers of intercellular material, which are concentrically clustered around an

osteonic canal, form a cylinder-shaped unit called an **osteon** (Haversian system). Many of these units cemented together form the substance of bone (see chapter 7).

Each osteonic canal contains a blood vessel, so that every bone cell is fairly close to a nutrient supply. In addition, the bone cells have many cytoplasmic processes that extend outward and pass through minute tubes in the matrix called *canaliculi.* Gap junctions attach these cellular processes to the membranes of nearby cells (see chapter 3). As a result, materials can move rapidly between blood vessels and bone cells. Thus, in spite of its inert appearance, bone is a very active tissue. Injured bone heals much more rapidly than does injured cartilage.

## Blood

**Blood,** another type of connective tissue, is composed of cells that are suspended in a fluid intercellular matrix called *blood plasma.* These cells include red blood cells (transport gases), white blood cells (fight disease), and cellular fragments called *platelets* (involved in blood clotting). Most blood cells form in special tissues (hematopoietic tissues) in red marrow within the hollow parts of certain bones (fig. 5.27). Blood is described in chapter 14.

Of the blood cells, only the red cells function entirely within the blood vessels. White blood cells typically migrate from the blood through capillary walls. They enter connective tissues where they carry on their major activities, and they usually reside there until they die. Table 5.7 lists the characteristics of the types of connective tissue.

Red blood cells

Intercellular fluid (plasma)

White blood cells

Platelet

**FIGURE 5.27**

Blood tissue consists of red blood cells, white blood cells, and platelets suspended in an intercellular fluid (200× micrograph enlarged to 800×).

| TABLE 5.7 | CONNECTIVE TISSUES | | |
|---|---|---|---|
| **Type** | **Description** | **Function** | **Location** |
| Loose connective tissue | Cells in fluid-gel matrix | Binds organs together, holds tissue fluids | Beneath the skin, between muscles, beneath epithelial tissues |
| Adipose tissue | Cells in fluid-gel matrix | Protects, insulates, and stores fat | Beneath the skin, around the kidneys, behind the eyeballs, on the surface of the heart |
| Reticular connective tissue | Cells in fluid-gel matrix | Supports | Walls of liver, spleen, and lymphatic organs |
| Dense connective tissue | Cells in fluid-gel matrix | Binds organs together | Tendons, ligaments, dermis |
| Elastic connective tissue | Cells in fluid-gel matrix | Provides elastic quality | Connecting parts of the spinal column, in walls of arteries and airways |
| Hyaline cartilage | Cells in solid-gel matrix | Supports, protects, provides framework | Ends of bones, nose, and rings in walls of respiratory passages |
| Elastic cartilage | Cells in solid-gel matrix | Supports, protects, provides flexible framework | Framework of external ear and part of larynx |
| Fibrocartilage | Cells in solid-gel matrix | Supports, protects, absorbs shock | Between bony parts of spinal column, parts of pelvic girdle, and knee |
| Bone | Cells in solid matrix | Supports, protects, provides framework | Bones of skeleton, middle ear |
| Blood | Cells and platelets in fluid matrix | Transports gases, defends against disease, clotting | Within blood vessels |

1. Describe the general characteristics of cartilage.

2. Explain why injured bone heals more rapidly than does injured cartilage.

3. What are the major components of blood?

## MUSCLE TISSUES

### General Characteristics

**Muscle tissues** are contractile; that is, their elongated cells, or *muscle fibers,* can shorten and thicken. As they contract, muscle fibers pull at their attached ends, which moves body parts. The three types of muscle tissue (skeletal, smooth, and cardiac) are discussed in chapter 9.

### Skeletal Muscle Tissue

**Skeletal muscle tissue** (fig. 5.28) forms muscles that usually attach to bones and that we control by conscious effort. For this reason, it is often called *voluntary* muscle tissue. The long and threadlike cells of skeletal muscle have alternating light and dark cross-markings called *striations.* Each cell has many nuclei (multinucleate). A message from a nerve cell can stimulate a muscle fiber to contract by causing protein filaments within the muscle fiber to slide past one another. Then, the muscle fiber relaxes. Skeletal muscles move the head, trunk, and limbs and enable us to make facial expressions, write, talk, and sing, as well as chew, swallow, and breathe.

### Smooth Muscle Tissue

**Smooth muscle tissue** (fig. 5.29) is called smooth because its cells lack striations. This tissue comprises the walls of hollow internal organs, such as the stomach, intestines, urinary bladder, uterus, and blood vessels. Unlike skeletal muscle, smooth muscle usually cannot be stimulated to contract by conscious efforts. Thus, it is a type of *involuntary* muscle tissue.

Smooth muscle cells are shorter than those of skeletal muscle, and each has a single, centrally located nucleus. The cells are spindle-shaped. Smooth muscle tissue moves food through the digestive tract, constricts blood vessels, and empties the urinary bladder.

Striations

Muscle fiber

Nuclei

**FIGURE 5.28**
Skeletal muscle tissue is composed of striated muscle fibers that contain many nuclei (250× micrograph enlarged to 750×).

Cell membrane

Cytoplasm

Nucleus

**FIGURE 5.29**
Smooth muscle tissue consists of spindle-shaped cells, each with a large nucleus (250× micrograph enlarged to 900×).

Striations

Nucleus

Intercalated disk

**FIGURE 5.30**

Cardiac muscle fibers are branched and interconnected, with a single nucleus each (400×).

## Cardiac Muscle Tissue

**Cardiac muscle tissue** (fig. 5.30) is only in the heart. Its cells, which are striated, are joined end-to-end. The resulting muscle fibers are branched and connected in complex networks. Each cell within a cardiac muscle fiber has a single nucleus. Where it touches another cell is a specialized intercellular junction called an *intercalated disk,* seen only in cardiac tissue.

Cardiac muscle, like smooth muscle, is controlled involuntarily and, in fact, can continue to function without being stimulated by nerve impulses. This tissue makes up the bulk of the heart and pumps blood through the heart chambers and into blood vessels.

The cells of different tissues vary greatly in their abilities to reproduce. Epithelial cells of the skin and the inner lining of the digestive tract, and the connective tissue cells that form blood cells in red bone marrow, continuously reproduce. However, striated and cardiac muscle cells and nerve cells do not usually reproduce at all after differentiating.

Fibroblasts respond rapidly to injuries by increasing in numbers and increasing fiber production. They are often the principal agents of repair in tissues that have limited abilities to regenerate. For instance, cardiac muscle tissue typically degenerates in the regions damaged by a heart attack. Such tissue may be replaced by connective tissue formed by fibroblasts, which later appears as a scar.

**1** List the general characteristics of muscle tissue.

**2** Distinguish among skeletal, smooth, and cardiac muscle tissues.

## NERVOUS TISSUES

**Nervous tissues** are found in the brain, spinal cord, and peripheral nerves. The basic cells are called *nerve cells,* or *neurons,* and they are among the more highly specialized body cells. Neurons sense certain types of changes in their surroundings. They respond by transmitting nerve impulses along cytoplasmic extensions (nerve fibers) to other neurons or to muscles or glands (fig. 5.31). As a result of the extremely complex patterns by which neurons connect with each other and with muscle and gland cells, they can coordinate, regulate, and integrate many body functions.

In addition to neurons, nervous tissue includes **neuroglial cells.** These cells support and bind the components of nervous tissue, carry on phagocytosis, and help supply nutrients to neurons by connecting them to blood vessels. They may also play a role in cell-to-cell communications. Nervous tissue is discussed in chapter 10. Table 5.8 summarizes the general characteristics of muscle and nervous tissues. Clinical Application 5.2 discusses bioengineered tissues.

**1** Describe the general characteristics of nervous tissue.

**2** Distinguish between neurons and neuroglial cells.

# Tissue Engineering

If an automobile or appliance part is damaged or malfunctions, replacing it is fairly simple. Not so for the human body, because the immune system presents a formidable barrier to introducing foreign tissue. Immunosuppressant drugs help, but they raise the risk of infection and cancer by dampening aspects of the immune response other than transplant rejection. A solution to the challenge of replacing body parts is tissue engineering, which combines synthetic materials with cells.

The basic recipe for a bioengineered tissue is to place cells in or on a scaffolding sculpted from a synthetic material that is accepted in the body. The cells secrete substances as they normally would, or they may be genetically altered to overproduce their natural secreted products or supply entirely different ones with therapeutic benefit, such as growth factors that might make the implant more acceptable to the body.

## REPLACEMENT BLOOD VESSELS

A semi-synthetic replacement blood vessel is a flexible yet strong tube of collagen (fig. 5A). An external layer of pig collagen provides strength and handling characteristics, and an inner layer of cow collagen provides a smooth surface. This biosynthetic structure does not evoke an immune response and serves as a scaffold for remodeling by the body's own cells. These bioengineered blood vessels can be manufactured to any length or width, and they withstand pressures comparable to that of the blood racing through the circulatory system. The vessels can be stitched next to natural blood vessels so smoothly that blood clots do not form nearly as easily as they do in completely synthetic grafts. The vessels can be used in cardiac bypass surgery, to replace damaged arteries in the legs, and to replace brain arteries damaged by a stroke.

**FIGURE 5A**

This replacement blood vessel is biosynthetic. It is fabricated from collagen, a natural biomaterial. It does not evoke an immune response when used to replace a damaged blood vessel. It is remodeled by the body's own cells and functions much more like a natural blood vessel than a synthetic graft.

*(continued)*

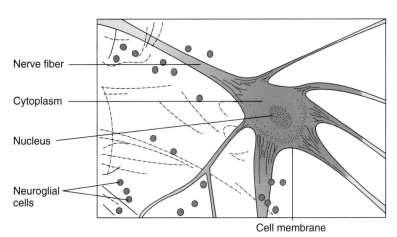

Nerve fiber

Cytoplasm

Nucleus

Neuroglial cells

Cell membrane

**FIGURE 5.31**

A nerve cell with nerve fibers extending into its surroundings (50× micrograph enlarged to 300×).

# Tissue Engineering (continued)

### NEW SKIN AND MORE

Burn patients can sometimes be helped by a bioengineered skin consisting of the patient's epidermal cells placed in sheets over dermal cells grown in culture. A nylon mesh framework supports both layers. This semisynthetic skin may also be useful for patients who have lost a great deal of skin in surgery to remove tattoos, cancers, and moles. Bioengineered skin is also used for in vitro toxicity testing. In many laboratories it has replaced live animals in testing cosmetic ingredients.

Other tissues on the bioengineering drawing board include liver, connective tissue, and bone marrow. A replacement cartilage similar to the skin recipe, consisting of chondrocytes in collagen, may help replace joints destroyed by arthritis.

A scaled-down version of an engineered tissue, called a cell implant, offers a new route to drug delivery, placing cells that naturally manufacture vital substances precisely where a patient needs them. The cells are "immunoisolated" because their packaging enables them to secrete without alerting the immune system. This is done by surrounding cells with a polymer membrane with holes small enough to allow nutrients in and the therapeutic biochemicals out, while excluding the larger molecules that trigger immune rejection.

Prime candidates for cell implants are pancreatic beta cells, which would secrete insulin to aid people with diabetes mellitus, and brain implants that would secrete dopamine, providing the biochemical that is missing in people who have Parkinson's disease (fig. 5B). ■

Leaked dopamine

Dopamine-secreting cell

Implant

**FIGURE 5B**

People who have Parkinson's disease lack the neurotransmitter dopamine in certain parts of their brains, but cannot receive dopamine directly because it does not cross the blood-brain barrier. A cell implant may one day help them. The dark, central portion of this micrograph shows a cross section of such an implant (100×). Containing thousands of dopamine-secreting cells, the implant continuously delivers therapeutic amounts of dopamine directly to the brain. The cells in the implant, as well as surrounding brain tissue, stain brown, indicating dopamine production. The implant shown here measures approximately 1 cm long and is about the diameter of a thin pencil lead.

**TABLE 5.8  MUSCLE AND NERVOUS TISSUES**

| Type | Description | Function | Location | Type | Description | Function | Location |
|------|-------------|----------|----------|------|-------------|----------|----------|
| Skeletal muscle tissue | Long, threadlike cells, striated, many nuclei | Voluntary movements of skeletal parts | Muscles usually attached to bones | Cardiac muscle tissue | Branched cells, striated, single nucleus | Heart movements | Heart muscle |
| Smooth muscle tissue | Shorter cells, single, central nucleus | Involuntary movements of internal organs | Walls of hollow internal organs | Nervous tissue | Cell with cytoplasmic extensions | Sensory reception and conduction of nerve impulses | Brain, spinal cord, and peripheral nerves |

# Introduction (page 136)

Cells are organized in layers or groups to form tissues. Intercellular materials, varying from solid to liquid, separate cells. The four major types of human tissue are epithelial tissues, connective tissues, muscle tissues, and nervous tissues.

# Epithelial Tissues (page 136)

1. General characteristics
   a. Epithelial tissue covers all free body surfaces and is the major tissue of glands.
   b. A basement membrane anchors epithelium to connective tissue. Epithelial tissue lacks blood vessels, contains little intercellular material, and is continuously replaced.
   c. It functions in protection, secretion, absorption, and excretion.
2. Simple squamous epithelium
   a. This tissue consists of a single layer of thin, flattened cells through which substances pass easily.
   b. It functions in the exchange of gases in the lungs and lines blood vessels, lymph vessels, and membranes within the thorax and abdomen.
3. Simple cuboidal epithelium
   a. This tissue consists of a single layer of cube-shaped cells.
   b. It carries on secretion and absorption in the kidneys and various glands.
4. Simple columnar epithelium
   a. This tissue is composed of elongated cells whose nuclei are near the basement membrane.
   b. It lines the uterus and digestive tract, where it functions in protection, secretion, and absorption.
   c. Absorbing cells often have microvilli.
   d. This tissue usually contains goblet cells that secrete mucus.
5. Pseudostratified columnar epithelium
   a. This tissue appears stratified because the nuclei are at two or more levels.
   b. Its cells may have cilia that move mucus or egg cells over the surface of the tissue.
   c. It lines tubes of the respiratory and reproductive systems.
6. Stratified squamous epithelium
   a. This tissue is composed of many layers of cells, the topmost of which are flattened.
   b. It protects underlying cells from harmful environmental effects.
   c. It covers the skin and lines the oral cavity, throat, vagina, and anal canal.
7. Stratified cuboidal epithelium
   a. This tissue is composed of two or three layers of cube-shaped cells.
   b. It lines the larger ducts of the sweat glands, salivary glands, and pancreas.
   c. It functions in protection.
8. Stratified columnar epithelium
   a. The top layer of cells in this tissue contains elongated columns. Cube-shaped cells make up the bottom layers.
   b. It is in the vas deferens, part of the male urethra, and parts of the pharynx.
   c. This tissue functions in protection and secretion.
9. Transitional epithelium
   a. This tissue is specialized to become distended.
   b. It is in the walls of organs of the urinary tract.
   c. It helps prevent the contents of the urinary passageways from diffusing out.
10. Glandular epithelium
    a. Glandular epithelium is composed of cells that are specialized to secrete substances.
    b. A gland consists of one or more cells.
       (1) Exocrine glands secrete into ducts.
       (2) Endocrine glands secrete into tissue fluid or blood.
    c. Glands are classified according to the organization of their cells.
       (1) Simple glands have unbranched ducts.
       (2) Compound glands have branched ducts.
       (3) Tubular glands consist of simple epithelial-lined tubes.
       (4) Alveolar glands consist of saclike dilations connected to the surface by narrowed ducts.
    d. Exocrine glands are classified according to composition of their secretions.
       (1) Merocrine glands secrete watery fluids without loss of cytoplasm. Most secretory cells are merocrine.
          (a) Serous cells secrete watery fluid with a high enzyme content.
          (b) Mucous cells secrete mucus.
       (2) Apocrine glands lose portions of their cells during secretion.
       (3) Holocrine glands release cells filled with secretions.

# Connective Tissues (page 144)

1. General characteristics
   a. Connective tissue connects, supports, protects, provides frameworks, fills spaces, stores fat, produces blood cells, protects against infection, and helps repair damaged tissues.
   b. Connective tissue cells usually have considerable intercellular material between them.
   c. This intercellular matrix consists of fibers and a ground substance.
2. Major cell types
   a. Fibroblasts produce collagenous and elastic fibers.
   b. Macrophages are phagocytes.
   c. Mast cells may release heparin and histamine, and usually are near blood vessels.
3. Connective tissue fibers
   a. Collagenous fibers are composed of collagen and have great tensile strength.
   b. Elastic fibers are composed of microfibrils embedded in elastin and are very elastic.
   c. Reticular fibers are very fine collagenous fibers.
4. Loose connective tissue
   a. This tissue forms thin membranes between organs and binds them.

b. It is beneath the skin and between muscles.
c. Its intercellular spaces contain tissue fluid.
5. Adipose tissue
   a. Adipose tissue is a specialized form of connective tissue that stores fat, cushions, and insulates.
   b. It is found beneath the skin, in certain abdominal membranes, and around the kidneys, heart, and various joints.
6. Reticular connective tissue
   a. This tissue largely consists of thin, branched collagenous fibers.
   b. It supports the walls of the liver, spleen, and lymphatic organs.
7. Dense connective tissue
   a. This tissue is largely composed of strong, collagenous fibers that bind structures.
   b. Regular dense connective tissue is found in tendons and ligaments, whereas irregular tissue is found in the dermis.
8. Elastic connective tissue
   a. This tissue is mainly composed of elastic fibers.
   b. It imparts an elastic quality to the walls of certain hollow internal organs such as the lungs and blood vessels.
9. Cartilage
   a. Cartilage provides a supportive framework for various structures.
   b. Its intercellular material is composed of fibers and a gel-like ground substance.
   c. It lacks a direct blood supply and is slow to heal.
   d. Cartilaginous structures are enclosed in a perichondrium, which contains blood vessels.
   e. Major types are hyaline cartilage, elastic cartilage, and fibrocartilage.
   f. Cartilage is at the ends of various bones, in the ear, in the larynx, and in pads between bones of the spinal column, pelvic girdle, and knees.
10. Bone
    a. The intercellular matrix of bone contains mineral salts and collagen.
    b. Its cells usually form concentric circles around osteonic canals. Canaliculi connect them.
    c. It is an active tissue that heals rapidly.
11. Blood
    a. Blood is composed of cells suspended in fluid.
    b. Blood cells are formed by special tissue in the hollow parts of certain bones.

## Muscle Tissues (page 153)
1. General characteristics
   a. Muscle tissue contracts, moving structures that are attached to it.
   b. Three types are skeletal, smooth, and cardiac muscle tissues.
2. Skeletal muscle tissue
   a. Muscles containing this tissue usually attach to bones and are controlled by conscious effort.
   b. Cells or muscle fibers are long and threadlike, containing several nuclei, with alternating light and dark cross-markings.
   c. Muscle fibers contract when stimulated by nerve impulses, then immediately relax.
3. Smooth muscle tissue
   a. This tissue of spindle-shaped cells, each with one nucleus, is in walls of hollow internal organs.
   b. Usually it is involuntarily controlled.
4. Cardiac muscle tissue
   a. This tissue is found only in the heart.
   b. Cells, each with a single nucleus, are joined by intercalated disks and form branched networks.
   c. Cardiac muscle tissue is involuntarily controlled.

## Nervous Tissues (page 154)
1. Nervous tissue is in the brain, spinal cord, and peripheral nerves.
2. Neurons
   a. Neurons sense changes and respond by transmitting nerve impulses to other neurons or to muscles or glands.
   b. They coordinate, regulate, and integrate body activities.
3. Neuroglial cells
   a. Some of these cells bind and support nervous tissue.
   b. Others carry on phagocytosis.
   c. Still others connect neurons to blood vessels.
   d. Some are involved in cell-to-cell communication.

## CRITICAL THINKING QUESTIONS

1. Joints such as the elbow, shoulder, and knee contain considerable amounts of cartilage and loose connective tissue. How does this explain the fact that joint injuries are often very slow to heal?
2. Disorders of collagen are characterized by deterioration of connective tissues. Why would you expect such diseases to produce widely varying symptoms?
3. Sometimes, in response to irritants, mucous cells secrete excess mucus. What symptoms might this produce if it occurred in (a) the respiratory passageways or (b) the digestive tract?
4. Tissue engineering combines living cells with synthetic materials to create functional substitutes for human tissues. What components would you use to engineer replacement (a) skin, (b) blood, (c) bone, and (d) muscle?
5. Collagen and elastin are added to many beauty products. What type of tissue are they normally part of?
6. In the lungs of smokers, a process called metaplasia occurs where the normal lining cells of the lung are replaced by squamous metaplastic cells (many layers of squamous epithelial cells). Functionally, why is this an undesirable body reaction to tobacco smoke?
7. Cancer-causing agents (carcinogens) usually act on cells that are reproducing. Which of the four tissues would carcinogens most influence? Least influence?

1. Define *tissue*.
2. Name the four major types of tissue found in the human body.
3. Describe the general characteristics of epithelial tissues.
4. Distinguish between simple epithelium and stratified epithelium.
5. Explain how the structure of simple squamous epithelium provides its function.
6. Name an organ that includes each of the following tissues, and give the function of the tissue:
   a. Simple squamous epithelium
   b. Simple cuboidal epithelium
   c. Simple columnar epithelium
   d. Pseudostratified columnar epithelium
   e. Stratified squamous epithelium
   f. Stratified cuboidal epithelium
   g. Stratified columnar epithelium
   h. Transitional epithelium
7. Define *gland*.
8. Distinguish between an exocrine gland and an endocrine gland.
9. Explain how glands are classified according to the structure of their ducts and the organization of their cells.
10. Explain how glands are classified according to the nature of their secretions.
11. Distinguish between a serous cell and a mucous cell.
12. Describe the general characteristics of connective tissue.
13. Describe three major types of connective tissue cells.
14. Distinguish between collagen and elastin.
15. Explain the difference between loose connective tissue and dense connective tissue.
16. Explain how the quantity of adipose tissue in the body reflects diet.
17. Distinguish between regular and irregular dense connective tissues.
18. Distinguish between elastic and reticular connective tissues.
19. Explain why injured loose connective tissue and cartilage are usually slow to heal.
20. Name the major types of cartilage, and describe their differences and similarities.
21. Describe how bone cells are organized in bone tissue.
22. Explain how bone cells receive nutrients.
23. Describe the solid components of blood.
24. Describe the general characteristics of muscle tissues.
25. Distinguish among skeletal, smooth, and cardiac muscle tissues.
26. Describe the general characteristics of nervous tissue.
27. Distinguish between neurons and neuroglial cells.

**For More Information Related to This Chapter, Please Visit** *Hole's Human Anatomy and Physiology* **Home Page**
http://www.mhhe.com/biosci/abio/

# SKIN AND THE INTEGUMENTARY SYSTEM

## UNDERSTANDING WORDS

**alb-**, white: *alb*inism—condition characterized by a lack of pigment.

**cut-**, skin: sub*cut*aneous—beneath the skin.

**derm-**, skin: *derm*is—inner layer of the skin.

**epi-**, upon, after, in addition: *epi*dermis—outer layer of the skin.

**follic-**, small bag: hair *follic*le—tubelike depression in which a hair develops.

**hol-**, entire, whole: *hol*ocrine gland—gland discharges the entire cell containing the secretion.

**kerat-**, horn: *kerat*in—protein produced as epidermal cells die and harden.

**melan-**, black: *melan*in—dark pigment produced by certain cells.

**por-**, passage, channel: *por*e—opening by which a sweat gland communicates to the skin's surface.

**seb-**, grease: *seb*aceous gland—gland that secretes an oily substance.

## CHAPTER OBJECTIVES

AFTER YOU HAVE STUDIED THIS CHAPTER, YOU SHOULD BE ABLE TO:

1. Describe the four major types of membranes.

2. Describe the structure of the layers of the skin.

3. List the general functions of each layer of the skin.

4. Describe the accessory organs associated with the skin.

5. Explain the functions of each accessory organ.

6. Explain how the skin helps regulate body temperature.

7. Summarize the factors that determine skin color.

## KEY TERMS

**conduction** (kon-duk'shun)
**convection** (kon-vek'shun)
**cutaneous membrane** (ku-ta'ne-us mem'brān)
**dermis** (der'mis)
**epidermis** (ep"i-der'mis)
**evaporation** (e-vap"o-ra'shun)
**hair follicle** (hār fol'i-kl)
**integumentary** (in-teg-u-men'tar-e)

**keratinization** (ker"ah-tin"ĭ-za'shun)
**melanin** (mel'ah-nin)
**mucous membrane** (mu'kus mem'brān)
**sebaceous gland** (se-ba'shus gland)
**serous membrane** (se'rus mem'brān)
**subcutaneous** (sub"ku-ta'ne-us)
**sweat gland** (swet gland)

We often do not appreciate an organ system until it malfunctions. Torn muscles, pinched nerves, and broken bones are common conditions that make us aware of our organ systems. Textbook sales representative Amy Porter survived a rather unusual experience that painfully revealed the importance of the integumentary system. She literally lost her skin, and as a result, nearly lost her life.

In October 1995, ten days after taking a course of antibiotic drugs, Amy developed a rash that quickly turned to painful blisters covering her entire body. In the emergency room, a dermatologist diagnosed a severe form of Stevens-Johnson syndrome called toxic epidermal necrolysis. In short, Amy's entire epidermis (outer skin layer) was peeling off, as if she had suffered devastating burns everywhere.

Amy survived, thanks to tremendous courage and a skin bank at the University of Texas at Galveston. For several weeks, thousands of staples fastened donor skin to her body. Amy endured painful daily baths. Gradually the donor skin grafts wore off as new skin grew from beneath. Sedation enabled Amy to withstand the pain.

Losing one's epidermis is life-threatening, providing a route to infection and disrupting regulation of body temperature. Only one in 2 million people develop Stevens-Johnson syndrome; of those, only 10% survive. Because of her severe case, Amy's odds were even worse. But she made it.

---

Two or more kinds of tissues grouped together and performing specialized functions constitute an **organ.** Thus membranes, which are thin, sheetlike structures that are usually composed of epithelial and connective tissues and covering body surfaces and lining body cavities, are organs (fig. 6.1). The cutaneous membrane (skin) together with various accessory organs make up the **integumentary organ system.**

## TYPES OF MEMBRANES

The four major types of membranes are *serous, mucous, synovial,* and *cutaneous.* Usually these structures are thin. Serous, mucous, and cutaneous membranes are composed of epithelial tissue and some underlying connective tissue; synovial membranes are composed entirely of connective tissues.

**Serous membranes** line the body cavities that lack openings to the outside. They form the inner linings of the thorax and abdomen, and they cover the organs within these cavities (see fig. 1.9). A serous membrane consists of a layer of simple squamous epithelium (mesothelium) and a thin layer of loose connective tissue. Cells of a serous membrane secrete watery *serous fluid,* which helps lubricate the surfaces of the membrane.

**Mucous membranes** line the cavities and tubes that open to the outside of the body. These include the oral and nasal cavities, and the tubes of the digestive, respiratory, urinary, and reproductive systems. A mucous membrane consists of epithelium overlying a layer of loose connective tissue; however, the type of epithelium varies with the location of the membrane. For example, stratified squamous epithelium lines the oral cavity, pseudostratified columnar epithelium lines part of the nasal cavity, and simple columnar epithelium lines the small intestine. Specialized cells within a mucous membrane secrete *mucus.*

**Synovial membranes** form the inner linings of joint cavities between the ends of bones at freely movable

**FIGURE** 6.1

An organ, such as the skin, is composed of several kinds of tissues (30×).

Stratified squamous epithelium

Irregular dense connective tissue

Glandular epithelium

Adipose tissue

joints (synovial joints). These membranes usually include loose connective tissue with elastic fibers and some adipose tissue. Cells of a synovial membrane secrete a thick, colorless *synovial fluid* into the joint cavity, which lubricates the ends of the bones within the joint. (Synovial joints are described in chapter 8.)

The **cutaneous membrane** is an organ of the integumentary organ system and is more commonly called *skin.* It is described in detail in this chapter.

## SKIN AND ITS TISSUES

The skin is one of the larger and more versatile organs of the body, and it is vital in maintaining homeostasis. It is a protective covering that prevents many harmful substances, including microorganisms, from entering the body. Skin also retards water loss by diffusion from

Epidermis

Dermis

Subcutaneous layer

Hair shaft

Sweat gland pore

Stratum corneum

Capillary

Stratum basale

Dermal papilla

Basement membrane

Touch receptor

Sweat gland duct

Sebaceous gland

Arrector pili muscle

Hair follicle

Sweat gland

Nerve fiber

Adipose cells

Blood vessels

**FIGURE 6.2**
A section of skin.

deeper tissues, and helps regulate body temperature. It houses sensory receptors, contains immune system cells, synthesizes various chemicals, and excretes small quantities of waste.

The skin includes two distinct tissue layers. The outer layer, called the **epidermis,** is composed of stratified squamous epithelium. The inner layer, or **dermis,** is thicker than the epidermis, and it is made up of connective tissue containing collagen and elastic fibers, epithelial tissue, smooth muscle tissue, nervous tissue, and blood. A basement membrane that is anchored to the dermis by short fibrils separates the two skin layers.

Beneath the dermis, masses of loose connective and adipose tissues bind the skin to underlying organs. These tissues form the **subcutaneous layer** or **hypodermis** (fig. 6.2).

**1** Name the four types of membranes, and explain how they differ.

**2** List the general functions of the skin.

**3** Name the tissue in the outer layer of the skin.

**4** Name the tissues in the inner layer of the skin.

An inherited condition called *epidermolysis bullosa* (EB) destroys the vital integrity of the skin's layered organization. Symptoms include very easy blistering and scarring. Different types of EB reflect the specific proteins affected. In EB simplex, blisters form only on the hands and feet, and usually only during warm weather. EB simplex is an abnormality in the protein keratin in epidermal cells. In the severe dystrophic form, collagen fibers that anchor the dermis to the epidermis are abnormal, causing the layers to separate, forming many blisters. The basement membrane form of EB is so severe that it causes death in infancy. It is a defect in epiligrin, a protein that anchors the epidermis to the basement membrane.

**FIGURE 6.3**

(*a*) The layers of the epidermis are distinguished by changes in cells as they are pushed toward the surface of the skin.
(*b*) Micrograph from the palm of the hand (50×).

## Epidermis

Since the epidermis is composed of stratified squamous epithelium, it lacks blood vessels. However, the deepest layer of epidermal cells, called the *stratum basale,* is close to the dermis, and is nourished by dermal blood vessels. Cells of the stratum basale can reproduce and grow. As new cells enlarge, they push the older epidermal cells away from the dermis toward the surface of the skin. The farther the cells travel, the poorer their nutrient supply becomes, and, in time, they die.

> In psoriasis, a chronic skin disease, cells in the epidermis divide seven times more frequently than normal. Excess cells accumulate, forming bright red patches covered with silvery scales, which are keratinized cells. The anticancer drug methotrexate is sometimes used to treat severe cases. Five million people in the United States and 2% of all people worldwide have psoriasis.

The cell membranes of the older cells (keratinocytes) thicken and develop many desmosomes that fasten them to adjacent cells (see chapter 3). At the same time, the cells begin to harden (keratinize), with strands of tough, fibrous, waterproof keratin protein synthesized and stored within the cell. As a result, many layers of tough, tightly packed dead cells accumulate in the epidermis, forming an outermost layer called the *stratum corneum.* The dead cells that compose it are eventually shed, such as when the skin is rubbed briskly with a towel.

The structural organization of the epidermis varies from region to region. It is thickest on the palms of the hands and the soles of the feet, where it may be 0.8–1.4 mm thick. In most areas, only four layers can be distinguished. They are the *stratum basale* (stratum germinativum, or basal cell layer), which is the deepest layer; the *stratum spinosum,* a thick layer; the *stratum granulosum,* a granular layer; and the *stratum corneum,* a fully keratinized layer (horny layer). An additional layer, the *stratum lucidum* (between the stratum granulosum and the stratum corneum) is in the thickened skin of the palms and soles. The cells of these layers change shape as they are pushed toward the surface (fig. 6.3).

**TABLE 6.1** | **LAYERS OF THE EPIDERMIS**

| Layer | Location | Characteristics |
|-------|----------|-----------------|
| Stratum corneum | Outermost layer | Many layers of keratinized, dead epithelial cells that are flattened and nonnucleated |
| Stratum lucidum | Between stratum corneum and stratum granulosum on soles and palms | Cells appear clear; nuclei, organelles, and cell membranes are no longer visible |
| Stratum granulosum | Beneath the stratum corneum | Three to five layers of flattened granular cells that contain shrunken fibers of keratin and shriveled nuclei |
| Stratum spinosum | Beneath the stratum granulosum | Many layers of cells with centrally located, large, oval nuclei and developing fibers of keratin; cells becoming flattened |
| Stratum basale (basal cell layer) | Deepest layer | A single row of cuboidal or columnar cells that reproduce and grow; this layer also includes melanocytes |

Contact dermatitis is superficial inflammation (redness and swelling), or irritation of the skin. In allergic contact dermatitis, the immune system reacts to an allergen (an innocuous substance recognized as foreign), causing a red scaliness. The rash resulting from exposure to oils in poison ivy is an example of allergic contact dermatitis; 50 to 70% of people with this allergy also react to poison oak and sumac, mango peel, gingko fruit, and an oil in cashew shells. Metals in jewelry, acids in fruits, and materials in shoes also trigger allergic contact dermatitis. It is also seen among workers in certain fields, such as hairdressers, butchers, furniture makers, shrimp peelers, and bakers.

Irritant contact dermatitis is damage caused by an irritating substance, not an immune system reaction (allergy). The skin becomes red and itchy, with small, oozing blisters. Babies are famous for skin irritations— caused by everything from the perpetual drool on their faces to their wet diapers. "Dishpan hands" and reactions to cosmetics are common examples of irritant contact dermatitis. Men with outbreaks on the left sides of their necks may use an aftershave lotion that reacts with sunlight when they drive.

In body regions other than the palms and soles, the epidermis is usually much thinner, averaging 0.07–0.12 mm. The stratum granulosum may be missing where the epidermis is thin. Table 6.1 describes the characteristics of each layer of the epidermis.

In healthy skin, production of epidermal cells is closely balanced with loss of dead cells from the stratum corneum, so that skin does not wear away completely. In fact, the rate of cellular reproduction increases where the skin is rubbed or pressed regularly, causing growth of thickened areas called *calluses* on the palms and soles, and keratinized conical masses on the toes called *corns*. Other changes in the skin include the common rashes described in table 6.2.

Specialized cells in the epidermis called **melanocytes** produce the dark pigment melanin that provides skin color, discussed further on page 175 (fig. 6.4). Melanin absorbs light energy and in this way helps protect deeper cells from the damaging effects of ultraviolet radiation in sunlight.

Because blood vessels in the dermis supply nutrients to the epidermis, interference with blood flow may kill epidermal cells. For example, when a person lies in one position for a prolonged period, the weight of the body pressing against the bed blocks the skin's blood supply. If cells die, the tissues begin to break down (necrosis), and a pressure ulcer (also called a decubitus ulcer or bedsore) may appear.

Pressure ulcers usually occur in the skin overlying bony projections, such as on the hip, heel, elbow, or shoulder. Frequently changing body position or massaging the skin to stimulate blood flow in regions associated with bony prominences can prevent ulcers. In the case of a paralyzed person who cannot feel pressure or respond to it by shifting position, caretakers must turn the body often to prevent pressure ulcers.

Melanocytes lie in the deepest portion of the epidermis and in the underlying connective tissue of the dermis. Although they are the only cells that can produce melanin, the pigment also may be present in nearby epidermal cells. This happens because melanocytes have long, pigment-containing cellular extensions that pass upward between neighboring epidermal cells, and the extensions can transfer granules of melanin into these other cells by a process called *cytocrine secretion*. Nearby epidermal cells may contain more melanin than the melanocytes (fig. 6.5). Clinical Application 6.1 discusses one consequence of excessive sun exposure—skin cancer.

TABLE 6.2 RASHES

| Illness | Description of Rash | Cause |
|---------|---------------------|-------|
| Chicken pox | Tiny pustules start on back, chest, or scalp and spread for three to four days. Pustules form blisters, then crust, then fall away. | *Herpes varicella* |
| Fifth disease | Begins with "slapped cheek" appearance, then red spots suddenly cover entire body, lasting up to two days. | *Human parvovirus B19* |
| Impetigo | Thin-walled blisters and thick, crusted lesions. | *Staphylococcus aureus, Streptococcus pyogenes* |
| Lyme disease | Large rash resembling a bull's-eye usually on thighs or trunk. | *Borrelia burgdorferi* |
| Rosacea | Flushing leads to sunburned appearance in center of face. Red pimples and then wavy red lines develop. | Unknown, but may be a microscopic mite living in hair follicles |
| Roseola infantum | Following high fever, red spots suddenly cover entire body, lasting up to two days. | *Herpesvirus 6* |
| Scarlet fever | Rash resembling sunburn with goose bumps begins below ears, on chest and underarms, and spreads to abdomen, limbs, and face. Skin may peel. | *Group A Streptococcus* |
| Shingles | Small, clear blisters appear on inflamed skin. Blisters enlarge, become cloudy, crust, then fall off. | The virus that causes chicken pox hides in peripheral nerves, affecting the area where the nerve endings reach the skin. |

**FIGURE 6.4**

Melanocytes (arrows) that are mainly in the stratum basale at the deepest layer of the epidermis produce the pigment melanin (160×).

(a)

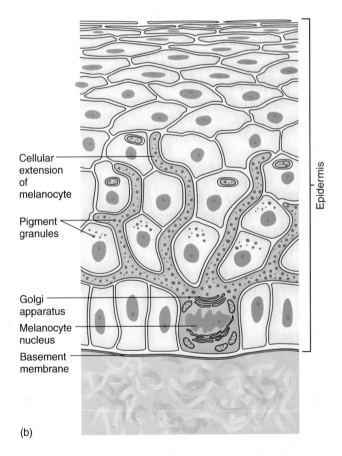

(b)

**FIGURE 6.5**

(a) Transmission electron micrograph of a melanocyte with pigment-containing granules (10,000×). (b) A melanocyte may have pigment-containing extensions that pass between epidermal cells.

# Skin Cancer

Like cigarette smoking, a deep, dark tan was once very desirable. A generation ago, a teenager might have spent hours on a beach, skin glistening with oil, maybe even using a reflecting device to concentrate sun exposure on the face. Today, as they lather on sunblock, many of these people realize that the tans of yesterday may cause cancer tomorrow.

Just four hours of unprotected sunbathing has immediate and lasting effects on the skin. The ultraviolet radiation alters collagen and elastin and dilates blood vessels in the dermis. A few days later, the outer skin layer may blister and peel. Cells that peel have undergone programmed cell death, which removes cells that have possibly turned cancerous from environmental exposure. People with the inherited disorder xeroderma pigmentosum lack DNA repair enzymes, and must be completely covered by clothing and sunblock to avoid developing skin cancers (fig. 6A).

Skin cancer usually arises from nonpigmented epithelial cells within the deep layer of the epidermis or from pigmented melanocytes. Skin cancers originating from epithelial cells are called *cutaneous carcinomas* (basal cell carcinoma or squamous cell carcinoma); those arising from melanocytes are *cutaneous melanomas* (melanocarcinomas or malignant melanomas) (fig. 6B).

Cutaneous carcinomas are the most common type of skin cancer, most frequently occurring in light-skinned people over forty years of age. These cancers usually appear in persons who are regularly exposed to sunlight, such as farmers, sailors, athletes, and sunbathers.

Cutaneous carcinoma often develops from hard, dry, scaly growths (lesions) that have reddish bases. Such lesions may be either flat or raised, and they firmly adhere to the skin, appearing most often on the neck, face, and scalp. Fortunately, cutaneous carcinomas are typically slow growing and can usually be cured completely by surgical removal or radiation treatment.

Cutaneous melanomas are pigmented with melanin, often with a variety of colored areas—variegated brown, black, gray, or blue. They usually have irregular rather than smooth outlines (fig. 6B).

**FIGURE 6A**

This child has xeroderma pigmentosum. Sun exposure causes extreme freckling, and skin cancer is likely to develop because he lacks DNA repair enzymes. The large lesion on his chin is a skin cancer.

---

**1** Explain how the epidermis is formed.

**2** What factors help prevent loss of body fluids through the skin?

**3** What is the function of melanin?

## Dermis

The surface between the epidermis and dermis is usually uneven. This is because the epidermis has ridges projecting inward and the dermis has conical *dermal papillae* passing into the spaces between the ridges (see fig. 6.2).

Fingerprints are due to these undulations of the skin seen at the distal end of the palmar surface of a finger. Fingerprints are used for purposes of identification, because they are individually unique. Fingerprint pattern is genetically determined, and the prints form during fetal existence. However, during a certain time early in development, when the fetus moves, the print pattern can change. Because no two fetuses move exactly alike, even the fingerprints of identical twins are slightly different.

Aging skin changes, particularly in regions exposed to sunlight. The rate and degree of change vary from person to person. Typically, the skin becomes dry, scaly, thin, and more translucent with age. Connective tissue fibers degenerate, leading to loss of elasticity and tensile strength, followed by wrinkling.

Cutaneous melanomas may appear in people of any age, and seem to be caused by short, intermittent exposure to high-intensity sunlight. Thus, risk of melanoma increases in persons who stay indoors but occasionally sustain blistering sunburns.

Cutaneous melanomas most often affect light-skinned people who burn rather than tan. The cancer usually appears in the skin of the trunk, especially the back, or the limbs, arising from normal-appearing skin or from a mole (nevus). The lesion spreads horizontally through the skin, but eventually may thicken and grow downward into the skin, invading deeper tissues. If the melanoma is surgically removed while it is in its horizontal growth phase, it may be arrested. Once it thickens and spreads into deeper tissues, unfortunately, it becomes difficult to treat. The survival rate with conventional treatment is very low. However, a form of experimental gene therapy injects genes directly into melanoma cells. These genes direct the cells to produce surface proteins that attract the immune system to attack the cancer.

Incidence of melanoma has been increasing rapidly for the past twenty years. To reduce the chances of occurrence, avoid exposure to high-intensity sunlight, use sunscreens and sunblocks, and examine the skin regularly. Report any unusual lesions—particularly those that change in color, shape, or surface texture—to a physician.

Replacements for natural suntanning may be ineffective or dangerous. "Sunless tanning agents" do not tan the skin at all, but merely dye it temporarily. "Tan accelerators," according to manufacturers, are nutrients that supposedly increase melanin synthesis in the sunlight, but there is no scientific proof that they work. Tanning booths may be dangerous. Even those claiming to be safe because they emit only partial ultraviolet radiation may cause skin cancer.

Enjoy the sun—but protect yourself! ■

**FIGURE 6B**
A cutaneous malignant melanoma.

The dermis binds the epidermis to the underlying tissues. It is largely composed of irregular dense connective tissue that includes tough collagenous fibers and elastic fibers in a gel-like ground substance. Networks of these fibers give the skin toughness and elasticity. On the average, the dermis is 1.0–2.0 mm thick; however, it may be as thin as 0.5 mm or less on the eyelids, or as thick as 3.0 mm on the soles of the feet.

The dermis also contains muscle fibers. Some regions, such as the skin that encloses the testes (scrotum), contain many smooth muscle cells that can wrinkle the skin when they contract. Other smooth muscles in the dermis are associated with accessory organs such as hair follicles and glands. Many striated muscle fibers are anchored to the dermis in the skin of the face. They help produce the voluntary movements associated with facial expressions.

Nerve fibers are scattered throughout the dermis. Motor fibers carry impulses to dermal muscles and glands, and sensory fibers carry impulses away from specialized sensory receptors, such as touch receptors (see fig. 6.2).

One type of dermal sensory receptor (Pacinian corpuscles) is stimulated by heavy pressure, whereas another type (Meissner's corpuscles) senses light touch. Still other receptors respond to temperature changes or to factors that can damage tissues. Sensory receptors are discussed in chapter 12. The dermis also contains blood vessels, hair follicles, sebaceous glands, and sweat glands, which are discussed later in the chapter.

## Subcutaneous Layer

The subcutaneous layer (hypodermis) beneath the dermis consists of loose connective and adipose tissues (see fig. 6.2). The collagenous and elastic fibers of this layer are continuous with those of the dermis. Most of these fibers parallel the surface of the skin, extending in all directions. As a result, no sharp boundary separates the dermis and the subcutaneous layer.

The adipose tissue of the subcutaneous layer insulates, helping to conserve body heat and impeding the entrance of heat from the outside. The amount of adipose tissue varies greatly with each individual's nutritional condition. It also varies in thickness from one region to another. For example, adipose tissue is usually thick over the abdomen, but absent in the eyelids.

The subcutaneous layer contains the major blood vessels that supply the skin. Branches of these vessels form a network (rete cutaneum) between the dermis and the subcutaneous layer. They, in turn, give off smaller vessels that supply the dermis above and the underlying adipose tissue.

---

Subcutaneous injections are administered through a hollow needle into the subcutaneous layer beneath the skin. Intradermal injections are injected within the skin. Subcutaneous injections and intramuscular injections (administered into muscles) are sometimes called hypodermic injections.

Some substances are administered through the skin by means of an adhesive transdermal patch that includes a small reservoir containing a drug. The drug passes from the reservoir through a permeable membrane at a known rate. It then diffuses into the epidermis and enters the blood vessels of the dermis. Transdermal patches are commonly used to protect against motion sickness, chest pain associated with heart disease, and elevated blood pressure. A transdermal patch that delivers nicotine is used to help people stop smoking.

---

**1** What kinds of tissues make up the dermis?

**2** What are the functions of these tissues?

**3** What are the functions of the subcutaneous layer?

---

## ACCESSORY ORGANS OF THE SKIN

Accessory organs of the skin extend downward from the epidermis and include hair follicles, nails, and skin glands. As long as accessory organs remain intact, severely burned or injured dermis can regenerate.

## Hair Follicles

Hair is present on all skin surfaces except the palms, soles, lips, nipples, and parts of the external reproductive organs; however, it is not always well developed. For example, hair on the forehead is usually very fine.

Each hair develops from a group of epidermal cells at the base of a tubelike depression called a **hair follicle.** This follicle extends from the surface into the dermis and contains the hair *root*. The epidermal cells at its base are nourished from dermal blood vessels in a projection of connective tissue (hair papilla) at the deep end of the follicle. As these epidermal cells reproduce and grow, older cells are pushed toward the surface. The cells that move upward and away from the nutrient supply become keratinized and die. Their remains constitute the structure of a developing hair, whose *shaft* extends away from the skin surface. In other words, a hair is composed of dead epidermal cells (figs. 6.6 and 6.7).

---

*Folliculitis* is an inflammation of the hair follicles in response to bacterial infection. The condition can be picked up in dirty swimming pools or hot tubs. One woman got a severe case by repeatedly using a loofah sponge containing bacteria.

---

Usually a hair grows for a time and then rests while it remains anchored in its follicle. Later a new hair begins to grow from the base of the follicle, and the old hair is pushed outward and drops off. Sometimes, however, the hairs are not replaced. When this occurs in the scalp, the result is baldness, described in Clinical Application 6.2.

Genes determine hair color by directing the type and amount of pigment epidermal melanocytes produce. For example, dark hair has much more melanin than blond hair. The white hair of a person with albinism lacks melanin altogether. Bright red hair contains an iron pigment (trichosiderin) that is not in hair of any other color. A mixture of pigmented and unpigmented hair usually appears gray.

A bundle of smooth muscle cells, forming the *arrector pili muscle* (see figs. 6.2 and 6.6), attaches to each hair follicle. This muscle is positioned so that a short hair within the follicle stands on end when the muscle contracts. If a person is upset or very cold, nerve impulses may stimulate the arrector pili muscles to contract, raising gooseflesh or goose bumps. Each hair follicle also has associated with it one or more sebaceous (oil-producing) glands, discussed later in the chapter.

---

Some interesting hair characteristics are inherited. The direction of a cowlick is inherited, with a clockwise whorl being more common than a counterclockwise whorl. A white forelock, and hairy ears, elbows, nose tip, or palms are also inherited.

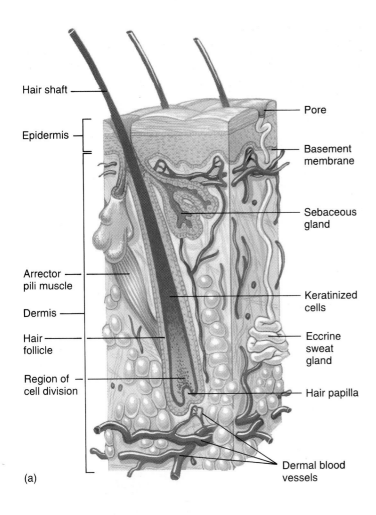

Hair shaft

Epidermis

Arrector pili muscle

Dermis

Hair follicle

Region of cell division

Pore

Basement membrane

Sebaceous gland

Keratinized cells

Eccrine sweat gland

Hair papilla

Dermal blood vessels

(a)

Dermal tissue

Hair follicle

Hair root

Region of cell division

Hair papilla

Adipose tissue

(b)

**FIGURE** 6.6

(a) A hair grows from the base of a hair follicle when epidermal cells reproduce and older cells move outward and become keratinized. (b) A light micrograph of a hair follicle (160×).

Keratinized cells of hair shaft

Squamous cells of epidermis

**FIGURE** 6.7

A scanning electron micrograph of a hair emerging from the epidermis (340× micrograph enlarged to 900×).

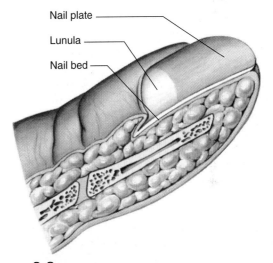

Nail plate

Lunula

Nail bed

**FIGURE** 6.8

Nails grow from epithelial cells that reproduce and become keratinized in the lunula.

## Nails

**Nails** are protective coverings on the ends of the fingers and toes. Each nail consists of a *nail plate* that overlies a surface of skin called the *nail bed*. Specialized epithelial cells that are continuous with the epithelium of the skin produce the nail bed. The whitish, thickened, half-moon–shaped region (lunula) at the base of a nail plate is the most active growing region. The epithelial cells here reproduce, and the newly formed cells are keratinized. This gives rise to tiny, keratinized scales that become part of the nail plate, pushing it forward over the nail bed. In time, the plate extends beyond the end of the nail bed and with normal use gradually wears away (fig. 6.8).

# Hair Loss

A healthy person loses from twenty to 100 hairs a day as part of the normal growth cycle of hair. A hair typically grows for two to six years, rests for two to three months, then falls out. A new hair grows in its place. At any time, 90% of hair is in the growth phase.

In the United States, about 57.5 million people have some degree of baldness. Pattern baldness, in which the top of the head loses hair, affects 35 million men and 20 million women. The women tend to be past menopause, when lowered amounts of the hormone estrogen contribute to hair loss, which is more even on the scalp than it is in men. Pattern baldness is called *androgenic alopecia*, because it is associated with testosterone, an androgenic (male) hormone. About 2.5 million people have an inherited condition called *alopecia areata,* in which the body manufactures antibodies that attack the hair follicles. This results in oval bald spots in mild cases, but complete loss of scalp and body hair in severe cases.

Various conditions can cause temporary hair loss. Lowered estrogen levels shortly before and after giving birth may cause a woman's hair to fall out in clumps. Taking birth control pills, cough medications, certain antibiotics, vitamin A derivatives, antidepressants, and many other medications can also cause temporary hair loss. A sustained high fever may prompt hair loss six weeks to three months later.

Many people losing their hair seek treatment (fig. 6C). One treatment is minoxidil (Rogaine), a drug originally used to lower high blood pressure. It was approved as a treatment for baldness in men in 1988 and in women in 1991. Rogaine causes new hair to grow in 10 to 14% of cases, but in 90% of people it slows hair loss. However, when a person stops taking it, any new hair falls out.

Hair transplants move hair follicles from a hairy part of a person's body to a bald part, and they are successful. Several other approaches, however, are potentially damaging—to the wallet as well as the scalp. Suturing on hair pieces often leads to scarring and infection. In 1984, the Food and Drug Administration banned hair implants of high-density artificial fibers, because they too become easily infected. Products called "thinning hair supplements" are conditioners, often found in ordinary shampoo, that merely make hair feel thicker. They are generally concoctions of herbs and the carbohydrate polysorbate. Labels claim the product "releases hairs trapped in the scalp." ■

**FIGURE 6C**
Being bald can be beautiful, but many people with hair loss seek ways to grow hair.

---

Nail appearance mirrors health. Bluish nail beds may reflect a circulatory problem. A white nail bed or oval depressions in a nail can indicate anemia. A pigmented spot under a nail that isn't caused by an injury may be a melanoma. Horizontal furrows may result from a period of serious illness, or indicate malnutrition. Certain disorders of the lungs, heart, or liver may cause extreme curvature of the nails. Red streaks in noninjured nails may be traced to rheumatoid arthritis, ulcers, or hypertension.

## Skin Glands

**Sebaceous glands** (see fig. 6.2) contain groups of specialized epithelial cells and are usually associated with hair follicles. They are holocrine glands (see chapter 5), and their cells produce globules of a fatty material that accumulate, swelling and bursting the cells. The resulting mixture of fatty material and cellular debris is called *sebum*.

Sebum is secreted into hair follicles through short ducts and helps keep the hairs and the skin soft, pliable, and waterproof (fig. 6.9). Acne results from excess sebum secretion (Clinical Application 6.3).

Hair  Hair follicle  Sebaceous gland

**FIGURE** 6.9

A sebaceous gland secretes sebum into a hair follicle (shown here in cross section: 175× micrograph enlarged to 300×).

Epidermis

Dermis

Eccrine sweat gland

**FIGURE** 6.10

Light micrograph of an eccrine sweat gland (30×).

Sebaceous glands are scattered throughout the skin, but are not on the palms and soles. In some regions, such as the lips, the corners of the mouth, and parts of the external reproductive organs, sebaceous glands open directly to the surface of the skin rather than being connected to hair follicles.

**Sweat glands** (sudoriferous glands) are widespread in the skin. Each gland consists of a tiny tube that originates as a ball-shaped coil in the deeper dermis or superficial subcutaneous layer. The coiled portion of the gland is closed at its deep end and is lined with sweat-secreting epithelial cells. The most numerous sweat glands, called **eccrine glands,** respond throughout life to body temperature elevated by environmental heat or physical exercise (figs. 6.10 and 6.11). These glands are common on the forehead, neck, and back, where they produce profuse sweat on hot days or during intense physical activity. They also cause the moisture that appears on the palms and soles when a person is emotionally stressed.

With advancing age, sweat gland activity decreases, and in the very old, fibrous tissues may replace sweat glands. As a result, the person becomes less able to control body temperature. Sebaceous gland activity also lessens with age, so that the skin of elderly people tends to be dry.

The fluid the eccrine sweat glands secrete is carried by a tube (duct) that opens at the surface as a *pore.* Sweat is mostly water, but it also contains small quantities of salts and wastes, such as urea and uric acid. Thus, sweating is also an excretory function.

The secretions of certain sweat glands, called **apocrine glands,** develop a scent as they are metabolized by

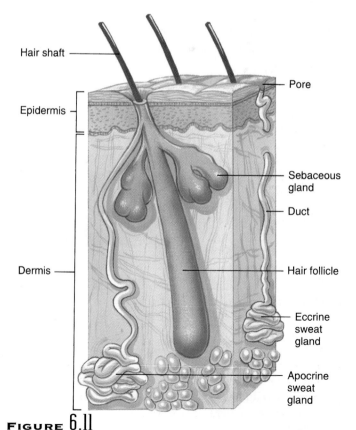

Hair shaft

Pore

Epidermis

Sebaceous gland

Duct

Dermis

Hair follicle

Eccrine sweat gland

Apocrine sweat gland

**FIGURE** 6.11

Note the difference in location of the ducts of the eccrine and apocrine sweat glands.

# Acne

Many young people are all too familiar with *acne vulgaris,* a disorder of the sebaceous glands. Excess sebum and squamous epithelial cells clog the glands, producing blackheads and whiteheads (comedones). The blackness is not dirt, but results from the accumulated cells blocking light. In addition, the clogged sebaceous gland provides an attractive environment for anaerobic bacteria that signals the immune system to trigger inflammation. The inflamed, raised area is a pimple (pustule).

## A HORMONAL PROBLEM

Acne is the most common skin disease, affecting 80% of people at some time between the ages of 11 and 30. It is largely hormonally induced. Just before puberty, the adrenal glands increase production of androgens, which stimulate increased secretion of sebum. At puberty, sebum production surges again. Acne usually develops because the sebaceous glands are extra responsive to androgens, but in some cases, androgens may be produced in excess.

Acne can cause skin blemishes far more serious than the perfect models in acne medication ads depict (fig. 6D). Scarring from acne can lead to emotional problems. Fortunately, several highly effective treatments are available.

## WHAT TO DO—AND NOT DO

Acne is not caused by uncleanliness or eating too much chocolate or greasy food. Although cleansing products containing soaps, detergents, or astringents can remove surface sebum, they do not stop the flow of oil that contributes to acne.

Abrasive products are actually harmful, because they irritate the skin and increase inflammation.

Several acne treatments are available, but most take weeks to months to work. Women with acne are sometimes prescribed birth control pills, because the estrogens counter androgen excess. Other drugs with estrogenic effects are available in Europe but not in the United States or Canada. Isotretinoin is a derivative of vitamin A that helps nearly all people achieve relief or even permanent cures, but has several side effects and causes birth defects. Certain antibiotics can treat acne by clearing bacteria from sebaceous glands. Topical treatments include tretinoin (another vitamin A derivative), salicylic acid (an aspirin solution), and benzoyl peroxide.

Treatment for severe acne requires a doctor's care. Drug combinations are tailored to the severity of the condition (table 6A). ■

**FIGURE** 6D

Acne can affect one's appearance.

| TABLE 6A | ACNE TREATMENTS (BY INCREASING SEVERITY) | |
|---|---|
| **Condition** | **Treatment** |
| Noninflammatory comedonal acne (blackheads and whiteheads) | Topical tretinoin or salicylic acid |
| Papular inflammatory acne | Topical antibiotic |
| Widespread blackheads and pustules | Topical tretinoin and systemic antibiotic |
| Severe cysts | Systemic isotretinoin |
| Explosive acne (ulcerated lesions, fever, joint pain) | Systemic corticosteroids |

TABLE 6.3  SKIN GLANDS

| Type | Description | Function | Location |
|---|---|---|---|
| Sebaceous glands | Groups of specialized epithelial cells | Keep hair soft, pliable, waterproof | Near or connected to hair follicles, everywhere but on palms and soles |
| Eccrine sweat glands | Abundant sweat glands with odorless secretion | Lower body temperature | Originate in deep dermis or subcutaneous layer and open to surface on forehead, neck, and back |
| Apocrine sweat glands | Less numerous sweat glands with secretions that develop odors | Wet skin during pain, fear, emotional upset, and sexual arousal | Near hair follicles in armpit, groin, around nipples |
| Ceruminous glands | Modified sweat glands | Secrete earwax | External ear canal |
| Mammary glands | Modified sweat glands | Secrete milk | Breasts |

skin bacteria (fig 6.11). (Although they are currently called apocrine, these glands secrete by the same mechanism as eccrine glands—see chapter 3.) Apocrine sweat glands become active at puberty, and can wet certain areas of the skin when a person is emotionally upset, frightened, or in pain. Apocrine sweat glands are also active during sexual arousal. In adults, the apocrine glands are most numerous in axillary regions, the groin, and the area around the nipples. Ducts of these glands open into hair follicles.

Other sweat glands are structurally and functionally modified to secrete specific fluids, such as the ceruminous glands of the external ear canal that secrete ear wax and the female mammary glands that secrete milk (see chapter 22). Table 6.3 summarizes skin glands.

1 Explain how a hair forms.

2 What causes gooseflesh?

3 What is the function of the sebaceous glands?

4 How does the composition of a fingernail differ from that of a hair?

5 Describe the locations of the sweat glands.

6 How do the functions of eccrine sweat glands and apocrine sweat glands differ?

## REGULATION OF BODY TEMPERATURE

The regulation of body temperature is vitally important because even slight shifts can disrupt the rates of metabolic reactions. Normally, the temperature of deeper body parts remains close to a set point of 37°C (98.6°F). The maintenance of a stable temperature requires that the amount of heat the body loses be balanced by the amount it produces. The skin plays a key role in the homeostatic mechanism that regulates body temperature.

## Heat Production and Loss

Because cellular metabolism releases heat, the most active cells are the major heat producers. These include skeletal and heart muscle cells, and the cells of certain glands, such as the liver.

In intense heat, nerve impulses stimulate structures in the skin and other organs to release heat. For example, during physical exercise, active muscles release heat, which the blood carries away. The warmed blood reaches the part of the brain (the hypothalamus) that controls the body's temperature set point, which signals muscles in the walls of specialized dermal blood vessels to relax. As the vessels dilate (vasodilation), more blood enters them and some of the heat the blood carries escapes to the outside. At the same time, deeper blood vessels contract (vasoconstriction), diverting blood to the surface, and the skin reddens. The heart is stimulated to beat faster, moving more blood out of the deeper regions.

The primary means of body heat loss is **radiation,** by which *infrared heat rays* escape from warmer surfaces to cooler surroundings. These rays radiate in all directions, much like those from the bulb of a heat lamp.

Conduction and convection release lesser amounts of heat. In **conduction,** heat moves from the body directly into the molecules of cooler objects in contact with its surface. For example, heat is lost by conduction into the seat of a chair when a person sits down. The heat loss continues as long as the chair is cooler than the body surface touching it.

Heat is also lost by conduction to the air molecules that contact the body. As air becomes heated, it moves away from the body, carrying heat with it, and is replaced by cooler air moving toward the body. This type of continuous circulation of air over a warm surface is **convection.**

Still another means of body heat loss is **evaporation.** When the body temperature rises above normal, the nervous system stimulates eccrine sweat glands to release sweat onto the surface of the skin. As this fluid evaporates, it carries heat away from the surface, cooling the skin.

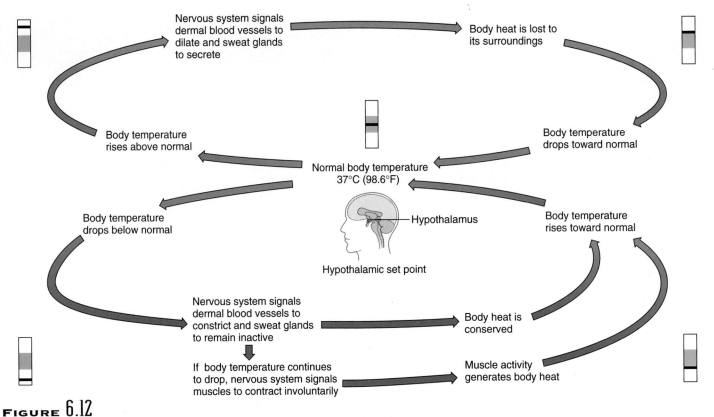

**FIGURE 6.12**

Body temperature regulation is an example of homeostasis.

*(Figure labels, clockwise from top:)*

Nervous system signals dermal blood vessels to dilate and sweat glands to secrete

Body heat is lost to its surroundings

Body temperature drops toward normal

Body temperature rises above normal

Normal body temperature 37°C (98.6°F)

Hypothalamus

Hypothalamic set point

Body temperature rises toward normal

Body temperature drops below normal

Nervous system signals dermal blood vessels to constrict and sweat glands to remain inactive

Body heat is conserved

Muscle activity generates body heat

If body temperature continues to drop, nervous system signals muscles to contract involuntarily

---

With excessive heat loss, the brain triggers a different set of responses in skin structures. Muscles in the walls of dermal blood vessels are stimulated to contract; this decreases the flow of heat-carrying blood through the skin, which tends to lose color, and helps reduce heat loss by radiation, conduction, and convection. At the same time, sweat glands remain inactive, decreasing heat loss by evaporation. If the body temperature continues to drop, the nervous system may stimulate muscle fibers in the skeletal muscles throughout the body to contract slightly. This action requires an increase in the rate of cellular respiration and produces heat as a by-product. If this response does not raise the body temperature to normal, small groups of muscles may contract rhythmically with still greater force, causing the person to shiver, and thus generating more heat. Figure 6.12 summarizes the body's temperature-regulating mechanism, and Clinical Application 6.4 examines two causes of elevated body temperature.

## Problems in Temperature Regulation

The body's temperature-regulating mechanism does not always operate satisfactorily, and the consequences may be dangerous. For example, air can hold only a limited amount of water vapor, so that on a hot, humid day the air may become nearly saturated with water. At such times, the sweat glands may be activated, but the sweat cannot quickly evaporate. The skin becomes wet, but the person remains hot and uncomfortable. Body temperature may rise, a condition called hyperthermia. In addi-tion, if the air temperature is high, heat loss by radiation is less effective. In fact, if the air temperature exceeds body temperature, the person may gain heat from the surroundings, elevating body temperature even higher.

> Because body temperature regulation depends, in large part, on evaporation of sweat from the skin's surface, and because high humidity hinders evaporation, athletes are advised to slow down their activities on hot, humid days. On these days, they should also stay out of the sunlight whenever possible and drink enough fluids to avoid dehydration. Such precautions can prevent the symptoms of *heat exhaustion,* which include fatigue, dizziness, headache, muscle cramps, and nausea.

Hypothermia, or lowered body temperature, can result from prolonged exposure to cold, or as part of an illness. It can be extremely dangerous. Hypothermia begins with shivering and a feeling of coldness, but if not treated, progresses to mental confusion, lethargy, loss of reflexes and consciousness, and, eventually, a shutting down of major organs. If the temperature in the body's core drops just a few degrees, fatal respiratory failure or heart arrhythmia may result. However, the extremities can withstand drops of 20 to 30°F below normal.

Certain people are at higher risk for developing hypothermia. These include the very young and the very old, very thin individuals, and the homeless. Hypother-

# Elevated Body Temperature

It was a warm June morning when the harried and hurried father strapped his 5-month-old son Bryan into the backseat of his car and headed for work. Tragically, the father forgot to drop his son off at the babysitter's. When his wife called him at work late that afternoon to inquire why the child was not at the sitter's, the shocked father realized his mistake and hurried down to his parked car. But it was too late—Bryan had died. Left for 10 hours in the car in the sun, all windows shut, the baby's temperature had quickly soared. Two hours after he was discovered, the child's temperature still exceeded 41°C (106°F).

Sarah L.'s case of elevated body temperature was more typical. She awoke with a fever of 40°C (104°F) and a terribly painful sore throat. At the doctor's office, a test revealed that Sarah had a *Streptococcus* infection. The fever was her body's attempt to fight the infection.

The true cases of young Bryan and Sarah illustrate two reasons why body temperature may rise— inability of the temperature homeostatic mechanism to handle an extreme environment, and an immune system response to infection. In the case of Bryan, sustained exposure to very high heat overwhelmed the temperature-regulating mechanism, resulting in hyperthermia. Body heat built up faster than it could dissipate, and body temperature rose, even though the set point of the thermostat was normal. His blood vessels dilated so greatly in an attempt to dissipate the excess heat that after a few hours his circulatory system collapsed.

Sarah's fever was a special case of hyperthermia, in which molecules on the surfaces of the infectious agents stimulated phagocytes to release a substance called interleukin-1 (also called endogenous pyrogen, meaning "fire maker from within"). The bloodstream carried IL-1 to the hypothalamus, where it raised the set point controlling temperature. In response, the brain signalled skeletal muscles to increase heat production, blood flow to the skin to decrease, and sweat glands to decrease secretion. As a result, body temperature rose to the new set point, and Sarah had a fever. The increased body temperature helped her immune system kill the pathogens—and made her quite uncomfortable for a while. This discomfort caused Sarah to be inactive, thus conserving energy for the immune response.

What should one do when body temperature rises? Hyperthermia in response to exposure to intense, sustained heat should be rapidly treated by administering liquids to replace lost body fluids and electrolytes, sponging the skin with water to increase cooling by evaporation, and covering the person with a refrigerated blanket. Fever can be lowered with ibuprofen or acetaminophen, or aspirin in adults. However, some health professionals believe that a slightly elevated temperature should not be reduced (with medication or cold baths), because it may be part of a normal immune response. ∎

---

mia can be prevented by dressing appropriately and staying active in the cold. A person suffering from hypothermia must be warmed gradually so that respiratory and cardiovascular functioning remain stable.

> Hypothermia is intentionally induced during certain surgical procedures involving the heart or central nervous system (brain or spinal cord). In heart surgery, for example, body temperature may be lowered to between 78°F (26°C) and 89°F (32°C). The cooling lowers the body's metabolic rate so that less oxygen is required. Hypothermia for surgery is accomplished by packing the patient in ice, or by removing blood, cooling it, and returning it to the body.

1. Why is regulation of body temperature so important?
2. How is body heat produced?
3. How does the body lose heat when heat production is excessive?
4. How does the skin help regulate body temperature?
5. What are the dangers of hypothermia?

## SKIN COLOR

Heredity and the environment determine skin color.

### Genetic Factors

Regardless of racial origin, all people have about the same number of melanocytes in their skin. Differences in skin color result from differences in the amount of melanin these cells produce, which is genetically controlled. The more melanin, the darker the skin. The distribution and the size of pigment granules within melanocytes also influence skin color. The granules in very dark skin are single and large; those in lighter skin occur in clusters of two to four granules and are smaller. People who inherit mutant melanin genes have nonpigmented skin, a condition called *albinism*. It affects people of all races and also many other species (fig. 6.13).

Among the general U.S. population, only 1 in 20,000 people has albinism. Among the native Hopi people in Arizona, however, the incidence is 1 in 200. The reason for this is as much sociological as it is biological. Men with albinism help the women rather than risk severe sunburn in the fields with the other men. They disproportionately contribute to the next generation because they have more frequent contact with women.

## Environmental Factors

Environmental factors such as sunlight, ultraviolet light from sunlamps, or X rays also affect skin color. These factors rapidly darken existing melanin, and they stimulate melanocytes to produce more pigment and transfer it to nearby epidermal cells within a few days. This is why sunbathing results in skin tanning. Unless exposure to sunlight continues, however, the tan fades as pigmented epidermal cells become keratinized and wear away.

## Physiological Factors

Blood in the dermal vessels adds color to the skin. For example, when blood is well oxygenated, the blood pigment hemoglobin is bright red, making the skins of light-complexioned people appear pinkish. On the other hand, when the blood oxygen concentration is low, hemoglobin is dark red, and the skin appears bluish—a condition called *cyanosis*.

The state of the blood vessels also affects skin color. If the vessels are dilated, more blood enters the dermis, and the skin of a light-complexioned person reddens. This may happen when a person is overheated, embarrassed, or under the influence of alcohol. Conversely, conditions that constrict blood vessels cause the skin to lose this reddish color. Thus, if body temperature drops abnormally or if a person is frightened, the skin may appear pale. Illnesses may also affect skin color.

A yellow-orange plant pigment called *carotene*, which is especially common in yellow vegetables, may give skin a yellowish cast if a person consumes too much of it. This results from accumulation of carotene in the adipose tissue of the subcutaneous layer.

**FIGURE** 6.13

Albinism is common among the Hopi people of Arizona because men with the condition father more children than men with pigmented skin. The little girl in the middle contrasts sharply with her playmates, who do not have albinism.

An unusual skin color can signal a medical problem. Some newborns develop yellowish skin shortly after birth, a condition called *jaundice*. A blood incompatibility or an immature liver can cause jaundice. An observant British hospital nurse discovered a treatment for newborn jaundice in 1958. She liked to take her tiny charges out in the sun, and she noticed that a child whose skin had a yellow pallor developed normal pigmentation when he lay in sunlight. However, the part of the child's body covered by a diaper and therefore not exposed to the sun remained yellow. Further investigation showed that sunlight enables the body to break down bilirubin, the liver substance that accumulates in the skin. Today, newborns who develop persistently yellowish skin may have to lie under artificial "bili lights," clad only in protective goggles.

1 What factors influence skin color?

2 Which of these factors are genetic? Which are environmental?

## HEALING OF WOUNDS AND BURNS

**Inflammation** is a normal response to injury or stress. Blood vessels in affected tissues dilate and become more permeable, forcing fluids to leave the blood vessels and enter the damaged tissues. Inflamed skin may become reddened, swollen, warm, and painful to touch. However, the dilated blood vessels provide the tissues with more nutrients and oxygen, which aids healing. The specific events in the healing process depend on the nature and extent of the injury.

## Cuts

If a break in the skin is shallow, epithelial cells along its margin are stimulated to reproduce more rapidly than usual. The newly formed cells simply fill the gap.

If the injury extends into the dermis or subcutaneous layer, blood vessels break and the escaping blood forms a clot in the wound. A clot consists mainly of a fibrous protein (fibrin) that forms from the blood plasma, blood cells, and platelets that become entrapped in the protein fibers. Tissue fluids seep into the area and dry. The blood clot and the dried fluids form a *scab* that covers and protects underlying tissues. Before long, fibroblasts migrate into the injured region and begin forming new collagenous fibers that bind the edges of the wound together. Suturing or otherwise closing a large break in the skin speeds this process. In addition, the connective tissue matrix releases *growth factors* that stimulate certain cells to proliferate and regenerate the damaged tissue.

As healing continues, blood vessels extend into the area beneath the scab. Phagocytic cells remove dead cells and other debris. Eventually, the damaged tissues are replaced, and the scab sloughs off. If the wound is extensive, the newly formed connective tissue may appear on the surface as a *scar*.

In large, open wounds, healing may be accompanied by formation of *granulations* that develop in the exposed tissues as small, rounded masses. A granulation consists of a new branch of a blood vessel and a cluster of collagen-secreting fibroblasts that the vessel nourishes. In time, some of the blood vessels are resorbed, and the fibroblasts migrate away, leaving a scar that is largely composed of collagenous fibers. Figure 6.14 shows the stages in the healing of a wound.

## Burns

Slightly burned skin, such as from a minor sunburn, may become warm and reddened (erythema) as dermal blood vessels dilate. This response may be accompanied by mild edema, and, in time, the surface layer of skin may shed. A burn injuring only the epidermis is called a *superficial partial-thickness* (first degree) *burn.*

Healing usually occurs within a few days to two weeks, with no scarring.

A burn that destroys some epidermis as well as some underlying dermis is a *deep partial-thickness* (second degree) *burn.* Fluid escapes from damaged dermal capillaries, and as it accumulates beneath the outer layer of epidermal cells, blisters appear. The injured region becomes moist and firm, and may vary in color from dark red to waxy white. Such a burn most commonly occurs as a result of exposure to hot objects, hot liquids, flames, or burning clothing.

The healing of a deep partial-thickness burn depends upon accessory organs of the skin that survive the injury because they are located deep in the dermis. These organs, which include hair follicles, sweat glands, and sebaceous glands, contain epithelial cells. During healing, these cells grow out onto the surface of the dermis, spread over it, and form a new epidermis. In time, the skin usually completely recovers, and scar tissue does not develop unless an infection occurs.

A burn that destroys the epidermis, dermis, and the accessory organs of the skin is called a *full-thickness* (third degree) *burn.* The injured skin becomes dry and leathery, and it may vary in color from red to black to white.

A full-thickness burn usually occurs as a result of immersion in hot liquids or prolonged exposure to hot objects, flames, or corrosive chemicals. Since most of the epithelial cells in the affected region are likely to be destroyed, spontaneous healing can occur only by growth of epithelial cells inward from the margin of the burn. If the injury is extensive, treatment may involve removing a thin layer of skin from an unburned region of the body and transplanting it in the injured area. This procedure is called an *autograft.*

If the burn is too extensive to replace with skin from other parts of the body, cadaveric skin from a skin bank may be used to cover the injury. In this case, the transplant, a *homograft,* is a temporary covering that decreases the size of the wound, helps prevent infection, and helps preserve deeper tissues. In time, after healing has begun, the temporary covering may be removed and replaced with an autograft, as skin becomes available in areas that have healed. However, skin grafts can leave extensive scars.

Various skin substitutes also may be used to temporarily cover extensive burns. These include amniotic membrane that previously surrounded a human fetus before its birth and artificial membranes composed of silicone, polyurethane, or nylon together with a network of collagenous fibers. Still another type of skin substitute can be prepared by culturing human epithelial cells. In a laboratory, a bit of human skin the size of a postage stamp can grow to the size of a bathmat in about three weeks.

The treatment of a burn patient requires estimating the extent of the body's surface that is affected. Physicians

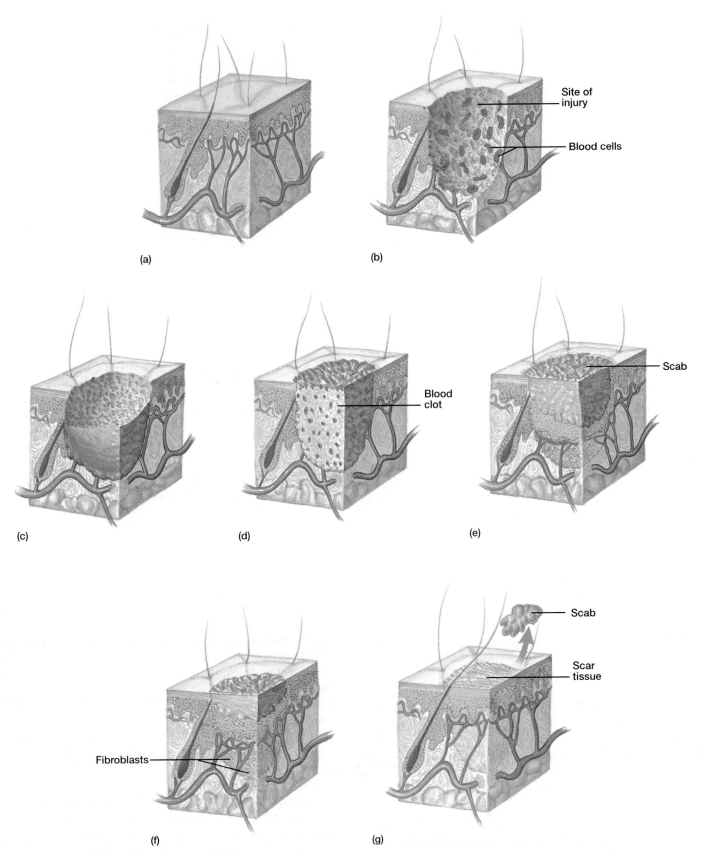

**FIGURE 6.14**

(a) If normal skin is (b) injured deeply, (c) blood escapes from dermal blood vessels, and (d) a blood clot soon forms. (e) The blood clot and dried tissue fluid form a scab that protects the damaged region. (f) Later, blood vessels send out branches, and fibroblasts migrate into the area. (g) The fibroblasts produce new connective tissue fibers, and when the skin is mostly repaired, the scab sloughs off.

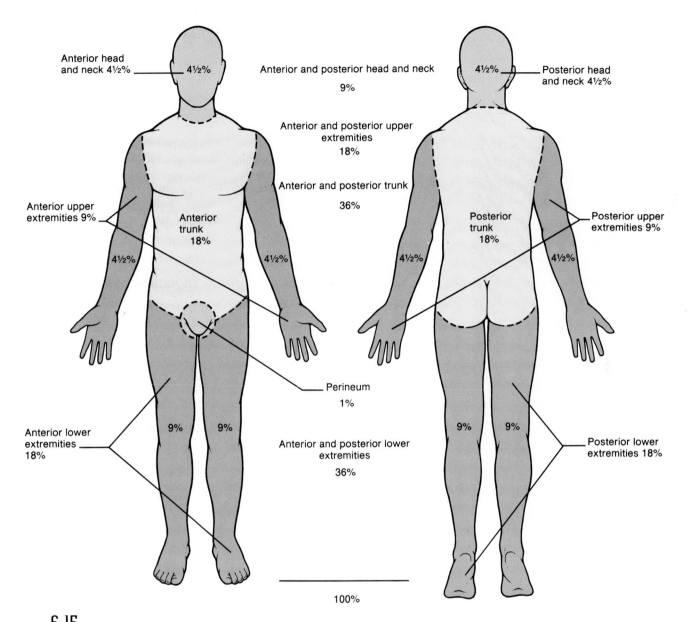

Anterior head and neck 4½%

4½%

Anterior and posterior head and neck
9%

4½%

Posterior head and neck 4½%

Anterior and posterior upper extremities
18%

Anterior upper extremities 9%

Anterior trunk
18%

Anterior and posterior trunk
36%

Posterior trunk
18%

Posterior upper extremities 9%

4½%

4½%

4½%

4½%

Perineum
1%

Anterior lower extremities
18%

9%

9%

Anterior and posterior lower extremities
36%

9%

9%

Posterior lower extremities 18%

100%

**FIGURE 6.15**

As an aid for estimating the extent of damage burns cause, the body is subdivided into regions, each representing 9% (or some multiple of 9%) of the total skin surface area.

use the "rule of nines," subdividing the skin's surface into regions, each accounting for 9% (or some multiple of 9%) of the total surface area (fig. 6.15). This estimate is important in planning to replace body fluids and electrolytes lost from injured tissues, and for covering the burned area with skin or skin substitutes.

**1** What is the tissue response to inflammation?

**2** In the healing of a wound, what occurs within the wound to cause the sloughing of the scab?

**3** Which type of burn is most likely to leave a scar? Why?

## COMMON SKIN DISORDERS

**athlete's foot** (ath'-lētz foot)   Fungus infection (*Tinea pedis*) usually in the skin of the toes and soles.

**birthmark** (berth'mark)   Congenital blemish or spot on the skin, visible at birth or soon after.

**boil** (boil)   Bacterial infection (furuncle) of the skin, produced when bacteria enter a hair follicle.

**carbuncle** (kar'bung-kl)   Bacterial infection, similar to a boil, that spreads into the subcutaneous tissues.

**cyst** (sist)   Liquid-filled sac or capsule.

**eczema** (ek'zĕ-mah)   Noncontagious skin rash often accompanied by itching, blistering, and scaling.

**INTEGUMENTARY SYSTEM**

The skin provides protection, contains sensory organs, and helps control body temperature.

## Skeletal System

Vitamin D activated by the skin helps provide calcium for bone matrix.

## Muscular System

Involuntary muscle contractions (shivering) work with the skin to control body temperature. Muscles act on facial skin to create expressions.

## Nervous System

Sensory receptors provide information about the outside world to the nervous system. Nerves control the activity of sweat glands.

## Endocrine System

Hormones help to increase skin blood flow during exercise. Other hormones stimulate either the synthesis or the decomposition of subcutaneous fat.

## Cardiovascular System

Skin blood vessels play a role in regulating body temperature.

## Lymphatic System

The skin provides an important first line of defense for the immune system.

## Digestive System

Excess calories may be stored as subcutaneous fat. Vitamin D activated by the skin stimulates dietary calcium absorption.

## Respiratory System

Stimulation of skin receptors may alter respiratory rate.

## Urinary System

The kidneys help compensate for water and electrolytes lost in sweat.

## Reproductive System

Sensory receptors play an important role in sexual activity and in the suckling reflex.

**erythema** (er″ĭ-the′mah)   Reddening of the skin due to dilation of dermal blood vessels in response to injury or inflammation.

**herpes** (her′pēz)   Infectious disease of the skin usually caused by the herpes simplex virus and characterized by recurring formations of small clusters of vesicles.

**keloid** (ke′loid)   Elevated, enlarging fibrous scar usually initiated by an injury.

**mole** (mol)   Fleshy skin tumor (nevus) that is usually pigmented; colors range from brown to black.

**pediculosis** (pĕ-dik″u-lo′sis)   Disease produced by an infestation of lice.

**pruritus** (proo-ri′tus)   Itching of the skin.

**pustule** (pus′tūl)   Elevated, pus-filled area.

**scabies** (ska′bēz)   Disease resulting from an infestation of mites.

**seborrhea** (seb″o-re′ah)   Hyperactivity of the sebaceous glands, accompanied by greasy skin and dandruff.

**ulcer** (ul′ser)   Open sore.

**urticaria** (ur″tĭ-ka′re-ah)   Allergic reaction of the skin that produces reddish, elevated patches (hives).

**wart** (wort)   Flesh-colored, raised area caused by a viral infection.

---

## CHAPTER SUMMARY

## Introduction  (page 161)

Organs are composed of two or more kinds of tissues. The skin and its accessory organs constitute the integumentary organ system.

## Types of Membranes  (page 161)

1. Serous membranes
   a. Serous membranes are organs that line body cavities lacking openings to the outside.
   b. They are composed of epithelium and loose connective tissue.
   c. Cells of serous membranes secrete watery serous fluid that lubricates membrane surfaces.
2. Mucous membranes
   a. Mucous membranes are organs that line cavities and tubes opening to the outside of the body.
   b. They are composed of epithelium and loose connective tissue.
   c. Cells of mucous membranes secrete mucus.
3. Synovial membranes
   a. Synovial membranes are organs that line joint cavities.
   b. They are composed of loose connective tissue with elastic fibers and some adipose tissue.
   c. Cells of synovial membranes secrete synovial fluid that lubricates the ends of bones at joints.
4. The cutaneous membrane is the external body covering commonly called the skin.

## Skin and Its Tissues  (page 161)

Skin is a protective covering, helps regulate body temperature, houses sensory receptors, synthesizes chemicals, and excretes wastes. It is composed of an epidermis and a dermis separated by a basement membrane. A subcutaneous layer lies beneath the dermis.

1. Epidermis
   a. The epidermis is a layer of stratified squamous epithelium that lacks blood vessels.
   b. The deepest layer, called stratum basale, contains cells reproducing and growing.
   c. Epidermal cells undergo keratinization as they are pushed toward the surface.
   d. The outermost layer, called stratum corneum, is composed of dead epidermal cells.
   e. Production of epidermal cells balances the rate at which they are lost at the surface.
   f. Epidermis protects underlying tissues against water loss, mechanical injury, and the effects of harmful chemicals.
   g. Melanin protects underlying cells from the effects of ultraviolet light.
   h. Melanocytes transfer melanin to nearby epidermal cells.
2. Dermis
   a. The dermis is a layer composed of irregular dense connective tissue that binds the epidermis to underlying tissues.
   b. It also contains muscle fibers, blood vessels, and nerve fibers.
   c. Dermal blood vessels supply nutrients to all skin cells and help regulate body temperature.
   d. Nervous tissue is scattered through the dermis.
      (1) Some dermal nerve fibers carry impulses to muscles and glands of the skin.
      (2) Other dermal nerve fibers are associated with sensory receptors in the skin.
3. Subcutaneous layer
   a. The subcutaneous layer is composed of loose connective tissue and adipose tissue.
   b. Adipose tissue helps conserve body heat.
   c. This layer contains blood vessels that supply the skin.

## Accessory Organs of the Skin (page 168)

1. Hair follicles
   a. Hair occurs in nearly all regions of the skin.
   b. Each hair develops from epidermal cells at the base of a tubelike hair follicle.
   c. As newly formed cells develop and grow, older cells are pushed toward the surface and undergo keratinization.
   d. A hair usually grows for a while, rests, and then is replaced by a new hair.
   e. Hair color is determined by genes that direct the type and amount of pigment in its cells.
   f. A bundle of smooth muscle cells and one or more sebaceous glands are attached to each hair follicle.

2. Nails
   a. Nails are protective covers on the ends of fingers and toes.
   b. They are produced by epidermal cells that undergo keratinization.
3. Skin glands
   a. Sebaceous glands secrete sebum, which helps keep skin and hair soft and waterproof.
   b. Usually associated with hair follicles, in some regions sebaceous glands open directly to the skin surface.
   c. Sweat glands are located in nearly all regions of the skin.
   d. Each sweat gland consists of a coiled tube.
   e. Eccrine sweat glands, located on the forehead, neck, back, palms, and soles, respond to elevated body temperature or emotional stress.
   f. Sweat is primarily water, but also contains salts and waste products.
   g. Apocrine sweat glands, located in the axillary regions, groin, and around the nipples, moisten the skin when a person is emotionally upset, scared, in pain, or sexually aroused.

## Regulation of Body Temperature (page 173)

Regulation of body temperature is vital because heat affects the rates of metabolic reactions. Normal temperature of deeper body parts is close to a set point of 37°C (98.6°F).

1. Heat production and loss
   a. Heat is a by-product of cellular respiration.
   b. When body temperature rises above normal, more blood enters dermal blood vessels and the skin reddens.
   c. Heat is lost to the outside by radiation, conduction, convection, and evaporation.
   d. Sweat gland activity increases heat loss by evaporation.
   e. If the body temperature drops below normal, dermal blood vessels constrict, causing the skin to lose color, and sweat glands become inactive.
   f. When heat is lost excessively, skeletal muscles involuntarily contract; this increases cellular respiration and produces additional heat.
2. Problems in temperature regulation
   a. Air can hold a limited amount of water vapor.

b. When the air is saturated with water, sweat may fail to evaporate, and body temperature may remain elevated.
c. Hypothermia is lowered body temperature. It causes shivering, mental confusion, lethargy, loss of reflexes and consciousness, and eventually major organ failure.

## Skin Color (page 175)

All humans have about the same concentration of melanocytes. Skin color is largely due to the amount of melanin in the epidermis.

1. Genetic factors
   a. Each person inherits genes for melanin production.
   b. Dark skin is due to genes that cause large amounts of melanin to be produced; lighter skin is due to genes that cause lesser amounts of melanin to form.
   c. Mutant genes may cause a lack of melanin in the skin.
2. Environmental factors
   a. Environmental factors that influence skin color include sunlight, ultraviolet light, and X rays.
   b. These factors darken existing melanin and stimulate additional melanin production.
3. Physiological factors
   a. The oxygen content of the blood in dermal vessels may cause the skin of light-complexioned persons to appear pinkish or bluish.
   b. Carotene in the subcutaneous layer may cause the skin to appear yellowish.
   c. Disease may affect skin color.

## Healing of Wounds and Burns (page 177)

Skin injuries trigger inflammation. The affected area becomes red, warm, swollen, and tender.

1. A cut in the epidermis is filled in by reproducing epithelial cells. Clots close deeper cuts, sometimes leaving a scar where connective tissue replaces skin. Granulations form as part of the healing process.
2. A superficial partial-thickness burn heals quickly with no scarring. The area is warm and red. A burn penetrating to the dermis is a deep partial-thickness burn. It blisters. Deeper skin structures help heal this more serious type of burn. A full-thickness burn is the most severe, and may require a skin graft.

## CRITICAL THINKING QUESTIONS

1. What special problems would result from the loss of 50% of a person's functional skin surface? How might this person's environment be modified to compensate partially for such a loss?
2. A premature infant typically lacks subcutaneous adipose tissue. Also, the surface area of an infant's body is relatively large compared to its volume. How do you think these factors affect the ability of an infant to regulate its body temperature?
3. As a rule, a superficial partial-thickness burn is more painful than one involving deeper tissues. How would you explain this observation?
4. Which of the following would result in the more rapid absorption of a drug: a subcutaneous injection or an intradermal injection? Why?
5. What methods might be used to cool the skin of a child experiencing a high fever? For each method you list, identify the means by which it promotes heat loss— radiation, conduction, convection, or evaporation.
6. How would you explain to an athlete the importance of keeping the body hydrated when exercising in warm weather?
7. Everyone's skin contains about the same number of melanocytes even though people come in many different colors. How is this possible?
8. How is skin peeling after a severe sunburn protective? How might a fever be protective?

1. Explain why a membrane is an organ.
2. Define *integumentary organ system.*
3. Distinguish between serous and mucous membranes.
4. Explain the function of synovial membranes.
5. List six functions of skin.
6. Distinguish between the epidermis and the dermis.
7. Describe the subcutaneous layer.
8. Explain what happens to epidermal cells as they undergo keratinization.
9. List the layers of the epidermis.
10. Describe the function of melanocytes.
11. Describe the structure of the dermis.
12. Review the functions of dermal nervous tissue.
13. Explain the functions of the subcutaneous layer.
14. Distinguish between a hair and a hair follicle.
15. Review how hair color is determined.
16. Describe how nails are formed.
17. Explain the function of sebaceous glands.
18. Distinguish between eccrine and apocrine sweat glands.
19. Explain the importance of body temperature regulation.
20. Describe the role of the skin in promoting the loss of excess body heat.
21. Explain how body heat is lost by radiation.
22. Distinguish between conduction and convection.
23. Describe the body's responses to decreasing body temperature.
24. Review how air saturated with water vapor may interfere with body temperature regulation.
25. Explain how environmental factors affect skin color.
26. Describe three physiological factors that affect skin color.
27. Distinguish between the healing of shallow and deeper breaks in the skin.
28. Distinguish among first-, second-, and third-degree burns.
29. Describe possible treatments for a third-degree burn.

**For More Information Related to This Chapter, Please Visit** *Hole's Human Anatomy and Physiology* **Home Page.**

**http://www.mhhe.com/biosci/abio/**

# SKELETAL SYSTEM

## UNDERSTANDING WORDS

**ax-,** axis: *ax*ial skeleton—upright portion of the skeleton that supports the head, neck, and trunk.

**-blast,** bud, a growing organism in early stages: osteo*blast*—cell that will form bone tissue.

**canal-,** channel: *canal*iculus—tubular passage.

**carp-,** wrist: *carp*als—wrist bones.

**-clast,** break: osteo*clast*—cell that breaks down bone tissue.

**clav-,** bar: *clav*icle—bone that articulates with the sternum and scapula.

**condyl-,** knob, knuckle: *condyl*e—rounded, bony process.

**corac-,** a crow's beak: *corac*oid process—beaklike process of the scapula.

**cribr-,** sieve: *cribr*iform plate—portion of the ethmoid bone with many small openings.

**crist-,** crest: *crist*a galli—bony ridge that projects upward into the cranial cavity.

**fov-,** pit: *fov*ea capitis—pit in the head of a femur.

**gladi-,** sword: *gladi*olus—middle portion of the bladelike sternum.

**glen-,** joint socket: *glen*oid cavity—depression in the scapula that articulates with the head of a humerus.

**inter-,** among, between: *inter*vertebral disk—structure located between vertebrae.

**intra-,** inside: *intra*membranous bone—bone that forms within sheetlike masses of connective tissue.

**lamell-,** thin plate: *lamell*a—thin bony plate.

**meat-,** passage: auditory *meat*us—canal of the temporal bone that leads inward to parts of the ear.

**odont-,** tooth: *odont*oid process—toothlike process of the second cervical vertebra.

**poie-,** make, produce: hemato*poie*sis—process by which blood cells are formed.

## CHAPTER OBJECTIVES

### AFTER YOU HAVE STUDIED THIS CHAPTER, YOU SHOULD BE ABLE TO:

1. Classify bones according to their shapes and name an example from each group.

2. Describe the general structure of a bone and list the functions of its parts.

3. Distinguish between intramembranous and endochondral bones, and explain how such bones grow and develop.

4. Describe the effects of sunlight, nutrition, hormonal secretions, and exercise on bone development.

5. Discuss the major functions of bones.

6. Distinguish between the axial and appendicular skeletons, and name the major parts of each.

7. Locate and identify the bones and the major features of the bones that comprise the skull, vertebral column, thoracic cage, pectoral girdle, upper limb, pelvic girdle, and lower limb.

## KEY TERMS

**articular cartilage** (ar-tik'u-lar kar'tĭ-lij)
**compact bone** (kom'pakt bōn)
**diaphysis** (di-af'ĭ-sis)
**endochondral bone** (en'do-kon'dral bōn)
**epiphyseal disk** (ep"ĭ-fiz'e-al disk)
**epiphysis** (e-pif'ĭ-sis)
**fontanel** (fon'tah-nel)
**hematopoiesis** (hem"ah-to-poi-e'sis)
**intramembranous bone** (in"trah-mem'brah-nus bōn)

**lever** (lev'er)
**marrow** (mar'o)
**meatus** (me-a'-tus)
**medullary cavity** (med'u-lār"e kav'ĭ-te)
**osteoblast** (os'te-o-blast)
**osteoclast** (os'te-o-klast)
**osteocyte** (os'te-o-sīt)
**osteon** (os'te-on)
**periosteum** (per"e-os'te-um)
**spongy bone** (spun'je bōn)
**trabeculae** (trah-bek'u-le)

C lifton Martin, a 64-year-old New York City business owner, had to wear a surgical mask whenever he left his house, to hide the bright red tissue protruding from his nostrils. The tissue was from a tumor that had been growing behind his nose and mouth for two decades. Although the tumor was benign, and hadn't spread, it had done enough damage where it was, weaving into the man's sinuses and throughout the area behind his face.

Martin's plight began with nosebleeds that gradually worsened. Over the years, many surgeons had refused to operate, citing the tumor's proximity to the brain, and the difficulty of reaching it without destroying the face. Two biopsies had led to extensive bleeding. In 1997, Martin found a surgeon who could remove such tumors. Dr. Ivo Janecka, of Harvard Medical School, had so thoroughly studied the bones of the face that he had compiled his own atlas to supplement existing texts. His specialty was the skull base, which is a thick basin of bone extending from the back of the head around the ears and nose. The skull base supports the brain, and also has minute passageways for nerves and blood vessels leading to the face.

For years surgeons had skillfully skirted the skull base. Neurosurgeons operated above it, head and neck surgeons beneath it. In contrast, Dr. Janecka's approach, called "facial translocation," creates flaps of bone in and around the face. He carefully divides the face into sectors, then gently moves aside pieces of skin, fat, and bone to get to the tumor. It is a little like slowly taking apart an intricate puzzle, then reassembling the pieces.

Facial translocation takes from 10 to 20 hours, and requires a large team of surgeons with different specialties who take turns as their area of expertise is revealed. The procedure is possible, Dr. Janecka says, because of this cooperation, coupled with more complete information on facial anatomy and the location of tumors derived from MRI and CT scans. The surgeons map out the path to a particular patient's tumor ahead of time, then practice on cadavers.

Clifton Martin's surgery took 10 hours. Dr. Janecka peeled back the left side of the face, including the lip area, cheek, nose, and eyelid, to reveal the tumor—the size of a small apple, behind the mouth and nose, snaking into the sinuses, and touching the skull base. After removing the tumor, Dr. Janecka reassembled Martin's facial bones, using small plates and screws made of the biocompatible metal titanium.

---

Nonliving material in the matrix of bone tissue makes the whole organ appear to be inert. A bone also contains very active, living tissues. An individual bone is composed of a variety of tissues: bone tissue, cartilage, dense connective tissue, blood, and nervous tissue.

## BONE STRUCTURE

The bones of the skeletal system differ greatly in size and shape, but they are similar in their structure, development, and functions.

### Bone Classification

Bones are classified according to their shapes—long, short, flat, or irregular (fig. 7.1).

**Long bones** have long longitudinal axes and expanded ends. Examples are the forearm and thigh bones.

**Short bones** are somewhat cubelike, with their lengths and widths roughly equal. The bones of the wrists and ankles are examples of this type.

**Flat bones** are platelike structures with broad surfaces, such as the ribs, scapulae, and some bones of the skull.

**Irregular bones** have a variety of shapes and are usually connected to several other bones. Irregular bones include the vertebrae that comprise the backbone and many facial bones.

In addition to these four groups of bones, some authorities recognize a fifth group called *sesamoid* or round bones. These bones are usually small and nodular, and are embedded within tendons adjacent to joints, where the tendons are compressed. The kneecap (patella) is a very large example of a sesamoid bone.

### Parts of a Long Bone

The femur, a long bone in the thigh, illustrates the structure of bone (fig. 7.2). At each end of such a bone is an expanded portion called an **epiphysis** (pl., *epiphyses*), which articulates (or forms a joint) with another bone. On its outer surface, the articulating portion of the epiphysis is coated with a layer of hyaline cartilage called **articular cartilage.** The shaft of the bone, which is located between the epiphyses, is called the **diaphysis.**

Except for the articular cartilage on its ends, the bone is completely enclosed by a tough, vascular covering of fibrous tissue called the **periosteum.** This membrane is firmly attached to the bone, and the periosteal fibers are continuous with ligaments and tendons that are connected to the membrane. The periosteum also functions in the formation and repair of bone tissue.

A bone's shape makes possible its functions. Bony projections called *processes,* for example, provide sites for attachment of ligaments and tendons; grooves and openings are passageways for blood vessels and nerves; a depression of one bone might articulate with a process of another.

The wall of the diaphysis is mainly composed of tightly packed tissue called **compact bone** (cortical bone). This type of bone has a continuous matrix with no gaps (fig. 7.3a).

The epiphyses, on the other hand, are largely composed of **spongy bone** (cancellous bone) with thin layers

**FIGURE 7.1** 🏃

(a) The femur of the thigh is a long bone, (b) a tarsal bone of the ankle is a short bone, (c) a parietal bone of the skull is a flat bone, (d) a vertebra of the backbone is an irregular bone, and (e) the patella of the knee is a round bone.

**FIGURE 7.2** 🏃
Major parts of a long bone.

most highly developed in the regions of the epiphyses that are subjected to compressive forces. Both compact and spongy bone are strong and resist bending.

A bone usually has both compact and spongy bone tissues. Short, flat, and irregular bones typically consist of a mass of spongy bone that is either covered by a layer of compact bone or sandwiched between plates of compact bone (fig. 7.3c).

Compact bone in the diaphysis of a long bone forms a semirigid tube with a hollow chamber called the **medullary cavity** that is continuous with the spaces of the spongy bone. A thin layer of squamous epithelial cells called **endosteum** lines these areas, and a specialized type of soft connective tissue called **marrow** fills them. Marrow exists in two forms, red marrow and yellow marrow, described later in the chapter (see also fig. 7.2).

## Microscopic Structure

Recall from chapter 5 that bone cells (osteocytes) are located in minute, bony chambers called lacunae, which form concentric circles around osteonic canals (Haversian

of compact bone on their surfaces (fig. 7.3b). Spongy bone consists of numerous branching bony plates called **trabeculae.** Irregular connecting spaces between these plates help reduce the bone's weight. The bony plates are

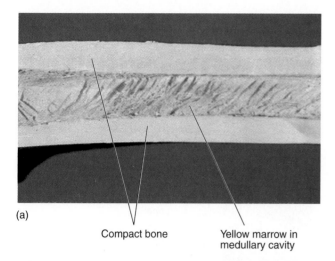

(a)

Compact bone

Yellow marrow in medullary cavity

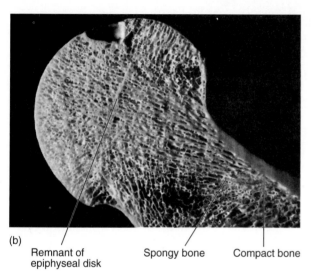

(b)

Remnant of epiphyseal disk

Spongy bone

Compact bone

(c)

Spongy bone

Compact bone

**FIGURE 7.3**

(*a*) In a femur, the wall of the diaphysis consists of compact bone. (*b*) The epiphyses of the femur contain spongy bone enclosed by a thin layer of compact bone. (*c*) This skull bone contains a layer of spongy bone sandwiched between plates of compact bone.

**FIGURE 7.4**

Scanning electron micrograph of a single osteon in compact bone (575×).

*Tissues and Organs: A Text-Atlas of Scanning Electron Microscopy*, by R. G. Kessel and R. H. Kardon. © 1979 W. H. Freeman and Company.

canals). Osteocytes communicate with nearby cells by means of cellular processes passing through canaliculi. The intercellular material of bone tissue is largely collagen and inorganic salts. Collagen gives bone its strength and resilience, and inorganic salts make it hard and resistant to crushing.

### Compact Bone

In compact bone, the osteocytes and layers of intercellular material concentrically clustered around an osteonic canal form a cylinder-shaped unit called an **osteon** (Haversian system). Many of these units cemented together form the substance of compact bone (fig. 7.4). The orientation of the osteons resists compressive forces.

Each osteonic canal contains blood vessels and nerve fibers surrounded by loose connective tissue. Blood in these vessels nourishes bone cells associated with the osteonic canal via gap junctions between osteocytes.

Osteonic canals travel longitudinally through bone tissue. They are interconnected by transverse *perforating canals* (Volkmann's canals). These canals contain larger blood vessels and nerves by which the smaller blood vessels and nerve fibers in the osteonic canals communicate with the surface of the bone and the medullary cavity (fig. 7.5).

### Spongy Bone

Spongy bone is also composed of osteocytes and intercellular material, but the bone cells do not aggregate around osteonic canals. Instead, the cells lie within the

**FIGURE 7.5**

Compact bone is composed of osteons cemented together by bone matrix.

trabeculae, and receive nourishment from substances diffusing into the canaliculi that lead to the surface of these thin, bony plates.

> Severe bone pain is a symptom of inherited sickle cell disease. Under low oxygen conditions, abnormal hemoglobin (an oxygen-carrying protein) bends the red blood cells that contain it into a sickle shape, which obstructs circulation. X rays can reveal blocked arterial blood flow in bones of sickle cell disease patients.

1. Explain how bones are classified.

2. List five major parts of a long bone.

3. How do compact and spongy bone differ in structure?

4. Describe the microscopic structure of compact bone.

## BONE DEVELOPMENT AND GROWTH

Parts of the skeletal system begin to form during the first few weeks of prenatal development, and bony structures continue to grow and develop into adulthood. Bones form by replacing existing connective tissue in one of two ways. Some bones originate within sheetlike layers of connective tissues; they are called intramembranous bones. Others begin as masses of cartilage that are later replaced by bone tissue; they are called endochondral bones (fig. 7.6).

### Intramembranous Bones

The broad, flat bones of the skull are intramembranous bones. During their development (osteogenesis), membranelike layers of unspecialized, or primitive, connective tissues appear at the sites of the future bones. Dense networks of blood vessels supply these connective tissue layers, which may form around the vessels. These primitive cells enlarge and differentiate into bone-forming cells called **osteoblasts,** which, in turn, deposit bony matrix

(a)

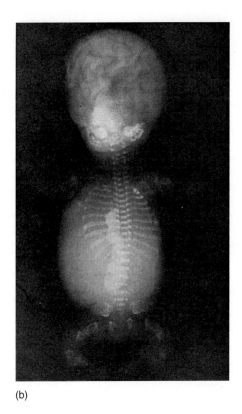

(b)

**FIGURE 7.6**

(a) Note the stained bones of this fourteen-week fetus. (b) Bones can fracture even before birth. This fetus has numerous broken bones because of an inherited defect in collagen called osteogenesis imperfecta. Often parents of such children are unfairly accused of child abuse when their children frequently break bones.

around themselves. As a result, spongy bone forms in all directions along blood vessels within the layers of primitive connective tissues. Later, some spongy bone may become compact bone as spaces fill with bone matrix.

As development continues, the osteoblasts may become completely surrounded by matrix, and in this manner they become secluded within lacunae. At the same time, matrix enclosing the cellular processes of the osteoblasts gives rise to canaliculi. Once isolated in lacunae, these cells are called **osteocytes** (fig. 7.7).

Cells of the primitive connective tissue that persist outside the developing bone give rise to the periosteum. Osteoblasts on the inside of the periosteum form a layer of compact bone over the surface of the newly formed spongy bone.

This process of replacing connective tissue to form an intramembranous bone is called *intramembranous ossification*. Table 7.1 lists the major steps of the process.

## Endochondral Bones

Most of the bones of the skeleton are **endochondral bones.** They develop from masses of hyaline cartilage shaped like future bony structures. These cartilaginous models grow rapidly for a time, and then begin to extensively change. For example, cartilage cells enlarge and

Lacuna          Osteocyte          Nucleus

**FIGURE 7.7**

Transmission electron micrograph of an osteocyte isolated within a lacuna (4,700×).

their lacunae grow. The surrounding matrix breaks down, and soon the cartilage cells die and degenerate.

About the same time, a periosteum forms from connective tissue that encircles the developing structure. As the cartilage decomposes, blood vessels and undifferentiated connective tissue cells invade the disintegrating tissue. Some of the invading cells differentiate into

TABLE 7.1

## MAJOR STEPS IN BONE DEVELOPMENT

| Intramembranous Ossification | Endochondral Ossification |
|---|---|
| **1.** Sheets of primitive connective tissue appear at sites of future bones. | **1.** Masses of hyaline cartilage form models of future bones. |
| **2.** Primitive connective tissue cells arrange around blood vessels in these layers. | **2.** Cartilage tissue breaks down. Periosteum develops. |
| **3.** Connective tissue cells differentiate into osteoblasts, which deposit spongy bone. | **3.** Blood vessels and differentiating osteoblasts from the periosteum invade the disintegrating tissue. |
| **4.** Osteoblasts become osteocytes when bony matrix completely surrounds them. | **4.** Osteoblasts form spongy bone in the space occupied by cartilage. |
| **5.** Connective tissue on the surface of each developing structure forms a periosteum. | **5.** Osteoblasts become osteocytes when bony matrix completely surrounds them. |
| **6.** Osteoblasts on the inside of the periosteum deposit compact bone over the spongy bone. | **6.** Osteoblasts beneath the periosteum deposit compact bone around spongy bone. |

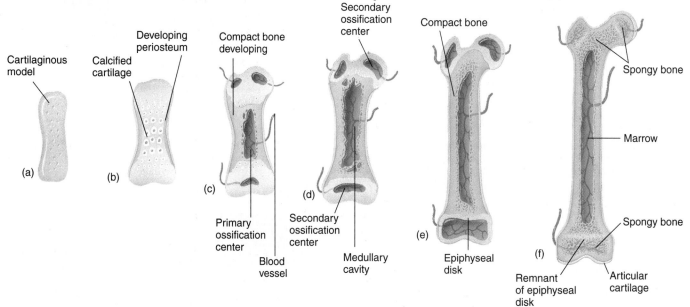

**FIGURE 7.8**

Major stages (a–f) in the development of an endochondral bone. (Relative bone sizes not to scale.)

osteoblasts and begin to form spongy bone in the spaces previously housing the cartilage. Once completely surrounded by the bony matrix, osteoblasts are called osteocytes. As ossification continues, osteoblasts beneath the periosteum deposit compact bone around the spongy bone.

The process of forming an endochondral bone by the replacement of hyaline cartilage is called *endochondral ossification.* Its major steps are listed in table 7.1 and illustrated in figure 7.8.

## Growth of an Endochondral Bone

In a long bone, bony tissue begins to replace hyaline cartilage in the center of the diaphysis. This region is called the *primary ossification center,* and bone develops from it toward the ends of the cartilaginous structure. Meanwhile, osteoblasts from the periosteum deposit a thin

layer of compact bone around the primary ossification center. The epiphyses of the developing bone remain cartilaginous and continue to grow. Later, *secondary ossification centers* appear in the epiphyses, and spongy bone forms in all directions from them. As spongy bone is deposited in the diaphysis and in the epiphysis, a band of cartilage, called the **epiphyseal disk** (metaphysis), remains between the two ossification centers (see figs. 7.2, 7.3b, and 7.8).

The cartilaginous cells of the epiphyseal disk occur in four layers, each of which may be several cells thick, as shown in figure 7.9. The first layer, closest to the end of the epiphysis, is composed of resting cells that do not actively participate in growth. This layer anchors the epiphyseal disk to the bony tissue of the epiphysis.

The second layer contains rows of many young cells undergoing mitosis. As new cells appear and as

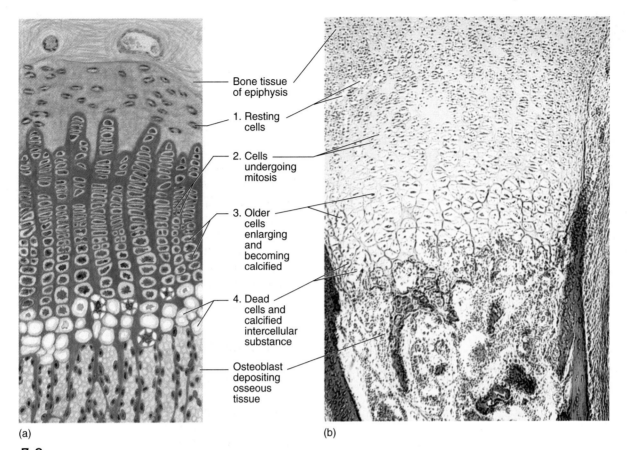

Bone tissue of epiphysis

1. Resting cells

2. Cells undergoing mitosis

3. Older cells enlarging and becoming calcified

4. Dead cells and calcified intercellular substance

Osteoblast depositing osseous tissue

(a)

(b)

**FIGURE 7.9**

(a) The cartilaginous cells of an epiphyseal disk lie in four layers, each of which may be several cells thick.
(b) A micrograph of an epiphyseal disk (100×).

intercellular material forms around them, the cartilaginous disk thickens.

The rows of older cells, which are left behind when new cells appear, form the third layer. These cells enlarge and thicken the epiphyseal disk still more. Consequently, the entire bone lengthens. At the same time, calcium salts accumulate in the intercellular matrix adjacent to the oldest cartilaginous cells, and as the matrix calcifies, the cells begin to die.

The fourth layer of the epiphyseal disk is quite thin. It is composed of dead cells and calcified intercellular substance.

In time, large, multinucleated cells called **osteoclasts** break down the calcified matrix. These large cells originate by the fusion of single-nucleated white blood cells called monocytes (see chapter 14). Osteoclasts secrete an acid that dissolves the inorganic component of the calcified matrix, and their lysosomal enzymes digest the organic components. After osteoclasts remove the matrix, bone-building osteoblasts invade the region and deposit bone tissue in place of the calcified cartilage.

A long bone continues to lengthen while the cartilaginous cells of the epiphyseal disks are active. However, once the ossification centers of the diaphysis and

Osteoclast

Bone tissue

**FIGURE 7.10**

Micrograph of a bone-resorbing osteoclast (135×).

epiphyses meet and the epiphyseal disks ossify, lengthening is no longer possible in that end of the bone.

A developing bone thickens as compact bone is deposited on the outside, just beneath the periosteum. As this compact bone forms on the surface, osteoclasts erode other bone tissue on the inside (fig. 7.10). The resulting

| TABLE 7.2 | OSSIFICATION TIMETABLE | | |
|-----------|------------------------|---|---|
| **Age** | **Occurrence** | **Age** | **Occurrence** |
| Third month of prenatal development | Ossification in long bones begins. | 15 to 18 years (females) 17 to 20 years (males) | Bones of the upper limbs and scapulae completely ossify. |
| Fourth month of prenatal development | Most primary ossification centers have appeared in the diaphyses of bones. | 16 to 21 years (females) 18 to 23 years (males) | Bones of the lower limbs and coxal bones completely ossify. |
| Birth to 5 years | Secondary ossification centers appear in the epiphyses. | 21 to 23 years (females) 23 to 25 years (males) | Bones of the sternum, clavicles, and vertebrae completely ossify. |
| 5 to 12 years in females, or 5 to 14 years in males | Ossification rapidly spreads from the ossification centers, and certain bones are ossifying. | By 23 years (females) By 25 years (males) | Nearly all bones completely ossify. |

space becomes the medullary cavity of the diaphysis, which later fills with marrow.

The bone in the central regions of the epiphyses and diaphysis remains spongy, and hyaline cartilage on the ends of the epiphyses persists throughout life as articular cartilage. Table 7.2 lists the ages at which various bones ossify.

## Homeostasis of Bone Tissue

After the intramembranous and endochondral bones form, the actions of osteoclasts and osteoblasts continually remodel them. Thus, throughout life, osteoclasts resorb bone tissue and osteoblasts replace the bone. These opposing processes of **resorption** and **deposition** are well regulated so that the total mass of bone tissue within an adult skeleton normally remains nearly constant, even though 3% to 5% of bone calcium is exchanged each year.

**FIGURE 7.11**

The presence of epiphyseal disks (arrows) in a child's bone indicates that the bone is still lengthening.

A child's long bones are still growing if a radiograph shows epiphyseal disks (fig. 7.11). If a disk is damaged as a result of a fracture before it ossifies, elongation of that long bone may prematurely cease, or if growth continues, it may be uneven. For this reason, injuries to the epiphyses of a young person's bones are of special concern. On the other hand, an epiphysis is sometimes altered surgically in order to equalize growth of bones that are developing at very different rates.

In bone cancers, abnormally active osteoclasts destroy bone tissue. Interestingly, cancer of the prostate gland can have the opposite effect. If such cancer cells reach the bone marrow, as they do in most cases of advanced prostatic cancer, they stimulate osteoblast activity. This promotes formation of new bone on the surfaces of the bony trabeculae.

## Factors Affecting Bone Development, Growth, and Repair

A number of factors influence bone development, growth, and repair. These include nutrition, exposure to sunlight, hormonal secretions, and physical exercise. For example, vitamin D is necessary for proper absorption of calcium in the small intestine. In the absence of this vitamin, calcium is poorly absorbed, the inorganic salt portion of bone matrix lacks calcium, and bones are deformed. In children, this condition is called *rickets,* and in adults it is called *osteomalacia.*

Vitamin D is relatively uncommon in natural foods, except for eggs. But it is readily available in milk and other dairy products fortified with vitamin D. Vitamin D also forms from a substance (dehydrocholesterol) produced by cells in the digestive tract or obtained in the diet. Dehydrocholesterol is carried by the blood to the skin, and when exposed to ultraviolet light from the sun, it is converted to a compound that becomes vitamin D.

Vitamins A and C are also required for normal bone development and growth. Vitamin A is necessary for bone resorption during normal development. Thus, deficiency of vitamin A may retard bone development. Vitamin C is required for collagen synthesis, so its lack also may inhibit bone development. In this case, osteoblasts produce less collagen in the intercellular material of the bone tissue, and the resulting bones are abnormally slender and fragile.

Hormones secreted by the pituitary gland, thyroid gland, parathyroid glands, and ovaries or testes affect bone growth and development. The pituitary gland, for instance, secretes **growth hormone,** which stimulates reproduction of cartilage cells in the epiphyseal disks. In the absence of this hormone, the long bones of the limbs fail to develop normally, and the child has *pituitary dwarfism.* Such a person is very short, but has normal body proportions. If excessive growth hormone is released before the epiphyseal disks are ossified, height may exceed 8 feet—a condition called *pituitary gigantism.* In an adult, excessive secretion of growth hormone causes a condition called *acromegaly,* in which the hands, feet, and jaw enlarge (see chapter 13).

> Pituitary dwarfism is treated with human growth hormone (HGH). Today, HGH is plentiful and pure, thanks to recombinant DNA technology. Bacteria given the human gene for HGH secrete the hormone. Previously, HGH was pooled from donors or cadavers. This introduced the risk of transmitting infection. A controversial use of HGH is to give it to children who are of short stature, but not abnormally so, or to use it to enhance height with the goal of improving athletic ability.

Thyroid hormone stimulates replacement of cartilage in the epiphyseal disks of long bones with bone tissue. Thyroid hormone can halt bone growth by causing premature ossification of the disks. Deficiency of thyroid hormone also may stunt growth, because without its stimulation, the pituitary gland does not secrete enough growth hormone (see chapter 13).

Both male and female sex hormones (called androgens and estrogens, respectively) from the testes, ovaries, and adrenal glands promote formation of bone tissue. Beginning at puberty, these hormones are abundant, causing the long bones to grow considerably. However, sex hormones also stimulate ossification of the epiphyseal disks, and consequently they stop bone lengthening at a relatively early age. The effect of estrogens on the disks is somewhat stronger than that of androgens. For this reason, females typically reach their maximum heights earlier than males.

Sites of muscle attachments

**FIGURE 7.12**

Note the increased amount of bone at the sites of muscle attachments in the femur on the left. The thickened bone is better able to withstand the force resulting from muscle contraction.

Physical stress also stimulates bone growth. For example, when skeletal muscles contract, they pull at their attachments on bones, and the resulting stress stimulates the bone tissue to thicken and strengthen (hypertrophy). Conversely, with lack of exercise, the same bone tissue wastes, becoming thinner and weaker (atrophy). This is why the bones of athletes are usually stronger and heavier than those of nonathletes (fig. 7.12). It is also why fractured bones immobilized in casts may shorten. Clinical Application 7.1 describes what happens when a bone breaks.

1. Describe the development of an intramembranous bone.
2. Explain how an endochondral bone develops.
3. List the steps in the growth of a long bone.
4. Explain how nutritional factors affect bone development.
5. What effects do hormones have on bone growth?
6. How does physical exercise affect bone structure?

## BONE FUNCTION

Bones shape, support, and protect body structures. They also act as levers that aid body movements, house tissues that produce blood cells, and store various inorganic salts.

## Support and Protection

Bones give shape to structures such as the head, face, thorax, and limbs. They also provide support and protection. For example, the bones of the lower limbs, pelvis, and backbone support the body's weight. The bones of

# Fractures

When 7-year-old Jacob fell from the tree limb he had been hanging from and held out his arm at an odd angle, it was obvious that he had broken a bone. An X ray at the hospital emergency room confirmed this, and Jacob spent the next six weeks with his broken arm immobilized in a cast.

Many of us have experienced fractured, or broken, bones. A fracture is classified by its cause and the nature of the break. For example, a break due to injury is a *traumatic* fracture, whereas one resulting from disease is a *spontaneous*, or *pathologic*, fracture.

A broken bone exposed to the outside by an opening in the skin is termed a *compound (open) fracture.* It has the added danger of infection, because microorganisms enter through the broken skin. A break protected by uninjured skin is a *closed fracture.* Figure 7A shows several types of traumatic fractures.

### REPAIR OF A FRACTURE

Whenever a bone breaks, blood vessels within it and its periosteum rupture, and the periosteum is likely to tear. Blood escaping from the broken vessels spreads through the damaged area and soon forms a blood clot, or *hematoma.* Vessels in surrounding tissues dilate, swelling and inflaming tissues.

Within days or weeks, developing blood vessels and large numbers of osteoblasts originating from the periosteum invade the hematoma. The osteoblasts rapidly multiply in the regions close to the new blood vessels, building spongy bone nearby. Granulation tissue develops, and in regions farther from a blood supply, fibroblasts produce masses of fibrocartilage.

Meanwhile, phagocytic cells begin to remove the blood clot as well as any dead or damaged cells in the affected area. Osteoclasts also appear and resorb bone fragments, aiding in "cleaning up" debris.

In time, fibrocartilage fills the gap between the ends of the broken bone. This mass, termed a cartilaginous *callus,* is later replaced by bone tissue in much the same way that the hyaline cartilage of a developing endochondral bone is replaced. That is, the cartilaginous callus breaks down, blood vessels and osteoblasts invade the area, and a bony callus fills the space.

A *greenstick* fracture is incomplete, and the break occurs on the convex surface of the bend in the bone.

A *fissured* fracture involves an incomplete longitudinal break.

A *comminuted* fracture is complete and fragments the bone.

A *transverse* fracture is complete, and the break occurs at a right angle to the axis of the bone.

An *oblique* fracture occurs at an angle other than a right angle to the axis of the bone.

A *spiral* fracture is caused by twisting a bone excessively.

**FIGURE 7A**
Various types of fractures.

Typically, more bone is produced at the site of a healing fracture than is necessary to replace the damaged tissues. Osteoclasts remove the excess, and the final result is a bone shaped very much like the original. Figure 7B shows the steps in the healing of a fracture.

The rate of fracture repair depends upon several factors. For instance, if the ends of the broken bone are close together, healing is more rapid than if they are far apart. Setting fractured bones and using casts or metal pins to keep the broken ends together help speed healing, as well as aligning the fractured parts. Also, some bones naturally heal more rapidly than others. The long bones of the upper limbs, for example, may heal in half the time required by the long bones of the lower limbs, as Jacob was happy to discover. He also healed quickly because of his young age. ■

(a) Blood escapes from ruptured blood vessels and forms a hematoma.

(b) Spongy bone forms in regions close to developing blood vessels, and fibrocartilage forms in more distant regions.

(c) A bony callus replaces fibrocartilage.

(d) Osteoclasts remove excess bony tissue, restoring new bone structure much like the original.

FIGURE 7B

Major steps in repair of a fracture.

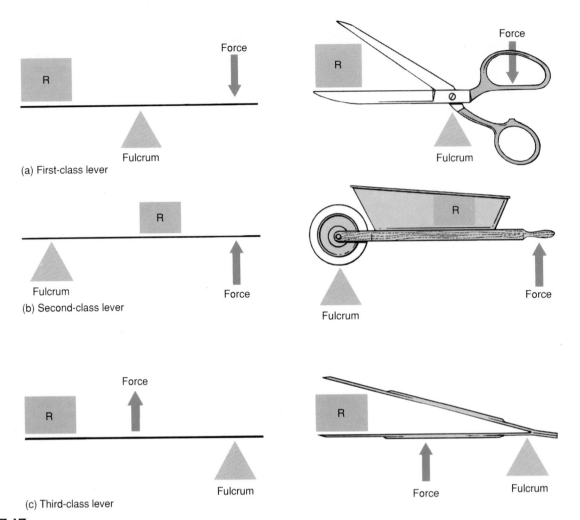

**FIGURE 7.13**

Three types of levers: (*a*) A first-class lever is used in a pair of scissors, (*b*) a second-class lever is used in a wheelbarrow, and (*c*) a third-class lever is used in a pair of forceps.

the skull protect the eyes, ears, and brain. Those of the rib cage and shoulder girdle protect the heart and lungs, whereas bones of the pelvic girdle protect the lower abdominal and internal reproductive organs.

## Body Movement

Whenever limbs or other body parts move, bones and muscles interact as simple mechanical devices called *levers*. A lever has four basic components: (1) a rigid bar or rod, (2) a pivot or fulcrum on which the bar turns, (3) an object that is moved against resistance, and (4) a force that supplies energy for the movement of the bar.

A pair of scissors is a lever. The handle and blade form a rigid bar that rocks on a pivot near the center (the screw). The material to be cut by the blades represents the resistance, while the person on the handle end supplies the force needed for cutting the material.

Figure 7.13 shows the three types of levers, which differ in their arrangements. A first-class lever's parts are like those of a pair of scissors. Its pivot is located between the resistance and the force, making the sequence of components resistance–pivot–force. Other examples of first-class levers are seesaws and hemostats (devices used to clamp blood vessels).

The parts of a second-class lever are in the sequence pivot–resistance–force, as in a wheelbarrow. The parts of a third-class lever are in the sequence resistance–force–pivot. Eyebrow tweezers or forceps used to grasp an object illustrate this type of lever.

The actions of bending and straightening the upper limb at the elbow illustrate bones and muscles functioning as levers (fig. 7.14). When the upper limb bends, the forearm bones represent the rigid bar; the elbow joint is the pivot; the hand is moved against the resistance provided by its weight; and the force is supplied by muscles

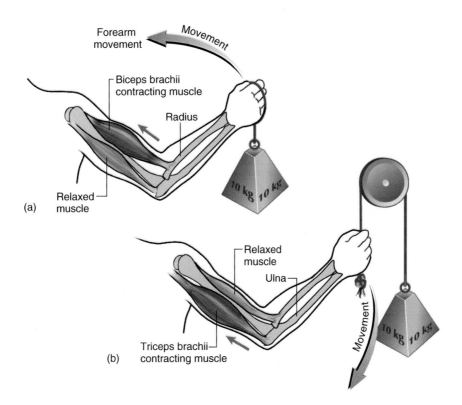

**FIGURE 7.14**

(a) When the forearm bends at the elbow or (b) when the forearm straightens at the elbow, the bones and muscles function as a lever.

on the anterior side of the arm. One of these muscles, the *biceps brachii,* is attached by a tendon to a projection (radial tuberosity) on the *radius* bone in the forearm, a short distance below the elbow. Since the parts of this lever are arranged in the sequence resistance–force–pivot, it is a third-class lever.

When the upper limb straightens at the elbow, the forearm bones again serve as the rigid bar, the hand as the resistance, and the elbow joint as the pivot. However, this time the *triceps brachii,* a muscle located on the posterior side of the arm, supplies the force. A tendon of this muscle is attached to a projection (olecranon process) of the ulna bone at the point of the elbow. Since the parts of the lever are arranged resistance–pivot–force, it is a first-class lever.

Levers provide a range of movements. Levers that move limbs, for example, are arranged in ways that produce rapid motions, whereas others, such as those that move the head, help maintain posture with minimal effort.

## Blood Cell Formation

Very early in life the process of blood cell formation, called **hematopoiesis,** begins in the yolk sac, which lies outside the human embryo (see chapter 23). Later in development, blood cells are manufactured in the liver and spleen, and still later they form in bone marrow.

Marrow is a soft, netlike mass of connective tissue within the medullary cavities of long bones, in the irreg-

ular spaces of spongy bone, and in the larger osteonic canals of compact bone tissue. There are two kinds of marrow—red marrow and yellow marrow. *Red marrow* functions in the formation of red blood cells (erythrocytes), white blood cells (leukocytes), and blood platelets (thrombocytes). It is red because of the red, oxygen-carrying pigment **hemoglobin** contained within the red blood cells.

Red marrow occupies the cavities of most bones in an infant. With increasing age, however, yellow marrow replaces much of it. *Yellow marrow* stores fat and is inactive in blood cell production.

In an adult, red marrow is primarily found in the spongy bone of the skull, ribs, sternum, clavicles, vertebrae, and pelvis. If the blood cell supply is deficient, some yellow marrow may change back into red marrow and produce blood cells. Chapter 14 discusses blood cell formation.

## Inorganic Salt Storage

Recall that the intercellular matrix of bone tissue contains collagen and inorganic mineral salts. The salts account for about 70% of the matrix by weight, and are mostly tiny crystals of a type of calcium phosphate called *hydroxyapatite.* Clinical Application 7.2 discusses osteoporosis, a condition that results from loss of bone mineral.

The human body requires calcium for a number of vital metabolic processes, including blood clot formation,

# Osteoporosis

It is an all-too-familiar scenario. The elderly woman pulls herself out of bed, reaches for the night table for support, and misses. She falls, landing on her hip. A younger woman would pull herself up, and maybe ache for a few minutes and develop a black-and-blue mark by the next day. But the 80-year-old, with weakened, brittle bones, suffers a broken hip. Each year in the United States, 200,000 senior citizens break their hips, more than 90% of the time as the result of an accident.

In **osteoporosis,** the skeletal system loses bone volume and mineral content. This disorder is associated with aging. Within affected bones, trabeculae are lost, and the bones develop spaces and canals, which enlarge and fill with fibrous and fatty tissues. Such bones easily fracture and may spontaneously break because they are no longer able to support body weight. For example, a person with osteoporosis may suffer a spontaneous fracture of the thigh bone (femur) at the hip or the collapse of sections of the backbone (vertebrae). Similarly, the distal portion of a forearm bone (radius) near the wrist may be fractured as a result of a minor stress.

Osteoporosis causes many fractures in persons over forty-five years of age. Although it may affect either gender, it is most common in thin, light-complexioned females after menopause (see chapter 22).

Factors that increase the risk of osteoporosis include low intake of dietary calcium and lack of physical exercise (particularly during the early growing years). However, excessively strenuous exercise in adolescence can delay puberty, which raises the risk of developing osteoporosis later in life for both sexes.

In females, declining levels of the hormone estrogen contribute to development of osteoporosis. The ovaries produce estrogen until menopause. Evidence of the estrogen–osteoporosis link comes from studies on women who have declining estrogen levels and increased risk of osteoporosis. These include young women who have had their ovaries removed, women who have anorexia nervosa (self-starvation) that stopped their menstrual cycles, and women past menopause. Drinking alcohol, smoking cigarettes, and inheriting certain genes may also increase a person's risk of developing osteoporosis.

Osteoporosis may be prevented if steps are taken early enough. Bone mass usually peaks at about age thirty-five. Thereafter, bone loss may exceed bone formation in both males and females. To reduce such loss, doctors suggest that people in their mid-twenties and older should take in 1,000–1,500 milligrams of calcium daily. An 8-ounce glass of nonfat milk, for example, contains about 275 milligrams of calcium. It is also recommended that people engage in regular exercise, especially walking or jogging, in which the bones support body weight. Additionally, post-menopausal women may require estrogen replacement therapy. As a rule, women have about 30% less bone mass than men; after menopause, women typically lose bone mass twice as fast as men do. Also, people with osteoporosis can slow progress of the disease by taking a drug that is a form of the hormone calcitonin, if they can tolerate the side effect of throat irritation.

Confirming osteoporosis is sometimes difficult. A radiograph may not reveal a decrease in bone density until 20% to 30% of the bone tissue is lost. Noninvasive diagnostic techniques, however, can detect rapid changes in bone mass. These include a *densitometer scanner* that measures the density of wrist bones, and *quantitative computed tomography,* which can visualize the density of other bones.

Alternatively, a physician may take a bone sample, usually from a hipbone, to directly assess the condition of the tissue. Such a biopsy may also be used to judge the effectiveness of treatment for bone disease. ■

nerve impulse conduction, and muscle cell contraction. When the blood is low in calcium, parathyroid hormone stimulates osteoclasts to break down bone tissue, releasing calcium salts from the intercellular matrix into the blood. On the other hand, very high blood calcium inhibits osteoclast activity, and calcitonin from the thyroid gland stimulates osteoblasts to form bone tissue, storing excess calcium in the matrix (fig. 7.15). The details of this homeostatic mechanism are presented in chapter 13.

In addition to storing calcium and phosphorus (as calcium phosphate), bone tissue contains lesser amounts of magnesium, sodium, potassium, and carbonate ions.

**Figure 7.15**
Hormonal regulation of bone calcium resorption and deposition.

Calcium is consumed, blood calcium levels increase.

↓

Thyroid gland releases calcitonin.

↓

Osteoblasts deposit calcium in bones.

↓

Blood calcium levels are returned to normal (homeostasis).

↑

Osteoclasts break down bone to release calcium.

↑

Parathyroid gland releases parathyroid hormone.

↑

Blood calcium levels are low.

Bones also accumulate certain harmful metallic elements such as lead, radium, and strontium, which are not normally present in the body but are sometimes accidentally ingested.

1. Name three major functions of bones.

2. Explain how parts of the upper limb form a first-class lever and a third-class lever.

3. Distinguish between the functions of red marrow and yellow marrow.

4. Explain regulation of the concentration of blood calcium.

5. List the substances normally stored in bone tissue.

## SKELETAL ORGANIZATION

### Number of Bones

The number of bones in a human skeleton is often reported to be 206, but the actual number varies from person to person. People may lack certain bones or have

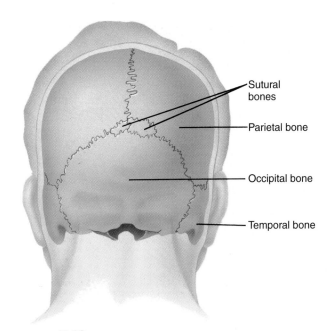

**Figure 7.16**
Sutural bones are extra bones that sometimes develop in sutures between the flat bones of the skull.

extra ones. For example, the flat bones of the skull usually grow together and tightly join along irregular lines called **sutures.** Occasionally, extra bones called sutural bones (wormian bones) develop in these sutures (fig. 7.16). Extra small, round sesamoid bones may develop in tendons, where they reduce friction in places where tendons pass over bony prominences (table 7.3).

### Divisions of the Skeleton

For purposes of study, it is convenient to divide the skeleton into two major portions—an axial skeleton and an appendicular skeleton (fig. 7.17). The **axial skeleton** consists of the bony and cartilaginous parts that support and protect the organs of the head, neck, and trunk. These parts include the following:

1. **Skull.** The skull is composed of the *cranium* (brain case) and the *facial bones.*

2. **Hyoid bone.** The hyoid (hi′oid) bone is located in the neck between the lower jaw and the larynx (fig. 7.18). It does not articulate with any other bones but is fixed in position by muscles and ligaments. The hyoid bone supports the tongue and is an attachment for certain muscles that help move the tongue during swallowing. It can be felt approximately a finger's width above the anterior prominence of the larynx.

3. **Vertebral column.** The vertebral column, or backbone, consists of many vertebrae separated by cartilaginous *intervertebral disks.* This column

| TABLE 7.3 | BONES OF THE ADULT SKELETON | |
|---|---|---|

**1. Axial Skeleton**

| | | |
|---|---|---|
| a. **Skull** | | 22 bones |
| | 8 cranial bones | |
| |   frontal 1 | |
| |   parietal 2 | |
| |   occipital 1 | |
| |   temporal 2 | |
| |   sphenoid 1 | |
| |   ethmoid 1 | |
| | 13 facial bones | |
| |   maxilla 2 | |
| |   palatine 2 | |
| |   zygomatic 2 | |
| |   lacrimal 2 | |
| |   nasal 2 | |
| |   vomer 1 | |
| |   inferior nasal concha 2 | |
| | 1 mandible | |
| b. **Middle ear bones** | | 6 bones |
| | malleus 2 | |
| | incus 2 | |
| | stapes 2 | |
| c. **Hyoid** | | 1 bone |
| | hyoid bone 1 | |
| d. **Vertebral column** | | 26 bones |
| | cervical vertebra 7 | |
| | thoracic vertebra 12 | |
| | lumbar vertebra 5 | |
| | sacrum 1 | |
| | coccyx 1 | |
| e. **Thoracic cage** | | 25 bones |
| | rib 24 | |
| | sternum 1 | |

**2. Appendicular Skeleton**

| | | |
|---|---|---|
| a. **Pectoral girdle** | | 4 bones |
| | scapula 2 | |
| | clavicle 2 | |
| b. **Upper limbs** | | 60 bones |
| | humerus 2 | |
| | radius 2 | |
| | ulna 2 | |
| | carpal 16 | |
| | metacarpal 10 | |
| | phalanx 28 | |
| c. **Pelvic girdle** | | 2 bones |
| | coxal bone 2 | |
| d. **Lower limbs** | | 60 bones |
| | femur 2 | |
| | tibia 2 | |
| | fibula 2 | |
| | patella 2 | |
| | tarsal 14 | |
| | metatarsal 10 | |
| | phalanx 28 | |
| | **Total** | **206 bones** |

forms the central axis of the skeleton. Near its distal end, several vertebrae fuse to form the **sacrum,** which is part of the pelvis. A small, rudimentary tailbone called the **coccyx** is attached to the end of the sacrum.

4. **Thorax.** The thorax protects the organs of the thoracic cavity and the upper abdominal cavity. It is composed of twelve pairs of **ribs,** which articulate posteriorly with thoracic vertebrae. It also includes the **sternum** (ster'num), or breastbone, to which most of the ribs are attached anteriorly.

The **appendicular skeleton** consists of the bones of the upper and lower limbs and the bones that anchor the limbs to the axial skeleton. It includes the following:

1. **Pectoral girdle.** The pectoral girdle is formed by a **scapula** (scap'u-lah), or shoulder blade, and a **clavicle** (klav'ĭ-k'l), or collarbone, on both sides of the body. The pectoral girdle connects the bones of the upper limbs to the axial skeleton and aids in upper limb movements.

2. **Upper limbs.** Each upper limb consists of a **humerus** (hu'mer-us), or arm bone; two forearm bones—a **radius** (ra'de-us) and an **ulna** (ul'nah); and a hand. The humerus, radius, and ulna articulate with each other at the elbow joint. At the distal end of the radius and ulna is the hand. The hand includes the wrist, palm, and fingers. There are eight **carpals** (kar'pals) or wrist bones. The five bones of the palm are called **metacarpals,** and the fourteen finger bones are called **phalanges** (fah-lan'jēz).

3. **Pelvic girdle.** The pelvic girdle is formed by two **coxal** (kok'sal) bones or hipbones, which are attached to each other anteriorly and to the sacrum posteriorly. They connect the bones of the lower limbs to the axial skeleton and, with the sacrum and coccyx, form the **pelvis,** which protects the lower abdominal and internal reproductive organs.

4. **Lower limbs.** Each lower limb consists of a **femur** (fe'mur), or thighbone; two leg bones—a large **tibia** (tib'e-ah), or shinbone, and a slender **fibula** (fib'u-lah), or calf bone; and a foot. The femur, tibia, and fibula articulate with each other at the knee joint, where the **patella** (pah-tel'ah), or kneecap, covers the anterior surface. At the distal ends of the tibia and fibula is the foot. The foot includes the ankle, instep, and toes. There are seven **tarsals** (tahr'sals) or anklebones. The five bones of the instep are called **metatarsals,** and the fourteen bones of the toes (like the fingers) are called **phalanges.** Table 7.4 defines some terms used to describe skeletal structures.

**1** Distinguish between the axial and appendicular skeletons.

**2** List the bones of the axial skeleton and of the appendicular skeleton.

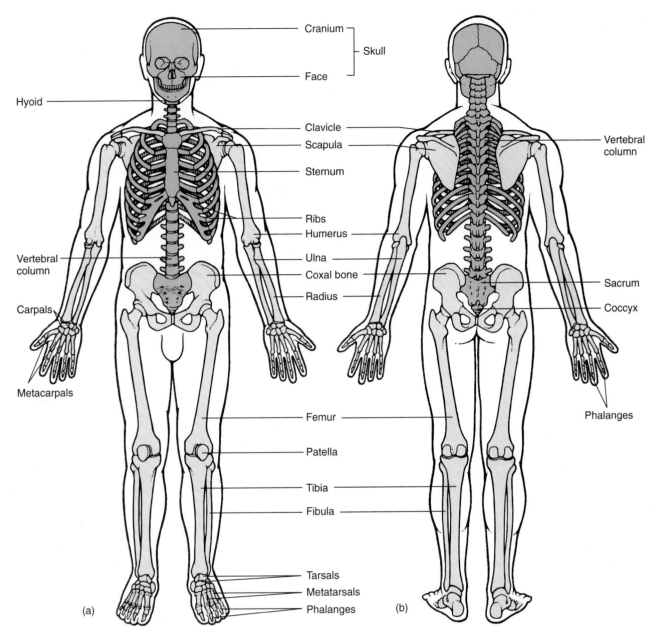

**FIGURE 7.17**

Major bones of the skeleton. (*a*) Anterior view. (*b*) Posterior view. The axial portions are shown in orange and the appendicular portions are shown in yellow.

## SKULL

A human skull usually consists of twenty-two bones that, except for the lower jaw, are firmly interlocked along lines called sutures. Eight of these interlocked bones make up the cranium, and thirteen form the facial skeleton. The **mandible** (man′dǐ-b′l), or lower jawbone, is a movable bone held to the cranium by ligaments (figs. 7.19, 7.21, and 7.22). Some facial and cranial bones together form the orbit of the eye (fig. 7.20). Plates 8–36 on pages 241–255 show a set of photographs of the human skull and its parts.

## Cranium

The **cranium** (kra′ne-um) encloses and protects the brain, and its surface provides attachments for muscles that make chewing and head movements possible. Some of the cranial bones contain air-filled cavities called *sinuses*, which are lined with mucous membranes and connect by passageways to the nasal cavity. Sinuses reduce the weight of the skull and increase the intensity of the voice by serving as resonant sound chambers.

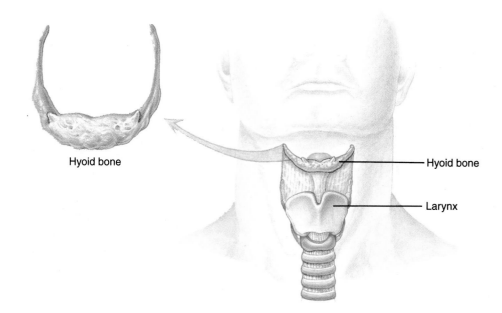

**FIGURE 7.18**

The hyoid bone supports the tongue and serves as an attachment for muscles that move the tongue and function in swallowing.

Hyoid bone

Hyoid bone

Larynx

**TABLE 7.4 TERMS USED TO DESCRIBE SKELETAL STRUCTURES**

| Term | Definition | Example |
|------|------------|---------|
| condyle (kon′dīl) | A rounded process that usually articulates with another bone | Occipital condyle of the occipital bone (fig. 7.22) |
| crest (krest) | A narrow, ridgelike projection | Iliac crest of the ilium (fig. 7.50) |
| epicondyle (ep″ĭ-kon′dīl) | A projection situated above a condyle | Medial epicondyle of the humerus (fig. 7.45) |
| facet (fas′et) | A small, nearly flat surface | Facet of a thoracic vertebra (fig. 7.38) |
| fissure (fish′ūr) | A cleft or groove | Inferior orbital fissure in the orbit of the eye (fig. 7.20) |
| fontanel (fon″tah-nel′) | A soft spot in the skull where membranes cover the space between bones | Anterior fontanel between the frontal and parietal bones (fig. 7.33) |
| foramen (fo-ra′men) | An opening through a bone that usually serves as a passageway for blood vessels, nerves, or ligaments | Foramen magnum of the occipital bone (fig. 7.22) |
| fossa (fos′ah) | A relatively deep pit or depression | Olecranon fossa of the humerus (fig. 7.45) |
| fovea (fo′ve-ah) | A tiny pit or depression | Fovea capitis of the femur (fig. 7.53) |
| head (hed) | An enlargement on the end of a bone | Head of the humerus (fig. 7.45) |
| linea (lin′e-ah) | A narrow ridge | Linea aspera of the femur (fig. 7.53) |
| meatus (me-a′tus) | A tubelike passageway within a bone | External auditory meatus of the ear (fig. 7.21) |
| process (pros′es) | A prominent projection on a bone | Mastoid process of the temporal bone (fig. 7.21) |
| ramus (ra′mus) | A branch or similar extension | Ramus of the mandible (fig. 7.31) |
| sinus (si′nus) | A cavity within a bone | Frontal sinus of the frontal bone (fig. 7.27) |
| spine (spīn) | A thornlike projection | Spine of the scapula (fig. 7.43) |
| suture (soo′cher) | An interlocking line of union between bones | Lambdoidal suture between the occipital and parietal bones (fig. 7.21) |
| trochanter (tro-kan′ter) | A relatively large process | Greater trochanter of the femur (fig. 7.53) |
| tubercle (tu′ber-kl) | A small, knoblike process | Tubercle of a rib (fig. 7.41) |
| tuberosity (tu″bĕ-ros′ĭ-te) | A knoblike process usually larger than a tubercle | Radial tuberosity of the radius (fig. 7.46) |

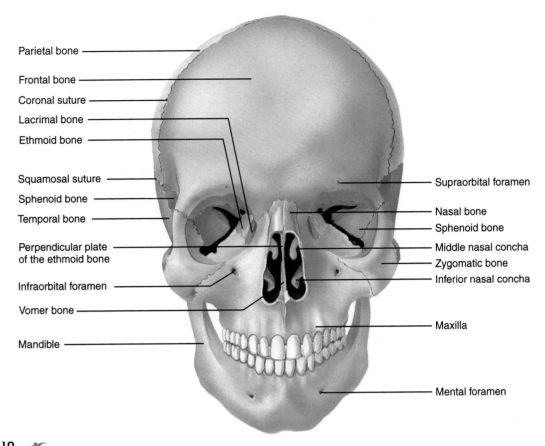

**FIGURE 7.19**

Anterior view of the skull.

Parietal bone

Frontal bone

Coronal suture

Lacrimal bone

Ethmoid bone

Squamosal suture

Sphenoid bone

Temporal bone

Perpendicular plate of the ethmoid bone

Infraorbital foramen

Vomer bone

Mandible

Supraorbital foramen

Nasal bone

Sphenoid bone

Middle nasal concha

Zygomatic bone

Inferior nasal concha

Maxilla

Mental foramen

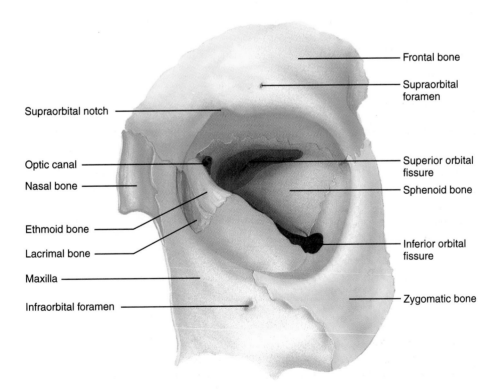

**FIGURE 7.20**

The orbit of the eye includes both cranial and facial bones.

Supraorbital notch

Optic canal

Nasal bone

Ethmoid bone

Lacrimal bone

Maxilla

Infraorbital foramen

Frontal bone

Supraorbital foramen

Superior orbital fissure

Sphenoid bone

Inferior orbital fissure

Zygomatic bone

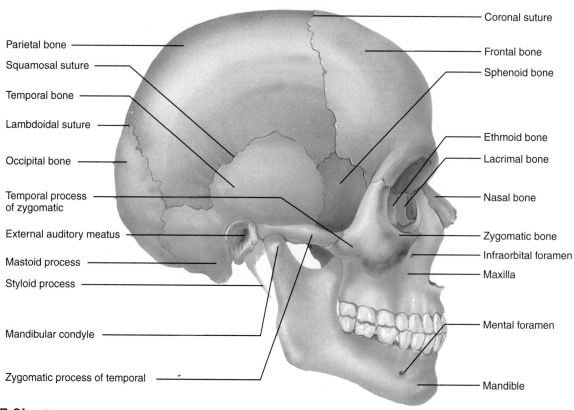

Parietal bone

Squamosal suture

Temporal bone

Lambdoidal suture

Occipital bone

Temporal process
of zygomatic

External auditory meatus

Mastoid process

Styloid process

Mandibular condyle

Zygomatic process of temporal

Coronal suture

Frontal bone

Sphenoid bone

Ethmoid bone

Lacrimal bone

Nasal bone

Zygomatic bone

Infraorbital foramen

Maxilla

Mental foramen

Mandible

**FIGURE 7.21**
Lateral view of the skull.

Zygomatic bone

Sphenoid bone

Vomer bone

Zygomatic arch

Mandibular fossa

Styloid process

External auditory meatus

Mastoid process

Occipital condyle

Temporal bone

Incisive foramen

Palatine process of
maxilla

Median palatine suture

Palatine bone

Greater palatine
foramen

Foramen lacerum

Foramen ovale

Foramen spinosum

Carotid canal

Jugular foramen

Stylomastoid foramen

Foramen magnum

Lambdoidal suture

Occipital bone

**FIGURE 7.22**
Inferior view of the skull.

TABLE 7.5 CRANIAL BONES

| Name and Number | Description | Special Features |
|---|---|---|
| Frontal (1) | Forms forehead, roof of nasal cavity, and roofs of orbits | Supraorbital foramen, frontal sinuses |
| Parietal (2) | Form side walls and roof of cranium | Fused at midline along sagittal suture |
| Occipital (1) | Forms back of skull and base of cranium | Foramen magnum, occipital condyles |
| Temporal (2) | Form side walls and floor of cranium | External auditory meatus, mandibular fossa, mastoid process, styloid process, zygomatic process |
| Sphenoid (1) | Forms parts of base of cranium, sides of skull, and floors and sides of orbits | Sella turcica, sphenoidal sinuses |
| Ethmoid (1) | Forms parts of roof and walls of nasal cavity, floor of cranium, and walls of orbits | Cribriform plates, perpendicular plate, superior and middle nasal conchae, ethmoidal sinuses, crista galli |

The eight bones of the cranium (table 7.5) are as follows:

1. **Frontal bone.**   The frontal (frun'tal) bone forms the anterior portion of the skull above the eyes, including the forehead, the roof of the nasal cavity, and the roofs of the orbits (bony sockets) of the eyes. On the upper margin of each orbit, the frontal bone is marked by a *supraorbital foramen* through which blood vessels and nerves pass to the tissues of the forehead. Within the frontal bone are two *frontal sinuses,* one above each eye near the midline. The frontal bone is a single bone in adults, but it develops in two parts (see fig. 7.33). These halves grow together and usually completely fuse by the fifth or sixth year of life.

2. **Parietal bones.**   One parietal (pah-ri'ĕ-tal) bone is located on each side of the skull just behind the frontal bone. Each is shaped like a curved plate and has four borders. Together, the parietal bones form the bulging sides and roof of the cranium. They are fused at the midline along the *sagittal suture,* and they meet the frontal bone along the *coronal suture.*

3. **Occipital bone.**   The occipital (ok-sip'ĭ-tal) bone joins the parietal bones along the *lambdoidal* (lam'doid-al) *suture.* It forms the back of the skull and the base of the cranium. A large opening on its lower surface, the *foramen magnum,* houses nerve fibers from the brain that pass through and enter the vertebral canal to become part of the spinal cord. Rounded processes called *occipital condyles,* located on each side of the foramen magnum,

articulate with the first vertebra (atlas) of the vertebral column.

4. **Temporal bones.**   A temporal (tem'por-al) bone on each side of the skull joins the parietal bone along a *squamosal* (skwa-mo'sal) *suture.* The temporal bones form parts of the sides and the base of the cranium. Located near the inferior margin is an opening, the *external auditory* (acoustic) *meatus,* which leads inward to parts of the ear. The temporal bones also house the internal ear structures and have depressions called the *mandibular fossae* (glenoid fossae) that articulate with condyles of the mandible. Below each external auditory meatus are two projections—a rounded *mastoid process* and a long, pointed *styloid process.* The mastoid process provides an attachment for certain muscles of the neck, whereas the styloid process anchors muscles associated with the tongue and pharynx. An opening near the mastoid process, the *carotid canal,* transmits the internal carotid artery. An opening between the temporal and occipital bones, the *jugular foramen,* accommodates the internal jugular vein (see fig. 7.22).

The mastoid process may become infected. The tissues in this region of the temporal bone contain a number of interconnected air cells lined with mucous membranes that communicate with the middle ear. These spaces sometimes become inflamed when microorganisms spread into them from an infected middle ear (*otitis media*). The resulting mastoid infection, called *mastoiditis,* is of particular concern because nearby membranes that surround the brain may become infected.

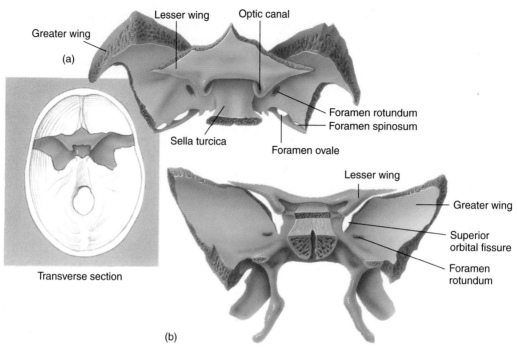

**FIGURE 7.23**

(a) The sphenoid bone viewed from above. (b) Posterior view. (The sphenoidal sinuses are within the bone and are not visible in this representation.)

A *zygomatic process* projects anteriorly from the temporal bone in the region of the external auditory meatus. It joins the *zygomatic bone* and helps form the prominence of the cheek.

5. **Sphenoid bone.** The sphenoid (sfe'noid) bone (fig. 7.23) is wedged between several other bones in the anterior portion of the cranium. It consists of a central part and two winglike structures that extend laterally toward each side of the skull. This bone helps form the base of the cranium, the sides of the skull, and the floors and sides of the orbits. Along the midline within the cranial cavity, a portion of the sphenoid bone indents to form the saddle-shaped *sella turcica* (sel'ah tur'si-ka). In this depression lies the pituitary gland, which hangs from the base of the brain by a stalk.

The sphenoid bone also contains two *sphenoidal sinuses.* These lie side by side and are separated by a bony septum that projects downward into the nasal cavity.

6. **Ethmoid bone.** The ethmoid (eth'moid) bone (fig. 7.24) is located in front of the sphenoid bone. It consists of two masses, one on each side of the nasal cavity, which are joined horizontally by thin *cribriform* (krib'rĭ-form) *plates.* These plates form part of the roof of the nasal cavity, and nerves associated with the sense of smell pass through tiny openings (*olfactory foramina*) in them. Portions of the ethmoid bone also form sections of the cranial floor, orbital walls, and nasal cavity walls. A *perpendicular plate* projects downward in the midline from the cribriform plates to form most of the nasal septum.

Delicate, scroll-shaped plates called the *superior nasal concha* (kong'kah) and the *middle nasal concha* project inward from the lateral portions of the ethmoid bone toward the perpendicular plate. These bony plates support mucous membranes that line the nasal cavity. The mucous membranes, in turn, begin moistening, warming, and filtering air as it enters the respiratory tract. The lateral portions of the ethmoid bone contain many small air spaces, the *ethmoidal sinuses.* Figure 7.25 shows various structures in the nasal cavity.

Projecting upward into the cranial cavity between the cribriform plates is a triangular process of the ethmoid bone called the *crista galli* (kris'tă gal'li; cock's comb). Membranes that enclose the brain attach to this process. Figure 7.26 shows a view of the cranial cavity.

Transverse section

**FIGURE 7.24**

The ethmoid bone viewed (*a*) from above and (*b*) from behind.

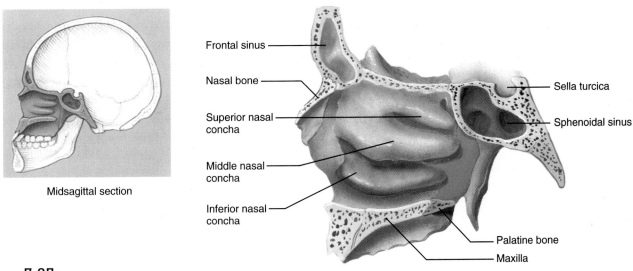

Midsagittal section

**FIGURE 7.25**

Lateral wall of the nasal cavity.

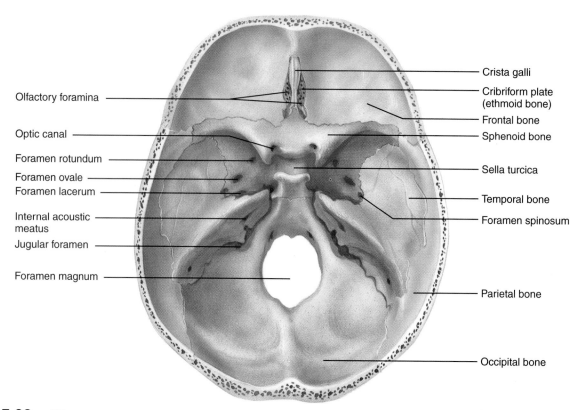

Olfactory foramina

Optic canal

Foramen rotundum

Foramen ovale

Foramen lacerum

Internal acoustic meatus

Jugular foramen

Foramen magnum

Crista galli

Cribriform plate (ethmoid bone)

Frontal bone

Sphenoid bone

Sella turcica

Temporal bone

Foramen spinosum

Parietal bone

Occipital bone

**FIGURE 7.26**

Floor of the cranial cavity viewed from above.

Trepanation was an early treatment in many cultures in the Old and New World. The procedure involves drilling holes in the skull and sometimes removing pieces of cranium. Trepanation was used for physical ailments such as fractures, or for spiritual ills, such as headaches, seizures, and mental disorders. Early evidence of trepanation is a male skull from about 7,000 years ago found in a Neolithic burial site in Alsace. The skull contained two drill holes, one in the frontal bone, the other covering both parietal bones. New bone had formed a thin shell over the smaller, anterior lesion, and had partially covered the larger hole. The fact that the man lived to the old age of 55 attests to the skill of the long-ago surgeon. Neanderthals, who lived 20,000 to 30,000 years ago, also used trepanation. Modern researchers used CT scans to identify the trepanation, which can superficially resemble fractures or mauling of a skull after death.

## Facial Skeleton

The **facial skeleton** consists of thirteen immovable bones and a movable lower jawbone. In addition to forming the basic shape of the face, these bones provide attachments for muscles that move the jaw and control facial expressions.

The bones of the facial skeleton are as follows:

1. **Maxillary bones.** The maxillary (mak′sĭ-ler″e) bones (pl., *maxillae,* mak-sĭl′e) form the upper jaw; together they form the keystone of the face, since all the other immovable facial bones articulate with them.

Portions of these bones comprise the anterior roof of the mouth (*hard palate*), the floors of the orbits, and the sides and floor of the nasal cavity. They also contain the sockets of the upper teeth. Inside the maxillae, lateral to the nasal cavity, are *maxillary sinuses*. These spaces are the largest of the sinuses, and they extend from the floor of the orbits to the roots of the upper teeth. Figure 7.27 shows the locations of the maxillary and other sinuses. Table 7.6 displays a summary of the sinuses.

During development, portions of the maxillary bones called *palatine processes* grow together and fuse along the midline, or median palatine suture. This forms the anterior section of the hard palate (see fig. 7.22).

The inferior border of each maxillary bone projects downward, forming an *alveolar* (al-ve′o-lar) *process*. Together these processes form a horseshoe-shaped *alveolar arch* (dental arch). Teeth occupy cavities in this arch (dental alveoli). Dense connective tissue binds teeth to the bony sockets (see chapter 17).

**FIGURE 7.27**

Locations of the sinuses.

| TABLE 7.6 | SINUSES OF THE CRANIAL AND FACIAL BONES | | |
|---|---|---|---|
| **Sinuses** | **Number** | **Location** | |
| Frontal sinuses | 2 | Frontal bone above each eye and near the midline | |
| Sphenoidal sinuses | 2 | Sphenoid bone above the posterior portion of the nasal cavity | |
| Ethmoidal sinuses | 2 groups of small spaces | Ethmoid bone on either side of the upper portion of the nasal cavity | |
| Maxillary sinuses | 2 | Maxillary bones lateral to the nasal cavity and extending from the floor of the orbits to the roots of the upper teeth | |

In cleft palate, the palatine processes of the maxillae are incompletely fused at birth. However, the defect in facial bone development actually occurs early in development. Infants with cleft palate may have trouble suckling because of the opening between the oral and nasal cavities. A temporary prosthetic device (artificial palate) may be inserted within the mouth, or a special type of rubber nipple can be used on bottles, until surgery can be performed.

2. **Palatine bones.** The L-shaped palatine (pal'ah-tīn) bones (fig. 7.28) are located behind the maxillae. The horizontal portions form the posterior section of the hard palate and the floor of the nasal cavity. The perpendicular portions help form the lateral walls of the nasal cavity.

3. **Zygomatic bones.** The zygomatic (zi″go-mat′ik) bones are responsible for the prominences of the cheeks below and to the sides of the eyes. These bones also help form the lateral walls and the floors of the orbits. Each bone has a *temporal process,* which extends posteriorly to join the zygomatic process of a temporal bone. Together these processes form a *zygomatic arch* (see figs. 7.21 and 7.22).

4. **Lacrimal bones.** A lacrimal (lak′rĭ-mal) bone is a thin, scalelike structure located in the medial wall of each orbit between the ethmoid bone and the maxilla (see fig. 7.21). A groove in its anterior portion leads from the orbit to the nasal cavity, providing a pathway for a channel that carries tears from the eye to the nasal cavity.

5. **Nasal bones.** The nasal (na′zal) bones are long, thin, and nearly rectangular (see fig. 7.19). They lie side by side and are fused at the midline, where they form the bridge of the nose. These bones are attachments for the cartilaginous tissues that form the shape of the nose.

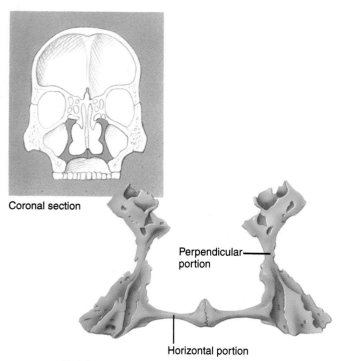

Coronal section

**FIGURE** 7.28

The horizontal portions of the palatine bones form the posterior section of the hard palate, and the perpendicular portions help form the lateral walls of the nasal cavity.

Perpendicular portion

Horizontal portion

6. **Vomer bone.** The thin, flat vomer (vo'mer) bone is located along the midline within the nasal cavity. Posteriorly, it joins the perpendicular plate of the ethmoid bone, and together they form the nasal septum (figs. 7.29 and 7.30).

7. **Inferior nasal conchae.** The inferior nasal conchae (kong'ke) are fragile, scroll-shaped bones attached to the lateral walls of the nasal cavity. They are the largest of the conchae and are positioned below the superior and middle nasal conchae of the ethmoid bone (see figs. 7.19 and 7.25). Like the ethmoidal conchae, the inferior conchae support mucous membranes within the nasal cavity.

8. **Mandible.** The mandible (man'dĭ-b'l), or lower jawbone, is a horizontal, horseshoe-shaped body with a flat *ramus* projecting upward at each end. The rami are divided into a posterior *mandibular condyle* and an anterior *coronoid* process (fig. 7.31). The mandibular condyles articulate with the mandibular fossae of the temporal bones, whereas the coronoid processes provide attachments for muscles used in chewing. Other large chewing muscles are inserted on the lateral surfaces of the rami. A curved bar of bone on the superior border of the mandible, the

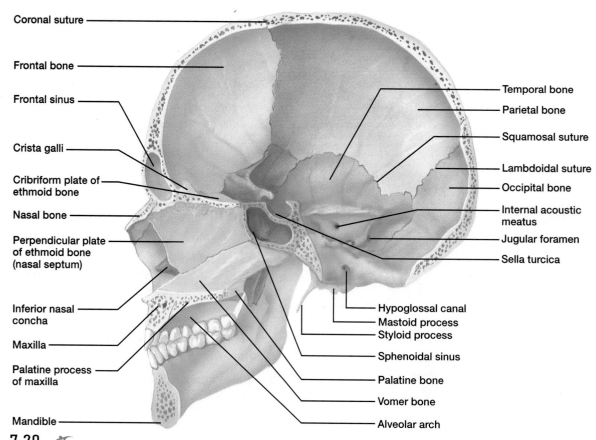

Coronal suture
Frontal bone
Frontal sinus
Crista galli
Cribriform plate of ethmoid bone
Nasal bone
Perpendicular plate of ethmoid bone (nasal septum)
Inferior nasal concha
Maxilla
Palatine process of maxilla
Mandible

Temporal bone
Parietal bone
Squamosal suture
Lambdoidal suture
Occipital bone
Internal acoustic meatus
Jugular foramen
Sella turcica
Hypoglossal canal
Mastoid process
Styloid process
Sphenoidal sinus
Palatine bone
Vomer bone
Alveolar arch

**FIGURE** 7.29

Sagittal section of the skull.

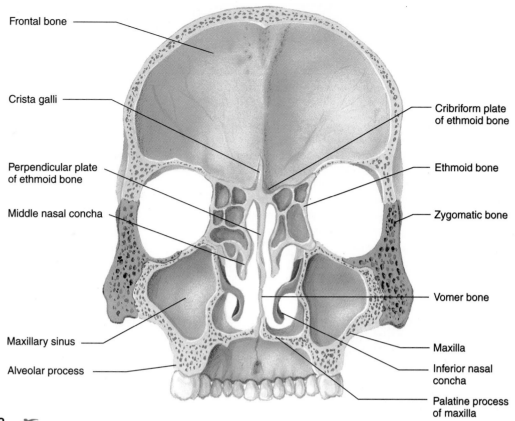

Frontal bone

Crista galli

Cribriform plate of ethmoid bone

Perpendicular plate of ethmoid bone

Ethmoid bone

Middle nasal concha

Zygomatic bone

Vomer bone

Maxillary sinus

Maxilla

Alveolar process

Inferior nasal concha

Palatine process of maxilla

**FIGURE 7.30**

Coronal section of the skull (posterior view).

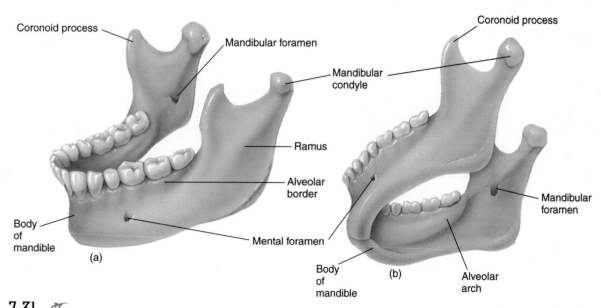

Coronoid process

Mandibular foramen

Coronoid process

Mandibular condyle

Ramus

Alveolar border

Mandibular foramen

Body of mandible

Mental foramen

Body of mandible

Alveolar arch

(a)

(b)

**FIGURE 7.31**

(*a*) Lateral view of the mandible. (*b*) Inferior view.

TABLE

7.7  BONES OF THE
FACIAL SKELETON

| Name and Number | Description | Special Features |
|---|---|---|
| Maxillary (2) | Form upper jaw, anterior roof of mouth, floors of orbits, and sides and floor of nasal cavity | Alveolar processes, maxillary sinuses, palatine process |
| Palatine (2) | Form posterior roof of mouth, and floor and lateral walls of nasal cavity | |
| Zygomatic (2) | Form prominences of cheeks, and lateral walls and floors of orbits | Temporal process |
| Lacrimal (2) | Form part of medial walls of orbits | Groove that leads from orbit to nasal cavity |
| Nasal (2) | Form bridge of nose | |
| Vomer (1) | Forms inferior portion of nasal septum | |
| Inferior nasal conchae (2) | Extend into nasal cavity from its lateral walls | |
| Mandible (1) | Forms lower jaw | Body, ramus, mandibular condyle, coronoid process, alveolar process, mandibular foramen, mental foramen |

(a)

(b)

**FIGURE 7.32**

Radiographs of the skull (a) from the front and (b) from the side.

*alveolar border,* contains the hollow sockets (dental alveoli) that bear the lower teeth.

On the medial side of the mandible, near the center of each ramus, is a *mandibular foramen.* This opening admits blood vessels and a nerve, which supply the roots of the lower teeth. Dentists inject anesthetic into the tissues near this foramen to temporarily block nerve impulse conduction and desensitize teeth on that side of the jaw. Branches

of the blood vessels and the nerve emerge from the mandible through the *mental foramen,* which opens on the outside near the point of the jaw. They supply the tissues of the chin and lower lip.

Table 7.7 describes the fourteen facial bones. Figure 7.32 shows features of these bones on radiographs. Table 7.8 lists the major openings (*foramina*) and passageways through bones of the skull, as well as their general locations and the structures they transmit.

TABLE 7.8

## PASSAGEWAYS THROUGH BONES OF THE SKULL

| Passageway | Location | Major Stuctures Transmitted |
|---|---|---|
| Carotid canal (fig. 7.22) | Inferior surface of the temporal bone | Internal carotid artery, veins, and nerves |
| Foramen lacerum (fig. 7.22) | Floor of cranial cavity between temporal and sphenoid bones | Branch of pharyngeal artery (in life, opening is largely covered by fibrocartilage) |
| Foramen magnum (fig. 7.26) | Base of skull in occipital bone | Nerve fibers passing between the brain and spinal cord as it exits from the base of the brain, also certain arteries |
| Foramen ovale (fig. 7.22) | Floor of cranial cavity in sphenoid bone | Mandibular division of trigeminal nerve and veins |
| Foramen rotundum (fig. 7.26) | Floor of cranial cavity in sphenoid bone | Maxillary division of trigeminal nerve |
| Foramen spinosum (fig. 7.26) | Floor of cranial cavity in sphenoid bone | Middle meningeal blood vessels and branch of mandibular nerve |
| Greater palatine foramen (fig. 7.22) | Posterior portion of hard palate in palatine bone | Palatine blood vessels and nerves |
| Hypoglossal canal (fig. 7.29) | Near margin of foramen magnum in occipital bone | Hypoglossal nerve |
| Incisive foramen (fig. 7.22) | Anterior portion of hard palate | Nasopalatine nerves, openings of vomeronasal organ |
| Inferior orbital fissure (fig. 7.20) | Floor of the orbit | Maxillary nerve and blood vessels |
| Infraorbital foramen (fig. 7.20) | Below the orbit in maxillary bone | Infraorbital blood vessels and nerves |
| Internal acoustic meatus (fig. 7.26) | Floor of cranial cavity in temporal bone | Branches of facial and vestibulocochlear nerves, and blood vessels |
| Jugular foramen (fig. 7.26) | Base of the skull between temporal and occipital bones | Glossopharyngeal, vagus and accessory nerves, and blood vessels |
| Mandibular foramen (fig. 7.31) | Inner surface of ramus of mandible | Inferior alveolar blood vessels and nerves |
| Mental foramen (fig. 7.31) | Near point of jaw in mandible | Mental nerve and blood vessels |
| Optic canal (fig. 7.20) | Posterior portion of orbit in sphenoid bone | Optic nerve and ophthalmic artery |
| Stylomastoid foramen (fig. 7.22) | Between styloid and mastoid processes | Facial nerve and blood vessels |
| Superior orbital fissure (fig. 7.20) | Lateral wall of orbit | Oculomotor, trochlear, and abducens nerves, and ophthalmic division of trigeminal nerve |
| Supraorbital foramen (fig. 7.19) | Upper margin or orbit in frontal bone | Supraorbital blood vessels and nerves |

## Infantile Skull

At birth, the skull is incompletely developed, with fibrous membranes connecting the cranial bones. These membranous areas are called **fontanels** (fon″tah-nel′z) or, more commonly, soft spots. They permit some movement between the bones, so that the developing skull is partially compressible and can slightly change shape. This action, called *molding,* enables an infant's skull to more easily pass through the birth canal. Eventually, the fontanels close as the cranial bones grow together. The posterior fontanel usually closes about two months after birth; the sphenoid fontanel closes at about three months; the mastoid fontanel closes near the end of the first year; and the anterior fontanel may not close until the middle or end of the second year.

Other characteristics of an infantile skull (fig. 7.33) include a relatively small face with a prominent forehead and large orbits. The jaw and nasal cavity are small, the paranasal sinuses are incompletely formed, and the frontal bone is in two parts. The skull bones are thin, but they are also somewhat flexible and thus are less easily fractured than adult bones.

In the infantile skull, a frontal suture (metopic suture) separates the two parts of the developing frontal bone in the midline. This suture usually closes before the sixth year; however, in a few adults, the frontal suture remains open.

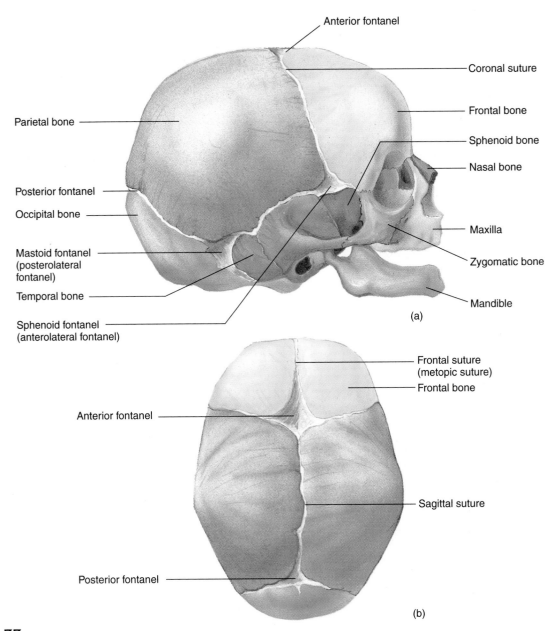

**FIGURE 7.33**

(*a*) Lateral view and (*b*) superior view of the infantile skull.

**1** Locate and name each of the bones of the cranium.

**2** Locate and name each of the facial bones.

**3** Explain how an adult skull differs from that of an infant.

## VERTEBRAL COLUMN

The **vertebral column** extends from the skull to the pelvis and forms the vertical axis of the skeleton (fig. 7.34). It is composed of many bony parts called **vertebrae** (ver'tĕ-bre) that are separated by masses of fibrocartilage called *intervertebral disks* and are connected to one another by ligaments. The vertebral column supports the head and the trunk of the body, yet is flexible enough to permit movements, such as bending forward, backward, or to the side, and turning or rotating on the central axis. It also protects the spinal cord, which passes through a *vertebral canal* formed by openings in the vertebrae.

An infant has thirty-three separate bones in the vertebral column. Five of these bones eventually fuse to form the sacrum, and four others join to become the coccyx. As a result, an adult vertebral column has twenty-six bones.

Normally, the vertebral column has four curvatures, which give it a degree of resiliency. The names of the curves correspond to the regions in which they occur, as shown in figure 7.34. The *thoracic* and *pelvic curvatures*

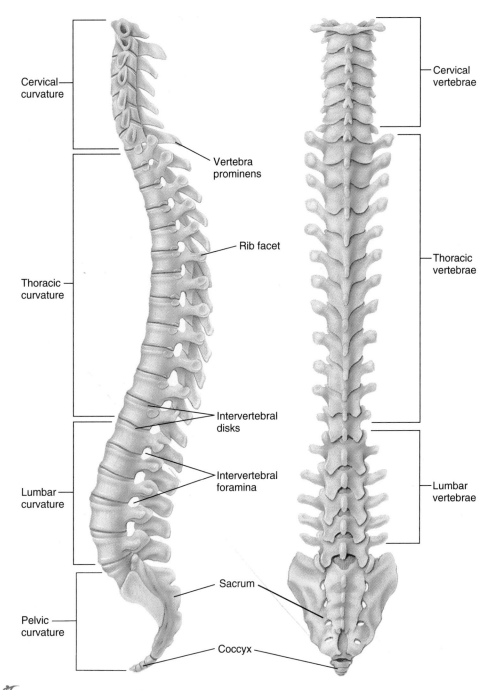

Cervical curvature

Vertebra prominens

Rib facet

Thoracic curvature

Intervertebral disks

Intervertebral foramina

Lumbar curvature

Sacrum

Pelvic curvature

Coccyx

Cervical vertebrae

Thoracic vertebrae

Lumbar vertebrae

**FIGURE** 7.34

The curved vertebral column consists of many vertebrae separated by intervertebral disks. (*a*) Left lateral view. (*b*) Posterior view.

are concave anteriorly and are called primary curves. The *cervical curvature* in the neck and the *lumbar curvature* in the lower back are convex anteriorly and are called secondary curves. The cervical curvature develops when a baby begins to hold up its head, and the lumbar curvature develops when the child begins to stand.

## A Typical Vertebra

Although the vertebrae in different regions of the vertebral column have special characteristics, they also have features in common. A typical vertebra (fig. 7.35) has a

drum-shaped *body* (centrum), which forms the thick, anterior portion of the bone. A longitudinal row of these vertebral bodies supports the weight of the head and trunk. The intervertebral disks, which separate adjacent vertebrae, are fastened to the roughened upper and lower surfaces of the vertebral bodies. These disks cushion and soften the forces caused by such movements as walking and jumping, which might otherwise fracture vertebrae or jar the brain. The bodies of adjacent vertebrae are joined on their anterior surfaces by *anterior ligaments* and on their posterior surfaces by *posterior ligaments*.

**FIGURE 7.35**

(*a*) Lateral view of a typical thoracic vertebra. (*b*) Adjacent vertebrae join at their articulating processes. (*c*) Superior view of a typical thoracic vertebra.

Projecting posteriorly from each vertebral body are two short stalks called *pedicles* (ped′ĭ-k'lz). They form the sides of the *vertebral foramen.* Two plates called *laminae* (lam′ĭ-ne) arise from the pedicles and fuse in the back to become a *spinous process.* The pedicles, laminae, and spinous process together complete a bony *vertebral arch* around the vertebral foramen, through which the spinal cord passes.

Between the pedicles and laminae of a typical vertebra is a *transverse process,* which projects laterally and posteriorly. Various ligaments and muscles are attached to the dorsal spinous process and the transverse processes. Projecting upward and downward from each vertebral arch are *superior* and *inferior articulating processes.* These processes bear cartilage-covered facets by which each vertebra is joined to the one above and the one below it.

On the lower surfaces of the vertebral pedicles are notches that align to help form openings called *intervertebral foramina* (in″ter-ver′tĕ-bral fo-ram′ĭ-nah). These openings provide passageways for spinal nerves that proceed between adjacent vertebrae and connect to the spinal cord.

Gymnasts, jumpers, pole vaulters, and other athletes who repeatedly land on hard surfaces sometimes break or crack their vertebrae, particularly the articulating processes. Such a vertebral fracture is called *spondylolysis.* Limiting hard landings and using padding and floor mats can prevent these injuries.

### Cervical Vertebrae

Seven **cervical vertebrae** comprise the bony axis of the neck. These are the smallest of the vertebrae, but their bone tissues are denser than those in any other region of the vertebral column.

The transverse processes of the cervical vertebrae are distinctive because they have *transverse foramina,* which are passageways for arteries leading to the brain. Also, the spinous processes of the second through the fifth cervical vertebrae are uniquely forked (bifid). These processes provide attachments for muscles.

The spinous process of the seventh vertebra is longer and protrudes beyond the other cervical spines. It is called the *vertebra prominens,* and because it can be

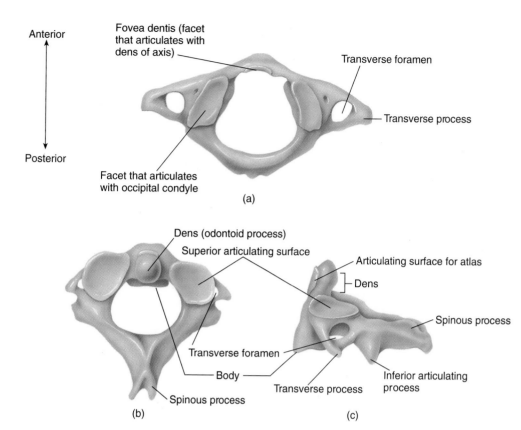

FIGURE 7.36

Superior view of the (a) atlas and (b) axis. (c) Lateral view of the axis.

felt through the skin, it is a useful landmark for locating other vertebral parts (see fig. 7.34).

Two of the cervical vertebrae, shown in figure 7.36, are of special interest. The first vertebra, or **atlas** (at'las), supports the head. It has practically no body or spine and appears as a bony ring with two transverse processes. On its superior surface, the atlas has two kidney-shaped *facets,* which articulate with the occipital condyles.

The second cervical vertebra, or **axis** (ak'sis), bears a toothlike *dens* (odontoid process) on its body. This process projects upward and lies in the ring of the atlas. As the head is turned from side to side, the atlas pivots around the dens (figs. 7.36 and 7.37).

## Thoracic Vertebrae

The twelve **thoracic vertebrae** are larger than those in the cervical region. Each vertebra has a long, pointed spinous process, which slopes downward, and facets on the sides of its body, which articulate with a rib.

Beginning with the third thoracic vertebra and moving inferiorly, the bodies of these bones increase in size. Thus, they are adapted to bear increasing loads of body weight.

FIGURE 7.37

Radiograph of the cervical vertebrae.

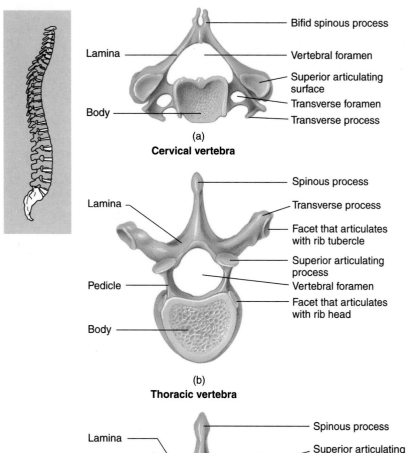

Lamina

Body

Bifid spinous process

Vertebral foramen

Superior articulating surface

Transverse foramen

Transverse process

(a)
**Cervical vertebra**

Lamina

Pedicle

Body

Spinous process

Transverse process

Facet that articulates with rib tubercle

Superior articulating process

Vertebral foramen

Facet that articulates with rib head

(b)
**Thoracic vertebra**

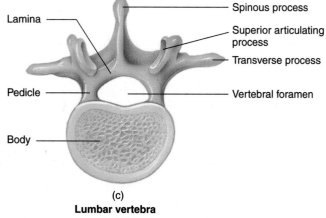

Lamina

Pedicle

Body

Spinous process

Superior articulating process

Transverse process

Vertebral foramen

(c)
**Lumbar vertebra**

**FIGURE 7.38**

Superior view of (a) a cervical vertebra, (b) a thoracic vertebra, and (c) a lumbar vertebra.

## Lumbar Vertebrae

The five **lumbar vertebrae** in the small of the back (loins) support more weight than the superior vertebrae, and have larger and stronger bodies. Compared to other types of vertebrae, the transverse processes of these vertebrae project posteriorly at sharp angles, whereas their short, thick spinous processes are nearly horizontal. Figure 7.38 compares the structures of the cervical, thoracic, and lumbar vertebrae.

Gymnasts, football players, and others who bend their vertebral columns excessively and forcefully may experience slipping of a vertebra over the one below it. This painful condition is called *spondylolisthesis*. Usually, the fifth lumbar vertebra slides forward over the body of the sacrum. Doing exercises to strengthen the back muscles associated with the vertebral column may prevent this condition.

## Sacrum

The **sacrum** (sa′krum) is a triangular structure at the base of the vertebral column. It is composed of five vertebrae that develop separately but gradually fuse between ages 18 and 30. The spinous processes of these fused bones form a ridge of tubercles, the *median sacral crest*. Nerves and blood vessels pass through rows of openings, called the *dorsal sacral foramina,* located to the sides of the tubercles (fig. 7.39).

The sacrum is wedged between the coxal bones of the pelvis and is united to them at its *auricular surfaces* by fibrocartilage of the *sacroiliac joints*. The pelvic girdle, at these joints, transmits the body's weight to the legs (see fig. 7.17).

The sacrum forms the posterior wall of the pelvic cavity. The upper anterior margin of the sacrum, which represents the body of the first sacral vertebra, is called the *sacral promontory* (sa′kral prom′on-to″re). During a vaginal examination, a physician can feel this projection and use it as a guide in determining the size of the pelvis. This measurement is helpful in estimating how easily an infant may be able to pass through a woman's pelvic cavity during childbirth.

The vertebral foramina of the sacral vertebrae form the *sacral canal,* which continues through the sacrum to an opening of variable size at the tip, called the *sacral hiatus* (hi-a′tus). This foramen exists because the laminae of the last sacral vertebra are not fused. On the ventral surface of the sacrum, four pairs of *pelvic sacral foramina* provide passageways for nerves and blood vessels.

## Coccyx

The **coccyx** (kok′siks), or tailbone, is the lowest part of the vertebral column and is usually composed of four vertebrae that fuse by the twenty-fifth year. Ligaments attach it to the margins of the sacral hiatus (see fig. 7.39). Sitting

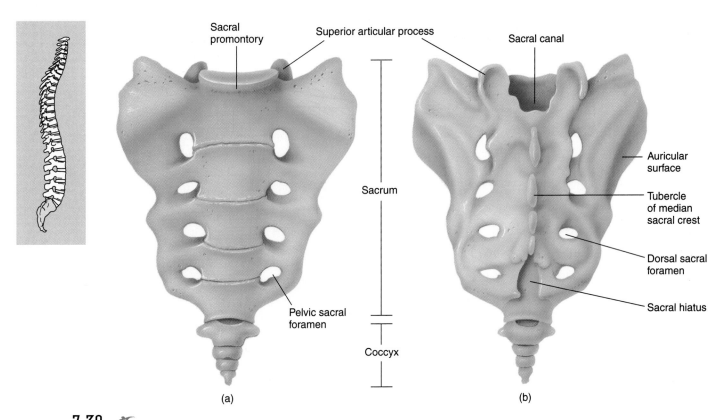

**FIGURE 7.39**

(a) Anterior view of the sacrum and coccyx. (b) Posterior view.

| | | | | | | | |
|---|---|---|---|---|---|---|---|
| **TABLE 7.9** | | **BONES OF THE VERTEBRAL COLUMN** | | | | | |
| **Bones** | **Number** | **Special Features** | | **Bones** | **Number** | **Special Features** | |
| Cervical vertebrae | 7 | Transverse foramina; facets of atlas articulate with occipital condyles of skull; dens of axis articulates with atlas; spinous processes of second through fifth vertebrae are bifid | | Lumbar vertebrae | 5 | Large bodies; transverse processes that project posteriorly at sharp angles; short, thick spinous processes directed nearly horizontally | |
| | | | | Sacrum | 5 vertebrae fused into 1 bone | Dorsal sacral foramina, auricular surfaces, sacral promontory, sacral canal, sacral hiatus, pelvic sacral foramina | |
| Thoracic vertebrae | 12 | Pointed spinous processes that slope downward; facets that articulate with ribs | | Coccyx | 4 vertebrae fused into 1 bone | Attached by ligaments to the margins of the sacral hiatus | |

presses on the coccyx; and it moves forward, acting like a shock absorber. Sitting down with great force can fracture or dislocate the coccyx. Table 7.9 summarizes the bones of the vertebral column, and Clinical Application 7.3 discusses disorders of the vertebral column.

**1** Describe the structure of the vertebral column.

**2** Explain the difference between the vertebral column of an adult and that of an infant.

**3** Describe a typical vertebra.

**4** How do the structures of cervical, thoracic, and lumbar vertebrae differ?

# Disorders of the Vertebral Column

Changes in the intervertebral disks may cause various problems. Each disk is composed of a tough, outer layer of fibrocartilage (annulus fibrosus) and an elastic central mass (nucleus pulposus). With age, these disks degenerate—the central masses lose firmness and the outer layers thin and weaken, developing cracks. Extra pressure, as when a person falls or lifts a heavy object, can break the outer layers of the disks, squeezing out the central masses. Such a rupture may press on the spinal cord or on spinal nerves that branch from it. This condition, called a *ruptured,* or *herniated, disk,* may cause back pain and numbness or loss of muscular function in the parts innervated by the affected spinal nerves.

A surgical procedure called a *laminectomy* may relieve the pain of a herniated disk by removing a portion of the posterior arch of a vertebra. This reduces the pressure on the affected nerve tissues. Alternatively, a protein-digesting enzyme (chymopapain) may be injected into the injured disk to shrink it.

Sometimes problems develop in the curvatures of the vertebral column because of poor posture, injury, or disease. An exaggerated thoracic curvature causes rounded shoulders and a hunchback. This condition, called *kyphosis,* occasionally develops in adolescents who undertake strenuous athletic activities. Unless corrected before

bone growth completes, the condition can permanently deform the vertebral column.

Sometimes the vertebral column develops an abnormal lateral curvature, so that one hip or shoulder is lower than the other. This may displace or compress the thoracic and abdominal organs. With unknown cause, this condition, called *scoliosis,* is most common in adolescent females. It also may accompany such diseases as poliomyelitis, rickets, or tuberculosis. An accentuated lumbar curvature is called *lordosis,* or swayback.

As a person ages, the intervertebral disks tend to shrink and become more rigid, and compression is more likely to fracture the vertebral bodies. Consequently, height may decrease, and the thoracic curvature of the vertebral column may be accentuated, bowing the back. ■

## THORACIC CAGE

The **thoracic cage** includes the ribs, the thoracic vertebrae, the sternum, and the costal cartilages that attach the ribs to the sternum. These bones support the shoulder girdle and upper limbs, protect the viscera in the thoracic and upper abdominal cavities, and play a role in breathing (fig. 7.40).

### Ribs

The usual number of rib pairs is twelve—one pair attached to each of the twelve thoracic vertebrae. Some individuals develop extra ribs associated with their cervical or lumbar vertebrae.

The first seven rib pairs, which are called the *true ribs* (vertebrosternal ribs), join the sternum directly by their costal cartilages. The remaining five pairs are called *false ribs,* because their cartilages do not reach the sternum directly. Instead, the cartilages of the upper three false ribs (vertebrochondral ribs) join the cartilages of the

seventh rib, whereas the last two rib pairs have no attachments to the sternum. These last two pairs (or sometimes the last three pairs) are called *floating ribs* (vertebral ribs).

A typical rib (fig. 7.41) has a long, slender shaft, which curves around the chest and slopes downward. On the posterior end is an enlarged *head* by which the rib articulates with a facet on the body of its own vertebra and with the body of the next higher vertebra. The neck of the rib is flattened, lateral to the head, where ligaments attach. A *tubercle,* close to the head of the rib, articulates with the transverse process of the vertebra.

The costal cartilages are composed of hyaline cartilage. They are attached to the anterior ends of the ribs and continue in line with them toward the sternum.

### Sternum

The **sternum** (ster′num), or breastbone, is located along the midline in the anterior portion of the thoracic cage. It is a flat, elongated bone that develops in three parts—an upper *manubrium* (mah-nu′bre-um), a middle *body,* and

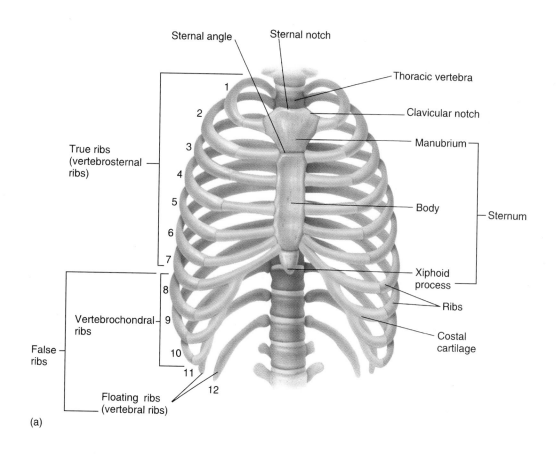

Sternal angle

Sternal notch

Thoracic vertebra

Clavicular notch

Manubrium

Body

Sternum

Xiphoid process

Ribs

Costal cartilage

True ribs (vertebrosternal ribs)

1
2
3
4
5
6
7

False ribs

Vertebrochondral ribs

8
9
10
11
12

Floating ribs (vertebral ribs)

(a)

**FIGURE 7.40**

(a) The thoracic cage includes the thoracic vertebrae, the sternum, the ribs, and the costal cartilages that attach the ribs to the sternum. (b) Radiograph of the thoracic cage viewed from the front. The light region behind the sternum and above the diaphragm is the heart.

(b)

a lower *xiphoid* (zif′oid) *process* that projects downward (see fig. 7.40).

The sides of the manubrium and the body are notched where they articulate with costal cartilages. The manubrium also articulates with the clavicles by facets on its superior border. It usually remains as a separate bone until middle age or later, when it fuses to the body of the sternum.

The manubrium and body of the sternum lie in different planes, so that the line of union between them projects slightly forward. This projection, which occurs at the level of the second costal cartilage, is called the *sternal angle* (angle of Louis). It is commonly used as a clinical landmark to locate a particular rib accurately (see fig. 7.40).

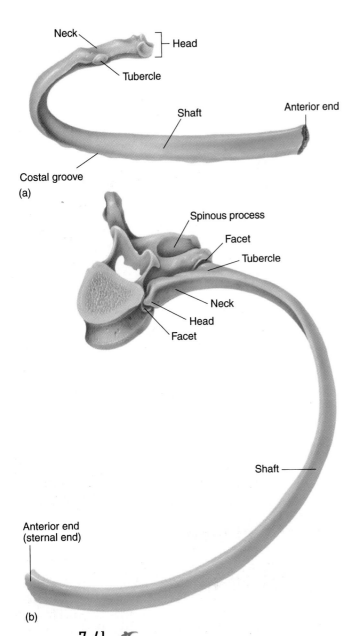

Neck

Head

Tubercle

Shaft

Anterior end

Costal groove

(a)

Spinous process

Facet

Tubercle

Neck

Head

Facet

Shaft

Anterior end
(sternal end)

(b)

**FIGURE 7.41**

(a) A typical rib (posterior view). (b) Articulations of a rib with a thoracic vertebra (superior view).

The xiphoid process begins as a piece of cartilage. It slowly ossifies, and by middle age it usually fuses to the body of the sternum also.

Red marrow within the spongy bone of the sternum produces blood cells into adulthood. Since the sternum has a thin covering of compact bone and is easy to reach, samples of its marrow may be removed to diagnose diseases. This procedure, a *sternal puncture*, suctions (aspirates) some marrow through a hollow needle. (Marrow may also be removed from the iliac crest of a coxal bone.)

1 Which bones comprise the thoracic cage?

2 Describe a typical rib.

3 What are the differences among true, false, and floating ribs?

## PECTORAL GIRDLE

The **pectoral** (pek'to-ral) **girdle** (shoulder girdle) is composed of four parts—two clavicles (collarbones) and two scapulae (shoulder blades). Although the word *girdle* suggests a ring-shaped structure, the pectoral girdle is an incomplete ring. It is open in the back between the scapulae, and the sternum separates its bones in front. The pectoral girdle supports the upper limbs and is an attachment for several muscles that move them (fig. 7.42).

### Clavicles

The **clavicles** are slender, rodlike bones with elongated S-shapes (fig. 7.42). Located at the base of the neck, they run horizontally between the sternum and the shoulders. The medial (or sternal) ends of the clavicles articulate with the manubrium, and the lateral (or acromial) ends join processes of the scapulae.

The clavicles brace the freely movable scapulae, helping to hold the shoulders in place. They also provide attachments for muscles of the upper limbs, chest, and back. Because of its elongated double curve, the clavicle is structurally weak. If compressed lengthwise due to abnormal pressure on the shoulder, it is likely to fracture.

### Scapulae

The **scapulae** are broad, somewhat triangular bones located on either side of the upper back. They have flat bodies with concave anterior surfaces. The posterior surface of each scapula is divided into unequal portions by a *spine*. Above the spine is the *supraspinous fossa*, and below the spine is the *infraspinous fossa*. This spine leads to a *head*, which bears two processes—an *acromion* (ah-kro'me-on) *process* that forms the tip of the shoulder and a *coracoid* (kor'ah-koid) *process* that curves anteriorly and inferiorly to the clavicle (fig. 7.43). The acromion process articulates with the clavicle and provides attachments for muscles of the upper limb and chest. The coracoid process also provides attachments for upper limb and chest muscles. On the head of the scapula between the processes is a depression called the *glenoid cavity* (glenoid fossa of the scapula). It articulates with the head of the arm bone (humerus).

The scapula has three borders. The *superior border* is on the superior edge. The *axillary* or *lateral border* is directed toward the upper limb. The *vertebral* or *medial border* is closest to the vertebral column, about 5 cm away.

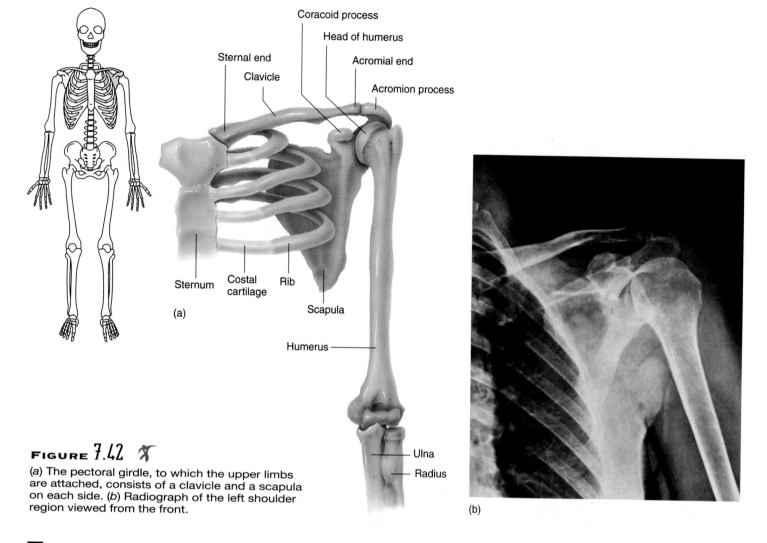

**FIGURE 7.42**

(*a*) The pectoral girdle, to which the upper limbs are attached, consists of a clavicle and a scapula on each side. (*b*) Radiograph of the left shoulder region viewed from the front.

**1** Which bones form the pectoral girdle?

**2** What is the function of the pectoral girdle?

In the epic poem the *Iliad,* Homer describes a man whose "shoulders were bent and met over his chest." The man probably had a rare inherited condition, called cleidocranial dysplasia, in which certain bones do not grow. The skull consists of small fragments joined by connective tissue, rather than large, interlocking hard bony plates. The scapulae are stunted or missing.

Cleidocranial dysplasia was first reported in a child in the huge Arnold family, founded by a Chinese immigrant to South Africa. The child had been kicked by a horse, and X rays revealed that the fontanels atop the head had never closed. The condition became known as "Arnold head." In 1997, researchers traced the condition to a malfunctioning gene that normally instructs certain cells to specialize as bone. Mice missing both copies of this gene develop a skeleton that is completely cartilage—bone never replaces the original cartilage model.

## UPPER LIMB

The bones of the upper limb form the framework of the arm, forearm, and hand. They also provide attachments for muscles, and they function in levers that move limb parts. These bones include a humerus, a radius, an ulna, carpals, metacarpals, and phalanges (fig. 7.44).

### Humerus

The **humerus** (fig. 7.45) is a heavy bone that extends from the scapula to the elbow. At its upper end is a smooth, rounded *head* that fits into the glenoid cavity of the scapula. Just below the head are two processes—a *greater tubercle* on the lateral side and a *lesser tubercle* on the anterior side. These tubercles provide attachments for muscles that move the upper limb at the shoulder. Between them is a narrow furrow, the *intertubercular groove,* through which a tendon passes from a muscle in the arm (biceps brachii) to the shoulder.

The narrow depression along the lower margin of the head that separates it from the tubercles is called the *anatomical neck.* Just below the head and the tubercles of the humerus is a tapering region called the *surgical*

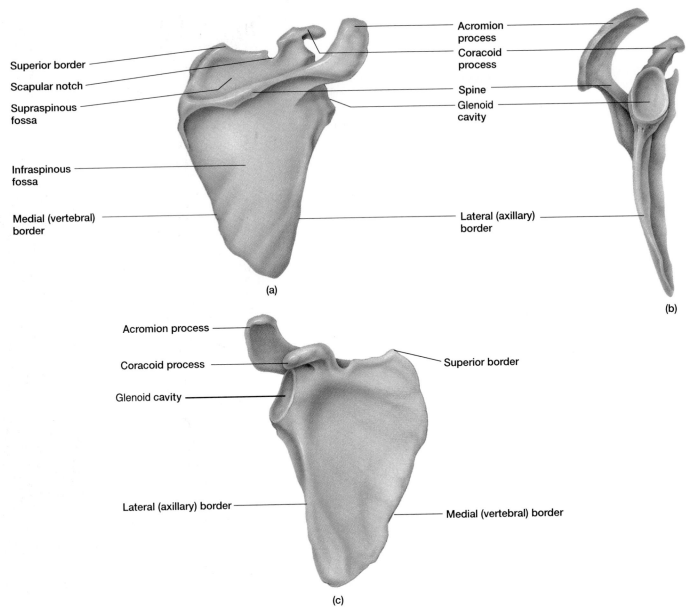

**FIGURE 7.43**

(a) Posterior surface of the right scapula. (b) Lateral view showing the glenoid cavity that articulates with the head of the humerus. (c) Anterior surface.

*neck,* so named because fractures commonly occur there. Near the middle of the bony shaft on the lateral side is a rough V-shaped area called the *deltoid tuberosity.* It provides an attachment for the muscle (deltoid) that raises the upper limb horizontally to the side.

At the lower end of the humerus are two smooth *condyles*—a knoblike *capitulum* (kah-pit'u-lum) on the lateral side and a pulley-shaped *trochlea* (trok'le-ah) on the medial side. The capitulum articulates with the radius at the elbow, whereas the trochlea joins the ulna.

Above the condyles on either side are *epicondyles,* which provide attachments for muscles and ligaments of the elbow. Between the epicondyles anteriorly is a depression, the *coronoid* (kor'o-noid) *fossa,* that receives a process of the ulna (coronoid process) when the elbow bends. An-

other depression on the posterior surface, the *olecranon* (o"lek'ra-non) *fossa,* receives an ulnar process (olecranon process) when the upper limb straightens at the elbow.

Many a thirtyish parent of a young little leaguer or softball player becomes tempted to join in. But if he or she has not pitched in many years, sudden activity may break the forearm. Forearm pain while pitching is a signal that a fracture could happen. Medical specialists advise returning to the pitching mound gradually. Start with twenty pitches, five days a week, for two to three months before regular games begin. By the season's start, 120 pitches per daily practice session should be painless.

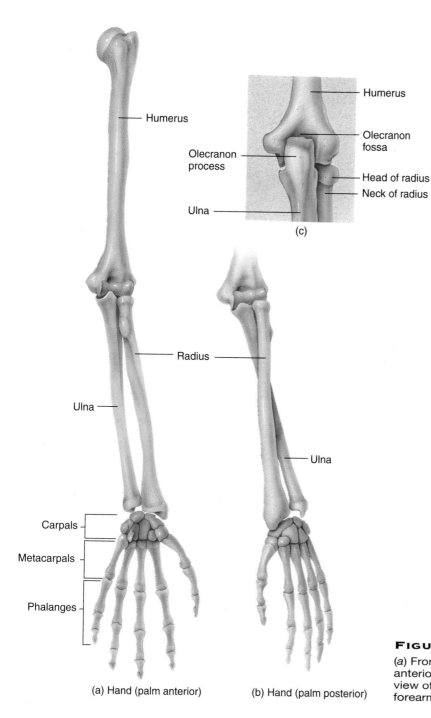

Humerus

Olecranon
process

Ulna

(c)

Humerus

Olecranon
fossa

Head of radius

Neck of radius

Humerus

Radius

Ulna

Ulna

Carpals

Metacarpals

Phalanges

(a) Hand (palm anterior)

(b) Hand (palm posterior)

(d)

**FIGURE** 7.44 🏃

(a) Frontal view of the left upper limb with the hand (palm anterior) and (b) with the hand (palm posterior). (c) Posterior view of the right elbow. (d) Radiograph of the left elbow and forearm, viewed anteriorly.

## Radius

The **radius,** located on the thumb side of the forearm, is somewhat shorter than its companion, the ulna (fig. 7.46). The radius extends from the elbow to the wrist and crosses over the ulna when the hand is turned so that the palm faces backward.

A thick, disklike *head* at the upper end of the radius articulates with the capitulum of the humerus and a notch of the ulna (radial notch). This arrangement allows the radius to rotate freely.

On the radial shaft just below the head is a process called the *radial tuberosity.* It is an attachment for a muscle

(biceps brachii) that bends the upper limb at the elbow. At the distal end of the radius, a lateral *styloid* (sti'loid) *process* provides attachments for ligaments of the wrist.

## Ulna

The **ulna** is longer than the radius and overlaps the end of the humerus posteriorly. At its proximal end, the ulna has a wrenchlike opening, the *trochlear notch* (semilunar notch), that articulates with the trochlea of the humerus. A process lies on either side of this notch. The *olecranon process,* located above the trochlear notch, provides an attachment for the muscle (triceps brachii) that straightens the upper limb at the elbow. During this movement, the olecranon process of the ulna fits into the olecranon fossa of the humerus. Similarly, the *coronoid process,* just below the trochlear notch, fits into the coronoid fossa of the humerus when the elbow bends.

At the distal end of the ulna, its knoblike *head* articulates laterally with a notch of the radius (ulnar notch) and with a disk of fibrocartilage inferiorly (see fig. 7.46). This disk, in turn, joins a wrist bone (triangular). A medial *styloid process* at the distal end of the ulna provides attachments for ligaments of the wrist.

## Hand

The hand is composed of a wrist, a palm, and five fingers (fig. 7.47). The skeleton of the wrist consists of eight small **carpal bones** that are firmly bound in two rows of four bones each. The resulting compact mass is called a *carpus* (kar′pus).

The carpus is rounded on its proximal surface, where it articulates with the radius and with the fibrocartilaginous disk on the ulnar side. The carpus is concave anteriorly, forming a canal through which tendons and nerves extend to the palm. Its distal surface articulates with the metacarpal bones. Figure 7.47 names the individual bones of the carpus.

Five **metacarpal** (met″ah-kar′pal) **bones,** one in line with each finger, form the framework of the palm. These bones are cylindrical, with rounded distal ends that form the knuckles of a clenched fist. The metacarpals articulate proximally with the carpals and distally with the phalanges. The metacarpal on the lateral side is the most freely movable; it permits the thumb to oppose the fingers when grasping something. These bones are numbered 1 to 5, beginning with the metacarpal of the thumb (fig. 7.47).

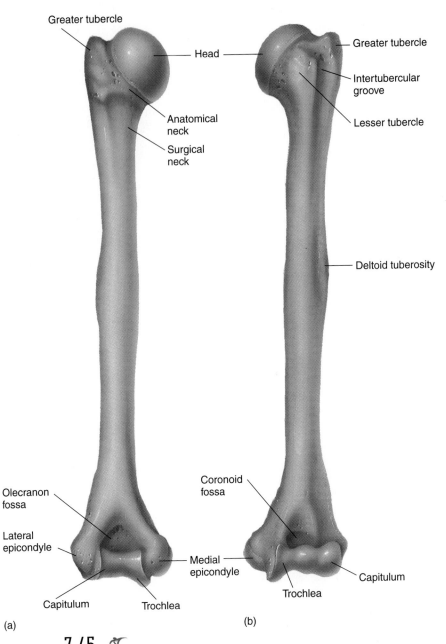

**FIGURE 7.45**

(a) Posterior surface and (b) anterior surface of the left humerus.

The **phalanges** are the finger bones. There are three in each finger—a proximal, a middle, and a distal phalanx—and two in the thumb. (The thumb lacks a middle phalanx.) Thus, each hand has fourteen finger bones. Table 7.10 summarizes the bones of the pectoral girdle and upper limbs.

**1** Locate and name each of the bones of the upper limb.

**2** Explain how the bones of the upper limb articulate with one another.

It is not uncommon for a baby to be born with an extra finger or toe, but since the extra digit is usually surgically removed early in life, hands like the ones in figure 7.48 are rare. Polydactyly ("many digits") is an inherited trait. It is common in cats. A lone but popular male cat brought the trait from England to colonial Boston.

(b)

**FIGURE 7.46**

(a) The head of the right radius articulates with the radial notch of the ulna, and the head of the ulna articulates with the ulnar notch of the radius. (b) Lateral view of the proximal end of the ulna.

## PELVIC GIRDLE

The **pelvic girdle** consists of the two coxal bones (hipbones), which articulate with each other anteriorly and with the sacrum posteriorly (fig. 7.49). The sacrum, coccyx, and pelvic girdle together form the bowl-shaped *pelvis*. The pelvic girdle supports the trunk of the body, provides attachments for the lower limbs, and protects the urinary bladder, the distal end of the large intestine,

and the internal reproductive organs. The body's weight is transmitted through the pelvic girdle to the lower limbs and then onto the ground.

### Coxal Bones

Each **coxal bone** develops from three parts—an ilium, an ischium, and a pubis. These parts fuse in the region of a cup-shaped cavity called the *acetabulum* (as″ĕ-tab′u-lum). This depression, on the lateral surface of the hipbone, receives the rounded head of the femur or thighbone (fig. 7.50).

The **ilium** (il′e-um), which is the largest and most superior portion of the coxal bone, flares outward, forming the prominence of the hip. The margin of this prominence is called the *iliac crest.* The smooth, concave surface on the anterior aspect of the ilium is the *iliac fossa.*

Posteriorly, the ilium joins the sacrum at the *sacroiliac* (sa″kro-il′e-ak) *joint.* Anteriorly, a projection of the ilium, the *anterior superior iliac spine,* can be felt lateral to the groin. This spine provides attachments for ligaments and muscles, and is an important surgical landmark.

A common injury in contact sports, such as football, is bruising the soft tissues and bone associated with the anterior superior iliac spine. Wearing protective padding can prevent this painful injury, called a *hip pointer.*

On the posterior border of the ilium is a *posterior superior iliac spine.* Below this spine is a deep indentation, the *greater sciatic notch,* through which a number of nerves and blood vessels pass.

The **ischium** (is′ke-um), which forms the lowest portion of the coxal bone, is L-shaped, with its angle, the *ischial tuberosity,* pointing posteriorly and downward. This tuberosity has a rough surface that provides attachments for ligaments and lower limb muscles. It also supports the weight of the body during sitting. Above the ischial tuberosity, near the junction of the ilium and ischium, is a sharp projection called the *ischial spine.* Like the sacral promontory, this spine, which can be felt during a vaginal examination, is used as a guide for determining pelvis size. The distance between the ischial spines is the shortest diameter of the pelvic outlet.

(a)

(b)

**FIGURE 7.47**

(a) The right hand posterior view. (b) Radiograph of the right hand. Note the small sesamoid bone associated with the joint at the base of the thumb (arrow).

**FIGURE 7.48**

A person with polydactyly has extra digits.

The **pubis** (pu′bis) constitutes the anterior portion of the coxal bone. The two pubic bones come together at the midline to form a joint called the *symphysis pubis* (sim′fĭ-sis pu′bis). The angle these bones form below the symphysis is the *pubic arch*.

A portion of each pubis passes posteriorly and downward to join an ischium. Between the bodies of these bones on either side is a large opening, the *obturator foramen*, which is the largest foramen in the skeleton. An obturator membrane covers and nearly closes this foramen (see figs. 7.49 and 7.50).

## Greater and Lesser Pelves

If a line were drawn along each side of the pelvis from the sacral promontory downward and anteriorly to the upper margin of the symphysis pubis, it would mark the *pelvic brim* (linea terminalis). This margin separates the lower, or lesser (true), pelvis from the upper, or greater (false), pelvis (fig. 7.51).

| TABLE 7.10 | BONES OF THE PECTORAL GIRDLE AND UPPER LIMBS | |
|---|---|---|
| **Name and Number** | **Location** | **Special Features** |
| Clavicle (2) | Base of neck between sternum and scapula | Sternal end, acromial end |
| Scapula (2) | Upper back, forming part of shoulder | Body, spine, head, acromion process, coracoid process, glenoid cavity |
| Humerus (2) | Arm, between scapula and elbow | Head, greater tubercle, lesser tubercle, intertubercular groove, surgical neck, deltoid tuberosity, capitulum, trochlea, medial epicondyle, lateral epicondyle, coronoid fossa, olecranon fossa |
| Radius (2) | Lateral side of forearm, between elbow and wrist | Head, radial tuberosity, styloid process, ulnar notch |
| Ulna (2) | Medial side of forearm, between elbow and wrist | Trochlear notch, olecranon process, head, styloid process, radial notch |
| Carpal (16) | Wrist | Arranged in two rows of four bones each |
| Metacarpal (10) | Palm | One in line with each finger |
| Phalanx (28) | Finger | Three in each finger; two in each thumb |

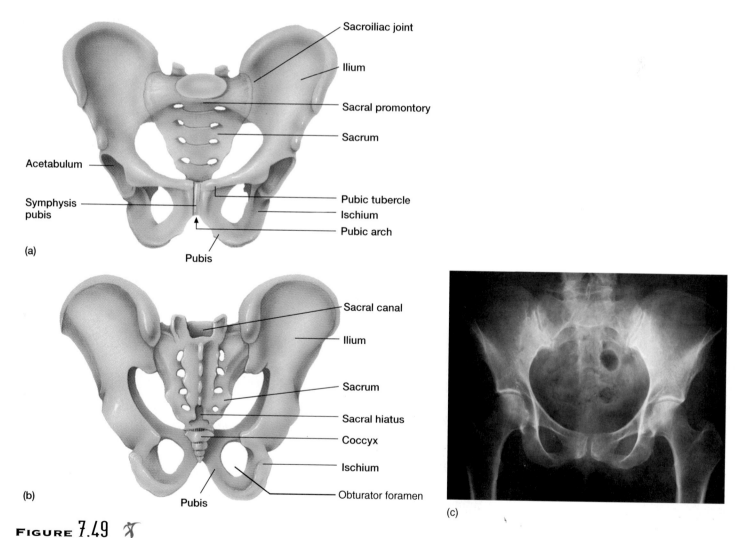

**FIGURE 7.49**

(a) Anterior view and (b) posterior view of the pelvic girdle. This girdle provides an attachment for the lower limbs, and together with the sacrum and coccyx forms the pelvis. (c) Radiograph of the pelvic girdle.

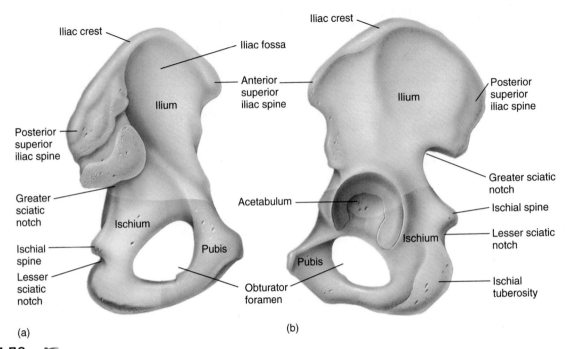

**FIGURE 7.50**

(a) Medial surface of the left coxal bone. (b) Lateral view.

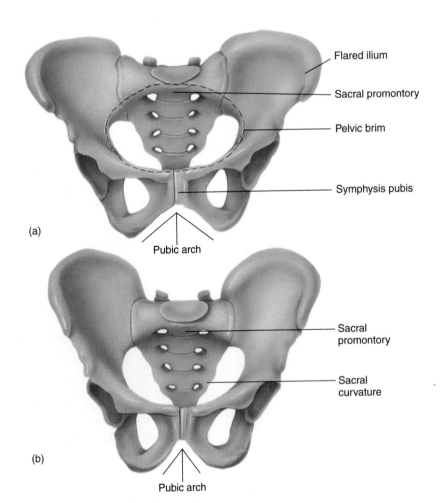

**FIGURE 7.51**

The female pelvis is usually wider in all diameters and roomier than that of the male. (a) Female pelvis. (b) Male pelvis.

| TABLE 7.11 | DIFFERENCES BETWEEN THE MALE AND FEMALE SKELETONS | |
|---|---|---|
| **Part** | **Differences** | |
| Skull | Male skull is larger and heavier, with more conspicuous muscular attachments. Male forehead is shorter, facial area is less round, jaw larger, and mastoid processes more prominent than those of a female. | |
| Pelvis | Male pelvic bones are heavier, thicker, and have more obvious muscular attachments. The obturator foramina and the acetabula are larger and closer together than those of a female. | |
| Pelvic cavity | Male pelvic cavity is narrower in all diameters, and is longer, less roomy, and more funnel-shaped. The distances between the ischial spines and between the ischial tuberosities are lesser than in a female. | |
| Sacrum | Male sacrum is narrower, sacral promontory projects forward to a greater degree, and sacral curvature is bent less sharply posteriorly than in a female. | |
| Coccyx | Male coccyx is less movable than that of a female. | |

The *greater pelvis* is bounded posteriorly by the lumbar vertebrae, laterally by the flared parts of the iliac bones, and anteriorly by the abdominal wall. The false pelvis helps support the abdominal organs.

The *lesser pelvis* is bounded posteriorly by the sacrum and coccyx, and laterally and anteriorly by the lower ilium, ischium, and pubis bones. This portion of the pelvis surrounds a short, canal-like cavity that has an upper inlet and a lower outlet. An infant passes through this cavity during childbirth.

## Differences between Male and Female Pelves

Some basic structural differences distinguish the male and the female pelves, even though it may be difficult to find all of the "typical" characteristics in any one individual. These differences arise from the function of the female pelvis as a birth canal. Usually, the female iliac bones are more flared than those of the male, and consequently, the female hips are usually broader than the male's. The angle of the female pubic arch may be greater, there may be more distance between the ischial spines and the ischial tuberosities, and the sacral curvature may be shorter and flatter. Thus, the female pelvic cavity is usually wider in all diameters than that of the male. Also, the bones of the female pelvis are usually lighter, more delicate, and show less evidence of muscle attachments (fig. 7.51). Table 7.11 summarizes some of the differences between the female and male skeletons.

**1** Locate and name each bone that forms the pelvis.

**2** Distinguish between the greater pelvis and the lesser pelvis.

**3** How are male and female pelves different?

## LOWER LIMB

The bones of the lower limb form the frameworks of the thigh, leg, and foot. They include a femur, a tibia, a fibula, tarsals, metatarsals, and phalanges (fig. 7.52).

## Femur

The **femur,** or thighbone, is the longest bone in the body and extends from the hip to the knee. A large, rounded *head* at its proximal end projects medially into the acetabulum of the coxal bone. On the head, a pit called the *fovea capitis* marks the attachment of a ligament. Just below the head are a constriction, or *neck,* and two large processes—a superior, lateral *greater trochanter* and an inferior, medial *lesser trochanter.* These processes provide attachments for muscles of the lower limbs and buttocks. On the posterior surface in the middle third of the shaft is a longitudinal crest called the *linea aspera.* This rough strip is an attachment for several muscles (fig. 7.53).

At the distal end of the femur, two rounded processes, the *lateral* and *medial condyles,* articulate with the tibia of the leg. A patella also articulates with the femur on its distal anterior surface.

On the medial surface at its distal end is a prominent *medial epicondyle,* and on the lateral surface is a *lateral epicondyle.* These projections provide attachments for muscles and ligaments.

## Patella

The **patella,** or kneecap, is a flat sesamoid bone located in a tendon that passes anteriorly over the knee (see fig. 7.52). Because of its position, the patella controls the angle at which this tendon continues toward the tibia, and so it functions in lever actions associated with lower limb movements.

As a result of a blow to the knee or a forceful unnatural movement of the leg, the patella sometimes slips to one side. This painful condition is called a *patellar dislocation.* Doing exercises that strengthen muscles associated with the knee and wearing protective padding can prevent knee displacement. Unfortunately, once the soft tissues that hold the patella in place are stretched, patellar dislocation tends to recur.

(a)

(b)

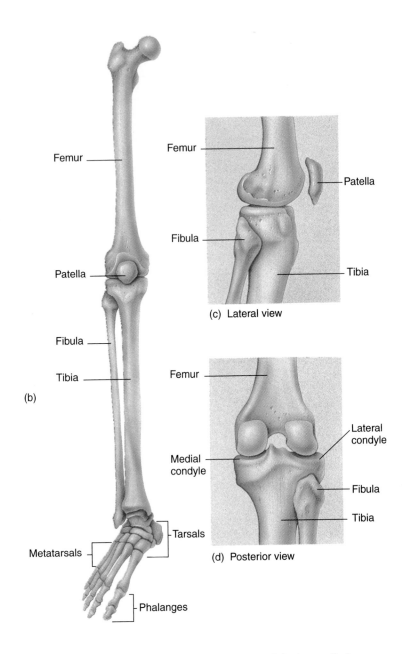

(c) Lateral view

(d) Posterior view

**FIGURE** 7.52

(*a*) Radiograph of the right knee, showing the ends of the femur, tibia, and fibula. (*b*) Anterior view of the right lower limb.
(*c*) Lateral view of the right knee. (*d*) Posterior view of the right knee.

## Tibia

The **tibia,** or shinbone, is the larger of the two leg bones and is located on the medial side. Its proximal end is expanded into *medial* and *lateral condyles,* which have concave surfaces and articulate with the condyles of the femur. Below the condyles, on the anterior surface, is a process called the *tibial tuberosity,* which provides an attachment for the *patellar ligament* (a continuation of the patella-bearing tendon). A prominent *anterior crest* extends downward from the tuberosity and attaches connective tissues in the leg.

At its distal end, the tibia expands to form a prominence on the inner ankle called the *medial malleolus* (mah-le′o-lus), which is an attachment for ligaments. On

its lateral side is a depression that articulates with the fibula. The inferior surface of the tibia's distal end articulates with a large bone (the talus) in the foot (fig. 7.54).

The skeleton is particularly vulnerable to injury during the turbulent teen years, when bones grow rapidly. Athletic teens sometimes develop Osgood-Schlatter disease, which is a painful swelling of a bony projection of the tibia below the knee. Overusing the thigh muscles to straighten the lower limb irritates the area, causing the swelling. Usually a few months of rest and no athletic activity allows the bone to heal on its own. Rarely, a cast must be used to immobilize the knee.

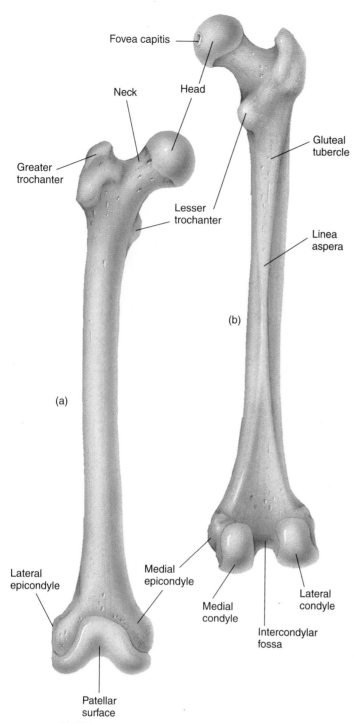

**FIGURE 7.53**

(*a*) Anterior surface and (*b*) posterior surface of the right femur.

**FIGURE 7.54**

Bones of the right leg viewed from the front.

## Fibula

The **fibula** is a long, slender bone located on the lateral side of the tibia. Its ends are slightly enlarged into a proximal *head* and a distal *lateral malleolus.* The head articulates with the tibia just below the lateral condyle; however, it does not enter into the knee joint and does not bear any body weight. The lateral malleolus articulates with the ankle and protrudes on the lateral side (fig. 7.54).

## Foot

The foot consists of a *tarsus* (tahr′sus), a *metatarsus* (met″ah-tar′sus), and five toes. The tarsus is composed of seven **tarsal bones** (ankle bones). These bones are arranged so that one of them, the **talus** (ta′lus), can move freely where it joins the tibia and fibula. The remaining tarsal bones are bound firmly together, forming a mass supporting the talus. Figures 7.55 and 7.56 name the individual bones of the tarsus.

(a)

Fibula

Tibia

Talus

Medial
cuneiform

Navicular

Calcaneus

Metatarsals

Phalanges

Tarsus

(b)

**FIGURE 7.55**

(a) Radiograph of the right foot viewed from the medial side. (b) The talus moves freely where it articulates with the tibia and fibula.

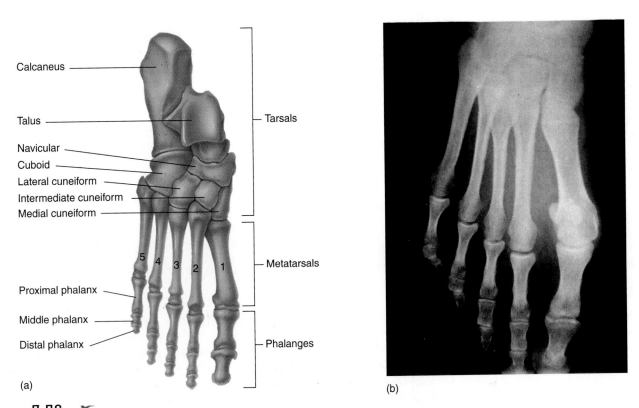

Calcaneus

Talus

Navicular

Cuboid

Lateral cuneiform

Intermediate cuneiform

Medial cuneiform

Tarsals

5 4 3 2 1

Metatarsals

Proximal phalanx

Middle phalanx

Distal phalanx

Phalanges

(a)

(b)

**FIGURE 7.56**

(a) The right foot viewed superiorly. (b) Radiograph of the right foot viewed superiorly.

TABLE 7.12

## TABLE 7.12 BONES OF THE PELVIC GIRDLE AND LOWER LIMBS

| Name and Number | Location | Special Features |
|---|---|---|
| Coxal bone (2) | Hip, articulating with the other coxal bone anteriorly and with the sacrum posteriorly | Ilium, iliac crest, anterior superior iliac spine, ischium, ischial tuberosity, ischial spine, obturator foramen, acetabulum, pubis |
| Femur (2) | Thigh, between hip and knee | Head, fovea capitis, neck, greater trochanter, lesser trochanter, linea aspera, lateral condyle, medial condyle, gluteal tuberosity, intercondylar fossa |
| Patella (2) | Anterior surface of knee | A flat sesamoid bone located within a tendon |
| Tibia (2) | Medial side of leg, between knee and ankle | Medial condyle, lateral condyle, tibial tuberosity, anterior crest, medial malleolus, intercondylar eminence |
| Fibula (2) | Lateral side of leg, between knee and ankle | Head, lateral malleolus |
| Tarsal (14) | Ankle | Freely movable talus that articulates with leg bones; six other tarsal bones bound firmly together |
| Metatarsal (10) | Instep | One in line with each toe, arranged and bound by ligaments to form arches |
| Phalanx (28) | Toe | Three in each toe, two in great toe |

The largest of the anklebones, the **calcaneus** (kal-ka′ne-us), or heel bone, is located below the talus where it projects backward to form the base of the heel. The calcaneus helps support the weight of the body and provides an attachment for muscles that move the foot.

The metatarsus consists of five elongated **metatarsal** (met″ah-tar′sal) **bones,** which articulate with the tarsus. They are numbered 1 to 5, beginning on the medial side (fig. 7.56). The heads at the distal ends of these bones form the ball of the foot. The tarsals and metatarsals are arranged and bound by ligaments to form the arches of the foot. A longitudinal arch extends from the heel to the toe, and a transverse arch stretches across the foot. These arches provide a stable, springy base for the body. Sometimes, however, the tissues that bind the metatarsals weaken, producing fallen arches, or flat feet.

An infant with two casts on her feet is probably being treated for clubfoot, a very common birth defect in which the foot twists out of its normal position, turning in, out, up, down, or some combination of these directions. Clubfoot probably results from arrested development during fetal existence, but the precise cause is not known. Clubfoot can almost always be corrected with special shoes, or surgery, followed by several months in casts to hold the feet in the correct position.

The **phalanges** of the toes are shorter, but otherwise similar to those of the fingers, and align and articulate with the metatarsals. Each toe has three phalanges—a proximal, a middle, and a distal phalanx—except the great toe, which has only two because it lacks the middle phalanx (fig. 7.56). Table 7.12 summarizes the bones of the pelvic girdle and lower limbs.

1. Locate and name each of the bones of the lower limb.

2. Explain how the bones of the lower limb articulate with one another.

3. Describe how the foot is adapted to support the body.

## CLINICAL TERMS RELATED TO THE SKELETAL SYSTEM

**achondroplasia** (a-kon″dro-pla′ze-ah)   Inherited condition that retards formation of cartilaginous bone. The result is a type of dwarfism.

**acromegaly** (ak″ro-meg′ah-le)   Abnormal enlargement of facial features, hands, and feet in adults as a result of overproduction of growth hormone.

**Colles fracture** (kol′ēz frak′tūre)   Fracture at the distal end of the radius that displaces the smaller fragment posteriorly.

**epiphysiolysis** (ep″ĭ-fiz″e-ol′ĭ-sis)   Separation or loosening of the epiphysis from the diaphysis of a bone.

**laminectomy** (lam″ĭ-nek′to-me)   Surgical removal of the posterior arch of a vertebra, usually to relieve symptoms of a ruptured intervertebral disk.

**lumbago** (lum-ba′go)   Dull ache in the lumbar region of the back.

**orthopedics** (or″tho-pe′diks)   Medical specialty that prevents, diagnoses, and treats diseases and abnormalities of the skeletal and muscular systems.

**ostealgia** (os″te-al′je-ah) )   Pain in a bone.

**ostectomy** (os-tek′to-me)   Surgical removal of a bone.

**osteitis** (os″te-i′tis)   Inflammation of bone tissue.

**osteochondritis** (os″te-o-kon-dri′tis)   Inflammation of bone and cartilage tissues.

**osteogenesis** (os″te-o-jen′ĕ-sis)   Bone development.

**SKELETAL SYSTEM**

Bones provide support, protection, and movement and also play a role in calcium balance.

## Integumentary System

Vitamin D, activated in the skin, plays a role in calcium availability for bone matrix.

## Lymphatic System

Cells of the immune system originate in the bone marrow.

## Muscular System

Muscles pull on bones to cause movement.

## Digestive System

Absorption of dietary calcium provides material for bone matrix.

## Nervous System

Proprioceptors sense the position of body parts. Pain receptors warn of trauma to bone. Bones protect the brain and spinal cord

## Respiratory System

Ribs and muscles work together in breathing.

## Endocrine System

Some hormones act on bone to help regulate blood calcium levels.

## Urinary System

The kidneys and bones work together to help regulate blood calcium levels.

## Cardiovascular System

Blood transports nutrients to bone cells. Bone helps regulate plasma calcium levels, important to heart function.

## Reproductive System

The pelvis helps support the uterus during pregnancy. Bone may provide a source of calcium during lactation.

**osteogenesis imperfecta** (os"te-o-jen'ĕ-sis im-per-fek'ta) Inherited condition of deformed and abnormally brittle bones.

**osteoma** (os"te-o'mah) Tumor composed of bone tissue.

**osteomalacia** (os"te-o-mah-la'she-ah) Softening of adult bone due to a disorder in calcium and phosphorus metabolism, usually caused by vitamin D deficiency.

**osteomyelitis** (os"te-o-mi"ĕ-li'tis) Bone inflammation caused by the body's reaction to bacterial or fungal infection.

**osteonecrosis** (os"te-o-ne-kro'sis) Death of bone tissue. This condition most commonly occurs in the femur head in elderly persons and may be due to obstructed arteries supplying the bone.

**osteopathology** (os"te-o-pah-thol'o-je) Study of bone diseases.

**osteopenia** (os"te-o-pe'ni-ah) Decrease in bone mass due to reduction in rate of bone tissue formation.

**osteoporosis** (os"te-o-po-ro'sis) Decreased bone mineral content.

**osteotomy** (os"te-ot'o-me) Cutting a bone.

## CHAPTER SUMMARY

## Introduction (page 185)

Individual bones are the organs of the skeletal system. A bone contains very active tissues.

## Bone Structure (page 185)

Bone structure reflects its function.

1. Bone classification
   Bones are grouped according to their shapes—long, short, flat, irregular, or sesamoid.
2. Parts of a long bone
   a. Epiphyses at each end are covered with articular cartilage and articulate with other bones.
   b. The shaft of a bone is called the diaphysis.
   c. Except for the articular cartilage, a bone is covered by a periosteum.
   d. Compact bone has a continuous matrix with no gaps.
   e. Spongy bone has irregular interconnecting spaces between bony plates.
   f. Both compact and spongy bone are strong and resist bending.
   g. The diaphysis contains a medullary cavity filled with marrow.
3. Microscopic structure
   a. Compact bone contains osteons cemented together.
   b. Osteonic canals contain blood vessels that nourish the cells of osteons.
   c. Perforating canals connect osteonic canals transversely and communicate with the bone's surface and the medullary cavity.
   d. Diffusion from the surface of thin bony plates nourishes cells of spongy bones.

## Bone Development and Growth (page 188)

1. Intramembranous bones
   a. Certain flat bones of the skull are intramembranous bones.
   b. They develop from layers of connective tissues.
   c. Osteoblasts within the membranous layers form bone tissue.
   d. Mature bone cells are called osteocytes.
   e. Primitive connective tissue gives rise to the periosteum.
2. Endochondral bones
   a. Most of the bones of the skeleton are endochondral.
   b. They develop as hyaline cartilage that is later replaced by bone tissue.
3. Growth of an endochondral bone
   a. Primary ossification center appears in the diaphysis, whereas secondary ossification centers appear in the epiphyses.
   b. An epiphyseal disk remains between the primary and secondary ossification centers.
   c. An epiphyseal disk consists of layers of cells: resting cells, young reproducing cells, older enlarging cells, and dying cells.
   d. The epiphyseal disks are responsible for lengthening.
   e. Long bones continue to lengthen until the epiphyseal disks are ossified.
   f. Growth in thickness is due to intramembranous ossification beneath the periosteum.
   g. The action of osteoclasts forms the medullary cavity.
4. Homeostasis of bone tissue
   a. Osteoclasts and osteoblasts continually remodel bone.
   b. The total mass of bone remains nearly constant.
5. Factors affecting bone development, growth, and repair
   a. Deficiencies of vitamin A, C, or D result in abnormal development.
   b. Insufficient secretion of pituitary growth hormone may result in dwarfism; excessive secretion may result in gigantism, or acromegaly.
   c. Deficiency of thyroid hormone delays bone growth.
   d. Male and female sex hormones promote bone formation and stimulate ossification of the epiphyseal disks.

## Bone Function (page 193)

1. Support and protection
   a. Bones shape and form body structures.
   b. Bones support and protect softer, underlying tissues.
2. Body movement
   a. Bones and muscles function together as levers.
   b. A lever consists of a rod, a pivot (fulcrum), a resistance, and a force that supplies energy.
   c. Parts of a first-class lever are arranged resistance–pivot–force; of a second-class lever pivot–resistance–force; of a third-class lever resistance–force–pivot.
3. Blood cell formation
   a. At different ages, hematopoiesis occurs in the yolk sac, the liver, the spleen, and the red bone marrow.
   b. Red marrow houses developing red blood cells, white blood cells, and blood platelets.

4. Inorganic salt storage
   a. The intercellular material of bone tissue contains large quantities of calcium phosphate in the form of hydroxyapatite.
   b. When blood calcium ion concentration is low, osteoclasts resorb bone, releasing calcium salts.
   c. When blood calcium ion concentration is high, osteoblasts are stimulated to form bone tissue and store calcium salts.
   d. Bone stores small amounts of sodium, magnesium, potassium, and carbonate ions.
   e. Bone tissues may accumulate lead, radium, or strontium.

## Skeletal Organization (page 199)

1. Number of bones
   a. Usually a human skeleton has 206 bones, but the number may vary.
   b. Extra bones in sutures are called sutural bones.
2. Divisions of the skeleton
   a. The skeleton can be divided into axial and appendicular portions.
   b. The axial skeleton consists of the skull, hyoid bone, vertebral column, and thoracic cage.
   c. The appendicular skeleton consists of the pectoral girdle, upper limbs, pelvic girdle, and lower limbs.

## Skull (page 201)

The skull consists of twenty-two bones, which include eight cranial bones, thirteen facial bones, and one mandible.

1. Cranium
   a. The cranium encloses and protects the brain, and provides attachments for muscles.
   b. Some cranial bones contain air-filled sinuses that help reduce the weight of the skull.
   c. Cranial bones include the frontal bone, parietal bones, occipital bone, temporal bones, sphenoid bone, and ethmoid bone.
2. Facial skeleton
   a. Facial bones form the basic shape of the face and provide attachments for muscles.
   b. Facial bones include the maxillary bones, palatine bones, zygomatic bones, lacrimal bones, nasal bones, vomer bone, inferior nasal conchae, and mandible.
3. Infantile skull
   a. Incompletely developed bones, connected by fontanels, enable the infantile skull to change shape slightly during childbirth.
   b. Proportions of the infantile skull are different from those of an adult skull, and its bones are less easily fractured.

## Vertebral Column (page 214)

The vertebral column extends from the skull to the pelvis and protects the spinal cord. It is composed of vertebrae separated by intervertebral disks. An infant has thirty-three vertebral bones and an adult has twenty-six. The vertebral column has four curvatures—cervical, thoracic, lumbar, and pelvic.

1. A typical vertebra
   a. A typical vertebra consists of a body, pedicles, laminae, spinous process, transverse processes, and superior and inferior articulating processes.
   b. Notches on the upper and lower surfaces of the pedicles on adjacent vertebrae form intervertebral foramina through which spinal nerves pass.

2. Cervical vertebrae
   a. Cervical vertebrae comprise the bones of the neck.
   b. Transverse processes have transverse foramina.
   c. The atlas (first vertebra) supports the head.
   d. The dens of the axis (second vertebra) provides a pivot for the atlas when the head is turned from side to side.
3. Thoracic vertebrae
   a. Thoracic vertebrae are larger than cervical vertebrae.
   b. Their long spinous processes slope downward, and facets on the sides of bodies articulate with the ribs.
4. Lumbar vertebrae
   a. Vertebral bodies of lumbar vertebrae are large and strong.
   b. Their transverse processes project posteriorly at sharp angles, and their spinous processes are directed horizontally.
5. Sacrum
   a. The sacrum, formed of five fused vertebrae, is a triangular structure that has rows of dorsal sacral foramina.
   b. It is united with the coxal bones at the sacroiliac joints.
   c. The sacral promontory provides a guide for determining the size of the pelvis.
6. Coccyx
   a. The coccyx, composed of four fused vertebrae, forms the lowest part of the vertebral column.
   b. It acts as a shock absorber when a person sits.

## Thoracic Cage (page 220)

The thoracic cage includes the ribs, thoracic vertebrae, sternum, and costal cartilages. It supports the pectoral girdle and upper limbs, protects viscera, and functions in breathing.

1. Ribs
   a. Twelve pairs of ribs are attached to the twelve thoracic vertebrae.
   b. Costal cartilages of the true ribs join the sternum directly; those of the false ribs join indirectly or not at all.
   c. A typical rib has a shaft, head, and tubercles that articulate with the vertebrae.
2. Sternum
   a. The sternum consists of a manubrium, body, and xiphoid process.
   b. It articulates with costal cartilages and clavicles.

## Pectoral Girdle (page 222)

The pectoral girdle is composed of two clavicles and two scapulae. It forms an incomplete ring that supports the upper limbs and provides attachments for muscles that move the upper limbs.

1. Clavicles
   a. Clavicles are rodlike bones that run horizontally between the sternum and shoulders.
   b. They hold the shoulders in place and provide attachments for muscles.
2. Scapulae
   a. The scapulae are broad, triangular bones with bodies, spines, heads, acromion processes, coracoid processes, glenoid cavities, supraspinous and infraspinous fossae, superior borders, axillary borders, and vertebral borders.
   b. They articulate with the humerus of each upper limb, and provide attachments for muscles of the upper limbs and chest.

## Upper Limb (page 223)

Limb bones provide the frameworks and attachments of muscles, and function in levers that move the limb and its parts.

1. Humerus
   a. The humerus extends from the scapula to the elbow.
   b. It has a head, greater tubercle, lesser tubercle, intertubercular groove, anatomical neck, surgical neck, deltoid tuberosity, capitulum, trochlea, epicondyles, coronoid fossa, and olecranon fossa.
2. Radius
   a. The radius is located on the thumb side of the forearm between the elbow and wrist.
   b. It has a head, radial tuberosity, styloid process, and ulnar notch.
3. Ulna
   a. The ulna is longer than the radius and overlaps the humerus posteriorly.
   b. It has a trochlear notch, olecranon process, coronoid process, head, styloid process, and radial notch.
   c. It articulates with the radius laterally and with a disk of fibrocartilage inferiorly.
4. Hand
   a. The hand is composed of a wrist, palm, and five fingers.
   b. It includes eight carpals that form a carpus, five metacarpals, and fourteen phalanges.

## Pelvic Girdle (page 227)

The pelvic girdle consists of two coxal bones that articulate with each other anteriorly and with the sacrum posteriorly. The sacrum, coccyx, and pelvic girdle form the pelvis. The girdle provides support for body weight and attachments for muscles, and protects visceral organs.

1. Coxal bones
   Each coxal bone consists of an ilium, ischium, and pubis, which are fused in the region of the acetabulum.
   a. Ilium
      (1) The ilium, the largest portion of the coxal bone, joins the sacrum at the sacroiliac joint.
      (2) It has an iliac crest with anterior and posterior superior iliac spines, and iliac fossae.
   b. Ischium
      (1) The ischium is the lowest portion of the coxal bone.
      (2) It has an ischial tuberosity and ischial spine.
   c. Pubis
      (1) The pubis is the anterior portion of the coxal bone.
      (2) Pubis bones are fused anteriorly at the symphysis pubis.
2. Greater and lesser pelves
   a. The lesser pelvis is below the pelvic brim; the greater pelvis is above it.
   b. The lesser pelvis functions as a birth canal; the greater pelvis helps support abdominal organs.
3. Differences between male and female pelves
   a. Differences between male and female pelves are related to the function of the female pelvis as a birth canal.
   b. Usually the female pelvis is more flared; pubic arch is broader; distance between the ischial spines and the ischial tuberosities is greater; and sacral curvature is shorter.

## Lower Limb (page 231)

Bones of the lower limb provide the frameworks of the thigh, leg, and foot.

1. Femur
   a. The femur extends from the hip to the knee.
   b. It has a head, fovea capitis, neck, greater trochanter, lesser trochanter, linea aspera, lateral condyle, and medial condyle.
2. Patella
   a. The patella is a flat sesamoid bone in the tendon that passes anteriorly over the knee.
   b. It controls the angle of this tendon and functions in lever actions associated with lower limb movements.
3. Tibia
   a. The tibia is located on the medial side of the leg.
   b. It has medial and lateral condyles, tibial tuberosity, anterior crest, and medial malleolus.
   c. It articulates with the talus of the ankle.
4. Fibula
   a. The fibula is located on the lateral side of the tibia.
   b. It has a head and lateral malleolus that articulates with the ankle but does not bear body weight.
5. Foot
   a. The foot consists of an ankle, an instep, and five toes.
   b. It includes seven tarsals that form the tarsus, five metatarsals, and fourteen phalanges.

## CRITICAL THINKING QUESTIONS

1. What steps do you think should be taken to reduce the chances of a person accumulating abnormal metallic elements such as lead, radium, and strontium in bones?
2. Why do you think incomplete, longitudinal fractures of bone shafts (greenstick fractures) are more common in children than in adults?
3. When a child's bone is fractured, growth may be stimulated at the epiphyseal disk. What problems might this extra growth cause in an upper or lower limb before the growth of the other limb compensates for the difference in length?
4. Why do elderly persons often develop bowed backs and appear shorter than they were in earlier years?
5. How might the condition of an infant's fontanels be used to evaluate skeletal development? How might the fontanels be used to estimate intracranial pressure?
6. Why are women more likely than men to develop osteoporosis? What steps can reduce the risk of developing this condition?
7. How does the structure of a bone make it strong yet lightweight?
8. Archeologists discover skeletal remains of humanlike animals in Ethiopia. Examination of the bones suggests that the remains represent four types of individuals. Two of the skeletons have bone densities that are 30% less than those of the other two skeletons. The skeletons with the lower bone mass also have broader front pelvic bones. Within the two groups defined by bone mass, smaller skeletons have bones with evidence of epiphyseal plates, but larger bones have only a thin line where the epiphyseal plates should be. Give the age group and gender of the individuals in this find.

# REVIEW EXERCISES

## Part A

1. List four groups of bones based upon their shapes, and name an example from each group.
2. Sketch a typical long bone, and label its epiphyses, diaphysis, medullary cavity, periosteum, and articular cartilages.
3. Distinguish between spongy and compact bone.
4. Explain how osteonic canals and perforating canals are related.
5. Explain how the development of intramembranous bone differs from that of endochondral bone.
6. Distinguish between osteoblasts and osteocytes.
7. Explain the function of an epiphyseal disk.
8. Explain how a bone grows in thickness.
9. Define *osteoclast*.
10. Explain how osteoclasts and osteoblasts regulate bone mass.
11. Describe the effects of vitamin deficiencies on bone development.
12. Explain the causes of pituitary dwarfism and gigantism.
13. Describe the effects of thyroid and sex hormones on bone development.
14. Explain the effects of exercise on bone structure.
15. Provide several examples to illustrate how bones support and protect body parts.
16. Describe a lever, and explain how its parts may be arranged to form first-, second-, and third-class levers.
17. Explain how upper limb movements function as levers.
18. Describe the functions of red and yellow bone marrow.
19. Explain the mechanism that regulates the concentration of blood calcium ions.
20. List three substances that may be abnormally stored in bone.
21. Distinguish between the axial and appendicular skeletons.
22. Name the bones of the cranium and the facial skeleton.
23. Explain the importance of fontanels.
24. Describe a typical vertebra.
25. Explain the differences among cervical, thoracic, and lumbar vertebrae.
26. Describe the locations of the sacroiliac joint, the sacral promontory, and the sacral hiatus.
27. Name the bones that comprise the thoracic cage.
28. List the bones that form the pectoral and pelvic girdles.
29. Name the bones of the upper limb.
30. Define *coxal bone*.
31. List the major differences that may occur between the male and female pelves.
32. List the bones of the lower limb.

## Part B

Match the parts listed in column I with the bones listed in column II.

| I | II |
|---|---|
| 1. Coronoid process | A. Ethmoid bone |
| 2. Cribriform plate | B. Frontal bone |
| 3. Foramen magnum | C. Mandible |
| 4. Mastoid process | D. Maxillary bone |
| 5. Palatine process | E. Occipital bone |
| 6. Sella turcica | F. Temporal bone |
| 7. Supraorbital notch | G. Sphenoid bone |
| 8. Temporal process | H. Zygomatic bone |
| 9. Acromion process | I. Femur |
| 10. Deltoid tuberosity | J. Fibula |
| 11. Greater trochanter | K. Humerus |
| 12. Lateral malleolus | L. Radius |
| 13. Medial malleolus | M. Scapula |
| 14. Olecranon process | N. Sternum |
| 15. Radial tuberosity | O. Tibia |
| 16. Xiphoid process | P. Ulna |

# HUMAN SKULL

■ *The following set of reference plates is presented to help you locate some of the more prominent features of the human skull. As you study these photographs, it is important to remember that individual human skulls vary in every characteristic. Also, the photographs in this set depict bones from several different skulls.*

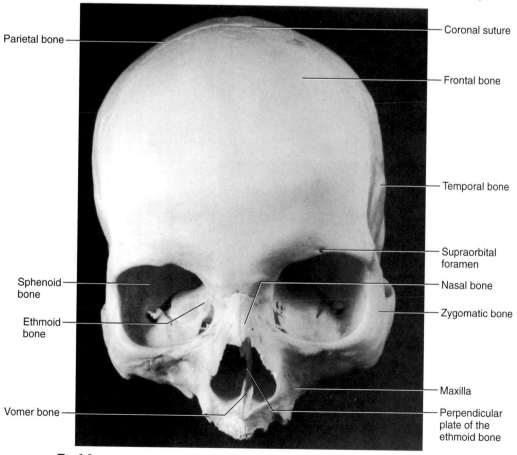

Parietal bone — 
Coronal suture
Frontal bone
Temporal bone
Supraorbital foramen
Sphenoid bone
Nasal bone
Ethmoid bone
Zygomatic bone
Vomer bone
Maxilla
Perpendicular plate of the ethmoid bone

**PLATE Eight**

The skull, frontal view.

**HUMAN SKULL**

Sagittal suture

Parietal bone

Coronal suture

Frontal
bone

Squamosal suture

Temporal bone

Sphenoid bone

Nasal bone

Lacrimal bone

External auditory meatus

Zygomatic arch

Ethmoid bone

Zygomatic bone

Maxilla

**PLATE Nine**

The skull, left anterolateral view.

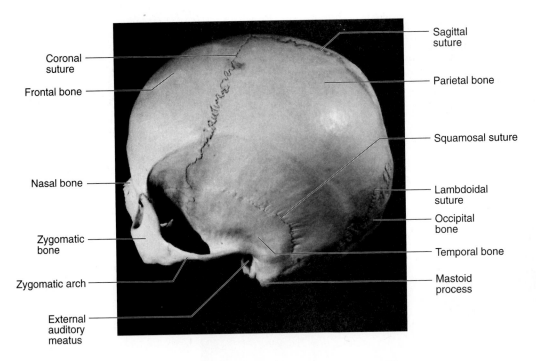

Sagittal
suture

Coronal
suture

Frontal bone

Parietal bone

Squamosal suture

Nasal bone

Lambdoidal
suture

Occipital
bone

Zygomatic
bone

Temporal bone

Zygomatic arch

Mastoid
process

External
auditory
meatus

**PLATE Ten**

The skull, left posterolateral view.

Frontal bone

Supraorbital foramen

Nasal bone

Lacrimal bone

Ethmoid bone

Zygomatic bone

Inferior orbital fissure

**PLATE** Eleven

Bones of the left orbital region.

Nasal bone

Lacrimal bone

Ethmoid bone

Superior orbital fissure

Perpendicular plate of ethmoid bone

Middle nasal concha

Inferior nasal concha

Infraorbital foramen

Maxilla

Vomer bone

**PLATE** Twelve

Bones of the anterior nasal region.

Supraorbital foramen

Frontal bone

Lacrimal bone

Zygomatic bone

Infraorbital foramen

Maxilla

Squamosal suture

Sphenoid bone

Temporal bone

Zygomatic arch

**PLATE Thirteen**

Bones of the left zygomatic region.

Sphenoid bone

Squamosal suture

Zygomatic bone

Temporal process of zygomatic bone

Temporal bone

Zygomatic process of temporal bone

External auditory meatus

Mandibular fossa

**PLATE Fourteen**

Bones of the left temporal region.

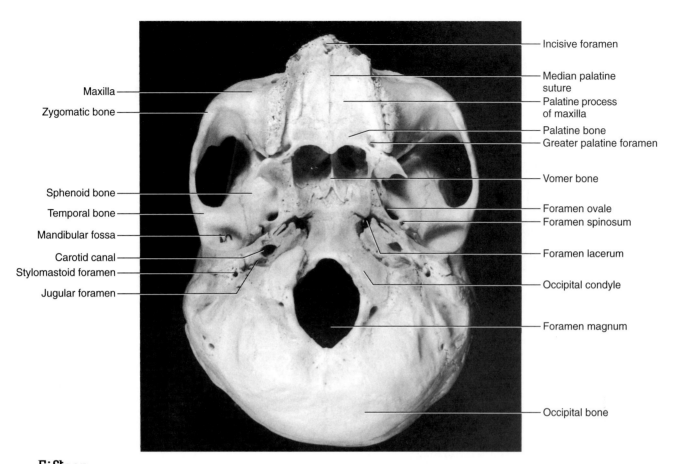

Maxilla

Zygomatic bone

Sphenoid bone

Temporal bone

Mandibular fossa

Carotid canal

Stylomastoid foramen

Jugular foramen

Incisive foramen

Median palatine suture

Palatine process of maxilla

Palatine bone

Greater palatine foramen

Vomer bone

Foramen ovale

Foramen spinosum

Foramen lacerum

Occipital condyle

Foramen magnum

Occipital bone

**PLATE Fifteen**
The skull, inferior view.

Temporal bone

Mandibular fossa

Foramen lacerum

Mastoid process

Occipital bone

Occipital condyle

Foramen magnum

Vomer bone

Sphenoid bone

Foramen ovale

Foramen spinosum

Carotid canal

Jugular foramen

Stylomastoid foramen

**PLATE Sixteen**
Base of the skull, sphenoidal region.

**HUMAN SKULL**

Foramen ovale

Foramen spinosum

Foramen lacerum

Carotid canal

Jugular foramen

Occipital condyle

Foramen magnum

Occipital bone

**PLATE** Seventeen

Base of the skull, occipital region.

Incisive foramen

Median palatine suture

Palatine process of maxilla

Palatine bone

Greater palatine foramen

Vomer bone

Foramen ovale

Foramen spinosum

Foramen lacerum

Carotid foramen

Jugular foramen

Stylomastoid foramen

Occipital condyle

Sphenoid bone

Occipital bone

Foramen magnum

PLATE Eighteen

Base of the skull, maxillary region.

Coronoid process

Mandibular condyle

Mandibular ramus

Body

Alveolar arch

Mental foramen

**PLATE Nineteen**

Mandible, lateral view.

Coronoid process

Mandibular condyle

Mandibular ramus

Mandibular foramen

**PLATE Twenty**

Mandible, medial surface of right ramus.

— Frontal
suture

— Supraorbital
notch

— Orbital portion

**PLATE Twenty-One**

Frontal bone, anterior view.

Occipital — 
condyle

Foramen — 
magnum

**PLATE Twenty-Two**

Occipital bone, inferior view.

— External auditory
meatus

— Mastoid process

Mandibular —
fossa

Zygomatic —
process

**PLATE Twenty-Three**

Temporal bone, left lateral view.

— Crista
galli

Orbital —
surface

— Ethmoidal sinus

— Middle nasal
concha

— Perpendicular
plate

**PLATE Twenty-Four**

Ethmoid bone, right lateral view.

Sphenoidal sinus

Greater wing

Lesser wing

Superior orbital fissure

Foramen rotundum

**PLATE Twenty-Five**

Sphenoid bone, anterior view.

Greater wing

Lesser wing

Foramen rotundum

Sella turcica

Foramen ovale

Foramen spinosum

**PLATE Twenty-Six**

Sphenoid bone, posterior view.

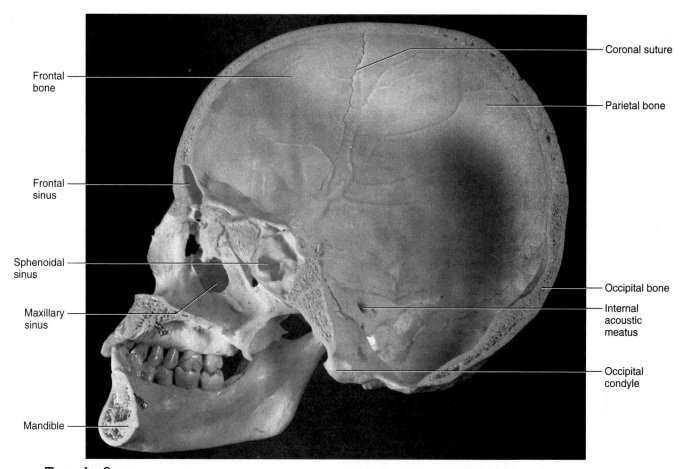

**PLATE Twenty-Seven**

The skull, sagittal section.

Frontal bone

Frontal sinus

Sphenoidal sinus

Maxillary sinus

Mandible

Coronal suture

Parietal bone

Occipital bone

Internal acoustic meatus

Occipital condyle

**PLATE Twenty-Eight**

Ethmoidal region, sagittal section.

Frontal bone

Frontal sinus

Ethmoidal sinus

Maxillary sinus

Ethmoid bone

Sphenoid bone

Frontal bone

Ethmoid bone

Ethmoidal sinus

Maxillary sinus

Parietal bone

Sella turcica

Sphenoidal sinus

Sphenoid bone

**PLATE Twenty-Nine**

Sphenoidal region, sagittal section.

Crista galli of ethmoid bone

Foramen ovale

Foramen spinosum

Foramen lacerum

Jugular foramen

Frontal sinus

Frontal bone

Sphenoid bone

Sella turcica

Parietal bone

Foramen magnum

Occipital bone

**PLATE Thirty**

The skull, floor of the cranial cavity.

Frontal
sinus

Frontal bone

Crista galli

Cribriform
plate of ethmoid
bone

Sphenoid
bone

**PLATE Thirty-One**

Frontal region, transverse section.

Optic canal

Superior
orbital
fissure

Sella turcica

Foramen rotundum

Foramen ovale

Foramen spinosum

Foramen lacerum

Jugular foramen

Foramen magnum

**PLATE Thirty-Two**

Sphenoidal region, floor of the cranial cavity.

**PLATE Thirty-Three**

Skull of a fetus, left anterolateral view.

**PLATE Thirty-Four**

Skull of a fetus, left superior view.

**PLATE Thirty-Five**
Skull of a child, right lateral view.

**PLATE Thirty-Six**
Skull of an aged person, left lateral view.

# JOINTS OF THE SKELETAL SYSTEM

## UNDERSTANDING WORDS

**acetabul-,** vinegar cup: *acetabul*um—depression of the coxal bone that articulates with the head of the femur.

**annul-,** ring: *annul*ar ligament—ring-shaped band of connective tissue within the elbow that encircles the head of the radius.

**arth-,** joint: *arth*rology—study of joints and ligaments.

**burs-,** bag, purse: prepatellar *burs*a—fluid-filled sac between the skin and the patella.

**condyl-,** knob: medial *condyl*e—rounded bony process at the distal end of the femur.

**fov-,** pit: *fov*ea capitis—pit in the head of the femur to which a ligament is attached.

**glen-,** joint socket: *glen*oid cavity—depression in the scapula that articulates with the head of the humerus.

**labr-,** lip: glenoidal *labr*um—rim of fibrocartilage attached to the margin of the glenoid cavity.

**ov-,** egglike: syn*ov*ial fluid—thick fluid within a joint cavity that resembles egg white.

**sutur-,** sewing: *sutur*e—type of joint in which flat bones are interlocked by a set of tiny bony processes.

**syndesm-,** binding together: *syndesm*osis—type of joint in which the bones are held together by long fibers of connective tissue.

## CHAPTER OBJECTIVES

### AFTER YOU HAVE STUDIED THIS CHAPTER, YOU SHOULD BE ABLE TO:

1. Explain how joints can be classified according to the type of tissue that binds the bones together.

2. Describe how bones of fibrous joints are held together.

3. Describe how bones of cartilaginous joints are held together.

4. Describe the general structure of a synovial joint.

5. List six types of synovial joints and name an example of each type.

6. Explain how skeletal muscles produce movements at joints and identify several types of joint movements.

7. Describe the shoulder joint and explain how its articulating parts are held together.

8. Describe the elbow, hip, and knee joints and explain how their articulating parts are held together.

## KEY TERMS

**articulation** (ar-tik'u-la"shun)
**bursa** (ber'sah)
**gomphosis** (gom-fo'sis)
**ligament** (lig'ah-ment)
**meniscus** (me-nis'kus)

**suture** (su'chur)
**symphysis** (sim'fi-sis)
**synchondrosis** (sin"kon-dro'sis)
**syndesmosis** (sin"des-mo'sis)
**synovial** (si-no've-al)

out is a metabolic disorder in which lack of an enzyme blocks recycling of two of the four DNA nucleotides called purines. As a result, uric acid crystals accumulate in joints, causing great pain.

In humans, gout mostly affects the small joints in the foot, usually those of the large toes. For many years gout was attributed solely to eating a great deal of red meat, which is rich in purines. Today we know that while such a diet may exacerbate gout, a genetic abnormality causes the illness. Yet researchers recently discovered evidence that is consistent with the association of gout with eating red meat—signs of the condition in *Tyrannosaurus rex!*

An arthritis specialist and two paleontologists examined a cast of the right forearm of a dinosaur named Sue, a long-ago resident of the Hell Creek Formation in South Dakota, whose fossilized remains were found in 1990 jutting from a butte. Although telltale uric acid crystals had long since decomposed, X rays revealed patterns of bone erosion that could only have resulted from gout. The researchers examined only Sue's forearm, however, because she had been discovered on native American land and had been illegally traded by a fossil dealer. As a result of this dubious background, the Federal Bureau of Investigation had confiscated Sue. So the researchers examined bones from 83 other dinosaurs, but only found evidence of gout in one other individual.

Sue had a hard life. Her facial bones and a lower limb bone were broken, and a tooth was found embedded in a rib, a legacy of an ancient battle. Whatever the reason for her injuries, Sue may have experienced the same kind of persistent pain that humans do.

Joints, or **articulations,** are functional junctions between bones. They bind parts of the skeletal system, make possible bone growth, permit parts of the skeleton to change shape during childbirth, and enable the body to move in response to skeletal muscle contractions.

## CLASSIFICATION OF JOINTS

Joints vary considerably in structure and function. However, they can be classified by the type of tissue that binds the bones at each junction. Three general groups are fibrous joints, cartilaginous joints, and synovial joints.

Joints can also be grouped according to the degree of movement possible at the bony junctions. In this scheme, joints are classified as immovable (synarthrotic), slightly movable (amphiarthrotic), and freely movable (diarthrotic). There is some overlap in the structural and functional classification schemes. Currently, structural classification is the one most commonly used.

### Fibrous Joints

**Fibrous joints** are so named because the dense connective tissue holding them together contains many collagenous fibers. They lie between bones that closely contact one another. As a rule, no appreciable movement occurs between these bones; sometimes, however, a very small degree of motion is possible. The three types of fibrous joints are:

1. **Syndesmosis.** In this type of joint, the bones are bound by long fibers of connective tissue that form an *interosseous ligament.* Because this ligament is flexible and may be twisted, the joint may permit slight movement and thus is amphiarthrotic (am″fe-ar-thro′tik). A syndesmosis is at the distal ends of the tibia and fibula, where they join to form the tibiofibular articulation (fig. 8.1).

**FIGURE 8.1**

The articulation between the distal ends of the tibia and fibula is an example of a syndesmosis.

2. **Suture.** Sutures are only between flat bones of the skull, where the broad margins of adjacent bones grow together and unite by a thin layer of dense connective tissue called a *sutural ligament.* Recall from chapter 7 that the infantile skull is incompletely developed, with several of the bones connected by membranous areas called *fontanels* (see fig. 7.33). These areas allow the skull to change shape slightly during childbirth, but as the bones continue to grow, the fontanels close and sutures replace them. With time, some of the bones at sutures interlock by tiny bony processes. Such a suture is in the adult human skull where the parietal and occipital bones meet to form the lambdoidal suture. Because they are immovable, sutures are synarthrotic (sin′ ar-thro′tik) joints (figs. 8.2 and 8.3).

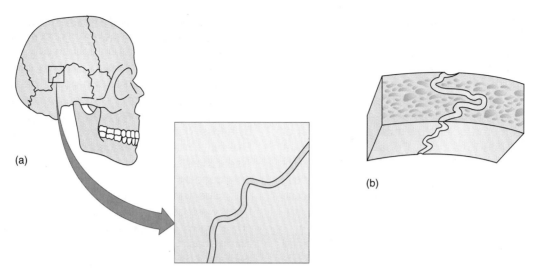

**FIGURE 8.2**

(a) The fibrous joints between the bones of the cranium are immovable and are called sutures. (b) A sutural ligament connects the bones at a suture.

Parietal bone

Margin of suture

Suture

Occipital bone

(a)

(b)

**FIGURE 8.3**

(a) Sutures between the parietal and occipital bones of the skull. (b) The inner margin of a parietal suture. The grooves on the inside of this parietal bone mark the paths of blood vessels located near the brain's surface.

3. **Gomphosis.** A gomphosis is a joint formed by the union of a cone-shaped bony process in a bony socket. The peglike root of a tooth fastened to a jawbone by a *periodontal ligament* is such a joint. This ligament surrounds the root and firmly attaches it to the jaw with bundles of thick collagenous fibers. A gomphosis is a synarthrotic joint (fig. 8.4).

**1** What is a joint?

**2** How are joints classified?

**3** Describe three types of fibrous joints.

**4** What is the function of the fontanels?

## Cartilaginous Joints

Hyaline cartilage or fibrocartilage connects the bones of **cartilaginous joints.** The two types are:

1. **Synchondrosis.** In a synchondrosis, bands of hyaline cartilage unite the bones. Many of these joints are temporary structures that disappear during growth. An example is an immature long bone where a band of hyaline cartilage (the epiphyseal disk) connects an epiphysis to a diaphysis. This cartilage band participates in bone lengthening, and in time is converted into bone. When this ossification is completed, usually before the age of 25 years, movement no longer occurs at the joint. Thus, the joint is synarthrotic (see fig. 7.11).

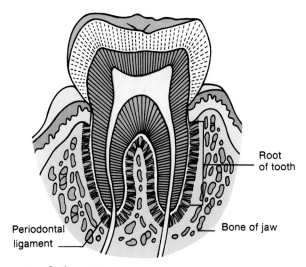

**FIGURE 8.4**

The articulation between the root of a tooth and the jawbone is a gomphosis.

Root of tooth

Periodontal ligament

Bone of jaw

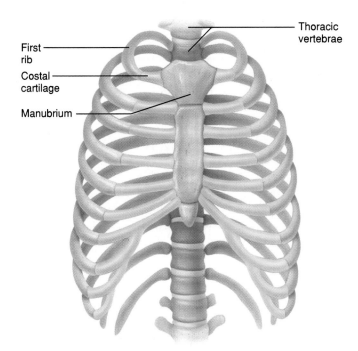

**FIGURE 8.5**

The articulation between the first rib and the manubrium is a synchondrosis.

Thoracic vertebrae

First rib

Costal cartilage

Manubrium

Another synchondrosis occurs between the manubrium (sternum) and the first rib, which are directly united by costal cartilage (fig. 8.5). This joint is also synarthrotic, but permanent. The joints between the costal cartilages and the sternum of ribs 2 through 7 are usually synovial joints.

2. **Symphysis.** The articular surfaces of the bones at a symphysis are covered by a thin layer of hyaline cartilage, and the cartilage, in turn, is attached to a pad of springy fibrocartilage. A limited amount of movement occurs at such a joint whenever forces compress or deform the cartilaginous pad. An example of this type of joint is the symphysis pubis in the pelvis, which allows maternal pelvic bones to shift as an infant passes through the birth canal (fig. 8.6*a*). The joint formed by the bodies of two adjacent vertebrae separated by an intervertebral disk is also a symphysis (fig. 8.6*b* and reference plate 51).

Each intervertebral disk is composed of a band of fibrocartilage (annulus fibrosus) that surrounds a gelatinous core (nucleus pulposus). The disk absorbs shocks and helps equalize pressure between the vertebrae when the body moves. Since each disk is slightly flexible, the combined movement of many of the joints in the vertebral column allows the back to bend forward or to the side or to twist. Because these joints allow slight movements, they are amphiarthrotic joints.

## Synovial Joints

Most joints of the skeletal system are **synovial joints,** and because they allow free movement, they are diarthrotic (di″ar-thro′tik). These joints are more complex structurally than fibrous or cartilaginous joints. They consist of articular cartilage, a joint capsule, and a synovial membrane.

Virtuoso violinist Niccolò Paganini (1782–1840) astounded concertgoers with his ability to reach three octaves across the bridge of his instrument. So lax were his joints that he could bend his thumb backward until the nail touched the back of his hand.

Paganini had "benign joint hypermobility syndrome," defined as a range of motion much greater than normal. Today the condition is studied in people whose professions make lax joints either a benefit or a liability. In athletes and dancers, for example, loose joints increase the risk of injury. Musicians are especially interesting. The nimble fingers, hands, and wrists of hypermobility syndrome help woodwind and string players, but lax joints also tend to cause back and knee problems. Rather than gaining strength from repetitive movements of playing instruments, these joints must bear weight from long hours of sitting in one position.

Perhaps rock guitarists make the best use of hypermobile joints. They stretch their fingers like Paganini while jumping about onstage to better distribute their weight on the other joints!

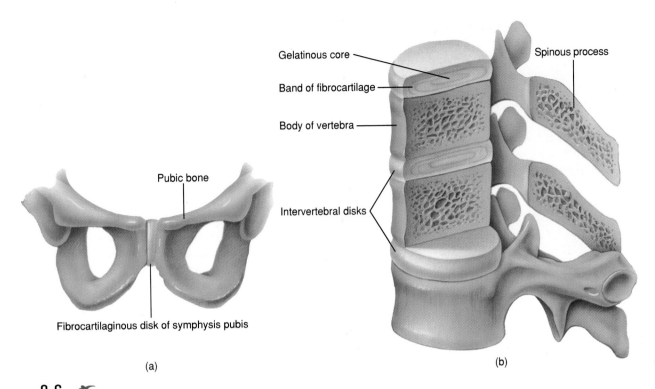

Gelatinous core

Band of fibrocartilage

Body of vertebra

Pubic bone

Intervertebral disks

Spinous process

Fibrocartilaginous disk of symphysis pubis

(a)

(b)

**FIGURE 8.6**

Fibrocartilage composes (a) the symphysis pubis that separates the pubic bones and (b) the intervertebral disks that separate vertebrae.

## GENERAL STRUCTURE OF A SYNOVIAL JOINT

The articular ends of the bones in a synovial joint are covered with a thin layer of hyaline cartilage (fig. 8.7). This layer, which is called the **articular cartilage,** resists wear and minimizes friction when it is compressed as the joint moves.

Typically, the bone beneath articular cartilage (subchondral plate) contains cancellous bone, which is somewhat elastic (fig. 8.7). This plate absorbs shocks, helping protect the joint from stresses caused by the load of body weight and by forces produced by contracting muscles.

Excessive mechanical stress due to obesity or certain athletic activities may fracture a subchondral plate. Although such fractures usually heal, the bone that regenerates may be less elastic than the original, reducing its protective function.

A tubular **joint capsule** (articular capsule) that has two distinct layers holds together the bones of a synovial joint. The outer layer largely consists of dense connective tissue, whose fibers attach to the periosteum around the circumference of each bone of the joint near its articular end. Thus, the outer fibrous layer of the capsule completely encloses the other parts of the joint. It is, however, flexible enough to permit movement and strong

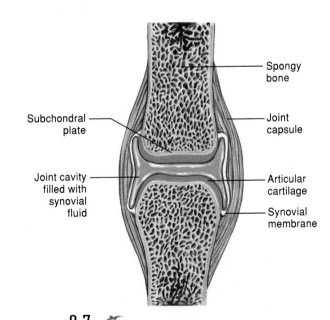

Spongy bone

Subchondral plate

Joint capsule

Joint cavity filled with synovial fluid

Articular cartilage

Synovial membrane

**FIGURE 8.7**

The generalized structure of a synovial joint.

enough to help prevent the articular surfaces from being pulled apart.

Bundles of strong, tough collagenous fibers called **ligaments** reinforce the joint capsule and help bind the articular ends of the bones. Some ligaments appear as thickenings in the fibrous layer of the capsule, whereas others are *accessory structures* located outside the capsule. In either case, these structures help prevent excessive movement at the joint. That is, the ligament is relatively inelastic, and it tightens when the joint is stressed.

The inner layer of the joint capsule consists of a shiny, vascular lining of loose connective tissue called the **synovial membrane.** This membrane, which is only a few cells thick, covers all of the surfaces within the joint capsule, except the areas the articular cartilage covers. The synovial membrane surrounds a closed sac called the *synovial cavity,* into which the membrane secretes a clear, viscous fluid called **synovial fluid.** In some regions, the surface of the synovial membrane has villi as well as larger folds and projections that extend into the cavity. Besides filling spaces and irregularities of the joint cavity, these extensions increase the surface area of the synovial membrane. The membrane may also store adipose tissue and form movable fatty pads within the joint. This multifunctional membrane also reabsorbs fluid, which is important when a joint cavity is injured or infected.

Synovial fluid has a consistency similar to uncooked egg white, and it moistens and lubricates the smooth cartilaginous surfaces within the joint. It also helps supply articular cartilage with nutrients that are obtained from blood vessels of the synovial membrane. Although the volume of synovial fluid in a joint cavity is usually just enough to cover the articulating surfaces with a thin film of fluid, it may provide a protective cushion. The amount of synovial fluid in the cavity of the knee is 0.5 mL or less.

A physician can determine the cause of joint inflammation or degeneration (arthritis) by aspirating a sample of synovial fluid from the affected joint using a procedure called arthrocentesis. Bloody fluid with lipid on top indicates a fracture extending into the joint. Clear fluid is found in osteoarthritis, a degeneration of collagen in the joint that is inherited or associated with aging. Cloudy, yellowish fluid may indicate the autoimmune disorder rheumatoid arthritis, and crystals in the synovial fluid signal the metabolic condition gout. If the fluid is cloudy but red-tinged and containing pus, a bacterial infection may be present that requires prompt treatment. Normal synovial fluid has 180 or fewer leukocytes (white blood cells) per cubic mm. If the fluid is infected, the leukocyte count exceeds 2,000.

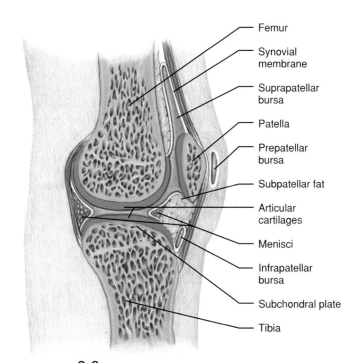

**FIGURE 8.8**

Menisci separate the articulating surfaces of the femur and tibia. Several bursae are associated with the knee joint.

Some synovial joints are partially or completely divided into two compartments by disks of fibrocartilage called **menisci** (sing., *meniscus*) located between the articular surfaces. Each meniscus attaches to the fibrous layer of the joint capsule peripherally, and its free surface projects into the joint cavity. In the knee joint, crescent-shaped menisci cushion the articulating surfaces and help distribute body weight onto these surfaces (fig. 8.8).

Certain synovial joints are also associated with fluid-filled sacs called **bursae.** Each bursa has an inner lining of synovial membrane, which may be continuous with the synovial membrane of a nearby joint cavity. These sacs contain synovial fluid and are commonly located between the skin and underlying bony prominences, as in the case of the patella of the knee or the olecranon process of the elbow. Bursae cushion and aid the movement of tendons that glide over bony parts or over other tendons. The names of bursae indicate their locations. Figure 8.8 shows a *suprapatellar bursa,* a *prepatellar bursa,* and an *infrapatellar bursa.*

1  Describe two types of cartilaginous joints.

2  What is the function of an intervertebral disk?

3  Describe the structure of a synovial joint.

4  What is the function of the synovial fluid?

Articular cartilage, like other cartilaginous structures, lacks a direct blood supply (see chapter 5). Surrounding synovial fluid supplies oxygen, nutrients, and other vital chemicals. Normal body movements force these substances into the joint cartilage. When a joint is immobilized or is not used for a long time, inactivity may cause degeneration of the articular cartilage. The degeneration may reverse when joint movements resume. However, it is important to avoid exercises that greatly compress the tissue during the period of regeneration. Otherwise, chondrocytes in the thinned cartilage may be injured, hindering repair.

## TYPES OF SYNOVIAL JOINTS

The articulating bones of synovial joints have a variety of shapes that allow different kinds of movement. Based upon their shapes and the movements they permit, these joints can be classified into six major types—ball-and-socket joints, condyloid joints, gliding joints, hinge joints, pivot joints, and saddle joints.

1. A **ball-and-socket joint** (spheroidal joint) consists of a bone with a globular or slightly egg-shaped head that articulates with the cup-shaped cavity of another bone. Such a joint allows a wider range of motion than does any other kind, permitting movements in all planes, as well as rotational movement around a central axis. The hip and shoulder contain joints of this type (fig. 8.9a).

2. In a **condyloid joint** (ellipsoidal joint), the ovoid condyle of one bone fits into the elliptical cavity of another bone, as in the joints between the metacarpals and phalanges. This type of joint permits a variety of movements in different planes; rotational movement, however, is not possible (fig. 8.9b).

3. The articulating surfaces of **gliding joints** (arthrodial joints) are nearly flat or slightly curved. These joints allow sliding or back-and-forth motion and twisting movements. Most of the joints within the wrist and ankle, as well as those between the articular processes of adjacent vertebrae, belong to this group (fig. 8.9c). The sacroiliac joints and the joints formed by ribs 2 through 7 connecting with the sternum are also gliding joints.

4. In a **hinge joint** (ginglymoidal joint), the convex surface of one bone fits into the concave surface of another, as in the elbow and the joints of the phalanges. Such a joint resembles the hinge of a door in that it permits movement in one plane only (fig. 8.9d).

5. In a **pivot joint** (trochoidal joint), the cylindrical surface of one bone rotates within a ring formed of bone and fibrous tissue of a ligament. Movement at such a joint is limited to rotation around a central axis. The joint between the proximal ends of the radius and the ulna, where the head of the radius rotates in a ring formed by the radial notch of the ulna and a ligament (annular ligament), is of this type. Similarly, a pivot joint functions in the neck as the head turns from side to side. In this case, the ring formed by a ligament (transverse ligament) and the anterior arch of the atlas rotates around the dens of the axis (fig. 8.9e).

6. A **saddle joint** (sellar joint) forms between bones whose articulating surfaces have both concave and convex regions. The surface of one bone fits the complementary surface of the other. This physical relationship permits a variety of movements, as in the case of the joint between the carpal (trapezium) and the metacarpal of the thumb (fig. 8.9f).

Table 8.1 summarizes the types of joints.

## TYPES OF JOINT MOVEMENTS

Skeletal muscle action produces movements at synovial joints. Typically, one end of a muscle is attached to a relatively immovable or fixed part on one side of a joint, and the other end of the muscle is fastened to a movable part on the other side. When the muscle contracts, its fibers pull its movable end (*insertion*) toward its fixed end (*origin*), and a movement occurs at the joint.

The following terms describe movements at joints that occur in different directions and in different planes (figs. 8.10, 8.11, and 8.12):

**flexion** (flek'shun)   Bending parts at a joint so that the angle between them decreases and the parts come closer together (bending the lower limb at the knee).

**extension** (ek-sten'shun)   Straightening parts at a joint so that the angle between them increases and the parts move farther apart (straightening the lower limb at the knee).

(a) **Ball-and-socket joint**

Hipbone

Head of femur in acetabulum

Femur

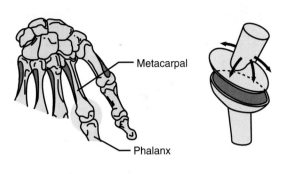

(b) **Condyloid joint**

Metacarpal

Phalanx

(c) **Gliding joint**

Carpals

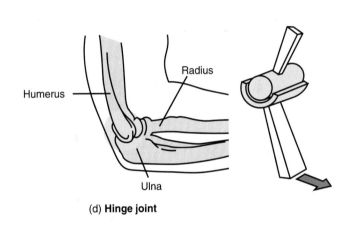

(d) **Hinge joint**

Humerus

Radius

Ulna

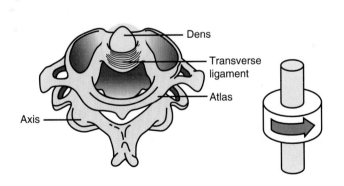

(e) **Pivot joint**

Dens

Transverse ligament

Atlas

Axis

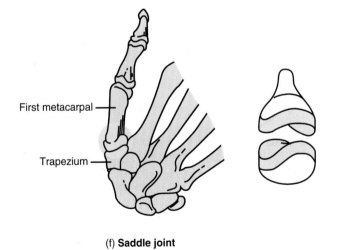

(f) **Saddle joint**

First metacarpal

Trapezium

**FIGURE 8.9**

Types and examples of synovial (freely movable) joints.

TABLE 8.1 TYPES OF JOINTS

| Type of Joint | Description | Possible Movements | Example |
|---|---|---|---|
| Fibrous | Articulating bones fastened together by thin layer of dense connective tissue containing many collagenous fibers | | |
| **1.** *Syndesmosis* (amphiarthrotic) | Bones bound by interosseous ligament | Joint flexible and may be twisted | Tibiofibular articulation |
| **2.** *Suture* (synarthrotic) | Flat bones united by sutural ligament | None | Parietal bones articulate at sagittal suture of skull |
| **3.** *Gomphosis* (synarthrotic) | Cone-shaped process fastened in bony socket by periodontal ligament | None | Root of tooth united with mandible |
| Cartilaginous | Articulating bones connected by hyaline cartilage or fibrocartilage | | |
| **1.** *Synchondrosis* (synarthrotic) | Bones united by bands of hyaline cartilage | Movement occurs during growth process until ossification occurs | Joint between epiphysis and diaphysis of a long bone |
| **2.** *Symphysis* (amphiarthrotic) | Articular surfaces separated by thin layers of hyaline cartilage attached to band of fibrocartilage | Limited movement, as when back is bent or twisted | Joints between bodies of vertebrae |
| Synovial (diarthrotic) | Articulating bones surrounded by a joint capsule of ligaments and synovial membranes; ends of articulating bones covered by hyaline cartilage and separated by synovial fluid | | |
| **1.** *Ball-and-socket* | Ball-shaped head of one bone articulates with cup-shaped socket of another | Movements in all planes; rotation | Shoulder, hip |
| **2.** *Condyloid* | Oval-shaped condyle of one bone articulates with elliptical cavity of another | Variety of movements in different planes, but no rotation | Joints between metacarpals and phalanges |
| **3.** *Gliding* | Articulating surfaces are nearly flat or slightly curved | Sliding or twisting | Joints between various bones of wrist and ankle |
| **4.** *Hinge* | Convex surface of one bone articulates with concave surface of another | Flexion and extension | Elbow and joints of phalanges |
| **5.** *Pivot* | Cylindrical surface of one bone articulates with ring of bone and fibrous tissue | Rotation | Joint between proximal ends of radius and ulna |
| **6.** *Saddle* | Articulating surfaces have both concave and convex regions; surface of one bone fits the complementary surface of another | Variety of movements | Joint between carpal and metacarpal of thumb |

**hyperextension** (hi″per-ek-sten′shun)   Excess extension of the parts at a joint, beyond the anatomical position (bending the head back beyond the upright position).

**dorsiflexion** (dor″sĭ-flek′shun)   Bending the foot at the ankle toward the shin (bending the foot upward).

**plantar flexion** (plan′tar flek′shun)   Bending the foot at the ankle toward the sole (bending the foot downward).

**abduction** (ab-duk′shun)   Moving a part away from the midline (lifting the upper limb horizontally to form a right angle with the side of the body).

**adduction** (ah-duk′shun)   Moving a part toward the midline (returning the upper limb from the horizontal position to the side of the body).

**rotation** (ro-ta′shun)   Moving a part around an axis (twisting the head from side to side). Medial rotation involves movement toward the midline, whereas lateral rotation involves movement in the opposite direction.

**FIGURE** 8.10 ✗
Joint movements illustrating adduction, abduction, dorsiflexion, plantar flexion, hyperextension, extension, and flexion.

**circumduction** (ser"kum-duk'shun)   Moving a part so that its end follows a circular path (moving the finger in a circular motion without moving the hand).

**supination** (soo"pĭ-na'shun)   Turning the hand so the palm is upward.

**pronation** (pro-na'shun)   Turning the hand so the palm is downward.

**eversion** (e-ver'zhun)   Turning the foot so the sole faces laterally.

**inversion** (in-ver'zhun)   Turning the foot so the sole faces medially.

**protraction** (pro-trak'shun)   Moving a part forward (thrusting the chin forward).

**retraction** (re-trak'shun)   Moving a part backward (pulling the chin backward).

**elevation** (el"ĕ-va'shun)   Raising a part (shrugging the shoulders).

**depression** (de-presh'un)   Lowering a part (drooping the shoulders).

Where movements of body parts are part of the definition, we will simply describe movements of those parts, for example, adduction of the lower limb or rotation of the head. Special cases also fall herein, as with plantar flexion of the foot. Other movements are described by the change in geometry at a joint, such as the action of the biceps brachii, flexion at the elbow. Here we will go with the more descriptive "flexion of the forearm at the elbow." Table 8.2 lists information on several joints.

**1.** Name six types of synovial joints.

**2.** Describe the structure of each type of synovial joint.

**3.** What terms describe movements that occur at synovial joints?

Supination    Pronation

Circumduction

Rotation

**FIGURE 8.11**

Joint movements illustrating rotation, circumduction, supination, and pronation.

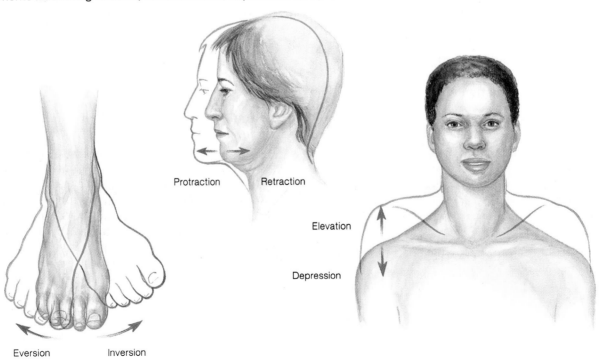

Protraction    Retraction

Elevation

Depression

Eversion    Inversion

**FIGURE 8.12**

Joint movements illustrating eversion, inversion, protraction, retraction, elevation, and depression.

## TABLE 8.2 JOINTS OF THE BODY

| Joint | Location | Type of Joint | Type of Movement |
|---|---|---|---|
| Skull | Cranial and facial bones | Suture, fibrous | Immovable, synarthrotic |
| Temporomandibular | Temporal bone, mandible | Modified hinge, synovial | Elevation, depression, protraction, retraction, diarthrotic |
| Atlantooccipital | Atlas, occipital bone | Condyloid, synovial | Flexion, extension, circumduction, diarthrotic |
| Intervertebral | Between adjacent vertebral bodies | Symphysis, cartilaginous | Slight movement, amphiarthrotic |
| Intervertebral | Between articular processes | Gliding, synovial | Flexion, extension, slight rotation, diarthrotic |
| Sacroiliac | Sacrum and hipbone | Gliding, synovial | Little to no movement, diarthrotic |
| Vertebrocostal | Vertebrae and ribs | Gliding, synovial | Slight movement during breathing, diarthrotic |
| Sternoclavicular | Sternum and clavicle | Gliding, synovial | Slight movement when shrugging shoulders, diarthrotic |
| Sternocostal | Sternum and rib 1 | Synchondrosis, cartilaginous | Immovable, synarthrotic |
| Sternocostal | Sternum and ribs 2–7 | Gliding, synovial | Slight movement during breathing, diarthrotic |
| Acromioclavicular | Scapula and clavicle | Gliding, synovial | Protraction, retraction, elevation, depression, diarthrotic |
| Shoulder (glenohumeral) | Humerus and scapula | Ball-and-socket, synovial | Flexion, extension, adduction, abduction, rotation, circumduction, diarthrotic |
| Elbow | Humerus and ulna | Hinge, synovial | Flexion, extension, diarthrotic |
| Proximal radioulnar | Radius and ulna | Pivot, synovial | Rotation, diarthrotic |
| Distal radioulnar | Radius and ulna | Syndesmosis, fibrous | Slight movement, amphiarthrotic |
| Wrist (radiocarpal) | Radius and carpals | Condyloid, synovial | Flexion, extension, adduction, abduction, circumduction, diarthrotic |
| Intercarpal | Adjacent carpals | Gliding, synovial | Slight movement, diarthrotic |
| Carpometacarpal | Carpal and metacarpal 1 | Saddle, synovial | Flexion, extension, adduction, abduction, diarthrotic |
| Carpometacarpal | Carpals and metacarpals 2–5 | Condyloid, synovial | Flexion, extension, adduction, abduction, diarthrotic |
| Metacarpophalangeal | Metacarpal and proximal phalanx | Condyloid, synovial | Flexion, extension, adduction, abduction, diarthrotic |
| Interphalangeal | Adjacent phalanges | Hinge, synovial | Flexion, extension, diarthrotic |
| Symphysis pubis | Pubic bones | Symphysis, cartilaginous | Slight movement, amphiarthrotic |
| Hip | Hipbone and femur | Ball-and-socket, synovial | Flexion, extension, adduction, abduction, rotation, circumduction, diarthrotic |
| Knee (tibiofemoral) | Femur and tibia | Modified hinge, synovial | Flexion, extension, slight rotation when flexed, diarthrotic |
| Knee (femoropatellar) | Femur and patella | Gliding, synovial | Slight movement, diarthrotic |
| Proximal tibiofibular | Tibia and fibula | Gliding, synovial | Slight movement, diarthrotic |
| Distal tibiofibular | Tibia and fibula | Syndesmosis, fibrous | Slight movement, amphiarthrotic |
| Ankle (talocrural) | Talus, tibia, and fibula | Hinge, synovial | Dorsiflexion, plantar flexion, slight circumduction, diarthrotic |
| Intertarsal | Adjacent tarsals | Gliding, synovial | Inversion, eversion, diarthrotic |
| Tarsometatarsal | Tarsals and metatarsals | Gliding, synovial | Slight movement, diarthrotic |
| Metatarsophalangeal | Metatarsal and proximal phalanx | Condyloid, synovial | Flexion, extension, adduction, abduction, diarthrotic |

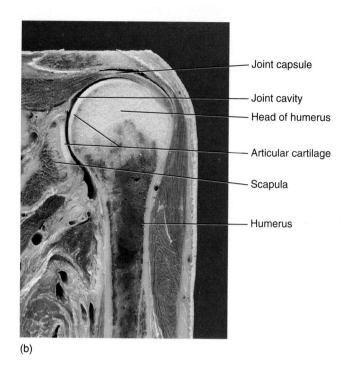

**FIGURE 8.13**

(a) The shoulder joint allows movements in all directions. Note that a bursa is associated with this joint. (b) Photograph of the shoulder joint (coronal section).

## EXAMPLES OF SYNOVIAL JOINTS

The shoulder, elbow, hip, and knee are large, freely movable joints. Although these joints have much in common, each has a unique structure that makes possible its specific function.

### Shoulder Joint

The **shoulder joint** is a ball-and-socket joint that consists of the rounded head of the humerus and the shallow glenoid cavity of the scapula. The coracoid and acromion processes of the scapula protect these parts, and dense connective tissue and muscle hold them together.

The joint capsule of the shoulder is attached along the circumference of the glenoid cavity and the anatomical neck of the humerus. Although it completely envelopes the joint, the capsule is very loose, and by itself is unable to keep the bones of the joint in close contact. However, muscles and tendons surround and reinforce the capsule, keeping together the articulating parts of the shoulder (fig. 8.13).

> The tendons of several muscles intimately blend with the fibrous layer of the shoulder joint capsule, forming the *rotator cuff*, which reinforces and supports the shoulder joint. Throwing a ball can create powerful centrifugal forces that injure the rotator cuff.

The ligaments that help prevent displacement of the articulating surfaces of the shoulder joint include the following (fig. 8.14):

1. **Coracohumeral** (kor″ah-ko-hu′mer-al) **ligament.** This ligament is composed of a broad band of connective tissue that connects the coracoid process of the scapula to the greater tubercle of the humerus. It strengthens the superior portion of the joint capsule.

2. **Glenohumeral** (gle″no-hu′mer-al) **ligaments.** These include three bands of fibers that appear as thickenings in the ventral wall of the joint capsule. They extend from the edge of the glenoid cavity to the lesser tubercle and the anatomical neck of the humerus.

3. **Transverse humeral ligament.** This ligament consists of a narrow sheet of connective tissue fibers that runs between the lesser and the greater tubercles of the humerus. Together with the intertubercular groove of the humerus, the ligament forms a canal (retinaculum) through which the long head of the biceps brachii muscle passes.

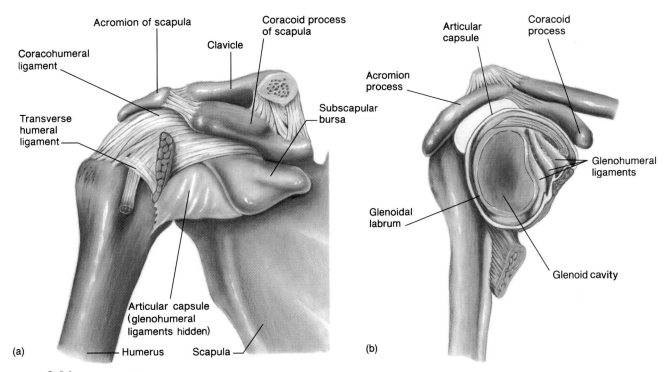

**FIGURE 8.14**

(a) Ligaments hold together the articulating surfaces of the shoulder. (b) The glenoidal labrum is a ligament composed of fibrocartilage.

4. **Glenoidal labrum** (gle′noid-al la′brum). This ligament is composed of fibrocartilage. It is attached along the margin of the glenoid cavity and forms a rim with a thin, free edge that deepens the cavity.

Several bursae are associated with the shoulder joint. The major ones include the *subscapular bursa* located between the joint capsule and the tendon of the subscapularis muscle, the *subdeltoid bursa* between the joint capsule and the deep surface of the deltoid muscle, the *subacromial bursa* between the joint capsule and the undersurface of the acromion process of the scapula, and the *subcoracoid bursa* between the joint capsule and the coracoid process of the scapula. Of these, the subscapular bursa is usually continuous with the synovial cavity of the joint cavity, and although the others do not communicate with the joint cavity, they may be connected to each other (see figs. 8.13 and 8.14).

Due to the looseness of its attachments and the relatively large articular surface of the humerus compared to the shallow depth of the glenoid cavity, the shoulder joint is capable of a very wide range of movement. These movements include flexion, extension, abduction, adduction, rotation, and circumduction. Motion occurring simultaneously in the joint formed between the scapula and the clavicle may also aid such movements.

Because the bones of the shoulder joint are mainly held together by supporting muscles rather than by bony structures and strong ligaments, the joint is somewhat weak. Consequently, the articulating surfaces may become displaced or dislocated easily. Such a *dislocation* most commonly occurs with a forceful impact during abduction, as when a person falls on an outstretched arm. This movement may press the head of the humerus against the lower part of the joint capsule where its wall is thin and poorly supported by ligaments. Dislocations commonly affect joints of the shoulders, knees, fingers, and jaw.

### Elbow Joint

The **elbow joint** is a complex structure that includes two articulations—a hinge joint between the trochlea of the humerus and the trochlear notch of the ulna, and a gliding joint between the capitulum of the humerus and a shallow depression (fovea) on the head of the radius. A joint capsule completely encloses and holds together these unions (fig. 8.15). Ulnar and radial collateral ligaments thicken the two joints, and fibers

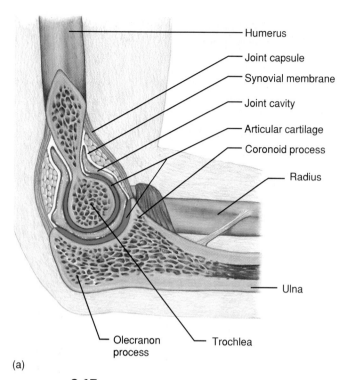

Humerus
Joint capsule
Synovial membrane
Joint cavity
Articular cartilage
Coronoid process
Radius
Ulna
Olecranon process
Trochlea

(a)

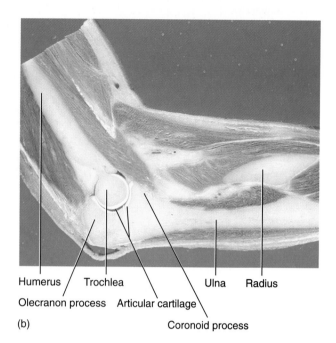

Humerus   Trochlea          Ulna   Radius
Olecranon process   Articular cartilage
Coronoid process

(b)

**FIGURE 8.15**

(a) The elbow joint allows hinge movements, as well as pronation and supination of the hand. (b) Photograph of the elbow joint (sagittal section).

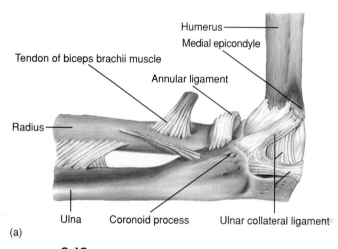

Tendon of biceps brachii muscle
Annular ligament
Humerus
Medial epicondyle
Radius
Ulna   Coronoid process   Ulnar collateral ligament

(a)

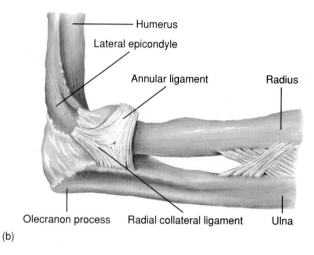

Humerus
Lateral epicondyle
Annular ligament
Radius
Olecranon process   Radial collateral ligament   Ulna

(b)

**FIGURE 8.16**

(a) The ulnar collateral ligament and (b) the radial collateral ligament strengthen the capsular wall of the elbow joint.

from a muscle (brachialis) in the arm reinforce its anterior surface.

The **ulnar collateral ligament,** which is a thick band of dense connective tissue, is located in the medial wall of the capsule. The anterior portion of this ligament connects the medial epicondyle of the humerus to the medial margin of the coronoid process of the ulna. Its posterior part is attached to the medial epicondyle of the humerus and to the olecranon process of the ulna (fig. 8.16a).

The **radial collateral ligament,** which strengthens the lateral wall of the joint capsule, is a fibrous band extending between the lateral epicondyle of the humerus and the *annular ligament* of the radius. The annular ligament, in turn, attaches to the margin of the trochlear notch of the ulna, and it encircles the head of the radius, keeping the head in contact with the radial notch of the ulna. The elbow joint capsule encloses the resulting radioulnar joint so that its function is closely associated with the elbow (fig. 8.16b).

The *synovial membrane* that forms the inner lining of the elbow capsule projects into the joint cavity between the radius and ulna, and partially divides the joint into humerus-ulnar and humerus-radial portions. Also, varying amounts of adipose tissue form fatty pads between the synovial membrane and the fibrous layer of the joint capsule. These pads help protect nonarticular bony areas during joint movements.

The only movements that can occur at the elbow between the humerus and ulna are hinge-type movements—flexion and extension. The head of the radius, however, is free to rotate in the annular ligament. This movement allows pronation and supination of the hand.

1.   Which parts help keep together the articulating surfaces of the shoulder joint?

2.   What factors allow an especially wide range of motion in the shoulder?

3.   Which structures form the hinge joint of the elbow?

4.   Which parts of the elbow permit pronation and supination of the hand?

Arthroscopy enables a surgeon to visualize the interior of a joint, and even perform diagnostic or therapeutic procedures, guided by the image on a video screen. An arthroscope is a thin, tubular instrument about 25 cm long containing optical fibers that transmit an image. The surgeon inserts the device through a small incision in the joint capsule. It is far less invasive than conventional surgery. Many runners have undergone arthroscopy and raced just weeks later.

Arthroscopy is combined with a genetic technique called the polymerase chain reaction (PCR) to rapidly diagnose infection. Guided by an arthroscope, the surgeon samples a small piece of the synovial membrane. PCR detects and amplifies specific DNA sequences, such as those of bacteria. For example, the technique can rapidly diagnose Lyme disease by detecting DNA from the causative bacterium *Borrelia burgdorferi*. This is valuable because a variety of bacteria can infect joints, and choosing the appropriate antibiotic, based on knowing the type of bacterium, is crucial for fast and complete recovery.

## Hip Joint

The **hip joint** is a ball-and-socket joint that consists of the head of the femur and the cup-shaped acetabulum of the coxal bone. A ligament (ligamentum capitis) attaches to a pit (fovea capitis) on the head of the femur and to connective tissue within the acetabulum. This attachment, however, seems to have little importance in holding the articulating bones together, but rather carries blood vessels to the head of the femur (fig. 8.17).

(a)

(b)

**FIGURE 8.17**

(a) The acetabulum provides the socket for the head of the femur in the hip joint. (b) The pit (fovea capitis) in the femur's head marks attachment of a ligament that carries blood vessels there.

A horseshoe-shaped ring of fibrocartilage (acetabular labrum) at the rim of the acetabulum deepens the cavity of the acetabulum. It encloses the head of the femur and helps hold it securely in place. In addition, a heavy, cylindrical joint capsule that is reinforced with still other ligaments surrounds the articulating structures and connects the neck of the femur to the margin of the acetabulum (fig. 8.18).

The major ligaments of the hip joint include the following (fig. 8.19):

1. **Iliofemoral** (il″e-o-fem′o-ral) **ligament.** This ligament consists of a Y-shaped band of very strong fibers that connects the anterior inferior iliac spine of the coxal bone to a bony line (intertrochanteric line) extending between the greater and lesser trochanters of the femur. The iliofemoral ligament is the strongest ligament in the body.

2. **Pubofemoral** (pu″bo-fem′o-ral) **ligament.** The pubofemoral ligament extends between the superior portion of the pubis and the iliofemoral ligament. Its fibers also blend with the fibers of the joint capsule.

Coxal bone

Joint cavity

Articular cartilage

Synovial membrane

Joint capsule

Ligamentum capitis

Femur

(a)

Coxal bone

Articular cartilage

Joint cavity

Head of femur

Joint capsule

Femur

(b)

**FIGURE 8.18**

(a) A ring of cartilage in the acetabulum and a ligament-reinforced joint capsule hold together the hip joint. (b) Photograph of the hip joint (coronal section).

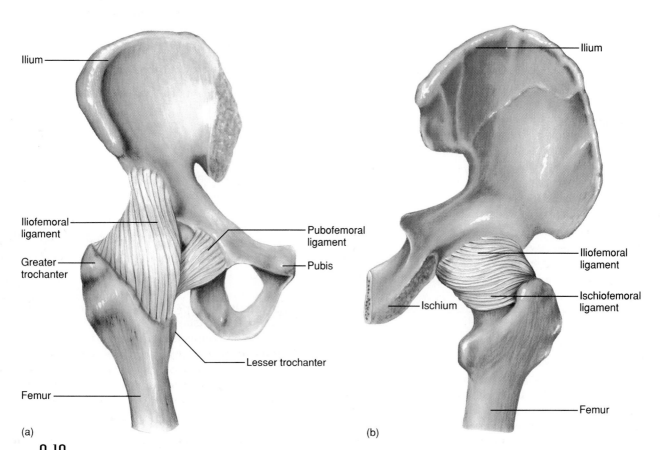

Ilium

Iliofemoral ligament

Greater trochanter

Femur

Pubofemoral ligament

Pubis

Lesser trochanter

(a)

Ilium

Iliofemoral ligament

Ischiofemoral ligament

Ischium

Femur

(b)

**FIGURE 8.19**

The major ligaments of the right hip joint. (a) Anterior view. (b) Posterior view.

# CLINICAL APPLICATION

## Replacing Joints

Surgeons use several synthetic materials to replace joints that are severely damaged by arthritis or injury. Metals such as cobalt-chrome and titanium alloys are used to replace larger joints, whereas silicone polymers are more commonly used to replace smaller joints. Such artificial joints must be durable yet not provoke immune system rejection. They must also allow normal healing to occur, and not move surrounding structures out of their normal positions.

Before the advent of joint replacements, surgeons removed damaged or diseased joint surfaces, hoping that scar tissue filling in the area would restore mobility. This type of surgery was rarely successful. In the 1950s, Alfred Swanson, an army surgeon in Grand Rapids, Michigan, invented the first joint implants using silicone polymers. By 1969, after much refinement, the first silicone-based joint implants hit the market.

These devices provided flexible hinges for joints of the toes, fingers, and wrists. Since then, more than two dozen joint replacement models have been developed, and more than a million people have them, mostly in the hip.

A surgeon inserts a joint implant in a procedure called implant resection arthroplasty. The surgeon first removes the surface of the joint bones and excess cartilage. Next, the centers of the tips of abutting bones are hollowed out,

and the stems of the implant are inserted here. The hinge part of the implant lies between the bones, aligning them yet allowing them to bend, as they would at a natural joint. Bone cement fixes the implant in place. Finally, the surgeon repairs the tendons, muscles, and ligaments. As the site of the implant heals, the patient must exercise the joint. A year of physical therapy may be necessary to fully benefit from replacement joints.

Newer joint replacements use materials that resemble natural body chemicals. Hip implants, for example, may bear a coat of hydroxylapatite, which interacts with natural bone. Instead of filling in spaces with bone cement, some investigators are testing a variety of porous coatings that allow bone tissue to grow into the implant area. ∎

3. **Ischiofemoral** (is″ke-o-fem′o-ral) **ligament.** This ligament consists of a band of strong fibers that originates on the ischium just posterior to the acetabulum and blends with the fibers of the joint capsule.

Muscles surround the joint capsule of the hip. The articulating parts of the hip are held more closely together than those of the shoulder, allowing considerably less freedom of movement. The structure of the hip joint, however, still permits a wide variety of movements, including extension, flexion, abduction, adduction, rotation, and circumduction. The hip is one of the joints most frequently replaced (Clinical Application 8.1).

## Knee Joint

The **knee joint** is the largest and most complex of the synovial joints. It consists of the medial and lateral condyles at the distal end of the femur, and the medial and lateral condyles at the proximal end of the tibia. In addition, the femur articulates anteriorly with the patella. Although the knee is sometimes considered a modified hinge joint, the articulations between the femur

and tibia are condyloid, and the joint between the femur and patella is a gliding joint.

The *joint capsule* of the knee is relatively thin, but ligaments and the tendons of several muscles greatly strengthen it. For example, the fused tendons of several muscles in the thigh cover the capsule anteriorly. Fibers from these tendons descend to the patella, partially enclose it, and continue downward to the tibia. The capsule attaches to the margins of the femoral and tibial condyles as well as between these condyles (fig. 8.20).

The ligaments associated with the joint capsule that help keep the articulating surfaces of the knee joint in contact include the following (fig. 8.21):

1. **Patellar** (pah-tel′ar) **ligament.** This ligament is a continuation of a tendon from a large muscle group in the thigh (quadriceps femoris). It consists of a strong, flat band that extends from the margin of the patella to the tibial tuberosity.

2. **Oblique popliteal** (ŏ′blēk pop-lit′e-al) **ligament.** This ligament connects the lateral condyle of the femur to the margin of the head of the tibia.

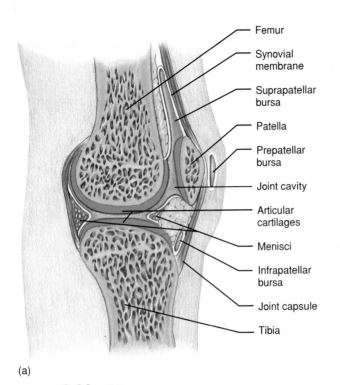

- Femur
- Synovial membrane
- Suprapatellar bursa
- Patella
- Prepatellar bursa
- Joint cavity
- Articular cartilages
- Menisci
- Infrapatellar bursa
- Joint capsule
- Tibia

(a)

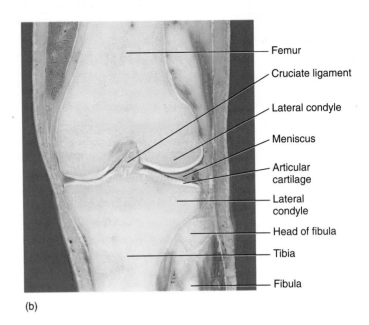

- Femur
- Cruciate ligament
- Lateral condyle
- Meniscus
- Articular cartilage
- Lateral condyle
- Head of fibula
- Tibia
- Fibula

(b)

**FIGURE 8.20**

(a) The knee joint is the most complex of the synovial joints (sagittal section). (b) Photograph of the knee joint (coronal section).

- Femur
- Posterior cruciate ligament
- Medial condyle
- Anterior cruciate ligament
- Lateral condyle
- Medial meniscus
- Lateral meniscus
- Medial condyle
- Lateral condyle
- Tibial collateral ligament
- Fibular collateral ligament
- Fibula
- Tibia
- Patellar ligament (cut)

(a)

- Femur
- Joint capsule
- Oblique popliteal ligament
- Fibular collateral ligament
- Arcuate popliteal ligament
- Tibial collateral ligament
- Fibula
- Tibia

(b)

**FIGURE 8.21**

Ligaments within the knee joint help to strengthen it. (a) Anterior view of right knee (patella removed). (b) Posterior view of left knee.

3. **Arcuate** (ar'ku-āt) **popliteal ligament.** This ligament appears as a Y-shaped system of fibers that extends from the lateral condyle of the femur to the head of the fibula.

4. **Tibial collateral** (tib'e-al kŏ-lat'er-al) **ligament** (medial collateral ligament). This ligament is a broad, flat band of tissue that connects the medial condyle of the femur to the medial condyle of the tibia.

5. **Fibular** (fib'u-lar) **collateral ligament** (lateral collateral ligament). This ligament consists of a strong, round cord located between the lateral condyle of the femur and the head of the fibula.

In addition to the ligaments that strengthen the joint capsule, two ligaments within the joint, called **cruciate** (kroo'she-āt) **ligaments,** help prevent displacement of the articulating surfaces. These strong bands of fibrous tissue stretch upward between the tibia and the femur, crossing each other on the way. They are named according to their positions of attachment to the tibia. Thus, the *anterior cruciate ligament* originates from the anterior intercondylar area of the tibia and extends to the lateral condyle of the femur. The *posterior cruciate ligament* connects the posterior intercondylar area of the tibia to the medial condyle of the femur.

---

The young soccer player, running at full speed, suddenly switches direction and is literally stopped in her tracks by a popping sound followed by a searing pain in her knee. Two hours after she veered toward the ball, her knee is swollen and painful, due to bleeding within the joint. She has torn the anterior cruciate ligament, a common serious knee injury.

---

Two fibrocartilaginous *menisci* separate the articulating surfaces of the femur and tibia. Each meniscus is roughly C-shaped, with a thick rim and a thinner center, and attaches to the head of the tibia. The medial and lateral menisci form depressions that fit the corresponding condyles of the femur (fig. 8.21).

Several bursae are associated with the knee joint. These include a large extension of the knee joint cavity called the *suprapatellar bursa* located between the anterior surface of the distal end of the femur and the muscle group (quadriceps femoris) above it; a large *prepatellar bursa* between the patella and the skin; and a smaller *infrapatellar bursa* between the proximal end of the tibia and the patellar ligament (see fig. 8.8).

As with a hinge joint, the basic structure of the knee joint permits flexion and extension. However, when the knee is flexed, rotation is also possible. Clinical Application 8.2 discusses some common joint disorders.

---

Tearing or displacing a meniscus is another common knee injury, usually resulting from forcefully twisting the knee when the leg is flexed. Since the meniscus is composed of fibrocartilage, this type of injury heals very slowly. Also, a torn and displaced portion of cartilage jammed between the articulating surfaces impedes movement of the joint. Following such a knee injury, the synovial membrane may become inflamed (acute synovitis) and secrete excess fluid, distending the joint cavity so that the knee swells above and on the sides of the patella.

---

**1** Which structures help keep the articulating surfaces of the hip together?

**2** What types of movement does the structure of the hip permit?

**3** What types of joints are within the knee?

**4** Which parts help hold together the articulating surfaces of the knee?

---

## CLINICAL TERMS RELATED TO JOINTS

**ankylosis** (ang″ kĭ-lo'sis)   Abnormal stiffness of a joint or fusion of bones at a joint, often due to damage of the joint membranes from chronic rheumatoid arthritis.

**arthralgia** (ar-thral'je-ah)   Pain in a joint.

**arthrodesis** (ar'thro-de'sis)   Surgery to fuse the bones at a joint.

**arthrogram** (ar'thro-gram)   Radiograph of a joint after an injection of radiopaque fluid into the joint cavity.

**arthrology** (ar-throl'o-je)   Study of joints and diseases of them.

**arthropathy** (ar-throp'ah-the)   Any joint disease.

**arthroplasty** (ar'thro-plas″te)   Surgery to make a joint more movable.

**arthrostomy** (ar-thros'to-me)   Surgical opening of a joint to allow fluid drainage.

**arthrotomy** (ar-throt'o-me)   Surgical incision of a joint.

**hemarthrosis** (hem″ar-thro'sis)   Blood in a joint cavity.

**hydrarthrosis** (hi″drar-thro'sis)   Accumulation of fluid within a joint cavity.

**luxation** (luk-sa'shun)   Dislocation of a joint.

**subluxation** (sub″luk-sa'shun)   Partial dislocation of a joint.

**synovectomy** (sin″o-vek'to-me)   Surgical removal of the synovial membrane of a joint.

# Joint Disorders

Joints have a tough job. They must support weight, provide a great variety of body movements, and are used very frequently. In addition to this normal wear and tear, these structures are sometimes subjected to injury from overuse, infection, an immune system launching a misplaced attack, or degeneration. Here is a look at some common joint problems.

## SPRAINS

*Sprains* result from overstretching or tearing the connective tissues, ligaments, and tendons associated with a joint, but they do not dislocate the articular bones. Usually forceful wrenching or twisting sprains the wrist or ankles. For example, inverting an ankle too far can sprain it by stretching the ligaments on its lateral surface. Severe injuries may pull these tissues loose from their attachments.

A sprained joint is painful and swollen, restricting movement. Immediate treatment of a sprain is rest; more serious cases require medical attention. However, immobilization of a joint, even for a brief period, causes bone resorption and weakens ligaments. Consequently, exercise may help strengthen the joint.

## BURSITIS

Overuse of a joint or stress on a bursa may cause *bursitis,* an inflammation of a bursa. The bursa between the heel bone (calcaneus) and the Achilles tendon may become inflamed as a result of a sudden increase in physical activity using the feet. Similarly, a form of bursitis called tennis elbow affects the bursa between the olecranon process and the skin. Bursitis is treated with rest. Medical attention may be necessary.

## ARTHRITIS

*Arthritis* is a disease that causes inflamed, swollen, and painful joints. More than a hundred different types of arthritis affect 50 million people in the United States. Arthritis can also be part of other syndromes (table 8A). The most common causes of arthritis are discussed below.

| TABLE 8A | DIFFERENT TYPES OF ARTHRITIS | | |
|---|---|---|---|
| **Some More-Common Forms of Arthritis** | | | |
| *Type* | *Incidence in the United States* | | |
| Osteoarthritis | 15.8 million | | |
| Rheumatoid arthritis | 2.9 million | | |
| Spondyloarthropathies | 2.5 million | | |
| **Some Less-Common Forms of Arthritis** | | | |
| *Type* | *Incidence in the United States* | *Age of Onset* | *Symptoms* |
| Gout | 1.6 million (85% male) | >40 | Sudden onset of extreme pain and swelling of a large joint |
| Juvenile rheumatoid arthritis | 250,000 | <18 | Joint stiffness, often in knee |
| Scleroderma | 300,000 | 30–50 | Skin hardens and thickens |
| Systemic lupus erythematosus | 300,000 (>90% female) | teens–50s | Fever, weakness, upper body rash, joint pain |
| Kawasaki disease | Hundreds of cases in local outbreaks | 6 months–11 years | Fever, joint pain, red rash on palms and soles, heart complications |
| Strep A infection | 100,000 | any age | Confusion, body aches, shock, low blood pressure, dizziness, arthritis, pneumonia |

Source: From "Arthritis: Modern Treatment for That Old Pain in the Joints" in *FDA Consumer,* July/August 1991.

## RHEUMATOID ARTHRITIS (RA)

This autoimmune disorder (a condition in which the immune system attacks the body's healthy tissues) is the most painful and debilitating form of arthritis. The synovial membrane of a joint becomes inflamed and thickens, forming a mass called a pannus. Then, the articular cartilage is damaged, and fibrous tissue infiltrates, interfering with joint movements. In time, the joint may ossify, fusing the articulating bones (bony ankylosis). Joints severely damaged by RA may be surgically replaced.

RA may affect many joints or only a few. It is usually a systemic illness, accompanied by fatigue, muscular atrophy, anemia, and osteoporosis, as well as changes in the skin, eyes, lungs, blood vessels, and heart. RA usually affects adults, but there is a juvenile form.

## OSTEOARTHRITIS

This degenerative disorder is the most common type of arthritis. It usually occurs with aging, but an in-herited form may appear as early as one's thirties. A person may first become aware of osteoarthritis when a blow to the affected joint produces pain that is much more intense than normal. Gradually, the area of the affected joint deforms (fig. 8A). For example, arthritic fingers take on a gnarled appearance, or a knee may bulge.

In osteoarthritis, articular cartilage softens and disintegrates gradually, roughening the articular surfaces. Joints become painful, with restricted movement. For example, arthritic fingers may lock into place while a person is playing the guitar or tying a shoelace. Osteoarthritis most often affects joints that are used the most over a lifetime, such as those of the fingers, hips, knees, and the lower parts of the vertebral column.

Fortunately, nonsteroidal anti-inflammatory drugs can usually control osteoarthritis symptoms. Exercise can keep stiff joints more flexible, and such simple measures as wearing gloves in the winter can alleviate symptoms.

## LYME ARTHRITIS

Lyme disease is a bacterial infection passed in a tick bite that causes inter-mittent arthritis of several joints, usually weeks after the initial symptoms of rash, fatigue, and flu-like aches and pains. Lyme arthritis was first observed in Lyme, Connecticut, where an astute woman kept a journal after noticing that many of her young neighbors had what appeared to be the very rare juvenile form of rheumatoid arthritis. Her observations led Allen Steere, a Yale University rheumatologist, to trace the illness to a tick-borne bacterial infection. Antibiotic treatment beginning as soon as the early symptoms of Lyme disease are recognized can prevent development of the associated arthritis.

Other types of bacteria can cause arthritis too. These include common *Staphylococcus* and *Streptococcus* species, *Neisseria gonorrhoeae* (which causes the sexually transmitted disease gonorrhea), and *Mycobacterium* (which causes tuberculosis). Arthritis may also be associated with AIDS, because the immune system breakdown raises the risk of infection by bacteria that can cause arthritis. ■

Femur

Patella

Cartilage

Synovial membrane lining

Damage to cartilage has occurred

Tibia

Synovial fluid

(a) Normal knee

(b) Osteoarthritic joint

**FIGURE 8A**

An inherited defect in collagen or prolonged wear and tear destroys joints in osteoarthritis.

# Introduction (page 257)

A joint forms wherever two or more bones meet.
Joints are the functional junctions between bones.

# Classification of Joints (page 257)

Joints are classified according to the type of tissue that binds the bones together.

1. Fibrous joints
   a. Bones at fibrous joints are fastened tightly together by a layer of dense connective tissue with many collagenous fibers.
   b. Little or no movement occurs at a fibrous joint.
   c. There are three types of fibrous joints.
      (1) A syndesmosis is characterized by bones bound by relatively long fibers of connective tissue.
      (2) A suture occurs where flat bones are united by a thin layer of connective tissue and become interlocked by a set of bony processes.
      (3) A gomphosis is formed by the union of a cone-shaped bony process to a bony socket.
2. Cartilaginous joints
   a. A layer of cartilage holds together bones of cartilaginous joints.
   b. There are two types of cartilaginous joints.
      (1) A synchondrosis is characterized by bones united by hyaline cartilage that may disappear as a result of growth.
      (2) A symphysis is a joint whose articular surfaces are covered by hyaline cartilage and attached to a pad of fibrocartilage.
3. Synovial joints
   a. Synovial joints have a more complex structure than other types of joints.
   b. These joints include articular cartilage, a joint capsule, and a synovial membrane.

# General Structure of a Synovial Joint (page 260)

Articular cartilage covers articular ends of bones.

1. A joint capsule strengthened by ligaments holds bones together .
2. A synovial membrane that secretes synovial fluid lines the inner layer of a joint capsule.
3. Synovial fluid moistens, provides nutrients, and lubricates the articular surfaces.
4. Menisci divide some synovial joints into compartments.
5. Some synovial joints have fluid-filled bursae.
   a. Bursae are usually located between the skin and underlying bony prominences.
   b. Bursae cushion and aid movement of tendons over bony parts.
   c. Bursae are named according to their locations.

# Types of Synovial Joints (page 262)

1. Ball-and-socket joints
   a. In a ball-and-socket joint, the globular head of a bone fits into the cup-shaped cavity of another bone.
   b. These joints permit a wide variety of movements.
   c. The hip and shoulder are ball-and-socket joints.
2. Condyloid joints
   a. A condyloid joint consists of an ovoid condyle of one bone fitting into an elliptical cavity of another bone.
   b. This joint permits a variety of movements.
   c. The joints between the metacarpals and phalanges are condyloid.
3. Gliding joints
   a. Articular surfaces of gliding joints are nearly flat.
   b. These joints permit the articular surfaces to slide back and forth.
   c. Most of the joints of the wrist and ankle are gliding joints.
4. Hinge joints
   a. In a hinge joint, the convex surface of one bone fits into the concave surface of another bone.
   b. This joint permits movement in one plane only.
   c. The elbow and the joints of the phalanges are the hinge type.
5. Pivot joints
   a. In a pivot joint, a cylindrical surface of one bone rotates within a ring of bone and fibrous tissue.
   b. This joint permits rotational movement.
   c. The articulation between the proximal ends of the radius and the ulna is a pivot joint.
6. Saddle joints
   a. A saddle joint is formed between bones that have complementary surfaces with both concave and convex regions.
   b. This joint permits a variety of movements.
   c. The articulation between the carpal and metacarpal of the thumb is a saddle joint.

# Types of Joint Movements (page 262)

1. Muscles acting at synovial joints produce movements in different directions and in different planes.
2. Joint movements include flexion, extension, hyperextension, dorsiflexion, plantar flexion, abduction, adduction, rotation, circumduction, supination, pronation, eversion, inversion, elevation, depression, protraction, and retraction.

# Examples of Synovial Joints (page 268)

1. Shoulder joint
   a. The shoulder joint is a ball-and-socket joint that consists of the head of the humerus and the glenoid cavity of the scapula.
   b. A cylindrical joint capsule envelops the joint.
      (1) The capsule is loose and by itself cannot keep the articular surfaces together.
      (2) It is reinforced by surrounding muscles and tendons.
   c. Several ligaments help prevent displacement of the bones.

d. Several bursae are associated with the shoulder joint.
e. Because its parts are loosely attached, the shoulder joint permits a wide range of movements.

2. Elbow joint
   a. The elbow has a hinge joint between the humerus and the ulna, and a gliding joint between the humerus and the radius.
   b. The joint capsule is reinforced by collateral ligaments.
   c. A synovial membrane partially divides the joint cavity into two portions.
   d. The joint between the humerus and the ulna permits flexion and extension only.

3. Hip joint
   a. The hip joint is a ball-and-socket joint between the femur and the coxal bone.
   b. A ring of fibrocartilage deepens the cavity of the acetabulum.

c. The articular surfaces are held together by a heavy joint capsule that is reinforced by ligaments.
d. The hip joint permits a wide variety of movements.

4. Knee joint
   a. The knee joint includes two condyloid joints between the femur and the tibia, and a gliding joint between the femur and the patella.
   b. Ligaments and tendons strengthen the relatively thin joint capsule.
   c. Several ligaments, some of which are within the joint capsule, bind articular surfaces.
   d. Two menisci separate the articulating surfaces of the femur and the tibia.
   e. Several bursae are associated with the knee joint.
   f. The knee joint permits flexion and extension; when the lower limb is flexed at the knee, some rotation is possible.

## CRITICAL THINKING QUESTIONS

1. How would you explain to an athlete why damaged joint ligaments and cartilages are so slow to heal following an injury?

2. Compared to the shoulder and hip joints, in what way is the knee joint poorly protected, and thus especially vulnerable to injuries?

3. Based upon your knowledge of joint structures, which do you think could be more satisfactorily replaced by a prosthetic device, a hip joint or a knee joint? Why?

4. If a patient's forearm and elbow were immobilized by a cast for several weeks, what changes would you expect to occur in the bones of the upper limb?

5. Why is it important to encourage an inactive patient to keep all joints mobile, even if it is necessary to have another person or a device move the joints (passive movement)?

6. How would you explain to a person with a dislocated shoulder that the shoulder is likely to become more easily dislocated in the future?

## REVIEW EXERCISES

## Part A

1. Define *joint.*
2. Explain how joints are classified.
3. Compare the structure of a fibrous joint with that of a cartilaginous joint.
4. Distinguish between a syndesmosis and a suture.
5. Describe a gomphosis, and name an example.
6. Compare the structures of a synchondrosis and a symphysis.
7. Explain how the joints between vertebrae permit movement.
8. Describe the general structure of a synovial joint.
9. Describe how a joint capsule may be reinforced.
10. Explain the function of the synovial membrane.
11. Explain the function of synovial fluid.
12. Define *meniscus.*
13. Define *bursa.*
14. List six types of synovial joints, and name an example of each type.
15. Describe the movements permitted by each type of synovial joint.
16. Name the parts that comprise the shoulder joint.
17. Name the major ligaments associated with the shoulder joint.
18. Explain why the shoulder joint permits a wide range of movements.
19. Name the parts that comprise the elbow joint.

20. Describe the major ligaments associated with the elbow joint.
21. Name the movements permitted by the elbow joint.
22. Name the parts that comprise the hip joint.
23. Describe how the articular surfaces of the hip joint are held together.
24. Explain why there is less freedom of movement in the hip joint than in the shoulder joint.
25. Name the parts that comprise the knee joint.
26. Describe the major ligaments associated with the knee joint.
27. Explain the function of the menisci of the knee.
28. Describe the locations of the bursae associated with the knee joint.

## Part B

Match the movements in column I with the descriptions in column II.

| I | II |
|---|---|
| 1. Rotation | A. Turning palm upward |
| 2. Supination | B. Decreasing angle between parts |
| 3. Extension | C. Moving part forward |
| 4. Eversion | D. Moving part around an axis |
| 5. Protraction | E. Turning sole of foot outward |
| 6. Flexion | F. Increasing angle between parts |
| 7. Pronation | G. Lowering a part |
| 8. Abduction | H. Turning palm downward |
| 9. Depression | I. Moving part away from midline |

 **For More Information Related to This Chapter, Please Visit** *Hole's Human Anatomy and Physiology* **Home Page**
http://www.mhhe.com/biosci/abio/

# MUSCULAR SYSTEM

## CHAPTER OBJECTIVES

AFTER YOU HAVE STUDIED THIS CHAPTER, YOU SHOULD BE ABLE TO:

1. Describe how connective tissue is part of the structure of a skeletal muscle.

2. Name the major parts of a skeletal muscle fiber and describe the function of each part.

3. Explain the major events that occur during muscle fiber contraction.

4. Explain how energy is supplied to the muscle fiber contraction mechanism, how oxygen debt develops, and how a muscle may become fatigued.

5. Distinguish between fast and slow muscle fibers.

6. Distinguish between a twitch and a sustained contraction.

7. Describe how exercise affects skeletal muscles.

8. Explain how various types of muscular contractions produce body movements and help maintain posture.

9. Distinguish between the structures and functions of a multiunit smooth muscle and a visceral smooth muscle.

10. Compare the contraction mechanisms of skeletal, smooth, and cardiac muscle fibers.

11. Explain how the locations of skeletal muscles help produce movements and how muscles interact.

12. Identify and locate the major skeletal muscles of each body region and describe the action of each muscle.

## KEY TERMS

**actin** (ak'tin)
**antagonist** (an-tag'o-nist)
**aponeurosis** (ap″o-nu-ro'sēz)
**fascia** (fash'e-ah)
**insertion** (in-ser'shun)
**motor neuron** (mo'tor nu'ron)
**motor unit** (mo'tor u'nit)
**muscle impulse** (mus'el im'puls)
**myofibril** (mi″o-fi'bril)
**myogram** (mi'o-gram)

**myosin** (mi'o-sin)
**neurotransmitter** (nu″ro-trans'mit-er)
**origin** (or'ĭ-jin)
**oxygen debt** (ok'sĭ-jen det)
**prime mover** (prīm mōōv'er)
**recruitment** (re-krōōt'ment)
**sarcomere** (sar'ko-mēr)
**synergist** (sin'er-jist)
**threshold stimulus** (thresh'old stim'u-lus)

## UNDERSTANDING WORDS

**calat-,** something inserted: inter*calat*ed disk—membranous band that connects cardiac muscle cells.

**erg-,** work: syn*erg*ist—muscle that works together with a prime mover to produce a movement.

**fasc-,** bundle: *fasc*iculus—bundle of muscle fibers.

**-gram,** something written: myo*gram*—recording of a muscular contraction.

**hyper-,** over, more: muscular *hyper*trophy—enlargement of muscle fibers.

**inter-,** between: *inter*calated disk—membranous band that connects cardiac muscle cells.

**iso-,** equal: *iso*tonic contraction—contraction during which the tension in a muscle remains unchanged.

**laten-,** hidden: *laten*t period—period between a stimulus and the beginning of a muscle contraction.

**myo-,** muscle: *myo*fibril—contractile fiber of a muscle cell.

**reticul-,** a net: sarcoplasmic *reticul*um—network of membranous channels within a muscle fiber.

**sarco-,** flesh: *sarco*plasm—substance (cytoplasm) within a muscle fiber.

**syn-,** together: *syn*ergist—muscle that works with a prime mover to produce a movement.

**tetan-,** stiff: *tetan*ic contraction—sustained muscular contraction.

**-tonic,** stretched: iso*tonic* contraction—contraction during which the tension of a muscle remains unchanged.

**-troph,** well fed: muscular hyper*troph*y—enlargement of muscle fibers.

**voluntar-,** of one's free will: *voluntar*y muscle—muscle that can be controlled by conscious effort.

ike many things in life, individual muscles aren't appreciated until we see what happens when they do not work. For children with Moebius syndrome, absence of the sixth and seventh cranial nerves, which carry impulses from the brain to the muscles of the face, leads to an odd collection of symptoms.

The first signs of Moebius syndrome are typically difficulty sucking, excessive drooling, and sometimes crossed eyes. The child has difficulty swallowing and chokes easily, cannot move the tongue well, and is very sensitive to bright light because he or she cannot squint or blink, or even avert the eyes. Special bottles and feeding tubes can help the child eat, and surgery can correct eye defects.

Children with Moebius syndrome are slow to reach developmental milestones, but do finally walk. As they get older, if they are lucky, they are left with only one symptom, but it is a rather obvious one—inability to form facial expressions.

A young lady named Chelsey Thomas called attention to this very rare condition when she underwent two surgeries that would enable her to smile. In 1995 and 1996, when she was 7 years old, Chelsey had two transplants of nerve and muscle tissue from her legs to either side of her mouth, supplying the missing "smile apparatus." Gradually, she acquired the subtle, and not-so-subtle, muscular movements of the mouth that make the human face so expressive. Chelsey inspired several other youngsters to undergo "smile surgery."

---

The three types of muscle tissues are skeletal, smooth, and cardiac, as described in chapter 5. This chapter focuses primarily on the skeletal muscles, which are usually attached to bones and are under conscious control.

## STRUCTURE OF A SKELETAL MUSCLE

A skeletal muscle is an organ of the muscular system. It is composed primarily of skeletal muscle tissue, nervous tissue, blood, and connective tissues.

### Connective Tissue Coverings

An individual skeletal muscle is separated from adjacent muscles and held in position by layers of dense connective tissue called **fascia.** This connective tissue surrounds each muscle and may project beyond the end of its muscle fibers to form a cordlike **tendon.** Fibers in a tendon intertwine with those in the periosteum of a bone, attaching the muscle to the bone. In other cases, the connective tissues associated with a muscle form broad, fibrous sheets called **aponeuroses,** which may be attached to the coverings of adjacent muscles (figs. 9.1 and 9.2).

A tendon, or the connective tissue sheath of a tendon (tenosynovium), may become painfully inflamed and swollen following an injury or the repeated stress of athletic activity. These conditions are called *tendinitis* and *tenosynovitis,* respectively. The tendons most commonly affected are those associated with the joint capsules of the shoulder, elbow, hip, and knee, and those involved with moving the wrist, hand, thigh, and foot.

The layer of connective tissue that closely surrounds a skeletal muscle is called the *epimysium.* Other layers of connective tissue, called the *perimysium,* extend inward from the epimysium and separate the muscle tissue into small sections. These sections contain bundles of skeletal muscle fibers called *fascicles* (fasciculi). Each muscle

Aponeuroses

Skeletal muscles

Tendons

**FIGURE 9.1**
Tendons attach muscles to bones, whereas aponeuroses attach muscles to other muscles.

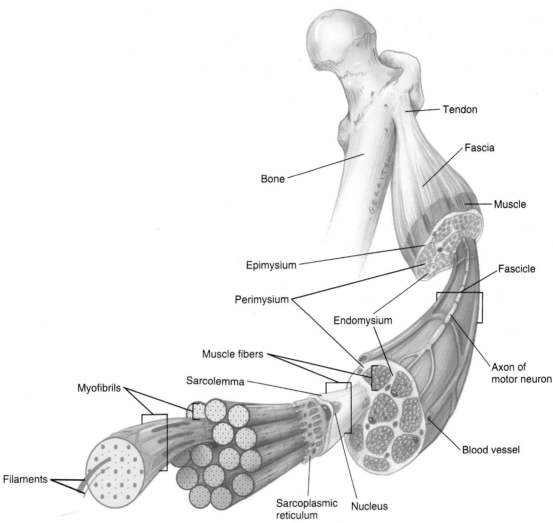

**FIGURE** 9.2

A skeletal muscle is composed of a variety of tissues, including layers of connective tissue. Fascia covers the surface of the muscle, epimysium lies beneath the fascia, and perimysium extends into the structure of the muscle where it separates muscle cells into fascicles. Endomysium separates individual muscle fibers.

fiber within a fascicle (fasciculus) is surrounded by a layer of connective tissue in the form of a thin covering called *endomysium* (figs. 9.2 and 9.3). Layers of connective tissue, therefore, enclose and separate all parts of a skeletal muscle. This arrangement allows the parts to move somewhat independently. Also, many blood vessels and nerves pass through these layers.

The fascia associated with each individual organ of the muscular system is part of a complex network of fasciae that extends throughout the body. The portion of the network that surrounds and penetrates the muscles is called *deep fascia.* It is continuous with the *subcutaneous fascia* that lies just beneath the skin, forming the subcutaneous layer described in chapter 6. The network is also continuous with the *subserous fascia* that forms the connective tissue layer of the serous membranes covering organs in various body cavities and lining those cavities (see chapter 6).

A *compartment* is the space that contains a particular group of muscles, blood vessels, and nerves, all tightly enclosed by fascia. The limbs have many such compartments. If an injury causes fluid, such as blood from an internal hemorrhage, to accumulate within a compartment, the pressure inside will rise. The increased pressure, in turn, may interfere with blood flow into the region, reducing the supply of oxygen and nutrients to the affected tissues. This condition, called *compartment syndrome,* often produces severe, unrelenting pain. Persistently elevated compartmental pressure may irreversibly damage the enclosed muscles and nerves. Treatment for compartment syndrome may require immediate intervention by a surgical incision through the fascia (fasciotomy) to relieve the pressure and restore circulation.

**FIGURE 9.3**

Scanning electron micrograph of a fascicle (fasciculus) surrounded by its connective tissue sheath, the perimysium. Muscle fibers within the fascicle are surrounded by endomysium (320×).

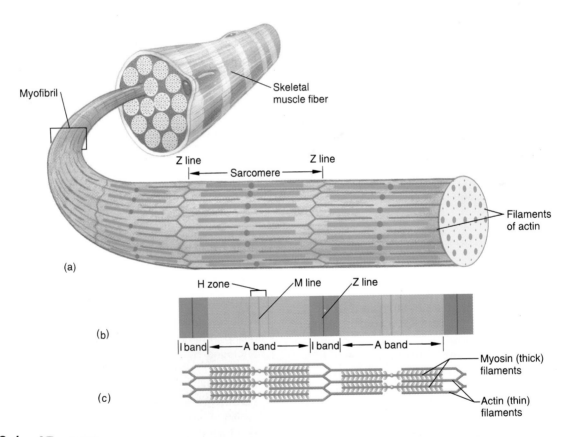

**FIGURE 9.4**

(a) A skeletal muscle fiber contains numerous myofibrils, each consisting of (b) repeating units called sarcomeres. (c) The characteristic striations of a sarcomere are due to the arrangement of actin and myosin filaments.

## Skeletal Muscle Fibers

Recall from chapter 5 that a skeletal muscle fiber represents a single cell of a muscle (see fig. 5.28). Each skeletal muscle fiber is a thin, elongated cylinder with rounded ends that are attached to connective tissues as-

sociated with a muscle. Just beneath the muscle cell membrane (*sarcolemma*), the cytoplasm (*sarcoplasm*) of the fiber contains many small, oval nuclei and mitochondria. The sarcoplasm also contains abundant, parallel threadlike **myofibrils** (fig. 9.4a).

The myofibrils play a fundamental role in the muscle contraction mechanism. They contain two kinds of protein filaments—thick filaments composed of the protein **myosin** and thin filaments composed of the protein **actin** (fig. 9.4a and c). The organization of these filaments produces the characteristic alternating light and dark *striations* of skeletal muscle fiber.

The striation pattern of skeletal muscle fibers has two main parts. The first, the *I bands* (the light bands), are composed of thin actin filaments held by direct attachment to structures called *Z lines*.

The second part of the striation pattern consists of the *A bands* (the dark bands), which are composed of thick myosin filaments overlapping thin actin filaments. The myosin filaments are also held in place by the Z lines, but are attached to them by a large protein called **titin** (connectin). Note that the A band consists not only of a region where the thick and thin filaments overlap, but also a central region (*H zone*) consisting only of thick filaments, plus a thickening known as the *M line* (fig. 9.4b). The segment of a myofibril that extends from one Z line to the next Z line is called a **sarcomere** (figs. 9.4 and 9.5).

Within the cytoplasm of a muscle fiber is a network of membranous channels that surrounds each myofibril and runs parallel to it. These membranes form the **sarcoplasmic reticulum,** which corresponds to the endoplasmic reticulum of other cells. Another set of membranous channels are the **transverse tubules** (T-tubules) that extend as invaginations from the fiber's membrane, and pass all the way through the fiber. Thus, each of these tubules opens to the outside of the muscle fiber and contains extracellular fluid. Furthermore, each transverse tubule lies between two enlarged portions of the sarcoplasmic reticulum called *cisternae* near the region where the actin and myosin filaments overlap. The sarcoplasmic reticulum and the transverse tubules activate muscle contraction when the fiber is stimulated (fig. 9.6).

**FIGURE** 9.5

A sarcomere (16,000×).

**FIGURE** 9.6

Within the sarcoplasm of a skeletal muscle fiber are a network of sarcoplasmic reticulum and a system of transverse tubules.

Although muscle fibers and the connective tissues associated with them are flexible, they can tear if overstretched. This type of injury is common in athletes and is called a *muscle strain*. The seriousness of the injury depends on the degree of damage the tissues sustain. In a mild strain, only a few muscle fibers are injured, the fascia remains intact, and little function is lost. In a severe strain, many muscle fibers as well as fascia tear, and muscle function may be lost completely. A severe strain is very painful and is accompanied by discoloration and swelling of tissues due to ruptured blood vessels. Surgery may be required to reconnect the separated tissues.

**1** Describe how connective tissue is associated with a skeletal muscle.

**2** Describe the general structure of a skeletal muscle fiber.

**3** Explain why skeletal muscle fibers appear striated.

**4** Explain the physical relationship between the sarcoplasmic reticulum and the transverse tubules.

## Neuromuscular Junction

Each skeletal muscle fiber is connected to an extension (a nerve fiber) of a **motor neuron** that passes outward from the brain or spinal cord. Normally a skeletal muscle fiber contracts only upon stimulation by a motor neuron.

The site where the nerve fiber and muscle fiber meet is called a **neuromuscular junction** (myoneural junction). There, the muscle fiber membrane is specialized to form a **motor end plate** where nuclei and mitochondria are abundant, and the sarcolemma is extensively folded (fig. 9.7).

The end of the motor nerve fiber branches, projecting into recesses of the muscle fiber membrane at the motor end plate. The membrane of the nerve fiber and the membrane of the muscle fiber are separated by a small gap called the **synaptic cleft.** The cytoplasm at the distal ends of the nerve fiber is rich in mitochondria and contains many tiny vesicles (synaptic vesicles) that store chemicals called **neurotransmitters.**

When a nerve impulse traveling from the brain or spinal cord reaches the end of a motor nerve fiber, some of the vesicles release a neurotransmitter into the synaptic cleft. This action stimulates the muscle fiber to contract.

## Motor Units

A muscle fiber usually has a single motor end plate. Motor nerve fibers, however, are densely branched. By means of these branches, one motor nerve fiber may connect to many muscle fibers. Furthermore, when a motor nerve fiber transmits an impulse, all of the muscle fibers connected to it are stimulated to contract simultaneously. Together, a motor neuron and the muscle fibers it controls constitute a **motor unit** (fig. 9.8).

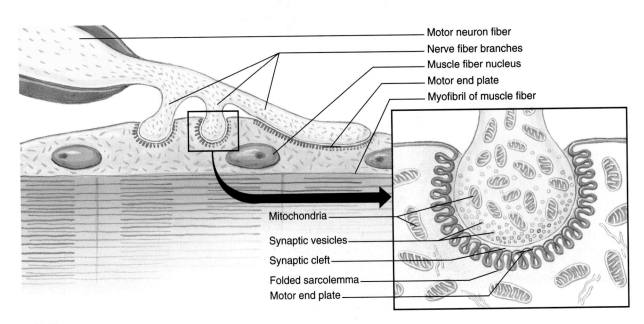

Motor neuron fiber
Nerve fiber branches
Muscle fiber nucleus
Motor end plate
Myofibril of muscle fiber

Mitochondria
Synaptic vesicles
Synaptic cleft
Folded sarcolemma
Motor end plate

**FIGURE** 9.7

A neuromuscular junction includes the end of a motor neuron and the motor end plate of a muscle fiber.

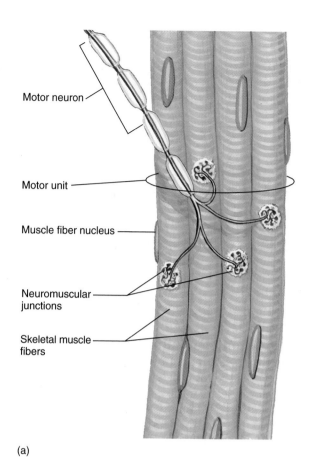

Motor neuron

Motor unit

Muscle fiber nucleus

Neuromuscular
junctions

Skeletal muscle
fibers

(a)

Motor nerve fiber

Muscle fiber

Neuromuscular junction

(b)

**FIGURE 9.8**

(*a*) A motor unit consists of one motor neuron and all the muscle fibers with which it communicates. (*b*) Micrograph of a neuromuscular junction (500×).

In the summer months of the early 1950s, parents in the United States lived in terror of their children contracting poliomyelitis, a viral infection that attacks nerve cells that stimulate skeletal muscles to contract. In half of the millions of affected children, fever, headache, and nausea rapidly progressed to a stiffened back and neck, drowsiness, and then the feared paralysis, usually of the lower limbs or muscles that control breathing or swallowing. Today, many a middle-aged person with a limp owes this slight disability to polio. Vaccines introduced in the middle 1950s ended the nightmare of polio—or so we thought.

Today, a third of the 1.6 million polio survivors in the United States are experiencing the fatigue, muscle weakness and atrophy, and difficulty breathing of post-polio syndrome. Researchers do not yet know whether this condition is a reactivation of the past infection, a new infection, or the effect of aging on nerve cells damaged in childhood.

The number of muscle fibers in a motor unit varies considerably. The fewer muscle fibers in the motor units, however, the finer the movements that can be produced in a particular muscle. For example, the motor units of the muscles that move the eyes may contain fewer than ten muscle fibers per motor unit and can produce very slight movements. Conversely, the motor units of the large muscles in the back may include a hundred or more muscle fibers. When these motor units are stimulated, the movements produced are coarse compared to those of the eye muscles.

## SKELETAL MUSCLE CONTRACTION

A muscle fiber contraction is a complex interaction of several cellular and chemical constituents. The final result is a movement within the myofibrils in which the filaments of actin and myosin slide past one another, shortening the sarcomeres. When this happens, the muscle fiber shortens and pulls on its attachments.

### Roles of Myosin and Actin

Myosin accounts for about two-thirds of the protein within skeletal muscles. A myosin molecule is composed of two twisted protein strands with globular parts called cross-bridges projecting outward along their lengths. Many of these molecules comprise a myosin filament. In the presence of calcium ions, the myosin cross-bridges

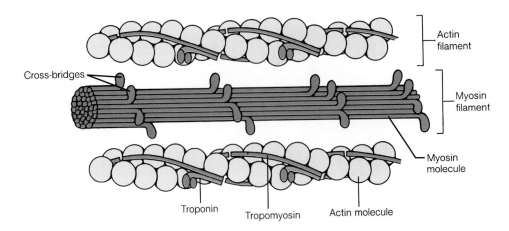

Actin filament

Myosin filament

Myosin molecule

Cross-bridges

Myosin molecule

Troponin    Tropomyosin    Actin molecule

**FIGURE 9.9**

Myosin molecules have cross-bridges that extend toward nearby actin filaments.

react with actin filaments and form linkages with them. This reaction between the myosin and actin filaments provides the force that shortens myofibrils during muscle contraction. Actin-myosin interactions are the basis for contraction in all three types of muscle.

> When adolescents or young adults die suddenly of heart failure, the cause is often familial hypertrophic cardiomyopathy. At the molecular level, this inherited heart disease is caused by a defect in the myosin cross-bridge at the point where it would normally contact an actin filament. As a result, actin and myosin cannot interact properly. The heart muscle overgrows (hypertrophies) to compensate for its inefficiency.

Actin accounts for about one-fourth of the total protein in skeletal muscle. An actin molecule is a globular structure with a binding site to which the cross-bridges of the myosin molecules can attach. Many of these actin molecules, aggregated in a double twisted strand (helix), form an actin filament.

Two other types of proteins, **tropomyosin** and **troponin,** associate with actin filaments. Tropomyosin molecules are rod-shaped and occupy the longitudinal grooves of the actin helix. Each tropomyosin has a troponin molecule attached to its surface, forming a tropomyosin-troponin complex (fig. 9.9).

When a muscle fiber is at rest, these tropomyosin-troponin complexes block the binding sites on the actin molecules and thus prevent formation of linkages. If the concentration of calcium ions is great, however, the calcium ions bind to the troponin, changing its shape

(conformation) and altering the position of the tropomyosin. The movement of the tropomyosin molecules exposes the binding sites on the actin filaments, and linkages form between the actin and myosin filaments.

The **sliding filament theory** of muscle contraction states that the head of a myosin cross-bridge can attach to an actin binding site and bend slightly, pulling the actin filament. Then the head can release, straighten, combine with another binding site farther down the actin filament, and pull again (fig. 9.10).

Myosin cross-bridges contain the enzyme **ATPase,** which catalyzes the breakdown of ATP to ADP and phosphate. This reaction releases energy (see chapter 4) that provides the force for muscle contraction. Breakdown of ATP puts the myosin cross-bridge in a "cocked" position. When a muscle is stimulated to contract, a cocked cross-bridge attaches to actin and pulls the actin filament toward the center of the sarcomere, shortening the sarcomere and thus the muscle. When another ATP binds, the cross-bridge first releases from the actin binding site, and then breaks down the ATP to return to the cocked position. According to this theory, if the muscle continues to be stimulated to contract, this cycle repeats over and over. If stimulation of contraction stops, binding sites are once again blocked by tropomyosin, and (as long as ATP is present) the cross-bridges return to the cocked position, but do not attach to actin (fig. 9.11).

Actin, myosin, troponin, and tropomyosin are abundant in muscle cells. Scarcer proteins are also vital to muscle function. This is the case for a rod-shaped muscle protein called **dystrophin.** It accounts for only 0.002% of total muscle protein in skeletal muscle, but its absence causes the devastating inherited disorder Duchenne muscular dystrophy, a disease that usually

Actin filament
"Cocked" cross-bridge
Myosin filament

ADP←—ATP

(a)

Linkage broken

ATP

(d)

Binding sites
Ca++        Ca++

Linkage forms

(b)

Cross-bridge pulls actin filament

(c)

**Figure 9.10**

According to the sliding filament theory, (a) ATP breakdown provides energy to "cock" the unattached myosin cross-bridge. When calcium ion concentration is low, the muscle remains relaxed. (b) When calcium ion concentration rises, binding sites on actin filaments open and cross-bridges attach. (c) Upon binding to actin, cross-bridges spring from the cocked position and pull on actin filaments. (d) ATP binds to the cross-bridge (but is not yet broken down), causing it to release from the actin filament. As long as ATP and calcium ions are present the cycle continues.

affects boys. Dystrophin binds to the inside face of muscle cell membranes, supporting them against the powerful force of contraction. Without even these minute amounts of dystrophin, muscle cells burst and die. Other forms of muscular dystrophy result from abnormalities of other proteins to which dystrophin attaches.

## Stimulus for Contraction

A skeletal muscle fiber normally does not contract until a neurotransmitter stimulates it. In skeletal muscle, the neurotransmitter is a compound called **acetylcholine (ACh).** This substance is synthesized in the cytoplasm of the motor neuron and is stored in vesicles near the distal end of its nerve fiber. When a nerve impulse (or *action potential,* described in chapter 10) reaches the end of the axon, some of these vesicles release acetylcholine into the gap between the nerve fiber and the motor end plate (see fig. 9.7).

Acetylcholine diffuses rapidly across the gap, combines with certain protein molecules (receptors) in the sarcolemma, and stimulates the muscle fiber membrane. The response is a **muscle impulse,** very much like a nerve impulse, which passes in all directions over the surface of the sarcolemma. It then travels through the transverse tubules, deep into the fiber, and reaches the sarcoplasmic reticulum.

The sarcoplasmic reticulum has a high concentration of calcium ions compared to the sarcoplasm. This is due to active transport of calcium ions (calcium pump) in the membrane of the sarcoplasmic reticulum. In response to a muscle impulse, the membranes of the cisternae become more permeable to these ions, and the calcium ions diffuse out of the cisternae into the sarcoplasm of the muscle fiber.

When a high concentration of calcium ions is in the sarcoplasm, linkages form between the actin and myosin filaments, and a muscle contracts. The contraction, which also requires ATP, continues as long as nerve impulses cause ACh release. When the nerve impulses cease, two events lead to muscle relaxation.

First, the acetylcholine that stimulated the muscle fiber in the first place is rapidly decomposed by an enzyme called **acetylcholinesterase.** This enzyme is present at the neuromuscular junction on the membranes of the motor end plate. The action of acetylcholinesterase prevents a single nerve impulse from continuously stimulating the muscle fiber.

Second, when acetylcholine is broken down, the stimulus to the sarcolemma and membranes within the muscle fiber ceases. The calcium pump (which requires ATP) quickly moves calcium ions back into the sarcoplasmic reticulum, decreasing the calcium ion concentration of the sarcoplasm. The cross-bridge linkages break, and, once again, the troponin and tropomyosin

Sarcomere

Cross-bridges | A band | Z line

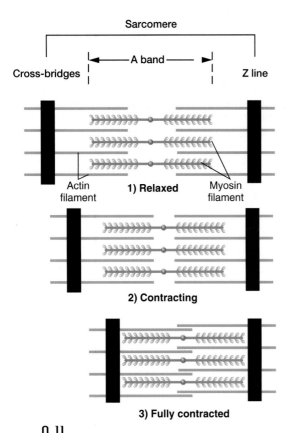

Actin filament
1) Relaxed
Myosin filament

2) Contracting

3) Fully contracted

**FIGURE 9.11**

When a skeletal muscle contracts, individual sarcomeres shorten as thick and thin filaments slide past one another (23,000×).

molecules inhibit the interaction between the filaments. Consequently, the muscle fiber relaxes. Table 9.1 summarizes the major events leading to muscle contraction and relaxation.

If acetylcholine receptors at the motor end plate are decreased in number, or blocked, muscles cannot receive the signal to contract. This may occur as the result of a disease, such as myasthenia gravis, or a poison, such as nerve gas. A drug called pyridostigmine bromide is used to treat myasthenia gravis. The drug inhibits the enzyme (acetylcholinesterase) that normally breaks down acetylcholine, keeping the neurotransmitter around longer. It was given to veterans of the Persian Gulf War who complained of muscle aches in the months following their military service. Health officials reasoned that the drug's effect on myasthenia gravis might also help restore muscle function if the veterans' symptoms arose from exposure to nerve gas during the war. Acetylcholinesterase inhibitors are also used as insecticides. The buildup of acetylcholine causes an insect to twitch violently, then die.

Sarcomere

Z line | A band | Z line

In September 1985, two teenage tourists from Hong Kong went to the emergency room at Montreal Children's Hospital complaining of extreme nausea and weakness. Although doctors released them when they could not identify a cause of the symptoms, the girls returned that night—far sicker. Now they were becoming paralyzed and had difficulty breathing. This time, physicians recognized symptoms of botulism.

Botulism occurs when the bacterium *Clostridium botulinum* grows in an anaerobic (oxygen-poor) environment, such as in a can of food. The bacteria produce a toxin that prevents the release of acetylcholine from nerve terminals. Symptoms include: nausea, vomiting, and diarrhea (digestive tract); headache, dizziness, and blurred or double vision (nervous system); and finally, weakness, hoarseness, and difficulty swallowing and, eventually, breathing (muscular system). Fortunately, physicians can administer an antitoxin substance that binds to and inactivates botulinum toxin in the bloodstream, stemming further symptoms, although not correcting damage already done.

Prompt treatment saved the touring teens, and astute medical detective work led to a restaurant in Vancouver where the sisters and thirty-four others had eaten roast beef sandwiches. The bread had been coated with a garlic-butter spread. The garlic was bottled with soybean oil and should have been refrigerated. It was not. With bacteria that the garlic had picked up in the soil where it grew, and eight months sitting outside of the refrigerator, conditions were just right for *C. botulinum* to produce its deadly toxin.

## TABLE 9.1 MAJOR EVENTS OF MUSCLE CONTRACTION AND RELAXATION

| Muscle Fiber Contraction | Muscle Fiber Relaxation |
|---|---|
| 1. The distal end of a motor neuron releases acetylcholine. | 1. Acetylcholinesterase decomposes acetylcholine, and the muscle fiber membrane is no longer stimulated. |
| 2. Acetylcholine diffuses across the gap at the neuromuscular junction. | 2. Calcium ions are actively transported into the sarcoplasmic reticulum. |
| 3. The sarcolemma is stimulated, and a muscle impulse travels over the surface of the muscle fiber and deep into the fiber through the transverse tubules and reaches the sarcoplasmic reticulum. | 3. ATP causes linkages between actin and myosin filaments to break without ATP breakdown. |
| 4. Calcium ions diffuse from the sarcoplasmic reticulum into the sarcoplasm and bind to troponin molecules. | 4. Cross-bridges recock. |
| 5. Tropomyosin molecules move and expose specific sites on actin filaments. | 5. Troponin and tropomyosin molecules inhibit the interaction between myosin and actin filaments. |
| 6. Actin and myosin filaments form linkages. | 6. Muscle fiber remains relaxed, yet ready until stimulated again. |
| 7. Actin filaments are pulled inward by myosin cross-bridges. | |
| 8. Muscle fiber shortens as a contraction occurs. | |

1 Describe a neuromuscular junction.

2 Define *motor unit.*

3 List four proteins associated with myofibrils, and explain their structural and functional relationships.

4 Explain how the filaments of a myofibril interact during muscle contraction.

5 Explain how a motor nerve impulse can trigger a muscle contraction.

## Energy Sources for Contraction

The energy used during muscle fiber contraction comes from ATP molecules. This energy makes possible the reaction between the actin and myosin filaments. However, a muscle fiber has only enough ATP to contract briefly. Therefore, when a fiber is active, ATP must be regenerated.

The initial source of energy available to regenerate ATP from ADP and phosphate is **creatine phosphate.** Like ATP, creatine phosphate contains a high-energy phosphate bond, and it is actually four to six times more abundant in muscle fibers than ATP. Creatine phosphate, however, cannot directly supply energy to a cell's energy-utilizing reactions. Instead, it stores excess energy released from mitochondria. Thus, whenever sufficient ATP is present, an enzyme in the mitochondria (creatine phosphokinase) promotes the synthesis of creatine phosphate, which stores the excess energy in its phosphate bond (fig. 9.12).

As ATP is decomposed to ADP, the energy from creatine phosphate molecules is transferred to these ADP

**FIGURE 9.12**

A muscle cell uses energy released in cellular respiration to synthesize ATP. ATP is then used to power muscle contraction, or to synthesize creatine phosphate. Later, creatine phosphate may be used to synthesize ATP.

molecules, quickly converting them back into ATP. The amount of ATP and creatine phosphate in a skeletal muscle, however, is usually not sufficient to support maximal muscle activity for more than a few seconds. As a result, the muscle fibers in an active muscle soon depend upon cellular respiration of glucose as a source of energy for synthesizing ATP. Typically, a muscle stores glucose in the form of glycogen.

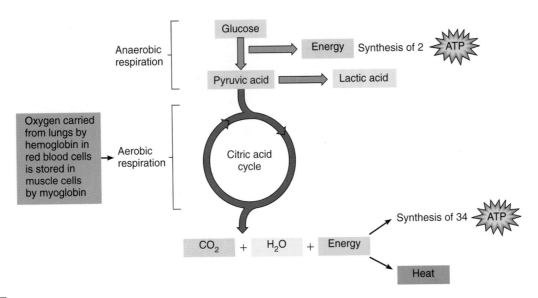

**FIGURE** 9.13

The oxygen required to support aerobic respiration is carried in the blood and stored in myoglobin. In the absence of sufficient oxygen, pyruvic acid is converted to lactic acid. The maximum number of ATPs generated per glucose molecule varies with cell type, and is 36 (2 + 34) in skeletal muscle.

## Oxygen Supply and Cellular Respiration

Recall from chapter 4 that glycolysis, the early phase of cellular respiration, occurs in the cytoplasm and is *anaerobic,* not dependent on oxygen. This phase only partially breaks down energy-supplying glucose, and releases only a few ATP molecules. The complete breakdown of glucose occurs in the mitochondria and is *aerobic,* requiring oxygen. This process, which includes the complex series of reactions of the *citric acid cycle,* produces many ATP molecules.

Blood carries the oxygen necessary to support aerobic respiration from the lungs to body cells. Oxygen is transported within the red blood cells loosely bound to molecules of hemoglobin, the pigment responsible for the red color of blood. In regions of the body where the oxygen concentration is relatively low, oxygen is released from hemoglobin and becomes available for cellular respiration.

Another pigment, **myoglobin,** is synthesized in muscle cells and imparts the reddish-brown color of skeletal muscle tissue. Like hemoglobin, myoglobin can loosely combine with oxygen, and in fact has a greater attraction for oxygen than does hemoglobin. Myoglobin can temporarily store oxygen in muscle tissue, which reduces a muscle's requirement for a continuous blood supply during contraction. This oxygen storage is important, because blood flow may decrease during muscular contraction when contracting muscle fibers compress blood vessels (fig. 9.13).

## Oxygen Debt

When a person is resting or moderately active, the respiratory and circulatory systems can usually supply sufficient oxygen to the skeletal muscles to support aerobic respiration. However, when skeletal muscles are used strenuously for even a minute or two, these systems usually cannot supply enough oxygen to sustain aerobic respiration. The muscle fibers must increasingly utilize the anaerobic phase of respiration for energy. The transition to anaerobically produced ATP is called *anaerobic threshold.*

Chapter 4 discussed how anaerobic respiration breaks glucose down to pyruvic acid and converts it to lactic acid, which diffuses out of the muscle fibers and is carried to the liver by the blood. Liver cells can react lactic acid to form *glucose,* but this requires energy from ATP (fig. 9.14). During strenuous exercise, available oxygen is primarily used to synthesize ATP for muscle contraction rather than to make ATP for converting lactic acid into glucose. Consequently, as lactic acid accumulates, a person develops an **oxygen debt** that must be repaid at a later time. The amount of oxygen debt equals the amount of oxygen liver cells require to convert the accumulated lactic acid into glucose, plus the amount the muscle cells require to resynthesize ATP and creatine phosphate, and restore their original concentrations.

The conversion of lactic acid back into glucose is relatively slow. It may require several hours to repay an oxygen debt following strenuous exercise.

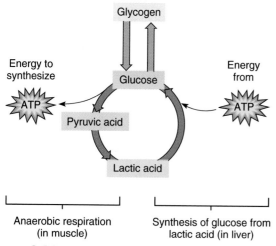

**FIGURE** 9.14

Liver cells can convert lactic acid, generated by muscles anaerobically, into glucose.

---

break the linkages between actin and myosin filaments before muscle fibers can relax. In a cramp, part of the muscle may be contracting uncontrollably while other parts may be rigid.

> A few hours after death, the skeletal muscles partially contract, fixing the joints. This condition, called *rigor mortis,* may continue for seventy-two hours or more. It results from an increase in membrane permeability to calcium ions, which promotes cross-bridge attachment, and a decrease in availability of ATP in the muscle fibers, which prevents cross-bridge release from actin. Thus, the actin and myosin filaments of the muscle fibers remain linked until the muscles begin to decompose.

> The runners are on the starting line, their muscles primed for a sprint. Glycogen will be broken down to release glucose, and creatine phosphate will supply high-energy phosphate groups to replenish ATP stores by phosphorylating ADP. The starting gun fires. Energy comes first from residual ATP, but almost instantaneously, creatine phosphate begins donating high energy phosphates to ADP, regenerating ATP. Meanwhile, oxidation of glucose ultimately produces more ATP. But because the runner cannot take in enough oxygen to meet the high demand, most ATP is generated in glycolysis. Formation of lactic acid causes fatigue and possibly leg muscle cramps as the runner crosses the finish line. Already, her liver is actively converting lactic acid back to pyruvate, and storing glycogen. In her muscles, creatine phosphate begins to build again.

## Muscle Fatigue

A muscle exercised strenuously for a prolonged period may lose its ability to contract, a condition called *fatigue.* This condition may result from an interruption in the muscle's blood supply or, rarely, from exhaustion of the supply of acetylcholine in its motor nerve fibers. However, muscle fatigue is most likely to arise from accumulation of lactic acid in the muscle as a result of anaerobic respiration (see chapter 4). The lowered pH from the lactic acid prevents muscle fibers from responding to stimulation.

Occasionally a muscle becomes fatigued and develops cramps at the same time. A cramp is a painful condition due to a lack of ATP, which is required to move calcium ions back into the sarcoplasmic reticulum and

Tolerance to lactic acid varies, so that some persons experience muscular fatigue more quickly than others. Resistance to the effects of lactic acid also varies among athletes, although in general, athletes can exercise and produce less lactic acid than nonathletes. In part, this is because the strenuous exercise of physical training stimulates new capillaries to grow within the muscles, supplying more oxygen and nutrients to the muscle fibers. Such physical training also causes muscle fibers to produce additional mitochondria, increasing their ability to carry on aerobic respiration. Some muscle fibers may be more likely to accumulate lactic acid than others, as described below.

## Fast and Slow Muscles

Muscles vary in contraction speed. Those that move the eyes contract about ten times faster than those that maintain posture, and the muscles that move the limbs contract at intermediate rates. Speed of contraction reflects the specialized function of the muscle.

The slow-contracting (slow-twitch) postural muscles, such as the long muscles of the back, are often called *red muscles* because most of their fibers contain the red, oxygen-storing pigment myoglobin. These fibers also are well supplied with oxygen-carrying blood. In addition, red muscle fibers contain many mitochondria, an adaptation for aerobic respiration. These fibers have a high respiratory capacity and can generate ATP fast enough to keep up with the ATP breakdown that occurs when they contract. For this reason, these fibers can contract for prolonged periods without fatiguing.

Fast-contracting (fast-twitch) muscles are called *white muscles* because they contain less myoglobin and have a poorer blood supply than red muscles. They include certain hand muscles as well as those that move the eyes. Most of the fibers of these muscles have fewer mitochondria and thus have a reduced respiratory capacity. However, the white fibers have a more extensive sarcoplasmic reticulum, which stores and reabsorbs calcium ions, and their ATPase has higher activity than that of red fibers. Because of these factors, white muscle fibers can contract rapidly, though they tend to fatigue as lactic acid accumulates and as ATP and the biochemicals to regenerate ATP are depleted.

Some muscles contain a predominate number of slow-contracting red fibers or fast-contracting white fibers, but most skeletal muscles have both types. Researchers have also identified muscle fibers with intermediate characteristics.

---

Birds that migrate long distances have abundant dark, slow-twitch muscles—this is why their meat is dark. In contrast, chickens that can only flap around the barnyard have abundant fast-twitch muscles, and mostly white meat.

World-class distance runners are the human equivalent of the migrating bird. Their muscles may contain over 90% slow-twitch fibers! In some European nations, athletic coaches measure slow-twitch to fast-twitch muscle fiber ratios to predict who will excel at long-distance events and who will fare better in sprints.

---

## Heat Production

Heat is a by-product of cellular respiration; all active cells generate heat. Since muscle tissue represents such a large proportion of the total body mass, it is a major source of heat.

Less than half of the energy released in cellular respiration is available for use in metabolic processes; the rest becomes heat. Active muscles release large amounts of heat. Blood transports this heat throughout the body, helping to maintain body temperature. Homeostatic mechanisms promote heat loss when the temperature of the internal environment begins to rise (see chapters 1 and 6).

1. What are the sources of energy used to regenerate ATP?

2. What are the sources of oxygen required for aerobic respiration?

3. How do lactic acid and oxygen debt relate to muscle fatigue?

4. Distinguish between fast-contracting and slow-contracting muscles.

5. What is the relationship between cellular respiration and heat production?

## MUSCULAR RESPONSES

One way to observe muscle contraction is to remove a single muscle fiber from a skeletal muscle and connect it to a device that senses and records changes in the fiber's length. An electrical stimulator is usually used to promote muscle contraction.

### Threshold Stimulus

When an isolated muscle fiber is exposed to a series of stimuli of increasing strength, the fiber remains unresponsive until a certain strength of stimulation is applied. This minimal strength required to cause contraction is called the **threshold stimulus.** An impulse in a motor neuron normally releases enough ACh to bring the muscle fibers in its motor unit to threshold.

### All-or-None Response

A muscle fiber exposed to a stimulus of threshold strength or above responds to its fullest extent. Increasing the strength of the stimulus does not affect the degree to which the fiber contracts. In other words, a muscle fiber does not partially contract—if it contracts at all, it contracts completely, even though it may not shorten completely. This phenomenon is called the **all-or-none response.**

### Recording a Muscle Contraction

To record how a whole muscle responds to stimulation, a skeletal muscle can be removed from a frog or other small animal and mounted in a special laboratory apparatus that stretches it to an optimal length. The muscle is then stimulated electrically, and when it contracts, it pulls on a lever. The lever's movement is recorded, and the resulting pattern is called a **myogram.**

If a muscle is exposed to a single stimulus of sufficient strength to activate some of its motor units, the muscle will contract and then relax. This action—a single contraction that lasts only a fraction of a second—is called a **twitch.** A twitch produces a myogram like that in figure 9.15. Note that there was a delay between the time the stimulus was applied and the time the muscle responded. This is the **latent period.** In a frog muscle, the latent period lasts for about 0.01 second; in a human muscle, it is even shorter. The latent period is followed by a *period of contraction* when the muscle pulls at its attachments, and a *period of relaxation* when the apparatus stretches it to its former length.

If a muscle is exposed to two stimuli (of threshold strength or above) too quickly, it may respond with a twitch to the first stimulus but not to the second. This is because it takes an instant following a contraction for muscle fibers to become responsive to further stimulation. Thus, for a very brief moment following stimulation, a muscle remains unresponsive. This time is called the **refractory period.**

**FIGURE 9.15**

A myogram of a single muscle twitch.

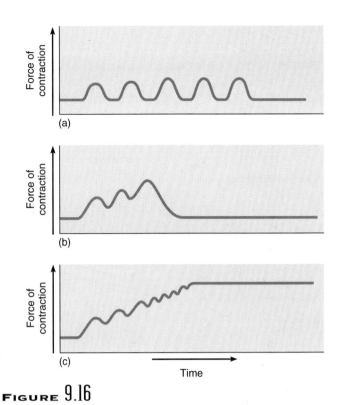

**FIGURE 9.16**

Myograms of (a) a series of twitches showing the staircase effect, (b) summation, and (c) a tetanic contraction. Note that stimulation frequency increases from one myogram to the next.

## Staircase Effect

The force a muscle fiber exerts in a twitch may depend on whether it has recently been stimulated to contract. A muscle fiber that has been inactive can be subjected to a series of stimuli, such that it undergoes a series of twitches with complete relaxation in between (fig. 9.16a). However, the strength of each successive contraction increases slightly, soon reaching a maximum. This phenomenon, called the **staircase effect** (treppe), is small and brief. Muscle fiber contraction is otherwise an all-or-none response.

The staircase effect seems to involve a net increase in the concentration of calcium ions available in the sarcoplasm of the muscle fibers. This increase might occur if each stimulus in the series caused the release of calcium ions and if the sarcoplasmic reticulum failed to recapture those ions immediately.

## Summation

The force that a muscle fiber can generate is not limited to the maximum force of a single twitch. A muscle fiber exposed to a series of stimuli of increasing frequency reaches a point when it is unable to completely relax before the next stimulus in the series arrives. When this happens, the individual twitches begin to combine and the muscle contraction becomes sustained. In such a sustained contraction, the force of individual twitches combines by the process of **summation** (fig. 9.16b). When the resulting forceful, sustained contraction lacks even partial relaxation, it is called a **tetanic** (te-tan-ik) **contraction** (tetanus) (fig. 9.16c).

## Recruitment of Motor Units

Since the muscle fibers within a muscle are organized into motor units, and each motor unit is controlled by a single motor neuron, all the muscle fibers in a motor unit are stimulated at the same time. Therefore, a motor unit also responds in an all-or-none manner. A whole muscle, however, does not behave like this, because it is composed of many motor units controlled by different motor neurons, some of which are more easily stimulated than others. Thus, if only the more easily stimulated motor neurons are involved, few motor units contract. At higher intensities of stimulation, other motor neurons respond and more motor units are activated. Such an increase in the number of activated motor units is called **recruitment.** As the intensity of stimulation increases, recruitment of motor units continues until finally all possible motor units are activated in that muscle.

## Sustained Contractions

At the same time that twitches combine, the strength of the contractions may increase due to the recruitment of motor units. The smaller motor units, which have finer fibers, tend to respond earlier in the series of stimuli. The larger motor units, which contain thicker fibers, respond later and more forcefully. The product is a sustained contraction of increasing strength called *multiple motor unit summation.*

Although twitches may occur occasionally in human skeletal muscles, as when an eyelid twitches, such contractions are of limited use. More commonly,

(a) Muscle contracts and shortens

Movement

(b) Muscle contracts but does not shorten

No movement

**FIGURE** 9.17

(a) Isotonic contractions occur when a muscle contracts and shortens. (b) Isometric contractions occur when a muscle contracts but does not shorten.

muscular contractions are sustained, and they are smooth rather than irregular or jerky because a mechanism within the spinal cord stimulates contractions in different sets of motor units at different moments. Thus, while some motor units are contracting, others are relaxing.

Tetanic contractions occur frequently in skeletal muscles during everyday activities. In many cases, the condition occurs in only a portion of a muscle. For example, when a person lifts a weight or walks, sustained contractions are maintained in the upper limb or lower limb muscles for varying lengths of time. These contractions are responses to a rapid series of stimuli transmitted from the brain and spinal cord on motor neuron fibers.

Even when a muscle appears to be at rest, a certain amount of sustained contraction is occurring in its fibers. This is called **muscle tone** (tonus), and it is a response to nerve impulses originating repeatedly in the spinal cord and traveling to a few muscle fibers. The result is a continuous state of partial contraction.

Muscle tone is particularly important in maintaining posture. Tautness in the muscles of the neck, trunk, and lower limbs enables a person to hold the head upright, stand, or sit. If tone is suddenly lost, such as when a person loses consciousness, the body will collapse. Muscle tone is maintained in health, but is lost if motor nerve fibers are cut or if diseases interfere with conduction of nerve impulses.

When skeletal muscles are contracted very forcefully, they may generate up to 50 pounds of pull for each square inch of muscle cross section. Consequently, large muscles such as those in the thigh can pull with several hundred pounds of force. Occasionally, this force is so great that the tendons of muscles tear away from their attachments to the bones.

## Types of Contractions

Sometimes muscles shorten when they contract. For example, if a person lifts an object, the muscles remain taut, their attached ends pull closer together, and the object is moved. This type of contraction is termed **isotonic.**

At other times, a skeletal muscle contracts, but the parts to which it is attached do not move. This happens, for instance, when a person pushes against the wall of a building. Tension within the muscles increases, but the wall does not move, and the muscles remain the same length. Contractions of this type are called **isometric.** Isometric contractions occur continuously in postural muscles that stabilize skeletal parts and hold the body upright. Figure 9.17 illustrates isotonic and isometric contractions.

A third type of contraction, called a lengthening or an **eccentric** contraction, occurs when the force a muscle

# Use and Disuse of Skeletal Muscles

Skeletal muscles are very responsive to use and disuse. Those that are forcefully exercised tend to enlarge. This phenomenon is called *muscular hypertrophy*. Conversely, a muscle that is not used *atrophies*—it decreases in size and strength.

The way a muscle responds to use also depends on the type of exercise. For instance, when a muscle contracts weakly, as during swimming and running, its slow, fatigue-resistant red fibers are most likely to be activated. As a result, these fibers develop more mitochondria, and more extensive capillary networks. Such changes increase the fibers' abilities to resist fatigue during prolonged exercise, although their sizes and strengths may remain unchanged.

Forceful exercise, such as weightlifting, in which a muscle exerts more than 75% of its maximum tension, uses the muscle's fast, fatigable white fibers. In response, existing muscle fibers develop new filaments of actin and myosin, and as their diameters increase, the entire muscle enlarges. However, no new muscle fibers are produced during hypertrophy.

Since the strength of a contraction is directly proportional to the diameter of the muscle fibers, an enlarged muscle can contract more strongly than before. However, such a change does not increase the muscle's ability to resist fatigue during activities such as running or swimming.

If regular exercise stops, capillary networks shrink and the number of mitochondria within the muscle fibers fall. Actin and myosin filaments diminish, and the entire muscle atrophies. Injured limbs immobilized in casts, or accidents or diseases that interfere with motor nerve impulses, commonly cause muscle atrophy. A muscle that cannot be exercised may shrink to less than one-half its usual size within a few months.

Muscle fibers whose motor neurons are severed not only shrink, but also may fragment and, in time, be replaced by fat or fibrous tissue. However, reinnervation of such a muscle within the first few months following an injury can restore function. ■

---

generates is less than that required to move or lift an object, as in laying a book down on a table. Even in such a contraction, cross-bridges are working to shorten the muscle.

Most body actions involve both isotonic and isometric contraction. In walking, for instance, certain leg and thigh muscles contract isometrically and keep the limb stiff as it touches the ground, while other muscles contract isotonically, bending the limb and lifting it. Similarly, walking down stairs involves eccentric contraction of certain thigh muscles. Clinical Application 9.1 discusses very noticeable effects of muscle use and disuse.

1 Define *threshold stimulus.*

2 What is an all-or-none response?

3 Distinguish between a twitch and a sustained contraction.

4 Define *muscle tone.*

5 Explain the differences among isometric, isotonic, and isokinetic contractions.

## SMOOTH MUSCLES

The contractile mechanisms of smooth and cardiac muscles are essentially the same as those of skeletal muscles. However, the cells of these tissues have important structural and functional differences.

### Smooth Muscle Fibers

As discussed in chapter 5, smooth muscle cells are shorter than the fibers of skeletal muscle, and they have single, centrally located nuclei. Smooth muscle cells are elongated with tapering ends and contain filaments of actin and myosin in myofibrils that extend throughout their lengths. However, the filaments are very thin and more randomly arranged than those in skeletal muscle fibers. As a result, smooth muscle cells lack striations. They also lack transverse tubules, and their sarcoplasmic reticula are not well developed.

The two major types of smooth muscles are multiunit and visceral. In **multiunit smooth muscle,** the muscle fibers are less well organized and occur as separate fibers rather than in sheets. Smooth muscle of this type is found in the irises of the eyes and in the walls of blood vessels. Typically, multiunit smooth muscle contracts only after stimulation by motor nerve impulses or certain hormones.

**Visceral smooth muscle** is composed of sheets of spindle-shaped cells held in close contact by gap junctions. The thick portion of each cell lies next to the thin parts of adjacent cells. Visceral smooth muscle is the more common type and is found in the walls of hollow organs, such as the stomach, intestines, urinary bladder, and uterus. Usually there are two thicknesses of smooth muscle in the walls of these organs. The fibers of the outer coats are directed longitudinally, whereas those of the inner coats are arranged circularly. These muscular layers change the sizes and shapes of these organs as they function.

Fibers of visceral smooth muscles can stimulate each other. When one fiber is stimulated, the impulse moving over its surface may excite adjacent fibers that, in turn, stimulate still others. Some visceral smooth muscle cells also display *rhythmicity*—a pattern of spontaneous repeated contractions.

These two features of visceral smooth muscle—transmission of impulses from cell to cell and rhythmicity—are largely responsible for the wavelike motion called **peristalsis** that occurs in certain tubular organs (see chapter 17). Peristalsis consists of alternate contractions and relaxations of the longitudinal and circular muscles. These movements help force the contents of a tube along its length. In the intestines, for example, peristaltic waves move masses of partially digested food and help to mix them with digestive fluids. Peristalsis in the ureters moves urine from the kidneys to the urinary bladder.

## Smooth Muscle Contraction

Smooth muscle contraction resembles skeletal muscle contraction in a number of ways. Both mechanisms reflect reactions of actin and myosin; both are triggered by membrane impulses and release of calcium ions; and both use energy from ATP molecules. There are, however, significant differences between smooth and skeletal muscle action. For example, smooth muscle fibers lack troponin, the protein that binds to calcium ions in skeletal muscle. Instead, smooth muscle uses a protein called *calmodulin,* which binds to calcium ions released when its fibers are stimulated, thus activating the actin-myosin contraction mechanism. In addition, much of the calcium necessary for smooth muscle contraction diffuses into the cell from the extracellular fluid.

Acetylcholine, the neurotransmitter in skeletal muscle, as well as *norepinephrine,* affect smooth muscle. Each of these neurotransmitters stimulates contractions in some smooth muscles and inhibits contractions in others. The discussion of the autonomic nervous system in chapter 11 describes these actions in greater detail.

Hormones affect smooth muscles by stimulating or inhibiting contraction in some cases and altering the degree of response to neurotransmitters in others. For example, during the later stages of childbirth, the hormone oxytocin stimulates smooth muscles in the wall of the uterus to contract (see chapter 22).

Stretching of smooth muscle fibers can also trigger contractions. This response is particularly important to the function of visceral smooth muscle in the walls of certain hollow organs, such as the urinary bladder and the intestines. For example, when partially digested food stretches the wall of the intestine, automatic contractions move the contents away.

Smooth muscle is slower to contract and slower to relax than skeletal muscle. On the other hand, smooth muscle can forcefully contract longer with the same amount of ATP. Unlike skeletal muscle, smooth muscle fibers can change length without changing tautness; because of this, smooth muscles in the stomach and intestinal walls can stretch as these organs fill, holding the pressure inside the organs constant.

**1** Describe the two major types of smooth muscle.

**2** What special characteristics of visceral smooth muscle make peristalsis possible?

**3** How is smooth muscle contraction similar to skeletal muscle contraction?

**4** How do the contraction mechanisms of smooth and skeletal muscles differ?

## CARDIAC MUSCLE

Cardiac muscle appears only in the heart. It is composed of striated cells joined end to end, forming fibers that are interconnected in branching, three-dimensional networks. Each cell contains a single nucleus and many filaments of actin and myosin similar to those in skeletal muscle. A cardiac muscle cell also has a well-developed sarcoplasmic reticulum, a system of transverse tubules, and many mitochondria. However, the cisternae of the sarcoplasmic reticulum of a cardiac muscle fiber are less developed and store less calcium than those of a skeletal muscle fiber. On the other hand, the transverse tubules of cardiac muscle fibers are larger than those in skeletal muscle, and they release many calcium ions into the sarcoplasm in response to a single muscle impulse.

The calcium ions in transverse tubules come from the fluid outside the muscle fiber. Thus, extracellular calcium partially controls the strength of cardiac muscle contraction and enables cardiac muscle fibers to contract longer than skeletal muscle fibers can.

Intercalated disk

Cardiac muscle cells

**FIGURE** 9.18

The intercalated disks of cardiac muscle, shown in this transmission electron micrograph, bind adjacent cells and allow ions to move between cells (12,500×).

| TABLE 9.2 | CHARACTERISTICS OF MUSCLE TISSUES | | |
|---|---|---|---|
| | **Skeletal** | **Smooth** | **Cardiac** |
| *Major location* | Skeletal muscles | Walls of hollow organs | Wall of the heart |
| *Major function* | Movement of bones at joints; maintenance of posture | Movement of walls of hollow organs; peristalsis | Pumping action of the heart |
| *Cellular characteristics* | | | |
| *Striations* | Present | Absent | Present |
| *Nucleus* | Multiple nuclei | Single nucleus | Single nucleus |
| *Special features* | Transverse tubule system is well developed | Lacks transverse tubules | Transverse tubule system is well developed; intercalated disks separate cells |
| *Mode of control* | Voluntary | Involuntary | Involuntary |
| *Contraction characteristics* | Contracts and relaxes relatively rapidly | Contracts and relaxes relatively slowly; some types self-exciting; rhythmic | Network of fibers contracts as a unit; self-exciting; rhythmic; remains refractory until contraction ends |

Drugs called calcium channel blockers are used to stop spasms of the heart muscle. They do this by blocking ion channels that admit extracellular calcium into cardiac muscle cells.

The opposing ends of cardiac muscle cells are connected by cross-bands called *intercalated disks.* These bands are actually complex membrane junctions. They help join cells and transmit the force of contraction from cell to cell. Intercellular junctions of the fused membranes of intercalated disks allow ions to diffuse between the cells. This allows muscle impulses to travel rapidly from cell to cell (see figs. 9.18 and 5.30).

When one portion of the cardiac muscle network is stimulated, the impulse passes to other fibers of the network, and the whole structure contracts as a unit (a *syncytium*); that is, the network responds to stimulation in an all-or-none manner. Cardiac muscle is also self-exciting and rhythmic. Consequently, a pattern of contraction and relaxation repeats again and again, causing the rhythmic contraction of the heart. Also, unlike skeletal muscle, cardiac muscle remains refractory until a contraction ends, so that sustained or tetanic contractions do not occur in the heart muscle. Table 9.2 summarizes characteristics of the three types of muscles.

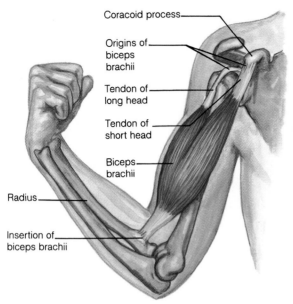

Coracoid process

Origins of biceps brachii

Tendon of long head

Tendon of short head

Biceps brachii

Radius

Insertion of biceps brachii

**FIGURE 9.19**

The biceps brachii has two heads that originate on the scapula. This muscle is inserted on the radius by a single tendon.

1  How is cardiac muscle similar to skeletal muscle?

2  How does cardiac muscle differ from skeletal muscle?

3  What is the function of intercalated disks?

4  What characteristic of cardiac muscle causes the heart to contract as a unit?

## SKELETAL MUSCLE ACTIONS

Skeletal muscles generate a great variety of body movements. The action of each muscle mostly depends upon the kind of joint it is associated with and the way the muscle is attached on either side of that joint.

### Origin and Insertion

Recall from chapter 7 that one end of a skeletal muscle is usually fastened to a relatively immovable or fixed part, and the other end is connected to a movable part on the other side of a joint. The immovable end is called the **origin** of the muscle, and the movable end is called its **insertion.** When a muscle contracts, its insertion is pulled toward its origin (fig. 9.19).

Some muscles have more than one origin or insertion. The *biceps brachii* in the arm, for example, has two origins. This is reflected in its name *biceps,* meaning "two heads." (*Note:* The head of a muscle is the part nearest its origin.) As figure 9.19 shows, one head of the muscle is attached to the coracoid process of the scapula, and the other head arises from a tubercle above the glen-

oid cavity of the scapula. The muscle extends along the anterior surface of the humerus and is inserted by a single tendon on the radial tuberosity of the radius. When the biceps brachii contracts, its insertion is pulled toward its origin, and the elbow bends.

### Interaction of Skeletal Muscles

Skeletal muscles almost always function in groups. As a result, when a particular body part moves, a person must do more than contract a single muscle; instead, after learning to make a particular movement, the person wills the movement to occur, and the nervous system stimulates the appropriate group of muscles.

By carefully observing body movements, it is possible to determine the roles of particular muscles. For instance, lifting the upper limb horizontally away from the side requires contracting the *deltoid* muscle, which is said to be the **prime mover.** However, while a prime mover is acting, certain nearby muscles also contract. When a deltoid muscle contracts, nearby muscles help hold the shoulder steady and in this way make the action of the prime mover more effective. Muscles that contract and assist a prime mover are called **synergists.**

Still other muscles act as **antagonists** to prime movers. These muscles can resist a prime mover's action and cause movement in the opposite direction—the antagonist of the prime mover that raises the upper limb can lower the upper limb, or the antagonist of the prime mover that bends the upper limb can straighten it. If both a prime mover and its antagonist contract simultaneously, the structure they act upon remains rigid. Similarly, smooth body movements depend upon the antagonists' relaxing and giving way to the prime movers whenever the prime movers contract. Once again, the nervous system controls these complex actions, as described in chapter 11.

The movements termed "flexion" and "extension" describe changes in the angle between bones that meet at a joint. For example, flexion at the elbow refers to a movement of the forearm that bends the elbow, or decreases the angle. Alternatively, one could say that flexion at the elbow results from the action of the biceps brachii on the radius of the forearm.

Since students often find it helpful to think of movements in terms of the specific actions of the muscles involved, we may also describe flexion and extension in these terms. Thus the action of the biceps brachii may be described as "flexion of the forearm at the elbow" and the action of the quadriceps group as "extension of the leg at the knee." We believe that this occasional departure from strict anatomical terminology eases understanding and learning.

**FIGURE** 9.20
Anterior view of superficial skeletal muscles.

**FIGURE** 9.21
Posterior view of superficial skeletal muscles.

**1** Distinguish between the origin and the insertion of a muscle.

**2** Define *prime mover.*

**3** What is the function of a synergist? An antagonist?

## MAJOR SKELETAL MUSCLES

This section concerns the locations, actions, origins, and insertions of some of the major skeletal muscles. The tables that summarize the information concerning groups of these muscles also include the names of nerves that supply the individual muscles within each group. Chapter 11 presents the origins and pathways of these nerves.

Figures 9.20 and 9.21 show the locations of superficial skeletal muscles—that is, those near the surface. Notice that the names of muscles often describe them. A name may indicate a muscle's size, shape, location, action, number of attachments, or the direction of its fibers, as in the following examples:

**pectoralis major**   A muscle of large size (*major*) located in the pectoral region (chest).
**deltoid**   Shaped like a delta or triangle.
**extensor digitorum**   Extends the digits (fingers or toes).
**biceps brachii**   A muscle with two heads (*biceps*), or points of origin, located in the brachium or arm.

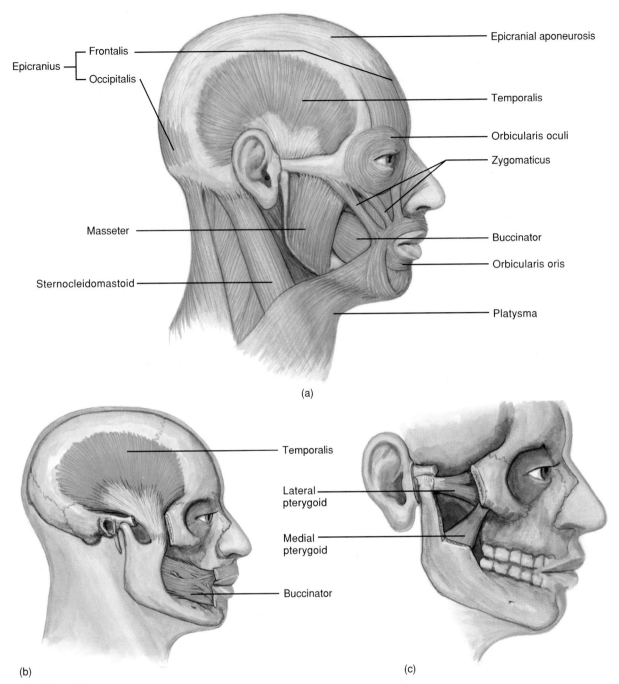

**FIGURE 9.22** 🏃

(a) Muscles of facial expression and mastication; isolated views of (b) the temporalis and buccinator muscles and (c) the lateral and medial pterygoid muscles.

**sternocleidomastoid**   Attached to the sternum, clavicle, and mastoid process.

**external oblique**   Located near the outside, with fibers that run obliquely or in a slanting direction.

## Muscles of Facial Expression

A number of small muscles beneath the skin of the face and scalp enable us to communicate feelings through facial expression. Many of these muscles are located around the eyes and mouth, and they make possible such expressions as surprise, sadness, anger, fear, disgust, and pain. As a group, the muscles of facial expression connect the bones of the skull to connective tissue in regions of the overlying skin. Figure 9.22 and reference plate 61 show these muscles, and Table 9.3 lists them. The muscles of facial expression include the following:

Epicranius                 Buccinator
Orbicularis oculi          Zygomaticus
Orbicularis oris           Platysma

TABLE 9.3

## MUSCLES OF FACIAL EXPRESSION

| Muscle | Origin | Insertion | Action | Nerve Supply |
|---|---|---|---|---|
| Epicranius | Occipital bone | Skin and muscles around eye | Raises eyebrow as when surprised | Facial n. |
| Orbicularis oculi | Maxillary and frontal bones | Skin around eye | Closes eye as in blinking | Facial n. |
| Orbicularis oris | Muscles near the mouth | Skin of central lip | Closes lips, protrudes lips as for kissing | Facial n. |
| Buccinator | Outer surfaces of maxilla and mandible | Orbicularis oris | Compresses cheeks inward as when blowing air | Facial n. |
| Zygomaticus | Zygomatic bone | Orbicularis oris | Raises corner of mouth as when smiling | Facial n. |
| Platysma | Fascia in upper chest | Lower border of mandible | Draws angle of mouth downward as when pouting | Facial n. |

The **epicranius** (ep″ĭ-kra′ne-us) covers the upper part of the cranium and consists of two muscular parts—the *frontalis* (frun-ta′lis), which lies over the frontal bone, and the *occipitalis* (ok-sip″ĭ-ta′lis), which lies over the occipital bone. These muscles are united by a broad, tendinous membrane called the *epicranial aponeurosis,* which covers the cranium like a cap. Contraction of the epicranius raises the eyebrows and horizontally wrinkles the skin of the forehead, as when a person expresses surprise. Headaches often result from sustained contraction of this muscle.

The **orbicularis oculi** (or-bik′u-la-rus ok′u-li) is a ringlike band of muscle, called a *sphincter muscle,* that surrounds the eye. It lies in the subcutaneous tissue of the eyelid and closes or blinks the eye. At the same time, it compresses the nearby tear gland, or *lacrimal gland,* aiding the flow of tears over the surface of the eye. Contraction of the orbicularis oculi also causes the folds, or crow's feet, that radiate laterally from the corner of the eye.

The **orbicularis oris** (or-bik′u-la-rus o′ris) is a sphincter muscle that encircles the mouth. It lies between the skin and the mucous membranes of the lips, extending upward to the nose and downward to the region between the lower lip and chin. The orbicularis oris is sometimes called the kissing muscle because it closes and puckers the lips.

The **buccinator** (buk′sĭ-na″tor) is located in the wall of the cheek. Its fibers are directed forward from the bones of the jaws to the angle of the mouth, and when they contract, the cheek is compressed inward. This action helps hold food in contact with the teeth when a person is chewing. The buccinator also aids in blowing air out of the mouth, and for this reason, it is sometimes called the trumpeter muscle.

The **zygomaticus** (zi″go-mat′ik-us) extends from the zygomatic arch downward to the corner of the mouth. When it contracts, the corner of the mouth is drawn upward, as in smiling or laughing.

The **platysma** (plah-tiz′mah) is a thin, sheetlike muscle whose fibers extend from the chest upward over the neck to the face. It pulls the angle of the mouth downward, as in pouting. The platysma also helps lower the mandible. The muscles that move the eye are described in chapter 12.

## Muscles of Mastication

Four pairs of muscles attached to the mandible produce chewing movements. Three pairs of these muscles close the lower jaw, as in biting; the fourth pair can lower the jaw, cause side-to-side grinding motions of the mandible, and pull the mandible forward, causing it to protrude. The muscles of mastication are shown in figure 9.22 and reference plate 61, and are listed in table 9.4. They include the following:

*Masseter*
*Temporalis*
*Medial pterygoid*
*Lateral pterygoid*

The **masseter** (mas-se′ter) is a thick, flattened muscle that can be felt just in front of the ear when the teeth are clenched. Its fibers extend downward from the zygomatic arch to the mandible. The masseter raises the jaw, but it can also control the rate at which the jaw falls open in response to gravity (fig. 9.22a).

The **temporalis** (tem-po-ra′lis) is a fan-shaped muscle located on the side of the skull above and in front of the ear. Its fibers, which also raise the jaw, pass downward beneath the zygomatic arch to the mandible (fig. 9.22a and b). Tensing this muscle is associated with temporomandibular joint syndrome, discussed in Clinical Application 9.2.

# CLINICAL APPLICATION

# TMJ Syndrome

Facial pain, headache, ringing in the ears, a clicking jaw, insomnia, teeth sensitive to heat or cold, backache, dizziness, and pain in front of the ears are aches and pains that may all result from temporomandibular joint (TMJ) syndrome. This condition is caused by a misaligned jaw or simply by a habit of grinding or clenching the teeth. These conditions may stress the temporomandibular joint, the articulation between the mandibular condyle of the mandible and the mandibular fossa of the temporal bone. Loss of coordination of these structures affects the nerves that pass through the neck and jaw region, causing the symptoms. In TMJ syndrome, tensing a muscle in the forehead can cause a headache, or a spasm in the muscle that normally opens the auditory tubes during swallowing can result in inability to clear the ears.

Doctors diagnose TMJ syndrome using an electromyograph, in which electrodes record muscle activity in four pairs of head and neck muscle groups. A form of treatment is transcutaneous electrical nerve stimulation (TENS), which stimulates the facial muscles for up to an hour. Another treatment is an orthotic device fitted by a dentist. Worn for 3 to 6 months, the device fine-tunes the action of jaw muscles to form a more comfortable bite. Finally, once the correct bite is determined, a dentist can use bonding materials to alter shapes of certain teeth to provide a more permanent treatment for TMJ syndrome. ■

| TABLE 9.4 MUSCLES OF MASTICATION | | | | |
|---|---|---|---|---|
| **Muscle** | **Origin** | **Insertion** | **Action** | **Nerve Supply** |
| Masseter | Lower border of zygomatic arch | Lateral surface of mandible | Elevates mandible | Trigeminal n. |
| Temporalis | Temporal bone | Coronoid process and anterior ramus of mandible | Elevates mandible | Trigeminal n. |
| Medial pterygoid | Sphenoid, palatine, and maxillary bones | Medial surface of mandible | Elevates mandible, and moves it from side to side | Trigeminal n. |
| Lateral pterygoid | Sphenoid bone | Anterior surface of mandibular condyle | Depresses and protracts mandible, and moves it from side to side | Trigeminal n. |

The **medial pterygoid** (ter'ĭ-goid) extends back and downward from the sphenoid, palatine, and maxillary bones to the ramus of the mandible. It closes the jaw (fig. 9.22c), and moves it from side to side.

The fibers of the **lateral pterygoid** extend forward from the region just below the mandibular condyle to the sphenoid bone. This muscle can open the mouth, pull the mandible forward to make it protrude, and move the mandible from side to side (fig. 9.22c).

When, in 1995, two dentists examined an eyeless cadaver's skull from an unusual perspective, they discovered an apparently newly seen muscle in the head. Named the sphenomandibularis, the muscle extends about an inch and a half from behind the eyes to the inside of the jawbone, and may assist chewing movements. In traditional dissection from the side, the new muscle's origin and insertion are not visible, so it may have appeared to be part of the larger and overlying temporalis muscle. Although the sphenomandibularis inserts on the inner side of the jawbone, as does the temporalis, it originates differently, on the sphenoid bone. The dentists then identified the sphenomandibularis in twenty-five other cadavers, and other researchers found it in live patients undergoing MRI scans.

TABLE 9.5

## MUSCLES THAT MOVE THE HEAD AND VERTEBRAL COLUMN

| Muscle | Origin | Insertion | Action | Nerve Supply |
|---|---|---|---|---|
| Sternocleidomastoid | Anterior surface of sternum and upper surface of clavicle | Mastoid process of temporal bone | Pulls head to one side, flexes neck or elevates sternum | Accessory, C2 and C3 cervical nerves |
| Splenius capitis | Spinous processes of lower cervical and upper thoracic vertebrae | Occipital bone | Rotates head, bends head to one side, or extends neck | Cervical nerves |
| Semispinalis capitis | Processes of lower cervical and upper thoracic vertebrae | Occipital bone | Extends head, bends head to one side, or rotates head | Cervical and thoracic spinal nerves |
| Erector spinae | | | | |
| *Iliocostalis (lateral) group* | | | | |
| Iliocostalis lumborum | Iliac crest | Lower six ribs | Extends lumbar region of vertebral column | Lumbar spinal nerves |
| Iliocostalis thoracis | Lower six ribs | Upper six ribs | Holds spine erect | Thoracic spinal nerves |
| Iliocostalis cervicis | Upper six ribs | Fourth through sixth cervical vertebrae | Extends cervical region of vertebral column | Cervical spinal nerves |
| *Longissimus (intermediate) group* | | | | |
| Longissimus thoracis | Lumbar vertebrae | Thoracic and upper lumbar vertebrae and ribs 9 and 10 | Extends thoracic region of vertebral column | Spinal nerves |
| Longissimus cervicis | Fourth and fifth thoracic vertebrae | Second through sixth cervical vertebrae | Extends cervical region of vertebral column | Spinal nerves |
| Longissimus capitis | Upper thoracic and lower cervical vertebrae | Mastoid process of temporal bone | Extends and rotates head | Cervical spinal nerves |
| *Spinalis (medial) group* | | | | |
| Spinalis thoracis | Upper lumbar and lower thoracic vertebrae | Upper thoracic vertebrae | Extends vertebral column | Spinal nerves |
| Spinalis cervicis | Ligamentum nuchae and seventh cervical vertebra | Axis | Extends vertebral column | Spinal nerves |
| Spinalis capitis | Upper thoracic and lower cervical vertebrae | Occipital bone | Extends vertebral column | Spinal nerves |

## Muscles that Move the Head and Vertebral Column

Paired muscles in the neck and back flex, extend, and rotate the head and hold the torso erect (figs. 9.23 and 9.25 and table 9.5). They include the following:

> Sternocleidomastoid
> Splenius capitis
> Semispinalis capitis
> Erector spinae

The **sternocleidomastoid** (ster″no-kli″do-mas′toid) is a long muscle in the side of the neck that extends upward from the thorax to the base of the skull behind the ear. When the sternocleidomastoid on one side contracts, the face turns to the opposite side. When both muscles contract, the head bends toward the chest. If other muscles fix the head in position, the sternocleidomastoids can raise the sternum, aiding forceful inhalation (fig. 9.25 and table 9.5).

The **splenius capitis** (sple′ne-us kap′ĭ-tis) is a broad, straplike muscle in the back of the neck. It connects the

Splenius capitis (cut)

Longissimus capitis

Semispinalis capitis (cut)

Longissimus cervicis

Iliocostalis cervicis

Longissimus thoracis

Semispinalis capitis

Spinalis capitis

Splenius capitis

Spinalis cervicis

Iliocostalis thoracis

Spinalis thoracis

Iliocostalis lumborum

**FIGURE 9.23**

Deep muscles of the back and the neck help move the head (posterior view) and hold the torso erect. The splenius capitis and semispinalis capitis are removed on the left to show underlying muscles.

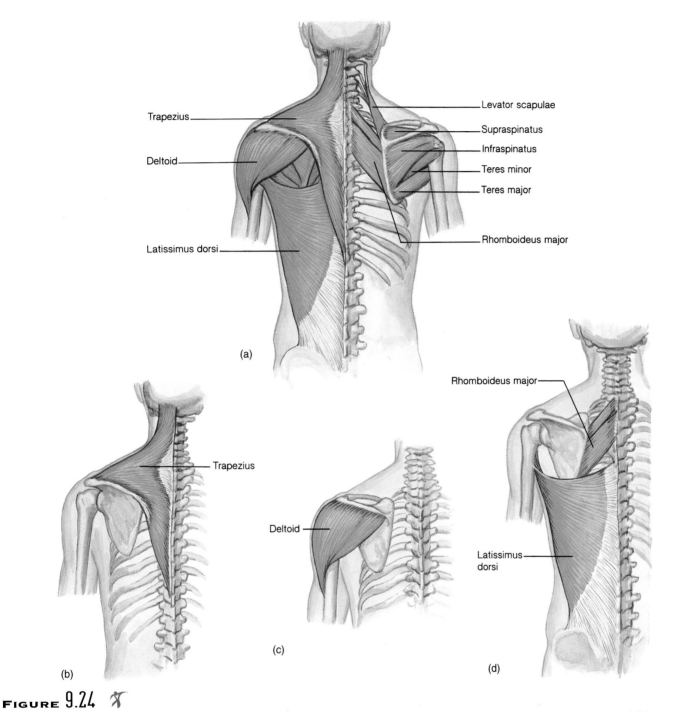

**FIGURE 9.24** ✗

(a) Muscles of the posterior shoulder. The right trapezius is removed to show underlying muscles. Isolated views of (b) trapezius, (c) deltoid, and (d) rhomboideus and latissimus dorsi muscles.

base of the skull to the vertebrae in the neck and upper thorax. A splenius capitis acting singly rotates the head and bends it toward one side. Acting together, these muscles bring the head into an upright position (fig. 9.23 and table 9.5).

The **semispinalis capitis** (sem″e-spi-na′lis kap′ĭ-tis) is a broad, sheetlike muscle extending upward from the vertebrae in the neck and thorax to the occipital bone. It

extends the head, bends it to one side, or rotates it (fig. 9.23 and table 9.5).

**Erector spinae** muscles run longitudinally along the back, with origins and insertions at many places on the axial skeleton. These muscles extend and rotate the head, and maintain the erect position of the vertebral column. Erector spinae can be subdivided into medial, intermediate, and lateral groups (table 9.5).

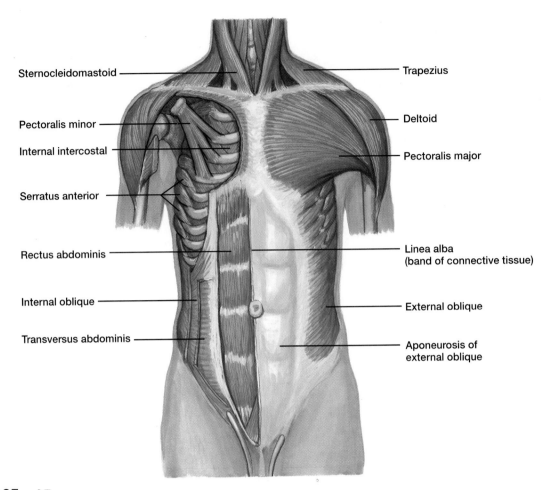

Sternocleidomastoid

Pectoralis minor

Internal intercostal

Serratus anterior

Rectus abdominis

Internal oblique

Transversus abdominis

Trapezius

Deltoid

Pectoralis major

Linea alba
(band of connective tissue)

External oblique

Aponeurosis of
external oblique

**FIGURE** 9.25

Muscles of the anterior chest and abdominal wall. The right pectoralis major is removed to show the pectoralis minor.

## Muscles that Move the Pectoral Girdle

The muscles that move the pectoral girdle are closely associated with those that move the arm. A number of these chest and shoulder muscles connect the scapula to nearby bones and move the scapula upward, downward, forward, and backward (figs. 9.24, 9.25, 9.26; reference plates 63, 64; table 9.6). Muscles that move the pectoral girdle include the following:

*Trapezius*
*Rhomboideus major*
*Levator scapulae*
*Serratus anterior*
*Pectoralis minor*

The **trapezius** (trah-pe′ze-us) is a large, triangular muscle in the upper back that extends horizontally from the base of the skull and the cervical and thoracic vertebrae to the shoulder. Its fibers are arranged into three

groups—upper, middle, and lower. Together these fibers rotate the scapula. The upper fibers acting alone raise the scapula and shoulder, as when the shoulders are shrugged to express a feeling of indifference. The middle fibers pull the scapula toward the vertebral column, and the lower fibers draw the scapula and shoulder downward. When other muscles fix the shoulder in position, the trapezius can pull the head backward or to one side (fig. 9.24).

The **rhomboideus** (rom-boid′-ē-us) **major** connects the upper thoracic vertebrae to the scapula. It raises the scapula and adducts it (fig. 9.24).

The **levator scapulae** (le-va′tor scap′u-lē) is a straplike muscle that runs almost vertically through the neck, connecting the cervical vertebrae to the scapula. It elevates the scapula (figs. 9.24 and 9.26).

The **serratus anterior** (ser-ra′tus an-te′re-or) is a broad, curved muscle located on the side of the chest. It arises as fleshy, narrow strips on the upper ribs and extends along the medial wall of the axilla to the ventral

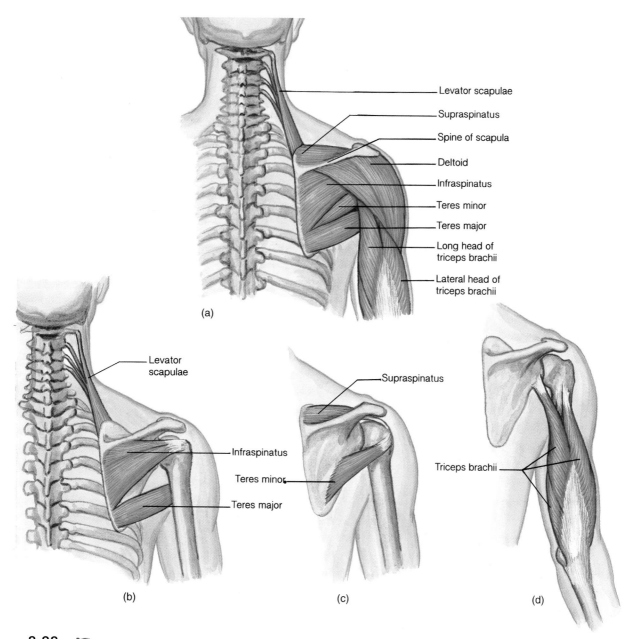

**FIGURE 9.26**

(a) Muscles of the posterior surface of the scapula and the arm. (b and c) Muscles associated with the scapula. (d) Isolated view of the triceps brachii.

surface of the scapula. It pulls the scapula downward and anteriorly, and is used to thrust the shoulder forward, as when pushing something (fig. 9.25).

The **pectoralis** (pek″to-ra′lis) **minor** is a thin, flat muscle that lies beneath the larger pectoralis major. It extends laterally and upward from the ribs to the scapula and pulls the scapula forward and downward. When other muscles fix the scapula in position, the pectoralis minor can raise the ribs and thus aid forceful inhalation (fig. 9.25).

A small, triangular region, called the *triangle of auscultation,* is located in the back where the trapezius overlaps the superior border of the latissimus dorsi and the underlying rhomboideus major. This area, which is near the medial border of the scapula, enlarges when a person bends forward with the arms folded across the chest. By placing the bell of a stethoscope within the triangle of auscultation, a physician can usually clearly hear the sounds of the respiratory organs.

TABLE 9.6

## MUSCLES THAT MOVE THE PECTORAL GIRDLE

| Muscle | Origin | Insertion | Action | Nerve Supply |
| --- | --- | --- | --- | --- |
| Trapezius | Occipital bone and spines of cervical and thoracic vertebrae | Clavicle, spine, and acromion process of scapula | Rotates scapula; various fibers raise scapula, pull scapula medially, or pull scapula and shoulder downward | Accessory n. |
| Rhomboideus major | Spines of upper thoracic vertebrae | Medial border of scapula | Raises and adducts scapula | Dorsal scapular n. |
| Levator scapulae | Transverse processes of cervical vertebrae | Medial margin of scapula | Elevates scapula | Dorsal scapular, and cervical nerves |
| Serratus anterior | Outer surfaces of upper ribs | Ventral surface of scapula | Pulls scapula anteriorly and downward | Long thoracic n. |
| Pectoralis minor | Sternal ends of upper ribs | Coracoid process of scapula | Pulls scapula forward and downward or raises ribs | Pectoral n. |

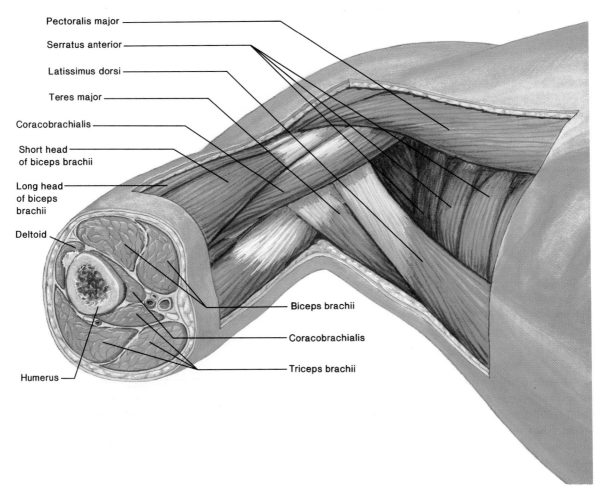

Pectoralis major
Serratus anterior
Latissimus dorsi
Teres major
Coracobrachialis
Short head of biceps brachii
Long head of biceps brachii
Deltoid
Humerus

Biceps brachii
Coracobrachialis
Triceps brachii

**FIGURE 9.27**

Cross section of the arm.

## Muscles that Move the Arm

The arm is one of the more freely movable parts of the body because muscles connect the humerus to regions of the pectoral girdle, ribs, and vertebral column. These muscles can be grouped according to their primary actions—flexion, extension, abduction, and rotation (figs. 9.26, 9.27, 9.28; reference plates 62, 63, 64; table 9.7). Muscles that move the arm include the following:

**Flexors**
*Coracobrachialis*
*Pectoralis major*

**Abductors**
*Supraspinatus*
*Deltoid*

**Extensors**
*Teres major*
*Latissimus dorsi*

**Rotators**
*Subscapularis*
*Infraspinatus*
*Teres minor*

### Flexors

The **coracobrachialis** (kor″ah-ko-bra′ke-al-is) extends from the scapula to the middle of the humerus along its medial surface. It flexes and adducts the arm (figs. 9.27 and 9.28).

The **pectoralis major** is a thick, fan-shaped muscle located in the upper chest. Its fibers extend from the

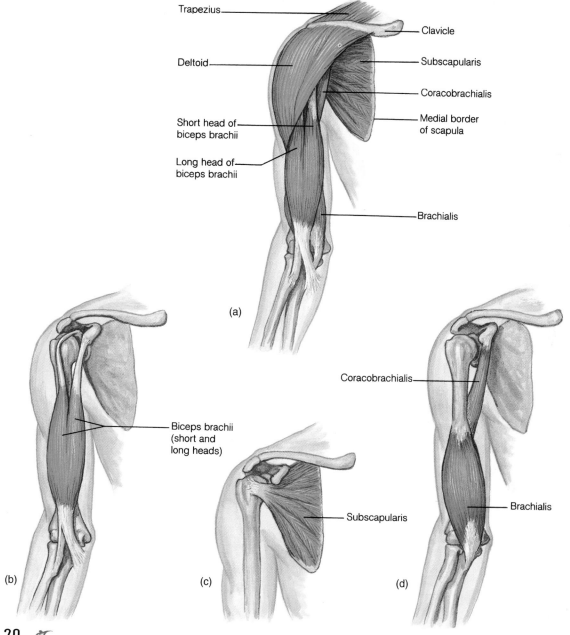

**FIGURE 9.28**

(a) Muscles of the anterior shoulder and the arm, with the rib cage removed. (b, c, and d) Isolated views of muscles associated with the arm.

TABLE 9.7 MUSCLES THAT MOVE THE ARM

| Muscle | Origin | Insertion | Action | Nerve Supply |
|---|---|---|---|---|
| Coracobrachialis | Coracoid process of scapula | Shaft of humerus | Flexes and adducts the arm | Musculocutaneus n. |
| Pectoralis major | Clavicle, sternum, and costal cartilages of upper ribs | Intertubercular groove of humerus | Flexes, adducts, and rotates arm medially | Pectoral n. |
| Teres major | Lateral border of scapula | Intertubercular groove of humerus | Extends, adducts, and rotates arm medially | Lower subscapular n. |
| Latissimus dorsi | Spines of sacral, lumbar, and lower thoracic vertebrae, iliac crest, and lower ribs | Intertubercular groove of humerus | Extends, adducts, and rotates the arm medially, or pulls the shoulder downward and back | Thoracodorsal n. |
| Supraspinatus | Posterior surface of scapula above spine | Greater tubercle of humerus | Abducts the arm | Suprascapular n. |
| Deltoid | Acromion process, spine of the scapula, and the clavicle | Deltoid tuberosity of humerus | Abducts, extends, and flexes arm | Axillary n. |
| Subscapularis | Anterior surface of scapula | Lesser tubercle of humerus | Rotates arm medially | Subscapular n. |
| Infraspinatus | Posterior surface of scapula below spine | Greater tubercle of humerus | Rotates arm laterally | Suprascapular n. |
| Teres minor | Lateral border of scapula | Greater tubercle of humerus | Rotates arm laterally | Axillary n. |

center of the thorax through the armpit to the humerus. This muscle primarily pulls the arm forward and across the chest. It can also rotate the humerus medially and adduct the arm from a raised position (fig. 9.25).

## Extensors

The **teres** (te′rēz) **major** connects the scapula to the humerus. It extends the humerus and can also adduct and rotate the arm medially (figs. 9.24 and 9.26).

The **latissimus dorsi** (lah-tis′ĭ-mus dor′si) is a wide, triangular muscle that curves upward from the lower back, around the side, and to the armpit. It can extend and adduct the arm and rotate the humerus medially. It also pulls the shoulder downward and back. This muscle is used to pull the arm back in swimming, climbing, and rowing (figs. 9.24 and 9.27).

## Abductors

The **supraspinatus** (su″prah-spi′na-tus) is located in the depression above the spine of the scapula on its posterior surface. It connects the scapula to the greater tubercle of the humerus and abducts the arm (figs. 9.24 and 9.26).

The **deltoid** (del′toid) is a thick, triangular muscle that covers the shoulder joint. It connects the clavicle and scapula to the lateral side of the humerus and abducts the arm. The deltoid's posterior fibers can extend the humerus, and its anterior fibers can flex the humerus (fig. 9.24).

A humerus fractured at its surgical neck may damage the nerve that supplies the deltoid muscle (axillary nerve) (see fig. 7.45). If this occurs, the muscle is likely to shrink and weaken. In order to test the deltoid for such weakness, a physician may ask a patient to abduct the arm against some resistance and maintain that posture for a time.

## Rotators

The **subscapularis** (sub-scap′u-lar-is) is a large, triangular muscle that covers the anterior surface of the scapula. It connects the scapula to the humerus and rotates the arm medially (fig. 9.28).

The **infraspinatus** (in″frah-spi′na-tus) occupies the depression below the spine of the scapula on its posterior surface. The fibers of this muscle attach the scapula to the humerus and rotate the arm laterally (fig. 9.26).

The **teres minor** is a small muscle connecting the scapula to the humerus. It rotates the arm laterally (figs. 9.24 and 9.26).

## Muscles that Move the Forearm

Most forearm movements are produced by muscles that connect the radius or ulna to the humerus or pectoral girdle. A group of muscles located along the anterior surface of the humerus flexes the forearm at the elbow, whereas a single posterior muscle extends this joint. Other muscles cause movements at the radioulnar joint and rotate the forearm.

The muscles that move the forearm are shown in figures 9.28, 9.29, 9.30, 9.31, and in reference plates 63, 65, and are listed in table 9.8, grouped according to their primary actions. They include the following:

| Flexors | Extensor | Rotators |
|---------|----------|----------|
| Biceps brachii | Triceps brachii | Supinator |
| Brachialis | | Pronator teres |
| Brachioradialis | | Pronator quadratus |

## Flexors

The **biceps brachii** (bi′seps bra′ke-i) is a fleshy muscle that forms a long, rounded mass on the anterior side of the arm. It connects the scapula to the radius and flexes the forearm at the elbow and rotates the hand laterally (supination), as when a person turns a doorknob or screwdriver (fig. 9.28).

The **brachialis** (bra′ke-al-is) is a large muscle beneath the biceps brachii. It connects the shaft of the humerus to the ulna and is the strongest flexor of the elbow (fig. 9.28).

The **brachioradialis** (bra″ke-o-ra″de-a′lis) connects the humerus to the radius. It aids in flexing the elbow (fig. 9.29).

## Extensor

The **triceps brachii** (tri′seps bra′ke-i) has three heads and is the only muscle on the back of the arm. It connects the humerus and scapula to the ulna, and is the primary extensor of the elbow (figs. 9.26 and 9.27).

## Rotators

The **supinator** (su′pǐ-na-tor) is a short muscle whose fibers run from the ulna and the lateral end of the humerus to the radius. It assists the biceps brachii in rotating the forearm laterally (supination) (fig. 9.29).

The **pronator teres** (pro-na′tor te′rēz) is a short muscle connecting the ends of the humerus and ulna to the radius. It rotates the arm medially, as when the hand is turned so the palm is facing downward (pronation) (fig. 9.29).

The **pronator quadratus** (pro-na′tor kwod-ra′tus) runs from the distal end of the ulna to the distal end of the radius. It assists the pronator teres in rotating the arm medially (fig. 9.29).

## Muscles that Move the Hand

Movements of the hand include movements of the wrist and fingers. Many muscles move the wrist, hand, and fingers. They originate from the distal end of the humerus and from the radius and ulna. The two major

groups of these muscles are flexors on the anterior side of the forearm and extensors on the posterior side. Figures 9.29, 9.30, 9.31, reference plate 65, and table 9.9 concern these muscles. The muscles that move the hand include the following:

**Flexors**
*Flexor carpi radialis*
*Flexor carpi ulnaris*
*Palmaris longus*
*Flexor digitorum profundus*
*Flexor digitorum superficialis*

**Extensors**
*Extensor carpi radialis longus*
*Extensor carpi radialis brevis*
*Extensor carpi ulnaris*
*Extensor digitorum*

## Flexors

The **flexor carpi radialis** (flek′sor kar-pi′ra″de-a′lis) is a fleshy muscle that runs medially on the anterior side of the forearm. It extends from the distal end of the humerus into the hand, where it is attached to metacarpal bones. The flexor carpi radialis flexes and abducts the hand at the wrist (fig. 9.29).

The **flexor carpi ulnaris** (flek′sor kar-pi′ ul-na′ris) is located along the medial border of the forearm. It connects the distal end of the humerus and the proximal end of the ulna to carpal and metacarpal bones. It flexes and adducts the hand at the wrist (fig. 9.29).

The **palmaris longus** (pal-ma′ris long′gus) is a slender muscle located on the medial side of the forearm between the flexor carpi radialis and the flexor carpi ulnaris. It connects the distal end of the humerus to fascia of the palm and flexes the hand at the wrist (fig. 9.29).

Some of the first signs of Parkinson's disease appear in the hands. In this disorder, certain brain cells degenerate and damage nerve cells that control muscles. Once called "shaking palsy," the disease often begins with a hand tremor that resembles the motion of rolling a marble between the thumb and forefinger. Another sign is called "cogwheel rigidity." When a doctor rotates the patient's hand in an arc, the hand resists the movement and then jerks, like the cogs in a gear.

The **flexor digitorum profundus** (flek′sor dij″ĭ-to′rum pro-fun′dus) is a large muscle that connects the ulna to the distal phalanges. It flexes the distal joints of the fingers, as when a fist is made (fig. 9.31).

The **flexor digitorum superficialis** (flek′sor dij″ĭ-to′rum su″per-fish″e-a′lis) is a large muscle located beneath the flexor carpi ulnaris. It arises by three

Biceps brachii

Brachialis

Supinator

Pronator teres

Brachioradialis

Extensor carpi radialis longus

Flexor carpi radialis

Palmaris longus

Flexor carpi ulnaris

Pronator quadratus

Flexor digitorum superficialis

(a)

Brachioradialis

Flexor carpi ulnaris

(b)

Pronator teres

Pronator quadratus

(c)

(d)

Flexor digitorum superficialis

Flexor carpi radialis

(e)

**FIGURE** 9.29

(a) Muscles of the anterior forearm. (b–e) Isolated views of muscles associated with the anterior forearm.

Triceps brachii

Brachioradialis

Extensor carpi radialis longus

Flexor carpi ulnaris

Extensor carpi radialis brevis

Extensor carpi ulnaris

Extensor digitorum

Extensor retinaculum

(a)

Extensor carpi radialis longus and brevis

(b)

Extensor carpi ulnaris

Extensor digitorum

(c)

**FIGURE 9.30**

(a) Muscles of the posterior forearm. (b and c) Isolated views of muscles associated with the posterior forearm.

heads—one from the medial epicondyle of the humerus, one from the medial side of the ulna, and one from the radius. It is inserted in the tendons of the fingers and flexes the fingers and, by a combined action, flexes the hand at the wrist (fig. 9.29).

## Extensors

The **extensor carpi radialis longus** (eks-ten′sor kar-pi′ ra″de-a′lis long′gus) runs along the lateral side of the forearm, connecting the humerus to the hand. It extends

the hand at the wrist and assists in abducting the hand (figs. 9.30 and 9.31).

The **extensor carpi radialis brevis** (eks-ten′sor kar-pi′ ra″de-a′lis brev′ĭs) is a companion of the extensor carpi radialis longus and is located medially to it. This muscle runs from the humerus to metacarpal bones and extends the hand at the wrist. It also assists in abducting the hand (figs. 9.30 and 9.31).

The **extensor carpi ulnaris** (eks-ten′sor kar-pi′ ul-na′ris) is located along the posterior surface of the ulna and

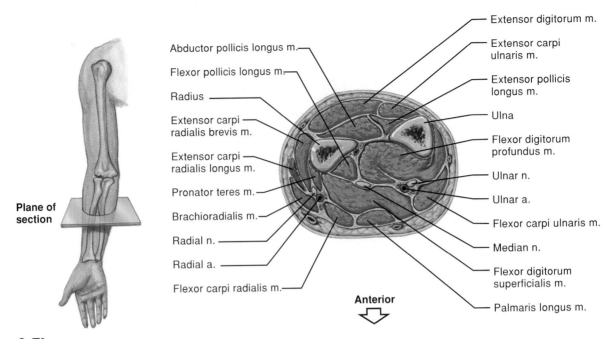

**FIGURE 9.31**
A cross section of the forearm.

| TABLE 9.8 | MUSCLES THAT MOVE THE FOREARM | | | |
|---|---|---|---|---|
| **Muscle** | **Origin** | **Insertion** | **Action** | **Nerve Supply** |
| Biceps brachii | Coracoid process and tubercle above glenoid cavity of scapula | Radial tuberosity of radius | Flexes forearm at elbow and rotates hand laterally | Musculocutaneous n. |
| Brachialis | Anterior shaft of humerus | Coronoid process of ulna | Flexes forearm at elbow | Musculocutaneous, median, and radial nerves |
| Brachioradialis | Distal lateral end of humerus | Lateral surface of radius above styloid process | Flexes forearm at elbow | Radial n. |
| Triceps brachii | Tubercle below glenoid cavity and lateral and medial surfaces of humerus | Olecranon process of ulna | Extends forearm at elbow | Radial n. |
| Supinator | Lateral epicondyle of humerus and crest of ulna | Lateral surface of radius | Rotates forearm laterally | Radial n. |
| Pronator teres | Medial epicondyle of humerus and coronoid process of ulna | Lateral surface of radius | Rotates forearm medially | Median n. |
| Pronator quadratus | Anterior distal end of ulna | Anterior distal end of radius | Rotates forearm medially | Median n. |

connects the humerus to the hand. It extends the hand at the wrist and assists in adducting it (figs. 9.30 and 9.31).

The **extensor digitorum** (eks-ten′sor dij″ĭ-to′rum) runs medially along the back of the forearm. It connects the humerus to the posterior surface of the phalanges and extends the fingers (figs. 9.30 and 9.31).

A structure called the *extensor retinaculum* consists of a group of heavy connective tissue fibers in the fascia of the wrist (fig. 9.30). It connects the lateral margin of the radius with the medial border of the styloid process

of the ulna and certain bones of the wrist. The retinaculum gives off branches of connective tissue to the underlying wrist bones, creating a series of sheathlike compartments through which the tendons of the extensor muscles pass to the wrist and fingers.

## Muscles of the Abdominal Wall

The walls of the chest and pelvic regions are supported directly by bone, but those of the abdomen are not.

TABLE 9.9

## TABLE 9.9 · MUSCLES THAT MOVE THE HAND

| Muscle | Origin | Insertion | Action | Nerve Supply |
|---|---|---|---|---|
| Flexor carpi radialis | Medial epicondyle of humerus | Base of second and third metacarpals | Flexes and abducts hand at the wrist | Median n. |
| Flexor carpi ulnaris | Medial epicondyle of humerus and olecranon process | Carpal and metacarpal bones | Flexes and adducts hand at the wrist | Ulnar n. |
| Palmaris longus | Medial epicondyle of humerus | Fascia of palm | Flexes hand at the wrist | Median n. |
| Flexor digitorum profundus | Anterior surface of ulna | Bases of distal phalanges in fingers 2–5 | Flexes distal joints of fingers | Median and ulnar nerves |
| Flexor digitorum superficialis | Medial epicondyle of humerus, coronoid process of ulna, and radius | Tendons of fingers | Flexes fingers and hand | Median n. |
| Extensor carpi radialis longus | Distal end of humerus | Base of second metacarpal | Extends and abducts hand at the wrist | Radial n. |
| Extensor carpi radialis brevis | Lateral epicondyle of humerus | Base of second and third metacarpals | Extends and abducts hand at the wrist | Radial n. |
| Extensor carpi ulnaris | Lateral epicondyle of humerus | Base of fifth metacarpal | Extends and adducts hand at the wrist | Radial n. |
| Extensor digitorum | Lateral epicondyle of humerus | Posterior surface of phalanges in fingers 2–5 | Extends fingers | Radial n. |

## TABLE 9.10 · MUSCLES OF THE ABDOMINAL WALL

| Muscle | Origin | Insertion | Action | Nerve Supply |
|---|---|---|---|---|
| External oblique | Outer surfaces of lower ribs | Outer lip of iliac crest and linea alba | Tenses abdominal wall and compresses abdominal contents | Intercostal nerves 7–12 |
| Internal oblique | Crest of ilium and inguinal ligament | Cartilages of lower ribs, linea alba, and crest of pubis | Same as above | Intercostal nerves 7–12 |
| Transversus abdominis | Costal cartilages of lower ribs, processes of lumbar vertebrae, lip of iliac crest, and inguinal ligament | Linea alba and crest of pubis | Same as above | Intercostal nerves 7–12 |
| Rectus abdominis | Crest of pubis and symphysis pubis | Xiphoid process of sternum and costal cartilages | Same as above; also flexes vertebral column | Intercostal nerves 7–12 |

Instead, the anterior and lateral walls of the abdomen are composed of layers of broad, flattened muscles. These muscles connect the rib cage and vertebral column to the pelvic girdle. A band of tough connective tissue, called the **linea alba,** extends from the xiphoid process of the sternum to the symphysis pubis. It is an attachment for some of the abdominal wall muscles.

Contraction of these muscles decreases the volume of the abdominal cavity and increases the pressure inside. This action helps press air out of the lungs during forceful exhalation, and also aids in defecation, urination, vomiting, and childbirth.

The abdominal wall muscles are shown in figure 9.32 and reference plate 62, and are listed in table 9.10. They include the following:

*External oblique*
*Internal oblique*
*Transversus abdominis*
*Rectus abdominis*

The **external oblique** (eks-ter'nal ŏ-blēk) is a broad, thin sheet of muscle whose fibers slant downward from the lower ribs to the pelvic girdle and the

Rectus abdominis

(a)

(b)

(c)

(d)

External oblique

Internal oblique

Transversus abdominis

Peritoneum
Transversus abdominis
Internal oblique
External oblique
Skin

Linea alba

Rectus abdominis

(e)

**FIGURE 9.32**

(*a–d*) Isolated muscles of the abdominal wall. (*e*) Transverse section through the abdominal wall.

linea alba. When this muscle contracts, it tenses the abdominal wall and compresses the contents of the abdominal cavity.

Similarly, the **internal oblique** (in-ter′nal ŏ-blēk) is a broad, thin sheet of muscle located beneath the external oblique. Its fibers run up and forward from the pelvic girdle to the lower ribs. Its function is similar to that of the external oblique.

The **transversus abdominis** (trans-ver′sus ab-dom′ĭ-nis) forms a third layer of muscle beneath the external and internal obliques. Its fibers run horizontally from the lower ribs, lumbar vertebrae, and ilium to the linea alba and pubic bones. It functions in the same manner as the external and internal obliques.

The **rectus abdominis** (rek′tus ab-dom′ĭ-nis) is a long, straplike muscle that connects the pubic bones to the ribs and sternum. Three or more fibrous bands cross the muscle transversely, giving it a segmented appearance. The muscle functions with other abdominal wall muscles to compress the contents of the abdominal cavity, and it also helps to flex the vertebral column.

## Muscles of the Pelvic Outlet

Two muscular sheets span the outlet of the pelvis—a deeper **pelvic diaphragm** and a more superficial **urogenital diaphragm.** The pelvic diaphragm forms the floor of the pelvic cavity, and the urogenital diaphragm fills the space within the pubic arch. Figure 9.33 and table 9.11 show the muscles of the male and female pelvic outlets. They include the following:

**Pelvic Diaphragm**
*Levator ani*
*Coccygeus*

**Urogenital Diaphragm**
*Superficial transversus perinei*
*Bulbospongiosus*
*Ischiocavernosus*
*Sphincter urethrae*

### Pelvic Diaphragm

The **levator ani** (le-va′tor ah-ni′) muscles form a thin sheet across the pelvic outlet. They are connected at the midline posteriorly by a ligament that extends from the tip of the coccyx to the anal canal. Anteriorly, they are separated in the male by the urethra and the anal canal, and in the female by the urethra, vagina, and anal canal. These muscles help support the pelvic viscera and provide sphincterlike action in the anal canal and vagina.

An *external anal sphincter* that is under voluntary control and an *internal anal sphincter* that is formed of involuntary muscle fibers of the intestine encircle the anal canal and keep it closed.

The **coccygeus** (kok-sij′e-us) is a fan-shaped muscle that extends from the ischial spine to the coccyx and sacrum. It aids the levator ani.

### Urogenital Diaphragm

The **superficial transversus perinei** (su″per-fish′al trans-ver′sus per″ĭ-ne′i) consists of a small bundle of muscle fibers that passes medially from the ischial tuberosity along the posterior border of the urogenital diaphragm. It assists other muscles in supporting the pelvic viscera.

In males, the **bulbospongiosus** (bul″bo-spon″je-o′sus) muscles are united surrounding the base of the penis. They assist in emptying the urethra. In females,

| TABLE 9.11 | MUSCLES OF THE PELVIC OUTLET | | | | |
|---|---|---|---|---|---|
| **Muscle** | **Origin** | **Insertion** | **Action** | **Nerve Supply** |
| Levator ani | Pubic bone and ischial spine | Coccyx | Supports pelvic viscera, and provides sphincterlike action in anal canal and vagina | Pudendal n. |
| Coccygeus | Ischial spine | Sacrum and coccyx | Same as above | S4 and S5 nerves |
| Superficial transversus perinei | Ischial tuberosity | Central tendon | Supports pelvic viscera | Pudendal n. |
| Bulbospongiosus | Central tendon | Males: Urogenital diaphragm and fascia of penis | Males: Assists emptying of urethra | Pudendal n. |
| | | Females: Pubic arch and root of clitoris | Females: Constricts vagina | |
| Ischiocavernosus | Ischial tuberosity | Pubic arch | Assists function of bulbospongiosus | Pudendal n. |
| Sphincter urethrae | Margins of pubis and ischium | Fibers of each unite with those from other side | Opens and closes urethra | Pudendal n. |

**FIGURE 9.33**

External view of muscles of (*a*) the male pelvic outlet, and (*b*) the female pelvic outlet. (*c*) Internal view of female pelvic and urogenital diaphragms.

these muscles are separated medially by the vagina and constrict the vaginal opening. They can also retard the flow of blood in veins, which helps maintain an erection in the penis of the male and in the clitoris of the female.

The **ischiocavernosus** (is″ke-o-kav″er-no′sus) muscle is a tendinous structure that extends from the ischial tuberosity to the margin of the pubic arch. It assists the bulbospongiosus muscle.

The **sphincter urethrae** (sfingk′ter u-re′thrē) are muscles that arise from the margins of the pubic and ischial bones. Each arches around the urethra and unites with the one on the other side. Together they act as a sphincter that

closes the urethra by compression and opens it by relaxation, thus helping control the flow of urine.

## Muscles that Move the Thigh

The muscles that move the thigh are attached to the femur and to some part of the pelvic girdle. (An important exception is the sartorius, described later.) They can be separated into anterior and posterior groups. The muscles of the anterior group primarily flex the thigh; those of the posterior group extend, abduct, or rotate it. The muscles in these groups are shown in figures 9.34, 9.35, 9.36, 9.37, and in reference plates 66 and 67, and are

**FIGURE 9.34**

(a) Muscles of the anterior right thigh. Isolated views of (b) the vastus intermedius, (c–e) adductors of the thigh, (f–g) flexors of the thigh.

listed in table 9.12. Muscles that move the thigh include the following:

**Anterior Group**
*Psoas major*
*Iliacus*

**Posterior Group**
*Gluteus maximus*
*Gluteus medius*
*Gluteus minimus*
*Tensor fasciae latae*

Still another group of muscles, attached to the femur and pelvic girdle, adducts the thigh. This group includes:

*Pectineus*
*Adductor longus*
*Adductor magnus*
*Gracilis*

## Anterior Group

The **psoas** (so'as) **major** is a long, thick muscle that connects the lumbar vertebrae to the femur. It flexes the thigh (fig. 9.34).

The **iliacus** (il'e-ak-us), a large, fan-shaped muscle, lies along the lateral side of the psoas major. The iliacus and the psoas major are the primary flexors of the thigh, and they advance the lower limb in walking movements (fig. 9.34).

## Posterior Group

The **gluteus maximus** (gloo'te-us mak'si-mus) is the largest muscle in the body and covers a large part of each buttock. It connects the ilium, sacrum, and coccyx to the femur by fascia of the thigh and extends the thigh. The gluteus maximus helps to straighten the lower limb at the hip when a person walks, runs, or climbs. It is also used to raise the body from a sitting position (fig. 9.35).

The **gluteus medius** (gloo'te-us me'de-us) is partly covered by the gluteus maximus. Its fibers extend from the ilium to the femur, and they abduct the thigh and rotate it medially (fig. 9.35).

The **gluteus minimus** (gloo'te-us min'ĭ-mus) lies beneath the gluteus medius and is its companion in attachments and functions (fig. 9.35).

The **tensor fasciae latae** (ten'sor fash'e-e lah-tē) connects the ilium to the fascia of the thigh, which continues downward to the tibia. This muscle abducts and flexes the thigh and rotates it medially (fig. 9.35).

The gluteus medius and gluteus minimus help support and maintain the normal position of the pelvis. If these muscles are paralyzed as a result of injury or disease, the pelvis tends to drop to one side whenever the foot on that side is raised. Consequently, the person walks with a waddling limp called the *gluteal gait*.

## Thigh Adductors

The **pectineus** (pek-tin'e-us) muscle runs from the spine of the pubis to the femur. It adducts and flexes the thigh (fig. 9.34).

The **adductor longus** (ah-duk'tor long'gus) is a long, triangular muscle that runs from the pubic bone to the femur. It adducts the thigh and assists in flexing and rotating it laterally (fig. 9.34).

The **adductor magnus** (ah-duk'tor mag'nus) is the largest adductor of the thigh. It is a triangular muscle that connects the ischium to the femur. It adducts the thigh and assists in extending and rotating it laterally (fig. 9.34).

The **gracilis** (gras'il-is) is a long, straplike muscle that passes from the pubic bone to the tibia. It adducts the thigh and flexes the leg at the knee (fig. 9.34).

## Muscles that Move the Leg

The muscles that move the leg connect the tibia or fibula to the femur or to the pelvic girdle. They fall into two major groups—those that flex the knee and those that extend it. The muscles of these groups are shown in figures 9.34, 9.35, 9.36, 9.37, in reference plates 66 and 67, and are listed in table 9.13. Muscles that move the leg include the following:

**Flexors**
*Biceps femoris*
*Semitendinosus*
*Semimembranosus*
*Sartorius*

**Extensor**
*Quadriceps femoris group*

### Flexors

As its name implies, the **biceps femoris** (bi'seps fem'or-is) has two heads, one attached to the ischium and the other attached to the femur. This muscle passes along the back of the thigh on the lateral side and connects to the proximal ends of the fibula and tibia. The biceps femoris is one of the hamstring muscles, and its tendon (hamstring) can be felt as a lateral ridge behind the knee. This muscle flexes and rotates the leg laterally and extends the thigh (figs. 9.35 and 9.36).

The **semitendinosus** (sem"e-ten'dĭ-no-sus) is another hamstring muscle. It is a long, bandlike muscle on the back of the thigh toward the medial side, connecting the ischium to the proximal end of the tibia. The semitendinosus is so named because it becomes tendinous in the middle of the thigh, continuing to its insertion as a long, cordlike tendon. It flexes and rotates the leg medially, and extends the thigh (fig. 9.36).

The **semimembranosus** (sem"e-mem'brah-no-sus) is the third hamstring muscle and is the most medially located muscle in the back of the thigh. It connects the

Gluteus medius

Gluteus maximus

Biceps femoris

Tensor fasciae latae

Sartorius

Rectus femoris

Vastus lateralis

Fascia

(a)

Gluteus medius

Gluteus maximus

Gluteus minimus

(b)          (c)          (d)

**FIGURE 9.35**
(a) Muscles of the lateral right thigh. (b–d) Isolated views of the gluteal muscles.

ischium to the tibia, and flexes and rotates the leg medially and extends the thigh (fig. 9.36).

The **sartorius** (sar-to′re-us) is an elongated, straplike muscle that passes obliquely across the front of the thigh and then descends over the medial side of the knee. It connects the ilium to the tibia, and flexes the leg and the thigh. It can also abduct the thigh and rotate it laterally (figs. 9.34 and 9.35).

The tendinous attachments of the hamstring muscles to the ischial tuberosity are sometimes torn as a result of strenuous running or kicking motions. This painful injury is commonly called "pulled hamstrings," and is usually accompanied by internal bleeding from damaged blood vessels that supply the muscles.

Gluteus medius

Gluteus maximus

Adductor magnus

Vastus lateralis covered by fascia

Gracilis

Semitendinosus

Biceps femoris

Semimembranosus

Sartorius

Gastrocnemius

(a)

Semimembranosus

Biceps femoris (short head)

(b)

Semitendinosus

Biceps femoris (long head)

(c)

FIGURE 9.36

(*a*) Muscles of the posterior right thigh. (*b* and *c*) Isolated views of muscles that flex the leg at the knee.

## Extensor

The large, fleshy muscle group called the **quadriceps femoris** (kwod'rĭ-seps fem'or-is) occupies the front and sides of the thigh, and is the primary extensor of the knee. It is composed of four parts—*rectus femoris, vastus lateralis, vastus medialis,* and *vastus intermedius* (figs. 9.34 and 9.37). These parts connect the ilium and femur to a common *patellar tendon,* which passes over the front of the knee and attaches to the patella. This tendon then continues as the *patellar ligament* to the tibia.

Occasionally, as a result of traumatic injury in which muscle such as the quadriceps femoris is compressed against an underlying bone, new bone tissue may begin to develop within the damaged muscle. This condition is called *myositis ossificans.* When the bone tissue matures several months following the injury, surgery can remove the newly formed bone.

FIGURE 9.37 🦅

A cross section of the thigh.

| Muscle | Origin | Insertion | Action | Nerve Supply |
|--------|--------|-----------|--------|--------------|
| Psoas major | Lumbar intervertebral disks; bodies and transverse processes of lumbar vertebrae | Lesser trochanter of femur | Flexes thigh | Branches of L1-3 nerves |
| Iliacus | Iliac fossa of ilium | Lesser trochanter of femur | Flexes thigh | Femoral n. |
| Gluteus maximus | Sacrum, coccyx, and posterior surface of ilium | Posterior surface of femur and fascia of thigh | Extends thigh at hip | Inferior gluteal n. |
| Gluteus medius | Lateral surface of ilium | Greater trochanter of femur | Abducts and rotates thigh medially | Superior gluteal n. |
| Gluteus minimus | Lateral surface of ilium | Greater trochanter of femur | Same as gluteus medius | Superior gluteal n. |
| Tensor fasciae latae | Anterior iliac crest | Fascia of thigh | Abducts, flexes, and rotates thigh medially | Superior gluteal n. |
| Pectineus | Spine of pubis | Femur distal to lesser trochanter | Adducts and flexes thigh | Obturator and femoral nerves |
| Adductor longus | Pubic bone near symphysis pubis | Posterior surface of femur | Adducts, flexes, and rotates thigh laterally | Obturator n. |
| Adductor magnus | Ischial tuberosity | Posterior surface of femur | Adducts, extends, and rotates thigh laterally | Obturator and branch of sciatic n. |
| Gracilis | Lower edge of symphysis pubis | Medial surface of tibia | Adducts thigh and flexes leg at the knee | Obturator n. |

TABLE 9.13

## MUSCLES THAT MOVE THE LEG

| Muscle | Origin | Insertion | Action | Nerve Supply |
|---|---|---|---|---|
| **Hamstring Group** | | | | |
| Biceps femoris | Ischial tuberosity and linea aspera of femur | Head of fibula and lateral condyle of tibia | Flexes and rotates leg laterally and extends thigh | Tibial n. |
| Semitendinosus | Ischial tuberosity | Medial surface of tibia | Flexes and rotates leg medially and extends thigh | Tibial n. |
| Semimembranosus | Ischial tuberosity | Medial condyle of tibia | Flexes and rotates leg medially and extends thigh | Tibial n. |
| *Sartorius* | Anterior superior iliac spine | Medial surface of tibia | Flexes leg and thigh, abducts and rotates thigh laterally | Femoral n. |
| **Quadriceps Femoris Group** | | | | |
| Rectus femoris | Spine of ilium and margin of acetabulum | | | |
| Vastus lateralis | Greater trochanter and posterior surface of femur | Patella by common tendon, which continues as patellar ligament to tibial tuberosity | Extends leg at knee | Femoral n. |
| Vastus medialis | Medial surface of femur | | | |
| Vastus intermedius | Anterior and lateral surfaces of femur | | | |

## Muscles that Move the Foot

Movements of the foot include movements of the ankle and toes. A number of muscles that move the foot are located in the leg. They attach the femur, tibia, and fibula to bones of the foot and are responsible for moving the foot upward (dorsiflexion) or downward (plantar flexion), and turning the foot so the toes are inward (inversion) or outward (eversion). These muscles are shown in figures 9.38, 9.39, 9.40, 9.41, in reference plates 68, 69, 70, and are listed in table 9.14. Muscles that move the foot include the following:

**Dorsal Flexors**
*Tibialis anterior*
*Peroneus tertius*
*Extensor digitorum longus*

**Plantar Flexors**
*Gastrocnemius*
*Soleus*
*Flexor digitorum longus*

**Invertor**
*Tibialis posterior*

**Evertor**
*Peroneus longus*

### Dorsal Flexors

The **tibialis anterior** (tib″e-a′lis an-te′re-or) is an elongated, spindle-shaped muscle located on the front of the leg. It arises from the surface of the tibia, passes medially over the distal end of the tibia, and attaches to bones of the ankle and foot. Contraction of the tibialis anterior causes dorsiflexion and inversion of the foot (fig. 9.38).

The **peroneus tertius** (per″o-ne′us ter′shus) is a muscle of variable size that connects the fibula to the lateral side of the foot. It functions in dorsiflexion and eversion of the foot (fig. 9.38).

The **extensor digitorum longus** (eks-ten′sor dij″ĭ-to′rum long′gus) is situated along the lateral side of the leg just behind the tibialis anterior. It arises from the proximal end of the tibia and the shaft of the fibula. Its tendon divides into four parts as it passes over the front of the ankle. These parts continue over the surface of the foot and attach to the four lateral toes. The actions of the extensor digitorum longus include dorsiflexion of the foot, eversion of the foot, and extension of the toes (figs. 9.38 and 9.39).

### Plantar Flexors

The **gastrocnemius** (gas″trok-ne′me-us) on the back of the leg forms part of the calf. It arises by two heads from the femur. The distal end of this muscle joins the

Patella

Patellar ligament

Tibialis anterior

Peroneus longus

Extensor digitorum
longus

Peroneus brevis

Gastrocnemius

Soleus

Tibia

Extensor
retinacula

(a)

Tibialis
anterior

Peroneus
tertius

Extensor
digitorum
longus

(b)                    (c)                    (d)

**FIGURE** 9.38

(a) Muscles of the anterior right leg. (b–d) Isolated views of muscles associated with the anterior leg.

strong *calcaneal tendon* (Achilles tendon), which descends to the heel and attaches to the calcaneus. The gastrocnemius is a powerful plantar flexor of the foot that aids in pushing the body forward when a person walks or runs. It also flexes the leg at the knee (figs. 9.39 and 9.40).

Strenuous athletic activity may partially or completely tear the calcaneal (Achilles) tendon. This injury occurs most frequently in middle-aged athletes who run or play sports that involve quick movements and directional changes. A torn calcaneal tendon usually requires surgical treatment.

**FIGURE 9.39**

(a) Muscles of the lateral right leg. Isolated views of (b) peroneus longus and (c) peroneus brevis.

The **soleus** (so'le-us) is a thick, flat muscle located beneath the gastrocnemius, and together these two muscles form the calf of the leg. The soleus arises from the tibia and fibula, and it extends to the heel by way of the calcaneal tendon. It acts with the gastrocnemius to cause plantar flexion of the foot (figs. 9.39 and 9.40).

The **flexor digitorum longus** (flek'sor dij"ĭ-to'rum long'gus) extends from the posterior surface of the tibia to the foot. Its tendon passes along the plantar surface of the foot. There the muscle divides into four parts that attach to the terminal bones of the four lateral toes. This muscle assists in plantar flexion of the foot, flexion of the four lateral toes, and inversion of the foot (fig. 9.40).

**FIGURE** 9.40 🏃

(a) Muscles of the posterior right leg. (b–e) Isolated views of muscles associated with the posterior right leg.

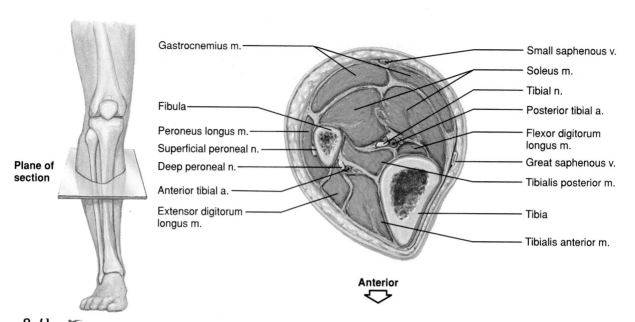

**FIGURE 9.41**

A cross section of the leg (superior view).

| TABLE 9.14 | MUSCLES THAT MOVE THE FOOT | | | | |
|---|---|---|---|---|---|
| **Muscle** | **Origin** | **Insertion** | **Action** | **Nerve Supply** |
| Tibialis anterior | Lateral condyle and lateral surface of tibia | Tarsal bone (cuneiform) and first metatarsal | Dorsiflexion and inversion of foot | Deep peroneal n. |
| Peroneus tertius | Anterior surface of fibula | Dorsal surface of fifth metatarsal | Dorsiflexion and eversion of foot | Deep peroneal n. |
| Extensor digitorum longus | Lateral condyle of tibia and anterior surface of fibula | Dorsal surfaces of second and third phalanges of four lateral toes | Dorsiflexion and eversion of foot and extension of toes | Deep peroneal n. |
| Gastrocnemius | Lateral and medial condyles of femur | Posterior surface of calcaneus | Plantar flexion of foot and flexion of leg at knee | Tibial n. |
| Soleus | Head and shaft of fibula and posterior surface of tibia | Posterior surface of calcaneus | Plantar flexion of foot | Tibial n. |
| Flexor digitorum longus | Posterior surface of tibia | Distal phalanges of four lateral toes | Plantar flexion and inversion of foot, and flexion of four lateral toes | Tibial n. |
| Tibialis posterior | Lateral condyle and posterior surface of tibia, and posterior surface of fibula | Tarsal and metatarsal bones | Plantar flexion and inversion of foot | Tibial n. |
| Peroneus longus | Lateral condyle of tibia, and head and shaft of fibula | Tarsal and metatarsal bones | Plantar flexion and eversion of foot; also supports arch | Superficial peroneal n. |

### Invertor

The **tibialis posterior** (tib″e-a′lis pos-tēr′e-or) is the deepest of the muscles on the back of the leg. It connects the fibula and tibia to the ankle bones by means of a tendon that curves under the medial malleolus. This muscle assists in inversion and plantar flexion of the foot (fig. 9.40).

### Evertor

The **peroneus longus** (per″o-ne′us long′gus) is a long, straplike muscle located on the lateral side of the leg. It connects the tibia and the fibula to the foot by means of a stout tendon that passes behind the lateral malleolus. It everts the foot, assists in plantar flexion, and helps support the arch of the foot (figs. 9.39 and 9.41).

As in the wrist, fascia in various regions of the ankle thicken to form retinacula. Anteriorly, for example, *extensor retinacula* connect the tibia and fibula as well as the calcaneus and fascia of the sole. These retinacula form sheaths for tendons crossing the front of the ankle (fig. 9.39).

Posteriorly, on the inside, a *flexor retinaculum* runs between the medial malleolus and the calcaneus, and forms sheaths for tendons passing beneath the foot (fig. 9.40). *Peroneal retinacula* connect the lateral malleolus and the calcaneus, providing sheaths for tendons on the lateral side of the ankle (fig. 9.39).

---

## CLINICAL TERMS RELATED TO THE MUSCULAR SYSTEM

**contracture** (kon-trak′tūr)  Condition in which there is great resistance to the stretching of a muscle.

**convulsion** (kun-vul′shun)  Series of involuntary contractions of various voluntary muscles.

**electromyography** (e-lek″tro-mi-og′rah-fe)  Technique for recording the electrical changes that occur in muscle tissues.

**fibrillation** (fi″bri-la′shun)  Spontaneous contractions of individual muscle fibers, producing rapid and uncoordinated activity within a muscle.

**fibrosis** (fi-bro′sis)  Degenerative disease in which connective tissue with many fibers replaces skeletal muscle tissue.

**fibrositis** (fi″bro-si′tis)  Inflammation of connective tissues with many fibers, especially in the muscle fascia. This disease is also called muscular rheumatism.

**muscular dystrophy** (mus′ku-lar dis′tro-fe)  Progressive muscle weakness and atrophy caused by deficient dystrophin protein.

**myalgia** (mi-al′je-ah)  Pain resulting from any muscular disease or disorder.

**myasthenia gravis** (mi″as-the′ne-ah grav′is)  Chronic disease characterized by muscles that are weak and easily fatigued. It results from the immune system's attack on neuromuscular junctions so that stimuli are not transmitted from motor neurons to muscle fibers.

**myokymia** (mi″o-ki′me-ah)  Persistent quivering of a muscle.

**myology** (mi-ol′o-je)  Study of muscles.

**myoma** (mi-o′mah)  Tumor composed of muscle tissue.

**myopathy** (mi-op′ah-the)  Any muscular disease.

**myositis** (mi″o-si′tis)  Inflammation of skeletal muscle tissue.

**myotomy** (mi-ot′o-me)  Cutting of muscle tissue.

**myotonia** (mi″o-to′ne-ah)  Prolonged muscular spasm.

**paralysis** (pah-ral′ĭ-sis)  Loss of ability to move a body part.

**paresis** (pah-re′sis)  Partial or slight paralysis of the muscles.

**shin splints** (shin′ splints)  Soreness on the front of the leg due to straining the flexor digitorum longus, often as a result of walking up and down hills.

**torticollis** (tor″tĭ-kol′is)  Condition in which the neck muscles, such as the sternocleidomastoids, contract involuntarily. It is more commonly called wryneck.

**MUSCULAR SYSTEM**
Muscles provide the force for moving body parts.

## Integumentary System

The skin increases heat loss during skeletal muscle activity. Sensory receptors function in the reflex contol of skeletal muscles.

## Skeletal System

Bones provide attachments that allow skeletal muscles to cause movement.

## Nervous System

Neurons control muscle contractions.

## Endocrine System

Hormones help increase blood flow to exercising skeletal muscles.

## Cardiovascular System

Blood flow delivers oxygen and nutrients and removes wastes.

## Lymphatic System

Muscle action pumps lymph through lymphatic vessels.

## Digestive System

Skeletal muscles are important in swallowing. The digestive system absorbs needed nutrients.

## Respiratory System

Breathing depends on skeletal muscles. The lungs provide oxygen for body cells and eliminate carbon dioxide.

## Urinary System

Skeletal muscles help control urine elimination.

## Reproductive System

Skeletal muscles are important in sexual activity.

# Introduction (page 282)

The three types of muscle tissue are skeletal, smooth, and cardiac.

# Structure of a Skeletal Muscle (page 282)

Skeletal muscles are composed of nervous, vascular, and various connective tissues, as well as skeletal muscle tissue.

1. Connective tissue coverings
   a. Fascia covers each skeletal muscle.
   b. Other connective tissues surround cells and groups of cells within the muscle's structure.
   c. Fascia is part of a complex network of connective tissue that extends throughout the body.
2. Skeletal muscle fibers
   a. Each skeletal muscle fiber is a single muscle cell, which is the unit of contraction.
   b. Muscle fibers are cylindrical cells with many nuclei.
   c. The cytoplasm contains mitochondria, sarcoplasmic reticulum, and myofibrils of actin and myosin.
   d. The arrangement of the actin and myosin filaments causes striations.
   e. Transverse tubules extend from the cell membrane into the cytoplasm and are associated with the cisternae of the sarcoplasmic reticulum.
3. Neuromuscular junction
   a. Motor neurons stimulate muscle fibers to contract.
   b. The motor end plate of a muscle fiber lies on one side of a neuromuscular junction.
   c. In response to a nerve impulse, the end of a motor nerve fiber secretes a neurotransmitter, which diffuses across the junction and stimulates the muscle fiber.
4. Motor units
   a. One motor neuron and the muscle fibers associated with it constitute a motor unit.
   b. Muscles whose motor units contain small numbers of muscle fibers produce finer movements.

# Skeletal Muscle Contraction (page 287)

Muscle fiber contraction results from a sliding movement of actin and myosin filaments.

1. Role of myosin and actin
   a. Cross-bridges of myosin filaments form linkages with actin filaments.
   b. The reaction between actin and myosin filaments provides the basis for contraction.
   c. When a fiber is at rest, troponin and tropomyosin molecules interfere with linkage formation.
   d. Calcium ions remove the inhibition.
2. Stimulus for contraction
   a. Muscle fiber is usually stimulated by acetylcholine released from the end of a motor nerve fiber.
   b. Acetylcholinesterase decomposes acetylcholine to prevent continuous stimulation.
   c. Stimulation causes muscle fiber to conduct an impulse that travels over the surface of the sarcolemma and reaches the deep parts of the fiber by means of the transverse tubules.

   d. A muscle impulse signals the sarcoplasmic reticulum to release calcium ions.
   e. Linkages form between myosin and actin, and the actin filaments move inward, shortening the sarcomere.
   f. The muscle fiber relaxes when calcium ions are transported back into the sarcoplasmic reticulum.
3. Energy sources for contraction
   a. ATP supplies the energy for muscle fiber contraction.
   b. Creatine phosphate stores energy that can be used to synthesize ATP as it is decomposed.
   c. Active muscles depend upon cellular respiration for energy.
4. Oxygen supply and cellular respiration
   a. Anaerobic respiration yields few ATP molecules, whereas aerobic respiration provides many ATP molecules.
   b. Hemoglobin in red blood cells carries oxygen from the lungs to body cells.
   c. Myoglobin in muscle cells stores some oxygen temporarily.
5. Oxygen debt
   a. During rest or moderate exercise, oxygen is sufficient to support aerobic respiration.
   b. During strenuous exercise, oxygen deficiency may develop, and lactic acid may accumulate as a result of anaerobic respiration.
   c. The amount of oxygen needed to convert accumulated lactic acid to glucose and to restore supplies of ATP and creatine phosphate is called oxygen debt.
6. Muscle fatigue
   a. A fatigued muscle loses its ability to contract.
   b. Muscle fatigue is usually due to the effects of accumulation of lactic acid.
   c. Athletes usually produce less lactic acid than nonathletes because of their increased ability to supply oxygen and nutrients to muscles.
7. Fast and slow muscles
   a. The speed of contraction is related to a muscle's specific function.
   b. Slow-contracting, or red, muscles can generate ATP fast enough to keep up with ATP breakdown and can contract for long periods.
   c. Fast-contracting, or white, muscles have reduced ability to carry on aerobic respiration and tend to fatigue relatively rapidly.
8. Heat production
   a. Muscles represent an important source of body heat.
   b. Most of the energy released by cellular respiration is lost as heat.

# Muscular Responses (page 294)

1. Threshold stimulus is the minimal stimulus needed to elicit a muscular contraction.
2. All-or-none response
   a. If a muscle fiber contracts at all, it will contract completely.
   b. Motor units respond in an all-or-none manner.
3. Recording a muscle contraction
   a. A myogram is a recording of an electrically stimulated isolated muscle pulling a lever.

b. A twitch is a single, short contraction reflecting stimulation of some motor units in a muscle.
c. The latent period is the time between stimulus and responding muscle contraction.
d. During the refractory period immediately following contraction, a muscle cannot respond.

4. Staircase effect
a. An inactive muscle undergoes a series of contractions of increasing strength when subjected to a series of stimuli.
b. This staircase effect seems to be due to failure to remove calcium ions from the sarcoplasm rapidly enough.

5. Summation
a. A rapid series of stimuli may produce summation of twitches and sustained contraction.
b. Forceful, sustained contraction without relaxation is a tetanic contraction.

6. Recruitment of motor units
a. At low intensity of stimulation, relatively small numbers of motor units contract.
b. At increasing intensities of stimulation, other motor units are recruited until the muscle contracts with maximal tension.

7. Sustained contractions
a. When contractions fuse, the strength of contraction may increase due to recruitment of fibers.
b. Even when a muscle is at rest, its fibers usually maintain tone—that is, remain partially contracted.

8. Types of contractions
a. When a muscle contracts and its ends are pulled closer together, the contraction is called isotonic.
b. When a muscle contracts but its attachments do not move, the contraction is called isometric.
c. When the force a muscle generates is less than that required to move or lift an object, the contraction is called isokinetic.
d. Most body movements involve both isometric and isotonic contractions.

## Smooth Muscles  (page 297)

The contractile mechanisms of smooth and cardiac muscles are similar to those of skeletal muscle.

1. Smooth muscle fibers
a. Smooth muscle cells contain filaments of myosin and actin.
b. They lack transverse tubules, and the sarcoplasmic reticula are not well developed.
c. Types include multiunit smooth muscle and visceral smooth muscle.
d. Visceral smooth muscle displays rhythmicity.
e. Peristalsis aids movement of material through hollow organs.

2. Smooth muscle contraction
a. In smooth muscles, calmodulin binds to calcium ions and activates the contraction mechanism.
b. Both acetylcholine and norepinephrine are neurotransmitters for smooth muscles.
c. Hormones and stretching affect smooth muscle contractions.
d. With a given amount of energy, smooth muscle can maintain a contraction for a longer time than can skeletal muscle.
e. Smooth muscles can change lengths without changing tautness.

## Cardiac Muscle  (page 298)

1. Cardiac muscle contracts for a longer time than skeletal muscle because transverse tubules supply extra calcium ions.
2. Intercalated disks connect the ends of adjacent cardiac muscle cells and hold the cells together.
3. A network of fibers contracts as a unit and responds to stimulation in an all-or-none manner.
4. Cardiac muscle is self-exciting, rhythmic, and remains refractory until a contraction is completed.

## Skeletal Muscle Actions  (page 300)

1. Origin and insertion
a. The movable end of attachment of a skeletal muscle to a bone is its insertion, and the immovable end is its origin.
b. Some muscles have more than one origin or insertion.
2. Interaction of skeletal muscles
a. Skeletal muscles function in groups.
b. A prime mover is responsible for most of a movement; synergists aid prime movers; antagonists can resist the movement of a prime mover.
c. Smooth movements depend upon antagonists giving way to the actions of prime movers.

## Major Skeletal Muscles  (page 301)

Muscle names often describe sizes, shapes, locations, actions, number of attachments, or direction of fibers.

1. Muscles of facial expression
a. These muscles lie beneath the skin of the face and scalp and are used to communicate feelings through facial expression.
b. They include the epicranius, orbicularis oculi, orbicularis oris, buccinator, zygomaticus, and platysma.
2. Muscles of mastication
a. These muscles are attached to the mandible and are used in chewing.
b. They include the masseter, temporalis, medial pterygoid, and lateral pterygoid.
3. Muscles that move the head and vertebral column
a. Muscles in the neck and back move the head.
b. They include the sternocleidomastoid, splenius capitis, semispinalis capitis, and erector spinae.
4. Muscles that move the pectoral girdle
a. Most of these muscles connect the scapula to nearby bones and are closely associated with muscles that move the arm.
b. They include the trapezius, rhomboideus major, levator scapulae, serratus anterior, and pectoralis minor.
5. Muscles that move the arm
a. These muscles connect the humerus to various regions of the pectoral girdle, ribs, and vertebral column.
b. They include the coracobrachialis, pectoralis major, teres major, latissimus dorsi, supraspinatus, deltoid, subscapularis, infraspinatus, and teres minor.
6. Muscles that move the forearm
a. These muscles connect the radius and ulna to the humerus and pectoral girdle.
b. They include the biceps brachii, brachialis, brachioradialis, triceps brachii, supinator, pronator teres, and pronator quadratus.

7. Muscles that move the hand
   a. These muscles arise from the distal end of the humerus and from the radius and ulna.
   b. They include the flexor carpi radialis, flexor carpi ulnaris, palmaris longus, flexor digitorum profundus, flexor digitorum superficialis, extensor carpi radialis longus, extensor carpi radialis brevis, extensor carpi ulnaris, and extensor digitorum.
   c. An extensor retinaculum forms sheaths for tendons of the extensor muscles.
8. Muscles of the abdominal wall
   a. These muscles connect the rib cage and vertebral column to the pelvic girdle.
   b. They include the external oblique, internal oblique, transversus abdominis, and rectus abdominis.
9. Muscles of the pelvic outlet
   a. These muscles form the floor of the pelvic cavity and fill the space of the pubic arch.
   b. They include the levator ani, coccygeus, superficial transversus perinei, bulbospongiosus, ischiocavernosus, and sphincter urethrae.
10. Muscles that move the thigh
    a. These muscles are attached to the femur and to some part of the pelvic girdle.
    b. They include the psoas major, iliacus, gluteus maximus, gluteus medius, gluteus minimus, tensor fasciae latae, pectineus, adductor longus, adductor magnus, and gracilis.
11. Muscles that move the leg
    a. These muscles connect the tibia or fibula to the femur or pelvic girdle.
    b. They include the biceps femoris, semitendinosus, semimembranosus, sartorius, and the quadriceps femoris group.
12. Muscles that move the foot
    a. These muscles attach the femur, tibia, and fibula to various bones of the foot.
    b. They include the tibialis anterior, peroneus tertius, extensor digitorum longus, gastrocnemius, soleus, flexor digitorum longus, tibialis posterior, and peroneus longus.
    c. Retinacula form sheaths for tendons passing to the foot.

## CRITICAL THINKING QUESTIONS

1. Why do you think athletes generally perform better if they warm up by exercising lightly before a competitive event?
2. Following childbirth, a woman may lose urinary control (incontinence) when sneezing or coughing. Which muscles of the pelvic floor should be strengthened by exercise to help control this problem?
3. What steps might be taken to minimize atrophy of skeletal muscles in patients who are confined to bed for prolonged times?
4. As lactic acid and other substances accumulate in an active muscle, they stimulate pain receptors, and the muscle may feel sore. How might the application of heat or substances that dilate blood vessels help relieve such soreness?
5. Several important nerves and blood vessels course through the muscles of the gluteal region. In order to avoid the possibility of damaging such parts, intramuscular injections are usually made into the lateral, superior portion of the gluteus medius. What landmarks would help you locate this muscle in a patient?
6. Following an injury to a nerve, the muscles it supplies with motor nerve fibers may become paralyzed. How would you explain to a patient the importance of moving the disabled muscles passively or contracting them with electrical stimulation?

## REVIEW EXERCISES

## Part A

1. List the three types of muscle tissue.
2. Distinguish between a tendon and an aponeurosis.
3. Describe the connective tissue coverings of a skeletal muscle.
4. Distinguish among deep fascia, subcutaneous fascia, and subserous fascia.
5. List the major parts of a skeletal muscle fiber, and describe the function of each part.
6. Describe a neuromuscular junction.
7. Explain the function of a neurotransmitter substance.
8. Define *motor unit,* and explain how the number of fibers within a unit affects muscular contractions.
9. Describe the major events that occur when a muscle fiber contracts.
10. Explain how ATP and creatine phosphate function in muscle contraction.
11. Describe how oxygen is supplied to skeletal muscles.
12. Describe how an oxygen debt may develop.
13. Explain how muscles may become fatigued and how a person's physical condition may affect tolerance to fatigue.
14. Distinguish between fast-contracting and slow-contracting muscles.
15. Explain how the actions of skeletal muscles affect maintenance of body temperature.
16. Define *threshold stimulus.*
17. Explain *all-or-none response.*
18. Explain *recruitment.*
19. Describe the staircase effect.

20. Explain how a skeletal muscle can be stimulated to produce a sustained contraction.
21. Distinguish between a tetanic contraction and muscle tone.
22. Distinguish between isometric and isotonic contractions, and explain how each is used in body movements.
23. Compare the structures of smooth and skeletal muscle fibers.
24. Distinguish between multiunit and visceral smooth muscles.
25. Define *peristalsis* and explain its function.
26. Compare the characteristics of smooth and skeletal muscle contractions.
27. Compare the structures of cardiac and skeletal muscle fibers.
28. Compare the characteristics of cardiac and skeletal muscle contractions.
29. Distinguish between a muscle's origin and its insertion.
30. Define *prime mover, synergist,* and *antagonist.*

## Part B

Match the muscles in column I with the descriptions and functions in column II.

| I | II |
|---|---|
| 1. Buccinator | A. Inserted on the coronoid process of the mandible. |
| 2. Epicranius | B. Draws the corner of the mouth upward. |
| 3. Lateral pterygoid | C. Can raise and adduct the scapula. |
| 4. Platysma | D. Can pull the head into an upright position. |
| 5. Rhomboideus major | E. Consists of two parts— the frontalis and the occipitalis. |
| 6. Splenius capitis | F. Compresses the cheeks. |
| 7. Temporalis | G. Extends over the neck from the chest to the face. |
| 8. Zygomaticus | H. Pulls the jaw from side to side. |
| 9. Biceps brachii | I. Primary extensor of the elbow. |
| 10. Brachialis | J. Pulls the shoulder back and downward. |
| 11. Deltoid | K. Abducts the arm. |
| 12. Latissimus dorsi | L. Rotates the arm laterally. |
| 13. Pectoralis major | M. Pulls the arm forward and across the chest. |
| 14. Pronator teres | N. Rotates the arm medially. |
| 15. Teres minor | O. Strongest flexor of the elbow. |
| 16. Triceps brachii | P. Strongest supinator of the forearm. |
| 17. Biceps femoris | Q. Inverts the foot. |
| 18. External oblique | R. A member of the quadriceps femoris group. |
| 19. Gastrocnemius | S. A plantar flexor of the foot. |
| 20. Gluteus maximus | T. Compresses the contents of the abdominal cavity. |
| 21. Gluteus medius | U. Largest muscle in the body. |
| 22. Gracilis | V. A hamstring muscle. |
| 23. Rectus femoris | W. Adducts the thigh. |
| 24. Tibialis anterior | X. Abducts the thigh. |

## Part C

Which muscles can you identify in the bodies of these models whose muscles are enlarged by exercise?

# SURFACE ANATOMY

■ *The following set of reference plates is presented to help you locate some of the more prominent surface features in various regions of the body. For the most part, the labeled structures are easily seen or palpated through the skin. As a review, you may want to locate as many of these features as possible on your own body.*

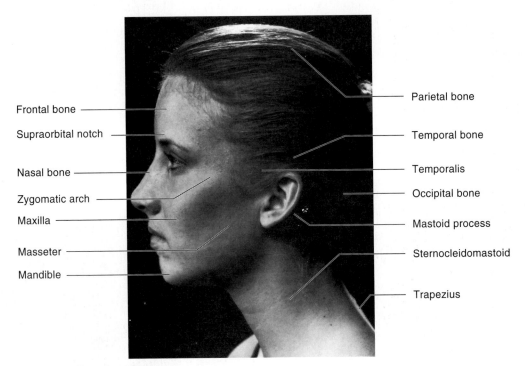

Frontal bone

Supraorbital notch

Nasal bone

Zygomatic arch

Maxilla

Masseter

Mandible

Parietal bone

Temporal bone

Temporalis

Occipital bone

Mastoid process

Sternocleidomastoid

Trapezius

**PLATE** Thirty-Seven

Surface anatomy of head and neck, lateral view.

Clavicle

Deltoid

Pectoralis major

Biceps brachii

Serratus anterior

Brachioradialis

Extensor carpi
radialis longus

Extensor
digitorum

Trapezius

Acromial process

Triceps brachii

Border of scapula

Latissimus dorsi

Tendon of
biceps brachii

Lateral epicondyle
of humerus

Olecranon process
of ulna

**PLATE Thirty-Eight**

Surface anatomy of upper limb and thorax, lateral view.

Trapezius

Teres major

Spinous processes
of vertebrae

Biceps brachii
Triceps brachii
Deltoid

Infraspinatus

Border of scapula

Vertebral spine

Latissimus dorsi

**PLATE Thirty-Nine**

Surface anatomy of back and upper limbs, posterior view.

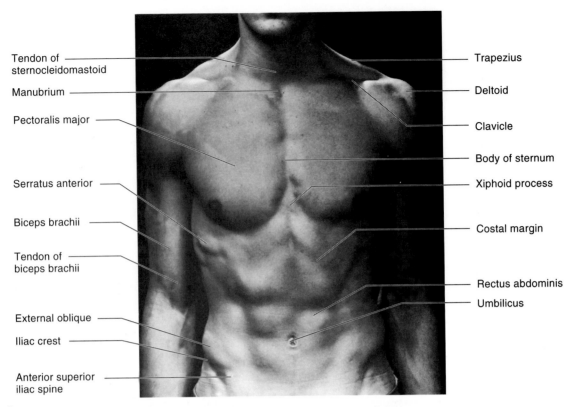

Tendon of sternocleidomastoid

Manubrium

Pectoralis major

Serratus anterior

Biceps brachii

Tendon of biceps brachii

External oblique

Iliac crest

Anterior superior iliac spine

Trapezius

Deltoid

Clavicle

Body of sternum

Xiphoid process

Costal margin

Rectus abdominis

Umbilicus

**PLATE Forty**
Surface anatomy of torso and arms, anterior view.

Latissimus dorsi

Gluteus medius

Gluteus maximus

Hamstring muscles

Tendon of biceps femoris

Tendon of semitendinosus

Triceps brachii

Lateral epicondyle of humerus

Olecranon process of ulna

Iliac crest

Posterior superior iliac spine

Sacrum

Styloid process of radius

Greater trochanter of femur

Ischial tuberosity

**PLATE Forty-One**
Surface anatomy of torso and thighs, posterior view.

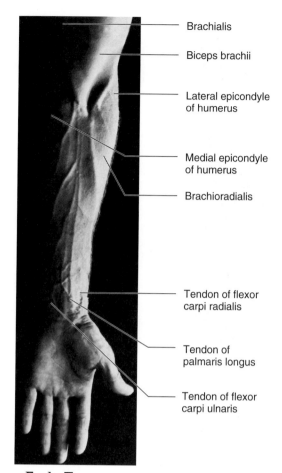

Brachialis

Biceps brachii

Lateral epicondyle
of humerus

Medial epicondyle
of humerus

Brachioradialis

Tendon of flexor
carpi radialis

Tendon of
palmaris longus

Tendon of flexor
carpi ulnaris

PLATE Forty-Two

Surface anatomy of forearm, anterior view.

Rectus femoris

Sartorius

Vastus lateralis

Vastus medialis

Patella

Patellar ligament

Tibial tuberosity

Tibialis anterior

Gastrocnemius

Peroneus longus

Anterior border
of tibia

PLATE Forty-Four

Surface anatomy of knee and surrounding area, anterior view.

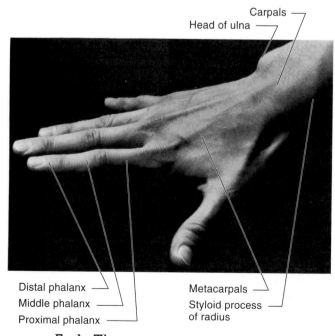

Carpals

Head of ulna

Distal phalanx

Middle phalanx

Proximal phalanx

Metacarpals

Styloid process
of radius

PLATE Forty-Three

Surface anatomy of the hand.

Rectus femoris

Biceps femoris

Vastus lateralis

Tendon of
biceps femoris

Patella

Gastrocnemius

Peroneus longus

Tibialis anterior

PLATE Forty-Five

Surface anatomy of knee and surrounding area, lateral view.

**SURFACE ANATOMY**

Gastrocnemius

Tibialis anterior

Tibia

Soleus

Tendon of tibialis anterior

Medial malleolus

Calcaneal tendon

Tendon of tibialis posterior

Calcaneus

Metatarsals

Phalanges

PLATE Forty-Six

Surface anatomy of ankle and leg, medial view.

Lateral malleolus

Medial malleolus

Tendon of tibialis anterior

Tarsals

Tendons of extensor digitorum longus

Metatarsals

Proximal phalanx

Middle phalanx

Distal phalanx

PLATE Forty-Seven

Surface anatomy of ankle and foot.

# NERVOUS SYSTEM I

## Basic Structure and Function

## CHAPTER OBJECTIVES

AFTER YOU HAVE STUDIED THIS CHAPTER, YOU SHOULD BE ABLE TO:

1. Explain the general functions of the nervous system.

2. Describe the general structure of a neuron.

3. Explain how neurons are classified.

4. Name four types of neuroglial cells and describe the functions of each.

5. Explain how an injured nerve fiber may regenerate.

6. Explain how a membrane becomes polarized.

7. Describe the events that lead to the conduction of a nerve impulse.

8. Explain how a nerve impulse is transmitted from one neuron to another.

9. Distinguish between excitatory and inhibitory postsynaptic potentials.

10. Explain two ways impulses are processed in neuronal pools.

### KEY TERMS

action potential (ak′shun po-ten′shal)
axon (ak′son)
central nervous system (sen′tral ner′vus sis′tem)
convergence (kon-ver′jens)
dendrite (den′drit)
divergence (di-ver′jens)
facilitation (fah-sil″ĭ-ta′shun)
myelin (mi′ĕ-lin)

neurilemma (nu″ri-lem′mah)
neuroglia (nu-rog′le-ah)
neuron (nu′ron)
neurotransmitter (nu″ro-trans-mit′er)
peripheral nervous system (pĕ-rif′er-al ner′vus sis′tem)
summation (sum-ma′shun)
synapse (sin′aps)
threshold (thresh′old)

## UNDERSTANDING WORDS

**astr-,** starlike: *astr*ocyte—star-shaped neuroglial cell.

**ax-,** axle: *ax*on:—cylindrical nerve fiber that carries impulses away from a neuron cell body.

**dendr-,** tree: *dendr*ite—branched nerve fiber that serves as the receptor surface of a neuron.

**ependym-,** tunic: *ependym*a—neuroglial cells that line spaces within the brain and spinal cord.

**-lemm,** rind or peel: neuri*lemm*a—sheath that surrounds the myelin of a nerve fiber.

**moto-,** moving: *mot*or neuron—neuron that stimulates a muscle to contract or a gland to release a secretion.

**multi-,** many: *multi*polar neuron—neuron with many fibers extending from the cell body.

**oligo-,** few: *oligo*dendrocyte—small neuroglial cell with few cellular processes.

**peri-,** all around: *peri*pheral nervous system—portion of the nervous system that consists of the nerves branching from the brain and spinal cord.

**saltator-,** a dancer: *saltator*y conduction—nerve impulse conduction in which the impulse seems to jump from node to node along the nerve fiber.

**sens-,** feeling: *sens*ory neuron—neuron that can be stimulated by a sensory receptor and conducts impulses into the brain or spinal cord.

**syn-,** together: *syn*apse—junction between two neurons.

**uni-,** one: *uni*polar—neuron with only one fiber extending from the cell body.

CHAPTER 10

UNIT THREE

When the five living U.S. presidents gathered at the funeral of former President Richard M. Nixon in April 1994, Gerald Ford, Jimmy Carter, George Bush, and Bill Clinton knew that all was not right with their compatriot Ronald Reagan. The former president was forgetful, responded inappropriately to questions, and, in the words of Gerald Ford, seemed "hollowed out." Reagan's memory continued to fade in and out, and 6 months later he penned a moving letter to the public confirming that he had Alzheimer's disease. The spells of forgetfulness and cloudy reasoning would come and go over the next several years, eventually increasing in frequency and severity to the point where he would not even recognize his wife Nancy.

Because Alzheimer's disease affects millions of families, finding a treatment or cure is a primary health care objective. Doing so requires pinpointing just what goes awry in the nervous system to trigger the cascade of destruction that ultimately strangles the brain in gummy protein plaques and tangles. Research is focusing on replenishing the messenger molecule that is deficient in the brain of a person with Alzheimer's disease, and supplying other biochemicals that stimulate nervous tissue to grow.

The ability to diagnose Alzheimer's disease, even before symptoms begin, will likely be possible long before we have treatments. This is already feasible for relatives of the 5% to 10% of affected individuals who have inherited the disorder. Inherited Alzheimer's disease tends to have an early onset, in one's forties or fifties. Three genes are known to directly cause the disease. One gene encodes amyloid, the gummy protein that forms intractable plaques in the brain. The other two genes, called presenilin 1 and presenilin 2, control amyloid production. A fourth gene, called apolipoprotein E, when abnormal increases the risk of developing Alzheimer's disease. Understanding how these genes malfunction to cause Alzheimer's disease will yield clues that can be used to help people with the noninherited form of the illness as well.

The ability to diagnose a disorder without a treatment may appear to be of dubious value. However, early knowledge may enable families to plan better for caring for ill relatives. It may also aid physicians in distinguishing Alzheimer's disease from other disorders. Finally, the ability to identify the condition early will enable researchers to test new treatments as the disease starts, when they may be more likely to slow or prevent symptoms.

## GENERAL FUNCTIONS OF THE NERVOUS SYSTEM

The nervous system is composed of some blood vessels and connective tissue but predominantly neural tissue, which consists of two cell types: nerve cells or **neurons,** and **neuroglia** (glial cells). Neurons, the structural and functional units of the nervous system, are specialized to react to physical and chemical changes in their surroundings. They transmit information, in the form of electrochemical changes called **nerve impulses,** along **nerve fibers,** to other neurons and to cells outside the nervous system (fig. 10.1). **Nerves** are bundles of nerve fibers. Neuroglia were once thought only to fill spaces and surround or support neurons. Today we know that they have many other functions, including nourishing neurons and perhaps even sending and receiving messages.

Perhaps the most important part of the nervous system at the cellular level is not a cell at all but the small spaces between neurons, called **synapses.** Much of the effort of the nervous system centers on sending and receiving electrochemical messages from neuron to neuron at synapses. The actual carriers of this information are biological messenger molecules called **neurotransmitters.**

The organs of the nervous system can be divided into two groups. One group, consisting of the brain and spinal cord, forms the **central nervous system** (CNS), and the other, composed of the nerves (peripheral nerves) that connect the central nervous system to other body parts, is called the **peripheral nervous system** (PNS)

(fig. 10.2). Together these systems provide three general functions—sensory, integrative, and motor.

Structures called **sensory receptors** at the ends of peripheral neurons provide the sensory function of the nervous system (see chapter 11). These receptors gather information by detecting changes inside and outside the body. They monitor external environmental factors such as light and sound intensities as well as the temperature, oxygen concentration, and other conditions of the body's internal environment.

Sensory receptors convert their information into nerve impulses, which are then transmitted over peripheral nerves to the central nervous system. There the signals are integrated—that is, they are brought together, creating sensations, adding to memory, or helping produce thoughts. Following integration, conscious or subconscious decisions are made and then acted upon by means of motor functions.

The motor functions of the nervous system employ peripheral neurons that carry impulses from the central nervous system to responsive structures called *effectors.* These effectors are outside the nervous system and include muscles that contract in response to nerve impulse stimulation, and glands that secrete when stimulated.

The nervous system can detect changes in the body, make decisions on the basis of the information received, and stimulate muscles or glands to respond. Typically, these responses counteract the effects of the changes, and in this way, the nervous system helps maintain homeostasis. Clinical Application 10.1 discusses a com-

**FIGURE 10.1**

Neurons are the structural and functional units of the nervous system (600×). Neuroglial cells surround the neuron, appearing as dark dots. Note the location of nerve fibers (dendrites and a single axon).

Labels: Dendrites, Cell body, Axon

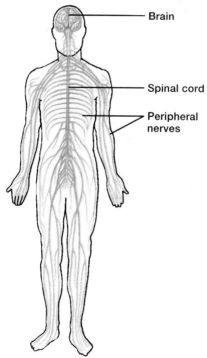

**FIGURE 10.2**

The nervous system consists of the brain and spinal cord (central nervous system) and peripheral nerves (peripheral nervous system).

Labels: Brain, Spinal cord, Peripheral nerves

mon medical problem attributed to the nervous system that may involve its blood supply as well as neurons.

## Neuron Structure

Neurons vary considerably in size and shape, but they share certain features. For example, every neuron has a **cell body** and nerve fibers (tubular membrane-bound processes containing cytoplasm) that conduct electrical signals to or from the cell body. Figure 10.3 shows some of the other structures common to neurons.

A neuron's cell body (soma or perikaryon) contains granular cytoplasm, mitochondria, lysosomes, a Golgi apparatus, and many microtubules. A network of fine threads called **neurofibrils** extends into the fibers and supports them. Scattered throughout the cytoplasm are many membranous packets of **chromatophilic substance** (Nissl bodies), which consist of rough endoplasmic reticulum. Cytoplasmic inclusions in neurons contain glycogen, lipids, or pigments such as melanin.

Near the center of the neuron cell body is a large, spherical nucleus with a conspicuous nucleolus. Mature neurons generally do not reproduce.

Neurons are notoriously difficult to culture in the laboratory, but cell biologists can do so by coaxing other cell types to express the genes that cause them to specialize as neurons. Researchers at one biotechnology company begin by culturing human olfactory epithelium, cells that provide our sense of smell. Under certain culture conditions, these cells gradually become immature neurons. By carefully manipulating the chemicals in the culture medium, researchers stimulate the cells to specialize into neurons characteristic of those in the central nervous system. This technique may one day provide replacement cell implants for treating such brain disorders as Parkinson's disease and Alzheimer's disease—if the cultured neurons can release the same neurotransmitters and form the same connections to other neurons as do the brain neurons that they would replace.

Two kinds of nerve fibers, **dendrites** and **axons,** extend from the cell bodies of most neurons. A neuron may have many dendrites but only one axon.

Dendrites are usually highly branched, providing receptive surfaces to which processes from other neurons communicate. (In some kinds of neurons, the cell body itself provides such a receptive surface.) Often the dendrites have tiny, thornlike spines (dendritic spines) on their surfaces, which are contact points for other neurons.

The axon, which often arises from a slight elevation of the cell body (axonal hillock), is a slender, cylindrical process with a nearly smooth surface and uniform

# Migraine

The signs of a migraine are unmistakable—a pounding head, waves of nausea, sometimes shimmering images in the peripheral visual field, and extreme sensitivity to light or sound. Inherited susceptibilities and environmental factors probably cause migraines. Environmental triggers include sudden exposure to bright light, eating a particular food (chocolate, red wine, nuts, and processed meats top the list), lack of sleep, stress, high altitude, stormy weather, and excessive caffeine or alcohol intake. Because 70% of the millions of people who suffer from migraine worldwide are women, hormonal influences may also be involved.

Although it is considered a headache, a migraine attack is actually a response to changes in the diameters of blood vessels in the face, head, and neck. Constriction followed by dilation of these vessels causes head pain (usually on one side), nausea and perhaps vomiting, and sensitivity to light.

## MIGRAINE TYPES

The two major variants of migraine are called "classic" and "common." Ten to 15% of sufferers experience classic migraine, which lasts 4 to 6 hours and begins with an "aura" of light in the peripheral vision. Common migraine usually lacks an aura, and may last for 3 to 4 days. A third, very rare type, familial hemiplegic migraine, may lead neurologists to finally understand precisely how all migraines occur.

Familial hemiplegic migraine runs in families. In addition to severe head pain, it paralyzes one side of the body for a few hours to a few days, and may cause loss of consciousness. This form of migraine results from a single amino acid change in a neuronal protein calcium channel in three brain regions (cerebellum, brain stem, and hippocampus). Interestingly, two other types of mutations in the responsible gene cause two different nervous system disorders. A shortened protein causes episodic ataxia, a movement disorder that makes a person periodically walk as if intoxicated. Extra copies of a particular amino acid in the protein cause spinocerebellar ataxia type 6, which causes chronic lack of coordination.

## TREATMENTS

Learning what goes awry in the rare, inherited form of migraine may shed light on how more common forms of the disease begin and progress. The researchers who discovered the gene for familial hemiplegic migraine are already using the clues in the calcium channel protein to develop a drug that can prevent migraine attacks.

Current drug treatments, although very effective, must be taken as soon as symptoms begin. The first drug to directly stop a migraine in its tracks was Imitrex (sumatriptan), which became available in 1996. Imitrex mimics the action of the neurotransmitter serotonin, levels of which fluctuate during an attack. The drug constricts blood vessels in the brain, decreasing blood flow to certain areas. Newer drugs, called triptans, more precisely target the neurons that are affected in a migraine attack—specifically, those in an area called the trigeminal nucleus. These neurons control cerebral blood vessel dilation. Imitrex can cause cardiac side effects because it also binds to serotonin receptors on blood vessels in the heart.

Drugs can help about 85% of migraine sufferers. With several new drugs in development, and a new understanding of how this painful condition develops, the future is bright for vanquishing migraine. ∎

diameter. It is specialized to conduct electrical signals called nerve impulses away from the cell body. Its cytoplasm includes many mitochondria, microtubules, and neurofibrils (ribosomes are found only in the cell body). An axon begins as a single fiber but may give off branches, called *collaterals.* Near its end, an axon may have many fine extensions, each with a specialized ending called an *axon terminal.* This terminates as a *synaptic knob* very close to the receptive surface of another cell, separated only by a space called the **synaptic cleft.**

In addition to conducting nerve impulses, an axon conveys biochemicals that are produced in the neuron cell body, which can be quite a task in these very long cells. This mechanism, called *axonal transport,* involves vesicles, mitochondria, ions, nutrients, and neurotransmitters that travel from the cell body to the ends of the axon.

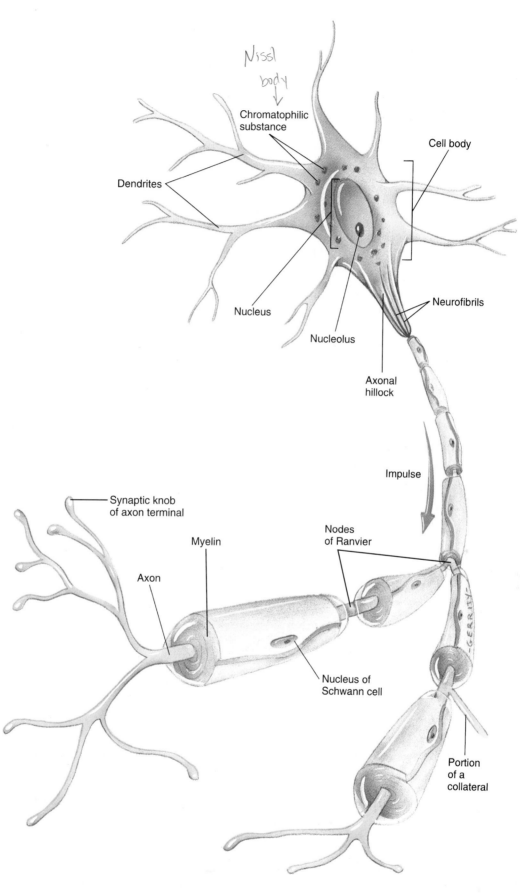

Nissl
body

Chromatophilic
substance

Cell body

Dendrites

Neurofibrils

Nucleus

Nucleolus

Axonal
hillock

Impulse

Synaptic knob
of axon terminal

Nodes
of Ranvier

Myelin

Axon

—GEARITY—

Nucleus of
Schwann cell

Portion
of a
collateral

**FIGURE 10.3**

A common neuron.

**FIGURE 10.4**

(a) The portion of a Schwann cell that winds tightly around an axon forms the myelin sheath. The cytoplasm and nucleus of the Schwann cell, remaining on the outside, form the neurilemmal sheath. (b) Light micrograph of a myelinated nerve fiber (longitudinal section) (300× micrograph enlarged to 650×). (c) An axon lying in a longitudinal groove of a Schwann cell lacks a myelin sheath.

Sheaths composed of many neuroglial cells called **Schwann cells** enclose larger axons of peripheral neurons. Schwann cells wind tightly around these axons, somewhat like a bandage wrapped around a finger. As a result, such fibers are coated with many layers of cell membranes that have little or no cytoplasm between them. These layers, called **myelin,** have a higher proportion of lipid than other surface membranes. Myelin forms a *myelin sheath* on the outside of an axon. In addition, the portions of the Schwann cells that contain most of the cytoplasm and the nuclei remain outside the myelin sheath and comprise a **neurilemma** or *neurilemmal sheath,* which surrounds the

myelin sheath (fig. 10.4). Narrow gaps in the myelin sheath between Schwann cells are called **nodes of Ranvier** (fig. 10.4).

Schwann cells also enclose, but do not wind around, the smallest axons of peripheral neurons. Consequently, these fibers lack myelin sheaths. Instead, the fiber or a group of fibers may lie in a longitudinal groove of Schwann cells.

Fibers that have myelin sheaths are called *myelinated* (medullated) nerve fibers, and those that lack these sheaths are *unmyelinated* nerve fibers (fig. 10.5). Groups of myelinated fibers appear white. Masses of such fibers

# Multiple Sclerosis

In 1964, at age 20, skier Jimmie Huega won the Olympic bronze medal in the slalom event. In 1967, his vision blurred and then his legs became slightly numb. Jimmie ignored these intermittent symptoms, and after a while they disappeared.

Three years later, the symptoms returned, and this time Jimmie sought medical help. On the basis of his symptoms, which affected more than one body part and occurred sporadically, physicians diagnosed multiple sclerosis (MS). Today, diagnosis also includes a magnetic resonance imaging scan (MRI), which can detect brain lesions.

Jimmie is still active. For many of the 300,000 people in the United States with MS, the progressive deterioration causes permanent paralysis. For some patients, taking a drug based on the immune system biochemical beta interferon can prevent flare-ups and mitigate symptoms.

In MS, the myelin coating in various sites through the brain and spinal cord forms hard scars, called scleroses, that block the underlying neurons from transmitting messages. Muscles that no longer receive input from motor neurons stop contracting, and eventually they atrophy. Symptoms reflect the specific neurons affected. Shortcircuiting in one part of the brain may affect fine coordination in one hand; if another brain part is affected, vision may be altered.

What might destroy myelin in MS? A virus may cause the body's immune system to attack the cells producing myelin. This would happen if viruses lay latent in nerve cells, then emerged years later bearing proteins also found on nerve cells. The immune system, interpreting the proteins as foreign, would attack the viruses as well as the neurons (an autoimmune response).

A virus is suspected for a few reasons: viral infections can strip neurons of their myelin sheaths; viral infections can cause repeated bouts of symptoms; and most compelling, MS is far more common in some geographical regions (the temperate zones of Europe, South America, and North America) than others, suggesting a pattern of infection. ■

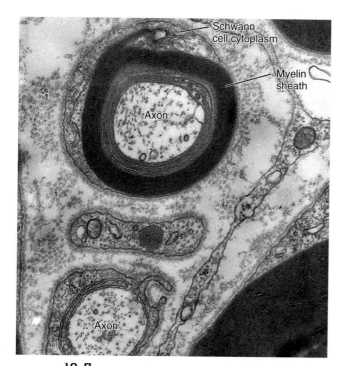

**FIGURE 10.5**

A transmission electron micrograph of myelinated and unmyelinated axons in cross section (30,000×).

impart color to the *white matter* in the brain and spinal cord, but here myelin is produced by another kind of neuroglial cell (oligodendrocyte) rather than by Schwann cells. Myelinated nerve fibers in the brain and spinal cord lack neurilemmal sheaths.

Unmyelinated nerve tissue appears gray. Thus, the *gray matter* within the brain and spinal cord contains an abundance of unmyelinated nerve fibers and neuron cell bodies. Clinical Application 10.2 discusses a disorder in which peripheral neurons lose their myelin.

**1** List the general functions of the nervous system.

**2** Describe a neuron.

**3** Explain how a fiber in the peripheral nervous system becomes myelinated.

## CLASSIFICATION OF NEURONS AND NEUROGLIA

Neurons vary in size and shape. They may differ in the length and size of their axons and dendrites, and in the number of processes by which they communicate with other neurons.

Neurons also vary in function. Some carry impulses into the brain or spinal cord; others carry impulses out from the brain or spinal cord; and still others conduct impulses from neuron to neuron within the brain or spinal cord.

## Classification of Neurons

On the basis of *structural differences,* neurons can be classified into three major groups, as figure 10.6 shows. Each type of neuron is specialized to send a nerve impulse in one direction, originating at a sensitive region called the **trigger zone.**

1. **Bipolar neurons.** The cell body of a bipolar neuron has only two nerve fibers, one arising from either end. Although these fibers are similar in structure, one is an axon and the other is a dendrite. Such neurons are found within specialized parts of the eyes, nose, and ears.

2. **Unipolar neurons.** Each unipolar neuron has a single nerve fiber extending from its cell body. A short distance from the cell body, this fiber divides into two branches which really function as a single axon: One branch (peripheral process) is associated with dendrites near a peripheral body part. The other branch (central process) enters the brain or spinal cord. The cell bodies of some unipolar neurons aggregate in specialized masses of nerve tissue called *ganglia,* which are located outside the brain and spinal cord.

3. **Multipolar neurons.** Multipolar neurons have many nerve fibers arising from their cell bodies. Only one fiber of each neuron is an axon; the rest are dendrites. Most neurons whose cell bodies lie within the brain or spinal cord are of this type. The neuron illustrated in figure 10.3 is multipolar.

Neurons can also be classified by *functional differences* into the following groups:

1. **Sensory neurons** (afferent neurons) carry nerve impulses from peripheral body parts into the brain or spinal cord. These neurons have specialized *receptor ends* at the tips of their dendrites, or they have dendrites that are near *receptor cells* in the skin or in certain sensory organs.

   Changes that occur inside or outside the body are likely to stimulate receptor ends or receptor cells, triggering sensory nerve impulses. The impulses travel on sensory neuron axons into the brain or spinal cord. Most sensory neurons are unipolar, although some are bipolar.

2. **Interneurons** (also called association or internuncial neurons) lie within the brain or spinal

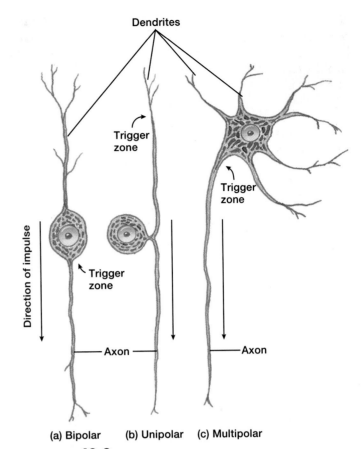

**FIGURE** 10.6

Structural types of neurons include (*a*) the bipolar neuron, (*b*) the unipolar neuron, and (*c*) the multipolar neuron.

cord. They are multipolar and form links between other neurons. Interneurons transmit impulses from one part of the brain or spinal cord to another. That is, they may direct incoming sensory impulses to appropriate regions for processing and interpreting. Other incoming impulses are transferred to motor neurons.

3. **Motor neurons** (efferent neurons) are multipolar and carry nerve impulses out of the brain or spinal cord to effectors—structures that respond, such as muscles or glands. For example, when motor impulses reach muscles, they contract; when motor impulses reach glands, they release secretions.

   Two specialized groups of motor neurons, accelerator and inhibitory neurons, innervate smooth and cardiac muscles. *Accelerator neurons* increase muscular activities, whereas *inhibitory neurons* decrease such actions.

   Motor neurons that control skeletal muscle are under voluntary (conscious) control. Other motor neurons, such as those that control glands and

TABLE 10.1 TYPES OF NEURONS

**A. Classified by Structure**

| Type | Structural Characteristics | Location |
|---|---|---|
| 1. Bipolar neuron | Cell body with a nerve fiber arising from each end | In specialized parts of the eyes, nose, and ears |
| 2. Unipolar neuron | Cell body with a single nerve fiber that divides into two branches | In ganglia outside the brain or spinal cord |
| 3. Multipolar neuron | Cell body with many nerve fibers, one of which is an axon | Most common type of neuron in the brain and spinal cord |

**B. Classified by Function**

| Type | Functional Characteristics | Structural Characteristics |
|---|---|---|
| 1. Sensory neuron | Conducts nerve impulses from receptors in peripheral body parts into the brain or spinal cord | Most unipolar; some bipolar |
| 2. Interneuron | Transmits nerve impulses between neurons within the brain and spinal cord | Multipolar |
| 3. Motor neuron | Conducts nerve impulses from the brain or spinal cord out to effectors—muscles or glands | Multipolar |

smooth and cardiac muscle, are largely under involuntary control.

Table 10.1 summarizes the classification of neurons.

## Classification of Neuroglia

Neuroglia (glial cells) were once thought to be mere by-standers to neural function, providing scaffolding, and controlling the sites at which neurons contact one another (figs. 10.7 and 10.8). These important cells have additional functions. In the embryo, neuroglia guide neurons to their positions and may stimulate them to specialize. Neuroglia also produce the growth factors that nourish neurons, and remove ions and neurotransmitters that accumulate between neurons, enabling them to continue transmitting information. Experiments in which neuroglia grow in culture suggest that some types of these cells may play a direct role in communicating with neurons.

Schwann cells are the neuroglia of the peripheral nervous system. The central nervous system contains the following types of neuroglia:

1. **Astrocytes.** As their name implies, astrocytes are star-shaped cells. They are commonly found between neurons and blood vessels, where they provide support and hold structures together by means of abundant cellular processes. Astrocytes aid metabolism of certain substances, such as glucose, and they may help regulate the concentrations of important ions, such as potassium ions, within the interstitial space of nervous tissue. Astrocytes also respond to injury of brain tissue and form a special type of scar tissue, which fills spaces and closes gaps in the CNS. These multifunctional cells may also have a nutritive function, regulating movement of substances from blood vessels to neurons and bathing nearby neurons in growth factors. Gap junctions link astrocytes to one another, forming protein-lined channels through which calcium ions travel, possibly stimulating neurons.

2. **Oligodendrocytes.** Oligodendrocytes resemble astrocytes but are smaller and have fewer processes. They commonly occur in rows along myelinated nerve fibers, and they form myelin in the brain and spinal cord.

    Unlike the Schwann cells of the peripheral nervous system, oligodendrocytes can send out a number of cellular processes, each of which forms a myelin sheath around a nearby axon. In this way, a single oligodendrocyte may provide myelin for many axons. However, these cells do not form neurilemmal sheaths.

3. **Microglia.** Microglial cells are small and have fewer processes than other types of neuroglia. These cells are scattered throughout the central nervous system, where they help support neurons and phagocytize bacterial cells and cellular debris. They usually increase in number whenever the brain or spinal cord is inflamed because of injury or disease.

**FIGURE** 10.7

Types of neuroglial cells in the central nervous system include (*a*) microglial cell, (*b*) oligodendrocyte, (*c*) astrocyte, and (*d*) ependymal cell.

**FIGURE** 10.8

A scanning electron micrograph of a neuron cell body and some of the neuroglial cells associated with it (1,000×).

(*Tissues and Organs: A Text-Atlas of Scanning Electron Microscopy*, by R. G. Kessel and R. H. Kardon. © 1979 W. H. Freeman and Company.)

4. **Ependyma.** Ependymal cells are cuboidal or columnar in shape and may have cilia. They form the inner lining of the *central canal* that extends downward through the spinal cord. Ependymal cells also form a one-cell-thick epithelial-like membrane that covers the inside of spaces within the brain called *ventricles* (see chapter 11). Throughout the ventricles, gap junctions join ependymal cells to one another. They form a porous layer through which substances diffuse freely between the interstitial fluid of the brain tissues and the fluid (cerebrospinal fluid) within the ventricles.

Ependymal cells also cover the specialized capillaries called *choroid plexuses* that are associated with the ventricles of the brain. Here they help regulate the composition of the cerebrospinal fluid.

Neuroglia form more than half of the volume of the brain. Table 10.2 summarizes characteristics of neuroglial cells.

TABLE 10.2

# TYPES OF NEUROGLIAL CELLS OF THE CENTRAL NERVOUS SYSTEM

| Type | Characteristics | Functions |
|------|-----------------|-----------|
| Astrocytes | Star-shaped cells found between neurons and blood vessels | Structural support, formation of scar tissue, transport of substances between blood vessels and neurons, communicate with one another and with neurons, mop up excess ions and neurotransmitters |
| Oligodendrocytes | Shaped like astrocytes, but with fewer cellular processes, and arranged in rows along nerve fibers | Form myelin sheaths within the brain and spinal cord; produce nerve growth factors |
| Microglia | Small cells with few cellular processes and found throughout the CNS | Structural support and phagocytosis (immune protection) |
| Ependyma | Cuboidal and columnar cells in the inner lining of the ventricles of the brain and the central canal of the spinal cord | Form a porous layer through which substances diffuse between the interstitial fluid of the brain and spinal cord, and the cerebrospinal fluid |

Abnormal neuroglia are associated with certain disorders. Most brain tumors, for example, consist of neuroglia that multiply too often. Gene therapy is being tested to treat such tumors. Researchers add genes that instruct the cancerous glia to bear cell surface proteins that attract the immune system or render them more sensitive to cancer-fighting drugs.

Another experimental medical approach utilizing neuroglia is to construct implants consisting of certain types of neuroglia. Once implanted, these cells would secrete substances that may

- replace neurochemicals whose absence causes degenerative diseases of the nervous system, such as Parkinson's disease, Alzheimer's disease, multiple sclerosis, and amyotrophic lateral sclerosis,
- repair damaged spinal cords,
- halt damage to delicate nervous tissues caused by AIDS or cancer chemotherapy.

Myelin begins to form on nerve fibers during the fourteenth week of prenatal development. By the time of birth, many nerve fibers are not completely myelinated. All myelinated fibers have begun to develop sheaths by the time a child starts to walk, and myelination continues into adolescence.

Excess myelin seriously impairs nervous system functioning. In Tay-Sachs disease, an inherited defect in a lysosomal enzyme causes myelin to accumulate, burying neurons in fat. The affected child begins to show symptoms by 6 months of age, gradually losing sight, hearing, and muscle function until death occurs by age 4. Thanks to genetic screening among people of eastern European descent who are most likely to carry this gene, fewer than ten children are born in the United States with this condition each year.

## Regeneration of Nerve Fibers

Injury to the cell body usually kills the neuron, and because mature neurons do not reproduce, it is not replaced. However, a damaged axon may regenerate. For example, if injury or disease separates an axon in a peripheral nerve from its cell body, the distal portion of the axon and its myelin sheath deteriorate within a few weeks. Macrophages remove the fragments of myelin and other cellular debris. The proximal end of the injured axon develops sprouts shortly after the injury. Influenced by nerve growth factors that nearby glia secrete, one of these sprouts may grow into a tube formed by remaining basement membrane and connective tissue. At the same time, any remaining Schwann cells proliferate along the length of the degenerating fiber, and form new myelin around the growing axon.

Growth of a regenerating fiber is slow (3 to 4 millimeters per day), but eventually the new fiber may reestablish the former connection (fig. 10.9). Nerve growth factors, secreted by glial cells, may help direct the growing fiber. However, the regenerating fiber may still end up in the wrong place, so full function often does not return.

If an axon of a nerve fiber within the central nervous system is separated from its cell body, the distal portion of the axon will degenerate, but more slowly than a separated axon in the peripheral nervous system. However, axons within the central nervous system lack a neurilemma, and the myelin-producing oligodendrocytes fail to proliferate following an injury. Consequently, if the proximal end of a damaged axon begins to grow, there is no tube of sheath cells to guide it. Therefore regeneration is unlikely.

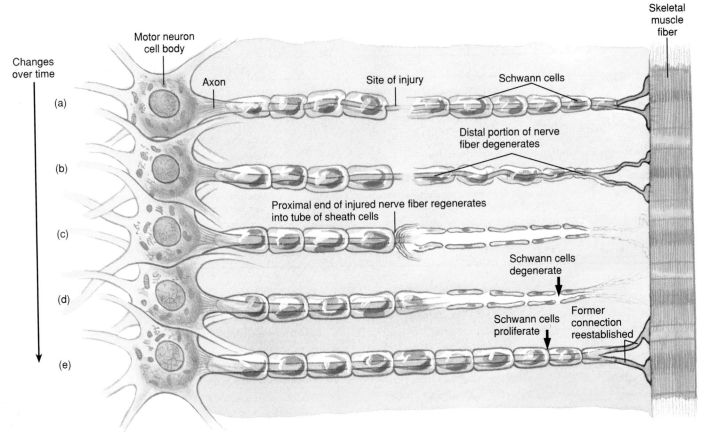

**FIGURE 10.9**

If a myelinated axon is injured, the following events may occur over several weeks to months: (a) The proximal portion of the fiber may survive, but (b) the portion distal to the injury degenerates. (c and d) In time, the proximal portion may develop extensions that grow into the tube of basement membrane and connective tissue cells that the fiber previously occupied, and (e) possibly reestablish the former connection. Nerve growth factors that neuroglial cells secrete assist in the regeneration process.

If a peripheral nerve is severed, it is very important that the two cut ends be connected as soon as possible so that the regenerating sprouts of the nerve fibers can more easily reach the tubes formed by the basement membranes and connective tissues on the distal side of the gap. When the gap exceeds 3 millimeters, the regenerating fibers may form a tangled mass called a *neuroma.* It is composed of sensory nerve fibers, and is painfully sensitive to pressure. Neuromas sometimes complicate a patient's recovery following limb amputation.

**1** What is a neuroglial cell?

**2** Name and describe four types of neuroglial cells.

**3** What are some functions of neuroglia?

**4** Explain how an injured peripheral nerve fiber might regenerate.

**5** Explain why functionally significant regeneration is unlikely in the central nervous system.

## CELL MEMBRANE POTENTIAL

A cell membrane is usually electrically charged, or *polarized,* so that the inside is negatively charged with respect to the outside. This polarization is due to an unequal distribution of positive and negative ions on either side of the membrane, and it is particularly important in the conduction of muscle and nerve impulses.

### Distribution of Ions

Potassium ions are the major intracellular positive ion (cation), and sodium ions are the major extracellular cation. The distribution of ions inside and outside of cell membranes is determined in part by channels in those membranes, as discussed in chapter 3. These channels, formed by membrane proteins, can be quite selective; that is, a particular channel may allow one kind of ion to pass through and exclude all other ions of different size and charge. Furthermore, some channels are always open, whereas others may be either open or closed, somewhat like a gate. As we shall see, both chemical and

Protein    Gatelike mechanism

Cell membrane

(a) Channel closed                    (b) Channel open

**FIGURE 10.10**

A gatelike mechanism can close (a) or open (b) some of the channels in cell membranes through which ions pass.

electrical factors can affect the opening and closing of these gatelike channels (fig. 10.10).

A resting nerve cell is one that is not being stimulated to send a nerve impulse. As a consequence of the selectivity of specific ion channels, potassium ions tend to pass through resting cell membranes much more easily than sodium ions. The relative ease with which potassium ions diffuse through membranes makes them a major contributor to membrane polarization. Calcium ions are less able to cross the resting cell membrane than either sodium ions or potassium ions, and have a special role in nerve function, described later.

## Resting Potential

The active transport of sodium and potassium ions maintains a greater concentration of sodium ions ($Na^+$) outside cells and a greater concentration of potassium ions ($K^+$) inside cells. The cytoplasm of these cells has many negatively charged ions called anions, which include phosphate ($PO_4^{-2}$), sulfate ($SO_4^{-2}$), and proteins, that cannot diffuse through the cell membranes (fig. 10.11a).

Sodium and potassium ions follow the laws of diffusion stated earlier and show a net movement from high concentration to low concentration across a membrane as permeabilities permit. The resting cell membrane is more permeable to potassium ions than to sodium ions. If we consider an imaginary cell, before a membrane potential has been established, we would expect potassium ions to diffuse out of the cell more rapidly than sodium could diffuse in. This means that every millisecond, more positive charges leave the cell than enter it (fig. 10.11a). As a result, the outside of the membrane gains a slight surplus of positive charges, the inside is left with a surplus of impermeable negative charges, and a membrane potential is established (fig. 10.11b). In this new situation, the negative condition inside tends to pull sodium ions into the cell, but resists potassium ions diffusing out. All this

time the cell continues to expend energy to actively transport sodium and potassium ions in the opposite directions, thus maintaining the concentration gradients for those ions responsible for their diffusion in the first place.

The difference in electrical charge between two points is called the *potential difference* because it represents stored electrical energy that can be used to do work. It is measured in units called *volts*. In the case of a resting nerve cell, the potential difference between the inside and the outside of the cell membrane is about −70 millivolts (negative because of the excess negative charges on the inside of the cell membrane). This separation of charge is called the **resting potential,** and the work it may do is to send the nerve impulse or **action potential.** To understand how neurons initiate action potentials, we must first understand how nerve cells respond to signals called stimuli.

## Local Potential Changes

Nerve cells are excitable; that is, they can respond to changes in their surroundings. Some nerve cells, for example, detect changes in temperature, light, or pressure outside the body, whereas others respond to signals from inside the body, often from other neurons. In either case, such changes or stimuli usually affect the resting potential in a particular local region of the cell membrane. If, in response to a stimulus, the membrane potential becomes more negative than the resting potential, the membrane is *hyperpolarized;* if the resting potential becomes less negative (more positive) than the resting potential, the membrane is *depolarized.*

Up to a point, the changes in the local potential of a membrane are *graded.* This means the amount of change in potential is directly proportional to the intensity of the stimulation. If a sufficiently strong depolarization occurs, a level called the *threshold potential* is reached.

**FIGURE 10.11**

Development of the resting membrane potential. (*a*) Active transport creates a concentration gradient across the cell membrane for sodium ions (Na$^+$) and potassium ions (K$^+$). K$^+$ diffuses out of the cell rather slowly but nonetheless faster than Na$^+$ can diffuse in. (*b*) This unequal diffusion results in a net loss of positive charge and a resultant excess of negative charge inside the membrane.

Reaching threshold opens certain ion channels, described in the next section, that are sensitive to changes in membrane potential. In many cases, a single stimulus is not sufficient to reach threshold potential.

If another stimulus of the same type arrives before the effect of the first one subsides, the change in local membrane potential is greater. This additive phenomenon is called *summation,* and as a result of summated potentials, the threshold potential may be reached. Thus, many subthreshold potential changes may combine to reach threshold. At threshold, an action potential is produced in a nerve fiber.

**FIGURE 10.12**

At rest (*a*), the membrane potential is about −70 millivolts. When the membrane reaches threshold (*b*), voltage-sensitive sodium channels open, some Na⁺ diffuses inward, and the membrane is depolarized. Soon afterward (*c*), voltage-sensitive potassium channels open, K⁺ diffuses out, and the membrane is repolarized. (Negative ions not shown.)

## Action Potentials

Some of the gatelike ion channels in the cell membrane are sensitive to membrane voltage. At the resting membrane potential they remain closed, but when threshold is reached, they open for an instant. The so-called trigger zone of a neuron contains a large number of voltage-sensitive sodium channels. As sodium channels in the trigger zone open, sodium ions diffuse in. This movement is aided by attraction of the sodium ions to the negative electrical condition on the inside of the membrane.

As the sodium ions rush inward, the membrane potential changes from its resting value (fig. 10.12*a*) and momentarily becomes positive on the inside (this is still considered depolarization). At the peak of the action potential, membrane potential may reach +30mV (fig. 10.12*b*). At almost the same time, other, slower, voltage-sensitive potassium channels open for a moment and allow some of the potassium ions to pass through. As potassium ions diffuse outward, the inside of the membrane becomes negatively charged once more. The membrane is therefore **repolarized** (note that it hyperpolarizes for an instant). Thus, the resting potential is quickly reestablished, and it remains in this resting state until it is stimulated again (fig. 10.12*c*).

This rapid sequence of changes, involving depolarization and repolarization, takes about 1/1,000th second or less—this is the action potential (fig. 10.13). Actually, only a small proportion of the sodium and potassium ions present move through the membrane during an action potential, so that many action potentials could occur before the concentrations of sodium ions and potassium ions on either side of the membrane would change significantly.

The active transport mechanism in the membrane works to maintain the original concentrations of sodium and potassium ions.

An action potential at the trigger zone causes an electric current to flow a short distance down the fiber, which stimulates the adjacent membrane to its threshold level, triggering another action potential. The second action potential causes another electric current to flow farther down the fiber. This sequence of events results in a series of action potentials occurring sequentially down the fiber, traveling to the end of the fiber without decreasing in amplitude, even if branches occur. The propagation of action potentials along a fiber is the **nerve impulse** (fig. 10.14).

A nerve impulse is similar to the muscle impulse mentioned in chapter 9. In the muscle fiber, stimulation at the motor end plate triggers an impulse to travel over the surface of the fiber and down into its transverse tubules. See table 10.3 for a summary of the events leading to the conduction of a nerve impulse.

(a)

(b)

(c)

**FIGURE 10.14** [icon]
(*a*) An action potential in one region stimulates the adjacent region, and (*b* and *c*) a wave of action potentials (a nerve impulse) moves along the fiber. (Negative ions not shown.)

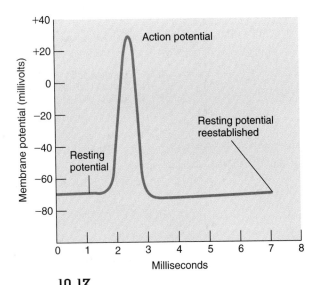

**FIGURE 10.13**
An oscilloscope records an action potential.

| TABLE 10.3 | EVENTS LEADING TO NERVE IMPULSE CONDUCTION |
|---|---|

1. Nerve fiber membrane maintains resting potential by diffusion of Na⁺ and K⁺ down their concentration gradients as the cell pumps them up the gradients.
2. Neurons receive stimulation, causing local potentials, which may sum to reach threshold.
3. Sodium channels in a local region of the membrane open.
4. Sodium ions diffuse inward, depolarizing the membrane.
5. Potassium channels in the membrane open.
6. Potassium ions diffuse outward, repolarizing the membrane.
7. The resulting action potential causes an electric current that stimulates adjacent portions of the membrane.
8. Series of action potentials occurs sequentially along the length of the nerve fiber as a nerve impulse.

## Refractory Period

For a very short time following passage of a nerve impulse, a threshold stimulus will not trigger another impulse on a nerve fiber. This brief period, called the **refractory period,** has two parts. During the *absolute refractory period,* which lasts about 1/2,500 of a second, the fiber's membrane is changing in sodium permeability and cannot be stimulated. This is followed by a *relative refractory period,* during which the membrane is reestablishing its resting potential. While the membrane is in the relative refractory period, even though repolarization is incomplete, a threshold stimulus of high intensity may trigger an impulse.

As time passes, the intensity of stimulation required to trigger an impulse decreases until the fiber's original excitability is restored. This return to the resting state usually takes from 10 to 30 milliseconds.

Because of the refractory period, a nerve fiber cannot be continuously stimulated. The refractory period limits the rate of nerve impulse conduction. The time between impulses cannot be less than the absolute refractory period, and the maximum rate of nerve impulse conduction is therefore about one impulse per millisecond.

## All-or-None Response

Like muscle fiber contraction, nerve impulse conduction is an all-or-none response. In other words, if a nerve fiber responds at all, it responds completely. Thus, a nerve impulse is conducted whenever a stimulus of threshold intensity or above is applied to a nerve fiber, and all impulses carried on that fiber will be of the same strength. A greater intensity of stimulation produces not a stronger impulse but rather more impulses per second.

## Impulse Conduction

An unmyelinated nerve fiber conducts an impulse over its entire surface. A myelinated fiber functions differently. Myelin contains a high proportion of lipid that excludes water and water-soluble substances. Thus, it serves as an insulator and prevents almost all flow of ions through the membrane that is enclosed in myelin.

It might seem that the myelin sheath would prevent conduction of a nerve impulse, and this would be true if the sheath were continuous. However, nodes of Ranvier between Schwann cells or oligodendrocytes interrupt the sheath (see fig. 10.3). At these nodes, the nerve fiber membrane contains channels for sodium and potassium ions that open during a threshold depolarization.

When a myelinated nerve fiber is stimulated to threshold, an action potential occurs at the trigger zone. This causes an electric current to flow away from the trigger zone through the cytoplasm of the axon. As this local current reaches the first node, it stimulates the membrane to its threshold level, and an action potential occurs there, sending an electric current to the next node. Consequently, a nerve impulse traveling along a myelinated nerve fiber involves action potentials occurring only at the nodes. Because the action potentials appear to jump from node to node, this type of impulse conduction is called **saltatory conduction.** Conduction on myelinated axons is many times faster than conduction on unmyelinated axons (fig. 10.15).

The speed of nerve impulse conduction is also determined by the diameter of the fiber—the greater the diameter, the faster the impulse. For example, an impulse on a thick, myelinated nerve fiber, such as a motor fiber associated with a skeletal muscle, might travel 120 meters

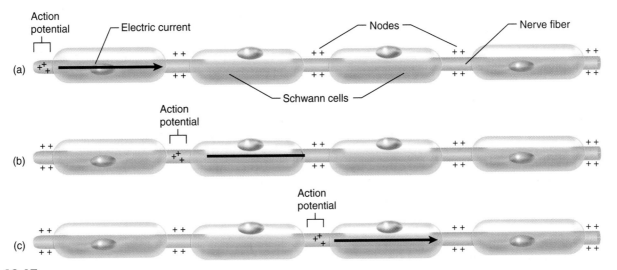

**FIGURE 10.15**

On a myelinated fiber, a nerve impulse appears to jump from node to node.

# Factors Affecting Impulse Conduction

A number of substances alter nerve fiber membrane permeability to ions. For example, calcium ions are required to close sodium channels in nerve fiber membranes during an action potential. Consequently, if calcium is deficient, sodium channels remain open, and sodium ions diffuse through the membrane again and again so that impulses are transmitted repeatedly. If these spontaneous impulses travel along nerve fibers to skeletal muscle fibers, the muscles continuously spasm (tetanus or tetany). This can occur in women during pregnancy as the developing fetus uses maternal calcium. Tetanic contraction may also occur when the diet lacks calcium or vitamin D, or when prolonged diarrhea depletes the body of calcium.

A small increase in the concentration of extracellular potassium ions causes the resting potential of nerve fibers to be less negative (partially depolarized). As a result, the threshold potential is reached with a less intense stimulus than usual. The affected fibers are very excitable, and the person may experience convulsions.

If the extracellular potassium ion concentration is greatly decreased, the resting potentials of the nerve fibers may become so negative that action potentials cannot occur. In this case, impulses are not triggered, and muscles become paralyzed.

Certain anesthetic drugs, such as procaine, decrease membrane permeability to sodium ions. In the tissue fluids surrounding a nerve fiber, these drugs prevent impulses from passing through the affected region. Consequently, the drugs keep impulses from reaching the brain, preventing perception of touch and pain. ∎

---

per second, whereas an impulse on a thin, unmyelinated nerve fiber, such as a sensory fiber associated with the skin, might move only 0.5 meter per second. Clinical Application 10.3 discusses factors that influence nerve impulse conduction.

1. Summarize how a resting potential is achieved.

2. Explain how a polarized nerve fiber responds to stimulation.

3. List the major events that occur during an action potential.

4. Define *refractory period*.

5. Explain how impulse conduction differs in myelinated and unmyelinated nerve fibers.

## THE SYNAPSE

Nerve impulses may pass from neuron to neuron at connections along complex nerve pathways called synapses (fig. 10.16). A **presynaptic neuron** brings the impulse to the synapse, and as a result stimulates or inhibits a **postsynaptic neuron.** A narrow space or synaptic cleft (synaptic gap) separates the two neurons (fig. 10.17). These are functional rather than physical connections. The process by which the impulse in the presynaptic neuron signals the postsynaptic neuron is called **synaptic transmission.**

A nerve impulse travels along the axon to the axon terminal. Axons usually have several rounded *synaptic knobs* (buttons) at their terminals, which dendrites lack. These knobs contain numerous membranous sacs, called synaptic vesicles. When a nerve impulse reaches a synaptic knob, voltage-sensitive calcium channels open, resulting in an inward diffusion of calcium from the extracellular fluid. The increased calcium concentration inside the cell initiates a series of events that causes the synaptic vesicles to fuse with the cell membrane, releasing their neurotransmitter by exocytosis.

## Synaptic Potentials

Released neurotransmitter molecules diffuse across the synaptic cleft and react with specific receptor molecules in or on the postsynaptic neuron membrane. Effects of neurotransmitters may vary. Some open ion channels, and others close them. Because these ion channels

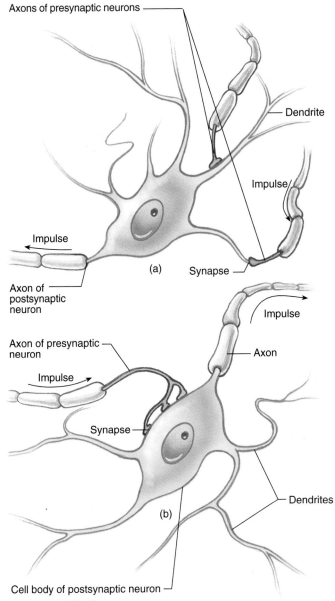

Axons of presynaptic neurons

Dendrite

Impulse

Impulse

(a)          Synapse

Axon of postsynaptic neuron

Axon of presynaptic neuron

Impulse

Impulse

Axon

Synapse

Dendrites

(b)

Cell body of postsynaptic neuron

**FIGURE 10.16**

For an impulse to continue from one neuron to another, it must cross the synaptic cleft at a synapse. A synapse usually occurs (a) between an axon and a dendrite, or (b) between an axon and a cell body.

respond to neurotransmitter molecules, they are called *chemically sensitive,* in contrast to the voltage sensitive ion channels involved in action potentials. Changes in chemically sensitive ion channels create local potentials, called **synaptic potentials,** which enable one neuron to influence another.

Synaptic potentials are graded and can depolarize or hyperpolarize the receiving cell membrane. For example, if a neurotransmitter binds to a postsynaptic receptor and opens sodium ion channels, the ions diffuse inward, depolarizing the membrane, possibly triggering an action

potential. This type of membrane change is called an **excitatory postsynaptic potential** (EPSP), and it lasts for about 15 milliseconds.

If a different neurotransmitter binds other receptors and increases membrane permeability to potassium ions, these ions diffuse outward, hyperpolarizing the membrane. Since an action potential is now less likely to occur, this change is called an **inhibitory postsynaptic potential** (IPSP).

Within the brain and spinal cord, each neuron may receive the synaptic knobs of a thousand or more nerve fibers on its dendrites and cell body. Furthermore, at any moment, some of the postsynaptic potentials are excitatory on a particular neuron, while others are inhibitory (fig. 10.18).

The integrated sum of the EPSPs and IPSPs determines whether an action potential results. If the net effect is more excitatory than inhibitory, threshold may be reached, and an action potential triggered. Conversely, if the net effect is inhibitory, no impulse will be transmitted.

This summation of the excitatory and inhibitory effects of the postsynaptic potentials commonly takes place at the trigger zone, usually in a proximal region of the axon, but found in some dendrites as well. This region has an especially low threshold for triggering an action potential; thus, it serves as a decision-making part of the neuron.

**1** Describe a synapse.

**2** Explain the function of a neurotransmitter.

**3** Distinguish between EPSP and IPSP.

**4** Describe the net effects of EPSPs and IPSPs.

## Neurotransmitters

The nervous system produces at least thirty different kinds of neurotransmitters. Some neurons release only one type; others produce two or three kinds. Neurotransmitters include *acetylcholine,* which stimulates skeletal muscle contractions (see chapter 9); a group of compounds called *monoamines* (such as epinephrine, norepinephrine, dopamine, and serotonin), which are formed by modifying amino acid molecules; a group of unmodified *amino acids* (such as glycine, glutamic acid, aspartic acid, and gamma-aminobutyric acid—GABA); and a large group of *peptides* (such as enkephalins and substance P), which are short chains of amino acids.

Most types of neurotransmitters are synthesized in the cytoplasm of the synaptic knobs and stored in synaptic vesicles. When an action potential passes along the membrane of a synaptic knob, it increases the membrane's

(a)

(b)

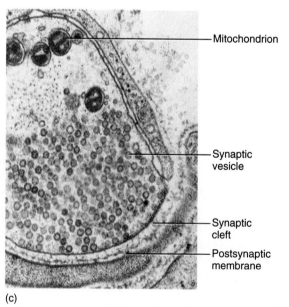

(c)

**FIGURE 10.17**

(*a*) When a nerve impulse reaches the synaptic knob at the end of an axon, (*b*) synaptic vesicles release a neurotransmitter substance that diffuses across the synaptic cleft. (*c*) A transmission electron micrograph of a synaptic knob filled with synaptic vesicles (37,500×).

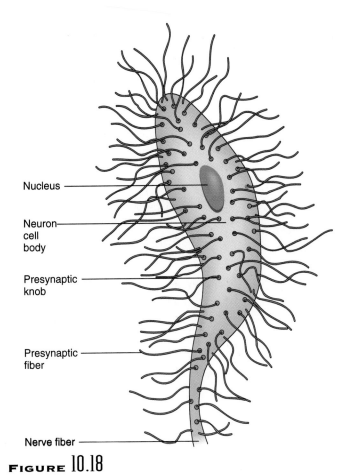

Nucleus

Neuron cell body

Presynaptic knob

Presynaptic fiber

Nerve fiber

**FIGURE** 10.18

The synaptic knobs of many axons may communicate with the cell body of a neuron.

permeability to calcium ions by opening its calcium ion channels. Calcium ions diffuse inward, and in response, some of the synaptic vesicles fuse with the presynaptic membrane and release their contents by exocytosis into the synaptic cleft. The more calcium that enters the synaptic knob, the more vesicles, and thus the more neurotransmitter released. Clinical Application 10.4 discusses myasthenia gravis, an autoimmune disorder that affects neuromuscular junctions. Tables 10.4 and 10.5 list disorders and drugs that alter neurotransmitter levels.

After a vesicle releases its neurotransmitter, it becomes part of the cell membrane. It eventually returns to the cytoplasm by endocytosis, and may then provide material to form new secretory vesicles. Table 10.6 summarizes this activity.

In order to keep signal duration short, enzymes in synaptic clefts and on postsynaptic membranes rapidly decompose some neurotransmitters. The enzyme acetylcholinesterase, for example, decomposes acetylcholine. This enzyme is present on postsynaptic membranes. Other neurotransmitters are transported back into the synaptic knob of the presynaptic neuron or into nearby neurons or neuroglial cells, a process called *reuptake.* The enzyme monoamine oxidase inactivates the monoamine neurotransmitters epinephrine and norepinephrine after reuptake. This enzyme is found in mitochondria in the synaptic knob. Destruction or removal of neurotransmitter prevents continuous stimulation of the postsynaptic neuron.

| TABLE 10.4 | DISORDERS ASSOCIATED WITH NEUROTRANSMITTER IMBALANCES | |
|---|---|---|
| **Condition** | **Symptoms** | **Imbalance of Neurotransmitter in Brain** |
| Alzheimer's disease | Memory loss, depression, disorientation, dementia, hallucinations, death | Deficient acetylcholine |
| Clinical depression | Debilitating, inexplicable sadness | Deficient norepinephrine and/or serotonin |
| Epilepsy | Seizures, loss of consciousness | Excess GABA leads to excess norepinephrine and dopamine |
| Huntington disease | Personality changes, loss of coordination, uncontrollable dancelike movements, death | Deficient GABA |
| Hypersomnia | Excessive sleeping | Excess serotonin |
| Insomnia | Inability to sleep | Deficient serotonin |
| Mania | Elation, irritability, overtalkativeness, increased movements | Excess norepinephrine |
| Myasthenia gravis | Progressive muscular weakness | Deficient acetylcholine receptors at neuromuscular junctions |
| Parkinson's disease | Tremors of hands, slowed movements, muscle rigidity | Deficient dopamine |
| Schizophrenia | Inappropriate emotional responses, hallucinations | Deficient GABA leads to excess dopamine |
| Sudden infant death syndrome ("crib death") | Baby stops breathing, dies if unassisted | Excess dopamine |
| Tardive dyskinesia | Uncontrollable movements of facial muscles | Deficient dopamine |

# Myasthenia Gravis

In an autoimmune disorder, the immune system attacks part of the body. In myasthenia gravis (MG), that part is the nervous system, particularly receptors for acetylcholine on muscle cells at neuromuscular junctions, where neuron meets muscle cell. People with MG have one-third the normal number of acetylcholine receptors at these junctions. On a whole-body level, this causes weak and easily fatigued muscles.

MG affects hundreds of thousands of people worldwide, usually women, beginning in their twenties or thirties and men in their sixties and seventies. The specific symptoms depend upon the site of attack. For 85% of patients, the disease causes generalized muscle weakness. Many people develop a characteristic flat smile and nasal voice and have difficulty chewing and swallowing due to affected facial and neck muscles. Many have limb weakness. About 15% of patients experience the illness only in the muscles surrounding their eyes. The disease reaches crisis level when respiratory muscles are affected, requiring a ventilator to support breathing. MG does not affect sensation or reflexes.

Until 1958, MG was a serious threat to health, with a third of patients dying, a third worsening, and only a third maintaining or improving their condition. Today, most people with MG can live near-normal lives, thanks to a combination of the following treatments:

- Drugs that inhibit acetylcholinesterase, which boosts availability of acetylcholine.
- Removing the thymus gland, which oversees much of the immune response.
- Immunosuppressant drugs.
- Intravenous antibodies to bind and inactivate the ones causing the damage.
- Plasma exchange, which rapidly removes the damaging antibodies from the circulation. This helps people in crisis. ∎

## TABLE 10.5 DRUGS THAT ALTER NEUROTRANSMITTER LEVELS

| Drug | Neurotransmitter Affected* | Mechanism of Action | Effect |
|------|---------------------------|---------------------|--------|
| Tryptophan | Serotonin | Stimulates neurotransmitter synthesis | Sleepiness |
| Reserpine | Norepinephrine | Decreases packaging neurotransmitter into vesicles | Decreases blood pressure |
| Curare | Acetylcholine | Decreases neurotransmitter in synaptic cleft | Muscle paralysis |
| Valium | GABA | Enhances receptor binding | Decreases anxiety |
| Nicotine | Acetylcholine | Stimulates synthesis of enzyme that degrades neurotransmitter | Increases alertness |
| Cocaine | Norepinephrine | Blocks reuptake | Euphoria |
| Tricyclic antidepressants | Norepinephrine | Blocks reuptake | Mood elevation |
| Monoamine oxidase inhibitors | Norepinephrine | Blocks enzymatic degradation of neurotransmitter in presynaptic cell | Mood elevation |
| Prozac and related drugs | Serotonin | Blocks reuptake | Mood elevation |

*Others may be affected as well.

**TABLE 10.6** EVENTS LEADING TO NEUROTRANSMITTER RELEASE

1. Action potential passes along a nerve fiber and over the surface of its synaptic knob.
2. Synaptic knob membrane becomes more permeable to calcium ions, and they diffuse inward.
3. In the presence of calcium ions, synaptic vesicles fuse to synaptic knob membrane.
4. Synaptic vesicles release their neurotransmitter by exocytosis into the synaptic cleft.
5. Synaptic vesicles become part of the membrane.
6. The added membrane provides material for endocytotic vesicles.

## Neuropeptides

Neurons in the brain or spinal cord synthesize **neuropeptides.** These peptides act as neurotransmitters or as *neuromodulators*—substances that alter a neuron's response to a neurotransmitter or block the release of a neurotransmitter.

Among the neuropeptides are the *enkephalins* that occur throughout the brain and spinal cord. Each enkephalin molecule is a chain of five amino acids. Synthesis of enkephalins increases during periods of painful stress, and they bind to the same receptors in the brain (opiate receptors) as the narcotic morphine. Enkephalins relieve pain sensations, and probably have other functions as well.

Another morphinelike peptide, called *beta endorphin,* is found in the brain and cerebrospinal fluid. It acts longer than enkephalins and is a much more potent pain reliever (Clinical Application 10.5).

Still another neuropeptide, which consists of eleven amino acids and is widely distributed throughout the nervous system, is *substance P.* It functions as a neurotransmitter (or perhaps as a neuromodulator) in the neurons that transmit pain impulses into the spinal cord and on to the brain. Enkephalins and endorphins may relieve pain by inhibiting the release of substance P from pain-transmitting neurons.

## IMPULSE PROCESSING

The way the nervous system processes nerve impulses and acts upon them reflects, in part, the organization of neurons and their nerve fibers within the brain and spinal cord.

## Neuronal Pools

The neurons within the central nervous system are organized into groups called **neuronal pools** that have varying numbers of cells. Each neuronal pool receives impulses from input (afferent) nerve fibers. These impulses are processed according to the special characteristics of the pool, and any resulting impulses are conducted away on output (efferent) fibers.

Each input fiber divides many times as it enters, and its branches spread over a certain region of the neuronal pool. The branches give off smaller branches, and their terminals form hundreds of synapses with the dendrites and cell bodies of the neurons in the pool.

## Facilitation

As a result of incoming impulses and neurotransmitter release, a particular neuron of the neuronal pool may receive excitatory and inhibitory stimulation. If the net effect of the stimulation is excitatory, threshold may be reached, and an outgoing impulse is triggered. If the net effect is excitatory but subthreshold, an impulse is not triggered, but the neuron is more excitable to incoming stimulation, a state called **facilitation.**

## Convergence

Any single neuron in a neuronal pool may receive impulses from two or more incoming fibers. Fibers originating from different parts of the nervous system leading to the same neuron exhibit **convergence.**

Convergence makes it possible for a neuron to summate impulses from different sources. For example, if a neuron receives subthreshold stimulation from one input fiber, it may reach threshold if it receives additional stimulation from a second input fiber. Thus, an output impulse triggered from this neuron reflects summation of impulses from two different sources. Such an output impulse may travel to a particular effector and evoke a response (fig. 10.19*a*).

Incoming impulses often represent information from various sensory receptors that detect changes. Convergence allows the nervous system to collect, process, and respond to information.

# Opiates in the Human Body

Opiate drugs, such as morphine, heroin, codeine, and opium, are potent painkillers derived from the poppy plant. These drugs alter pain perception, making it easier to tolerate, and elevate mood.

The human body produces its own opiates, called endorphins (for "endogenous morphine"), that are peptides. Like the poppy-derived opiates that they structurally resemble, endorphins influence mood and perception of pain.

The discovery of endorphins began in 1971 in research laboratories at Stanford University and the Johns Hopkins School of Medicine, where researchers exposed pieces of brain tissue from experimental mammals to morphine. The morphine was radioactively labeled (some of the atoms were radioactive isotopes) so researchers could follow its destination in the brain.

The morphine indeed bound to receptors on the membranes of certain nerve cells, particularly in the neurons that transmit pain. Why, the investigators wondered, would an animal's brain contain receptors for a chemical made by a poppy? Could a mammal's body manufacture its own opiates? The opiate receptor, then, would normally bind the body's own opiates (the endorphins) but would also be able to bind the chemically similar compounds made by the poppy. Over the next few years, researchers identified several types of endorphins in the human brain and associated their release with situations involving pain relief, such as acupuncture and analgesia to mother and child during childbirth.

The existence of endorphins explains why some people who are addicted to opiate drugs such as heroin experience withdrawal pain when they stop taking the drug. Initially, the body interprets the frequent binding of heroin to its endorphin receptors as an excess of endorphins. To bring the level down, the body slows its own production of endorphins. Then, when the addict stops taking the heroin, the body is caught short of opiates (heroin and endorphins). The result is pain.

Opiate drugs can be powerfully addicting when abused—that is, taken repeatedly by a person who is not in pain. These same drugs, however, are extremely useful in dulling severe pain, particularly in terminal illnesses. ■

## Divergence

Impulses leaving a neuron of a neuronal pool often exhibit **divergence** by passing into several other output fibers. For example, an impulse from one neuron may stimulate two others; each of these, in turn, may stimulate several others, and so forth. Such a pattern of diverging nerve fibers can amplify an impulse—that is, spread it to increasing numbers of neurons within the pool (fig. 10.19b).

As a result of divergence, an impulse originating from a single neuron in the central nervous system may be amplified so that enough impulses reach the motor units within a skeletal muscle to cause forceful contraction. Similarly, an impulse originating from a sensory receptor may diverge and reach several different regions of the central nervous system, where the resulting impulses can be processed and acted upon.

The nervous system enables us to experience the world, and to think and feel emotion. This organ system is also very sensitive to outside influences. Clinical Application 10.6 discusses one way that an outside influence can affect the nervous system—drug addiction.

1. Define *neuropeptide*.

2. What is a neuronal pool?

3. Define *facilitation*.

4. What is convergence?

5. What is the relationship between divergence and amplification?

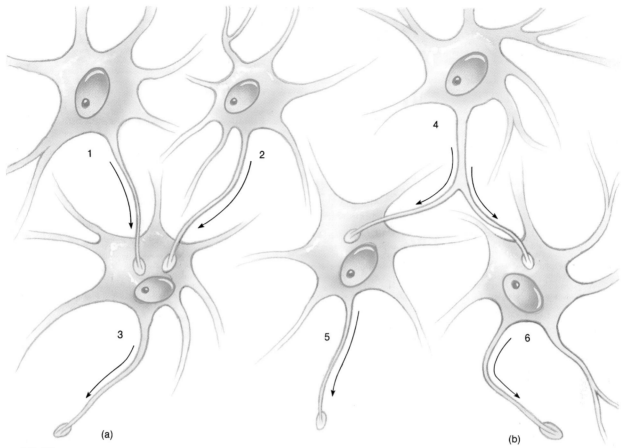

**FIGURE** 10.19   ▭

(a) Nerve fibers of neurons 1 and 2 converge to the cell body of neuron 3. (b) The nerve fiber of neuron 4 diverges to the cell bodies of neurons 5 and 6.

10.6

## CLINICAL APPLICATION

# Drug Addiction

Drug abuse and addiction are ancient as well as contemporary problems. A 3,500-year-old Egyptian document decries that society's reliance on opium. In the 1600s, a smokable form of opium enslaved many Chinese, and the Japanese and Europeans discovered the addictive nature of nicotine. During the American Civil War, morphine was a widely used painkiller; cocaine was introduced a short time later to relieve veterans addicted to morphine. Today, we continue to abuse drugs intended for medical use. LSD was originally used in psychotherapy but was abused in the 1960s as a hallucinogen. PCP was an anesthetic before being abused in the 1980s.

Why do people become addicted to certain drugs? Answers lie in the complex interactions of neurons, drugs, and individual behaviors.

### THE ROLE OF RECEPTORS

Eating hot fudge sundaes is highly enjoyable, but we usually don't feel driven to consume them repeatedly. Why do certain drugs compel a

*(continued)*

# Drug Addiction (continued)

person to repeatedly use them, even when knowing that doing so can be dangerous—the definition of addiction? The biology of neurotransmission helps to explain how we, and other animals, become addicted to certain drugs.

Understanding how neurotransmitters fit receptors can explain the actions of certain drugs. When a drug alters the activity of a neurotransmitter on a postsynaptic neuron, it either halts or enhances synaptic transmission. A drug that binds to a receptor, blocking a neurotransmitter from binding there, is called an antagonist. A drug that activates the receptor, triggering an action potential, or that helps a neurotransmitter to bind, is called an agonist. The effect of a drug depends upon whether it is an antagonist or an agonist; on the particular behaviors the affected neurotransmitter normally regulates; and in which parts of the brain drugs affect neurotransmitters and their binding to receptors. Many addictive substances bind to receptors for the neurotransmitter dopamine, in a brain region called the nucleus accumbens.

With repeated use of an addictive substance, the number of receptors it targets can decline. This means that the person must use more of the drug to feel the same effect. Neural pathways that use the neurotransmitter noradrenaline control arousal, dreaming, and mood. Amphetamine enhances norepinephrine activity, thereby heightening alertness and

mood. Amphetamine's structure is so similar to that of norepinephrine that it binds to norepinephrine receptors and triggers the same changes in the postsynaptic membrane.

Cocaine has a complex mechanism of action, both blocking reuptake of noradrenaline and binding to molecules that transport dopamine to postsynaptic cells. Cocaine's rapid and short-lived "high" reflects its rapid stay in the brain—its uptake takes just 4 to 6 minutes, and within 20 minutes the drug loses half its activity.

GABA is an inhibitory neurotransmitter used in a third of the brain's synapses. The drug valium causes relaxation and inhibits seizures and anxiety by helping GABA bind to receptors on postsynaptic neurons. Valium is therefore a GABA agonist.

## NICOTINE ADDICTION

Many medical professionals agree that cigarette smoking is highly addictive (figs. 10A and 10B). According to the Diagnostic & Statistical Manual of Mental Disorders, a person addicted to tobacco

1. must smoke more to attain the same effects (tolerance) over time;
2. experiences withdrawal symptoms when smoking stops, including weight gain, difficulty concentrating, insomnia, restlessness, anxiety, depression, slowed metabolism, and lowered heart rate;
3. smokes more often and for longer than intended;
4. spends considerable time obtaining cigarettes;

5. devotes less time to other activities;
6. continues to smoke despite knowing it is unhealthy;
7. wants to stop, but cannot easily do so.

Nicotine causes addiction, and the addiction supplies enough of the other chemicals in cigarette smoke to destroy health. The site of nicotine's activity is the neuron.

An activated form of nicotine binds protein receptors, called nicotinic receptors, that are parts of cell membranes of certain brain neurons. These receptors normally receive the neurotransmitter acetylcholine. When sufficient nicotine binds, a channel within the receptor opens, allowing positive ions to enter the neuron (fig. 10C). When a certain number of positive ions enter, the neuron is stimulated to release (by exocytosis) the neurotransmitter dopamine from its other end. The dopamine provides the pleasurable feelings associated with smoking. Addiction stems from two sources, researchers hypothesize—seeking the good feelings of sending off all that dopamine, and avoiding painful withdrawal symptoms.

Binding nicotinic receptors isn't the only effect of nicotine on the brain. When a smoker increases the number of cigarettes smoked, the number of nicotinic receptors on the brain cells increases. This happens because of the way that the nicotine binding impairs the recycling of receptor proteins by endocytosis, so receptors are produced faster than they are taken apart. However, after a period of steady nicotine exposure,

**FIGURE 10A**

Celebrities helped glamorize smoking in the past, and some still do.

SURGEON GENERAL'S WARNING: Smoking Causes Lung Cancer, Heart Disease, Emphysema, And May Complicate Pregnancy

**FIGURE 10B**

This photograph, published by the American Medical Association in 1944, shows a man using a prosthetic hand to light a cigarette—indicating that even the medical community then accepted smoking as a routine part of life. Today physicians feel quite differently, as the Surgeon General's warning on cigarette packages and advertisements indicates.

many of the receptors malfunction and no longer admit the positive ions that trigger the nerve impulse. This may be why as time goes on it takes more nicotine to produce the same effects.

Many questions remain concerning the biological effects of tobacco smoking. Why don't lab animals experience withdrawal? Why do people who have successfully stopped smoking often start again 6 months later, even though withdrawal eases within 2 weeks of quitting? Why do some people become addicted easily, yet others smoke only a few cigarettes a day and can stop anytime? While scientists try to answer these questions, society must deal with questions of rights and responsibilities that cigarette smoking causes. ■

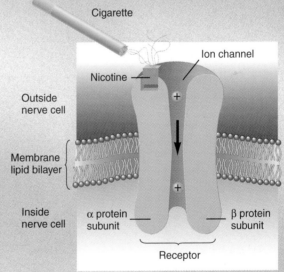

**FIGURE 10C**

The seat of tobacco addiction lies in nicotine's binding to nerve cell surface receptors that normally bind the neurotransmitter acetylcholine. Not only does nicotine alter the receptor so that positive ions enter the cell, triggering dopamine release, but the chemical's repeated presence in a heavy smoker stimulates excess receptors to accumulate—although they soon become nonfunctional. Nicotine's effects are complex.

**NERVOUS SYSTEM**

Neurons carry impulses that allow body systems to communicate.

## Integumentary System

Sensory receptors provide the nervous system with information about the outside world.

## Lymphatic System

Stress may impair the immune response.

## Skeletal System

Bones protect the brain and spinal cord and help maintain plasma calcium, which is important to neuron function.

## Digestive System

The nervous system can influence digestive function.

## Muscular System

Nerve impulses control movement and carry information about the position of body parts.

## Respiratory System

The nervous system alters respiratory activity to control oxygen levels and blood pH.

## Endocrine System

The hypothalamus controls secretion of many hormones.

## Urinary System

Nerve impulses affect urine production and elimination.

## Cardiovascular System

Nerve impulses help control blood flow and blood pressure.

## Reproductive System

The nervous system plays a role in egg and sperm formation, sexual pleasure, childbirth, and nursing.

## Introduction (page 344)

Organs of the nervous system are divided into the central and peripheral nervous systems. These divisions provide sensory, integrative, and motor functions.

## General Functions of the Nervous System (page 344)

1. Sensory receptors detect changes in internal and external body conditions.
2. Integrative functions bring sensory information together and make decisions that motor functions act upon.
3. Motor impulses stimulate effectors to respond.
4. Neuron structure
   a. A neuron includes a cell body, cell processes, and the organelles usually found in cells.
   b. Dendrites and the cell body provide receptive surfaces.
   c. A single axon arises from the cell body and may be enclosed in a myelin sheath and a neurilemma.

## Classification of Neurons and Neuroglia (page 349)

Neurons differ in structure and function.

1. Classification of neurons
   a. On the basis of structure, neurons are classified as bipolar, unipolar, or multipolar.
   b. On the basis of function, neurons are classified as sensory neurons, interneurons, or motor neurons.
2. Classification of neuroglia
   a. Neuroglial cells make up a large portion of the nervous system and have several functions.
   b. They fill spaces, support neurons, hold nervous tissue together, play a role in the metabolism of glucose, help regulate potassium ion concentration, produce myelin, carry on phagocytosis, rid synapses of excess ions and neurotransmitters, nourish neurons, and may directly communicate with neurons.
   c. They include Schwann cells in the peripheral nervous system and astrocytes, oligodendrocytes, microglia, and ependymal cells in the central nervous system.
3. Regeneration of nerve fibers
   a. If a neuron cell body is injured, the neuron is likely to die.
   b. If a peripheral nerve fiber is severed, its distal portion will die, but under the influence of nerve growth factors, the proximal portion may regenerate and reestablish its former connections, provided a tube of connective tissue guides it.
   c. Significant regeneration is not likely in the central nervous system.

## Cell Membrane Potential (page 354)

A cell membrane is usually polarized as a result of an unequal distribution of ions on either side. Pores and channels in membranes that allow passage of some ions but not others control ion distribution.

1. Distribution of ions
   a. Membrane pores and channels, formed by proteins, may be always open or sometimes open and sometimes closed.
   b. Potassium ions pass more readily through resting neuron cell membranes than do sodium and calcium ions.
2. Resting potential
   a. A high concentration of sodium ions is on the outside of the membrane and a high concentration of potassium ions is on the inside.
   b. Large numbers of negatively charged ions are inside the cell.
   c. In a resting cell, more positive ions leave the cell than enter it, so the inside of the cell membrane develops a negative charge with respect to the outside.
3. Local potential changes
   a. Stimulation of a membrane affects its resting potential in a local region.
   b. The membrane is depolarized if it becomes less negative; it is hyperpolarized if it becomes more negative.
   c. Local potential changes are graded and subject to summation.
   d. Reaching threshold potential triggers an action potential.
4. Action potentials
   a. At threshold, sodium channels open and sodium ions diffuse inward, depolarizing the membrane.
   b. About the same time, potassium channels open and potassium ions diffuse outward, repolarizing the membrane.
   c. This rapid change in potential is an action potential.
   d. Many action potentials can occur before active transport reestablishes the original resting potential.
   e. The propagation of action potentials along a nerve fiber is an impulse.
5. Refractory period
   a. The refractory period is a brief time following passage of a nerve impulse when the membrane is unresponsive to an ordinary stimulus.
   b. During the absolute refractory period, the membrane cannot be stimulated; during the relative refractory period, the membrane can be stimulated with a high intensity stimulus.
6. All-or-none response
   a. A nerve impulse is an all-or-none response to a stimulus of threshold intensity applied to a fiber.
   b. All the impulses conducted on a fiber are the same.
7. Impulse conduction
   a. Unmyelinated fibers conduct impulses that travel over their entire surfaces.
   b. Myelinated fibers conduct impulses that travel from node to node.
   c. Impulse conduction is more rapid on myelinated fibers with large diameters.

# The Synapse (page 360)

A synapse is a junction between two cells. A synaptic cleft is the gap between parts of two cells at a synapse. Synapses can occur between two neurons, a receptor cell and a neuron, or a neuron and an effector.

1. Synaptic transmission
   a. Impulses usually travel from dendrite or cell body, then along the axon to a synapse.
   b. Axons have synaptic knobs at their distal ends that secrete neurotransmitters.
   c. The neurotransmitter is released when a nerve impulse reaches the end of an axon, and the neurotransmitter diffuses across the synaptic cleft.
   d. A neurotransmitter reaching the nerve fiber or cell body on the distal side of the cleft triggers a nerve impulse.

2. Synaptic potentials
   a. Some neurotransmitters can depolarize postsynaptic membranes, triggering an action potential. This is an excitatory postsynaptic potential (EPSP).
   b. Others hyperpolarize the membranes, inhibiting action potentials. This is an inhibitory postsynaptic potential (IPSP).
   c. EPSPs and IPSPs are summed in a trigger zone of the neuron.

3. Neurotransmitters
   a. The nervous system produces at least thirty types of neurotransmitters.
   b. Calcium ions diffuse into synaptic knobs in response to action potentials, releasing neurotransmitters.
   c. Neurotransmitters are quickly decomposed or removed from synaptic clefts.

4. Neuropeptides
   a. Neuropeptides are chains of amino acids.
   b. Some neuropeptides are neurotransmitters or neuromodulators.
   c. They include enkephalins, endorphins, and substance P.

# Impulse Processing (page 365)

The way impulses are processed reflects the organization of neurons in the brain and spinal cord.

1. Neuronal pools
   a. Neurons are organized into pools within the central nervous system.
   b. Each pool receives, processes, and conducts away impulses.

2. Facilitation
   a. Each neuron in a pool may receive excitatory and inhibitory stimuli.
   b. A neuron is facilitated when it receives subthreshold stimuli and becomes more excitable.

3. Convergence
   a. Impulses from two or more incoming fibers may converge on a single neuron.
   b. Convergence enables a neuron to summate impulses from different sources.

4. Divergence
   a. Impulses leaving a pool may diverge by passing onto several output fibers.
   b. Divergence amplifies impulses.

# CRITICAL THINKING QUESTIONS

1. A drug called tacrine slows breakdown of acetylcholine in synaptic clefts. Which illness discussed in the chapter might tacrine theoretically treat?

2. Is Imitrex, the drug used to treat migraine, an agonist or an antagonist?

3. How would you explain the following observations?
   a. When motor nerve fibers in the leg are severed, the muscles they innervate become paralyzed; however, in time, control over the muscles often returns.
   b. When motor nerve fibers in the spinal cord are severed, the muscles they control become permanently paralyzed.

4. People who inherit familial periodic paralysis often develop very low blood potassium concentrations. How would you explain the fact that the paralysis may disappear quickly when potassium ions are administered intravenously?

5. What might be deficient in the diet of a pregnant woman who is complaining of leg muscle cramping? How would you explain this to her?

6. Why are rapidly growing cancers that originate in nervous tissue more likely to be composed of neuroglial cells than of neurons?

7. How are multiple sclerosis and Tay-Sachs disease opposite one another?

# REVIEW EXERCISES

1. Explain the relationship between the central nervous system and the peripheral nervous system.
2. List three general functions of the nervous system.
3. Distinguish between neurons and neuroglial cells.
4. Describe the generalized structure of a neuron.
5. Define *myelin*.
6. Distinguish between myelinated and unmyelinated nerve fibers.
7. Describe how an injured nerve fiber may regenerate.
8. Discuss the functions of each type of neuroglial cell.
9. Explain how a membrane may become polarized.
10. Define *resting potential*.

11. Distinguish between depolarizing and hyperpolarizing.
12. List the changes that occur during an action potential.
13. Distinguish between action potentials and nerve impulses.
14. Define *refractory period.*
15. Define *saltatory conduction.*
16. Define *synapse.*
17. Explain how a nerve impulse is transmitted from one neuron to another.
18. Explain the role of calcium in the release of neurotransmitters.
19. Define *neuropeptide.*

20. Distinguish between excitatory and inhibitory postsynaptic potentials.
21. Describe the "trigger zone" of a neuron.
22. Describe the relationship between an input nerve fiber and its neuronal pool.
23. Define *facilitation.*
24. Distinguish between convergence and divergence.
25. Explain how nerve impulses are amplified.
26. Explain how neurons are classified on the basis of their structure.
27. Explain how neurons are classified on the basis of their function.

**For More Information Related to This Chapter, Please Visit** *Hole's Human Anatomy and Physiology* **Home Page**

http://www.mhhe.com/biosci/abio/

# NERVOUS SYSTEM II

## Divisions of the Nervous System

**cephal-,** head: en*cephal*itis—inflammation of the brain.

**chiasm-,** cross: optic *chiasm*a—X-shaped structure produced by the crossing over of optic nerve fibers.

**flacc-,** flabby: *flacc*id paralysis—paralysis characterized by loss of tone in muscles innervated by damaged nerve fibers.

**funi-,** small cord or fiber: *funi*culus—major nerve tract or bundle of myelinated nerve fibers within the spinal cord.

**gangli-,** swelling: *gangli*on—mass of neuron cell bodies.

**mening-,** membrane: *mening*es—membranous coverings of the brain and spinal cord.

**plex-,** interweaving: choroid *plex*us—mass of specialized capillaries associated with spaces in the brain.

### CHAPTER OBJECTIVES

#### AFTER YOU HAVE STUDIED THIS CHAPTER, YOU SHOULD BE ABLE TO:

1. Describe the coverings of the brain and spinal cord.
2. Describe the structure of the spinal cord and its major functions.
3. Describe a reflex arc.
4. Define reflex behavior.
5. Name the major parts of the brain and describe the functions of each.
6. Distinguish among motor, sensory, and association areas of the cerebral cortex.
7. Explain hemisphere dominance.
8. Explain the stages in memory storage.
9. Describe the formation and function of cerebrospinal fluid.
10. Explain the functions of the limbic system and the reticular formation.
11. List the major parts of the peripheral nervous system.
12. Describe the structure of a peripheral nerve and how its fibers are classified.
13. Name the cranial nerves and list their major functions.
14. Explain how spinal nerves are named and their functions.
15. Describe the general characteristics of the autonomic nervous system.
16. Distinguish between the sympathetic and the parasympathetic divisions of the autonomic nervous system.
17. Describe a sympathetic and a parasympathetic nerve pathway.
18. Explain how the autonomic neurotransmitters differently affect visceral effectors.

### KEY TERMS

**adrenergic** (ad″ren-er′jik)

**autonomic nervous system** (aw″to-nom′ik ner′vus sis′tem)

**brain stem** (brān stem)

**cerebellum** (ser″ĕ-bel′um)

**cerebral cortex** (ser′ĕ-bral kor′teks)

**cerebral hemisphere** (ser′ĕ-bral hem′i-sfĕr)

**cerebrospinal fluid** (ser″ĕ-bro-spi′nal floo′id)

**cerebrum** (ser′ĕ-brum)

**cholinergic** (ko″lin-er′jik)

**choroid plexus** (ko′roid plek′sus)

**diencephalon** (di″en-sef′ah-lon)

**effector** (e-fek′tor)

**hypothalamus** (hi″po-thal′ah-mus)

**medulla oblongata** (mĕ-dul′ah ob″long-ga′tah)

**midbrain** (mid′brān)

**parasympathetic** (par″ah-sim″pah-thet′ik)

**postganglionic** (pōst″gang-gle-on′ik)

**preganglionic** (pre″gang-gle-on′ik)

**receptor** (re-sep′tor)

**reflex** (re′fleks)

**reticular formation** (rĕ-tik′u-lar fōr-ma′shun)

**sympathetic** (sim″pah-thet′ik)

**thalamus** (thal′ah-mus)

**ventricle** (ven′trĭ-kl)

September 13, 1848, was a momentous day for Phineas Gage, a young man who worked in Vermont evening out terrain for railroad tracks. To blast away rock, he would drill a hole, fill it with gunpowder, cover that with sand, insert a fuse, and then press down with an iron rod called a tamping iron. The explosion would go down into the rock.

But on that fateful September day, Gage began pounding on the tamping iron before his coworker had put down the sand. The gunpowder exploded outward, slamming the inch-thick, 40-inch-long iron rod straight through Gage's skull. It pierced his brain like an arrow propelled through a soft melon, shooting out the other side of his head. Remarkably, Gage stood up just a few moments later, fully conscious and apparently unharmed by the hole just blasted through his head.

As it turned out, Gage was harmed in the freak accident, but in ways so subtle that they were not at first evident. His friends reported that "Gage was no longer Gage." Although retaining his intellect and abilities to move, speak, learn, and remember, Gage's personality dramatically changed. Once a trusted, honest, and dedicated worker, the 25-year-old became irresponsible, shirking work, cursing, and pursuing what his doctor termed "animal propensities."

Researchers as long ago as 1868 hypothesized that the tamping iron had ripped out a part of Gage's brain controlling personality. In 1994, computer analysis more precisely pinpointed the damage to the famous brain of Phineas Gage, which, along with the tamping iron, wound up in a museum at Harvard University. Researchers reconstructed the trajectory of the tamping iron, localizing two small areas in the front of the brain that control rational decision making and processing of emotion.

More than a hundred years after Gage's accident, in 1975, 21-year-old Karen Ann Quinlan drank alcohol after taking a prescription sedative, and her heart and lungs stopped. When she was found, Karen had no pulse, was not breathing, had dilated pupils, and was unresponsive. Cardiopulmonary resuscitation restored her pulse, but once at the hospital she was placed on a ventilator. Within 12 hours, some functions returned—her pupils constricted, she moved, gagged, grimaced, and even opened her eyes. Within a few months, she could even breathe unaided for short periods.

Because Karen's responses were random and not purposeful, and she was apparently unaware of herself and her environment, she was said to be in a *persistent vegetative state.* Her basic life functions were intact, but she had to be fed and given water intravenously. Fourteen months after Karen took the fateful pills and alcohol, her parents made a request that was to launch the right-to-die movement. They asked that Karen be taken off of life support. Doctors removed Karen's ventilator, and she lived for 9 more years in a nursing home before dying of infection. She never regained awareness.

A rod blasted through the head of a young railway worker has taught us much about the biology of personality.

Throughout Karen and her family's ordeal, researchers tried to fathom what had happened to her. A CAT scan performed 5 years after the accident showed atrophy in two major brain regions, the cerebrum and the cerebellum. But when researchers analyzed Karen Ann Quinlan's brain in 1993, they were surprised. The most severely damaged part of her brain was the thalamus, an area thought to function merely as a relay station to higher brain structures. Karen's tragic case revealed that the thalamus is also important in processing thoughts, in providing the awareness and responsiveness that makes a person a conscious being.

The cases of Phineas Gage and Karen Ann Quinlan dramatically illustrate the function of the human brain by revealing what can happen when it is damaged. Nearly every aspect of our existence depends upon the brain and other parts of the nervous system, from thinking and feeling; to sensing, perceiving, and responding to the environment; to carrying out vital functions such as breathing and heartbeat. This chapter describes how the billions of neurons comprising the nervous system interact to enable us to survive and to enjoy the world around us.

Bones, membranes, and fluid surround the organs of the central nervous system (CNS). More specifically, the brain lies within the cranial cavity of the skull, whereas the spinal cord, which is continuous with the brain, occupies the vertebral canal within the vertebral column. Beneath these bony coverings, membranes called **meninges,** located between the bone and the soft tissues of the nervous system, protect the brain and spinal cord (fig. 11.1a).

## MENINGES

The meninges (sing., *meninx*) have three layers—dura mater, arachnoid mater, and pia mater (fig. 11.1b). The **dura mater** is the outermost layer. It is primarily composed of tough, white dense connective tissue and contains many blood vessels and nerves. It attaches to the inside of the cranial cavity and forms the internal periosteum of the surrounding skull bones (see reference plate 53).

In some regions, the dura mater extends inward between lobes of the brain and forms supportive and protective partitions (table 11.1). In other areas, the dura mater splits into two layers, forming channels called *dural sinuses,* shown in figure 11.1b. Venous blood flows through these channels as it returns from the brain to vessels leading to the heart.

The dura mater continues into the vertebral canal as a strong, tubular sheath that surrounds the spinal cord. It is attached to the cord at regular intervals by a band of pia mater (denticulate ligaments) that extends the length of the spinal cord on either side. The dural sheath termi-

| **TABLE 11.1** | **PARTITIONS OF THE DURA MATER** |
|---|---|
| **Partition** | **Location** |
| Falx cerebelli | Separates the right and left cerebellar hemispheres |
| Falx cerebri | Extends downward into the longitudinal fissure, and separates the right and left cerebral hemispheres (fig. 11.1b) |
| Tentorium cerebelli | Separates the occipital lobes of the cerebrum from the cerebellum (fig. 11.1a) |

nates as a blind sac at the level of the second sacral vertebra, below the end of the spinal cord. The sheath around the spinal cord is not attached directly to the vertebrae but is separated by an *epidural space,* which lies between the dural sheath and the bony walls (fig. 11.2). This space contains blood vessels, loose connective tissue, and adipose tissue that provide a protective pad around the spinal cord.

A blow to the head may rupture some blood vessels associated with the brain, and the escaping blood may collect in the space beneath the dura mater. This condition, called *subdural hematoma,* can increase pressure between the rigid bones of the skull and the soft tissues of the brain. Unless the accumulating blood is promptly evacuated, compression of the brain may lead to functional losses or even death.

**FIGURE 11.1**

(a) Membranes called meninges enclose the brain and spinal cord. (b) The meninges include three layers: dura mater, arachnoid mater, and pia mater.

376

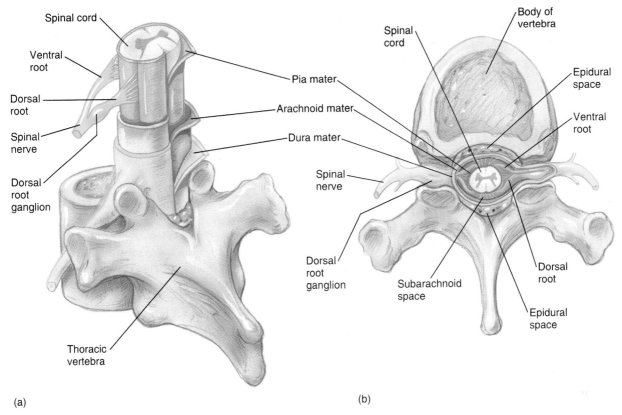

**FIGURE 11.2**

(a) The dura mater ensheaths the spinal cord. (b) Tissues forming a protective pad around the cord fill the epidural space between the dural sheath and the bone of the vertebra.

The **arachnoid mater** is a thin, weblike membrane that lacks blood vessels and is located between the dura and pia maters. It spreads over the brain and spinal cord but generally does not dip into the grooves and depressions on their surfaces. Many thin strands extend from its undersurface and are attached to the pia mater. Between the arachnoid and pia maters is a *subarachnoid space,* which contains the clear, watery **cerebrospinal fluid.**

The **pia mater** is very thin and contains many nerves, as well as blood vessels that nourish the underlying cells of the brain and spinal cord. The pia mater is attached to the surfaces of these organs and follows their irregular contours, passing over the high areas and dipping into the depressions.

---

*Meningitis* is an inflammation of the meninges. Bacteria or viruses that invade the cerebrospinal fluid are the usual causes of this condition. Meningitis may affect the dura mater, but it is more commonly limited to the arachnoid and pia maters.

Meningitis occurs most often in infants and children and is considered a serious childhood infection. Possible complications of this disease include loss of vision or hearing, paralysis, mental retardation, and death.

---

**1** Describe the meninges.

**2** Name the layers of the meninges.

**3** Explain where cerebrospinal fluid is located.

---

## SPINAL CORD

The **spinal cord** is a slender column of nerve fibers that is continuous with the brain and extends downward through the vertebral canal. The spinal cord is said to begin where nervous tissue leaves the cranial cavity at the level of the foramen magnum (see reference plate 55). The cord tapers to a point and terminates near the intervertebral disk that separates the first and second lumbar vertebrae (fig. 11.3*a*).

### Structure of the Spinal Cord

The spinal cord consists of thirty-one segments, each of which gives rise to a pair of **spinal nerves.** These nerves branch to various body parts and connect them with the central nervous system.

In the neck region, a thickening in the spinal cord, called the *cervical enlargement,* supplies nerves to the upper limbs. A similar thickening in the lower back, the *lumbar enlargement,* gives off nerves to the lower limbs.

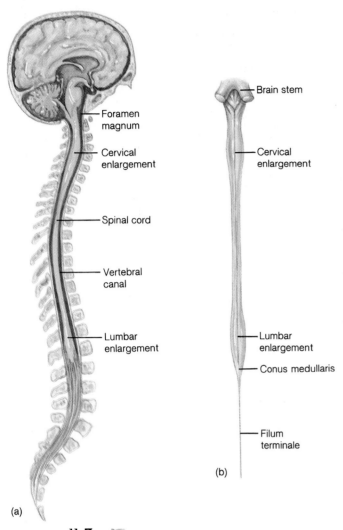

**FIGURE 11.3**

(a) The spinal cord begins at the level of the foramen magnum. (b) Posterior view of the spinal cord with the spinal nerves removed.

*Labels in figure (a):*
- Foramen magnum
- Cervical enlargement
- Spinal cord
- Vertebral canal
- Lumbar enlargement

*Labels in figure (b):*
- Brain stem
- Cervical enlargement
- Lumbar enlargement
- Conus medullaris
- Filum terminale

with relatively large cell bodies in the anterior horns (anterior horn cells) give rise to motor fibers that pass out through spinal nerves to various skeletal muscles. However, the majority of neurons in the gray matter are interneurons (see chapter 10).

A horizontal bar of gray matter in the middle of the spinal cord, the *gray commissure,* connects the wings of the gray matter on the right and left sides. This bar surrounds the **central canal,** which is continuous with the ventricles of the brain and contains cerebrospinal fluid. The central canal is prominent during embryonic development, but it becomes almost microscopic in an adult.

The gray matter divides the white matter of the spinal cord into three regions on each side—the *anterior, lateral,* and *posterior columns* (or *funiculi*). Each column consists of longitudinal bundles of myelinated nerve fibers that comprise major nerve pathways called **nerve tracts.**

## Functions of the Spinal Cord

The spinal cord has two major functions. First, it is a conduit for nerve impulses to and from the brain. Second, it is a center for spinal reflexes.

The nerve tracts of the spinal cord together with the spinal nerves provide a two-way communication system between the brain and body parts outside the nervous system. The tracts that conduct sensory impulses to the brain are called **ascending tracts;** those that conduct motor impulses from the brain to motor neurons reaching muscles and glands are called **descending tracts.**

The nerve fibers within these tracts are axons. Typically, all the axons within a given tract originate from neuron cell bodies located in the same part of the nervous system and end together in some other part. The names that identify nerve tracts often reflect these common origins and terminations. For example, a *spinothalamic tract* begins in the spinal cord and carries sensory impulses associated with the sensations of pain and touch to the thalamus of the brain. A *corticospinal tract* originates in the cortex of the brain and carries motor impulses downward through the spinal cord and spinal nerves. These impulses control skeletal muscle movements.

### Ascending Tracts

Among the major ascending tracts of the spinal cord are the following (fig. 11.5):

1. **Fasciculus gracilis** (fah-sik′u-lus gras′il-is) and **fasciculus cuneatus** (ku′ne-at-us). These tracts are located in the posterior funiculi of the spinal cord. Their fibers conduct sensory impulses from the skin, muscles, tendons, and joints to the brain, where they are interpreted as sensations of touch, pressure, and body movement.

Just inferior to the lumbar enlargement, the spinal cord tapers to a structure called the *conus medullaris.* From this tip, fibers extend downward to become spinal nerves at the remaining lumbar and sacral levels. Originating from among them, a thin cord of connective tissue descends to the upper surface of the coccyx. This cord is called the *filum terminale* (fig. 11.3*b*).

Two grooves, a deep *anterior median fissure* and a shallow *posterior median sulcus,* extend the length of the spinal cord, dividing it into right and left halves. A cross section of the cord (fig. 11.4) reveals that it consists of white matter surrounding a core of gray matter. The pattern the gray matter produces roughly resembles a butterfly with its wings outspread. The upper and lower wings of gray matter are called the *posterior horns* and the *anterior horns,* respectively. Between them on either side is a protrusion of gray matter called the *lateral horn.* Neurons

(a)

(b)

**FIGURE** 11.4

(a) A cross section of the spinal cord. (b) Identify the parts of the spinal cord in this micrograph (7.5×).

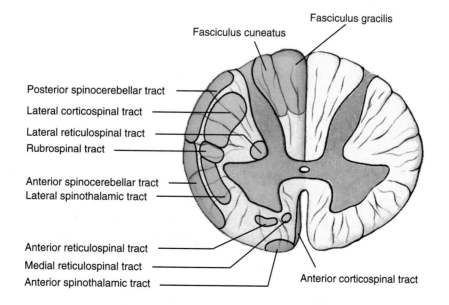

**FIGURE** 11.5

Major ascending and descending tracts within a cross section of the spinal cord. Ascending tracts are in orange, descending tracts in tan. (Tracts are shown only on one side.)

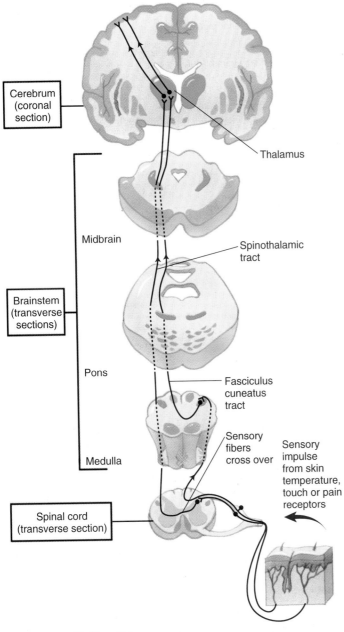

**FIGURE 11.6** 🏃

Sensory impulses originating in skin touch receptors ascend in the fasciculus cuneatus tract and cross over in the medulla of the brain. Pain and temperature information ascends in the lateral spinothalamic tract, which crosses over in the spinal cord.

At the base of the brain in an area called the *medulla oblongata,* most of the fasciculus gracilis and fasciculus cuneatus fibers cross over (decussate) from one side to the other—that is, those ascending on the left side of the spinal cord pass across to the right side and vice versa. As a result, the impulses originating from sensory receptors on the left side of the body reach the right side of the brain, and those originating on the right side of the body reach the left side of the brain (fig. 11.6).

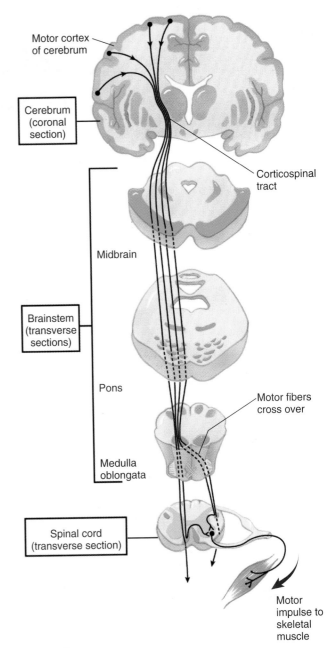

**FIGURE 11.7** 🏃

Most motor fibers of the corticospinal tract begin in the cerebral cortex, cross over in the medulla, and descend in the spinal cord, where they synapse with neurons whose fibers lead to spinal nerves supplying skeletal muscles. Some fibers cross over in the spinal cord.

2. **Spinothalamic** (spi″no-thah-lam′ik) **tracts.** The *lateral* and *anterior spinothalamic tracts* are located in the lateral and anterior funiculi, respectively. Impulses in these tracts cross over in the spinal cord. The lateral tracts conduct impulses from various body regions to the brain and give rise to sensations of pain and temperature (fig. 11.6). Impulses carried on fibers of the anterior tracts are interpreted as touch and pressure.

TABLE 11.2 NERVE TRACTS OF THE SPINAL CORD

| Tract | Location | Function |
|---|---|---|
| **Ascending Tracts** | | |
| 1. Fasciculus gracilis and fasciculus cuneatus | Posterior funiculi | Conduct sensory impulses associated with the senses of touch, pressure, and body movement from skin, muscles, tendons, and joints to the brain |
| 2. Spinothalamic tracts (lateral and anterior) | Lateral and anterior funiculi | Conduct sensory impulses associated with the senses of pain, temperature, touch, and pressure from various body regions to the brain |
| 3. Spinocerebellar tracts (posterior and anterior) | Lateral funiculi | Conduct sensory impulses required for the coordination of muscle movements from muscles of the lower limbs and trunk to the cerebellum |
| **Descending Tracts** | | |
| 1. Corticospinal tracts (lateral and anterior) | Lateral and anterior funiculi | Conduct motor impulses associated with voluntary movements from the brain to skeletal muscles |
| 2. Reticulospinal tracts (lateral, anterior, and medial) | Lateral and anterior funiculi | Conduct motor impulses associated with the maintenance of muscle tone and the activity of sweat glands from the brain |
| 3. Rubrospinal tracts | Lateral funiculi | Conduct motor impulses associated with muscular coordination and the maintenance of posture from the brain |

3. **Spinocerebellar** (spi″no-ser″ĕ-bel′ar) **tracts.** The *posterior* and *anterior spinocerebellar tracts* lie near the surface in the lateral funiculi of the spinal cord. Fibers in the posterior tracts remain uncrossed, whereas those in the anterior tracts cross over in the medulla. Impulses conducted on their fibers originate in the muscles of the lower limbs and trunk and then travel to the cerebellum of the brain. These impulses coordinate muscular movements.

## Descending Tracts

The major descending tracts of the spinal cord are shown in figure 11.5. They include the following:

1. **Corticospinal** (kor″tĭ-ko-spi′nal) **tracts.** The *lateral* and *anterior corticospinal tracts* occupy the lateral and anterior funiculi, respectively. Most of the fibers of the lateral tracts cross over in the lower portion of the medulla oblongata. Some fibers of the anterior tracts cross over at various levels of the spinal cord. The corticospinal tracts conduct motor impulses from the brain to spinal nerves and outward to various skeletal muscles. Thus, they help control voluntary movements (fig. 11.7).

   The corticospinal tracts are sometimes called *pyramidal tracts* after the pyramid-shaped regions in the medulla oblongata through which they pass. Other descending tracts are called *extrapyramidal tracts,* and they include the reticulospinal and rubrospinal tracts.

2. **Reticulospinal** (rĕ-tik″u-lo-spi′nal) **tracts.** The *lateral reticulospinal tracts* are located in the lateral funiculi, whereas the *anterior* and *medial reticulospinal tracts* are in the anterior funiculi. Some fibers in the lateral tracts cross over, whereas others remain uncrossed. Those of the anterior and medial tracts remain uncrossed. Motor impulses transmitted on the reticulospinal tracts originate in the brain and control muscular tone and activity of sweat glands.

3. **Rubrospinal** (roo″bro-spi′nal) **tracts.** The fibers of the rubrospinal tracts cross over in the brain and pass through the lateral funiculi. They carry motor impulses from the brain to skeletal muscles, and coordinate muscles and control posture.

A hemi-lesion of the spinal cord (severed on only one side) affecting the corticospinal and spinothalamic tracts can cause Brown-Séquard's syndrome. Because ascending tracts cross over at different levels, the injured side of the body becomes paralyzed and loses touch sensation. The other side of the body retains movement, but loses sensations of pain and temperature.

Table 11.2 summarizes the nerve tracts of the spinal cord. Clinical Application 11.1 describes injuries to the spinal cord.

# Spinal Cord Injuries

Christopher Reeve's life changed forever in a split second on May 27, 1995. Reeve, best known for his portrayal of Superman in four films, was one of 300 equestrians competing on that bright Saturday in Culpeper County, Virginia. He and his horse, Buck, were poised to clear the third of fifteen hurdles in a 2-mile event. The horse's front legs went over the hurdle, and Reeve's back arched as he propelled himself forward. But Buck stopped, his back legs never clearing the fence. Reeve hurled forward, striking his head on the fence. He landed on the grass—unconscious, not moving or breathing.

Reeve had broken the first and second cervical vertebrae, between the neck and the brain stem. Someone performed mouth-to-mouth resuscitation until paramedics inserted a breathing tube and then stabilized him on a board. At a nearby hospital, Reeve received methylprednisolone, a drug that diminishes the extremely damaging swelling that occurs as the immune system responds to the injury. If given within 8 hours of the accident, this drug can save a fifth of the damaged neurons. Reeve was then flown to a larger medical center, where he was sedated and fluid was suctioned from his lungs.

The first few days following a spinal cord injury are devastating. At first, the vertebrae are compressed and may break, which sets off action potentials in neurons, many of which soon die. The massive neuron death releases calcium ions, which activate tissue-degrading enzymes. Then white blood cells arrive and produce inflammation that can destroy healthy as well as damaged neurons. Axons tear, myelin coatings are stripped off, and vital connections between nerves and muscles are cut. The tissue cannot regenerate.

Thousands of people sustain spinal cord injuries each year. The consequences depend on the extent of damage the cord sustains and where the damage occurs.

Normal spinal reflexes depend on two-way communication between the spinal cord and the brain. Injuring nerve pathways depresses the cord's reflex activities in sites below the injury. At the same time, sensations and muscular tone in the parts the affected fibers innervate lessen. This condition, *spinal shock,* may last for days or weeks, al-though normal reflex activity may eventually return. However, if nerve fibers are severed, some of the cord's functions may be permanently lost.

Less severe injuries to the spinal cord, as from a blow to the head, whiplash, or rupture of an intervertebral disk, compress or distort the cord. Pain, weakness, and muscular atrophy in the regions the damaged nerve fibers supply may occur.

Among the more common causes of severe direct injury to the spinal cord are gunshot wounds, stabbings, and fractures or dislocations of vertebrae during vehicular accidents (fig. 11A). Regardless of the cause, if nerve fibers in ascending tracts are cut, sensations arising from receptors below the level of the injury are lost. Damage to descending tracts results in loss of motor functions.

For example, if the right lateral corticospinal tract is severed in the neck near the first cervical vertebra, control of the voluntary muscles in the right upper and lower limbs is lost, paralyzing them (hemiplegia). Problems of this type in fibers of the descending tracts produce *upper motor neuron syndrome,* characterized by *spastic paralysis* in which muscle tone increases, with very little atrophy of the muscles. However, uncoordinated reflex activity (hyperreflexia) usually occurs, when the flexor and extensor muscles of affected limbs alternately spasm.

Injury to motor neurons or their fibers in the horns of the spinal cord results in *lower motor neuron syndrome.* It is characterized by *flaccid paralysis,* a total loss of muscle tone and reflex activity, and the muscles atrophy. ■

**FIGURE 11A**

A dislocation of the atlas may cause a compression injury to the spinal cord.

Spinal cord

Dens

Atlas

Axis

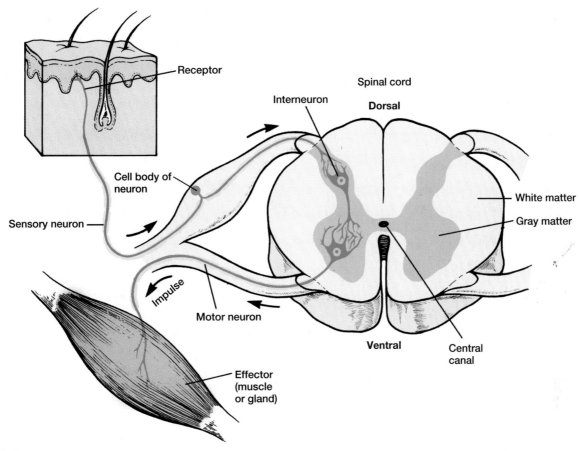

**FIGURE** 11.8 ▭ ✗

A reflex arc usually includes a receptor, a sensory neuron, interneurons, a motor neuron, and an effector.

**1** Describe the structure of the spinal cord.

**2** What are ascending and descending tracts?

**3** What is the consequence of fibers crossing over?

**4** Name the major tracts of the spinal cord, and list the kinds of impulses each conducts.

## REFLEX ARCS

Nerve impulses follow nerve pathways as they travel through the nervous system. The simplest of these pathways, including only a few neurons, constitutes a **reflex arc.** Reflex arcs carry out the simplest responses—reflexes.

A reflex arc begins with a receptor at the end of a sensory nerve fiber. This fiber usually leads to several interneurons within the central nervous system, which serve as a processing center, or *reflex center*. Fibers from these interneurons may connect with interneurons in other parts of the nervous system. They also communicate with motor neurons, whose fibers pass outward from the central nervous system to effectors. Reflexes whose arcs pass through the spinal cord are called **spinal reflexes** (fig. 11.8).

## Reflex Behavior

**Reflexes** are automatic, subconscious responses to changes (stimuli) within or outside the body. They help maintain homeostasis by controlling many involuntary processes such as heart rate, breathing rate, blood pressure, and digestion. Reflexes also carry out the automatic actions of swallowing, sneezing, coughing, and vomiting.

The *knee-jerk reflex* (patellar tendon reflex) is an example of a simple monosynaptic reflex so-called because it uses only two neurons—a sensory neuron communicating directly to a motor neuron. Striking the patellar ligament just below the patella initiates this reflex. The quadriceps femoris muscle group, which is attached to the patella by a tendon, is pulled slightly, stimulating stretch receptors within the muscle group. These receptors, in turn, trigger impulses that pass along the fiber of a sensory neuron into the lumbar region of the spinal cord. Within the spinal cord, the sensory axon synapses with a motor neuron. The impulse then continues along the axon of the motor neuron and travels back to the quadriceps femoris. The muscles respond by contracting, and the reflex is completed as the leg extends (fig. 11.9).

This reflex helps maintain an upright posture. For example, if a person is standing still and the knee begins to

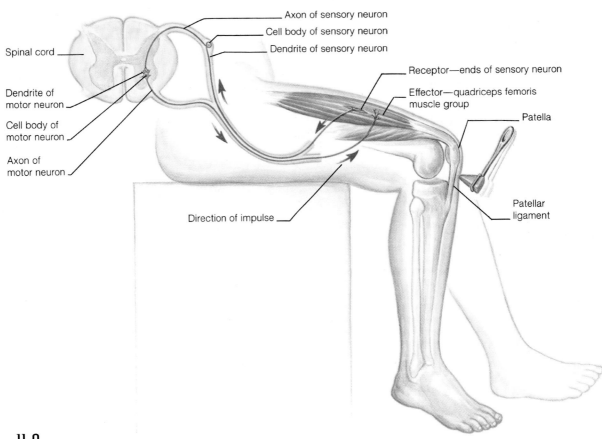

**FIGURE 11.9**

The knee-jerk reflex involves two neurons—a sensory neuron and a motor neuron. It is an example of a monosynaptic reflex.

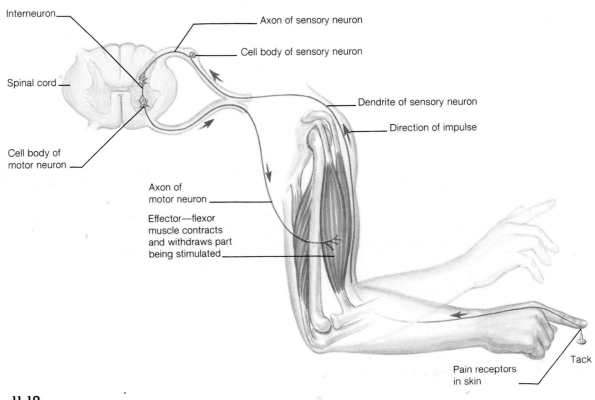

**FIGURE 11.10**

A withdrawal reflex involves a sensory neuron, an interneuron, and a motor neuron.

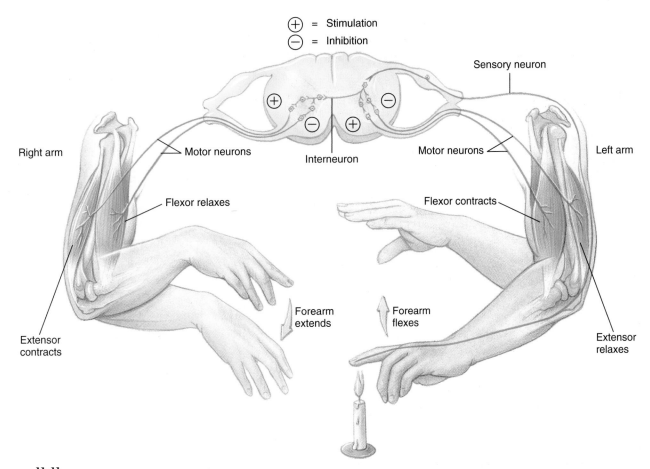

+ = Stimulation
− = Inhibition

Sensory neuron

Right arm
Motor neurons
Interneuron
Motor neurons
Left arm

Flexor relaxes
Flexor contracts

Forearm extends
Forearm flexes

Extensor contracts
Extensor relaxes

**FIGURE 11.11**

When the flexor muscle on one side is stimulated to contract in a withdrawal reflex, the extensor muscle on the opposite side also contracts. This helps to maintain balance.

bend as a result of gravity, the quadriceps femoris is stretched, the reflex is triggered, and the leg straightens again. Adjustments within the stretch receptors themselves keep the reflex responsive at different muscle lengths.

Another type of reflex, called a *withdrawal reflex* (fig. 11.10), occurs when a person unexpectedly touches a finger to something painful, activating skin receptors and sending sensory impulses to the spinal cord. There the impulses pass on to interneurons of a reflex center and are directed to motor neurons. The motor neurons transmit signals to the flexor muscles in the arm, which contract in response, pulling the part away from the painful stimulus.

At the same time, some of the incoming impulses stimulate interneurons that inhibit the action of the antagonistic extensor muscles (reciprocal innervation). This inhibition allows flexor muscles to effectively withdraw the hand.

While flexor muscles of the stimulated side (ipsilateral side) contract, the flexor muscles of the other arm (contralateral side) are inhibited. Furthermore, the extensor muscles on this other side contract. This phenome-

non, called a *crossed extensor reflex,* is due to interneuron pathways within the reflex center of the spinal cord that allow sensory impulses arriving on one side of the cord to pass across to the other side and produce an opposite effect (fig. 11.11).

Concurrent with the withdrawal reflex, other interneurons in the spinal cord carry sensory impulses upward to the brain. The person becomes aware of the experience and may feel pain.

A withdrawal reflex is protective because it prevents or limits tissue damage when a body part touches something potentially harmful. Table 11.3 summarizes the components of a reflex arc. Clinical Application 11.2 discusses some familiar reflexes.

**1** What is a nerve pathway?

**2** Describe a reflex arc.

**3** Define *reflex.*

**4** Describe the actions that occur during a withdrawal reflex.

11.2

# Uses of Reflexes

Since normal reflexes depend on normal neuron functions, reflexes are commonly used to obtain information concerning the condition of the nervous system. An anesthesiologist, for instance, may try to initiate a reflex in a patient who is being anesthetized in order to determine how the anesthetic drug is affecting nerve functions. Also, in the case of injury to some part of the nervous system, observing reflexes may reveal the location and extent of damage.

Injury to any component of a reflex arc alters its function. For example, a *plantar reflex* is normally initiated by stroking the sole of the foot, and the usual response is flexion of the foot and toes. However, damage to certain nerve pathways (corticospinal tract) may trigger an abnormal response called the *Babinski reflex,* which is a dorsiflexion, extending the great toe upward and fanning apart the smaller toes. If the injury is minor, the response may consist of plantar flexion with failure of the great toe to flex, or plantar flexion followed by dorsiflexion. The Babinski reflex is normally present in infants up to the age of twelve months and is thought to reflect immaturity in their corticospinal tracts.

Other reflexes that may be tested during a neurological examination include the following:

1. *Biceps-jerk reflex.* Extending a person's forearm at the elbow elicits this reflex. The examiner's finger is placed on the inside of the extended elbow over the tendon of the biceps muscle, and the finger is tapped. The biceps contracts in response, and the forearm flexes at the elbow.
2. *Triceps-jerk reflex.* Flexing a person's forearm at the elbow and tapping the short tendon of the triceps muscle close to its insertion near the tip of the elbow elicits this reflex. The muscle contracts in response, and the forearm extends slightly.
3. *Abdominal reflexes.* These reflexes occur when the examiner strokes the skin of the abdomen. For example, a dull pin drawn from the sides of the abdomen upward toward the midline and above the umbilicus causes the abdominal muscles underlying the skin to contract, and the umbilicus moves toward the stimulated region.
4. *Ankle-jerk reflex.* Tapping the calcaneal tendon just above its insertion on the calcaneus elicits this reflex. The response is plantar flexion, produced by contraction of the gastrocnemius and soleus muscles.
5. *Cremasteric reflex.* This reflex is elicited in males by stroking the upper inside of the thigh. In response, the testis on the same side is elevated by contracting muscles. ■

TABLE 11.3 — PARTS OF A REFLEX ARC

| Part | Description | Function |
| --- | --- | --- |
| Receptor | The receptor end of a dendrite or a specialized receptor cell in a sensory organ | Sensitive to a specific type of internal or external change |
| Sensory neuron | Dendrite, cell body, and axon of a sensory neuron | Transmits nerve impulse from the receptor into the brain or spinal cord |
| Interneuron | Dendrite, cell body, and axon of a neuron within the brain or spinal cord | Serves as processing center; conducts nerve impulse from the sensory neuron to a motor neuron |
| Motor neuron | Dendrite, cell body, and axon of a motor neuron | Transmits nerve impulse from the brain or spinal cord out to an effector |
| Effector | A muscle or gland | Responds to stimulation by the motor neuron and produces the reflex or behavioral action |

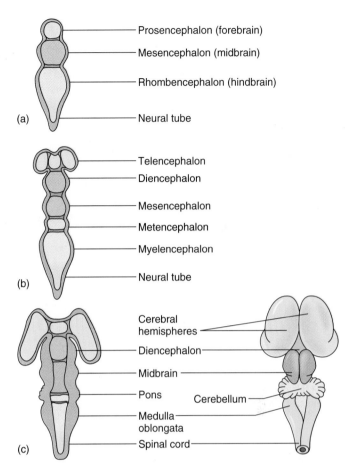

**FIGURE 11.12**

(a) The brain develops from a tubular structure with three cavities. (b) The cavities persist as the ventricles and their interconnections. (c) The wall of the tube gives rise to various regions of the brain, brain stem, and spinal cord.

## BRAIN

The **brain** is the largest and most complex part of the nervous system. It occupies the cranial cavity and is composed of about one hundred billion ($10^{11}$) multipolar neurons and innumerable nerve fibers, by which these neurons communicate with one another and with neurons in other parts of the system.

The brain contains nerve centers associated with sensory functions and is responsible for sensations and perceptions. It issues motor commands to skeletal muscles and carries on higher mental functions, such as memory and reasoning. It also contains centers that coordinate muscular movements, as well as centers and nerve pathways that regulate visceral activities. In addition to overseeing the function of the entire body, the brain also provides characteristics such as personality, as evidenced by the strange case of Phineas Gage discussed at the chapter's start.

### Brain Development

The basic structure of the brain reflects the way it forms during early (embryonic) development. It begins as the neural tube that gives rise to the central nervous system. The portion that becomes the brain has three major cavities, or vesicles, at one end—the *forebrain* (prosencephalon), *midbrain* (mesencephalon), and *hindbrain* (rhombencephalon) (fig. 11.12). Later, the forebrain divides into anterior and posterior portions (telencephalon and diencephalon, respectively), and the hindbrain partially divides into two parts (metencephalon and myelencephalon). The resulting five cavities persist in the mature brain as fluid-filled spaces called **ventricles** and the tubes that connect them. The tissue surrounding the spaces differentiates into the structural and functional regions of the brain.

The wall of the anterior portion of the forebrain gives rise to the *cerebrum* and *basal ganglia,* whereas the posterior portion forms a section of the brain called the *diencephalon.* The region the midbrain produces continues to be called the *midbrain* in the adult structure, and the hindbrain gives rise to the *cerebellum, pons,* and *medulla oblongata* (fig. 11.13 and table 11.4). Together, the midbrain, pons, and medulla oblongata comprise the **brain stem,** which attaches the brain to the spinal cord.

On a cellular level, the brain develops as specific neurons attract others by secreting growth hormones. Apoptosis (programmed cell death) destroys excess neural connections.

### Structure of the Cerebrum

The cerebrum, which develops from the anterior portion of the forebrain, is the largest part of the mature brain. It consists of two large masses or **cerebral hemispheres,**

Skull

Meninges

Cerebrum

Diencephalon

Midbrain

Brain stem — Pons

Medulla oblongata

Convolution

Sulcus

Corpus callosum

Transverse fissure

Cerebellum

Spinal cord

**FIGURE 11.13**

The major portions of the brain include the cerebrum, the diencephalon, the cerebellum, and the brain stem (see also reference plate 76).

| TABLE 11.4 | STRUCTURAL DEVELOPMENT OF THE BRAIN | | |
|---|---|---|---|
| **Embryonic Vesicle** | **Spaces Produced** | **Regions of the Brain Produced** | |
| Forebrain (prosencephalon) | | | |
| Anterior portion (telencephalon) | Lateral ventricles | Cerebrum Basal ganglia | |
| Posterior portion (diencephalon) | Third ventricle | Thalamus Hypothalamus Posterior pituitary gland Pineal gland | |
| Midbrain (mesencephalon) | Cerebral aqueduct | Midbrain | |
| Hindbrain (rhombencephalon) | | | |
| Anterior portion (metencephalon) | Fourth ventricle | Cerebellum Pons | |
| Posterior portion (myelencephalon) | Fourth ventricle | Medulla oblongata | |

A fetus with *anencephaly* has a face and lower brain structures, but lacks most higher brain structures. On an ultrasound scan, the head looks like little more than a face. Rarely, an infant with anencephaly survives to be born but usually dies within 24 hours. Sometimes the parents donate its organs.

Anencephaly is a type of neural tube defect (NTD). It occurs at about the twenty-eighth day of prenatal development, when a sheet of tissue that normally folds to form a neural tube, which develops into the central nervous system, remains open at the top. A less-serious NTD is *spina bifida,* in which an opening farther down the neural tube causes a lesion in the spine. The most serious form of this condition results in paralysis from that point downward. Sometimes spina bifida can be improved or even corrected with surgery.

The precise cause of neural tube defects is not known, but it involves folic acid; taking supplements of this vitamin sharply cuts the recurrence risk among women who have had an affected child. Most pregnant women take a blood test at the fifteenth week of pregnancy to detect fluid leaking from an NTD.

which are essentially mirror images of each other (fig. 11.14 and reference plate 49). A deep bridge of nerve fibers called the **corpus callosum** connects the cerebral hemispheres. A layer of dura mater called the *falx cerebri* separates them.

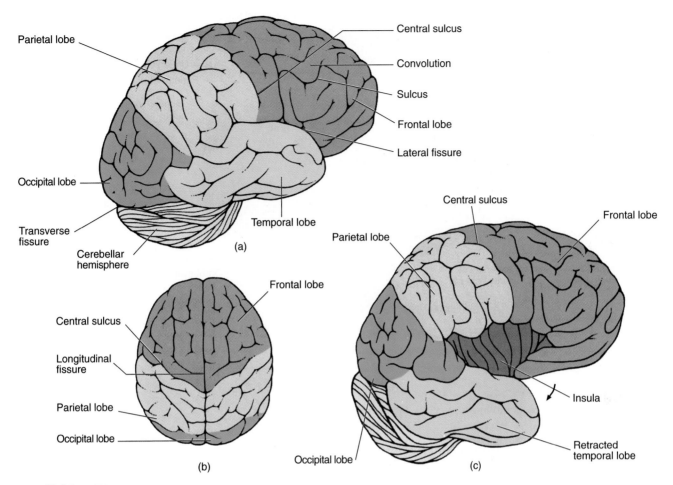

**FIGURE 11.14**

Colors in this figure distinguish the lobes of the cerebral hemispheres. (*a*) Lateral view of the right hemisphere. (*b*) Hemispheres viewed from above. (*c*) Lateral view of the right hemisphere with the insula exposed.

Many ridges or **convolutions** (sing., *gyrus*; pl., *gyri*), separated by grooves, mark the cerebrum's surface. Generally, a shallow groove is called a **sulcus,** and a very deep groove is called a **fissure.** The pattern of these elevations and depressions is complex, and it is distinct in all normal brains. For example, a *longitudinal fissure* separates the right and left cerebral hemispheres; a *transverse fissure* separates the cerebrum from the cerebellum; and sulci divide each hemisphere into lobes (figs. 11.13 and 11.14).

In a disorder called *lissencephaly* ("smooth brain"), a newborn has a smooth cerebral cortex, completely lacking the characteristic convolutions. Absence of a protein early in prenatal development prevents certain neurons from migrating within the brain, which blocks formation of convolutions. The child is profoundly mentally retarded, with frequent seizures and other neurological problems.

The lobes of the cerebral hemispheres (fig. 11.14) are named after the skull bones that they underlie. They include the following:

1. **Frontal lobe.** The frontal lobe forms the anterior portion of each cerebral hemisphere. It is bordered posteriorly by a *central sulcus* (fissure of Rolando), which passes out from the longitudinal fissure at a right angle, and inferiorly by a *lateral sulcus* (fissure of Sylvius), which exits the undersurface of the brain along its sides.

2. **Parietal lobe.** The parietal lobe is posterior to the frontal lobe and is separated from it by the central sulcus.

3. **Temporal lobe.** The temporal lobe lies inferior to the frontal and parietal lobes and is separated from them by the lateral sulcus.

4. **Occipital lobe.** The occipital lobe forms the posterior portion of each cerebral hemisphere and is separated from the cerebellum by a shelflike

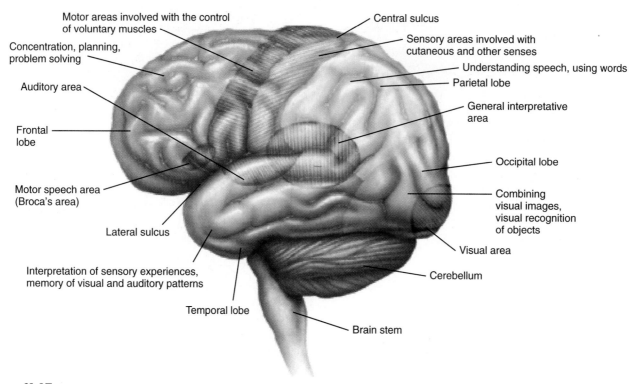

**FIGURE 11.15**
Some motor, sensory, and association areas of the left cerebral cortex.

extension of dura mater called the *tentorium cerebelli.* The occipital lobe and the parietal and temporal lobes have no distinct boundary.

5. **Insula.** The insula (island of Reil) is located deep within the lateral sulcus and is covered by parts of the frontal, parietal, and temporal lobes. A *circular sulcus* separates it from them.

A thin layer of gray matter (2 to 5 millimeters thick) called the **cerebral cortex** constitutes the outermost portion of the cerebrum. It covers the convolutions, dipping into the sulci and fissures. The cerebral cortex contains nearly 75% of all the neuron cell bodies in the nervous system.

Just beneath the cerebral cortex is a mass of white matter that makes up the bulk of the cerebrum. This mass contains bundles of myelinated nerve fibers that connect neuron cell bodies of the cortex with other parts of the nervous system. Some of these fibers pass from one cerebral hemisphere to the other by way of the corpus callosum, and others carry sensory or motor impulses from the cortex to nerve centers in the brain or spinal cord.

## Functions of the Cerebrum

The cerebrum provides higher brain functions: interpreting impulses from sense organs, initiating voluntary muscular movements, storing information as memory, and retrieving this information in reasoning. The cerebrum is also the seat of intelligence and personality.

### Functional Regions of the Cortex

The regions of the cerebral cortex that perform specific functions have been located using a variety of techniques. Persons who have suffered brain disease or injury, such as Karen Ann Quinlan and Phineas Gage, or have had portions of their brains removed surgically, have provided clues as to the functions of the impaired brain regions.

In other studies, areas of cortices have been exposed surgically and stimulated mechanically or electrically, with researchers observing the responses in certain muscles or the specific sensations that result. As a result of such investigations, researchers have divided the cerebral cortex into sections known as motor, sensory, and association areas. They overlap somewhat.

### Motor Areas

The *primary motor areas* of the cerebral cortex lie in the frontal lobes just in front of the central sulcus (precentral gyrus), and in the anterior wall of this sulcus (fig. 11.15). The nervous tissue in these regions contains many large *pyramidal cells,* named for their pyramid-shaped cell bodies.

Impulses from these pyramidal cells travel downward through the brain stem and into the spinal cord on the *corticospinal tracts.* Most of the nerve fibers in these tracts cross over from one side of the brain to the other within the brain stem and descend as lateral corticospinal tracts. Other fibers, in the anterior corticospinal tracts, cross over at various levels of the spinal cord (see fig. 11.7).

**FIGURE 11.16**

(a) Motor areas that control voluntary muscles (only right hemisphere shown). (b) Sensory areas involved with cutaneous and other senses (only left hemisphere shown).

Within the spinal cord, the corticospinal fibers synapse with motor neurons in the gray matter of the anterior horns. Axons of the motor neurons lead outward through peripheral nerves to voluntary muscles. Impulses transmitted on these pathways in special patterns and frequencies are responsible for fine movements in skeletal muscles. More specifically, as figure 11.16 shows, cells in the upper portions of the motor areas send impulses to muscles in the thighs and legs; those in the middle portions control muscles in the arms and forearms; and those in lower portions activate muscles of the head, face, and tongue.

The *reticulospinal* and *rubrospinal tracts* coordinate and control motor functions that maintain balance and posture. Many of these fibers pass into the basal ganglia on the way to the spinal cord. Some of the impulses conducted on these pathways normally inhibit muscular actions.

In addition to the primary motor areas, certain other regions of the frontal lobe control motor functions. For example, a region called *Broca's area* is just anterior to the primary motor cortex and superior to the lateral sulcus (see fig. 11.15), usually in the left cerebral hemisphere. It coordinates the complex muscular actions of the mouth, tongue, and larynx, which make speech possible. A person with an injury to this area may be able to understand spoken words, but may be unable to speak.

Above Broca's area is a region called the *frontal eye field*. The motor cortex in this area controls voluntary movements of the eyes and eyelids. Nearby is the cortex responsible for movements of the head that direct the eyes. Another region just in front of the primary motor area controls the muscular movements of the hands and fingers that make such skills as writing possible (see fig. 11.15).

An injury to the motor system may impair the ability to produce purposeful muscular movements. Such a condition that affects use of the upper and lower limbs, head, or eyes is called *apraxia*. When apraxia affects the speech muscles, disrupting speaking ability, it is called *aphasia*.

## Sensory Areas

Sensory areas, in several lobes of the cerebrum, interpret impulses from sensory receptors, producing feelings or sensations. For example, the sensations of temperature, touch, pressure, and pain in the skin arise in the anterior portions of the parietal lobes along the central sulcus (postcentral gyrus) and in the posterior wall of this sulcus (see fig. 11.15). The posterior parts of the occipital lobes provide vision, whereas the posterior, dorsal portions of the temporal lobes contain the centers for hearing. The sensory areas for taste are near the bases of the central sulci along the lateral sulci, and the sense of smell arises from centers deep within the cerebrum.

Like motor fibers, sensory fibers, such as those in the *fasciculus cuneatus tract,* cross over in the spinal cord or the brain stem (see fig. 11.6). Thus, the centers in the right central hemisphere interpret impulses originating from the left side of the body, and vice versa. However, the sensory areas concerned with vision receive impulses from both eyes, and those concerned with hearing receive impulses from both ears.

## Association Areas

Association areas are regions of the cerebral cortex that are not primarily sensory or motor in function, and interconnect with each other and with other brain structures. These areas occupy the anterior portions of the frontal lobes and are widespread in the lateral portions of the parietal, temporal, and occipital lobes. They analyze and interpret sensory experiences and help provide memory, reasoning, verbalizing, judgment, and emotions (see fig. 11.15).

The association areas of the frontal lobes provide higher intellectual processes, such as concentrating, planning, and complex problem solving. The anterior and inferior portions of these lobes (prefrontal areas) control emotional behavior and produce awareness of the possible consequences of behavior.

The parietal lobes have association areas that help interpret sensory information, and aid in understanding speech and choosing words to express thoughts and feelings. Awareness of the form of objects, including one's own body parts, stems from the posterior regions of these lobes.

The association areas of the temporal lobes and the regions at the posterior ends of the lateral fissures interpret complex sensory experiences, such as those needed to understand speech and to read. These regions also store memories of visual scenes, music, and other complex sensory patterns.

The occipital lobes have association areas adjacent to the visual centers. These are important in analyzing visual patterns and combining visual images with other sensory experiences—as when one recognizes another person.

Of particular importance is the region where the parietal, temporal, and occipital association areas join near the posterior end of the lateral sulcus. This region, called the *general interpretative area*, plays the primary role in complex thought processing. It makes it possible for a person to recognize words and arrange them to express a thought, and to read and understand ideas presented in writing. Table 11.5 summarizes the functions of the cerebral lobes.

| TABLE 11.5 | FUNCTIONS OF THE CEREBRAL LOBES | | |
|---|---|---|---|
| **Lobe** | **Functions** | **Lobe** | **Functions** |
| Frontal lobes | Motor areas control movements of voluntary skeletal muscles. | Temporal lobes | Sensory areas are responsible for hearing. |
| | Association areas carry on higher intellectual processes for concentrating, planning, complex problem solving and judging the consequences of behavior. | | Association areas interpret sensory experiences and remember visual scenes, music, and other complex sensory patterns. |
| Parietal lobes | Sensory areas provide sensations of temperature, touch, pressure, and pain involving the skin. | Occipital lobes | Sensory areas are responsible for vision. |
| | Association areas function in understanding speech and in using words to express thoughts and feelings. | | Association areas combine visual images with other sensory experiences. |

A person with *dyslexia* sees letters separately and must be taught to read in a different way than people whose nervous systems allow them to group letters into words. Three to 10% of people have dyslexia. The condition probably has several causes, with inborn visual and perceptual skills interacting with the way the child learns to read. Dyslexia has nothing to do with intelligence—many brilliant thinkers were "slow" in school because educators had not yet learned how to help them.

**1** How does the brain form during early development?

**2** Describe the cerebrum.

**3** List the general functions of the cerebrum.

**4** Where in the brain are the primary motor and sensory regions located?

**5** Explain the functions of association areas.

## Hemisphere Dominance

Both cerebral hemispheres participate in basic functions, such as receiving and analyzing sensory impulses, controlling skeletal muscles on opposite sides of the body, and storing memory. However, in most persons one side acts as a **dominant hemisphere** for certain other functions.

In over 90% of the population, for example, the left hemisphere is dominant for the language-related activities of speech, writing, and reading. It is also dominant for complex intellectual functions requiring verbal, analytical, and computational skills. In other persons, the right hemisphere is dominant, and in some, the hemispheres are equally dominant.

Tests indicate that the left hemisphere is dominant in 90% of right-handed adults and in 64% of left-handed ones. The right hemisphere is dominant in 10% of right-handed adults and in 20% of left-handed ones. The hemispheres are equally dominant in the remaining 16% of left-handed persons. As a consequence of hemisphere dominance, Broca's area on one side almost completely controls the motor activities associated with speech. For this reason, over 90% of patients with language impairment stemming from problems in the cerebrum have disorders in the left hemisphere.

In addition to carrying on basic functions, the nondominant hemisphere specializes in nonverbal functions, such as motor tasks that require orientation of the body in space, understanding and interpreting musical patterns, and visual experiences. It also provides emotional and intuitive thought processes. For example, although the region in the nondominant hemisphere that corresponds to Broca's area does not control speech, it influences the emotional aspects of spoken language.

Nerve fibers of the *corpus callosum,* which connect the cerebral hemispheres, enable the dominant hemisphere to control the motor cortex of the nondominant hemisphere. These fibers also transfer sensory information reaching the nondominant hemisphere to the general interpretative area of the dominant one, where the information can be used in decision making.

## Memory

Memory, one of the most astonishing capabilities of the brain, is the consequence of learning. Whereas learning is the acquisition of new knowledge, memory is the persistence of that learning, with the ability to access it at a later time. Two types of memory, short term and long term, have been recognized for many years, and researchers are now beginning to realize that they differ in characteristics other than duration.

Short-term or "working" memories are thought to be electrical in nature. Neurons may be connected in a circuit so that the last in the series stimulates the first. As long as the pattern of stimulation continues, the thought is remembered. When the electrical events cease, so does the memory—unless it enters long-term memory.

Long-term memory probably changes the structure or function of neurons in ways that enhance synaptic transmission, perhaps by establishing certain patterns of synaptic connections. Synaptic patterns fulfill two requirements of long-term memory. First, there are enough synapses to encode an almost limitless number of memories—each of the 10 billion neurons in the cortex can make tens of thousands of synaptic connections to other neurons, forming 60 trillion synapses. Second, a certain pattern of synapses can remain unchanged for years.

Understanding how neurons in different parts of the brain encode memories, and how short-term memories are converted to long-term memories (a process called **memory consolidation**) is at the forefront of research into the functioning of the human brain. According to one theory called **long-term synaptic potentiation**, primarily in an area of the cerebral cortex called the **hippocampus,** frequent, nearly simultaneous, and repeated stimulation of the same neurons strengthens their synaptic connections. This strengthening results in more frequent action potentials triggered in postsynaptic cells in response to the repeated stimuli. Clinical Application 11.3 discusses some common causes of damage to the cerebrum.

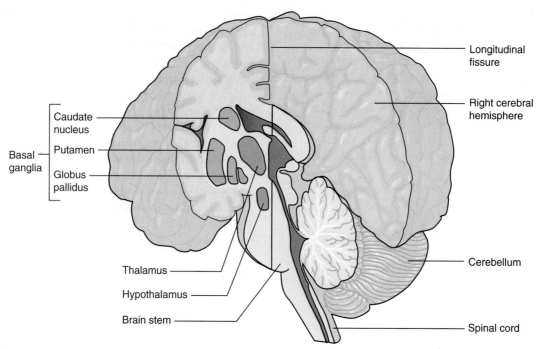

Basal ganglia
- Caudate nucleus
- Putamen
- Globus pallidus

Longitudinal fissure

Right cerebral hemisphere

Cerebellum

Thalamus

Hypothalamus

Brain stem

Spinal cord

**FIGURE 11.17**

A coronal section of the left cerebral hemisphere reveals some of the basal ganglia.

---

Like much of our knowledge about the human brain, we know about the role of the hippocampus by observing the unusual behaviors and skills of animals or people in whom these structures have been damaged. In 1953, a surgeon removed parts of the hippocampus and another area called the amygdala of a young man called H. M., thinking this drastic action might relieve his severe epilepsy. His seizures indeed became less frequent, but H. M. suffered a profound loss in the ability to consolidate short-term memories into long-term ones. As a result, events in H. M.'s life fade from memory as quickly as they occur. He is unable to recall any events that took place since surgery, living today as if it was the 1950s. He can read the same magazine article repeatedly with renewed interest each time. With practice, he improves skills that require procedural memory, such as puzzle solving. But, since factual memory is impossible, he insists that he has never seen the puzzle before!

## Basal Ganglia

The **basal ganglia** (basal nuclei) are masses of gray matter located deep within the cerebral hemispheres. They are called the *caudate nucleus,* the *putamen,* and the *globus pallidus,* and they develop from the anterior portion of the forebrain (fig. 11.17). The neuron cell bodies that the basal ganglia contain relay motor impulses originating in the cerebral cortex and passing into the brain stem and spinal cord. The basal ganglia produce most of the inhibitory neurotransmitter *dopamine.* Impulses from the basal ganglia normally inhibit motor functions, thus controlling certain muscular activities. Clinical Application 11.4 discusses Parkinson's disease, in which neurons in the basal ganglia degenerate.

1. What is hemisphere dominance?

2. What are the functions of the nondominant hemisphere?

3. Distinguish between short-term and long-term memory.

4. What is the function of the basal ganglia?

## Ventricles and Cerebrospinal Fluid

A series of interconnected cavities called *ventricles* are located within the cerebral hemispheres and brain stem (fig. 11.18 and reference plates 53 and 54). These spaces are continuous with the central canal of the spinal cord and are filled with cerebrospinal fluid.

The largest ventricles are the *lateral ventricles,* which are the first and second ventricles (the first ventricle in the left cerebral hemisphere and the second ventricle in the right cerebral hemisphere). They extend into

# Cerebral Injuries and Abnormalities

The specific symptoms associated with a cerebral injury or abnormality depend upon the areas and extent of damage. A person with damage to the association areas of the frontal lobes may have difficulty concentrating on complex mental tasks, appearing disorganized and easily distracted.

If the general interpretative area of the dominant hemisphere is injured, the person may be unable to interpret sounds as words or to understand written ideas. However, the dominance of one hemisphere usually does not become established until after five or six years of age. Consequently, if the general interpretative area is destroyed in a child, the corresponding region of the other side of the brain may be able to take over the functions, and the child's language abilities may develop normally. If such an injury occurs in an adult, the non-dominant hemisphere may develop only limited interpretative functions, producing a severe intellectual disability. Following are three common cerebral abnormalities.

- In a *concussion,* the brain is jarred against the cranium, usually as a result of a blow to the head, causing loss of consciousness. Short-term memory loss, mental cloudiness, difficulty concentrating and remembering, and a fierce headache may occur in the days after a concussion, but recovery is usually complete.

- *Cerebral palsy* (CP) is motor impairment at birth, often stemming from a brain anomaly occurring during prenatal development. Until recently, most cases of CP were blamed on "birth trauma," but recently researchers determined that the most common cause is a blocked cerebral blood vessel, which leads to atrophy of the brain region deprived of its blood supply. Birth trauma and brain infection cause some cases.

CP affects about 1 in every 1,000 births and is especially prevalent among premature babies. One-half to two-thirds of affected babies improve and can even outgrow the condition by age 7. Sometimes seizures or learning disabilities are present. Clinicians classify CP by the number of limbs and the types of neurons affected.

- In a "stroke," or *cerebrovascular accident* (CVA), a sudden interruption in blood flow in a vessel supplying brain tissues damages the cerebrum. The affected blood vessel may rupture, bleeding into the brain, or be blocked by a clot. In either case, brain tissues downstream from the vascular accident die or permanently lose function. Temporary interruption in cerebral blood flow, perhaps by a clot that quickly breaks apart, produces a much less serious *transient ischemic attack* (TIA). ∎

---

the cerebral hemispheres and occupy portions of the frontal, temporal, and occipital lobes.

A narrow space that constitutes the *third ventricle* is located in the midline of the brain, beneath the corpus callosum. This ventricle communicates with the lateral ventricles through openings (*interventricular foramina*) in its anterior end.

The *fourth ventricle* is located in the brain stem just in front of the cerebellum. A narrow canal, the *cerebral aqueduct* (aqueduct of Sylvius), connects it to the third ventricle and passes lengthwise through the brain stem. This ventricle is continuous with the central canal of the spinal cord and has openings in its roof that lead into the subarachnoid space of the meninges.

Tiny, reddish cauliflowerlike masses of specialized capillaries from the pia mater, called **choroid plexuses,** secrete *cerebrospinal fluid* (CSF). These structures project into the cavities of the ventricles (fig. 11.19). A single layer of specialized ependymal cells (see chapter 10) joined closely by tight junctions covers the choroid plexuses. In much the same way that astrocytes provide a barrier between the blood and the brain interstitial fluid (blood-brain barrier), these cells block passage of water-soluble substances between the blood and the

# Parkinson's Disease

Parkinson's disease affects millions of people worldwide. In this disorder, neurons in the basal ganglia that synthesize the neurotransmitter dopamine degenerate. The resulting decline in dopamine levels in the striatum causes slowed movements, difficulty in initiating voluntary muscular actions, rigidity, loss of balance, and tremors.

The standard treatment for many years has been to give levo-dopa, which is a precursor to dopamine that can cross the blood-brain barrier. Once in the brain, levo-dopa is converted to dopamine, which cannot cross the barrier. However, with prolonged use, the brain becomes dependent on the external supply and decreases its own production of dopamine even further. Eventually levo-dopa only alleviates symptoms on some days. A new drug, pramipexole, seems to even out the supply of dopamine by mimicking the action of dopamine. If given with levo-dopa, the new drug improves symptoms and even helps lift depression associated with advanced Parkinson's disease. Prolonged use of levo-dopa can also cause tardive dyskinesia, a condition that produces uncontrollable facial tics, perhaps by abnormally increasing the dopamine level in brain areas not implicated in Parkinson's disease.

A controversial and experimental treatment for Parkinson's disease is to transplant fetal brain tissue into patients' brains, where the tissue produces dopamine. These procedures have been done experimentally since the 1980s. At first, results were mixed, because often the patient's immune system destroyed the foreign tissue. By 1995, giving immune-suppressant drugs to transplant recipients greatly improved the success of the procedure. One pa-

tient, treated in 1995, experienced a return of function. He could once again feed and dress himself, and could lower his daily dosage of levo-dopa. After his death from other causes 18 months after the transplant, his brain was found to contain dense groupings of neurons in the striatum that had actively secreted dopamine—proof that the transplant had a least partially corrected the disease.

The cause of Parkinson's disease isn't known. It has been attributed to exposure to certain chemicals in pesti-

cides and designer drugs, and to severe and frequent injury such as occurred in heavyweight champ Mohammed Ali, who suffers from the disorder (fig. 11B). The chemical exposures and injury may increase the rate of oxygen free radical formation, which damages brain tissue.

Clues to the cause of Parkinson's disease may come from genetics. Researchers have identified a mutant gene that accounts for the disease in several well-studied, large families. The gene encodes a protein called alpha-synuclein, which is found in the basal ganglia. When abnormal, the protein folds improperly, which may cause deposits to form in the brain. The next step in the research will be to determine whether the brain deposits characteristic of Parkinson's disease, called Lewy bodies (fig. 11C), are comprised of misfolded alpha-synuclein.

Only a small percentage of cases of Parkinson's disease are inherited. However, understanding how the symptoms arise in these individuals will provide clues that may eventually help the many others with the disease too. ■

**FIGURE 11B**

Professional boxers are at higher risk of developing Parkinson's disease from repeated blows to the head. Muhammed Ali has Parkinson's disease as a result of many years of receiving blows to the head.

**FIGURE 11C**

The chemical composition of Lewy bodies, which are characteristic of the brains of people with Parkinson's disease, may provide clues to the cause of the condition.

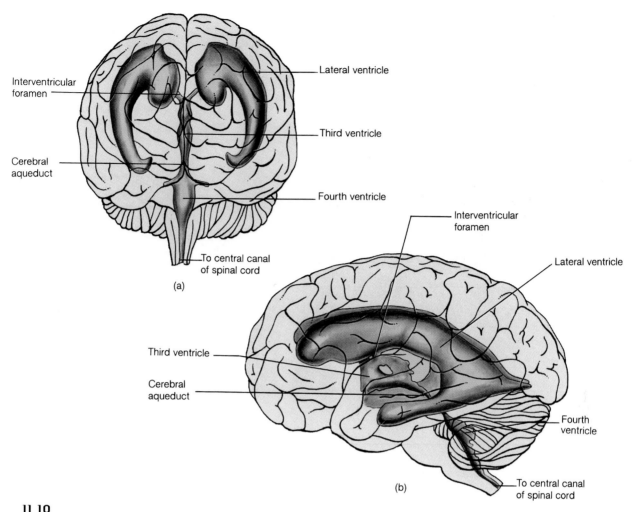

**FIGURE 11.18**

(a) Anterior view of the ventricles within the cerebral hemispheres and brain stem. (b) Lateral view.

cerebrospinal fluid. At the same time, the cells selectively transfer certain substances from the blood into the cerebrospinal fluid by facilitated diffusion and transfer other substances by active transport (see chapter 3), thus regulating the composition of the cerebrospinal fluid.

Most of the cerebrospinal fluid arises in the lateral ventricles, from where it slowly circulates into the third and fourth ventricles and into the central canal of the spinal cord. It also enters the subarachnoid space of the meninges by passing through the wall of the fourth ventricle near the cerebellum.

Humans secrete nearly 500 milliliters of cerebrospinal fluid daily. However, only about 140 milliliters are in the nervous system at any time, because cerebrospinal fluid is continuously reabsorbed into the blood. The CSF is reabsorbed through tiny, fingerlike structures called *arachnoid granulations* that project from the subarachnoid space into the blood-filled dural sinuses (fig. 11.19).

Cerebrospinal fluid is a clear, somewhat viscid liquid that differs in composition from the fluid that leaves the capillaries in other parts of the body. Specifically, it contains a greater concentration of sodium and lesser concentrations of glucose and potassium than do other extracellular fluids. Its function is nutritive as well as protective. Cerebrospinal fluid helps maintain a stable ionic concentration in the central nervous system and provides a pathway to the blood for waste. The cerebrospinal fluid may also supply information about the internal environment to autonomic centers in the hypothalamus and brain stem, because the fluid forms from blood plasma and therefore its composition reflects changes in body fluids. Clinical Application 11.5 discusses the pressure that cerebrospinal fluid generates.

Because it occupies the subarachnoid space of the meninges, cerebrospinal fluid completely surrounds the brain and spinal cord. In effect, these organs float in the

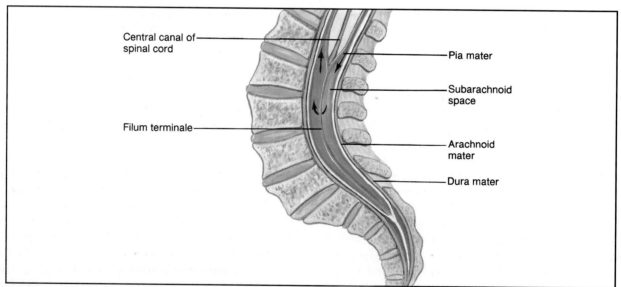

**FIGURE 11.19**

Choroid plexuses in ventricle walls secrete cerebrospinal fluid. The fluid circulates through the ventricles and central canal, enters the subarachnoid space, and is reabsorbed into the blood of the dural sinuses through arachnoid granulations.

fluid. The CSF protects them by absorbing forces that might otherwise jar and damage their delicate tissues.

**1** Where are the ventricles of the brain located?

**2** How does cerebrospinal fluid form?

**3** Describe the pattern of cerebrospinal fluid circulation.

## Diencephalon

The **diencephalon** develops from the posterior forebrain and is located between the cerebral hemispheres and above the brain stem (see figs. 11.13 and 11.20). It surrounds the third ventricle and is largely composed of gray matter. Within the diencephalon, a dense mass, called the **thalamus,** bulges into the third ventricle

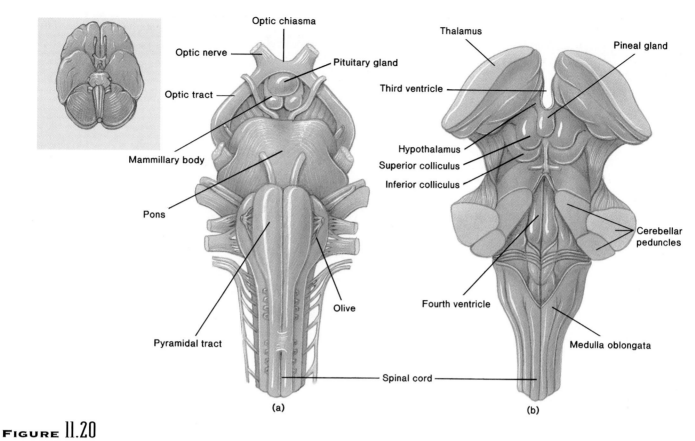

Optic chiasma
Optic nerve
Pituitary gland
Optic tract
Mammillary body
Pons
Olive
Pyramidal tract
Spinal cord

Thalamus
Pineal gland
Third ventricle
Hypothalamus
Superior colliculus
Inferior colliculus
Cerebellar peduncles
Fourth ventricle
Medulla oblongata

(a)

(b)

**FIGURE 11.20**

(*a*) Ventral view of the brain stem. (*b*) Dorsal view of the brain stem with the cerebellum removed, exposing the fourth ventricle.

from each side. Another region of the diencephalon that includes many nuclei is the **hypothalamus.** It lies below the thalamic nuclei and forms the lower walls and floor of the third ventricle (see reference plates 49 and 53).

Other parts of the diencephalon include (1) the **optic tracts** and the **optic chiasma** that is formed by the optic nerve fibers crossing over; (2) the **infundibulum,** a conical process behind the optic chiasma to which the pituitary gland is attached; (3) the **posterior pituitary gland,** which hangs from the floor of the hypothalamus; (4) the **mammillary** (mam′ĭ-ler″e) **bodies,** which are two rounded structures behind the infundibulum; and (5) the **pineal gland,** which forms as a cone-shaped evagination from the roof of the diencephalon (see chapter 13).

The thalamus is a selective gateway for sensory impulses ascending from other parts of the nervous system to the cerebral cortex. It receives all sensory impulses (except those associated with the sense of smell) and channels them to appropriate regions of the cortex for interpretation. In addition, all regions of the cerebral cortex

can communicate with the thalamus by means of descending fibers.

The thalamus seems to transmit sensory information by synchronizing action potentials. Consider vision. An image on the retina stimulates the *lateral geniculate nucleus* (LGN) region of the thalamus, which then sends action potentials to a part of the visual cortex. Researchers have observed that those action potentials are synchronized—that is, fired simultaneously—by the LGN's neurons only if the stimuli come from a single object, such as a bar. If the stimulus is two black dots, the resulting thalamic action potentials are not synchronized. The synchronicity of action potentials, therefore, may be a way that the thalamus selects which stimuli to relay to higher brain structures. Therefore, the thalamus is not only a messenger but also an editor.

Nerve fibers connect the hypothalamus to the cerebral cortex, thalamus, and parts of the brain stem so that it can receive impulses from them and send impulses to them. The hypothalamus maintains homeostasis by regulating a variety of visceral activities and by linking the nervous and endocrine systems.

# Cerebrospinal Fluid Pressure

Because cerebrospinal fluid (CSF) is secreted and reabsorbed continuously, the fluid pressure in the ventricles remains relatively constant. However, infection, a tumor, or a blood clot can interfere with the fluid's circulation, increasing pressure within the ventricles (intracranial pressure). This can collapse cerebral blood vessels, retarding blood flow. Brain tissues may be injured by being forced against the skull.

A *lumbar puncture* (spinal tap) measures CSF pressure. A physician inserts a fine, hollow needle into the subarachnoid space between the third and fourth or between the fourth and fifth lumbar vertebrae—below the end of the spinal cord (fig. 11D). An instrument called a *manometer* measures the pressure of the fluid, which is usually about

**FIGURE 11D**

A lumbar puncture is performed by inserting a fine needle between the third and fourth lumbar vertebrae and withdrawing a sample of cerebrospinal fluid from the subarachnoid space.

Spinal cord

Conus medullaris

Subarachnoid space

Third lumbar vertebra

Dura mater

Arachnoid mater

Skin

Sacrum

Filum terminale

Coccyx

The hypothalamus regulates:

1. Heart rate and arterial blood pressure.
2. Body temperature.
3. Water and electrolyte balance.
4. Control of hunger and body weight.
5. Control of movements and glandular secretions of the stomach and intestines.
6. Production of neurosecretory substances that stimulate the pituitary gland to release hormones that help regulate growth, control various glands, and influence reproductive physiology.
7. Sleep and wakefulness.

Structures in the region of the diencephalon also are important in controlling emotional responses. For example, portions of the cerebral cortex in the medial parts of the frontal and temporal lobes connect with the hypo-

130 millimeters of water (10 millimeters of mercury). At the same time, samples of CSF may be withdrawn and tested for the presence of abnormal constituents. Red blood cells in the CSF, for example, may indicate a hemorrhage in the central nervous system.

A temporary drain inserted into the subarachnoid space between the fourth and fifth lumbar vertebrae can relieve pressure. In a fetus or infant whose cranial sutures have not yet united, increasing intracranial pressure (ICP) may cause an enlargement of the cranium called *hydrocephalus,* or "water on the brain." A shunt to relieve hydrocephalus drains fluid away from the cranial cavity and into the digestive tract, where it is either reabsorbed into the blood or excreted (fig. 11E). ■

(a)

Ventricles
(b)

**FIGURE** 11E
CT scans of the human brain. (*a*) Normal ventricles. (*b*) Ventricles enlarged by accumulated fluid.

thalamus, thalamus, basal ganglia, and other deep nuclei. Together, these structures comprise a complex called the limbic system.

The **limbic system** controls emotional experience and expression and can modify the way a person acts. It produces such feelings as fear, anger, pleasure, and sorrow. The limbic system seems to recognize upsets in a person's physical or psychological condition that might threaten life. By causing pleasant or unpleasant feelings about experiences, the limbic system guides a person into behavior that may increase the chance of survival. In addition, portions of the limbic system interpret sensory impulses from the receptors associated with the sense of smell (olfactory receptors).

### Brain Stem

The brain stem connects the brain to the spinal cord. It consists of the midbrain, pons, and medulla oblongata. These structures include many tracts of nerve fibers and masses of gray matter called *nuclei* (see figs. 11.13 and 11.20).

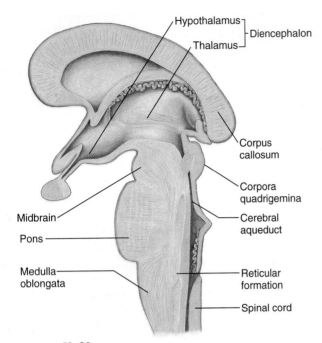

**FIGURE** 11.21

The reticular formation (shown in green) extends from the superior portion of the spinal cord into the diencephalon.

## Midbrain

The **midbrain** (mesencephalon) is a short section of the brain stem between the diencephalon and the pons. It contains bundles of myelinated nerve fibers that join lower parts of the brain stem and spinal cord with higher parts of the brain. The midbrain includes several masses of gray matter that serve as reflex centers. It also contains the *cerebral aqueduct* that connects the third and fourth ventricles (fig. 11.21).

Two prominent bundles of nerve fibers on the underside of the midbrain comprise the *cerebral peduncles.* These fibers include the corticospinal tracts and are the main motor pathways between the cerebrum and lower parts of the nervous system (see fig. 11.20). Beneath the cerebral peduncles are some large bundles of sensory fibers that carry impulses upward to the thalamus.

Two pairs of rounded knobs on the superior surface of the midbrain mark the location of four nuclei, known collectively as *corpora quadrigemina.* The upper masses (superior colliculi) contain the centers for certain visual reflexes, such as those responsible for moving the eyes to view something as the head turns. The lower ones (inferior colliculi) contain the auditory reflex centers that operate when it is necessary to move the head to hear sounds more distinctly (see fig. 11.20).

Near the center of the midbrain is a mass of gray matter called the *red nucleus.* This nucleus communicates with the cerebellum and with centers of the spinal cord, and it provides reflexes that maintain posture.

## Pons

The **pons** appears as a rounded bulge on the underside of the brain stem where it separates the midbrain from the medulla oblongata (see fig. 11.20). The dorsal portion of the pons largely consists of longitudinal nerve fibers, which relay impulses to and from the medulla oblongata and the cerebrum. Its ventral portion contains large bundles of transverse nerve fibers, which transmit impulses from the cerebrum to centers within the cerebellum.

Several nuclei of the pons relay sensory impulses from peripheral nerves to higher brain centers. Other nuclei function with centers of the medulla oblongata to regulate the rate and depth of breathing.

## Medulla Oblongata

The **medulla oblongata** is an enlarged continuation of the spinal cord extending from the level of the foramen magnum to the pons (see fig. 11.20). Its dorsal surface flattens to form the floor of the fourth ventricle, and its ventral surface is marked by the corticospinal tracts, most of whose fibers cross over at this level. On each side of the medulla oblongata is an oval swelling called the *olive,* from which a large bundle of nerve fibers arises and passes to the cerebellum.

Because of the medulla oblongata's location, all the ascending and descending nerve fibers connecting the brain and spinal cord must pass through it. As in the spinal cord, the white matter of the medulla surrounds a central mass of gray matter. Here, however, the gray matter breaks up into nuclei that are separated by nerve fibers. Some of these nuclei relay ascending impulses to the other side of the brain stem and then on to higher brain centers. The *nucleus gracilis* and the *nucleus cuneatus,* for example, receive sensory impulses from fibers of the fasciculus gracilis and the fasciculus cuneatus, and pass them on to the thalamus or the cerebellum.

Other nuclei within the medulla oblongata control vital visceral activities. These centers include the following:

1. **Cardiac center.** Peripheral nerves transmit impulses originating in the cardiac center to the heart, where they increase or decrease heart rate.

2. **Vasomotor center.** Certain cells of the vasomotor center initiate impulses that travel to smooth muscles in the walls of blood vessels and stimulate them to contract, constricting the vessels (vasoconstriction) and thereby increasing blood pressure. A decrease in the activity of these cells can produce the opposite effect—dilation of the blood vessels (vasodilation) and a consequent drop in blood pressure.

3. **Respiratory center.** The respiratory center acts with centers in the pons to regulate the rate, rhythm, and depth of breathing.

Some nuclei within the medulla oblongata are centers for certain nonvital reflexes, such as those associated with coughing, sneezing, swallowing, and vomiting. However, since the medulla also contains vital reflex centers, injuries to this part of the brain stem are often fatal.

## Reticular Formation

Scattered throughout the medulla oblongata, pons, and midbrain is a complex network of nerve fibers associated with tiny islands of gray matter. This network, the **reticular formation** (reticular activating system), extends from the superior portion of the spinal cord into the diencephalon (fig. 11.21). Its intricate system of nerve fibers connects centers of the hypothalamus, basal ganglia, cerebellum, and cerebrum with fibers in all the major ascending and descending tracts.

When sensory impulses reach the reticular formation, it responds by activating the cerebral cortex into a state of wakefulness. Without this arousal, the cortex remains unaware of stimulation and cannot interpret sensory information or carry on thought processes. Thus, decreased activity in the reticular formation results in sleep. If the reticular formation is injured and ceases to function, the person remains unconscious, even with strong stimulation. This is called a comatose state.

---

A person in a persistent vegetative state is occasionally awake, but not aware; a person in a coma is not awake or aware. Sometimes following a severe injury a person will become comatose and then gradually enter a persistent vegetative state. Coma and persistent vegetative states are also seen in the end stage of neurodegenerative disorders such as Alzheimer's disease; when there is an untreatable mass in the brain, such as a blood clot or tumor; or in anencephaly, when a newborn lacks higher brain structures.

---

The reticular formation also filters incoming sensory impulses. Impulses judged to be important, such as those originating in pain receptors, are passed on to the cerebral cortex, while others are disregarded. This selective action of the reticular formation frees the cortex from what would otherwise be a continual bombardment of sensory stimulation, and allows it to concentrate on more significant information. The cerebral cortex can also activate the reticular system, so that intense cerebral activity tends to keep a person awake. In addition, the reticular formation regulates motor activities so that various skeletal muscles move together evenly, and it inhibits or enhances certain spinal reflexes.

## Types of Sleep

The two types of normal sleep are *slow wave* and *rapid eye movement* (REM). Slow-wave sleep (also called non-REM sleep) occurs when a person is very tired, and reflects decreasing activity of the reticular formation. It is restful, dreamless, and accompanied by reduced blood pressure and respiratory rate. Slow-wave sleep may range from light to heavy and is usually described in four stages. It may last from 70 to 90 minutes. Slow-wave and REM sleep alternate.

REM sleep is also called "paradoxical sleep" because some areas of the brain are active. As its name implies, the eyes can be seen rapidly moving beneath the eyelids. Cats and dogs in REM sleep sometimes twitch their limbs. In humans, REM sleep usually lasts from 5 to 15 minutes. This "dream sleep" is apparently very important. If a person lacks REM sleep for just one night, sleep on the next night makes up for it! During REM sleep, heart and respiratory rates are irregular. Certain drugs, such as marijuana and alcohol, interfere with REM sleep. Table 11.6 describes several disorders of sleep.

---

The person to go the longest without sleep was a 17-year-old student who stayed awake, in a sleep laboratory under medical supervision, for 264 hours. Fortunately he suffered no ill effects, but during his ordeal he was irritable and had blurred vision, slurred speech, and memory lapses. Toward the end, he seemed confused about his identity.

Rats experimentally deprived of sleep do not fare as well. In a study conducted at the University of Chicago, rats kept awake by a moving floor developed skin sores and hormonal and metabolic changes, dying within 11 to 32 days. Control rats allowed to nap survived.

---

**1** What are the major functions of the thalamus? Of the hypothalamus?

**2** How may the limbic system influence a person's behavior?

**3** Which vital reflex centers are located in the brain stem?

**4** What is the function of the reticular formation?

**5** Describe two types of sleep.

## Cerebellum

The **cerebellum** is a large mass of tissue located inferior to the occipital lobes of the cerebrum and posterior to the pons and medulla oblongata (see figs. 11.13 and 11.22). It consists of two lateral hemispheres partially separated by a layer of dura mater called the *falx cerebelli*. A structure called the *vermis* connects the cerebellar hemispheres at the midline.

TABLE 11.6  SLEEP DISORDERS

| Disorder | Symptoms | Percent of Population |
|---|---|---|
| Fatal familial insomnia | Inability to sleep. Emotional instability, hallucinations, stupor, coma, death within 13 months of onset around age 50. Both slow-wave and REM sleep abolished. | Very rare, inherited |
| Insomnia | Inability to fall or remain asleep. | 10% |
| Narcolepsy | Abnormal REM sleep causes extreme daytime sleepiness. Begins between ages of 15 and 25. | 0.02–0.06% |
| Obstructive sleep apnea syndrome | Upper airway collapses repeatedly during sleep, blocking breathing. Snoring and daytime sleepiness. | 4–5% |
| Parasomnias | Sleepwalking, sleeptalking, and night terrors. Outgrown. | <5% of children |
| REM-sleep behavior disorder | Excessive motor activity during REM sleep, which disturbs continuous sleep. | Very rare |
| Restless-leg syndrome | Brief, repetitive leg jerks during sleep. Leg pain forces person to get up several times a night. | Very rare, inherited |
| Sleep paralysis | Inability to move for up to a few minutes after awakening or when falling asleep. | Very rare, inherited |

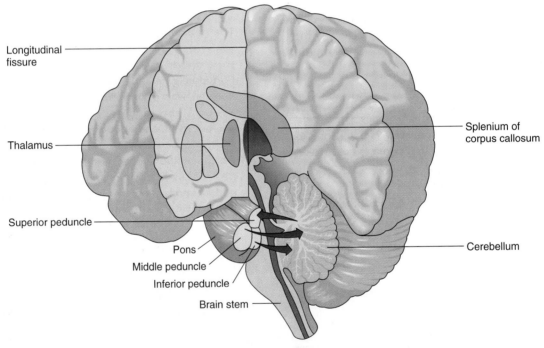

**FIGURE 11.22**

The cerebellum, which is located below the occipital lobes of the cerebrum, communicates with other parts of the nervous system by means of the cerebellar peduncles.

Like the cerebrum, the cerebellum is primarily composed of white matter with a thin layer of gray matter, the **cerebellar cortex,** on its surface. This cortex doubles over on itself in a series of complex folds that have myelinated nerve fibers branching into them. A cut into the cerebellum reveals a treelike pattern of white matter, called the *arbor vitae,* that is surrounded by gray matter. A number of nuclei lie deep within each cerebellar

hemisphere. The largest and most important is the *dentate nucleus.*

The cerebellum communicates with other parts of the central nervous system by means of three pairs of nerve tracts called **cerebellar peduncles** (ser″ĕ-bel′ar pe-dung′k'ls). One pair, the *inferior peduncles,* brings sensory information concerning the actual position of body parts such as limbs and joints to the cerebellum via the spinal cord and medulla oblongata. The *middle peduncles* transmit information from the cerebral cortex about the desired position of these body parts. After integrating and analyzing the information from these two sources, the cerebellum sends correcting impulses from the dentate nucleus via the *superior peduncles* to the midbrain (fig. 11.22). These corrections are incorporated into motor impulses that travel downward through the pons, medulla oblongata, and spinal cord in the appropriate patterns to move the body in the desired way.

Thus, the cerebellum integrates sensory information concerning the position of body parts and coordinates skeletal muscle activity and maintains posture. It receives sensory impulses from receptors in muscles, tendons, and joints (proprioceptors) and from special sense organs, such as the eyes and ears (see chapter 12). For example, the cerebellum uses sensory information from the semicircular canals of the inner ears concerning the motion and position of the head to help maintain equilibrium. Damage to the cerebellum is likely to result in tremors, inaccurate movements of voluntary muscles, loss of muscle tone, a reeling walk, and loss of equilibrium (fig. 11.23).

Table 11.7 summarizes the characteristics and functions of the major parts of the brain. Clinical Application 11.6 discusses how brain waves reflect brain activity.

**1** Where is the cerebellum located?

**2** What are the major functions of the cerebellum?

**3** What kinds of receptors provide information to the cerebellum?

**FIGURE 11.23**

A sagittal section of the brain showing the cerebrum, diencephalon, brain stem, and cerebellum (see also plate 76).

| TABLE 11.7 | MAJOR PARTS OF THE BRAIN | |
| --- | --- | --- |
| **Part** | **Characteristics** | **Functions** |
| 1. Cerebrum | Largest part of the brain; two hemispheres connected by the corpus callosum | Controls higher brain functions, including interpreting sensory impulses, initiating muscular movements, storing memory, reasoning, and determining intelligence |
| 2. Basal ganglia | Masses of gray matter deep within the cerebral hemispheres | Relay stations for motor impulses originating in the cerebral cortex and passing into the brain stem and spinal cord |
| 3. Diencephalon | Includes masses of gray matter (thalamus and hypothalamus) | The thalamus is a relay station for sensory impulses ascending from other parts of the nervous system to the cerebral cortex; the hypothalamus helps maintain homeostasis by regulating visceral activities and by linking the nervous and endocrine systems |
| 4. Brain stem | Connects the cerebrum to the spinal cord | |
|   a. Midbrain | Contains masses of gray matter and bundles of nerve fibers that join the spinal cord to higher regions of the brain | Contains reflex centers that move the eyes and head, and maintains posture |
|   b. Pons | A bulge on the underside of the brain stem that contains masses of gray matter and nerve fibers | Relays nerve impulses to and from the medulla oblongata and cerebrum; helps regulate rate and depth of breathing |
|   c. Medulla oblongata | An enlarged continuation of the spinal cord that extends from the foramen magnum to the pons, and contains masses of gray matter and nerve fibers | Conducts ascending and descending impulses between the brain and spinal cord; contains cardiac, vasomotor, and respiratory control centers and various nonvital reflex control centers |
| 5. Cerebellum | A large mass of tissue located below the cerebrum and posterior to the brain stem; includes two lateral hemispheres connected by the vermis | Communicates with other parts of the CNS by nerve tracts; integrates sensory information concerning the position of body parts; and coordinates muscle activities and maintains posture |

# Brain Waves

*Brain waves* are recordings of fluctuating electrical changes in the brain. To obtain such a recording, electrodes are positioned on the surface of a surgically exposed brain (an electrocorticogram, ECoG) or on the outer surface of the head (an electroencephalogram, EEG). These electrodes detect electrical changes in the extracellular fluid of the brain in response to changes in potential among large groups of neurons. The resulting signals from the electrodes are amplified and recorded. Brain waves originate from the cerebral cortex, but also reflect activities in other parts of the brain that influence the cortex, such as the reticular formation. Because the intensity of electrical changes is proportional to the degree of neuronal activity, brain waves vary markedly in amplitude and frequency between sleep and wakefulness.

Brain waves are classified as alpha, beta, theta, and delta waves. *Alpha waves* are recorded most easily from the posterior regions of the head and have a frequency of 8–13 cycles per second. They occur when a person is awake but resting, with the eyes closed. These waves disappear during sleep, and if the wakeful person's eyes open, higher frequency beta waves replace the alpha waves.

*Beta waves* have a frequency of more than 13 cycles per second and are usually recorded in the anterior region of the head. They occur when a person is actively engaged in mental activity or is under tension.

*Theta waves* have a frequency of 4–7 cycles per second and occur mainly in the parietal and temporal regions of the cerebrum. They are normal in children, but do not usually occur in adults. However, some adults produce theta waves in early stages of sleep or at times of emotional stress.

*Delta waves* have a frequency below 4 cycles per second and occur during sleep. They originate from the cerebral cortex when it is not being activated by the reticular formation (fig. 11F).

Brain wave patterns can be useful for diagnosing disease conditions, such as distinguishing types of seizure disorders (epilepsy) and locating brain tumors. Brain waves are also used to determine when *brain death* has occurred. Brain death, characterized by the cessation of neuronal activity, may be verified by an EEG that lacks waves (isoelectric EEG). However, drugs that greatly depress brain functions must be excluded as the cause of the flat EEG pattern before brain death can be confirmed. ∎

**FIGURE 11F**

Brain waves record fluctuating electrical changes in the brain.

Alpha waves

Beta waves

Theta waves

Delta waves

50 µv

1 sec.

| TABLE 11.8 | SUBDIVISIONS OF THE NERVOUS SYSTEM |
|---|---|

1. Central nervous system (CNS)
   a. Brain
   b. Spinal cord
2. Peripheral nervous system (PNS)
   a. Cranial nerves arising from the brain
      (1) Somatic fibers connecting to the skin and skeletal muscles
      (2) Autonomic fibers connecting to viscera
   b. Spinal nerves arising from the spinal cord
      (1) Somatic fibers connecting to the skin and skeletal muscles
      (2) Autonomic fibers connecting to viscera

## PERIPHERAL NERVOUS SYSTEM

The **peripheral nervous system** (PNS) consists of the nerves that branch from the central nervous system (CNS), connecting it to other body parts. The PNS includes the *cranial nerves* that arise from the brain and the *spinal nerves* that arise from the spinal cord.

The peripheral nervous system can also be subdivided into somatic and autonomic nervous systems. Generally, the **somatic nervous system** consists of the cranial and spinal nerve fibers that connect the CNS to the skin and skeletal muscles, so it oversees conscious activities. The **autonomic nervous system** includes fibers that connect the CNS to viscera such as the heart, stomach, intestines, and various glands. Thus, the autonomic nervous system controls unconscious actions. Table 11.8 outlines the subdivisions of the nervous system.

### Structure of Peripheral Nerves

As described in chapter 10, a peripheral nerve consists of connective tissue surrounding bundles of nerve fibers. The outermost layer of the connective tissue, called the *epineurium,* is dense and includes many collagenous fibers. Each bundle of nerve fibers (fascicle) is, in turn, enclosed in a sleeve of looser connective tissue called the *perineurium.* A small amount of loose connective tissue called *endoneurium* surrounds individual nerve fibers (figs. 11.24 and 11.25). Blood vessels in the epineurium and perineurium give rise to a network of capillaries in the endoneurium.

### Nerve Fiber Classification

Like nerve fibers, nerves that conduct impulses into the brain or spinal cord are called **sensory nerves,** and those that carry impulses to muscles or glands are termed **motor nerves.** Most nerves, however, include both sensory and motor fibers, and they are called **mixed nerves.**

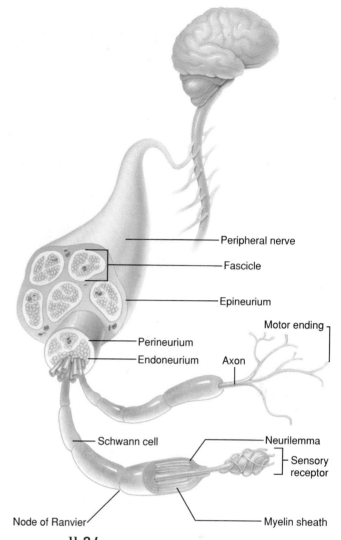

**FIGURE 11.24**

The structure of a peripheral mixed nerve.

Nerves originating from the brain that communicate with other body parts are called **cranial nerves,** whereas those originating from the spinal cord that communicate with other body parts are called **spinal nerves.** The nerve fibers within these structures can be subdivided further into four groups as follows:

1. **General somatic efferent fibers** carry motor impulses outward from the brain or spinal cord to skeletal muscles and stimulate them to contract.

2. **General visceral efferent fibers** carry motor impulses outward to various smooth muscles and glands associated with internal organs, causing certain muscles to contract or glands to secrete.

3. **General somatic afferent fibers** carry sensory impulses inward to the brain or spinal cord from receptors in the skin and skeletal muscles.

**CHAPTER ELEVEN**   Nervous System II

407

Epineurium

Perineurium

Endoneurium

Blood vessel

Fascicle

Nerve fiber

**FIGURE 11.25**

Scanning electron micrograph of a peripheral nerve in cross section (350×). Note the bundles or fascicles of nerve fibers. Copyright by R. G. Kessel and R. H. Kardon, *Tissues and Organs: A Text-Atlas of Scanning Electron Microscopy.* All rights reserved.

4. **General visceral afferent fibers** carry sensory impulses to the central nervous system from blood vessels and internal organs.

The term *general* in each of these categories indicates that the fibers are associated with general structures such as the skin, skeletal muscles, glands, and viscera. Three other groups of fibers, found only in cranial nerves, are associated with more specialized, or *special,* structures:

1. **Special visceral efferent fibers** carry motor impulses outward from the brain to the muscles used in chewing, swallowing, speaking, and forming facial expressions.

2. **Special visceral afferent fibers** carry sensory impulses inward to the brain from the olfactory and taste receptors.

3. **Special somatic afferent fibers** carry sensory impulses inward from the receptors of sight, hearing, and equilibrium.

## Cranial Nerves

Twelve pairs of **cranial nerves** arise from the underside of the brain. Except for the first pair, which begins within the cerebrum, these nerves originate from the brain stem. They pass from their sites of origin through foramina of the skull and lead to areas of the head, neck, and trunk.

Although most cranial nerves are mixed nerves, some of those associated with special senses, such as smell and vision, contain only sensory fibers. Others that innervate muscles and glands are primarily composed of motor fibers and have only limited sensory functions.

Neuron cell bodies to which the sensory fibers in the cranial nerves attach are located outside the brain and are usually in groups called *ganglia* (sing., *ganglion*). On the other hand, motor neuron cell bodies are typically located within the gray matter of the brain.

Cranial nerves are designated by numbers or name. The numbers indicate the order in which the nerves arise from the brain, from anterior to posterior. The names describe primary functions or the general distribution of their fibers (fig. 11.26).

The first pair of cranial nerves, the **olfactory nerves (I),** are associated with the sense of smell and contain only sensory neurons. These neurons synapse with bipolar neurons, located in the lining of the upper nasal cavity, that serve as *olfactory receptor cells.* Axons from these receptors pass upward through the cribriform plates of the ethmoid bone. The synapses occur in the *olfactory bulbs,* which are extensions of the cerebral cortex, located just beneath the frontal lobes. Sensory impulses travel from the olfactory bulbs along *olfactory tracts* to cerebral centers where they are interpreted. The result of this interpretation is the sensation of smell.

The second pair, the **optic nerves (II),** lead from the eyes to the brain and are associated with vision. The sensory cell bodies of these nerve fibers form ganglion cell layers within the eyes, and their axons pass through the *optic foramina* of the orbits and continue into the visual nerve pathways of the brain (see chapter 12).

The third pair, the **oculomotor** (ok"u-lo-mo'tor) **nerves (III),** arise from the midbrain and pass into the orbits of the eyes. One component of each nerve connects to a number of voluntary muscles, including those that raise the eyelids and four of the six muscles that move the eye.

Olfactory bulb

Olfactory tract

Optic tract

Vestibulocochlear (VIII)

Hypoglossal (XII)

Vagus (X)

Olfactory (I)

Optic (II)

Oculomotor (III)

Trochlear (IV)

Trigeminal (V)

Abducens (VI)

Facial (VII)

Glossopharyngeal (IX)

Accessory (XI)

**FIGURE 11.26**

The cranial nerves, except for the first pair, arise from the brain stem. They are identified either by numbers indicating their order, their function, or the distribution of their fibers.

A second portion of each oculomotor nerve is part of the autonomic nervous system, supplying involuntary muscles inside the eyes. These muscles help adjust the amount of light that enters the eyes and help focus the lenses of the eyes.

The fibers of the oculomotor nerves are primarily motor, with a few sensory fibers. These transmit sensory information to the brain concerning the condition of various eye muscles.

The fourth pair, the **trochlear** (trok'le-ar) **nerves (IV),** are the smallest cranial nerves. They arise from the midbrain and carry motor impulses to a fifth pair of external eye muscles, the *superior oblique muscles,* which are not supplied by the oculomotor nerves. The trochlear nerves also contain sensory fibers that transmit information about the condition of certain eye muscles.

The fifth pair, the **trigeminal** (tri-jem'i-nal) **nerves (V),** are the largest of the cranial nerves and arise from the pons. They are mixed nerves, with the sensory portions more extensive than the motor portions. Each sensory component includes three large branches, called the ophthalmic, maxillary, and mandibular divisions (fig. 11.27).

The *ophthalmic division* consists of sensory fibers that bring impulses to the brain from the surface of the eye, the tear gland, and the skin of the anterior scalp, forehead, and upper eyelid. The fibers of the *maxillary division* carry sensory impulses from the upper teeth, upper gum, and upper lip, as well as from the mucous lining of the palate and facial skin. The *mandibular division* includes both motor and sensory fibers. The sensory branches transmit impulses from the scalp behind the ears, the skin of the jaw, the lower teeth, the lower gum, and the lower lip. The motor branches supply the muscles of mastication and certain muscles in the floor of the mouth.

The sixth pair, the **abducens** (ab-du'senz) **nerves (VI),** are quite small, and originate from the pons near the medulla oblongata. They enter the orbits of the eyes and supply motor impulses to the remaining pair of external eye muscles, the *lateral rectus muscles.* The sensory fibers of these nerves provide the brain with information concerning the condition of certain muscles.

A disorder of the trigeminal nerve called *trigeminal neuralgia* (tic douloureux) causes severe recurring pain in the face and forehead on the affected side. If drugs cannot control the pain, surgery may be used to sever the sensory portion of the nerve. Although this procedure may relieve the pain, the patient loses sensations in other body regions that the sensory branch supplies. Consequently, after such surgery, care must be taken when eating or drinking hot foods or liquids, and the mouth must be inspected daily for food particles or damage to the cheeks from biting.

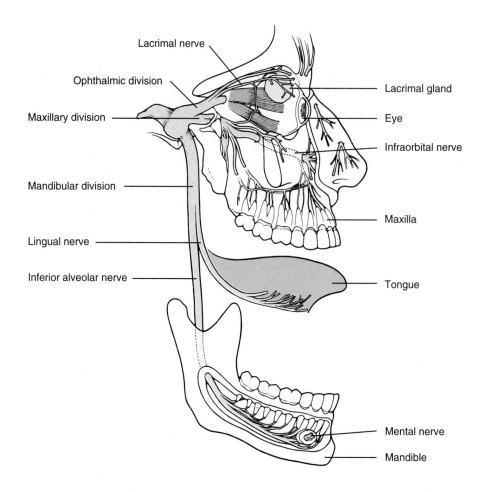

**FIGURE 11.27**

Each trigeminal nerve has three large branches that supply various regions of the head and face: the ophthalmic division, the maxillary division, and the mandibular division.

The seventh pair of cranial nerves, the **facial** (fa'shal) **nerves (VII),** arise from the lower part of the pons and emerge on the sides of the face. Their sensory branches are associated with taste receptors on the anterior two-thirds of the tongue, and some of their motor fibers transmit impulses to muscles of facial expression. Still other motor fibers of these nerves function in the autonomic nervous system by stimulating secretions from tear glands and certain salivary glands (submandibular and sublingual glands) (fig. 11.28).

The eighth pair, the **vestibulocochlear** (ves-tib"u-lo-kok'le-ar) **nerves (VIII,** acoustic, or auditory, nerves), are sensory nerves that arise from the medulla oblongata. Each of these nerves has two distinct parts—a vestibular branch and a cochlear branch.

The neuron cell bodies of the *vestibular branch* fibers are located in ganglia near the vestibule and semicircular canals of the inner ear. These structures contain receptors that sense changes in the position of the head, and in response initiate and send impulses to the cerebellum, where they are used in reflexes that maintain equilibrium.

The neuron cell bodies of the *cochlear branch* fibers are located in a ganglion of the cochlea, a part of the inner ear that houses the hearing receptors. Impulses from this branch pass through the medulla oblongata and midbrain on their way to the temporal lobe, where they are interpreted.

The ninth pair, the **glossopharyngeal** (glos"o-fah-rin'je-al) **nerves (IX),** are associated with the tongue and pharynx. These nerves arise from the medulla oblongata, and, although they are mixed nerves, their predominant fibers are sensory. These fibers carry impulses from the lining of the pharynx, tonsils, and posterior third of the tongue to the brain. Fibers in the motor component of the glossopharyngeal nerves innervate constrictor muscles in the wall of the pharynx that function in swallowing.

The tenth pair, the **vagus** (va'gus) **nerves (X),** originate in the medulla oblongata and extend downward through the neck into the chest and abdomen. These nerves are mixed, containing both somatic and autonomic branches, with the autonomic fibers predominant.

Temporal nerve

Zygomatic nerve

Buccal nerve

Facial nerve

Posterior auricular nerve

Parotid salivary gland

Mandibular nerve

Cervical nerve

**FIGURE** 11.28

The facial nerves are associated with taste receptors on the tongue and with muscles of facial expression.

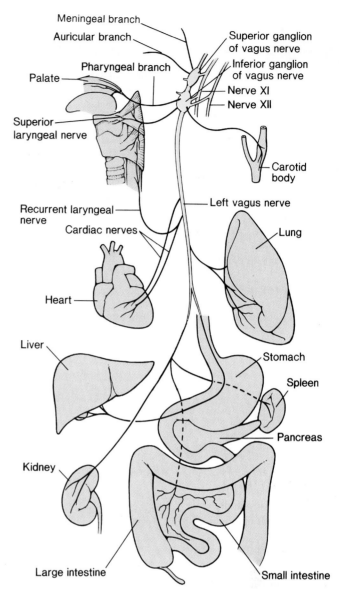

Meningeal branch

Auricular branch

Pharyngeal branch

Palate

Superior ganglion of vagus nerve

Inferior ganglion of vagus nerve

Nerve XI

Nerve XII

Superior laryngeal nerve

Carotid body

Recurrent laryngeal nerve

Left vagus nerve

Cardiac nerves

Lung

Heart

Liver

Stomach

Spleen

Pancreas

Kidney

Large intestine

Small intestine

**FIGURE** 11.29

The vagus nerves extend from the medulla oblongata downward into the chest and abdomen to supply many organs.

Among the somatic components of the vagus nerves are motor fibers that carry impulses to muscles of the larynx. These fibers are associated with speech and swallowing reflexes that employ muscles in the soft palate and pharynx. Vagal sensory fibers carry impulses from the linings of the pharynx, larynx, and esophagus, and from the viscera of the thorax and abdomen to the brain. Autonomic motor fibers of the vagus nerves supply the heart and many smooth muscles and glands in the viscera of the thorax and abdomen (fig. 11.29).

The eleventh pair, the **accessory** (ak-ses'o-re) **nerves (XI,** spinal accessory), originate in the medulla oblongata and the spinal cord; thus, they have both cranial and spinal branches. Each *cranial branch* of an accessory nerve joins a vagus nerve and carries impulses to muscles of the soft palate, pharynx, and larynx. The *spinal branch* descends into the neck and supplies motor fibers to the trapezius and sternocleidomastoid muscles.

The twelfth pair, the **hypoglossal** (hi"po-glos'al) **nerves (XII),** arise from the medulla oblongata and pass into the tongue. They primarily consist of motor fibers that carry impulses to muscles that move the tongue in

speaking, chewing, and swallowing. Table 11.9 summarizes the functions of the cranial nerves.

1 Define *peripheral nervous system.*

2 Distinguish between somatic and autonomic nerve fibers.

3 Describe the structure of a peripheral nerve.

4 Distinguish among sensory, motor, and mixed nerves.

5 Name the cranial nerves, and list the major functions of each.

TABLE **11.9**

## FUNCTIONS OF CRANIAL NERVES

| Nerve | | Type | Function |
|---|---|---|---|
| I | Olfactory | Sensory | Sensory fibers transmit impulses associated with the sense of smell. |
| II | Optic | Sensory | Sensory fibers transmit impulses associated with the sense of vision. |
| III | Oculomotor | Primarily motor | Motor fibers transmit impulses to muscles that raise the eyelids, move the eyes, adjust the amount of light entering the eyes, and focus the lenses. |
| | | | Some sensory fibers transmit impulses associated with the condition of muscles. |
| IV | Trochlear | Primarily motor | Motor fibers transmit impulses to muscles that move the eyes. |
| | | | Some sensory fibers transmit impulses associated with the condition of muscles. |
| V | Trigeminal | Mixed | |
| | Ophthalmic division | | Sensory fibers transmit impulses from the surface of the eyes, tear glands, scalp, forehead, and upper eyelids. |
| | Maxillary division | | Sensory fibers transmit impulses from the upper teeth, upper gum, upper lip, lining of the palate, and skin of the face. |
| | Mandibular division | | Sensory fibers transmit impulses from the scalp, skin of the jaw, lower teeth, lower gum, and lower lip. |
| | | | Motor fibers transmit impulses to muscles of mastication and to muscles in the floor of the mouth. |
| VI | Abducens | Primarily motor | Motor fibers transmit impulses to muscles that move the eyes. |
| | | | Some sensory fibers transmit impulses associated with the condition of muscles. |
| VII | Facial | Mixed | Sensory fibers transmit impulses associated with taste receptors of the anterior tongue. |
| | | | Motor fibers transmit impulses to muscles of facial expression, tear glands, and salivary glands. |
| VIII | Vestibulocochlear | Sensory | |
| | Vestibular branch | | Sensory fibers transmit impulses associated with the sense of equilibrium. |
| | Cochlear branch | | Sensory fibers transmit impulses associated with the sense of hearing. |
| IX | Glossopharyngeal | Mixed | Sensory fibers transmit impulses from the pharynx, tonsils, posterior tongue, and carotid arteries. |
| | | | Motor fibers transmit impulses to salivary glands and to muscles of the pharynx used in swallowing. |
| X | Vagus | Mixed | Somatic motor fibers transmit impulses to muscles associated with speech and swallowing; autonomic motor fibers transmit impulses to the viscera of the thorax and abdomen. |
| | | | Sensory fibers transmit impulses from the pharynx, larynx, esophagus, and viscera of the thorax and abdomen. |
| XI | Accessory | Primarily motor | |
| | Cranial branch | | Motor fibers transmit impulses to muscles of the soft palate pharynx, and larynx. |
| | Spinal branch | | Motor fibers transmit impulses to muscles of the neck and back. |
| XII | Hypoglossal | Primarily motor | Motor fibers transmit impulses to muscles that move the tongue. |

## Spinal Nerves

Thirty-one pairs of spinal nerves originate from the spinal cord. They are mixed nerves, and they provide two-way communication between the spinal cord and parts of the upper and lower limbs, neck, and trunk.

Spinal nerves are not named individually, but are grouped by the level from which they arise, with each nerve numbered in sequence (fig. 11.30). Thus, there are eight pairs of *cervical nerves* (numbered C1 to C8), twelve pairs of *thoracic nerves* (numbered T1 to T12),

five pairs of *lumbar nerves* (numbered L1 to L5), five pairs of *sacral nerves* (numbered S1 to S5), and one pair of *coccygeal nerves* (Co).

The nerves arising from the superior part of the spinal cord pass outward almost horizontally, whereas those from the inferior portions of the spinal cord descend at sharp angles. This arrangement is a consequence of growth. In early life, the spinal cord extends the entire length of the vertebral column, but with age, the column grows more rapidly than the cord. Thus, the adult spinal

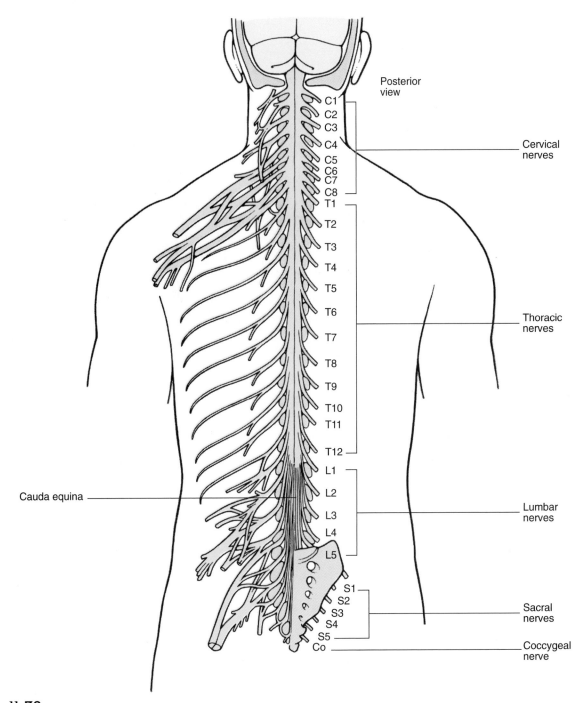

**FIGURE 11.30**

The 31 pairs of spinal nerves are grouped according to the level from which they arise, and are numbered in sequence.

cord ends at the level between the first and second lumbar vertebrae, so the lumbar, sacral, and coccygeal nerves descend to their exits beyond the end of the cord. These descending nerves form a structure called the *cauda equina* (horse's tail) (fig. 11.30).

Each spinal nerve emerges from the cord by two short branches, or roots, which lie within the vertebral column. The **dorsal root** (posterior, or sensory, root) can be identified by an enlargement called the *dorsal root ganglion.* This ganglion contains the cell bodies of the sensory neurons whose dendrites conduct impulses inward from the peripheral body parts. The axons of these neurons extend through the dorsal root and into the spinal cord, where they form synapses with dendrites of other neurons.

The **ventral root** (anterior, or motor, root) of each spinal nerve consists of axons from the motor neurons whose cell bodies are located within the gray matter of

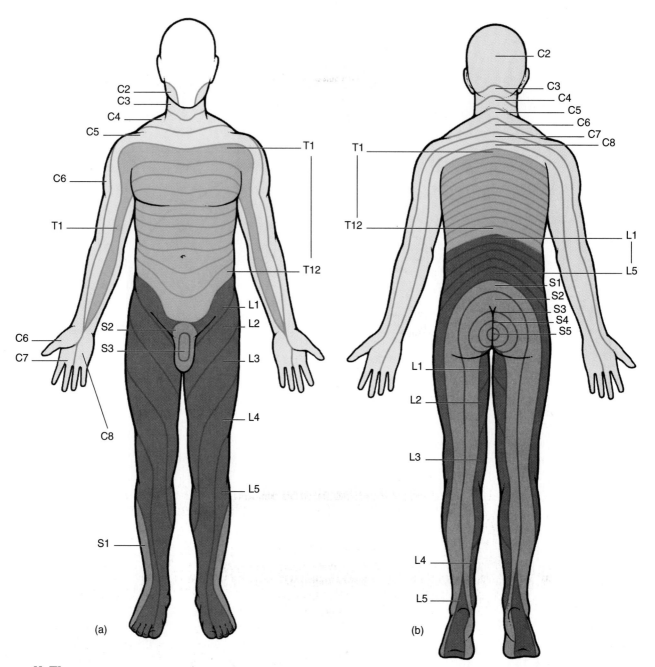

FIGURE 11.31

**FIGURE** 11.31

(a) Dermatomes on the anterior body surface and (b) on the posterior surface. Note that spinal nerve C1 does not supply any skin area.

An area of skin that the sensory nerve fibers of a particular spinal nerve innervate is called a *dermatome.* Dermatomes are highly organized, but they vary considerably in size and shape, as figure 11.31 indicates. A map of the dermatomes is often useful in localizing the sites of injuries to dorsal roots or the spinal cord.

the cord. A ventral root and a dorsal root unite to form a spinal nerve, which extends outward from the vertebral canal through an *intervertebral foramen.* Just beyond its foramen, each spinal nerve branches. One of these parts, the small *meningeal branch,* reenters the vertebral canal through the intervertebral foramen and supplies the meninges and blood vessels of the cord, as well as the intervertebral ligaments and the vertebrae.

**FIGURE 11.32**

(a) Each spinal nerve has a posterior and an anterior branch. (b) The thoracic and lumbar spinal nerves also have a visceral branch.

As figure 11.32 shows, a *posterior branch* (posterior ramus) of each spinal nerve turns posteriorly and innervates the muscles and skin of the back. The main portion of the nerve, the *anterior branch* (anterior ramus), continues forward to supply muscles and skin on the front and sides of the trunk and limbs.

The spinal nerves in the thoracic and lumbar regions have a fourth, or *visceral branch,* which is part of the autonomic nervous system. Except in the thoracic region, anterior branches of the spinal nerves combine to form complex networks called **plexuses** instead of continuing directly to the peripheral body parts. In a plexus, the fibers of various spinal nerves are sorted and recombined, so that fibers associated with a particular peripheral body part reach it in the same nerve, even though the fibers originate from different spinal nerves (fig. 11.33).

### Cervical Plexuses

The **cervical plexuses** lie deep in the neck on either side. They are formed by the anterior branches of the first four cervical nerves. Fibers from these plexuses supply the muscles and skin of the neck. In addition, fibers from the third, fourth, and fifth cervical nerves pass into the right and left **phrenic** (fren′ik) **nerves,** which conduct motor impulses to the muscle fibers of the diaphragm.

### Brachial Plexuses

The anterior branches of the lower four cervical nerves and the first thoracic nerve give rise to **brachial plexuses.** These networks of nerve fibers are located deep within the shoulders between the neck and the axillae (armpits). The major branches emerging from the brachial plexuses include the following (fig. 11.34):

1. *Musculocutaneous nerves* supply muscles of the arms on the anterior sides, and the skin of the forearms.

2. *Ulnar nerves* supply muscles of the forearms and hands, and the skin of the hands.

3. *Median nerves* supply muscles of the forearms, and muscles and skin of the hands.

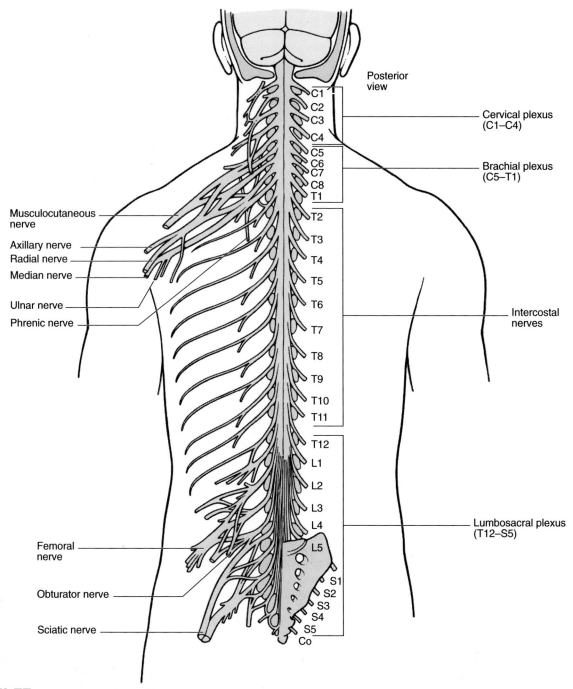

**FIGURE 11.33**

The anterior branches of the spinal nerves in the thoracic region give rise to intercostal nerves. Those in other regions combine to form complex networks called plexuses.

4. *Radial nerves* supply muscles of the arms on the posterior sides, and the skin of the forearms and hands.

5. *Axillary nerves* supply muscles and skin of the superior, lateral, and posterior regions of the arm.

Other nerves associated with the brachial plexus that innervate various skeletal muscles include:

1. The *lateral* and *medial pectoral nerves* supply the pectoralis major and pectoralis minor muscles.

2. The *dorsal scapular nerve* supplies the rhomboideus major and levator scapulae muscles.

3. The *lower subscapular nerve* supplies the subscapularis and teres major muscles.

Humerus

Axillary n.

Musculocutaneous n.

Radial n.

Radius

C5
C6
C7
C8
T1

Median n.

Ulnar n.

Ulna

**FIGURE 11.34**

Nerves of the brachial plexus.

4. The *thoracodorsal nerve* supplies the latissimus dorsi muscle.

5. The *suprascapular nerve* supplies the supraspinatus and infraspinatus muscles.

## Lumbosacral Plexuses

The **lumbosacral** (lum″bo-sa′kral) **plexuses** are formed by the last thoracic nerve and the lumbar, sacral, and coccygeal nerves. These networks of nerve fibers extend from the lumbar region of the back into the pelvic cavity, giving rise to a number of motor and sensory fibers associated with the lower abdominal wall, external genitalia, buttocks, thighs, legs, and feet. The major branches of these plexuses include the following (fig. 11.35):

1. The *obturator nerves* supply the adductor muscles of the thighs.

2. The *femoral nerves* divide into many branches, supplying motor impulses to muscles of the thighs and legs, and receiving sensory impulses from the skin of the thighs and legs.

3. The *sciatic nerves* are the largest and longest nerves in the body. They pass downward into the buttocks and descend into the thighs, where they divide into *tibial* and *common peroneal nerves.* The many branches of these nerves supply muscles and skin in the thighs, legs, and feet.

Other nerves associated with the lumbosacral plexus that innervate various skeletal muscles include:

1. The *pudendal nerve* supplies the muscles of the perineum.

2. The *inferior* and *superior gluteal nerves* supply the gluteal muscles and the tensor fasciae latae muscle.

The anterior branches of the thoracic spinal nerves do not enter a plexus. Instead, they travel into spaces between the ribs and become **intercostal** (in″ter-kos′tal) **nerves.** These nerves supply motor impulses to the intercostal muscles and the upper abdominal wall muscles. They also receive sensory impulses from the skin of the thorax and abdomen. Clinical Application 11.7 discusses injuries to the spinal nerves.

**1** How are spinal nerves grouped?

**2** Describe how a spinal nerve joins the spinal cord.

**3** Name and locate the major nerve plexuses.

**FIGURE 11.35**

Nerves of the lumbosacral plexus. (*a*) Anterior view. (*b*) Posterior view.

## AUTONOMIC NERVOUS SYSTEM

The autonomic nervous system is the part of the peripheral nervous system that functions independently (autonomously) and continuously, without conscious effort. This system controls visceral activities by regulating the actions of smooth muscles, cardiac muscles, and various glands. It oversees heart rate, blood pressure, breathing rate, body temperature, and other visceral activities that aid in maintaining homeostasis. Portions of the autonomic nervous system also respond during times of emotional stress and prepare the body to meet the demands of strenuous physical activity.

## General Characteristics

Reflexes in which sensory signals originate from receptors within the viscera and the skin regulate autonomic activities. Afferent nerve fibers transmit these signals to nerve centers within the brain or spinal cord. In response, motor impulses travel out from these centers on efferent nerve fibers within cranial and spinal nerves.

Typically, these efferent fibers lead to ganglia outside the central nervous system. The impulses they carry are integrated within the ganglia and are relayed to various organs (muscles or glands) that respond by contracting, secreting, or being inhibited. The integrative function of the ganglia provides the autonomic nervous system with some degree of independence from the brain and spinal cord,

# CLINICAL APPLICATION

## Spinal Nerve Injuries

Birth injuries, dislocations, vertebral fractures, stabs, gunshot wounds, and pressure from tumors can all injure spinal nerves. Suddenly bending the neck, called whiplash, can compress the nerves of the cervical plexuses, causing persistent headache and pain in the neck and skin, which the cervical nerves supply. If a broken or dislocated vertebra severs or damages the phrenic nerves associated with the cervical plexuses, partial or complete paralysis of the diaphragm may result.

Intermittent or constant pain in the neck, shoulder, or upper limb may result from prolonged abduction of the upper limb, as in painting or typing. This is due to too much pressure on the brachial plexus. This condition, called *thoracic outlet syndrome,* may also result from a con-genital skeletal malformation that com-presses the plexus during upper limb and shoulder movements.

Degenerative changes may com-press an intervertebral disk in the lum-bar region, producing *sciatica,* which causes pain in the lower back and gluteal region that can radiate to the thigh, calf, ankle, and foot. Sciatica is most common in middle-aged people, particularly distance runners. It usually compresses spinal nerve roots between L2 and S1, some of which contain fibers of the sciatic nerve. Rest, drugs, or surgery are used to treat sciatica.

In *carpal tunnel syndrome,* re-peated hand movements, such as typing, inflame the tendons that pass through the carpal tunnel, which is a space between bones in the wrist. The swollen tendons compress the median nerve in the wrist, causing pain to shoot up the upper limb. Surgery or avoiding repetitive hand movements can re-lieve symptoms. ■

---

and the visceral efferent nerve fibers associated with these ganglia comprise the autonomic nervous system.

The autonomic nervous system includes two divi-sions, called the **sympathetic** and **parasympathetic divi-sions,** that act in conjunction with one another in various situations. For example, many organs have nerve fibers from each of the divisions. Impulses on one set of fibers may activate an organ, whereas impulses on the other set inhibit it. Thus, the divisions may function antagonisti-cally, regulating the actions of some organs by alternately activating or inhibiting them.

The functions of the autonomic divisions are var-ied; that is, each activates some organs and inhibits oth-ers. This reveals that the divisions have important functional differences. The sympathetic division primar-ily prepares the body for energy-expending, stressful, or emergency situations. Conversely, the parasympathetic division is most active under ordinary, restful condi-tions. It also counterbalances the effects of the sympa-thetic division, and restores the body to a resting state following a stressful experience. For example, during an emergency, the sympathetic division increases heart and breathing rates; following the emergency, the parasympa-thetic division decreases these activities.

## Autonomic Nerve Fibers

All of the nerve fibers of the autonomic nervous system are efferent, or motor, fibers. In the motor pathways of the somatic nervous system, a single neuron typically links the central nervous system and a skeletal muscle. In the autonomic system, motor pathways include two neurons, as figure 11.36 shows. The cell body of one neu-ron is located in the brain or spinal cord. Its axon, the **preganglionic fiber,** leaves the CNS and synapses with one or more nerve fibers whose cell bodies are housed within an autonomic ganglion. The axon of such a sec-ond neuron is called a **postganglionic fiber,** and it ex-tends to a visceral effector.

### Sympathetic Division

Within the sympathetic division (thoracolumbar divi-sion), the preganglionic fibers originate from neurons within the lateral horn of the spinal cord. These neurons are found in all of the thoracic segments and in the upper two or three lumbar segments of the cord. Their axons exit through the ventral roots of spinal nerves along with various somatic motor fibers.

After traveling a short distance, preganglionic fibers leave the spinal nerves through branches called *white rami* (sing., *ramus*) and enter sympathetic ganglia. Two groups of such ganglia, called **paravertebral ganglia,** are located in chains along the sides of the vertebral column. These ganglia, with the fibers that connect them, com-prise the **sympathetic trunks** (fig. 11.37).

The paravertebral ganglia lie just beneath the parietal pleura in the thorax and beneath the parietal

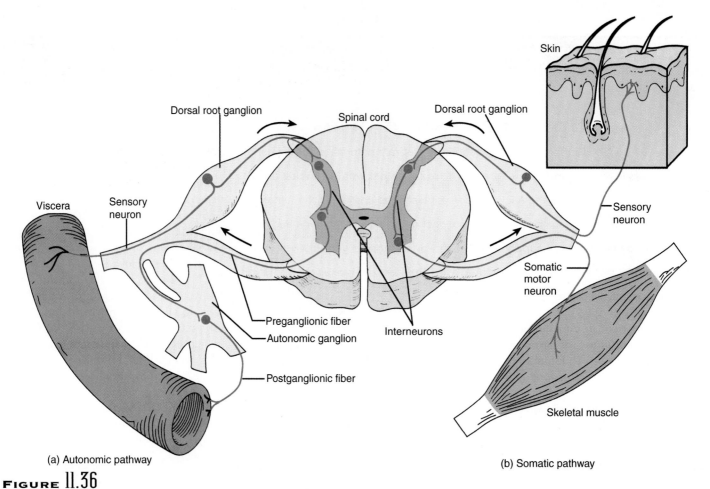

**FIGURE 11.36**

(a) Autonomic pathways include two neurons between the central nervous system and an effector. (b) Somatic pathways usually have a single neuron between the central nervous system and an effector.

**FIGURE 11.37**

A chain of paravertebral ganglia extends along each side of the vertebral column.

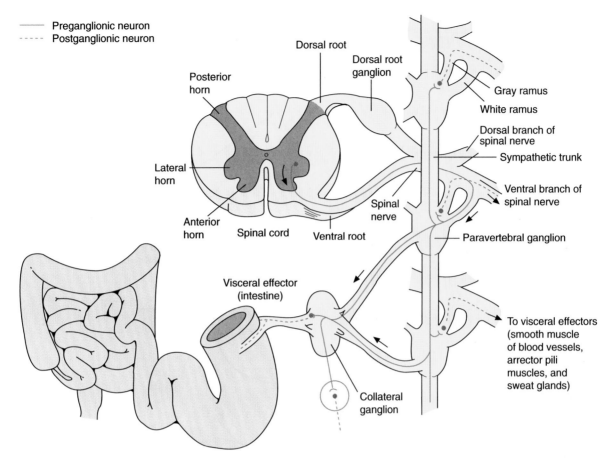

**FIGURE** 11.38

Sympathetic fibers leave the spinal cord in the ventral roots of spinal nerves, enter sympathetic ganglia, and synapse with other neurons that extend to visceral effectors.

peritoneum in the abdomen (see chapter 1). Although these ganglia are located some distance from the viscera they help control, other sympathetic ganglia are nearer to the viscera. The *collateral ganglia,* for example, are found within the abdomen, closely associated with certain large blood vessels (fig. 11.38).

Some of the preganglionic fibers that enter paravertebral ganglia synapse with neurons within these ganglia. Other fibers extend through the ganglia and pass up or down the sympathetic trunk and synapse with neurons in ganglia at higher or lower levels within the chain. Still other fibers pass through to collateral ganglia before they synapse. Typically, a preganglionic axon will synapse with several other neurons within a sympathetic ganglion.

The axons of the second neurons in sympathetic pathways, the postganglionic fibers, extend out from the sympathetic ganglia to visceral effectors. Those leaving paravertebral ganglia usually pass through branches called **gray rami** and return to a spinal nerve before proceeding to an effector (fig. 11.38). These branches appear gray because the postganglionic axons generally are unmyelinated, whereas the preganglionic axons in the white rami are nearly all myelinated.

An important exception to the usual arrangement of sympathetic fibers occurs in a set of preganglionic fibers that pass through the sympathetic ganglia and extend out to the medulla of each adrenal gland. These fibers terminate within the glands on special hormone-secreting cells that release **norepinephrine** (20%) and **epinephrine** (80%) when they are stimulated. Chapter 13 discusses the functions of the adrenal medulla and its hormones. Figure 11.39 shows the sympathetic division.

## Parasympathetic Division

The preganglionic fibers of the parasympathetic division (craniosacral division) arise from neurons in the midbrain, pons, and medulla oblongata of the brain stem and from the sacral region of the spinal cord (fig. 11.40). From there, they lead outward on cranial or sacral nerves to ganglia located near or within various organs (terminal ganglia). The short postganglionic fibers continue from the ganglia to specific muscles or glands within these organs (fig. 11.41). Parasympathetic preganglionic axons are usually myelinated, and the parasympathetic postganglionic fibers are unmyelinated.

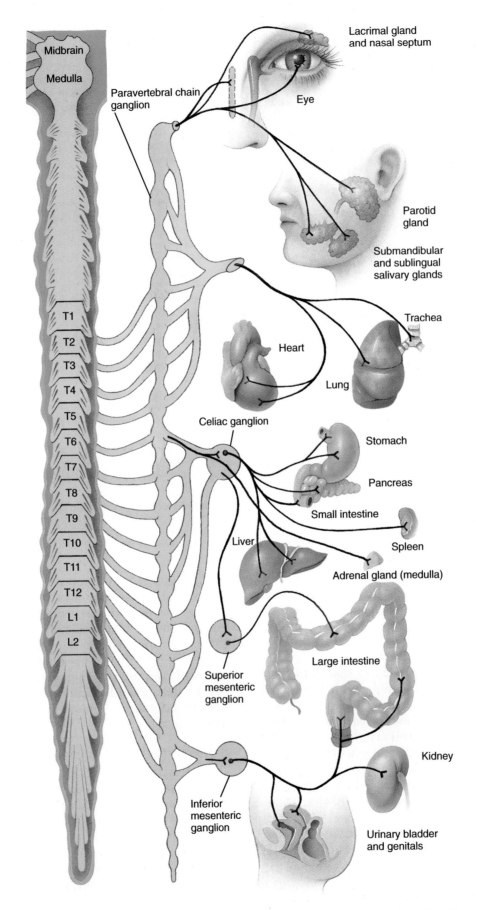

**FIGURE** 11.39

The preganglionic fibers of the sympathetic division of the autonomic nervous system arise from the thoracic and lumbar regions of the spinal cord. Note that the adrenal medulla is directly innervated by a preganglionic fiber, as described in the text.

The parasympathetic preganglionic fibers associated with parts of the head are included in the oculomotor, facial, and glossopharyngeal nerves. Those fibers that innervate organs of the thorax and upper abdomen are parts of the vagus nerves. (The vagus nerves carry about 75% of all parasympathetic fibers.) Preganglionic fibers arising from the sacral region of the spinal cord lie within the branches of the second through the fourth sacral spinal nerves, and they carry impulses to viscera within the pelvic cavity (see fig. 11.40).

**1** What is the general function of the autonomic nervous system?

**2** How are the divisions of the autonomic system distinguished?

**3** Describe a sympathetic nerve pathway and a parasympathetic nerve pathway.

## Autonomic Neurotransmitters

The preganglionic fibers of the sympathetic and parasympathetic divisions all secrete acetylcholine, and for this reason they are called **cholinergic fibers.** The parasympathetic postganglionic fibers are also cholinergic fibers. Most sympathetic postganglionic fibers, however, secrete norepinephrine (noradrenalin) and are called **adrenergic fibers** (see fig. 11.41). Exceptions to this include the sympathetic postganglionic fibers which stimulate sweat glands and a few sympathetic fibers to blood vessels in skin (which cause vasodilation); these fibers secrete acetylcholine and therefore are cholinergic (adrenergic sympathetic fibers to blood vessels cause vasoconstriction).

The different postganglionic neurotransmitters (mediators) are responsible for the different effects that the sympathetic and parasympathetic divisions have on organs. Since each division can activate some effectors and inhibit others, it is not surprising that the divisions usually are antagonistic. For example, the sympathetic nervous system increases heart rate and dilates pupils, whereas parasympathetic stimulation decreases heart rate and constricts pupils. However, this is not always the case. For example, the diameters of most blood vessels lack parasympathetic innervation, and are thus regulated by the sympathetic division. Smooth muscles in the walls of these vessels are continuously stimulated by sympathetic impulses; they are thereby maintained in a

**FIGURE** 11.40

The preganglionic fibers of the parasympathetic division of the autonomic nervous system arise from the brain and sacral region of the spinal cord.

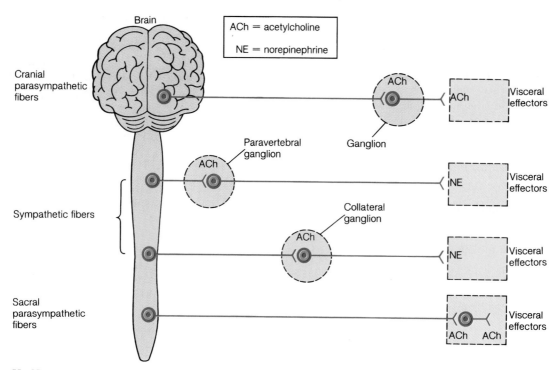

Brain

ACh = acetylcholine

NE = norepinephrine

Cranial parasympathetic fibers

ACh

ACh

ACh

Visceral effectors

Paravertebral ganglion

Ganglion

ACh

Sympathetic fibers

NE

Visceral effectors

Collateral ganglion

ACh

NE

Visceral effectors

Sacral parasympathetic fibers

ACh    ACh

Visceral effectors

**FIGURE 11.41**

Most sympathetic fibers are adrenergic and secrete norepinephrine at the ends of the postganglionic fibers; parasympathetic fibers are cholinergic and secrete acetylcholine at the ends of the postganglionic fibers. The two arrangements of parasympathetic postganglionic fibers are seen in both cranial and sacral portions. Similarly, sympathetic paravertebral and collateral ganglia are seen in both the thoracic and lumbar portions of the nervous system.

state of partial contraction called *sympathetic tone*. Decreasing sympathetic stimulation allows the muscular walls of such blood vessels to relax, increasing their diameters (vasodilation). Conversely, increasing sympathetic stimulation vasoconstricts vessels. Table 11.10 summarizes the effects of adrenergic and cholinergic fibers on various visceral effectors.

## Actions of Autonomic Neurotransmitters

As in the case of stimulation at neuromuscular junctions (see chapter 9) and synapses (see chapter 10), the actions of autonomic neurotransmitters result from their binding to protein receptors in the membranes of effector cells. Receptor binding alters the membrane. For example, the membrane's permeability to certain ions may increase, and in smooth muscle cells, an action potential followed by muscular contraction may result. Similarly, a gland cell may respond to a change in its membrane by secreting a product.

Acetylcholine can combine with two types of cholinergic receptors, called *muscarinic receptors* and *nicotinic receptors.* These receptor names come from *muscarine,* a toxin from a fungus that can activate muscarinic receptors, and *nicotine,* the toxin of tobacco that can activate nicotinic receptors. The muscarinic receptors are located in the membranes of effector cells at the

ends of all postganglionic parasympathetic nerve fibers and at the ends of the cholinergic sympathetic fibers. Responses from these receptors are excitatory and occur relatively slowly. The nicotinic receptors occur in the synapses between the preganglionic and postganglionic neurons of the parasympathetic and sympathetic pathways. They produce rapid, excitatory responses (fig. 11.42). (Receptors at neuromuscular junctions of skeletal muscles are nicotinic.)

Epinephrine and norepinephrine are the two chemical mediators of the sympathetic nervous system. The adrenal gland releases both as hormones, but only norepinephrine is released as a neurotransmitter by the sympathetic nervous system. These biochemicals can then combine with adrenergic receptors of effector cells.

The two major types of adrenergic receptors are termed *alpha* and *beta.* Exciting them elicits different responses in the effector organs. For example, stimulation of the alpha receptors in vascular smooth muscle causes vasoconstriction, whereas stimulation of the beta receptors in bronchial smooth muscle causes relaxation leading to bronchodilation. Furthermore, although norepinephrine has a somewhat stronger effect on alpha receptors, both of these mediators can stimulate both kinds of receptors. Consequently, the way each of these adrenergic substances influences effector cells depends on the relative numbers of alpha and beta receptors in the cell membranes.

TABLE 11.10

## EFFECTS OF AUTONOMIC STIMULATION ON VARIOUS EFFECTORS

| Effector Location | Response to Sympathetic Stimulation | Response to Parasympathetic Stimulation |
|---|---|---|
| **Integumentary system** | | |
| Apocrine glands | Increased secretion | No action |
| Eccrine glands | Increased secretion (cholinergic effect) | No action |
| **Special senses** | | |
| Iris of eye | Dilation | Constriction |
| Tear gland | Slightly increased secretion | Greatly increased secretion |
| **Endocrine system** | | |
| Adrenal cortex | Increased secretion | No action |
| Adrenal medulla | Increased secretion | No action |
| **Digestive system** | | |
| Muscle of gallbladder wall | Relaxation | Contraction |
| Muscle of intestinal wall | Decreased peristaltic action | Increased peristaltic action |
| Muscle of internal anal sphincter | Contraction | Relaxation |
| Pancreatic glands | Reduced secretion | Greatly increased secretion |
| Salivary glands | Reduced secretion | Greatly increased secretion |
| **Respiratory system** | | |
| Muscles in walls of bronchioles | Dilation | Constriction |
| **Cardiovascular system** | | |
| Blood vessels supplying muscles | Constriction (alpha adrenergic) Dilation (beta adrenergic) Dilation (cholinergic) | No action |
| Blood vessels supplying skin | Constriction | No action |
| Blood vessels supplying heart (coronary arteries) | Dilation (beta adrenergic) Constriction (alpha adrenergic) | Dilation |
| Muscles in wall of heart | Increased contraction rate | Decreased contraction rate |
| **Urinary system** | | |
| Muscle of bladder wall | Relaxation | Contraction |
| Muscle of internal urethral sphincter | Contraction | Relaxation |
| **Reproductive systems** | | |
| Blood vessels to clitoris and penis | No action | Dilation leading to erection of clitoris and penis |
| Muscles associated with internal reproductive organs | Male ejaculation, female orgasm | |

The enzyme acetylcholinesterase rapidly decomposes the acetylcholine that cholinergic fibers release. (Recall that this decomposition also occurs at the neuromuscular junctions of skeletal muscle.) Thus, acetylcholine usually produces an effect on the postsynaptic membrane for only a fraction of a second.

Much of the norepinephrine released from adrenergic fibers is removed from the synapse by active transport back into the nerve endings. The enzyme monoamine oxidase found in mitochondria then inactivates norepinephrine. This may take a few seconds, during which some molecules may diffuse into nearby tissues or the bloodstream, where other enzymes decompose them. On the other hand, some norepinephrine molecules may escape decomposition and remain active for awhile. For these reasons, norepinephrine is likely to produce a more prolonged effect than acetylcholine. In fact, when the adrenal medulla releases norepinephrine and epinephrine into the blood in response to sympathetic stimulation, these substances may trigger sympathetic responses in organs throughout the body that last up to thirty seconds.

Many drugs influence autonomic functions. Some, like ephedrine, enhance sympathetic effects by stimulating release of norepinephrine from postganglionic sympathetic nerve endings. Others, like reserpine, inhibit sympathetic activity by preventing norepinephrine synthesis. Another group of drugs, which includes pilocarpine, produces parasympathetic effects, and some, like atropine, block the action of acetylcholine on visceral effectors.

## Control of Autonomic Activity

Although the autonomic nervous system has some independence resulting from impulse integration within its ganglia, it is largely controlled by the brain and spinal cord. For example, recall the control centers in the medulla oblongata for cardiac, vasomotor, and respiratory activities. These reflex centers receive sensory impulses from viscera by means of vagus nerve fibers, and use autonomic nerve pathways to stimulate motor

(a) Parasympathetic fiber

(b) Sympathetic fiber (cholinergic)

(c) Somatic motor fiber

**FIGURE** 11.42

(*a, b,* and *c*) Muscarinic receptors occur in the membranes of effector cells at the ends of autonomic cholinergic fibers. Nicotinic receptors are found in the membranes of postganglionic autonomic neurons and skeletal muscle fibers.

responses in various muscles and glands. Thus, they control the autonomic nervous system. Similarly, the hypothalamus helps regulate body temperature, hunger, thirst, and water and electrolyte balance by influencing autonomic pathways.

Still higher levels within the brain, including the limbic system and the cerebral cortex, control the autonomic nervous system during emotional stress. In this way, the autonomic pathways can affect emotional expression and behavior. Subsequent chapters that deal with individual organs and organ systems discuss regulation of particular organs.

1 Distinguish between cholinergic and adrenergic fibers.

2 Explain how the fibers of one autonomic division can control the actions of a particular organ.

3 Which neurotransmitters are used in the autonomic nervous system?

4 Describe two types of cholinergic receptors and two types of adrenergic receptors.

## CLINICAL TERMS RELATED TO THE NERVOUS SYSTEM

**analgesia** (an″al-je′ze-ah)   Loss or reduction in the ability to sense pain, but without loss of consciousness.

**analgesic** (an″al-je′sik)   Pain-relieving drug.

**anesthesia** (an″es-the′ze-ah)   Loss of feeling.

**aphasia** (ah-fa′ze-ah)   Disturbance or loss in the ability to use words or to understand them, usually due to damage to cerebral association areas.

**apraxia** (ah-prak′se-ah)   Impairment in a person's ability to make correct use of objects.

**ataxia** (ah-tak′se-ah)   Partial or complete inability to coordinate voluntary movements.

**cordotomy** (kor-dot′o-me)   Surgical procedure severing a nerve tract within the spinal cord, usually to relieve intractable pain.

**craniotomy** (kra″ne-ot′o-me)   Surgical procedure opening a part of the skull.

**encephalitis** (en″sef-ah-li′tis)   Inflammation of the brain and meninges characterized by drowsiness and apathy.

**epilepsy** (ep′ĭ-lep″se)   Disorder of the central nervous system that is characterized by temporary disturbances in normal brain impulses; it may be accompanied by convulsive seizures and loss of consciousness.

**Huntington disease** (hunt′ing-tun di-zez′)   Hereditary disorder of the brain producing progressively worsening, uncontrollable dancelike movements and personality changes.

**laminectomy** (lam″i-nek′to-me)   Surgical removal of the posterior arch of a vertebra, usually to relieve symptoms of a ruptured intervertebral disk that is pressing on a spinal nerve.

**neuralgia** (nu-ral′je-ah)   Sharp, recurring pain associated with a nerve, usually caused by inflammation or injury.

**neuritis** (nu-ri′tis)   Inflammation of a nerve.

**vagotomy** (va-got′o-me)   Surgical severing of a vagus nerve.

---

## CHAPTER SUMMARY

## Introduction  (page 376)

Bone and protective membranes called meninges surround the brain and spinal cord.

## Meninges  (page 376)

1. The meninges consist of a dura mater, arachnoid mater, and pia mater.
2. Cerebrospinal fluid occupies the space between the arachnoid and pia maters.

## Spinal Cord  (page 377)

The spinal cord is a nerve column that extends from the brain into the vertebral canal. It terminates at the level between the first and second lumbar vertebrae.

1. Structure of the spinal cord
   a. The spinal cord is composed of thirty-one segments, each of which gives rise to a pair of spinal nerves.
   b. It is characterized by a cervical enlargement, a lumbar enlargement, and two deep longitudinal grooves that divide it into right and left halves.
   c. White matter surrounds a central core of gray matter.
   d. The white matter is composed of bundles of myelinated nerve fibers.
2. Functions of the spinal cord
   a. The cord provides a two-way communication system between the brain and structures outside the nervous system.
   b. Ascending tracts carry sensory impulses to the brain; descending tracts carry motor impulses to muscles and glands.
   c. Many of the fibers in the ascending and descending tracts cross over in the spinal cord or brain.

## Reflex Arcs  (page 383)

A reflex arc usually includes a sensory neuron, a reflex center composed of interneurons, and a motor neuron. The reflex arc is the behavioral unit of the nervous system.

1. Reflex behavior
   a. Reflexes are automatic, subconscious responses to changes.
   b. They help maintain homeostasis.
   c. The knee-jerk reflex employs only two neurons.
   d. Withdrawal reflexes are protective actions.

## Brain  (page 387)

The brain is the largest and most complex part of the nervous system. It contains nerve centers that are associated with sensations. The brain issues motor commands and carries on higher mental functions.

1. Brain development
   a. Brain structure reflects the way it forms.
   b. The brain develops from a neural tube with three cavities—the forebrain, midbrain, and hindbrain.
   c. The cavities persist as ventricles, and the walls give rise to structural and functional regions.
2. Structure of the cerebrum
   a. The cerebrum consists of two cerebral hemispheres connected by the corpus callosum.
   b. Its surface is marked by ridges and grooves; sulci divide each hemisphere into lobes.
   c. The cerebral cortex is a thin layer of gray matter near the surface.
   d. White matter consists of myelinated nerve fibers that interconnect neurons within the nervous system and communicate with other body parts.
3. Functions of the cerebrum
   a. The cerebrum is concerned with higher brain functions, such as thought, reasoning, interpretation of sensory impulses, control of voluntary muscles, and memory storage.
   b. The cerebral cortex has sensory, motor, and association areas.
   c. The primary motor regions lie in the frontal lobes near the central sulcus. Other areas of the frontal lobes control special motor functions.
   d. Areas that interpret sensory impulses from the skin are located in the parietal lobes near the central sulcus; other specialized sensory areas are in the temporal and occipital lobes.
   e. Association areas analyze and interpret sensory impulses and provide memory, reasoning, verbalizing, judgment, and emotions.
   f. One cerebral hemisphere usually dominates for certain intellectual functions.
   g. Short-term memory is probably electrical. Long-term memory is thought to be encoded in patterns of synaptic connections.
4. Basal ganglia
   a. Basal ganglia are masses of gray matter located deep within the cerebral hemispheres.
   b. They relay motor impulses originating in the cerebral cortex, and aid in controlling motor activities.

5. Ventricles and cerebrospinal fluid
   a. Ventricles are connected cavities within the cerebral hemispheres and brain stem.
   b. Cerebrospinal fluid fills the ventricles.
   c. Choroid plexuses in the walls of the ventricles secrete cerebrospinal fluid.
   d. Ependymal cells of the choroid plexus regulate the composition of cerebrospinal fluid.
   e. Cerebrospinal fluid circulates through the ventricles and is reabsorbed into the blood of the dural sinuses.
6. Diencephalon
   a. The diencephalon contains the thalamus and hypothalamus.
   b. The thalamus selects incoming sensory impulses and relays them to the cerebral cortex.
   c. The hypothalamus is important in maintaining homeostasis.
   d. The limbic system produces emotional feelings and modifies behavior.
7. Brain stem
   a. The brain stem extends from the base of the brain to the spinal cord.
   b. The brain stem consists of the midbrain, pons, and medulla oblongata.
   c. The midbrain contains reflex centers associated with eye and head movements.
   d. The pons transmits impulses between the cerebrum and other parts of the nervous system and contains centers that help regulate rate and depth of breathing.
   e. The medulla oblongata transmits all ascending and descending impulses and contains several vital and nonvital reflex centers.
   f. The reticular formation filters incoming sensory impulses, arousing the cerebral cortex into wakefulness in response to meaningful impulses.
   g. Normal sleep results from decreasing activity of the reticular formation, and paradoxical sleep occurs when activating impulses are received by some parts of the brain, but not by others.
8. Cerebellum
   a. The cerebellum consists of two hemispheres connected by the vermis.
   b. A thin cortex of gray matter surrounds the white matter of the cerebellum.
   c. The cerebellum functions primarily as a reflex center, coordinating skeletal muscle movements and maintaining equilibrium.

# Peripheral Nervous System (page 407)

The peripheral nervous system consists of cranial and spinal nerves that branch out from the brain and spinal cord to all body parts. It can be subdivided into somatic and autonomic portions.

1. Structure of peripheral nerves
   a. A nerve consists of a bundle of nerve fibers surrounded by connective tissues.
   b. The connective tissues form an outer epineurium, a perineurium enclosing bundles of nerve fibers, and an endoneurium surrounding each fiber.
2. Nerve fiber classification
   a. Nerves are cordlike bundles of nerve fibers. Nerves can be classified as sensory nerves, motor nerves, or mixed nerves, depending on which type of fibers they contain.

   b. Nerve fibers within the central nervous system can be subdivided into groups with general and special functions.
3. Cranial nerves
   a. Twelve pairs of cranial nerves connect the brain to parts in the head, neck, and trunk.
   b. Although most cranial nerves are mixed, some are pure sensory, and others are primarily motor.
   c. The names of cranial nerves indicate their primary functions or the general distributions of their fibers.
   d. Some cranial nerve fibers are somatic and others are autonomic.
4. Spinal nerves
   a. Thirty-one pairs of spinal nerves originate from the spinal cord.
   b. These mixed nerves provide a two-way communication system between the spinal cord and the upper limbs, lower limbs, neck, and trunk.
   c. Spinal nerves are grouped according to the levels from which they arise, and they are numbered sequentially.
   d. Each nerve emerges by a dorsal and a ventral root.
      (1) A dorsal root contains sensory fibers and has a dorsal root ganglion.
      (2) A ventral root contains motor fibers.
   e. Just beyond its foramen, each spinal nerve divides into several branches.
   f. Most spinal nerves combine to form plexuses that direct nerve fibers to a particular body part.

# Autonomic Nervous System (page 418)

The autonomic nervous system functions without conscious effort. It is concerned primarily with regulating visceral activities that maintain homeostasis.

1. General characteristics
   a. Autonomic functions are reflexes controlled from centers in the hypothalamus, brain stem, and spinal cord.
   b. Autonomic nerve fibers are associated with ganglia where impulses are integrated before distribution to effectors.
   c. The integrative function of the ganglia provides a degree of independence from the central nervous system.
   d. The autonomic nervous system consists of the visceral efferent fibers associated with these ganglia.
   e. The autonomic nervous system is subdivided into two divisions—sympathetic and parasympathetic.
   f. The sympathetic division prepares the body for stressful and emergency conditions.
   g. The parasympathetic division is most active under ordinary conditions.
2. Autonomic nerve fibers
   a. The autonomic fibers are efferent, or motor.
   b. Sympathetic fibers leave the spinal cord and synapse in ganglia.
      (1) Preganglionic fibers pass through white rami to reach paravertebral ganglia.
      (2) Paravertebral ganglia and interconnecting fibers comprise the sympathetic trunks.
      (3) Preganglionic fibers synapse within paravertebral or collateral ganglia.
      (4) Postganglionic fibers usually pass through gray rami to reach spinal nerves before passing to effectors.
      (5) A special set of sympathetic preganglionic fibers passes through ganglia and extends to the adrenal medulla.

c. Parasympathetic fibers begin in the brain stem and sacral region of the spinal cord, and synapse in ganglia near various organs or in the organs themselves.
3. Autonomic neurotransmitters
   a. Sympathetic and parasympathetic preganglionic fibers secrete acetylcholine.
   b. Most sympathetic postganglionic fibers secrete norepinephrine and are adrenergic; postganglionic parasympathetic fibers secrete acetylcholine and are cholinergic.
   c. The different effects of the autonomic divisions are due to the different neurotransmitters the postganglionic fibers release.
4. Actions of autonomic neurotransmitters
   a. Neurotransmitters combine with receptors and alter cell membranes.

b. There are two types of cholinergic receptors and two types of adrenergic receptors.
   c. How cells respond to neurotransmitters depends upon the number and type of receptors in their membranes.
   d. Acetylcholine acts very briefly; norepinephrine and epinephrine may have more prolonged effects.
5. Control of autonomic activity
   a. The central nervous system largely controls the autonomic nervous system.
   b. The medulla oblongata uses autonomic fibers to regulate cardiac, vasomotor, and respiratory activities.
   c. The hypothalamus uses autonomic fibers in regulating visceral functions.
   d. The limbic system and cerebral cortex control emotional responses through the autonomic nervous system.

## CRITICAL THINKING QUESTIONS

1. If a physician plans to obtain a sample of spinal fluid from a patient, what anatomical site can be safely used, and how should the patient be positioned to facilitate this procedure?
2. What functional losses would you expect to observe in a patient who has suffered injury to the right occipital lobe of the cerebral cortex? To the right temporal lobe?
3. The Brown-Séquard syndrome is due to an injury on one side of the spinal cord. It is characterized by paralysis below the injury and on the same side as the injury, and by loss of sensations of temperature and pain on the opposite side. How would you explain these symptoms?
4. The biceps-jerk reflex employs motor neurons that exit from the spinal cord in the 5th spinal nerve (C5), that is, fifth from the top of the cord. The triceps-jerk reflex involves motor neurons in the 7th spinal nerve (C7). How

might these reflexes be used to help locate the site of damage in a patient with a neck injury?
5. Substances used by intravenous drug abusers are sometimes obtained in tablet form and are crushed and dissolved before they are injected. Such tablets may contain fillers, such as talc or cornstarch, that may obstruct tiny blood vessels in the cerebrum. What problems might these obstructions create?
6. In planning treatment for a patient who has had a cerebrovascular accident (CVA), why would it be important to know whether the CVA was caused by a ruptured or obstructed blood vessel?
7. What symptoms might the sympathetic division of the autonomic nervous system produce in a patient who is experiencing stress?
8. How would you distinguish between a patient in a coma and one in a persistent vegetative state?

## REVIEW EXERCISES

1. Name the layers of the meninges, and explain their functions.
2. Describe the location of cerebrospinal fluid within the meninges.
3. Describe the structure of the spinal cord.
4. Name the major ascending and descending tracts of the spinal cord, and list the functions of each.
5. Explain the consequences of nerve fibers crossing over.
6. Describe a reflex arc.
7. Define reflex.
8. Describe a withdrawal reflex.
9. Describe how the brain develops.
10. Describe the structure of the cerebrum.
11. Define cerebral cortex.
12. Describe the location and function of the primary motor areas of the cortex.
13. Describe the location and function of Broca's area.

14. Describe the location and function of the sensory areas of the cortex.
15. Explain the function of the association areas of the lobes of the cerebrum.
16. Define hemisphere dominance.
17. Explain the function of the corpus callosum.
18. Distinguish between short-term and long-term memory.
19. Describe the location and function of the basal ganglia.
20. Describe the location of the ventricles of the brain.
21. Explain how cerebrospinal fluid is produced and how it functions.
22. Name the parts of the diencephalon, and describe the general functions of each.
23. Define the limbic system, and explain its functions.
24. Name the parts of the midbrain, and describe the general functions of each.
25. Describe the pons and its functions.

26. Describe the medulla oblongata and its functions.
27. Describe the location and function of the reticular formation.
28. Distinguish between normal and paradoxical sleep.
29. Describe the functions of the cerebellum.
30. Distinguish between the somatic and autonomic nervous systems.
31. Describe the structure of a peripheral nerve.
32. Distinguish among sensory, motor, and mixed nerves.
33. List four general types of nerve fibers.
34. Name, locate, and describe the major functions of each pair of cranial nerves.
35. Explain how the spinal nerves are grouped and numbered.
36. Define *cauda equina.*
37. Describe the structure of a spinal nerve.
38. Define *plexus,* and locate the major plexuses of the spinal nerves.

39. Distinguish between the sympathetic and parasympathetic divisions of the autonomic nervous system.
40. Explain how autonomic ganglia provide a degree of independence from the central nervous system.
41. Distinguish between a preganglionic fiber and a postganglionic fiber.
42. Define *paravertebral ganglion.*
43. Trace a sympathetic nerve pathway through a ganglion to an effector.
44. Explain why the effects of the sympathetic and parasympathetic autonomic divisions differ.
45. Distinguish between cholinergic and adrenergic nerve fibers.
46. Define *sympathetic tone.*
47. Explain how autonomic neurotransmitters influence the actions of effector cells.
48. Distinguish between alpha adrenergic and beta adrenergic receptors.
49. Describe three examples in which the central nervous system employs autonomic nerve pathways.

 **For More Information Related to This Chapter, Please Visit** *Hole's Human Anatomy and Physiology* **Home Page.**
http://www.mhhe.com/biosci/abio/

# SOMATIC AND SPECIAL SENSES

## CHAPTER OBJECTIVES

### AFTER YOU HAVE STUDIED THIS CHAPTER, YOU SHOULD BE ABLE TO:

1. Name five kinds of receptors and explain the function of each.

2. Explain how receptors stimulate sensory impulses.

3. Explain how a sensation is produced.

4. Distinguish between somatic and special senses.

5. Describe the receptors associated with the senses of touch and pressure, temperature, and pain.

6. Describe how the sense of pain is produced.

7. Explain the importance of stretch receptors in muscles and tendons.

8. Explain the relationship between the senses of smell and taste.

9. Name the parts of the ear and explain the function of each part.

10. Distinguish between static and dynamic equilibrium.

11. Name the parts of the eye and explain the function of each part.

12. Explain how the eye refracts light.

13. Explain how depth and distance are perceived.

14. Describe the visual nerve pathway.

## KEY TERMS

**accommodation** (ah-kom″o-da′shun)

**ampulla** (am-pul′lah)

**auditory** (aw′di-to″re)

**chemoreceptor** (ke″mo-re-sep′tor)

**cochlea** (kok′le-ah)

**cornea** (kor′ne-ah)

**dynamic equilibrium** (di-nam′ik e″kwĭ-lib′re-um)

**labyrinth** (lab′i-rinth)

**macula** (mak′u-lah)

**mechanoreceptor** (mek″ah-no-re-sep′tor)

**olfactory** (ol-fak′to-re)

**optic** (op′tik)

**photoreceptor** (fo″to-re-sep′tor)

**projection** (pro-jek′shun)

**proprioceptor** (pro″pre-o-sep′tor)

**referred pain** (re-furd′ pān)

**refraction** (re-frak′shun)

**retina** (ret′i-nah)

**rhodopsin** (ro-dop′sin)

**sclera** (skle′rah)

**sensory adaptation** (sen′so-re ad″ap-ta′shun)

**static equilibrium** (stat′ik e″kwĭ-lib′re-um)

**thermoreceptor** (ther″mo-re-sep′tor)

## UNDERSTANDING WORDS

**aud-,** to hear: *aud*itory—pertaining to hearing.

**choroid,** skinlike: *choroid* coat—middle, vascular layer of the eye.

**cochlea,** snail: *cochlea*—coiled tube within the inner ear.

**corn-,** horn: *corn*ea—transparent outer layer in the anterior portion of the eye.

**iris,** rainbow: *iris*—colored, muscular part of the eye.

**labyrinth,** maze: *labyrinth*—complex system of connecting chambers and tubes of the inner ear.

**lacri-,** tears: *lacri*mal gland—tear gland.

**lut-,** yellow: macula *lut*ea—yellowish spot on the retina.

**macula,** spot: *macula* lutea—yellowish spot on the retina.

**malle-,** hammer: *malle*us—one of the three bones in the middle ear.

**oculi-,** eye: orbicularis *oculi*—muscle associated with the eyelid.

**olfact-,** to smell: *olfact*ory—pertaining to the sense of smell.

**palpebra,** eyelid: levetor *palpebra*e superioris—muscle associated with the eyelid.

**photo-,** light: *photo*receptor—specialized structures in the eye responsive to light.

**scler-,** hard: *scler*a—tough, outer protective layer of the eye.

**therm-,** heat: *therm*oreceptor—receptor sensitive to changes in temperature.

**tympan-,** drum: *tympan*ic membrane—eardrum.

**vitre-,** glass: *vitre*ous humor—clear, jellylike substance within the eye.

ohn Dalton, a famous English chemist, saw things differently than most people. In a 1794 lecture, he described his visual world. Sealing wax that appeared red to other people was as green as a leaf to Dalton and his brother. Pink wildflowers were blue, and Dalton perceived the cranesbill plant as "sky blue" in daylight, but "very near yellow, but with a tincture of red," in candlelight. He concluded, ". . . that part of the image which others call red, appears to me little more than a shade, or defect of light." The Dalton brothers, like 7 percent of males and 0.4 percent of females today, had the inherited trait of colorblindness.

Dalton was very curious about the cause of his colorblindness, so he made arrangements with his personal physician, Joseph Ransome, to dissect his eyes after he died. Ransome snipped off the back of one eye, removing the retina, where the cone cells that provide color vision are nestled among the more abundant rod cells that impart black-and-white vision. Because Ransome could see red and green normally when he peered through the back of his friend's eyeball, he concluded that it was not an abnormal filter in front of the eye that altered color vision.

Fortunately, Ransome stored the eyes in dry air, where they remained relatively undamaged. In 1994, Dalton's eyes underwent DNA analysis at London's Institute of Ophthalmology. The research showed that Dalton's remaining retina lacked one of three types of pigments, called photopigments, that enable cone cells to capture certain incoming wavelengths of light.

Although people have studied colorblindness for centuries, we are still learning more about it. Recently, researchers investigated why colorblind men lacking cones that capture green light are affected to different degrees. They discovered that colorblind men who can discern a few shades of green have red cone cells that can detect some wavelengths of light that fall within the green region of the spectrum. Color vision may be more complex than we had thought!

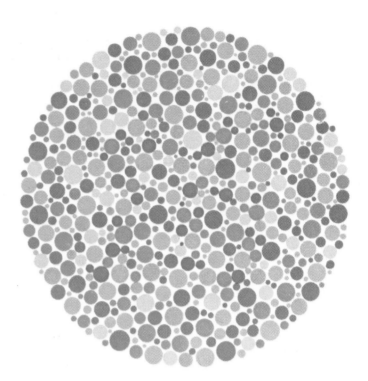

People who are colorblind must get along in a multicolored world. To help them overcome the disadvantage of not seeing important color differences, researchers have developed computer algorithms that convert colored video pictures into shades they can see. This circle of dots is a test to determine whether someone is colorblind. Affected individuals cannot see a different color in certain of the circles in such a drawing. As a result, their brains cannot perceive the embedded pattern that forms the number 16 that others can see.

The above has been reproduced from *Ishihara's Tests for Colour Blindness* published by Kanehara & Co. Ltd, Tokyo, Japan, but tests for colour blindness cannot be conducted with this material. For accurate testing, the original plates should be used.

---

All senses work in basically the same way. Special sensory receptors collect information from the environment and stimulate neurons to send a message to the brain. There the cerebral cortex forms a perception, a person's particular view of the stimulus. Table 12.1 outlines the pathways from sensation to perception that describe an apple.

## RECEPTORS AND SENSATIONS

Sensory receptors are diverse, but share certain features. For example, each type of receptor is particularly sensitive to a distinct kind of environmental change and is much less sensitive to other forms of stimulation.

## Receptor Types

Five general groups of sensory receptors are known, based on their sensitivities to changes in one of the following factors:

1. *Chemical concentration.* Receptors that are stimulated by changes in the concentration of chemical substances are called **chemoreceptors.** Receptors associated with the senses of smell and taste are of this type. Chemoreceptors in internal organs, for example, can detect changes in the blood concentrations of substances including oxygen, hydrogen ions, and glucose.

TABLE 12.1

## INFORMATION FLOW FROM THE ENVIRONMENT THROUGH THE NERVOUS SYSTEM

| Information Flow | Smell | Taste | Sight | Hearing |
|---|---|---|---|---|
| Sense receptors ↓ | Olfactory cells in nose ↓ | Taste bud receptor cells ↓ | Rods and cones in retina ↓ | Hair cells in cochlea ↓ |
| Stimulation of nerve fibers ↓ | Olfactory nerve fibers ↓ | Sensory fibers in various cranial nerves ↓ | Optic nerve fibers ↓ | Auditory nerve fibers ↓ |
| Impulse transmission to CNS | Cortex ↓ | Cortex ↓ | Midbrain and cortex ↓ | Midbrain and cortex ↓ |
| Sensation (memory, experience) ↓ | A pleasant smell ↓ | A sweet taste ↓ | A small, round, red object ↓ | A crunching sound ↓ |
| Perception | The smell of an apple | The taste of an apple | The sight of an apple | Biting into an apple |

2. *Tissue damage.* Tissue damage stimulates **pain receptors** (nociceptors). Triggering factors include exposure to excessive mechanical, electrical, thermal, or chemical energy.

3. *Temperature change.* Receptors sensitive to temperature change are called **thermoreceptors.**

4. *Mechanical forces.* A number of sensory receptors sense mechanical forces, such as changes in the pressure or movement of fluids. As a rule, these **mechanoreceptors** detect changes that deform the receptors. For example, **proprioceptors** sense changes in the tensions of muscles and tendons, whereas **baroreceptors** (pressoreceptors) in certain blood vessels can detect changes in blood pressure. Similarly, **stretch receptors** in the lungs sense degree of inflation.

5. *Light intensity.* Light receptors, or **photoreceptors,** are only in the eyes. They respond to light energy of sufficient intensity.

## Sensory Impulses

Sensory receptors can be ends of nerve fibers or other kinds of cells located close to them. In either case, stimulation causes local changes in their membrane potentials (receptor potentials), generating a graded electric current that reflects the intensity of stimulation (see chapter 10).

If a receptor is a nerve fiber and the change in membrane potential reaches threshold, an action potential is generated, and a sensory impulse travels on the nerve fiber. However, if the receptor is another type of cell, its receptor potential must be transferred to a nerve fiber to trigger an action potential. Peripheral nerves transmit such sensory impulses to the central nervous system, where they are analyzed and interpreted within the brain (see chapter 11).

## Sensations

A **sensation** (perception) is a feeling that occurs when the brain interprets sensory impulses. Because all the nerve impulses that travel away from sensory receptors into the central nervous system are alike, the resulting sensation depends on which region of the cerebral cortex receives the impulse. For example, impulses reaching one region are always interpreted as sounds, and those reaching another portion are always sensed as touch.

It makes little difference how receptors are stimulated. Pain receptors, for example, can be stimulated by heat, cold, or pressure, but the sensation is always the same because in each case the same part of the brain interprets the resulting nerve impulses. Similarly, factors other than light, such as a sharp blow to the head, may trigger nerve impulses in visual receptors. When this happens, the person may "see" lights, even though no light is entering the eye, since any impulses reaching the visual cortex are interpreted as light. Normally receptors only respond to specific stimuli, so the brain creates the correct sensation.

At the same time that a sensation forms, the cerebral cortex interprets it to seem to come from the receptors being stimulated. This process is called **projection,** because the brain projects the sensation back to its apparent source. Projection allows a person to pinpoint the region of stimulation. Thus, we are aware that the eyes see an apple, the nose smells it, and the ears hear the teeth crunch into it.

## Sensory Adaptation

When sensory receptors are continuously stimulated, many of them undergo an adjustment called **sensory adaptation.** As the receptors adapt, impulses leave them at decreasing rates, until finally these receptors may completely fail to send signals. Once receptors adapt, impulses are triggered only if the strength of the stimulus changes.

Imagine a person walking into a fish market. At first the fishy odor seems intense, but it becomes less and less noticeable as the smell (olfactory) receptors adapt. Soon, the once offensive odor is barely noticeable, but if the person leaves and reenters, he or she again smells the stench.

**1** List the five general types of sensory receptors.

**2** What do all types of receptors have in common?

**3** Explain how a sensation occurs.

**4** What is sensory adaptation?

## SOMATIC SENSES

Receptors associated with the skin, muscles, joints, and viscera provide somatic senses. These senses can be divided into three groups:

1. Senses associated with changes at the body surface (exteroceptive senses), which include the senses of touch, pressure, temperature, and pain.

2. Senses associated with changes in muscles and tendons and in body position (proprioceptive senses).

3. Senses associated with changes in viscera (visceroceptive senses). Except for visceral pain, these senses are discussed in subsequent chapters.

### Touch and Pressure Senses

The senses of touch and pressure derive from three kinds of receptors (fig. 12.1). As a group, these receptors sense mechanical forces that deform or displace tissues. The touch and pressure receptors include:

1. **Sensory nerve fibers.** These receptors are common in epithelial tissues where their free ends lie between epithelial cells. They are associated with sensations of touch and pressure.

2. **Meissner's corpuscles.** These are small, oval masses of flattened connective tissue cells in connective tissue sheaths. Two or more sensory nerve fibers branch into each corpuscle and end within it as tiny knobs.

   Meissner's corpuscles are abundant in the hairless portions of the skin, such as the lips, fingertips, palms, soles, nipples, and external genital organs. They sense motion of objects that barely contact the skin, interpreting impulses from them as the sensation of light touch. They are also used when a person touches something to judge its texture (fig. 12.2a).

**FIGURE 12.1**

Touch and pressure receptors include free ends of sensory nerve fibers, Meissner's corpuscles, and Pacinian corpuscles.

3. **Pacinian corpuscles.** These sensory bodies are relatively large, ellipsoidal structures composed of connective tissue fibers and cells. They are common in the deeper subcutaneous tissues of the hands, feet, penis, clitoris, urethra, and breasts, and also in tendons of muscles and ligaments of joints (fig. 12.2b).

   Heavy pressure stimulates Pacinian corpuscles. They may also detect vibrations in tissues.

### Temperature Senses

Temperature receptors (thermoreceptors) include two groups of *free nerve endings* located in the skin. Those that respond to warmer temperatures are called *heat receptors* and those that respond to colder temperatures are called *cold receptors*. The heat receptors are most sensitive to temperatures above 25°C (77°F) and become unresponsive at temperatures above 45°C (113°F). As 45°C is approached, pain receptors are also triggered, producing a burning sensation.

Cold receptors are most sensitive to temperatures between 10°C (50°F) and 20°C (68°F). If the temperature drops below 10°C, pain receptors are stimulated, and the person feels a freezing sensation.

At intermediate temperatures, the brain interprets sensory input from different combinations of these receptors as a particular temperature sensation. Both heat and cold receptors rapidly adapt, so that within about a minute of continuous stimulation, the sensation of heat or cold be-

(a)

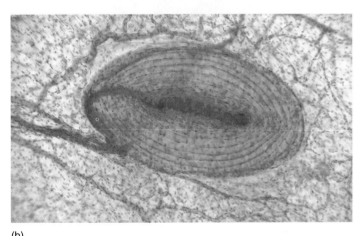

(b)

**FIGURE 12.2**

(a) Light micrograph of a Meissner's corpuscle from the skin of the palm (250×). (b) Light micrograph of a Pacinian corpuscle (25× enlarged to 100×).

gins to fade. Imagine stepping into a bathtub containing "hot" water that soon feels pleasantly warm, or diving into a "freezing" pool of water that soon becomes refreshingly cool. These are familiar temperature adaptations.

## Sense of Pain

Other receptors that consist of *free nerve endings* sense pain. These receptors are widely distributed throughout the skin and internal tissues, except in the nervous tissue of the brain, which lacks pain receptors.

### Pain Receptors

The pain receptors are protective in that they are stimulated when tissues are damaged. Pain sensation is usu-

ally perceived as unpleasant, signaling that action be taken to remove the source of the stimulation.

Although most pain receptors can be stimulated by more than one type of change, some are most sensitive to mechanical damage whereas others are particularly sensitive to extremes in temperature. Still other pain receptors are most responsive to chemicals, such as hydrogen ions, potassium ions, or specific breakdown products of proteins, histamine, and acetylcholine. A deficiency of blood (ischemia) and thus a deficiency of oxygen (hypoxia) in a tissue, or stimulation of mechanical-sensitive receptors, also triggers the pain sensation. The pain elicited during a muscle cramp, for example, results from interruption of blood flow that occurs as the sustained contraction squeezes capillaries, as well as to stimulation of mechanical-sensitive pain receptors. Also, when blood flow is interrupted, pain-stimulating chemicals accumulate. Increasing blood flow through the sore tissue may relieve the resulting pain, and this is why heat is sometimes applied to reduce muscle soreness. The heat dilates blood vessels and thus promotes blood flow, which helps reduce the concentration of the pain-stimulating substances. In some conditions, accumulating chemicals lower the thresholds of pain receptors, making inflamed tissues more sensitive to heat or pressure than before.

Pain receptors adapt very little, if at all. Once such a receptor is activated, even by a single stimulus, it may continue to send impulses into the central nervous system for some time.

### Visceral Pain

As a rule, pain receptors are the only receptors in viscera whose stimulation produces sensations. Pain receptors in these organs respond differently to stimulation than those associated with surface tissues. For example, localized damage to intestinal tissue during surgical procedures may not elicit any pain sensations, even in a conscious person. However, when visceral tissues are subjected to more widespread stimulation, as when intestinal tissues are stretched or when the smooth muscles in the intestinal walls undergo spasms, a strong pain sensation may follow. Once again, the resulting pain results from stimulation of mechanical-sensitive receptors and to decreased blood flow accompanied by a lower tissue oxygen concentration and accumulation of pain-stimulating chemicals.

Visceral pain may feel as if it is coming from some part of the body other than the part being stimulated—a phenomenon called **referred pain.** For example, pain originating in the heart may be referred to the left shoulder or the inside of the left arm. Pain from the lower esophagus, stomach, or small intestine may seem to be coming from the upper central (epigastric) region of the abdomen. Pain from the urogenital tract may be referred to the lower central (hypogastric) region of the abdomen or to the sides between the ribs and the hip (fig. 12.3).

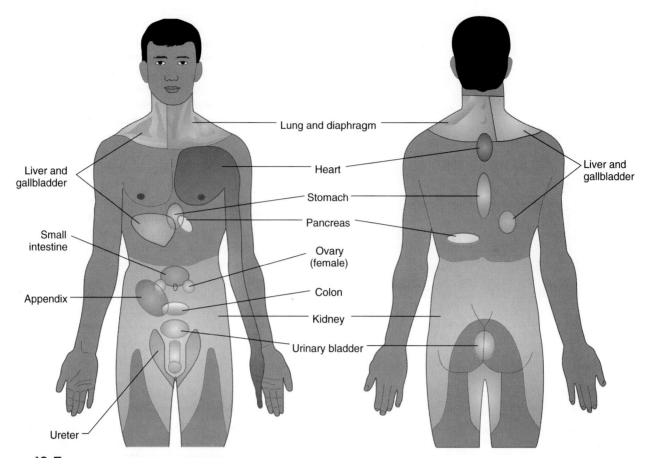

**FIGURE 12.3**

Surface regions to which visceral pain may be referred.

Referred pain may derive from *common nerve pathways* that sensory impulses coming both from skin areas and from internal organs use (see chapter 11). Pain impulses from the heart seem to be conducted over the same nerve pathways as those from the skin of the left shoulder and the inside of the left arm, as shown in figure 12.4. Consequently, the cerebral cortex may incorrectly interpret the source of the impulses as the shoulder and the inside of the left arm, rather than the heart.

Pain originating in the parietal layers of thoracic and abdominal membranes—parietal pleura, parietal pericardium, or parietal peritoneum—is usually not referred; instead, such pain is felt directly over the area being stimulated.

## Pain Nerve Pathways

The nerve fibers that conduct impulses away from pain receptors are of two main types: acute pain fibers and chronic pain fibers.

The *acute pain fibers* (also known as A-delta fibers) are thin, myelinated nerve fibers. They conduct nerve impulses rapidly, at velocities up to 30 meters per second. These impulses are associated with the sensation of sharp pain, which typically seems to originate in a local

area of skin. This type of pain seldom continues after the pain-producing stimulus stops.

The *chronic pain fibers* (C fibers) are thin, unmyelinated nerve fibers. They conduct impulses more slowly than acute pain fibers, at velocities up to 2 meters per second. These impulses cause the dull, aching pain sensation that may be widespread and difficult to pinpoint. Such pain may continue for some time after the original stimulus ceases. Although acute pain is usually sensed as coming from the surface, chronic pain is likely to be felt in deeper tissues as well as in the skin. Visceral pain impulses are usually carried on C fibers.

Commonly, an event that stimulates pain receptors will trigger impulses on both types of pain fibers. This causes a dual sensation—a sharp, pricking pain, then a dull, aching one. The aching pain is usually more intense and may worsen over time. Chronic pain that resists relief and control can be debilitating.

Pain impulses that originate from tissues of the head reach the brain on sensory fibers of the fifth, seventh, ninth, and tenth cranial nerves. All other pain impulses travel on sensory fibers of spinal nerves, and they pass into the spinal cord by way of the dorsal roots of these spinal nerves.

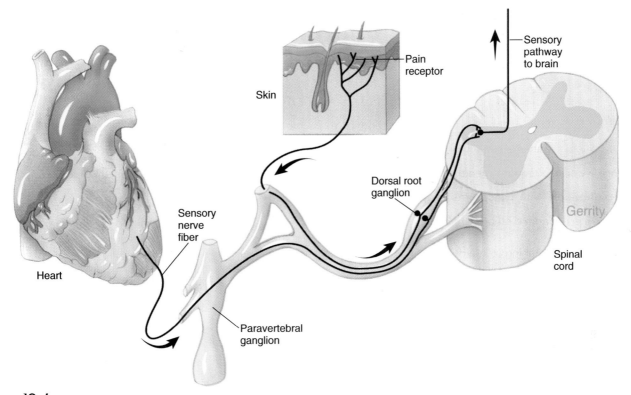

**FIGURE** 12.4

Pain originating in the heart may feel as if it is coming from the skin because sensory impulses from those two regions follow common nerve pathways to the brain.

Upon reaching the spinal cord, pain impulses enter the gray matter of the dorsal horn, where they are processed (see chapters 10 and 11). The fast-conducting fibers synapse with long nerve fibers that cross over to the opposite side of the spinal cord and ascend in the lateral spinothalamic tracts. The impulses carried on the slow-conducting fibers pass through one or more interneurons before reaching the long fibers that cross over and ascend to the brain.

Within the brain, most of the pain fibers terminate in the reticular formation (see chapter 11), and from there are conducted on fibers of still other neurons to the thalamus, hypothalamus, and cerebral cortex.

### Regulation of Pain Impulses

Awareness of pain occurs when pain impulses reach the level of the thalamus—that is, even before they reach the cerebral cortex. However, the cerebral cortex must judge the intensity of pain and locate its source. The cerebral cortex is also responsible for emotional and motor responses to pain.

Still other parts of the brain regulate the flow of pain impulses from the spinal cord, including areas of gray matter in the midbrain, pons, and medulla oblongata. Impulses from special neurons in these areas descend in the lateral funiculus (see chapter 11) to various levels within the spinal cord. The impulses stimulate the ends of certain nerve fibers to release biochemicals that can block pain signals by inhibiting presynaptic nerve fibers in the posterior horn of the spinal cord.

Among the inhibiting substances released in the posterior horn are neuropeptides called *enkephalins* and the monoamine *serotonin* (see chapter 10). Enkephalins can suppress both acute and chronic pain impulses; thus they can relieve strong pain sensations, much as morphine and other opiate drugs do. In fact, enkephalins were discovered because they bind to the same receptor sites on neuron membranes as does morphine. Serotonin stimulates other neurons to release enkephalins.

Another group of neuropeptides with pain-suppressing, morphinelike actions are the *endorphins*. They are found in the pituitary gland and in regions of the nervous system, such as the hypothalamus, that transmit pain impulses. Enkephalins and endorphins are released in response to extreme pain impulses, providing natural pain control. Clinical Application 12.1 discusses treatments for severe pain.

1.   Describe three types of touch and pressure receptors.
2.   Describe thermoreceptors.
3.   What types of stimuli excite pain receptors?
4.   What is referred pain?
5.   Explain how neuropeptides control pain.

# Cancer Pain and Chronic Pain

One of the shortcomings of modern medicine is that many patients suffering from the pain of advanced, untreatable cancer do not receive adequate medication to ease their final days. Studies show that 50% of people with cancer are in pain at the time of their diagnosis, and 90% of those in advanced stages are in pain.

Doctors are beginning to advocate pain relief for all people with cancer who require it, not only for humanitarian reasons but for clinical ones. Studies on rats with lung cancer show that tumors spread more rapidly in animals not given pain medication. Plus, the widespread belief that giving narcotics would addict cancer patients has been shown to be unfounded. Narcotics are much more likely to be addicting when they are taken to provide euphoria than when they are taken to relieve severe pain.

Drugs used to help cancer patients include:

- Anti-inflammatory agents such as aspirin and ibuprofen.

- Weak narcotics such as codeine.
- Strong narcotics such as morphine.
- Opiates delivered directly to the spine via an implanted reservoir.

Many patients are offered the option of patient-controlled analgesia, in which they determine their own dosage schedule.

Chronic pain is of three types: lower back pain, migraine, and myofascial syndrome (inflammation of muscles and their fascia). A variety of approaches treat chronic pain:

- Biofeedback. A mechanical device detects and amplifies an autonomic body function, such as blood pressure or heart rate. A person made aware of the measurement can concentrate to attempt to alter it to within a normal range.
- Anti-inflammatory drugs.
- Stretching exercises.
- Trigger point injections of local anesthetic drugs into cramping muscles.
- The antidepressant amitriptyline raises serotonin levels in the central nervous system, relieving some chronic pain.
- Transcutaneous electrical nerve stimulation (TENS) places electrodes on nerves causing pain. The patient feels a tingling sensation, then pain relief.
- A nerve block is invasive and interrupts a pain signal by freezing (cryotherapy) or introducing an anesthetic drug (neurolysis). Side effects include numbness or paralysis.
- A dorsal column stimulator consists of electrodes implanted near the spinal cord. ■

## Stretch Receptors

*Stretch receptors* are proprioceptors that send information to the spinal cord and brain concerning the lengths and tensions of muscles. The two main kinds of stretch receptors are muscle spindles and Golgi tendon organs; however, no sensation results when they are stimulated.

**Muscle spindles** are located in skeletal muscles near their junctions with tendons. Each spindle consists of one or more small, modified skeletal muscle fibers (intrafusal fibers) enclosed in a connective tissue sheath. Near its center, each intrafusal fiber has a specialized nonstriated region with the end of a sensory nerve fiber wrapped around it, (fig. 12.5a).

The striated portions of the intrafusal fiber contract to keep the spindle taut at different muscle lengths.

Thus, if the whole muscle is stretched, the muscle spindle is also stretched, triggering sensory nerve impulses on its nerve fiber. Such sensory impulses travel into the spinal cord and onto motor fibers leading back to the same muscle, contracting it. This action, called a **stretch reflex,** opposes the lengthening of the muscle and helps maintain the desired position of a limb in spite of gravitational or other forces tending to move it (see chapter 10).

**Golgi tendon organs** are found in tendons close to their attachments to muscles. Each is connected to a set of skeletal muscle fibers and is innervated by a sensory neuron (fig. 12.5b). These receptors have high thresholds and are stimulated by increased tension. Sensory impulses from them produce a reflex that inhibits contraction of the muscle whose tendon they occupy. Thus, the

Sensory nerve fiber

Nerve endings

Intrafusal fiber

Skeletal muscle fiber

Muscle spindle

Connective tissue sheath

(a)

Golgi tendon organ

Sensory nerve fiber

Tendon

Skeletal muscle fiber

(b)

**FIGURE 12.5**

(a) Increased muscle length stimulate muscle spindles, which stimulate muscle contraction. (b) Golgi tendon organs occupy tendons, where they inhibit muscle contraction.

| TABLE 12.2 SOMATIC RECEPTORS | | |
|---|---|---|
| **Type** | **Function** | **Sensation** |
| Free nerve endings (mechanoreceptors) | Detect changes in pressure | Touch, pressure |
| Meissner's corpuscles (mechanoreceptors) | Detect objects moving over the skin | Touch, texture |
| Pacinian corpuscles (mechanoreceptors) | Detect changes in pressure | Deep pressure, vibrations |
| Free nerve endings (thermoreceptors) | Detect changes in temperature | Heat, cold |
| Free nerve endings (pain receptors) | Detect tissue damage | Pain |
| Muscle spindle (mechanoreceptors) | Detect changes in muscle length | None |
| Golgi tendon organ (mechanoreceptors) | Detect changes in muscle tension | None |

Golgi tendon organs stimulate a reflex with an effect opposite that of a stretch reflex. This reflex also helps maintain posture, and it protects muscle attachments from being pulled away from their insertions by exces-

sive tension. Table 12.2 summarizes the somatic receptors and their functions.

1 Describe a muscle spindle.

2 Explain how muscle spindles help maintain posture.

3 Where are Golgi tendon organs located?

4 What is the function of Golgi tendon organs?

## SPECIAL SENSES

Special senses are those whose sensory receptors are within large, complex sensory organs in the head. These organs and their respective senses include the following:

- olfactory organs ⟶ smell
- taste buds ⟶ taste
- ears ⟶ hearing;

  static equilibrium;

  dynamic equilibrium
- eyes ⟶ sight

# Mixed-up Senses—Synesthesia

"His name was purple."
"The song was full of shimmering green triangles."
"The paint smelled red."
"The sunset was salty."

To 1 in 500,000 of us, senses become unusually associated in a little-understood condition called synesthesia. A sensation involving one sense is perceived in terms of another, so that, most commonly, visions that take on characteristic smells or sounds are associated with particular colors. Although synesthetic people are unlikely to share the same associations, they do share some aspects of their talents—their particular quirks are involuntary, remain constant over their lifetimes (a certain name or song is always a certain color), and they are absolutely convinced that what they perceive is reality, not something in their "mind's eye."

We do not know what causes synesthesia, although it does seem to be inherited. The condition has been sheepishly reported to psychologists and physicians for at least 200 years. Various theories (all unproven) have attributed the condition to an immature nervous system that cannot sort out sensory stimuli; altered brain circuitry; and simply an exaggerated use of metaphor—taking such descriptions as a sharp flavor, warm color, and sweet person too literally.

In 1980, measurements of cerebral blood flow linked synesthesia to a brain abnormality. A person who "tasted" in geometric shapes inhaled a harmless radioactive chemical, xenon-

133, which her tissues absorbed. The rate at which the chemical left certain parts of her brain indicated how metabolically active that part was. When brain activity was assessed during a synesthetic experience, blood flow in her left hemisphere, particularly in the temporal lobe, plummeted 18%—a decrease seen only when tissue dies as a result of a stroke. But the woman was perfectly healthy. The left hemisphere is the site of the language center. Another clue to the origin of synesthesia is that people with temporal lobe epilepsy are often synesthetic. Researchers believe that synesthesia reflects a breakdown in the translation of a perception into language—but we have much to learn about this fascinating mixing of the senses. ■

## SENSE OF SMELL

The ability to detect the strong scent of a fish market, the antiseptic odor of a hospital, the aroma of a ripe melon—and thousands of other smells—is possible thanks to a yellowish patch of tissue the size of a quarter high up in the nose. This fabric of sensation is actually a layer of 12 million specialized cells.

## Olfactory Receptors

Smell, or olfactory, receptors are similar to those for taste in that they are chemoreceptors sensitive to chemicals dissolved in liquids. These two chemical senses function closely together and aid in food selection, because we smell food at the same time we taste it. In fact, it is often difficult to tell what part of a food sensation is due to smell and what part is due to taste. For this reason, an onion tastes quite different when sampled with the nostrils closed, because much of the usual onion sensation is due to odor. Similarly, if copious mucous secretions from an upper respiratory infection (such as a cold) cover the olfactory receptors, food may seem to lose its taste. Clinical Application 12.2

discusses an unusual and disturbing type of sensory abnormality.

## Olfactory Organs

The **olfactory organs,** which contain the olfactory receptors, also include epithelial supporting cells. These organs appear as yellowish brown masses surrounded by pinkish mucous membrane. They cover the upper parts of the nasal cavity, the superior nasal conchae, and a portion of the nasal septum (fig. 12.6).

The **olfactory receptor cells** are bipolar neurons surrounded by columnar epithelial cells. These neurons have knobs at the distal ends of their dendrites covered with hairlike cilia. The cilia project into the nasal cavity and are thought to be the sensitive portions of the receptors (fig. 12.7). A person's 12 million olfactory receptor cells each have ten to twenty cilia.

Chemicals that stimulate olfactory receptors enter the nasal cavity as gases, but they must dissolve at least partially in the watery fluids that surround the cilia before they can be detected. In some cases, these chemicals bind to specific plasma membrane receptors on the olfactory receptor cells, depolarizing them and thereby generating a nerve impulse.

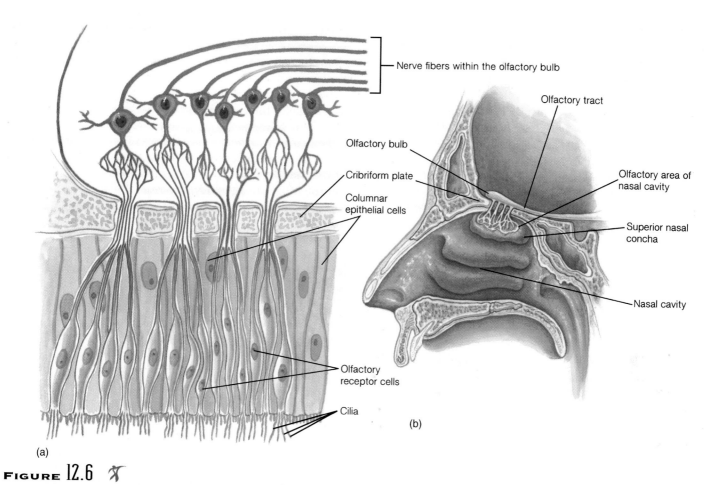

**FIGURE 12.6**

(a) Columnar epithelial cells support olfactory receptor cells, which have cilia at their distal ends. (b) The olfactory area is associated with the superior nasal concha.

**FIGURE 12.7**

Light micrograph of the olfactory epithelium (250×).

The expert nose of the bloodhound is due to its 4 billion olfactory cells. A bloodhound's olfactory epithelium spread out covers 59 square inches, compared to 1 1/2 square inches in a human. Still, the human sense of smell is nothing to sneeze at—people can detect one molecule of green pepper smell in a gaseous sea of 3 trillion other molecules. Our 12 million smell cells and their many million more receptors allow us to discern some 10,000 scents. But, without air, we cannot smell, as early astronauts could attest; eating through tubes, early astronauts could not smell their food. Currently, astronauts eat in pressurized cabins, unaffected by the vacuum in which they travel.

Sensory receptors are not the same as membrane receptors. Sensory receptors may be as small as individual cells or as large as complex organs such as the eye or ear. They respond to sensory stimuli. Membrane receptors are molecules such as proteins and glycoproteins located on the plasma membranes of cells. They allow cells, such as neurons and olfactory receptor cells, to respond to specific molecules reaching their membranes. Thus, the olfactory receptors are cells that respond to chemical stimuli, but they depend on plasma membrane receptors to do so.

## Olfactory Nerve Pathways

Once olfactory receptors are stimulated, nerve impulses travel along their axons through tiny openings in the cribriform plates of the ethmoid bone. These fibers synapse with neurons located in the enlargements of the **olfactory bulbs,** structures that lie on either side of the crista galli of the ethmoid bone (see figs. 7.26 and 12.6). Within the olfactory bulbs, the sensory impulses are analyzed, and as a result, additional impulses travel along the **olfactory tracts** to portions of the limbic system (see chapter 11). The main interpreting areas for the olfactory impulses (olfactory cortex) are located deep within the temporal lobes and at the bases of the frontal lobes, anterior to the hypothalamus.

## Olfactory Stimulation

Biologists are not certain how stimulated receptors encode specific smells, but a leading hypothesis is that each odor stimulates a distinct set of receptor subtypes. The brain then recognizes the combination of receptors as an *olfactory code.* For example, perhaps there are ten types of odor receptors. Banana might stimulate receptors 2, 4, and 7; garlic receptors 1, 5, and 9. Some investigators have proposed seven primary odors, but others hypothesize that the number is much higher and may reflect the functioning of hundreds of genes.

Because the olfactory organs are high in the nasal cavity above the usual pathway of inhaled air, sniffing and forcing air over the receptor areas may be necessary to smell a faint odor. Olfactory receptors undergo sensory adaptation rather rapidly, so the intensity of a smell drops about 50% within a second following the stimulation. Within a minute, the receptors may become almost insensitive to a given odor, but even though they have adapted to one scent, their sensitivity to other odors persists.

The olfactory receptor neurons are the only nerve cells in direct contact with the outside environment. Because of their exposed positions, these neurons are subject to damage; they are the only example of damaged neurons that are regularly replaced. Despite replacement, progressive diminishing of olfactory sense usually occurs with age. In fact, some investigators estimate that about 1% of the olfactory receptors are lost every year.

**1** Where are the olfactory receptors located?

**2** Trace the pathway of an olfactory impulse from a receptor to the cerebrum.

**3** Why is the sense of smell likely to decrease with age?

Genes may control which smells we like and dislike. One study asked young women to smell tee shirts that young men had worn for two days, without benefit of deodorant. The women tended to prefer the scents of the men who were least like them genetically. In an evolutionary sense, preferring to mate with a genetically dissimilar partner may lessen the chance of inherited diseases being transmitted to a child. This is because more closely related individuals are more likely to have the same disease-causing genes.

## SENSE OF TASTE

**Taste buds** are the special organs of taste. They resemble orange sections and are primarily located on the surface of the tongue where they are associated with tiny elevations called **papillae** (figs. 12.8 and 12.9). Taste buds are also scattered in the roof of the mouth, the linings of the cheeks, and the walls of the pharynx.

## Taste Receptors

Each taste bud includes a group of modified epithelial cells, which are the **taste cells** (gustatory cells) that function as receptors. Each of our 10,000 taste buds houses 60 to 100 taste cells. The taste bud also includes a number of epithelial supporting cells. The entire structure is somewhat spherical, with an opening, the **taste pore,** on its free surface. Tiny projections (microvilli), called **taste hairs,** protrude from the outer ends of the taste cells and jut out through the taste pore. These taste hairs are the sensitive parts of the receptor cells.

Interwoven among and wrapped around the taste cells is a network of nerve fibers. The ends of these fibers

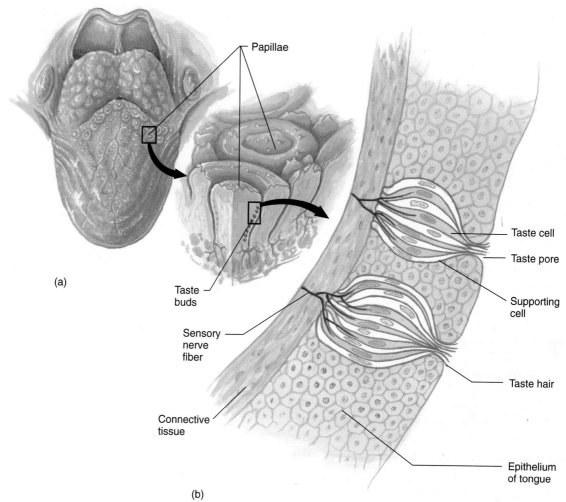

Papillae

Taste cell

Taste pore

Supporting cell

Taste buds

Sensory nerve fiber

Taste hair

Connective tissue

Epithelium of tongue

(a)

(b)

**FIGURE** 12.8

(a) Taste buds on the surface of the tongue are associated with nipplelike elevations called papillae. (b) A taste bud contains taste cells and has an opening, the taste pore, at its free surface.

**FIGURE** 12.9

A light micrograph of some taste buds (arrows) (225×).

closely contact the receptor cell membranes. A stimulated receptor cell triggers an impulse on a nearby nerve fiber, which travels into the brain.

A chemical to be tasted must dissolve in the watery fluid surrounding the taste buds. The salivary glands supply this fluid. To demonstrate its importance, blot your tongue and try to taste some dry food; then repeat the test after moistening your tongue with saliva.

As is the case for smell, the mechanism of tasting is not well understood, but may involve combinations of chemicals binding specific receptor sites on taste hair surfaces, altering membrane polarization, and thereby generating sensory impulses on nearby nerve fibers. The amount of change is directly proportional to the concentration of the stimulating substance.

Although the taste cells in all taste buds appear very much alike microscopically, there are at least four types, corresponding to a particular kind of chemical stimulus. This produces at least four primary taste (gustatory) sensations.

**Figure 12.10**

Color indicates patterns of taste receptors: (*a*) sweet receptors, (*b*) sour receptors, (*c*) salt receptors, and (*d*) bitter receptors.

## Taste Sensations

The four *primary taste sensations* are *sweet, sour, salty,* and *bitter.* Some investigators recognize two other taste sensations, *alkaline* and *metallic.* Each of the many flavors we experience daily is believed to result from one of the primary sensations or from some combination of two or more of them. The way we experience flavors may also reflect the concentration of chemicals as well as the sensations of smell, texture (touch), and temperature. Furthermore, chemicals in some foods—chili peppers and ginger, for instance—may stimulate pain receptors that cause a burning sensation.

The tongue is often depicted as having regions in which taste buds specialize in a particular sensation—the tip tastes sweetness, the front saltiness, the sides sour, and the back bitter. Although taste buds with all specificities are scattered everywhere, and a single taste bud can have receptors for all four general types of tastes, each of the four major types of taste receptors is somewhat concentrated in these regions of the tongue's surface (fig. 12.10).

*Sweet receptors* are most plentiful near the tip of the tongue. This is why a child may prefer to lick a lollipop rather than chew it. The chemicals that stimulate these receptors are usually carbohydrates, but a few inorganic substances, including some salts of lead and beryllium, also elicit sweet sensations.

*Sour receptors* primarily occur along the margins of the tongue, and acids stimulate them. The intensity of a sour sensation is roughly proportional to the concentration of the hydrogen ions in the substance being tasted.

*Salt receptors* are most abundant in the tip and the upper front portion of the tongue. Ionized inorganic salts mainly stimulate salt receptors. The quality of the sensation that each salt produces depends upon the kind of positively charged ion, such as $Na^+$ from table salt, that it releases into solution.

For many years biology teachers have given students strips of paper impregnated with a chemical called PTC to study a "bitterness tasting" gene. Typically, seven in ten students make a horrible face upon tasting the paper, because they taste PTC as extremely bitter. The other students look around in amazement, for they cannot taste a thing! PTC-tasters also find artificial sweeteners offensive and may taste cheese as bitter rather than pleasantly tart. Researchers hypothesize that some infants who fail to gain weight may simply not like the taste of milk. Genes may also control whether one likes the taste of broccoli, or can distinguish diet from regular soft drinks.

*Bitter receptors* are located toward the back of the tongue. A variety of chemicals stimulates them, including many organic compounds. Inorganic salts of magnesium and calcium produce bitter sensations, too.

One group of bitter compounds of particular interest are the *alkaloids,* which include a number of poisons such as strychnine, nicotine, and morphine. The fact that we spit out bitter substances may be a protective mechanism to avoid poisonous alkaloids in foods.

Taste receptors, like olfactory receptors, rapidly undergo sensory adaptation. One can avoid the resulting loss of taste by moving bits of food over the surface of the tongue to stimulate different receptors at different moments.

Although taste cells are very close to the surface of the tongue and are therefore exposed to environmental wear and tear, the sense of taste is not as likely to diminish with age as is the sense of smell. This is because taste cells are modified epithelial cells and reproduce continually. A taste cell functions for only about three days before it is replaced.

# CLINICAL APPLICATION

## Smell and Taste Disorders

Imagine a spicy slice of pizza, or freshly brewed coffee, and your mouth waters in anticipation. But for millions of people, the senses of smell and taste are dulled, distorted, or gone altogether. Many more of us get some idea of their plight when a cold temporarily stifles these senses.

Compared to the loss of hearing or sight, being unable to taste or smell normally may seem more an oddity than an illness. But those with such ailments would probably disagree. In some situations, a poor or lacking sense of smell can even be dangerous. One patient died in a house fire because he did not smell the smoke in time to escape.

The direct connection between the outside environment and the brain makes the sense of smell very vulnerable to damage. Smell and taste disorders can be triggered by colds and flu, allergies, nasal polyps (swollen mucous membranes inside the nose), a head injury, chemical exposure, a nutritional or metabolic problem, or a disease. In many cases, a cause cannot be identified.

Drugs can alter taste and smell in many ways, affecting cell turnover, the neural conduction system, the status of receptors, and changes in nutritional status. Consider what happened to twelve hikers touring Peru and Bolivia. A day before a long hike, three of them had begun taking acetazolamide (Diamox), a drug that prevents acute mountain sickness. The night after the climb, the group went out for beer. To three of the people, the brew tasted unbearably bitter, and a drink of cola to wash away the taste was equally offensive. At fault: acetazolamide.

Drugs containing sulfur atoms squelch taste. They include the anti-inflammatory drug penicillamine, the antihypertensive drug captopril (Capoten), and transdermal (patch) nitroglycerin to treat chest pain. The antibiotic tetracycline and the antiprotozoan metronidazole (Flagyl) cause a metallic taste. Cancer chemotherapy and radiation treatment often alter taste and smell.

Exposure to toxic chemicals can affect taste and smell, too. A 45-year-old woman from Altoona, Pennsylvania, suddenly found that once-pleasant smells had become offensive. Her doctor traced her problem to inhaling a paint stripper. Hydrocarbon solvents in the product—toluene, methanol, and methylene chloride—were responsible for her *cacosmia,* the association of an odor of decay with normally inoffensive stimuli. ∎

## Taste Nerve Pathways

Sensory impulses from taste receptors located in the anterior two-thirds of the tongue travel on fibers of the facial nerve (VII); impulses from receptors in the posterior one-third of the tongue and the back of the mouth pass along the glossopharyngeal nerve (IX); and impulses from receptors at the base of the tongue and the pharynx travel on the vagus nerve (X) (see chapter 11).

These cranial nerves conduct the impulses into the medulla oblongata. From there, the impulses ascend to the thalamus and are directed to the gustatory cortex of the cerebrum, located in the parietal lobe along a deep portion of the lateral sulcus.

Sensory information from olfactory receptors is also conducted to the limbic system, a brain center for memory and emotions. This is why we may become nostalgic over a scent from the past. A whiff of the perfume that grandma used to wear may bring back a flood of memories. The input to the limbic system also explains why odors can alter mood so easily. For example, the scent of new-mown hay or rain on a summer's morning generally makes us feel good. Clinical Application 12.3 and table 12.3 discuss disorders of smell and taste.

| TABLE 12.3 | TYPES OF SMELL AND TASTE DISORDERS | |
|---|---|---|
| Doctors use these terms when discussing taste and smell disorders: | | |
| | **Smell** | **Taste** |
| Loss of sensation | Anosmia | Ageusia |
| Diminished sensation | Hyposmia | Hypogeusia |
| Heightened sensation | Hyperosmia | Hypergeusia |
| Distorted sensation | Dysosmia | Dysgeusia |

1   Why is saliva necessary to taste?

2   Name the four primary taste sensations.

3   What characteristic of taste receptors helps maintain a sense of taste with age?

4   Trace a sensory impulse from a taste receptor to the cerebral cortex.

## SENSE OF HEARING

The organ of hearing, the **ear,** has external, middle, and inner sections. In addition to making hearing possible, the ear provides the sense of equilibrium.

### External Ear

The external ear consists of two parts: an outer, funnel-like structure called the **auricle** (pinna), and an S-shaped tube, the **external auditory meatus** (external auditory canal) that leads inward for about 2.5 centimeters (fig. 12.11).

The external auditory meatus passes into the temporal bone. Near its opening, hairs guard the tube. It is lined with skin that contains many modified sweat glands called *ceruminous glands,* which secrete wax (cerumen). The hairs and wax help keep large foreign objects, such as insects, out of the ear.

Vibrations are transmitted through matter as sound waves. Just as the sounds of some musical instruments are produced by vibrating strings or reeds, the sounds of the human voice are caused by vibrating vocal folds in the larynx. The auricle of the ear helps collect sound waves traveling through air and directs them into the external auditory meatus.

After entering the meatus, the sound waves pass to the end of the tube and alter the pressure on the **tympanic membrane** (eardrum). The tympanic membrane moves back and forth in response, reproducing the vibrations of the sound wave source.

### Middle Ear

The tympanic membane is thought of as the boundary between the outer ear and the middle ear, but it is technically part of the middle ear. The middle ear includes the tympanic membrane, the tympanic cavity, and three small bones called auditory ossicles.

The **tympanic cavity** is an air-filled space in the temporal bone that separates the external and internal ears. The tympanic membrane is a semitransparent membrane covered by a thin layer of skin on its outer surface and by mucous membrane on the inside. It has an oval margin and is cone-shaped, with the apex of the cone directed inward. One of the auditory ossicles (malleus) maintains its cone shape.

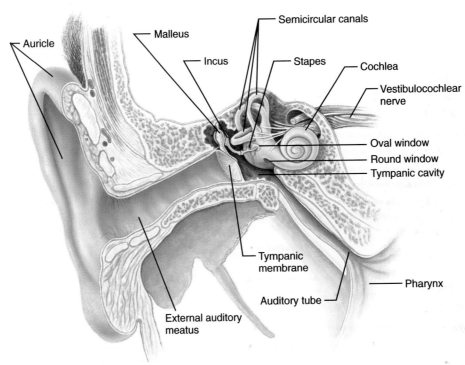

**FIGURE** 12.11

Major parts of the ear.

The three **auditory ossicles,** called the *malleus,* the *incus,* and the *stapes,* are attached to the wall of the tympanic cavity by tiny ligaments and are covered by mucous membrane. These bones bridge the tympanic membrane and the inner ear, transmitting vibrations between these parts. Specifically, the malleus is attached to the tympanic membrane, and when the tympanic membrane vibrates, the malleus vibrates in unison with it. The malleus vibrates the incus, and the incus passes the movement on to the stapes. Ligaments hold the stapes to an opening in the wall of the tympanic cavity called the **oval window.** Vibration of the stapes, which acts like a piston at the oval window, moves a fluid within the inner ear. These vibrations of the fluid stimulate the hearing receptors (fig. 12.11).

In addition to transmitting vibrations, the auditory ossicles form a lever system that helps increase (amplify) the force of the vibrations as they pass from the tympanic membrane to the oval window. Also, because the ossicles transmit vibrations from the large surface of the tympanic membrane to a much smaller area at the oval window, the vibrational force concentrates as it travels from the external to the inner ear. As a result of these two factors, the pressure (per square millimeter) that the stapes applies at the oval window is about twenty-two times greater than that which sound waves exert on the tympanic membrane.

The middle ear also contains two small skeletal muscles that are attached to the auditory ossicles and are controlled involuntarily. One of them, the *tensor tympani,* is inserted on the medial surface of the malleus and is anchored to the cartilaginous wall of the auditory tube. When it contracts, it pulls the malleus inward. The other muscle, the *stapedius,* is attached to the posterior side of the stapes and the inner wall of the tympanic cavity. It pulls the stapes outward (fig. 12.12). These muscles are the effectors in the **tympanic reflex,** which

**FIGURE 12.12**

Two small muscles attached to the (*a*) malleus and (*b*) stapes, the tensor tympani and the stapedius, are effectors in the tympanic reflex. Figure 12.11 does not show these muscles.

is elicited in about one-tenth second following a long, external sound. When the reflex occurs, the muscles contract, and the malleus and stapes move. As a result, the bridge of ossicles in the middle ear becomes more rigid, reducing its effectiveness in transmitting vibrations to the inner ear.

The tympanic reflex reduces pressure from loud sounds that might otherwise damage the hearing receptors. The tympanic reflex is also elicited by ordinary vocal sounds, as when a person speaks or sings, and this action muffles the lower frequencies of such sounds, improving the hearing of higher frequencies, which are common in human vocal sounds. In addition, the tensor tympani muscle also steadily pulls on the tympanic membrane. This is important because a loose tympanic membrane would not be able to transmit vibrations effectively to the auditory ossicles.

> The middle ear muscles take 100 to 200 milliseconds to contract. For this reason, the tympanic reflex cannot protect the hearing receptors from the effects of loud sounds that occur very rapidly, such as those from an explosion or a gunshot. On the other hand, this protective mechanism can reduce the effects of intense sounds that arise slowly, such as the roar of thunder.

## Auditory Tube

An **auditory tube** (eustachian tube) connects each middle ear to the throat. This tube allows air to pass between the tympanic cavity and the outside of the body by way of the throat (nasopharynx) and mouth. It helps maintain equal air pressure on both sides of the eardrum, which is necessary for normal hearing (see fig. 12.11).

The function of the auditory tube becomes noticeable during rapid change in altitude. As a person moves from a high altitude to a lower one, the air pressure on the outside of the tympanic membrane steadily increases. As a result, the tympanic membrane may be pushed inward, out of its normal position, impairing hearing.

When the air pressure difference is great enough, some air may force its way up through the auditory tube into the middle ear. This equalizes the pressure on both sides of the tympanic membrane, which moves back into its regular position, causing a popping sound as normal hearing returns. A reverse movement of air ordinarily occurs when a person moves from a low altitude to a higher one.

The auditory tube is usually closed by valvelike flaps in the throat, which may inhibit air movements into the middle ear. Swallowing, yawning, or chewing aid in opening the valves, and can hasten equalization of air pressure.

> Signs of an ear infection in an infant or toddler are hard to miss—irritability, screaming incessantly for no apparent reason, perhaps fever or tugging on the affected ear. A doctor viewing the painful ear with an instrument called an otoscope sees a red and bulging eardrum. The diagnosis: *otitis media*, or a middle ear infection.
>
> Ear infections occur because the mucous membranes that line the auditory tubes are continuous with the linings of the middle ears, creating a conduit for bacteria infecting the throat or nasal passages to travel to the ear. This route to infection is greater in young children because their auditory tubes are shorter than they are in adults. Half of all children in the United States have an ear infection by the first birthday, and 90% have one by age six.
>
> Physicians treat acute otitis media with antibiotics. Because recurrent infections may cause hearing loss and interfere with learning, children with recurrent otitis media are often fitted with tympanostomy tubes, which are inserted into affected ears during a brief surgical procedure. The tubes form a small tunnel through the tympanic membrane so the ears can drain. By the time the tubes fall out, the child has usually outgrown the susceptibility to ear infections.

## Inner Ear

The inner ear is a complex system of intercommunicating chambers and tubes called a **labyrinth.** Each ear has two such regions—the osseous labyrinth and the membranous labyrinth.

The *osseous labyrinth* is a bony canal in the temporal bone; the *membranous labyrinth* is a tube that lies within the osseous labyrinth and has a similar shape (fig. 12.13a). Between the osseous and membranous labyrinths is a fluid called *perilymph,* which cells in the wall of the bony canal secrete. Within the membranous labyrinth is a slightly different fluid called *endolymph.*

The parts of the labyrinths include a **cochlea** that functions in hearing, and three **semicircular canals** that provide a sense of equilibrium. A bony chamber called the **vestibule,** located between the cochlea and the semicircular canals, houses membranous structures that serve both hearing and equilibrium.

The cochlea is shaped like a snail shell, coiled around a bony core (modiolus) with a thin, bony shelf (spiral lamina) that wraps around the core like a spiral staircase (fig. 12.13b). The shelf divides the bony labyrinth of the cochlea into upper and lower compartments. The upper compartment, called the *scala vestibuli,* leads from the oval window to the apex of the spiral. The lower compartment, the *scala tympani,* extends from the apex of the cochlea to a membrane-covered opening in the wall of the inner ear called the **round window.** These compartments constitute the bony labyrinth of the cochlea, and they are filled with perilymph. At the apex

(a)

Osseous labyrinth

Membranous labyrinth

Perilymph

Semicircular canals

Endolymph

Cochlear duct

Cochlea

Utricle

Saccule

Ampulla

Vestibule

Round window

Cochlear duct

Scala vestibule

Scala tympani

(b)

Modiolus

Spiral ganglion

Cochlear nerve

Vestibular membrane

Cochlear duct (scala media)

Helicotrema

Spiral lamina

**FIGURE 12.13**

Within the inner ear (a) perilymph separates the osseous labyrinth of the inner ear from the membranous labyrinth, which contains endolymph. (b) The spiral lamina coils around a bony core, the modiolus.

of the cochlea, the fluids in the chambers can flow together through a small opening (helicotrema) (fig. 12.14).

A portion of the membranous labyrinth within the cochlea, called the *cochlear duct* (scala media), lies between the two bony compartments, and is filled with endolymph. The cochlear duct ends as a closed sac at the apex of the cochlea. The duct is separated from the scala vestibuli by a *vestibular membrane* (Reissner's membrane) and from the scala tympani by a *basilar membrane* (fig. 12.14).

The basilar membrane extends from the bony shelf of the cochlea and forms the floor of the cochlear duct. It contains many thousands of stiff, elastic fibers whose lengths vary, becoming progressively longer from the base of the cochlea to its apex. Vibrations entering the perilymph at the oval window travel along the scala

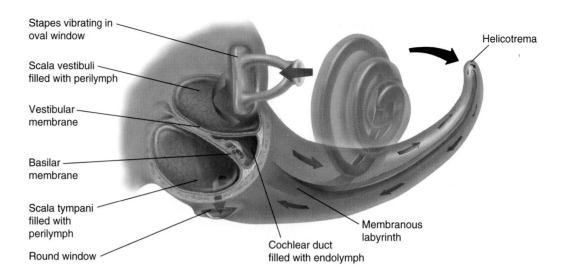

**FIGURE 12.14**

The cochlea is a coiled, bony canal with a membranous tube (labyrinth) inside. If the cochlea could be unwound, the membranous labyrinth would be seen ending as a closed sac at the apex where the bony canal makes a u-turn.

Stapes vibrating in oval window

Scala vestibuli filled with perilymph

Vestibular membrane

Basilar membrane

Scala tympani filled with perilymph

Round window

Cochlear duct filled with endolymph

Membranous labyrinth

Helicotrema

vestibuli and pass through the vestibular membrane to enter the endolymph of the cochlear duct, where they move the basilar membrane. After passing through the basilar membrane, the vibrations enter the perilymph of the scala tympani, and their forces are dissipated into the air in the tympanic cavity by movement of the membrane covering the round window.

The **organ of Corti,** which contains about 16,000 hearing receptor cells, is located on the upper surface of the basilar membrane and stretches from the apex to the base of the cochlea. The receptor cells, called **hair cells,** are arranged in four parallel rows, and they possess numerous hairlike processes (stereocilia) that extend into the endolymph of the cochlear duct. Above these hair cells is a *tectorial membrane,* which is attached to the bony shelf of the cochlea and passes like a roof over the receptor cells, contacting the tips of their hairs (figs. 12.15 and 12.16).

Because of its structure, different portions of the basilar membrane are moved by different frequencies of vibration. Thus, a particular sound frequency causes the hairs of a specific group of receptor cells to shear back and forth against the tectorial membrane more intensely, while each of the other frequencies deflects another set of receptor cells.

These receptor cells are epithelial cells, but they respond to stimuli somewhat like neurons (see chapter 10). For example, when a receptor cell is at rest, its membrane is polarized. When its hairs move in a certain direction, selective ion channels open, and its cell membrane depolarizes. The membrane then becomes more permeable specifically to calcium ions. The receptor cell has no axon or dendrites, but it does have neurotransmitter-containing vesicles in the cytoplasm near its base. In the presence of calcium ions, some of these vesicles fuse with the cell membrane and release neurotransmitter to the outside. The neurotransmitter stimulates the ends of nearby sensory nerve fibers, and in response they transmit nerve impulses along the cochlear branch of the vestibulocochlear nerve (cranial nerve VIII) to the brain.

Although the ear of a young person with normal hearing is able to detect sound waves with frequencies varying from about 20 to 20,000 or more vibrations per second, the range of greatest sensitivity is between 2,000 and 3,000 vibrations per second (fig. 12.17).

## Auditory Nerve Pathways

The cochlear branches of the vestibulocochlear nerves enter the auditory nerve pathways that extend into the medulla oblongata and proceed through the midbrain to the thalamus. From there they pass into the auditory cortices of the temporal lobes of the cerebrum, where they are interpreted. On the way, some of these fibers cross over, so that impulses arising from each ear are interpreted on both sides of the brain. Consequently, damage to a temporal lobe on one side of the brain is not necessarily accompanied by complete hearing loss in the ear on that side (fig. 12.18).

Table 12.4 summarizes the pathway of vibrations through the parts of the middle and inner ears. Clinical Application 12.4 examines types of hearing loss.

Units called *decibels* (dB) measure sound intensity. The decibel scale begins at 0 dB, which is the intensity of the sound that is least perceptible by a normal human ear. The decibel scale is logarithmic, so that a sound of 10 dB is 10 times as intense as the least perceptible sound; a sound of 20 dB is 100 times as intense; and a sound of 30 dB is 1,000 times as intense.

On this scale, a whisper has an intensity of about 40 dB, normal conversation measures 60–70 dB, and heavy traffic produces about 80 dB. A sound of 120 dB, such as that commonly produced by the amplified sound at a rock concert, produces discomfort, and a sound of 140 dB, such as that emitted by a jet plane at takeoff, causes pain. Frequent or prolonged exposure to sounds with intensities above 90 dB is likely to cause permanent hearing loss.

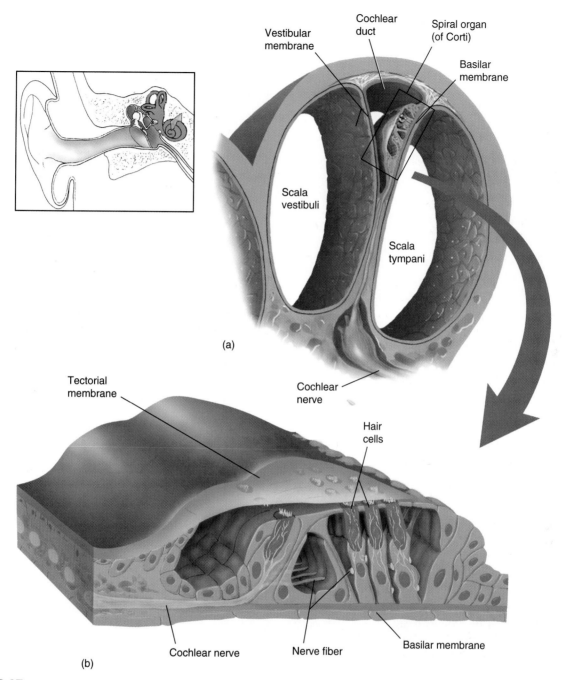

Vestibular membrane
Cochlear duct
Spiral organ (of Corti)
Basilar membrane
Scala vestibuli
Scala tympani
Cochlear nerve

(a)

Tectorial membrane
Hair cells
Cochlear nerve
Nerve fiber
Basilar membrane

(b)

**FIGURE 12.15**

(*a*) Cross section of the cochlea. (*b*) Organ of Corti and the tectorial membrane.

1  Describe the external, middle, and inner ears.

2  Explain how sound waves are transmitted through the parts of the ear.

3  Describe the tympanic reflex.

4  Distinguish between the osseous and membranous labyrinths.

5  Explain the function of the organ of Corti.

## SENSE OF EQUILIBRIUM

The feeling of equilibrium derives from two senses—*static equilibrium* and *dynamic equilibrium*. Different sensory organs provide these two components of equilibrium. The organs associated with static equilibrium sense the position of the head, maintaining stability and posture when the head and body are still. When the head and body suddenly move or rotate, the organs of dynamic equilibrium detect such motion and aid in maintaining balance.

(a)

(b)

**FIGURE 12.16**

(*a*) A micrograph of the organ of Corti and the tectorial membrane (75× micrograph enlarged to 300×). (*b*) A scanning electron micrograph of hair cells in the organ of Corti (6,700×).

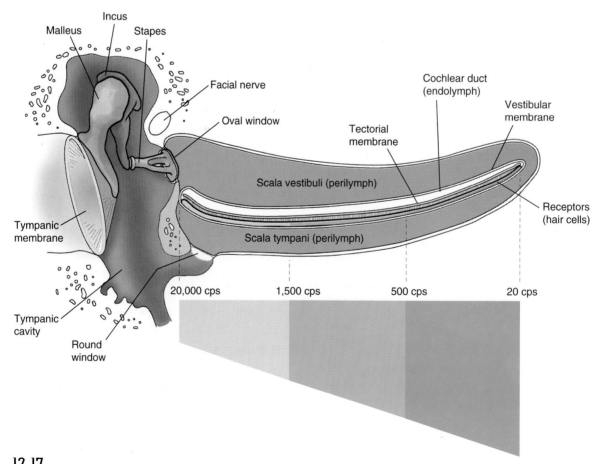

**FIGURE 12.17**

Receptors in regions of the cochlear duct sense different frequencies of vibration, expressed in cycles per second (cps).

Auditory cortex
(temporal lobe)

Thalamus

Medial geniculate
body of thalamus

Midbrain

Pons

Cochlear
nuclei

Vestibulococlear
nerve

Superior
olivary
nucleus

Medulla
oblongata

**FIGURE** 12.18

The auditory nerve pathway extends into the medulla
oblongata, proceeds through the midbrain to the thalamus,
and passes into the auditory cortex of the cerebrum.

## Static Equilibrium

The organs of **static equilibrium** are located within the
vestibule, a bony chamber between the semicircular
canals and the cochlea. More specifically, the membranous labyrinth inside the vestibule consists of two expanded chambers—a **utricle** and a **saccule.** The larger
utricle communicates with the saccule and the membranous portions of the semicircular canals; the saccule, in
turn, communicates with the cochlear duct (fig. 12.19).

Each of these chambers has a small patch of hair
cells and supporting cells called a **macula** on its wall.
When the head is upright, the hairs of the macula in the
utricle project vertically, while those in the saccule project horizontally. In each case, the hairs contact a sheet of
gelatinous material (otolithic membrane) that has crystals
of calcium carbonate (otoliths) embedded on its surface.
These particles add weight to the gelatinous sheet, making it more responsive to changes in position. The hair
cells, which are sensory receptors, have nerve fibers
wrapped around their bases. These fibers are associated
with the vestibular portion of the vestibulocochlear
nerve.

Gravity is the stimulus that causes these hairs to respond. This usually occurs when the head bends forward, backward, or to one side. Such movements tilt the
gelatinous mass of one or more maculae, and as the
gelatinous material sags in response to gravity, the hairs
projecting into it bend. This action stimulates the hair
cells, and they signal the nerve fibers associated with
them. The resulting nerve impulses travel into the central nervous system by means of the vestibular branch of
the vestibulocochlear nerve, informing the brain of the
head's position. The brain responds to this information
by sending motor impulses to skeletal muscles, and they
may contract or relax appropriately to maintain balance
(figs. 12.20 and 12.21).

| TABLE 12.4 | STEPS IN THE GENERATION OF SENSORY IMPULSES FROM THE EAR |
|---|---|

1. Sound waves enter the external auditory meatus.
2. Waves of changing pressures cause the tympanic membrane to reproduce the vibrations coming from the sound wave source.
3. Auditory ossicles amplify and transmit vibrations to the end of the stapes.
4. Movement of the stapes at the oval window transmits vibrations to the perilymph in the scala vestibuli.
5. Vibrations pass through the vestibular membrane and enter the endolymph of the cochlear duct.
6. Different frequencies of vibration in endolymph move specific regions of the basilar membrane, thus stimulating specific sets of receptor cells.

7. A receptor cell becomes depolarized; its membrane becomes more permeable to calcium ions.
8. In the presence of calcium ions, vesicles at the base of the receptor cell release neurotransmitter.
9. Neurotransmitter stimulates the ends of nearby sensory neurons.
10. Sensory impulses are triggered on fibers of the cochlear branch of the vestibulocochlear nerve.
11. The auditory cortex of the temporal lobe interprets the sensory impulses.

# Hearing Loss

Several factors can impair hearing, including interference with transmission of vibrations to the inner ear (*conductive deafness*) or damage to the cochlea or the auditory nerve and its pathways (*sensorineural deafness*). Disease, injury, and heredity all can impair hearing. There are more than 100 forms of inherited deafness, many of which are part of other syndromes. About 8% of people have hearing loss.

About 95% of cases of hearing loss are conductive. One cause is accumulated dry wax or a foreign object in the ear, which plugs the auditory meatus. Changes in the tympanic membrane or auditory ossicles can also block hearing. The tympanic membrane may harden as a result of disease, becoming less responsive to sound waves, or an injury may tear or perforate it.

A common disorder of the auditory ossicles is *otosclerosis,* in which new bone is deposited abnormally around the base of the stapes. This interferes with the ossicles' movement, which is necessary to transmit vibrations to the inner ear. Surgery often can restore some hearing to a person with otosclerosis by chipping away the bone that holds the stapes in position, or replacing the stapes with a wire or plastic substitute.

Two tests used to diagnose conductive deafness are the Weber test and the Rinne test. In the Weber test, the handle of a vibrating tuning fork is pressed against the forehead. A person with normal hearing perceives the sound coming from directly in front, whereas a person with sound conduction blockage in one middle ear hears the sound coming from the impaired side.

In the Rinne test, a vibrating tuning fork is held against the bone behind the ear. After the sound is no longer heard by conduction through the bones of the skull, the fork is moved to just in front of the external auditory meatus. In middle ear conductive deafness, the vibrating fork can no longer be heard, but a normal ear will continue to hear its tone.

Very loud sounds can cause sensorineural deafness. If exposure is brief, hearing loss may be temporary, but when exposure is repeated and prolonged, such as occurs in foundries, near jackhammers, or on a firing range, impairment may be permanent. Pete Townshend, of the rock group The Who, suffers from hearing loss from many years of performing loud concerts. President Clinton was also recently diagnosed with hearing loss, which often becomes noticeable in one's fifties.

Other causes of sensorineural deafness include tumors in the central nervous system, brain damage as a result of vascular accidents, and the use of certain drugs.

Because hearing loss and other ear problems can begin gradually, it is important to be aware of their signs, which may include:

- difficulty hearing people talking softly
- inability to understand speech when there is background noise

- ringing in the ears
- dizziness
- loss of balance

New parents should notice whether their infant responds to sounds in a way that indicates normal hearing. Before 1993, 50% of hearing impaired infants were not diagnosed until age 2. Since then, the federal government has advised hearing exams as part of a well-baby visit to a doctor. If the baby's responses indicate a possible problem, the next step is to see an audiologist, who identifies and measures hearing loss.

Often a hearing aid can help people with conductive hearing loss. A hearing aid has a tiny microphone that picks up sound waves and converts them to electrical signals, which are then amplified so that the person can hear them. An ear mold holds the device in place, either behind the outer ear, in the outer ear, or in the ear canal.

A cochlear implant enables people with sensorineural hearing loss to detect some sounds, although it usually remains difficult to discern distinct words. The device converts sound waves to electrical signals, which stimulate neurons in the cochlea.

Many persons over age 65 lose the ability to hear high-frequency sounds and can no longer discriminate speech sounds as well as they once did—a condition called *presbycusis.* Thus, it is important to use lower voice tones in speaking to an older person. Also, the speaker should face the listener so that lipreading is possible. ■

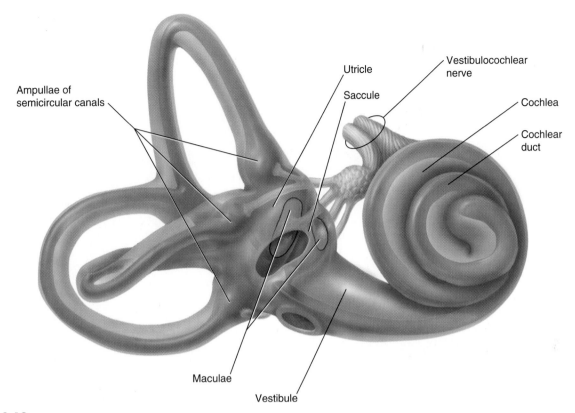

Ampullae of
semicircular canals

Utricle

Saccule

Vestibulocochlear
nerve

Cochlea

Cochlear
duct

Maculae

Vestibule

**FIGURE 12.19**

The saccule and utricle, which are expanded portions of the membranous labyrinth, are located within the bony chamber of the vestibule.

The maculae also participate in the sense of dynamic equilibrium. For example, if the head or body is thrust forward or backward abruptly, the gelatinous mass of the maculae lags slightly behind, and the hair cells are stimulated. In this way, the maculae aid the brain in detecting movements such as falling, and in maintaining posture while walking.

## Dynamic Equilibrium

Each semicircular canal follows a circular path about 6 millimeters in diameter. The three bony semicircular canals lie at right angles to each other and occupy three different planes in space. Two of them, the *superior canal* and the *posterior canal,* stand vertically, whereas the third, the *lateral canal,* is horizontal. Their orientations closely approximate the three body planes (see chapter 1).

Suspended in the perilymph of each bony canal is a membranous semicircular canal that ends in a swelling called an **ampulla.** The ampullae communicate with the utricle of the vestibule.

An ampulla contains a septum that crosses the tube and houses a sensory organ. Each of these organs, called a **crista ampullaris,** contains a number of sensory hair cells and supporting cells. As in the maculae, the hairs of the hair cells extend upward into a dome-shaped gelatinous mass called the *cupula.* Also, the hair cells are connected at their bases to nerve fibers that make up part of the vestibular branch of the vestibulocochlear nerve (fig. 12.22).

Rapid turns of the head or body stimulate the hair cells of the crista. At such times, the semicircular canals move with the head or body, but the fluid inside the membranous canals tends to remain stationary because of inertia. This bends the cupula in one or more of the canals in a direction opposite that of the head or body movement, and the hairs embedded in it also bend. This bending of the hairs stimulates the hair cells to signal their associated nerve fibers, and, as a result, impulses travel to the brain (fig. 12.23).

Parts of the cerebellum are particularly important in interpreting impulses from the semicircular canals. Analysis of such information allows the brain to predict the consequences of rapid body movements, and by modifying signals to appropriate skeletal muscles, the cerebellum can maintain balance.

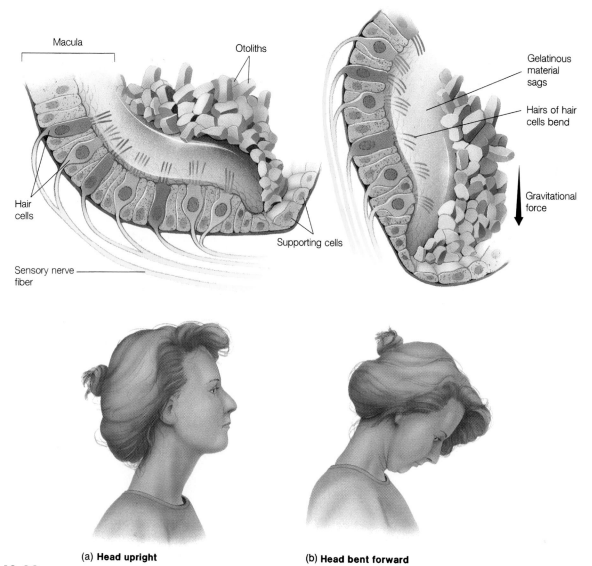

Macula

Otoliths

Gelatinous material sags

Hairs of hair cells bend

Gravitational force

Hair cells

Supporting cells

Sensory nerve fiber

(a) **Head upright**

(b) **Head bent forward**

**FIGURE** 12.20 ▭ ✗

The macula responds to changes in position of the head. (*a*) Macula with the head in an upright position. (*b*) Macula with the head bent forward.

Other sensory structures aid in maintaining equilibrium. For example, various proprioceptors, particularly those associated with the joints of the neck, inform the brain about the position of body parts. In addition, the eyes detect changes in posture that result from body movements. Such visual information is so important that even if the organs of equilibrium are damaged, keeping the eyes open and moving slowly is sufficient to maintain normal balance.

1 Distinguish between the senses of static and dynamic equilibrium.

2 Which structures provide the sense of static equilibrium? Of dynamic equilibrium?

3 How does sensory information from other receptors help maintain equilibrium?

Motion sickness is a disturbance of the inner ear's sensation of balance. Nine out of 10 people have experienced this nausea and vomiting, usually when riding in a car or on a boat. Astronauts began reporting a form of motion sickness called space adaptation syndrome in 1968, when spacecraft were made roomy enough for astronauts to move about while in flight.

Although the cause of motion sickness is not known, one theory is that it results when visual information contradicts the inner ear's sensation that one is motionless. Consider a woman riding in a car. Her inner ears tell her that she is not moving, but the passing scenery tells her eyes that she is moving. The problem is compounded if she tries to read. The brain reacts to these seemingly contradictory sensations by signaling a "vomiting center" in the medulla oblongata.

**FIGURE 12.21**

Scanning electron micrograph of hairs of hair cells, such as those found in the utricle and saccule (8,000×).

## SENSE OF SIGHT

A number of *accessory organs* assist the visual receptors, which are in the eyes. These include the eyelids and lacrimal apparatus that help protect the eyes, and a set of extrinsic muscles that move them.

### Visual Accessory Organs

Each eye, lacrimal gland, and associated extrinsic muscles are housed within the pear-shaped orbital cavity of the skull. The orbit, which is lined with the periosteums of various bones, also contains fat, blood vessels, nerves, and connective tissues.

Each **eyelid** (palpebra) is composed of four layers—skin, muscle, connective tissue, and conjunctiva. The skin of the eyelid, which is the thinnest skin of the body, covers the lid's outer surface and fuses with its inner lining near the margin of the lid (fig. 12.24).

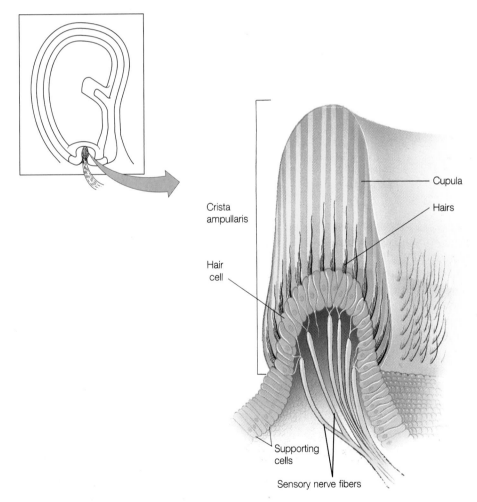

**FIGURE 12.22**

A crista ampullaris is located within the ampulla of each semicircular canal.

**(a) Head in still position**

**(b) Head rotating**

FIGURE 12.23

(a) When the head is stationary, the cupula of the crista ampullaris remains upright. (b) When the head is moving rapidly, the cupula is bent, opposite the motion of the head, stimulating sensory receptors.

The muscles that move the eyelids include the *orbicularis oculi* and the *levator palpebrae superioris.* Fibers of the orbicularis oculi encircle the opening between the lids and spread out onto the cheek and forehead. This muscle acts as a sphincter that closes the lids when it contracts.

Fibers of the levator palpebrae superioris muscle arise from the roof of the orbit and are inserted in the connective tissue of the upper lid. When these fibers contract, the upper lids are raised and the eye opens.

The connective tissue layer of the eyelid, which helps give it form, contains many modified sebaceous glands (tarsal glands). Ducts carry the oily secretions of these glands to openings along the borders of the lids. This secretion helps keep the lids from sticking together.

The **conjunctiva** is a mucous membrane that lines the inner surfaces of the eyelids and folds back to cover the anterior surface of the eyeball, except for its central portion (cornea). Although the tissue that lines the eye-

lids is relatively thick, the conjunctiva that covers the eyeball is very thin. It is also freely movable and quite transparent, so that blood vessels are clearly visible beneath it.

A child in school with "pinkeye" is usually sent straight home. Bacteria cause this highly contagious form of inflammation of the conjunctiva, or *conjunctivitis.* Viral conjunctivitis is not usually contagious. Allergy or exposure to an irritating chemical may also cause conjunctivitis.

The *lacrimal apparatus* consists of the **lacrimal gland,** which secretes tears, and a series of *ducts,* which carry the tears into the nasal cavity (fig. 12.25). The gland is located in the orbit, above and lateral to the eye. It secretes tears continuously, and they pass out through

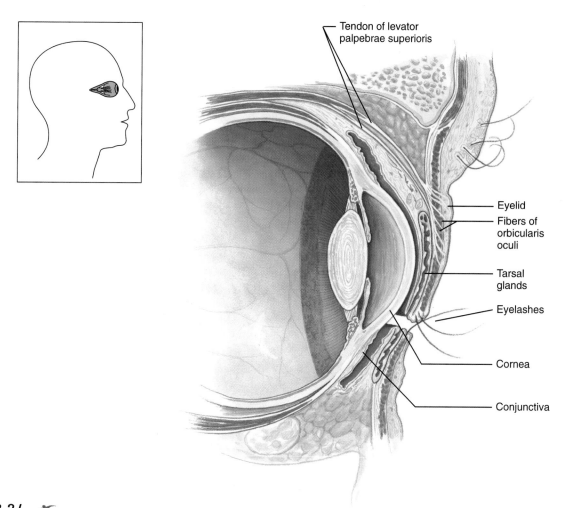

**FIGURE 12.24**

Sagittal section of the closed eyelids and the anterior portion of the eye.

Tendon of levator palpebrae superioris, Eyelid, Fibers of orbicularis oculi, Tarsal glands, Eyelashes, Cornea, Conjunctiva

tiny tubules and flow downward and medially across the eye.

Two small ducts (superior and inferior canaliculi) collect tears, and their openings (puncta) can be seen on the medial borders of the eyelids. From these ducts, the fluid moves into the *lacrimal sac,* which lies in a deep groove of the lacrimal bone, and then into the *naso-lacrimal duct,* which empties into the nasal cavity.

Glandular cells of the conjunctiva also secrete a tearlike liquid that, together with the secretion of the lacrimal gland, moistens and lubricates the surface of the eye and the lining of the lids. Tears contain an enzyme, called *lysozyme,* that has antibacterial properties, reducing the chance of eye infections.

Tear glands secrete excessively when a person is upset, or the conjunctiva is irritated. Tears spill over the edges of the eyelids, and the nose fills with fluid. When a person cries, parasympathetic nerve fibers carry motor impulses to the lacrimal glands.

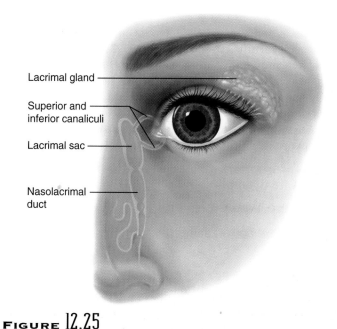

Labels: Lacrimal gland, Superior and inferior canaliculi, Lacrimal sac, Nasolacrimal duct

**FIGURE 12.25**

The lacrimal apparatus consists of a tear-secreting gland and a series of ducts.

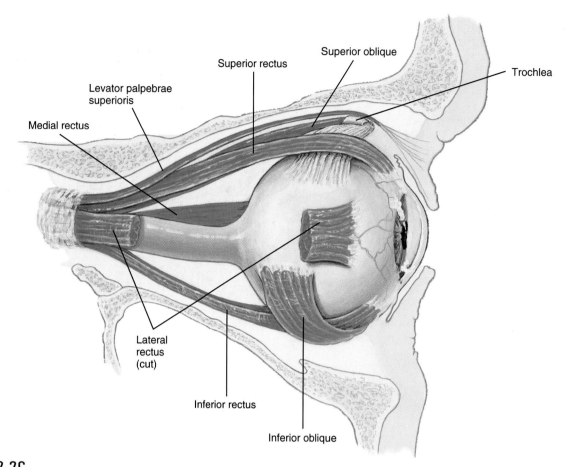

**FIGURE 12.26**

The extrinsic muscles of the right eye (lateral view). Note the pulley-like trochlea that the superior oblique passes through.

The **extrinsic muscles** of the eye arise from the bones of the orbit and are inserted by broad tendons on the eye's tough outer surface. Six such muscles move the eye in various directions (fig. 12.26). Although any given eye movement may use more than one of them, each muscle is associated with one primary action, as follows:

1.  **Superior rectus**—rotates the eye upward and toward the midline.

2.  **Inferior rectus**—rotates the eye downward and toward the midline.

3.  **Medial rectus**—rotates the eye toward the midline.

4.  **Lateral rectus**—rotates the eye away from the midline.

5.  **Superior oblique**—rotates the eye downward and away from the midline.

6.  **Inferior oblique**—rotates the eye upward and away from the midline.

The motor units of the extrinsic eye muscles contain the fewest muscle fibers (five to ten) of any muscles in the body, so they can move the eyes with great precision. Also, the eyes move together so that they align when looking at something. Such alignment is the result of complex motor adjustments that contract certain eye muscles while relaxing their antagonists. For example, when the eyes move to the right, the lateral rectus of the right eye and the medial rectus of the left eye must contract. At the same time, the medial rectus of the right eye and the lateral rectus of the left eye must relax. A person whose eyes are not coordinated well enough to align has *strabismus*. Table 12.5 summarizes the muscles associated with the eyelids and eye.

**1** Explain how the eyelid is moved.

**2** Describe the conjunctiva.

**3** What is the function of the lacrimal apparatus?

**4** Describe the function of each extrinsic eye muscle.

TABLE 12.5

## MUSCLES ASSOCIATED WITH THE EYELIDS AND EYES

| Skeletal Muscles | | | Smooth Muscles | | |
|---|---|---|---|---|---|
| **Name** | **Innervation** | **Function** | **Name** | **Innervation** | **Function** |
| Muscles of the eyelids | | | Ciliary muscles | Oculomotor nerve (III) parasympathetic fibers | Relax suspensory ligaments |
| Orbicularis oculi | Facial nerve (VII) | Closes eye | | | |
| Levator palpebrae superioris | Oculomotor nerve (III) | Opens eye | Iris, circular muscles | Oculomotor nerve (III) parasympathetic fibers | Constrict pupil |
| Extrinsic muscles of the eyes | | | Iris, radial muscles | Sympathetic fibers | Dilate pupil |
| Superior rectus | Oculomotor nerve (III) | Rotates eye upward and toward midline | | | |
| Inferior rectus | Oculomotor nerve (III) | Rotates eye downward and toward midline | | | |
| Medial rectus | Oculomotor nerve (III) | Rotates eye toward midline | | | |
| Lateral rectus | Abducens nerve (VI) | Rotates eye away from midline | | | |
| Superior oblique | Trochlear nerve (IV) | Rotates eye downward and away from midline | | | |
| Inferior oblique | Oculomotor nerve (III) | Rotates eye upward and away from midline | | | |

When one eye deviates from the line of vision, the person has double vision (diplopia). If this condition persists, the brain may eventually suppress the image from the deviated eye. As a result, the turning eye may become blind (suppression amblyopia). Treating the eye deviation early in life with exercises, eyeglasses, and surgery can prevent such monocular blindness. For this reason, vision screening programs for preschool children are very important.

## Structure of the Eye

The eye is a hollow, spherical structure about 2.5 centimeters in diameter. Its wall has three distinct layers—an outer *fibrous tunic*, a middle *vascular tunic*, and an inner *nervous tunic*. The spaces within the eye are filled with fluids that support its wall and internal structures, and help maintain its shape. Figure 12.27 shows the major parts of the eye.

### The Outer Tunic

The anterior sixth of the outer tunic bulges forward as the transparent **cornea**, which is the window of the eye and helps focus entering light rays. It is largely composed of connective tissue with a thin layer of epithelium on its surface. The cornea is transparent because it contains few cells and no blood vessels. Also, the cells and collagenous fibers occur in unusually regular patterns.

On the other hand, the cornea is well supplied with nerve fibers that enter its margin and radiate toward its center. These fibers are associated with many pain receptors that have very low thresholds. Cold receptors are also abundant in the cornea, but heat and touch receptors are not.

Along its circumference, the cornea is continuous with the **sclera,** the white portion of the eye. The sclera makes up the posterior five-sixths of the outer tunic and is opaque due to many large, haphazardly arranged collagenous and elastic fibers. The sclera protects the eye and is an attachment for the extrinsic muscles.

In the back of the eye, the **optic nerve** and blood vessels pierce the sclera. The dura mater that encloses these structures is continuous with the sclera.

### The Middle Tunic

The middle, or vascular, tunic of the eyeball (uveal layer) includes the choroid coat, the ciliary body, and the iris.

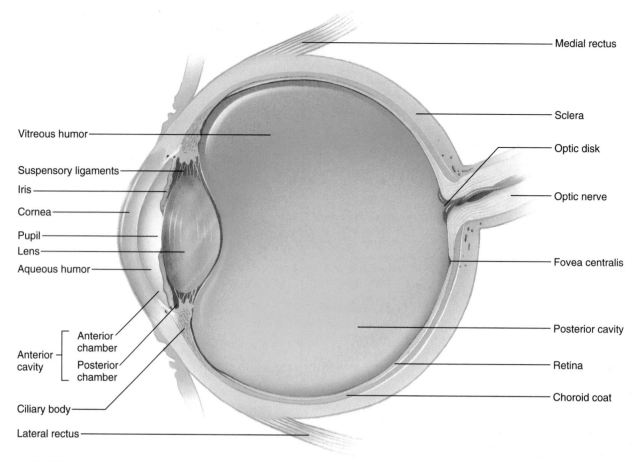

Vitreous humor

Suspensory ligaments

Iris

Cornea

Pupil

Lens

Aqueous humor

Anterior cavity
- Anterior chamber
- Posterior chamber

Ciliary body

Lateral rectus

Medial rectus

Sclera

Optic disk

Optic nerve

Fovea centralis

Posterior cavity

Retina

Choroid coat

**FIGURE 12.27**

Transverse section of the left eye (superior view).

The **choroid coat,** in the posterior five-sixths of the globe of the eye, loosely joins the sclera. Blood vessels pervade the choroid coat and nourish surrounding tissues. The choroid coat also contains numerous pigment-producing melanocytes that give it a brownish-black appearance. The melanin of these cells absorbs excess light and helps keep the inside of the eye dark.

In 1905, doctors transplanted the cornea of an 11-year-old boy who lost his eye in an accident into a man whose cornea had been destroyed by a splash of a caustic chemical, marking one of the first successful human organ transplants. Today, corneal transplants are commonly used to treat corneal disease, the most common cause of blindness worldwide. In this procedure, called a *penetrating keratoplasty,* a piece of donor cornea replaces the central two-thirds of the defective cornea. These transplants are highly successful because the cornea lacks blood vessels, and therefore the immune system does not have direct access to the new, "foreign" tissue. Unfortunately, as is the case for many transplantable body parts, donor tissue is in short supply.

The **ciliary body,** which is the thickest part of the middle tunic, extends forward from the choroid coat and forms an internal ring around the front of the eye. Within the ciliary body are many radiating folds called *ciliary processes* and two distinct groups of muscle fibers that constitute the *ciliary muscles.* Figure 12.28 shows these structures.

Many strong but delicate fibers, called *suspensory ligaments* (zonular fibers), extend inward from the ciliary processes and hold the transparent lens in position. The distal ends of these fibers are attached along the margin of a thin capsule that surrounds the lens. The body of the lens, which lacks blood vessels, lies directly behind the iris and pupil and is composed of specialized epithelial cells.

These epithelial cells originate from a single layer of epithelium located beneath the anterior portion of the lens capsule. The cells multiply, and the newly formed cells on the surface of the lens capsule differentiate into columnar cells called *lens fibers.* These lens fibers constitute the substance of the lens. Lens fiber production continues slowly throughout life, thickening the lens from front to back. Simultaneously, the deeper lens fibers are compressed toward the center of the structure (fig. 12.29).

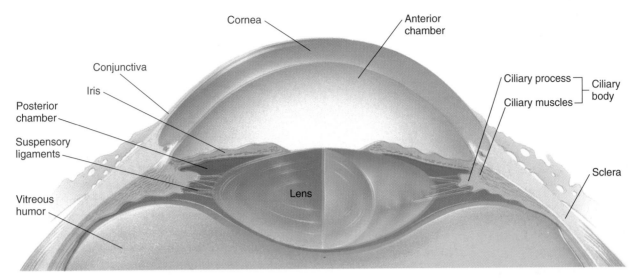

**FIGURE 12.28**
Anterior portion of the eye.

**FIGURE 12.29**
A scanning electron micrograph of the long, flattened lens fibers (2,650×). Note the fingerlike junctions where one fiber joins another.

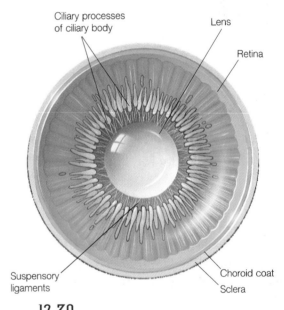

**FIGURE 12.30**
The lens and ciliary body viewed from behind.

The lens capsule is a clear, membranelike structure largely composed of intercellular material. It is quite elastic, a quality that keeps it under constant tension. As a result, the lens can assume a globular shape. However, the suspensory ligaments attached to the margin of the capsule are also under tension, and they pull outward, flattening the capsule and the lens (fig. 12.30).

If the tension on the suspensory ligaments relaxes, the elastic capsule rebounds, and the lens surface becomes more convex. This change occurs in the lens when the eye focuses to view a close object and is called **accommodation.**

The ciliary muscles relax the suspensory ligaments during accommodation. One set of these muscle fibers forms a circular sphincterlike structure around the ciliary processes. The fibers of the other set extend back from fixed points in the sclera to the choroid coat. When the circular muscle fibers contract, the diameter of the ring formed by the ciliary processes decreases; when the other fibers contract, the choroid coat is pulled forward and the ciliary body shortens. Both of these actions relax

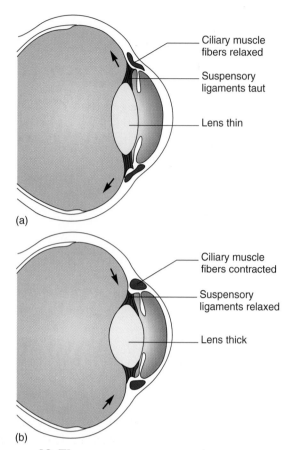

**Figure 12.31**

(*a*) The lens thins as the ciliary muscle fibers relax. (*b*) The lens thickens as the ciliary muscle fibers contract.

Labels for figure (a):
- Ciliary muscle fibers relaxed
- Suspensory ligaments taut
- Lens thin

Labels for figure (b):
- Ciliary muscle fibers contracted
- Suspensory ligaments relaxed
- Lens thick

the suspensory ligaments, thickening the lens. In this thickened state, the lens is focused for viewing closer objects than before (fig. 12.31).

To focus on a distant object, the ciliary muscles relax, increasing tension on the suspensory ligaments. The lens thins again.

A common eye disorder, particularly in older people, is *cataract.* The lens or its capsule slowly becomes cloudy, opaque, and discolored. As a result, clear images cannot focus on the retina. In time, the person may become blind.

Cataract is often treated by surgically removing the lens and replacing it with an artificial one (intraocular lens implant). Sometimes eyeglasses or contact lenses can correct vision following removal of the lens.

**1** Describe the outer and middle tunics of the eye.

**2** What factors contribute to the transparency of the cornea?

**3** How does the shape of the lens change during accommodation?

The **iris** is a thin diaphragm mostly composed of connective tissue and smooth muscle fibers. It is seen from the outside as the colored portion of the eye. The iris extends forward from the periphery of the ciliary body and lies between the cornea and the lens. It divides the space separating these parts, which is called the *anterior cavity,* into an *anterior chamber* (between the cornea and the iris) and a *posterior chamber* (between the iris and the vitreous humor, occupied by the lens).

The epithelium on the inner surface of the ciliary body continuously secretes a watery fluid called **aqueous humor** into the posterior chamber. The fluid circulates from this chamber through the **pupil,** a circular opening in the center of the iris, and into the anterior chamber (fig. 12.32). Aqueous humor fills the space between the cornea and the lens, providing nutrients and maintaining the shape of the front of the eye. It subsequently leaves the anterior chamber through veins and a special drainage canal, the scleral venous sinus (canal of Schlemm), located in its wall at the junction of the cornea and the sclera.

A disorder called *glaucoma* sometimes develops in the eyes as a person ages when the rate of aqueous humor formation exceeds the rate of its removal. Fluid accumulates in the anterior chamber of the eye, and the fluid pressure rises.

Because liquids cannot be compressed, the increasing pressure from the anterior chamber is transmitted to all parts of the eye, and, in time, the blood vessels that supply the receptor cells of the retina may squeeze shut. If this happens, cells that fail to receive nutrients and oxygen may die, causing permanent blindness.

Drugs, laser therapy, or surgery to promote the outflow of aqueous humor can treat glaucoma if it is diagnosed early. However, since glaucoma in its early stages typically produces no symptoms, discovery of the condition usually depends on measuring the intraocular pressure using an instrument called a *tonometer.*

The smooth muscle fibers of the iris form two groups, a *circular set* and a *radial set.* These muscles control the size of the pupil, which is the opening that light passes through as it enters the eye. The circular set of muscle fibers acts as a sphincter, and when it contracts, the pupil gets smaller (constricts) and the intensity of the light entering decreases. When the radial muscle fibers contract, the diameter of the pupil in-

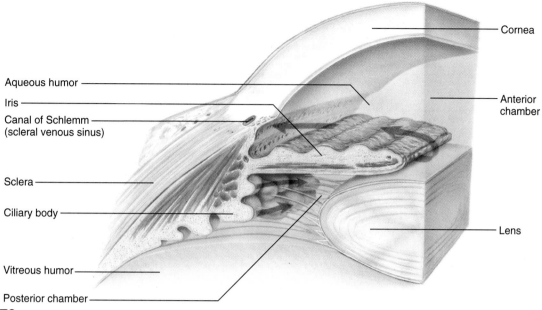

Aqueous humor
Iris
Canal of Schlemm
(scleral venous sinus)

Sclera

Ciliary body

Vitreous humor

Posterior chamber

Cornea

Anterior
chamber

Lens

**FIGURE 12.32**

Aqueous humor (blue arrows), which is secreted into the posterior chamber, circulates into the anterior chamber and leaves it through the canal of Schlemm (scleral venous sinus).

creases (dilates) and the intensity of the light entering increases.

The sizes of the pupils change constantly in response to pupillary reflexes triggered by such factors as light intensity, gaze, accommodation, and variations in emotional state. For example, bright light elicits a reflex, and impulses travel along parasympathetic nerve fibers to the *circular muscles* of the irises. The pupils constrict in response. Conversely, in dim light, impulses travel on sympathetic nerve fibers to the *radial muscles* of the irises, and the pupils dilate (fig. 12.33).

The amount and distribution of melanin in the irises and the density of the tissue within the body of the iris determine eye color. If melanin is present only in the epithelial cells on the iris's posterior surface, more wavelengths of light are scattered by the iris, causing it to look blue or even green. When the same distribution of melanin occurs with denser-than-usual tissue within the body of the iris, it appears gray. When melanin is within the body of the iris as well as in the epithelial covering, the iris appears brown.

### The Inner Tunic

The inner tunic of the eye consists of the **retina,** which contains the visual receptor cells (photoreceptors). This nearly transparent sheet of tissue is continuous with the optic nerve in the back of the eye and extends forward as the inner lining of the eyeball. It ends just behind the margin of the ciliary body.

The retina is thin and delicate, but its structure is quite complex. It has a number of distinct layers, including pigmented epithelium, neurons, nerve fibers, and limiting membranes (figs. 12.34 and 12.35).

There are five major groups of retinal neurons. The nerve fibers of three of these groups—the *receptor cells, bipolar neurons,* and *ganglion cells*—provide a direct pathway for impulses triggered in the receptors to the optic nerve and brain. The nerve fibers of the other two groups of retinal cells, called *horizontal cells* and *amacrine cells,* pass laterally between retinal cells (see fig. 12.34). The horizontal and amacrine cells modify the impulses transmitted on the fibers of the direct pathway.

In the central region of the retina is a yellowish spot called the **macula lutea** that occupies about 1 square millimeter. A depression in its center, called the **fovea centralis,** is in the region of the retina that produces the sharpest vision.

Just medial to the fovea centralis is an area called the **optic disk** (fig. 12.36). Here the nerve fibers from the retina leave the eye and become parts of the optic nerve. A central artery and vein also pass through at the optic disk. These vessels are continuous with capillary networks of the retina, and together with vessels in the underlying choroid coat, they supply blood to the cells of the inner tunic. Because the optic disk lacks receptor cells, it is commonly referred to as the *blind spot* of the eye.

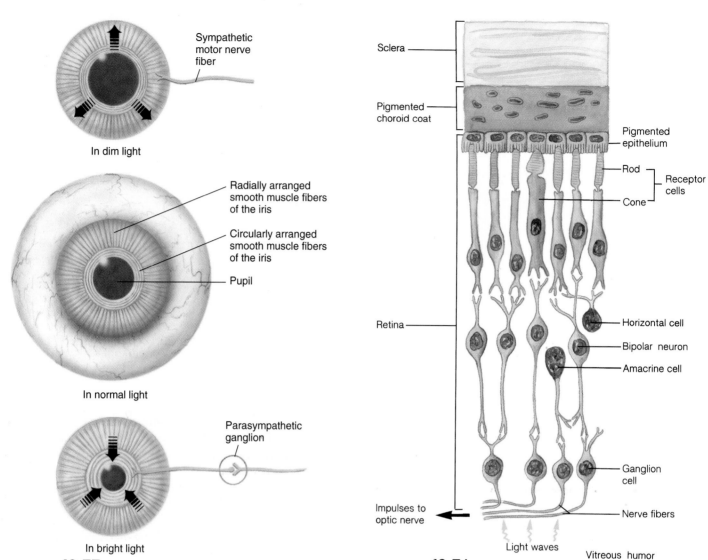

**FIGURE 12.33**

In dim light, the radial muscles of the iris are stimulated to contract, and the pupil dilates. In bright light, the circular muscles of the iris are stimulated to contract, and the pupil constricts.

**FIGURE 12.34**

The retina consists of several cell layers.

**FIGURE 12.35**

Note the layers of cells and nerve fibers in this light micrograph of the retina (75×).

(a)

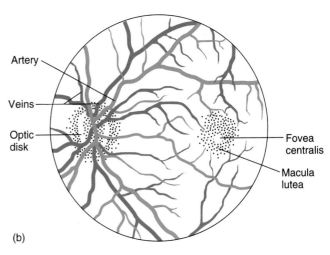

(b)

**FIGURE 12.36**

(a) Nerve fibers leave the eye in the area of the optic disk (arrow) to form the optic nerve (53×). (b) Major features of the retina.

**TABLE 12.6 LAYERS OF THE EYE**

| Layer/Tunic | Posterior Portion | Function | Anterior Portion | Function |
|---|---|---|---|---|
| Outer layer | Sclera | Protection | Cornea | Light transmission and refraction |
| Middle layer | Choroid coat | Blood supply, pigment prevents reflection | Ciliary body, iris | Accommodation; controls light intensity |
| Inner layer | Retina | Photoreception, impulse transmission | None | |

The space enclosed by the lens, ciliary body, and retina is the largest compartment of the eye and is called the *posterior cavity*. It is filled with a transparent, jelly-like fluid called **vitreous humor,** which together with some collagenous fibers comprise the **vitreous body.** The vitreous body supports the internal structures of the eye and helps maintain its shape.

In summary, light waves entering the eye must pass through the cornea, aqueous humor, lens, vitreous humor, and several layers of the retina before they reach the photoreceptors (see fig. 12.34). Table 12.6 summarizes the layers of the eye.

1. Explain the origin of aqueous humor and trace its path through the eye.

2. How is the size of the pupil regulated?

3. Describe the structure of the retina.

With age, tiny dense clumps of gel or deposits of crystal-like substances form in the vitreous humor. These clumps may cast shadows on the retina, and as a result, the person sees small moving specks in the field of vision. These specks, or *floaters,* are most apparent when looking at a plain background, such as the sky or a blank wall.

Also with age, the vitreous humor may shrink and pull away from the retina. This may mechanically stimulate receptor cells of the retina, and the person may see flashes of light. The presence of floaters or light flashes usually does not indicate a serious eye problem; however, if they develop suddenly or seem to be increasing in number or frequency, an ophthalmologist should examine the eye.

# Light Refraction

When a person sees something, either the object is giving off light, or light waves are reflected from it. These light waves enter the eye, and an image of what is seen focuses upon the retina. The light rays must bend to be focused, a phenomenon called **refraction.**

Refraction occurs when light waves pass at an oblique angle from a medium of one optical density into a medium of a different optical density. For example, as figure 12.37 shows, when light passes obliquely from a less dense medium such as air into a denser medium such as glass, or from air into the cornea of the eye, the light is bent toward a line perpendicular to the surface between these substances. When the surface between such refracting media is curved, a lens is formed. A lens with a *convex* surface causes light waves to converge, and a lens with a *concave* surface causes light waves to diverge (fig. 12.38). Clinical Application 12.5 discusses some familiar problems with refraction.

The convex surface of the cornea refracts light waves from objects outside the eye, providing about 75% of the total refractive power of the eye. The light is refracted again by the convex surface of the lens and to a lesser extent by the surfaces of the fluids within the eye chambers.

If the shape of the eye is normal, light waves are focused sharply upon the retina, much as a motion picture image is focused on a screen for viewing. Unlike the motion picture image, however, the one formed on the retina is upside down and reversed from left to right (fig. 12.39). When the visual cortex of the cerebrum interprets such an image, it somehow corrects this, and things are seen in their proper positions.

Light waves coming from objects more than 20 feet away are traveling in nearly parallel lines, and they are focused on the retina by the cornea and by the lens in its more flattened or "at rest" condition. Light waves arriving from objects less than 20 feet away, however, reach the eye along more divergent lines—in fact, the closer the object, the more divergent the lines.

Divergent light waves tend to focus behind the retina unless something increases the refracting power of the eye. Accommodation accomplishes this increase, thickening the lens. As the lens thickens, light waves converge more strongly, so that diverging light waves coming from close objects focus on the retina.

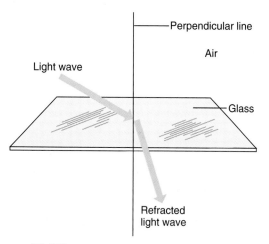

**FIGURE 12.37**

When light passes at an oblique angle from air into glass, the light waves bend toward a line perpendicular to the surface of the glass.

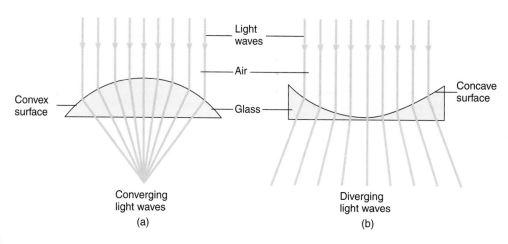

**FIGURE 12.38**

(*a*) A lens with a convex surface causes light waves to converge. (*b*) A lens with a concave surface causes them to diverge.

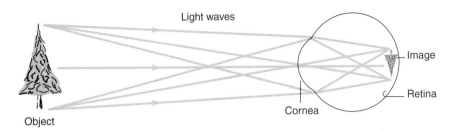

Light waves

Image

Retina

Cornea

Object

**FIGURE 12.39**

The image of an object forms upside down on the retina.

**1** What is refraction?

**2** What parts of the eye provide refracting surfaces?

**3** Why is it necessary to accommodate for viewing close objects?

## Visual Receptors

The photoreceptors of the eye are modified neurons of two distinct kinds. One group of receptor cells, called **rods,** have long, thin projections at their terminal ends. The cells of the other group, called **cones,** have short, blunt projections. The retina contains about 100 million rods and three million cones.

Instead of being located in the surface layer of the retina, rods and cones are found in a deep portion, closely associated with a layer of pigmented epithelium (see figs. 12.34 and 12.35). The projections from the receptors extend into the pigmented layer and contain light-sensitive visual pigments.

The epithelial pigment of the retina absorbs light waves that are not absorbed by the receptor cells, and together with the pigment of the choroid coat, it keeps light from reflecting off the surfaces inside the eye. The pigment layer also stores vitamin A, which the receptor cells use to synthesize visual pigments.

Researchers can grow retinal epithelial cells in laboratory cultures, and the cells retain their pigment. This means that someday scientists may be able to grow tissue that can be implanted into a person's eye to treat some forms of blindness.

The visual receptors are only stimulated when light reaches them. Thus, when a light image is focused on an area of the retina, some receptors are stimulated, and impulses travel from them to the brain. However, the impulse leaving each activated receptor provides only a fragment of the information required for the brain to interpret a total scene.

*Albinism* is an inherited condition in which a missing enzyme blocks pigment synthesis, causing very pale, highly sun-sensitive skin. More severe forms of albinism also affect the eyes, making vision blurry and intolerant to light. A person may squint even in very faint light. The extrasensitivity is due to the fact that light reflects inside the lenses, excessively stimulating visual receptors. The eyes of many people with albinism also dart about uncontrollably, a condition called *nystagmus.*

Rods and cones function differently. Rods are hundreds of times more sensitive to light than are cones, and as a result, rods provide vision in dim light. In addition, rods produce colorless vision, whereas cones can detect colors.

Still another difference is that cones provide sharp images, whereas rods produce more general outlines of objects. This difference is due to the fact that nerve fibers from many rods may converge, and their impulses may be transmitted to the brain on the same nerve fiber (see chapter 10). Thus, if a point of light stimulates a rod, the brain cannot tell which one of many receptors has actually been stimulated. Such a convergence of impulses occurs to a much lesser degree among cones, so when a cone is stimulated, the brain is able to pinpoint the stimulation more accurately (fig. 12.40).

The area of sharpest vision, the fovea centralis in the macula lutea, lacks rods, but contains densely packed cones with few or no converging fibers. Also, the overlying layers of the retina, as well as the retinal blood vessels, are displaced to the sides in the fovea, which more fully exposes the receptors to incoming light. Consequently, to view something in detail, a person moves the eyes so that the important part of an image falls upon the fovea centralis.

The concentration of cones decreases in areas farther away from the macula lutea, whereas the concentration of rods increases in these areas. Also, the degree of convergence among the rods and cones increases toward the periphery of the retina. As a result, the visual sensations from images focused on the sides of the retina tend to be blurred compared with those focused on the central portion of the retina.

# Refraction Disorders

The elastic quality of the lens capsule lessens with time. People over 45 years of age are often unable to accommodate sufficiently to read the fine print in books and newspapers or on medicine bottles. Their eyes remain focused for distant vision. This condition is termed *presbyopia,* or farsightedness of age. Eyeglasses or contact lenses can usually make up for the eye's loss of refracting power.

Other visual problems result from eyeballs that are too short or too long for sharp focusing. If an eye is too short, light waves are not focused sharply on the retina because their point of focus lies behind it. A person with this condition may be able to bring the image of distant objects into focus by accommodation, but this requires contraction of the ciliary muscles at times when these muscles are at rest in a normal eye. Still more accommodation is necessary to view closer objects, and the person may suffer from ciliary muscle fatigue, pain, and headache when doing close work.

Since people with short eyeballs are usually unable to accommodate enough to focus on the very close objects, they are said to be *farsighted.* Eyeglasses or contact lenses with *convex* surfaces can remedy this condition (hyperopia) by focusing images closer to the front of the eye.

If an eyeball is too long, light waves tend to be focused in front of the retina, and the image produced on it is blurred. In other words, the refracting power of the eye, even when the lens is flattened, is too great. Although a person with this problem may be able to focus on close objects by ac-

commodation, distance vision is invariably poor. For this reason, the person is said to be *nearsighted.* Eyeglasses or contact lenses with *concave* surfaces that focus images farther from the front of the eye treat nearsightedness (myopia) (figs. 12A and 12B).

Still another refraction problem, *astigmatism,* reflects a defect in the curvature of the cornea or the lens. The normal cornea has a spherical curvature, like the inside of a ball; an astigmatic cornea usually has an elliptical curvature, like the bowl of a spoon. As a result, some portions of an image are in focus on the retina, but other portions are blurred, and vision is distorted.

Without corrective lenses, astigmatic eyes tend to accommodate back and forth reflexly in an attempt to sharpen focus. The consequence of this continual action is often ciliary muscle fatigue and headache. ■

The bony orbit usually protects the eye, but a forceful blow can cause damage. If the eye is jarred sufficiently, some of its contents may be displaced. The suspensory ligaments may be torn and the lens dislocated into the posterior cavity, or the retina may pull away from the underlying vascular choroid coat. Once the retina is detached, photoreceptor cells may die because of lack of oxygen and nutrients. Unless a *detached retina* is repaired surgically, this injury may cause visual loss or blindness.

## Visual Pigments

Both rods and cones contain light-sensitive pigments that decompose when they absorb light energy. The light-sensitive pigment in rods is called **rhodopsin,** or visual purple, and it is embedded in membranous disks that are stacked within these receptor cells (fig. 12.41). In the presence of light, rhodopsin molecules break down into molecules of a colorless protein called *opsin* and a yellowish organic molecule called *retinal* (retinene) that is synthesized from vitamin A.

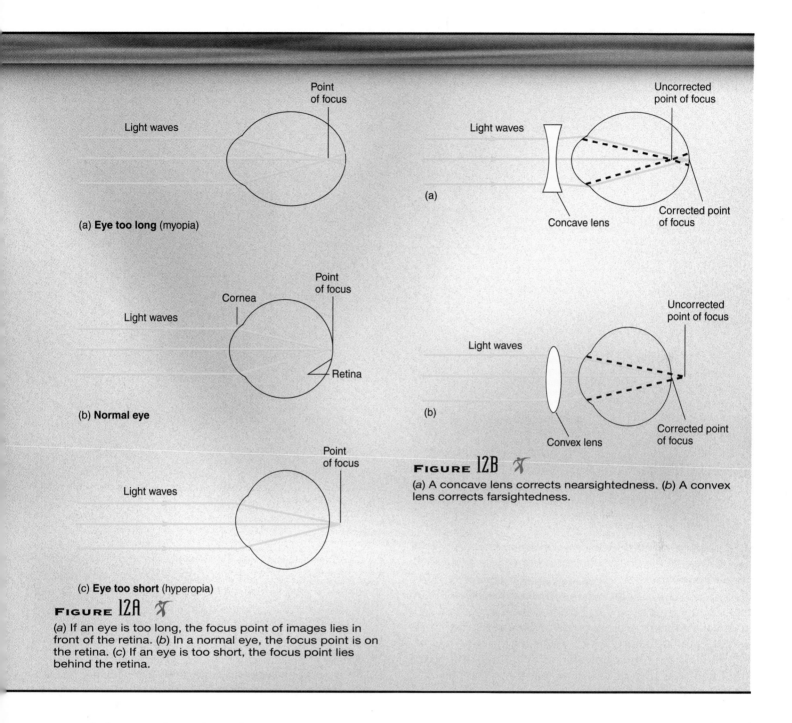

(a) **Eye too long** (myopia)

Cornea

Point of focus

Light waves

Retina

(b) **Normal eye**

Point of focus

Light waves

(c) **Eye too short** (hyperopia)

**FIGURE 12A** ⚘

(*a*) If an eye is too long, the focus point of images lies in front of the retina. (*b*) In a normal eye, the focus point is on the retina. (*c*) If an eye is too short, the focus point lies behind the retina.

Uncorrected point of focus

Light waves

(a)

Concave lens

Corrected point of focus

Uncorrected point of focus

Light waves

(b)

Convex lens

Corrected point of focus

**FIGURE 12B** ⚘

(*a*) A concave lens corrects nearsightedness. (*b*) A convex lens corrects farsightedness.

In darkness, sodium channels in portions of the receptor cell membranes are kept open by a nucleotide called *cyclic guanosine monophosphate* (cGMP). When rhodopsin molecules absorb light, they change shape and release opsin, in mere trillionths of a second. The released opsin then becomes an active enzyme. This enzyme activates a G-protein linked enzyme (transducin), which in turn activates still another enzyme (phosphodiesterase). The third enzyme of this series breaks down cGMP, and as the concentration of cGMP decreases, sodium channels close, and the receptor cell membrane hyperpolarizes (see chapter 10). The degree of hyperpolarization is directly proportional to the intensity of the light stimulating the receptor cells.

The hyperpolarization reaches the synaptic end of the cell, inhibiting release of neurotransmitter. Through a complex mechanism, decreased release of neurotransmitter by photoreceptor cells either stimulates or inhibits nerve impulses (action potentials) in nearby retinal neurons. Consequently, complex patterns of nerve impulses travel away from the retina, through the optic nerve, and into the brain, where they are interpreted as vision.

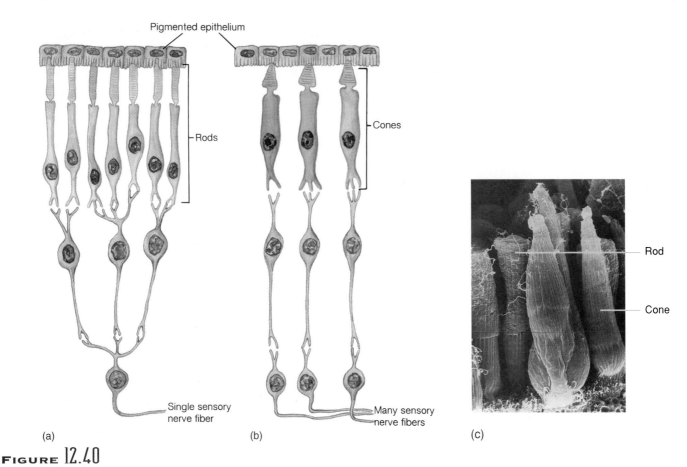

Pigmented epithelium

Rods

Cones

Rod

Cone

Single sensory
nerve fiber

Many sensory
nerve fibers

(a)  (b)  (c)

**FIGURE 12.40**

(a) A single sensory nerve fiber transmits impulses from several rods to the brain. (b) Separate sensory nerve fibers transmit impulses from cones to the brain. (c) A scanning electron micrograph of rods and cones (1,350×).

In bright light, nearly all of the rhodopsin in the rods decomposes, sharply reducing the sensitivity of these receptors. The cones continue to function, however, and in bright light we therefore see in color. In dim light, rhodopsin can be regenerated from opsin and retinal faster than it is broken down. This regeneration requires cellular energy, which ATP provides (see chapter 4). Under these conditions, the rods continue to function and the cones remain unstimulated. Hence, we see only shades of gray in dim light.

The light sensitivity of an eye whose rods have converted the available opsin and retinal to rhodopsin increases about 100,000 times, and the eye is said to be *dark adapted.* A person needs a dark-adapted eye to see in dim light. For example, when going from daylight into a darkened theater, it may be difficult to see well enough to locate a seat, but soon the eyes adapt to the dim light and vision improves. Later, leaving the theater and entering the sunlight may cause discomfort or even pain. This occurs at the moment that most of the rhodopsin decomposes in response to the bright light. At the same time, the light sensitivity of the eyes decreases greatly, and they become *light adapted.*

Too little vitamin A in the diet reduces the quantity of retinal, impairing rhodopsin production and sensitivity of the rods. The result is night blindness, or poor vision in dim light.

The light-sensitive pigments of cones, called *iodopsins,* are similar to rhodopsin in that they are composed of retinal combined with a protein; the protein, however, differs from the protein in the rods. In fact, there are three sets of cones within the retina, each containing an abundance of one of three different visual pigments.

The wavelength of a particular kind of light determines the color perceived from it. For example, the shortest wavelengths of visible light are perceived as violet, whereas the longest wavelengths of visible light are seen as red. One type of cone pigment (erythrolabe) is most sensitive to red light waves, another (chlorolabe) to green light waves, and a third (cyanolabe) to blue light waves. The sensitivities of these pigments do overlap somewhat. For example, both red and green light pigments are sensi-

**FIGURE** 12.41

Rhodopsin is embedded in disks of membrane that are stacked within the rod cells.

Disks of membrane within cell

Mitochondria

Nucleus

Synaptic ending

Rod cell

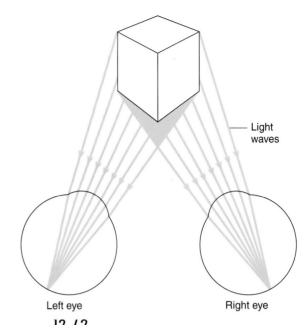

**FIGURE** 12.42

Stereoscopic vision results from formation of two slightly different retinal images.

Light waves

Left eye

Right eye

Insects, fish, and birds have more acute color vision than mammals. Among the mammals, primates—including humans—have the most complex color vision, with red, green, and blue cones. Most other mammals have only yellow and blue cones.

## Stereoscopic Vision

**Stereoscopic vision** (stereopsis) simultaneously perceives distance, depth, height, and width of objects. Such vision is possible because the pupils are 6–7 centimeters apart. Consequently, objects that are close (less than 20 feet away) produce slightly different retinal images. That is, the right eye sees a little more of one side of an object, while the left eye sees a little more of the other side. These two images are somehow superimposed and interpreted by the visual cortex of the brain. The result is the perception of a single object in three dimensions (fig. 12.42).

Because this type of depth perception depends on vision with two eyes (binocular vision), it follows that a one-eyed person is less able to judge distance and depth accurately. To compensate, a person with one eye can use the relative sizes and positions of familiar objects as visual clues.

## Visual Nerve Pathways

As mentioned in chapter 11, the axons of the ganglion cells in the retina leave the eyes to form the *optic nerves.*

tive to orange light waves. On the other hand, red pigment absorbs orange light waves more effectively.

The color perceived depends upon which sets of cones the light in a given image stimulates. If all three types of sets of cones are stimulated, the light is perceived as white, and if none are stimulated, it is seen as black.

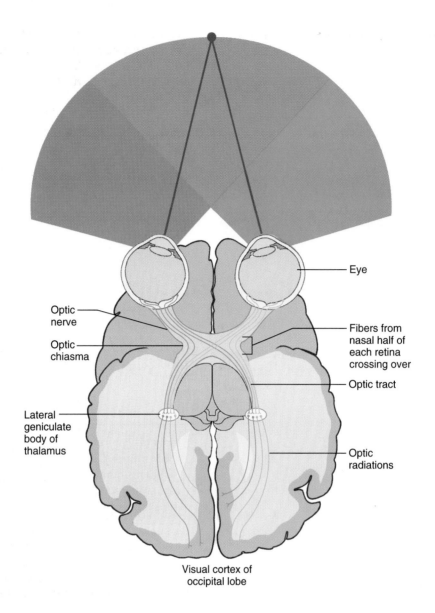

**FIGURE 12.43**

The visual pathway includes the optic nerve, optic chiasma, optic tract, and optic radiations.

Just anterior to the pituitary gland, these nerves give rise to the X-shaped *optic chiasma,* and within the chiasma some of the fibers cross over. More specifically, the fibers from the nasal (medial) half of each retina cross over, whereas those from the temporal (lateral) sides do not. Thus, fibers from the nasal half of the left eye and the temporal half of the right eye form the right *optic tract;* and fibers from the nasal half of the right eye and the temporal half of the left eye form the left optic tract.

The nerve fibers continue in the optic tracts, and just before they reach the thalamus, a few of them leave to enter nuclei that function in various visual reflexes. Most of the fibers, however, enter the thalamus and synapse in its posterior portion (lateral geniculate body). From this region the visual impulses enter nerve pathways called *optic radiations,* and the pathways lead to the visual cortex of the occipital lobes (fig. 12.43).

Since each visual cortex receives impulses from each eye, a person may develop partial blindness in both eyes if either visual cortex is injured. For example, if the right visual cortex (or the right optic tract) is injured, sight may be lost in the temporal side of the right eye and the nasal side of the left eye. Similarly, damage to the central portion of the optic chiasma, where fibers from the nasal sides of the eyes cross over, blinds the nasal sides of both eyes.

Other fibers conducting visual impulses pass downward into various regions of the brain stem. These impulses are important for controlling head and eye movements associated with tracking an object visually, for controlling the simultaneous movements of both eyes, and for controlling certain visual reflexes, such as those that move the iris muscles.

**1** Distinguish between the rods and the cones of the retina.

**2** Explain the roles of visual pigments.

**3** What factors make stereoscopic vision possible?

**4** Trace the pathway of visual impulses from the retina to the occipital cortex.

## CLINICAL TERMS RELATED TO THE SENSES

**amblyopia** (am″ble-o′pe-ah)  Dim vision due to a cause other than a refractive disorder or lesion.

**anopia** (an-o′pe-ah)  Absence of an eye.

**audiometry** (aw″de-om′ĕ-tre)  Measurement of auditory acuity for various frequencies of sound waves.

**blepharitis** (blef″ah-ri′tis)  Inflammation of the margins of the eyelids.

**causalgia** (kaw-zal′je-ah)  Persistent, burning pain usually associated with injury to a limb.

**conjunctivitis** (kon-junk″tĭ-vi′tis)  Inflammation of the conjunctiva.

**diplopia** (di-plo′pe-ah)  Double vision, or the sensation of seeing two objects when only one is viewed.

**emmetropia** (em″ĕ-tro′pe-ah)  Normal condition of the eyes; eyes with no refractive defects.

**enucleation** (e-nu″kle-a′shun)  Removal of the eyeball.

**exophthalmos** (ek″sof-thal′mos)  Condition in which the eyes protrude abnormally.

**hemianopsia** (hem″e-an-op′se-ah)  Defective vision affecting half of the visual field.

**hyperalgesia** (hi″per-al-je′ze-ah)  Abnormally increased sensitivity to pain.

**iridectomy** (ir″ĭ-dek′to-me)  Surgical removal of part of the iris.

**iritis** (i-ri′tis)  Inflammation of the iris.

**keratitis** (ker″ah-ti′tis)  Inflammation of the cornea.

**labyrinthectomy** (lab″i-rin-thek′to-me)  Surgical removal of the labyrinth.

**labyrinthitis** (lab″i-rin-thi′tis)  Inflammation of the labyrinth.

**Ménière's disease** (men″e-ārz′ dĭ-zez)  Inner ear disorder characterized by ringing in the ears, increased sensitivity to sounds, dizziness, and loss of hearing.

**neuralgia** (nu-ral′je-ah)  Pain resulting from inflammation of a nerve or a group of nerves.

**neuritis** (nu-ri′tis)  Inflammation of a nerve.

**nystagmus** (nis-tag′mus)  Involuntary oscillation of the eyes.

**otitis media** (o-ti′tis me′de-ah)  Inflammation of the middle ear.

**otosclerosis** (o″to-skle-ro′sis)  Formation of spongy bone in the inner ear, which often causes deafness by fixing the stapes to the oval window.

**pterygium** (tĕ-rij′e-um)  Abnormally thickened patch of conjunctiva that extends over part of the cornea.

**retinitis pigmentosa** (ret″ĭ-ni′tis pig″men-to′sa)  Inherited progressive retinal sclerosis that deposits pigment in the retina and causes retinal atrophy.

**retinoblastoma** (ret″i-no-blas-to′mah)  Inherited, highly malignant tumor arising from immature retinal cells.

**tinnitus** (tĭ-ni′tus)  Ringing or buzzing in the ears.

**tonometry** (to-nom′e-tre)  Measurement of fluid pressure within the eyeball.

**trachoma** (trah-ko′mah)  Bacterial disease of the eye, characterized by conjunctivitis, which may lead to blindness.

**tympanoplasty** (tim″pah-no-plas′te)  Surgical reconstruction of the middle ear bones and establishment of continuity from the tympanic membrane to the oval window.

**uveitis** (u″ve-i′tis)  Inflammation of the uvea, the region of the eye that includes the iris, the ciliary body, and the choroid coat.

**vertigo** (ver′tĭ-go)  Sensation of dizziness.

## CHAPTER SUMMARY

## Introduction  (page 432)

Sensory receptors are sensitive to environmental changes and initiate impulses to the brain and spinal cord.

## Receptors and Sensations  (page 432)

1. Receptor types
   a. Each type of receptor is sensitive to a distinct type of stimulus.
   b. The major types of receptors include the following:

(1) Chemoreceptors, sensitive to changes in chemical concentration.
(2) Pain receptors, sensitive to tissue damage.
(3) Thermoreceptors, sensitive to temperature changes.
(4) Mechanoreceptors, sensitive to mechanical forces.
(5) Photoreceptors, sensitive to light.

2. Sensory impulses
   a. When receptors are stimulated, changes occur in their membrane potentials.

b. Receptor potentials are transferred to nerve fibers, triggering action potentials.
3. Sensations
   a. Sensations are feelings resulting from sensory stimulation.
   b. A particular part of the sensory cortex interprets every impulse that reaches it in the same way.
   c. The cerebral cortex projects a sensation back to the region of stimulation.
4. Sensory adaptations are adjustments of sensory receptors to continuous stimulation. Impulses are triggered at slower and slower rates.

## Somatic Senses  (page 434)

Somatic senses receive information from receptors in skin, muscles, joints, and viscera. They can be grouped as exteroceptive, proprioceptive, and visceroceptive senses.

1. Touch and pressure senses
   a. Free ends of sensory nerve fibers are the receptors for the sensations of touch and pressure.
   b. Meissner's corpuscles are the receptors for the sensations of light touch.
   c. Pacinian corpuscles are the receptors for the sensations of heavy pressure and vibrations.
2. Thermoreceptors include two sets of free nerve endings that are heat and cold receptors.
3. Sense of pain
   a. Pain receptors
      (1) Pain receptors are free nerve endings that tissue damage stimulates.
      (2) Pain receptors provide protection, do not adapt rapidly, and can be stimulated by changes in temperature, mechanical force, and chemical concentration.
   b. The only receptors in viscera that provide sensations are pain receptors. These receptors are most sensitive to certain chemicals and lack of blood flow. The sensations they produce feel as if they come from some other part of the body (referred pain).
   c. Pain nerve pathways
      (1) The two main types of pain fibers are acute pain fibers and chronic pain fibers.
      (2) Acute pain fibers are fast conducting; chronic pain fibers are slower conducting.
      (3) Pain impulses are processed in the dorsal horn of the spinal cord, and they ascend in the spinothalamic tracts.
      (4) Within the brain, pain impulses pass through the reticular formation before being conducted to the cerebral cortex.
   d. Regulation of pain impulses
      (1) Awareness of pain occurs when impulses reach the thalamus.
      (2) The cerebral cortex judges the intensity of pain and locates its source.
      (3) Impulses descending from the brain cause neurons to release pain-relieving substances, such as enkephalins and serotonin.
      (4) Endorphin is a pain-relieving biochemical produced in the brain.
   e. Certain neuropeptides synthesized in the brain and spinal cord inhibit pain impulses.
4. Stretch receptors
   a. Stretch receptors provide information about the condition of muscles and tendons.

b. Muscle spindles are stimulated when a muscle is relaxed, and they initiate a reflex that contracts the muscle.
c. Golgi tendon organs are stimulated when muscle tension increases, and they initiate a reflex that relaxes the muscle.

## Special Senses  (page 439)

Special senses are those whose receptors occur in relatively large, complex sensory organs of the head.

## Sense of Smell  (page 440)

1. Olfactory receptors
   a. Olfactory receptors are chemoreceptors that chemicals dissolved in liquid stimulate.
   b. Olfactory receptors function together with taste receptors and aid in food selection.
2. Olfactory organs
   a. The olfactory organs consist of receptors and supporting cells in the nasal cavity.
   b. Olfactory receptors are neurons with cilia that sense lipid-soluble chemicals.
3. Olfactory nerve pathways.
   a. Nerve impulses travel from the olfactory receptors through the olfactory nerves, olfactory bulbs, and olfactory tracts.
   b. They go to interpreting centers in the limbic system of the brain.
4. Olfactory stimulation
   a. Olfactory impulses may result when various gaseous molecules combine with specific sites on the cilia of the receptor cells.
   b. Olfactory receptors adapt rapidly.
   c. Olfactory receptors are often damaged by environmental factors, but are not replaced.

## Sense of Taste  (page 442)

1. Taste receptors
   a. Taste buds consist of receptor cells and supporting cells.
   b. Taste cells have taste hairs that are sensitive to particular chemicals dissolved in water.
   c. Taste hair surfaces have receptor sites to which chemicals combine and trigger impulses to the brain.
   d. Each of the four primary kinds of taste cells is particularly sensitive to a certain group of chemicals.
2. Taste sensations
   a. The four primary taste sensations are sweet, sour, salty, and bitter.
   b. Various taste sensations result from the stimulation of one or more sets of taste receptors.
   c. Sweet receptors are most plentiful near the tip of the tongue, sour receptors along the margins, salt receptors in the tip and upper front, and bitter receptors toward the back.
3. Taste nerve pathways
   a. Sensory impulses from taste receptors travel on fibers of the facial, glossopharyngeal, and vagus nerves.
   b. These impulses are carried to the medulla and ascend to the thalamus, and then to the gustatory cortex in the parietal lobes.

## Sense of Hearing  (page 446)

1. The external ear collects sound waves created by vibrating objects.
2. Middle ear
   a. Auditory ossicles of the middle ear conduct sound waves from the tympanic membrane to the oval window of the inner ear. They also increase the force of these waves.
   b. Skeletal muscles attached to the auditory ossicles provide the tympanic reflex, which protects the inner ear from the effects of loud sounds.
3. Auditory tubes connect the middle ears to the throat and help maintain equal air pressure on both sides of the tympanic membranes.
4. Inner ear
   a. The inner ear consists of a complex system of connected tubes and chambers—the osseous and membranous labyrinths. It includes the cochlea, which in turn houses the organ of Corti.
   b. The organ of Corti contains the hearing receptors that vibrations in the fluids of the inner ear stimulate.
   c. Different frequencies of vibrations stimulate different sets of receptor cells; the human ear can detect sound frequencies from about 20 to 20,000 vibrations per second.
5. Auditory nerve pathways
   a. The nerve fibers from hearing receptors travel in the cochlear branch of the vestibulocochlear nerves.
   b. Auditory impulses travel into the medulla oblongata, midbrain, and thalamus, and are interpreted in the temporal lobes of the cerebrum.

## Sense of Equilibrium  (page 451)

1. Static equilibrium maintains the stability of the head and body when they are motionless. The organs of static equilibrium are located in the vestibule.
2. Dynamic equilibrium balances the head and body when they are moved or rotated suddenly. The organs of this sense are located in the ampullae of the semicircular canals.
3. Other structures that help maintain equilibrium include the eyes and the proprioceptors associated with certain joints.

## Sense of Sight  (page 457)

1. Visual accessory organs include the eyelids and lacrimal apparatus that protect the eye, and the extrinsic muscles that move the eye.
2. Structure of the eye

   a. The wall of the eye has an outer, a middle, and an inner tunic that function as follows:
      (1) The outer layer (sclera) is protective, and its transparent anterior portion (cornea) refracts light entering the eye.
      (2) The middle layer (choroid coat) is vascular and contains pigments that help keep the inside of the eye dark.
      (3) The inner layer (retina) contains the visual receptor cells.
   b. The lens is a transparent, elastic structure. The ciliary muscles control its shape.
   c. The iris is a muscular diaphragm that controls the amount of light entering the eye; the pupil is an opening in the iris.
   d. Spaces within the eye are filled with fluids (aqueous and vitreous humors) that help maintain its shape.
3. Light refraction
   a. Light waves are primarily refracted by the cornea and lens to focus an image on the retina.
   b. The lens must thicken to focus on close objects.
4. Visual receptors
   a. The visual receptors are rods and cones.
   b. Rods are responsible for colorless vision in relatively dim light, and cones provide color vision.
5. Visual pigments
   a. A light-sensitive pigment in rods (rhodopsin) decomposes in the presence of light and triggers a complex series of reactions that initiate nerve impulses on the optic nerve.
   b. Three sets of cones provide color vision. Each set contains a different light-sensitive pigment, and each set is sensitive to a different wavelength of light; the color perceived depends on which set or sets of cones are stimulated.
6. Stereoscopic vision
   a. Stereoscopic vision provides perception of distance and depth.
   b. Stereoscopic vision occurs because of the formation of two slightly different retinal images that the brain superimposes and interprets as one image in three dimensions.
   c. A one-eyed person uses relative sizes and positions of familiar objects to judge distance and depth.
7. Visual nerve pathways
   a. Nerve fibers from the retina form the optic nerves.
   b. Some fibers cross over in the optic chiasma.
   c. Most of the fibers enter the thalamus and synapse with others that continue to the visual cortex of the occipital lobes.
   d. Other impulses pass into the brain stem and function in various visual reflexes.

## CRITICAL THINKING QUESTIONS

1. How would you interpret the following observation? A person enters a tub of water and reports that it is uncomfortably warm, yet a few moments later says the water feels comfortable, even though the water temperature is unchanged.

2. Why are some serious injuries, such as those produced by a bullet entering the abdomen, relatively painless, whereas others, such as those that crush the skin, are quite painful?

3. Labyrinthitis is an inflammation of the tissues of the inner ear. What symptoms would you expect to observe in a patient with this disorder?

4. Sometimes, as a result of an injury to the eye, the retina detaches from its pigmented epithelium. Assuming that the retinal tissues remain functional, what is likely to happen to the person's vision if the retina moves unevenly toward the interior of the eye?

5. The auditory tubes of a child are shorter and directed more horizontally than those of an adult. How might this explain the greater prevalence of middle ear infections in children compared to adults?

6. A patient with heart disease experiences pain at the base of the neck and in the left shoulder and arm after exercise. How would you explain to the patient the origin of this pain?

## REVIEW EXERCISES

1. List five groups of sensory receptors, and name the kind of change to which each is sensitive.
2. Explain how sensory receptors stimulate sensory impulses.
3. Define *sensation*.
4. Explain the projection of a sensation.
5. Define *sensory adaptation.*
6. Explain how somatic senses can be grouped.
7. Describe the functions of free nerve endings, Meissner's corpuscles, and Pacinian corpuscles.
8. Explain how thermoreceptors function.
9. Compare pain receptors with other types of somatic receptors.
10. List the factors that are likely to stimulate visceral pain receptors.
11. Define *referred pain.*
12. Explain how neuropeptides relieve pain.
13. Distinguish between muscle spindles and Golgi tendon organs.
14. Explain how the senses of smell and taste function together to create the perception of the flavors of foods.
15. Describe the olfactory organ and its function.
16. Trace a nerve impulse from the olfactory receptor to the interpreting centers of the brain.
17. Explain how an olfactory code distinguishes odor stimuli.
18. Explain how the salivary glands aid the taste receptors.
19. Name the four primary taste sensations, and describe the patterns in which the taste receptors are roughly distributed on the tongue.
20. Explain why taste sensation is less likely to diminish with age than olfactory sensation.
21. Trace the pathway of a taste impulse from the receptor to the cerebral cortex.
22. Distinguish among the external, middle, and inner ears.
23. Trace the path of a sound vibration from the tympanic membrane to the hearing receptors.

24. Describe the functions of the auditory ossicles.
25. Describe the tympanic reflex, and explain its importance.
26. Explain the function of the auditory tube.
27. Distinguish between the osseous and the membranous labyrinths.
28. Describe the cochlea and its function.
29. Describe a hearing receptor.
30. Explain how a hearing receptor stimulates a sensory neuron.
31. Trace a nerve impulse from the organ of Corti to the interpreting centers of the cerebrum.
32. Describe the organs of static and dynamic equilibrium and their functions.
33. Explain how the sense of vision helps maintain equilibrium.
34. List the visual accessory organs, and describe the functions of each.
35. Name the three layers of the eye wall, and describe the functions of each.
36. Describe how accommodation is accomplished.
37. Explain how the iris functions.
38. Distinguish between aqueous humor and vitreous humor.
39. Distinguish between the macula lutea and the optic disk.
40. Explain how light waves focus on the retina.
41. Distinguish between rods and cones.
42. Explain why cone vision is generally more acute than rod vision.
43. Describe the function of rhodopsin.
44. Explain how the eye adapts to light and dark.
45. Describe the relationship between light wavelengths and color vision.
46. Define *stereoscopic vision.*
47. Explain why a person with binocular vision is able to judge distance and depth of close objects more accurately than a one-eyed person.
48. Trace a nerve impulse from the retina to the visual cortex.

**For More Information Related to This Chapter, Please Visit** *Hole's Human Anatomy and Physiology* **Home Page**
http://www.mhhe.com/biosci/abio/

# ENDOCRINE SYSTEM

## CHAPTER OBJECTIVES

AFTER YOU HAVE STUDIED THIS CHAPTER, YOU SHOULD BE ABLE TO:

1. Distinguish between endocrine and exocrine glands.

2. Describe how hormones can be classified according to their chemical composition.

3. Explain how steroid and nonsteroid hormones affect target cells.

4. Discuss how negative feedback mechanisms regulate hormonal secretion.

5. Explain how the nervous system controls hormonal secretion.

6. Name and describe the locations of the major endocrine glands, and list the hormones they secrete.

7. Describe the general functions of the various hormones.

8. Explain how the secretion of each hormone is regulated.

9. Distinguish between physical and psychological stress.

10. Describe the general stress response.

## KEY TERMS

**adenylate cyclase** (ah-den′ĭ-lat si′klās)

**adrenal cortex** (ah-dre′nal kor′teks)

**adrenal medulla** (ah-dre′nal me-dul′ah)

**anterior pituitary** (an-ter′e-or pĭ-tu′ĭ-tar″e)

**cyclic AMP** (sik′lik AMP)

**hormone** (hor′mōn)

**kinase** (ki′nās)

**metabolic rate** (met″ah-bol′ik rāt)

**negative feedback** (neg′ah-tiv fēd′bak)

**pancreas** (pan′kre-as)

**parathyroid gland** (par″ah-thi′roid gland)

**pineal gland** (pin′e-al gland)

**posterior pituitary** (pos-ter′e-or pĭ-tu′ĭ-tar″e)

**prostaglandin** (pros″tah-glan′din)

**steroid** (ste′roid)

**target cell** (tar′get sel)

**thymus gland** (thi′mus gland)

**thyroid gland** (thi′roid gland)

## UNDERSTANDING WORDS

**cort-,** bark, rind: adrenal *cort*ex—outer portion of an adrenal gland.

**-crin,** to secrete: endo*crine*—pertaining to internal secretions.

**diuret-,** to pass urine: *diuret*ic—substance that promotes the production of urine.

**endo-,** within: *endo*crine gland—gland that releases its secretion internally into a body fluid.

**exo-,** outside: *exo*crine gland—gland that releases its secretion to the outside through a duct.

**horm-,** impetus, impulse: *horm*one—substance that a cell secretes that affects another cell.

**hyper-,** above: *hyper*thyroidism—condition resulting from an above-normal secretion of thyroid hormone.

**hypo-,** below: *hypo*thyroidism—condition resulting from a below-normal secretion of thyroid hormone.

**lact-,** milk: pro*lact*in—hormone that promotes milk production.

**med-,** middle: adrenal *med*ulla—middle section of an adrenal gland.

**para-,** beside: *para*thyroid glands—set of glands located near the surface of the thyroid gland.

**toc-,** birth: oxy*toc*in—hormone that stimulates the uterine muscles to contract during childbirth.

**-tropic,** influencing: adrenocortico*tropic* hormone—a hormone secreted by the anterior pituitary gland that stimulates the adrenal cortex.

**vas-,** vessel: *vas*opressin—substance that causes blood vessel walls to contract.

The sweet-smelling urine that is the hallmark of insulin-dependent diabetes mellitus was noted as far back as an Egyptian papyrus from 1500 B.C. In A.D. 96 in Greece, Aretaeus of Cappadocia described the condition as a "melting down of limbs and flesh into urine." A look at the 3-year-old boy on the previous page illustrates how apt a description that was. In December 1922, the boy weighed only 15 pounds. But he was far luckier than people who had suffered with this form of diabetes before him. He was among the first to receive as a drug insulin, a hormone that his body could not produce. As the inset shows, the boy rapidly improved, doubling his weight in just two months of insulin treatment.

Insulin and the gland that produces it—the pancreas—are familiar components of the endocrine system. Understanding insulin-dependent diabetes mellitus provides a fascinating glimpse into the evolution of medical technology that continues today.

In 1921, Canadian physiologists Sir Frederick Grant Banting and Charles Herbert Best discovered the link between lack of insulin and diabetes. They induced diabetes

symptoms in a dog by removing its pancreas, then cured it by administering insulin from another dog's healthy pancreas. Just a year later, people with diabetes—such as the starving 3-year-old—began to receive insulin extracted from pigs or cattle.

And so it went until 1982, when pure human insulin became available by genetically engineering bacteria to produce the human protein, helping those who were allergic to the product from pigs or cows. Today, people with this form of diabetes receive insulin in several injections a day, or with an infusion pump inserted beneath the skin. A pump more closely approximates the pancreas's natural release of insulin, but it is not perfect. Alternative approaches to improve insulin delivery include a pancreas transplant or an implant of insulin-producing cell clusters.

Although a person with insulin-dependent diabetes mellitus today is certainly far better off than the boy on the brink of the insulin discovery, we still cannot exactly duplicate the function of the pancreas in secreting insulin. Better understanding of the endocrine system will lead to better treatment of this and other hormonal disorders.

## GENERAL CHARACTERISTICS OF THE ENDOCRINE SYSTEM

The **endocrine system** includes the cells, tissues, and organs that secrete hormones into body fluids (the internal environment). In contrast, **exocrine** structures secrete into tubes or ducts, which lead to internal or external body surfaces. The thyroid and parathyroid glands, for example, secrete hormones into the blood and are therefore endocrine glands, but sweat glands and sebaceous (oil) glands in the skin are exocrine (fig. 13.1). Two other terms are used to describe patterns of secretion. **Paracrine** secretions are restricted to the interstitial fluid and affect only neighboring cells. **Autocrine** secretions affect the secreting cell itself. Paracrine and autocrine secretions are sometimes called "local hormones."

> The interrelationships of the glands of the endocrine system, although not well understood, are vividly obvious in families that have an inherited cancer syndrome called *multiple endocrine neoplasia* (MEN). One family member might have a tumor of the adrenal glands called pheochromocytoma; another might have thyroid cancer; yet a third relative might have parathyroid hyperplasia, a precancerous condition.

Endocrine glands and their hormones help regulate metabolic processes. They control the rates of certain chemical reactions, aid in transporting substances through membranes, and help regulate water balance, electrolyte balance, and blood pressure. They also play vital roles in reproduction, development, and growth.

Specialized small groups of cells produce some hormones. However, the larger endocrine glands—the pituitary gland, thyroid gland, parathyroid glands, adrenal glands, and pancreas—are the subject of this chapter (fig. 13.2). Subsequent chapters discuss several other hormone-secreting glands and tissues.

## HORMONE ACTION

A **hormone** is a biochemical secreted by a cell that affects the functions of another cell. Hormones are released into the extracellular spaces surrounding secretory cells. Paracrine hormones travel only short distances and have local actions. Other hormones are carried by the blood to all parts of the body and may produce general effects. In either case, the physiological action of a particular hormone is restricted to its **target cells,** because only those cells have surface receptors for the particular hormone.

### Chemistry of Hormones

Chemically, most hormones are either steroids (or steroidlike substances) that are synthesized from cholesterol (see chapter 2), or they are amines, peptides, proteins, or glycoproteins that are synthesized from amino acids. Hormones are organic compounds. They can stimulate changes in **target cells** even if present in extremely low concentrations.

*Steroids* are lipids that include complex rings of carbon and hydrogen atoms. Steroids differ by the types and numbers of atoms attached to these rings and the ways they are joined (see chapter 2). All steroid hormones are derived from cholesterol, including sex hor-

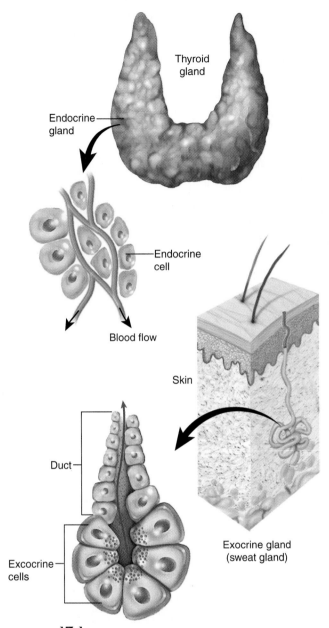

**FIGURE 13.1**

Endocrine glands release hormones into the internal environment (body fluids). Exocrine glands secrete to the outside environment, through ducts that lead to body surfaces.

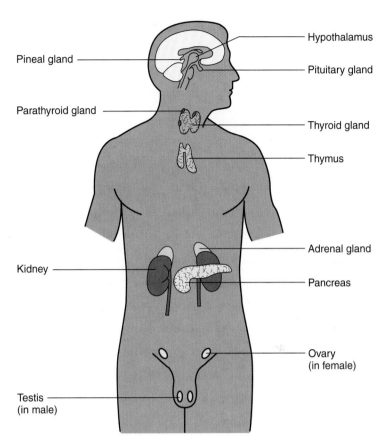

**FIGURE 13.2** 🕱

Locations of major endocrine glands. Some organs that secrete hormones are not shown, such as the heart and small intestine.

The *peptide* hormones are short chains of amino acids. This group includes hormones associated with the posterior pituitary gland and some produced in the hypothalamus.

*Protein* hormones are composed of long chains of amino acids, linked to form intricate molecular structures (see chapter 2). They include the hormone secreted by the parathyroid gland and some of those secreted by the anterior pituitary gland (fig. 13.3). Certain other hormones from the anterior pituitary gland are *glycoproteins,* which consist of a protein joined to a carbohydrate.

Another group of compounds, called prostaglandins, are paracrine substances. They regulate neighboring cells. Prostaglandins are lipids (20-carbon fatty acids that include 5-carbon rings) and are synthesized from a type of fatty acid (arachidonic acid) in cell membranes. Prostaglandins occur in a wide variety of cells, including those of the liver, kidneys, heart, lungs, thymus gland, pancreas, brain, and reproductive organs. (Their cousins, the leukotrienes, affect white blood cells and are discussed in chapter 16.)

Table 13.1 lists the names and abbreviations of some of the hormones discussed in this chapter. Table 13.2 and figure 13.3 summarize the chemical composition of hormones.

mones (such as estrogens and testosterone) and those the adrenal cortex secretes (including aldosterone and cortisol). Vitamin D is a modified steroid and can be converted into a hormone, as is discussed later in this chapter (see also chapter 18).

Some neurons produce hormones called *amines* that are derived from amino acids, including norepinephrine and epinephrine. These hormones are also synthesized in the adrenal medulla (the inner portion of the adrenal gland) from the amino acid tyrosine (see chapter 11).

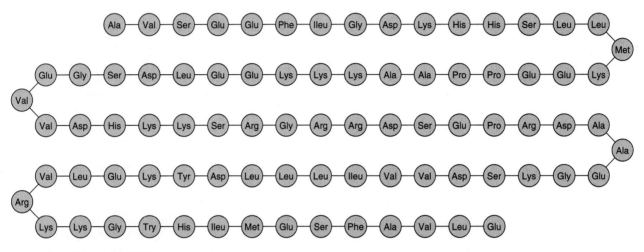

(a) Cortisol

(b) Norepinephrine

(c) Parathyroid hormone (PTH)

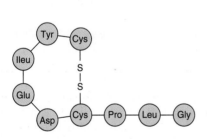

(d) Oxytocin

(e) Prostaglandin PGE₂

**FIGURE 13.3**

(*a* and *b*) Structural formulas of a steroid hormone (cortisol) and an amine hormone (norepinephrine). (*c* and *d*) Amino acid sequence of a protein hormone (PTH) and a peptide hormone (oxytocin). (*e*) A prostaglandin (PGE₂).

1. How do endocrine glands and exocrine glands differ?

2. What is a hormone?

3. How are hormones chemically classified?

## Actions of Hormones

Hormones exert their effects by altering metabolic processes. For example, a hormone might alter activity of an enzyme necessary for synthesizing a particular substance, or alter the rate at which particular chemicals are transported through cell membranes.

Hormones bind to specific receptors in order to exert effects on their target cells. Each hormone receptor is a protein or glycoprotein molecule that has a *binding site* for a specific hormone. A hormone delivers its message to a cell by uniting with the binding site of its receptor. The more receptors the hormone binds, the greater the cellular response.

### Steroid Hormones

Steroid hormones are soluble in the lipids that make up the bulk of cell membranes. These hormones can easily

TABLE 13.1

## HORMONE NAMES AND ABBREVIATIONS

| Source | Name | Abbreviation | Synonym |
|--------|------|--------------|---------|
| Hypothalamus | Corticotropin-releasing hormone | CRH | |
| | Gonadotropin-releasing hormone | GnRH | Luteinizing hormone releasing hormone (LHRH) |
| | Somatostatin | SS | Growth hormone release-inhibiting hormone (GIH) |
| | Growth hormone-releasing hormone | GHRH | |
| | Prolactin release-inhibiting hormone | PIH | |
| | Prolactin-releasing hormone | PRH | |
| | Thyrotropin-releasing hormone | TRH | |
| Anterior pituitary gland | Adrenocorticotropic hormone | ACTH | Corticotropin |
| | Follicle-stimulating hormone | FSH | Follitropin |
| | Growth hormone | GH | Somatotropin (STH) |
| | Luteinizing hormone | LH | Lutropin, interstitial cell-stimulating hormone (ICSH) |
| | Prolactin | PRL | |
| | Thyroid-stimulating hormone | TSH | Thyrotropin |
| Posterior pituitary gland | Antidiuretic hormone | ADH | Vasopressin |
| | Oxytocin | OT | |
| Thyroid gland | Calcitonin | | |
| | Thyroxine | $T_4$ | |
| | Triiodothyronine | $T_3$ | |
| Parathyroid gland | Parathyroid hormone | PTH | Parathormone |
| Adrenal medulla | Epinephrine | EPI | Adrenalin |
| | Norepinephrine | NE | Noradrenalin |
| Adrenal cortex | Aldosterone | | |
| | Cortisol | | Hydrocortisone |
| Pancreas | Glucagon | | |
| | Insulin | | |
| | Somatostatin | SS | |

TABLE 13.2

## TYPES OF HORMONES

| Types of Compound | Formed From | Examples |
|-------------------|-------------|----------|
| Amines | Amino acids | Norepinephrine, epinephrine |
| Peptides | Amino acids | ADH, OT, TRH, SS, GnRH |
| Proteins | Amino acids | PTH, GH, PRL |
| Glycoproteins | Protein and carbohydrate | FSH, LH, TSH |
| Steroids | Cholesterol | Estrogen, testosterone, aldosterone, cortisol |

The activated genes, in turn, are transcribed into messenger RNA (mRNA), which enters the cytoplasm, where it directs synthesis of specific proteins. These proteins bring about the cellular changes associated with the particular hormone (fig. 13.4, table 13.3, and Clinical Application 13.1).

### Nonsteroid Hormones

A nonsteroid hormone, such as an amine, peptide, or protein, combines with specific receptor molecules on the target cell membrane. Each receptor molecule is a protein that has a *binding site* and an *activity site*. The hormone combines with the binding site of the receptor. This causes the receptor's activity site to interact with other membrane proteins. Receptor binding may alter the function of enzymes or membrane transport mechanisms, changing the concentrations of still other cellular components. The hormone that triggers this cascade of biochemical activity is considered a *first messenger*. The biochemicals in the cell that induce the changes that are recognized as responses to the hormone are called *second messengers.*

diffuse into target cells. Once inside a target cell, steroid hormones combine (usually within the nucleus) with specific protein molecules—the receptors. The resulting *hormone-receptor complex* binds within the nucleus to a particular region of the DNA and activates specific genes.

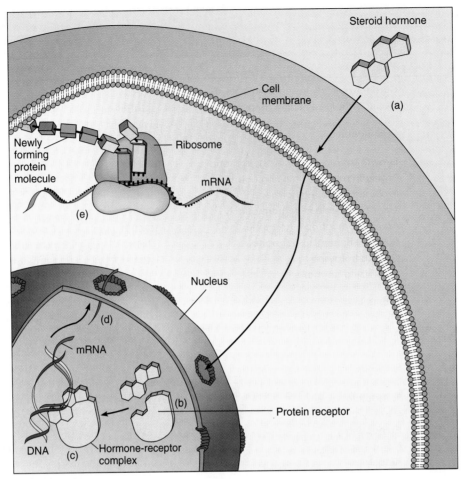

**FIGURE 13.4**

(a) A steroid hormone passes through a cell membrane and (b) combines with a protein receptor, usually in the nucleus. (c) The hormone-receptor complex activates messenger RNA synthesis. (d) The mRNA leaves the nucleus and (e) guides protein synthesis. *Note:* In the bloodstream, most lipid-soluble steroids are bound to proteins. Only the few that are not bound are free to enter cells, as shown here.

### TABLE 13.3 — SEQUENCE OF STEROID HORMONE ACTION

1. Endocrine gland secretes steroid hormone.
2. Steroid hormone diffuses through target cell membrane and enters nucleus.
3. Hormone combines with a receptor molecule.
4. Steroid hormone-receptor complex binds to DNA and promotes synthesis of messenger RNA.
5. Messenger RNA enters the cytoplasm and directs protein synthesis.
6. Newly synthesized proteins produce hormone's specific effects.

Many hormones use **cyclic adenosine monophosphate** (cyclic AMP, or cAMP) as a second messenger. In this mechanism, a hormone binds to its receptor, and the resulting hormone-receptor complex activates a protein called a **G protein,** which then activates an enzyme called **adenylate cyclase** that is bound to the inside of the cell membrane. The activated enzyme removes two phosphates from ATP and circularizes it, forming cyclic AMP (fig. 13.5). Cyclic AMP, in turn, activates another set of enzymes called **protein kinases.** Protein kinases transfer phosphate groups from ATP molecules to protein substrate molecules. This phosphorylation alters the shapes of the substrate molecules and converts some of them from inactive forms into active ones.

The activated protein molecules then induce changes in various cellular processes (fig. 13.6). Thus, the response of any particular cell to such a hormone is determined not only by the type of membrane receptors present, but also by the kinds of protein substrate molecules in the cell. Table 13.4 summarizes these actions. Cellular responses to second messenger activation include altering membrane permeabilities, activating enzymes, promoting synthesis of certain proteins, stimulating or inhibiting specific metabolic pathways, promoting cellular movements, and initiating secretion of hormones and other substances.

# CLINICAL APPLICATION

# Steroids and Athletes—An Unhealthy Combination

Canadian track star Ben Johnson flew past his competitors in the 100-meter run at the 1988 Summer Olympics in Seoul (fig. 13A). But 72 hours later, the gold medal awarded for his record-breaking time of 9.79 seconds was rescinded after traces of the drug stanozolol were detected in his urine.

Stanozolol is one of several synthetic versions of the steroid hormone testosterone. Like testosterone, these drugs promote signs of masculinity (their androgenic effect) and increased synthesis of muscle proteins (their anabolic effect). Used in the past to treat medical conditions such as anemia and breast cancer, steroids are used by professional and amateur athletes alike, all looking to build muscle tissue easily.

Steroid users may improve their performances and physiques in the short term, but in the long run they may suffer for it. Steroids hasten adulthood, stunting height and causing early hair loss, and in males lead to breast development and in females to a deepened voice, hairiness, and a male physique. The kidney, liver, and heart may be damaged, and atherosclerosis may develop because steroids raise LDL and lower HDL—the opposite of a healthy cholesterol profile. In males, the body mistakes the synthetic

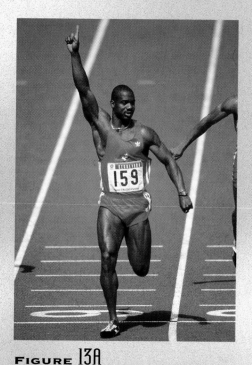

**FIGURE 13A**

Canadian track star Ben Johnson ran away with the gold medal in the 100-meter race at the 1988 Summer Olympics—but then had to return the award when traces of a steroid drug showed up in his urine.

steroids for the natural hormone and lowers its own production of testosterone. The price of athletic prowess today may be infertility later.

In 1976 the International Olympic Committee banned the use of steroids by athletes and required urine tests to detect illicit drug use. The Anabolic Steroids Act of 1990 made possession with intent to distribute the drugs a federal offense in the United States.

Ben Johnson was caught in his tracks by a urine test able to detect part-per-billion traces of synthetic steroids even weeks after they are taken. Johnson at first claimed the stanozolol in his urine was the result of a drink spiked with an approved anti-inflammatory drug used on his ankle, but a test of his natural testosterone showed it to be only 15% of normal—a sure sign that this athlete had been taking steroids for a long time.

Unfortunately, athletes' use of drugs to pump up their muscles continues. At the 1992 Summer Olympics in Barcelona, athletes from Germany, the United States, China, Britain, and the former Soviet Union were turned away because of evidence they had used nonsteroid drugs that they thought would have steroidlike effects. ■

Because many hormones utilize the cAMP-mediated second messenger system to exert their effects, an abnormality in this signaling system can lead to symptoms from many endocrine glands. In McCune-Albright syndrome, for example, a defect in the G protein that activates adenylate cyclase results in the conversion of ATP to cAMP even without hormonal stimulation. As a result, cells in the pituitary, thyroid, gonads, and adrenal glands secrete hormones in excess. These conditions are inherited because the components of the second messenger system are proteins.

Another enzyme, phosphodiesterase, quickly inactivates cAMP, so that its action is short-lived. For this reason, a continuing response in a target cell depends upon a continuing signal from hormone molecules combining with receptors in the target-cell membrane. Since enzymes continuously degrade hormones, a sustained response depends on continued hormone secretion.

Hormones whose actions depend upon cyclic AMP include releasing hormones from the hypothalamus; TSH, ACTH, FSH, and LH from the anterior pituitary gland; ADH from the posterior pituitary gland; PTH from the parathyroid glands; norepinephrine and epinephrine

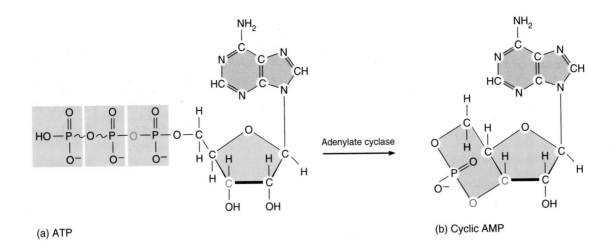

(a) ATP

(b) Cyclic AMP

**FIGURE 13.5**

Adenylate cyclase catalyzes conversion of (a) ATP molecules into cyclic AMP (b). The atoms forming the new bond are shown in red.

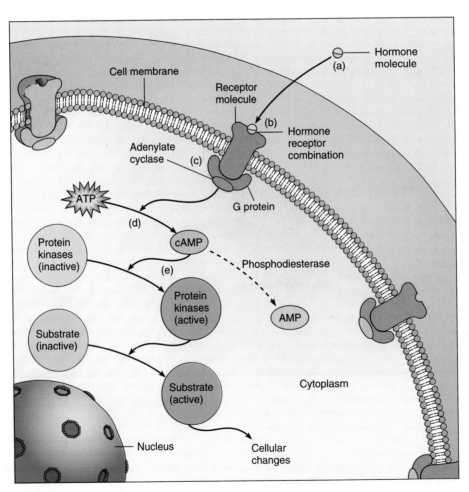

**FIGURE 13.6**

(a) Body fluids carry nonsteroid hormone molecules to the target cell, where they combine (b) with receptor sites on the cell membrane. (c) This may activate molecules of adenylate cyclase, which (d) catalyze conversion of ATP into cyclic AMP. (e) Cyclic AMP promotes a series of reactions leading to the cellular changes associated with the hormone's action. The dashed arrow indicates breakdown of cAMP, which turns off the cellular response to the hormone.

| TABLE 13.4 | SEQUENCE OF NONSTEROID HORMONE ACTIONS |

| | |
|---|---|
| **1.** Endocrine gland secretes nonsteroid hormone. | **5.** Adenylate cyclase converts ATP into cyclic AMP. |
| **2.** Body fluid carries hormone to its target cell. | **6.** Cyclic AMP activates protein kinases. |
| **3.** Hormone combines with receptor site on membrane of its target cell, activating G protein. | **7.** These enzymes activate protein substrates in the cell that change metabolic processes. |
| **4.** Adenylate cyclase molecules are activated within target cell's membrane. | **8.** Cellular changes produce the hormone's effects. |

from the adrenal glands; calcitonin from the thyroid gland; and glucagon from the pancreas.

Other nonsteroid hormones employ second messengers other than cAMP. For example, a second messenger called diacylglycerol (DAG), like cAMP, activates a protein kinase leading to a cellular response.

In another mechanism, a hormone binding to its receptor increases calcium ion concentration within the target cell. Such a hormone may stimulate transport of calcium ions inward through the cell membrane or induce release of calcium ions from cellular storage sites via a second messenger called inositol triphosphate ($IP_3$). The calcium ions combine with the protein *calmodulin* (see chapter 9), altering its molecular structure in a way that activates the molecule. Activated calmodulin can then interact with enzymes, altering their activities and thus producing diverse responses.

Still another hormonal mechanism uses *cyclic guanosine monophosphate* (cyclic GMP, or cGMP). Like cAMP, cGMP is a nucleotide derivative, and functions in much the same manner as a second messenger.

Cellular response to a steroid hormone is directly proportional to the number of hormone-receptor complexes that form. In contrast, response to a hormone operating through a second messenger is greatly amplified. This is possible because many second messenger molecules can be activated in response to just a few hormone-receptor complexes. Because of such amplification, cells are highly sensitive to changes in the concentrations of nonsteroid hormones.

|   |   |
|---|---|
| **1** | How does a steroid hormone act on its target cells? |
| **2** | How does a nonsteroid hormone act on its target cells? |
| **3** | What is a second messenger? |

## Prostaglandins

Prostaglandins are paracrine substances, acting locally, that are very potent and present in very small quantities. They are not stored in cells, but are synthesized just before they are released. They are rapidly inactivated.

Some prostaglandins regulate cellular responses to hormones. For example, different prostaglandins can either activate or inactivate adenylate cyclase in cell membranes, thereby controlling production of cAMP and altering the cell's response to a hormone.

Prostaglandins produce a variety of effects. Some prostaglandins can relax smooth muscle in the airways of the lungs and in the blood vessels, dilating these passageways. Yet other prostaglandins can contract smooth muscle in the walls of the uterus, causing menstrual cramps and labor contractions. They stimulate secretion of hormones from the adrenal cortex and inhibit secretion of hydrochloric acid from the wall of the stomach. Prostaglandins also influence movements of sodium ions and water in the kidneys, help regulate blood pressure, and have powerful effects on both male and female reproductive physiology. When tissues are injured, prostaglandins promote inflammation (see chapter 5).

Understanding prostaglandin function has medical applications. Drugs such as aspirin and certain steroids that relieve the joint pain of rheumatoid arthritis inhibit production of prostaglandins in the synovial fluid of affected joints. Daily doses of aspirin may reduce the risk of heart attack by altering prostaglandin activity. Prostaglandins may be used as drugs to dilate constricted blood vessels to relieve hypertension.

|   |   |
|---|---|
| **1** | What are prostaglandins? |
| **2** | Describe one possible function of prostaglandins. |
| **3** | What kinds of effects do prostaglandins produce? |

## CONTROL OF HORMONAL SECRETIONS

Hormones are continually excreted in the urine and broken down by various enzymes, primarily in the liver. Therefore, increasing or decreasing blood levels of a hormone requires increased or decreased secretion. Hormone secretion is precisely regulated.

## Control Sources

Hormone secretion must be precisely controlled so that these biochemicals can effectively maintain the internal environment. Control of hormone secretion occurs in three ways:

- The hypothalamus controls the anterior pituitary gland's release of **tropic hormones,** which then stimulate other endocrine glands to release hormones. The hypothalamus constantly receives information about the internal environment from neural connections and cerebrospinal fluid, made possible by its location near the thalamus and the third ventricle (fig. 13.7).

- Another group of glands responds directly to changes in the composition of the internal environment. For example, when the blood glucose level rises, the pancreas secretes insulin, and when the blood glucose level falls, it secretes glucagon, as we shall see later in the chapter.

- The nervous system stimulates some glands directly. The adrenal medulla, for example, secretes its hormones (epinephrine and norepinephrine) in response to preganglionic sympathetic nerve impulses. The secretory cells replace the postganglionic sympathetic neurons, which would normally secrete norepinephrine alone as a neurotransmitter.

## Negative Feedback Systems

Commonly, **negative feedback systems** control hormonal secretions (see chapter 1). In such a system, an endocrine gland or the system controlling it is sensitive to the concentration of a substance the gland secretes or to the concentration of a product from a process it controls. Whenever this concentration reaches a certain level, the endocrine gland is inhibited (a negative effect), and its secretory activity decreases (fig. 13.8). Then, as the concentration of the gland's hormone decreases, the concentration of the regulated substance or process decreases too, and the inhibition of the gland is diminished. When the gland is less inhibited, it increases secretion of its hormone again. Such negative feedback systems work to keep the concentrations of hormones relatively stable (fig. 13.9).

**1** How does the nervous system help regulate hormonal secretions?

**2** What is a feedback system?

**3** How does a negative feedback system control hormonal secretions?

## PITUITARY GLAND

The **pituitary gland** (hypophysis) is about 1 centimeter in diameter and is located at the base of the brain. It is attached to the hypothalamus by the pituitary stalk, or *infundibulum,* and lies in the sella turcica of the sphenoid bone, as figure 13.7 shows.

The pituitary gland consists of two distinct portions: an *anterior lobe* (adenohypophysis) and a *posterior lobe* (neurohypophysis). The anterior lobe secretes a number of hormones, including **growth hormone** (GH),

Anterior cerebral artery

Cerebral cortex

Optic nerve

Pituitary stalk

Anterior lobe of pituitary gland

Sphenoidal sinus

Sphenoid bone

Third ventricle

Hypothalamus

Optic chiasma

Oculomotor nerve

Trochlear nerve

Posterior lobe of pituitary gland

Basilar artery

Sella turcica

**FIGURE 13.7**

The pituitary gland is attached to the hypothalamus and lies in the sella turcica of the sphenoid bone.

thyroid-stimulating hormone (TSH), adrenocorticotropic hormone (ACTH), follicle-stimulating hormone (FSH), luteinizing hormone (LH), and prolactin (PRL). Although the cells of the posterior lobe (pituicytes) do not synthesize any hormones, two important hormones, antidiuretic hormone (ADH) and oxytocin (OT), are secreted by special neurons called neurosecretory cells whose nerve endings secrete into the bloodstream within the posterior lobe. The cell bodies of these neurosecretory cells are in the hypothalamus.

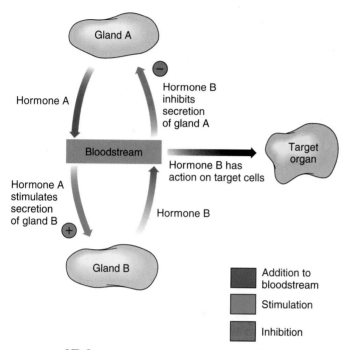

**FIGURE 13.8**

An example of a negative feedback system: (1) Gland A secretes a hormone that stimulates gland B to increase secretion of another hormone. (2) The hormone from gland B alters its target cells and inhibits activity of gland A.

During fetal development, a narrow region appears between the anterior and posterior lobes of the pituitary gland. Called the *intermediate lobe* (pars intermedia), this region produces melanocyte-stimulating hormone (MSH), which regulates the formation of melanin—the pigment found in the skin, and in portions of the eyes and brain. The region atrophies during prenatal development, and appears only as a vestige in adults.

The brain controls most of the pituitary gland's activities (fig. 13.10). For example, the pituitary gland's posterior lobe releases hormones into the bloodstream in response to nerve impulses from the hypothalamus. However, releasing hormones from the hypothalamus primarily control secretions from the anterior lobe. These releasing hormones are carried in the blood via a capillary net associated with the hypothalamus. The vessels merge to form the **hypophyseal portal veins** that pass downward along the pituitary stalk and give rise to a capillary net in the anterior lobe. Thus, substances released into the blood from the hypothalamus are carried directly to the anterior lobe. The hypothalamus, therefore, is an endocrine gland itself, yet it also controls other endocrine glands. This is also true of the anterior pituitary.

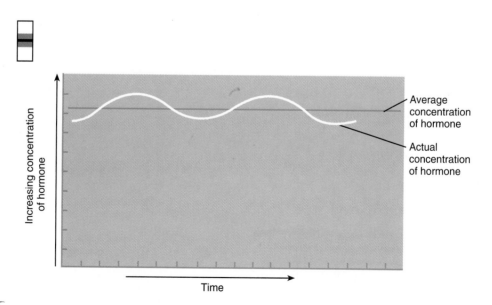

**FIGURE 13.9**

As a result of negative feedback, hormone concentrations remain relatively stable, although they may fluctuate slightly above and below average concentrations.

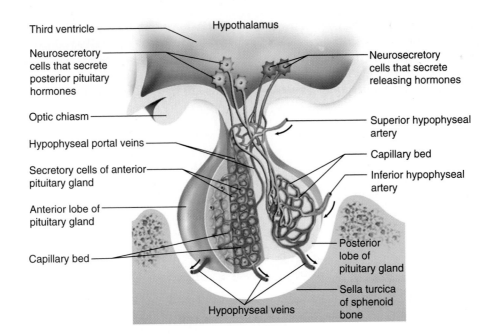

**FIGURE 13.10**

Hypothalamic releasing hormones stimulate cells of the anterior lobe to secrete hormones. Nerve impulses originating in the hypothalamus stimulate nerve endings in the posterior lobe of the pituitary gland to release hormones.

Upon reaching the anterior lobe of the pituitary, each of the hypothalamic releasing hormones acts on a specific population of cells in the anterior pituitary. Some of the resulting actions are inhibitory (prolactin inhibitory hormone and somatostatin), but most stimulate the anterior pituitary to release hormones that stimulate the secretions of peripheral endocrine glands. In many of these cases, important negative feedback relationships regulate hormone levels in the bloodstream. Figure 13.11 shows this general relationship.

The following sections discuss the hormones of the anterior pituitary as a group. The hormones of the peripheral endocrine glands will then be discussed individually.

**1** Where is the pituitary gland located?

**2** List the hormones that the anterior and posterior lobes of the pituitary gland secrete.

**3** Explain how the hypothalamus controls the actions of the pituitary gland.

## Anterior Pituitary Hormones

The anterior lobe of the pituitary gland is enclosed in a dense capsule of collagenous connective tissue and largely consists of epithelial tissue arranged in blocks around many thin-walled blood vessels. Within the epithelial tissue are five types of secretory cells. They are *somatotropes* that secrete GH, *mammatropes* that secrete PRL, *thyrotropes* that secrete TSH, *corticotropes* that secrete ACTH, and *gonadotropes* that secrete FSH and LH

(figs. 13.12 and 13.13). In males, LH (luteinizing hormone) is known as ICSH (interstitial cell-stimulating hormone) because it affects the interstitial cells of the testes (see chapter 22).

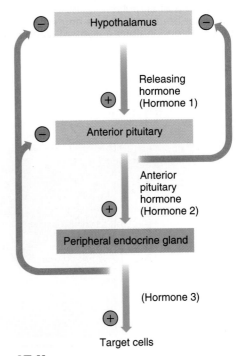

**FIGURE 13.11**

Hypothalamic control of the peripheral endocrine glands may utilize as many as three types of hormones, with multiple negative feedback controls, indicated by ⊖.

Secretory cells

**FIGURE 13.12**

Light micrograph of the anterior pituitary gland (100× micrograph enlarged to 240×).

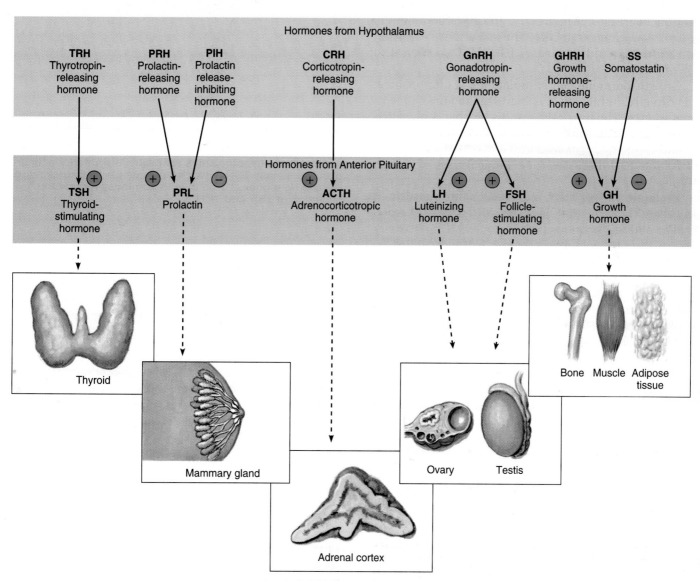

Hormones from Hypothalamus

| **TRH** Thyrotropin-releasing hormone | **PRH** Prolactin-releasing hormone | **PIH** Prolactin release-inhibiting hormone | **CRH** Corticotropin-releasing hormone | **GnRH** Gonadotropin-releasing hormone | **GHRH** Growth hormone-releasing hormone | **SS** Somatostatin |

Hormones from Anterior Pituitary

| **TSH** Thyroid-stimulating hormone | **PRL** Prolactin | **ACTH** Adrenocorticotropic hormone | **LH** Luteinizing hormone | **FSH** Follicle-stimulating hormone | **GH** Growth hormone |

Thyroid

Mammary gland

Adrenal cortex

Ovary    Testis

Bone   Muscle   Adipose tissue

**FIGURE 13.13**

Hormones released from the hypothalamus, the corresponding hormones released from the anterior lobe of the pituitary gland, and their target organs.

**Growth hormone,** which is also called *soma-totropin* (STH), is a protein that stimulates cells to increase in size and more rapidly reproduce. It enhances the movement of amino acids through the cell membranes and increases the rate of protein synthesis. GH also decreases the rate at which cells utilize carbohydrates and increases the rate at which they use fats.

Growth hormone is secreted in pulses, especially during sleep. Two biochemicals from the hypothalamus control its secretion. They are released alternately, exerting opposite effects. *Growth hormone-releasing hormone* (GHRH) stimulates secretion of GH, and *somatostatin* (SS) inhibits secretion.

Nutritional state seems to play a role in control of GH. For example, more GH is released during periods of protein deficiency and abnormally low blood glucose concentration. Conversely, when blood protein and glucose concentrations increase, growth hormone secretion decreases. Apparently, the hypothalamus is able to sense changes in the concentrations of certain blood nutrients, and it releases (GHRH) in response to some of them.

Growth hormone can stimulate elongation of bone tissue directly, but its effect on cartilage requires a mediator substance. The liver releases this biochemical, called *somatomedin* (insulinlike growth factor), in response to GH. Somatomedin promotes growth of cartilage. Clinical Application 13.2 discusses some clinical uses of growth hormone.

**Prolactin** is a protein, and as its name suggests, it promotes milk production. In males, PRL decreases secretion of luteinizing hormone (LH). Because LH is necessary for production of male sex hormones (androgens), excess prolactin secretion may cause a male to produce too few sex hormones and become infertile (see chapter 22).

Two biochemicals from the hypothalamus regulate prolactin secretion. One of these, *prolactin release-inhibiting hormone* (PIH), restrains secretion of prolactin, whereas the other, *prolactin-releasing hormone* (PRH), stimulates its secretion.

**Thyroid-stimulating hormone,** also called *thyrotropin,* is a glycoprotein. It controls secretion of certain hormones from the thyroid gland. TSH can also stimulate growth of the gland, and abnormally high TSH levels may lead to an enlarged thyroid gland, or *goiter.*

The hypothalamus partially regulates TSH secretion by producing thyrotropin-releasing hormone (TRH). Circulating thyroid hormones help regulate TSH secretion by inhibiting release of TRH and TSH; therefore, as the blood concentration of thyroid hormones increases, secretions of TRH and TSH decline (fig. 13.14).

External factors influence release of TRH and TSH. These include exposure to extreme cold, which is accompanied by increased hormonal secretion, and emotional stress, which sometimes increases hormonal secretion and other times decreases secretion.

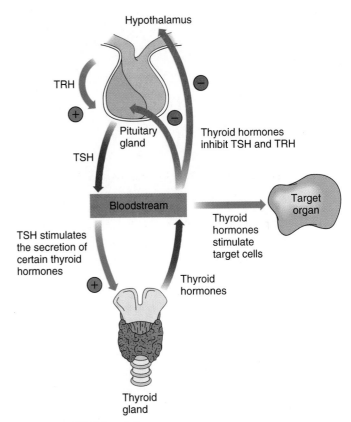

**FIGURE 13.14**

TRH from the hypothalamus stimulates the anterior pituitary gland to release TSH, which stimulates the thyroid gland to release hormones. These thyroid hormones reduce the secretion of TSH and TRH.

**1** How does growth hormone affect the cellular metabolism of carbohydrates, fats, and proteins?

**2** What are the functions of prolactin in females? In males?

**3** How is TSH secretion regulated?

**Adrenocorticotropic hormone** is a peptide that controls the manufacture and secretion of certain hormones from the outer layer (cortex) of the adrenal gland. The secretion of ACTH is regulated in part by *corticotropin-releasing hormone* (CRH), which is released from the hypothalamus in response to decreased concentrations of adrenal cortical hormones. Stress can increase secretion of ACTH by stimulating release of CRH.

Both **follicle-stimulating hormone** and **luteinizing hormone** are glycoproteins and are called *gonadotropins,* which means they exert their actions on the gonads or reproductive organs. FSH, for example, is responsible for growth and development of egg-cell–containing follicles in the ovaries. It also stimulates the follicular cells to secrete a group of female sex hormones, collectively called *estrogen.*

# Growth Hormone Ups and Downs

Insufficient secretion of growth hormone during childhood produces *hypopituitary dwarfism.* Body proportions and mental development are normal, but because secretions of other anterior pituitary hormones are also below normal amounts, additional hormone deficiency symptoms may appear. For example, a child with this condition often fails to develop adult sexual features unless he or she receives hormone therapy.

Human growth hormone, manufactured using recombinant DNA technology, is a valuable drug in treating pituitary dwarfism, although treatment must begin before the bones completely ossify. The hormone also has some controversial uses. Some people want to use it to increase height in children who are short, but not abnormally so. A few years ago, growth hormone was given experimentally to older individuals to see if it would slow aging. Although muscle tone improved and the participants reported feeling well, side effects arose, and the value of such treatment was not confirmed. Bovine growth hormone is given to dairy cows to increase their milk production.

Oversecretion of growth hormone in childhood may result in *gigantism,* in which height may eventually exceed 8 feet. Gigantism is usually accompanied by a tumor of the pituitary gland, which secretes excess pituitary hormones and GH. As a result, a giant often suffers from other metabolic disturbances.

If growth hormone is oversecreted in an adult after the epiphyses of the long bones have ossified, the person does not grow taller. The soft tissues, however, continue to enlarge and the bones thicken, producing a large tongue, nose, hands and feet, and a protruding jaw. This condition, *acromegaly,* is also often associated with a pituitary tumor (fig. 13B). ■

(a)

(b)

(c)

(d)

**FIGURE 13B**
Oversecretion of growth hormone in adulthood causes acromegaly. Note the changes in this woman's facial features at ages (*a*) 9, (*b*) 16, (*c*) 33, and (*d*) 52.

In males, FSH stimulates the initial production of sperm cells in the testes at puberty. LH promotes secretion of sex hormones in both males and females and is essential for release of egg cells from the ovaries. Other functions of the gonadotropins and their interactions are discussed in chapter 22.

The mechanism that regulates secretion of gonadotropins is not well understood. However, it is known that the hypothalamus secretes a *gonadotropin-releasing hormone* (GnRH). The hypothalamus does not secrete this hormone in significant amounts until puberty. We know this because gonadotropins are virtually absent in the body fluids of infants and children.

1. What is the function of ACTH?

2. Describe the functions of FSH and LH in a male and in a female.

3. What is a gonadotropin?

## Posterior Pituitary Hormones

Unlike the anterior lobe of the pituitary gland, which is primarily composed of glandular epithelial cells, the posterior lobe largely consists of nerve fibers and neuroglial cells (*pituicytes*). The neuroglial cells support the nerve fibers that originate in the hypothalamus. The hypothalamic cells that give rise to these fibers are called **neurosecretory cells** because their secretions function not as neurotransmitters but as hormones.

Specialized neurons in the hypothalamus produce the two hormones associated with the posterior pituitary—**antidiuretic hormone** and **oxytocin** (see fig. 13.10). These hormones travel down axons through the pituitary stalk to the posterior pituitary and are stored in vesicles (secretory granules) near the ends of the axons. The hormones are released into the blood in response to nerve impulses coming from the hypothalamus.

Antidiuretic hormone and oxytocin are short polypeptides with similar sequences (fig. 13.15). A *diuretic* is a substance that increases urine production. An *antidiuretic,* then, is a chemical that decreases urine formation. ADH produces its antidiuretic effect by causing the kidneys to reduce the amount of water they excrete. In this way, ADH plays an important role in regulating the concentration of body fluids (see chapter 20).

> Drinking alcohol is often followed by frequent and copious urination. This is because alcohol (ethyl alcohol) inhibits ADH secretion. A person must replace the lost body fluid to maintain normal water balance. Although it seems counterintuitive, drinking too much beer can lead to dehydration because the body loses more fluid than it replaces.

ADH present in sufficient concentrations contracts certain smooth muscles, including those in the walls of blood vessels. As a result, vascular resistance and blood pressure may increase. For this reason, ADH is also called *vasopressin.* Although ADH is seldom present in quantities sufficient to cause high blood pressure, its secretion increases following severe blood loss. Blood pressure may drop as a consequence of profuse bleeding.

Oxytocin

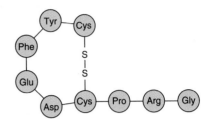

Antidiuretic hormone

**FIGURE 13.15**

The structure of oxytocin differs from that of ADH by only two amino acids, yet they function differently.

In this situation, ADH's vasopressor effect may help to minimize the drop and return blood pressure toward normal.

> ADH's two effects—vasoconstriction and water retention—are possible because the hormone binds two different receptors on target cells. V1 receptors, when ADH binds, increase the concentration of the second messenger inositol triphosphate, which increases calcium ion concentration, leading to vasoconstriction. The second receptor, V2, is found on parts of the kidneys' microscopic tubules called collecting ducts. There ADH binding activates the cAMP second messenger system, which ultimately causes collecting duct cells to reabsorb water that would otherwise be excreted as urine.

The hypothalamus regulates secretion of ADH. Certain neurons in this part of the brain, called *osmoreceptors,* sense changes in the concentration of body fluids. For example, if a person is dehydrating due to a lack of water intake, the solutes in blood become more and more concentrated. The osmoreceptors, sensing the resulting increase in osmotic pressure, signal the posterior pituitary to release ADH, which causes the kidneys to retain water.

On the other hand, if a person drinks a large amount of water, body fluids become more dilute, inhibiting the release of ADH. Consequently, the kidneys excrete more dilute urine until the concentration of body fluids returns to normal.

Blood volume also affects ADH secretion. Increased blood volume stretches the walls of certain blood vessels, stimulating volume receptors that signal the hypothalamus to inhibit release of ADH. However, if hemorrhage decreases blood volume, these receptors are stretched less, and therefore send fewer inhibiting impulses. As a result, ADH secretion increases, and as before, ADH causes the kidneys to conserve water. This helps prevent further volume loss.

Oxytocin also has an antidiuretic action, but less so than ADH. In addition, oxytocin can contract smooth muscles in the uterine wall, playing a role in the later stages of childbirth. The uterus becomes more and more sensitive to oxytocin's effects during pregnancy. Stretching of uterine and vaginal tissues late in pregnancy, caused by the growing fetus, initiates nerve impulses to the hypothalamus, which then signals the posterior pituitary to release oxytocin, which in turn stimulates the uterine contractions of labor.

> A baby first displayed symptoms at 5 months of age—he drank huge amounts of water. By 13 months, he had become severely dehydrated, although he drank nearly continuously. His parents were constantly changing his wet diapers. Doctors finally diagnosed a form of the condition *diabetes insipidus*, which impairs ADH regulation of water balance. The boy was drinking sufficient fluids, but his kidneys could not absorb them. ADH V2 receptors on the kidney tubules' collecting ducts were defective. The hormone binds, but the receptor fails to trigger cAMP formation. The boy's ADH was still able to constrict blood vessels because the V1 receptors were unaffected. A high-calorie diet and providing lots of water preserved the boy's mental abilities, but he remained small for his age. Tumors and injury affecting the hypothalamus and posterior pituitary can also cause diabetes insipidus.

In the breasts, oxytocin contracts certain cells near the milk-producing glands and their ducts. In lactating breasts, this action forces liquid from the milk glands into the milk ducts and ejects the milk from the breasts.

The mechanical stimulation provided by suckling the nipple initiates nerve impulses that travel to the mother's hypothalamus. The hypothalamus responds by signaling the posterior pituitary to release oxytocin, which in turn stimulates milk release. Thus, milk is normally not ejected from the milk glands and ducts until the baby sucks. Oxytocin has no known function in males, although it is present in the male posterior pituitary. Table 13.5 reviews the hormones of the pituitary gland.

> If the uterus is not sufficiently contracting to expel a fully developed fetus, oxytocin is sometimes given intravenously to stimulate uterine contractions, thus inducing labor. Oxytocin is also administered to the mother following childbirth to ensure that the uterine muscles contract enough to squeeze broken blood vessels closed, minimizing the danger of hemorrhage.

1. What is the function of ADH?
2. How is the secretion of ADH controlled?
3. What effects does oxytocin produce in females?

## THYROID GLAND

The **thyroid gland,** as figure 13.16 shows, is a very vascular structure that consists of two large lateral lobes connected by a broad isthmus. It is located just below the larynx on either side and anterior to the trachea. It has a special ability to remove iodine from the blood.

### Structure of the Gland

A capsule of connective tissue covers the thyroid gland, which is made up of many secretory parts called *follicles.* The cavities within these follicles are lined with a single layer of cuboidal epithelial cells, and are filled with a clear, viscous glycoprotein called *colloid.* The follicular cells produce and secrete hormones that may either be stored in the colloid or released into the blood of nearby capillaries (fig. 13.17). Other hormone-secreting cells, called *extrafollicular cells* (C cells), lie outside the follicles.

### Thyroid Hormones

The thyroid gland produces three important hormones. The follicular cells synthesize two of these, which have marked effects on the metabolic rates of body cells. The extrafollicular cells produce the third type of hormone, which influences blood concentrations of calcium and phosphate ions.

The two important thyroid hormones that affect cellular metabolic rates are **thyroxine,** or tetraiodothyronine (also called $T_4$, because it includes four atoms of iodine), and **triiodothyronine** (also called $T_3$, because it includes three atoms of iodine). These hormones help regulate the metabolism of carbohydrates, lipids, and proteins. Specifically, thyroxine and triiodothyronine increase the rate at which cells release energy from carbohydrates, enhance the rate of protein synthesis, and stimulate breakdown and mobilization of lipids. These hormones are essential for normal growth and development and for maturation of the nervous system (fig. 13.18). TSH from the anterior pituitary gland controls them.

TABLE 13.5 HORMONES OF THE PITUITARY GLAND

## Anterior Lobe

| Hormone | Action | Source of Control |
|---|---|---|
| Growth hormone (GH) | Stimulates increase in size and rate of reproduction of body cells; enhances movement of amino acids through membranes; promotes growth of long bones | Secretion inhibited by somatostatin (SS) and stimulated by growth hormone-releasing hormone (GHRH) from the hypothalamus |
| Prolactin (PRL) | Sustains milk production after birth; amplifies effect of LH in males | Secretion inhibited by prolactin release-inhibiting hormone (PIH) and stimulated by prolactin-releasing hormone (PRH) from the hypothalamus |
| Thyroid-stimulating hormone (TSH) | Controls secretion of hormones from the thyroid gland | Thyrotropin-releasing hormone (TRH) from the hypothalamus |
| Adrenocorticotropic hormone (ACTH) | Controls secretion of certain hormones from the adrenal cortex | Corticotropin-releasing hormone (CRH) from the hypothalamus |
| Follicle-stimulating hormone (FSH) | Development of egg-containing follicles in ovaries; stimulates follicular cells to secrete estrogen; in males, stimulates production of sperm cells | Gonadotropin-releasing hormone (GnRH) from the hypothalamus |
| Luteinizing hormone (LH or ICSH in males) | Promotes secretion of sex hormones; releases egg cell in females | Gonadotropin-releasing hormone (GnRH) from the hypothalamus |

## Posterior Lobe

| Hormone | Action | Source of Control |
|---|---|---|
| Antidiuretic hormone (ADH) | Causes kidneys to reduce water excretion; in high concentration, raises blood pressure | Hypothalamus in response to changes in blood water concentration and blood volume |
| Oxytocin (OT) | Contracts muscles in uterine wall and those associated with milk-secreting glands | Hypothalamus in response to stretch in uterine and vaginal walls, and stimulation of breasts |

(a) **Anterior view**

**FIGURE 13.16**

(a) The thyroid gland consists of two lobes connected anteriorly by an isthmus. (b) Follicular cells secrete thyroid hormones.

**FIGURE 13.17**

A light micrograph of thyroid gland tissue (40× micrograph enlarged to 240×). The open spaces that follicular cells surround are filled with colloid.

Follicular cells

Colloid

Extrafollicular cells

Thyroxine (T₄)          Triiodothyronine (T₃)

**FIGURE 13.18**

The hormones thyroxine and triiodothyronine have very similar molecular structures.

Follicular cells require iodine salts (iodides) to produce thyroxine and triiodothyronine. Such salts are normally obtained from foods, and after they have been absorbed from the intestine, the blood carries some of them to the thyroid gland. An efficient active transport mechanism called the *iodine pump* moves the iodides into the follicular cells, where they are concentrated. The iodides, together with an amino acid (tyrosine), are used to synthesize these thyroid hormones.

Follicular cells also secrete a protein called *thyroglobulin,* which is the main ingredient of thyroid colloid. Thyroglobulin stores thyroid hormones whenever they are produced in excess. The stored hormones are bound to the thyroglobulin until the hormone concentration of the body fluids drops below a certain level. Then enzymes release the hormones from the colloid, and they diffuse into the blood. Once in the blood, thyroid hormones combine with blood proteins (alpha globulins) and are transported to body cells. Triiodothyronine is nearly five times more potent, but thyroxine accounts for at least 95% of circulating thyroid hormones.

The thyroid gland produces **calcitonin,** but it is sometimes not considered to be a thyroid hormone because it is synthesized by the C cells, which are distinct from the gland's follicles. Calcitonin plays a role in the control of blood calcium and phosphate ion concentrations. It helps lower concentrations of calcium and phosphate ions by inhibiting the rate at which they leave the bones and enter extracellular fluids by action of osteoclasts (see chapter 7). At the same time, calcitonin increases the rate at which calcium and phosphate ions are deposited in bone matrix by stimulating activity of osteoblasts. It also increases the excretion of calcium ions and phosphate ions by the kidneys.

Calcitonin secretion is stimulated by a high blood calcium ion concentration, as may occur following absorption of calcium ions from a recent meal. Certain hormones also prompt its secretion, such as gastrin, which is released from active digestive organs. Calcitonin helps prevent prolonged elevation of blood calcium ion concentration after eating.

Table 13.6 summarizes the actions and sources of control of the thyroid hormones. Clinical Application 2.1, table 13.7, and figures 13.19, 13.20, and 13.21 discuss disorders of the thyroid gland.

**1** Where is the thyroid gland located?

**2** What hormones of the thyroid gland affect carbohydrate metabolism and protein synthesis?

**3** What substance is essential for the production of thyroxine and triiodothyronine?

**4** How does calcitonin influence the concentrations of blood calcium and phosphate ions?

On a spring morning in 1991, U.S. President George Bush set out on his daily run. After only a few hundred yards, he stopped, winded and experiencing an irregular heartbeat. Over the preceding two months, his wife, Barbara, had inexplicably lost 18 pounds, and her eyesight had become blurry and her eyes dry. Both Bushes were feeling unusually fidgety. Blood tests found that they had hyperthyroidism—and so did their dog, Millie. Because of the three cases, the White House environment was carefully studied, but no cause was ever identified.

## TABLE 13.6  HORMONES OF THE THYROID GLAND

| Hormone | Action | Source of Control |
|---|---|---|
| Thyroxine ($T_4$) | Increases rate of energy release from carbohydrates; increases rate of protein synthesis; accelerates growth; stimulates activity in the nervous system | TSH from the anterior pituitary gland |
| Triiodothyronine ($T_3$) | Same as above, but five times more potent than thyroxine | Same as above |
| Calcitonin | Lowers blood calcium and phosphate ion concentrations by inhibiting release of calcium and phosphate ions from bones, and by increasing the rate at which calcium and phosphate ions are deposited in bones | Elevated blood calcium ion concentration, digestive hormones |

## TABLE 13.7  DISORDERS OF THE THYROID GLAND

| Condition | Mechanism/Symptoms |
|---|---|
| *Hyperthyroid* | |
| Grave's disease | Autoantibodies (against self) bind TSH receptors on thyroid cell membranes, mimicking action of TSH, overstimulating gland (hyperthyroidism); this is an exothalmic goiter |
| Hyperthyroidism | High metabolic rate, sensitivity to heat, restlessness, hyperactivity, weight loss, protruding eyes, goiter |
| *Hypothyroid* | |
| Hashimoto's disease | Autoantibodies (against self) attack thyroid cells, producing hypothyroidism |
| Hypothyroidism (infantile) | Cretinism—stunted growth, abnormal bone formation, mental retardation, low body temperature, sluggishness |
| Hypothyroidism (adult) | Myxedema—low metabolic rate, sensitivity to cold, sluggishness, poor appetite, swollen tissues, mental dullness |
| Simple goiter | Deficiency of thyroid hormones due to iodine deficiency; because no thyroid hormones inhibit pituitary release of TSH, thyroid is overstimulated and enlarges, but functions below normal (hypothyroidism) |

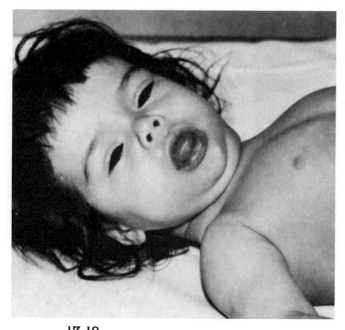

**FIGURE 13.19**

Cretinism is due to an underactive thyroid gland during infancy and childhood.

**FIGURE 13.20**

Hyperthyroidism may cause the eyes to protrude.

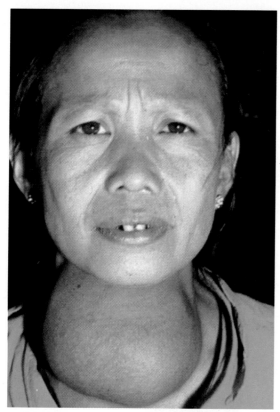

**FIGURE 13.21**

An iodine deficiency causes simple (endemic) goiter, and results in high levels of TSH.

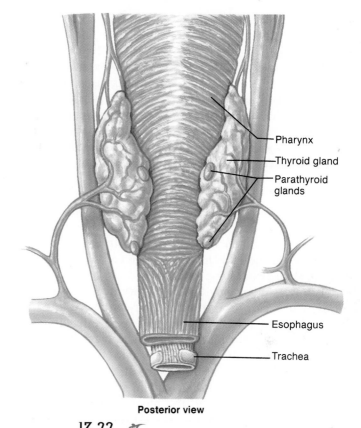

**Posterior view**

**FIGURE 13.22**

The parathyroid glands are embedded in the posterior surface of the thyroid gland.

## PARATHYROID GLANDS

The **parathyroid glands** are located on the posterior surface of the thyroid gland, as figure 13.22 shows. Usually there are four of them—a superior and an inferior gland associated with each of the thyroid's lateral lobes. The parathyroid glands secrete a hormone that regulates the concentrations of calcium and phosphate ions in the blood.

### Structure of the Glands

Each parathyroid gland is a small, yellowish brown structure covered by a thin capsule of connective tissue. The body of the gland consists of many tightly packed secretory cells that are closely associated with capillary networks (fig. 13.23).

### Parathyroid Hormone

The parathyroid glands secrete a protein, **parathyroid hormone** (PTH) or *parathormone* (see fig. 13.3), which increases blood calcium ion concentration and decreases blood phosphate ion concentration through actions in the bones, kidneys, and intestines.

**FIGURE 13.23**

Light micrograph of the parathyroid gland (540×).

The intercellular matrix of bone tissue contains a considerable amount of calcium phosphate and calcium carbonate. PTH stimulates bone resorption by osteocytes and osteoclasts, and inhibits the activity of osteoblasts (see chapter 7). As bone resorption increases, calcium and phosphate ions are released into the blood. At the same time, PTH causes the kidneys to conserve blood calcium ions and to excrete more phosphate ions in the urine. It also indirectly stimulates absorption of calcium ions from food in the intestine by influencing metabolism of vitamin D.

Vitamin D (cholecalciferol) is synthesized from dietary cholesterol, which intestinal enzymes convert into provitamin D (7-dehydrocholesterol). This provitamin is largely stored in the skin, and exposure to the ultraviolet wavelengths of sunlight changes it to vitamin D. Vitamin D is also obtained from some foods.

The liver changes vitamin D to hydroxycholecalciferol, which is carried in the bloodstream or is stored in tissues. When PTH is present, hydroxycholecalciferol can be changed in the kidneys into an active form of vitamin D (dihydroxycholecalciferol), which controls absorption of calcium ions from the intestine (fig. 13.24).

A negative feedback mechanism operating between the parathyroid glands and the blood calcium ion concentration regulates secretion of PTH (fig. 13.25). As the concentration of blood calcium ions rises, less PTH is secreted; as the concentration of blood calcium ions drops, more PTH is released.

Cancer cells in nonendocrine tissues sometimes secrete hormones, producing symptoms of an *endocrine paraneoplastic syndrome*. A patient with this unusual syndrome in 1994 led to discovery of a new hormone. In a middle-aged woman with oncogenic osteomalacia, a small tumor in the soft tissue of her thigh disrupted phosphate homeostasis, causing her bones to demineralize, leading to her symptoms of bone aching. To the surprise of medical researchers, the tumor was not producing parathyroid hormone but a new hormone, tentatively named phosphaturic factor, which increases excretion of phosphate in urine, weakening the bones. The woman recovered when her tumor was removed, but it grew back and symptoms returned.

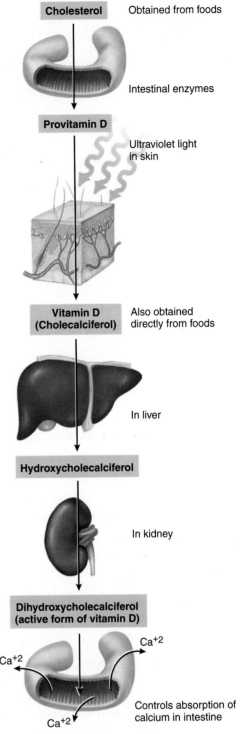

**FIGURE 13.24**

Mechanism by which PTH promotes calcium absorption in the intestine.

The opposite effects of calcitonin and PTH maintain calcium ion homeostasis. This is important in a number of physiological processes. For example, as the blood calcium ion concentration drops (hypocalcemia), the nervous system becomes abnormally excitable, and impulses may be triggered spontaneously. As a result, muscles, including the respiratory muscles, may undergo tetanic contractions, and the person may die due to a failure to breathe. An abnormally high concentration of blood calcium ions (hypercalcemia) depresses the nervous system. Consequently, muscle contractions are weak and reflexes are sluggish. Table 13.8 lists parathyroid disorders.

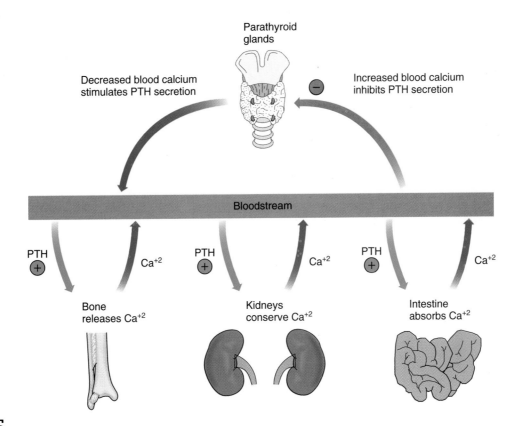

**FIGURE** 13.25

Parathyroid hormone stimulates bone to release calcium and the kidneys to conserve calcium. It indirectly stimulates the intestine to absorb calcium. The resulting increase in blood calcium concentration inhibits secretion of parathyroid hormone.

| TABLE 13.8 | DISORDERS OF THE PARATHYROID GLANDS | | |
|---|---|---|---|
| **Condition** | **Symptoms/Mechanism** | **Cause** | **Treatment** |
| Hyperparathyroidism | Fatigue, muscular weakness, painful joints, altered mental functions, depression, weight loss, bone weakening. Increased PTH secretion overstimulates osteoclasts. | Tumor | Remove tumor, correct bone deformities |
| Hypoparathyroidism | Muscle cramps and seizures. Decreased PTH secretion reduces osteoclast activity, diminishing blood calcium ion concentration. | Inadvertent surgical removal; injury | Calcium salt injections, massive doses of vitamin D |

1  Where are the parathyroid glands located?

2  How does parathyroid hormone help regulate the concentrations of blood calcium and phosphate ions?

3  How does the negative feedback system of the parathyroid glands differ from that of the thyroid gland?

## ADRENAL GLANDS

The **adrenal glands** (suprarenal glands) are closely associated with the kidneys. A gland sits atop each kidney like a cap and is embedded in the mass of adipose tissue that encloses the kidney.

## Structure of the Glands

The adrenal glands are shaped like pyramids. Each adrenal gland is very vascular and consists of two parts (fig. 13.26). The central portion is the adrenal medulla, and the outer part is the adrenal cortex. These regions are not sharply divided, but they are distinct glands that secrete different hormones.

The **adrenal medulla** consists of irregularly shaped cells grouped around blood vessels. These cells are intimately connected with the sympathetic division of the autonomic nervous system. In fact, these adrenal medullary cells are modified postganglionic neurons, and preganglionic autonomic nerve fibers lead to them directly from the central nervous system (see chapter 11).

**FIGURE** 13.26

(a) An adrenal gland consists of an outer cortex and an inner medulla. (b) The cortex has three layers, or zones, of cells.

**FIGURE** 13.27

Light micrograph of the adrenal medulla and the adrenal cortex (75×).

Norepinephrine

Epinephrine

**FIGURE** 13.28

Epinephrine and norepinephrine have similar molecular structures and similar functions.

The **adrenal cortex** makes up the bulk of the adrenal gland. It is composed of closely packed masses of epithelial layers that form an outer, a middle, and an inner zone of the cortex—the zona glomerulosa, the zona fasciculata, and the zona reticularis, respectively (fig. 13.27).

## Hormones of the Adrenal Medulla

The cells of the adrenal medulla (chromaffin cells) produce, store, and secrete two closely related hormones, **epinephrine** (adrenalin) and **norepinephrine** (noradrena-

lin). Both of these substances are a type of amine called a *catecholamine*, and they have similar molecular structures and physiological functions (fig. 13.28). In fact, epinephrine is synthesized from norepinephrine.

The synthesis of these hormones begins with the amino acid *tyrosine*. In the first step of the process, an enzyme (tyrosine hydroxylase) in the secretory cells converts tyrosine into a substance called *dopa*. A second enzyme (dopa decarboxylase) converts dopa to dopamine, and a third enzyme (dopamine betahydroxylase) converts

TABLE 13.9 COMPARATIVE EFFECTS OF EPINEPHRINE AND NOREPINEPHRINE

| Structure or Function Affected | Epinephrine | Norepinephrine |
|---|---|---|
| Heart | Rate increases | Rate increases |
| | Force of contraction increases | Force of contraction increases |
| Blood vessels | Vessels in skeletal muscle vasodilate, decreasing resistance to blood flow | Blood flow to skeletal muscles increases, resulting from constriction of blood vessels in skin and viscera |
| Systemic blood pressure | Some increase due to increased cardiac output | Great increase due to vasoconstriction, counteracted in muscle blood vessels during exercise |
| Airways | Dilated | Some dilation |
| Reticular formation of brain | Activated | Little effect |
| Liver | Promotes breakdown of glycogen to glucose, increasing blood sugar level | Little effect on blood sugar |
| Metabolic rate | Increases | Increases |

dopamine to norepinephrine. Still another enzyme (phenylethanolamine N-methyltransferase) converts the norepinephrine to epinephrine. About 15% of the norepinephrine is not converted but is stored unchanged. The hormones are stored in tiny vesicles (chromaffin granules), much like neurotransmitters are stored in vesicles in neurons.

The effects of the adrenal medullary hormones generally resemble those that result when sympathetic nerve fibers stimulate their effectors. These effects include increased heart rate and increased force of cardiac muscle contraction, elevated blood pressure, increased breathing rate, and decreased activity in the digestive system (see table 11.10). The hormonal effects last up to ten times longer than the neurotransmitter effects because the hormones are removed from the tissues relatively slowly.

The ratio of the two hormones in the adrenal medullary secretion varies with different physiological conditions, but usually the secretion is about 80% epinephrine and 20% norepinephrine. Although these hormones' effects are generally similar, certain effectors respond differently to them. These differences are due to the relative numbers of alpha and beta receptors in the membranes of the effector cells. As described in chapter 11, both hormones can stimulate both classes of receptors, although norepinephrine has a greater effect on alpha receptors. Table 13.9 compares some of the differences in the effects of these hormones.

Impulses arriving on sympathetic nerve fibers stimulate the adrenal medulla to release its hormones at the same time sympathetic impulses stimulate other effectors. As a rule, these impulses originate in the hypothalamus in response to stress. Thus, the medullary secretions function together with the sympathetic division of the autonomic nervous system in preparing the body for energy-expending action—"fight or flight."

1. Describe the location and structure of the adrenal glands.

2. Name the hormones the adrenal medulla secretes.

3. What general effects do hormones secreted by the adrenal medulla produce?

4. What usually stimulates release of hormones from the adrenal medulla?

## Hormones of the Adrenal Cortex

The cells of the adrenal cortex produce more than thirty different steroids, including several hormones (corticosteroids). Unlike the adrenal medullary hormones, without which a person can survive, some of those released by the cortex are vital. In fact, in the absence of adrenal cortical secretions, a person usually dies within a week if he or she does not receive extensive electrolyte therapy. The most important adrenal cortical hormones are aldosterone, cortisol, and certain sex hormones.

### Aldosterone

Cells in the outer zone (zona glomerulosa) of the adrenal cortex synthesize **aldosterone.** This hormone is called a *mineralocorticoid* because it helps regulate the concentration of mineral electrolytes, such as sodium and potassium ions. More specifically, aldosterone causes the kidney to conserve sodium ions and to excrete potassium ions. The cells that secrete aldosterone respond directly to changes in the composition of blood plasma. However, whereas an increase in plasma potassium strongly stimulates these cells, a decrease in plasma sodium only slightly stimulates them. Control of aldosterone secretion is indirectly linked to plasma sodium level by the **renin-angiotensin system.**

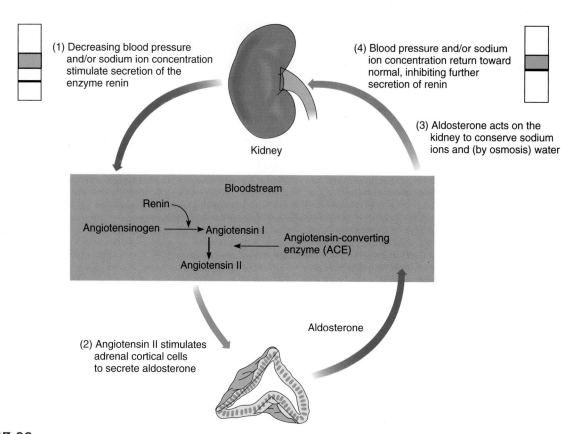

(1) Decreasing blood pressure and/or sodium ion concentration stimulate secretion of the enzyme renin

(4) Blood pressure and/or sodium ion concentration return toward normal, inhibiting further secretion of renin

Kidney

(3) Aldosterone acts on the kidney to conserve sodium ions and (by osmosis) water

Bloodstream

Renin

Angiotensinogen ⟶ Angiotensin I

Angiotensin-converting enzyme (ACE)

Angiotensin II

Aldosterone

(2) Angiotensin II stimulates adrenal cortical cells to secrete aldosterone

**FIGURE 13.29**

Aldosterone increases blood volume and pressure by promoting conservation of sodium ions and water (steps 1–4).

Groups of specialized kidney cells (juxtaglomerular cells) are able to respond to changes in blood pressure and the plasma sodium ion concentration. If either of these factors decreases, the cells release an enzyme called **renin.** Renin decomposes a blood protein called **angiotensinogen,** which releases a peptide called **angiotensin I.** Another enzyme (angiotensin-converting enzyme, or ACE) in the lungs converts angiotensin I into another form, **angiotensin II,** which is carried in the bloodstream (fig. 13.29). When angiotensin II reaches the adrenal cortex, it stimulates the release of aldosterone. ACTH also stimulates secretion of aldosterone, and is necessary for aldosterone secretion to respond to other stimuli.

Aldosterone, in conserving sodium ions, indirectly retains water by osmosis. This helps maintain blood sodium ion concentration and blood volume. Angiotensin II helps maintain systemic blood pressure by constricting blood vessels (fig. 13.29).

ACE inhibitors are a class of drugs used to treat some forms of high blood pressure (hypertension). They work by competing with angiotensin-converting enzyme, blocking formation of angiotensin II and preventing inactivation of a substance called bradykinin, a vasodilator. Both effects lead to dilation of blood vessels, lowering blood pressure.

### Cortisol

**Cortisol** (hydrocortisone) is a *glucocorticoid,* which means it affects glucose metabolism. It is produced in the middle zone (zona fasciculata) of the adrenal cortex and has a molecular structure similar to aldosterone (fig. 13.30). In addition to affecting glucose, cortisol influences protein and fat metabolism. Among the more important actions of cortisol are the following:

1. It inhibits the synthesis of protein in various tissues, increasing blood concentration of amino acids.

2. It promotes the release of fatty acids from adipose tissue, increasing the use of fatty acids as an energy source and decreasing the use of glucose as an energy source.

3. It stimulates liver cells to synthesize glucose from noncarbohydrates (gluconeogenesis), such as circulating amino acids and glycerol, thus increasing blood glucose concentration.

Cortisol's actions help keep the blood glucose concentration within the normal range between meals. These actions are important because just a few hours without food can exhaust liver glycogen, another major source of glucose.

A negative feedback mechanism much like that controlling the thyroid hormones $T_3$ and $T_4$ regulates cortisol

Cortisol

Aldosterone

**FIGURE 13.30**

Cortisol and aldosterone are steroids with similar molecular structures.

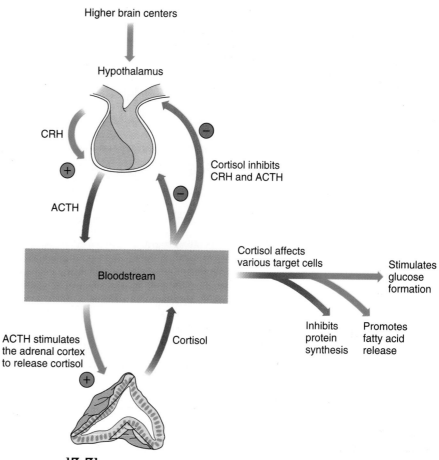

**FIGURE 13.31**

Negative feedback regulates cortisol secretion, similar to the regulation of thyroid hormone secretion (see fig. 13.14).

release. It involves the hypothalamus, anterior pituitary gland, and adrenal cortex. The hypothalamus secretes CRH (corticotropin-releasing hormone) into the hypophyseal portal veins, which carry the CRH to the anterior pituitary gland, stimulating it to secrete ACTH. In turn, ACTH stimulates the adrenal cortex to release cortisol. Cortisol inhibits release of both CRH and ACTH. As concentration of these substances falls, cortisol production drops.

---

Cortisol and related compounds are often used as drugs to reduce inflammation. They relieve pain by:

- decreasing permeability of capillaries, preventing leakage of fluids that swell surrounding tissues
- stabilizing lysosomal membranes, preventing release of their enzymes, which destroy tissue
- inhibiting prostaglandin synthesis

Unfortunately, the concentration of cortisol compounds necessary to stifle inflammation is toxic, so these drugs can only be used for a short time. They are used to treat autoimmune disorders, allergies, asthma, and patients who have received organ transplants or tissue grafts.

---

The set point of the feedback loop controlling cortisol secretion changes from time to time, enabling hormone output to alter to meet the demands of changing conditions. For example, under stress—injury, disease, extreme temperature, or emotional upset—nerve impulses send the brain information concerning the stressful condition. In response, brain centers signal the hypothalamus to release more CRH, leading to a higher concentration of cortisol until the stress subsides (fig. 13.31).

### Sex Hormones

Cells in the inner zone (zona reticularis) of the adrenal cortex produce sex hormones. These hormones are male (adrenal androgens), but some of them are converted into female hormones (estrogen) by the skin, liver, and adipose tissues. These hormones may supplement the supply of sex hormones from the gonads and stimulate early development of the reproductive organs. Also, adrenal androgens may play a role in controlling the female sex drive. Table 13.10 summarizes the actions of the cortical hormones. Clinical Application 13.3 discusses some of the effects of a malfunctioning adrenal gland on health.

# Disorders of the Adrenal Cortex

John F. Kennedy's beautiful bronze complexion may have resulted not from sunbathing, but from a disorder of the adrenal glands. When he ran for the presidency in 1960, Kennedy knew he had *Addison's disease,* but his staff kept his secret, for fear it would affect his career. Kennedy had almost no adrenal tissue but was able to function by receiving mineralocorticoids and glucocorticoids, the standard treatment.

In Addison's disease, the adrenal cortex does not secrete hormones sufficiently, due to immune system attack (autoimmunity) or an infection such as tuberculosis. Signs and symptoms include decreased blood sodium, increased blood potassium, low blood glucose level (hypoglycemia), dehydration, low blood pressure, frequent infections, fatigue, nausea and vomiting, loss of appetite, and increased skin pigmentation. Some sufferers experience salt cravings—one woman reported eating bowls and bowls of salty chicken noodle soup, with pickles and briny pickle juice added! Without treatment, death comes within days from severe disturbances in electrolyte balance.

An adrenal tumor or oversecretion of ACTH by the anterior pituitary causes hypersecretion of glucocorticoids (primarily cortisol), resulting in *Cushing's syndrome.* It may also result from taking corticosteroid drugs for many years, such as to treat asthma or rheumatoid arthritis. Tissue protein level plummets, due to muscle wasting and loss of bone tissue. Blood glucose level remains elevated, and excess sodium is retained. As a result, tissue fluid increases, blood pressure rises, and the skin appears puffy and thin, and may bruise easily.

Adipose tissue deposited in the face and back produce a characteristic "moon face" and "buffalo hump." Increase in adrenal sex hormone secretion may masculinize a female, causing growth of facial hair and a deepening voice. Other symptoms include extreme fatigue, sleep disturbances, skin rashes, headache, and leg muscle cramps.

Treatment of Cushing's syndrome attempts to reduce ACTH secretion. This may entail removing a tumor in the pituitary gland or partially or completely removing the adrenal glands.

Both Addison's disease and Cushing's syndrome are rare, and for this reason, they are often misdiagnosed, or, in early stages, the patient's report of symptoms is not taken seriously. Addison's disease affects thirty-nine to sixty people of every million, and Cushing's syndrome affects five to twenty-five people per million. ■

---

**TABLE 13.10 HORMONES OF THE ADRENAL CORTEX**

| Hormone | Action | Factors Regulating Secretion |
|---|---|---|
| Aldosterone | Helps regulate the concentration of extracellular electrolytes by conserving sodium ions and excreting potassium ions | Electrolyte concentrations in body fluids, and renin-angiotensin mechanism |
| Cortisol | Decreases protein synthesis, increases fatty acid release, and stimulates glucose synthesis from noncarbohydrates | CRH from the hypothalamus, and ACTH from the anterior pituitary gland |
| Adrenal androgens | Supplement sex hormones from the gonads; may be converted into estrogen | |

---

1. Name the important hormones of the adrenal cortex.

2. What is the function of aldosterone?

3. What does cortisol do?

4. How are blood concentrations of aldosterone and cortisol regulated?

# PANCREAS

The **pancreas** contains two major types of secretory tissues. This reflects its dual function as an exocrine gland that secretes digestive juice through a duct, and an endocrine gland that releases hormones into body fluids.

## Structure of the Gland

The pancreas is an elongated, somewhat flattened organ that is posterior to the stomach and behind the parietal peritoneum (fig. 13.32). It is attached to the first section of the small intestine (duodenum) by a duct, which transports its digestive juice into the intestine. The digestive functions of the pancreas are discussed in chapter 17.

The endocrine portion of the pancreas consists of cells grouped around blood vessels. These groups, called *islets of Langerhans,* include three distinct types of hormone-secreting cells—*alpha cells,* which secrete glucagon; *beta cells,* which secrete insulin; and *delta cells,* which secrete somatostatin (fig. 13.33).

## Hormones of the Islets of Langerhans

**Glucagon** is a protein that stimulates the liver to break down glycogen into glucose (glycogenolysis) and to convert noncarbohydrates, such as amino acids, into glucose (gluconeogenesis). Glucagon also stimulates breakdown of fats into fatty acids and glycerol.

In a negative feedback system, a low concentration of blood glucose stimulates release of glucagon from the alpha cells. When blood glucose concentration returns toward normal, glucagon secretion decreases. This mechanism prevents hypoglycemia from occurring at times when glucose concentration is relatively low, such as between meals, or when glucose is being used rapidly—during periods of exercise, for example.

The hormone **insulin** is also a protein, and its main effect is exactly opposite that of glucagon. Insulin stimulates the liver to form glycogen from glucose and inhibits conversion of noncarbohydrates into glucose. Insulin also has the special effect of promoting the facilitated diffusion (see chapter 3) of glucose through the membranes of cells bearing insulin receptors. These cells include those of cardiac muscle, adipose tissues, and resting skeletal muscles (glucose uptake by exercising skeletal muscles is not dependent on insulin). Insulin action decreases the concentration of blood glucose, promotes transport of amino acids into cells, and increases protein synthesis. It also stimulates adipose cells to synthesize and store fat.

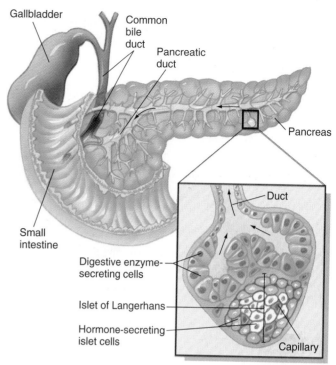

**FIGURE** 13.32

The hormone-secreting cells of the pancreas are grouped in clusters, or islets, that are closely associated with blood vessels. Other pancreatic cells secrete digestive enzymes into ducts.

**FIGURE** 13.33

Light micrograph of an islet of Langerhans within the pancreas (50× micrograph enlarged to 200×).

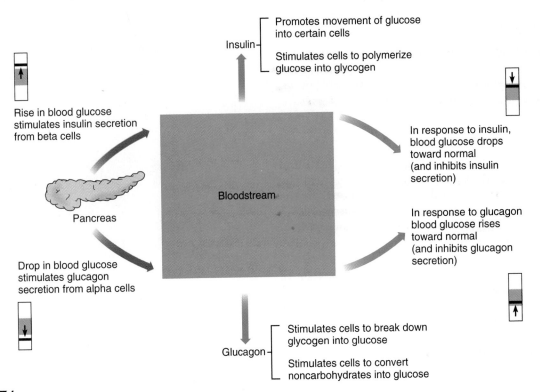

Insulin
- Promotes movement of glucose into certain cells
- Stimulates cells to polymerize glucose into glycogen

Rise in blood glucose stimulates insulin secretion from beta cells

Pancreas

Bloodstream

In response to insulin, blood glucose drops toward normal (and inhibits insulin secretion)

In response to glucagon blood glucose rises toward normal (and inhibits glucagon secretion)

Drop in blood glucose stimulates glucagon secretion from alpha cells

Glucagon
- Stimulates cells to break down glycogen into glucose
- Stimulates cells to convert noncarbohydrates into glucose

**FIGURE 13.34**

Insulin and glucagon function together to help maintain a relatively stable blood glucose concentration. Negative feedback responding to blood glucose concentration controls the levels of both hormones.

An enzyme called glucokinase enables beta cells to "sense" glucose level, important information in determining rates of synthesis of glucagon and insulin. One form of a rare type of diabetes mellitus, maturity-onset diabetes of the young (MODY), is caused by a mutation in a gene encoding glucokinase—the beta cells cannot accurately assess when they must produce insulin. Other genetic mutations that cause MODY alter insulin's structure, secretion, or cell surface receptors, or the ability of liver cells to form glycogen in response to insulin. MODY is treated with drugs or dietary modification.

A negative feedback system sensitive to the concentration of blood glucose regulates insulin secretion. When glucose concentration is relatively high, as may occur following a meal, the beta cells release insulin. By promoting formation of glycogen in the liver and entrance of glucose into adipose and muscle cells, insulin helps prevent excessive rise in the blood glucose concentration (hyperglycemia). Then, when the glucose concentration falls, between meals or during the night, insulin secretion decreases (fig. 13.34).

As insulin concentration decreases, less and less glucose enters the adipose and muscle cells, and the glucose remaining in the blood is available for cells to use that lack insulin receptors, such as nerve cells. Neurons readily use a continuous supply of glucose to produce ATP.

Nerve cells, including those of the brain, obtain glucose by a facilitated-diffusion mechanism that is not dependent on insulin, but rather only on the blood glucose concentration. For this reason, nerve cells are particularly sensitive to changes in the blood glucose concentration. Conditions that cause such changes—excessive insulin secretion, for example—are likely to affect brain functions.

At the same time that insulin concentration is decreasing, glucagon secretion is increasing. Therefore, these hormones function together to maintain a relatively constant blood glucose concentration, despite great variations in the amounts of ingested carbohydrates.

**Somatostatin** (similar to the hypothalamic hormone), which the delta cells release, helps regulate glucose metabolism by inhibiting secretion of glucagon and insulin. Table 13.11 summarizes the hormones of the islets of Langerhans, and Clinical Application 13.4 discusses diabetes mellitus, a derangement of the control of glucose metabolism.

**1** What is the name of the endocrine portion of the pancreas?

**2** What is the function of glucagon?

**3** What is the function of insulin?

**4** How are the secretions of glucagon and insulin controlled?

**5** Why are nerve cells particularly sensitive to changes in blood glucose concentration?

# Diabetes Mellitus

Life for a person with insulin-dependent *diabetes mellitus* means constant awareness of the illness—usually several insulin injections a day, frequent finger punctures to monitor blood glucose level, a restrictive diet, and concern over complications, which include loss of vision, leg ulcers, and kidney damage. The many symptoms of this form of diabetes mellitus reflect disturbances in carbohydrate, protein, and fat metabolism.

In Latin, *diabetes* means "increased urine output," and *mellitus* means "honey," referring to urine's sugar content. Lack of insulin decreases movement of glucose into skeletal muscle and adipose cells, inhibiting glycogen formation. As a result, blood glucose concentration rises (hyperglycemia). When blood glucose reaches a certain level, the kidneys begin to excrete the excess, and glucose appears in the urine (glycosuria). Water follows the glucose by osmosis, causing excess urinary water loss. As a result, the person becomes dehydrated and very thirsty.

Diabetes mellitus decreases protein synthesis, causing tissues to waste away as glucose-starved cells use protein as an energy source. Weight falls and wounds cannot heal. Fatty acids accumulate in the blood as a result of decreased fat synthesis and storage. Ketone bodies, a by-product of fat metabolism, also increase in the blood. They are excreted in the urine as sodium salts, and large quantities of water follow by osmosis, intensifying dehydration and lowering sodium ion concentration in the blood. Accumulation of ketones and loss of sodium ions lead to metabolic acidosis, a condition that lowers the pH of body fluids. Acidosis and dehydration adversely affect brain neurons. Without treatment (insulin replacement), the person becomes disoriented and may enter a diabetic coma and die.

Insulin-dependent diabetes mellitus (IDDM, also called type I, or juvenile-onset diabetes) usually appears before age 20. About 15% of people with diabetes mellitus have this form. It is an autoimmune disorder in which the immune system attacks pancreatic beta cells, ultimately destroying them and halting insulin secretion. Treatment for IDDM is administering enough insulin to control carbohydrate metabolism. Insulin cannot be taken in oral form because it is a peptide, and digestive enzymes break it down.

A milder, second form of diabetes, noninsulin-dependent diabetes mellitus (NIDDM), begins gradually, in people over 40. Usually in NIDDM, cells lose insulin receptors, and therefore cannot respond to insulin, even if it is abundant. Heredity and a lifestyle of overeating and underexercising are risk factors for developing NIDDM. People who develop it are often overweight. Treatment for NIDDM includes careful control of diet to avoid foods that stimulate insulin production, exercising, and maintaining desirable body weight.

The glucose-tolerance test is used to diagnose diabetes mellitus. The patient ingests a known quantity of glucose, and blood glucose concentration is measured at intervals to determine glucose utilization. If the person has diabetes, blood glucose concentration rises excessively and remains elevated for several hours. In a healthy person, glucose rise is less dramatic and the level returns to normal in about one-and-a-half hours. ■

| TABLE 13.11 HORMONES OF THE ISLETS OF LANGERHANS | | |
|---|---|---|
| **Hormone** | **Action** | **Source of Control** |
| Glucagon | Stimulates the liver to break down glycogen and convert noncarbohydrates into glucose; stimulates breakdown of fats | Blood glucose concentration |
| Insulin | Promotes formation of glycogen from glucose, inhibits conversion of noncarbohydrates into glucose, and enhances movement of glucose through adipose and muscle cell membranes, decreasing blood glucose concentration; promotes transport of amino acids into cells; enhances synthesis of proteins and fats | Blood glucose concentration |
| Somatostatin | Helps regulate carbohydrates | Not determined |

## OTHER ENDOCRINE GLANDS

Other organs that produce hormones and, therefore, are parts of the endocrine system include the pineal gland, thymus gland, reproductive glands, and certain glands of the digestive tract, the heart, and the kidneys.

### Pineal Gland

The **pineal gland** is a small, oval structure located deep between the cerebral hemispheres, where it attaches to the upper portion of the thalamus near the roof of the third ventricle. It largely consists of specialized *pineal cells* and supportive *neuroglial cells* (see fig. 11.20b).

The pineal gland secretes a hormone, **melatonin,** which is synthesized from serotonin. Varying patterns of light and dark outside the body control the gland, with environmental information reaching the gland by means of nerve impulses. In the presence of light, nerve impulses originating in the retinas of the eyes travel to the hypothalamus. From the hypothalamus, they enter the reticular formation and then pass downward into the spinal cord. In the spinal cord, the impulses travel along sympathetic nerve fibers back into the brain, and finally they reach the pineal gland. In response to impulses that light triggers, melatonin secretion from the pineal gland decreases.

In the absence of light, nerve impulses from the eyes are decreased, and secretion of melatonin increases. Melatonin secretion is involved in the regulation of **circadian rhythms,** which are patterns of repeated activity associated with the environmental cycles of day and night (Clinical Application 13.5). Circadian rhythms responding to light and dark include sleep/wake rhythms and seasonal cycles of fertility in many mammals.

Melatonin inhibits secretion of gonadotropins from the anterior pituitary gland and helps regulate the female reproductive cycle (menstrual cycle). It may also control onset of puberty (see chapter 22).

### Thymus Gland

The **thymus gland,** which lies in the mediastinum posterior to the sternum and between the lungs, is large in young children but diminishes in size with age. This gland secretes a group of hormones, called **thymosins,** that affect production and differentiation of certain white blood cells (T lymphocytes). The thymus gland plays an important role in immunity and is discussed in chapter 16.

### Reproductive Glands

The reproductive organs that secrete important hormones include the **ovaries,** which produce estrogen and progesterone; the **placenta,** which produces estrogens, progesterone, and a gonadotropin, and the **testes,** which produce testosterone. These glands and their secretions are discussed in chapter 22.

### Digestive Glands

The digestive glands that secrete hormones are generally associated with the linings of the stomach and small intestine. These structures and their secretions are described in chapter 17.

### Other Hormone-Producing Organs

Other organs that produce hormones include the heart, which secretes *atrial natriuretic peptide* (chapter 15), and the kidneys, which secrete a red blood cell growth hormone called *erythropoietin* (chapter 14).

1. Where is the pineal gland located?
2. What seems to be the function of the pineal gland?
3. Where is the thymus gland located?

# Misrepresenting Melatonin

If something sounds too good to be true, it usually is. So it may be with melatonin, a hormone that the pineal gland of vertebrates (animals with backbones) produces, and millions of people take in pill form. Table 13A lists the conditions that melatonin has been claimed to treat—so far, insomnia is the only valid entry. Taking melatonin is widely reported to overcome the fatigue of jet lag, but in fact, melatonin only resets the body's clock by an hour. In species that are seasonal breeders, such as deer and hamsters, additional melatonin disturbs reproduction—we don't yet know it's effects on human reproduction.

Melatonin was discovered, as a pineal gland hormone, in the 1950s, but research stalled until interest in circadian rhythms resurged in the 1970s. The pineal gland synthesizes melatonin during periods of darkness, but not in the light. Biologists hypothesize that animals use melatonin levels to distinguish day from night, and to tell the season by determining relative lengths of day and night. The hormone may cue species other than humans when to molt, mate, migrate, and hibernate.

Molecular biologists recently discovered that melatonin binds to two types of receptors on brain neurons, one that is very abundant and one that is relatively uncommon. The major receptors are found on cells of the suprachiasmatic nucleus, a region that regulates the circadian clock. Binding to the second, less-abundant receptors, however, induces sleepiness.

Endocrinologists still have much to learn about melatonin's role in the human body. Compelling questions that remain include:

- What are melatonin's effects on a human fetus?
- What are long-term effects of taking the hormone?
- Do different people react differently to the same dosage of melatonin?
- Does melatonin interact with any drugs?
- What are the consequences of taking too much melatonin, or taking it at the wrong time of day? ■

| TABLE 13.A | MELATONIN HAS BEEN CLAIMED TO TREAT: | |
|---|---|---|
| AIDS | | Influenza |
| Alzheimer's disease | | Insomnia |
| Autism | | Jet lag |
| Cancer | | Parkinson's disease |
| Cataracts | | Schizophrenia |
| Depression | | Seizure disorders |
| Diabetes mellitus | | Sudden infant death syndrome |
| Heart disease | | |

## STRESS AND ITS EFFECTS

Because survival depends upon maintaining homeostasis, factors that change the body's internal environment are potentially life threatening. When such dangers are sensed, nerve impulses are directed to the hypothalamus, triggering physiological responses that resist a loss of homeostasis. These responses often include increased activity in the sympathetic division of the autonomic nervous system and increased secretion of adrenal hormones. A factor capable of stimulating such a response is called a **stressor,** and the condition it produces in the body is called **stress.**

## Types of Stress

Stressors may be physical, psychological, or a combination of both.

*Physical stress* threatens tissues. This includes extreme heat or cold, decreased oxygen concentration, infections, injuries, prolonged heavy exercise, and loud sounds. Often physical stress is accompanied by unpleasant or painful sensations.

*Psychological stress* results from thoughts about real or imagined dangers, personal losses, unpleasant social interactions (or lack of social interactions), or any factors that threaten a person. Feelings of anger, fear, grief, anxiety, depression, and guilt cause psychological stress. In other instances, psychological stress may stem from pleasant stimuli, such as friendly social contact, feelings of joy or happiness, or sexual arousal. The factors that produce psychological stress vary greatly from person to person. A situation that is stressful to one person may not affect another, and what is stressful at one time may not be at another time.

## Responses to Stress

The hypothalamus controls response to stress, termed the *general stress* (or *general adaptation*) *syndrome.* This response, evoked to stress of any kind, maintains homeostasis.

> Inflammation is the immune system's generalized response to limit the effects of injury or infection. However, inflammation is painful and possibly destructive. The endocrine system keeps the immune system in check by increasing secretion by the pituitary and adrenal glands to temper inflammation. This is an example of how homeostasis operates between organ systems as well as within them.

Recall that the hypothalamus receives information from nearly all body parts, including visceral receptors, the cerebral cortex, the reticular formation, and limbic system. At times of stress, the hypothalamus responds to incoming impulses by activating the "fight or flight" response. More specifically, sympathetic impulses from the hypothalamus raise blood glucose concentration, the level of blood glycerol and fatty acids, heart rate, blood pressure, and breathing rate, and dilate the air passages. The response also shunts blood from the skin and digestive organs into the skeletal muscles, and increases secretion of epinephrine from the adrenal medulla. The epinephrine, in turn, intensifies these sympathetic responses and prolongs their effects (fig. 13.35).

At the same time that sympathetic activity increases, the hypothalamus's release of corticotropin-releasing hormone (CRH) stimulates the anterior pituitary gland to secrete ACTH, which increases the adrenal cortex's secretion of cortisol. Cortisol supplies cells with amino acids and extra energy sources, and diverts glucose from skeletal muscles to brain tissue (fig. 13.36). Stress can also trigger release of glucagon from the pancreas, growth hormone (GH) from the anterior pituitary, and antidiuretic hormone (ADH) from the posterior pituitary gland. Secretion of renin from the kidney may also be stimulated.

Glucagon and growth hormone help mobilize energy sources, such as glucose, glycerol, and fatty acids, and stimulate cells to take up amino acids, facilitating repair of injured tissues. ADH stimulates the kidneys to retain water. This action decreases urine output and helps to maintain blood volume—particularly important if a person is bleeding excessively or sweating heavily. Renin, by increasing angiotensin II levels, helps stimulate the kidneys to retain sodium (through aldosterone), and through the vasoconstrictor action of angiotension II contributes to maintaining blood pressure. Table 13.12 summarizes the body's reactions to stress.

> The 17-year-old was visiting her family physician for the third time. She had recurrent stomach pains and seemed to have a constant respiratory infection. Unlike her previous visits, this time the woman seemed noticeably upset. Suspecting that her physical symptoms stemmed from a struggle to deal with a stressful situation, the doctor looked for signs—increased heart and respiratory rate, elevated blood pressure, and excessive sweating. He took a blood sample to measure levels of epinephrine and cortisol, while prodding her to talk about whatever was bothering her.
>
> The young woman's cortisol was indeed elevated, which could account for her gastrointestinal pain and high blood pressure, as well as her impaired immunity. On the doctor's advice, she began seeing a psychologist. Her symptoms began to abate when she was able to discuss the source of her stress—a family member's illness and anxiety about beginning college.

**1** What is stress?

**2** Distinguish between physical stress and psychological stress.

**3** Describe the general stress syndrome.

## CLINICAL TERMS RELATED TO THE ENDOCRINE SYSTEM

**adrenalectomy** (ah-dre″nah-lek′to-me)  Surgical removal of the adrenal glands.

**adrenogenital syndrome** (ah-dre″no-jen′ĭ-tal sin′drōm)  Group of symptoms associated with changes in sexual characteristics as a result of increased secretion of adrenal androgens.

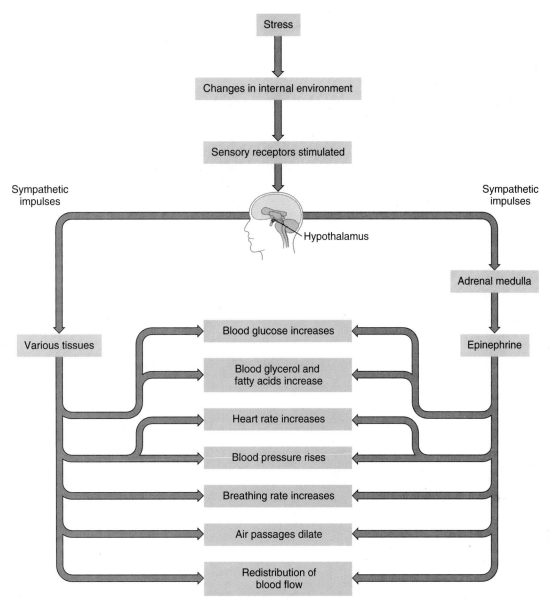

**FIGURE 13.35**

During stress, the hypothalamus helps prepare the body for fight or flight by triggering sympathetic impulses to various organs. It also stimulates release of epinephrine, intensifying the sympathetic responses.

**exophthalmos** (ek″sof-thal′mos) Abnormal protrusion of the eyes.

**hirsutism** (her′sut-izm) Excessive growth of hair, especially in women.

**hypercalcemia** (hi″per-kal-se′me-ah) Excess of blood calcium.

**hyperglycemia** (hi″per-gli-se′me-ah) Excess of blood glucose.

**hypocalcemia** (hi″po-kal-se′me-ah) Deficiency of blood calcium.

**hypophysectomy** (hi-pof″ĭ-sek′to-me) Surgical removal of the pituitary gland.

**parathyroidectomy** (par″ah-thi″roi-dek′to-me) Surgical removal of the parathyroid glands.

**pheochromocytoma** (fe-o-kro″mo-si-to′mah) Type of tumor found in the adrenal medulla and usually accompanied by high blood pressure.

**polyphagia** (pol″e-fa′je-ah) Excessive eating.

**thymectomy** (thi-mek′to-me) Surgical removal of the thymus gland.

**thyroidectomy** (thi″roi-dek′to-me) Surgical removal of the thyroid gland.

**thyroiditis** (thi″roi-di′tis) Inflammation of the thyroid gland.

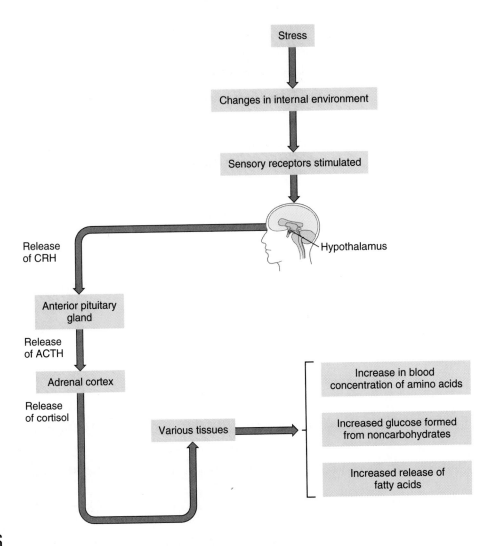

**FIGURE 13.36**

As a result of stress, the hypothalamus stimulates the adrenal cortex to release cortisol, which promotes responses that resist the effects of stress. Compare this to figure 13.31.

| TABLE 13.12 | MAJOR EVENTS IN THE GENERAL STRESS SYNDROME |
|---|---|

1. As a result of stress, nerve impulses are transmitted to the hypothalamus.

2. Sympathetic impulses arising from the hypothalamus increase blood glucose concentration, blood glycerol concentration, blood fatty acid concentration, heart rate, blood pressure, and breathing rate. They dilate air passages, shunt blood into skeletal muscles, and increase secretion of epinephrine from the adrenal medulla.

3. Epinephrine intensifies and prolongs sympathetic actions.

4. The hypothalamus secretes CRH, which stimulates secretion of ACTH by the anterior pituitary gland.

5. ACTH stimulates release of cortisol by the adrenal cortex.

6. Cortisol increases the concentration of blood amino acids, releases fatty acids, and forms glucose from noncarbohydrate sources.

7. Secretion of glucagon from the pancreas and growth hormone from the anterior pituitary increase.

8. Glucagon and growth hormone aid mobilization of energy sources and stimulate uptake of amino acids by cells.

9. Secretion of ADH from the posterior pituitary increases.

10. ADH promotes the retention of water by the kidneys, which increases blood volume.

11. Renin increases blood levels of angiotensin II, which acts as a vasoconstrictor and also stimulates aldosterone secretion by the adrenal cortex.

12. Aldosterone stimulates sodium retention by the kidneys.

## ENDOCRINE SYSTEM

Glands secrete hormones that have a variety of effects on cells, tissues, organs, and organ systems.

### Integumentary System

Melanocytes produce skin pigment in response to hormonal stimulation.

### Skeletal System

Hormones act on bones to control calcium balance.

### Muscular System

Hormones help increase blood flow to exercising muscles.

### Nervous System

Neurons control the secretions of the anterior and posterior pituitary glands and the adrenal medulla.

### Cardiovascular System

Hormones are carried in the boodstream; some have direct actions on the heart and blood vessels.

### Lymphatic System

Hormones stimulate lymphocyte production.

### Digestive System

Hormones help control digestive system activity.

### Respiratory System

Decreased oxygen causes hormonal stimulation of red blood cell production; red blood cells transport oxygen and carbon dioxide.

### Urinary System

Hormones act on the kidneys to help control water and electrolyte balance.

### Reproductive System

Sex hormones play a major role in development of secondary sex characteristics, egg, and sperm.

## Introduction (page 480)

Endocrine glands secrete their products into body fluids (the internal environment); exocrine glands secrete their products into ducts that lead to the outside of the body.

## General Characteristics of the Endocrine System (page 480)

As a group, endocrine glands regulate metabolic processes.

## Hormone Action (page 480)

Endocrine glands secrete hormones that affect target cells possessing specific receptors.

1. Chemistry of hormones
   a. Each kind of hormone has a special molecular structure and is very potent.
   b. Chemically, hormones are steroids, amines, peptides, proteins, or glycoproteins.
2. Actions of hormones
   a. Steroid hormones
      (1) Steroid hormones enter target cells and combine with receptors to form complexes.
      (2) These complexes activate specific genes in the nucleus, which direct synthesis of specific proteins.
      (3) The degree of cellular response is proportional to the number of hormone-receptor complexes formed.
   b. Nonsteroid hormones
      (1) Nonsteroid hormones combine with receptors in the target-cell membrane.
      (2) A hormone-receptor complex stimulates membrane proteins, such as adenylate cyclase, to induce the formation of second messenger molecules.
      (3) A second messenger, such as cAMP, activates protein kinases.
      (4) Protein kinases activate certain protein substrate molecules, which, in turn, change cellular processes.
      (5) The cellular response to a nonsteroid hormone is amplified because the enzymes induced by a small number of hormone-receptor complexes can catalyze formation of a large number of second messenger molecules.
3. Prostaglandins
   a. Prostaglandins are paracrine substances present in small quantities that have powerful hormonelike effects.
   b. Prostaglandins modulate hormones that regulate formation of cyclic AMP.

## Control of Hormonal Secretions (page 487)

The concentration of each hormone in the body fluids is precisely regulated.

1. Control sources
   a. Some endocrine glands secrete in response to nerve impulses.
   b. Other glands secrete hormones in response to releasing hormones the hypothalamus secretes.
2. Negative feedback systems
   a. In a negative feedback system, a gland is sensitive to the concentration of a substance it regulates.
   b. When the concentration of the regulated substance reaches a certain concentration, it inhibits the gland.
   c. As the gland secretes less hormone, the controlled substance also decreases.

## Pituitary Gland (page 488)

The pituitary gland, which is attached to the base of the brain, has an anterior lobe and a posterior lobe. Releasing hormones from the hypothalamus control most pituitary secretions.

1. Anterior pituitary hormones
   a. The anterior pituitary consists largely of epithelial cells, and it secretes GH, PRL, TSH, ACTH, FSH, and LH.
   b. Growth hormone (GH)
      (1) Growth hormone stimulates body cells to grow and reproduce.
      (2) Growth hormone-releasing hormone and somatostatin from the hypothalamus control GH secretion.
   c. Prolactin (PRL)
      (1) PRL promotes breast development and stimulates milk production.
      (2) In males, prolactin decreases secretion of LH (ICSH).
      (3) Prolactin release-inhibiting hormone from the hypothalamus restrains secretion of prolactin, whereas prolactin-releasing hormone promotes its secretion.
   d. Thyroid-stimulating hormone (TSH)
      (1) TSH controls secretion of hormones from the thyroid gland.
      (2) The hypothalamus, by secreting thyrotropin-releasing hormone, regulates TSH secretion.
   e. Adrenocorticotropic hormone (ACTH)
      (1) ACTH controls the secretion of certain hormones from the adrenal cortex.
      (2) The hypothalamus, by secreting corticotropin-releasing hormone, regulates ACTH secretion.
   f. Follicle-stimulating hormone (FSH) and luteinizing hormone (LH) are gonadotropins that influence the reproductive organs.
2. Posterior pituitary hormones
   a. The posterior lobe of the pituitary gland largely consists of neuroglial cells and nerve fibers that originate in the hypothalamus.
   b. The two hormones of the posterior pituitary are produced in the hypothalamus.
   c. Antidiuretic hormone (ADH)
      (1) ADH causes the kidneys to reduce the amount of water they excrete.
      (2) In high concentration, ADH constricts blood vessel walls, raising blood pressure.
      (3) The hypothalamus regulates ADH secretion.

        d. Oxytocin (OT)
            (1) Oxytocin has an antidiuretic effect and can
                contract muscles in the uterine wall.
            (2) OT also contracts certain cells associated with
                production and ejection of milk from the milk
                glands of the breasts.

# Thyroid Gland (page 495)

The thyroid gland is located in the neck and consists of two
lateral lobes.

1. Structure of the gland
    a. The thyroid gland consists of many hollow secretory
       parts called follicles.
    b. The follicles are fluid filled and store the hormones
       the follicle cells secrete.
    c. Extrafollicular cells secrete calcitonin.
2. Thyroid hormones
    a. Thyroxine and triiodothyronine
        (1) These hormones increase the rate of metabolism,
            enhance protein synthesis, and stimulate lipid
            breakdown.
        (2) These hormones are needed for normal growth
            and development, and for maturation of the
            nervous system.
    b. Calcitonin
        (1) Calcitonin lowers blood calcium and phosphate
            ion concentrations.
        (2) This hormone prevents prolonged elevation of
            calcium after a meal.

# Parathyroid Glands (page 499)

The parathyroid glands are located on the posterior surface of
the thyroid.

1. Structure of the glands
    a. Each gland is small and yellow-brown, within a thin
       connective tissue capsule.
    b. Each gland consists of secretory cells that are well
       supplied with capillaries.
2. Parathyroid hormone (PTH)
    a. PTH increases blood calcium ion concentration and
       decreases blood phosphate ion concentration.
    b. PTH stimulates resorption of bone tissue, causes the
       kidneys to conserve calcium ions and excrete
       phosphate ions, and indirectly stimulates absorption
       of calcium ions from the intestine.
    c. A negative feedback mechanism operating between the
       parathyroid glands and the blood regulates these glands.

# Adrenal Glands (page 501)

The adrenal glands are located atop the kidneys.

1. Structure of the glands
    a. Each adrenal gland consists of a medulla and a cortex.
    b. The adrenal medulla and adrenal cortex are distinct
       glands that secrete different hormones.
2. Hormones of the adrenal medulla
    a. The adrenal medulla secretes epinephrine and
       norepinephrine.
    b. These hormones are synthesized from tyrosine and are
       closely related to each other chemically.

        c. These hormones produce effects similar to those of the
           sympathetic nervous system.
        d. Sympathetic impulses originating from the
           hypothalamus stimulate secretion of these hormones.
3. Hormones of the adrenal cortex
    a. The cortex produces several types of steroids that
       include hormones.
    b. Aldosterone
        (1) It causes the kidneys to conserve sodium ions and
            water, and to excrete potassium ions.
        (2) It is secreted in response to increased potassium
            ion concentration, or presence of angiotensin II.
        (3) By conserving sodium ions and water, it helps
            maintain blood volume and pressure.
    c. Cortisol
        (1) It inhibits protein synthesis, releases fatty acids,
            and stimulates glucose formation from
            noncarbohydrates.
        (2) A negative feedback mechanism involving
            secretion of CRH from the hypothalamus and
            ACTH from the anterior pituitary gland controls
            its level.
    d. Adrenal sex hormones
        (1) These hormones are of the male type although
            some can be converted into female hormones.
        (2) They supplement the sex hormones produced by
            the gonads.

# Pancreas (page 507)

The pancreas secretes digestive juices as well as hormones.

1. Structure of the gland
    a. The pancreas is posterior to the stomach and is
       attached to the small intestine.
    b. The endocrine portion, which is called the islets of
       Langerhans, secretes glucagon, insulin, and
       somatostatin.
2. Hormones of the islets of Langerhans
    a. Glucagon stimulates the liver to produce glucose,
       increasing concentration of blood glucose. It also
       breaks down fat.
    b. Insulin moves glucose through cell membranes,
       stimulates its storage, promotes protein synthesis, and
       stimulates fat storage.
    c. Nerve cells lack insulin receptors and depend upon
       diffusion for a glucose supply.
    d. Somatostatin inhibits insulin and glucagon release.

# Other Endocrine Glands (page 510)

1. Pineal gland
    a. The pineal gland is attached to the thalamus near the
       roof of the third ventricle.
    b. Postganglionic sympathetic nerve fibers innervate it.
    c. It secretes melatonin, which inhibits secretion of
       gonadotropins from the anterior pituitary gland.
    d. It may help regulate the female reproductive cycle.
2. Thymus gland
    a. The thymus gland lies posterior to the sternum and
       between the lungs.
    b. Its size diminishes with age.
    c. It secretes thymosin, which affects the production of
       certain lymphocytes that, in turn, provide immunity.

3. Reproductive glands
   a. The ovaries secrete estrogen and progesterone.
   b. The placenta secretes estrogen, progesterone, and a gonadotropin.
   c. The testes secrete testosterone.
4. The digestive glands include certain glands of the stomach and small intestine that secrete hormones.
5. Other hormone-producing organs include the heart and kidneys.

## Stress and Its Effects (page 511)

Stress occurs when the body responds to stressors that threaten the maintenance of homeostasis. Stress responses include increased activity of the sympathetic nervous system and increased secretion of adrenal hormones.

1. Types of stress
   a. Physical stress results from environmental factors that are harmful or potentially harmful to tissues.
   b. Psychological stress results from thoughts about real or imagined dangers.
   c. Factors that produce psychological stress vary with the individual and the situation.
2. Responses to stress
   a. Responses to stress maintain homeostasis.
   b. The hypothalamus controls a general stress syndrome.

## CRITICAL THINKING QUESTIONS

1. Based on your understanding of the actions of glucagon and insulin, would a person with diabetes mellitus be likely to require more insulin or more sugar following strenuous exercise? Why?
2. What problems might result from the prolonged administration of cortisol to a person with a severe inflammatory disease?
3. How might the environment of a patient with hyperthyroidism be modified to minimize the drain on body energy resources?
4. Which hormones should be administered to an adult whose anterior pituitary gland has been removed? Why?
5. A patient who has lost a large volume of blood will secrete excess aldosterone from the adrenal cortex. What effect will this increased secretion have on the patient's blood concentrations of sodium and potassium ions?
6. Both growth hormone and growth hormone-releasing hormone have been successfully used to promote growth in children with short statures. What is the difference in the ways these hormones produce their effects?

## REVIEW EXERCISES

1. What is an endocrine gland?
2. Define *hormone* and *target cell*.
3. Explain how hormones can be grouped on the basis of their chemical composition.
4. Explain how steroid hormones influence cells.
5. Distinguish between the binding site and the activity site of a receptor molecule.
6. Explain how nonsteroid hormones may function through the formation of cAMP.
7. Explain how nonsteroid hormones may function through an increase in intracellular calcium ion concentration.
8. Explain how the cellular response to a hormone operating through a second messenger is amplified.
9. Define *prostaglandins,* and explain their general function.
10. Describe a negative feedback system.
11. Define *releasing hormone,* and provide an example of one.
12. Describe the location and structure of the pituitary gland.
13. List the hormones the anterior pituitary gland secretes.
14. Explain how the brain controls pituitary gland activity.
15. Explain how growth hormone produces its effects.
16. List the major factors that affect growth hormone secretion.
17. Summarize the functions of prolactin.
18. Describe regulation of concentrations of circulating thyroid hormones.
19. Explain the control of secretion of ACTH.
20. List the major gonadotropins, and explain the general functions of each.
21. Compare the cellular structures of the anterior and posterior lobes of the pituitary gland.
22. Name the hormones associated with the posterior pituitary, and explain their functions.
23. Explain how the release of ADH is regulated.
24. Describe the location and structure of the thyroid gland.
25. Name the hormones the thyroid gland secretes, and list the general functions of each.
26. Define *iodine pump.*
27. Describe the location and structure of the parathyroid glands.
28. Explain the general functions of parathyroid hormone.
29. Describe mechanisms that regulate the secretion of parathyroid hormone.
30. Distinguish between the adrenal medulla and the adrenal cortex.

31. List the hormones produced by the adrenal medulla, and describe their general functions.

32. List the steps in the synthesis of adrenal medullary hormones.

33. Name the most important hormones of the adrenal cortex, and describe the general functions of each.

34. Describe the regulation of the secretion of aldosterone.

35. Describe how the secretion of cortisol is regulated.

36. Describe the location and structure of the pancreas.

37. List the hormones the islets of Langerhans secrete, and describe the general functions of each.

38. Summarize how the secretion of hormones from the pancreas is regulated.

39. Describe the location and general function of the pineal gland.

40. Describe the location and general function of the thymus gland.

41. Distinguish between a stressor and stress.

42. List several factors that cause physical and psychological stress.

43. Describe the general stress syndrome.

 For More Information Related to This Chapter, Please Visit *Hole's Human Anatomy and Physiology* Home Page
http://www.mhhe.com/biosci/abio/

# BLOOD

## UNDERSTANDING WORDS

**agglutin-,** to glue together: *agglutin*ation—clumping together of red blood cells.

**bil-,** bile: *bil*irubin—pigment excreted in the bile.

**-crit,** to separate: hemato*crit*—percentage by volume of cells in a blood sample, determined by separating the cells from the plasma.

**embol-,** stopper: *embol*ism—obstruction in a blood vessel.

**erythr-,** red: *erythr*ocyte—red blood cell.

**hem-,** blood: *hem*oglobin—red pigment responsible for the color of blood.

**hepar-,** liver: *hepar*in—anticoagulant secreted by liver cells.

**leuko-,** white: *leuko*cyte—white blood cell.

**-lys,** to break up: fibrino*lys*in—protein-splitting enzyme that can digest fibrin.

**macro-,** large: *macro*phage—large phagocytic cell.

**-osis,** abnormal condition: leukocyt*osis*—condition in which white blood cells are overproduced.

**-poie,** make, produce: erythro*poie*tin—hormone that stimulates the production of red blood cells.

**poly-,** many: *poly*cythemia—condition in which red blood cells are overproduced.

**-sta,** halt, make stand: hemo*sta*sis—arrest of bleeding from damaged blood vessels.

**thromb-,** clot: *thromb*ocyte—blood platelet involved in the formation of a blood clot.

## CHAPTER OBJECTIVES

AFTER YOU HAVE STUDIED THIS CHAPTER, YOU SHOULD BE ABLE TO:

1. Describe the general characteristics of the blood and discuss its major functions.

2. Distinguish among the types of blood cells.

3. Explain how blood cell counts are made and how they are used.

4. Discuss the life cycle of a red blood cell.

5. Explain control of red blood cell production.

6. List the major components of blood plasma and describe the functions of each.

7. Define *hemostasis,* and explain the mechanisms that help to achieve it.

8. Review the major steps in blood coagulation.

9. Explain how coagulation can be prevented.

10. Explain the basis for blood typing.

11. Describe how blood reactions may occur between fetal and maternal tissues.

## KEY TERMS

**albumin** (al-bu'min)
**antibody** (an'ti-bod"e)
**antigen** (an'-ti-jen)
**basophil** (ba'-so-fil)
**coagulation** (ko-ag"u-la'-shun)
**embolus** (em'-bo-lus)
**eosinophil** (e"o-sin'o-fil)
**erythrocyte** (ĕ-rith'ro-sīt)
**erythropoietin** (e-rith"ro-poi'ĕ-tin)
**fibrinogen** (fi-brin'o-jen)

**globulin** (glob'u-lin)
**hemostasis** (he"mo-sta'sis)
**leukocyte** (lu'ko-sīt)
**lymphocyte** (lim'fo-sīt)
**macrophage** (mak'ro-fāj)
**monocyte** (mon'o-sīt)
**neutrophil** (nu'tro-fil)
**plasma** (plaz'mah)
**platelet** (plāt'let)
**thrombus** (throm'bus)

mbilical cord blood was once regarded as medical waste. Today, it is recognized as a rich, abundant source of stem cells that give rise to specialized blood cells. Because it can replenish a person's blood supply, umbilical cord blood is used to treat certain tumors, immunodeficiency syndromes, anemias, leukemias, and other blood disorders.

Using stem cells from umbilical cords to treat these conditions offers many advantages over the existing approach, bone marrow transplantation. Unlike the stem cells that are exceedingly rare in bone marrow, those in umbilical cord blood are abundant. Collecting blood from the umbilical cord does not hurt the donor, who is at the time yowling as he or she enters the world. Compared to bone marrow stem cells, those in the umbilical cord blood are less likely to attack a recipient's tissues, a deadly condition called graft-versus-host disease. Plus, transplants from cord blood are more likely to succeed, and less likely to transmit infection. Because cells fresh from a fetus are less able to stimulate an immune response, matching between cord blood donor and recipient may not have to be as close as it is for a bone marrow transplant.

The first umbilical cord blood transplant was performed in 1988, on a child with Fanconi anemia, an inherited severe lack of red blood cells. In 1990, the technique cured a child of leukemia. In 1992, the National Heart, Lung and Blood Institute established a placental blood bank, which treated two dozen children—more than half did extremely well. Soon the agency was sponsoring a network of similar facilities, and commercial umbilical cord storage services are now available too. These banks screen donated cord blood for infectious agents and genetic disease, standardize collection procedures, and measure the percentage of stem cells per sample. A unit of donor cord blood is usually sufficient to treat a child or a small adult.

New parents today are routinely asked to donate the cord blood that was not too long ago thrown away—either to treat their own child should he or she develop a blood or immune disorder some day, or to help another person.

Blood is often considered a type of connective tissue whose cells are suspended in a liquid intercellular material. Blood is vital in transporting substances between body cells and the external environment, thereby helping to maintain a stable internal environment.

## BLOOD AND BLOOD CELLS

Whole blood is slightly heavier and three to four times more viscous than water. Its cells, which form mostly in red bone marrow, include red blood cells and white blood cells. Blood also contains cellular fragments called blood platelets (fig. 14.1).

### Blood Volume and Composition

Blood volume varies with body size, changes in fluid and electrolyte concentrations, and the amount of adipose tissue. An average-sized adult has a blood volume of about 5 liters.

If a blood sample is allowed to stand in a tube for a while and prevented from clotting, the cells separate from the liquid portion of the blood and settle to the bottom. Centrifuging the sample quickly packs the cells into the lower part of the centrifuge tube, as figure 14.2 shows. The percentage of cells and liquid in the blood sample can then be calculated.

A blood sample is usually about 45% cells by volume. This percentage is called the **hematocrit** (HCT), or **packed cell volume** (PCV). The remaining 55% of a blood sample is clear, straw-colored **plasma**. Appendix C lists values for the hematocrit and other common blood tests in healthy individuals.

**Blood Sample**

- Plasma
- "Buffy coat"
- Red blood cells

**Peripheral Blood Smear**

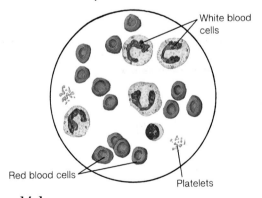

White blood cells

Red blood cells

Platelets

**FIGURE 14.1**

Blood consists of a liquid portion called plasma and a solid portion that includes red blood cells, white blood cells, and platelets. When blood components are separated, the white blood cells and platelets form a thin layer, called the "buffy coat," between the plasma and the red blood cells.

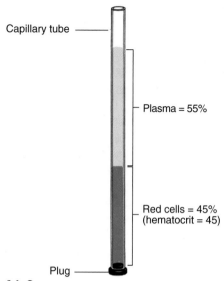

Capillary tube

Plasma = 55%

Red cells = 45%
(hematocrit = 45)

Plug

**FIGURE** 14.2

If a blood-filled capillary tube is centrifuged, the red cells pack in the lower portion, and the percentage of red cells (hematocrit) can be determined.

In addition to red blood cells, which comprise over 99% of the blood cells, the formed elements of the blood include white blood cells and blood platelets. Plasma is a complex mixture that includes water, amino acids, proteins, carbohydrates, lipids, vitamins, hormones, electrolytes, and cellular wastes.

## The Origin of Blood Cells

Blood cells originate in bone marrow from **hemocytoblasts,** a type of stem cell (fig. 14.3). A stem cell can differentiate into any of a number of specialized cell types. As hemocytoblasts reproduce, the new cells respond to different secreted growth factors, called **colony stimulating factors,** that turn on some genes and turn off others. This exposure to growth factors ultimately sculpts the distinctive formed elements of blood, including the cellular components of the immune system. A protein called *thrombopoietin* (TPO) stimulates large cells called **megakaryocytes** to proliferate.

1. What are the major components of blood?
2. What factors affect blood volume?
3. How is hematocrit determined?
4. How do blood cells form?

## Characteristics of Red Blood Cells

**Red blood cells,** or **erythrocytes,** are tiny, biconcave disks that are thin near their centers and thicker around their rims (fig. 14.4). This special shape is an adaptation for the red cell's function of transporting gases; it increases the surface area through which gases can diffuse. The shape also places the cell membrane closer to oxygen-carrying *hemoglobin* within the cell. Because of

its shape, a red blood cell can readily squeeze through the narrow passages of capillaries.

Each red blood cell is about one-third hemoglobin by volume. This protein is responsible for the color of the blood. The rest of the cell mainly consists of membrane, water, electrolytes, and enzymes. When the hemoglobin combines with oxygen, the resulting *oxyhemoglobin* is bright red, and when the oxygen is released, the resulting *deoxyhemoglobin* is darker. Blood rich in deoxyhemoglobin may appear bluish when it is viewed through blood vessel walls.

A person experiencing prolonged oxygen deficiency (hypoxia) may become *cyanotic.* The skin and mucous membranes appear bluish due to an abnormally high blood concentration of deoxyhemoglobin. Exposure to low temperature may also result in cyanosis. Such exposure constricts superficial blood vessels, which slows blood flow and removes more oxygen than usual from blood flowing through the vessels.

Red blood cells have nuclei during their early stages of development, but extrude them as the cells mature, providing more space for hemoglobin. Since they lack nuclei, red blood cells cannot synthesize messenger RNA or reproduce. Because they also lack mitochondria, red blood cells produce ATP through glycolysis only, and use none of the oxygen they carry. As long as cytoplasmic enzymes function, these can carry on vital energy-releasing processes. With time, however, red blood cells become less and less active.

In *sickle cell disease,* a single DNA base change causes hemoglobin to crystallize in a low-oxygen environment. This bends the red blood cells containing the hemoglobin into a sickle shape, which blocks circulation in small vessels, causing excruciating joint pain and damaging many organs. As the spleen works overtime to recycle the short-lived red blood cells, infection becomes likely.

Most children with sickle cell disease are diagnosed at birth and receive antibiotics daily for years to prevent infection. Hospitalization for blood transfusions may be necessary if the person experiences painful sickling "crises" of blocked circulation.

A bone marrow transplant can completely cure sickle cell disease but has a 15% risk of causing death. A new treatment is an old drug, used to treat cancer, called hydroxyurea. It activates dormant genes that produce a slightly different form of hemoglobin in the fetus. Because of the presence of the functional fetal hemoglobin, the sickle hemoglobin cannot crystallize as quickly as it otherwise would. Sickling becomes delayed, which enables red blood cells carrying sickled hemoglobin to reach the lungs faster—where fresh oxygen restores the cells' normal shapes.

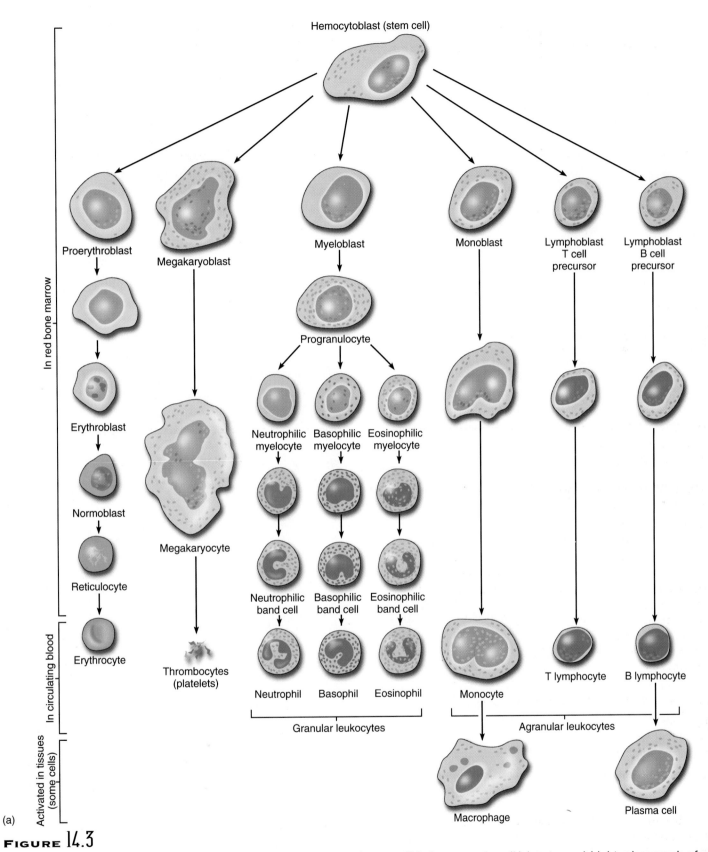

Hemocytoblast (stem cell)

In red bone marrow

Proerythroblast

Erythroblast

Normoblast

Reticulocyte

Megakaryoblast

Megakaryocyte

Myeloblast

Progranulocyte

Neutrophilic myelocyte

Basophilic myelocyte

Eosinophilic myelocyte

Neutrophilic band cell

Basophilic band cell

Eosinophilic band cell

Monoblast

Lymphoblast T cell precursor

Lymphoblast B cell precursor

In circulating blood

Erythrocyte

Thrombocytes (platelets)

Neutrophil

Basophil

Eosinophil

Granular leukocytes

Monocyte

T lymphocyte

B lymphocyte

Agranular leukocytes

Activated in tissues (some cells)

Macrophage

Plasma cell

(a)

## Figure 14.3

(a) Origin and development of blood cells from a hemocytoblast (stem cell) in bone marrow. (b) (next page) Light micrograph of a hemocytoblast (arrow) in red bone marrow (500×).

(b)

**FIGURE** $14.3$ *Continued*

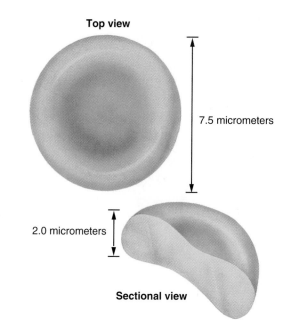

Top view

7.5 micrometers

2.0 micrometers

Sectional view

(a)

1. Describe a red blood cell.

2. How does the biconcave shape of a red blood cell make possible its function?

3. What is the function of hemoglobin?

4. What changes occur in a red blood cell as it matures?

## Red Blood Cell Counts

The number of red blood cells in a cubic millimeter (mm³) of blood is called the *red blood cell count* (RBCC or RCC). Although this number varies from time to time even in healthy individuals, the typical range for adult males is 4,600,000–6,200,000 cells per mm³, and that for adult females is 4,200,000–5,400,000 cells per mm³. For children, the average range is 4,500,000–5,100,000 cells per mm³. These values may vary slightly with the hospital, physician, and type of equipment used to make blood cell counts. The number of red blood cells generally increases following exercise, a large meal, a rise in temperature, or an increase in altitude.

> The equivalent units of microliters (μL) are sometimes used in place of mm³ in describing blood cell counts. For example, 4,600,000 cells per mm³ is equivalent to 4,600,000 cells per μL.

Since an increasing number of circulating red blood cells increases the blood's *oxygen-carrying capacity,* changes in this number may affect health. For this reason, red cell counts are routinely consulted to help diagnose and evaluate the courses of various diseases.

## Destruction of Red Blood Cells

Red blood cells are quite elastic and flexible, and they readily bend as they pass through small blood vessels. With age, however, these cells become more fragile, and

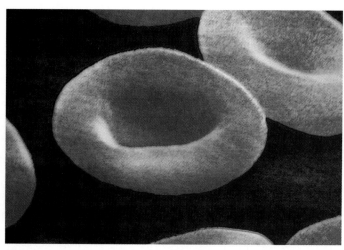

(b)

**FIGURE** $14.4$

(*a*) Note the biconcave shape of a red blood cell. (*b*) Scanning electron micrograph of human red blood cells (falsely colored) (7,500×).

they are frequently damaged simply by passing through capillaries, particularly those in active muscles.

Damaged or worn red cells rupture as they pass through the spleen or liver. In these organs, macrophages (see chapter 5) phagocytize and destroy damaged red cells and their contents. Hemoglobin molecules liberated from the red blood cells break down into their four component polypeptide "globin" chains, each surrounding a heme group. Heme further decomposes into iron and a greenish pigment called **biliverdin**. The iron, combined with a protein called *transferrin,* may be carried by the blood to the hematopoietic (red blood cell–forming) tissue in the red bone marrow and reused in synthesizing new hemoglobin. About 80% of the iron is stored in the

Polypeptide chain ($\beta_2$)  Polypeptide chain ($\beta_1$)

Iron-containing
heme groups

Polypeptide chain ($\alpha_2$)  Polypeptide chain ($\alpha_1$)

(a) Hemoglobin molecule

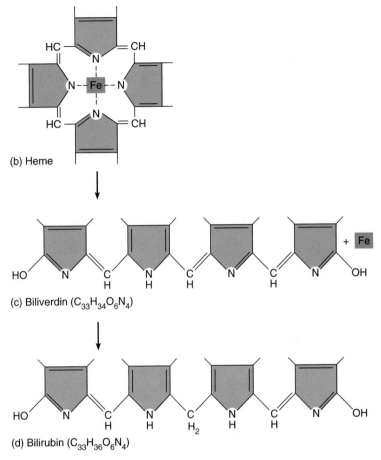

(b) Heme

(c) Biliverdin ($C_{33}H_{34}O_6N_4$)

(d) Bilirubin ($C_{33}H_{36}O_6N_4$)

**FIGURE 14.5**

(a) When a hemoglobin molecule decomposes, (b) the heme groups break down into (c) iron (Fe) and biliverdin. (d) Most of the biliverdin is then converted to bilirubin.

liver cells in the form of an iron-protein complex called *ferritin.* In time, the biliverdin is converted to an orange pigment called **bilirubin.** Biliverdin and bilirubin are excreted in the bile as bile pigments (fig. 14.5). Table 14.1 summarizes the process of red blood cell destruction.

In an inherited condition called *hemochromatosis,* the small intestine absorbs too much iron from food. The metal accumulates in the liver, pancreas, heart, and endocrine glands and makes the skin appear bronze. A symptom is increased susceptibility to bacterial infection, because bacteria thrive in the iron-rich tissues, but the condition often goes undiagnosed for many years. Women do not develop symptoms until after menopause, when their menstrual periods no longer relieve the condition each month. Treatment for hemochromatosis is simple—periodically removing blood. Without treatment, premature death results from chronic liver disease or heart failure.

Hemochromatosis is common. It affects 0.3 to 0.5% of many populations, and carriers may be as prevalent as 1 in 10 to 1 in 15 individuals.

**TABLE 14.1  MAJOR EVENTS IN RED BLOOD CELL DESTRUCTION**

1. Squeezing through the capillaries of active tissues damages red blood cells.
2. Macrophages in the liver and spleen phagocytize damaged red blood cells.
3. Hemoglobin from the red blood cells is decomposed into heme and globin.
4. Heme is decomposed into iron and biliverdin.
5. Iron is made available for reuse in the synthesis of new hemoglobin or is stored in the liver as ferritin.
6. Some biliverdin is converted into bilirubin.
7. Biliverdin and bilirubin are excreted in bile as bile pigments.

1 What is the typical red blood cell count for an adult male? For an adult female?

2 What happens to damaged red blood cells?

3 What are the end products of hemoglobin breakdown?

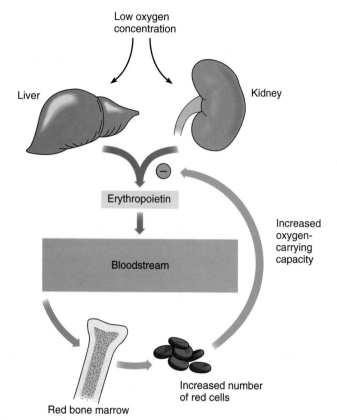

Low oxygen
concentration

Liver

Kidney

Erythropoietin

Bloodstream

Increased
oxygen-
carrying
capacity

Red bone marrow

Increased number
of red cells

**FIGURE** 14.6

Low oxygen concentration causes the kidneys and liver to release erythropoietin, which stimulates production of red blood cells that carry oxygen to tissues.

## Red Blood Cell Production and Its Control

As mentioned in chapter 7, red blood cell formation (hematopoiesis) initially occurs in the yolk sac, liver, and spleen. After an infant is born, these cells are produced almost exclusively by tissue lining the spaces in bones, filled with red bone marrow.

Within the red marrow, hemocytoblasts give rise to **erythroblasts** that can synthesize hemoglobin molecules at the rate of 2 million to 3 million per second. The erythroblasts also reproduce and give rise to many new cells. The nuclei of these newly formed cells soon shrink and are extruded by being pinched off in thin coverings of cytoplasm and cell membrane. The resulting cells are erythrocytes. Some of these young red cells may contain a netlike structure (reticulum) for a day or two. This network is the remainder of the endoplasmic reticulum, and such cells are called **reticulocytes.** This is the stage that exits the bone marrow to enter the blood. When the reticulum degenerates, the cells are fully mature and considered to be erythrocytes.

The average life span of a red blood cell is 120 days. During that time, a red blood cell travels through the body about 75,000 times. Many of these cells are removed from the circulation each day, yet the number of

cells in the circulating blood remains relatively stable. This observation suggests a *homeostatic* control of the rate of red cell production.

A *negative feedback mechanism* utilizing the hormone **erythropoietin** controls the rate of red blood cell formation. In response to prolonged oxygen deficiency, erythropoietin is released, primarily from the kidneys and to a lesser extent from the liver. (In a fetus, the liver is the main site of erythropoietin production.)

At high altitudes, for example, the percentage of oxygen in the air remains the same, but the $PO_2$ decreases. The amount of oxygen delivered to the tissues initially decreases. As figure 14.6 shows, this drop in oxygen concentration triggers release of erythropoietin, which travels via the blood to the red bone marrow and stimulates increased erythrocyte production.

After a few days, large numbers of newly formed red cells begin to appear in the circulating blood. The increased rate of production continues as long as the kidney and liver tissues experience oxygen deficiency. Eventually, the number of erythrocytes in the circulation may be sufficient to supply these tissues with their oxygen requirements. When this happens, or when the oxygen level in the air returns to normal, erythropoietin release decreases, and the rate of red cell production falls.

Several members of a large Danish family have inherited a condition called erythrocytosis. Their reticulocytes are extra-sensitive to erythropoietin, and as a result, these individuals produce about 25% more red blood cells than normal. The result: great physical endurance. One man from this family has won three Olympic gold medals and two world championships in cross-country skiing.

Other conditions produce a low oxygen concentration and stimulate erythropoietin release. These include loss of blood, which decreases the oxygen-carrying capacity of the circulatory system, and chronic lung diseases, which decrease the respiratory surface area available for gas exchange. Table 14.2 summarizes the steps in the regulation of red blood cell production.

**1** Where are red blood cells produced?

**2** How is the production of red blood cells controlled?

## Dietary Factors Affecting Red Blood Cell Production

The availability of two B-complex vitamins—vitamin $B_{12}$ and folic acid—significantly influences red blood cell production. These vitamins are required for DNA synthesis, so they are necessary for growth and reproduction of all cells. Since cellular reproduction occurs at a particularly

TABLE 14.2

## MAJOR EVENTS IN THE HORMONAL CONTROL OF RED BLOOD CELL PRODUCTION

1. The kidney and liver tissues experience an oxygen deficiency.
2. These tissues release the hormone erythropoietin.
3. Erythropoietin travels to the red bone marrow and stimulates an increase in production of red cells.
4. As increasing numbers of red blood cells are released into the circulation, the oxygen-carrying capacity of the blood rises.
5. The oxygen concentration in the kidney and liver tissues increases, and the release of erythropoietin decreases.

high rate in hematopoietic tissue, this tissue is especially vulnerable to deficiency of either of these vitamins. Lack of vitamin $B_{12}$ is usually due to a disorder in the stomach lining rather than to a dietary deficiency, because certain cells in the stomach secrete a substance called *intrinsic factor* that promotes absorption of vitamin $B_{12}$.

In the absence of intrinsic factor, vitamin $B_{12}$ absorption decreases, causing the red bone marrow to form abnormally large, irregularly shaped, thin-membraned fragile cells. This condition, called *pernicious anemia,* can cause permanent brain damage if not treated promptly with vitamin $B_{12}$ injections. Taking excess folic acid—as some pregnant women do to prevent certain birth defects—can mask a vitamin $B_{12}$ deficiency. These women must be careful to follow a balanced diet and get enough vitamin $B_{12}$.

Iron is required for hemoglobin synthesis. Although much of the iron released during the decomposition of hemoglobin is available for reuse, some iron is lost each day and must be replaced. Iron is slowly absorbed from food in the small intestine. Only a small proportion of iron is absorbed even when food contains an abundance of this mineral. The rate of iron absorption varies somewhat with the total amount of iron in the body. Thus, when the amount of iron is low, absorption rate increases, and when the tissues are becoming saturated with iron, the rate is greatly decreased. Figure 14.7 summarizes the life cycle of a red blood cell. The dietary factors that affect red blood cell production are summarized in table 14.3.

Vitamin C increases absorption of iron in the digestive tract. Drinking orange juice with a meat meal is a good way to boost iron intake.

A deficiency of red blood cells or a reduction in the quantity of the hemoglobin they contain results in a condition called **anemia.** This reduces the oxygen-carrying capacity of the blood, and the affected person may appear pale and lack energy (fig. 14.8 and table 14.4). A pregnant woman may become anemic if she doesn't eat iron-rich foods, because her blood volume increases due to fluid retention to accommodate the requirements of the fetus. This increased blood volume decreases the hematocrit. Clinical Application 14.1 discusses another disorder of red blood cells that may have had an impact on American history.

## Types of White Blood Cells

**White blood cells,** or **leukocytes,** protect against disease. Although they work mostly outside the circulatory system, they are transported in the blood to sites of infection.

Normally, five types of white cells are in circulating blood, distinguished by size, the nature of the cytoplasm, the shape of the nucleus, and staining characteristics. Some types of leukocytes have granular cytoplasm and make up a group called **granulocytes,** whereas others lack cytoplasmic granules and are called **agranulocytes.**

A typical granulocyte is about twice the size of a red cell. The members of this group include **neutrophils, eosinophils,** and **basophils.** These cells develop in the red bone marrow from hemocytoblasts in much the same manner as red cells. However, they have a relatively short life span, averaging about 12 hours.

A defect in the ability of granulocytes to destroy pathogenic bacteria can make infection deadly. In *chronic granulomatous disease,* granulocytes engulf bacteria and bring them into the cell, but once there, cannot produce superoxide, a chemical that kills the bacteria. As a result, certain infections rage out of control.

**Neutrophils** have fine cytoplasmic granules that stain light purple with a combination of acid and base stains. The nucleus of an older neutrophil is lobed and consists of two to five sections connected by thin strands of chromatin, and for this reason, they are also called *polymorphonuclear leukocytes.* Younger neutrophils are also called *bands* because their nucleus is C-shaped. Neutrophils are the first white blood cells to arrive at the infection site. These cells phagocytize bacteria, fungi, and

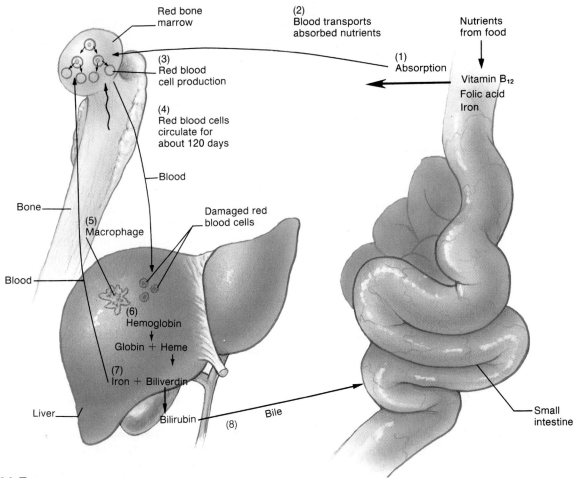

**FIGURE 14.7**

Life cycle of a red blood cell. (1) The intestine absorbs essential nutrients. (2) Blood transports nutrients to red bone marrow. (3) In the marrow, red blood cells arise from reproduction of less specialized cells. (4) Mature red blood cells are released into the bloodstream, where they circulate for about 120 days. (5) Macrophages destroy damaged red blood cells in the spleen and liver. (6) Hemoglobin liberated from red blood cells breaks down into heme and globin. (7) Iron from heme returns to red bone marrow and is reused. (8) Biliverdin and bilirubin are excreted in the bile.

| TABLE 14.3 DIETARY FACTORS AFFECTING RED BLOOD CELL PRODUCTION | | |
|---|---|---|
| **Substance** | **Source** | **Function** |
| Vitamin $B_{12}$ (requires intrinsic factor for absorption via small intestine) | Absorbed from small intestine | DNA synthesis |
| Iron (requires vitamin C for absorption in small intestine) | Absorbed from small intestine; conserved during red cell destruction and made available for reuse | Hemoglobin synthesis |
| Folic acid | Absorbed from small intestine | DNA synthesis |

some viruses. Neutrophils account for 54% to 62% of the leukocytes in a typical adult blood sample (fig. 14.9).

**Eosinophils** contain coarse, uniformly sized cytoplasmic granules that stain deep red in acid stain. The nucleus usually has only two lobes (bilobed). Eosinophils moderate allergic reactions and defend against parasitic worm infestation. These cells make up 1% to

3% of the total number of circulating leukocytes (fig. 14.10).

**Basophils** are similar to eosinophils in size and in the shape of their nuclei. However, they have fewer, more irregularly shaped cytoplasmic granules that stain deep blue in basic stain. These granules often obscure a view of the nucleus. Basophils migrate to damaged tissues where

# King George III and Porphyria

Some historians blame the American Revolution on a blood disease—an abnormality of hemoglobin that afflicted King George III, who ruled England at the time. So puzzling were his symptoms that not until this century did medical researchers discover the underlying disorder, called *porphyria*.

At age 50, the king first experienced abdominal pain and constipation, followed by weak limbs, fever, a rapid pulse, hoarseness, and dark red urine. Next, nervous system symptoms began, including insomnia, headaches, visual problems, restlessness, delirium, convulsions, and stupor. His confused and racing thoughts, combined with his ripping off his wig and running about naked while at the peak of a fever, convinced court observers that the king was mad. Just as Parliament was debating his ability to rule, he mysteriously recovered.

But George's plight was far from over. He suffered a relapse 13 years later, then again 3 years after that. Always the symptoms appeared in the same order, beginning with abdominal pain, fever, and weakness and progressing to the nervous system symptoms. Finally, an attack in 1811 placed him in an apparently permanent stupor, and he was dethroned by the Prince of Wales. He lived for several more years, experiencing further episodes of his odd affliction.

In George III's time, physicians were permitted to do very little to the royal body, basing diagnoses on what the patient told them. Twentieth-century researchers found that George's red urine was caused by an inborn error of metabolism. In porphyria, because of the absence of an enzyme, part of the blood pigment hemoglobin, called a porphyrin ring, is routed into the urine instead of being broken down and metabolized by cells. Porphyrin builds up and attacks the nervous system, causing many of the other symptoms. Examination of the medical records of King George III's descendants reveal several of them also had symptoms of porphyria. The underlying defect in red blood cell recycling had appeared in its various guises as different problems. ■

(a)

(b)

## FIGURE 14.8

(a) Light micrograph of normal human erythrocytes (400× micrograph enlarged to 1,200×). (b) Light micrograph of erythrocytes from a person with hypochromic anemia (280× micrograph enlarged to 900×).

| TABLE **14.4** | TYPES OF ANEMIA | |
|---|---|---|
| **Type** | **Cause** | **Defect** |
| Aplastic anemia | Toxic chemicals, radiation | Damaged bone marrow |
| Hemolytic anemia | Toxic chemicals | Red blood cells destroyed |
| Iron deficiency anemia | Dietary lack of iron | Hemoglobin deficient |
| Pernicious anemia | Inability to absorb vitamin $B_{12}$ | Excess of immature cells |
| Sickle cell disease | Defective gene | Red blood cells abnormally shaped |
| Thalassemia | Defective gene | Hemoglobin deficient; red blood cells short-lived |

they release *histamine* that promotes inflammation, and *heparin* that inhibits blood clotting, thus increasing blood flow to injured tissues. This type of leukocyte usually accounts for less than 1% of the leukocytes (fig. 14.11).

The leukocytes of the agranulocyte group include **monocytes** and **lymphocytes.** Monocytes generally arise from red bone marrow. Lymphocytes are formed in the organs of the lymphatic system as well as in the red bone marrow.

**Monocytes** are the largest cells in blood, two to three times greater in diameter than red blood cells. Their nuclei vary in shape and are round, kidney-shaped, oval, or lobed. Monocytes leave the bloodstream and become *macrophages* that phagocytize bacteria, dead cells, and other debris in the tissues. They usually make up 3% to 9% of the leukocytes in a blood sample and live for several weeks or even months (fig. 14.12).

**Lymphocytes** are usually only slightly larger than erythrocytes. A typical lymphocyte contains a large, round nucleus surrounded by a thin rim of cytoplasm. The major types of lymphocytes are *T cells* and *B cells,* both important in *immunity.* T cells directly attack microorganisms, tumor cells, and transplant cells. B cells produce **antibodies** (see chapter 16) that attack foreign cells or proteins. Lymphocytes account for 25% to 33% of the circulating leukocytes. They may live for years (fig. 14.13).

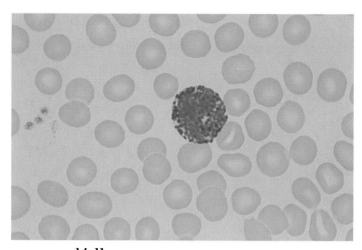

**FIGURE** 14.11

A basophil has cytoplasmic granules that stain deep blue (400× micrograph enlarged to 950×).

**FIGURE** 14.9

A neutrophil has a lobed nucleus with two to five components (400× micrograph enlarged to 950×).

**FIGURE** 14.12

A monocyte may leave the bloodstream and become a macrophage (500× micrograph enlarged to 1,250×).

**FIGURE** 14.10

An eosinophil has red-staining cytoplasmic granules (400× micrograph enlarged to 950×).

**FIGURE** 14.13

The lymphocyte contains a large, round nucleus (400× micrograph enlarged to 950×).

1. Distinguish between granulocytes and agranulocytes.

2. List five types of white blood cells, and explain how they differ from one another.

3. Describe the function of each type of white blood cell.

## White Blood Cells Fight Infection

Leukocytes protect against infection in various ways. Some leukocytes phagocytize bacterial cells in the body, and others produce proteins (*antibodies*) that destroy or disable foreign particles.

Leukocytes can squeeze between the cells that form blood vessel walls. This movement, called **diapedesis,** allows the white cells to leave the circulation (fig. 14.14). Once outside the blood, they move through interstitial spaces using a form of self-propulsion called *ameboid motion.*

The most mobile and active phagocytic leukocytes are neutrophils and monocytes. Although neutrophils are unable to ingest particles much larger than bacterial cells, monocytes can engulf relatively large objects. Monocytes contain numerous lysosomes, which are filled with digestive enzymes that break down organic molecules in captured bacteria. Neutrophils and monocytes often become so engorged with digestive products and bacterial toxins that they also die.

When microorganisms invade human tissues, certain body cells respond by releasing biochemicals that dilate local blood vessels. One such substance, **hista-mine,** dilates arterioles, allowing more blood to flood the capillaries. The tissues redden and copious fluids leak into the interstitial spaces. The swelling this *inflammatory reaction* produces delays the spread of invading microorganisms into other regions (see chapter 5). At the same time, damaged cells release chemicals that attract leukocytes. This phenomenon is called **positive chemotaxis,** and when combined with diapedesis brings many white blood cells into inflamed areas quickly (fig. 14.15).

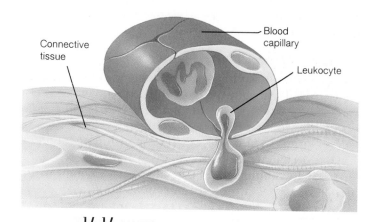

**FIGURE** 14.14

Leukocytes can squeeze between the cells of a capillary wall and enter the tissue space outside the bloodstream.

**FIGURE** 14.15

When bacteria invade the tissues, leukocytes migrate into the region and destroy the microbes by phagocytosis.

As bacteria, leukocytes, and damaged cells accumulate in an inflamed area, a thick fluid called *pus* often forms, and remains while the invading microorganisms are active. If the pus cannot escape to the outside of the body or into a body cavity, it may remain trapped in the tissues for some time. Eventually surrounding cells absorb it.

**1** How do white blood cells fight infection?

**2** How do white blood cells reach microorganisms that are outside blood vessels?

## White Blood Cell Counts

The procedure used to count white blood cells is similar to that used for counting red cells. However, before a *white cell count* is made, the red cells in the blood sample are destroyed so they will not be mistaken for white cells. Normally, a cubic millimeter of blood includes 5,000 to 10,000 white cells.

The total number and percentages of different white blood cell types are of clinical interest. A rise in the number of circulating white cells may indicate infection. A total number of white cells exceeding 10,000 per mm³ of blood constitutes **leukocytosis,** indicating acute infection, such as appendicitis. It may also appear following vigorous exercise, emotional disturbances, or great loss of body fluids.

A total white cell count below 5,000 per mm³ of blood is called **leukopenia.** Such a deficiency may accompany typhoid fever, influenza, measles, mumps, chicken pox, AIDS, or poliomyelitis. It may also result from anemia or from lead, arsenic, or mercury poisoning.

A *differential white blood cell count* (DIFF) lists percentages of the types of leukocytes in a blood sample. This test is useful because the relative proportions of white cells may change in particular diseases. Neutrophils, for instance, usually increase during bacterial infections, and eosinophils may increase during certain parasitic infections and allergic reactions. In HIV infection and AIDS, the numbers of a type of lymphocyte called helper T cells plummet.

Table 14.5 lists some disorders that alter the numbers of particular types of white blood cells. Clinical Application 14.2 examines leukemia, cancer of white blood cells.

**1** What is the normal human white blood cell count?

**2** Distinguish between leukocytosis and leukopenia.

**3** What is a differential white blood cell count?

## Blood Platelets

**Platelets,** or **thrombocytes,** are not complete cells. They arise from very large cells in the red bone marrow, called

| TABLE 14.5 | WHITE BLOOD CELL ALTERATIONS |
|---|---|
| **White Blood Cell Population Change** | **Illness** |
| Elevated lymphocytes | Hairy cell leukemia, whooping cough, mononucleosis |
| Elevated eosinophils | Tapeworm infestation, hookworm infestation |
| Elevated monocytes | Typhoid fever, malaria, tuberculosis |
| Too few helper T cells | AIDS |

**megakaryocytes,** that fragment a little like a shattered plate, releasing small sections of cytoplasm—platelets—into the circulation. The larger fragments of the megakaryocytes shrink and become platelets as they pass through the blood vessels of the lungs.

Each platelet lacks a nucleus and is less than half the size of a red blood cell. It is capable of ameboid movement and may live for about ten days. In normal blood, the *platelet count* varies from 130,000 to 360,000 platelets per mm³.

Platelets help repair damaged blood vessels by sticking to broken surfaces. They release **serotonin,** which contracts smooth muscles in the vessel walls, reducing blood flow. Table 14.6 summarizes the characteristics of blood cells and platelets.

**1** What is the normal human blood platelet count?

**2** What is the function of blood platelets?

## BLOOD PLASMA

Plasma is the clear, straw-colored, liquid portion of the blood in which the cells and platelets are suspended. It is approximately 92% water and contains a complex mixture of organic and inorganic biochemicals. Functions of plasma constituents include transporting nutrients, gases, and vitamins; helping to regulate fluid and electrolyte balance; and maintaining a favorable pH. Figure 14.16 shows the chemical makeup of plasma.

### Plasma Proteins

**Plasma proteins** are the most abundant dissolved substances (solutes) in plasma. These proteins remain in the blood and interstitial fluids, and ordinarily are not used as energy sources. The three main plasma protein groups are albumins, globulins, and fibrinogen. The groups differ in chemical composition and physiological function.

# Leukemia

The young woman had noticed symptoms for several months before she finally went to the doctor. At first it was just fatigue and headaches, which she attributed to studying for final exams. She had frequent colds and bouts of fever, chills, and sweats that she thought were just minor infections. When she developed several bruises and bone pain and noticed that her blood did not clot very quickly after cuts and scrapes, she consulted her physician, who examined her and took a blood sample. One glance at a blood smear under a microscope alarmed the doctor—there were far too few red blood cells and platelets, and too many white blood cells. She sent the sample to a laboratory to diagnose the type of *leukemia,* or cancer of the white blood cells, that was causing her patient's symptoms.

The young woman had *myeloid leukemia.* Her red bone marrow was producing too many granulocytes, but they were immature cells, unable to fight infection. This explained the frequent illnesses. The leukemic cells were crowding out red blood cells and their precursors in the red marrow, causing her anemia and resulting fatigue. Platelet deficiency (thrombocytopenia) led to increased tendency to bleed. Finally, spread of the cancer cells outside the marrow painfully weakened the surrounding bone. Eventually, if she wasn't treated, the cancer cells would spread outside the circulatory system, causing other tissues that would normally not produce white blood cells to do so.

A second type of leukemia, distinguished by the source of the cancer cells, is *lymphoid leukemia.* These cancer cells are lymphocytes, produced in lymph nodes. Many of the symptoms are similar to those of myeloid leukemia. Sometimes a person has no leukemia symptoms at all, and a routine blood test detects the condition (fig. 14A).

Leukemia is also classified as acute or chronic. An acute condition appears suddenly, symptoms progress rapidly, and death occurs in a few months without treatment. Chronic forms begin more slowly and may remain undetected for months or even years or, in rare cases, decades. Without treatment, life expectancy after symptoms develop is about 3 years. With treatment, 50% to 80% of patients enter remission, a period of stability that may become a cure. Chemotherapy may be necessary for a year or two to increase the chances of long remission.

Leukemia treatment includes correcting symptoms by giving blood transfusions, treating infections, and using drugs that kill cancer cells. Several drugs in use for many years have led to spectacular increases in cure rates, particularly for acute lymphoid leukemia in children. Some newer treatments offer hope for other types of leukemia, too. For example, alpha interferon treats a chronic form called hairy cell leukemia that mostly affects adult males, and a type of retinoic acid treats acute promyelocytic leukemia. A bone marrow transplant can cure leukemia, but it is a very risky procedure. ■

(a)

(b)

**FIGURE 14A**

(*a*) Normal blood cells (400× micrograph enlarged to 800×). (*b*) Blood cells from a person with granulocytic leukemia (400× micrograph enlarged to 800×). Note the increased number of leukocytes.

TABLE 14.6

## CELLULAR COMPONENTS OF BLOOD

| Component | Description | Number Present | Function |
|---|---|---|---|
| Red blood cell (erythrocyte) | Biconcave disk without a nucleus, about one-third hemoglobin | 4,200,000 to 6,200,000 per $mm^3$ | Transports oxygen and carbon dioxide |
| White blood cell (leukocyte) | | 5,000 to 10,000 per $mm^3$ | Destroys pathogenic microorganisms and parasites and removes worn cells |
| *Granulocytes* | About twice the size of red cells; cytoplasmic granules are present | | |
| 1. Neutrophil | Nucleus with two to five lobes; cytoplasmic granules stain light purple in combined acid and base stains | 54%–62% of white cells present | Phagocytizes small particles |
| 2. Eosinophil | Nucleus bilobed; cytoplasmic granules stain red in acid stain | 1%–3% of white cells present | Kills parasites and helps control inflammation and allergic reaction |
| 3. Basophil | Nucleus lobed; cytoplasmic granules stain blue in basic stain | Less than 1% of white cells present | Releases heparin and histamine |
| *Agranulocytes* | Cytoplasmic granules are absent | | |
| 1. Monocyte | Two to three times larger than a red cell; nuclear shape varies from round to lobed | 3%–9% of white cells present | Phagocytizes large particles |
| 2. Lymphocyte | Only slightly larger than a red cell; its nucleus nearly fills cell | 25%–33% of white cells present | Provides immunity |
| Platelet (thrombocyte) | Cytoplasmic fragment | 130,000 to 360,000 per $mm^3$ | Helps control blood loss from broken vessels |

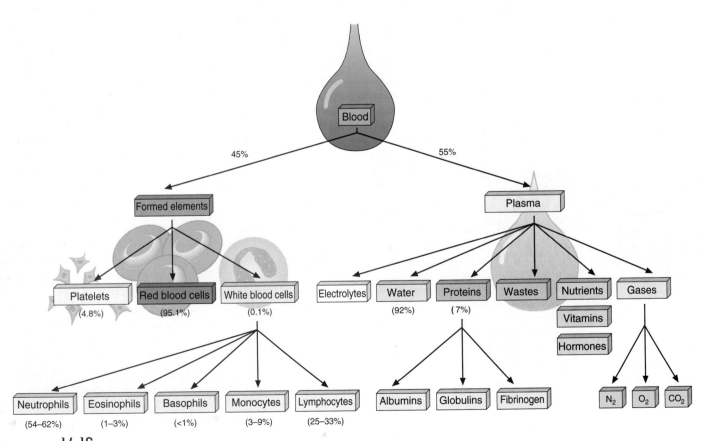

**FIGURE 14.16**
Blood composition.

**Albumins** are the smallest of the plasma proteins, yet account for about 60% of these proteins by weight. They are synthesized in the liver, and because they are so plentiful, albumins are an important determinant of the *osmotic pressure* of the blood.

As explained in chapter 3, whenever the concentration of solutes changes on either side of a cell membrane, water moves through the membrane toward the region where the solutes are in higher concentration. For this reason, it is important that the concentration of solutes in the plasma remains relatively stable. Otherwise, water tends to leave the blood and enter the tissues, or leave the tissues and enter the blood, by osmosis. Because the presence of albumins (and other plasma proteins) adds to the osmotic pressure of the plasma, albumins help regulate water movement between the blood and the tissues. In doing so, they help control blood volume, which, in turn, directly affects blood pressure. Albumins also bind and transport certain molecules, such as bilirubin and free fatty acids.

> If the concentration of plasma proteins falls, tissues swell, a condition called *edema*. This may result from starvation or a protein-deficient diet, either of which requires the body to use protein for energy, or from an impaired liver that cannot synthesize plasma proteins. As blood concentration of plasma proteins drops, so does the blood osmotic pressure, sending fluids into the intercellular spaces.

**Globulins,** which make up about 36% of the plasma proteins, can be further subdivided into *alpha, beta,* and *gamma globulins.* The liver synthesizes alpha and beta globulins. They have a variety of functions, including transport of lipids and fat-soluble vitamins. Lymphatic tissues produce the gamma globulins, which are a type of antibody (see chapter 16).

**Fibrinogen,** which constitutes about 4% of the plasma protein, plays a primary role in blood coagulation. Synthesized in the liver, it is the largest of the plasma proteins. The function of fibrinogen is discussed later in this chapter. Table 14.7 summarizes the characteristics of the plasma proteins.

1. List three types of plasma proteins.

2. How do albumins help maintain water balance between the blood and the tissues?

3. Which of the globulins functions in immunity?

4. What is the role of fibrinogen?

## Gases and Nutrients

The most important *blood gases* are oxygen and carbon dioxide. Plasma also contains a considerable amount of

| TABLE 14.7 PLASMA PROTEINS | | | |
|---|---|---|---|
| **Protein** | **Percentage of Total** | **Origin** | **Function** |
| Albumin | 60% | Liver | Helps maintain blood osmotic pressure |
| Globulin | 36% | | |
|   Alpha globulins | | Liver | Transports lipids and fat-soluble vitamins |
|   Beta globulins | | Liver | Transports lipids and fat-soluble vitamins |
|   Gamma globulins | | Lymphatic tissues | Constitute the antibodies of immunity |
| Fibrinogen | 4% | Liver | Plays a key role in blood coagulation |

dissolved nitrogen, which ordinarily has no physiological function. Chapter 21 discusses blood gases and their transport.

> As a rule, blood gases are evaluated using a fresh sample of whole blood obtained from an artery. This blood is cooled to decrease the rates of metabolic reactions, and an anticoagulant is added to prevent clotting. In the laboratory, the partial pressures of oxygen and carbon dioxide of the blood are determined, the blood pH is measured, and the plasma bicarbonate concentration is calculated. Such information is used to diagnose and treat disorders of circulation, respiration, and electrolyte balance. Appendix C lists average values for these laboratory tests.

The *plasma nutrients* include amino acids, simple sugars, nucleotides, and lipids absorbed from the digestive tract. For example, plasma transports glucose from the small intestine to the liver, where it may be stored as glycogen or altered to form fat. If blood glucose concentration drops below the normal range, glycogen may be broken down into glucose, as described in chapter 13.

Recently absorbed amino acids are also carried to the liver. Here they may be used to manufacture proteins, or deaminated and used as an energy source (see chapter 4).

Plasma lipids include fats (triglycerides), phospholipids, and cholesterol. Because lipids are not water soluble and plasma is almost 92% water, these lipids combine with proteins in **lipoprotein** complexes. Lipoprotein molecules are relatively large and consist of a surface layer of phospholipid, cholesterol, and protein

TABLE

## 14.8 PLASMA LIPOPROTEINS

| Lipoprotein | Characteristics | Functions |
|---|---|---|
| Chylomicron | High concentration of triglycerides | Transports dietary fats to muscle and adipose cells |
| Very low-density lipoprotein (VLDL) | Relatively high concentration of triglycerides; produced in the liver | Transports triglycerides synthesized in the liver from carbohydrates to adipose cells |
| Low-density lipoprotein (LDL) | Relatively high concentration of cholesterol; formed from remnants of VLDL molecules that have given up their triglycerides | Delivers cholesterol to various cells, including liver cells |
| High-density lipoprotein (HDL) | Relatively high concentration of protein and low concentration of lipids | Transports to the liver remnants of chylomicrons that have given up their triglycerides |

surrounding a triglyceride core. *Apoproteins* in the outer layer can combine with receptors on the membranes of specific target cells. Lipoprotein molecules vary in the proportions of the lipids they contain.

Because lipids are less dense than proteins, as the proportion of lipids in a lipoprotein increases, the density of the particle decreases. Conversely, as the proportion of lipids decreases, the density increases. Lipoproteins are classified on the basis of their densities, which reflect their composition. *Chylomicrons* mainly consist of triglycerides absorbed from the small intestine (see chapter 17). *Very low-density lipoproteins* (VLDL) have a relatively high concentration of triglycerides. *Low-density lipoproteins* (LDL) have a relatively high concentration of cholesterol and are the major cholesterol-carrying lipoproteins. *High-density lipoproteins* (HDL) have a relatively high concentration of protein and a lower concentration of lipids.

Many residents of Tangier Island in the Chesapeake Bay have inherited a deficiency in a particular blood lipoprotein from the original settlers, who arrived in 1686. A low blood level of alpha lipoprotein, which normally transports cholesterol in the blood, leads to low blood cholesterol. However, excess cholesterol accumulates elsewhere, such as in the thymus gland, and in scavenging macrophages.

Chylomicrons, discussed further in chapter 17, transport dietary fats to muscle and adipose cells. Similarly, very low-density lipoproteins, produced in the liver, transport triglycerides synthesized from excess dietary carbohydrates. After VLDL molecules deliver their loads of triglycerides to adipose cells, an enzyme, lipoprotein lipase, converts their remnants to low-density lipoproteins. Because most of the triglycerides have been removed, LDL molecules have a higher cholesterol content than do the original VLDL molecules. Various cells, including liver cells, have surface receptors that combine with apoproteins associated with LDL molecules. These cells slowly remove LDL from plasma by receptor-mediated endocytosis, supplying cells with cholesterol (see chapter 3).

After chylomicrons deliver their triglycerides to cells, their remnants are transferred to high-density lipoproteins. These HDL molecules, which form in the liver and small intestine, transport chylomicron remnants to the liver, where they rapidly enter cells by receptor-mediated endocytosis. The liver disposes of the cholesterol it obtains in this manner by secreting it into bile or by using it to synthesize bile salts. Table 14.8 summarizes the characteristics and functions of these lipoproteins.

Much of the cholesterol and bile salts in bile are later reabsorbed by the small intestine and transported back to the liver, and the secretion-reabsorption cycle repeats. During each cycle, some of the cholesterol and bile salts escape reabsorption, reach the large intestine, and are eliminated with the feces.

1. Which gases are in plasma?

2. Which nutrients are in blood plasma?

3. How are triglycerides transported in plasma?

4. How is cholesterol eliminated from the liver?

## Nonprotein Nitrogenous Substances

Molecules that contain nitrogen atoms but are not proteins comprise a group called **nonprotein nitrogenous substances.** In plasma, this group includes amino acids, urea, uric acid, creatine, and creatinine. Amino acids come from protein digestion and amino acid absorption. Urea and uric acid are products of protein and nucleic acid catabolism, respectively, and creatinine results from the metabolism of creatine. As discussed in chapter 9, creatine occurs as **creatine phosphate** in muscle and brain tissues as well as in the blood, where it stores high-energy phosphate bonds, much like those of ATP molecules.

Normally, the concentration of nonprotein nitrogenous (NPN) substances remains relatively stable because protein intake and utilization are balanced with excretion of nitrogenous wastes. Because about half of the

NPN substances is urea, which the kidneys ordinarily excrete, a rise in the blood urea nitrogen (BUN) may suggest a kidney disorder. It may also result from excessive protein catabolism or infection.

## Plasma Electrolytes

Blood plasma contains a variety of *electrolytes*. Recall that electrolytes release ions when dissolved in water. Plasma electrolytes are absorbed from the intestine or released as by-products of cellular metabolism. They include sodium, potassium, calcium, magnesium, chloride, bicarbonate, phosphate, and sulfate ions. Of these, sodium and chloride ions are the most abundant. Such ions are important in maintaining the osmotic pressure and the pH of plasma, and like other plasma constituents, they are regulated so that their blood concentrations remain relatively stable. These electrolytes are discussed in chapter 21 in connection with water and electrolyte balance.

1    What is a nonprotein nitrogenous substance?

2    Why does kidney disease increase the blood concentration of these substances?

3    What are the sources of plasma electrolytes?

## HEMOSTASIS

The term **hemostasis** refers to the stoppage of bleeding, which is vitally important when blood vessels are damaged. Following an injury to the blood vessels, several actions may help prevent excessive blood loss, including blood vessel spasm, platelet plug formation, and blood coagulation.

These mechanisms are most effective in minimizing blood losses from small vessels. Injury to a larger vessel may result in a severe hemorrhage that requires special treatment.

### Blood Vessel Spasm

Cutting or breaking an arteriole or a venule stimulates the smooth muscles in its wall to contract, and blood loss lessens almost immediately. In fact, the ends of a severed vessel may close completely by such a *vasospasm*. This effect results from direct stimulation of the vessel wall as well as from reflexes elicited by pain receptors in the injured tissues.

Although the reflex response may last only a few minutes, the effect of the direct stimulation usually continues for about thirty minutes. By then, a *platelet plug* has formed and blood coagulation is occurring. Also, platelets release serotonin, which contracts smooth muscles in the blood vessel walls. This vasoconstriction further helps to reduce blood loss.

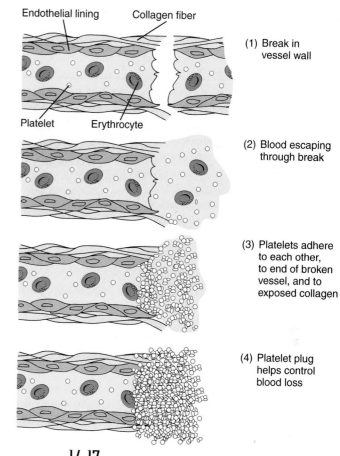

Endothelial lining    Collagen fiber

Platelet    Erythrocyte

(1) Break in vessel wall

(2) Blood escaping through break

(3) Platelets adhere to each other, to end of broken vessel, and to exposed collagen

(4) Platelet plug helps control blood loss

**FIGURE 14.17**
Steps in platelet plug formation.

## Platelet Plug Formation

Platelets stick to exposed ends of injured blood vessels. They adhere to any rough surface, particularly to the collagen in connective tissue underlying the endothelial lining of blood vessels.

When platelets contact collagen, their shapes change drastically, and numerous spiny processes begin to protrude from their membranes. At the same time, they tend to stick to each other, forming a platelet plug in the vascular break. A plug may control blood loss from a small break, but a larger one may require a blood clot to halt bleeding. Figure 14.17 shows the steps in platelet plug formation.

1    What is hemostasis?

2    How does a blood vessel spasm help control bleeding?

3    Describe the formation of a platelet plug.

## Blood Coagulation

**Coagulation,** the most effective hemostatic mechanism, causes formation of a *blood clot* by a series of reactions, each one activating the next in a chain reaction or *cascade*.

| TABLE 14.9 | HEMOSTATIC MECHANISMS | |
| --- | --- | --- |
| Mechanism | Stimulus | Effect |
| Blood vessel spasm | Direct stimulus to vessel wall or to pain receptors; platelets release serotonin | Smooth muscles in vessel wall contract reflexly; vasoconstriction helps maintain prolonged vessel spasm |
| Platelet plug formation | Exposure of platelets to rough surfaces or to collagen of connective tissue | Platelets adhere to rough surfaces and to each other, forming a plug |
| Blood coagulation | Cellular damage and blood contact with foreign surfaces result in activation of factors that favor coagulation | Blood clot forms as a result of a series of reactions, terminating in conversion of fibrinogen into fibrin |

**FIGURE 14.18**
A scanning electron micrograph of fibrin threads (2,800×).

Coagulation may occur by an extrinsic or an intrinsic clotting mechanism. Release of biochemicals from broken blood vessels or damaged tissues triggers the **extrinsic clotting mechanism.** Blood contact with foreign surfaces in the absence of tissue damage stimulates the **intrinsic clotting mechanism.**

Blood coagulation is very complex, and utilizes many biochemicals called *clotting factors* (designated by Roman numerals indicating the order of their discovery). Vitamin K is necessary for some clotting factors to function. Some of these factors promote coagulation (procoagulants) and others inhibit it (anticoagulants). Whether or not the blood coagulates depends on the balance between these two groups of factors. Normally, the anticoagulants prevail, and the blood does not clot. However, as a result of injury (trauma), biochemicals that favor coagulation may increase in concentration, and the blood may coagulate.

The major event in blood clot formation is conversion of the soluble plasma protein *fibrinogen* (factor I) into insoluble threads of the protein **fibrin.** Activation of certain plasma proteins by still other protein factors triggers conversion of fibrinogen to fibrin. Table 14.9 summarizes the three primary hemostatic mechanisms.

## Extrinsic Clotting Mechanism

The extrinsic clotting mechanism is triggered when blood contacts damaged blood vessel walls or tissues outside blood vessels. Such damaged tissues release a complex of substances called *tissue thromboplastin* (factor III) that is associated with disrupted cell membranes. The presence of tissue thromboplastin activates factor VII, which combines with and activates factor X. Further, factor X combines with and activates factor V. These reactions, which

also depend on the presence of calcium ions (factor IV), lead to production of *prothrombin activator.*

**Prothrombin** (factor II) is an alpha globulin that the liver continually produces and is thus a normal constituent of plasma. In the presence of calcium ions, prothrombin activator converts prothrombin into **thrombin** (factor IIa). Thrombin, in turn, catalyzes a reaction that fragments fibrinogen (factor I). The fibrinogen fragments join, forming long threads of fibrin. Fibrinogen is a soluble plasma protein, but fibrin is insoluble. Thrombin also activates factor XIII, which strengthens and stabilizes fibrin threads.

Once fibrin threads form, they stick to exposed surfaces of damaged blood vessels, creating a meshwork that entraps blood cells and platelets (fig. 14.18). The resulting mass is a blood clot, which may block a vascular break and prevent further blood loss.

The amount of prothrombin activator that appears in the blood is directly proportional to the degree of tissue damage. Once a blood clot begins to form, it promotes still more clotting, because thrombin also acts directly on blood clotting factors other than fibrinogen, causing prothrombin to form still more thrombin. This type of self-initiating action is an example of a **positive feedback system,** in which the original action stimulates more of the same type of action. Such a mechanism produces unstable conditions and can operate for only a short time in a living system, because life depends on the maintenance of a stable internal environment (see chapter 1).

Normally, blood flow throughout the body prevents formation of a massive clot within the cardiovascular system by rapidly carrying excess thrombin away and keeping its concentration too low to enhance further clotting. Also, a naturally occurring substance called *an-*

*tithrombin,* present in the blood and on the surfaces of endothelial cells that line blood vessels, limits thrombin formation. Consequently, blood coagulation is usually limited to blood that is standing still (or moving relatively slowly), and clotting ceases where a clot contacts circulating blood.

In *disseminated intravascular clotting,* coagulation is abnormally activated in widespread regions of the circulatory system. This condition is usually associated with bacteria or bacterial toxins in the blood or with a disorder causing widespread tissue damage. As a result, many small clots may appear and obstruct blood flow into various tissues and organs, particularly the kidneys. As plasma clotting factors and platelets are depleted, the patient may develop serious bleeding.

## Intrinsic Clotting Mechanism

Unlike extrinsic clotting, all of the components necessary for intrinsic clotting are in the blood. Activation of a substance called the *Hageman factor* (factor XII) initiates intrinsic clotting. This happens when blood is exposed to a foreign surface such as collagen or when blood is stored in a glass container. Activated factor XII activates factor XI, which activates factor IX. Factor IX then joins with factor VIII and platelet phospholipids to activate factor X. These reactions, which also depend on the presence of calcium ions, lead to the production of prothrombin activator. The subsequent steps of blood clot formation are the same as those described for the extrinsic mechanism (fig. 14.19). Table 14.10 compares extrinsic and intrinsic clotting mechanisms. Table 14.11 lists the clotting factors, their sources, and mechanisms.

Laboratory tests commonly used to evaluate blood coagulation mechanisms include *prothrombin time* (PT) and *partial thromboplastin time* (PTT). These tests measure the time it takes for fibrin threads to form in a sample of blood plasma. However, the prothrombin time test checks the extrinsic clotting mechanism, whereas the partial thromboplastin test evaluates intrinsic clotting.

## Fate of Blood Clots

After a blood clot forms, it soon begins to retract as the tiny processes extending from the platelet membranes adhere to strands of fibrin within the clot, and contract. The blood clot shrinks, pulling the edges of the broken vessel closer together and squeezing a fluid called **serum** from the clot. Serum is essentially plasma minus all of its fibrinogen and most of the other clotting factors. Platelets associated with a blood clot also release *platelet-derived*

*growth factor* (PDGF), which stimulates smooth muscle cells and fibroblasts to repair damaged blood vessel walls.

Fibroblasts soon invade blood clots that form in ruptured vessels, producing connective tissue with numerous fibers throughout the clots, which helps strengthen and seal vascular breaks. Many clots, including those that form in tissues as a result of blood leakage (hematomas), disappear in time. In clot dissolution, fibrin threads absorb a plasma protein called *plasminogen* (profibrinolysin). Then a substance called plasminogen activator released from the lysosomes of damaged tissue cells converts plasminogen to *plasmin.* Plasmin is a protein-splitting enzyme that can digest fibrin threads and other proteins associated with blood clots. Plasmin formation may dissolve a whole clot; however, clots that fill large blood vessels are seldom removed naturally.

A blood clot abnormally forming in a vessel is a **thrombus.** If the clot dislodges or if a fragment of it breaks loose and is carried away by the blood flow, it is called an **embolus.** Generally, emboli continue to move until they reach narrow places in vessels where they may lodge and block blood flow, causing an **embolism.**

Abnormal clot formations are often associated with conditions that change the endothelial linings of vessels. For example, in *atherosclerosis,* accumulations of fatty deposits change arterial linings, sometimes initiating inappropriate clotting. This is the most common cause of thrombosis in medium-sized arteries (fig. 14.20).

Coagulation may also occur in blood that is flowing too slowly. The concentration of clot-promoting substances may increase to a critical level instead of being carried away by more rapidly moving blood, and a clot may form. This mechanism is the usual cause of thrombosis in veins.

A blood clot forming in a vessel that supplies a vital organ, such as the heart (coronary thrombosis) or the brain (cerebral thrombosis), kills tissues the vessel serves (*infarction*) and may be fatal. A blood clot that travels and then blocks a vessel that supplies a vital organ, such as the lungs (pulmonary embolism), affects the portion of the organ that the blocked blood vessel supplies. Plasminogen activators break up abnormal blood clots and are used as drugs to treat heart attacks and strokes.

Drugs based on "clot-busting" biochemicals can be lifesavers. *Tissue plasminogen activator* (tPA) may restore blocked coronary circulation if given within 4 hours of a heart attack. A drug derived from bacteria called *streptokinase* may also be successful, for a fraction of the cost. Another plasminogen activator used as a drug is *urokinase,* an enzyme produced in certain kidney cells. Heparin and coumadin are drugs that interfere with clot formation.

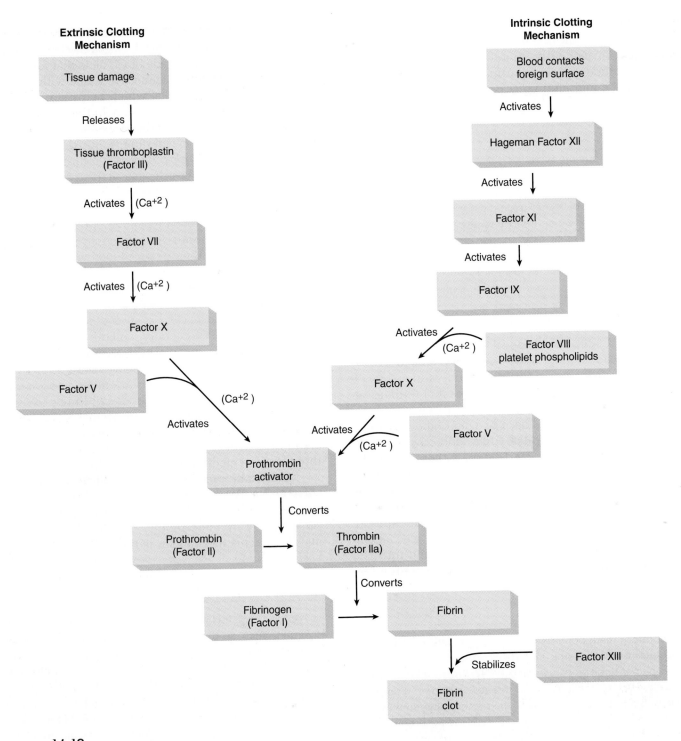

**FIGURE** 14.19

Schematic of blood-clotting mechanisms.

**1** Distinguish between extrinsic and intrinsic clotting mechanisms.

**2** What is the basic event in blood clot formation?

**3** What factors initiate the formation of fibrin?

**4** What prevents the formation of massive clots throughout the cardiovascular system?

**5** Distinguish between a thrombus and an embolus.

**6** How might atherosclerosis promote the formation of blood clots?

TABLE 14.10

## BLOOD COAGULATION

| Steps | Extrinsic Clotting Mechanism | Intrinsic Clotting Mechanism |
|---|---|---|
| Trigger | Damage to vessel or tissue | Blood contacts foreign surface |
| Initiation | Tissue thromboplastin | Hageman factor |
| Series of reactions involving several clotting factors and calcium ions ($Ca^{+2}$) lead to the production of | Prothrombin activator | Prothrombin activator |
| Prothrombin activator and calcium ions cause the conversion of | Prothrombin to thrombin | Prothrombin to thrombin |
| Thrombin causes fragmentation, then joining of | Fibrinogen to fibrin | Fibrinogen to fibrin |

TABLE 14.11

## CLOTTING FACTORS

| Clotting Factor | Source | Mechanism(s) |
|---|---|---|
| I (fibrinogen) | Synthesized in liver | Extrinsic and intrinsic |
| II (prothrombin) | Synthesized in liver, requires vitamin K | Extrinsic and intrinsic |
| III (tissue thromboplastin) | Damaged tissue | Extrinsic |
| IV (calcium ions) | Diet, bone | Extrinsic and intrinsic |
| V (proaccelerin) | Synthesized in liver, released by platelets | Extrinsic and intrinsic |
| VII (serum prothrombin conversion accelerator) | Synthesized in liver, requires vitamin K | Extrinsic |
| VIII (antihemophilic factor) | Released by platelets and endothelial cells | Intrinsic |
| IX (plasma thromboplastin component) | Synthesized in liver, requires vitamin K | Intrinsic |
| X (Stuart-Prower factor) | Synthesized in liver, requires vitamin K | Extrinsic and intrinsic |
| XI (plasma thromboplastin antecedent) | Synthesized in liver | Intrinsic |
| XII (Hageman factor) | Synthesized in liver | Intrinsic |
| XIII (fibrin-stabilizing factor) | Synthesized in liver, released by platelets | Extrinsic and intrinsic |

(a)  Lumen    Artery wall

## FIGURE 14.20

(a) Light micrograph of a normal artery (55×). (b) The inner wall of an artery changed as a result of atherosclerosis (55×).

Fatty deposit    Lumen filled with blood    Artery wall

(b)

TABLE 14.12

## FACTORS THAT INHIBIT BLOOD CLOT FORMATION

| Factor | Action | Factor | Action |
|---|---|---|---|
| Smooth lining of blood vessel | Prevents activation of intrinsic blood clotting mechanism | Antithrombin in plasma | Interferes with the action of thrombin |
| Prostacyclin | Inhibits adherence of platelets to blood vessel wall | Heparin from mast cells and basophils | Interferes with the formation of prothrombin activator |
| Fibrin threads | Adsorbs thrombin | | |

## Prevention of Coagulation

In a healthy vascular system, the endothelium of the blood vessels partly prevents spontaneous blood clot formation. This smooth lining discourages the accumulation of platelets and clotting factors. Endothelial cells also produce a prostaglandin (see chapter 13) called *prostacyclin* (PGI₂), which inhibits the adherence of platelets to the inner surface of healthy blood vessel walls.

When a clot is forming, fibrin threads latch onto or *adsorb* thrombin, thus helping prevent the spread of the clotting reaction. A plasma alpha globulin, *antithrombin*, inactivates additional thrombin by binding to it and blocking its action on fibrinogen.

In addition, basophils and mast cells in the connective tissue surrounding capillaries secrete the anticoagulant *heparin*. This substance interferes with formation of prothrombin activator, prevents the action of thrombin on fibrinogen, and promotes removal of thrombin by antithrombin and fibrin adsorption.

Heparin-secreting cells are particularly abundant in the liver and lungs, where capillaries trap small blood clots that commonly form in the slow-moving blood of veins. These cells secrete heparin continually, preventing additional clotting in the circulatory system. Table 14.12 summarizes clot-inhibiting factors. Clinical Application 14.3 discusses an ancient anticlotting treatment that is in use again—biochemicals in the saliva of leeches. Clinical Application 14.4 discusses clotting disorders.

---

*Thrombocytopenia* is a tendency to bleed because of a platelet count that drops below 100,000 platelets per cubic millimeter of blood. Symptoms include bleeding easily, capillary hemorrhages throughout the body, and small, bruiselike spots on the skin. Thrombocytopenia is a common side effect of cancer chemotherapy and radiation treatments, and can also develop as a complication of pregnancy, leukemia, bone marrow transplantation, infectious disease, cardiac surgery, or anemia.

Transfusion of platelets is the conventional treatment for thrombocytopenia. A new treatment is thrombopoietin (TPO), which stimulates formation and maturation of megakaryocytes, and thereby boosts platelet levels.

---

**1** How does the lining of a blood vessel help prevent blood clot formation?

**2** What is the function of antithrombin?

**3** How does heparin help prevent blood clot formation?

---

## BLOOD GROUPS AND TRANSFUSIONS

Experiments transfusing blood, which date from the late 1600s, first used lamb blood. By the 1800s, human blood was being used. Results were unpredictable—some recipients were cured, but some were killed when their kidneys failed under the strain of handling red blood cells that clumped when blood types were incompatible. So poor was the success rate that, by the late 1800s, many nations banned transfusions.

Around this time, Austrian physician Karl Landsteiner began investigating why transfusions sometimes worked and sometimes did not. In 1900, he determined that blood was of differing types, and that only certain combinations of them were compatible. In 1910, identification of the ABO blood antigen gene explained the observed blood type incompatibilities. Today, twenty different genes are known to contribute to the surface features of red blood cells, which determine compatibility between blood types.

### Antigens and Antibodies

The clumping of red blood cells when testing blood compatibility or resulting from a transfusion reaction is called **agglutination.** This phenomenon is due to a reaction between red blood cell surface molecules called **antigens** (formerly called *agglutinogens*), and protein **antibodies** (formerly called *agglutinins*) carried in the plasma. Although many different antigens are associated with human erythrocytes, only a few of them are likely to produce serious transfusion reactions. These include the antigens of the ABO group and those of the Rh group. Avoiding the mixture of certain kinds of antigens and antibodies prevents adverse transfusion reactions.

# The Return of the Medicinal Leech

It had taken surgeon Joseph Upton ten hours to sew the 5-year-old's ear back on, after a dog had bitten it off. At first the operation appeared to be a success, but after four days, trouble began. Blood flow in the ear was blocked. Close examination showed that the arteries that the surgeon had repaired were fine, but the smaller veins were becoming congested. So Dr. Upton tried an experimental technique—he applied twenty-four leeches to the wound area.

The leeches latched on for up to an hour each, drinking the boy's blood. Leech saliva contains several biochemicals, one of which is a potent anticoagulant called hirudin in honor of its source, the medicinal leech *Hirudo medicinalis.* Unlike conventional anticlotting agents such as heparin, which are short-acting, hirudin works for up to 24 hours after the leech has drunk its fill and dropped off. Hirudin specifically blocks thrombin in veins. The long-acting leech biochemical gave the boy's ear time to heal.

Leeches have long been part of medical practice, with references hailing back to the ancient Egyptians 2,500 years ago (fig. 14B). The leech's popularity peaked in Europe in the nineteenth century, when French physicians alone used more than a billion a year, to drain "bad humours" from the body to cure nearly every ill. Use of leeches fell in the latter half of the nineteenth century. They were rediscovered by Yugoslav plastic surgeons in 1960 and by French microsurgeons in the early 1980s. In 1985 Dr. Upton made headlines and brought leeches into the limelight by saving the boy's ear at Children's Hospital in Boston.

A leech's bite does not hurt, patients say. But for those unwilling to have one or more 3-inch long, slimy green-gray invertebrates picnicking on a wound, hirudin is also available as a drug called hirulog, produced by recombinant DNA technology (fig. 14C). ■

**FIGURE 14B**
For centuries, bloodletting with leeches was believed to cure many ills. This woman in seventeenth-century Belgium applies a medicinal leech to her arm.

**FIGURE 14C**
Microsurgeons sometimes use leeches to help maintain blood flow through veins in patients after reattaching severed ears or digits. An anticoagulant in the leech's saliva keeps the blood thin enough to flow.

© Biopharm (USA) Limited 1994.

# Coagulation Disorders

### HEMOPHILIA

In 1962, 5-year-old Bob Massie developed uncontrollable bleeding in his left knee, a symptom of his *hemophilia A,* an inherited clotting disorder. It took thirty transfusions of plasma over the next three months to stop the bleeding. Because the knee joint had swelled and locked into place during that time, Bob was unable to walk for the next seven years. Today, Bob still suffers from painful joint bleeds, but he injects himself with factor VIII, the coagulation protein that his body cannot make. Factor VIII controls his bleeding.

Hemophilia has left its mark on history. One of the earliest descriptions is in the Talmud, a second century B.C. Jewish document, which reads "If she circumcised her first child and he died, and a second one also died, she must not circumcise her third child." Queen Victoria (1819–1901) passed the hemophilia gene to several of her children, eventually spreading the condition to the royal families of England, Russia, Germany, and Spain. Hemophilia achieved notoriety when pooled blood donations containing factor VIII were discovered to transmit HIV in 1985. Ninety percent of people with severe hemophilia who used such pooled factor VIII prior to then have developed AIDS.

Different forms of hemophilia are caused by abnormalities of different clotting factors, but hemophilia A is the most common. Symptoms of the hemophilias include tendency to hemorrhage severely following minor injuries, frequent nosebleeds, large intramuscular hematomas, and blood in the urine.

### VON WILLEBRAND DISEASE

Easy bruising and a tendency to bleed easily are signs of *von Willebrand disease,* another inherited clotting disorder that is usually far less severe than hemophilia. Affected persons lack a plasma protein, von Willebrand factor, that is secreted by endothelial cells lining blood vessels and enables platelets to adhere to damaged blood vessel walls, a key step preceding actual clotting. Sometimes the condition can cause spontaneous bleeding from the mucous membranes of the gastrointestinal and urinary tracts. ■

## ABO Blood Group

The *ABO blood group* is based on the presence (or absence) of two major antigens on red cell membranes—antigen *A* and antigen *B.* A person's erythrocytes contain one of four antigen combinations: only A, only B, both A and B, or neither A nor B. A person with only antigen A has *type A blood;* a person with only antigen B has *type B blood;* one with both antigen A and B has *type AB blood;* and one with neither antigen A nor B has *type O blood.*

A person who has received mismatched blood quickly feels the effects of agglutination—anxiety, breathing difficulty, facial flushing, headache, and severe pain in the neck, chest, and lumbar area. Red blood cells burst, releasing free hemoglobin. Macrophages phagocytize the hemoglobin, converting it to bilirubin, which may sufficiently accumulate to cause the yellow skin of jaundice. Free hemoglobin in the kidneys may ultimately cause them to fail.

| TABLE 14.13 | ANTIGENS AND ANTIBODIES OF THE ABO BLOOD GROUP | |
| --- | --- | --- |
| **Blood Type** | **Antigen** | **Antibody** |
| A | A | anti-B |
| B | B | anti-A |
| AB | A and B | Neither anti-A nor anti-B |
| O | Neither A nor B | Both anti-A and anti-B |

Certain antibodies are synthesized in the plasma about two to eight months following birth. Specifically, whenever antigen A is absent in the red blood cells, an antibody called *anti-A* develops, and whenever antigen B is absent, an antibody called *anti-B* develops. Therefore, persons with type A blood also have antibody *anti-B* in their plasma; those with type B blood have antibody anti-A; those with type AB blood have neither antibody; and those with type O blood have both anti-A and anti-B (fig. 14.21 and table 14.13). The antibodies

Type A blood

Type B blood

Type AB blood

Type O blood

**FIGURE 14.21**

Each ABO blood type is distinguished by a different set of antigens and antibodies.

anti-A and anti-B are large and do not cross the placenta. Thus, a pregnant woman and her fetus may be of different ABO blood types, and agglutination in the fetus will not occur.

> The percentage of blood types in human populations reflects history and migration patterns. Type B blood is found in 5% to 10% of the English and Irish, but gradually increases eastward, reaching 25% to 30% in the former Soviet Union. Some Native American populations in the United States today do not have type B or AB blood, yet the Asian populations from whom they descend do. This suggests that the original group of settlers lacked individuals with type B or AB blood.

Because an antibody of one kind will react with an antigen of the same kind and clump red blood cells, such combinations must be avoided. The major concern in blood transfusion procedures is that the cells in the donated blood not clump due to antibodies present in the recipient's plasma. For this reason, a person with type A (anti-B) blood must not receive blood of type B or AB, either of which would clump in the presence of anti-B in the recipient's type A blood. Likewise, a person with type B (anti-A) blood must not be given type A or AB blood, and a person with type O (anti-A and anti-B) blood must not be given type A, B, or AB blood (fig. 14.22).

Because type AB blood lacks both anti-A and anti-B antibodies, an AB person can receive a transfusion of blood of any other type. For this reason, type AB persons are sometimes called *universal recipients*. However, type A (anti-B) blood, type B (anti-A) blood, and type O (anti-A and anti-B) blood still contain antibodies (either anti-A and/or anti-B) that could agglutinate type AB cells. Consequently, even for AB individuals, it is always best to use donor blood of the same type as the recipient blood. If the matching type is not available and type A, B, or O is used, it should be transfused slowly so that the donor blood is well diluted by the recipient's larger blood volume. This precaution usually avoids serious reactions between the donor's antibodies and the recipient's antigens.

Similarly, because type O blood lacks antigens A and B, this type could theoretically be transfused into persons with blood of any other type. Therefore, persons with type O blood are sometimes called *universal donors*. Type O blood, however, does contain both anti-A and anti-B antibodies and if it is given to a person with blood type A, B, or AB, it too should be transfused slowly to minimize the chance of an adverse reaction. When type O blood is given to blood types A, B, or AB, it is generally transfused as "packed cells," meaning the plasma has been removed. This also minimizes adverse reactions due to the anti-A and anti-B antibodies found

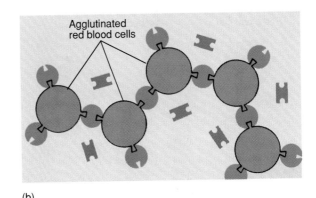

a)

b)

FIGURE 14.22

(*a*) If red blood cells with antigen A are added to blood containing antibody anti-A, (*b*) the antibodies react with the antigens, causing clumping (agglutination).

| TABLE 14.14 | PREFERRED AND PERMISSIBLE BLOOD TYPES FOR TRANSFUSIONS | | |
|---|---|---|---|
| **Blood Type of Recipient** | | **Preferred Blood Type of Donor** | **Permissible Blood Type of Donor (in an Extreme Emergency)** |
| A | | A | A, O |
| B | | B | B, O |
| AB | | AB | AB, A, B, O |
| O | | O | O |

in the plasma of type O blood. Table 14.14 summarizes preferred blood types for normal transfusions and permissible blood types for emergency transfusions.

1 Distinguish between antigens and antibodies.

2 What is a blood type?

3 What is the main concern when blood is transfused from one individual to another?

4 Why is a type AB person called a universal recipient?

5 Why is a type O person called a universal donor?

When is type O blood not really type O blood? Blood typing involves adding known antibodies to a blood sample to see if the blood will clump, demonstrating the presence of corresponding antigens on the red blood cell membranes. When a person has two copies of a rare gene, called the "h" gene, such a person lacks an enzyme that inserts a particular sugar onto red blood cell surfaces. Without that sugar, the A and B antigens cannot bind. The result—blood that tests as O (because it lacks A and B antigens) but can genetically be of any ABO type—A, B, AB, or O. Although this does not affect health, it can lead to questions when a child's ABO type cannot be derived from those of the parents.

## Rh Blood Group

The *Rh blood group* was named after the rhesus monkey in which it was first studied. In humans, this group includes several Rh antigens (factors). The most important of these is *antigen D;* however, if any of the antigen D and other Rh antigens are present on the red cell membranes, the blood is said to be *Rh-positive.* Conversely, if the red cells lack the Rh antigens, the blood is called *Rh-negative.*

As in the case of antigens A and B, the presence (or absence) of Rh antigens is an inherited trait. Unlike anti-A and anti-B, antibodies for Rh (*anti-Rh*) do not appear spontaneously. Instead, they form only in Rh-negative persons in response to special stimulation.

If an Rh-negative person receives a transfusion of Rh-positive blood, the recipient's antibody-producing cells are stimulated by the presence of the Rh antigens and will begin producing *anti-Rh antibodies.* Generally, no serious consequences result from this initial transfusion, but if the Rh-negative person—who is now sensitized to Rh-positive blood—receives another transfusion of Rh-positive blood some months later, the donated red cells are likely to agglutinate.

A related condition may occur when an Rh-negative woman is pregnant with an Rh-positive fetus for the first time. Such a pregnancy may be uneventful; however, at the time of this infant's birth (or if a miscarriage occurs), the placental membranes that separated the maternal

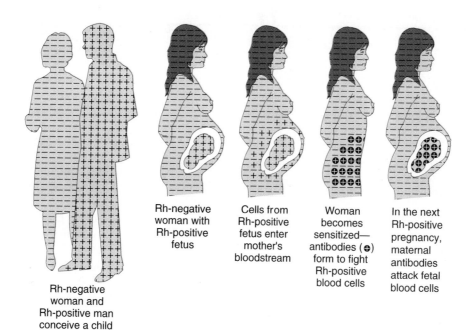

**FIGURE 14.23**

If a man who is Rh⁺ and a woman who is Rh⁻ conceive a child who is Rh⁺, the woman's body may manufacture antibodies that attack future Rh⁺ offspring.

blood from the fetal blood during the pregnancy tear, and some of the infant's Rh-positive blood cells may enter the maternal circulation. These Rh-positive cells may then stimulate the maternal tissues to begin producing anti-Rh antibodies (fig. 14.23).

If a woman who has already developed anti-Rh antibodies becomes pregnant with a second Rh-positive fetus, these anti-Rh antibodies, called hemolysins, cross the placental membrane and destroy the fetal red cells. The fetus then develops a condition called **erythroblastosis fetalis** (hemolytic disease of the newborn). Clinical Application 14.5 discusses human blood substitutes.

**1** What is the Rh blood group?

**2** What are two ways that Rh incompatibility can arise?

---

Erythroblastosis fetalis is extremely rare today, because obstetricians carefully track Rh status. An Rh⁻ woman who might carry an Rh⁺ fetus is given an injection of a drug called Rhogam. This is actually anti-Rh antibodies, which bind to and shield any Rh⁺ fetal cells that might contact the woman's cells, sensitizing her immune system. Rhogam must be given within 72 hours of possible contact with Rh⁺ cells—such situations include giving birth, terminating a pregnancy, miscarrying, or undergoing amniocentesis (a prenatal test in which a needle is inserted into the uterus).

## CLINICAL TERMS RELATED TO THE BLOOD

**anisocytosis** (an-i″so-si-to′sis)   Abnormal variation in the size of erythrocytes.

**antihemophilic plasma** (an″ti-he″mo-fil′ik plaz′mah)   Normal blood plasma that has been processed to preserve an antihemophilic factor.

**citrated whole blood** (sit′rāt-ed hōl blud)   Normal blood to which a solution of acid citrate has been added to prevent coagulation.

**dried plasma** (drīd plaz′mah)   Normal blood plasma that has been vacuum dried to prevent the growth of microorganisms.

**hemorrhagic telangiectasia** (hem″o-raj′ik tel-an″je-ek-ta′ze-ah)   Inherited tendency to bleed from localized lesions of the capillaries.

**heparinized whole blood** (hep′er-i-nīzed″ hōl blud)   Normal blood to which a solution of heparin has been added to prevent coagulation.

**macrocytosis** (mak″ro-si-to′sis)   Abnormally large erythrocytes.

**microcytosis** (mi″kro-si-to′sis)   Abnormally small erythrocytes.

**neutrophilia** (nu″tro-fil′e-ah)   Increase in the number of circulating neutrophils.

**packed red cells**   Concentrated suspension of red blood cells from which the plasma has been removed.

**pancytopenia** (pan″si-to-pe′ne-ah)   Abnormal depression of all the cellular components of blood.

# Replacing Blood

The medical need to replace blood is great. Trauma victims, surgical patients, and those suffering from chronic conditions that deplete specific blood components all require blood. The World Health Organization estimates that global demand for blood exceeds 100 million units a year.

For most uses, people in need of blood receive transfusions of donated blood. The field of blood substitutes develops chemicals that have some of blood's properties for temporary use until a blood transfusion is possible.

## TRANSFUSIONS

In the past, a blood transfusion used whole blood. Today, whole blood is often separated into its component parts, and only those needed are used.

A patient with anemia or an acute blood loss might receive concentrated (packed) red blood cells; a person with too few platelets might be given platelets; a person with cancer whose treatment has depleted the white blood cell count might be given a white blood cell preparation. Similarly, blood plasma might be used to replace lost blood volume, or to provide clotting factors.

Before donor blood is used to obtain blood components, it is tested for particular viruses, including those that cause hepatitis B and AIDS (see chapter 16). Then the blood compo-nents must be stored properly. For example, packed red blood cells can be stored for several years if they are frozen, but can be used for only about a month if they are not frozen. After thawing, however, such cells must be used within about a day. Platelet preparations must be used within five days, and white blood cell concentrates must be used immediately.

## BLOOD SUBSTITUTES

Synthetic bloods try to duplicate blood's oxygen-carrying ability. They are variations on chemicals that bind oxygen, mimicking the hemoglobin carried in red blood cells.

The first such synthetic blood was silicone oil. A mouse immersed into a beaker of silicone oil, for example, can "breathe" the oxygen in the oil. Researchers have tried related chemicals containing fluorine and carbon, which similarly bind oxygen to silicone oil. In 1990, the first red blood cell substitute was approved for use to maintain localized blood flow during certain surgical procedures. The product consists of two fluorine compounds, a mild detergent, and lipid from egg yolk.

In another approach, hemoglobin molecules are linked together and administered alone, rather than in cells as occurs in the circulatory system. Yet another type of artificial blood consists of hemoglobin packaged into synthetic microcapsules. However, these "neohemocytes" so far provoke an immune attack from the animals in which they have been tested. Artificial bloods based on hemoglobin do not offer the freedom from contamination possible with a completely synthetic product.

Typing and crossmatching blood takes precious time. Another approach to blood substitutes "cloaks" red blood cells, so that they can be given to any individual, making crossmatching unnecessary. In this technique, polyethylene glycol, a chemical used as an antifreeze, forms a fuzzy coat on the cells, which masks the ABO antigens. Therefore, a recipient with an incompatible blood type does not reject these cells. Such cloaked red blood cells are being developed for use in combat, to treat chronic anemias, and for the veterinary market, which has no dog and cat blood banks.

Perhaps the most unusual experimental source of a human blood substitute is tobacco. Researchers in France have added human hemoglobin genes to tobacco plants, which produce the proteins. Human hemoglobin produced in tobacco would not harbor infectious agents that could be present in donated blood. ■

**poikilocytosis** (poi″ki-lo-si-to′sis)   Irregularly shaped erythrocytes.

**purpura** (per′pu-rah)   Spontaneous bleeding into the tissues and through the mucous membranes.

**septicemia** (sep″ti-se′me-ah)   Presence of disease-causing microorganisms or their toxins in the blood.

**spherocytosis** (sfēr″o-si-to′sis)   Hemolytic anemia caused by defective proteins supporting the cell membranes of red blood cells. The cells are abnormally spherical.

**thalassemia** (thal″ah-se′me-ah)   Group of hereditary hemolytic anemias resulting from very thin, fragile erythrocytes.

## CHAPTER SUMMARY

## Introduction  (page 521)

Blood is often considered a type of connective tissue whose cells are suspended in a liquid intercellular material. It transports substances between the body cells and the external environment, and helps maintain a stable internal environment.

## Blood and Blood Cells  (page 521)

Blood contains red blood cells, white blood cells, and platelets.

1. Blood volume and composition
   a. Blood volume varies with body size, fluid and electrolyte balance, and adipose tissue content.
   b. Blood can be separated into formed elements and liquid portions.
      (1) The formed elements portion is mostly red blood cells.
      (2) The liquid plasma includes water, nutrients, hormones, electrolytes, and cellular wastes.
2. The origin of blood cells
   a. Blood cells develop from hemocytoblasts in bone marrow.
   b. Cells descended from hemocytoblasts respond to colony stimulating factors to specialize.
   c. Thrombopoietin stimulates megakaryocytes to give rise to platelets, much as erythropoietin stimulates formation of red blood cells.
3. Characteristics of red blood cells
   a. Red blood cells are biconcave disks with shapes that provide increased surface area and place their cell membranes close to internal structures.
   b. They contain hemoglobin, which combines loosely with oxygen.
   c. The mature form lacks nuclei, but contains enzymes needed for energy-releasing processes.
4. Red blood cell counts
   a. The red blood cell count equals the number of cells per mm$^3$ of blood.
   b. The average count may range from approximately 4,000,000 to 6,000,000 cells per mm$^3$.
   c. Red cell count is related to the oxygen-carrying capacity of the blood and is used in diagnosing and evaluating the courses of diseases.
5. Destruction of red blood cells
   a. Red cells are fragile and are damaged while moving through capillaries.
   b. Macrophages in the liver and spleen phagocytize damaged red blood cells.
   c. Hemoglobin molecules are decomposed, and the iron they contain is recycled.
   d. Biliverdin and bilirubin are pigments released from hemoglobin.
6. Red blood cell production and its control
   a. During fetal development, red cells form in the yolk sac, liver, and spleen; later, red cells are produced by the red bone marrow.
   b. The number of red blood cells remains relatively stable.
   c. A negative feedback mechanism involving erythropoietin from the kidneys and liver controls rate of red cell production.
      (1) Erythropoietin is released in response to low oxygen levels.
      (2) High altitude, loss of blood, or chronic lung disease can lower oxygen concentration.
7. Dietary factors affecting red blood cell production
   a. The availability of vitamin B$_{12}$, iron, and folic acid affects red blood cell production.
   b. The rate of iron absorption varies with the amount of iron in the body.
8. Types of white blood cells
   a. White blood cells control infection.
   b. Granulocytes include neutrophils, eosinophils, and basophils.
   c. Agranulocytes include monocytes and lymphocytes.
9. White blood cells fight infection
   a. Neutrophils and monocytes phagocytize foreign particles.
   b. Chemicals released by damaged cells attract and stimulate leukocytes.
   c. Eosinophils kill parasites and help control inflammation and allergic reactions.
   d. Basophils release heparin, which inhibits blood clotting, and histamine to increase blood flow to injured tissues.
   e. Lymphocytes (both T cells and B cells) are involved in immunity and produce antibodies that attack specific foreign antigens.
10. White blood cell counts
   a. Normal total white blood cell counts vary from 5,000 to 10,000 cells per mm$^3$ of blood.
   b. The number of white cells may change in abnormal conditions such as infections, emotional disturbances, or excessive loss of body fluids.
   c. A differential white cell count indicates the percentages of various types of leukocytes present.

11. Blood platelets
    a. Blood platelets are fragments of giant cells that detach and enter the circulation.
    b. The normal count varies from 130,000 to 360,000 platelets per mm$^3$.
    c. Platelets help close breaks in blood vessels.

# Blood Plasma (page 532)

Plasma is the liquid part of the blood that is composed of water and a mixture of organic and inorganic substances. It transports nutrients and gases, helps regulate fluid and electrolyte balance, and helps maintain stable pH.

1. Plasma proteins
   a. Plasma proteins remain in blood and interstitial fluids, and are not normally used as energy sources.
   b. Three major groups exist.
      (1) Albumins help maintain the osmotic pressure of blood.
      (2) Globulins include antibodies. They provide immunity and transport lipids and fat-soluble vitamins.
      (3) Fibrinogen functions in blood clotting.
2. Gases and nutrients
   a. Gases in plasma include oxygen, carbon dioxide, and nitrogen.
   b. Plasma nutrients include amino acids, simple sugars, and lipids.
      (1) Glucose is stored in the liver as glycogen and is released whenever the blood glucose concentration falls.
      (2) Amino acids are used to synthesize proteins and are deaminated for use as energy sources.
      (3) Lipoproteins function in the transport of lipids.
3. Nonprotein nitrogenous substances
   a. Nonprotein nitrogenous substances are composed of molecules that contain nitrogen atoms but are not proteins.
   b. They include amino acids, urea, uric acid, creatine, and creatinine.
      (1) Urea and uric acid are products of catabolism.
      (2) Creatinine results from the metabolism of creatine.
   c. These substances usually remain stable; an increase may indicate a kidney disorder.
4. Plasma electrolytes
   a. Plasma electrolytes are obtained by absorption from the intestines and are released as by-products of cellular metabolism.
   b. They include ions of sodium, potassium, calcium, magnesium, chloride, bicarbonate, phosphate, and sulfate.
   c. They are important in the maintenance of osmotic pressure and pH.

# Hemostasis (page 537)

Hemostasis refers to the stoppage of bleeding. Hemostatic mechanisms are most effective in controlling blood loss from small vessels.

1. Blood vessel spasm (vasospasm)
   a. Smooth muscles in walls of arterioles and arteries reflexly contract following injury.

   b. Platelets release serotonin that stimulates vasoconstriction and helps maintain vessel spasm.
2. Platelet plug formation
   a. Platelets adhere to rough surfaces and exposed collagen.
   b. Platelets stick together at the sites of injuries and form platelet plugs in broken vessels.
3. Blood coagulation
   a. Blood clotting, the most effective means of hemostasis, involves a series of reactions wherein each reaction stimulates the next reaction (cascade), which may be initiated by extrinsic or intrinsic mechanisms.
   b. The extrinsic clotting mechanism is triggered when blood contacts damaged tissue.
   c. The intrinsic clotting mechanism is triggered when blood contacts a foreign surface.
   d. Clot formation depends on the balance between clotting factors that promote clotting and those that inhibit clotting.
   e. The basic event of coagulation is the conversion of soluble fibrinogen into insoluble fibrin.
   f. After forming, the clot retracts and pulls the edges of a broken vessel closer together.
   g. A thrombus is an abnormal blood clot in a vessel; an embolus is a clot or fragment of a clot that moves in a vessel.
   h. Fibroblasts invade a clot, forming connective tissue throughout.
   i. Protein-splitting enzymes may eventually destroy a clot.
4. Prevention of coagulation
   a. The smooth lining of blood vessels discourages the accumulation of platelets.
   b. As a clot forms, fibrin adsorbs thrombin and prevents the reaction from spreading.
   c. Antithrombin interferes with the action of excessive thrombin.
   d. Some cells secrete heparin, an anticoagulant.

# Blood Groups and Transfusions (page 542)

Blood can be typed on the basis of the surface structures of its cells.

1. Antigens and antibodies
   a. Red blood cell membranes may contain antigens, and blood plasma may contain antibodies.
   b. Blood typing uses known antibodies to identify antigens on red cell membranes.
2. ABO blood group
   a. Blood can be grouped according to the presence or absence of antigens A and B.
   b. Whenever antigen A is absent, antibody anti-A is present; whenever antigen B is absent, antibody anti-B is present.
   c. Adverse transfusion reactions are avoided by preventing the mixing of red cells that contain an antigen with plasma that contains the corresponding antibody.
   d. Adverse reactions are due to agglutination (clumping) of the red blood cells.

3. Rh blood group
   a. Rh antigens are present on the red cell membranes of Rh-positive blood; they are absent in Rh-negative blood.
   b. If an Rh-negative person is exposed to Rh-positive blood, anti-Rh antibodies are produced in response.
   c. Mixing Rh-positive red cells with plasma that contains anti-Rh antibodies agglutinates the positive cells.
   d. If an Rh-negative female is pregnant with an Rh-positive fetus, some of the positive cells may enter the maternal blood at the time of birth and stimulate the maternal tissues to produce anti-Rh antibodies.
   e. Anti-Rh antibodies in maternal blood may pass through the placental tissues and react with the red cells of an Rh-positive fetus.

## CRITICAL THINKING QUESTIONS

1. What change would you expect to occur in the hematocrit of a person who is dehydrated? Why?
2. Erythropoietin is available as a drug. Why would athletes abuse it?
3. If a patient with an inoperable cancer is treated using a drug that reduces the rate of cell reproduction, what changes might occur in the blood cell count? How might the patient's environment be modified to compensate for the effects of these changes?
4. Hypochromic (iron-deficiency) anemia is common among aging persons who are admitted to hospitals for other conditions. What environmental and sociological factors might promote this form of anemia?
5. How would you explain to a patient with leukemia, who has a greatly elevated white blood cell count, the importance of avoiding bacterial infections?
6. If a woman whose blood is Rh-negative and contains anti-Rh antibodies is carrying a fetus with Rh-negative blood, will the fetus be in danger of developing erythroblastosis fetalis? Why?
7. In the United States, between 1977 and 1985, more than 10,000 men with hemophilia contracted the human immunodeficiency virus (HIV) from contaminated factor VIII that they received to treat the hemophilia. What are two abnormalities in the blood of these men?
8. Why do patients with liver diseases commonly develop blood-clotting disorders?

## REVIEW EXERCISES

1. List the major components of blood.
2. Define *hematocrit*, and explain how it is determined.
3. Describe a red blood cell.
4. Distinguish between oxyhemoglobin and deoxyhemoglobin.
5. Explain what is meant by a *red blood cell count*.
6. Describe the life cycle of a red blood cell.
7. Distinguish between biliverdin and bilirubin.
8. Define *erythropoietin*, and explain its function.
9. Explain how vitamin $B_{12}$ and folic acid deficiencies affect red blood cell production.
10. List two sources of iron that can be used for the synthesis of hemoglobin.
11. Distinguish between granulocytes and agranulocytes.
12. Name five types of leukocytes, and list the major functions of each type.
13. Explain the significance of white blood cell counts as aids to diagnosing diseases.
14. Describe a blood platelet, and explain its functions.
15. Name three types of plasma proteins, and list the major functions of each type.
16. Define *lipoprotein*.
17. Define *apoprotein*.
18. Distinguish between low-density lipoprotein and high-density lipoprotein.
19. Name the sources of VLDL, LDL, HDL, and chylomicrons.
20. Describe how lipoproteins are removed from plasma.
21. Explain how cholesterol is eliminated from plasma and from the body.
22. Define *nonprotein nitrogenous substances*, and name those commonly present in plasma.
23. Name several plasma electrolytes.
24. Define *hemostasis*.
25. Explain how blood vessel spasms are stimulated following an injury.
26. Explain how a platelet plug forms.
27. List the major steps leading to the formation of a blood clot.
28. Indicate the trigger and outline the steps for extrinsic clotting and for intrinsic clotting.
29. Distinguish between fibrinogen and fibrin.
30. Describe a positive feedback system that operates during blood clotting.
31. Define *serum*.

32. Distinguish between a thrombus and an embolus.

33. Explain how a blood clot may be removed naturally from a blood vessel.

34. Describe how blood coagulation may be prevented.

35. Name a vitamin required for blood clotting.

36. Distinguish between antigen and antibody.

37. Explain the basis of ABO blood types.

38. Explain why a person with blood type AB is sometimes called a universal recipient.

39. Explain why a person with blood type O is sometimes called a universal donor.

40. Distinguish between Rh-positive and Rh-negative blood.

41. Describe how a person may become sensitized to Rh-positive blood.

42. Describe *erythroblastosis fetalis*, and explain how this condition may develop.

 **For More Information Related to This Chapter, Please Visit *Hole's Human Anatomy and Physiology* Home Page**
http://www.mhhe.com/biosci/abio/

## CHAPTER 15

# CARDIOVASCULAR SYSTEM

## CHAPTER OBJECTIVES

**AFTER YOU HAVE STUDIED THIS CHAPTER, YOU SHOULD BE ABLE TO:**

1. Name the organs of the cardiovascular system and discuss their functions.

2. Name and describe the locations of the major parts of the heart and discuss the function of each part.

3. Trace the pathway of the blood through the heart and the vessels of the coronary circulation.

4. Discuss the cardiac cycle and explain how it is controlled.

5. Compare the structures and functions of the major types of blood vessels.

6. Describe the mechanisms that aid in returning venous blood to the heart.

7. Explain how blood pressure is produced and controlled.

8. Compare the pulmonary and systemic circuits of the cardiovascular system.

9. Identify and locate the major arteries and veins of the pulmonary and systemic circuits.

## KEY TERMS

**arteriole** (ar-te′re-ol)
**atrium** (a′tre-um)
**cardiac conduction system** (kar′de-ak kon-duk′shun sis′tem)
**cardiac cycle** (kar′de-ak si′kl)
**cardiac output** (kar′de-ak owt′poot)
**diastole** (di-as′to-le)
**electrocardiogram (ECG)** (e-lek″tro-kar′de-o-gram″)
**endocardium** (en″do-kar′de-um)
**epicardium** (ep″i-kar′de-um)
**functional syncytium** (funk′shun-al sin-sish′e-um)
**myocardium** (mi″o-kar′de-um)

**pacemaker** (pas′māk-er)
**pericardium** (per″i-kar′de-um)
**peripheral resistance** (pe-rif′er-al re-zis′tans)
**pulmonary circuit** (pul′mo-ner″e sur′kit)
**sphygmomanometer** (sfig″mo-mah-nom′ĕ-ter)
**systemic circuit** (sis-tem′ik sur′kit)
**systole** (sis′-to-le)
**vasoconstriction** (vas″o-kon-strik′shun)
**vasodilation** (vas″o-di-la′shun)
**ventricle** (ven′tri-kl)
**venule** (ven′ūl)
**viscosity** (vis-kos′i-te)

## UNDERSTANDING WORDS

**angio-,** vessel: *angio*tensin—substance that constricts blood vessels.

**ather-,** porridge: *ather*osclerosis—deposits of plaque in arteries.

**brady-,** slow: *brady*cardia—abnormally slow heartbeat.

**diastol-,** dilation: *diastol*ic pressure—blood pressure when the ventricle of the heart is relaxed.

**ectop-,** out of place: *ectop*ic beat—heartbeat that occurs before it is expected in a normal series of cardiac cycles.

**edem-,** swelling: *edem*a—condition in which fluids accumulate in the tissues and cause them to swell.

**-gram,** letter, drawing: electrocardio*gram*—recording of electrical changes in the myocardium during a cardiac cycle.

**lun-,** moon: semi*lun*ar valve—valve with crescent-shaped flaps.

**my-,** muscle: *my*ocardium—muscle tissue within the wall of the heart.

**papill-,** nipple: *papill*ary muscle—small mound of muscle within a ventricle of the heart.

**phleb-,** vein: *phleb*itis—inflammation of a vein.

**scler-,** hard: *scler*osis—condition in which a blood vessel wall loses its elasticity and hardens.

**syn-,** together: *syn*cytium—mass of merging cells that act together.

**systol-,** contraction: *systol*ic pressure—blood pressure resulting from a ventricular contraction.

**tachy-,** rapid: *tachy*cardia—abnormally fast heartbeat.

Massachusetts physician John T. Harrington's history of heart trouble began when he was 8 years old and had rheumatic fever. He was ill for several weeks but recovered completely—or so he thought for many years.

Rheumatic heart disease is an inflammation that follows an infection by *Streptococcus* bacteria. Because the cell surfaces of these bacteria resemble those of heart valve cells, the alerted immune system sometimes attacks the valve cells, mistaking them for bacteria. (The valves route blood through the heart.) The resulting valve damage can take many years to produce symptoms.

(a)

(b)

Harrington's aortic valve—the one between the heart and the aorta—was indeed damaged. It kept him out of sports in high school, then out of Vietnam, and became an object of much curiosity and amusement to his medical school friends. Although the valve damage produced an interesting sound and abnormalities in various diagnostic tests, he felt no ill effects until 1983.

At that time, Harrington began jogging, just a little, a half mile down a hill from his house, and the half mile back. But late that year, he started feeling pressure in his chest upon completing the workout, which would vanish when he rested. Finally, in early 1984, at age 47, he saw a cardiologist, who said he needed a "valve job."

Harrington's first replacement valve was bioprosthetic, meaning that it came from an animal—specifically, the outer heart muscle layer of a cow. It wouldn't last more than 10 years, but the human body is less likely to react to a bioprosthetic valve by forming dangerous blood clots than it is to a valve made of synthetic materials.

The cow valve lasted 8 years. Just after Christmas in 1992, while sitting at the breakfast table with his young grandsons, Harrington experienced crushing chest pain. The valve had torn. Waking up in the hospital some time later, Harrington learned that he now had a synthetic valve. This one should last about 30 years.

A heart valve is only a small component of the cardiovascular system, but it is vitally important, as Harrington's reliance on a replacement heart valve indicates.

**Rheumatic fever progresses to rheumatic heart disease when heart valves become scarred. (*a*) The valve on the top is healthy. (*b*) The thickened and scarred valve leaflets on the bottom are the result of rheumatic heart disease, a misplaced immune attack triggered by a bacterial infection.**

A functional cardiovascular system is vital for survival, because without circulation the tissues lack a supply of oxygen and nutrients, and wastes accumulate. Under such conditions the cells soon begin irreversible change, which quickly leads to death. Figure 15.1 shows the general pattern of the cardiovascular system.

# STRUCTURE OF THE HEART

The heart is a hollow, cone-shaped, muscular pump located within the mediastinum of the thorax and resting upon the diaphragm.

## Size and Location of the Heart

Heart size varies with body size. However, an average adult's heart is generally about 14 centimeters long and 9 centimeters wide (fig. 15.2).

The heart is bordered laterally by the lungs, posteriorly by the spinal cord, and anteriorly by the sternum (fig. 15.3 and reference plates 50, 56, 71, and 72). Its *base,* which is attached to several large blood vessels, lies beneath the second rib. Its distal end extends downward and to the left, terminating as a bluntly pointed *apex* at the level of the fifth intercostal space. For this reason, it is possible to sense the *apical heartbeat* by feeling or listening to the chest wall between the fifth and sixth ribs, about 7.5 centimeters to the left of the midline.

**FIGURE 15.1** ▭

The cardiovascular system transports blood between the body cells and organs such as the lungs, intestines, and kidneys that communicate with the external environment.

**FIGURE 15.2**

Anterior view of a human heart.

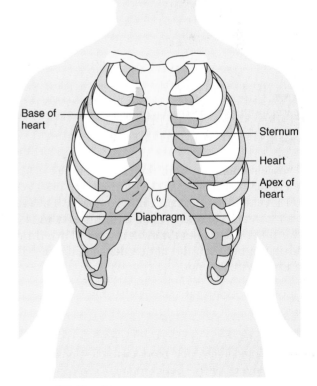

**FIGURE 15.3** 🐦

The heart is located posterior to the sternum, where it lies upon the diaphragm.

## Coverings of the Heart

The **pericardium** encloses the heart and the proximal ends of the large blood vessels to which it attaches. The pericardium consists of an outer fibrous bag, the *fibrous pericardium,* that surrounds a more delicate, double-layered sac. The inner layer of this sac, the *visceral pericardium* (epicardium), covers the heart. At the base of the heart, the visceral pericardium turns back upon itself to become the *parietal pericardium.* The parietal pericardium, in turn, forms the inner lining of the fibrous pericardium.

The fibrous pericardium is a tough, protective sac largely composed of white fibrous connective tissue. It is attached to the central portion of the diaphragm, the posterior of the sternum, the vertebral column, and the large

Right lung
Left lung
Superior vena cava
Aorta
Diaphragm
Right auricle
Right atrium
Atrioventricular sulcus
Right ventricle
Pulmonary trunk
Left auricle
Fibrous pericardium
Cut edge of parietal pericardium
Heart (covered by visceral pericardium)
Left ventricle
Interventricular sulcus
Pericardial cavity

**FIGURE 15.4**

The heart is within the mediastinum and is enclosed by a layered pericardium.

blood vessels emerging from the heart (see figs. 1.9, 15.4 and reference plates 56 and 57). Between the parietal and visceral layers of the pericardium is a space, the *pericardial cavity,* that contains a small amount of serous fluid that the pericardial membranes secrete. This fluid reduces friction between the pericardial membranes as the heart moves within them.

In *pericarditis,* inflammation of the pericardium due to viral or bacterial infection produces adhesions that attach the layers of the pericardium. This condition is very painful and interferes with heart movements.

**1** Where is the heart located?

**2** Where would you listen to hear the apical heartbeat?

**3** Distinguish between the visceral pericardium and the parietal pericardium.

**4** What is the function of the fluid in the pericardial cavity?

## Wall of the Heart

The wall of the heart is composed of three distinct layers: an outer epicardium, a middle myocardium, and an inner endocardium (fig. 15.5).

The **epicardium,** which corresponds to the visceral pericardium, is protective. It is a serous membrane that consists of connective tissue covered by epithelium, and it includes blood capillaries, lymph capillaries, and nerve fibers. The deeper portion of the epicardium often contains fat, particularly along the paths of coronary arteries and cardiac veins that provide blood flow through the myocardium.

The middle layer, or **myocardium,** is thick and largely consists of the cardiac muscle tissue that pumps blood out of the heart chambers. The muscle fibers are arranged in planes, separated by connective tissues that are richly supplied with blood capillaries, lymph capillaries, and nerve fibers.

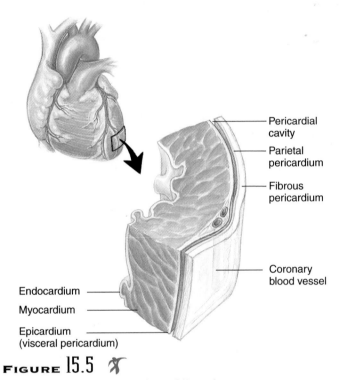

Endocardium
Myocardium
Epicardium
(visceral pericardium)

Pericardial cavity
Parietal pericardium
Fibrous pericardium
Coronary blood vessel

**FIGURE 15.5**

The wall of the heart consists of three layers: an endocardium, a myocardium, and an epicardium.

The inner layer, or **endocardium,** consists of epithelium and connective tissue that contains many elastic and collagenous fibers. The endocardium also contains blood vessels and some specialized cardiac muscle fibers called *Purkinje fibers,* described later in this chapter.

The endocardium lines all of the heart chambers and covers the structures, such as the heart valves, that project into them. This inner lining is also continuous with the inner lining of the blood vessels (endothelium) attached to the heart. Table 15.1 summarizes the characteristics of the three layers of the heart wall.

## Heart Chambers and Valves

Internally, the heart is divided into four hollow chambers, two on the left and two on the right. The upper chambers, called **atria** (sing., *atrium*), have thin walls and receive blood returning to the heart. Small, earlike projections, called **auricles,** extend anteriorly from the atria (see fig. 15.4). The lower chambers, the **ventricles,** force the blood out of the heart into arteries.

A structure called the *interatrial septum* separates the right from the left atrium. An *interventricular septum* separates the two ventricles. The atrium on each side communicates with its corresponding ventricle through an opening called the **atrioventricular orifice,** guarded by an *atrioventricular valve* (*A-V valve*).

Grooves on the surface of the heart mark the divisions between its chambers, and they also contain major blood vessels that supply the heart tissues. The deepest of these grooves is the **atrioventricular** (coronary) **sulcus,** which encircles the heart between the atria and ventricles. Two **interventricular** (anterior and posterior) **sulci** mark the septum that separates the right and left ventricles (see fig. 15.4).

| TABLE 15.1 | WALL OF THE HEART | | |
|---|---|---|---|
| **Layer** | **Composition** | **Function** | |
| Epicardium (visceral pericardium) | Serous membrane of connective tissue covered with epithelium and including blood capillaries, lymph capillaries, and nerve fibers | Forms a protective outer covering; secretes serous fluid | |
| Myocardium | Cardiac muscle tissue separated by connective tissues and including blood capillaries, lymph capillaries, and nerve fibers | Contracts to pump blood from the heart chambers | |
| Endocardium | Membrane of epithelium and connective tissues, including elastic and collagenous fibers, blood vessels, and specialized muscle fibers | Forms a protective inner lining of the chambers and valves | |

When increasing blood volume stretches muscle cells associated with the atria, the cells secrete a peptide hormone called *atrial natriuretic peptide* (ANP). ANP inhibits release of renin from the kidneys and of aldosterone from the adrenal cortex. The overall result is increased excretion of sodium ions and water from the kidneys and lowered blood volume and blood pressure. Researchers are investigating use of ANP to treat high blood pressure.

**1** Describe the layers of the heart wall.

**2** Name and locate the four chambers of the heart.

**3** Name the orifices between the upper and the lower chambers of the heart.

The right atrium receives blood from two large veins: the *superior vena cava* and the *inferior vena cava.* These veins return blood that is low in oxygen from tissues. A smaller vein, the *coronary sinus,* also drains blood into the right atrium from the myocardium of the heart itself.

A large **tricuspid valve** guards the atrioventricular orifice between the right atrium and the right ventricle. It is composed of three leaflets, or cusps, as its name implies. This valve permits the blood to move from the right atrium into the right ventricle and prevents it from moving in the opposite direction. The cusps fold passively out of the way against the ventricular wall when the blood pressure is greater on the atrial side, and they close passively when the pressure is greater on the ventricular side (figs. 15.6, 15.7, 15.8, and 15.9).

Strong, fibrous strings, called *chordae tendineae,* are attached to the cusps on the ventricular side. These strings originate from small mounds of cardiac muscle tissue, the **papillary muscles,** that project inward from the walls of the ventricle. The papillary muscles contract when the ventricle contracts. As the tricuspid valve closes, these muscles pull on the chordae tendineae and prevent the cusps from swinging back into the atrium.

The right ventricle has a thinner muscular wall than the left ventricle. This right chamber pumps the blood a fairly short distance to the lungs against a relatively low resistance to blood flow. The left ventricle, on the other hand, must force the blood to all the other parts of the body against a much greater resistance to flow.

When the muscular wall of the right ventricle contracts, the blood inside its chamber is put under increasing pressure, and the tricuspid valve closes passively. As a result, the only exit is through the *pulmonary trunk,* which divides to form the left and right *pulmonary arteries.* At the base of this trunk is a **pulmonary valve** (pulmonary semilunar valve), which consists of three cusps (see fig. 15.8). This valve opens as the right ventricle contracts. However, when the ventricular muscles relax, the

blood begins to back up in the pulmonary trunk. This closes the pulmonary valve, preventing a return flow into the ventricle. Unlike the tricuspid valve, the pulmonary valve does not have chordae tendineae or papillary muscles attached to its cusps.

The left atrium receives the blood from the lungs through four *pulmonary veins*—two from the right lung and two from the left lung. The blood passes from the left atrium into the left ventricle through the atrioventricular orifice, which a valve guards. This valve consists of two leaflets, and it is appropriately named the **bicuspid,** or **mitral** (shaped like a miter, a turbanlike headdress), **valve.** It prevents the blood from flowing back into the left atrium from the ventricle when the ventricle contracts. As with the tricuspid valve, the papillary muscles and the chordae tendineae prevent the cusps of the bicuspid valve from swinging back.

When the left ventricle contracts, the bicuspid valve closes passively, and the only exit is through a large artery called the *aorta.* Its branches distribute blood to all parts of the body.

At the base of the aorta is an **aortic valve** (aortic semilunar valve) that consists of three cusps (see fig. 15.8). It opens and allows blood to leave the left ventricle as it contracts. When the ventricular muscles relax, this valve closes and prevents blood from backing up into the ventricle. Table 15.2 summarizes the heart valves.

*Mitral valve prolapse* (MVP) is common, affecting up to 6% of some populations. In this condition, one (or both) of the cusps of the mitral valve stretches and bulges into the left atrium during ventricular contraction. The valve usually continues to function adequately, but sometimes, blood regurgitates into the left atrium. Through a stethoscope, a regurgitating MVP sounds like a click at the end of ventricular contraction, then a murmur as blood goes back through the valve into the left atrium.

Symptoms of MVP include chest pain, palpitations, fatigue, and anxiety. The mitral valve can be damaged by certain species of *Streptococcus* bacteria (see opening vignette). Endocarditis, an inflammation of the endocardium due to infection, appears as a plantlike growth on the valve. Individuals with MVP must take antibiotics before undergoing dental work to prevent infection by *Streptococcus* bacteria in the mouth. Also, people with MVP are particularly susceptible to endocarditis.

**1** Which blood vessels carry blood into the right atrium?

**2** Where does blood go after it leaves the right ventricle?

**3** Which blood vessels carry blood into the left atrium?

**4** What prevents blood from flowing back into the ventricles when they are relaxed?

Superior vena cava

Pulmonary valve

Right pulmonary artery

Branches of right pulmonary veins

Right atrium

Tricuspid valve

Right ventricle

Inferior vena cava

Aorta

Left pulmonary artery

Pulmonary trunk

Left pulmonary veins

Left atrium

Bicuspid valve

Chordae tendineae

Papillary muscle

Left ventricle

Interventricular septum

(a)

Superior vena cava

Pulmonary valve

Aortic valve

Right pulmonary artery

Branches of right pulmonary veins

Right atrium

Opening of coronary sinus

Tricuspid valve

Right ventricle

Inferior vena cava

Aorta

Left pulmonary artery

Pulmonary trunk

Left pulmonary veins

Left atrium

Bicuspid valve

Chordae tendineae

Papillary muscle

Left ventricle

Interventricular septum

(b)

**FIGURE 15.6**

Coronal sections of the heart (a) showing the connection between the right ventricle and the pulmonary trunk and (b) showing the connection between the left ventricle and the aorta.

Right atrium

Cusps of tricuspid valve

Chordae tendineae

Interventricular septum

Papillary muscles

Muscular ridges

**FIGURE 15.7**

Photograph of a human tricuspid valve.

**FIGURE 15.8**

Photograph of the pulmonary and aortic valves of the heart (superior view).

## Skeleton of the Heart

Rings of dense connective tissue surround the pulmonary trunk and aorta at their proximal ends. These rings are continuous with others that encircle the atrioventricular orifices. They provide firm attachments for the heart valves and for muscle fibers, and prevent the outlets of the atria and ventricles from dilating during contraction. The fibrous rings, together with other masses of dense connective tissue in the portion of the septum between the ventricles (interventricular septum), constitute the *skeleton of the heart* (see fig. 15.9).

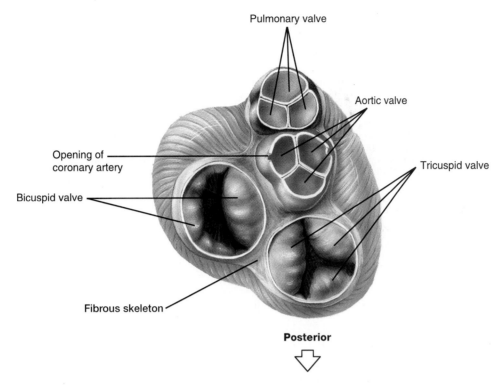

Pulmonary valve

Aortic valve

Opening of coronary artery

Tricuspid valve

Bicuspid valve

Fibrous skeleton

**Posterior**

**FIGURE** 15.9

The skeleton of the heart consists of fibrous rings to which the heart valves are attached (superior view).

| TABLE 15.2 VALVES OF THE HEART | | | | | |
|---|---|---|---|---|---|
| **Valve** | **Location** | **Function** | **Value** | **Location** | **Function** |
| Tricuspid valve | Right atrioventricular orifice | Prevents blood from moving from right ventricle into right atrium during ventricular contraction | Bicuspid (mitral) valve | Left atrioventricular orifice | Prevents blood from moving from left ventricle into left atrium during ventricular contraction |
| Pulmonary valve | Entrance to pulmonary trunk | Prevents blood from moving from pulmonary trunk into right ventricle during ventricular relaxation | Aortic valve | Entrance to aorta | Prevents blood from moving from aorta into left ventricle during ventricular relaxation |

## Path of Blood through the Heart

Blood that is low in oxygen concentration and high in carbon dioxide concentration enters the right atrium through the venae cavae and the coronary sinus. As the right atrial wall contracts, the blood passes through the right atrioventricular orifice and enters the chamber of the right ventricle (fig. 15.10).

When the right ventricular wall contracts, the tricuspid valve closes the right atrioventricular orifice, and the blood moves through the pulmonary valve into the pulmonary trunk and its branches (pulmonary arteries). From these vessels, blood enters the capillaries associated with the alveoli of the lungs. Gas exchanges occur between the blood in the capillaries and the air in the alveoli. The freshly oxygenated blood, which is now relatively low in carbon dioxide, returns to the heart through the pulmonary veins that lead to the left atrium.

The left atrial wall contracts, and the blood moves through the left atrioventricular orifice and into the chamber of the left ventricle. When the left ventricular

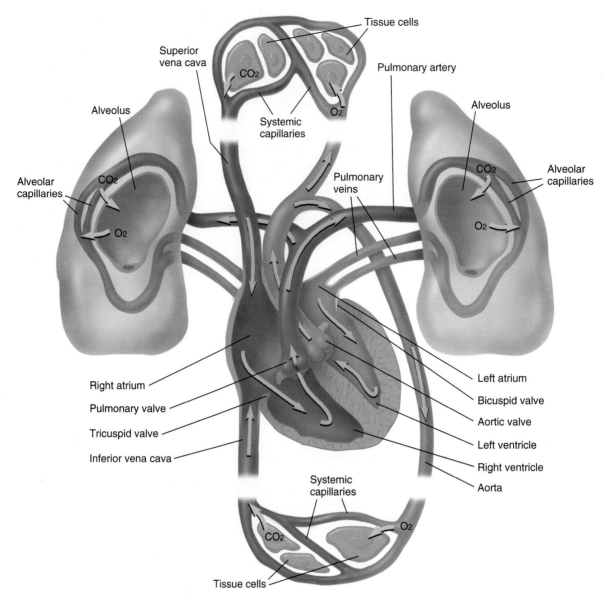

**FIGURE 15.10**

The right ventricle forces blood to the lungs, whereas the left ventricle forces blood to all other body parts. (Structures are not drawn to scale.)

wall contracts, the bicuspid valve closes the left atrioventricular orifice, and the blood passes through the aortic valve into the aorta and its branches. Figure 15.11 summarizes the path the blood takes as it moves through the heart and the pulmonary circuit.

---

*Magnetic resonance imaging* (MRI) can noninvasively image coronary arteries. Blood flow appears as a bright signal, and areas of diminished or absent blood flow, or blood turbulence, appear as blank areas. This approach is less invasive than the standard procedure of *coronary angiography,* in which a catheter is snaked through a blood vessel into the heart and a contrast image is used to show heart structure.

---

## Blood Supply to the Heart

The first two branches of the aorta, called the right and left **coronary arteries,** supply blood to the tissues of the heart. Their openings lie just beyond the aortic valve (figs. 15.12 and 15.13).

One branch of the left coronary artery, the *circumflex artery,* follows the atrioventricular sulcus between the left atrium and the left ventricle. Its branches supply blood to the walls of the left atrium and the left ventricle. Another branch of the left coronary artery, the *anterior interventricular artery* (or *left anterior descending artery*), travels in the anterior interventricular sulcus, and its branches supply the walls of both ventricles.

The right coronary artery passes along the atrioventricular sulcus between the right atrium and the right

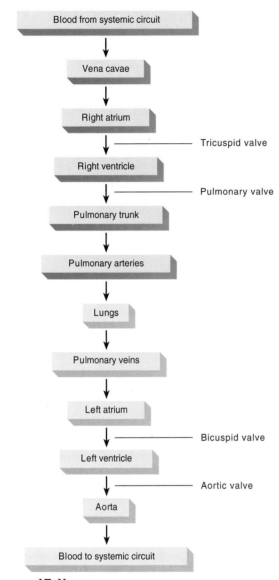

**FIGURE 15.11**

Path of blood through the heart and pulmonary circuit.

In most body parts, blood flow in arteries peaks during ventricular contraction. However, blood flow in the vessels of the myocardium is poorest during ventricular contraction. This is because the muscle fibers of the myocardium compress nearby vessels as they contract, interfering with blood flow. Also, the openings into the coronary arteries are partially blocked as the flaps of the aortic valve open. Conversely, during ventricular relaxation, the myocardial vessels are no longer compressed, and the orifices of the coronary arteries are not blocked by the aortic valve. This increases blood flow into the myocardium.

Blood that has passed through the capillaries of the myocardium is drained by branches of the **cardiac veins,** whose paths roughly parallel those of the coronary arteries. As figure 15.13b shows, these veins join the **coronary sinus,** an enlarged vein on the posterior surface of the heart in the atrioventricular sulcus. The coronary sinus empties into the right atrium. Figure 15.15 summarizes the path of blood that supplies the tissues of the heart.

If a thrombus or embolus blocks or narrows a coronary artery branch, myocardial cells are deprived of oxygen, producing ischemia and a painful condition called *angina pectoris.* The pain usually occurs during physical activity, when the myocardial cells' oxygen requirements exceed their oxygen supply. Pain lessens with rest. Emotional disturbance may also trigger angina pectoris.

Angina pectoris may cause a sensation of heavy pressure, tightening, or squeezing in the chest. The pain is usually felt in the region behind the sternum or in the anterior portion of the upper thorax, but may radiate to the neck, jaw, throat, shoulder, upper limb, back, or upper abdomen. Other symptoms include profuse perspiration (diaphoresis), difficulty breathing (dyspnea), nausea, or vomiting.

A blood clot that completely obstructs a coronary artery or one of its branches (coronary thrombosis) kills part of the heart. This is a *myocardial infarction* (MI), more commonly known as a heart attack.

ventricle. It gives off two major branches—a *posterior interventricular artery,* which travels along the posterior interventricular sulcus and supplies the walls of both ventricles, and a *marginal artery,* which passes along the lower border of the heart. Branches of the marginal artery supply the walls of the right atrium and the right ventricle (figs. 15.12, 15.13, and 15.14).

Because the heart must beat continually to supply blood to the tissues, myocardial cells require a constant supply of freshly oxygenated blood. The myocardium contains many capillaries fed by branches of the coronary arteries. The smaller branches of these arteries usually have connections (anastomoses) between vessels that provide alternate pathways for blood, called collateral circulation.

**1** Which structures make up the skeleton of the heart?

**2** Review the path of blood through the heart.

**3** How does blood composition differ in the right and left ventricle?

**4** Which vessels supply blood to the myocardium?

**5** How does blood return from the cardiac tissues to the right atrium?

Superior vena cava

Right pulmonary artery

Branches of right pulmonary veins

Right auricle

Right atrium

Right coronary artery

Inferior vena cava

(a)

Aorta

Ligamentum arteriosum

Left pulmonary artery

Pulmonary trunk

Left pulmonary veins

Left auricle

Left coronary artery

Circumflex artery

Cardiac vein

Left ventricle

Anterior interventricular artery

Apex of the heart

Marginal artery

Right ventricle

Aorta

Left pulmonary artery

Left pulmonary veins

Left auricle

Circumflex artery

Cardiac vein

Left ventricle

Apex of the heart

(b)

Superior vena cava

Right pulmonary arteries

Right pulmonary veins

Left atrium

Right atrium

Inferior vena cava

Coronary sinus

Middle cardiac vein

Posterior interventricular artery

Right ventricle

**FIGURE** 15.12 ✕

Blood vessels associated with the surface of the heart. (a) Anterior view. (b) Posterior view.

564

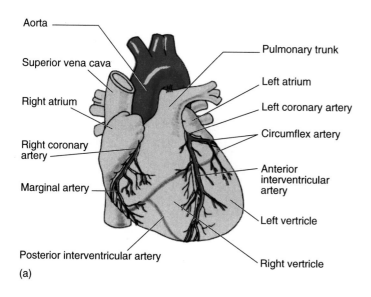

Aorta
Superior vena cava
Right atrium
Right coronary artery
Marginal artery
Posterior interventricular artery
(a)

Pulmonary trunk
Left atrium
Left coronary artery
Circumflex artery
Anterior interventricular artery
Left vertricle
Right vertricle

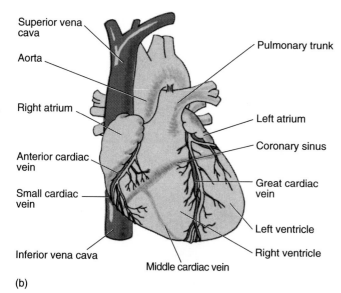

Superior vena cava
Aorta
Right atrium
Anterior cardiac vein
Small cardiac vein
Inferior vena cava
(b)

Pulmonary trunk
Left atrium
Coronary sinus
Great cardiac vein
Left ventricle
Right ventricle
Middle cardiac vein

**FIGURE 15.13**

(a) Branches of the coronary arteries supply blood to heart tissues. (b) Branches of the cardiac veins drain blood from heart tissues. (Faded color indicates vessels on posterior surface.)

**FIGURE 15.14**

An angiogram (radiograph) of the coronary arteries is a diagnostic procedure used to examine specific blood vessels.

A significant percentage of cases of heart failure in adults of African descent may be due to an inherited condition called familial amyloidosis. A gummy protein called amyloid forms deposits in the heart, causing angina (chest pain), failure of cardiac muscle function (cardiomyopathy), blockage of conduction of electrical impulses, and disturbed heart rhythm (arrhythmia). The enlarging amyloid deposits thicken the ventricular walls, which can be detected with an ultrasound technique called echocardiography. It is important to distinguish amyloidosis from other forms of arrhythmias, because drug treatments are different.

## HEART ACTIONS

Heart chambers function in a coordinated, interacting manner, so that the atrial walls contract while the ventricular walls relax, and ventricular walls contract while the atrial walls relax. Then, the atria and the ventricles relax for a moment. Such a series of events constitutes a complete heartbeat, or **cardiac cycle.**

**FIGURE 15.15**

Path of blood through the coronary circulation.

## Cardiac Cycle

During a cardiac cycle, the pressure within the chambers of the heart rises and falls. When the atria are relaxed, the blood flows into them from the large, attached veins. As these chambers fill, the pressure inside gradually increases. When the pressure rises above the pressure in the ventricles, the A-V valves are forced open. About 70% of the entering blood flows directly into the ventricles through the atrioventricular orifices before the atrial walls contract. Then, during atrial contraction (atrial systole), the atrial pressure rises suddenly, forcing the remaining 30% of the atrial contents into the ventricles. Atrial relaxation (atrial diastole) follows (fig. 15.16).

As the ventricles contract (ventricular systole), the A-V valves guarding the atrioventricular orifices are pushed closed by the pressure from the blood by the contracting ventricles; they begin to bulge back into the atria, greatly increasing atrial pressure. At the same time, the papillary muscles contract, and by pulling on the chordae

tendineae, they prevent the cusps of the A-V valves from bulging too far into the atria, allowing the atrial pressure to return to its previous level. During ventricular contraction, the A-V valves remain closed, and the atrial pressure increases again as the atria fill with blood. When the ventricles relax (ventricular diastole), the A-V valves open passively. When the pressure in the ventricles drops below the pressure in the atria, the blood flows through them into the ventricles, and the atrial pressure drops to a low point.

Pressure in the ventricles is low while they are filling, but when the atria contract, the ventricular pressure increases slightly. Then, as the ventricles contract, the ventricular pressure rises sharply, and as soon as the pressure exceeds that in the atria, the A-V valves close. The ventricular pressure continues to increase until it exceeds the pressure in the pulmonary trunk and aorta. Then, the pulmonary and aortic valves are pushed open, and blood is ejected from each valve's respective ventricle into these arteries. As blood flows out

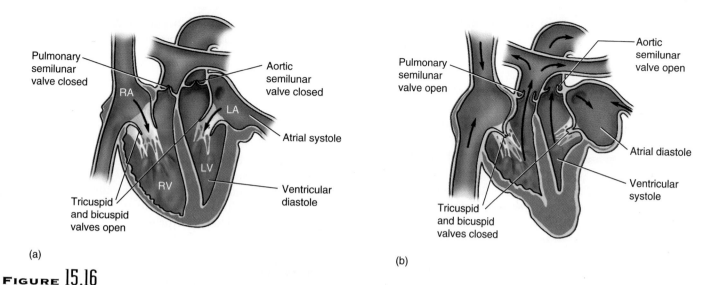

(a)                                                                    (b)

**FIGURE 15.16**

(a) The atria empty during atrial systole and (b) fill with blood during atrial diastole.

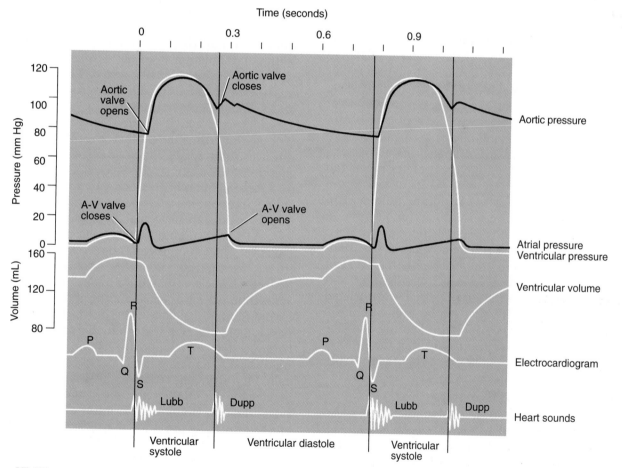

**FIGURE 15.17** 🏃 ▭

A graph of some of the changes that occur in the left ventricle during a cardiac cycle.

of the ventricles, the ventricular pressure begins to drop, and it continues to drop as the ventricles relax. When the ventricular pressure is less than that in the arteries, the pulmonary and aortic valves are closed by arterial blood flowing back toward the ventricles. As soon as the ventricular pressure falls below that of the atria, the A-V valves open, and the ventricles begin to fill once more. The graph in figure 15.17 summarizes some of the changes that occur in the left ventricle during a cardiac cycle.

## Heart Sounds

A heartbeat heard through a stethoscope sounds like "lubb–dupp." These sounds are due to vibrations in the heart tissues produced as the blood flow is suddenly speeded or slowed with the contraction and relaxation of the heart chambers, and with the opening and closing of the valves.

The first part of a heart sound (*lubb*) occurs during the ventricular contraction, when the A-V valves are closing. The second part (*dupp*) occurs during ventricular relaxation, when the pulmonary and aortic valves are closing (fig. 15.17).

Sometimes during inspiration, the interval between the closure of the pulmonary and the aortic valves is long enough that a sound coming from each of these events can be heard. In this case, the second heart sound is said to be *split*.

Heart sounds are of particular interest because they can indicate the condition of the heart valves. For example, inflammation of the endocardium (endocarditis) may change the shapes of the valvular cusps (valvular stenosis). Then, the cusps may close incompletely, and some blood may leak back through the valve. This produces an abnormal sound called a *murmur*. The seriousness of a murmur depends on the extent of valvular damage. Many heart murmurs are harmless. Fortunately for those who have serious problems, it is often possible to repair the damaged valves or to replace them, as happened to the doctor described at the chapter's beginning. Clinical Application 15.1 describes more severe heart problems.

Using a stethoscope, it is possible to hear sounds associated with the aortic and pulmonary valves by listening from the second intercostal space on either side of the sternum. The *aortic sound* comes from the right, and the *pulmonic sound* from the left. The sound associated with the bicuspid (mitral) valve can be heard from the fifth intercostal space at the nipple line on the left. The sound of the tricuspid valve can be heard at the fifth intercostal space just to the right of the sternum (fig. 15.18).

## Cardiac Muscle Fibers

Recall that cardiac muscle fibers function like those of skeletal muscles, but the fibers connect in branching networks. Stimulation to any part of the network sends impulses throughout the heart, which contracts as a unit.

A mass of merging cells that act as a unit is called a **functional syncytium.** Two such structures are in the heart—in the atrial walls and in the ventricular walls. These masses of cardiac muscle fibers are separated from each other by portions of the heart's fibrous skeleton, except for a small area in the right atrial floor. In this region, the *atrial syncytium* and the *ventricular syncytium* are connected by fibers of the **cardiac conduction system.**

1. Describe the pressure changes that occur in the atria and ventricles during a cardiac cycle.

2. What causes heart sounds?

3. What is a functional syncytium?

4. Where are the functional syncytia of the heart located?

## Cardiac Conduction System

Throughout the heart are clumps and strands of specialized cardiac muscle tissue whose fibers contain only a few myofibrils. Instead of contracting, these areas initiate and distribute impulses (cardiac impulses) throughout the myocardium. They comprise the cardiac conduction system, which coordinates the events of the cardiac cycle.

A key portion of this conduction system is the **sinoatrial node** (S-A node), a small, elongated mass of specialized cardiac muscle tissue just beneath the epicardium. It is located in the right atrium near the opening of the superior vena cava, and its fibers are continuous with those of the atrial syncytium.

The cells of the S-A node can reach threshold on their own, and their membranes contact one another. Without stimulation from nerve fibers or any other outside agents, the nodal cells initiate impulses that spread into the surrounding myocardium and stimulate the cardiac muscle fibers to contract.

S-A node activity is rhythmic. The S-A node initiates one impulse after another, seventy to eighty times a minute in an adult. Because it generates the heart's rhythmic contractions, the S-A node is often called the **pacemaker.** From the S-A node, bundles of atrial muscle, called *internodal atrial muscle,* preferentially conduct impulses along tracts to specific regions of the heart.

As a cardiac impulse travels from the S-A node into the atrial syncytium, it goes from cell to cell via gap junctions. The right and left atria contract almost simultaneously. Instead of passing directly into the ventricular syncytium, which is separated from the atrial syncytium by the fibrous skeleton of the heart, the cardiac impulse passes along fibers of the conduction system that are continuous with atrial muscle fibers. These conducting fibers lead to a mass of specialized cardiac muscle tissue called the **atrioventricular node** (A-V node). This node is located in the inferior portion of the septum that separates the atria (interatrial septum) and just beneath the endocardium. It provides the only normal conduction pathway between the atrial and ventricular syncytia, because the fibrous skeleton does not conduct the impulse.

The fibers that conduct the cardiac impulse into the A-V node (junctional fibers) have very small diameters, and because small fibers conduct impulses slowly, they delay transmission of the impulse. The impulse is delayed further as it moves through the A-V node, allowing time for the atria to contract completely so they empty all their blood into the ventricles prior to ventricular contraction.

# Heart Transplants

When Tina Orbacz was pregnant with her second child in 1990, she attributed her increasing fatigue to her pregnant state. But a month after her son's birth, she was even more exhausted. Plus, she had lost weight during her pregnancy, not gained it. Finally, cardiologists discovered that she was suffering from heart failure due to a birth defect, called an *atrial septal defect,* that weakened the tissue between the atria. Because her heart was failing, and a complication of this disorder is lung failure, Tina Orbacz, at the age of 24 years, found herself waiting for a heart-lung transplant.

In a *heart transplant,* the recipient's failing heart is removed, except for the posterior walls of the right and left atria and their connections to the venae cavae and pulmonary veins. The donor heart is similarly prepared and is attached to the atrial cuffs remaining in the recipient's thorax. Finally, the recipient's aorta and pulmonary arteries are connected to those of the donor heart.

Tina was lucky. Her condition stabilized, and she was able to survive the three-year wait until a 16-year-old who died in an accident donated a heart and lungs. Because she was young and her cell surface molecules closely matched those of the donated heart and lungs,

Tina has done extraordinarily well. Today, she exercises regularly and leads a normal life.

Because of the shortage of donor hearts, and the severity of the illnesses that lead to heart failure, many potential recipients die waiting for a transplant. A mechanical half-heart, called *a left ventricular assist device* (LVAD), can often maintain cardiac function long enough for a donor heart to become available.

This happened to Mike Dorsey, a 41-year-old father of six. Although he had to stay in the hospital to wear his LVAD, the device worked so well that he was able to exercise and help with office work while awaiting his transplant. The LVAD enabled him to increase his physical fitness, which contributed to the success of his eventual heart transplant six months later. ■

**FIGURE 15.18**
Thoracic regions where the sounds of each heart valve are most easily heard.

**FIGURE 15.19**

The cardiac conduction system.

**FIGURE 15.20**

Components of the cardiac conduction system.

Once the cardiac impulse reaches the distal side of the A-V node, it passes into a group of large fibers that make up the **A-V bundle** (atrioventricular bundle or bundle of His), and the impulse moves rapidly through them. The A-V bundle enters the upper part of the interventricular septum, and divides into right and left branches that lie just beneath the endocardium. About halfway down the septum, the branches give rise to enlarged **Purkinje fibers.** These larger fibers carry the impulse to distant regions of the ventricular myocardium much faster than cell-to-cell conduction could. Thus, the massive ventricular myocardium will also contract as a functioning unit.

The base of the aorta, which contains the aortic valves, is enlarged and protrudes somewhat into the interatrial septum close to the A-V bundle. Consequently, inflammatory conditions, such as bacterial endocarditis affecting the aortic valves (aortic valvulitis), may also affect the A-V bundle.

If a portion of the bundle is damaged, it may no longer conduct impulses normally. As a result, cardiac impulses may reach the two ventricles at different times so that they fail to contract together. This condition is called a *bundle branch block.*

The Purkinje fibers spread from the interventricular septum into the papillary muscles, which project inward from the ventricular walls, and then continue downward to the apex of the heart. There they curve around the tips of the ventricles and pass upward over the lateral walls of these chambers. Along the way, the Purkinje fibers give off many small branches, which become continuous with cardiac muscle fibers. These parts of the conduction system are shown in figure 15.19 and are summarized in figure 15.20.

The muscle fibers in the ventricular walls form irregular whorls. When impulses on the Purkinje fibers stimulate these muscle fibers, the ventricular walls contract with a twisting motion (fig. 15.21). This action squeezes blood out of the ventricular chambers and forces it into the aorta and pulmonary trunk.

Another property of the conduction system is that the Purkinje fibers transmit the impulse to the apex of the heart first. As a result, contraction begins at the apex and pushes the blood superiorly toward the aortic and pulmonary semilunar valves, rather than having the impulse begin superiorly and push blood toward the apex, as it would if the impulse traveled from cell to cell.

1 What is the function of the cardiac conduction system?

2 What kinds of tissues make up the cardiac conduction system?

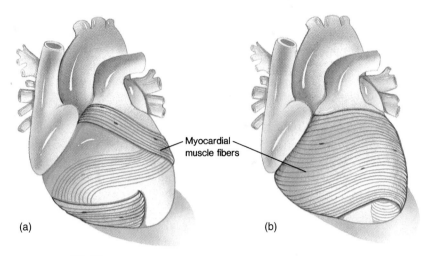

(a)

(b)

Myocardial muscle fibers

**FIGURE 15.21**

The muscle fibers within the ventricular walls are arranged in patterns of whorls. The fibers of groups (a) and (b) surround both ventricles in these anterior views of the heart.

walls are of much greater mass than those of the atria, the amount of electrical change is greater, and the pen deflects to a greater degree than before. When the electrical change finishes, the pen returns to the baseline, leaving a mark called the *QRS complex,* which usually consists of a *Q wave,* an *R wave,* and an *S wave.* This complex appears due to depolarization of the ventricular fibers just prior to the contraction of the ventricular walls.

The electrical changes occurring as the ventricular muscle fibers repolarize slowly produces a *T wave* as the pen deflects again, ending the ECG pattern. The record of the atrial repolarization seems to be missing from the pattern because the atrial fibers repolarize at the same time that the ventricular fibers depolarize. Thus, the QRS complex obscures the recording of the atrial repolarization.

**3** How is a cardiac impulse initiated?

**4** How is a cardiac impulse transmitted from the right atrium to the other heart chambers?

## Electrocardiogram

An *electrocardiogram* (ECG) is a recording of the electrical changes that occur in the myocardium during a cardiac cycle. (This pattern occurs as action potentials stimulate cardiac muscle fibers to contract, but it is not the same as individual action potentials.) Because body fluids can conduct electrical currents, such changes can be detected on the surface of the body.

To record an ECG, electrodes are placed on the skin and connected by wires to an instrument that responds to very weak electrical changes by moving a pen or stylus on a moving strip of paper. Up-and-down movements of the pen correspond to electrical changes in the myocardium. Because the paper moves past the pen at a known rate, the distance between pen deflections indicates time elapsing between phases of the cardiac cycle.

As figure 15.22a illustrates, a normal ECG pattern includes several deflections, or *waves,* during each cardiac cycle. Between cycles, the muscle fibers remain polarized, with no detectable electrical changes. Consequently, the pen does not move and simply marks along the baseline. When the S-A node triggers a cardiac impulse, the atrial fibers depolarize, producing an electrical change. The pen moves, and at the end of the electrical change, returns to the base position. This first pen movement produces a *P wave,* corresponding to depolarization of the atrial fibers that will lead to contraction of the atria (fig. 15.22b).

When the cardiac impulse reaches the ventricular fibers, they rapidly depolarize. Because the ventricular

Physicians use ECG patterns to assess the heart's ability to conduct impulses. For example, the time period between the beginning of a P wave and the beginning of a QRS complex (*P-Q interval,* or if the initial portion of the QRS wave is upright, *P-R interval*) indicates the time for the cardiac impulse to travel from the S-A node through the A-V node. Ischemia or other problems affecting the fibers of the A-V conduction pathways can increase this P-Q interval. Similarly, injury to the A-V bundle can increase duration of the QRS complex, because it may take longer for an impulse to spread throughout the ventricular walls (fig. 15.23).

**1** What is an electrocardiogram?

**2** Which cardiac events do the P wave, QRS complex, and T wave represent?

## Regulation of the Cardiac Cycle

The quantity of blood pumped changes to accommodate cellular requirements. For example, during strenuous exercise, the amount of blood the skeletal muscles require increases greatly, and heart rate increases in response. Since the S-A node normally controls heart rate, changes in this rate often involve factors that affect the pacemaker, such as the motor impulses carried on the parasympathetic and sympathetic nerve fibers (see figs. 11.39, 11.40, 15.24, and 15.39).

The parasympathetic fibers that innervate the heart arise from neurons in the medulla oblongata and make up parts of the *vagus nerves.* Most of these fibers branch to the S-A and A-V nodes. When the nerve impulses

(a)

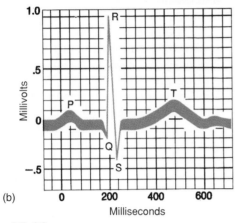

(b)

**FIGURE 15.22**

(a) A normal ECG. (b) In an ECG pattern, the P wave results from a depolarization of the atria, the QRS complex results from a depolarization of the ventricles, and the T wave results from a repolarization of the ventricles.

**FIGURE 15.23**

A prolonged QRS complex may result from damage to the A-V bundle fibers.

reach nerve fiber endings, they secrete acetylcholine, which decreases S-A and A-V nodal activity. As a result, heart rate decreases.

The vagus nerves continually carry impulses to the S-A and A-V nodes, braking heart action. Consequently, parasympathetic activity can change heart rate in either direction. An increase in the impulses slows the heart, and a decrease in the impulses releases the parasympathetic "brake" and increases heart rate.

Sympathetic fibers reach the heart by means of the *accelerator nerves,* whose branches join the S-A and A-V

nodes as well as other areas of the atrial and ventricular myocardium. The endings of these fibers secrete norepinephrine in response to nerve impulses, which increases the rate and force of myocardial contractions.

The *cardiac control center* of the medulla oblongata maintains balance between the inhibitory effects of the parasympathetic fibers and the excitatory effects of the sympathetic fibers. In this region of the brain, masses of neurons function as *cardioinhibitor* and *cardioaccelerator reflex centers.* These centers receive sensory impulses from throughout the circulatory system and relay motor impulses to the heart in response.

For example, receptors that are sensitive to stretch are located in certain regions of the aorta (aortic arch) and in the carotid arteries (carotid sinuses). These receptors, called *baroreceptors* (pressoreceptors), can detect changes in blood pressure. Rising pressure stretches the receptors, and they signal the cardioinhibitor center in the medulla. In response, the medulla sends parasympathetic motor impulses to the heart, decreasing the heart rate and force of contraction. This action helps lower blood pressure toward normal (fig. 15.24).

Another regulatory reflex involves stretch receptors in the venae cavae near the entrances to the right atrium. If venous blood pressure increases abnormally in these vessels, the receptors signal the cardioaccelerator center, and sympathetic impulses flow to the heart. As a result, heart rate and force of contraction increase, and the venous pressure is reduced.

Impulses from the cerebrum or hypothalamus also influence the cardiac control center. Such impulses may decrease heart rate, as occurs when a person faints following an emotional upset, or they may increase heart rate during a period of anxiety.

Two other factors that influence heart rate are temperature change and certain ions. Rising body temperature increases heart action, which is why heart rate usually increases during fever. On the other hand, abnormally low body temperature decreases heart action. As a result, a patient's body temperature is sometimes deliberately lowered (hypothermia) to slow the heart during surgery.

Of the ions that influence heart action, the most important are potassium ($K^+$) and calcium ($Ca^{++}$). Potassium affects the electrical potential of the cell membrane, altering its ability to reach the threshold for conducting an impulse (see chapter 10). Calcium affects the cardiac muscle fiber because its sarcoplasmic reticulum contains less calcium reserve than skeletal muscle, and therefore is more dependent on extracellular (bloodborne) calcium (see chapter 9). Although homeostatic mechanisms normally maintain the concentrations of these ions within narrow ranges, these mechanisms sometimes fail, and the consequences can be serious or even fatal. Clinical Application 15.2 examines abnormal heart rhythms.

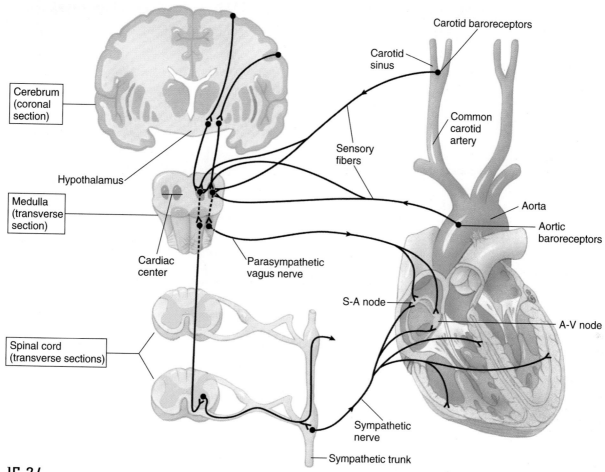

**FIGURE 15.24**

Autonomic nerve impulses alter the activities of the S-A and A-V nodes.

An excess of potassium ions (hyperkalemia) alters the usual polarized state of the cardiac muscle fibers, decreasing the rate and force of contractions. Very high potassium ion concentration may block conduction of cardiac impulses, and heart action may suddenly stop (cardiac arrest). Conversely, if the potassium concentration drops below normal (hypokalemia), the heart may develop a potentially life-threatening abnormal rhythm (arrhythmia).

Excess calcium ions (hypercalcemia) increase heart action, introducing danger that the heart will undergo a prolonged contraction. Conversely, low calcium ion concentration (hypocalcemia) depresses heart action because these ions help initiate muscle contraction.

**1** Which nerves supply parasympathetic fibers to the heart? Which nerves supply sympathetic fibers?

**2** How do parasympathetic and sympathetic impulses help control heart rate?

**3** How do changes in body temperature affect heart rate?

## BLOOD VESSELS

The blood vessels are organs of the cardiovascular system, and they form a closed circuit of tubes that carries blood from the heart to cells and back again. These vessels include **arteries, arterioles, capillaries, venules,** and **veins.** The arteries and arterioles conduct blood away from the ventricles of the heart and lead to the capillaries. The capillaries are sites of exchange of substances between blood and the body cells, and the venules and veins return blood from the capillaries to the atria. (Figure 5A shows a semisynthetic blood vessel substitute.)

### Arteries and Arterioles

**Arteries** are strong, elastic vessels that are adapted for carrying the blood away from the heart under high pressure. These vessels subdivide into progressively thinner tubes and eventually give rise to the finer branched **arterioles.**

The wall of an artery consists of three distinct layers, or *tunics,* shown in figure 15.25. The innermost layer, tunica interna (intima), is composed of a layer of simple squamous epithelium, called *endothelium,* that

# Arrhythmias

Each year, thousands of people die from a fast or irregular heartbeat. These are types of altered heart rhythm called *arrhythmia*. John Thomas was almost one of them—several times.

In a crowded Boston train in December 1990, 36-year-old Thomas lost consciousness for about 8 seconds, as his heart *fibrillated,* small areas of the myocardium contracting in an uncoordinated fashion (fig. 15A). As a result, the myocardium failed to contract as a whole, and blood was no longer pumped.

Fibrillation is cardiac chaos. Atrial fibrillation is not life threatening because the ventricles still pump blood, but ventricular fibrillation, which struck Thomas, is often deadly. Ventricular fibrillation can be caused by an obstructed coronary artery, toxic drug exposure, electric shock, or traumatic injury to the heart or chest wall.

Fortunately, Thomas had had a device called a defibrillator surgically implanted a year earlier, after he suffered a cardiac arrest. The device sends out an electrical jolt during fibrillation, which interrupts the abnormal heart rhythm, allowing the heart to resume a normal beat. Thomas' defibrillator saves his life several times a year.

An abnormally fast heartbeat, usually more than one hundred beats per minute, is called *tachycardia.* Increase in body temperature, nodal stimulation by sympathetic fibers, certain drugs or hormones, heart disease, excitement, exercise, anemia, or shock can all cause tachycardia. Figure 15B shows the ECG of a tachycardic heart.

*Bradycardia* means a slow heart rate, usually fewer than sixty beats per minute. Decreased body temperature, nodal stimulation by parasympathetic impulses, or certain drugs may cause bradycardia. It also may occur during sleep. Figure 15C shows the ECG of a bradycardic heart. Athletes sometimes have unusually slow heartbeats because their hearts have developed the ability to pump a greater-than-normal volume of blood with each beat. The slowest heartbeat recorded in a healthy athlete was 25 beats per minute!

A *premature beat* occurs before it is expected in a normal series of cardiac cycles. Cardiac impulses originating from unusual (ectopic) regions of the heart probably cause a premature beat. That is, the impulse originates from a site other than the S-A node. Cardiac impulses may arise from ischemic tissues or from muscle fibers irritated by disease or drugs.

A heart chamber *flutters* when it contracts regularly, but very rapidly, such as 250–350 times per minute. Although normal hearts may flutter occasionally, this condition is more likely to be due to damage to the myocardium (fig. 15D).

Any interference or block in cardiac impulse conduction may cause arrhythmia, the type varying with the location and extent of the block. Such arrhythmias arise because certain cardiac tissues other than the S-A node can function as pacemakers.

**FIGURE 15A**

Ventricular fibrillation is rapid, uncoordinated depolarization of the ventricles.

rests on a connective tissue membrane that is rich in elastic and collagenous fibers.

In addition to separating the flowing blood from the blood vessel wall, endothelium helps prevent blood clotting by secreting biochemicals that inhibit platelet aggregation (see chapter 14). The endothelial lining of an artery provides a smooth surface that allows blood cells and platelets to flow through without being damaged.

Endothelium also may help regulate local blood flow by secreting substances that either dilate or constrict blood vessels.

The middle layer, tunica media, makes up the bulk of the arterial wall. It includes smooth muscle fibers, which encircle the tube, and a thick layer of elastic connective tissue. The connective tissues give the vessel a tough elasticity that enables it to withstand the force of

The S-A node usually initiates seventy to eighty heartbeats per minute, called a sinus rhythm. If the S-A node is damaged, impulses originating in the A-V node may travel upward into the atrial myocardium and downward into the ventricular walls, stimulating them to contract. Under the influence of the A-V node acting as a *secondary pacemaker,* the heart may continue to pump blood, but at a rate of forty to sixty beats per minute, called a nodal rhythm. Similarly, the Purkinje fibers can initiate cardiac impulses, contracting the heart fifteen to forty times per minute.

An *artificial pacemaker* can treat a disorder of the cardiac conduction system. This device includes an electrical pulse generator and a lead wire that communicates with a portion of the myocardium. The pulse generator contains a permanent battery that provides energy and a microprocessor that can sense the cardiac rhythm and signal the heart to alter its contraction rate.

An artificial pacemaker is surgically implanted beneath the patient's skin in the shoulder. An external programmer adjusts its functions from the outside. The first pacemakers, made in 1958, were crude. Today, thanks to telecommunications advances, a physician can check a patient's pacemaker over the phone! A device called a pacemaker-cardioverter-defibrillator can correct both abnormal heart rhythm and cardiac arrest. ■

**FIGURE 15B**
Tachycardia is a rapid heartbeat.

**FIGURE 15C**
Bradycardia is a slow heartbeat.

**FIGURE 15D**
Atrial flutter is an abnormally rapid rate of atrial depolarization.

blood pressure and, at the same time, to stretch and accommodate the sudden increase in blood volume that accompanies ventricular contraction.

The outer layer, tunica externa (adventitia), is thin and chiefly consists of connective tissue with irregular elastic and collagenous fibers. This layer attaches the artery to the surrounding tissues. It also contains minute vessels (vasa vasorum) that give rise to capillar-

ies and provide blood to the more external cells of the artery wall.

The sympathetic branches of the autonomic nervous system innervate smooth muscle in artery and arteriole walls. Impulses on these *vasomotor fibers* contract the smooth muscles, reducing the diameter of the vessel. This is called **vasoconstriction.** If vasomotor impulses are inhibited, the muscle fibers relax and the diameter of

**Artery**

**Vein**

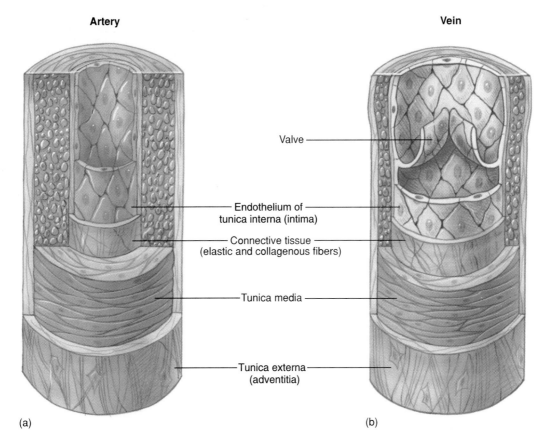

Valve

Endothelium of
tunica interna (intima)

Connective tissue
(elastic and collagenous fibers)

Tunica media

Tunica externa
(adventitia)

(a)

(b)

**FIGURE 15.25**

(a) The wall of an artery and (b) the wall of a vein.

Arteriole

Smooth muscle cell

Endothelium

Capillary

**FIGURE 15.26**

Small arterioles have smooth muscle fibers in their walls;
capillaries lack these fibers.

the vessel increases. This is called **vasodilation.** Changes
in the diameters of arteries and arterioles greatly influence blood flow and pressure.

Although the walls of the larger arterioles have
three layers similar to those of arteries, these walls thin
as the arterioles approach the capillaries. The wall of a
very small arteriole consists only of an endothelial lining and some smooth muscle fibers, surrounded by a
small amount of connective tissue (figs. 15.26 and 15.27).

**FIGURE 15.27**

Scanning electron micrograph of an arteriole cross section
(2,200×).

© R. G. Kessel and R. H. Kardon. *Tissues and Organs: A Text-Atlas of Scanning Electron
Microscopy,* 1979.

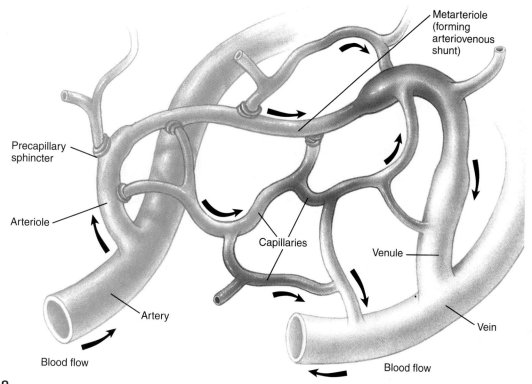

**FIGURE 15.28**

Some metarterioles provide arteriovenous shunts by connecting arterioles directly to venules.

Arterioles, which are microscopic continuations of arteries, give off branches called *metarterioles* that, in turn, join capillaries.

The arteriole and metarteriole walls are adapted for vasoconstriction and vasodilation in that their muscle fibers respond to impulses from the autonomic nervous system by contracting or relaxing. Thus, these vessels help control the flow of blood into the capillaries.

Sometimes metarterioles connect directly to venules, and blood entering them can bypass the capillaries. These connections between arteriole and venous pathways, shown in figure 15.28, are called *arteriovenous shunts.*

**1** Describe the wall of an artery.

**2** What is the function of the smooth muscle in the arterial wall?

**3** How is the structure of an arteriole different from that of an artery?

## Capillaries

**Capillaries** are the smallest blood vessels. They connect the smallest arterioles and the smallest venules. Capillaries are extensions of the inner linings of arterioles in that their walls are endothelium—a single layer of squamous

epithelial cells (fig. 15.29a). These thin walls form the semipermeable layer through which substances in the blood are exchanged for substances in the tissue fluid surrounding body cells.

### Capillary Permeability

The openings or intercellular channels in the capillary walls are thin slits where endothelial cells overlap. The sizes of these openings, and consequently the permeability of the capillary wall, vary from tissue to tissue. For example, the openings are relatively small in the capillaries of smooth, skeletal, and cardiac muscle, whereas those in capillaries associated with endocrine glands, the kidneys, and the lining of the small intestine are larger.

Capillaries with the largest openings include those of the liver, spleen, and red bone marrow. These capillaries are discontinuous, and the distance between their cells appears as little cavities (sinusoids) in the organ. Discontinuous capillaries allow large proteins and even intact cells to pass through as they enter or leave the circulation (fig. 15.29b and c). Clinical Application 3.2 discusses the blood-brain barrier, the protective tight capillaries in the brain. The barrier is not present in the pituitary and pineal glands and parts of the hypothalamus.

(a)

(b)

(c)

**FIGURE 15.29**

(a) Substances are exchanged between the blood and tissue fluid through openings (slits) separating endothelial cells. (b) Transmission electron micrograph of a capillary cross section (11,500×). (c) Note the narrow slitlike openings at the cell junctions (arrow) (micrograph b enlarged to 62,500×).

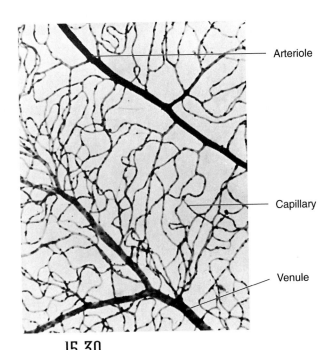

**FIGURE 15.30**

Light micrograph of a capillary network (100×).

## Capillary Arrangement

The higher a tissue's rate of metabolism, the denser its capillary networks. Muscle and nerve tissues, which use large quantities of oxygen and nutrients, are richly supplied with capillaries; cartilaginous tissues, the epidermis, and the cornea, where metabolism is slow, lack capillaries.

If the capillaries of an adult were unwound and spread end to end, they would cover from 25,000 to 60,000 miles.

The patterns of capillary arrangement also differ in various body parts. For example, some capillaries pass directly from arterioles to venules, but others lead to highly branched networks (fig. 15.30). Such physical arrangements make it possible for the blood to follow different pathways through a tissue that meet the demands of its cells.

Blood flow can be varied among tissues as well. During exercise, for example, blood can be directed into the capillary networks of the skeletal muscles, where the cells have increased requirements for oxygen and nutrients. At the same time, the blood can bypass some of the capillary networks in the tissues of the digestive tract, where demand for blood is less critical. Conversely, when a person is relaxing after a meal, blood can be shunted from the inactive skeletal muscles into the capillary networks of the digestive organs.

## Regulation of Capillary Blood Flow

The distribution of blood in the various capillary pathways is mainly regulated by the smooth muscles that encircle the capillary entrances. As figure 15.28 shows, these muscles form *precapillary sphincters,* which may close a capillary by contracting, or open it by relaxing. The precapillary sphincters respond to the demands of the cells. When the cells have low concentrations of oxygen and nutrients, the sphincter relaxes and blood flow increases; when cellular requirements have been met, the sphincter may contract again.

**1** Describe a capillary wall.

**2** What is the function of a capillary?

**3** What controls blood flow into capillaries?

## Exchanges in the Capillaries

The vital function of exchanging gases, nutrients, and metabolic by-products between the blood and the tissue fluid surrounding body cells occurs in the capillaries. The biochemicals exchanged move through the capillary walls by diffusion, filtration, and osmosis (see chapter 3). Diffusion is the most important means of transfer.

Because blood entering certain capillaries carries high concentrations of oxygen and nutrients, these substances diffuse through the capillary walls and enter the tissue fluid. Conversely, the concentrations of carbon dioxide and other wastes are generally greater in the tissues, and such wastes tend to diffuse into the capillary blood.

The paths these substances follow depend primarily on their solubilities in lipids. Substances that are soluble in lipid, such as oxygen, carbon dioxide, and fatty acids, can diffuse through most areas of the cell membranes that make up the capillary wall because the membranes are largely lipid. Lipid-insoluble substances, such as water, sodium ions, and chloride ions, diffuse through pores in the cell membranes and through the slitlike openings between the endothelial cells that form the capillary wall (see fig. 15.29). Plasma proteins generally remain in the blood because they are not soluble in the lipid portions of the capillary membranes, and they are too large to diffuse through the membrane pores or slitlike openings between the endothelial cells of most capillaries.

Recall from chapter 3 that in *filtration, hydrostatic pressure* forces molecules through a membrane. In the capillaries, the blood pressure generated when ventricle walls contract provides the force for filtration.

Blood pressure also moves blood through the arteries and arterioles. This pressure decreases as distance

from the heart increases, because of friction (peripheral resistance) between the blood and the vessel walls. For this reason, blood pressure is greater in the arteries than in the arterioles, and greater in the arterioles than in the capillaries. It is similarly greater at the arteriolar end of a capillary than at the venular end.

The walls of arteries and arterioles are too thick to allow blood components to pass through. However, the hydrostatic pressure of the blood pushes small molecules through capillary walls by filtration. This effect primarily occurs at the arteriolar ends of capillaries, whereas diffusion takes place along their entire lengths.

Plasma proteins, which remain in the capillaries, help make the *osmotic pressure* of the blood greater (hypertonic) than that of the tissue fluid. Although capillary blood has a greater osmotic attraction for water than does tissue fluid, the greater force of the blood pressure overcomes this attraction at the arteriolar end of the capillary. As a result, the net movement of water and dissolved substances is outward at the arteriolar end of the capillary by filtration.

More specifically, the force, including hydrostatic pressure, that moves fluid outward at the arteriolar end of a capillary is 41.3 millimeters of mercury (mm Hg), whereas the net osmotic pressure, which tends to move fluid inward, is only 28 mm Hg. Consequently, the net movement of water and dissolved substances is outward. However, blood pressure decreases as the blood moves through the capillary, and, at the venular end, the outward force equals 21.3 mm Hg, whereas the net osmotic pressure of the blood remains essentially unchanged at 28 mm Hg. Thus, there is a net movement of fluid into the venular end of the capillary by osmosis. Figure 15.31 shows this process.

Normally, more fluid leaves the capillaries than returns to them. *Lymphatic vessels* collect the excess fluid and return it to the venous circulation. This mechanism is discussed in chapter 16.

Sometimes unusual events increase blood flow to capillaries, and excess fluid enters spaces between tissue cells (interstitial spaces). This may occur, for instance, in response to certain chemicals such as *histamine* that vasodilate the metarterioles and increase capillary permeability. Enough fluid may leak out of the capillaries to overwhelm lymphatic drainage, and affected tissues become swollen (edematous) and painful.

**1** What forces are responsible for the exchange of substances between blood and the tissue fluid?

**2** Why is the fluid movement out of a capillary greater at its arteriolar end than at its venular end?

**3** Since more fluid leaves the capillary than returns to it, how is the remainder returned to the vascular system?

FIGURE 15.31

Water and other substances leave the capillaries because of a net outward filtration pressure at the arteriolar end of the capillaries. Water enters at the venular end of the capillaries because of a net inward force of osmotic pressure. Substances move in and out along the length of the capillaries according to their respective diffusion gradients.

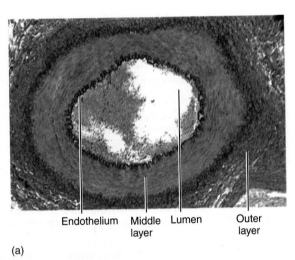

Endothelium   Middle   Lumen   Outer
              layer             layer

(a)

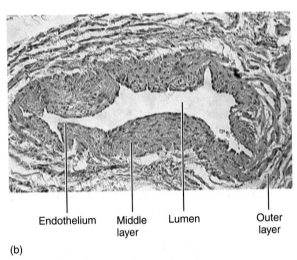

Endothelium   Middle   Lumen   Outer
              layer             layer

(b)

FIGURE 15.32

Note the structural differences in these cross sections of (a) an artery (100×) and (b) a vein (160×).

If the right ventricle of the heart is unable to pump blood out as rapidly as it enters, other parts of the body may develop edema, because the blood backs up into the veins, venules, and capillaries, increasing blood pressure in these vessels. As a result of this increased *back pressure,* osmotic pressure of the blood in the venular ends of the capillaries is less effective in attracting water from tissue fluid, and the tissues swell. This is true particularly in the lower extremities if the person is upright, or in the back if the person is supine. In the terminal stages of heart failure, edema is widespread, and fluid accumulates in the peritoneal cavity of the abdomen. This condition is called *ascites.*

## Venules and Veins

**Venules** are the microscopic vessels that continue from the capillaries and merge to form **veins.** The veins, which carry blood back to the atria, follow pathways that roughly parallel those of the arteries.

The walls of veins are similar to those of arteries in that they are composed of three distinct layers. However, the middle layer of the venous wall is poorly developed. Consequently, veins have thinner walls that contain less smooth muscle and less elastic tissue than those of comparable arteries but their lumens have a greater diameter (figs. 15.25 and 15.32).

Many veins, particularly those in the upper and lower limbs, contain flaplike *valves,* which project in-

**FIGURE 15.33**

(a) Venous valves allow blood to move toward the heart, but (b) prevent blood from moving backward away from the heart.

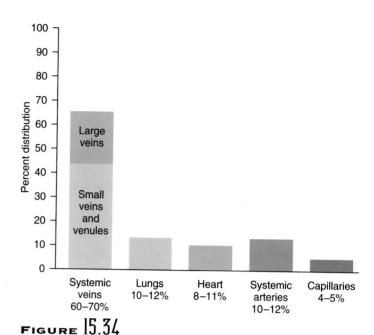

**FIGURE 15.34**

Most of the blood volume is contained within the veins and venules.

| TABLE 15.3 | CHARACTERISTICS OF BLOOD VESSELS | |
|---|---|---|
| **Vessel** | **Type of Wall** | **Function** |
| Artery | Thick, strong wall with three layers—an endothelial lining, a middle layer of smooth muscle and elastic tissue, and an outer layer of connective tissue | Carries relatively high pressure blood from the heart to arterioles |
| Arteriole | Thinner wall than an artery, but with three layers; smaller arterioles have an endothelial lining, some smooth muscle tissue, and a small amount of connective tissue | Connects an artery to a capillary, helps control the blood flow into a capillary by vasoconstricting or vasodilating |
| Capillary | Single layer of squamous epithelium | Provides a membrane through which nutrients, gases, and wastes are exchanged between the blood and tissue cells; connects an arteriole to a venule |
| Venule | Thinner wall, less smooth muscle and elastic tissue than in an arteriole | Connects a capillary to a vein |
| Vein | Thinner wall than an artery but with similar layers; the middle layer is more poorly developed; some with flaplike valves | Carries relatively low pressure blood from a venule to the heart; valves prevent a backflow of blood; serves as blood reservoir |

ward from their linings. Valves, shown in figure 15.33, are usually composed of two leaflets that are pushed closed if the blood begins to back up in a vein. These valves aid in returning blood to the heart because they are open as long as the flow is toward the heart, but close if it is in the opposite direction.

Veins also function as *blood reservoirs,* useful in times of blood loss. For example, if a hemorrhage accompanied by a drop in arterial blood pressure occurs, the muscular walls of the veins are reflexly stimulated

by sympathetic nerve impulses. The resulting venous constrictions help maintain blood pressure by returning more blood to the heart. This mechanism ensures a nearly normal blood flow even when as much as 25% of the blood volume is lost. Figure 15.34 illustrates the relative volumes of blood in the veins and other blood vessels.

Table 15.3 summarizes the characteristics of blood vessels. Clinical Application 15.3 examines disorders of blood vessels.

# CLINICAL APPLICATION

# Blood Vessel Disorders

In the arterial disease *atherosclerosis,* soft masses of fatty materials, particularly cholesterol, accumulate on the insides of the arterial walls. Such deposits, called *plaque,* protrude into the lumens of the vessels and interfere with blood flow (fig. 15E). Furthermore, plaque often forms a surface texture that can initiate formation of a blood clot, increasing the risk of developing thrombi or emboli that cause blood deficiency (*ischemia*) or tissue death (*necrosis*) downstream from the obstruction.

The walls of affected arteries may degenerate, losing their elasticity and becoming hardened or *sclerotic.* In this stage of the disease, called *arteriosclerosis,* a sclerotic vessel may rupture under the force of blood pressure.

Risk factors for developing atherosclerosis include a fatty diet, elevated blood pressure, cigarette smoking, obesity, and lack of physical exercise. Emotional and genetic factors may also increase susceptibility to atherosclerosis (see chapter 18).

If atherosclerosis so weakens the wall of an artery that blood pressure dilates a region of it, a pulsating sac called an *aneurysm* may form. Aneurysms tend to grow. If the resulting sac develops by a longitudinal splitting of the middle layer of the arterial wall, it is called a *dissecting aneurysm.* An aneurysm may cause symptoms by pressing on nearby organs, or it may rupture and produce a great loss of blood.

Aneurysms may also result from trauma, high blood pressure, infections, inherited disorders such as Marfan syndrome, or congenital defects in blood vessels. Common sites of aneurysms include the thoracic and abdominal aorta, and an arterial circle at the base of the brain (circle of Willis).

*Phlebitis,* or inflammation of a vein, is relatively common. It may occur in association with an injury or infection or after surgery, or it may develop for no apparent reason.

If inflammation is restricted to a superficial vein, such as the greater or lesser saphenous veins, blood flow may be rechanneled through other vessels. But if it occurs in a deep vein, such as the tibial, peroneal, popliteal, or femoral veins, the consequences can be quite serious, particularly if the blood within the affected vessel clots and blocks normal circulation. This condition, called *thrombophlebitis,* introduces a risk that a blood clot within a vein will detach, move with the venous blood, pass through the heart, and lodge in the pulmonary arterial system within a lung. Such an obstruction is called a *pulmonary embolism.*

*Varicose veins* are abnormal and irregular dilations in superficial veins, particularly in the legs. This condition is usually associated with prolonged, increased back pressure within the affected vessels due to gravity, as occurs when a person stands. Crossing the legs or sitting in a chair so that its edge presses against the area behind the knee can obstruct venous blood flow and aggravate varicose veins.

Increased venous back pressure stretches the veins and increases their diameters. Because the valves within these vessels do not change size, they soon lose their abilities to block the backward flow

1. How does the structure of a vein differ from that of an artery?

2. What are the functions of veins and venules?

3. How does venous circulation help to maintain blood pressure when hemorrhaging causes blood loss?

# BLOOD PRESSURE

Blood pressure is the force the blood exerts against the inner walls of the blood vessels. Although this force occurs throughout the vascular system, the term *blood pressure* most commonly refers to pressure in arteries supplied by branches of the aorta (systemic arteries).

## Arterial Blood Pressure

The arterial blood pressure rises and falls in a pattern corresponding to the phases of the cardiac cycle. That is, when the ventricles contract (ventricular systole), their walls squeeze the blood inside their chambers and force it into the pulmonary trunk and aorta. As a result, the pressures in these arteries increase sharply. The maximum pressure achieved during ventricular contraction is called of blood, and blood tends to accumulate in the enlarged regions.

Increased venous pressure is also accompanied by rising pressure within the venules and capillaries that supply the veins. Consequently, tissues in affected regions typically become edematous and painful.

Heredity, pregnancy, obesity, and standing for long periods raise the risk of developing varicose veins. Elevating the legs above the level of the heart or putting on support hosiery before arising in the morning can relieve discomfort. Intravenous injection of a substance that destroys veins (a sclerosing agent) or surgical removal of the affected veins may be necessary. ■

(a)

(b)

(c)

(d)

**FIGURE** 15E

As atherosclerosis develops, masses of fatty materials accumulate beneath the inner linings of certain arteries and arterioles and protrude into their lumens. (*a*) Normal arteriole. (*b, c,* and *d*) Accumulation of plaque on the inner wall of the arteriole (a–d 60×).

the **systolic pressure.** When the ventricles relax (ventricular diastole), the arterial pressure drops, and the lowest pressure that remains in the arteries before the next ventricular contraction is termed the **diastolic pressure.**

The surge of blood entering the arterial system during a ventricular contraction distends the elastic walls of the arteries, but the pressure drops almost immediately as the contraction ends, and the arterial walls recoil. This alternate expanding and recoiling of the arterial wall can be felt as a *pulse* in an artery that runs close to the surface. Figure 15.35 shows several sites where a pulse can be detected. The radial artery, for example, courses near the surface at the wrist and is commonly used to sense a person's radial pulse.

The radial pulse rate is equal to the rate at which the left ventricle contracts, and for this reason, it can be used to determine heart rate. A pulse can also reveal something about blood pressure, because an elevated pressure produces a pulse that feels strong and full, whereas a low pressure produces a pulse that is weak and easily compressed. Clinical Application 15.4 describes how to measure arterial blood pressure.

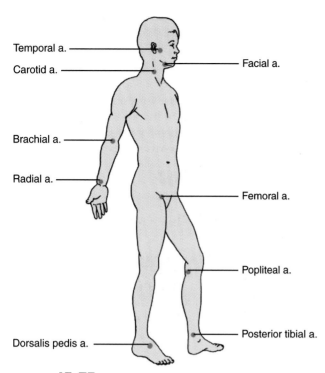

Temporal a.
Carotid a.
Facial a.
Brachial a.
Radial a.
Femoral a.
Popliteal a.
Posterior tibial a.
Dorsalis pedis a.

**FIGURE 15.35**

Sites where an arterial pulse is most easily detected (*a.* stands for *artery*).

1. Distinguish between systolic and diastolic blood pressure.

2. Which cardiac event causes systolic pressure? Diastolic pressure?

3. What causes a pulse in an artery?

## Factors that Influence Arterial Blood Pressure

Arterial pressure depends on a variety of factors. These include heart action, blood volume, resistance to flow, and blood viscosity (fig. 15.36).

### Heart Action

In addition to producing blood pressure by forcing blood into the arteries, heart action determines how much blood enters the arterial system with each ventricular contraction. The volume of blood discharged from the ventricle with each contraction is called the **stroke volume** and equals about 70 milliliters at rest. The volume discharged from the ventricle per minute is called the **cardiac output.** It is calculated by multiplying the stroke volume by the heart rate in beats per minute. (Cardiac output = stroke volume × heart rate.) Thus, if the stroke volume is 70 milliliters and the heart rate is 72 beats per minute, the cardiac output is 5,040 milliliters per minute.

Blood pressure varies with the cardiac output. If either the stroke volume or the heart rate increases, so does

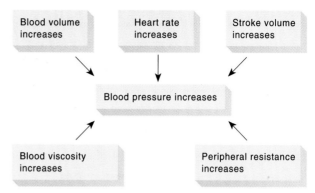

Blood volume increases → Heart rate increases → Stroke volume increases

→ Blood pressure increases ←

Blood viscosity increases ↗ Peripheral resistance increases ↖

**FIGURE 15.36**

Some of the factors that influence arterial blood pressure.

the cardiac output, and, as a result, blood pressure rises. Conversely, if the stroke volume or the heart rate decreases, the cardiac output and blood pressure also decrease.

### Blood Volume

**Blood volume** equals the sum of the formed elements and plasma volumes in the vascular system. Although the blood volume varies somewhat with age, body size, and sex, it is usually about 5 liters for adults.

Blood volume can be determined by injecting a known volume of an indicator, such as radioactive iodine, into the blood. After a time that allows for thorough mixing, a blood sample is withdrawn, and the concentration of the indicator measured. The total blood volume is calculated using the formula: blood volume = amount of indicator injected/concentration of indicator in blood sample.

Blood pressure is normally directly proportional to the volume of the blood within the cardiovascular system. Thus, any changes in the blood volume alter the blood pressure. For example, if a hemorrhage reduces blood volume, blood pressure drops. If a transfusion restores normal blood volume, normal pressure may be reestablished. Blood volume can also fall if the fluid balance is upset, as happens in dehydration. Fluid replacement can reestablish normal blood volume and pressure. Clinical Application 15.5 describes how the unusual conditions of microgravity in outer space affect the distribution of blood volume and control of blood pressure.

### Peripheral Resistance

Friction between blood and the walls of the blood vessels produces a force called **peripheral resistance,** which hinders blood flow. Blood pressure must overcome this

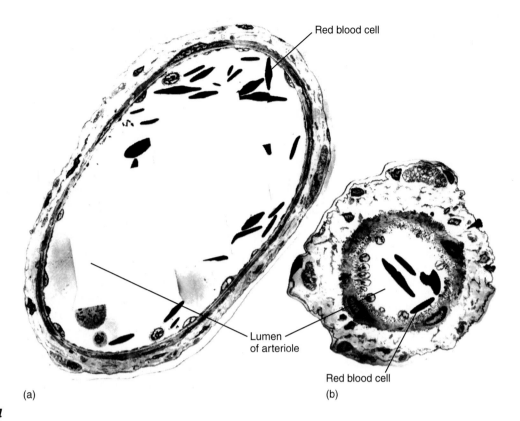

Red blood cell

Lumen of arteriole

Red blood cell

(a)                    (b)

**FIGURE 15.37**

(a) Relaxation of smooth muscle in the arteriole wall produces dilation, whereas (b) contraction of the smooth muscle causes constriction (a and b 1,500×).

force if the blood is to continue flowing. Therefore, factors that alter the peripheral resistance change blood pressure. For example, smooth muscles in the walls of contracting arterioles increases the peripheral resistance of these constricted vessels. Blood tends to back up into the arteries supplying the arterioles, and the arterial pressure rises. Dilation of the arterioles has the opposite effect—peripheral resistance lessens, and the arterial blood pressure drops in response (fig. 15.37).

Because arterial walls are quite elastic, when the ventricles discharge a surge of blood, arteries swell. Almost immediately, the elastic tissues recoil, and the vessel walls press against the blood inside. This action helps force the blood onward against the peripheral resistance in arterioles and capillaries. It also converts the intermittent flow of blood, which is characteristic of the arterial system, into a more continuous movement through the capillaries.

### Viscosity

The **viscosity** of a fluid is a physical property that derives from the ease with which its molecules flow past one another. The greater the viscosity, the greater the resistance to flowing.

Blood cells and plasma proteins increase blood viscosity. Since the greater the blood's resistance to flowing,

the greater the force needed to move it through the vascular system, it is not surprising that blood pressure rises as blood viscosity increases, and drops as viscosity decreases.

Although the viscosity of blood normally remains stable, any condition that alters the concentrations of blood cells or plasma proteins may alter viscosity. For example, anemia may decrease viscosity and consequently lower blood pressure. Excess red blood cells increase viscosity and blood pressure.

**1** How is cardiac output calculated?

**2** What is the relationship between cardiac output and blood pressure?

**3** How does blood volume affect blood pressure?

**4** What is the relationship between peripheral resistance and blood pressure? Between viscosity and blood pressure?

## Control of Blood Pressure

Blood pressure (BP) is determined by cardiac output (CO) and peripheral resistance (PR) according to this relationship: BP = CO × PR. Maintenance of normal blood pressure therefore involves the regulation of these two factors.

# Measurement of Arterial Blood Pressure

Systemic arterial blood pressure usually is measured using an instrument called a *sphygmomanometer* (fig. 15F). This device consists of an inflatable rubber cuff connected by tubing to a compressible bulb and a glass tube containing a column of mercury. The bulb is used to pump air into the cuff, and a rise in the mercury column indicates the pressure produced. Thus, the pressure in the cuff can be expressed in millimeters of mercury (mm Hg). A pressure of 100 mm Hg, for example, would be enough to force the mercury column upward for a distance of 100 mm.

To measure arterial blood pressure, the cuff of the sphygmomanometer is usually wrapped around the arm so that it surrounds the brachial artery. Air is pumped into the cuff until the cuff pressure exceeds the pressure in that artery. As a result, the vessel is squeezed closed, and its blood flow stopped. At this moment, if the diaphragm of a stethoscope is placed over the brachial artery at the distal border of the cuff, no sounds can be heard from the vessel because the blood flow is interrupted. As air is slowly released from the cuff, the air pressure inside it decreases. When the cuff pressure is approximately equal to the systolic blood pressure within the brachial artery, the artery opens enough for a small amount of blood to spurt through. This movement produces a sharp sound (Korotkoff's sound) that can be heard through the stethoscope. The height of the mercury column when this first tapping sound is heard represents the *arterial systolic pressure* (SP).

As the cuff pressure continues to drop, a series of increasingly louder sounds can be heard. Then, when the cuff pressure is approximately equal to that within the fully opened artery, the sounds become abruptly muffled and disappear. The height of the mercury column when this happens represents the *arterial diastolic pressure* (DP). The sound results from turbulence that occurs when the artery narrows.

The results of a blood pressure measurement are reported as a fraction, such as 120/80. In this notation, the upper number indicates the systolic pressure in mm Hg (SP), and the lower number indicates the diastolic pressure in mm Hg (DP). Figure 15G shows how these pressures decrease as distance from the left ventricle increases.

The difference between the systolic and diastolic pressures (SP–DP), which is called the *pulse pressure* (PP), is generally about 40 mm Hg.

---

*Cardiac output* depends on the volume of blood discharged from the left ventricle with each contraction (stroke volume) and heart rate. Mechanical, neural, and chemical factors affect these actions.

The blood volume entering the ventricle, for example, affects the stroke volume. As blood enters, myocardial fibers in the ventricular wall are mechanically stretched. Within limits, the greater the length of these fibers, the greater the force with which they contract. This relationship between fiber length (due to stretching of the cardiac muscle cell just before contraction) and force of contraction is called *Starling's law of the heart.* Because of it, the heart can respond to the immediate demands placed on it by the varying quantities of blood that return from the venous system. In other words, the more blood that enters the heart from the veins, the greater the ventricular distension, the stronger the ventricular contraction, the greater the stroke volume, and the greater the cardiac output (fig. 15.38). The less blood that returns from the veins, the lesser the ventricular distension, the weaker the ventricular contraction, the lesser the stroke volume, and the lesser the cardiac output.

Starling's law of the heart ensures that the volume of blood discharged from the heart is equal to the volume entering its chambers. Consequently, the volume of blood that enters the right atrium from the venae cavae is normally equal to the volume that leaves the left ventricle and enters the aorta.

Recall that baroreceptors in the walls of the aortic arch and carotid sinuses sense changes in blood pressure. If arterial pressure increases, nerve impulses travel from

The average pressure in the arterial system is also of interest because it represents the force that is effective throughout the cardiac cycle for driving blood to the tissues. This force, called the *mean arterial pressure*, is approximated by adding the diastolic pressure and one-third of the pulse pressure (DP + 1/3PP). ■

**FIGURE 15F**

A sphygmomanometer is used to measure arterial blood pressure.

**FIGURE 15G**

Blood pressure decreases as the distance from the left ventricle increases.

the receptors to the *cardiac center* of the medulla oblongata. This center relays parasympathetic impulses to the S-A node in the heart, and heart rate decreases in response. As a result of this *cardioinhibitor reflex*, cardiac output falls, and blood pressure decreases toward the normal level. Figure 15.39 summarizes this mechanism.

Conversely, decreasing arterial blood pressure initiates the *cardioaccelerator reflex*, which involves sympathetic impulses to the S-A node. As a result, the heart beats faster. This response increases cardiac output, increasing arterial pressure.

Recall that epinephrine increases heart rate, and consequently alters cardiac output and blood pressure. Other factors that increase heart rate and blood pressure include emotional responses, such as fear and anger; physical exercise; and a rise in body temperature.

Changes in arteriole diameters regulate peripheral resistance. Because blood vessels with smaller diameters offer a greater resistance to blood flow, factors that cause arteriole vasoconstriction increase peripheral resistance, and factors causing vasodilation decrease resistance.

The *vasomotor center* of the medulla oblongata continually sends sympathetic impulses to the smooth muscles in the arteriole walls, keeping them in a state of tonic contraction, which helps maintain the peripheral resistance associated with normal blood pressure. Because the vasomotor center responds to changes in blood pressure, it can increase peripheral resistance by increasing its outflow of sympathetic impulses, or it can decrease such resistance by decreasing its sympathetic outflow. In the latter case, the vessels undergo vasodilation as sympathetic stimulation decreases.

# Space Medicine

When the rescue team approached the space shuttle *Atlantis* just after it landed on September 26, 1996, they brought a stretcher, fully expecting to carry off Mission Specialist Shannon Lucid, Ph.D. The 53-year-old biochemist had just spent 188 days aboard the Russian *Mir* space station, as part of the NASA/*Mir* Science Program (fig. 15H). After all, Lucid had spent more time in space than any other U.S. astronaut, and it was widely known that about 70% of astronauts cannot stand at all upon reencountering gravity. But Shannon walked. She was a little wobbly, but she said, "No, I can stand up," and walked the 25 feet to the crew transporter, where she joyously received a gift from President Clinton—a gift-wrapped big box of M&M chocolate candies. After six months of having to consciously direct her hands to make contact with objects because of gravity changes, it must have been a great pleasure to grab the M&M box!

Shannon Lucid had expected to be in space for 140 days, but problems with the space shuttle and a fierce hurricane extended her stay. Because she spent the longest continuous time in space, her visit is supplying vital information for planning a trip to Mars, which would require that astronauts be in space at least two years, round-trip. ■

The human body evolved under conditions of constant gravity. So when a body is exposed to microgravity (very low gravity) or weightlessness for extended periods, some very obvious changes occur. The field of space medicine examines anatomic and physiologic responses to conditions in space. Shannon Lucid was expected to require the stretcher because of decreased muscle mass, mineral-depleted bones, and low blood volume. The 400 hours that she logged on the *Mir's* treadmill and stationary bicycle may have helped her stay in terrific physical shape.

From the moment she landed, Lucid was poked and prodded, monitored and tested, as medical researchers attempted to learn how six months in space affects cardiovascular functioning, respiratory capacity, mood, blood chemistry, circadian rhythms, muscular strength, body fluid composition, and many other aspects of anatomy and physiology. Just minutes after greeting her family, Lucid entered a magnetic resonance imaging tube to record the state of her various parts as they adapted to return to gravity.

The tendency to feel wobbly and be unable to stand upon returning to earth is one of the better-studied physiologic responses to experiencing low-gravity conditions. It is called orthostatic intolerance. Normally, gravity helps blood circulate in the lower limbs. In microgravity or no gravity, blood tends to pool in blood vessels in the center of the body, registering on receptors there. The body interprets this as excess blood, and in response, signals the kidneys to excrete more fluid. But there really isn't a large blood volume. On return to earth, the body actually has a pint to a quart less blood than it should, up to a 10 to 20% decrease in total blood volume. If blood vessels cannot constrict enough to counter the plummeting blood pressure, the weakness and wobbliness of orthostatic intolerance results. To minimize the effect, astronauts wear lower-body suction suits, which apply a vacuum force that helps draw blood into the blood vessels of the lower limbs. Maintaining fluid intake helps prevent dehydration.

## FIGURE 15H

Shannon Lucid's 188-day stay in space revealed to researchers much about the body's responses to microgravity conditions. While aboard the space station *Mir*, Lucid conducted experiments on quail embryos and growth of protein crystals.

**FIGURE** 15.38

Cardiac output is related to the volume of blood entering the heart.

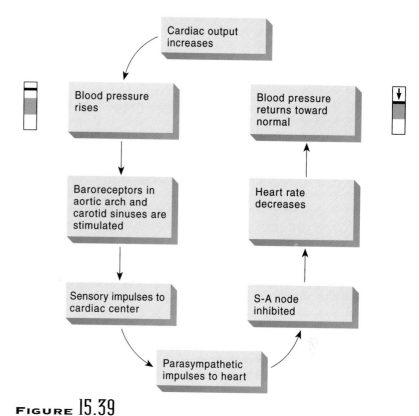

**FIGURE** 15.39

This mechanism helps regulate blood pressure by inhibiting the S-A node.

For instance, as figure 15.40 illustrates, whenever arterial blood pressure suddenly increases, baroreceptors in the aortic arch and carotid sinuses signal the vasomotor center, and the sympathetic outflow to the arteriole walls falls. The resulting vasodilation decreases peripheral resistance, and blood pressure decreases toward the normal level.

Similarly, if blood pressure drops, as following a hemorrhage, the vasomotor center increases sympathetic outflow. The resulting release of epinephrine and norepinephrine vasoconstrict most systemic vessels, increasing peripheral resistance. This helps return blood pressure toward normal.

The vasomotor center's control of vasoconstriction and vasodilation is especially important in the arterioles of the *abdominal viscera* (splanchnic region). These vessels, if fully dilated, could accept nearly all the blood of the body and cause the arterial pressure to approach zero. Thus, control of their diameters is essential in regulating normal peripheral resistance.

Certain chemicals, including carbon dioxide, oxygen, and hydrogen ions, also influence peripheral resistance by affecting precapillary sphincters and smooth muscles in arteriole and metarteriole walls. For example, increasing carbon dioxide, decreasing oxygen, and lowering pH relaxes these muscles in the systemic circulation. This increases local blood flow to tissues with high metabolic

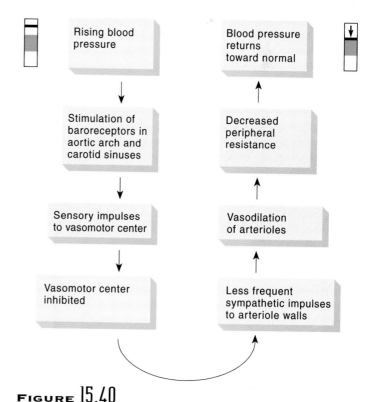

**FIGURE** 15.40

This mechanism helps regulate blood pressure by dilating arterioles.

# Hypertension

Hypertension, or high blood pressure, is persistently elevated arterial pressure. It is one of the more common diseases of the cardiovascular system in industrialized nations.

High blood pressure with unknown cause is called *essential* (also primary or idiopathic) *hypertension*. Elevated blood pressure that is a consequence of another problem, such as arteriosclerosis or kidney disease, is called *secondary hypertension*.

Arteriosclerosis is accompanied by decreased elasticity of the arterial walls and narrowed vessel lumens, which raise blood pressure. Kidney diseases often produce changes that interfere with blood flow to kidney cells. In response, the affected tissues may release an enzyme called *renin* that leads to the production of *angiotensin II*, a powerful vasoconstrictor that increases peripheral resistance in the arterial system, raising arterial pressure. Angiotensin II also stimulates the adrenal cortex to release *aldosterone*, which stimulates the kidneys to retain sodium ions and water. The resulting increase in blood volume contributes to increased blood pressure (fig. 15I).

Normally, this mechanism ensures that a decrease in blood flow to the kidneys is followed by an increase in arterial pressure, which, in turn, restores blood flow to the kidneys. If the decreased blood flow is the result of disease, such as atherosclerosis, the mechanism may cause high blood pressure and promote further deterioration of the arterial system.

In some individuals, high sodium intake leads to vasoconstriction, raising blood pressure. Obesity also is a risk factor for hypertension because it tends to increase peripheral resistance. Psychological stress, which activates sympathetic nerve impulses that cause generalized vasoconstriction, may also lead to hypertension. Yet another cause of hypertension may be an inability of endothelium to respond to a relaxing factor, leading to vasoconstriction.

Hypertension is called a "silent killer" because it may not have direct symptoms yet can set the stage for serious cardiovascular complications. For example, as the left ventricle

**FIGURE 15I**
This mechanism elevates blood pressure.

---

rates, such as exercising skeletal muscles. Clinical Application 15.6 discusses high blood pressure.

**1** What factors affect cardiac output?

**2** Explain Starling's law of the heart.

**3** What is the function of the baroreceptors in the walls of the aortic arch and carotid sinuses?

**4** How does the vasomotor center control peripheral resistance?

## Venous Blood Flow

Blood pressure decreases as the blood moves through the arterial system and into the capillary networks, so that little pressure remains at the venular ends of capillaries (see fig. 15G). Instead, blood flow through the venous system is only partly the direct result of heart action and depends on other factors, such as skeletal muscle contraction, breathing movements, and vasoconstriction of veins.

For example, contracting skeletal muscles thicken and press on nearby vessels, squeezing the blood in-works harder to pump blood at a higher pressure, the myocardium thickens, enlarging the heart. If the coronary blood vessels cannot support this overgrowth, parts of the heart muscle die and become replaced with fibrous tissue. Eventually, the enlarged and weakened heart dies.

Hypertension also contributes to the development of atherosclerosis. As arteries accumulate plaque, a *coronary thrombosis* or a *coronary embolism* may occur. Similar changes in the arteries of the brain increase the chances of a *cerebral vascular accident* (CVA), or stroke, due to a cerebral thrombosis, embolism, or hemorrhage.

When an embolus or hemorrhage causes a stroke, paralysis and other functional losses appear suddenly. A thrombus-caused stroke is slower. It may begin with clumsiness, progress to partial visual loss, then affect speech. One arm becomes paralyzed, then a day later, perhaps an entire side of the body is affected. Table 15A lists risk factors for a stroke.

A *transient ischemic attack* (TIA, or "ministroke") is a temporary block in a small artery. Symptoms include difficulty in speaking or understanding speech; numbness or weakness in the face, upper limb, lower limb, or one side; dizziness; falling; an unsteady gait; blurred vision; or blindness. These symptoms typically resolve within 24 hours with no lasting effects but may be a warning of an impending, more serious stroke.

Treatment of hypertension varies among patients and may include exercising regularly, controlling body weight, reducing stress, and limiting the diet to foods that are low in sodium. Drugs, such as diuretics and/or inhibitors of sympathetic nerve activity, may help control blood pressure. Diuretics increase urinary excretion of sodium and water, reducing the volume of body fluids. Sympathetic inhibitors block the synthesis of neurotransmitters, such as norepinephrine, or block receptor sites of effector cells. Table 15B summarizes drugs that treat hypertension. ∎

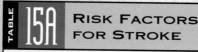

| TABLE 15A | RISK FACTORS FOR STROKE |
|---|---|

Alcohol consumption
Diabetes
Elevated serum cholesterol
Family history of cardiovascular disease
Hypertension
Smoking
Transient ischemic attacks

| TABLE 15B DRUGS TO TREAT HYPERTENSION | |
|---|---|
| **Type of Drug** | **Mechanism of Action** |
| Angiotensin-converting enzyme (ACE) inhibitors | Block formation of angiotensin II, preventing vasoconstriction |
| Beta blockers | Lower heart rate |
| Calcium channel blockers | Dilate blood vessels by keeping calcium ions out of muscle cells in vessel walls |
| Diuretics | Increase urine output, lowering blood volume |

side. As skeletal muscles press on veins with valves (called semilunar valves), some blood moves from one valve section to another. This massaging action of contracting skeletal muscles helps push the blood through the venous system toward the heart (fig. 15.41).

Respiratory movements also move venous blood. During inspiration, the pressure within the thoracic cavity is reduced as the diaphragm contracts and the rib cage moves upward and outward. At the same time, the pressure within the abdominal cavity is increased as the diaphragm presses downward on the abdominal viscera. Consequently, the blood tends to be squeezed out of the abdominal veins and forced into thoracic veins. During exercise, these respiratory movements act with skeletal muscle contractions to increase return of venous blood to the heart.

The veins also provide a blood reservoir that can adapt its capacity to changes in blood volume (see fig. 15.34). If some blood is lost and blood pressure falls, vein constriction (*venoconstriction*) can force blood out

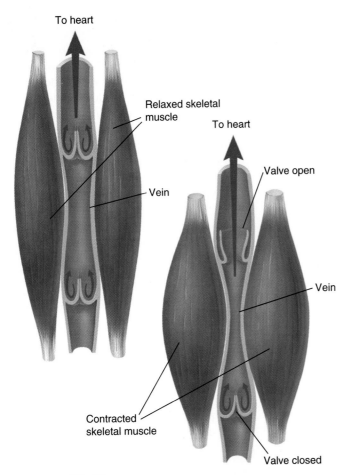

Relaxed skeletal muscle

To heart

To heart

Vein

Valve open

Vein

Contracted skeletal muscle

Valve closed

**FIGURE 15.41**

The massaging action of skeletal muscles helps move blood through the venous system toward the heart.

of this reservoir, returning venous blood to the heart. By maintaining venous return, venoconstriction helps to maintain blood pressure.

## Central Venous Pressure

Because all the veins, except those of the pulmonary circuit, drain into the right atrium, the pressure within this heart chamber is called *central venous pressure.* This pressure is of special interest because it affects the pressure within the peripheral veins. For example, if the heart is beating weakly, the central venous pressure increases, and blood backs up in the venous network, raising its pressure too. However, if the heart is beating forcefully, the central venous pressure and the pressure within the venous network decrease.

Other factors that increase the flow of blood into the right atrium, and thus elevate the central venous pressure, include increase in blood volume or widespread venoconstriction. An increase in central venous pressure can lead to peripheral edema because the resulting

higher capillary hydrostatic pressure favors movement of fluid into the tissues (see chapter 21). Clinical Application 15.7 discusses the effects of exercise on the heart and blood vessels.

As a result of disease or injury, blood or tissue fluid may accumulate in the pericardial cavity where it increases pressure. This condition, called *acute cardiac tamponade,* can be life threatening. As the pressure around the heart increases, it may compress the heart, interfere with the flow of blood into its chambers, and prevent pumping action. An early symptom of acute cardiac tamponade may be increased central venous pressure, with visible engorgement of the veins in the neck.

1. What is the function of the venous valves?

2. How do skeletal muscles affect venous blood flow?

3. How do respiratory movements affect venous blood flow?

4. What factors stimulate venoconstriction?

## PATHS OF CIRCULATION

The blood vessels can be divided into two major pathways. The **pulmonary circuit** consists of vessels that carry blood from the heart to the lungs and back to the heart. The **systemic circuit** carries blood from the heart to all other parts of the body and back again (fig. 15.42).

The circulatory pathways described in the following sections are those of an adult. The fetal pathways, which are somewhat different, are described in chapter 23.

## Pulmonary Circuit

Blood enters the pulmonary circuit as it leaves the right ventricle through the pulmonary trunk. The pulmonary trunk extends upward and posteriorly from the heart, and about 5 centimeters above its origin, it divides into the right and left pulmonary arteries. These branches penetrate the right and left lungs, respectively. Within the lungs, they divide into *lobar branches* (three on the right side and two on the left) that accompany the main divisions of the bronchi (airways) into the lobes of the lungs. After repeated divisions, the lobar branches give rise to arterioles that continue into the capillary networks associated with the walls of the alveoli (air sacs) (fig. 15.43).

# CLINICAL APPLICATION

## Exercise and the Cardiovascular System

We all know that exercise is good for the heart. Yet each year, a few individuals die of sudden cardiac arrest while shoveling snow, running, or engaging in some other strenuous activity. The explanation for this apparent paradox is that exercise *is* good for the heart—but only if it is a regular part of life.

Physiological responses to intense aerobic exercise generally increase blood flow, and therefore oxygen delivery, to active muscles. In muscles, vasodilation opens more capillaries. At the same time, vasoconstriction diminishes blood flow where it is not immediately needed, such as to the digestive tract. Blood flow, however, is maintained in the brain and kidneys, which require a steady stream of oxygen and nutrients to function. Respiratory movements and skeletal muscle activity increase venous return to the heart. As venous return to the heart increases, ventricular walls stretch, stimulating them to contract with greater force. Heart rate increases as well.

The cardiovascular system adapts to exercise as a way of life. The conditioned athlete experiences increases in heart pumping efficiency, blood volume, blood hemoglobin concentration, and the number of mitochondria in muscle fibers. All of these adaptations improve oxygen delivery to, and utilization by, muscle tissue.

An athlete's heart typically changes in response to these increased demands, and may enlarge 40% or more. Myocardial mass increases, the ventricular cavities expand, and the ventricle walls thicken. Stroke volume increases and heart rate decreases, as does blood pressure. To a physician unfamiliar with a conditioned cardiovascular system, a trained athlete may appear to be abnormal!

The cardiovascular system responds beautifully to a slow, steady buildup in exercise frequency and intensity. It does not react well to sudden demands—such as a person who never exercises suddenly shoveling snow or running 3 miles. Although sedentary people have a two- to sixfold increased risk of cardiac arrest while exercising than when not, people in shape have little or no excess risk while exercising.

How much exercise is enough to benefit the circulatory system? To achieve the benefits of exercise, the heart rate must be elevated to 70% to 85% of its "theoretical maximum" for at least half an hour three times a week. You can calculate your theoretical maximum by subtracting your age from 220. If you are 18 years old, your theoretical maximum is 202 beats per minute. Seventy to 85% of this value is 141 to 172 beats per minute. Some good activities for raising the heart rate are tennis, skating, skiing, handball, vigorous dancing, hockey, basketball, biking, and fast walking.

It is wise to consult a physician before starting an exercise program. People over the age of 30 are advised to have a stress test, which is an electrocardiogram taken while the subject is exercising. (The standard electrocardiogram is taken at rest.) An arrhythmia that appears only during exercise may indicate heart disease that has not yet produced symptoms. ■

---

The blood in the arteries and arterioles of the pulmonary circuit has a low concentration of oxygen and a high concentration of carbon dioxide. Gases are exchanged between the blood and the air as the blood moves through the *alveolar capillaries,* discussed in chapter 19.

Because the right ventricle contracts with less force than the left ventricle, the arterial pressure in the pulmonary circuit is less than that in the systemic circuit. Therefore, the alveolar capillary pressure is low.

The force that moves fluid out of an alveolar capillary is 23 mm Hg; the force pulling fluid into it is 22 mm Hg. Thus, such a capillary has a net filtration pressure of 1 mm Hg. This pressure causes a slight, continuous flow of fluid into the narrow interstitial space between the alveolar capillary and the alveolus.

The epithelial cells of the alveolar membranes are so tightly joined that sodium, chloride, and potassium ions, as well as glucose and urea, enter the interstitial space but usually fail to enter the alveoli. This helps maintain a high osmotic pressure in the interstitial fluid. Consequently, osmosis rapidly moves any water that gets into the alveoli back into the interstitial space. Although the alveolar surface must be moist to allow diffusion of oxygen and carbon dioxide, this mechanism prevents excess water from entering the alveoli and helps keep the alveoli from filling with fluid (fig. 15.44).

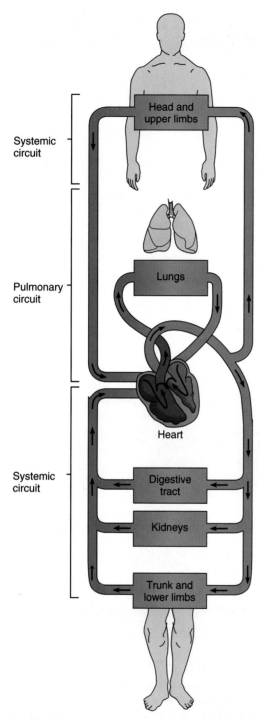

**FIGURE** 15.42

The pulmonary circuit consists of the vessels that carry blood between the heart and the lungs; all other vessels are included in the systemic circuit.

Fluid in the interstitial space may be drawn back into the alveolar capillaries by the somewhat higher osmotic pressure of the blood. Alternatively, lymphatic vessels (see chapter 16) may return it to the circulation.

As a result of the gas exchanges between the blood and the alveolar air, blood entering the venules of the pulmonary circuit is rich in oxygen and low in carbon

dioxide. These venules merge to form small veins, and they, in turn, converge to form larger veins. Four *pulmonary veins,* two from each lung, return blood to the left atrium, and this completes the vascular loop of the pulmonary circuit.

---

The ductus arteriosus is a fetal blood vessel that joins the aorta and the pulmonary artery. It normally closes shortly after birth, forcing blood to circulate through the lungs, which it did not do during fetal existence. Failure of this vessel to close, a condition called *patent ductus arteriosus,* can result in heart failure.

In a condition called *hypoplastic left heart syndrome,* physicians keep the ductus arteriosus open. This severe underdevelopment of the left side of the heart is responsible for 8% of fatal heart problems before the age of a year. Sometimes surgery can correct the condition, but in some cases a heart transplant is the only option. In order to transplant a heart, surgeons must keep the ductus arteriosus open. They do this by giving the small patient a prostaglandin drug, or by inserting a mold, called a stent, that keeps the vessel open. Several years ago, researchers attempted to treat hypoplastic left heart syndrome by transplanting a chimpanzee heart into a newborn named Baby Fae, to keep her alive until a human heart could be found. The experiment did not work, her body rejected the transplant.

---

*Pulmonary edema,* in which lungs fill with fluid, can accompany a failing left ventricle or a damaged bicuspid valve. A weak left ventricle may be unable to move the normal volume of blood into the systemic circuit. Blood backing up into the pulmonary circuit increases pressure in the alveolar capillaries, flooding the interstitial spaces with fluid. Increasing pressure in the interstitial fluid may rupture the alveolar membranes, and fluid may enter the alveoli more rapidly than it can be removed. This reduces the alveolar surface available for gas exchange, and the person may suffocate.

---

## Systemic Circuit

Contraction of the left ventricle forces the freshly oxygenated blood entering the left atrium into the systemic circuit. This circuit includes the aorta and its branches that lead to all of the body tissues, as well as the companion system of veins that returns blood to the right atrium.

**1** Distinguish between the pulmonary and systemic circuits of the cardiovascular system.

**2** Trace a drop of blood through the pulmonary circuit from the right ventricle.

**3** Explain why the alveoli normally do not fill with fluid.

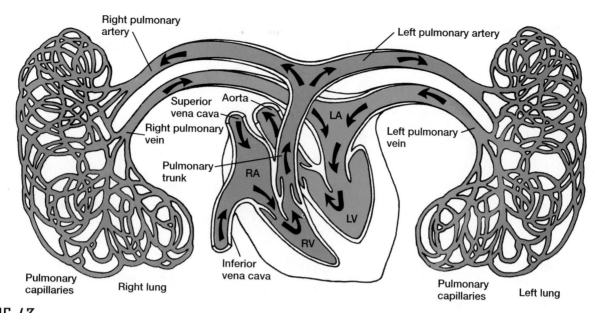

**FIGURE** 15.43

Blood is carried to the lungs through branches of the pulmonary arteries, and it returns to the heart through pulmonary veins.

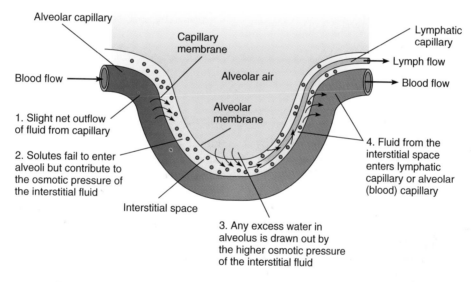

**FIGURE** 15.44

Cells of the alveolar wall are tightly joined. The relatively high osmotic pressure of the interstitial fluid draws water out of them.

## ARTERIAL SYSTEM

The **aorta** is the largest artery in the body. It extends upward from the left ventricle, arches over the heart to the left, and descends just anterior and to the left of the vertebral column.

### Principal Branches of the Aorta

The first portion of the aorta is called the *ascending aorta.* Located at its base are the three cusps of the aortic valve, and opposite each cusp is a swelling in the aortic wall called an **aortic sinus.** The right and left *coronary arteries* spring from two of these sinuses. Blood flow into these arteries is intermittent and is driven by the elastic recoil of the aortic wall following contraction of the left ventricle.

Several small structures called **aortic bodies** are located within the epithelial lining of the aortic sinuses. These bodies contain chemoreceptors that sense blood concentrations of oxygen and carbon dioxide.

Three major arteries originate from the *arch of the aorta* (aortic arch). They are the brachiocephalic (innominate) artery, the left common carotid artery, and the left subclavian artery. The aortic arch contains baroreceptors that detect changes in blood pressure.

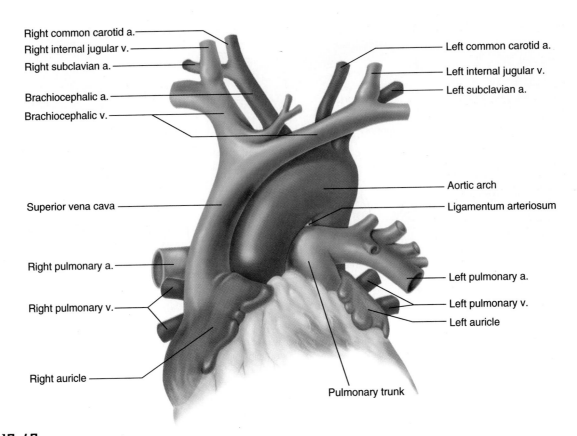

**FIGURE 15.45**
The major blood vessels associated with the heart.

Labels (clockwise from upper left):
Right common carotid a.
Right internal jugular v.
Right subclavian a.
Brachiocephalic a.
Brachiocephalic v.
Superior vena cava
Right pulmonary a.
Right pulmonary v.
Right auricle
Left common carotid a.
Left internal jugular v.
Left subclavian a.
Aortic arch
Ligamentum arteriosum
Left pulmonary a.
Left pulmonary v.
Left auricle
Pulmonary trunk

The **brachiocephalic** (brak″e-o-sĕ-fal′ik) **artery** supplies blood to the tissues of the upper limb and head, as its name suggests. It is the first branch from the aortic arch, and rises upward through the mediastinum to a point near the junction of the sternum and the right clavicle. There it divides, giving rise to the right **common carotid** (kah-rot′id) **artery,** which carries blood to the right side of the neck and head, and the right **subclavian** (sub-kla′ve-an) **artery,** which leads into the right arm. Branches of the subclavian artery also supply blood to parts of the shoulder, neck, and head.

The left *common carotid artery* and the left *subclavian artery* are respectively the second and third branches of the aortic arch. They supply blood to regions on the left side of the body corresponding to those supplied by their counterparts on the right (see fig. 15.45 and reference plates 71, 72, and 73).

Although the upper part of the *descending aorta* is positioned to the left of the midline, it gradually moves medially and finally lies directly in front of the vertebral column at the level of the twelfth thoracic vertebra. The portion of the descending aorta above the diaphragm is the **thoracic aorta,** and it gives off numerous small branches to the thoracic wall and the thoracic visceral organs. These branches, the *bronchial, pericardial,* and *esophageal arteries,* supply blood to the structures for which they were named. Other branches become *mediastinal arteries,* supplying various tissues within the mediastinum, and *posterior intercostal arteries* that pass into the thoracic wall.

Below the diaphragm, the descending aorta becomes the **abdominal aorta,** and it gives off branches to the abdominal wall and various abdominal organs. These branches include the following:

1. **Celiac** (se′le-ak) **artery.** This single vessel gives rise to the left *gastric, splenic,* and *hepatic arteries,* which supply upper portions of the digestive tract, the spleen, and the liver, respectively. (*Note:* The hepatic artery supplies the liver with about one-third of its blood flow, and this blood is oxygen-rich. The remaining two-thirds of the liver's blood flow arrives by means of the hepatic portal vein and is oxygen-poor.)

2. **Phrenic** (fren′ik) **arteries.** These paired arteries supply blood to the diaphragm.

3. **Superior mesenteric** (mes″en-ter′ik) **artery.** The superior mesenteric is a large, unpaired artery that branches to many parts of the intestinal tract, including the jejunum, ileum, cecum, ascending colon, and transverse colon.

4. **Suprarenal** (soo″prah-re′nal) **arteries.** This pair of vessels supplies blood to the adrenal glands.

5. **Renal** (re′nal) **arteries.** The renal arteries pass laterally from the aorta into the kidneys. Each artery then divides into several lobar branches within the kidney tissues.

6. **Gonadal** (go′nad-al) **arteries.** In a female, paired ovarian arteries arise from the aorta and pass into the pelvis to supply the ovaries. In a male, *spermatic arteries* originate in similar locations. They course downward and pass through the body wall by way of the *inguinal canal* to supply the testes.

7. **Inferior mesenteric artery.** Branches of this single artery lead to the descending colon, the sigmoid colon, and the rectum.

8. **Lumbar arteries.** Three or four pairs of lumbar arteries arise from the posterior surface of the aorta in the region of the lumbar vertebrae. These arteries supply muscles of the skin and the posterior abdominal wall.

9. **Middle sacral artery.** This small, single vessel descends medially from the aorta along the anterior surfaces of the lower lumbar vertebrae. It carries blood to the sacrum and coccyx.

The abdominal aorta terminates near the brim of the pelvis, where it divides into right and left *common iliac arteries*. These vessels supply blood to lower regions of the abdominal wall, the pelvic organs, and the lower extremities (fig. 15.46). Table 15.4 summarizes the main branches of the aorta.

## Arteries to the Neck, Head, and Brain

Branches of the subclavian and common carotid arteries supply blood to structures within the neck, head, and brain (figs. 15.47 and 15.48). The main divisions of the subclavian artery to these regions are the vertebral, thyrocervical, and costocervical arteries. The common carotid artery communicates with these regions by means of the internal and external carotid arteries.

The **vertebral arteries** arise from the subclavian arteries in the base of the neck near the tips of the lungs. They pass upward through the foramina of the transverse processes of the cervical vertebrae and enter the skull by

(a)

(b)

**FIGURE** 15.46

(a) Abdominal aorta and its major branches. (b) Angiogram (radiograph) of the abdominal aorta.

## TABLE 15.4 THE AORTA AND ITS PRINCIPAL BRANCHES

| Portion of Aorta | Major Branch | General Regions or Organs Supplied | Portion of Aorta | Major Branch | General Regions or Organs Supplied |
|---|---|---|---|---|---|
| Ascending aorta | Right and left coronary arteries | Heart | Abdominal aorta | Celiac artery | Organs of upper digestive tract |
| Arch of aorta | Brachiocephalic artery | Right upper limb, right side of head | | Phrenic artery | Diaphragm |
| | | | | Superior mesenteric artery | Portions of small and large intestines |
| | Left common carotid artery | Left side of head | | Suprarenal artery | Adrenal gland |
| | Left subclavian artery | Left upper limb | | Renal artery | Kidney |
| | | | | Gonadal artery | Ovary or testis |
| Descending aorta | | | | Inferior mesenteric artery | Lower portions of large intestine |
| Thoracic aorta | Bronchial artery | Bronchi | | Lumbar artery | Posterior abdominal wall |
| | Pericardial artery | Pericardium | | | |
| | Esophageal artery | Esophagus | | Middle sacral artery | Sacrum and coccyx |
| | Mediastinal artery | Mediastinum | | Common iliac artery | Lower abdominal wall, pelvic organs, and lower limb |
| | Posterior intercostal artery | Thoracic wall | | | |

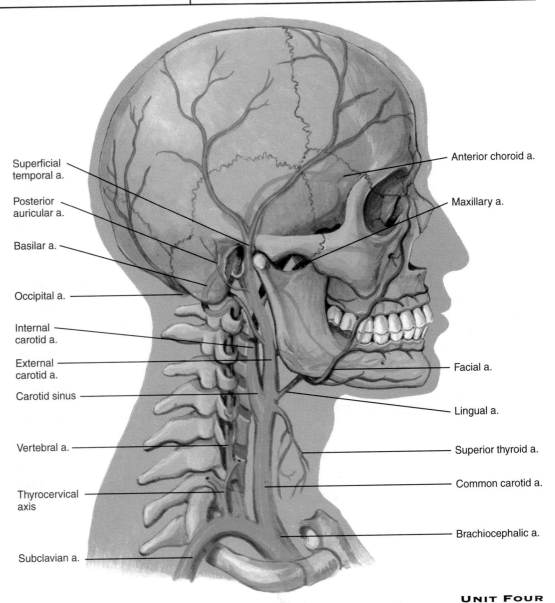

**FIGURE 15.47**

The main arteries of the head and neck. (Note that the clavicle has been removed.)

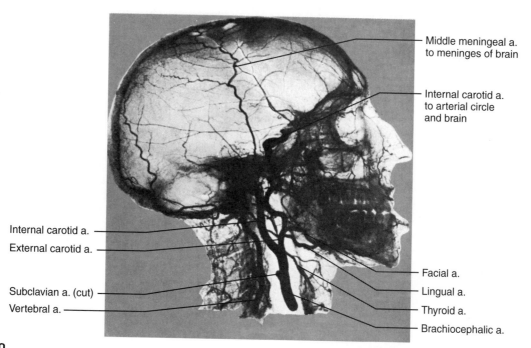

Middle meningeal a.
to meninges of brain

Internal carotid a.
to arterial circle
and brain

Internal carotid a.

External carotid a.

Subclavian a. (cut)

Vertebral a.

Facial a.

Lingual a.

Thyroid a.

Brachiocephalic a.

**FIGURE** 15.48

An angiogram of the arteries associated with the head.

way of the foramen magnum. Along their paths, these vessels supply blood to vertebrae and to their associated ligaments and muscles.

Within the cranial cavity, the vertebral arteries unite to form a single *basilar artery.* This vessel passes along the ventral brain stem and gives rise to branches leading to the pons, midbrain, and cerebellum. The basilar artery terminates by dividing into two *posterior cerebral arteries* that supply portions of the occipital and temporal lobes of the cerebrum. The posterior cerebral arteries also help form an arterial circle at the base of the brain, the **circle of Willis,** which connects the vertebral artery and internal carotid artery systems (fig. 15.49). The union of these systems provides alternate pathways through which blood can reach brain tissues in the event of an arterial occlusion.

The **thyrocervical** (thi″ro-ser′vĭ-kal) **arteries** are short vessels that give off branches at the thyrocervical axis to the thyroid gland, parathyroid glands, larynx, trachea, esophagus, and pharynx, as well as to various muscles in the neck, shoulder, and back. The **costocervical** (kos″to-ser′vĭ-kal) **arteries,** which are the third vessels to branch from the subclavians, carry blood to muscles in the neck, back, and thoracic wall.

The left and right *common carotid arteries* ascend deeply within the neck on either side. At the level of the upper laryngeal border, they divide to form the internal and external carotid arteries.

The **external carotid artery** courses upward on the side of the head, giving off branches to structures in the neck, face, jaw, scalp, and base of the skull. The

main vessels that originate from this artery include the following:

1. *Superior thyroid artery* to the hyoid bone, larynx, and thyroid gland.

2. *Lingual artery* to the tongue, muscles of the tongue, and salivary glands beneath the tongue.

3. *Facial artery* to the pharynx, palate, chin, lips, and nose.

4. *Occipital artery* to the scalp on the back of the skull, the meninges, the mastoid process, and various muscles in the neck.

5. *Posterior auricular artery* to the ear and the scalp over the ear.

The external carotid artery terminates by dividing into *maxillary* and *superficial temporal arteries.* The maxillary artery supplies blood to the teeth, gums, jaws, cheek, nasal cavity, eyelids, and meninges. The temporal artery extends to the parotid salivary gland and to various surface regions of the face and scalp.

The **internal carotid artery** follows a deep course upward along the pharynx to the base of the skull. Entering the cranial cavity, it provides the major blood supply to the brain. The major branches of the internal carotid artery include the following:

1. *Ophthalmic artery* to the eyeball and to various muscles and accessory organs within the orbit.

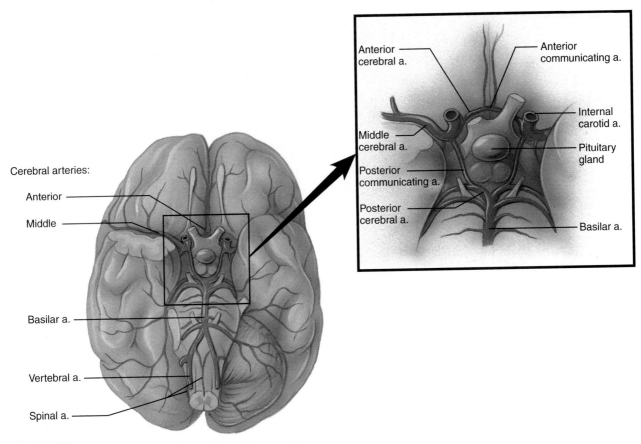

Cerebral arteries:

Anterior

Middle

Basilar a.

Vertebral a.

Spinal a.

Anterior cerebral a.

Anterior communicating a.

Middle cerebral a.

Internal carotid a.

Posterior communicating a.

Pituitary gland

Posterior cerebral a.

Basilar a.

**FIGURE 15.49**

View of inferior surface of the brain. The circle of Willis is formed by the anterior cerebral arteries, which are connected by the anterior communicating artery, and by the posterior vertebral arteries, which are connected to the internal carotid arteries by the posterior communicating arteries.

2. *Posterior communicating artery* that forms part of the circle of Willis.

3. *Anterior choroid artery* to the choroid plexus within the lateral ventricle of the brain and to nerve structures within the brain.

The internal carotid artery terminates by dividing into *anterior* and *middle cerebral arteries.* The middle cerebral artery passes through the lateral sulcus and supplies the lateral surface of the cerebrum, including the primary motor and sensory areas of the face and upper limbs, the optic radiations, and the speech area (see chapter 11). The anterior cerebral artery extends anteriorly between the cerebral hemispheres and supplies the medial surface of the brain.

Near the base of each internal carotid artery is an enlargement called a **carotid sinus.** Like the aortic sinuses, these structures contain baroreceptors that control blood pressure. A number of small epithelial masses, called **carotid bodies,** also occur in the wall of the carotid sinus. These bodies are very vascular and contain chemoreceptors that act with those of the aortic bodies to regulate circulation and respiration.

## Arteries to the Shoulder and Upper Limb

The subclavian artery, after giving off branches to the neck, continues into the arm (fig. 15.50). It passes between the clavicle and the first rib, and becomes the axillary artery.

The **axillary artery** supplies branches to structures in the axilla and the chest wall, including the skin of the shoulder, part of the mammary gland, the upper end of the humerus, the shoulder joint, and muscles in the back, shoulder, and chest. As this vessel leaves the axilla, it becomes the brachial artery.

The **brachial artery** courses along the humerus to the elbow. It gives rise to a *deep brachial artery* that curves posteriorly around the humerus and supplies the triceps muscle. Shorter branches pass into the muscles on the anterior side of the arm, whereas others descend on each side to the elbow and connect with arteries in the forearm. The resulting arterial network allows blood to reach the forearm even if a portion of the distal brachial artery becomes obstructed.

Within the elbow, the brachial artery divides into an ulnar artery and a radial artery. The **ulnar artery** leads

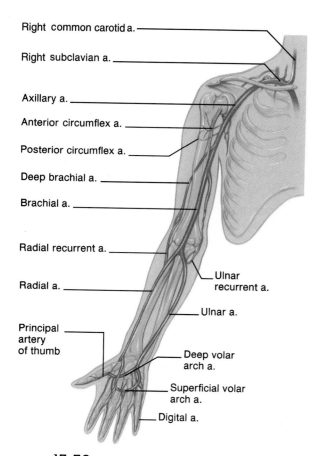

Right common carotid a.

Right subclavian a.

Axillary a.

Anterior circumflex a.

Posterior circumflex a.

Deep brachial a.

Brachial a.

Radial recurrent a.

Radial a.

Principal artery of thumb

Ulnar recurrent a.

Ulnar a.

Deep volar arch a.

Superficial volar arch a.

Digital a.

**FIGURE** 15.50

The main arteries to the shoulder and upper limb.

downward on the ulnar side of the forearm to the wrist. Some of its branches join the anastomosis around the elbow joint, whereas others supply blood to flexor and extensor muscles in the forearm.

The **radial artery,** a continuation of the brachial artery, travels along the radial side of the forearm to the wrist. As it nears the wrist, it comes close to the surface and provides a convenient vessel for taking the pulse (radial pulse). Branches of the radial artery join the anastomosis of the elbow and supply the lateral muscles of the forearm.

At the wrist, the branches of the ulnar and radial arteries join to form a network of vessels. Arteries arising from this network supply blood to structures in the wrist, hand, and fingers.

## Arteries to the Thoracic and Abdominal Walls

Blood reaches the thoracic wall through several vessels. These include branches from the subclavian artery and the thoracic aorta (fig. 15.51).

The subclavian artery contributes to this supply through a branch called the **internal thoracic artery.** This vessel originates in the base of the neck and passes downward on the pleura and behind the cartilages of the upper six ribs. It gives off two *anterior intercostal arteries* to each of the upper six intercostal spaces; these two arteries supply the intercostal muscles, other intercostal tissues, and the mammary glands.

Vertebral body

Posterior intercostal a.

Internal intercostal m.

Thoracic aorta

Internal thoracic a.

External intercostal m.

Sternum

Anterior intercostal a.

Costal cartilage

**FIGURE** 15.51

Arteries that supply the thoracic wall.

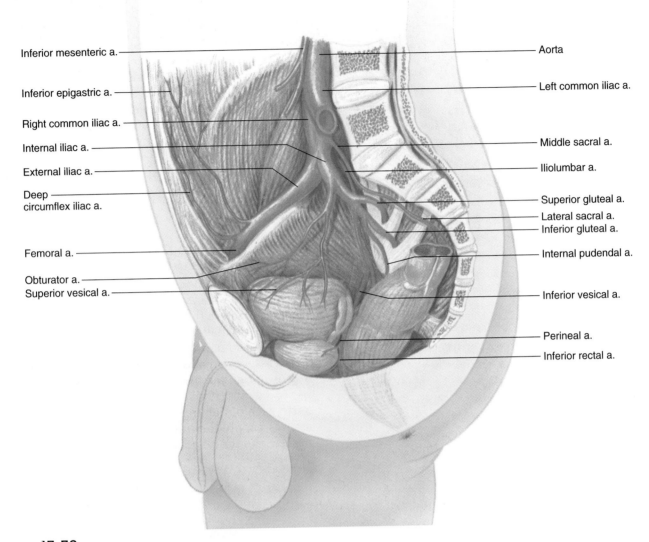

Inferior mesenteric a.

Inferior epigastric a.

Right common iliac a.

Internal iliac a.

External iliac a.

Deep circumflex iliac a.

Femoral a.

Obturator a.
Superior vesical a.

Aorta

Left common iliac a.

Middle sacral a.

Iliolumbar a.

Superior gluteal a.
Lateral sacral a.
Inferior gluteal a.

Internal pudendal a.

Inferior vesical a.

Perineal a.
Inferior rectal a.

**FIGURE 15.52**
Arteries that supply the pelvic region.

The *posterior intercostal arteries* arise from the thoracic aorta and enter the intercostal spaces between the third through the eleventh ribs. These arteries give off branches that supply the intercostal muscles, the vertebrae, the spinal cord, and deep muscles of the back.

Branches of the *internal thoracic* and *external iliac arteries* provide blood to the anterior abdominal wall. Paired vessels originating from the abdominal aorta, including the *phrenic* and *lumbar arteries,* supply blood to structures in the posterior and lateral abdominal wall.

## Arteries to the Pelvis and Lower Limb

The abdominal aorta divides to form the **common iliac arteries** at the level of the pelvic brim. These vessels provide blood to the pelvic organs, gluteal region, and lower limbs.

Each common iliac artery descends a short distance and divides into an internal (hypogastric) branch and an external branch. The **internal iliac artery** gives off numerous branches to various pelvic muscles and visceral structures, as well as to the gluteal muscles and the external genitalia. Figure 15.52 shows important branches of this vessel, including the following:

1. *Iliolumbar artery* to the ilium and muscles of the back.

2. *Superior* and *inferior gluteal arteries* to the gluteal muscles, pelvic muscles, and skin of the buttocks.

3. *Internal pudendal artery* to muscles in the distal portion of the alimentary canal, the external genitalia, and the hip joint.

4. *Superior* and *inferior vesical arteries* to the urinary bladder. In males, these vessels also supply the seminal vesicles and the prostate gland.

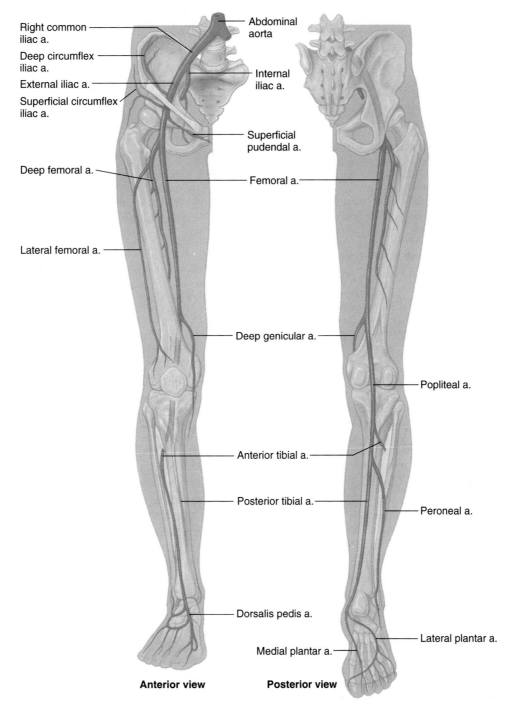

Right common iliac a.

Deep circumflex iliac a.

External iliac a.

Superficial circumflex iliac a.

Abdominal aorta

Internal iliac a.

Superficial pudendal a.

Deep femoral a.

Femoral a.

Lateral femoral a.

Deep genicular a.

Popliteal a.

Anterior tibial a.

Posterior tibial a.

Peroneal a.

Dorsalis pedis a.

Lateral plantar a.

Medial plantar a.

**Anterior view**

**Posterior view**

**FIGURE 15.53**

Main branches of the external iliac artery.

5. *Middle rectal artery* to the rectum.

6. *Uterine artery* to the uterus and vagina.

The **external iliac artery** provides the main blood supply to the lower limbs (fig. 15.53). It passes downward along the brim of the pelvis and gives off two large branches—an *inferior epigastric artery* and a *deep circumflex iliac artery*. These vessels supply the muscles and skin in the lower abdominal wall. Midway between the symphysis pubis and the anterior superior iliac spine of the ilium, the external iliac artery becomes the femoral artery.

The **femoral artery,** which passes fairly close to the anterior surface of the upper thigh, gives off many branches to muscles and superficial tissues of the thigh. These branches also supply the skin of the groin and the lower abdominal wall. Important subdivisions of the femoral artery include the following:

1. *Superficial circumflex iliac artery* to the lymph nodes and skin of the groin.

2. *Superficial epigastric artery* to the skin of the lower abdominal wall.

3. *Superficial* and *deep external pudendal arteries* to the skin of the lower abdomen and external genitalia.

4. *Profunda femoris artery* (the largest branch of the femoral artery) to the hip joint and muscles of the thigh.

5. *Deep genicular artery* to distal ends of thigh muscles and to an anastomosis around the knee joint.

As the femoral artery reaches the proximal border of the space behind the knee (popliteal fossa), it becomes the **popliteal artery.** Branches of this artery supply blood to the knee joint and to certain muscles in the thigh and calf. Also, many of its branches join the anastomosis of the knee and help provide alternate pathways for blood in the case of arterial obstructions. At the lower border of the popliteal fossa, the popliteal artery divides into the anterior and posterior tibial arteries.

The **anterior tibial** (tib'e-al) **artery** passes downward between the tibia and the fibula, giving off branches to the skin and muscles in anterior and lateral regions of the leg. It also communicates with the anastomosis of the knee and with a network of arteries around the ankle. This vessel continues into the foot as the *dorsalis pedis artery,* which supplies blood to the foot and toes.

The **posterior tibial artery,** the larger of the two popliteal branches, descends beneath the calf muscles, giving off branches to the skin, muscles, and other tissues of the leg along the way. Some of these vessels join the anastomoses of the knee and ankle. As it passes between the medial malleolus and the heel, the posterior tibial artery divides into the *medial* and *lateral plantar arteries.* Branches from these arteries supply blood to tissues of the heel, foot, and toes.

The largest branch of the posterior tibial artery is the *peroneal artery,* which travels downward along the fibula and contributes to the anastomosis of the ankle. The major vessels of the arterial system are shown in figure 15.54.

**1** Describe the structure of the aorta.

**2** Name the vessels that arise from the aortic arch.

**3** Name the branches of the thoracic and abdominal aorta.

**4** Which vessels supply blood to the head? To the upper limb? To the abdominal wall? To the lower limb?

## VENOUS SYSTEM

Venous circulation returns blood to the heart after gases, nutrients, and wastes are exchanged between the blood and body cells.

## Characteristics of Venous Pathways

The vessels of the venous system begin with the merging of capillaries into venules, venules into small veins, and small veins into larger ones. Unlike the arterial pathways, however, those of the venous system are difficult to follow. This is because the vessels commonly connect in irregular networks, so that many unnamed tributaries may join to form a relatively large vein.

On the other hand, the larger veins typically parallel the courses of named arteries, and these veins often have the same names as their companions in the arterial system. For example, the renal vein parallels the renal artery, and the common iliac vein accompanies the common iliac artery.

The veins that carry the blood from the lungs and myocardium back to the heart have already been described. The veins from all the other parts of the body converge into two major pathways, the *superior* and *inferior venae cavae,* which lead to the right atrium.

## Veins from the Head, Neck, and Brain

The **external jugular** (jug'u-lar) **veins** drain blood from the face, scalp, and superficial regions of the neck. These vessels descend on either side of the neck, passing over the sternocleidomastoid muscles and beneath the platysma. They empty into the *right* and *left subclavian veins* in the base of the neck (fig. 15.55).

The **internal jugular veins,** which are somewhat larger than the external jugular veins, arise from numerous veins and venous sinuses of the brain and from deep veins in various parts of the face and neck. They pass downward through the neck beside the common carotid arteries and also join the subclavian veins. These unions of the internal jugular and subclavian veins form large **brachiocephalic** (innominate) veins on each side. These vessels then merge in the mediastinum and give rise to the **superior vena cava,** which enters the right atrium.

---

A lung cancer, enlarged lymph node, or an aortic aneurysm can compress the superior vena cava, interfering with return of blood from the upper body to the heart. This produces pain, shortness of breath, distension of veins draining into the superior vena cava, and swelling of tissues in the face, head, and lower limbs. Restriction of blood flow to the brain may threaten life.

---

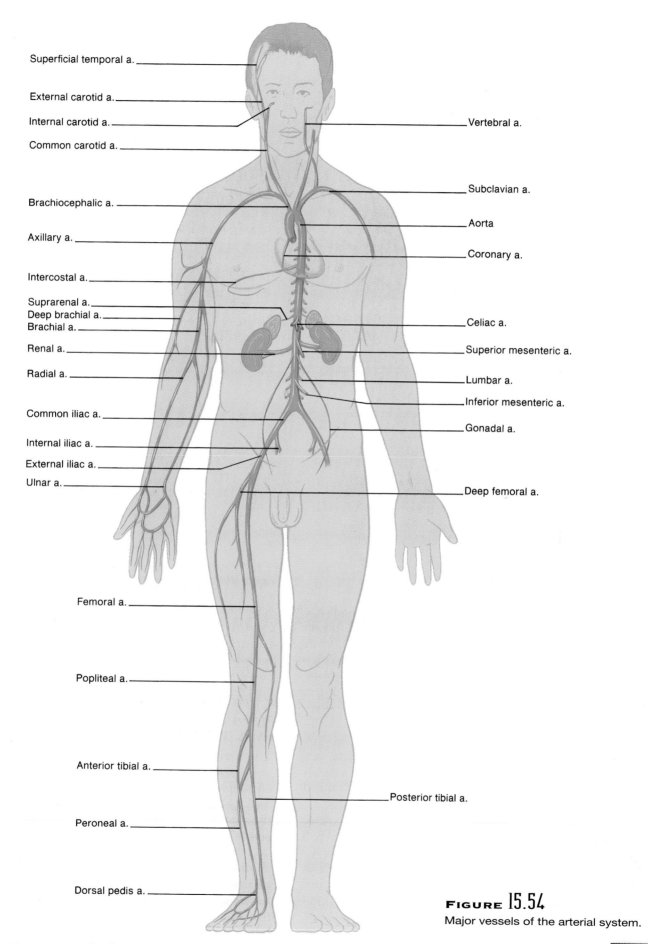

Superficial temporal a.

External carotid a.

Internal carotid a.

Common carotid a.

Vertebral a.

Brachiocephalic a.

Subclavian a.

Axillary a.

Aorta

Coronary a.

Intercostal a.

Suprarenal a.
Deep brachial a.
Brachial a.

Celiac a.

Renal a.

Superior mesenteric a.

Radial a.

Lumbar a.

Inferior mesenteric a.

Common iliac a.

Gonadal a.

Internal iliac a.

External iliac a.

Ulnar a.

Deep femoral a.

Femoral a.

Popliteal a.

Anterior tibial a.

Posterior tibial a.

Peroneal a.

Dorsal pedis a.

**FIGURE 15.54**

Major vessels of the arterial system.

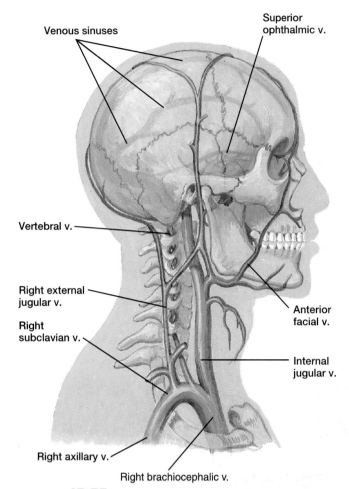

**FIGURE** 15.55

The main veins of the head and neck (*v.* stands for *vein*). (Note that the clavicle has been removed.)

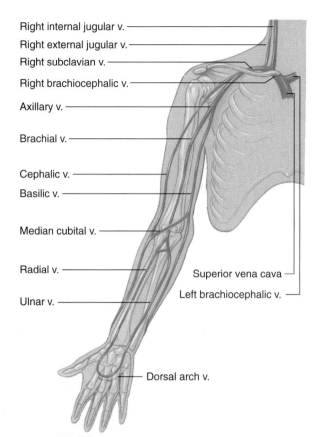

**FIGURE** 15.56

The main veins of the upper limb and shoulder.

## Veins from the Upper Limb and Shoulder

A set of deep veins and a set of superficial ones drain the upper limb. The deep veins generally parallel the arteries in each region and are given similar names, such as the *radial vein, ulnar vein, brachial vein,* and *axillary vein.* The superficial veins connect in complex networks just beneath the skin. They also communicate with the deep vessels of the upper limb, providing many alternate pathways through which the blood can leave the tissues (fig. 15.56).

The main vessels of the superficial network are the basilic and cephalic veins. They arise from anastomoses in the hand and wrist on the ulnar and radial sides, respectively.

The **basilic vein** passes along the back of the forearm on the ulnar side for a distance and then curves forward to the anterior surface below the elbow. It continues ascending on the medial side until it reaches the middle of the arm. There it penetrates the tissues deeply and joins the *brachial vein.* As the basilic and brachial veins merge, they form the *axillary vein.*

The **cephalic vein** courses upward on the lateral side of the upper limb from the hand to the shoulder. In the shoulder it pierces the tissues and empties into the axillary vein. Beyond the axilla, the axillary vein becomes the *subclavian vein.*

In the bend of the elbow, a *median cubital vein* ascends from the cephalic vein on the lateral side of the forearm to the basilic vein on the medial side. This large vein is usually visible. It is often used as a site for *venipuncture,* when it is necessary to remove a sample of blood for examination or to add fluids to the blood.

## Veins from the Abdominal and Thoracic Walls

Tributaries of the brachiocephalic and azygos veins drain the abdominal and thoracic walls. For example, the *brachiocephalic vein* receives blood from the *internal thoracic vein,* which generally drains the tissues the internal thoracic artery supplies. Some *intercostal veins* also empty into the brachiocephalic vein (fig. 15.57).

The **azygos** (az′ĭ-gos) **vein** originates in the dorsal abdominal wall and ascends through the mediastinum on the right side of the vertebral column to join the superior vena cava. It drains most of the muscular tissue in the abdominal and thoracic walls.

**FIGURE 15.57**

Veins that drain the thoracic wall.

Tributaries of the azygos vein include the *posterior intercostal veins* on the right side, which drain the intercostal spaces, and the *superior* and *inferior hemiazygos veins,* which receive blood from the posterior intercostal veins on the left. The right and left *ascending lumbar veins,* with tributaries that include vessels from the lumbar and sacral regions, also connect to the azygos system.

## Veins from the Abdominal Viscera

Although veins usually carry the blood directly to the atria of the heart, those that drain the abdominal viscera are exceptions (fig. 15.58). They originate in the capillary networks of the stomach, intestines, pancreas, and spleen, and carry blood from these organs through a **hepatic portal** (por'tal) **vein** to the liver. There the blood enters capillarylike **hepatic sinusoids** (hĕ-pat'ik si'nŭ-soids). This unique venous pathway is called the **hepatic portal system** (fig. 15.59).

The tributaries of the hepatic portal vein include the following vessels:

1. Right and left *gastric veins* from the stomach.

2. *Superior mesenteric vein* from the small intestine, ascending colon, and transverse colon.

3. *Splenic vein* from a convergence of several veins draining the spleen, the pancreas, and a portion of the stomach. Its largest tributary, the *inferior mesenteric vein,* brings blood upward from the descending colon, sigmoid colon, and rectum.

About 80% of the blood flowing to the liver in the hepatic portal system comes from the capillaries in the stomach and intestines and is oxygen-poor but rich in nutrients. As discussed in chapter 17, the liver handles these nutrients in a variety of ways. It regulates blood glucose concentration by polymerizing excess glucose into glycogen for storage, or by breaking down glycogen into glucose when blood glucose concentration drops below normal.

Similarly, the liver helps regulate blood concentrations of recently absorbed amino acids and lipids by modifying their molecules into forms cells can use, by oxidizing them, or by changing them into storage forms. The liver also stores certain vitamins and detoxifies harmful substances.

Blood in the hepatic portal vein nearly always contains bacteria that have entered through intestinal capillaries. Large *Kupffer cells* lining the hepatic sinusoids phagocytize these microorganisms, removing them from the portal blood before it leaves the liver.

After passing through the hepatic sinusoids of the liver, the blood in the hepatic portal system travels through a series of merging vessels into **hepatic veins.** These veins empty into the *inferior vena cava,* returning the blood to the general circulation.

Other veins empty into the inferior vena cava as it ascends through the abdomen. They include the *lumbar, gonadal, renal, suprarenal,* and *phrenic veins.* These vessels drain regions that arteries with corresponding names supply.

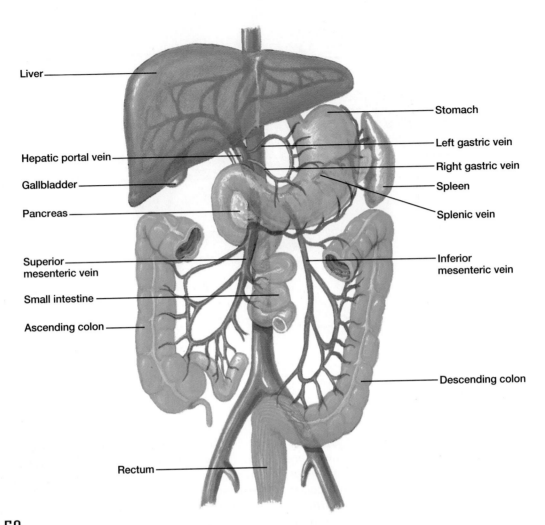

**FIGURE** 15.58

Veins that drain the abdominal viscera.

## Veins from the Lower Limb and Pelvis

As in the upper limb, veins that drain the blood from the lower limb can be divided into deep and superficial groups (fig. 15.60). The deep veins of the leg, such as the *anterior* and *posterior tibial veins,* have names that correspond to the arteries they accompany. At the level of the knee, these vessels form a single trunk, the **popliteal vein.** This vein continues upward through the thigh as the **femoral vein,** which, in turn, becomes the **external iliac vein.**

The superficial veins of the foot, leg, and thigh connect to form a complex network beneath the skin. These vessels drain into two major trunks: the small and great saphenous veins.

The small **saphenous** (sah-fe′nus) **vein** begins in the lateral portion of the foot and passes upward behind the lateral malleolus. It ascends along the back of the calf, enters the popliteal fossa, and joins the popliteal vein.

The **great saphenous vein,** which is the longest vein in the body, originates on the medial side of the foot. It ascends in front of the medial malleolus and extends upward along the medial side of the leg and thigh. In the thigh just below the inguinal ligament, it penetrates deeply and joins the femoral vein. Near its termination, the great saphenous vein receives tributaries from a number of vessels that drain the upper thigh, groin, and lower abdominal wall.

In addition to communicating freely with each other, the saphenous veins communicate extensively with the deep veins of the leg and thigh. Blood can thus return to the heart from the lower extremities by several routes.

In the pelvic region, vessels leading to the **internal iliac vein** carry blood away from organs of the reproductive, urinary, and digestive systems. This vein is formed by tributaries corresponding to the branches of the internal iliac artery, such as the *gluteal, pudendal, vesical, rectal, uterine,* and *vaginal veins.* Typically, these veins have many connections and form complex networks (plexuses) in the regions of the rectum, urinary bladder, and prostate gland (in the male) or uterus and vagina (in the female).

Liver

Hepatic portal vein

Gallbladder

Pancreas

Superior mesenteric vein

Small intestine

Ascending colon

Rectum

Stomach

Left gastric vein

Right gastric vein

Spleen

Splenic vein

Inferior mesenteric vein

Descending colon

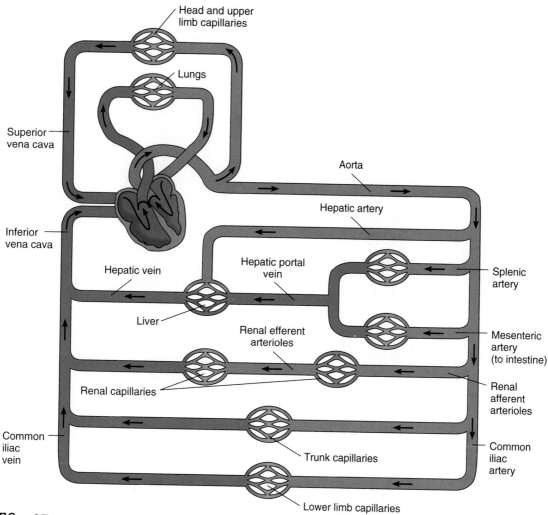

**FIGURE 15.59**

In this schematic drawing of the circulatory system, note how the hepatic portal vein drains one set of capillaries and leads to another set. A similar relationship exists in the kidneys.

The internal iliac veins originate deep within the pelvis and ascend to the pelvic brim. There they unite with the right and left external iliac veins to form the **common iliac veins.** These vessels, in turn, merge to produce the *inferior vena cava* at the level of the fifth lumbar vertebra. Figure 15.61 shows the major vessels of the venous system.

The chapter concludes with Clinical Application 15.8 which looks at molecular explanations of certain cardiovascular disorders, and Clinical Application 15.9, which discusses coronary artery disease.

## CLINICAL TERMS RELATED TO THE CARDIOVASCULAR SYSTEM

**anastomosis** (ah-nas″to-mo′sis)  Connection between two blood vessels, sometimes produced surgically.

**angiospasm** (an′je-o-spazm″)  Muscular spasm in the wall of a blood vessel.

**arteriography** (ar″te-re-og′rah-fe)  Injection of radiopaque solution into the vascular system for an X-ray examination of arteries.

**asystole** (a-sis′to-le)  Condition in which the myocardium fails to contract.

**cardiac tamponade** (kar′de-ak tam″po-nād′)  Compression of the heart by fluid accumulating within the pericardial cavity.

**congestive heart failure** (kon-jes′tiv hart fāl′yer)  Inability of the left ventricle to pump adequate blood to cells.

**cor pulmonale** (kor pul-mo-na′le)  Heart-lung disorder of pulmonary hypertension and hypertrophy of the right ventricle.

**embolectomy** (em″bo-lek′to-me)  Removal of an embolus through an incision in a blood vessel.

**endarterectomy** (en″dar-ter-ek′to-me)  Removal of the inner wall of an artery to reduce an arterial occlusion.

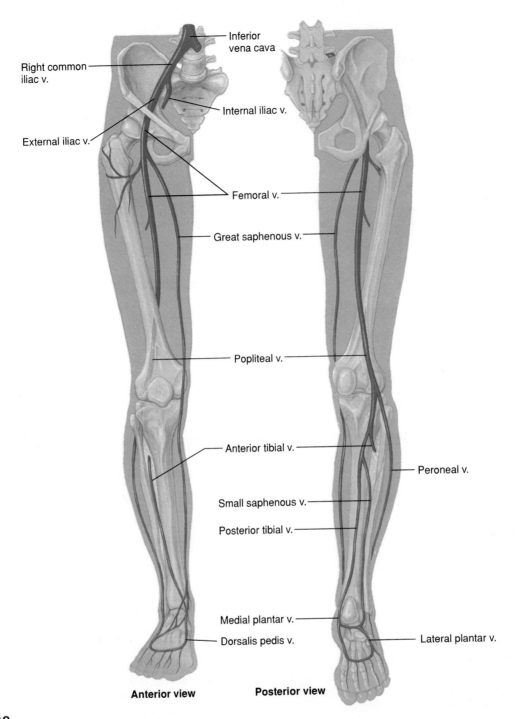

**FIGURE 15.60**

The main veins of the lower limb and pelvis.

**palpitation** (pal″pi-ta′shun)   Awareness of a heartbeat that is unusually rapid, strong, or irregular.

**pericardiectomy** (per″i-kar″de-ek′to-me)   Excision of the pericardium.

**phlebitis** (flĕ-bi′tis)   Inflammation of a vein, usually in the lower limbs.

**phlebotomy** (flĕ-bot′o-me)   Incision or puncture of a vein to withdraw blood.

**sinus rhythm** (si′nus rithm)   The normal cardiac rhythm regulated by the S-A node.

**thrombophlebitis** (throm″bo-fle-bi′tis)   Formation of a blood clot in a vein in response to inflammation of the venous wall.

**valvotomy** (val-vot′o-me)   Incision of a valve.

**venography** (ve-nog′rah-fe)   Injection of radiopaque solution into the vascular system for X-ray examination of veins.

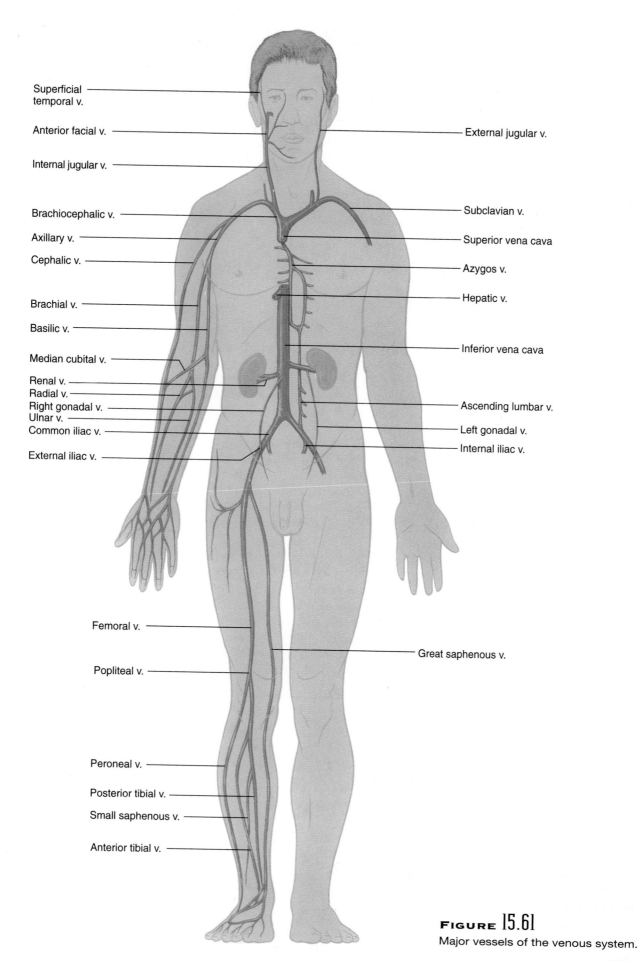

Superficial
temporal v.

Anterior facial v.

Internal jugular v.

Brachiocephalic v.

Axillary v.

Cephalic v.

Brachial v.

Basilic v.

Median cubital v.

Renal v.

Radial v.

Right gonadal v.

Ulnar v.

Common iliac v.

External iliac v.

External jugular v.

Subclavian v.

Superior vena cava

Azygos v.

Hepatic v.

Inferior vena cava

Ascending lumbar v.

Left gonadal v.

Internal iliac v.

Femoral v.

Great saphenous v.

Popliteal v.

Peroneal v.

Posterior tibial v.

Small saphenous v.

Anterior tibial v.

**FIGURE** 15.61

Major vessels of the venous system.

# Molecular Causes of Cardiovascular Disease

### A CONNECTIVE TISSUE DEFECT

In January 1986, volleyball champion Flo Hyman left the court during a game in Japan, collapsed, and suddenly died. Her aorta had burst, and death was instant. Hyman had *Marfan syndrome,* an inherited condition that also caused the characteristics that led her to excel in her sport—her great height and long fingers (fig. 15J).

In Marfan syndrome, an abnormal form of a connective tissue protein called fibrillin weakens the aorta wall. After Flo Hyman died, her siblings were examined, and her brother Michael was found to have a weakened aorta. By surgically repairing his aorta and giving him drugs to control his blood pressure and heart rate, physicians enabled him to avoid the sudden death that claimed his famous sister. Testing for the causative gene can alert physicians to affected individuals before the dangerous swelling in the aorta begins.

### A MYOSIN DEFECT

Each year, one or two seemingly healthy young people die suddenly during a sports event, usually basketball. The culprit is often *familial hypertrophic cardiomyopathy,* an inherited overgrowth of the heart muscle. The defect in this disorder is different than that behind Marfan syndrome. It is an abnormality in one of the myosin chains that comprise cardiac muscle. Again, detecting the responsible gene can alert affected individuals to their increased risk of sudden death. They can adjust the type of exercise they do to avoid stressing the cardiovascular system.

### A METABOLIC GLITCH

Sometimes inherited heart disease strikes very early in life. Jim D. died at 4 days of age, two days after suffering cardiac arrest. Two years later, his parents had another son. Like Jim, Kerry seemed normal at birth, but at 36 hours of age his heart rate plummeted, he had a seizure, and he stopped breathing. He was resuscitated. A blood test revealed excess long-chain fatty acids, indicating a metabolic disorder, an inability to utilize fatty acids. Lack of food triggered the symptoms because the boys could not use fatty acids for energy, as healthy people do. Kerry was able to survive for three years by following a diet low in fatty acids and eating frequently. Once he became comatose because he missed a meal. Eventually, he died of respiratory failure.

Kerry and Jim had inherited a deficiency of a mitochondrial enzyme that processes long-chain fatty acids. Because this is a primary energy source for cardiac muscle, their tiny hearts failed.

### CONTROLLING CHOLESTEROL

Low-density lipoprotein (LDL) receptors on liver cells admit cholesterol into the cells, keeping the fatty substance from building up in the bloodstream and forming artery-clogging plaques. When LDL receptors bind cholesterol, they activate a negative feedback system that temporarily halts the cell's production of cholesterol. In the severe form of familial hypercholesterolemia, a person inherits two defective copies of the gene encoding the LDL receptors. Yellowish lumps of cholesterol can be seen behind the knees and elbows, and heart failure usually causes death in childhood. People who inherit one defective gene have a milder form of the illness. They tend to develop coronary artery disease in young adulthood, but can delay symptoms by following a heart-healthy diet and regularly exercising.

These people have half the normal number of LDL receptors.

In Niemann-Pick type C disease, a defective protein disturbs the fate of cholesterol inside cells. Normally, the protein escorts cholesterol out of a cell's lysosomes (a type of organelle), which triggers the negative feedback mechanism that shuts off cholesterol synthesis. When the protein is absent or malfunctions, the cell keeps producing cholesterol and LDL receptors. Coronary artery disease develops, and the disease is typically fatal in childhood. ■

(a)

Normal heart

(b)

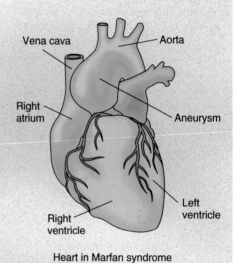

Heart in Marfan syndrome

**FIGURE 15J**

(a) Two years after she led the 1984 U.S. women's volleyball team to a silver medal in the Olympics, Flo Hyman died suddenly when her aorta burst, a symptom of Marfan syndrome. (b) Note the swelling (aneurysm) of the aorta in the heart on the right. A burst aneurysm here is fatal.

# CLINICAL APPLICATION

# Coronary Artery Disease

Dave R., a 52-year-old overweight accountant, had been having occasional chest pains for several months. The mild pain occurred during his usual weekend tennis match, and he attributed it to indigestion. The discomfort almost always diminished after the game, but recently the pain seemed more severe and prolonged. Dave asked his physician about the problem.

The physician explained that Dave was probably experiencing *angina pectoris,* a symptom of *coronary artery disease* (CAD), and suggested that he undergo an *exercise stress test.* Dave walked on a treadmill, speed and incline of which were increased while he exercised. During the test, an ECG recorded continuously, and Dave's blood pressure was monitored. Near the end of the test, when Dave's heart reached the desired rate, a small quantity of radioactive thallium-201 was injected into a vein. A *scintillation counter* scanned Dave's heart to determine if the blood carrying the thallium was uniformly distributed to the myocardium by branches of his coronary arteries (see fig. 15.13a).

The test revealed that Dave was developing CAD. In addition, he had hypertension and high blood cholesterol. The physician advised Dave to stop smoking, to reduce his intake of foods high in saturated fats, cholesterol, and sodium, and to exercise regularly. He was given medications to lower his blood pressure and to relieve the pain of angina. Dave was also cautioned to avoid stressful situations and to lose weight.

Six months later, in spite of faithful compliance with medical advice, Dave suffered a heart attack—a sign that blood flow to part of his myocardium had been obstructed, producing oxygen deficiency (ischemia). He was at home at the time of the attack, which began as severe, crushing chest pain, shortness of

breath, and sweating. Paramedics stabilized Dave's condition and transported him to a hospital.

At the hospital, a cardiologist concluded from an ECG that Dave's heart attack (acute myocardial infarction) was caused by a blood clot obstructing a coronary artery (occlusive coronary thrombosis). The cardiologist administered a "clot-busting" (thrombolytic) drug, tissue plasminogen activator (t-PA) intravenously.

After some time, the ECG showed that the blood vessel remained partially obstructed, so the cardiologist ordered a *coronary angiogram.* In this X-ray procedure, which was conducted in the cardiac catheterization laboratory, a thin plastic catheter was passed through a guiding sheath inserted into the femoral artery of Dave's right inguinal area. From there, the catheter was pushed into the aorta until it reached the region of the opening to the left coronary artery, and then near the opening to the right coronary artery.

The progress of the catheter was monitored with *X-ray fluoroscopy.* Each time the catheter was in proper position, a radiopaque dye (contrast medium) was released from its distal end into the blood. X-ray images that revealed the path of the dye as it entered a coronary artery and its branches were recorded on videotape and on motion picture film, which were later analyzed frame by frame. A single severe narrowing was discovered near the origin of Dave's left anterior descending artery. The cardiologist decided to perform *percutaneous transluminal coronary angioplasty*

(PTCA) in order to enlarge the opening (lumen) of that vessel.

The PTCA was performed by passing another plastic catheter through the guiding sheath used for the angiogram. This second tube had a tiny deflated balloon at its tip, and when the balloon was located in the region of the arterial narrowing, it was inflated for a short time with relatively high pressure. The inflating balloon compressed the atherosclerotic plaque (atheroma) that was obstructing the arterial wall. The expanding balloon also stretched the blood vessel wall, thus widening its lumen (recanalization). Blood flow to the myocardial tissue downstream from the obstruction improved immediately.

About 50% of the time, a vessel opened with PTCA becomes occluded again, because the underlying disease persists. To prevent this *restenosis,* the doctor inserted a *coronary stent,* which is an expandable tube or coil that holds the vessel wall open. The cardiologist had two other options that have a slightly higher risk of causing damage. He might have vaporized the plaque obstructing the vessel with an excimer laser pulse delivered along optical fibers threaded through the catheter. Or, he could have performed atherectomy, in which a cutting device attached to the balloon inserted into the catheter spins, removing plaque by withdrawing it on the catheter tip.

Should the coronary stent fail, or an obstruction block another heart vessel, Dave might benefit from *coronary bypass surgery.* A portion of his internal mammary artery inside his chest wall would be removed and stitched between the aorta and the blocked coronary artery at a point beyond the obstruction, restoring circulation through the heart. ∎

**CARDIOVASCULAR SYSTEM**

The heart pumps blood through as many as 60,000 miles of blood vessels delivering nutrients to, and removing wastes from, all body cells.

## Integumentary System

Changes in skin blood flow are important in temperature control.

## Skeletal System

Bones help control plasma calcium levels.

## Muscular System

Blood flow increases to exercising skeletal muscle, delivering oxygen and nutrients and removing wastes. Muscle actions help the blood circulate.

## Nervous System

The brain depends on blood flow for survival. The nervous system helps control blood flow and blood pressure.

## Endocrine System

Hormones are carried in the bloodstream. Some hormones directly affect the heart and blood vessels.

## Lymphatic System

The lymphatic system returns tissue fluids to the bloodstream.

## Digestive System

The digestive system breaks down nutrients into forms readily absorbed by the bloodstream.

## Respiratory System

The respiratory system oxygenates the blood and removes carbon dioxide. Respiratory movements help the blood circulate.

## Urinary System

The kidneys clear the blood of wastes and substances present in excess. The kidneys help control blood pressure and blood volume.

## Reproductive System

Blood pressure is important in normal function of the sex organs.

## Introduction (page 554)

The cardiovascular system provides oxygen and nutrients to tissues, and removes wastes.

## Structure of the Heart (page 554)

1. Size and location of the heart
   a. The heart is about 14 centimeters long and 9 centimeters wide.
   b. It is located within the mediastinum and rests on the diaphragm.
2. Coverings of the heart
   a. A layered pericardium encloses the heart.
   b. The pericardial cavity is a space between the visceral and parietal layers of the pericardium.
3. Wall of the heart
   a. The wall of the heart has three layers.
   b. These layers include an epicardium, a myocardium, and an endocardium.
4. Heart chambers and valves
   a. The heart is divided into four chambers—two atria and two ventricles—that communicate through atrioventricular orifices on each side.
   b. Right chambers and valves
      (1) The right atrium receives blood from the venae cavae and coronary sinus.
      (2) The tricuspid valve guards the right atrioventricular orifice.
      (3) The right ventricle pumps blood into the pulmonary trunk.
      (4) A pulmonary valve guards the base of the pulmonary trunk.
   c. Left chambers and valves
      (1) The left atrium receives blood from the pulmonary veins.
      (2) The bicuspid valve guards the left atrioventricular orifice.
      (3) The left ventricle pumps blood into the aorta.
      (4) An aortic valve guards the base of the aorta.
5. Skeleton of the heart
   a. The skeleton of the heart consists of fibrous rings that enclose the bases of the pulmonary artery, aorta, and atrioventricular orifices.
   b. The fibrous rings provide attachments for valves and muscle fibers, and prevent the orifices from excessively dilating during ventricular contractions.
6. Path of blood through the heart
   a. Blood that is relatively low in oxygen concentration and high in carbon dioxide concentration enters the right side of the heart from the venae cavae and coronary sinus, and then is pumped into the pulmonary circulation.
   b. After the blood is oxygenated in the lungs and some of its carbon dioxide is removed, it returns to the left side of the heart through the pulmonary veins.
   c. From the left ventricle, it moves into the aorta.
7. Blood supply to the heart
   a. The coronary arteries supply blood to the myocardium.
   b. It is returned to the right atrium through the cardiac veins and coronary sinus.

## Heart Actions (page 565)

1. Cardiac cycle
   a. The atria contract while the ventricles relax; the ventricles contract while the atria relax.
   b. Pressure within the chambers rises and falls in repeated cycles.
2. Heart sounds
   a. Heart sounds can be described as *lubb-dupp*.
   b. Heart sounds are due to the vibrations that the valve movements produce.
   c. The first part of the sound occurs as A-V valves close, and the second part is associated with the closing of pulmonary and aortic valves.
3. Cardiac muscle fibers
   a. Cardiac muscle fibers connect to form a functional syncytium.
   b. If any part of the syncytium is stimulated, the whole structure contracts as a unit.
   c. Except for a small region in the floor of the right atrium, the fibrous skeleton separates the atrial syncytium from the ventricular syncytium.
4. Cardiac conduction system
   a. This system, composed of specialized cardiac muscle tissue, initiates and conducts depolarization waves through the myocardium.
   b. Impulses from the S-A node pass slowly to the A-V node; impulses travel rapidly along the A-V bundle and Purkinje fibers.
   c. Muscle fibers in the ventricular walls are arranged in whorls that squeeze blood out of the contracting ventricles.
5. Electrocardiogram (ECG)
   a. An ECG records electrical changes in the myocardium during a cardiac cycle.
   b. The pattern contains several waves.
      (1) The P wave represents atrial depolarization.
      (2) The QRS complex represents ventricular depolarization.
      (3) The T wave represents ventricular repolarization.
6. Regulation of the cardiac cycle
   a. Physical exercise, body temperature, and concentration of various ions affect heartbeat.
   b. Branches of sympathetic and parasympathetic nerve fibers innervate the S-A and A-V nodes.
      (1) Parasympathetic impulses decrease heart action; sympathetic impulses increase heart action.
      (2) The cardiac center in the medulla oblongata regulates autonomic impulses to the heart.

## Blood Vessels (page 573)

The blood vessels form a closed circuit of tubes that transport blood between the heart and body cells. The tubes include arteries, arterioles, capillaries, venules, and veins.

1. Arteries and arterioles
   a. The arteries are adapted to carry relatively high pressure blood away from the heart.
   b. The arterioles are branches of arteries.
   c. The walls of arteries and arterioles consist of layers of endothelium, smooth muscle, and connective tissue.

d. Autonomic fibers that can stimulate vasoconstriction or vasodilation innervate smooth muscles in vessel walls.
2. Capillaries
Capillaries connect arterioles and venules. The capillary wall is a single layer of cells that forms a semipermeable membrane.
   a. Capillary permeability
      (1) Openings in the capillary walls are thin slits between endothelial cells.
      (2) The sizes of the openings vary from tissue to tissue.
      (3) Endothelial cells of brain capillaries are tightly fused, forming a blood-brain barrier through which substances move by facilitated diffusion.
   b. Capillary arrangement
      Capillary density varies directly with tissue metabolic rates.
   c. Regulation of capillary blood flow
      (1) Precapillary sphincters regulate capillary blood flow.
      (2) Precapillary sphincters open when cells are low in oxygen and nutrients, and close when cellular needs are met.
3. Exchanges in the capillaries
   a. Gases, nutrients, and metabolic by-products are exchanged between the capillary blood and the tissue fluid.
   b. Diffusion provides the most important means of transport.
   c. Diffusion pathways depend on lipid solubilities.
   d. Plasma proteins generally remain in the blood.
   e. Filtration, which is due to the hydrostatic pressure of blood, causes a net outward movement of fluid at the arteriolar end of a capillary.
   f. Osmosis causes a net inward movement of fluid at the venular end of a capillary.
   g. Some factors cause fluids to accumulate in the tissues.
4. Venules and veins
   a. Venules continue from capillaries and merge to form veins.
   b. Veins carry blood to the heart.
   c. Venous walls are similar to arterial walls, but are thinner and contain less muscle and elastic tissue.

# Blood Pressure  (page 582)

Blood pressure is the force blood exerts against the insides of blood vessels.

1. Arterial blood pressure
   a. The arterial blood pressure is produced primarily by heart action; it rises and falls with phases of the cardiac cycle.
   b. Systolic pressure occurs when the ventricle contracts; diastolic pressure occurs when the ventricle relaxes.
2. Factors that influence arterial blood pressure
   a. Heart action, blood volume, resistance to flow, and blood viscosity influence arterial blood pressure.
   b. Arterial pressure increases as cardiac output, blood volume, peripheral resistance, or blood viscosity increases.
3. Control of blood pressure
   a. Blood pressure is controlled in part by the mechanisms that regulate cardiac output and peripheral resistance.

b. Cardiac output depends on the volume of blood discharged from the ventricle with each beat and on the heart rate.
   (1) The more blood that enters the heart, the stronger the ventricular contraction, the greater the stroke volume, and the greater the cardiac output.
   (2) The cardiac center of the medulla oblongata regulates heart rate.
c. Changes in the diameter of arterioles, controlled by the vasomotor center of the medulla oblongata, regulate peripheral resistance.
4. Venous blood flow
   a. Venous blood flow is not a direct result of heart action; it depends on skeletal muscle contraction, breathing movements, and venoconstriction.
   b. Many veins contain flaplike valves that prevent blood from backing up.
   c. Venous constriction can increase venous pressure and blood flow.
5. Central venous pressure
   a. Central venous pressure is the pressure in the right atrium.
   b. Factors that influence it alter the flow of blood into the right atrium.
   c. It affects pressure within the peripheral veins.

# Paths of Circulation  (page 592)

1. Pulmonary circuit
   a. The pulmonary circuit consists of vessels that carry blood from the right ventricle to the lungs, alveolar capillaries, and vessels that lead back to the left atrium.
   b. Alveolar capillaries exert less pressure than those of the systemic circuit.
   c. Tightly joined epithelial cells of alveoli walls prevent most substances from entering the alveoli.
   d. Osmotic pressure rapidly draws water out of alveoli into the interstitial fluid, so alveoli do not fill with fluid.
2. Systemic circuit
   a. The systemic circuit is composed of vessels that lead from the heart to the body cells and back to the heart.
   b. It includes the aorta and its branches as well as the system of veins that return blood to the right atrium.

# Arterial System  (page 595)

1. Principal branches of the aorta
   a. The branches of the ascending aorta include the right and left coronary arteries.
   b. The branches of the aortic arch include the brachiocephalic, left common carotid, and left subclavian arteries.
   c. The branches of the descending aorta include the thoracic and abdominal groups.
   d. The abdominal aorta terminates by dividing into right and left common iliac arteries.
2. Arteries to the neck, head, and brain include branches of the subclavian and common carotid arteries.
3. Arteries to the shoulder and upper limb
   a. The subclavian artery passes into the arm, and in various regions is called the axillary and brachial artery.
   b. Branches of the brachial artery include the ulnar and radial arteries.

4. Arteries to the thoracic and abdominal walls
   a. Branches of the subclavian artery and thoracic aorta supply the thoracic wall.
   b. Branches of the abdominal aorta and other arteries supply the abdominal wall.
5. Arteries to the pelvis and lower limb
   The common iliac artery supplies the pelvic organs, gluteal region, and lower limb.

## Venous System (page 604)

1. Characteristics of venous pathways
   a. The veins return blood to the heart.
   b. Larger veins usually parallel the paths of major arteries.
2. Veins from the head, neck, and brain
   a. The jugular veins drain these regions.
   b. Jugular veins unite with subclavian veins to form the brachiocephalic veins.
3. Veins from the upper limb and shoulder
   a. Sets of superficial and deep veins drain the upper limb.
   b. The major superficial veins are the basilic and cephalic veins.
   c. The median cubital vein in the bend of the elbow is often used as a site for venipuncture.
4. Veins from the abdominal and thoracic walls include tributaries of the brachiocephalic, and azygos veins drain these walls.
5. Veins from the abdominal viscera
   a. The blood from the abdominal viscera generally enters the hepatic portal system and is carried to the liver.
   b. The blood in the hepatic portal system is rich in nutrients.
   c. The liver helps regulate the blood concentrations of glucose, amino acids, and lipids.
   d. Phagocytic cells in the liver remove bacteria from the portal blood.
   e. From the liver, hepatic veins carry blood to the inferior vena cava.
6. Veins from the lower limb and pelvis
   a. Sets of deep and superficial veins drain these regions.
   b. The deep veins include the tibial veins, and the superficial veins include the saphenous veins.

## CRITICAL THINKING QUESTIONS

1. Given the way capillary blood flow is regulated, do you think it is wiser to rest or to exercise following a heavy meal? Explain.
2. If a patient develops a blood clot in the femoral vein of the left lower limb, and a portion of the clot breaks loose, where is the blood flow likely to carry the embolus? What symptoms are likely?
3. When a person strains to lift a heavy object, intrathoracic pressure increases. What do you think will happen to the rate of venous blood returning to the heart during such lifting? Why?
4. Why is a ventricular fibrillation more likely to be life threatening than an atrial fibrillation?
5. Cirrhosis of the liver, a disease commonly associated with alcoholism, obstructs blood flow through the hepatic blood vessels. As a result, the blood backs up, and the capillary pressure greatly increases in the organs drained by the hepatic portal system. What effects might this increasing capillary pressure produce, and which organs would it affect?
6. If a cardiologist inserted a catheter into a patient's right femoral artery, which arteries would the tube have to pass through in order to reach the entrance of the left coronary artery?
7. How might the results of a cardiovascular exam differ for an athlete in top condition and a sedentary, overweight individual?
8. Cigarette smoke contains thousands of chemicals, including nicotine and carbon monoxide. Nicotine constricts blood vessels. Carbon monoxide prevents oxygen from binding to hemoglobin. How do these two components of smoke affect the cardiovascular system?
9. What structures and properties should an artificial heart have?

## REVIEW EXERCISES

1. Describe the general structure, function, and location of the heart.
2. Describe the pericardium.
3. Compare the layers of the cardiac wall.
4. Identify and describe the locations of the chambers and the valves of the heart.
5. Describe the skeleton of the heart, and explain its function.
6. Trace the path of the blood through the heart.
7. Trace the path of the blood through the coronary circulation.
8. Describe a cardiac cycle.
9. Describe the pressure changes that occur in the atria and ventricles during a cardiac cycle.
10. Explain the origin of heart sounds.
11. Describe the arrangement of the cardiac muscle fibers.
12. Distinguish between the roles of the S-A node and the A-V node.
13. Explain how the cardiac conduction system controls the cardiac cycle.
14. Describe and explain the normal ECG pattern.
15. Discuss how the nervous system regulates the cardiac cycle.

16. Describe two factors other than the nervous system that affect the cardiac cycle.
17. Distinguish between an artery and an arteriole.
18. Explain control of vasoconstriction and vasodilation.
19. Describe the structure and function of a capillary.
20. Describe the function of the blood-brain barrier.
21. Explain control of blood flow through a capillary.
22. Explain how diffusion functions in the exchange of substances between blood plasma and tissue fluid.
23. Explain why water and dissolved substances leave the arteriolar end of a capillary and enter the venular end.
24. Describe the effect of histamine on a capillary.
25. Distinguish between a venule and a vein.
26. Explain how veins function as blood reservoirs.

27. Distinguish between systolic and diastolic blood pressures.
28. Name several factors that influence the blood pressure, and explain how each produces its effect.
29. Describe the control of blood pressure.
30. List the major factors that promote the flow of venous blood.
31. Define central venous pressure.
32. Distinguish between the pulmonary and systemic circuits of the cardiovascular system.
33. Trace the path of blood through the pulmonary circuit.
34. Explain why the alveoli normally do not fill with fluid.
35. Describe the aorta, and name its principal branches.
36. Describe the relationship between the major venous pathways and the major arterial pathways.

 **For More Information Related to This Chapter, Please Visit** *Hole's Human Anatomy and Physiology* **Home Page**
http://www.mhhe.com/biosci/abio/

# 16
### CHAPTER

# LYMPHATIC SYSTEM AND IMMUNITY

## UNDERSTANDING WORDS

**auto-,** self: *auto*immune disease—condition in which the immune system attacks the body's own tissues.

**-gen,** become, be produced: aller*gen*—substance that stimulates an allergic response.

**humor-,** moisture, fluid: *humor*al immunity—immunity resulting from antibodies in body fluids.

**immun-,** free, exempt: *immun*ity—resistance to (freedom from) a specific disease.

**inflamm-,** to set on fire: *inflamm*ation—condition characterized by localized redness, heat, swelling, and pain in the tissues.

**nod-,** knot: *nod*ule—small mass of lymphocytes surrounded by connective tissue.

**path-,** disease, sickness: *path*ogen—disease-causing agent.

## CHAPTER OBJECTIVES

AFTER YOU HAVE STUDIED THIS CHAPTER, YOU SHOULD BE ABLE TO:

1. Describe the general functions of the lymphatic system.

2. Describe the location of the major lymphatic pathways.

3. Describe how tissue fluid and lymph form and explain the function of lymph.

4. Explain how lymphatic circulation is maintained and describe the consequence of lymphatic obstruction.

5. Describe a lymph node and its major functions.

6. Describe the location of the major chains of lymph nodes.

7. Discuss the functions of the thymus and spleen.

8. Distinguish between specific and nonspecific defenses, and provide examples of each.

9. Explain how two major types of lymphocytes are formed, activated, and how they function in immune mechanisms.

10. Name the major types of immunoglobulins and discuss their origins and actions.

11. Distinguish between primary and secondary immune responses.

12. Distinguish between active and passive immunity.

13. Explain how allergic reactions, tissue rejection reactions, and autoimmunity arise from immune mechanisms.

## KEY TERMS

**allergen** (al'er-jen)
**antibody** (an'ti-bod"e)
**antigen** (an'ti-jen)
**clone** (klōn)
**complement** (kom'plĕ-ment)
**hapten** (hap'ten)
**immunity** (i-mu'ni-te)
**immunoglobulin** (im"u-no-glob'u-lin)
**interferon** (in"ter-fēr'on)
**lymph** (limf)

**lymphatic pathway** (lim-fat'ik path'wa)
**lymph node** (limf nōd)
**lymphocyte** (lim'fo-sīt)
**macrophage** (mak'ro-fāj)
**pathogen** (path'o-jen)
**reticuloendothelial tissue** (rĕ-tik"u-lo-en"do-the'le-al tish'u)
**spleen** (splēn)
**thymus** (thi'mus)
**vaccine** (vak'sēn)

J im Hancock gave his daughter Leslie the gift of life when she was born. Six years later, he did it again. When Leslie's liver began to fail, Jim donated part of his own, saving her life.

Leslie was born with cystic fibrosis, a hereditary disorder that usually impairs the lungs and pancreas. When in the summer of 1995 doctors discovered that Leslie's digestive difficulties were not directly due to cystic fibrosis but to a hardening liver, they suggested that a liver transplant might be necessary in a few years. The drugs they prescribed to help Leslie's liver were not effective, and Leslie's condition rapidly worsened. She needed a new liver soon. The statistics were terrifying: A shortage of donor organs results in a third of the children awaiting donor livers dying before a transplant can be performed.

The Hancocks, however, were offered an alternative: Part of Jim's liver could be removed and transplanted into his daughter. "When the subject of a living-relative transplant came up, I never had to think about it. It was just something I wanted to do," said Jim. Transplants involving blood relatives have about a 90% success rate, compared to 80% for transplants involving nonrelatives. Leslie's immune system was not likely to reject her father's liver because the cell surfaces of this father and daughter are particularly similar. Leslie also took drugs to suppress her immune system, decreasing the chances of rejection even more.

The surgery was a success. Two months later, Leslie was back in her kindergarten classroom in Dubuque, Iowa.

Jim Hancock, of Dubuque, Iowa, gave his 6-year-old daughter Leslie part of his liver and a new chance at life. The transplant worked because the cell surfaces of this father and daughter are very similar. The little girl's immune system accepted the new liver as part of her body.

---

The lymphatic system is closely associated with the cardiovascular system because it includes a network of vessels that assist in circulating body fluids. Lymphatic vessels transport excess fluid away from interstitial spaces between cells in most tissues and return it to the bloodstream (fig. 16.1). Special lymphatic vessels (*lacteals*) located in the lining of the small intestine absorb fats resulting from digestion, then transport the fats to venous circulation. The organs of the lymphatic system also help defend the body against invasion by disease-causing agents.

## LYMPHATIC PATHWAYS

The lymphatic pathways begin as lymphatic capillaries that merge to form lymphatic vessels. These, in turn, lead to larger vessels that unite with the veins in the thorax.

### Lymphatic Capillaries

**Lymphatic capillaries** are microscopic, closed-ended tubes. They extend into the interstitial spaces, forming complex networks that parallel the networks of the blood capillaries (fig. 16.2). The walls of the lymphatic capillaries, like those of the blood capillaries, consist of single layers of squamous epithelial cells. These thin walls make it possible for tissue fluid (interstitial fluid) from the interstitial space to enter the lymphatic capillaries. Fluid inside a lymphatic capillary is called **lymph.**

### Lymphatic Vessels

The walls of **lymphatic vessels** are similar to those of veins, but thinner. That is, their walls are composed of three layers: an endothelial lining, a middle layer of smooth muscle and elastic fibers, and an outer layer of connective tissue. Also like veins, the lymphatic vessels have semilunar valves, which help prevent backflow of lymph. Figure 16.3 shows one of these valves.

The larger lymphatic vessels lead to specialized organs called **lymph nodes.** After leaving the nodes, the vessels merge to form still larger lymphatic trunks.

### Lymphatic Trunks and Collecting Ducts

The **lymphatic trunks,** which drain lymph, are named for the regions they serve. For example, the *lumbar trunk* drains lymph from the lower limbs, lower abdominal

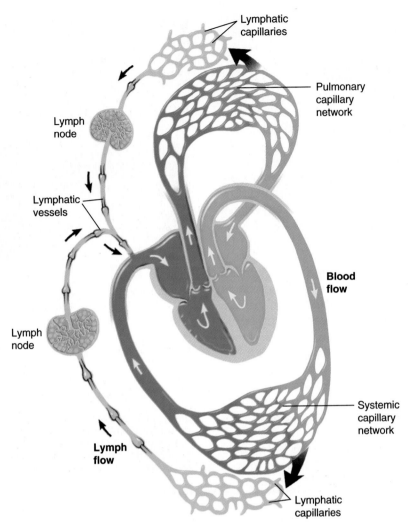

**FIGURE** 16.1

Schematic representation of lymphatic vessels transporting fluid from interstitial spaces to the bloodstream.

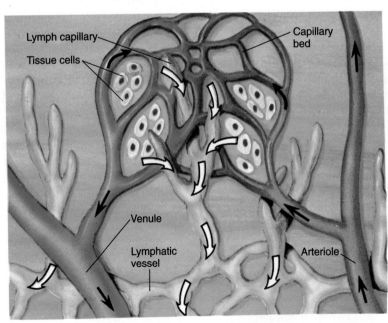

**FIGURE** 16.2

Lymph capillaries are microscopic, closed-ended tubes that begin in the interstitial spaces of most tissues.

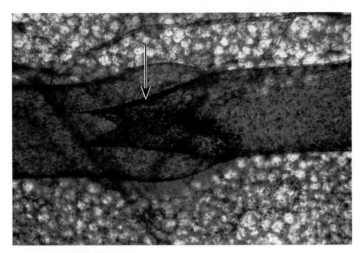

**FIGURE 16.3**

A light micrograph of the flaplike valve (arrow) within a lymphatic vessel (25×).

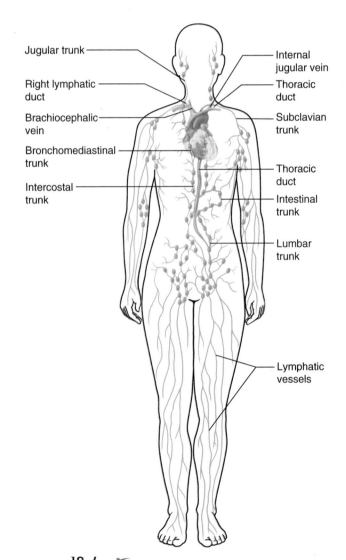

Jugular trunk

Right lymphatic duct

Brachiocephalic vein

Bronchomediastinal trunk

Intercostal trunk

Internal jugular vein

Thoracic duct

Subclavian trunk

Thoracic duct

Intestinal trunk

Lumbar trunk

Lymphatic vessels

**FIGURE 16.4**

Lymphatic vessels merge into larger lymphatic trunks, which in turn drain into collecting ducts.

**FIGURE 16.5**

A lymphangiogram (radiograph) of the lymphatic vessels and lymph nodes of the pelvic region.

wall, and pelvic organs; the *intestinal trunk* drains the abdominal viscera; the *intercostal* and *bronchomediastinal trunks* drain lymph from portions of the thorax; the *subclavian trunk* drains the upper limb; and the *jugular trunk* drains portions of the neck and head. These lymphatic trunks then join one of two **collecting ducts**—the thoracic duct or the right lymphatic duct. Figure 16.4 shows the location of the major lymphatic trunks and collecting ducts, and figure 16.5 shows a lymphangiogram, or radiograph, of some lymphatic vessels and lymph nodes.

The **thoracic duct** is the larger and longer of the two collecting ducts. It begins in the abdomen, passes upward through the diaphragm beside the aorta, ascends in front of the vertebral column through the mediastinum, and empties into the left subclavian vein near the junction of the left jugular vein. This duct drains lymph from the intestinal, lumbar, and intercostal trunks, as well as from the left subclavian, left jugular, and left bronchomediastinal trunks.

The **right lymphatic duct** originates in the right thorax at the union of the right jugular, right subclavian, and

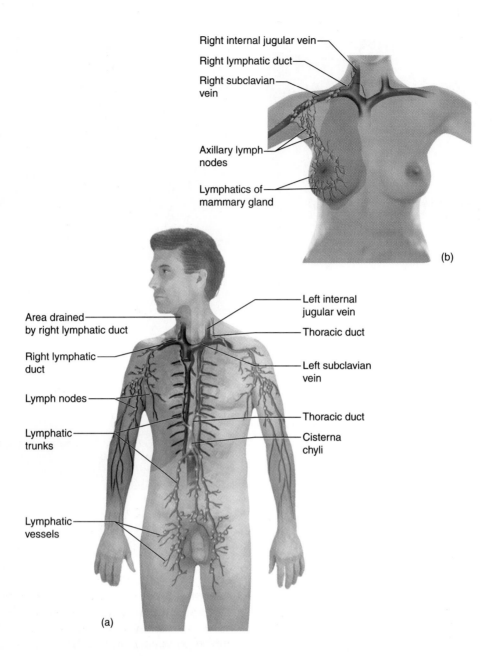

**FIGURE 16.6**

(a) The right lymphatic duct drains lymph from the upper right side of the body, whereas the thoracic duct drains lymph from the rest of the body. (b) Lymph drainage of the right breast.

right bronchomediastinal trunks. It empties into the right subclavian vein near the junction of the right jugular vein.

After leaving the two collecting ducts, lymph enters the venous system and becomes part of the plasma just before the blood returns to the right atrium. Thus, lymph from the lower body regions, the left upper limb, and the left side of the head and neck enters the thoracic duct; lymph from the right side of the head and neck, the right upper limb, and the right thorax enters the right lymphatic duct (fig. 16.6). Figure 16.7 summarizes the lymphatic pathway.

The skin is richly supplied with lymphatic capillaries. Consequently, if the skin is broken, or if something is injected into it (such as venom from a stinging insect), foreign substances rapidly enter the lymphatic system.

**1** What are the general functions of the lymphatic system?

**2** Through which lymphatic structures would lymph pass in traveling from a lower limb to the bloodstream?

**FIGURE 16.7**

The lymphatic pathway.

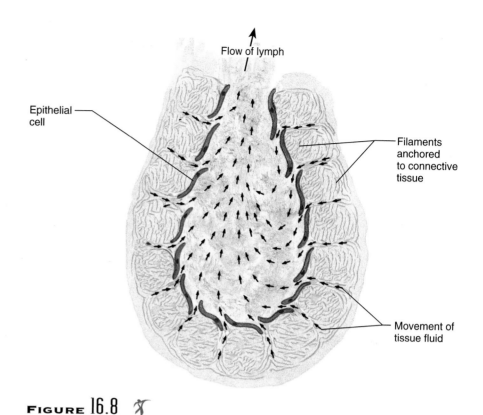

**FIGURE 16.8** 🏃

Tissue fluid enters lymphatic capillaries through flaplike valves between epithelial cells.

## TISSUE FLUID AND LYMPH

Lymph is essentially tissue fluid that has entered a lymphatic capillary. Thus, lymph formation depends upon tissue fluid formation.

### Tissue Fluid Formation

Recall from chapter 15 that tissue fluid originates from blood plasma. Tissue fluid is composed of water and dissolved substances that leave the blood capillaries by filtration and diffusion.

Tissue fluid contains nutrients and gases found in plasma and lacks large proteins. Some smaller proteins do leak out of the blood capillaries and enter the interstitial spaces. Usually, these smaller proteins are not reabsorbed when water and dissolved substances move back into the venular ends of these capillaries by osmosis and diffusion. As a result, the protein concentration of the tissue fluid tends to rise, raising the *osmotic pressure* of the fluid.

### Lymph Formation

Rising osmotic pressure of the tissue fluid interferes with the blood capillaries' osmotic reabsorption of water. The volume and hydrostatic pressure of fluid in the interstitial spaces then tends to increase. This increasing interstitial pressure forces some of the tissue fluid into the lymphatic capillaries, where it becomes lymph (see fig. 16.2).

### Lymph Function

Lymph returns to the bloodstream most of the proteins that leak out of the blood capillaries. At the same time, lymph transports foreign particles, such as bacteria or viruses, to lymph nodes.

Although these proteins and foreign particles cannot easily enter blood capillaries, the lymphatic capillaries are adapted to receive them. Specifically, the epithelial cells that form the walls of lymphatic vessels overlap each other but are not attached. This configuration, shown in figure 16.8, creates flaplike valves in the lymphatic capillary wall, which are pushed inward when the pressure is greater on the outside of the capillary, but close when the pressure is greater on the inside.

The epithelial cells of the lymphatic capillary wall are also attached to surrounding connective tissue cells by thin filaments. As a result, the lumen of a lymphatic capillary remains open even when the outside pressure increases.

**1** What is the relationship between tissue fluid and lymph?

**2** How do proteins in tissue fluid affect lymph formation?

**3** What are the major functions of lymph?

(a)

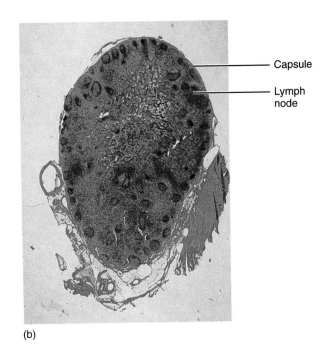

(b)

**FIGURE 16.9**

(a) A section of a lymph node. (b) Light micrograph of a lymph node (2.5× micrograph enlarged to 5×).

## LYMPH MOVEMENT

The hydrostatic pressure of tissue fluid drives the entry of lymph into lymphatic capillaries. However, muscular activity largely influences the movement of lymph through the lymphatic vessels.

### Lymph Flow

Lymph, like venous blood, is under relatively low hydrostatic pressure and may not flow readily through the lymphatic vessels without help. These forces include contraction of the skeletal muscles, pressure changes due to the action of the breathing muscles, and contraction of smooth muscles in the walls of the larger lymphatic trunks. Lymph flow peaks during physical exercise, due to the actions of skeletal and breathing muscles.

Contracting skeletal muscles compress lymphatic vessels. This squeezing action moves the lymph inside a vessel, but because the lymphatic vessels contain valves that prevent backflow, the lymph can only move toward a collecting duct. Additionally, the smooth muscles in the walls of the larger lymphatic trunks may contract and compress the lymph inside, forcing the fluid onward.

The breathing muscles aid lymph circulation by creating a relatively low pressure in the thorax during inhalation. At the same time, the contracting diaphragm increases the pressure in the abdominal cavity. Consequently, lymph is squeezed out of the abdominal vessels and forced into the thoracic vessels. Once again, valves within the lymphatic vessels prevent lymph backflow.

### Obstruction of Lymph Movement

The continuous movement of fluid from interstitial spaces into blood capillaries and lymphatic capillaries stabilizes the volume of fluid in these spaces. Conditions that interfere with lymph movement cause tissue fluids to accumulate in the interstitial spaces, producing *edema*. This may happen when surgery removes lymphatic tissue, obstructing certain lymphatic vessels. For example, surgery to remove a cancerous breast also usually removes nearby axillary lymph nodes, because associated lymphatic vessels can transport cancer cells to other sites (metastasis). Removing the lymphatic tissue can obstruct drainage from the upper limb, causing edema.

**1** What factors promote lymph flow?

**2** What is the consequence of lymphatic obstruction?

## LYMPH NODES

Lymph nodes (lymph glands) are located along the lymphatic pathways. They contain large numbers of *lymphocytes* and *macrophages*, which fight invading microorganisms.

### Structure of a Lymph Node

Lymph nodes vary in size and shape, but are usually less than 2.5 centimeters long and are somewhat bean-shaped. Figure 16.9 illustrates a section of a typical lymph node.

**FIGURE 16.10**

Lymph enters and leaves a lymph node through lymphatic vessels.

The indented region of a bean-shaped node is called the **hilum,** and it is the portion through which the blood vessels and nerves connect with the structure. The lymphatic vessels leading to a node (afferent vessels) enter separately at various points on its convex surface, but the lymphatic vessels leaving the node (efferent vessels) exit from the hilum.

A capsule of white connective tissue with numerous fibers encloses each lymph node. The capsule extends into the node and partially subdivides it into compartments that contain dense masses of actively dividing lymphocytes and macrophages, called **lymph nodules** with lighter staining germinal centers. These nodules are the structural units of the lymph node.

Nodules also occur singly or in groups associated with the mucous membranes of the respiratory and digestive tracts. The *tonsils,* described in chapter 17, are composed of partially encapsulated lymph nodules. Also, aggregations of nodules called *Peyer's patches* are scattered throughout the mucosal lining of the ileum of the small intestine.

The spaces within the node, called **lymph sinuses,** provide a complex network of chambers and channels through which lymph circulates as it passes through. Lymph enters a lymph node through *afferent lymphatic vessels,* moves slowly through the lymph sinuses, and leaves through *efferent lymphatic vessels* (fig. 16.10).

> Superficial lymphatic vessels inflamed by bacterial infection appear as red streaks beneath the skin, a condition called *lymphangitis.* Inflammation of the lymph nodes, called *lymphadenitis,* often follows. Affected nodes enlarge and may be quite painful.

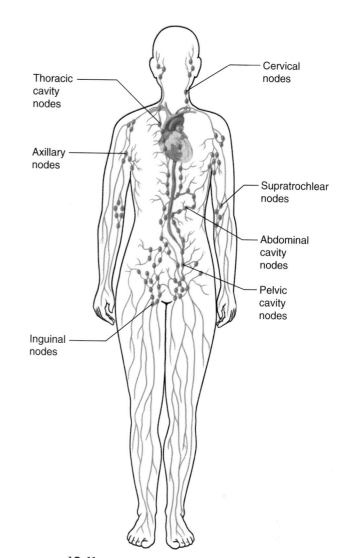

**FIGURE 16.11**

Major locations of lymph nodes.

## Locations of Lymph Nodes

The lymph nodes generally occur in groups or chains along the paths of the larger lymphatic vessels throughout the body, but are absent in the central nervous system. The major locations of the lymph nodes, shown in figure 16.11, are as follows:

1. **Cervical region.** Nodes in the cervical region follow the lower border of the mandible, in front of and behind the ears, and deep within the neck along the paths of the larger blood vessels. These nodes are associated with the lymphatic vessels that drain the skin of the scalp and face, as well as the tissues of the nasal cavity and pharynx.

2. **Axillary region.** Nodes in the underarm region receive lymph from vessels that drain the upper limbs, the wall of the thorax, the mammary glands (breasts), and the upper wall of the abdomen.

3. **Inguinal region.** Nodes in the inguinal region receive lymph from the lower limbs, the external genitalia, and the lower abdominal wall.

4. **Pelvic cavity.** Within the pelvic cavity, nodes primarily occur along the paths of the iliac blood vessels. They receive lymph from the lymphatic vessels of the pelvic viscera.

5. **Abdominal cavity.** Nodes within the abdominal cavity form chains along the main branches of the mesenteric arteries and the abdominal aorta. These nodes receive lymph from the abdominal viscera.

6. **Thoracic cavity.** Nodes of the thoracic cavity occur within the mediastinum and along the trachea and bronchi. They receive lymph from the thoracic viscera and from the internal wall of the thorax.

7. **Supratrochlear region.** These nodes are located superficially on the medial side of the elbow. They often enlarge in children as a result of infections from cuts and scrapes on the hands.

---

The illness described as "swollen glands" actually refers to enlarged cervical lymph nodes associated with throat or respiratory infection.

---

## Functions of Lymph Nodes

Lymph nodes have two primary functions: filtering potentially harmful particles from lymph before returning it to the bloodstream, and immune surveillance, provided by lymphocytes and macrophages. Along with the red bone marrow, the lymph nodes are centers for lymphocyte production. These cells attack infecting viruses, bacteria, and other parasitic cells that lymphatic vessels bring to the nodes. Macrophages in the nodes engulf and destroy foreign substances, damaged cells, and cellular debris.

1. Distinguish between a lymph node and a lymph nodule.

2. What factors promote the flow of lymph through a node?

3. In what body regions are lymph nodes most abundant?

4. What are the major functions of lymph nodes?

---

## THYMUS AND SPLEEN

Two other lymphatic organs, whose functions are similar to those of the lymph nodes, are the thymus and the spleen.

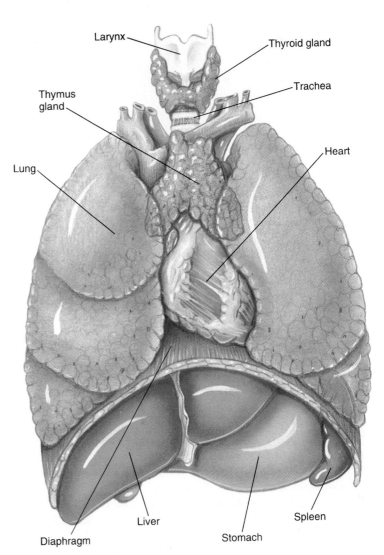

**FIGURE 16.12**

The thymus gland is bilobed and located between the lungs and above the heart. The spleen is located inferior to the diaphragm, posterior and lateral to the stomach.

## Thymus

The **thymus** gland is a soft, bilobed structure enclosed in a connective tissue capsule (fig. 16.12). It is located within the mediastinum, anterior to the aortic arch and posterior to the upper part of the sternum, and extends from the root of the neck to the pericardium. The thymus varies in size from person to person, and it is usually larger during infancy and early childhood. After puberty, the thymus shrinks, and in an adult, it may be quite small. In elderly persons, adipose and connective tissues replace the normal lymphatic tissue.

Connective tissues extend inward from the thymus's surface, subdividing it into lobules (fig. 16.13). The lobules contain many lymphocytes that developed from precursor cells in the bone marrow. The majority of these cells (thymocytes) are inactivated; however, some

**FIGURE 16.13**

A cross section of the thymus gland (10× micrograph enlarged to 20×). Note how the gland is subdivided into lobules.

mature into **T lymphocytes,** which leave the thymus and provide immunity. Epithelial cells within the thymus secrete a protein hormone called *thymosin,* which stimulates maturation of T lymphocytes after they leave the thymus and migrate to other lymphatic tissues.

## Spleen

The **spleen** is the largest of the lymphatic organs. It is located in the upper left portion of the abdominal cavity, just inferior to the diaphragm, posterior and lateral to the stomach (see fig. 16.12 and reference plates 4, 5, and 6).

The spleen resembles a large lymph node in that it is enclosed in connective tissue extending inward from the surface and partially subdividing the organ into chambers, or lobules. The organ also has a hilum on one surface through which blood vessels and nerves enter. However, unlike the sinuses of a lymph node, the spaces (venous sinuses) within the chambers of the spleen are filled with blood instead of lymph.

The tissues within the lobules of the spleen are of two types. The *white pulp* is distributed throughout the spleen in tiny islands. This tissue is composed of nodules (splenic nodules), which are similar to those in lymph nodes and contain many lymphocytes. The *red pulp,* which fills the remaining spaces of the lobules, surrounds the venous sinuses. This pulp contains numerous red blood cells, which impart its color, plus many lymphocytes and macrophages (figs. 16.14 and 16.15).

Blood capillaries within the red pulp are quite permeable. Red blood cells can squeeze through the pores in these capillary walls and enter the venous sinuses. The older, more fragile red blood cells may rupture as they make this passage, and the resulting cellular debris is removed by phagocytic macrophages within the splenic sinuses.

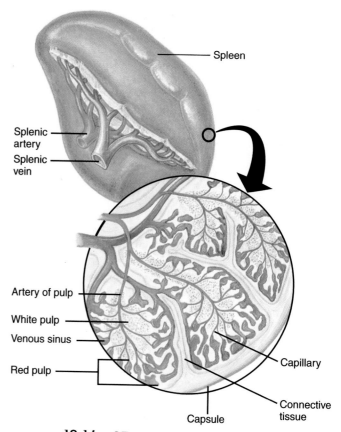

**FIGURE 16.14**

The spleen resembles a large lymph node.

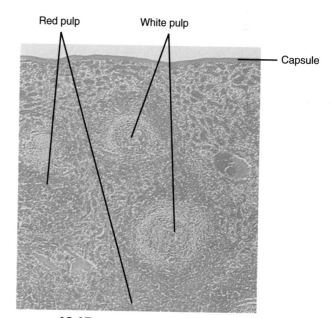

**FIGURE 16.15**

Light micrograph of the spleen (15×).

During fetal development, pulp cells of the spleen produce blood cells, much as red bone marrow cells do after birth. As the time of birth approaches, this splenic function ceases. However, in certain diseases, such as *erythroblastosis fetalis,* in which large numbers of red blood cells are destroyed, the splenic pulp cells may resume their hematopoietic activity.

The macrophages also engulf and destroy foreign particles, such as bacteria, that may be carried in the blood as it flows through the sinuses. The lymphocytes of the spleen, like those of the thymus, lymph nodes, and nodules, also help defend the body against infections. Thus, the spleen filters blood much as the lymph nodes filter lymph. Table 16.1 summarizes the characteristics of the major organs of the lymphatic system.

**1** Why are the thymus and spleen considered organs of the lymphatic system?

**2** What are the major functions of the thymus and the spleen?

## BODY DEFENSES AGAINST INFECTION

The presence and multiplication of a disease-causing agent, or **pathogen,** causes an **infection.** Pathogens include simple microorganisms (bacteria), complex microorganisms such as protozoa, and even spores of multicellular organisms, such as fungi. Viruses are pathogens, but they are considered to be infectious agents and not organisms, because their structure is far simpler than that of a living cell and they must infect a living cell to reproduce. An infection may be present even though an individual feels well. People who are infected with the human immunodeficiency virus (HIV), which causes AIDS, often live for many years in good health before becoming ill.

The human body can prevent entry of pathogens, or destroy them if they enter. Some mechanisms are quite general and protect against many types of pathogens, providing **nonspecific defense.** These mechanisms include species resistance, mechanical barriers, enzyme action, interferon, fever, inflammation, and phagocytosis. Other defense mechanisms are very precise, targeting specific pathogens. These mechanisms provide **specific defense,** or immunity. They are carried out by specialized lymphocytes that recognize foreign molecules (nonself antigens) in the body and respond to them.

| TABLE 16.1 | MAJOR ORGANS OF THE LYMPHATIC SYSTEM | |
|---|---|---|
| **Organ** | **Location** | **Function** |
| Lymph nodes | In groups or chains along the paths of larger lymphatic vessels | Filter foreign particles and debris from lymph; produce and house lymphocytes that destroy foreign particles in lymph; house macrophages that engulf and destroy foreign particles and cellular debris carried in lymph |
| Thymus | Within the mediastinum posterior to the upper portion of the sternum | Houses lymphocytes; differentiates thymocytes into T lymphocytes |
| Spleen | In upper left portion of abdominal cavity inferior to the diaphragm, posterior and lateral to the stomach | Houses macrophages that remove foreign particles, damaged red blood cells, and cellular debris from the blood; contains lymphocytes |

## NONSPECIFIC DEFENSES

### Species Resistance

**Species resistance** refers to the fact that a given kind of organism, or species (such as the human species, *Homo sapiens*), develops diseases that are unique to it. A species may be resistant to diseases that affect other species, because its tissues somehow fail to provide the temperature or chemical environment that a particular pathogen requires. For example, humans are subject to infections by the infectious agents that cause measles, mumps, gonorrhea, and syphilis, but other animal species are not. Similarly, humans are resistant to certain forms of malaria and tuberculosis that affect birds. However, new influenza strains that affect humans may come from birds, especially poultry.

In San Francisco in 1982, Simon Guzman became one of the very first recorded individuals to succumb to AIDS. He was the first person whose death was attributed to a parasitic infection previously seen only in sheep—*cryptosporidiosis.* This infection, which causes relentless diarrhea, illustrates one of the most bizarre effects of AIDS: alteration of species resistance.

| | |
|---|---|
| **TABLE 16.2** | **MAJOR ACTIONS THAT MAY OCCUR DURING AN INFLAMMATION RESPONSE** |

*Blood vessels dilate.*
*Capillary and venule permeability increase.*
   Tissues become red, swollen, warm, and painful.
*White blood cells invade the region.*
   Pus may form as white blood cells, bacterial cells, and cellular debris accumulate.
*Body fluids seep into the area.*
   A clot containing threads of fibrin may form.
*Fibroblasts appear.*
   A connective tissue sac may form around the injured tissues.
*Phagocytes are active.*
   Bacteria, dead cells, and other debris are removed.
*Cells reproduce.*
   Newly formed cells replace injured ones.

## Mechanical Barriers

The skin and mucous membranes lining the passageways of the respiratory, digestive, urinary, and reproductive systems create **mechanical barriers** that prevent the entrance of some infectious agents. As long as these barriers remain intact, many pathogens are unable to penetrate them. Another protection is that the epidermis sloughs off, removing superficial bacteria with it. In addition, the mucus-coated ciliated epithelium, described in chapter 19, that lines the respiratory passages entraps particles and sweeps them out of the airways and into the pharynx, where they are swallowed.

## Chemical Barriers

Enzymes in body fluids provide a **chemical barrier** to pathogens. Gastric juice, for example, contains the protein-splitting enzyme pepsin, and has a low pH due to the presence of hydrochloric acid. The combined effect of pepsin and hydrochloric acid kills many pathogens that enter the stomach. Similarly, tears contain the enzyme lysozyme, which has an antibacterial action against certain pathogens that may get onto the surfaces of the eyes. The accumulation of salt from perspiration also kills certain bacteria on the skin.

    **Interferons** are hormonelike peptides produced by certain cells, including lymphocytes and fibroblasts. Once released from the virus-infected cell, interferon binds to receptors on uninfected cells, stimulating them to synthesize antiviral proteins that block viral replication of a variety of viruses. Thus, the effect of interferon is nonspecific. Interferons also stimulate phagocytosis and enhance the activity of certain other cells that help to resist infections and the growth of tumors.

## Fever

A **fever** offers powerful protection. A viral or bacterial infection stimulates lymphocytes to proliferate, producing cells that secrete a substance called *endogenous pyrogen* ("fire maker from within"), also called IL-1. Endogenous pyrogen raises the thermoregulatory set point in the brain's hypothalamus to maintain a higher body temperature. Fever indirectly counters microbial growth because higher body temperature causes the liver and spleen to sequester iron, which reduces the level of iron in the blood. Since bacteria and fungi require more iron as temperature rises, their growth and reproduction in a fever-ridden body slows and may cease. Plus, phagocytic cells attack more vigorously when the temperature rises.

## Inflammation

**Inflammation** is a tissue response to pathogen invasion (see chapter 5) that produces localized redness, swelling, heat, and pain. The redness is a result of blood vessel dilation and the consequent increase of blood flow and volume within the affected tissues (hyperemia). This effect, coupled with an increase in permeability of nearby capillaries and venules, swells tissues (edema). The heat comes from blood from deeper body parts, which is generally warmer than that near the surface. Pain results from stimulation of nearby pain receptors.

    White blood cells accumulate at the sites of inflammation, where some of them help control pathogens by phagocytosis. In bacterial infections, the resulting mass of white blood cells, bacterial cells, and damaged tissue may form a thick fluid called *pus.*

    Body fluids (exudate) also collect in inflamed tissues. These fluids contain fibrinogen and other clotting factors that may stimulate a network of fibrin threads to form within the affected region. Later, fibroblasts may arrive and form fibers around the area until it is enclosed in a sac of fibrous connective tissue. This action inhibits the spread of pathogens and toxins to adjacent tissues.

    Once an infection is controlled, phagocytic cells remove dead cells and other debris from the site of inflammation. Cell reproduction replaces lost cells. Table 16.2 summarizes the process of inflammation.

TABLE 16.3

## TYPES OF NONSPECIFIC DEFENSES

| Type | Description |
|------|-------------|
| Species resistance | A species or organism is resistant to certain diseases to which other species are susceptible. |
| Mechanical barriers | Unbroken skin and mucous membranes prevent the entrance of some infectious agents. |
| Chemical barriers | Enzymes in various body fluids kill pathogens. pH and high salt concentration also harm pathogens. Interferons induce production of other proteins that block reproduction of viruses, stimulate phagocytosis, and enhance the activity of cells that resist infections and the growth of tumors. |
| Fever | Elevated body temperature aids recovery from bacterial infections. |
| Inflammation | A tissue response to injury that helps prevent the spread of infectious agents into nearby tissues. |
| Phagocytosis | Neutrophils, monocytes, and macrophages engulf and destroy foreign particles and cells. |

## Phagocytosis

Recall from chapter 14 that the most active phagocytic cells of the blood are *neutrophils* and *monocytes.* These cells can leave the bloodstream by squeezing between the cells of the blood vessel walls (diapedesis). Chemicals released from injured tissues attract these cells (chemotaxis). Neutrophils engulf and digest smaller particles; monocytes phagocytize larger ones.

Monocytes give rise to *macrophages* (histiocytes), which become fixed in various tissues and attach to the inner walls of blood and lymphatic vessels. Monocytes are relatively nonmotile and phagocytic, can produce new macrophages, and occupy lymph nodes, the spleen, liver, and lungs. A macrophage can engulf up to 100 bacteria, compared to the twenty or so bacteria that a neutrophil can engulf. Monocytes, macrophages, and neutrophils constitute the **mononuclear phagocytic system** (reticuloendothelial system).

Phagocytosis removes foreign particles from the lymph as it moves from the interstitial spaces to the bloodstream. Phagocytes in the blood vessels and in the tissues of the spleen, liver, or bone marrow usually remove particles that reach the blood. Table 16.3 summarizes the types of nonspecific defenses.

1 What is an infection?

2 Explain six nonspecific defense mechanisms.

## SPECIFIC DEFENSES (IMMUNITY)

**Immunity** is resistance to particular pathogens or to their toxins or metabolic by-products. Lymphocytes and macrophages that recognize specific foreign molecules (nonself antigens) carry out immune responses.

## Antigens

Before birth, cells inventory the proteins and other large molecules in the body, learning to identify these as "self." The lymphatic system responds to nonself, or foreign antigens, but not normally to self antigens. Receptors on T and B cell surfaces enable the cells to recognize foreign antigens.

**Antigens** may be proteins, polysaccharides, glycoproteins, or glycolipids. The antigens that are most effective in eliciting an immune response are large and complex, with few repeating parts. Sometimes, a smaller molecule that cannot by itself stimulate an immune response combines with a larger one, which makes it able to do so (antigenic). Such a small molecule is called a **hapten.** Stimulated lymphocytes react either to the hapten or to the larger molecule of the combination. Haptens are found in certain drugs, such as penicillin; in household and industrial chemicals; in dust particles; and in products of animal skins (dander).

## Lymphocyte Origins

During fetal development, red bone marrow releases undifferentiated lymphocytes into the circulation. About half of these cells reach the thymus, where they remain for a time. Here, these thymocytes specialize into T lymphocytes, or **T cells.** ("T" refers to *thymus-derived* lymphocytes.) Later, the blood transports T cells, where they comprise 70% to 80% of the circulating lymphocytes. T cells reside in lymphatic organs and are particularly abundant in the lymph nodes, the thoracic duct, and the white pulp of the spleen.

Other lymphocytes are thought to remain in the red bone marrow until they differentiate into B lymphocytes or **B cells.** (Historically, the "B" stands for *bursa of Fabricius,* an organ in the chicken where these cells were first identified.) The blood distributes B cells, which constitute 20% to 30% of circulating lymphocytes. B cells settle in lymphatic organs along with T cells and are abundant in lymph nodes, the spleen, bone marrow, and intestinal lining (figs. 16.16 and 16.17).

1 What is immunity?

2 What is the difference between an antigen and a hapten?

3 How do T cells and B cells originate?

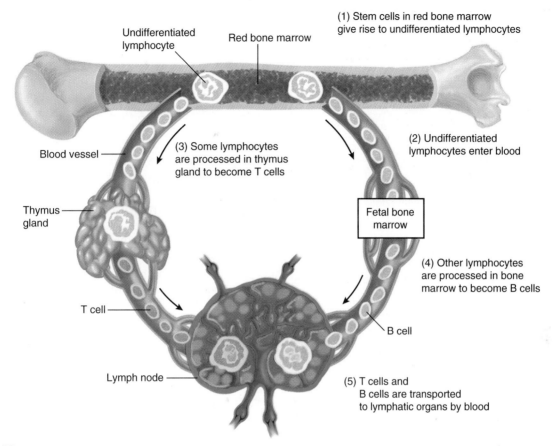

**FIGURE 16.16**

Bone marrow releases undifferentiated lymphocytes, which after processing become T lymphocytes (T cells) or B lymphocytes (B cells). Note that the diaphysis contains red marrow in the fetus.

**FIGURE 16.17**

Scanning electron micrograph of a human circulating lymphocyte (8,600×).

## T Cells and Cell-Mediated Immunity

T cells attach to foreign, antigen-bearing cells, such as bacterial cells, and interact directly—that is, by cell-to-cell contact. This type of response is called **cell-mediated immunity** (CMI).

T cells (and some macrophages) also synthesize and secrete polypeptides called **cytokines** (or more specifically, lymphokines) that enhance certain cellular responses to antigens. For example, *interleukin-1* and *interleukin-2* stimulate synthesis of several cytokines from other T cells. In addition, interleukin-1 helps activate T cells, whereas interleukin-2 causes T cells to proliferate and activates a certain type of T cell (cytotoxic T cells). Other cytokines called *colony-stimulating factors* (CSFs) stimulate production of leukocytes in the red bone marrow, cause B cells to grow and mature, and activate macrophages. Table 16.4 summarizes several cytokine types.

T cells may also secrete toxins that kill their antigen-bearing target cells; growth-inhibiting factors that prevent target-cell growth, or interferon that inhibits proliferation of viruses and tumor cells.

## TABLE 16.4 TYPES OF CYTOKINES

| Cytokine | Function |
|---|---|
| Colony-stimulating factors | Stimulate bone marrow to produce lymphocytes |
| Interferons | Block viral replication, stimulate macrophages to engulf viruses, stimulate B cells to produce antibodies, attack cancer cells |
| Interleukins | Control lymphocyte differentiation and growth |
| Tumor necrosis factor | Stops tumor growth, releases growth factors, causes fever that accompanies bacterial infection, stimulates lymphocyte differentiation |

## T Cell Types and Activation

T cell activation requires the presence of processed fragments of antigen attached to the surface of another kind of cell, called an **antigen-presenting cell** (or accessory cell). Macrophages, B cells, and several other cell types can serve as antigen-presenting cells.

The three major types of T cells are helper cells, memory cells, and cytotoxic cells. *Helper T cells* mobilize the immune system to stop a bacterial infection through a complex series of steps. First, a macrophage phagocytizes a bacterium, digesting it in its lysosomes. Some bacterial antigens exit the lysosomes and move to the macrophage's surface. Here, they are displayed on the cell membrane near certain protein molecules that are part of the "self" topography of the cell's surface. These self proteins belong to a group called the *major histocompatibility complex* (MHC). Then, a helper T cell contacts a displayed foreign antigen. If the displayed antigen fits and combines with the helper T cell's antigen receptors, the helper cell becomes activated. The role of the helper T cell in B cell activation is discussed later in the chapter.

A type of helper T cell called a CD4 cell is the prime target of HIV, the virus that causes AIDS. (CD4 stands for the "cluster-of-differentiation" antigen it bears that enables it to recognize a macrophage displaying a foreign antigen.) Considering the role of CD4 helper T cells as key players in establishing immunity—they stimulate B cells and secrete cytokines—it is no wonder that harming them destroys immunity.

*Memory T cells* are among the many T cells produced upon initial exposure to an antigen, but include only those cells not responding to the antigen at that time. These cells provide for a no-delay response to any future exposure to the same antigen, with immediate differentiation into cytotoxic T cells. This response generally vanquishes the invading organism before it has a chance to cause the body to produce signs and symptoms of disease.

A third major type of T cell is a *cytotoxic T cell*, which recognizes nonself antigens that cancerous cells or virally infected cells display on their surfaces near certain MHC proteins. A cytotoxic T cell becomes activated when it combines with an antigen that fits its receptors. Then, the T cell proliferates, enlarging its **clone** of cells (identical cells originating from reproduction of a single cell). Cytotoxic T cells then bind to the surfaces of antigen-bearing cells, where they release a protein called **perforin** that cuts porelike openings, destroying these cells. Another type of cell, called a **natural killer cell**, also uses perforin to destroy tumor cells. Both cytotoxic T cells and natural killer cells have other ways to lyse cells too. Cytotoxic T cells continually monitor the body's cells, recognizing and eliminating tumor cells and cells infected with viruses.

1. How do T cells become activated?

2. What is the function of cytokines?

3. How do cytotoxic T cells destroy antigen-bearing cells?

## B Cells and Antibody-Mediated Immunity

B cells attack foreign antigens in a different way than T cells. They differentiate into *plasma* cells, which produce and secrete large globular proteins called **antibodies,** which are also called **immunoglobulins.** A plasma cell is a virtual antibody factory, as evidenced by its characteristically huge Golgi apparatus. At the peak of an infection, a plasma cell may produce and secrete 2,000 antibody molecules a second! Body fluids carry antibodies, which then react in various ways to destroy specific antigens or antigen-bearing particles. This type of response is called **antibody-mediated immunity** (AMI), or humoral immunity. ("Humoral" refers to fluid.)

### B Cell Activation

A lymphocyte must be activated before it can respond to an antigen. A B cell, for example, may become activated when it encounters an antigen whose molecular shape fits the shape of the B cell's antigen receptors. In response to the receptor-antigen combination, the B cell reproduces repeatedly, expanding its clone. However, most antigens require T cell "help" to activate B cells.

When an activated helper T cell encounters a B cell that has already combined with an identical foreign antigen, the helper cell releases certain cytokines. These cytokines stimulate the B cell to proliferate, thus enlarging its clone of antibody-producing cells (figs. 16.18 and 16.19). T cells can also suppress antibody formation by releasing cytokines that inhibit B cell function.

Some members of the activated B cell's clone differentiate further into plasma cells, which secrete antibodies. These antibodies are similar in structure to the antigen-receptor molecules present on the original B cell's surface.

FIGURE 16.18

When a B cell encounters an antigen that fits its antigen receptor, it becomes activated and proliferates, thus enlarging its clone. Note that all cells in the clone have the same antigen receptor.

Thus, antibodies can combine with the antigen-bearing agent that has invaded the body, and react against it. Table 16.5 summarizes the steps leading to antibody production as a result of B and T cell activities.

A single type of B cell carries information to produce a single type of antibody. However, different B cells respond to different antigens on a pathogen's surface. Therefore, an immune response may include several types of antibodies manufactured against a single microbe or virus. This is called a **polyclonal response.** Clinical Application 16.1 discusses how researchers use clones of single B cells to produce single, or monoclonal, antibodies.

The human body can manufacture an apparently limitless number of different antibodies, but we have a limited number of antibody genes. The great diversity of antibody types is possible because different antibody protein subunits combine. During the early development of B cells, sections of their antibody genes randomly move to other chromosomal locations, creating new genetic instructions for antibodies. Antibody diversity is like using the limited number of words in a language to compose an infinite variety of stories.

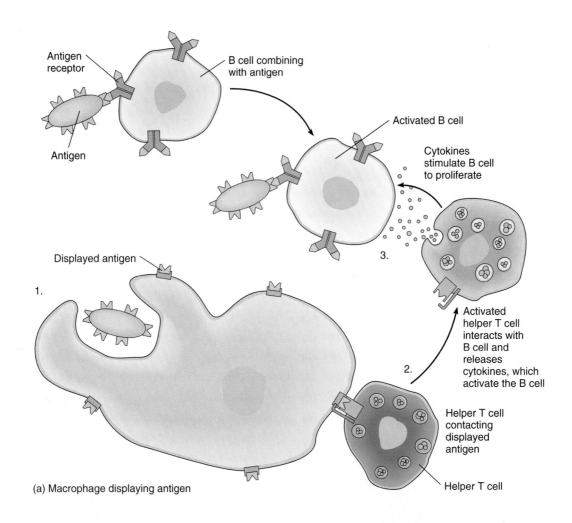

Antigen receptor

B cell combining with antigen

Antigen

Activated B cell

Cytokines stimulate B cell to proliferate

Displayed antigen

1.

3.

Activated helper T cell interacts with B cell and releases cytokines, which activate the B cell

2.

Helper T cell contacting displayed antigen

Helper T cell

(a) Macrophage displaying antigen

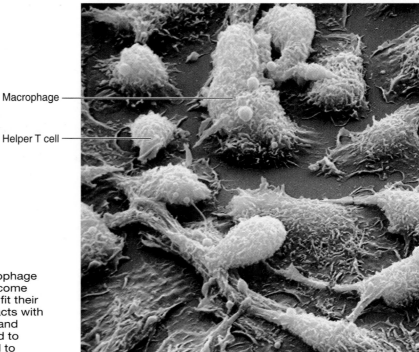

Macrophage

Helper T cell

(b)

## FIGURE 16.19

(a) 1. After digesting antigen-bearing agents, a macrophage displays antigens on its surface. 2. Helper T cells become activated when they contact displayed antigens that fit their antigen receptors. 3. An activated helper T cell interacts with a B cell that has combined with an identical antigen and causes the B cell to proliferate. (b) Macrophages bind to lymphocytes. During an infection, macrophages bind to helper T cells, activating them to trigger other immune defenses. Here, the round cells are helper T cells and the cells bearing projections are macrophages (1,040×).

## TABLE 16.5 STEPS IN ANTIBODY PRODUCTION

### B Cell Activity

1. Antigen-bearing agents enter tissues.
2. B cell becomes activated when it encounters an antigen that fits its antigen receptors, either alone or more often in conjunction with helper T cells.
3. Activated B cell proliferates, enlarging its clone.
4. Some of the newly formed B cells differentiate further to become plasma cells.
5. Plasma cells synthesize and secrete antibodies whose molecular structure is similar to the activated B cell's antigen receptors.
6. Antibodies combine with antigen-bearing agents, helping to destroy them.

### T Cell Activity

1. Antigen-bearing agents enter tissues.
2. Accessory cell, such as a macrophage, phagocytizes antigen-bearing agent, and the macrophage's lysosomes digest the agent.
3. Antigens from the digested antigen-bearing agents are displayed on the surface membrane of the accessory cell.
4. Helper T cell becomes activated when it encounters a displayed antigen that fits its antigen receptors.
5. Activated helper T cell releases cytokines when it encounters a B cell that has previously combined with an identical antigen-bearing agent.
6. Cytokines stimulate the B cell to proliferate.
7. Some of the newly formed B cells differentiate into antibody-secreting plasma cells.
8. Antibodies combine with antigen-bearing agents, helping to destroy them.

## Antibody Molecules

Antibodies are soluble, globular proteins that constitute the *gamma globulin* fraction of plasma proteins (see chapter 14). They are also called immunoglobulins.

Each immunoglobulin molecule is composed of four chains of amino acids that are linked together by pairs of sulfur atoms (disulfide bonds). Two of these chains are identical **light chains** (L-chains), and two are identical **heavy chains** (H-chains). The heavy chains contain about twice as many amino acids as the light chains. The five major types of immunoglobulin molecules each contain a particular kind of heavy chain.

As with other proteins, the sequences of amino acids of the heavy and light chains confer the unique, three-dimensional structure of each immunoglobulin. This special conformation, in turn, imparts the physiological properties of the molecule. For example, one end of each of the heavy and light chains consists of variable sequences of amino acids (variable regions). These regions are specialized to react with the shape of a specific antigen molecule.

Antibodies can bind to certain antigens because of the three-dimensional shape of the variable regions. The antibody contorts to form a pocket around the antigen.

**FIGURE 16.20**

An immunoglobulin molecule consists basically of two identical light chains of amino acids and two identical heavy chains of amino acids.

These specialized ends of the antibody molecule are called **antigen-binding sites,** and the particular parts that actually bind the antigen are called **idiotypes.**

The remaining portions of the chains are termed constant regions because their amino acid sequences are very similar from molecule to molecule. Constant regions impart other properties of the immunoglobulin molecule, such as its ability to bond to cellular structures or to combine with certain chemicals (fig. 16.20).

### Types of Immunoglobulins

Of the five major types of immunoglobulins, three constitute the bulk of the circulating antibodies. They are immunoglobulin G, which accounts for about 80% of the antibodies; immunoglobulin A, which makes up about 13%; and immunoglobulin M, which is responsible for about 6%. Immunoglobulin D and immunoglobulin E account for the remainder of the antibodies.

*Immunoglobulin G* (IgG) is in plasma and tissue fluids, and is particularly effective against bacteria, viruses, and toxins. IgG also activates a group of enzymes called **complement,** which is described in the following section.

*Immunoglobulin A* (IgA) is commonly found in exocrine gland secretions. It is in breast milk, tears, nasal fluid, gastric juice, intestinal juice, bile, and urine.

---

A newborn does not yet have its own antibodies, but does retain IgG that passed through the placenta from the mother. These maternal antibodies protect the infant against some illnesses to which the mother is immune. Just as the maternal antibody supply falls, the infant begins to manufacture its own. The newborn receives IgA from colostrum, a substance secreted from the mother's breasts for the first few days after birth. Antibodies in colostrum protect against certain digestive and respiratory infections.

---

# Immunotherapy

At the turn of the last century, German bacteriologist Paul Ehrlich developed the concept of the "magic bullet"—a substance that could enter the body and destroy diseased cells, yet spare the healthy ones. The biochemicals and cells of the immune system, with their great specificity for attacking nonself tissue, would be ideal magic bullets. *Immunotherapy* uses immune system components to fight disease—both humoral immunity (antibodies) and cell-mediated immunity (cytokines).

## MONOCLONAL ANTIBODIES— TARGETING IMMUNITY

Tapping the specificity of a single B cell, and using its single type, or *monoclonal,* antibody to target a specific antigen (such as on a cancer or bacterial cell) awaited finding a way to entice the normally short-lived mature B cells into persisting in culture. In 1975, British researchers Cesar Milstein and Georges Kohler devised *monoclonal antibody* (MAb) technology, an ingenious way to capture the antibody-making capacity of a single B cell.

Milstein and Kohler injected a mouse with antigen-laden red blood cells from a sheep. They then isolated a single B cell from the mouse's spleen, and fused it with a cancerous white blood cell from a mouse. The result was a fused cell, or *hybridoma,* with a valuable pair of talents: like the B cell, it produces large amounts of a single antibody type; like the cancer cell, it reproduces continuously (fig. 16A).

MAbs are used in basic research, veterinary and human health care, and agriculture. Cell biologists use pure antibodies to localize and isolate proteins. Diagnostic MAb "kits" detect tiny amounts of a single molecule. Most kits consist of a paper strip impregnated with a MAb, to which the user adds a body fluid. For example, a woman who suspects she is pregnant places drops of her urine onto the paper. A color change ensues if the MAb binds to human chorionic gonadotropin (see chapter 22), indicating pregnancy.

MAbs can highlight cancer before it can be detected by other means. The MAb is attached to a radioactive chemical, which is then detected when the MAb binds an antigen unique to the cancer cell surfaces. Detecting a cancer's recurrence with a MAb requires only an injection followed by a painless imaging procedure.

MAbs can ferry conventional cancer treatments to where they are needed and limit their toxicity by sparing healthy tissue. Drugs or radioactive chemicals are attached to MAbs that deliver them to antigens on cancer cells. When injected into a patient, the MAb and its cargo are engulfed by the cancer cells, which are destroyed. Although MAbs were originally derived from mice, human versions that prevent allergic reactions are now available .

## CYTOKINES

Immunotherapy experiments were difficult to do in the late 1960s because cytokines and antibodies could only be obtained in small amounts from cadavers. In the 1970s, recombinant DNA and monoclonal antibody technologies made it possible to make unlimited amounts of pure proteins—just as the AIDS

---

*Immunoglobulin M* (IgM) is a type of antibody that develops in the blood plasma, in response to contact with certain antigens in foods or bacteria. The antibodies anti-A and anti-B, described in chapter 14, are examples of IgM. IgM also activates complement.

*Immunoglobulin D* (IgD) is found on the surfaces of most B cells, especially those of infants. IgD is important in activating B cells (see figure 16.18).

*Immunoglobulin E* (IgE) appears in exocrine secretions along with IgA. It is associated with allergic reactions that are described later in this chapter. Table 16.6 summarizes the major immunoglobulins and their functions.

**1** What is an immunoglobulin?

**2** Describe the structure of an immunoglobulin molecule.

**3** Name the five major types of immunoglobulins.

## Antibody Actions

In general, antibodies react to antigens in three ways. Antibodies directly attack antigens, activate a set of enzymes epidemic was making it essential to find a purer source of biochemicals than cadavers!

Interferon was the first cytokine to be tested on a large scale. Although it did not live up to early expectations of being a wonder drug, it does treat a dozen or so conditions, including a type of leukemia, multiple sclerosis, hepatitis, and genital warts.

In another cytokine-based cancer treatment, certain T cells are removed from tumor samples and incubated with interleukin-2, which the patient also receives intravenously. The removed T cells, activated by the cytokine, are injected into the patient. Interleukin-2 is also used to treat kidney cancer. In another approach, colony-stimulating factors can boost the white blood cell supply in people whose immune systems are temporarily suppressed, such as those receiving drugs to treat cancer or AIDS, or transplant recipients.

Increasingly, cancer treatment consists of combinations of immune system cells or biochemicals, plus standard therapies. Immunotherapy can enable a patient to withstand higher doses of a conventional drug, or destroy cancer cells remaining after standard treatment.

The idea of using the immune system to treat cancer is not new. In 1893, a New York surgeon, William Coley, intentionally infected cancer patients with killed *Streptococcus* bacteria, after noticing that some patients with bacterial infections spontaneously recovered from the cancer. Sometimes "Coley's toxins" worked, although he did not understand their mechanism.

Today we know that an immune system biochemical called *tumor necrosis factor* often underlies the body's success in overcoming cancer. However, researchers do not yet understand all of this molecule's effects sufficiently to safely use it as an anticancer drug. ■

**FIGURE 16A**

Monoclonal antibodies are produced by a type of artificial cell combination called a hybridoma. It consists of a cancer cell (the flat blue cell) fused with a B cell (the round green cell). The cancer cell contributes rapid and continuous division; the B cell secretes a single antibody type (7,000×).

| TABLE 16.6 | CHARACTERISTICS OF MAJOR IMMUNOGLOBULINS | |
|---|---|---|
| **Type** | **Occurrence** | **Major Function** |
| IgG | Tissue fluid and plasma | Defends against bacterial cells, viruses, and toxins; activates complement |
| IgA | Exocrine gland secretions | Defends against bacterial cells and viruses |
| IgM | Plasma | Reacts with antigens occurring naturally on some red blood cell membranes following certain blood transfusions; activates complement |
| IgD | Surface of most B lymphocytes | B cell activation |
| IgE | Exocrine gland secretions | Promotes allergic reactions |

**FIGURE 16.21** ▭

Actions of the complement system.

| TABLE 16.7 | ACTIONS OF ANTIBODIES | |
|---|---|---|
| **General Action** | **Type of Effect** | **Description** |
| ***Direct Attack*** | | |
| | Agglutination | Antigens clump |
| | Precipitation | Antigens become insoluble |
| | Neutralization | Antigens lose toxic properties |
| ***Activation of Complement*** | | |
| | Opsonization | Alters cell membranes so cells are more susceptible to phagocytosis |
| | Chemotaxis | Attracts macrophages and neutrophils into region |
| | Inflammation | Promotes local tissue changes that help prevent spread of antigens |
| | Lysis | Cell membranes rupture |

(*complement*) that attack the antigens, or stimulate changes in local areas that help prevent spread of the antigens.

In a direct attack, antibodies combine with antigens and cause them to clump together (agglutinate) or to form insoluble substances (precipitation). Such actions make it easier for phagocytic cells to engulf the antigen-bearing

agents and eliminate them. In other instances, antibodies cover the toxic portions of antigen molecules and neutralize their effects (neutralization). However, under normal conditions, complement activation is more important in protecting against infection than direct antibody attack.

Complement is a group of proteins (complement system) in plasma and other body fluids. When certain IgG or IgM antibodies combine with antigens, they expose reactive sites on the constant regions of antibody molecules. This triggers a series of reactions leading to activation of the complement proteins, which in turn produce a variety of effects, including: coating the antigen-antibody complexes (opsonization), making the complexes more susceptible to phagocytosis; attracting macrophages and neutrophils into the region (chemotaxis); clumping antigen-bearing agents; rupturing membranes of foreign cells; and altering the molecular structure of viruses, making them harmless (fig. 16.21). Other proteins promote inflammation, which helps prevent the spread of infectious agents.

Immunoglobulin E promotes inflammation that may be so intense that it damages tissues. This antibody is usually attached to the membranes of widely distributed mast cells (see chapter 5). When antigens combine with the antibodies, the resulting antigen-antibody complexes stimulate mast cells to release biochemicals, such as histamine, that cause the changes associated with inflammation, such as vasodilation and edema.

Table 16.7 summarizes the actions of antibodies.

**1** In what general ways do antibodies function?

**2** What is the function of complement?

**3** How is complement activated?

## Immune Responses

When B cells or T cells become activated after first encountering the antigens for which they are specialized to react, their actions constitute a **primary immune response.** During such a response, plasma cells release antibodies (IgM, followed by IgG) into the lymph (fig. 16.22). The antibodies are transported to the blood and then throughout the body, where they help destroy antigen-bearing agents. Production and release of antibodies continues for several weeks.

Following a primary immune response, some of the B cells produced during proliferation of the clone remain dormant and serve as *memory cells* (fig. 16.22). If the identical antigen is encountered in the future, the clones of these memory cells enlarge, and they can respond rapidly with IgG to the antigen to which they

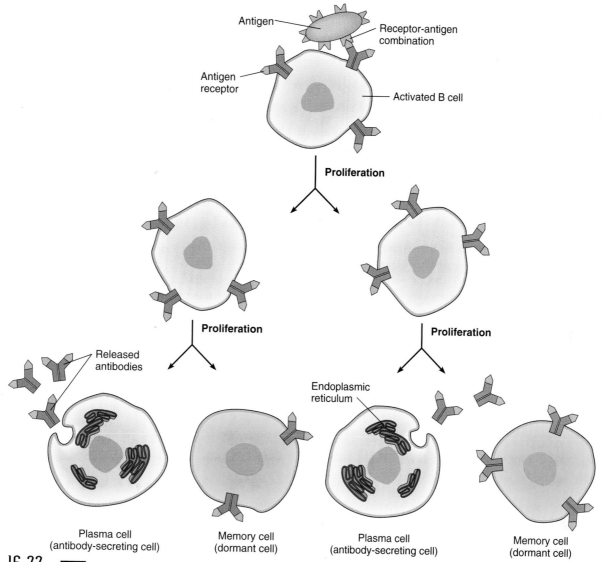

Antigen

Receptor-antigen combination

Antigen receptor

Activated B cell

**Proliferation**

**Proliferation**

**Proliferation**

Released antibodies

Endoplasmic reticulum

Plasma cell (antibody-secreting cell)

Memory cell (dormant cell)

Plasma cell (antibody-secreting cell)

Memory cell (dormant cell)

**FIGURE** 16.22

Activated B cells proliferate, giving rise to antibody-secreting plasma cells and dormant memory cells.

were previously sensitized. These memory B cells along with the previously discussed memory T cells produce a **secondary immune response.** Cells in lymph nodes called **follicular dendritic cells** may help memory by harboring and slowly releasing viral antigens after an initial infection. This constantly stimulates memory B cells, which present the antigens to T cells, maintaining immunity.

As a result of a primary immune response, detectable concentrations of antibodies usually appear in the body fluids within five to ten days following an exposure to antigens. If the identical antigen is encountered some time later, a secondary immune response may produce additional antibodies within a day or two (fig. 16.23). Although newly formed antibodies may

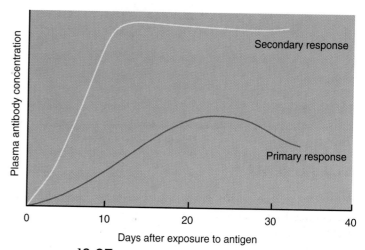

**FIGURE** 16.23

A primary immune response produces a lesser concentration of antibodies than does a secondary immune response.

persist in the body for only a few months, or perhaps a few years, memory cells live much longer. Consequently, the ability to produce a secondary immune response may be long-lasting.

Each year, a few children are born defenseless against infection due to *severe combined immune deficiency* (SCID), in which neither T nor B cells function. David Vetter was one such youngster. Born in Texas in 1971, he had no thymus gland and spent the 12 years of his life in a vinyl bubble, awaiting a treatment that never came. As David reached adolescence, he wanted to leave his bubble. An experimental bone marrow transplant was unsuccessful—soon afterwards David began vomiting and developed diarrhea, which are signs of infection. He left the bubble but died within days of a massive infection. Today, some children with SCID caused by an inherited enzyme deficiency can be treated with enzyme replacement therapy or experimental gene therapy.

## Types of Immunity

A generation ago, it was common at certain times of the year for grade school classrooms to be nearly empty, due to several infectious "childhood diseases," including measles, mumps, rubella, and chicken pox. However, each child usually suffered each illness only once, thanks to *naturally acquired active immunity.* This form of immunity develops after a primary immune response and is a response to exposure to a live pathogen and development of symptoms.

Today, most children in developed countries do not contract measles, mumps, or rubella because they develop another type of active immunity, produced in response to receiving a preparation called a **vaccine.** A vaccine contains an antigen that can stimulate a primary immune response against a particular disease-causing agent, but does not produce the severe symptoms of that disease.

A vaccine might contain bacteria or viruses that have been killed or attenuated (weakened) so that they cannot cause a serious infection, or it may contain a toxoid (toxin from an infectious organism that has been chemically altered to destroy its toxic effects). A vaccine may even consist of a single glycoprotein or similar large molecule from the pathogen's surface. This may be sufficient evidence of a foreign antigen to alert the immune system. A vaccine causes a person to develop *artificially acquired active immunity.*

Vaccines stimulate active immunity against a variety of diseases, including typhoid fever, cholera, whooping cough, diphtheria, tetanus, polio, chicken pox, measles (rubeola), German measles (rubella), mumps, influenza, hepatitis B, and bacterial pneumonia. Vaccines have virtually eliminated smallpox from the world, and greatly diminished the number of cases of poliomyelitis and infection by *Haemophilus influenzae* type B, which causes meningitis in preschool children. Unfortunately, distribution of vaccines is not equitable worldwide. Many thousands of people living in underdeveloped countries die of infectious diseases for which vaccines are widely available in other nations.

Viruses that rapidly mutate (change) present a great challenge to vaccine development, because their surfaces, which serve as antigens, become altered. It is a little like fighting an enemy who is constantly changing disguises. For this reason, a new vaccine against influenza must be developed each year. HIV, the virus that causes AIDS, is particularly changeable, which has severely hampered efforts to produce a vaccine.

Sometimes a person who has been exposed to infection requires protection against a disease-causing microorganism, but lacks the time necessary to develop active immunity. This happens with hepatitis, a viral infection of the liver. In such a case, it may be possible to inject the person with antiserum (ready-made antibodies). These antibodies may be obtained from gamma globulin separated from the blood plasma of persons who have already developed immunity against the disease.

An injection of gamma globulin provides *artificially acquired passive immunity.* This type of immunity is called passive because the recipient's cells do not produce the antibodies. Such immunity is short-term, seldom lasting more than a few weeks. Furthermore, because the recipient's lymphocytes might not have time to react to the pathogens for which protection was needed, susceptibility to those pathogens may persist in the future.

During pregnancy, certain antibodies (IgG) pass from the maternal blood into the fetal bloodstream. Receptor-mediated endocytosis (see chapter 3) utilizing receptor sites on cells of the fetal yolk sac accomplishes the transfer (see chapter 23). These receptor sites bind to a region common to the structure of IgG molecules. After entering the fetal cells, the antibodies are secreted into the fetal blood. As a result, the fetus acquires limited immunity against the pathogens for which the pregnant woman has developed active immunities. Thus, the fetus has *naturally acquired passive immunity,* which may persist for six months to a year after birth. Table 16.8 summarizes the types of immunity.

TABLE 16.8  TYPES OF IMMUNITY

| Type | Mechanism | Result |
|------|-----------|--------|
| Naturally acquired active immunity | Exposure to live pathogens | Symptoms of a disease and stimulation of an immune response |
| Artificially acquired active immunity | Exposure to a vaccine containing weakened or dead pathogens or their components | Stimulation of an immune response without the severe symptoms of a disease |
| Artificially acquired passive immunity | Injection of gamma globulin containing antibodies | Immunity for a short time without stimulating an immune response |
| Naturally acquired passive immunity | Antibodies passed to fetus from pregnant woman with active immunity | Short-term immunity for infant, without stimulating an immune response |

Competence of the immune system declines with age. Elderly persons have a higher incidence of tumors and an increased susceptibility to infections, such as influenza, pneumonia, and tuberculosis. This decline in effectiveness is primarily due to a loss of T cells. B cell activity changes little with age.

**1** Distinguish between a primary and a secondary immune response.

**2** Explain the difference between active and passive immunities.

## Allergic Reactions

Both allergic reactions and immune responses entail sensitizing of lymphocytes or combining of antigens with antibodies. An allergic reaction, however, is an immune attack against a nonharmful substance, and can damage tissues. An allergy is also called a hypersensitivity reaction. One form of allergic reaction can occur in almost anyone, but another form affects only people who have inherited an ability to produce exaggerated immune responses. The antigens that trigger allergic responses are called **allergens.**

A *delayed-reaction allergy* (type IV) may affect anyone. It results from repeated exposure of the skin to certain chemicals—commonly, household or industrial chemicals or some cosmetics. After repeated contacts, the presence of the foreign substance activates T cells, many of which collect in the skin. The T cells and the macrophages they attract release chemical factors, which, in turn, cause eruptions and inflammation of the skin (dermatitis). This reaction is called *delayed* because it usually takes about 48 hours to occur.

Hypersensitivities that take 1 to 3 hours to develop include *antibody-dependent cytotoxic reactions* (type II) and *immune complex reactions* (type III). In an antibody-dependent cytotoxic reaction, an antigen binds to a specific body cell, stimulating phagocytosis and complement-mediated lysis of the antigen. A transfusion reaction to mismatched blood is a type II hypersensitivity reaction. In an immune complex reaction, widespread antigen-antibody complexes cannot be cleared. **Autoimmunity,** the loss of the ability to tolerate self antigens, illustrates this type of hypersensitivity reaction. It is discussed later in the chapter.

The *tuberculin skin test* is used to detect individuals who have tuberculosis (TB) or who have had it (or a closely related infection) or been exposed to it in the past. The test uses a tuberculin preparation called *purified protein derivative* (PPD), which is introduced into the superficial layers of the test subject's skin (Mantoux test). If the person's T cells have been sensitized to the antigens of the mycobacteria that cause tuberculosis, an allergic reaction (positive test result) occurs within 48 to 72 hours. In a positive reaction, a localized region of the skin and subcutaneous tissue hardens (indurates). The absence of this reaction (negative result) signifies that the person's T cells have not previously been exposed to the mycobacterial antigens.

An *immediate-reaction* (type I or anaphylactic) *allergy* occurs within minutes after contact with an allergen. Persons with this type of allergy have an inherited tendency to overproduce IgE antibodies in response to certain antigens. IgE normally comprises a minute fraction of plasma proteins.

An immediate-reaction allergy activates B cells, which become sensitized when the allergen is first encountered. Subsequent exposures to the allergen trigger allergic reactions. In the initial exposure, IgE attaches to the membranes of widely distributed mast cells and basophils. When a subsequent allergen-antibody reaction occurs, these cells release allergy mediators such as *histamine, prostaglandin D₂,* and *leukotrienes* (fig. 16.24). These substances cause a variety of physiological effects, including dilation of blood vessels, increased vascular permeability that swells tissues, contraction of bronchial and intestinal smooth muscles, and increased mucus production. The result is a severe inflammation reaction that is responsible for the symptoms of the allergy, such as hives, hay fever, asthma, eczema, or gastric disturbances.

Anaphylactic shock is a severe form of immediate-reaction allergy, in which mast cells release allergy mediators throughout the body. The person may at first feel an inexplicable apprehension, and then suddenly the entire body itches and breaks out in red hives. Vomiting and diarrhea may follow. The face, tongue, and larynx begin to swell, and breathing becomes difficult. Unless the person receives an injection of epinephrine (adrenaline) and sometimes a tracheotomy (an incision into the windpipe so that breathing is restored), he or she will lose consciousness and may die within 5 minutes. Anaphylactic shock most often results from an allergy to penicillin or insect stings. Fortunately, thanks to prompt medical attention and people who know they have allergies avoiding the allergens, fewer than 100 people a year actually die of anaphylactic shock.

---

Anaphylactic shock seems to be too severe a response to something as innocuous as an insect bite or penicillin. One theory of the origin of allergies, particularly anaphylactic shock, is that they evolved at a time when insect bites and the natural substances from which antibiotics such as penicillin are made threatened human survival. The observation that IgE protects against roundworm and flatworm infections, in addition to taking part in allergic reactions, supports the idea that this antibody class is a holdover from times past, when immunity requirements might have been different from what they are today.

---

## Transplantation and Tissue Rejection

When a car breaks down, replacing the damaged or malfunctioning part often fixes the trouble. The same is sometimes true for the human body. Transplanted tissues and organs include corneas, kidneys, lungs, pancreases, bone marrow, pieces of skin, livers, and hearts. The danger the immune system poses to transplanted tissue is that the recipient's cells may recognize the donor's tissues as foreign and attempt to destroy the transplanted tissue. Such a response is called a **tissue rejection reaction.** The transplanted tissue may also produce substances that harm the recipient's tissue, a response called graft-versus-host disease.

Tissue rejection resembles the cell-mediated response against a nonself antigen. The greater the antigenic difference between the cell surface molecules of the recipient tissues and the donor tissues, the more rapid and severe the rejection reaction. Matching donor and recipient tissues can minimize the rejection reaction. This means locating a donor whose tissues are antigenically similar to those of the person needing a transplant—a procedure much like matching the blood of a donor with that of a recipient before giving a blood transfusion.

The four major varieties of grafts (transplant tissue) include:

- *Isograft.* Tissue is taken from a genetically identical twin.

- *Autograft.* Tissue is taken from elsewhere in a person's body, as in the case of skin grafts. (Technically, this is not a transplant, because it occurs within an individual.)

- *Allograft.* Tissue comes from an individual who is not genetically identical to the recipient, but of the same species.

- *Xenograft.* Tissue comes from a different species, such as pigs and baboons.

Table 16.9 presents examples of transplants.

---

*Major histocompatibility complex* or MHC is a general term for a large group of cell surface proteins that determine "self" in many species. The human version of the MHC produces human leukocyte antigens, or HLA types, which are patterns of proteins first discovered on white blood cells, but now known to be present on all cells that have nuclei. HLA antigens are "typed" to determine the chances of acceptance of transplanted tissue. The more HLA proteins two individuals have in common, the greater the chances of a successful transplant between them. HLA typing is also used in forensics to match the pattern of cell surface proteins between a suspect and tissue left at a crime scene, and to establish biological relationships, such as between a child and several men who may be the father.

---

*Immunosuppressive drugs* are used to reduce rejection of transplanted tissues. These drugs interfere with the recipient's immune response by suppressing formation of antibodies or production of T cells, thereby reducing the humoral and cellular responses. Unfortunately, the use of immunosuppressive drugs can leave a recipient unprotected against infections. It is not uncommon

(a)

(b)

## FIGURE 16.24

(a) Immediate-reaction allergy. 1. B cells are activated when they contact an allergen. 2. An activated B cell differentiates into an antibody-secreting plasma cell. 3. Antibodies attach to mast cells. 4. When allergens are encountered, they combine with the antibodies on the mast cells. 5. The mast cells release allergy mediators, which cause the symptoms of the allergy attack. (b) A mast cell releases histamine granules (3,000×).

## TABLE 16.9 TRANSPLANT TYPES

| Type | Donor | Example |
|------|-------|---------|
| Isograft | Identical twin | Bone marrow transplant from healthy twin to twin who has leukemia. |
| Autograft | Self | Skin graft from one part of body to replace burned skin. |
| Allograft | Same species | Kidney transplant from relative or closely matched donor. |
| Xenograft | Different species | Heart valves from a pig. |

for a patient to survive a transplant but die of infection because of a weakened immune system. The first immunosuppressive drug, cyclosporin, was discovered in a soil sample from Switzerland in the early 1980s. New drugs are more effective at selectively suppressing only those parts of the immune response that target transplanted tissue.

---

Less drastic than an organ transplant is a cell implant, which consists of small pieces of tissue. Implants of liver cells may treat cirrhosis; pancreas cells may treat diabetes; skeletal muscle cells may replace heart muscle damaged in a heart attack or treat muscular dystrophy; adrenal gland cells may treat Parkinson's disease; and brain cells may treat Alzheimer's and Huntington diseases.

Several biotechnology companies produce skin, blood vessel, and other tissue replacements that consist of biocompatible polymers seeded with cells that secrete tissue-specific biochemicals. One company manufactures blood vessels from three cell layers, with no synthetic materials. Antigens are removed from the cells so that they do not stimulate the immune system. Many people store their own blood for future use. Stem cells taken from a newborn's umbilical cord can be stored and then cultured later to provide healthy bone marrow stem cells, either to the donor or a compatible recipient.

Healthy fetal tissue is less likely to evoke immune rejection than adult tissue and has already been shown to relieve symptoms of Parkinson's disease when used in transplants. Researchers are developing ways to culture cells from early embryos and coax them to form specialized tissues. Once such cultures persist, embryos would no longer be required to generate tissues for transplant.

## Autoimmunity

Sometimes the immune system fails to distinguish self from nonself, producing antibodies, called **autoantibodies,** and cytotoxic T cells that attack and damage tissues and organs of the body. This attack against self is called **autoimmunity.** The specific nature of an autoimmune disorder depends upon the cell types that are the target of the immune attack. Table 16.10 lists some autoimmune disorders.

Why might the immune system attack body tissues? Perhaps a virus, while replicating within a human cell, "borrows" proteins from the host cell's surface and incorporates them onto its own surface. When the immune system "learns" the surface of the virus to destroy it, it also learns to attack the human cells that normally bear the particular protein. Another explanation of autoimmunity is that somehow T cells escape their "education" in the thymus, never learning to distinguish self from nonself.

A third possible route of autoimmunity is when a nonself antigen coincidentally resembles a self antigen. This may explain juvenile diabetes, which is a deficiency of insulin, the hormone required to transport glucose in the blood into cells that use it. Juvenile diabetes strikes 1 in 500 people under the age of 20—they must inject insulin several times a day.

Part of a protein on insulin-producing cells matches part of bovine serum albumin (BSA), which is a protein in cow's milk. Might children with an allergy to cow's milk develop antibodies against BSA, which later on attack the similar-appearing pancreas cells, causing juvenile diabetes? An ongoing study is tracking the health of children with family histories of juvenile diabetes who have avoided drinking cow's milk, compared to similarly high-risk children who have drunk cow's milk. Clinical Application 16.2 discusses the immunity breakdown of AIDS.

---

Chronic fatigue syndrome is a poorly understood immune system imbalance. The condition begins suddenly, producing fatigue so great that getting out of bed is an effort. Chills, fever, sore throat, swollen glands, muscle and joint pain, and headaches are also symptoms. The various disabling aches and pains reflect an overactive immune system. Affected people have up to forty times the normal amount of interleukin-2, and too many cytotoxic T cells, yet too little interferon. It is as if the immune system mounts a defense, and then doesn't know when to shut it off. The cause of chronic fatigue syndrome is not known.

TABLE 16.10

## AUTOIMMUNE DISORDERS

| Disorder | Symptoms | Antibodies Against |
|---|---|---|
| Glomerulonephritis | Lower back pain | Kidney cell antigens that resemble streptococcal bacteria antigens |
| Graves disease | Restlessness, weight loss, irritability, increased heart rate and blood pressure | Thyroid gland antigens near thyroid-stimulating hormone receptor, causing overactivity |
| Juvenile diabetes | Thirst, hunger, weakness, emaciation | Pancreatic beta cells |
| Hemolytic anemia | Fatigue and weakness | Red blood cells |
| Myasthenia gravis | Muscle weakness | Receptors for nerve messages on skeletal muscle |
| Pernicious anemia | Fatigue and weakness | Binding site for vitamin B on cells lining stomach |
| Rheumatic fever | Weakness, shortness of breath | Heart cell antigens that resemble streptococcal bacteria antigens |
| Rheumatoid arthritis | Joint pain and deformity | Cells lining joints |
| Scleroderma | Thick, hard, pigmented skin patches | Connective tissue cells |
| Systemic lupus erythematosus | Red rash on face, prolonged fever, weakness, kidney damage | DNA, neurons, blood cells |
| Ulcerative colitis | Lower abdominal pain | Colon cells |

**1** How are allergic reactions and immune reactions similar yet different?

**2** How does a tissue rejection reaction involve an immune response?

**3** How is autoimmunity an abnormal functioning of the immune response?

## CLINICAL TERMS RELATED TO THE LYMPHATIC SYSTEM AND IMMUNITY

**asplenia** (ah-sple′ne-ah)   Absence of a spleen.

**immunocompetence** (im″u-no-kom′pe-tens)   Ability to produce an immune response to the presence of antigens.

**immunodeficiency** (im″u-no-de-fish′en-se)   Inability to produce an immune response.

**lymphadenectomy** (lim-fad″ĕ-nek′to-me)   Surgical removal of lymph nodes.

**lymphadenopathy** (lim-fad″ĕ-nop′ah-the)   Enlargement of the lymph nodes.

**lymphadenotomy** (lim-fad″ĕ-not′o-me)   Incision of a lymph node.

**lymphocytopenia** (lim″fo-si″to-pe′ne-ah)   Too few lymphocytes in the blood.

**lymphocytosis** (lim″fo-si″to′sis)   Too many lymphocytes in the blood.

**lymphoma** (lim-fo′mah)   Tumor composed of lymphatic tissue.

**lymphosarcoma** (lim″fo-sar-ko′mah)   Cancer within the lymphatic tissue.

**splenectomy** (sple-nek′to-me)   Surgical removal of the spleen.

**splenitis** (sple-ni′tis)   Inflammation of the spleen.

**splenomegaly** (sple″no-meg′ah-le)   Abnormal enlargement of the spleen.

**splenotomy** (sple-not′o-me)   Incision of the spleen.

**thymectomy** (thi-mek′to-me)   Surgical removal of the thymus.

**thymitis** (thi-mi′tis)   Inflammation of the thymus.

# CLINICAL APPLICATION

# Immunity Breakdown: AIDS

## NATURAL HISTORY OF A MODERN PLAGUE

In late 1981 and early 1982, physicians from large cities in the United States began reporting to the Centers for Disease Control and Prevention cases of formerly rare infections in otherwise healthy young men. Some of the infections were prevalent in the general population, such as herpes simplex and cytomegalovirus, but in these young men were unusually severe. Oddly, some of the infections were caused by organisms known to infect only nonhuman animals. Other infections, particularly pneumonia caused by the microorganism *Pneumocystis carinii* and a cancer, Kaposi's sarcoma, were known only in individuals whose immune systems were suppressed (fig. 16B).

The bodies of the sick young men had become nesting places for all types of infectious agents, including viruses, bacteria, protozoans, and fungi. The infections were *opportunistic*, which means that they take advantage of a weakened immune system.

As the infections spread, a portrait of a lethal disease emerged. *Acquired immunodeficiency syndrome,* or AIDS, starts with recurrent fever, weakness, and weight loss. Then usually after a relatively healthy period, infections begin. Many patients die within 2 or 3 years of symptom onset, but the virus that causes AIDS can be present for a decade or longer before a person feels ill. Five percent of infected people have remained healthy for more than 15 years .

Scientists and physicians united to identify the source of the outbreak of infections. Attention turned first to the sufferers. What did they have in common? All had had blood contact with another AIDS patient. Many in the United States and Europe were homosexual men who transferred the infections through fragile rectal tissues during anal intercourse. Intravenous drug users or people who inject anabolic steroids spread AIDS by sharing needles. A growing AIDS population includes children who acquired the disease before or during birth from infected mothers. Today, worldwide, the most common mode of transmission is heterosexual intercourse.

Some people acquired AIDS from HIV-tainted blood transfused before 1985, when stricter blood banking precautions began. Many people with hemophilia were infected when they received blood products pooled from many donors. The toll on people with hemophilia has been staggering—90% of those who used clotting factors several times a month have AIDS, as do 70% of those who have ever received clotting factors.

## HIV

By 1983, researchers linked the human immunodeficiency virus, HIV, to AIDS. It targets lymphocytes in lymph nodes, and CD4 helper T cells. In most cases, HIV is transmitted through blood or semen.

HIV is a *retrovirus,* which means that its genetic material is RNA. The virus attaches to CD4 receptors on helper T cells, and sends in its RNA. A viral enzyme, *reverse transcriptase,* catalyzes construction of a DNA strand complementary to the viral RNA. The initial viral DNA strand replicates to form a DNA double helix, which enters the T cell's nucleus. Viral DNA can remain in immune system cells, such as memory cells, even long after drug therapy has diminished the detectable number of viruses in the bloodstream to near zero.

Using the invaded cell's protein synthetic machinery, HIV replicates, filling the cell with its RNA and proteins. Not only can the dying T cell no longer release cytokines or stimulate B cells to manufacture antibodies, but it bursts, unleashing many new HIV particles. HIV replicates at an astounding rate, producing 3 billion new particles a day, from the start of infection. Immune system cells can remove only 100 million to a billion per day, so soon the system becomes overwhelmed. Opportunistic infections set in. A danger sign in HIV-infected people is a falling CD4 T cell count, signaling collapse of the coordination of the immune response (table 16A). Initial diagnosis and extent of infection are assessed by measuring the amount of HIV RNA in the bloodstream, called viral load.

## TREATMENT

Several new drugs target HIV infection at varous stages. Drugs called protease inhibitors prevent HIV from processing its proteins to a functional size, crippling the virus. Older drugs, such as AZT, ddI, ddC, and 3TC, block viral reproduction. Combining drugs can keep viral load low and delay symptom onset and progression. The goal is to enable infected people to live normal life spans in relatively good health. Some 200 drugs are also available to treat AIDS-associated opportunistic infections and cancers.

## PROGRESS

Three groups of people are providing the clues that may lead to conquering HIV infection:

1. Infected individuals who never develop symptoms ("long-term nonprogressors"). These people have a weakened strain of HIV that lacks part of a gene that HIV normally uses to replicate. Enough of the virus remains to alert the immune system to protect against other strains. A vaccine might be based on this weakened strain.

2. People repeatedly exposed who never become infected. About 1% of the population has a gene variant that protects them from becoming infected with HIV. Cells of these individuals lack either of two receptor molecules that HIV requires to enter. Such "people who can't get AIDS" are described at the start of chapter 3.

3. Infected individuals who apparently become uninfected. Several infants infected at birth have apparently lost the virus as their immune systems matured.

Researchers are identifying how these people survive, evade, or vanquish HIV infection, and will use this knowledge to develop prevention and treatment strategies. Better understanding of the biology of HIV infection, plus new drug weapons and clues from survivors, are providing what has long been lacking in the global fight against AIDS—hope. ∎

**FIGURE 16B**

Prior to the appearance of AIDS, Kaposi's sarcoma was a rare cancer seen only in elderly Jewish and Italian men and in people with suppressed immune systems. In these groups it produces purplish patches on the lower limbs, but in AIDS patients, Kaposi's sarcoma patches appear all over the body and sometimes internally too. These lower limbs display characteristic lesions.

**TABLE 16A WALTER REED CLASSIFICATION\* OF HIV INFECTION**

| Stage | Characteristics | Approximate Time for Progression to Next Stage |
|---|---|---|
| 0 | Exposure to HIV | Months to years |
| 1 | Documented presence of HIV | 6 weeks–1 year |
| 2 | Chronically swollen lymph nodes (lymphadenopathy) | 3–5 years |
| 3 | Lymphadenopathy may disappear; helper T cell count falls below 400/mm3 | 18 months |
| 4 | Impaired cell-mediated immunity, as indicated by failure to respond to three of four skin tests measuring delayed hypersensitivity | 1 year |
| 5 | Total lack of delayed hypersensitivity; thrush and other viral or fungal infections of skin and mucous membranes | 7 months |
| 6 | Systemic immune deficiency, as indicated by rampant opportunistic infections; neurological impairment | 1 year |

\*Since the Walter Reed classification was developed in 1986, several cases have arisen of individuals who remain in the early stages for extended periods, perhaps even permanently. Most individuals, however, follow the stages outlined here.

**LYMPHATIC SYSTEM**

The lymphatic system is an important link between the interstitial fluid and the plasma; it also plays a major role in the response to infection.

## Integumentary System

The skin is a first line of defense against infection.

## Cardiovascular System

The lymphatic system returns tissue fluid to the bloodstream. Lymph originates as interstitial fluid, formed by the action of blood pressure.

## Skeletal System

Cells of the immune system originate in the bone marrow.

## Digestive System

Lymph plays a major role in the absorption of fats.

## Muscular System

Muscle action helps pump lymph through the lymphatic vessels.

## Respiratory System

Cells of the immune system patrol the respiratory system to defend against infection.

## Nervous System

Stress may impair the immune response.

## Urinary System

The kidneys control the volume of extracellular fluid, including lymph.

## Endocrine System

Hormones stimulate lymphocyte production.

## Reproductive System

Special mechanisms inhibit the female immune system in its attack of sperm as foreign invaders.

## Introduction (page 621)

The lymphatic system is closely associated with the cardiovascular system. It transports excess fluid to the bloodstream, absorbs fats, and helps defend the body against disease-causing agents.

## Lymphatic Pathways (page 621)

1. Lymphatic capillaries
   a. Lymphatic capillaries are microscopic, closed-ended tubes that extend into interstitial spaces.
   b. They receive lymph through their thin walls.
   c. Lacteals are lymphatic capillaries in the villi of the small intestine.
2. Lymphatic vessels
   a. Lymphatic vessels are formed by the merging of lymphatic capillaries.
   b. They have walls similar to veins, only thinner, and possess valves that prevent backflow of lymph.
   c. Larger lymphatic vessels lead to lymph nodes and then merge into lymphatic trunks.
3. Lymphatic trunks and collecting ducts
   a. Lymphatic trunks drain lymph from large body regions.
   b. Trunks lead to two collecting ducts within the thorax.
   c. Collecting ducts join the subclavian veins.

## Tissue Fluid and Lymph (page 625)

1. Tissue fluid formation
   a. Tissue fluid originates from blood plasma and includes water and dissolved substances that have passed through the capillary wall.
   b. It generally lacks large proteins, but some smaller proteins leak into interstitial spaces.
   c. As the protein concentration of tissue fluid increases, osmotic pressure increases.
2. Lymph formation
   a. Rising osmotic pressure in tissue fluid interferes with return of water to the blood capillaries.
   b. Increasing pressure within interstitial spaces forces some tissue fluid into lymphatic capillaries, and this fluid becomes lymph.
3. Lymph function
   a. Lymph returns protein molecules to the bloodstream.
   b. It transports foreign particles to the lymph nodes.

## Lymph Movement (page 626)

1. Lymph flow
   a. Lymph is under low pressure and may not flow readily without external aid.
   b. Lymph is moved by the squeezing of skeletal muscles and low pressure in the thorax created by breathing movements.
2. Obstruction of lymph movement
   a. Any condition that interferes with the flow of lymph results in edema.
   b. Obstruction of lymphatic vessels due to surgery causes edema.

## Lymph Nodes (page 626)

1. Structure of a lymph node
   a. Lymph nodes are usually bean-shaped, with blood vessels, nerves, and efferent lymphatic vessels attached to the indented region: afferent lymphatic vessels enter at points on the convex surface.
   b. Lymph nodes are enclosed in connective tissue that extends into the nodes and subdivides them into nodules.
   c. Nodules contain masses of lymphocytes and macrophages, as well as spaces through which lymph flows.
2. Locations of lymph nodes
   a. Lymph nodes aggregate in groups or chains along the paths of larger lymphatic vessels.
   b. They primarily occur in cervical, axillary, and inguinal regions, and within the pelvic, abdominal, and thoracic cavities.
3. Functions of lymph nodes
   a. Lymph nodes filter potentially harmful foreign particles from the lymph before it is returned to the bloodstream.
   b. Lymph nodes are centers for production of lymphocytes that act against foreign particles.
   c. They contain macrophages that remove foreign particles from lymph.

## Thymus and Spleen (page 628)

1. Thymus
   a. The thymus is a soft, bilobed organ located within the mediastinum.
   b. It slowly shrinks after puberty.
   c. It is composed of lymphatic tissue subdivided into lobules.
   d. Lobules contain lymphocytes, most of which are inactive, that develop from precursor cells in bone marrow.
   e. Some lymphocytes leave the thymus and provide immunity.
   f. The thymus secretes thymosin, which stimulates lymphocytes that have migrated to other lymphatic tissues.
2. Spleen
   a. The spleen is located in the upper left portion of the abdominal cavity.
   b. It resembles a large lymph node that is encapsulated and subdivided into lobules by connective tissue.
   c. Spaces within lobules are filled with blood.
   d. The spleen contains many macrophages and lymphocytes, which filter foreign particles and damaged red blood cells from the blood.

## Body Defenses Against Infection (page 630)

The presence and reproduction of pathogens cause infection. Pathogens include bacteria, complex single-celled organisms, fungi, and viruses. An infection may be present without immediately causing symptoms. The body has nonspecific and specific defenses against infection.

## Nonspecific Defenses  (page 630)

1. Species resistance
   Each species is resistant to certain diseases that may affect other species but is susceptible to diseases other species may resist.
2. Mechanical barriers
   a. Mechanical barriers include skin and mucous membranes.
   b. Intact mechanical barriers prevent entrance of some pathogens.
3. Chemical barriers
   a. Enzymes in gastric juice and tears kill some pathogens.
   b. Low Ph in the stomach prevents growth of some bacteria.
   c. High salt concentration in perspiration kills some bacteria.
   d. Interferons stimulate uninfected cells to synthesize antiviral proteins that block proliferation of viruses; stimulate phagocytosis; and enhance activity of cells that help resist infections and stifle tumor growth.
4. Fever
   a. Viral or bacterial infection stimulates certain lymphocytes to secrete endogenous pyrogen, which temporarily raises body temperature.
   b. Higher body temperature and the resulting decrease in blood iron level and increased phagocytic activity hamper infection.
5. Inflammation
   a. Inflammation is a tissue response to damage, injury, or infection.
   b. The response includes localized redness, swelling, heat, and pain.
   c. Chemicals released by damaged tissues attract white blood cells to the site.
   d. Clotting may occur in body fluids that accumulate in affected tissues.
   e. Fibrous connective tissue may form a sac around the injured tissue and thus prevent the spread of pathogens.
6. Phagocytosis
   a. The most active phagocytes in blood are neutrophils and monocytes; monocytes give rise to macrophages, which remain fixed in tissues.
   b. Phagocytic cells associated with the linings of blood vessels in the bone marrow, liver, spleen, and lymph nodes constitute the mononuclear phagocytic system.
   c. Phagocytes remove foreign particles from tissues and body fluids.

## Specific Defenses (Immunity) (page 632)

1. Antigens
   a. Before birth, body cells inventory "self" proteins and other large molecules.
   b. After inventory, lymphocytes develop receptors that allow them to differentiate between nonself (foreign) and self antigens.
   c. Nonself antigens combine with T cell and B cell surface receptors and stimulate these cells to cause an immune reaction.
   d. Haptens are small molecules that can combine with larger ones, becoming antigenic.

2. Lymphocyte origins
   a. Lymphocytes originate in red bone marrow and are released into the blood before they differentiate.
   b. Some reach the thymus where they mature into T cells.
   c. Others, the B cells, mature in the bone marrow.
   d. Both T cells and B cells reside in lymphatic tissues and organs.
3. T cells and cell-mediated immunity
   a. T cells respond to antigens by cell-to-cell contact.
   b. T cells secrete cytokines to enhance other cell responses to antigens.
   c. T cell types and activation.
      (1) T cells are activated when an antigen-presenting cell displays a foreign antigen.
      (2) When a macrophage acts as an accessory cell, it phagocytizes an antigen-bearing agent, digests the agent, and displays the antigens on its surface membrane in association with certain MHC proteins.
      (3) A helper T cell becomes activated when it encounters displayed antigens for which it is specialized to react.
      (4) CD4 helper T cells stimulate antibody-mediated and cell-mediated immunity. HIV cripples these cells.
      (5) Memory T cells allow for immediate response to second and subsequent antigen exposures.
      (6) Cytotoxic T cells recognize foreign antigens on tumor cells and cells whose surfaces indicate that they are infected by viruses. Stimulated cytotoxic T cells secrete perforin to destroy its target cells.
4. B cells and antibody-mediated immunity
   a. B cell activation
      (1) A B cell is activated when it encounters an antigen that fits its antigen receptors.
      (2) An activated B cell proliferates (especially when stimulated by a T cell), enlarging its clone.
      (3) Some activated B cells specialize into antibody-producing plasma cells.
      (4) Antibodies react against the antigen-bearing agent that stimulated their production.
      (5) An individual's diverse B cells defend against a very large number of pathogens.
   b. Antibody molecules
      (1) Antibodies are proteins called immunoglobulins.
      (2) They constitute the gamma globulin fraction of plasma.
      (3) Each immunoglobulin molecule consists of four chains of amino acids linked together.
      (4) Variable regions at the ends of these chains are specialized into antigen binding sites to react with antigens.
   c. Types of immunoglobulins
      (1) The five major types of immunoglobulins are IgG, IgA, IgM, IgD, and IgE.
      (2) IgG, IgA, and IgM make up most of the circulating antibodies.
   d. Antibody actions
      (1) Antibodies directly attack antigens, activate complement, or stimulate local tissue changes that are unfavorable to antigen-bearing agents.

(2) Direct attacks occur by means of agglutination, precipitation, or neutralization.

(3) Activated proteins of complement attract phagocytes, alter cells so they become more susceptible to phagocytosis, and lyse foreign cell membranes.

5. Immune responses
   a. B cells or T cells first encountering an antigen for which they are specialized to react, constitutes a primary immune response.
   (1) During this response, antibodies are produced for several weeks.
   (2) Some B cells remain dormant as memory cells, aided by follicular dendritic cells.
   b. A secondary immune response occurs rapidly as a result of memory cell response if the same antigen is encountered later.

6. Types of immunity
   a. A person who encounters a pathogen and has a primary immune response develops naturally acquired active immunity.
   b. A person who receives a vaccine containing a dead or weakened pathogen, or part of it, develops artificially acquired active immunity.
   c. A person who receives an injection of antibodies has artificially acquired passive immunity.
   d. When antibodies pass through a placental membrane from a pregnant woman to her fetus, the fetus develops naturally acquired passive immunity.
   e. Active immunity lasts much longer than passive immunity.

7. Allergic reactions
   a. Allergic or hypersensitivity reactions are excessive immune responses and may damage tissue.
   b. Delayed-reaction allergy, which can occur in anyone and inflame the skin, results from repeated exposure to allergens.
   c. Antibody-dependent cytotoxic allergic reactions occur when blood transfusions are mismatched.
   d. Immune complex allergic reactions involve autoimmunity, which is an immune reaction against self antigens.
   e. Immediate-reaction allergy is an inborn ability to overproduce IgE.
   f. Allergic reactions result from mast cells bursting and releasing allergy mediators such as histamine and serotonin.
   g. In anaphylactic shock, allergy mediators flood the body, causing severe symptoms.

8. Transplantation and tissue rejection
   a. A transplant recipient's immune system may react against the donated tissue in a tissue rejection reaction.
   b. Matching donor and recipient tissues and using immunosuppressive drugs can minimize tissue rejection.
   c. Immunosuppressive drugs may increase susceptibility to infection.
   d. Transplants may take place between genetically identical twins, from one body part to another, between unrelated individuals of the same species, or between individuals of different species.

9. Autoimmunity
   a. In autoimmune disorders, the immune system manufactures autoantibodies that attack one's own body tissues.
   b. Autoimmune disorders may result from a previous viral infection, faulty T cell development, or reaction to a nonself antigen that resembles a self antigen.

## CRITICAL THINKING QUESTIONS

1. How can removal of enlarged lymph nodes for microscopic examination aid in diagnosing certain diseases?

2. Why is an injection into the skin like an injection into the lymphatic system?

3. Why does vaccination provide long-lasting protection against a disease, whereas gamma globulin provides only short-term protection?

4. When a breast is surgically removed to treat breast cancer, the lymph nodes in the nearby axillary region are sometimes excised also. Why is this procedure likely to cause swelling of the upper limb on the treated side?

5. How would being born without a thymus affect susceptibility to infections?

6. A patient sees a doctor because of fatigue, night sweats (awakening drenched in perspiration), and weight loss.
   a. What questions should the physician ask this patient?
   b. What other symptoms might the physician look for or ask about?
   c. When the physician orders a blood test, what, specifically, should be examined?

7. Some parents keep their preschoolers away from other children to prevent them from catching illnesses. How might these well-meaning parents actually be harming their children?

8. A xenograft is tissue from nonhuman animals used to replace body parts in humans. For example, pigs are being developed to provide cardiovascular spare parts because their hearts and blood vessels are similar to ours. To increase the likelihood of such a xenotransplant working, researchers genetically engineer pigs to produce human antigens on their cell surfaces. How can this improve the chances of a human body not rejecting such a transplant?

1. Explain how the lymphatic system is related to the cardiovascular system.
2. Trace the general pathway of lymph from the interstitial spaces to the bloodstream.
3. Identify and describe the locations of the major lymphatic trunks and collecting ducts.
4. Distinguish between tissue fluid and lymph.
5. Describe the primary functions of lymph.
6. Explain why physical exercise promotes lymphatic circulation.
7. Explain how a lymphatic obstruction leads to edema.
8. Describe the structure of a lymph node, and list its major functions.
9. Locate the major body regions occupied by lymph nodes.
10. Describe the structure and functions of the thymus.
11. Describe the structure and functions of the spleen.
12. Distinguish between specific and nonspecific body defenses against infection.
13. Explain species resistance.
14. Name two mechanical barriers to infection.
15. Describe how enzymatic actions function as defense mechanisms.
16. Define *interferon*, and explain its action.
17. List the major effects of inflammation, and explain why each occurs.
18. Identify the major phagocytic cells in the blood and other tissues.
19. Distinguish between an antigen and a hapten.
20. Review the origin of T cells and B cells.
21. Explain cell-mediated immunity.
22. Define *cytokine*.
23. List three types of T cells and describe the function of each in the immune response.
24. Explain antibody-mediated immunity.
25. Define *clone of lymphocytes*.
26. Explain how a B cell is activated.
27. Explain the function of plasma cells.
28. Describe an immunoglobulin molecule.
29. Distinguish between the variable region and the constant region of an immunoglobulin molecule.
30. List the major types of immunoglobulins, and describe their main functions.
31. Explain four mechanisms by which antibodies may directly attack antigens.
32. Explain the function of complement.
33. Describe how T cells become activated.
34. Distinguish between a primary and a secondary immune response.
35. Distinguish between active and passive immunity.
36. Define *vaccine*.
37. Explain how a vaccine produces its effect.
38. Describe how a fetus may obtain antibodies from the maternal blood.
39. Explain the relationship between an allergic reaction and an immune response.
40. Distinguish between an antigen and an allergen.
41. List the major events leading to a delayed-reaction allergic response.
42. Describe how an immediate-reaction allergic response may occur.
43. Explain the relationship between tissue rejection and an immune response.
44. Describe two methods used to reduce the severity of a tissue rejection reaction.
45. How do immunosuppressant drugs increase the likelihood of success of a transplant, yet place a patient at a higher risk of developing infections?
46. Explain the relationship between autoimmunity and an immune response.

**For More Information Related to This Chapter, Please Visit** *Hole's Human Anatomy and Physiology* **Home Page**

http://www.mhhe.com/biosci/abio/

## CHAPTER 17

# DIGESTIVE SYSTEM

## CHAPTER OBJECTIVES

### AFTER YOU HAVE STUDIED THIS CHAPTER, YOU SHOULD BE ABLE TO:

1. Name and describe the locations and major parts of the organs of the digestive system.

2. Describe the general functions of each digestive organ.

3. Describe the structure of the wall of the alimentary canal.

4. Explain how the contents of the alimentary canal are mixed and moved.

5. List the enzymes the digestive organs and glands secrete and describe the function of each.

6. Describe how digestive secretions are regulated.

7. Explain how digestive reflexes control movement of material through the alimentary canal.

8. Describe the mechanisms of swallowing, vomiting, and defecating.

9. Explain how the products of digestion are absorbed.

## KEY TERMS

**absorption** (ab-sorp'shun)
**accessory organ** (ak-ses'o-re or'gan)
**alimentary canal** (al"i-men'tar-e kah-nal')
**bile** (bīl)
**chyme** (kīm)
**circular muscle** (ser'ku-lar mus'el)
**deciduous** (de-sid'u-us)
**feces** (fe'sēz)
**gastric juice** (gas'trik jōōs)

**intestinal juice** (in-tes'ti-nal jōōs)
**intrinsic** (in-trin'sik)
**longitudinal muscle** (lon"ji-tu'di-nal mus'el)
**mesentery** (mes'en-ter"e)
**mucous membrane** (mu'kus mem'brān)
**pancreatic juice** (pan"kre-at'ik jōōs)
**peristalsis** (per"i-stal'sis)
**serous layer** (se'rus la'er)
**sphincter muscle** (sfingk'ter mus'el)
**villus** (vil'us)

## UNDERSTANDING WORDS

**aliment-,** food: *aliment*ary canal—tubelike portion of the digestive system.

**cari-,** decay: dental *cari*es—tooth decay.

**cec-,** blindness: *cec*um—blind-ended sac at the beginning of the large intestine.

**chym-,** juice: *chym*e—semifluid paste of food particles and gastric juice formed in the stomach.

**decidu-,** falling off: *decidu*ous teeth—teeth that are shed during childhood.

**frenul-,** bridle, restraint: *frenul*um—membranous fold that anchors the tongue to the floor of the mouth.

**gastr-,** stomach: *gastr*ic gland—portion of the stomach that secretes gastric juice.

**hepat-,** liver: *hepat*ic duct—duct that carries bile from the liver to the common bile duct.

**hiat-,** opening: esophageal *hiat*us—opening through which the esophagus penetrates the diaphragm.

**lingu-,** tongue: *lingu*al tonsil—mass of lymphatic tissue at the root of the tongue.

**peri-,** around: *peri*stalsis—wavelike ring of contraction that moves material along the alimentary canal.

**pyl-,** gatekeeper, door: *pyl*oric sphincter—muscle that serves as a valve between the stomach and small intestine.

**rect-,** straight: *rect*um—distal portion of the large intestine.

**sorpt-,** to soak up: ab*sorpt*ion—uptake of substances.

**vill-,** hair: *vill*i—tiny projections of mucous membrane in the small intestine.

In late 1982, two young Australian physicians, Barry J. Marshall and J. Robin Warren, performed a daring experiment. They had been studying gastritis (stomach inflammation) and peptic ulcers (sores in the lining of the esophagus, stomach or small intestine) and had proposed that a bacterium, *Helicobacter pylori,* causes these problems. But because the medical community had for many decades been convinced that ulcers were the result of excess stomach acid due to stress, Marshall and Warren's idea was not very popular.

So, Marshall took a drastic step. He drank a culture of the suspect bacteria—and developed gastritis.

In the decade following Marshall and Warren's publication of the bacteria–gastritis link in early 1983, abundant evidence accumulated that they were right about gastritis. The medical community soon realized that this microbe that thrives in the highly acidic environs of the human stomach also causes a majority of peptic ulcers.

By 1994, treatment for gastritis and ulcers finally began to change. Instead of taking drugs for many years to lower stomach acid, sufferers can now often take a combination of already available antibiotic and acid-lowering or neutralizing drugs for just a few weeks.

By 1996, diagnosis for *H. pylori*–caused ulcers became far less invasive. Instead of undergoing a sampling of stomach tissue, sufferers can now take a breath test. A patient swallows urea (an organic compound) that includes a heavy isotope of carbon. If *H. pylori* is present, a bacterial enzyme, urease, breaks down the labeled urea, releasing carbon dioxide that incorporates the "heavy" carbon. An instrument called a mass spectrometer detects the telltale carbon, establishing presence of the "ulcer bug." A once debilitating digestive problem may now, for many people, be a short-lived pain and inconvenience.

---

**Digestion** is the mechanical and chemical breakdown of foods into forms that cell membranes can absorb. Mechanical digestion breaks large pieces into smaller ones without altering their chemical composition. Chemical digestion breaks food into simpler chemicals. The organs of the **digestive system** carry out these processes.

The digestive system consists of the **alimentary canal** extending from the mouth to the anus, and several accessory organs, which release secretions into the canal. The alimentary canal includes the mouth, pharynx, esophagus, stomach, small intestine, large intestine, and anal canal. The accessory organs include the salivary glands, liver, gallbladder, and pancreas. Figure 17.1 shows the major organs of the digestive system.

The digestive system originates from the middle layer (endoderm) of the embryo, which folds to form the tube of the alimentary canal. The accessory organs develop as buds from the tube.

## GENERAL CHARACTERISTICS OF THE ALIMENTARY CANAL

The alimentary canal is a muscular tube about 9 meters long that passes through the body's ventral cavity. The structure of its wall, how it moves food, and its innervation are similar throughout its length (fig. 17.2).

### Structure of the Wall

The wall of the alimentary canal consists of four distinct layers that are developed to different degrees from region to region. Although the four-layered structure persists throughout the alimentary canal, certain regions are specialized for particular functions. Beginning with the innermost tissues, these layers, shown in figure 17.3, include the following:

1.  **Mucosa (mucous membrane).** This layer is formed of surface epithelium, underlying connective tissue (lamina propria), and a small amount of smooth muscle. In some regions, it develops folds and tiny projections, which extend into the lumen of the digestive tube and increase its absorptive surface area. It may also contain glands that are tubular invaginations into which the lining cells secrete mucus and digestive enzymes. The mucosa protects the tissues beneath it and carries on secretion and absorption.

2.  **Submucosa.** The submucosa contains considerable loose connective tissue as well as glands, blood vessels, lymphatic vessels, and nerves. Its vessels nourish the surrounding tissues and carry away absorbed materials.

3.  **Muscular layer.** This layer, which provides movements of the tube, consists of two coats of smooth muscle tissue. The fibers of the inner coat encircle the tube. When these *circular fibers* (they are actually closed spirals) contract, the diameter of the tube decreases. The fibers of the outer muscular coat run lengthwise. When these *longitudinal fibers* (open spirals) contract, the tube shortens.

4.  **Serosa (serous layer).** The serous layer, or outer covering of the tube, is composed of the *visceral peritoneum,* which is formed of epithelium on the outside and connective tissue beneath. The cells of the serosa protect underlying tissues and secrete serous fluid, which moistens and lubricates the tube's outer surface so that the organs within the abdominal cavity slide freely against one another.

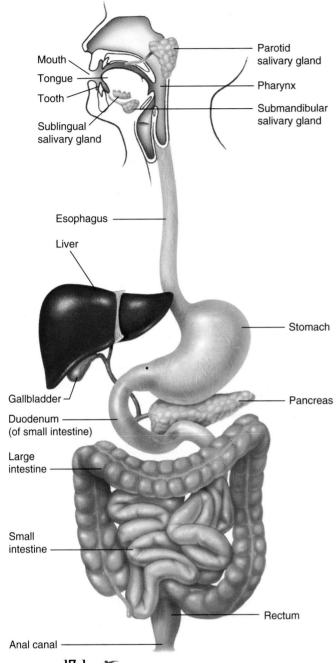

**FIGURE 17.1**

Major organs of the digestive system.

Table 17.1 summarizes the characteristics of these layers.

## Movements of the Tube

The motor functions of the alimentary canal are of two basic types—*mixing movements* and *propelling movements* (fig. 17.4). Mixing occurs when smooth muscles in small segments of the tube contract rhythmically. For example, when the stomach is full, waves of muscular con-

tractions move along its wall from one end to the other. These waves occur every twenty seconds or so, and their action mixes foods with the digestive juices that the mucosa secretes.

Propelling movements include a wavelike motion called **peristalsis.** When peristalsis occurs, a ring of contraction appears in the wall of the tube. At the same time, the muscular wall just ahead of the ring relaxes—a phenomenon called *receptive relaxation.* As the wave moves along, it pushes the tubular contents ahead of it. Peristalsis begins when food expands the tube. It causes the sounds that can be heard through a stethoscope applied to the abdominal wall.

An isotopic radiogastrogram is a multicolored scan that illustrates peristalsis in the stomach, what the technique's inventor calls the "stomachbeat." A patient complaining of severe indigestion eats a scrambled egg sandwich containing a radioactive isotope, which allows a special camera to detect the movement of food through the alimentary canal. A computer assembles hundreds of camera shots into an image.

## Innervation of the Tube

Branches of the sympathetic and parasympathetic divisions of the autonomic nervous system extensively innervate the alimentary canal. These nerve fibers, mainly associated with the tube's muscular layer, maintain muscle tone and regulate the strength, rate, and velocity of muscular contractions. Many of the postganglionic fibers synapse with a network or plexus of neurons within the wall of the canal (see figure 17.3). The *submucosal plexus* is important in controlling secretions by the gastrointestinal tract. The *myenteric plexus* of the muscular layer controls gastrointestinal motility.

Parasympathetic impulses generally increase the activities of the digestive system. Some of these impulses originate in the brain and are conducted through branches of the vagus nerves to the esophagus, stomach, pancreas, gallbladder, small intestine, and proximal half of the large intestine. Other parasympathetic impulses arise in the sacral region of the spinal cord and supply the distal half of the large intestine.

Sympathetic nerve impulses' effects on digestive actions usually are opposite those of the parasympathetic division. That is, sympathetic impulses inhibit certain digestive actions. For example, such impulses contract sphincter muscles in the wall of the alimentary canal, blocking movement of materials through the tube.

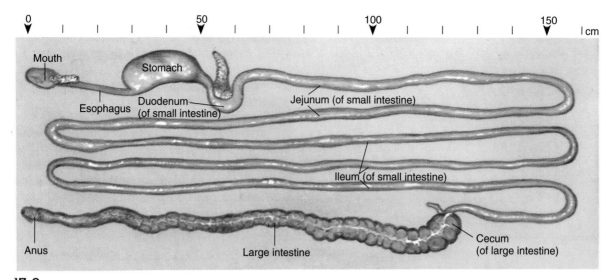

**FIGURE 17.2**

The alimentary canal is a muscular tube about 9 meters long.

So extensive are the nerve plexuses of the gastrointestinal tract that it is sometimes said to have a "second brain." The small intestine, for example, has at least 100 million neurons, many glial cells, and abundant and diverse neurotransmitters, neuropeptides, and growth factors.

**1** What are the general functions of the digestive system?

**2** Which organs constitute the digestive system?

**3** Describe the wall of the alimentary canal.

**4** Name the two types of movements that occur in the alimentary canal.

**5** How do parasympathetic nerve impulses affect digestive actions? What effect do sympathetic nerve impulses have?

## MOUTH

The **mouth,** which is the first portion of the alimentary canal, receives food and begins digestion by mechanically breaking up the solid particles into smaller pieces and mixing them with saliva (mastication). The mouth also functions as an organ of speech and sensory reception. It is surrounded by the lips, cheeks, tongue, and palate, and includes a chamber between the palate and tongue called the *oral cavity,* as well as a narrow space between the teeth, cheeks, and lips called the *vestibule* (fig. 17.5 and reference plate 49).

## Cheeks and Lips

The **cheeks** form the lateral walls of the mouth. They consist of outer layers of skin, pads of subcutaneous fat, muscles associated with expression and chewing, and inner linings of moist, stratified squamous epithelium.

Because cheek cells are easily removed, many biology laboratory classes have students collect them and use them to study cell structure. A similar "cheekbrush test" is used to identify people who are carriers of the inherited disorder cystic fibrosis. The patient simply swishes a brush on the inside of the cheek, then the doctor sends the brush to a laboratory. Here, cheek cells are removed from the brush, and the DNA is extracted and analyzed for the presence of the gene variant that causes cystic fibrosis.

The **lips** are highly mobile structures that surround the mouth opening. They contain skeletal muscles and sensory receptors useful in judging the temperature and texture of foods. Their normal reddish color is due to the many blood vessels near their surfaces. The external borders of the lips mark the boundaries between the skin of the face and the mucous membrane that lines the alimentary canal.

## Tongue

The **tongue** is a thick, muscular organ that occupies the floor of the mouth and nearly fills the oral cavity when the mouth is closed. It is covered by mucous membrane

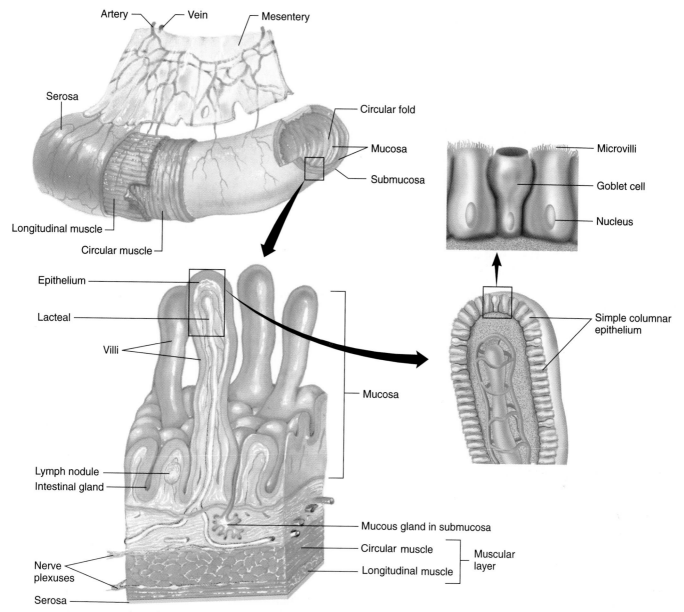

**FIGURE 17.3**

The wall of the small intestine, as in other portions of the alimentary canal, includes four layers: an inner mucosa, a submucosa, a muscular layer, and an outer serosa.

TABLE 17.1

## LAYERS OF THE WALL OF THE ALIMENTARY CANAL

| Layer | Composition | Function |
|-------|-------------|----------|
| Mucosa | Epithelium, connective tissue, smooth muscle | Protection, secretion, absorption |
| Submucosa | Loose connective tissue, blood vessels, lymphatic vessels, nerves | Nourishes surrounding tissues, transports absorbed materials |
| Muscular layer | Smooth muscle fibers arranged in circular and longitudinal groups | Movements of the tube and its contents |
| Serosa | Epithelium, connective tissue | Protection, lubrication |

**FIGURE 17.4** 🔲

(a) Mixing movements occur when small segments of the muscular wall of the alimentary canal rhythmically contract.
(b) Peristaltic waves move the contents along the canal.

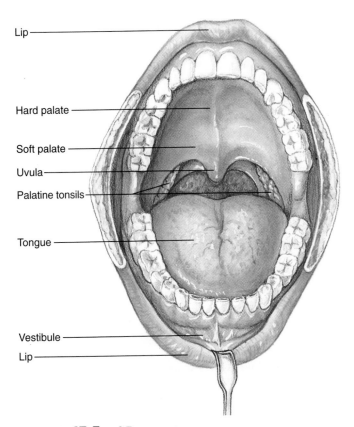

Lip ————————————————

Hard palate ————————

Soft palate ————————

Uvula ————————

Palatine tonsils ————————

Tongue ————————

Vestibule ————————

Lip ————————

**FIGURE 17.5** 🏃

The mouth is adapted for ingesting food and preparing it for digestion.

and is connected in the midline to the floor of the mouth by a membranous fold called the **frenulum.**

The *body* of the tongue is largely composed of skeletal muscle fibers that run in several directions. These muscles mix food particles with saliva during chewing and move food toward the pharynx during swallowing. Rough projections, called **papillae,** on the surface of the tongue provide friction, which helps handle

food. Between these papillae are most of the taste buds. Some taste buds are in the papillae (fig. 17.6).

The posterior region, or *root,* of the tongue is anchored to the hyoid bone. It is covered with rounded masses of lymphatic tissue called **lingual tonsils.**

## Palate

The **palate** forms the roof of the oral cavity and consists of a hard anterior part and a soft posterior part. The *hard palate* is formed by the palatine processes of the maxillary bones in front and the horizontal portions of the palatine bones in back. The *soft palate* forms a muscular arch, which extends posteriorly and downward as a cone-shaped projection called the **uvula.**

During swallowing, muscles draw the soft palate and the uvula upward. This action closes the opening between the nasal cavity and the pharynx, preventing food from entering the nasal cavity.

In the back of the mouth, on either side of the tongue and closely associated with the palate, are masses of lymphatic tissue called **palatine tonsils.** These structures lie beneath the epithelial lining of the mouth and, like other lymphatic tissues, help protect the body against infections (see chapter 16).

The palatine tonsils are common sites of infection, and when inflamed, produce *tonsillitis.* Infected tonsils may swell so greatly that they block the passageways of the pharynx and interfere with breathing and swallowing. Because the mucous membranes of the pharynx, auditory tubes, and middle ears are continuous, such an infection can travel from the throat into the middle ears (otitis media).

When tonsillitis occurs repeatedly and does not respond to antibiotic treatment, the tonsils are sometimes surgically removed (*tonsillectomy*). Tonsillectomies are done less often today than they were a generation ago, because the tonsils' role in immunity is now recognized.

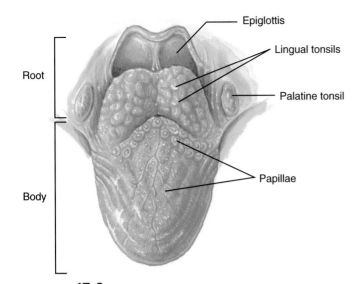

**FIGURE 17.6**

The surface of the tongue, superior view.

Still other masses of lymphatic tissue, called **pharyngeal tonsils,** or *adenoids,* are on the posterior wall of the pharynx, above the border of the soft palate. If the adenoids enlarge and block the passage between the pharynx and the nasal cavity, they also may be surgically removed (fig. 17.7).

**1** What is the function of the mouth?

**2** How does the tongue function as part of the digestive system?

**3** What is the role of the soft palate in swallowing?

**4** Where are the tonsils located?

## Teeth

The **teeth** are the hardest structures in the body, but they are not part of the skeletal system. They develop in sockets within the alveolar processes of the mandibular and

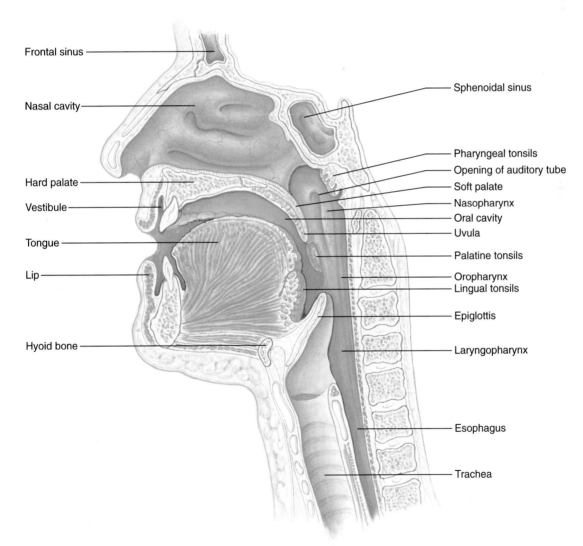

**FIGURE 17.7**

A sagittal section of the mouth, nasal cavity, and pharynx.

maxillary bones. Teeth are unique structures in that two sets form during development (fig. 17.8). The members of the first set, the *primary teeth* (deciduous teeth), usually erupt through the gums (gingiva) at regular intervals between the ages of 6 months and 2 to 4 years. The ten primary teeth in each jaw are located from the midline toward the sides in the following sequence: central incisor, lateral incisor, cuspid (canine), first molar, and second molar.

The primary teeth are usually shed in the same order they appeared. Before this happens, though, their roots are resorbed. The teeth are then pushed out of their sockets by pressure from the developing *secondary teeth* (permanent teeth). This secondary set consists of thirty-two teeth—sixteen in each jaw—and they are arranged from the midline as follows: central incisor, lateral incisor, cuspid (canine), first bicuspid (premolar), second bicuspid (premolar), first molar, second molar, and third molar (fig. 17.9). Table 17.2 summarizes the types and numbers of primary and secondary teeth.

The permanent teeth usually begin to appear at 6 years, but the set may not be completed until the third molars appear between 17 and 25 years. Sometimes these third molars, which are also called wisdom teeth, become wedged in abnormal positions within the jaws and fail to erupt. Such teeth are said to be *impacted*.

The teeth break pieces of food into smaller pieces, beginning mechanical digestion. This action increases the surface area of the food particles and thus makes it possible for digestive enzymes to react more effectively with the food molecules.

Different teeth are adapted to handle food in different ways. The *incisors* are chisel-shaped, and their sharp edges bite off large pieces of food. The *cuspids* are cone-shaped, and they grasp and tear food. The *bicuspids* and

*molars* have somewhat flattened surfaces and are specialized for grinding food particles.

Each tooth consists of two main portions—the *crown,* which projects beyond the gum, and the *root,* which is anchored to the alveolar process of the jaw. The region where these portions meet is called the *neck* of

(a)

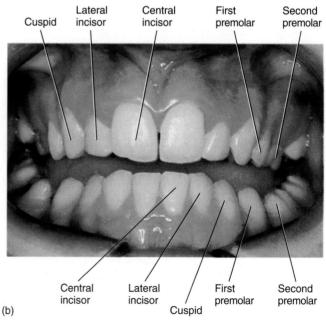

(b)

**FIGURE 17.9**

(a) The secondary teeth of the upper and lower jaws.
(b) Anterior view of the secondary teeth.

**FIGURE 17.8**

This child's skull reveals the primary and secondary teeth developing in the maxilla and mandible.

the tooth. Glossy, white *enamel* covers the crown. Enamel mainly consists of calcium salts and is the hardest substance in the body. Unfortunately, if abrasive action or injury damages enamel, it is not replaced. Enamel also tends to wear away with age.

The bulk of a tooth beneath the enamel is composed of *dentin,* a substance much like bone, but somewhat harder. Dentin, in turn, surrounds the tooth's central cavity (pulp cavity), which contains a combination of blood vessels, nerves, and connective tissue called pulp. Blood vessels and nerves reach this cavity through tubular *root canals,* which extend upward into the root. Tooth loss is most often associated with diseases of the gums (gingivitis) and the dental pulp (endodontitis).

The root is enclosed by a thin layer of bonelike material called *cementum,* which is surrounded by a *periodontal ligament* (periodontal membrane). This ligament contains bundles of thick collagenous fibers, which pass between the cementum and the bone of the alveolar process, firmly attaching the tooth to the jaw. The ligament also contains blood vessels and nerves near the surface of the cementum-covered root (fig. 17.10).

The mouth parts and their functions are summarized in table 17.3. Clinical Application 17.1 describes the effect of bacteria on teeth.

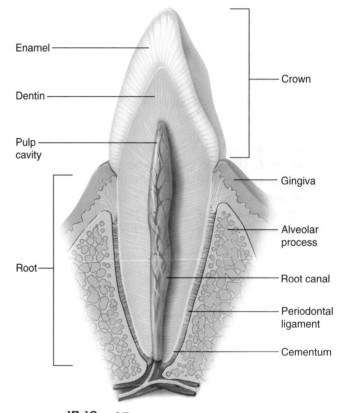

**FIGURE 17.10**

A section of a cuspid tooth.

| TABLE 17.2 | PRIMARY AND SECONDARY TEETH | | |
|---|---|---|---|
| **Primary Teeth (Deciduous)** | | **Secondary Teeth (Permanent)** | |
| *Type* | *Number* | *Type* | *Number* |
| Incisor | | Incisor | |
| Central | 4 | Central | 4 |
| Lateral | 4 | Lateral | 4 |
| Cuspid | 4 | Cuspid | 4 |
| | | Bicuspid | |
| | | First | 4 |
| | | Second | 4 |
| Molar | | Molar | |
| First | 4 | First | 4 |
| Second | 4 | Second | 4 |
| | | Third | 4 |
| **Total** | 20 | **Total** | 32 |

| TABLE 17.3 | MOUTH PARTS AND THEIR FUNCTIONS | | | | |
|---|---|---|---|---|---|
| **Part** | **Location** | **Function** | **Part** | **Location** | **Function** |
| Cheeks | Form lateral walls of mouth | Hold food in mouth; muscles function in chewing | Tongue | Occupies floor of mouth | Aids in mixing food with saliva; moves food toward pharynx; contains taste receptors |
| Lips | Surround mouth opening | Contain sensory receptors used to judge characteristics of foods | Palate | Forms roof of mouth | Holds food in mouth; directs food to pharynx |
| | | | Teeth | In sockets of mandibular and maxillary bones | Break food particles into smaller pieces; help mix food with saliva during chewing |

# Dental Caries

Sticky foods, such as caramel, lodge between the teeth and in the crevices of molars, feeding bacteria such as *Actinomyces, Streptococcus mutans,* and *Lactobacillus.* These microbes metabolize carbohydrates in the food, producing acid by-products that destroy tooth enamel and dentin (fig. 17A). The bacteria also produce sticky substances that hold them in place.

If a person eats candy bars but does not brush the teeth soon afterward, the actions of the acid-forming bacteria will produce decay, called *dental caries.* Unless a dentist cleans and fills the resulting cavity that forms where enamel is destroyed, the damage will spread to the underlying dentin. The tooth becomes very sensitive.

Dental caries can be prevented in several ways:

1. Brush and floss teeth regularly.
2. Have regular dental exams and cleanings.
3. Drink fluoridated water or receive a fluoride treatment. Fluoride is actually incorporated into the enamel's chemical structure, strengthening it.
4. Apply a sealant to children and adolescents' teeth that have crevices that might hold onto decay-causing bacteria. The sealant is a coating that keeps acids from eating away at tooth enamel.

One dental researcher took an unconventional approach to preventing dental caries that, understandably, was never commercialized. He invented a mouthwash consisting of mutant bacteria that would replace *Streptococcus mutans* but would not decay enamel. Consumer acceptance of a mutant bacterial brew was an obstacle! ■

**FIGURE 17A**

*Actinomyces* bacteria clinging to teeth release acids that decay tooth enamel (1,250×).

---

1. How do primary teeth differ from secondary teeth?

2. How are types of teeth adapted to provide specialized functions?

3. Describe the structure of a tooth.

4. Explain how a tooth is attached to the bone of the jaw.

**FIGURE** 17.11

Locations of the major salivary glands.

## SALIVARY GLANDS

The **salivary glands** secrete saliva. This fluid moistens food particles, helps bind them, and begins the chemical digestion of carbohydrates. Saliva is also a solvent, dissolving foods so that they can be tasted, and it helps cleanse the mouth and teeth. Bicarbonate ions ($HCO_3^-$) in saliva help buffer its acid concentration so that the pH of saliva usually remains near neutral, between 6.5 and 7.5. This is a favorable range for the action of the salivary enzyme and protects the teeth from dissolving in a highly acidic environment.

Many minor salivary glands are scattered throughout the mucosa of the tongue, palate, and cheeks. They continuously secrete fluid so that the lining of the mouth remains moist. There are also three pairs of major salivary glands: the parotid glands, the submandibular glands, and the sublingual glands.

### Salivary Secretions

Within a salivary gland are two types of secretory cells, *serous cells* and *mucous cells*, that occur in varying proportions within different glands. Serous cells produce a

watery fluid that contains a digestive enzyme called **amylase.** This enzyme splits starch and glycogen molecules into disaccharides—the first step in the chemical digestion of carbohydrates. Mucous cells secrete a thick liquid called **mucus,** which binds food particles and acts as a lubricant during swallowing.

Like other digestive structures, the salivary glands are innervated by branches of both sympathetic and parasympathetic nerves. Impulses arriving on sympathetic fibers stimulate the gland cells to secrete a small quantity of viscous saliva. Parasympathetic impulses, on the other hand, elicit the secretion of a large volume of watery saliva. Such parasympathetic impulses are activated reflexly when a person sees, smells, tastes, or even thinks about pleasant foods. Conversely, if food looks, smells, or tastes unpleasant, parasympathetic activity is inhibited, so that less saliva is produced and swallowing may become difficult.

### Major Salivary Glands

The **parotid glands** are the largest of the major salivary glands. Each gland lies anterior to and somewhat inferior to each ear, between the skin of the cheek and the masseter muscle. A *parotid duct* (Stensen's duct) passes from the gland inward through the buccinator muscle, entering the mouth just opposite the upper second molar on either side of the jaw. The parotid glands secrete a clear, watery fluid that is rich in amylase (fig. 17.11).

The **submandibular glands** are located in the floor of the mouth on the inside surface of the lower jaw. The secretory cells of these glands are predominantly serous, with a few mucous cells. The submandibular glands secrete a more viscous fluid than the parotid glands. The ducts of the submandibular glands (Wharton's ducts) open inferior to the tongue, near the frenulum (see fig. 17.11).

The **sublingual glands** are the smallest of the major salivary glands. They are found on the floor of the mouth inferior to the tongue. Because their cells are primarily the mucous type, their secretions, which enter the mouth through many separate ducts (Rivinus's ducts) tend to be thick and stringy (fig. 17.12). Table 17.4 summarizes the characteristics of the major salivary glands.

**1** What is the function of saliva?

**2** What stimulates the salivary glands to secrete saliva?

**3** Where are the major salivary glands located?

(a)  (b)  (c)

**FIGURE 17.12**

Light micrographs of (*a*) the parotid salivary gland (75×), (*b*) the submandibular salivary gland (180×), and (*c*) the sublingual salivary gland (80×).

| TABLE 17.4 | THE MAJOR SALIVARY GLANDS | | |
|---|---|---|---|
| **Gland** | **Location** | **Duct** | **Type of Secretion** |
| Parotid glands | Anterior to and somewhat inferior to the ears between the skin of the cheeks and the masseter muscles | Parotid ducts pass through the buccinator muscles and enter the mouth opposite the upper second molars | Clear, watery serous fluid, rich in amylase |
| Submandibular glands | In the floor of the mouth on the inside surface of the mandible | Ducts open inferior to the tongue near the frenulum | Primarily serous fluid, but with some mucus; more viscous than parotid secretion |
| Sublingual glands | In the floor of the mouth inferior to the tongue | Many separate ducts | Primarily thick, stringy mucus |

## PHARYNX AND ESOPHAGUS

The pharynx is a cavity posterior to the mouth from which the tubular esophagus leads to the stomach. The pharynx and the esophagus do not digest food, but both are important passageways, and their muscular walls function in swallowing.

### Structure of the Pharynx

The **pharynx** connects the nasal and oral cavities with the larynx and esophagus (see fig. 17.7). It can be divided into the following parts:

1. The **nasopharynx** is located superior to the soft palate. It communicates with the nasal cavity and provides a passageway for air during breathing. The auditory tubes, which connect the pharynx with the middle ears, open through the walls of the nasopharynx (see chapter 12).

2. The **oropharynx** is posterior to the mouth. It opens posterior to the soft palate into the nasopharynx and projects downward to the upper border of the epiglottis. This portion is a passageway for food moving downward from the mouth, and for air moving to and from the nasal cavity.

3. The **laryngopharynx** is located just inferior to the oropharynx. It extends from the upper border of the epiglottis downward to the lower border of the cricoid cartilage of the larynx and is a passageway to the esophagus.

The muscles in the walls of the pharynx form inner circular and outer longitudinal groups (fig. 17.13). The circular muscles, called *constrictor muscles,* pull the

Mandible

Superior constrictor muscles

Middle constrictor muscles

Inferior constrictor muscles

Skull

Tongue

Epiglottis

Larynx

Esophagus

**FIGURE 17.13**

Muscles of the pharyngeal wall, posterior view.

walls inward during swallowing. The *superior constrictor muscles,* which are attached to bony processes of the skull and mandible, curve around the upper part of the pharynx. The *middle constrictor muscles* arise from projections on the hyoid bone and fan around the middle of the pharynx. The *inferior constrictor muscles* originate from cartilage of the larynx and pass around the lower portion of the pharyngeal cavity. Some of the lower inferior constrictor muscle fibers contract most of the time, which prevents air from entering the esophagus during breathing.

Although the pharyngeal muscles are skeletal muscles, they are under voluntary control only in the sense that swallowing (deglutition) can be voluntarily initiated. Complex reflexes control the precise actions of these muscles during swallowing.

## Swallowing Mechanism

Swallowing reflexes can be divided into three stages. In the first stage, which is initiated voluntarily, food is chewed and mixed with saliva. Then, the tongue rolls this mixture into a mass (bolus) and forces it into the pharynx. The second stage begins as food reaches the pharynx and stimulates sensory receptors around the pharyngeal opening. This triggers the swallowing re-

flexes, illustrated in figure 17.14, which includes the following actions:

1. The soft palate raises, preventing food from entering the nasal cavity.

2. The hyoid bone and the larynx are elevated; the epiglottis closes off the top of the trachea so that food is less likely to enter.

3. The tongue is pressed against the soft palate, sealing off the oral cavity from the pharynx.

4. The longitudinal muscles in the pharyngeal wall contract, pulling the pharynx upward toward the food.

5. The lower portion of the inferior constrictor muscles relaxes, opening the esophagus.

6. The superior constrictor muscles contract, stimulating a peristaltic wave to begin in other pharyngeal muscles. This wave forces the food into the esophagus.

As the swallowing reflexes occur, breathing is momentarily inhibited. Then, during the third stage of swallowing, peristalsis transports the food in the esophagus to the stomach.

## Esophagus

The **esophagus** is a straight, collapsible tube about 25 centimeters long. It provides a passageway for food from the pharynx to the stomach. The esophagus descends through the thorax posterior to the trachea, passing through the mediastinum. It penetrates the diaphragm through an opening, the *esophageal hiatus,* and is continuous with the stomach on the abdominal side of the diaphragm (figs. 17.15, 17.16, and reference plates 57, 73).

Mucous glands are scattered throughout the submucosa of the esophagus. Their secretions moisten and lubricate the inner lining of the tube.

In a *hiatal hernia,* a portion of the stomach protrudes through a weakened area of the diaphragm, through the esophageal hiatus and into the thorax. As a result of a hiatal hernia, regurgitation (reflux) of gastric juice into the esophagus may inflame the esophageal mucosa, causing heartburn, difficulty in swallowing, or ulceration and blood loss. In response to the destructive action of gastric juice, columnar epithelium may replace the squamous epithelium that normally lines the esophagus (see chapter 5). This condition, called *Barrett's esophagus,* increases the risk of developing esophageal cancer.

**FIGURE** 17.14

Steps in the swallowing reflex. (*a*) The tongue forces food into the pharynx. (*b*) The soft palate, hyoid bone, and larynx are raised, the tongue is pressed against the palate, the epiglottis closes, and the inferior constrictor muscles relax so that the esophagus opens. (*c*) Superior constrictor muscles contract and force food into the esophagus. (*d*) Peristaltic waves move food through the esophagus to the stomach.

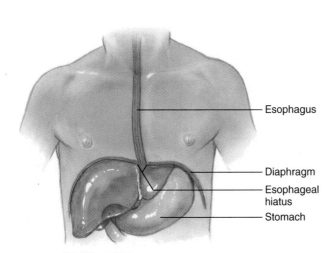

**FIGURE** 17.15

The esophagus functions as a passageway between the pharynx and the stomach.

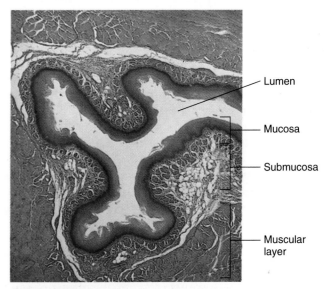

**FIGURE** 17.16

This cross section of the esophagus shows its muscular wall (10×).

(a)

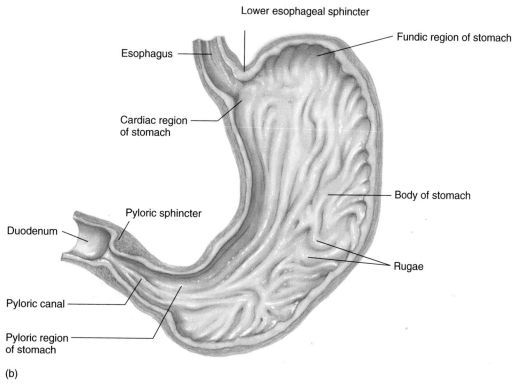

(b)

**FIGURE 17.17**

(*a*) Some parts of the stomach have three layers of muscle fibers. (*b*) Major regions of the stomach.

Just superior to the point where the esophagus joins the stomach, some of the circular muscle fibers have increased sympathetic muscle tone, forming the **lower esophageal sphincter** (cardiac sphincter) (fig. 17.17). These fibers usually remain contracted and they close the entrance to the stomach. In this way, they help prevent regurgitation of the stomach contents into the esophagus. When peristaltic waves reach the stomach, the muscle fibers that guard its entrance relax and allow the swallowed food to enter.

**1** Describe the regions of the pharynx.

**2** List the major events that occur during swallowing.

**3** What is the function of the esophagus?

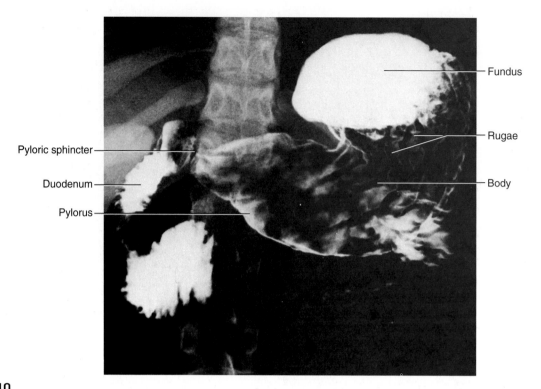

Pyloric sphincter

Duodenum

Pylorus

Fundus

Rugae

Body

**FIGURE 17.18**

Radiograph of a stomach. (*Note:* A radiopaque compound the patient swallowed appears white in the radiograph.)

## STOMACH

The **stomach** is a J-shaped, pouchlike organ, about 25–30 centimeters long, which hangs inferior to the diaphragm in the upper left portion of the abdominal cavity. It has a capacity of about one liter or more, and its inner lining is marked by thick folds (rugae) of the mucosal and submucosal layers that tend to disappear when its wall is distended. The stomach receives food from the esophagus, mixes it with gastric juice, initiates the digestion of proteins, carries on a limited amount of absorption, and moves food into the small intestine.

In addition to the two layers of smooth muscle—an inner circular layer and an outer longitudinal layer—found in other regions of the alimentary canal, some parts of the stomach have another inner layer of oblique fibers. This third innermost muscular layer is most highly developed near the opening of the esophagus and in the body of the stomach (fig. 17.17).

### Parts of the Stomach

The stomach, shown in figures 17.17 and 17.18 and reference plate 51, can be divided into the cardiac, fundic, body, and pyloric regions. The *cardiac region* is a small area near the esophageal opening (cardia). The *fundic region,* which balloons superior to the cardiac portion, is a temporary storage area and sometimes fills with swallowed air. This produces a gastric air bubble, which may

be used as a landmark on an radiograph of the abdomen. The dilated *body region,* which is the main part of the stomach, is located between the fundic and pyloric portions. The *pyloric region* (antrum) is a funnel-shaped portion that narrows and becomes the *pyloric canal* as it approaches the small intestine.

At the end of the pyloric canal, the circular layer of fibers in its muscular wall thickens, forming a powerful muscle, the **pyloric sphincter.** This muscle is a valve that controls gastric emptying.

*Hypertrophic pyloric stenosis* is a birth defect in which muscle overgrowth blocks the pyloric canal. The newborn vomits, with increasing force. To diagnose the condition, a radiograph is taken of the area after the infant drinks formula containing a radiopaque barium compound. Surgical splitting of the muscle blocking the passageway from stomach to small intestine is necessary to enable the infant to eat normally. Pyloric stenosis can occur later in life as a result of ulcers or cancer.

### Gastric Secretions

The mucous membrane that forms the inner lining of the stomach is thick, its surface studded with many small

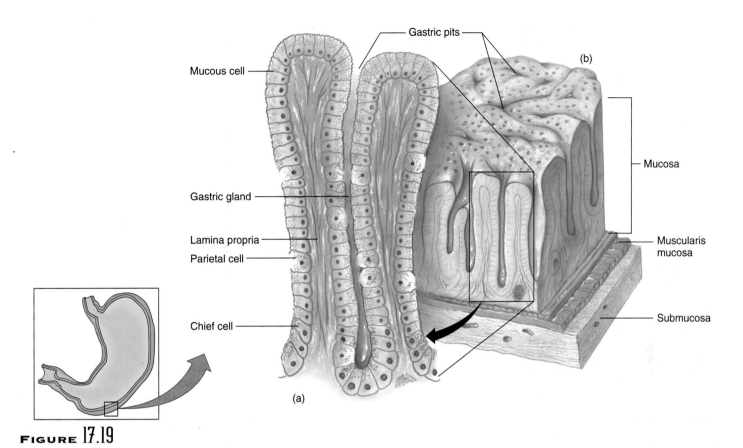

**FIGURE** 17.19

(*a*) Gastric glands include mucous cells, parietal cells, and chief cells. (*b*) The mucosa of the stomach is studded with gastric pits that are the openings of the gastric glands.

openings. These openings, called *gastric pits,* are located at the ends of tubular **gastric glands** (oxyntic glands) (fig. 17.19). Although their structure and the composition of their secretions vary in different parts of the stomach, gastric glands generally contain three types of secretory cells. One type, the *mucous cell* (goblet cell), occurs in the necks of the glands near the openings of the gastric pits. The other types, *chief cells* (peptic cells) and *parietal cells* (oxyntic cells), are found in the deeper parts of the glands (figs. 17.19 and 17.20). The chief cells secrete digestive enzymes, and the parietal cells release a solution containing hydrochloric acid. The products of the mucous cells, chief cells, and parietal cells together form **gastric juice.**

Much of what we know about the stomach's functioning comes from a French-Canadian explorer, Alexis St. Martin, who in 1822 accidentally shot himself in the abdomen. His extensive injuries eventually healed, but a hole, called a fistula, was left, allowing observers to look at his stomach in action. A U.S. Army surgeon, William Beaumont, spent eight years watching food digesting in the stomach, noting how the stomach lining changed in response to stress.

**FIGURE** 17.20

A light micrograph of cells associated with the gastric glands (50× micrograph enlarged to 100×).

TABLE 17.5

## MAJOR COMPONENTS OF GASTRIC JUICE

| Component | Source | Function |
|---|---|---|
| Pepsinogen | Chief cells of the gastric glands | Inactive form of pepsin |
| Pepsin | Formed from pepsinogen in the presence of hydrochloric acid | A protein-splitting enzyme that digests nearly all types of dietary protein |
| Hydrochloric acid | Parietal cells of the gastric glands | Provides the acid environment needed for the conversion of pepsinogen into pepsin and for the action of pepsin |
| Mucus | Goblet cells and mucous glands | Provides a viscous, alkaline protective layer on the stomach wall |
| Intrinsic factor | Parietal cells of the gastric glands | Aids in absorption of vitamin $B_{12}$ |

Of the several digestive enzymes in gastric juice, **pepsin** is by far the most important. The chief cells secrete it as an inactive, nonerosive enzyme precursor called **pepsinogen.** When pepsinogen contacts the hydrochloric acid from the parietal cells, however, it changes rapidly into pepsin. Pepsin, in turn, can also convert pepsinogen into more pepsin.

Pepsin begins the digestion of nearly all types of dietary protein. This enzyme is most active in an acid environment, which the hydrochloric acid in gastric juice provides.

Gastric juice also contains small quantities of a fat-splitting enzyme, *gastric lipase.* However, its action is weak due in part to the low pH of gastric juice. Gastric lipase acts mainly on butterfat.

The mucous cells of the gastric glands secrete large quantities of thin mucus. In addition, the cells of the mucous membrane, associated with the inner lining of the stomach and between the gastric glands, release a more viscous and alkaline secretion, which coats the inside of the stomach wall. This coating is especially important because pepsin can digest the proteins of stomach tissues, as well as those in foods. Thus, the coating normally prevents the stomach from digesting itself.

Still another component of gastric juice is **intrinsic factor.** The parietal cells of the gastric glands secrete intrinsic factor, which is required for vitamin $B_{12}$ absorption from the small intestine. Table 17.5 summarizes the components of gastric juice.

1. Where is the stomach located?

2. What are the secretions of the chief cells and parietal cells?

3. Which is the most important digestive enzyme in gastric juice?

4. Why doesn't the stomach digest itself?

## Regulation of Gastric Secretions

Gastric juice is produced continuously, but the rate varies considerably from time to time and is controlled both neurally and hormonally. More specifically, within the gastric glands, specialized cells closely associated with the parietal cells secrete the hormone *somatostatin,* which inhibits acid secretion. However, acetylcholine (ACh) released from nerve endings in response to parasympathetic impulses arriving on the vagus nerves suppresses the secretion of somatostatin and stimulates the gastric glands to secrete large amounts of gastric juice, which is rich in hydrochloric acid and pepsinogen. These parasympathetic impulses also stimulate certain stomach cells, mainly in the pyloric region, to release a peptide hormone called **gastrin,** which increases the secretory activity of gastric glands (fig. 17.21). Furthermore, parasympathetic impulses and gastrin promote release of *histamine* from gastric mucosal cells, which, in turn, stimulates additional gastric secretion.

Histamine is very effective in promoting secretion of gastric acid. Consequently, drugs that block the histamine receptors of gastric mucosal cells ($H_2$-blockers) are often used to inhibit excess gastric acid secretion.

Three stages of gastric secretion are recognized—the cephalic, gastric, and intestinal phases. The *cephalic phase* of gastric secretion begins before any food reaches the stomach and possibly even before eating. In this stage, parasympathetic reflexes operating through the vagus nerves stimulate gastric secretion whenever a person tastes, smells, sees, or even thinks about food. Furthermore, the hungrier the person is, the greater the amount of gastric secretion. The cephalic phase of secretion is responsible for 30%–50% of the secretory response to a meal.

1. Parasympathetic preganglionic nerve fiber

Vagus nerve fiber

4. Gastrin stimulates glands to release more gastric juice

Bloodstream

2. Parasympathetic postganglionic impulses stimulate the release of gastric juice from gastric glands

3. Impulses stimulate the release of gastrin

**FIGURE 17.21**

The secretion of gastric juice is regulated in part by parasympathetic nerve impulses that stimulate the release of gastric juice and gastrin.

| TABLE 17.6 | PHASES OF GASTRIC SECRETION |
| --- | --- |
| **Phase** | **Action** |
| Cephalic phase | The sight, taste, smell, or thought of food triggers parasympathetic reflexes. Gastric juice is secreted in response. |
| Gastric phase | Food in stomach chemically and mechanically stimulates release of gastrin, which, in turn, stimulates secretion of gastric juice; reflex responses also stimulate gastric juice secretion. |
| Intestinal phase | As food enters the small intestine, it stimulates intestinal cells to release intestinal gastrin, which, in turn, promotes the secretion of gastric juice from the stomach wall. |

The *gastric phase* of gastric secretion, which accounts for 40%–50% of the secretory activity, starts when food enters the stomach. The presence of food and the distension of the stomach wall trigger the stomach to release gastrin, which stimulates production of still more gastric juice.

As food enters the stomach and mixes with gastric juice, the pH of the contents rises, which enhances gastrin secretion. Consequently, the pH of the stomach contents drops. As the pH approaches 3.0, secretion of gastrin is inhibited. When the pH reaches 1.5, gastrin secretion ceases.

> Gastrin stimulates cell growth in the mucosa of the stomach and intestines, except where gastrin is produced. This effect helps repair mucosal cells damaged by normal stomach function, disease, or medical treatments.

For the stomach to secrete hydrochloric acid, hydrogen ions are removed from the blood, and an equivalent amount of alkaline bicarbonate ions is released into the blood. Consequently, following a meal, the blood concentration of bicarbonate ions increases, and the urine excretes excess bicarbonate ions. This phenomenon is called the *alkaline tide.*

The *intestinal phase* of gastric secretion, which accounts for about 5% of the total secretory response to a meal, begins when food leaves the stomach and enters the small intestine. When food first contacts the intestinal wall, it stimulates intestinal cells to release a hormone that again enhances gastric gland secretion. Many investigators believe this hormone is identical to gastrin, and they call it *intestinal gastrin.*

As more food moves into the small intestine, secretion of gastric juice from the stomach wall is inhibited, due to a sympathetic reflex triggered by acid in the upper part of the small intestine. Also, the presence of proteins and fats in this region of the intestine causes the release of the peptide hormone **cholecystokinin** from the intestinal wall, which decreases gastric motility. Similarly, fats in the small intestine stimulate intestinal cells to release *intestinal somatostatin,* which inhibits release of gastric juice. Overall, these actions decrease gastric secretion and motility, as the small intestine fills with food. Table 17.6 summarizes the phases of gastric secretion, and Clinical Application 17.2 discusses a common problem—indigestion.

**1** What controls gastric juice secretion?

**2** Distinguish between the cephalic, gastric, and intestinal phases of gastric secretion.

**3** What is the function of cholecystokinin?

## Gastric Absorption

Although gastric enzymes begin breaking down proteins, the stomach wall is not well-adapted to absorb digestive products. The stomach absorbs only small quantities of water and certain salts, as well as alcohol, and some lipid-soluble drugs.

> Most nutrients are absorbed in the small intestine. Alcohol, which is not a nutrient, is absorbed both in the small intestine and stomach. This is why the intoxicating effects of alcohol are felt soon after consuming alcoholic beverages.

# Oh, My Aching Stomach!

The food at the barbecue was so terrific that Perry W. didn't notice how much he was consuming—two burgers, three hot dogs, beans in a spicy sauce, loads of chips, several beers, and ice cream for dessert. Later, he paid for it, as a feeling of fullness became abdominal pain, then heartburn, as stomach contents backed up into his esophagus.

Perry found relief with an over-the-counter antacid product. Antacids raise the pH of the stomach within minutes. They usually include a compound containing either sodium, calcium, magnesium, or aluminum. Another ingredient in some products is simethicone, which breaks up gas bubbles in the digestive tract. If antacids do not help within a few minutes, or they are used for longer than two weeks, a doctor should be consulted. The problem may be more serious than overeating.

Avoiding acid indigestion and heartburn is a more healthful approach than gorging and then reaching for medication—or even taking products that lower acid production before a large or spicy meal. Some tips:

- Avoid large meals. The more food, the more acid the stomach produces.
- Eat slowly, so that stomach acid secretion is more gradual.

- Do not lie down immediately after eating, so that gravity helps food along the alimentary canal.
- If prone to indigestion or heartburn, avoid caffeine, which increases stomach acid secretion.
- Cigarettes and alcohol irritate the stomach lining and relax the sphincter at the junction between the stomach and the esophagus. This makes it easier for food to return to the esophagus, causing heartburn.
- Do not eat acidic foods, such as citrus fruits and tomatoes, unless it is at least three hours before bedtime.
- Use a pillow that elevates the head six to eight inches above the stomach. ■

## Mixing and Emptying Actions

Food entering the stomach stretches the smooth muscles in its wall. The stomach may enlarge, but its muscles maintain their tone, and internal pressure of the stomach normally is unchanged. A person may eat more than the stomach can comfortably hold, and when this happens, the internal pressure may rise enough to stimulate pain receptors. The result is a stomachache.

Following a meal, the mixing movements of the stomach wall aid in producing a semifluid paste of food particles and gastric juice called **chyme.** Peristaltic waves push the chyme toward the pyloric region of the stomach, and as chyme accumulates near the pyloric sphincter, this muscle begins to relax. Stomach contractions push chyme a little (5–15 milliliters) at a time into the small intestine. These contraction waves push most of the chyme backward into the stomach, mixing it further. Figure 17.22 illustrates this process.

The rate at which the stomach empties depends on the fluidity of the chyme and the type of food. Liquids usually pass through the stomach quite rapidly, but solids remain until they are well mixed with gastric juice. Fatty foods may remain in the stomach three to six hours; foods high in proteins move through more quickly; carbohydrates usually pass through more rapidly than either fats or proteins.

As chyme fills the duodenum, its internal pressure increases, and the intestinal wall stretches. These actions stimulate sensory receptors in the wall, triggering an **enterogastric reflex.** The name of this reflex, like those of other digestive reflexes, describes the origin and termination of reflex impulses. Thus, the enterogastric reflex begins in the small intestine (*entero*) and ends in the stomach (*gastro*).

As a result of the enterogastric reflex, fewer parasympathetic impulses arrive at the stomach, inhibiting peristalsis, and the intestine fills less rapidly (fig. 17.23). Also, if chyme entering the intestine is fatty, the intestinal wall releases the hormone cholecystokinin, which further inhibits peristalsis.

As chyme enters the duodenum (the first portion of the small intestine), accessory organs add secretions. These organs include the pancreas, the liver, and the gallbladder.

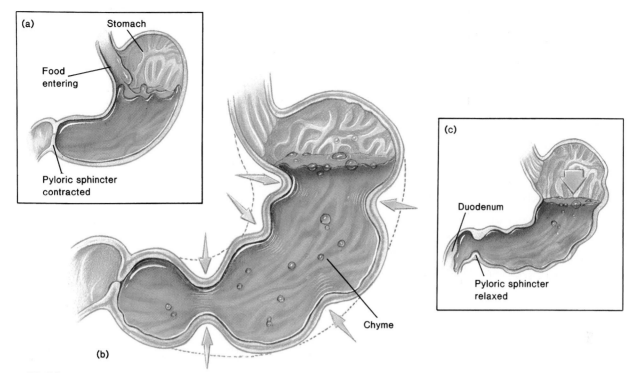

FIGURE 17.22

Stomach movements. (*a*) As the stomach fills, its muscular wall stretches, but the pyloric sphincter remains closed. (*b*) Mixing movements combine food and gastric juice, creating chyme. (*c*) Peristaltic waves move the chyme toward the pyloric sphincter, which relaxes and admits some chyme into the duodenum.

(4) Nerve impulses inhibit peristalsis in stomach wall

From CNS

Vagus nerve

To CNS

(1) Duodenum fills with chyme

(3) Sensory nerve impulses travel to central nervous system

(2) Sensory stretch receptors are stimulated

FIGURE 17.23

The enterogastric reflex partially regulates the rate at which chyme leaves the stomach.

*Vomiting* results from a complex reflex that empties the stomach another way. Irritation or distension in the stomach or intestines can trigger vomiting. Sensory impulses travel from the site of stimulation to the *vomiting center* in the medulla oblongata, and motor responses follow. These include taking a deep breath, raising the soft palate and thus closing the nasal cavity, closing the opening to the trachea (glottis), relaxing the circular muscle fibers at the base of the esophagus, contracting the diaphragm so it presses downward over the stomach, and contracting the abdominal wall muscles to increase pressure inside the abdominal cavity. As a result, the stomach is squeezed from all sides, forcing its contents upward and out through the esophagus, pharynx, and mouth.

Activity in the vomiting center can be stimulated by drugs (emetics), by toxins in contaminated foods, and sometimes by rapid changes in body motion. In this last situation, sensory impulses from the labyrinths of the inner ears reach the vomiting center, and can produce motion sickness. The vomiting center can also be activated by stimulation of higher brain centers through sights, sounds, odors, tastes, emotions, or mechanical stimulation of the back of the pharynx.

*Nausea* emanates from activity in the vomiting center or in nerve centers near it. During nausea, stomach movements usually are diminished or absent, and duodenal contents may move back into the stomach.

Duodenum

Common bile duct

Pancreatic duct

Tail of pancreas

Hepatopancreatic ampulla

Hepatopancreatic sphincter

Head of pancreas

Pancreatic duct

Common bile duct

Duodenal papilla

Intestinal lumen

Hepatopancreatic ampulla

Hepatopancreatic sphincter

**FIGURE 17.24** 🦅

The pancreas is closely associated with the duodenum.

1. How is chyme produced?

2. What factors influence how quickly chyme leaves the stomach?

3. Describe the enterogastric reflex.

4. Describe the vomiting reflex.

5. What factors may stimulate the vomiting reflex?

## PANCREAS

The **pancreas,** discussed as an endocrine gland in chapter 13, also has an exocrine function—secretion of a digestive juice called **pancreatic juice.**

### Structure of the Pancreas

The pancreas is closely associated with the small intestine and is located posterior to the parietal peritoneum. It extends horizontally across the posterior abdominal wall, with its head in the C-shaped curve of the duodenum and its tail against the spleen (fig. 17.24 and reference plate 59).

The cells that produce pancreatic juice, called *pancreatic acinar cells,* make up the bulk of the pancreas. These cells cluster around tiny tubes, into which they release their secretions. The smaller tubes unite to form larger ones, which, in turn, give rise to a *pancreatic duct* extending the length of the pancreas. The pancreatic duct usually connects with the duodenum at the same place where the bile duct from the liver and gallbladder joins the duodenum, although other connections may be present (see figs. 13.32 and 17.24).

The pancreatic and bile ducts join at a short, dilated tube called the *hepatopancreatic ampulla* (ampulla of Vater). A band of smooth muscle, called the *hepatopancreatic sphincter* (sphincter of Oddi), surrounds this ampulla.

### Pancreatic Juice

Pancreatic juice contains enzymes that digest carbohydrates, fats, proteins, and nucleic acids. The carbohydrate-

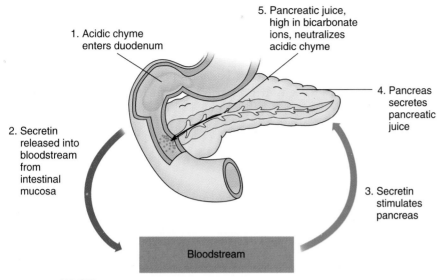

**FIGURE** 17.25

Acidic chyme entering the duodenum from the stomach stimulates release of secretin, which in turn stimulates release of pancreatic juice.

1. Acidic chyme enters duodenum
2. Secretin released into bloodstream from intestinal mucosa
3. Secretin stimulates pancreas
4. Pancreas secretes pancreatic juice
5. Pancreatic juice, high in bicarbonate ions, neutralizes acidic chyme

Bloodstream

digesting enzyme, **pancreatic amylase,** splits molecules of starch or glycogen into double sugars (disaccharides). The fat-digesting enzyme, **pancreatic lipase,** breaks triglyceride molecules into fatty acids and monoglycerides. (A monoglyceride molecule consists of one fatty acid bound to glycerol.)

The protein-splitting (proteolytic) enzymes are **trypsin, chymotrypsin,** and **carboxypeptidase.** Each of these enzymes splits the bonds between particular combinations of amino acids in proteins. Because no single enzyme can split all possible combinations of amino acids, several enzymes are necessary to completely digest protein molecules.

The proteolytic enzymes are stored in inactive forms within tiny cellular structures called *zymogen granules.* These enzymes, like gastric pepsin, are secreted in inactive forms and must be activated by other enzymes after they reach the small intestine. For example, the pancreatic cells release inactive **trypsinogen,** which is activated to trypsin when it contacts the enzyme *enterokinase,* which the mucosa of the small intestine secretes. Chymotrypsin and carboxypeptidase are activated, in turn, by trypsin. This mechanism prevents enzymatic digestion of proteins within the secreting cells and the pancreatic ducts.

A painful condition called *acute pancreatitis* results from a blockage in the release of pancreatic juice. Trypsinogen, activated as pancreatic juice builds up, digests parts of the pancreas. Alcoholism, gallstones, certain infections, traumatic injuries, or the side effects of some drugs can cause pancreatitis. The beginning of chapter 1 describes a hereditary form of pancreatitis.

Pancreatic juice also contains two **nucleases,** which are enzymes that break down nucleic acid molecules into nucleotides, as well as a high concentration of bicarbonate ions that makes it alkaline. The alkaline content of pancreatic juice provides a favorable environment for the actions of the digestive enzymes and helps neutralize the acidic chyme as it arrives from the stomach. At the same time, the alkaline condition in the small intestine blocks the action of pepsin, which might otherwise damage the duodenal wall.

## Regulation of Pancreatic Secretion

As with gastric and small intestinal secretions, the nervous and endocrine systems regulate release of pancreatic juice. For example, during the cephalic and gastric phases of gastric secretion, parasympathetic impulses reach the pancreas and stimulate it to release digestive enzymes. A peptide hormone, **secretin,** stimulates the pancreas to secrete a large quantity of fluid when acidic chyme enters the duodenum. Secretin is released into the blood from the duodenal mucous membrane in response to the acid in chyme. The pancreatic juice secreted at this time contains few, if any, digestive enzymes, but has a high concentration of bicarbonate ions that neutralizes the acid in chyme (fig. 17.25).

In cystic fibrosis, abnormal chloride channels in cells in various tissues cause water to be drawn into the cells from interstitial spaces. This dries out secretions in the lungs and pancreas, leaving a very sticky mucus that impairs the functioning of these organs. When the pancreas is plugged with mucus, its secretions, containing digestive enzymes, cannot reach the duodenum. Individuals with cystic fibrosis must take digestive enzyme supplements—usually as a powder mixed with a soft food such as applesauce—to prevent malnutrition.

The presence of proteins and fats in chyme within the duodenum also stimulates the release of cholecystokinin from the intestinal wall. As in the case of secretin, cholecystokinin reaches the pancreas by way of the bloodstream. Pancreatic juice secreted in response to cholecystokinin has a high concentration of digestive enzymes.

1. Where is the pancreas located?

2. List the enzymes in pancreatic juice.

3. What are the functions of the enzymes in pancreatic juice?

4. What regulates secretion of pancreatic juice?

## LIVER

The **liver,** the largest internal organ, is located in the upper right quadrant of the abdominal cavity, just inferior to the diaphragm. It is partially surrounded by the ribs, and extends from the level of the fifth intercostal space to the lower margin of the ribs. It is reddish-brown in color and well supplied with blood vessels (figs. 17.26, 17.27, 17.28 and reference plates 51, 57, 74).

### Liver Structure

A fibrous capsule encloses the liver, and connective tissue divides the organ into a large *right lobe* and a smaller *left lobe.* The *falciform ligament* is a fold of visceral peritoneum that separates the lobes and fastens the liver to the abdominal wall anteriorly. The liver also has two minor lobes, the *quadrate lobe* near the gallbladder, and the *caudate lobe* close to the vena cava (fig. 17.28).

A fold of visceral peritoneum called the *coronary ligament* attaches the liver to the diaphragm on its superior surface. Each lobe is separated into many tiny

**hepatic lobules,** which are the liver's functional units (fig. 17.29). A lobule consists of many *hepatic cells* radiating outward from a *central vein.* Vascular channels called **hepatic sinusoids** separate platelike groups of these cells from each other. Blood from the digestive tract, which is carried in the *hepatic portal vein* (see

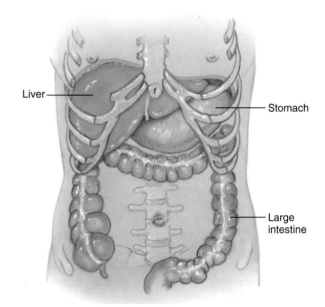

**FIGURE 17.26**

The ribs partially surround the liver.

**FIGURE 17.27**

This transverse section of the abdomen reveals the liver and other organs within the upper portion of the abdominal cavity.

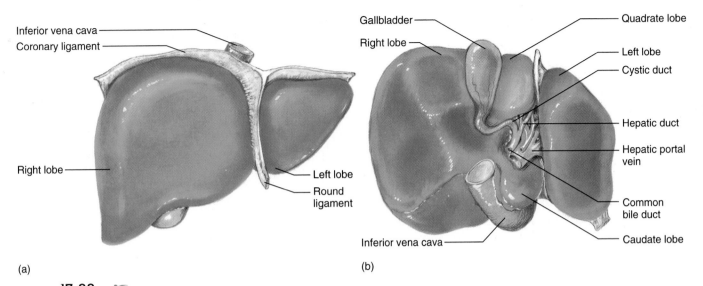

**FIGURE 17.28**

Lobes of the liver, viewed (*a*) anteriorly and (*b*) inferiorly.

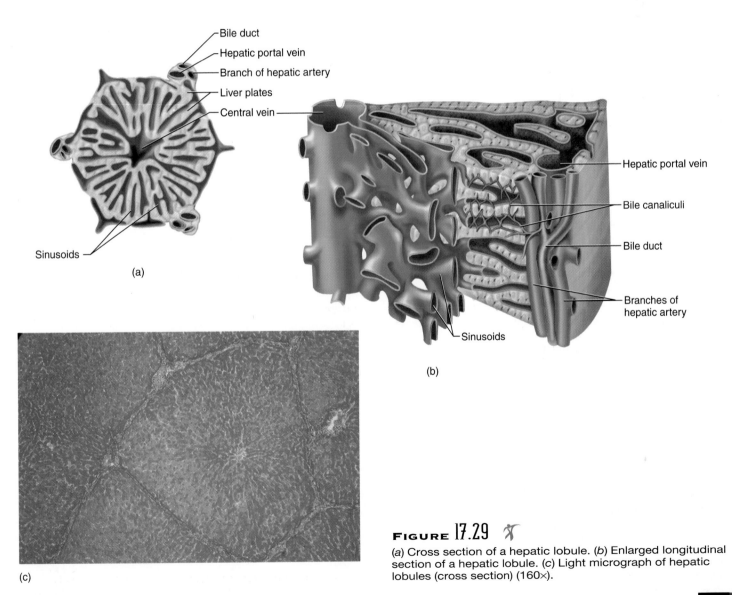

**FIGURE 17.29**

(*a*) Cross section of a hepatic lobule. (*b*) Enlarged longitudinal section of a hepatic lobule. (*c*) Light micrograph of hepatic lobules (cross section) (160×).

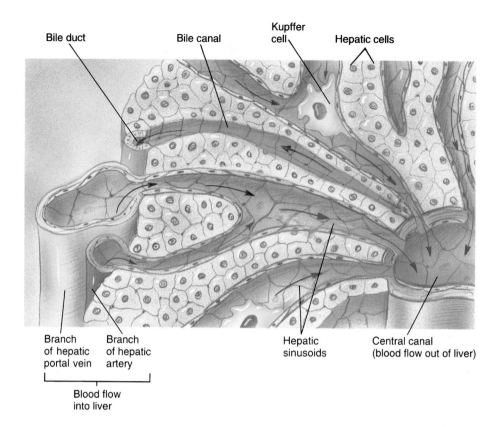

Bile duct     Bile canal     Kupffer cell     Hepatic cells

Branch of hepatic portal vein     Branch of hepatic artery     Hepatic sinusoids     Central canal (blood flow out of liver)

Blood flow into liver

**FIGURE 17.30**

The paths of blood and bile within a hepatic lobule.

chapter 15), brings newly absorbed nutrients into the sinusoids (fig. 17.30). At the same time, oxygenated blood from the hepatic artery mixes freely with the blood containing nutrients, then flows through the liver sinusoids and nourishes the hepatic cells.

Often blood in the portal veins contains some bacteria that have entered through the intestinal wall. However, large **Kupffer cells,** which are fixed to the inner lining (endothelium) of the hepatic sinusoids, remove most of the bacteria from the blood by phagocytosis. Then the blood passes into the *central veins* of the hepatic lobules and moves out of the liver.

Within the liver lobules are many fine *bile canals* (or canaliculi), which receive secretions from the hepatic cells. The canals of neighboring lobules unite to form larger ducts, and then converge to become the **hepatic ducts.** These ducts merge, in turn, to form the *common hepatic duct.*

## Liver Functions

The liver carries on many important metabolic activities. Recall from chapter 13 that the liver plays a key role in carbohydrate metabolism by helping maintain the normal concentration of blood glucose. Liver cells that respond to hormones such as insulin and glucagon lower

the blood glucose level by polymerizing glucose to glycogen and raise the blood glucose level by breaking down glycogen to glucose or by converting noncarbohydrates into glucose.

The liver's effects on lipid metabolism include oxidizing fatty acids at an especially high rate (see chapter 4); synthesizing lipoproteins, phospholipids, and cholesterol; and converting portions of carbohydrate and protein molecules into fat molecules. The blood transports fats synthesized in the liver to adipose tissue for storage.

The most vital liver functions are probably those related to protein metabolism. They include deaminating amino acids; forming urea (see chapter 20); synthesizing plasma proteins, such as clotting factors (see chapter 14); and converting certain amino acids to other amino acids.

Bacteria in the intestine produce ammonia, which is carried in the blood to the liver, where it is converted to urea. When this liver function fails, concentration of blood ammonia sharply rises, causing *hepatic coma,* a condition that can lead to death.

| **TABLE 17.7** | **MAJOR FUNCTIONS OF THE LIVER** | | |
|---|---|---|---|
| **General Function** | **Specific Function** | **General Function** | **Specific Function** |
| Carbohydrate metabolism | Polymerizes glucose to glycogen, breaks down glycogen to glucose, and converts noncarbohydrates to glucose | Protein metabolism | Deaminates amino acids; forms urea; synthesizes plasma proteins; interconverts amino acids |
| Lipid metabolism | Oxidizes fatty acids; synthesizes lipoproteins, phospholipids, and cholesterol; converts portions of carbohydrate and protein molecules into fats | Storage | Stores glycogen, vitamins A, D, and $B_{12}$, iron, and blood |
| | | Blood filtering | Removes damaged red blood cells and foreign substances by phagocytosis |
| | | Detoxification | Removes toxins from the blood |
| | | Secretion | Secretes bile |

The liver also stores many substances, including glycogen, iron, and vitamins A, D, and $B_{12}$. Extra iron from the blood combines with a protein (apoferritin) in liver cells, forming *ferritin*. The iron is stored in this form until blood iron concentration falls, when some of the iron is released. Thus, the liver is important in iron homeostasis.

Liver cells help destroy damaged red blood cells and phagocytize foreign antigens. The liver removes toxic substances such as alcohol from the blood (detoxification). The liver can also serve as a blood reservoir, storing 200 to 400 milliliters of blood. The liver's role in digestion is to secrete bile. Table 17.7 summarizes the major functions of the liver. Clinical Application 17.3 discusses hepatitis, an inflammation of the liver.

1 Describe the location of the liver.

2 Describe a hepatic lobule.

3 Review liver functions.

4 Which liver function participates in digestion?

## Composition of Bile

**Bile** is a yellowish-green liquid that hepatic cells continuously secrete. In addition to water, it contains *bile salts, bile pigments, cholesterol,* and *electrolytes.* Of these, bile salts are the most abundant. They are the only bile substances that have a digestive function.

Hepatic cells use cholesterol to produce bile salts, and in secreting these salts, they release some cholesterol into the bile. Cholesterol by itself has no special function in bile or in the alimentary canal.

Bile pigments (bilirubin and biliverdin) are breakdown products of hemoglobin from red blood cells. These pigments are normally excreted in the bile (see chapter 14). The yellowish skin, sclerae, and mucous membranes of jaundice result from excess deposition of bile pigments.

Jaundice can have several causes. In *obstructive jaundice*, bile ducts are blocked. In *hepatocellular jaundice*, the liver is diseased. In *hemolytic jaundice*, red blood cells are destroyed too rapidly.

## Gallbladder

The **gallbladder** is a pear-shaped sac located in a depression on the inferior surface of the liver. It is connected to the **cystic duct,** which, in turn, joins the hepatic duct (see fig. 17.28 and reference plate 59). The gallbladder has a capacity of 30–50 milliliters, is lined with columnar epithelial cells, and has a strong muscular layer in its wall. It stores bile between meals, concentrates bile by reabsorbing water, and releases bile into the duodenum when stimulated by cholecystokinin from the small intestine.

The **common bile duct** is formed by the union of the common hepatic and cystic ducts. It leads to the duodenum, where the hepatopancreatic sphincter muscle guards its exit (see fig. 17.24). This sphincter normally remains contracted, so that bile collects in the common bile duct and backs up into the cystic duct. When this happens, the bile flows into the gallbladder, where it is stored.

Although the bile is in the gallbladder, bile salts, bile pigments, and cholesterol become increasingly concentrated as the gallbladder lining reabsorbs some of the water and electrolytes. Although the cholesterol normally remains in solution, under certain conditions it may precipitate and form solid crystals. If cholesterol continues to come out of solution, these crystals become larger and larger, forming *gallstones* (fig. 17.31).

Gallstones may form if the bile is too concentrated, if the hepatic cells secrete too much cholesterol, or if the gallbladder is inflamed (cholecystitis). If such stones get

# Hepatitis

Television news reports in April 1997 showed sad children lining up to receive immunizations against hepatitis A. The viral infection had struck about 150 students and teachers in Michigan. The tainted frozen strawberries in their school lunches that caused the illness had also been distributed in six other states.

## LIVER INFLAMMATION CAUSES DISTINCT SYMPTOMS

Hepatitis is an inflammation of the liver caused by viral infection, or, more rarely, from reaction to a drug, alcoholism, or autoimmunity. The type of hepatitis that struck the schoolchildren—hepatitis A—is one of the least severe forms of this common illness. For the first few days, symptoms of hepatitis A resemble the flu, and include mild headache, low fever, fatigue, lack of appetite, nausea and vomiting, and sometimes stiff joints. By the end of the first week, more distinctive symptoms arise: a rash, pain in the upper right quadrant of the abdomen, dark and foamy urine, and pale feces. The skin and sclera of the eyes begin to turn yellow from accumulating bile pigments (jaundice). Great fatigue may continue for two or three weeks, and then gradually the person begins to feel better.

At the other end of the hepatitis spectrum is fulminant hepatitis, which is rare and can be caused by any of several viruses. Symptoms start suddenly and severely, and behavior and personality may change. Without medical attention, the condition progresses to kidney and liver failure, or coma.

Hepatitis B produces chronic symptoms that persist for more than 6 months. Perhaps as many as 300 million people worldwide are carriers of hepatitis B. They do not have symptoms, but can infect others. Five percent of such carriers eventually develop liver cancer.

## AN ALPHABET OF VIRAL CAUSES

Several types of viruses can cause hepatitis. Viral types are distinguished by the route of infection and by biochemical differences, such as gene sequences and surface proteins.

At the beginning of the twentieth century, before investigators knew how to study viruses, two types of hepatitis were distinguished epidemiologically. So-called "infectious hepatitis," which was transmitted from person to person, was later attributed to the hepatitis A virus. "Serum hepatitis" was transmitted by blood, and was later found to be caused by the hepatitis B virus. Hepatitis A often arose from food handlers who did not properly wash after using the bathroom, establishing a fecal-oral route of transmission. Hepatitis B was more often passed sexually.

By the mid-1970s, technology enabled physicians to identify either hepatitis A or hepatitis B virus. But then a problem arose: many cases of what appeared to be hepatitis were not caused by either of the known viral types. These were called "non-A non-B" hepatitis—which just meant that researchers did not know what caused them. Then in the 1980s, the "non-A and non-B" viruses were identified.

The hepatitis C virus causes more than 90% of cases of hepatitis that arise after a person receives a blood transfusion. Hepatitis D is bloodborne, usually associated with blood transfusion and intravenous drug use, and occurs in people already infected with hepatitis B. It kills about 20% of the people it infects. Hepatitis E infection is more common in developing nations, where it often severely affects pregnant women.

But some cases of hepatitis still could not be linked to a specific virus, and these became known as "non-A-E" hepatitis. Since 1994, two new hepatitis viruses have been identified and associated with some of the non-A-E cases. Very little is known about the hepatitis F virus, but it can pass from human feces to infect other primates. Hepatitis G is very rare, but seems to account for a significant percentage of cases of the fulminant form of hepatitis. However, in people with healthy immune systems, it produces symptoms so mild that they may not even be noticed.

Because hepatitis is most often caused by a virus, treatment with antibiotic drugs, which target bacteria, are ineffective. Often the person must simply endure the symptoms. Hepatitis C, however, sometimes responds to a form of interferon, an immune system biochemical. ■

into the bile duct, they may block the flow of bile, causing obstructive jaundice and considerable pain. Clinical Application 17.4 discusses disorders of the gallbladder.

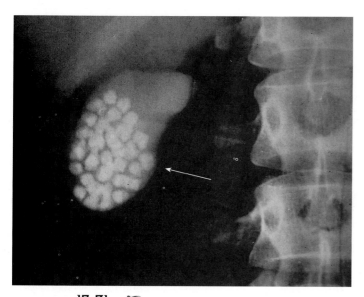

### Regulation of Bile Release

Normally, bile does not enter the duodenum until cholecystokinin stimulates the gallbladder to contract. The intestinal mucosa releases this hormone in response to proteins and fats in the small intestine. (*Note:* Recall its action to stimulate pancreatic enzyme secretion.) The hepatopancreatic sphincter usually remains contracted until a peristaltic wave in the duodenal wall approaches it. Then, just before the wave reaches it, the sphincter relaxes and a squirt of bile enters the duodenum (fig. 17.32). The hormones that help control digestive functions are summarized in table 17.8.

**FIGURE 17.31**

Radiograph of a gallbladder that contains gallstones (arrow).

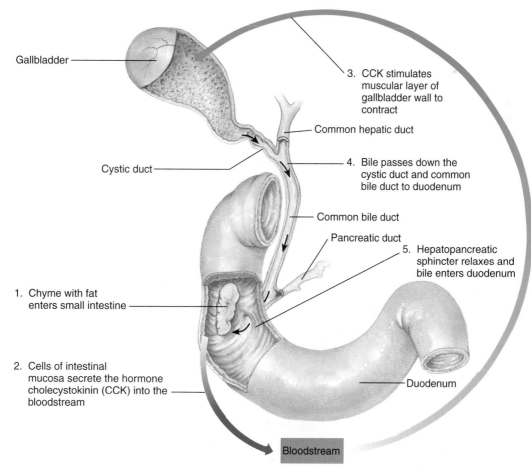

Gallbladder

3. CCK stimulates muscular layer of gallbladder wall to contract

Common hepatic duct

Cystic duct

4. Bile passes down the cystic duct and common bile duct to duodenum

Common bile duct

Pancreatic duct

5. Hepatopancreatic sphincter relaxes and bile enters duodenum

1. Chyme with fat enters small intestine

2. Cells of intestinal mucosa secrete the hormone cholecystokinin (CCK) into the bloodstream

Duodenum

Bloodstream

**FIGURE 17.32**

Fatty chyme entering the duodenum stimulates the gallbladder to release bile.

# Gallbladder Disease

Molly G., an overweight, 47-year-old college administrator and mother of four, had been feeling healthy until recently. Then she regularly began to feel pain in the upper right quadrant of her abdomen (see fig. 1.22). Sometimes the discomfort seemed to radiate around to her back and move upward into her right shoulder. Most commonly, she felt this pain after her evening meal; occasionally it also occurred during the night, awakening her. After an episode of particularly severe pain accompanied by sweating (diaphoresis) and nausea, Molly approached her physician.

During an examination of Molly's abdomen, the physician discovered tenderness in the epigastric region (see fig. 1.21). She decided that Molly might be experiencing the symptoms of *acute cholecystitis*—an inflammation of the gallbladder. The physician recommended that Molly have a *cholecystogram*—an X ray of the gallbladder.

Molly took tablets containing a contrast medium the night before the X-ray procedure. This schedule allowed time for the small intestine to absorb the substance, which was carried to the liver and excreted into the bile. Later, the bile and contrast medium would be stored and concentrated in the gallbladder and would make the contents of the gallbladder opaque to X rays.

Molly's cholecystogram (see fig. 17.31) revealed several stones (calculi) in her gallbladder, a condition called *cholelithiasis.* Because Molly's symptoms of gallbladder disease were worsening, her physician recommended that she consult with an abdominal surgeon about undergoing a *cholecystectomy*—surgical removal of the gallbladder.

During the surgical procedure, an incision was made in Molly's right subcostal region. Her gallbladder was excised from the liver. Then the cystic duct (see fig. 17.28) and hepatic ducts were explored for the presence of stones, but none were found.

Unfortunately, following her recovery from surgery, Molly's symptoms persisted. Her surgeon ordered a *cholangiogram*—an X-ray series of the bile ducts. This study showed a residual stone at the distal end of Molly's common bile duct (see fig. 17.24).

The surgeon extracted the residual stone using a *fiber-optic endoscope,* a long, flexible tube that can be passed through the patient's esophagus and stomach and into the duodenum. This instrument enables a surgeon to observe features of the gastrointestinal tract by viewing them directly through the eyepiece of the endoscope or by watching an image on the screen of a television monitor. A surgeon can also perform manipulations using specialized tools that are passed through the endoscope to its distal end.

In Molly's case, the surgeon performed an *endoscopic papillotomy*—an incision of the hepatopancreatic sphincter (see fig. 17.24). This was done by applying an electric current to a wire extending from the end of the endoscope. She then removed the exposed stone by manipulating a tiny basket at the tip of the endoscope. ■

## Functions of Bile Salts

Bile salts aid digestive enzymes and enhance absorption of fatty acids and certain fat-soluble vitamins. Molecules of fats tend to clump into *fat globules.* Bile salts affect fat globules much like a soap or detergent would, reducing surface tension and breaking fat globules into smaller droplets. This action is called **emulsification.** Monoglycerides that form from the action of pancreatic lipase on triglyceride molecules aid emulsification. This emulsification greatly increases the total surface area of the fatty substance, and the tiny droplets mix with water. Lipases can then digest the fat molecules more effectively.

Bile salts aid in absorption of fatty acids and cholesterol by forming complexes (micelles) that are very soluble in chyme and that epithelial cells can more easily absorb. The fat-soluble vitamins A, D, E, and K are also absorbed. Lack of bile salts results in poor lipid absorption and vitamin deficiencies.

The mucous membrane of the small intestine reabsorbs nearly all of the bile salts, along with fatty acids. The bile salts are carried in the blood to the liver, where hepatic cells resecrete them into the bile ducts. Liver cells synthesize bile salts, which replace the small quantities that are lost in the feces.

TABLE 17.8

## HORMONES OF THE DIGESTIVE TRACT

| Hormone | Source | Function |
|---|---|---|
| Gastrin | Gastric cells, in response to the presence of food | Causes gastric glands to increase their secretory activity |
| Intestinal gastrin | Cells of small intestine, in response to the presence of chyme | Causes gastric glands to increase their secretory activity |
| Somatostatin | Gastric cells | Inhibits secretion of acid by parietal cells |
| Intestinal somatostatin | Intestinal wall cells in response to the presence of fats | Inhibits secretion of acid by parietal cells |
| Cholecystokinin | Intestinal wall cells, in response to the presence of proteins and fats in the small intestine | Causes gastric glands to decrease their secretory activity and inhibits gastric motility; stimulates pancreas to secrete fluid with a high digestive enzyme concentration; stimulates gallbladder to contract and release bile |
| Secretin | Cells in the duodenal wall, in response to acidic chyme entering the small intestine | Stimulates pancreas to secrete fluid with a high bicarbonate ion concentration |

1. Explain how bile originates.

2. Describe the function of the gallbladder.

3. How is secretion of bile regulated?

4. How does bile function in digestion?

## SMALL INTESTINE

The **small intestine** is a tubular organ that extends from the pyloric sphincter to the beginning of the large intestine. With its many loops and coils, it fills much of the abdominal cavity. Although the small intestine is 5.5–6.0 meters (18–20 feet) long in a cadaver when the muscular wall lacks tone, it may be only half this long in a living person.

The small intestine receives secretions from the pancreas and liver. It also completes digestion of the nutrients in chyme, absorbs the products of digestion, and transports the remaining residues to the large intestine.

### Parts of the Small Intestine

The small intestine, shown in figures 17.33 and 17.34, and in reference plates 51, 58, 74, and 75, consists of three portions: the duodenum, the jejunum, and the ileum.

The **duodenum,** which is about 25 centimeters long and 5 centimeters in diameter, lies posterior to the parietal peritoneum (retroperitoneal). It is the shortest and most fixed portion of the small intestine. The duodenum follows a C-shaped path as it passes anterior to the right kidney and the upper three lumbar vertebrae.

The remainder of the small intestine is mobile and lies free in the peritoneal cavity. The proximal two-fifths of this portion is the **jejunum,** and the remainder is the

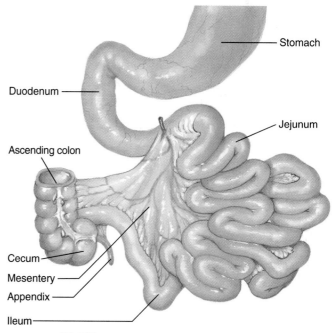

**FIGURE 17.33**

The three parts of the small intestine are the duodenum, the jejunum, and the ileum.

**ileum.** There is no distinct separation between the jejunum and ileum, but the diameter of the jejunum is usually greater, and its wall is thicker, more vascular, and more active than that of the ileum.

The jejunum and ileum are suspended from the posterior abdominal wall by a double-layered fold of peritoneum called **mesentery** (fig. 17.35). The mesentery supports the blood vessels, nerves, and lymphatic vessels that supply the intestinal wall.

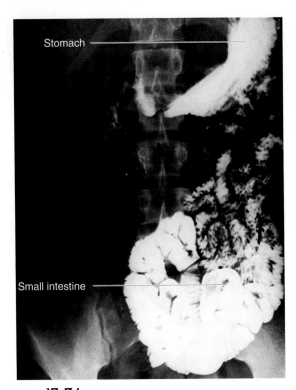

**FIGURE** 17.34

Radiograph showing a normal small intestine.

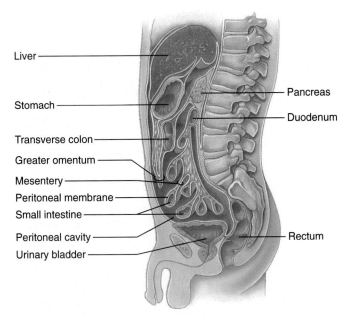

**FIGURE** 17.35

Mesentery formed by folds of the peritoneal membrane suspends portions of the small intestine from the posterior abdominal wall.

A filmy, double fold of peritoneal membrane called the *greater omentum* drapes like an apron from the stomach over the transverse colon and the folds of the small intestine. If infections occur in the wall of the alimentary canal, cells from the omentum may adhere to the in-

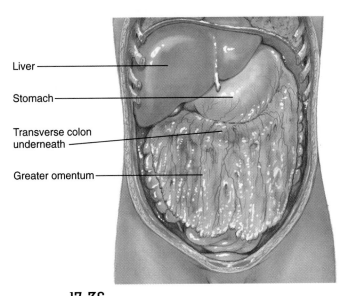

**FIGURE** 17.36

The greater omentum hangs like an apron over the abdominal organs.

flamed region and help wall it off so that the infection is less likely to enter the peritoneal cavity (fig. 17.36).

## Structure of the Small Intestinal Wall

Throughout its length, the inner wall of the small intestine has a velvety appearance. This is due to innumerable tiny projections of mucous membrane called **intestinal villi** (figs. 17.37 and 17.38). These structures are most numerous in the duodenum and the proximal portion of the jejunum. They project into the passageway, or **lumen,** of the alimentary canal, contacting the intestinal contents. Villi greatly increase the surface area of the intestinal lining, aiding absorption of digestive products.

Each villus consists of a layer of simple columnar epithelium and a core of connective tissue containing blood capillaries, a lymphatic capillary called a **lacteal,** and nerve fibers. At their free surfaces, the epithelial cells possess many fine extensions called *microvilli* that form a brushlike border and greatly increase the surface area of the intestinal cells, enhancing absorption (see figs. 17.3 and 17.39). The blood and lymph capillaries carry away absorbed nutrients, and impulses transmitted by the nerve fibers can stimulate or inhibit activities of the villus.

Between the bases of adjacent villi are tubular **intestinal glands** (crypts of Lieberkühn) that extend downward into the mucous membrane. The deeper layers of the small intestinal wall are much like those of other parts of the alimentary canal in that they include a submucosa, a muscular layer, and a serosa.

The lining of the small intestine also has many circular folds of mucosa, called *plicae circulares,* that are

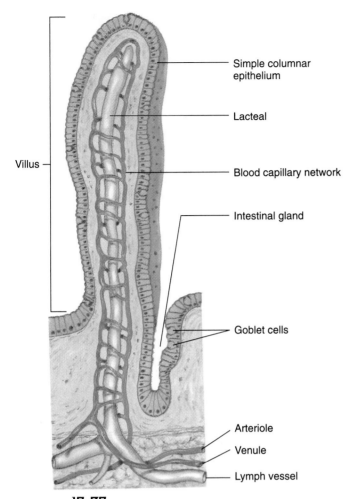

**FIGURE 17.37**

Structure of a single intestinal villus.

Simple columnar epithelium

Lacteal

Blood capillary network

Intestinal gland

Goblet cells

Villus

Arteriole

Venule

Lymph vessel

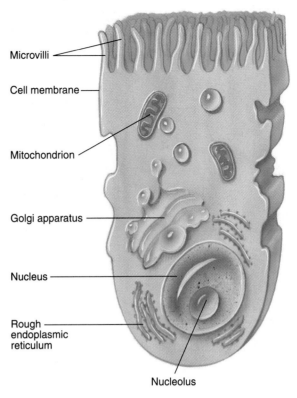

Microvilli

Cell membrane

Mitochondrion

Golgi apparatus

Nucleus

Rough endoplasmic reticulum

Nucleolus

(a) Intestinal epithelial cell

**FIGURE 17.38** 🏃

Light micrograph of intestinal villi from the wall of the duodenum (63×).

Lumen

Villus

Intestinal gland

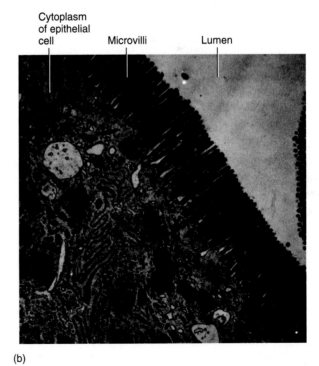

Cytoplasm of epithelial cell

Microvilli

Lumen

(b)

**FIGURE 17.39**

(a) Microvilli increase the surface area of intestinal epithelial cells. (b) Transmission electron micrograph of microvilli (16,000×).

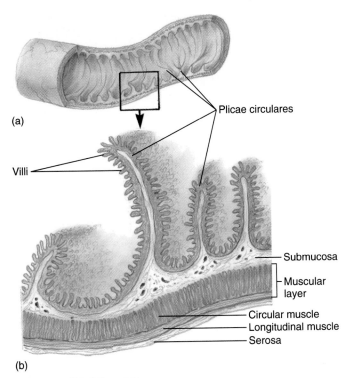

(a)

Plicae circulares

Villi

Submucosa

Muscular layer

Circular muscle
Longitudinal muscle
Serosa

(b)

**FIGURE 17.40**

(a) The inner lining of the small intestine contains many circular folds, the plicae circulares. (b) A longitudinal section through some of these folds.

especially well developed in the lower duodenum and upper jejunum. Together with the villi and microvilli, these folds help increase the surface area of the intestinal lining (fig. 17.40).

## Secretions of the Small Intestine

In addition to the mucus-secreting goblet cells, which are abundant throughout the mucosa of the small intestine, many specialized *mucus-secreting glands* (Brunner's glands) are in the submucosa within the proximal portion of the duodenum. These glands secrete large quantities of viscid, alkaline mucus in response to certain stimuli.

The epithelial cells that form the lining of the small intestine are continually replaced. New cells form within the intestinal glands by mitosis and migrate outward onto the villus surface. When the migrating cells reach the tip of the villus, they are shed. This *cellular turnover* renews the small intestine's epithelial lining every three to six days.

The intestinal glands at the bases of the villi secrete large amounts of a watery fluid (see fig. 17.37). The villi rapidly reabsorb this fluid, which carries digestive products into the villi. The fluid the intestinal glands secrete has a pH that is nearly neutral (6.5–7.5), and it lacks di-

gestive enzymes. However, the epithelial cells of the intestinal mucosa have digestive enzymes embedded in the membranes of the microvilli on their luminal surfaces. These enzymes break down food molecules just before absorption takes place. They include **peptidases,** which split peptides into their constituent amino acids; **sucrase, maltase,** and **lactase,** which split the double sugars (disaccharides) sucrose, maltose, and lactose into the simple sugars (monosaccharides) glucose, fructose, and galactose; and **intestinal lipase,** which splits fats into fatty acids and glycerol. Table 17.9 summarizes the sources and actions of the major digestive enzymes.

Many adults do not produce sufficient lactase to adequately digest lactose, or milk sugar. In this *lactose intolerance,* lactose remains undigested, increasing osmotic pressure of the intestinal contents and drawing water into the intestines. At the same time, intestinal bacteria metabolize undigested sugar, producing organic acids and gases. The overall result of lactose intolerance is bloating, intestinal cramps, and diarrhea. To avoid these unpleasant symptoms, people with lactose intolerance can take lactase in pill form before eating dairy products. Infants with lactose intolerance can drink formula based on soybeans rather than milk.

## Regulation of Small Intestinal Secretions

Since mucus protects the intestinal wall in the same way it protects the stomach lining, it is not surprising that mucus secretion increases in response to mechanical stimulation and the presence of irritants, such as gastric juice. Consequently, stomach contents entering the small intestine stimulate the duodenal mucous glands to release large quantities of mucus.

Goblet cells and intestinal glands secrete their products when they are stimulated by direct contact with chyme, which provides both chemical and mechanical stimulation. Distension of the intestinal wall activates the nerve plexuses within the wall and stimulates parasympathetic reflexes that also trigger release of small intestine secretions.

1. Describe the parts of the small intestine.

2. What is the function of an intestinal villus?

3. Distinguish between intestinal villi and microvilli.

4. How is surface area maximized in the small intestine?

5. What is the function of the intestinal glands?

6. List intestinal digestive enzymes.

TABLE 17.9

## SUMMARY OF THE MAJOR DIGESTIVE ENZYMES

| Enzyme | Source | Digestive Action |
|---|---|---|
| **Salivary Enzyme** | | |
| Amylase | Salivary glands | Begins carbohydrate digestion by breaking down starch and glycogen to disaccharides |
| **Gastric Enzymes** | | |
| Pepsin | Gastric glands | Begins protein digestion |
| Lipase | Gastric glands | Begins butterfat digestion |
| **Pancreatic Enzymes** | | |
| Amylase | Pancreas | Breaks down starch and glycogen into disaccharides |
| Lipase | Pancreas | Breaks down fats into fatty acids and glycerol |
| **Proteolytic Enzymes** <br> a. Trypsin <br> b. Chymotrypsin | Pancreas | Breaks down proteins or partially digested proteins into peptides |
| c. Carboxypeptidase | | Breaks down peptides into amino acids |
| Nucleases | Pancreas | Breaks down nucleic acids into nucleotides |
| **Intestinal Enzymes** | | |
| Peptidase | Mucosal cells | Breaks down peptides into amino acids |
| Sucrase, maltase, lactase | Mucosal cells | Breaks down disaccharides into monosaccharides |
| Lipase | Mucosal cells | Breaks down fats into fatty acids and glycerol |
| Enterokinase | Mucosal cells | Converts trypsinogen into trypsin |

**FIGURE 17.41**

Digestion breaks down complex carbohydrates into disaccharides, which are then broken down into monosaccharides, which are small enough for intestinal villi to absorb. The monosaccharides then enter the bloodstream.

## Absorption in the Small Intestine

Because villi greatly increase the surface area of the intestinal mucosa, the small intestine is the most important absorbing organ of the alimentary canal. In fact, the small intestine is so effective in absorbing digestive products, water, and electrolytes, that very little absorbable material reaches its distal end.

Carbohydrate digestion begins in the mouth with the activity of salivary amylase, and is completed in the small intestine by enzymes from the intestinal mucosa and pancreas. The resulting monosaccharides are absorbed by the villi and enter blood capillaries (fig. 17.41). Simple sugars are absorbed by active transport or facilitated diffusion (see chapter 3).

Protein digestion begins in the stomach as a result of pepsin activity, and is completed in the small intestine by enzymes from the intestinal mucosa and the pancreas. During this process, large protein molecules are broken down into amino acids, which are then absorbed into the villi by active transport and are carried away by the blood (fig. 17.42).

Fat molecules are digested almost entirely by enzymes from the intestinal mucosa and pancreas (fig. 17.43). The resulting fatty acid molecules are absorbed in the following steps: (1) The fatty acid molecules dissolve in the epithelial cell membranes of the villi and diffuse through them. (2) The endoplasmic reticula of the cells use the fatty acids to resynthesize fat molecules similar to those previously digested. (3) These fats collect in clusters that become encased in protein. (4) The resulting large molecules of lipoprotein are called *chylomicrons,* and they make their way to the lacteal of the villus. (5) Periodic contractions of smooth muscles in the villus help empty the lacteal into the cysterna chyli, which empties into the thoracic duct. The lymph carries the chylomicrons to the bloodstream (fig. 17.44).

**FIGURE 17.42**

The amino acids that result from dipeptide digestion are absorbed by intestinal villi and enter the blood.

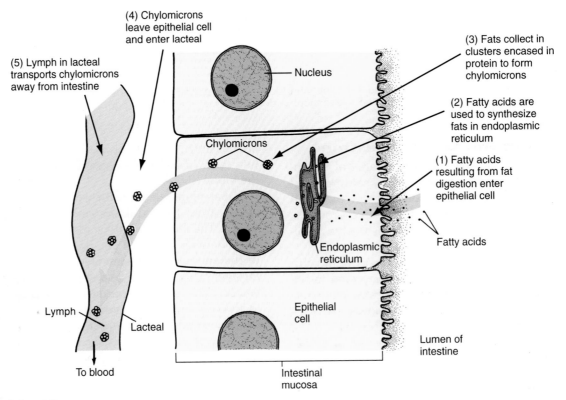

**FIGURE 17.43**

Fatty acids and glycerol result from fat digestion. Intestinal villi absorb them, and most are resynthesized into fat molecules before they enter the blood or lymph.

(4) Chylomicrons leave epithelial cell and enter lacteal

(5) Lymph in lacteal transports chylomicrons away from intestine

(3) Fats collect in clusters encased in protein to form chylomicrons

Nucleus

(2) Fatty acids are used to synthesize fats in endoplasmic reticulum

Chylomicrons

(1) Fatty acids resulting from fat digestion enter epithelial cell

Endoplasmic reticulum

Fatty acids

Lymph

Lacteal

Epithelial cell

Lumen of intestine

To blood

Intestinal mucosa

**FIGURE 17.44**

Fat absorption has several steps.

| TABLE | 17.10 | INTESTINAL ABSORPTION OF NUTRIENTS |

| Nutrient | Absorption Mechanism | Means of Transport |
| --- | --- | --- |
| Monosaccharides | Facilitated diffusion and active transport | Blood in capillaries |
| Amino acids | Active transport | Blood in capillaries |
| Fatty acids and glycerol | Facilitated diffusion of glycerol; diffusion of fatty acids into cells | |
| | (a) Most fatty acids are resynthesized back into fats and incorporated in chylomicrons for transport. | Lymph in lacteals |
| | (b) Some fatty acids with relatively short carbon chains are transported without being changed back into fats. | Blood in capillaries |
| Electrolytes | Diffusion and active transport | Blood in capillaries |
| Water | Osmosis | Blood in capillaries |

Chylomicrons are transported by the blood to capillaries of muscle and adipose tissues. A specific type of protein (apoprotein) associated with the surface of chylomicrons activates an enzyme (lipoprotein lipase) that is attached to the inner lining of such capillaries. As a result of the enzyme's action, fatty acids and monoglycerides are released from the chylomicrons, and they enter muscle or adipose cells for use as energy sources or to be stored. The remnants of the chylomicrons travel in the blood to the liver, where they bind to receptors on the surface of liver cells. The remnants quickly enter liver cells by receptor-mediated endocytosis (see chapter 3), and lysosomes degrade them. Some fatty acids with short carbon chains may be absorbed directly into the blood capillary of the villus without being converted back into fat.

> Be sure to read the labels on cakes claiming to be fat free. Often these products contain diglycerides, rather than the triglycerides found in dietary fats. Once diglycerides are broken down to their constituent fatty acids during digestion and enter the small intestine epithelium, they are reformed into fat. So yes, these cakes are technically no-fat, but the result after digestion may be the same as eating the real thing.

In addition to absorbing the products of carbohydrate, protein, and fat digestion, the intestinal villi absorb electrolytes and water. Certain ions, such as those of sodium, potassium, chloride, nitrate, and bicarbonate, are readily absorbed; but others, including ions of calcium, magnesium, and sulfate, are poorly absorbed.

Electrolytes are usually absorbed by active transport, and water by osmosis. Thus, even though the intestinal contents may be hypertonic to the epithelial cells at first, as nutrients and electrolytes are absorbed, they become slightly hypotonic to the cells. Then, water follows the nutrients and electrolytes into the villi by osmosis. The absorption process is summarized in table 17.10.

1 Which substances resulting from digestion of carbohydrate, protein, and fat molecules does the small intestine absorb?

2 Which ions does the small intestine absorb?

3 What transport mechanisms do intestinal villi use?

4 Describe how fatty acids are absorbed and transported.

> In *malabsorption* the small intestine digests, but does not absorb, some nutrients. Causes of malabsorption include surgical removal of a portion of the small intestine, obstruction of lymphatic vessels due to a tumor, or interference with the production and release of bile as a result of liver disease.
>
> Another cause of malabsorption is a reaction to *gluten*, found in certain grains, especially wheat and rye. This condition is called *celiac disease*. Microvilli are damaged and, in severe cases, villi may be destroyed. Both of these effects reduce the absorptive surface of the small intestine, preventing absorption of some nutrients. Symptoms of malabsorption include diarrhea, weight loss, weakness, vitamin deficiencies, anemia, and bone demineralization.

## Movements of the Small Intestine

Like the stomach, the small intestine carries on mixing movements and peristalsis. The major mixing movement is called *segmentation*, in which small, ringlike contractions occur periodically, cutting the chyme into segments and moving it back and forth. Segmentation also slows the movement of chyme through the small intestine.

Peristaltic waves propel chyme through the small intestine. These waves are usually weak, and they stop after pushing the chyme a short distance. Consequently, chyme moves slowly through the small intestine, taking from three to ten hours to travel its length.

As might be expected, parasympathetic impulses enhance both mixing and peristaltic movements, and sympathetic impulses inhibit them. Reflexes involving parasympathetic impulses to the small intestine sometimes originate in the stomach. For example, food filling the stomach distends its wall, triggering a reflex (gastroenteric reflex) that greatly increases peristaltic activity in the small intestine. Another reflex is initiated when the duodenum fills with chyme, stretching its wall. This reflex speeds movement through the small intestine.

If the small intestine wall becomes overdistended or irritated, a strong *peristaltic rush* may pass along the entire length of the organ, sweeping chyme into the large intestine so quickly that water, nutrients, and electrolytes that would normally be absorbed are not. The result is *diarrhea,* a condition in which defecation becomes more frequent and the stools become watery. Prolonged diarrhea causes imbalances in water and electrolyte concentrations.

The **ileocecal sphincter** joins the small intestine's ileum to the large intestine's cecum. Normally, this sphincter remains constricted, preventing the contents of the small intestine from entering the large intestine, and at the same time keeping the contents of the large intestine from backing up into the ileum. However, after a meal, a gastroileal reflex is elicited that increases peristalsis in the ileum and relaxes the sphincter, forcing some of the contents of the small intestine into the cecum.

1. Describe the movements of the small intestine.

2. How are the movements of the small intestine initiated?

3. What is a peristaltic rush?

4. What stimulus relaxes the ileocecal sphincter?

## LARGE INTESTINE

The **large intestine** is so named because its diameter is greater than that of the small intestine. This portion of the alimentary canal is about 1.5 meters long, and it begins in the lower right side of the abdominal cavity where the ileum joins the cecum. From there, the large intestine travels upward on the right side, crosses obliquely to the left, and descends into the pelvis. At its distal end, it opens to the outside of the body as the anus.

The large intestine absorbs water and electrolytes from the chyme remaining in the alimentary canal. It also reabsorbs and recycles water and remnants of digestive secretions. The large intestine also forms and stores feces.

## Parts of the Large Intestine

The large intestine consists of the cecum, the colon, the rectum, and the anal canal. Figures 17.45 and 17.46, and reference plates 51, 52, 58, and 75 depict the large intestine.

The **cecum,** at the beginning of the large intestine, is a dilated, pouchlike structure that hangs slightly inferior to the ileocecal opening. Projecting downward from it is a narrow tube with a closed end called the **vermiform (wormlike) appendix.** The human appendix has no known digestive function. However, it contains lymphatic tissue.

In *appendicitis,* the appendix becomes inflamed and infected. Surgery is required to prevent the appendix from rupturing. If it breaks open, the contents of the large intestine may enter the abdominal cavity and cause a serous infection of the peritoneum called *peritonitis.*

The **colon** is divided into four portions—the ascending, transverse, descending, and sigmoid colons. The **ascending colon** begins at the cecum and travels upward against the posterior abdominal wall to a point just inferior to the liver. There it turns sharply to the left (as the right colic, or hepatic, flexure) and becomes the **transverse colon.** The transverse colon is the longest and most movable part of the large intestine. It is suspended by a fold of peritoneum and sags in the middle below the stomach. As the transverse colon approaches the spleen, it turns abruptly downward (as the left colic, or splenic, flexure) and becomes the **descending colon.** At the brim of the pelvis, the descending colon makes an S-shaped curve, called the **sigmoid colon,** and then becomes the rectum.

The **rectum** lies next to the sacrum and generally follows its curvature. It is firmly attached to the sacrum by the peritoneum, and it ends about 5 centimeters inferior to the tip of the coccyx, where it becomes the anal canal (fig. 17.47).

The **anal canal** is formed by the last 2.5 to 4.0 centimeters of the large intestine. The mucous membrane in the canal is folded into a series of six to eight longitudinal *anal columns.* At its distal end, the canal opens to the outside as the **anus.** Two sphincter muscles guard the anus—an *internal anal sphincter muscle* composed of smooth muscle under involuntary control, and an *external anal sphincter muscle* composed of skeletal muscle under voluntary control.

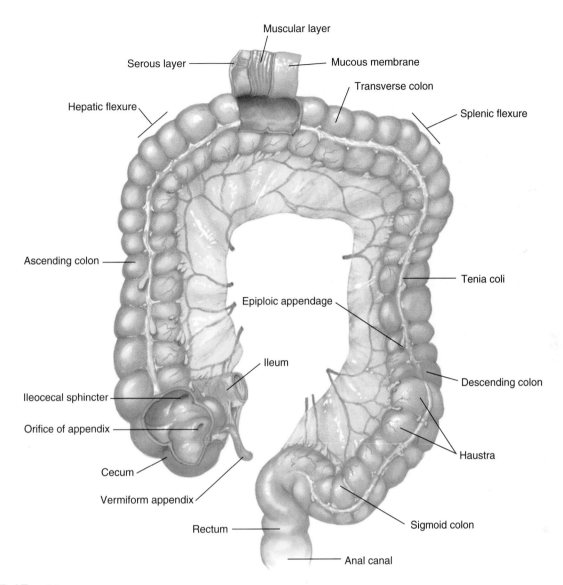

**FIGURE 17.45**

Parts of the large intestine (anterior view).

---

*Hemorrhoids* are, literally, a pain in the rear. Enlarged and inflamed branches of the rectal vein in the anal columns causes intense itching, sharp pain, and sometimes bright red bleeding. The hemorrhoids may be internal (which do not produce symptoms) or bulge out of the anus. Causes of hemorrhoids include anything that puts prolonged pressure on the delicate rectal tissue, including obesity, pregnancy, constipation, diarrhea, and liver disease.

Eating more fiber-rich foods and drinking lots of water can usually prevent or cure hemorrhoids. Warm soaks in the tub, cold packs, and careful wiping of painful areas also helps, as do external creams and ointments. Surgery—with a scalpel or a laser—can remove severe hemorrhoids.

**1** What is the general function of the large intestine?

**2** Describe the parts of the large intestine.

**3** Distinguish between the internal sphincter muscle and the external sphincter muscle of the anus.

## Structure of the Large Intestinal Wall

Although the wall of the large intestine includes the same types of tissues found in other parts of the alimentary canal, it has some unique features. The large intestinal wall lacks the villi and plicae circularis characteristic of the small intestine. Also, the layer of longitudinal muscle fibers does not uniformly cover its wall; the fibers are in three distinct bands (teniae coli) that extend the entire length of the colon. These bands exert tension

**FIGURE 17.46**

Radiograph of the large intestine.

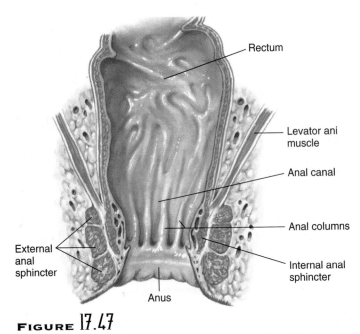

**FIGURE 17.47**

The rectum and the anal canal are located at the distal end of the alimentary canal.

- Rectum
- Levator ani muscle
- Anal canal
- Anal columns
- Internal anal sphincter
- External anal sphincter
- Anus

**FIGURE 17.48**

Light micrograph of the large intestinal wall (64×).

- Lumen
- Mucosa
- Submucosa
- Muscular layer
- Serosa

lengthwise on the wall, creating a series of pouches (haustra). The large intestinal wall also has small collections of fat (epiploic appendages) in the serosa on its outer surface (fig. 17.48).

## Functions of the Large Intestine

Unlike the small intestine, which secretes digestive enzymes and absorbs the products of digestion, the large intestine has little or no digestive function. However, the mucous membrane that forms the inner lining of the large intestine contains many tubular glands. Structurally, these glands are similar to those of the small intestine, but they are composed almost entirely of goblet cells. Consequently, mucus is the only significant secretion of this portion of the alimentary canal (fig. 17.49).

Mechanical stimulation from chyme and parasympathetic impulses control the rate of mucus secretion. In both cases, the goblet cells respond by increasing mucus production, which, in turn, protects the intestinal wall against the abrasive action of materials passing through it. Mucus also holds particles of fecal matter together, and, because it is alkaline, mucus helps control the pH of the large intestinal contents. This is important because acids are sometimes released from the feces as a result of bacterial activity.

Aspirin may reduce the death rate from cancers of the esophagus, stomach, colon, and rectum by up to 40%. These digestive cancers kill 81,000 people a year. Researchers do not know why aspirin seems to protect against these cancers. One suggestion is that it causes bleeding of the alimentary canal, as evidenced by rectal bleeding, which may send people to the doctor earlier, leading to earlier diagnosis, when treatment is more likely to be successful. Alternatively, aspirin may protect against these cancers by interfering with prostaglandin activity.

Lumen of
large
intestine

Goblet
cells

**FIGURE 17.49**

Light micrograph of the large intestinal mucosa (250× micrograph enlarged to 560×).

Chyme entering the large intestine usually has few nutrients remaining in it, and mostly consists of materials not digested or absorbed in the small intestine. It also contains water, electrolytes, mucus, and bacteria.

Absorption in the large intestine is normally limited to water and electrolytes, and this usually occurs in the proximal half of the tube. Electrolytes, such as sodium ions, can be absorbed by active transport, while the water follows, passively, entering the mucosa by osmosis. As a result, about 90% of the water that enters the large intestine is absorbed, and little sodium or water is lost in the feces.

The many bacteria that normally inhabit the large intestine, called *intestinal flora,* break down some of the molecules that escape the actions of human digestive enzymes. For instance cellulose, a complex carbohydrate in food of plant origin, passes through the alimentary canal almost unchanged, but colon bacteria can break down cellulose and use it as an energy source. This is why we can utilize dietary fiber. These bacteria also synthesize certain vitamins, such as K, $B_{12}$, thiamine, and riboflavin, which the intestinal mucosa absorbs. Bacterial actions in the large intestine may produce intestinal gas (flatus).

**1** How does the structure of the large intestine differ from that of the small intestine?

**2** What substances does the large intestine absorb?

**3** What useful substances do bacteria inhabiting the large intestine produce?

Hirschsprung disease causes extreme, chronic constipation and abdominal distension. The part of the large intestine distal to the distension lacks innervation. As a result, the person does not feel the urge to defecate. The problem begins in the embryo, when an abnormal gene prevents neurons from migrating to this portion of the gastrointestinal tract. Surgery may be used to treat Hirschsprung disease, which was once lethal.

## Movements of the Large Intestine

The movements of the large intestine—mixing and peristalsis—are similar to those of the small intestine, although usually slower. The mixing movements break the fecal matter into segments and turn it so that all portions are exposed to the intestinal mucosa. This helps in water and electrolyte absorption.

The peristaltic waves of the large intestine are different from those of the small intestine. Instead of occurring frequently, they happen only two or three times each day. These waves produce *mass movements* in which a large section of the intestinal wall constricts vigorously, forcing the intestinal contents to move toward the rectum. Typically, mass movements follow a meal, as a result of the gastrocolic reflex. Abnormal irritations of the intestinal mucosa can also trigger such movements. For instance, a person suffering from an inflamed colon (colitis) may experience frequent mass movements.

When it is appropriate to defecate, a person usually can initiate a *defecation reflex* by holding a deep breath and contracting the abdominal wall muscles. This action increases the internal abdominal pressure and forces feces into the rectum. As the rectum fills, its wall is distended and the defecation reflex is triggered, stimulating peristaltic waves in the descending colon, and the internal anal sphincter relaxes. At the same time, other reflexes involving the sacral region of the spinal cord strengthen the peristaltic waves, lower the diaphragm, close the glottis, and contract the abdominal wall muscles. These actions additionally increase the internal abdominal pressure and squeeze the rectum. The external anal sphincter is signaled to relax, and the feces are forced to the outside. A person can voluntarily inhibit defecation by contracting the external anal sphincter.

## Feces

Feces are composed of materials that were not digested or absorbed, with water, electrolytes, mucus, and bacteria. Usually the feces are about 75% water, and their color derives from bile pigments altered by bacterial action.

# CLINICAL APPLICATION

# Disorders of the Large Intestine

The large intestine is the source of many medical problems, from such familiar digestive discomforts as gas to more serious disorders.

## INTESTINAL GAS

People do not often talk about intestinal gas, but this common evidence of digestion is a source of pain and sometimes embarrassment. What exactly is in intestinal gas? Most of it is nitrogen and oxygen gulped in while breathing and eating. Undigested food fermented by bacteria contributes methane ($CH_4$), carbon dioxide ($CO_2$), and hydrogen. These five gases, which do not have odors, account for 99% of intestinal gas. The other 1% of compounds in intestinal gas do have distinctive odors, and result from the actions of intestinal bacteria on the nitrogen and sulfur in proteins, yielding the pungent-smelling ammonia ($NH_3$) and the foul hydrogen sulfide (HS).

Most people release a half liter of intestinal gas a day. That amount may increase if the diet is rich in hard-to-digest foods, such as beans. Foods rich in sulfur-containing amino acids make intestinal gas more foul. These foods include beans, bagels, bran, broccoli, brussels sprouts, cabbage, cauliflower, and onions. Eating slowly and avoiding milk if lactose intolerant can minimize intestinal gas production.

## DIARRHEA

Because the large intestine absorbs water from material within it, the rate of movement through it determines the consistency of feces. *Diarrhea,* the frequent and too-rapid passage of loose feces, results when material moves along so quickly that too little water is absorbed. The condition may reflect poisoning, infection, a diet too high in fiber, or nervousness. When the large intestine is the site of an infection or chemical irritation, diarrhea protects by flushing toxins out of the body.

## CONSTIPATION

*Constipation,* the infrequent passage of hard feces, is caused by abnormally slow movement of fecal matter through the large intestine. Because

Diverticula

## FIGURE 17B

Diverticulosis of the large intestine. The large intestine on the right is healthy. The organ on the left shows diverticula on the left side.

The pungent odor of the feces results from a variety of compounds that bacteria produce. These compounds include phenol, hydrogen sulfide, indole, skatole, and ammonia. Clinical Application 17.5 examines conditions affecting the large intestine.

1 How does peristalsis in the large intestine differ from peristalsis in the small intestine?

2 List the major events that occur during defecation.

3 Describe the composition of feces.

the feces remain in the large intestine longer than usual, excess water is absorbed. Constipation can be caused by a failure of the sensory cells in the rectum to signal the spinal cord to defecate or by the conscious suppression of defecation, both of which can be a result of emotional stress. A diet low in fiber can also cause constipation by slowing fecal movement through the large intestine. Eating foods high in fiber, drinking at least eight 8-ounce glasses of water a day, and regular exercise can prevent constipation.

## More Serious Disorders

In *diverticulosis,* parts of the intestinal wall weaken, and the inner mucous membrane protrudes through (fig. 17B). Many times there are no symptoms, but if the outpouching becomes blocked with chyme and then infected (a condition called diverticulitis), antibiotics or surgery may be required. It is not known exactly what causes diverticulosis, but the fact that it is nonexistent in populations whose diets are high in fiber and that it was not known in the United States before refined foods were introduced earlier this century suggests that lack of dietary fiber may be a cause. Fiber speeds movement of material through the digestive system, hastening exit of toxic substances.

In *ulcerative colitis,* inflammation of the inner lining of the colon and rectum produces abdominal and rectal pain, bloody diarrhea, and weight loss. Drugs can often treat the symptoms, but sometimes removal of the colon is necessary. Again, the cause of this disorder is not known.

Cancer of the large intestine and rectum, known as *colorectal cancer,* is the second most prevalent cancer in the United States, with 152,000 new cases and nearly 60,000 deaths yearly. Symptoms include a change in the frequency or consistency of bowel movements, bloody feces, and abdominal pain. A *hemoccult* (hidden blood) *test* can detect intestinal bleeding that may signal the presence of cancer. Blood in feces is often black and not visible. Follow-up at a doctor's office entails use of a fiber-optic colonoscope that searches and samples colorectal tissue for cancer. Physicians can sometimes diagnose colon cancer by performing genetic tests on cells in feces.

Some colon cancers may take many years to develop. A sequence of genetic changes causes the cells lining the large intestine first to divide more frequently than normal, then to enter a precancerous stage. Next, the growths form benign polyps, then, perhaps years later, cancer. Still other genetic changes control the cancer's spread. A susceptibility gene may trigger the entire process. By understanding the steps that lead to inherited types of colon cancer, researchers may learn how other colon malignancies form and progress as well (fig. 17C).

Cancer or other severe maladies may require that the large intestine be removed. A new opening for feces to exit the body is then needed. The free end of the intestine can be surgically attached to an opening created through the skin of the abdomen, and a bag attached to the opening to collect the fecal matter. This procedure is called a *colostomy.* ∎

**Figure 17C** ✗

To detect colon cancer, a patient receives an enema containing barium. The barium highlights the lower digestive tract, revealing an obstruction caused by a tumor. The arrows indicate the obstructed area.

## Clinical Terms Related to the Digestive System

**achalasia** (ak″ah-la′ze-ah)  Failure of the smooth muscle to relax at some junction in the digestive tube, such as that between the esophagus and stomach.

**achlorhydria** (ah″klōr-hi′dre-ah)  Lack of hydrochloric acid in gastric secretions.

**aphagia** (ah-fa′je-ah)  Inability to swallow.

**cholecystitis** (ko″le-sis-ti′tis)  Inflammation of the gallbladder.

**cholelithiasis** (ko″le-li-thi′ah-sis)  Stones in the gallbladder.

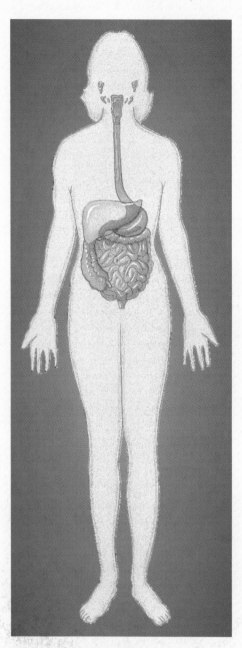

**DIGESTIVE SYSTEM**

The digestive system ingests, digests, and absorbs nutrients for use by all body cells.

## Integumentary System

Vitamin D activated in the skin plays a role in absorption of calcium from the digestive tract.

## Skeletal System

Bones are important in mastication.

## Muscular System

Muscles are important in mastication, swallowing, and the mixing and moving of digestion products through the gastrointestinal tract.

## Nervous System

The nervous system can influence digestive system activity.

## Endocrine System

Hormones can influence digestive system activity.

## Cardiovascular System

The bloodstream carries absorbed nutrients to all body cells.

## Lymphatic System

The lymphatic system plays a major role in the absorption of fats.

## Respiratory System

The digestive system and the respiratory system share common anatomical structures.

## Urinary System

The kidneys and liver work together to activate vitamin D.

## Reproductive System

Adequate availability of nutrients, including fats, is essential for conception and normal development.

**cholestasis** (ko″le-sta′sis)   Blockage in bile flow from the gallbladder.

**cirrhosis** (si-ro′sis)   Liver condition in which the hepatic cells degenerate and the surrounding connective tissues thicken.

**diverticulitis** (di″ver-tik″u-li′tis)   Inflammation of small pouches (diverticula) that sometimes form in the lining and wall of the colon.

**dumping syndrome** (dum′ping sin′drōm)   Symptoms, including diarrhea, that often occur following a gastrectomy.

**dysentery** (dis′en-ter″e)   Intestinal infection, caused by viruses, bacteria, or protozoans, that causes diarrhea and cramps.

**dyspepsia** (dis-pep′se-ah)   Indigestion; difficulty in digesting a meal.

**dysphagia** (dis-fa′je-ah)   Difficulty in swallowing.

**enteritis** (en″tĕ-ri′tis)   Inflammation of the intestine.

**esophagitis** (e-sof″ah-ji′tis)   Inflammation of the esophagus.

**gastrectomy** (gas-trek′to-me)   Partial or complete removal of the stomach.

**gastrostomy** (gas-tros′to-me)   The creation of an opening in the stomach wall through which food and liquids may be administered when swallowing is not possible.

**glossitis** (glŏ-si′tis)   Inflammation of the tongue.

**ileitis** (il″e-i′tis)   Inflammation of the ileum.

**ileus** (il′e-us)   Obstruction of the intestine due to an inhibition of motility or a mechanical cause.

**pharyngitis** (far″in-ji′tis)   Inflammation of the pharynx.

**pylorospasm** (pi-lor′o-spazm)   Spasm of the pyloric portion of the stomach or of the pyloric sphincter.

**pyorrhea** (pi″o-re′ah)   Inflammation of the dental periosteum with pus formation.

**stomatitis** (sto″mah-ti′tis)   Inflammation of the lining of the mouth.

---

## CHAPTER SUMMARY

## Introduction  (page 656)

Digestion is the process of mechanically and chemically breaking down foods so that they can be absorbed. The digestive system consists of an alimentary canal and several accessory organs.

## General Characteristics of the Alimentary Canal  (page 656)

Regions of the alimentary canal perform specific functions.

1.  Structure of the wall
    a. The wall consists of four layers.
    b. These layers include the mucosa, submucosa, muscular layer, and serosa.
2.  Movements of the tube
    a. Motor functions include mixing and propelling movements.
    b. Peristalsis is responsible for propelling movements.
    c. The wall of the tube undergoes receptive relaxation just ahead of a peristaltic wave.
3.  Innervation of the tube
    a. The tube is innervated by branches of the sympathetic and parasympathetic divisions of the autonomic nervous system.
    b. Parasympathetic impulses generally increase digestive activities; sympathetic impulses generally inhibit digestive activities.
    c. Sympathetic impulses contract certain sphincter muscles, controlling movement through the alimentary canal.

## Mouth  (page 658)

The mouth is adapted to receive food and begin preparing it for digestion. It also serves as an organ of speech and sensory perception.

1.  Cheeks and lips
    a. Cheeks form the lateral walls of the mouth.
    b. Lips are highly mobile and possess a variety of sensory receptors useful in judging the characteristics of food.
2.  Tongue
    a. The tongue is a thick, muscular organ that mixes food with saliva and moves it toward the pharynx.
    b. The rough surface of the tongue handles food and contains taste buds.
    c. Lingual tonsils are located on the root of the tongue.
3.  Palate
    a. The palate comprises the roof of the mouth and includes hard and soft portions.
    b. The soft palate closes the opening to the nasal cavity during swallowing.
    c. Palatine tonsils are located on either side of the tongue in the back of the mouth.
    d. Tonsils consist of lymphatic tissues.
4.  Teeth
    a. Two sets of teeth develop in sockets of the mandibular and maxillary bones.
    b. There are twenty primary and thirty-two secondary teeth.
    c. Teeth mechanically break food into smaller pieces, increasing the surface area exposed to digestive actions.
    d. Different kinds of teeth are adapted to handle foods in different ways, such as biting, grasping, or grinding.
    e. Each tooth consists of a crown and root, and is composed of enamel, dentin, pulp, nerves, and blood vessels.
    f. A tooth is attached to the alveolar process by collagenous fibers of the periodontal ligament.

## Salivary Glands  (page 665)

Salivary glands secrete saliva, which moistens food, helps bind food particles, begins chemical digestion of carbohydrates, makes taste possible, helps cleanse the mouth, and regulates pH in the mouth.

1.  Salivary secretions
    a. Salivary glands include serous cells that secrete digestive enzymes and mucous cells that secrete mucus.
    b. Parasympathetic impulses stimulate the secretion of serous fluid.

2. Major salivary glands
   a. The parotid glands are the largest, and they secrete saliva that is rich in amylase.
   b. The submandibular glands in the floor of the mouth produce viscous saliva.
   c. The sublingual glands in the floor of the mouth primarily secrete mucus.

# Pharynx and Esophagus  (page 666)

The pharynx and esophagus serve as passageways.

1. Structure of the pharynx
   a. The pharynx is divided into a nasopharynx, oropharynx, and laryngopharynx.
   b. The muscular walls of the pharynx contain fibers arranged in circular and longitudinal groups.
2. Swallowing mechanism
   a. Swallowing occurs in three stages.
      (1) Food is mixed with saliva and forced into the pharynx.
      (2) Involuntary reflex actions move the food into the esophagus.
      (3) Peristalsis transports food to the stomach.
3. Esophagus
   a. The esophagus passes through the mediastinum and penetrates the diaphragm.
   b. Circular muscle fibers at the distal end of the esophagus help prevent regurgitation of food from the stomach.

# Stomach  (page 670)

The stomach receives food, mixes it with gastric juice, carries on a limited amount of absorption, and moves food into the small intestine.

1. Parts of the stomach
   a. The stomach is divided into cardiac, fundic, body, and pyloric regions.
   b. The lower esophageal sphincter serves as a valve between the esophagus and the stomach.
   c. The pyloric sphincter serves as a valve between the stomach and the small intestine.
2. Gastric secretions
   a. Gastric glands secrete gastric juice.
   b. Gastric juice contains pepsin, hydrochloric acid, lipase, and intrinsic factor.
3. Regulation of gastric secretions
   a. Parasympathetic impulses and the hormone gastrin enhance gastric secretion.
   b. The three stages of gastric secretion are the cephalic, gastric, and intestinal phases.
   c. The presence of food in the small intestine reflexly inhibits gastric secretions.
4. Gastric absorption
   a. The stomach is not well adapted for absorption.
   b. A few substances such as water and other small molecules may be absorbed through the stomach wall.
5. Mixing and emptying actions
   a. As the stomach fills, its wall stretches, but its internal pressure remains unchanged.
   b. Mixing movements aid in producing chyme; peristaltic waves move the chyme into the pyloric region.
   c. The muscular wall of the pyloric region regulates chyme movement into the small intestine.
   d. The rate of emptying depends on the fluidity of the chyme and the type of food present.
   e. The upper part of the small intestine fills, and an enterogastric reflex inhibits peristalsis in the stomach.
   f. Vomiting results from a complex reflex that has many stimuli.

# Pancreas  (page 676)

The pancreas is closely associated with the duodenum.

1. Structure of the pancreas
   a. It produces pancreatic juice that is secreted into a pancreatic duct.
   b. The pancreatic duct leads to the duodenum.
2. Pancreatic juice
   a. Pancreatic juice contains enzymes that can split carbohydrates, proteins, fats, and nucleic acids.
   b. Pancreatic juice has a high bicarbonate ion concentration that helps neutralize chyme and causes the intestinal contents to be alkaline.
3. Regulation of pancreatic secretion
   a. Secretin from the duodenum stimulates the release of pancreatic juice that contains few digestive enzymes but has a high bicarbonate ion concentration.
   b. Cholecystokinin from the intestinal wall stimulates the release of pancreatic juice that has a high concentration of digestive enzymes.

# Liver  (page 678)

The liver is located in the upper right quadrant of the abdominal cavity.

1. Liver structure
   a. The liver is a highly vascular organ, enclosed in a fibrous capsule, and divided into lobes.
   b. Each lobe consists of hepatic lobules, the functional units of the liver.
   c. Bile from the lobules is carried by bile canals to hepatic ducts that unite to form the common bile duct.
2. Liver functions
   a. The liver has many functions. It metabolizes carbohydrates, lipids, and proteins; stores some substances; filters blood; destroys toxins; and secretes bile.
   b. Bile is the only liver secretion that directly affects digestion.
3. Composition of bile
   a. Bile contains bile salts, bile pigments, cholesterol, and electrolytes.
   b. Only the bile salts have digestive functions.
   c. Bile pigments are products of red blood cell breakdown.
4. Gallbladder
   a. The gallbladder stores bile between meals.
   b. A sphincter muscle controls release of bile from the common bile duct.
   c. Gallstones may sometimes form within the gallbladder.
5. Regulation of bile release
   a. Cholecystokinin from the small intestine stimulates bile release.

b. The sphincter muscle at the base of the common bile duct relaxes as a peristaltic wave in the duodenal wall approaches.
6. Functions of bile salts
    a. Bile salts emulsify fats and aid in the absorption of fatty acids, cholesterol, and certain vitamins.
    b. Bile salts are reabsorbed in the small intestine.

## Small Intestine (page 685)

The small intestine extends from the pyloric sphincter to the large intestine. It receives secretions from the pancreas and liver, completes digestion of nutrients, absorbs the products of digestion, and transports the residues to the large intestine.

1. Parts of the small intestine
    a. The small intestine consists of the duodenum, jejunum, and ileum.
    b. The small intestine is suspended from the posterior abdominal wall by mesentery.
2. Structure of the small intestinal wall
    a. The wall is lined with villi that greatly increase the surface area and aid in mixing and absorption.
    b. Microvilli on the free ends of epithelial cells increase the surface area even more.
    c. Intestinal glands are located between the villi.
    d. Circular folds in the lining of the intestinal wall also increase its surface area.
3. Secretions of the small intestine
    a. Intestinal glands secrete a watery fluid that lacks digestive enzymes but provides a vehicle for moving chyme to the villi.
    b. Digestive enzymes embedded in the surfaces of microvilli split molecules of sugars, proteins, and fats.
4. Regulation of small intestinal secretions
    a. Secretion is stimulated by gastric juice, chyme, and reflexes stimulated by distension of the small intestinal wall.
5. Absorption in the small intestine
    a. Villi absorb monosaccharides, amino acids, fatty acids, and glycerol.
    b. Villi also absorb water and electrolytes.
    c. Fat molecules with longer chains of carbon atoms enter the lacteals of the villi; fatty acids with relatively short carbon chains enter the blood capillaries of the villi.

6. Movements of the small intestine
    a. Movements include mixing by segmentation and peristalsis.
    b. Overdistension or irritation may stimulate a peristaltic rush and result in diarrhea.
    c. The ileocecal sphincter controls movement of the intestinal contents from the small intestine into the large intestine.

## Large Intestine (page 692)

The large intestine absorbs water and electrolytes, and forms and stores feces.

1. Parts of the large intestine
    a. The large intestine consists of the cecum, colon, rectum, and anal canal.
    b. The colon is divided into ascending, transverse, descending, and sigmoid portions.
2. Structure of the large intestinal wall
    a. The large intestine wall resembles the wall in other parts of the alimentary canal.
    b. The large intestine wall has a unique layer of longitudinal muscle fibers arranged in distinct bands.
3. Functions of the large intestine
    a. The large intestine has little or no digestive function, although it secretes mucus.
    b. Mechanical stimulation and parasympathetic impulses control the rate of mucus secretion.
    c. The large intestine absorbs water and electrolytes.
    d. Many bacteria inhabit the large intestine, where they break down certain undigestible substances and synthesize certain vitamins.
4. Movements of the large intestine
    a. Movements are similar to those in the small intestine.
    b. Mass movements occur two to three times each day.
    c. Defecation is stimulated by a reflex.
5. Feces
    a. Feces are formed and stored in the large intestine.
    b. Feces largely consist of water, undigested material, mucus, and bacteria.
    c. The color of feces is due to bile pigments that have been altered by bacterial action.

## CRITICAL THINKING QUESTIONS

1. If a patient has 95% of the stomach removed (subtotal gastrectomy) as treatment for severe ulcers or cancer, how would the digestion and absorption of foods be affected? How would the patient's eating habits have to be altered? Why?
2. Why may a person with inflammation of the gallbladder (cholecystitis) also develop an inflammation of the pancreas (pancreatitis)?
3. What effect is a before-dinner cocktail likely to have on digestion? Why are such beverages inadvisable for persons with ulcers?

4. What type of acid-alkaline disorder is likely to develop if the stomach contents are repeatedly lost by vomiting over a prolonged period? What acid-alkaline disorder is likely to develop as a result of prolonged diarrhea?
5. Several years ago, an extract from kidney beans was sold in health food stores as a "starch blocker." Advertisements claimed that one could eat a plate of spaghetti, yet absorb none of it, because starch-digesting enzyme function would be blocked. The kidney bean product indeed kept salivary amylase from functioning. However, people who took the starch blocker developed abdominal pain, bloating, and gas. Suggest a reason for the ill effects of the supposed starch blocker.

1. List and describe the locations of the major parts of the alimentary canal.
2. List and describe the locations of the accessory organs of the digestive system.
3. Name the four layers of the wall of the alimentary canal.
4. Distinguish between mixing movements and propelling movements.
5. Define *peristalsis.*
6. Explain the relationship between peristalsis and receptive relaxation.
7. Describe the general effects of parasympathetic and sympathetic impulses on the alimentary canal.
8. Discuss the functions of the mouth and its parts.
9. Distinguish among lingual, palatine, and pharyngeal tonsils.
10. Compare the primary and secondary teeth.
11. Explain how the various types of teeth are adapted to perform specialized functions.
12. Describe the structure of a tooth.
13. Explain how a tooth is anchored in its socket.
14. List and describe the locations of the major salivary glands.
15. Explain how the secretions of the salivary glands differ.
16. Discuss the digestive functions of saliva.
17. Name and locate the three major regions of the pharynx.
18. Describe the mechanism of swallowing.
19. Explain the function of the esophagus.
20. Describe the structure of the stomach.
21. List the enzymes in gastric juice, and explain the function of each enzyme.
22. Explain how gastric secretions are regulated.
23. Describe the mechanism that controls the emptying of the stomach.
24. Describe the enterogastric reflex.
25. Explain the mechanism of vomiting.
26. Describe the locations of the pancreas and the pancreatic duct.
27. List the enzymes in pancreatic juice, and explain the function of each enzyme.
28. Explain how pancreatic secretions are regulated.
29. Describe the structure of the liver.
30. List the major functions of the liver.
31. Describe the composition of bile.
32. Trace the path of bile from a bile canal to the small intestine.
33. Explain how gallstones form.
34. Define *cholecystokinin.*
35. Explain the functions of bile salts.
36. List and describe the locations of the parts of the small intestine.
37. Name the enzymes of the intestinal mucosa, and explain the function of each enzyme.
38. Explain regulation of the secretions of the small intestine.
39. Describe the functions of the intestinal villi.
40. Summarize how each major type of digestive product is absorbed.
41. Explain control of the movement of the intestinal contents.
42. List and describe the locations of the parts of the large intestine.
43. Explain the general functions of the large intestine.
44. Describe the defecation reflex.

 **For More Information Related to This Chapter, Please Visit** *Hole's Human Anatomy and Physiology* **Home Page**
http://www.mhhe.com/biosci/abio/

## CHAPTER 18

# NUTRITION AND METABOLISM

## CHAPTER OBJECTIVES

AFTER YOU HAVE STUDIED THIS CHAPTER, YOU SHOULD BE ABLE TO:

1. Define *nutrition, nutrients,* and *essential nutrients.*

2. List the major sources of carbohydrates, lipids, and proteins.

3. Describe how cells utilize carbohydrates.

4. Describe how cells utilize lipids.

5. Describe how cells utilize amino acids.

6. Define *nitrogen balance.*

7. Explain how the energy values of foods are determined.

8. Explain the factors that affect an individual's energy requirements.

9. Define *energy balance.*

10. Explain what is meant by desirable weight.

11. List the fat-soluble and water-soluble vitamins.

12. Describe the general functions of each vitamin.

13. Distinguish between a vitamin and a mineral.

14. List the major minerals and trace elements.

15. Describe the general functions of each mineral and trace element.

16. Describe an adequate diet.

17. Distinguish between primary and secondary malnutrition.

## KEY TERMS

acetyl coenzyme A (as′ĕ-til ko-en′zīm)

antioxidant (an″te-ok′sĭ-dant)

basal metabolic rate (ba′sal met″ah-bol′ik rāt)

calorie (kal′o-re)

calorimeter (kal″o-rim′ĕ-ter)

dynamic equilibrium (di-nam′ik e″kwĭ-lib′re-um)

energy balance (en′er-je bal′ans)

malnutrition (mal″nu-trish′un)

mineral (min′er-al)

nitrogen balance (ni′tro-jen bal′ans)

nutrient (nu′tre-ent)

triglyceride (tri-glis′er-īd)

vitamin (vi′tah-min)

## UNDERSTANDING WORDS

**bas-,** base: *bas*al metabolic rate—metabolic rate of body under resting (basal) conditions.

**calor-,** heat: *calor*ie—unit used in measurement of heat or energy content of foods.

**carot-,** carrot: *carot*ene—yellowish plant pigment responsible for the color of carrots and other yellowish plant parts.

**lip-,** fat: *lip*ids—fat or fatlike substance insoluble in water.

**mal-,** bad, abnormal: *mal*nutrition—poor nutrition resulting from lack of food or failure to use available foods to best advantage.

**-meter,** measure: calori*meter*—instrument used to measure the caloric content of food.

**nutri-,** nourish: *nutri*ent—substance needed to nourish body cells.

**obes-,** fat: *obes*ity—condition in which the body contains excess fat.

**pell-,** skin: *pell*agra—vitamin deficiency condition characterized by inflammation of the skin and other symptoms.

When gymnast Christy Henrich was buried on a Friday morning in July 1994, she weighed 61 pounds. Three weeks earlier, she had weighed an unbelievable 47 pounds. Between those two dates, she celebrated her twenty-second birthday.

Christy suffered from *anorexia nervosa,* a psychological disorder common among professional athletes in which the patient, usually a girl, perceives herself as fat—even though she is obviously not. A year before her death, Christy told a newspaper reporter, "My life is a horrifying nightmare. It feels like there's a beast inside me, like a monster."

Christy's decline into self-starvation began in 1988, when a judge at a gymnastics competition told her she'd have to lose weight if she wanted to make the U.S. Olympic team. Christy weighed 90 pounds. From then on, with only a few respites, she ate nearly nothing, exercised many hours each day, and used laxatives to drop her weight lower and lower. Finally, it dropped so low that her vital organs could not function.

World-class gymnast Christy Henrich died of complications of the self-starvation eating disorder anorexia nervosa in July 1994. In this photo, taken 11 months before her death, she weighed under 60 pounds. Concern over weight gain propelled her down the path of this deadly nutritional illness.

Anorexia nervosa is a killer. We will return to it, other eating disorders, and starvation at the chapter's end. First, we consider the human body's nutritional requirements.

## WHY WE EAT

The human body requires a daily input of calories, distributed among food substances needed in large amounts as well as some needed in small amounts. The **macronutrients,** needed in bulk, include carbohydrates, proteins, and fats. **Micronutrients** are essential in small daily doses, and include vitamins and minerals. The body also requires water.

In countries with adequate food supplies, most healthy individuals can obtain nourishment by eating a variety of foods and limiting fat intake. People who do not eat meat products can also receive adequate nutrition but must pay more attention to food choices to avoid developing nutrient deficiencies. For example, eliminating red meat also means eliminating an excellent source of iron, copper, zinc, and vitamin $B_{12}$. The fiber that often makes up much of a vegetarian's diet, although very healthful in many ways, also decreases absorption of iron. Therefore, a vegetarian must be careful to obtain sufficient iron from nonmeat sources. This is easily done. Fortified foods, green leafy vegetables, and especially whole grains provide many of the nutrients also present in meat. Table 18.1 lists the different types of vegetarian diets.

Nutrients are obtained from foods, and digestion breaks them down to sizes that can be absorbed and transported in the bloodstream. **Metabolism** refers to the ways that nutrients are altered chemically and used in anabolism (building up or synthesis) and catabolism (breaking down) to support the activities of life. (Chapter 4 discusses the metabolic pathways of carbohydrates, lipids, and proteins in detail.) Nutrients that human cells cannot synthesize, such as certain amino acids, are particularly important and are therefore called **essential nutrients.**

In an overall sense, we eat to obtain the nutrients that power the activities of life. In a more literal sense, eating is a complex, finely tuned homeostatic mechanism that balances nutrient utilization with nutrient procurement. At the center of this control are parts of the hypothalamus called the arcuate and paraventricular nuclei. Here, a molecule called neuropeptide Y links biochemical messages concerning nutrient use (levels of leptin, insulin, cholecystokinin, and glucocorticoids) to regulators of food intake. These regulators include release of peptides that control appetite, pituitary hormones that oversee endocrine control of nutrient use and storage, and insulin release.

| TABLE 18.1 | TYPES OF VEGETARIAN DIETS | |
|---|---|---|
| **Type** | **Food Restrictions** | |
| Vegan | No animal foods | |
| Ovo-vegetarian | Eggs allowed; no dairy or meat | |
| Lacto-vegetarian | Dairy allowed; no eggs or meat | |
| Lacto-ovo-vegetarian | Dairy and eggs allowed; no meat | |
| Pesco-vegetarian | Dairy, eggs, and fish allowed; no other meat | |
| Semivegetarian | Dairy, eggs, chicken, and fish allowed; no other meat | |

## CARBOHYDRATES

**Carbohydrates** are organic compounds, and include the sugars and starches. The energy held in their chemical bonds is used to power cellular processes.

### Carbohydrate Sources

Carbohydrates are ingested in a variety of forms. Complex carbohydrates include starch from grains and vegetables, and glycogen from meats. Foods containing starch and glycogen usually have many other nutrients, including valuable vitamins and minerals. The simple carbohydrates include *disaccharides* from cane sugar, beet sugar, and molasses, and *monosaccharides* from honey and fruits. During digestion, complex carbohydrates are broken down to monosaccharides, which are small enough to be absorbed.

Cellulose is a complex carbohydrate that is abundant in our food—it provides the crunch to celery and the crispness to lettuce. However, we cannot digest cellulose, and most of it passes through the alimentary canal largely unchanged. However, cellulose provides bulk (also called fiber or roughage) against which the muscular wall of the digestive system can push, facilitating the movement of food. *Hemicellulose*, *pectin*, and *lignin* are other plant carbohydrates that provide fiber.

### Carbohydrate Utilization

The monosaccharides that are absorbed from the digestive tract include *fructose*, *galactose*, and *glucose*. Liver enzymes convert fructose and galactose into glucose (fig. 18.1). Recall that glucose is the form of carbohydrate that is most commonly oxidized in glycolysis for cellular fuel.

Sugar substitutes provide concentrated sweetness, so that fewer calories are needed to sweeten a food. Aspartame, which is a dipeptide (two joined amino acids), is two hundred times as sweet as table sugar (sucrose). This means that just a pinch of it equals the sweetness of a teaspoonful or more of sugar. Saccharin, another sugar substitute, is three hundred times as sweet as sugar.

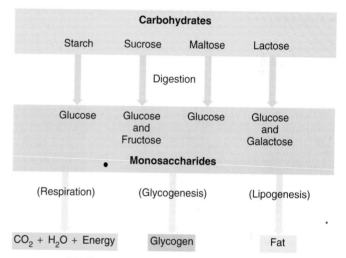

**FIGURE 18.1**

The liver converts the monosaccharides fructose and galactose into glucose.

**FIGURE 18.2**

Monosaccharides from foods are used for energy, stored as glycogen, or converted to fat.

Some excess glucose is polymerized to form *glycogen* (glycogenesis) and stored as a glucose reserve in the liver and muscles. Glucose can be rapidly mobilized from glycogen (glycogenolysis) when it is required to supply energy. However, only a certain amount of glycogen can be stored. Excess glucose beyond what is stored as glycogen is usually converted into fat and stored in adipose tissue (fig. 18.2).

Cells also use carbohydrates as starting materials for such vital biochemicals as the 5-carbon sugars *ribose* and *deoxyribose*, which are required for the production of the nucleic acids RNA and DNA, and the disaccharide *lactose* (milk sugar), which is synthesized when the breasts are actively secreting milk.

Many cells can also obtain energy by oxidizing fatty acids. However, some cells, such as neurons, normally depend on a continuous supply of glucose for survival. (Under some conditions, such as prolonged starvation, neurons may use other fuel sources.) Even a temporary decrease in the glucose supply may seriously impair nervous system function. Consequently, the body requires a minimum amount of carbohydrates. If adequate carbohydrates are not supplied in foods, the liver may convert some noncarbohydrates, such as amino acids from proteins or glycerol from fats, into glucose—a process called *gluconeogenesis*. Thus, the requirement for glucose has physiological priority over the need to synthesize certain other substances, such as proteins, from available amino acids.

1  List several common sources of carbohydrates.

2  In what form are carbohydrates utilized as a cellular fuel?

3  Explain what happens to excess glucose in the body.

4  Name two uses of carbohydrates other than supplying energy.

5  How does the body obtain glucose when its food supply of carbohydrates is insufficient?

---

An adult liver stores about 100 grams of glycogen, and muscle tissue stores another 200 grams, providing enough reserve to meet energy demands for about 12 hours when the person is resting. Whether these stores are filled depends, of course, on diet. People live on widely varying amounts of carbohydrates, often reflecting economic conditions. In the United States, a typical adult's diet supplies about 50% of total body energy from carbohydrates. In Asian countries where rice comprises much of the diet, carbohydrates contribute even more to the diet.

---

## Carbohydrate Requirements

Because carbohydrates provide the primary source of fuel for cellular processes, the need for carbohydrates varies with individual energy requirements. Therefore, physically active individuals require more carbohydrates than those who are sedentary. The minimal requirement for carbohydrates in the human diet is unknown. It is estimated, however, that an intake of at least 125–175 grams daily is necessary to spare protein (that is, to avoid protein breakdown) and to avoid metabolic disorders resulting from excess fat utilization. An average diet includes 200–300 grams of carbohydrates daily.

1  Why do the daily requirements for carbohydrates vary from person to person?

2  What is the daily minimum requirement for carbohydrates?

## LIPIDS

**Lipids** are organic compounds that include fats, oils, and fatlike substances (see chapter 2). They supply energy for cellular processes and to build structures, such as cell membranes. Lipids include fats, phospholipids, and cholesterol. The most common dietary lipids are the fats called *triglycerides* (see fig. 2.12).

## Lipid Sources

Triglycerides are found in plant- and animal-based foods. Saturated fats (which should comprise no more than 10% of the diet) are mainly found in foods of animal origin, such as meat, eggs, milk, and lard, as well as in palm and coconut oil. Unsaturated fats are contained in seeds, nuts, and plant oils.

Cholesterol is abundant in liver and egg yolk, and is present in smaller amounts in whole milk, butter, cheese, and meats. Foods of plant origin do not contain cholesterol. This is why a label on a plant-based food stating that it is "cholesterol-free" states the obvious.

---

Before 1993, "lite" and similar markings on prepared food packages could, and did, mean almost anything: A bottle of vegetable oil labeled "lite" referred to the product's color; a "lite" cheesecake referred to its texture! Now, "lite," or any similar spelling, has a distinct meaning: The product must have a third fewer calories than the "real thing," or half the fat calories.

Be wary of claims that a food product is "99% fat-free." This usually refers to percentage by weight—not calories, which is what counts. A creamy concoction that is 99% fat-free may be largely air and water, and therefore in that form, fat comprises very little of it. But when the air is compressed and the water absorbed, as happens in the stomach, the fat percentage may skyrocket.

---

## Lipid Utilization

Digestion breaks triglycerides down into fatty acids and glycerol. After being absorbed, these products are transported in the lymph and blood to tissues. In the cells, the process of beta oxidation breaks fatty acids down into 2-carbon units, which are further catabolized in the citric acid cycle to release the energy held in their bonds. Glycerol becomes an intermediate within the glycolytic pathway for further ATP production. The liver and adipose tissues control metabolism of lipids (fig. 18.3).

The liver can convert fatty acids from one form to another. However, it cannot synthesize certain of the fatty acids, such as *linoleic acid*. This **essential fatty acid** is required for the synthesis of phospholipids, which, in turn, are necessary for the formation of cell membranes and transport of circulating lipids. Good sources of linoleic acid include corn oil, cottonseed oil, and soy oil. Other essential fatty acids are *linolenic acid* and *arachidonic acid*.

The liver uses free fatty acids to synthesize triglycerides, phospholipids, and lipoproteins that may then be released into the blood. Thus, the liver regulates circulating lipids (see fig. 18.3). It also controls the total amount of cholesterol in the body by synthesizing cholesterol and releasing it into the blood, or by removing cholesterol from the blood and excreting it into the bile. The liver uses cholesterol to produce bile salts. Cholesterol is not used as an energy source, but it does provide structural material for cell and organelle membranes, and it furnishes starting materials for the synthesis of certain sex hormones and hormones produced by the adrenal cortex.

Excess triglycerides are stored in adipose tissue. If the blood lipid concentration drops (in response to fasting, for example), some of these triglycerides are hydrolyzed into free fatty acids and glycerol, and then released into the blood.

## Lipid Requirements

As with carbohydrates, the lipid content of human diets varies widely. A person who lives on burgers, fries, and shakes may consume 50% or more of total daily calories from fat. For a vegetarian, the percentage may be far lower. The American Heart Association advises that the diet not exceed 30% of total daily calories from fat.

The amounts and types of fats required for health are unknown. Since linoleic acid is an essential fatty acid, to prevent deficiency conditions from developing, nutritionists recommend that infants fed formula receive 3% of the energy intake in the form of linoleic acid. A typical adult diet consisting of a variety of foods usually provides adequate fats. Fat intake must also supply the required amounts of fat-soluble vitamins.

What tastes like fat, feels like fat, but isn't fat? Fake fat, or more formally, a fat substitute. Because it is really a carbohydrate or protein, a fat substitute offers half the calories of fat. The first fake fats, introduced in the early 1990s, were made from egg, milk, or whey proteins, mixed and broken down into tiny particles that impart the texture of fat.

In 1996, a product called Olestra became available. Olestra consists of a molecule of the sugar sucrose bound to six or more triglyceride molecules. It passes through the alimentary canal without being absorbed. Olestra is added to crackers, tortilla chips, and potato chips, and its uses may broaden if it proves to be safe. Using Olestra cuts calories in snack products approximately in half.

Although it took Olestra 25 years from its discovery in 1971 to be approved for marketing, the product is still under scrutiny. This is because it binds fat-soluble substances, including vitamins and carotenoids, taking them out of the body. Olestra may also cause abdominal cramps and "anal leakage." Olestra-laden potato chips taste and feel just like real chips—but the government-required label may discourage consumers from trying them: *"This product contains Olestra. Olestra may cause abdominal cramping and loose stools. Olestra inhibits the absorption of some vitamins and other nutrients. Vitamins A, D, E, and K have been added."*

1 What foods commonly supply lipids?

2 Which fatty acid is an essential nutrient?

3 What is the role of the liver in the utilization of lipids?

4 What is the function of cholesterol?

## PROTEINS

**Proteins** are polymers of amino acids with a wide variety of functions. Proteins include enzymes that control the rates of metabolic reactions; clotting factors; the keratin of skin and hair; elastin and collagen of connective tissue; plasma proteins that regulate water balance; the muscle components actin and myosin; certain hormones; and the antibodies that protect against infection (fig. 18.4).

Amino acids are also potential sources of energy. They are oxidized to release energy when they are present in excess or when supplies of carbohydrates and fats are insufficient. (Fig. 4.16 describes how the three macronutrients

**FIGURE 18.3**

The liver uses fatty acids to synthesize a variety of lipids.

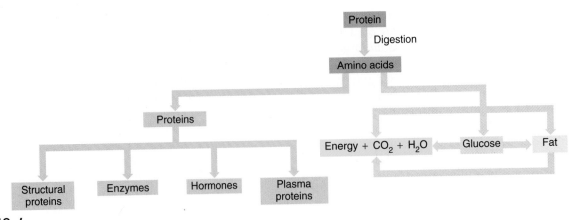

**FIGURE 18.4**

Proteins are digested to their constituent amino acids. These amino acids are then linked together, following genetic instructions, to build new proteins. Free amino acids are also used to supply energy under certain conditions.

enter the energy pathways.) When carbohydrate intake is sufficient, proteins remain available for tissue building and repair, rather than being broken down and reassembled into carbohydrates for use as an energy source. Using structural proteins to generate energy causes the tissue-wasting characteristic of starvation.

## Protein Sources

Foods rich in proteins include meats, fish, poultry, cheese, nuts, milk, eggs, and cereals. Legumes, including beans and peas, contain lesser amounts of protein. Digestion breaks proteins down into their component amino acids, and these smaller molecules are absorbed by intestinal cells and transported in the blood.

The human body can synthesize many amino acids (nonessential amino acids). However, eight amino acids the adult body needs (ten required for growing children) cannot be synthesized sufficiently or at all, and they are called **essential amino acids.** This term refers only to dietary intake since all amino acids are required for normal protein synthesis. Table 18.2 lists the amino acids found in foods, and indicates those that are essential.

All of the amino acids must be present in the body at the same time for growth and tissue repair to occur. In other words, if one essential amino acid is missing from the diet, protein synthesis cannot take place. Since essential amino acids are not stored, those present but not used in protein synthesis are oxidized as energy sources or are converted into carbohydrates or fats.

Proteins are classified as complete or incomplete on the basis of the amino acid types they provide. The **complete proteins,** which include those available in milk, meats, and eggs, contain adequate amounts of the essential amino acids to maintain human body tissues and promote normal growth and development. **Incomplete proteins,** such as *zein* in corn, which has too little of the essential amino acids tryptophan and lysine, are unable by themselves to maintain human tissues or to support normal growth and development.

| TABLE 18.2 | AMINO ACIDS IN FOODS |
|---|---|
| Alanine | Leucine (e) |
| Arginine (ch) | Lysine (e) |
| Asparagine | Methionine (e) |
| Aspartic acid | Phenylalanine (e) |
| Cysteine | Proline |
| Glutamic acid | Serine |
| Glutamine | Threonine (e) |
| Glycine | Tryptophan (e) |
| Histidine (ch) | Tyrosine |
| Hydroxyproline | Valine (e) |
| Isoleucine (e) | |

Eight essential amino acids (e) cannot be synthesized by human cells and must be provided in the diet. Two additional amino acids (ch) are essential in growing children.

Plant proteins typically contain too little of one or more essential amino acids to provide adequate nutrition for a person. Combining appropriate plant foods can provide a diversity of dietary amino acids. For example, beans are low in methionine, but have enough lysine. Rice lacks lysine, but has enough methionine. A meal of beans and rice provides enough of both types of amino acids.

Researchers can alter plants to make their protein more "complete." For example, genetic instructions for producing the amino acid tryptophan are inserted into corn cells, compensating for the low levels of this nutrient normally found in corn.

A protein in wheat called *gliadin* is an example of a *partially complete protein.* Although it does not contain enough lysine to promote human growth, it contains enough to maintain life.

TABLE 18.3

## CARBOHYDRATE, LIPID, AND PROTEIN NUTRIENTS

| Nutrient | Sources and RDA* for Adults | Calories per Gram | Utilization | Conditions Associated with | |
|---|---|---|---|---|---|
| | | | | *Excesses* | *Deficiencies* |
| Carbohydrate | Primarily from starch and sugars in foods of plant origin, and from glycogen in meats 0.8g/kg body weight | 4.1 | Oxidized for energy; used in production of ribose, deoxyribose, and lactose; stored in liver and muscles as glycogen; converted to fats and stored in adipose tissue | Obesity, dental caries, nutritional deficits | Metabolic acidosis |
| Lipid | Meats, eggs, milk, lard, plant oils 80–100 g | 9.5 | Oxidized for energy; production of triglycerides, phospholipids, lipoproteins, and cholesterol, stored in adipose tissue; glycerol portions of fat molecules may be used to synthesize glucose | Obesity, increased serum cholesterol, increased risk of heart disease | Weight loss, skin lesions |
| Protein | Meats, cheese, nuts, milk, eggs, cereals, legumes 125–175 g | 4.1 | Production of protein molecules used to build cell structure and to function as enzymes or hormones; used in the transport of oxygen, regulation of water balance, control of pH, formation of antibodies; amino acids may be broken down and oxidized for energy or converted to carbohydrates or fats for storage | Obesity | Extreme weight loss, wasting, anemia, growth retardation |

*RDA = recommended dietary allowance.

 Which foods provide rich sources of protein?

 Why are some amino acids called essential?

3 Distinguish between a complete protein and an incomplete protein.

## Nitrogen Balance

In a healthy adult, proteins are continuously built up and broken down. This occurs at different rates in different tissues, but the overall gain of body proteins equals the loss, producing a state of *dynamic equilibrium*. Since proteins contain a relatively high percentage of nitrogen, dynamic equilibrium also brings **nitrogen balance**—a condition in which the amount of nitrogen taken in is equal to the amount excreted.

A person who is starving has a *negative nitrogen balance* because the amount of nitrogen excreted as a result of amino acid oxidation exceeds the amount the diet replaces. Conversely, a growing child, a pregnant woman, or an athlete in training is likely to have a *positive nitrogen balance*, because more protein is being built into new tissue and less is being excreted or used for energy.

## Protein Requirements

In addition to supplying essential amino acids, proteins provide nitrogen and other elements for the synthesis of nonessential amino acids and certain nonprotein nitrogenous substances. The amount of protein individuals require varies according to body size, metabolic rate, and nitrogen balance condition.

For an average adult, nutritionists recommend a daily protein intake of about 0.8 grams per kilogram (0.4 grams per pound) of body weight. For a pregnant woman, who needs to maintain a positive nitrogen balance, the recommendation adds 30 grams of protein per day. Similarly, a nursing mother requires an additional 20 grams of protein per day to maintain a high level of milk production.

In addition to tissue wasting, protein deficiency may also decrease the level of plasma proteins, which decreases the osmotic pressure of the blood. As a result, fluids collect in the tissues, producing a condition called *nutritional edema*. Table 18.3 summarizes the sources, uses, and requirements for carbohydrate, lipid, and protein nutrients.

1. What are the physiological functions of proteins?

2. What is a negative nitrogen balance? A positive nitrogen balance?

3. How can inadequate nutrition cause edema?

## ENERGY EXPENDITURES

Carbohydrates, fats, and proteins supply energy. Since energy is required for all metabolic processes, energy availability is of prime importance to cell survival. If the diet is deficient in energy-supplying nutrients, structural molecules may gradually be consumed, leading to death.

### Energy Values of Foods

The amount of potential energy a food contains can be expressed as **calories,** which are units of heat. Although a calorie is defined as the amount of heat required to raise the temperature of a gram of water by 1 degree Celsius (° C), the calorie used to measure food energy is 1,000 times greater. This *large calorie* (Cal.) equals the amount of heat required to raise the temperature of a kilogram (1,000 grams) of water by 1° C (actually from 14.5° C to 15.5° C). This unit is also called a *kilocalorie*, but it is customary in nutritional studies to refer to it simply as a calorie.

Figure 18.5 shows a bomb calorimeter, which is used to measure the caloric contents of foods. It consists of a metal chamber submerged in a known volume of water. A food sample is dried, weighed, and placed in a nonreactive dish inside the chamber. The chamber is filled with oxygen gas and submerged in the water. Then, the food is ignited and allowed to oxidize completely. Heat released from the food raises the temperature of the surrounding water, and the change in temperature is measured. Since the volume of the water is known, the amount of heat released from the food can be calculated in calories.

Caloric values determined this way are somewhat higher than the amount of energy that metabolic oxidation actually releases, because nutrients generally are not completely absorbed from the digestive tract. Also, the body does not completely oxidize amino acids but excretes parts of them in urea or uses them to synthesize other nitrogenous substances. When such losses are considered, cellular oxidation yields on the average about 4.1 calories from 1 gram of carbohydrate, 4.1 calories from 1 gram of protein, and 9.5 calories from 1 gram of fat. The fact that more than twice as much energy is derived from equal amounts by weight of fats as from either proteins or carbohydrates is one reason why avoiding fatty foods helps weight loss. Fats also encourage weight gain because they add flavor to food, which can cause overeating.

1. What term designates the potential energy in a food?

2. How can the energy value of a food be determined?

### FIGURE 18.5

A bomb calorimeter measures the caloric content of a food sample.

3. What is the energy value of a gram of carbohydrate? A gram of protein? A gram of fat?

### Energy Requirements

The amount of energy required to support metabolic activities for 24 hours varies from person to person. The factors that influence energy needs include the individual's basal metabolic rate, degree of muscular activity, body temperature, and rate of growth.

The **basal metabolic rate** (BMR) measures the rate at which the body expends energy under *basal conditions*—when a person is awake and at rest, after an overnight fast, and in a comfortable, controlled environment. Tests of thyroid function can be used to estimate a person's BMR.

The amount of oxygen the body consumes is directly proportional to the amount of energy cellular respiration releases. The BMR, therefore, reveals the total amount of energy expended in a given time period to support the activities of such organs as the brain, heart, lungs, liver, and kidneys.

The average adult BMR indicates a requirement for approximately one Calorie of energy per hour for each kilogram of body weight. However, this requirement varies with such factors as sex, body size, body temperature, and level of endocrine gland activity. For example, since heat loss is directly proportional to the body surface area, and a smaller person has a greater surface area, he or she will have a higher BMR. Males tend to have higher metabolic rates than females. As body temperature increases, BMR increases, and as the blood level of thyroxine or epinephrine increases, so does the BMR.

Maintaining the basal metabolic rate usually requires the body's greatest expenditure of energy. The energy required to support voluntary muscular activity comes next, though this amount varies greatly with the type of activity (table 18.4). For example, energy needed to maintain posture while sitting at a desk might amount to 100 calories per hour above the basal need, whereas running or swimming might require 500–600 calories per hour.

| TABLE 18.4 | CALORIES USED DURING VARIOUS ACTIVITIES | |
|---|---|---|
| **Activity** | **Calories (per Hour)** | |
| Walking up stairs | 1,100 | |
| Running (jogging) | 570 | |
| Swimming | 500 | |
| Vigorous exercise | 450 | |
| Slow walking | 200 | |
| Dressing and undressing | 118 | |
| Sitting at rest | 100 | |

Maintenance of body temperature may require additional energy expenditure, particularly in cold weather. In this case, extra energy may be expended by involuntary muscular contractions, such as shivering, or through voluntary muscular actions, such as walking. Growing children and pregnant women, because their bodies are actively producing new tissues, also require more calories.

## Energy Balance

A state of **energy balance** exists when caloric intake in the form of foods equals caloric output resulting from the basal metabolic rate and muscular activities. Under these conditions, body weight remains constant, except perhaps for slight variations due to changes in water content.

If, on the other hand, caloric intake exceeds the output, a *positive energy balance* occurs, and tissues store excess nutrients. This increases body weight, since an excess of 3,500 calories can be stored as a pound of fat. Conversely, if caloric output exceeds input, the energy balance is negative, and stored materials are mobilized from the tissues for oxidation, causing weight loss.

**1** What is basal metabolic rate?

**2** What factors influence the BMR?

**3** What is *energy balance?*

## Desirable Weight

The most obvious and common nutritional disorders reflect calorie imbalances, which may result from societal and geographic factors. Obesity is a common problem in nations where food is plentiful and diverse. The tendency to become obese may be a holdover from thousands of years ago, when the ability to store energy in the form of fat was a survival advantage when food supplies were scarce or erratic. In many African nations, natural famines combined with political unrest cause mass starvation. Obesity and starvation are considered later in the chapter.

It is difficult to determine a desirable body weight. In the past, weight standards were based on average weights and heights within a certain population, and the degrees of underweight and overweight were expressed as percentage deviations from these averages. These stan-

(a)

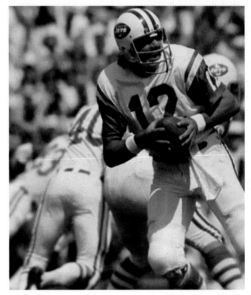

(b)

**FIGURE 18.6**

(a) An obese person is overweight and has excess adipose tissue. (b) An athlete may be overweight due to muscle overgrowth but is not considered obese. In fact, many athletes have very low percentages of body fat.

dards reflected the gradual gain in weight that usually occurs with age. Later, medical researchers recognized that such an increase in weight after the age of 25 to 30 years is not necessary and may not be conducive to health. Consequently, standards of *desirable weights* were prepared. More recent height–weight guidelines are based upon the characteristics of people who live the longest. These weights are somewhat more lenient than those in the desirable weight charts.

*Overweight* is defined as exceeding desirable weight by 10% to 20%. A person who is more than 20% above the desired weight is considered to be *obese*, although **obesity** is more correctly defined as excess adipose tissue. Therefore, overweight and obesity are not the same. For example, as figure 18.6 shows, an athlete or a person

# Obesity

The U.S. National Institutes of Health considers obesity a "killer disease," and for good reason. A person who is obese—defined as 20% above "ideal" weight based on population statistics considering age, sex, and build—is at higher risk for diabetes, digestive disorders, heart disease, kidney failure, hypertension, stroke, and cancers of the female reproductive organs and the gallbladder. The body is enormously strained to support the extra weight—miles of blood vessels are needed to nourish the additional pounds. In the United States, obesity is the second leading cause of preventable death, following cigarette smoking.

Obesity refers specifically to extra pounds of fat. The proportion of fat in a human body ranges from 5% to more than 50%, with "normal" for males falling between 12% and 23% and for females between 16% and 28%. An elite athlete may have a body fat level as low as 4%. Fat distribution also affects health. Excess poundage above the waist is linked to increased risk of heart disease, diabetes, hypertension, and lipid disorders. Figure 18A shows a do-it-yourself way to estimate obesity using a measurement called body mass index.

Both heredity and the environment contribute to obesity. We inherit genes that control metabolism, but the fact that identical twins reared in different households can grow into adults of vastly different weights indicates that environment influences weight too. Studies comparing body mass index between adult identical twins reared apart indicate that weight is about 70% influenced by genes and 30% by the environment.

## TREATMENT FOR OBESITY

### Diet and Exercise

A safe goal for weight loss using dietary restriction and exercise is 1 pound of fat per week. A pound of fat contains 3,500 calories of energy, so that pound can be shed by an appropriate combination of calorie cutting and exercise. This might mean eating 500 calories less per day or exercising off 500 calories each day. Actually more than a pound of weight will drop because water is lost as well as fat.

Dieting should apply to the energy-providing nutrients (carbohydrates, proteins, and fats) but never to the vitamins and minerals. A rule of thumb is to leave the proportion of protein calories about the same or slightly increased, cut fat calories in half, and cut carbohydrates by a third. Choose foods that you like, and distribute them into three or four balanced meals of 250 to 500 calories each.

Ideally, weight loss can be accomplished by changing diet and exercise habits. However, realistically two-thirds of those who lose weight regain it within five years. Physicians are increasingly regarding obesity as a chronic illness that may require more drastic measures than dieting and exercising.

### Drug Therapy

Physicians recommend drug therapy for obesity if the body mass index exceeds 30, or if it exceeds 27 and hypertension, diabetes mellitus, or hyperlipidemia are also present. Two very popular "diet drugs," fenfluramine and phentermine, were prescribed to millions of people in 1996 and early 1997, but were then withdrawn from the market when they were found to damage heart valves.

New obesity drugs are in development. Sibutramine is both a serotonin reuptake inhibitor and increases

---

whose work involves heavy muscular activity may be overweight, but not obese. Clinical Application 18.1 discusses obesity.

When a person needs to gain weight, diet can be altered to include more calories and to emphasize particular macronutrients. For example, a person recovering from a debilitating illness might stress carbohydrates, whereas a bodybuilder might consume extra protein to spur muscle development. An infant also needs to gain weight rapidly, which is best accomplished by consumption of human milk. The high fat content of human milk is important for the rapid growth of the infant's brain.

1    What is *desirable weight*?

2    Distinguish between overweight and obesity.

3    Under what conditions is weight gain desirable?

catecholamines in the brain. Tetrahydrolipstatin is an experimental drug that prevents intestinal absorption of fats. Finding drugs that block the proteins (adipocyte transcription factors) that enable fibroblasts to specialize into fat cells (adipocytes) is another approach to treating obesity. Yet other experimental strategies alter the homeostatic control of body weight, by stimulating production of cholecystokinin or leptin (so-called satiety factors), or by blocking the appetite stimulant neuropeptide Y.

*Surgery*

Surgery is recommended for people whose BMI exceeds 40, or if it exceeds 35 and medical problems are present. In an intestinal bypass operation, part of the small intestine is removed and the cut ends rejoined. This decreases the surface area through which nutrients are absorbed. Gastric reduction, also known as stomach stapling, similarly removes from digestive action a portion of the stomach. Some patients may undergo both operations. Side effects include intestinal blockage and electrolyte imbalances. ■

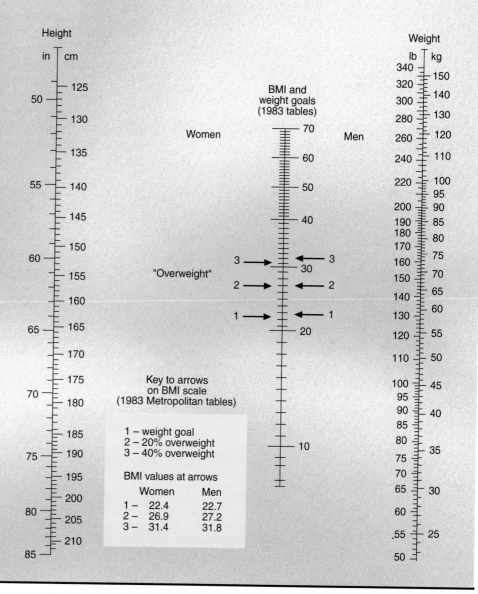

**FIGURE 18.A**

Body mass index (BMI) is a measurement that estimates obesity. BMI equals weight in kilograms divided by height in meters, squared (BMI = weight [kg]/height [m]$^2$). To calculate your BMI, draw a line from your height on the left side of the figure to your weight on the right side. The point where the line crosses the scale in the middle is your BMI. A value above 27.2 for men and 26.9 for women indicates obesity. A value above 30 is cause for concern; a value above 35 may be life threatening.

Key to arrows
on BMI scale
(1983 Metropolitan tables)

1 – weight goal
2 – 20% overweight
3 – 40% overweight

BMI values at arrows

| | Women | Men |
|---|---|---|
| 1 – | 22.4 | 22.7 |
| 2 – | 26.9 | 27.2 |
| 3 – | 31.4 | 31.8 |

## VITAMINS

**Vitamins** are organic compounds (other than carbohydrates, lipids, and proteins) required in small amounts for normal metabolic processes, but that body cells cannot synthesize in adequate amounts. Thus, vitamins are essential nutrients that must come from foods.

Vitamins can be classified on the basis of solubility, since some are soluble in fats (or fat solvents) and others are soluble in water. Those that are *fat-soluble* include vitamins A, D, E, and K; the *water-soluble* group includes the B vitamins and vitamin C. Table 18.5 lists, and corrects, some common misconceptions about vitamins.

> Different species have different vitamin requirements. For example, ascorbic acid is a vitamin (C) in humans, guinea pigs, and Indian fruit bats, but not in other animals, which can manufacture their own ascorbic acid.

## Fat-Soluble Vitamins

Because fat-soluble vitamins dissolve in fats, they associate with lipids and are influenced by the same factors that affect lipid absorption. For example, the presence of bile salts in the intestine promotes absorption of these vitamins. As a group, the fat-soluble vitamins are stored in moderate quantities within various tissues, and because they are fairly resistant to heating, they are usually not destroyed by cooking or food processing.

| | |
|---|---|
| TABLE | **18.5 VITAMIN FALLACIES AND FACTS** |

| Fallacy | Fact |
|---|---|
| Vitamins from foods are superior to those in tablet form | A vitamin molecule is chemically the same, whether it comes from a bean or a bottle |
| The more vitamins, the better | Too much of a water-soluble vitamin results in excretion of the vitamin through urination; too much of a fat-soluble vitamin can harm one's health |
| A varied diet provides all needed vitamins | Many people do need vitamin supplements, particularly pregnant and breast-feeding women |
| Vitamins provide energy | Vitamins do not directly supply energy; they aid in the release of energy from carbohydrates, fats, and proteins |

**1** What are vitamins?

**2** Distinguish between fat-soluble vitamins and water-soluble vitamins.

**3** How do bile salts affect the absorption of vitamins?

**Vitamin A** occurs in several forms, including retinol and retinal (retinene). Body cells synthesize this vitamin from a group of yellowish plant pigments called *carotenes*. Excess vitamin A or its precursors are mainly stored in the liver, which regulates their concentration in the body (fig. 18.7). An adult's liver stores enough vitamin A to supply body requirements for a year. Infants and children usually lack such reserves, and are therefore more likely to develop vitamin A deficiencies if their diets are inadequate.

Vitamin A is relatively stable to the effects of heat, acids, and alkalis. However, it is readily destroyed by oxidation and is unstable in light.

Vitamin A is important in vision. Retinal is used to synthesize *rhodopsin* (visual purple) in the rods of the retina, and may be required for production of light-sensitive pigments in the cones as well. The vitamin also functions in the synthesis of mucoproteins and mucopolysaccharides of mucus, in development of normal bones and teeth, and in maintenance of epithelial cells in skin and mucous membranes. Clinical Application 18.2 examines the ongoing controversy concerning the value of vitamins A and E in preventing illness.

Only foods of animal origin such as liver, fish, whole milk, butter, and eggs are sources of vitamin A. However, the vitamin's precursor, carotene, is widespread in leafy green vegetables and in yellow or orange vegetables and fruits.

**FIGURE 18.7**

A molecule of beta-carotene can be converted into two molecules of retinal, which in turn can be changed into retinol.

## CLINICAL APPLICATION

# Do Vitamins Protect Against Heart Disease and Cancer?

Vitamins A and E are popularly believed to protect against cardiovascular disease and cancer. Although studies have shown that people who eat diets rich in fruits and vegetables containing these vitamins have slightly lower incidences of these disorders, these investigations do not prove that the vitamins, and not something else in these foods, provide protection. A more meaningful study would carefully follow the result of taking a specific vitamin supplement. The Finnish Alpha-Tocopherol, Beta-Carotene Cancer Prevention Study did just that—with surprising results.

Researchers monitored the health of 29,000 adult male smokers over a 6-year period. Each man received vitamin A or E, both vitamins, or neither. The "endpoint" of the study was to determine which men developed lung cancer. Not only did these vitamins fail to protect against this specific cancer, but the men who received beta-carotene actually had an increased incidence of lung cancer! Neither vitamin lowered incidences of death from stroke or coronary heart disease.

The Finnish study has several limitations. Some critics claim that it was not conducted over a long-enough time period. Also, the study considers only one type of person who engages in a dangerous activity—smoking—and may not account for all possible contributing factors to disease development. Unraveling the connections between specific nutrients and health is extremely complex, because of the many variables. Overall, a diet rich in fruits and vegetables is healthful. But the Finnish study shows that claims that certain vitamins protect against cancer or heart disease may be premature. ■

Excess vitamin A produces peeling skin, hair loss, nausea, headache, and dizziness, a condition called *hypervitaminosis A*. Chronic overdoses of the vitamin may inhibit growth and cause the bones and joints to degenerate. "Megadosing" on fat-soluble vitamins is particularly dangerous during pregnancy. Excess vitamin A, for example, can cause birth defects.

A deficiency of vitamin A causes *night blindness*, in which a person cannot see normally in dim light. Vitamin A deficiency also causes degenerative changes in certain epithelial tissues, and the body becomes more susceptible to infection.

1 What biochemical do body cells use to synthesize vitamin A?

2 What conditions destroy vitamin A?

3 Which foods are good sources of vitamin A?

**Vitamin D** is a group of steroids that have similar properties. One of these substances, vitamin $D_3$ (cholecalciferol), is found in foods such as milk, egg yolk, and fish liver oils. Vitamin $D_2$ (ergocalciferol) is produced commercially by exposing a steroid obtained from yeasts (ergosterol) to ultraviolet light. Vitamin D can also be synthesized from dietary cholesterol that has been converted to provitamin D by intestinal enzymes, then stored in the skin and exposed to ultraviolet light (see chapter 13).

Like other fat-soluble vitamins, vitamin D resists the effects of heat, oxidation, acids, and alkalis. It is primarily stored in the liver and is less abundant in the skin, brain, spleen, and bones.

As it is needed, vitamin D stored in the form of hydroxycholecalciferol is released into the blood. When parathyroid hormone is present, this form of vitamin D is converted in the kidneys into an active form of the vitamin (dihydroxycholecalciferol). This substance, in turn, is carried as a hormone in the blood to the intestines. Here, it promotes absorption of calcium and phosphorus, ensuring that adequate amounts of these minerals are available in the blood for tooth and bone formation and metabolic processes.

Because natural foods are often poor sources of vitamin D, it is often added to food during processing. For example, homogenized, nonfat, and evaporated milk are typically fortified with vitamin D. *Fortified* means essential nutrients have been added to a food where they originally were absent or scarce. *Enriched* means essential nutrients have been partially replaced in a food that has lost nutrients during processing.

Excess vitamin D, or *hypervitaminosis D*, produces diarrhea, nausea, and weight loss. Over time, it may also cause calcification of certain soft tissues and irreversible kidney damage.

**FIGURE 18.8**
Vitamin D deficiency causes rickets, in which the bones and teeth do not develop normally.

In children, vitamin D deficiency results in *rickets*, in which the bones and teeth fail to develop normally (fig. 18.8). In adults or in the elderly who have little exposure to sunlight, such a deficiency may lead to *osteomalacia*, in which the bones decalcify and weaken due to disturbances in calcium and phosphorus metabolism.

1. Where is vitamin D stored?
2. What are the functions of vitamin D?
3. Which foods are good sources of vitamin D?
4. What are symptoms of vitamin D excess and deficiency?

**Vitamin E** includes a group of compounds, the most active of which is *alphatocopherol*. This vitamin is resistant to the effects of heat, acids, and visible light, but is unstable in alkalis and in the presence of ultraviolet light or oxygen. In fact, it combines so readily with oxygen that it may prevent oxidation of other compounds. For this reason, vitamin E is termed an **antioxidant.**

Vitamin E is found in all tissues, but is primarily stored in the muscles and adipose tissue. It is also highly concentrated in the pituitary and adrenal glands.

The precise functions of vitamin E are unknown, but is thought to act as an antioxidant by inhibiting the oxidation of vitamin A in the digestive tract and of polyunsaturated fatty acids in the tissues. It may also help maintain the stability of cell membranes. Antioxidants prevent formation of free radicals, which build up in certain diseases and aging.

Although vitamin E is widely distributed among foods, its richest sources are oils from cereal seeds such as wheat germ. Other good sources are salad oils, margarine, shortenings, fruits, nuts, and vegetables. Since this vitamin is so easily obtained, deficiency conditions are rare.

1. Where is vitamin E stored?
2. What are the functions of vitamin E?
3. Which foods are good sources of vitamin E?

**Vitamin K,** like the other fat-soluble vitamins, occurs in several chemical forms. One of these, vitamin $K_1$ (phylloquinone), is found in foods, whereas another, vitamin $K_2$, is produced by bacteria (*Escherichia coli*) that normally inhabit the human intestinal tract. These vitamins resist the effects of heat, but are destroyed by oxidation or by exposure to acids, alkalis, or light. The liver stores them to a limited degree.

Vitamin K primarily functions in the liver, where it is necessary for the formation of several proteins needed for blood clotting, including *prothrombin* (see chapter 14). Consequently, deficiency of vitamin K causes prolonged blood clotting time and a tendency to hemorrhage.

The richest sources of vitamin K are leafy green vegetables. Other good sources are egg yolk, pork liver, soy oil, tomatoes, and cauliflower. Table 18.6 summarizes the fat-soluble vitamins and their properties.

About 1 in every 200 to 400 newborns develops vitamin K deficiency because of an immature liver, poor transfer of vitamin K through the placenta, or lack of intestinal bacteria that can synthesize this vitamin. The deficiency causes "hemorrhagic disease of the newborn," which typically appears during the second to fifth day of life, producing abnormal bleeding. To prevent it, infants in many nations routinely receive injections of vitamin K shortly after birth. Adults may develop vitamin K deficiency if they are treated with antibiotic drugs that kill the intestinal bacteria that manufacture the vitamin. People with cystic fibrosis may develop vitamin K deficiency because they cannot digest fats well.

1. Where in the body is vitamin K synthesized?
2. What is the function of vitamin K?
3. Which foods are good sources of vitamin K?

TABLE
18.6  FAT-SOLUBLE VITAMINS

| Vitamin | Characteristics | Functions | Sources and RDA* for Adults | Conditions Associated with | |
|---------|----------------|-----------|----------------------------|-------------------------|---|
| | | | | Excesses | Deficiencies |
| Vitamin A | Occurs in several forms; synthesized from carotenes; stored in liver; stable in heat, acids, and alkalis; unstable in light | Necessary for synthesis of visual pigments, mucoproteins, and mucopolysaccharides; for normal development of bones and teeth; and for maintenance of epithelial cells | Liver, fish, whole milk, butter, eggs, leafy green vegetables, yellow and orange vegetables and fruits 4,000–5,000 IU** | Nausea, headache, dizziness, hair loss | Night blindness, degeneration of epithelial tissues |
| Vitamin D | A group of steroids; resistant to heat, oxidation, acids, and alkalis; stored in liver, skin, brain, spleen, and bones | Promotes absorption of calcium and phosphorus; promotes development of teeth and bones | Produced in skin exposed to ultraviolet light; in milk, egg yolk, fish liver oils, fortified foods 400 IU | Diarrhea, calcification of soft tissues, renal damage | Rickets, bone decalcification and weakening |
| Vitamin E | A group of compounds; resistant to heat and visible light; unstable in presence of oxygen and ultraviolet light; stored in muscles and adipose tissue | An antioxidant; prevents oxidation of vitamin A and polyunsaturated fatty acids; may help maintain stability of cell membranes | Oils from cereal seeds, salad oils, margarine, shortenings, fruits, nuts, and vegetables 30 IU | Hypertension | Rare, uncertain effects |
| Vitamin K | Occurs in several forms; resistant to heat, but destroyed by acids, alkalis, and light; stored in liver | Needed for synthesis of prothrombin, which functions in blood clotting | Leafy green vegetables, egg yolk, pork liver, soy oil, tomatoes, cauliflower 55–70 µg | None known | Easy bruising and bleeding |

*RDA = recommended dietary allowance.
**IU = international unit.

## Water-Soluble Vitamins

The water-soluble vitamins include the B vitamins and vitamin C. The **B vitamins** are several compounds that are essential for normal cellular metabolism. They help oxidize carbohydrates, lipids, and proteins. Since the B vitamins often occur together in foods, they are usually referred to as the *vitamin B complex*. Members of this group differ chemically and functionally.

The B-complex vitamins include the following:

1. **Thiamine**, or **vitamin B1**. In its pure form, thiamine is a crystalline compound called thiamine hydrochloride. It is destroyed by exposure to heat and oxygen, especially in alkaline environments. Figure 18.14 shows its molecular structure.

Thiamine is part of a coenzyme called *cocarboxylase*, which oxidizes carbohydrates. More specifically, thiamine is required for pyruvic acid to enter the citric acid cycle (see chapter 4); in the absence of this vitamin, pyruvic acid accumulates in the blood. Thiamine also functions as a coenzyme in the synthesis of the sugar ribose, which is part of the nucleic acid RNA.

Thiamine is primarily absorbed through the wall of the duodenum and is transported by the blood to body cells. Only small amounts are stored in the tissues, and excesses are excreted in the urine.

Since this vitamin functions in the oxidation of carbohydrates, the quantity cells require varies with caloric intake. It is recommended that an adult diet contain 0.5 milligram (mg) of thiamine for every 1,000 calories ingested daily.

Good sources of thiamine are lean pork, other lean meats, liver, eggs, whole-grain cereals, leafy green vegetables, and legumes.

A mild deficiency of thiamine produces loss of appetite, fatigue, and nausea. Prolonged deficiency leads to a disease called *beriberi*, which causes gastrointestinal disturbances, mental confusion, muscular weakness and paralysis, and heart enlargement. In severe cases, the heart may fail.

In developed nations, beriberi mainly occurs in people with chronic alcoholism who have substituted alcohol for foods. Moreover, since thiamine is required for the metabolic oxidation of alcohol, people with alcoholism are particularly likely to develop a thiamine deficiency.

2. **Riboflavin,** or **vitamin B2.** Riboflavin is a yellowish-brown crystalline substance that is relatively stable to the effects of heat, acids, and oxidation, but is destroyed by exposure to bases and ultraviolet light. This vitamin is part of several enzymes and coenzymes that are known as *flavoproteins.* One such coenzyme, FAD, is an electron carrier in the citric acid cycle and electron transport chain of aerobic respiration. Flavoproteins are essential for the oxidation of glucose and fatty acids, and for cellular growth. The absorption of riboflavin seems to be regulated by an active transport system that controls the amount taken in by the intestinal mucosa. Riboflavin is carried in the blood combined with blood proteins called *albumins.* Excess riboflavin in the blood is excreted in the urine, and any that remains unabsorbed in the intestine is lost in the feces.

The amount of riboflavin the body requires varies with caloric intake. About 0.6 mg of riboflavin per 1,000 calories is sufficient to meet daily cellular requirements.

Riboflavin is widely distributed in foods, and rich sources include meats and dairy products. Leafy green vegetables, whole-grain cereals, and enriched cereals provide lesser amounts. Vitamin B$_2$ deficiency produces dermatitis and blurred vision.

3. **Niacin.** Niacin, which is also known as *nicotinic acid,* occurs in plant tissues, and is stable in the presence of heat, acids, and bases. After ingestion, it is converted to a physiologically active form called *niacinamide* (fig. 18.9). Niacinamide is the form of niacin that is present in foods of animal origin.

Niacin functions as part of two coenzymes (coenzyme I, also called NAD [fig. 18.10], and coenzyme II, called NADP) that play essential roles in the oxidation of glucose, acting as electron carriers in glycolysis, the citric acid cycle, and the electron transport chain, as well as in the synthesis of proteins and fats. These coenzymes are also required for the synthesis of the sugars (ribose and deoxyribose) that are part of nucleic acids.

**FIGURE 18.9**

Body cells convert niacin from foods into physiologically active niacinamide.

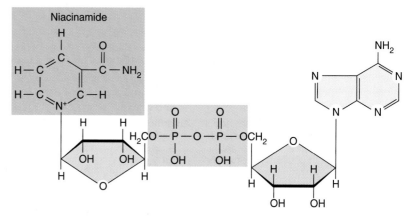

Coenzyme 1 (NAD or nicotinamide adenine dinucleotide)

**FIGURE 18.10**

Niacinamide is incorporated into molecules of coenzyme I.

Niacin is readily absorbed from foods, and human cells synthesize it from the essential amino acid *tryptophan.* Consequently, the daily requirement for niacin varies with tryptophan intake. Nutritionists recommend a daily niacin (or niacin equivalent) intake of 6.6 mg per 1,000 calories.

Rich sources of niacin (and tryptophan) include liver, lean meats, peanut butter, and legumes. Milk is a poor source of niacin, but a good source of tryptophan.

Historically, niacin deficiencies have been associated with diets largely consisting of corn and corn products, which are very low in niacin and lack adequate tryptophan. Such a deficiency causes a disease called *pellagra* that produces dermatitis, inflammation of the digestive tract, diarrhea, and mental disorders.

Although pellagra is relatively rare today, it was a serious problem in the rural South of the United States in the early 1900s. It sometimes occurs in people with chronic alcoholism who have substituted alcohol for foods.

4. **Pantothenic acid,** or **vitamin B5.** Pantothenic acid is a yellowish oil that is destroyed by heat, acids, and alkalis. It functions as part of a complex

Niacin in large doses is prescribed as a drug to lower serum cholesterol.

Vitamin B₆

**FIGURE 18.11**

Vitamin $B_6$ includes three closely related chemical compounds.

Vitamin B₁₂ (cyanocobalamin)

**FIGURE 18.12**

Vitamin $B_{12}$, which has the most complex molecular structure of the vitamins, contains cobalt (Co).

molecule called *coenzyme A*, which, in turn, reacts with intermediate products of carbohydrate and fat metabolism to become *acetyl coenzyme A*, which enters the citric acid cycle. Pantothenic acid is therefore essential to cellular energy release.

A daily adult intake of 4–7 mg of pantothenic acid seems to be adequate. Most diets apparently provide sufficient amounts, since deficiencies are rare, and no clearly defined set of deficiency

symptoms is known. Good sources of pantothenic acid include meats, whole-grain cereals, legumes, milk, fruits, and vegetables.

5. **Vitamin B6.** Vitamin $B_6$ is a group of three compounds that are chemically similar, as figure 18.11 shows. They are called *pyridoxine*, *pyridoxal*, and *pyridoxamine*. These compounds have similar actions and are fairly stable in the presence of heat and acids. Oxidation or exposure to bases or ultraviolet light destroys them. The vitamin $B_6$ compounds function as coenzymes that are essential in a wide variety of metabolic pathways, including those that synthesize proteins, amino acids, antibodies, and nucleic acids, as well as the conversion of tryptophan to niacin.

Since vitamin $B_6$ functions in the metabolism of nitrogen-containing substances, the requirement for this vitamin varies with the protein content of the diet rather than with caloric intake. The recommended daily allowance of vitamin $B_6$ is 2.0 mg, but because it is so widespread in foods, deficiency conditions are quite rare. Good sources of vitamin $B_6$ include liver, meats, bananas, avocados, beans, peanuts, whole-grain cereals, and egg yolk.

6. **Cyanocobalamin, or vitamin B12.** Cyanocobalamin has a complex molecular structure that contains a single atom of the element *cobalt* (fig. 18.12). In its pure form, this vitamin is red. It is stable to the effects of heat, but is inactivated by exposure to light or strong acids or bases.

Secretion of *intrinsic factor* from the parietal cells of the gastric glands regulates cyanocobalamin absorption. Intrinsic factor is thought to combine with cyanocobalamin and to facilitate its transfer through the epithelial lining of the small intestine and into the blood. Calcium ions must be present for the process to take place.

Cyanocobalamin is stored in various tissues, particularly those of the liver. An average adult has a reserve sufficient to supply cells for three to five years.

This vitamin is essential for the functions of all cells. It is part of coenzymes required for the synthesis of nucleic acids and the metabolism of carbohydrates and fats. Cyanocobalamin also seems to help form myelin in the central nervous system.

Only foods of animal origin contain cyanocobalamin. Good sources include liver, meats, milk, cheese, and eggs. In most countries, dietary lack of this vitamin is rare, although strict vegetarians may develop a deficiency.

When the gastric glands of some individuals fail to secrete adequate amounts of intrinsic factor, cyanocobalamin is poorly absorbed. This leads to *pernicious anemia*, in which abnormally large red blood cells called macrocytes are produced when bone marrow cells do not reproduce properly because of defective DNA synthesis.

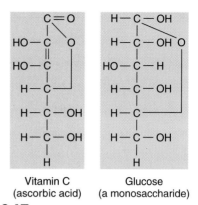

**Figure 18.13**

Vitamin C is chemically similar to some 6-carbon monosaccharides.

7. **Folacin,** or **folic acid.** Folacin is a yellow crystalline compound that exists in several forms. It is easily oxidized in an acid environment and is destroyed by heat in alkaline solutions; consequently, this vitamin may be lost in foods that are stored or cooked.

Folacin is readily absorbed from the digestive tract and is stored in the liver, where it is converted to a physiologically active substance called *folinic acid*. Folinic acid functions as a coenzyme that is necessary for the metabolism of certain amino acids and for the synthesis of DNA. It also acts with cyanocobalamin in promoting production of normal red blood cells.

Good sources of folacin include liver, leafy green vegetables, whole-grain cereals, and legumes. A deficiency of folacin leads to *megaloblastic anemia*, which is characterized by a reduced number of normal red blood cells and the presence of large, nucleated red cells. Folacin deficiency has been linked to neural tube defects, in which the tube that becomes the central nervous system in a fetus fails to close entirely. Neural tube defects include spina bifida and anencephaly. Taking folacin supplements just before and during pregnancy can greatly reduce the risk of a neural tube defect.

8. **Biotin.** Biotin is a relatively simple compound that is stable to the effects of heat, acids, and light, but may be destroyed by oxidation or bases. Figure 18.14 shows the molecular structure of biotin.

Biotin is a coenzyme in metabolic pathways for amino acids and fatty acids. It also plays a role in the synthesis of the purine nitrogenous bases of nucleic acids.

Metabolically active organs such as the brain, liver, and kidneys store small quantities of biotin. Bacteria that inhabit the intestinal tract synthesize biotin. The vitamin is widely distributed in foods, and dietary deficiencies are rare. Good sources include liver, egg yolk, nuts, legumes, and mushrooms.

**1** Which biochemicals comprise the vitamin B complex?

**2** Which foods are good sources of vitamin B complex?

**3** Which of the B-complex vitamins can be synthesized from tryptophan?

**4** What is the general function of each member of the B complex?

**Vitamin C,** or **ascorbic acid.** Ascorbic acid is a crystalline compound that contains six carbon atoms. Chemically, it is similar to the monosaccharides (fig. 18.13). Vitamin C is one of the least stable of the vitamins, in that oxidation, heat, light, or bases destroy it. However, vitamin C is fairly stable in acids.

Ascorbic acid is necessary for the production of the connective tissue protein *collagen*, for conversion of folacin to folinic acid, and in the metabolism of certain amino acids. It also promotes iron absorption and synthesis of certain hormones from cholesterol.

Although vitamin C is not stored in any great amount, tissues of the adrenal cortex, pituitary gland, and intestinal glands contain high concentrations. Excess vitamin C is excreted in the urine or oxidized.

Individual requirements for ascorbic acid may vary. Ten mg per day is sufficient to prevent deficiency symptoms, and 80 mg per day will saturate the tissues within a few weeks. Many nutritionists recommend a daily adult intake of 60 mg, which is enough to replenish normal losses and to provide a satisfactory level for cellular requirements.

Ascorbic acid is fairly widespread in plant foods; particularly high concentrations are found in citrus fruits and tomatoes. Leafy green vegetables are also good sources.

Prolonged deficiency of ascorbic acid leads to *scurvy*, which occurs more frequently in infants and children. Scurvy produces abnormal bone development and swollen, painful joints. Because of a tendency for cells to pull apart in scurvy, the gums may swell and bleed easily, resistance to infection is lowered, and wounds heal slowly. If a woman takes large doses of ascorbic acid during pregnancy, the newborn may develop symptoms of scurvy when it receives a far lower daily dose of the vitamin after birth than it did before. Table 18.7 summarizes the water-soluble vitamins and their characteristics.

TABLE 18.7  WATER-SOLUBLE VITAMINS

721

| Vitamin | Characteristics | Functions | Sources and RDA* for Adults | Conditions Associated with | |
|---|---|---|---|---|---|
| | | | | Excesses | Deficiencies |
| Thiamine (Vitamin B₁) | Destroyed by heat and oxygen, especially in alkaline environment | Part of coenzyme needed for oxidation of carbohydrates; coenzyme needed in synthesis of ribose | Lean meats, liver, eggs, whole-grain cereals, leafy green vegetables, legumes 1.5 mg | None known | Beriberi, muscular weakness, heart enlarges |
| Riboflavin (Vitamin B₂) | Stable to heat, acids, and oxidation; destroyed by bases and ultraviolet light | Part of enzymes and coenzymes, such as FAD, needed for oxidation of glucose and fatty acids and for cellular growth | Meats, dairy products, leafy green vegetables, whole-grain cereals 1.7 mg | None known | Dermatitis, blurred vision |
| Niacin (Nicotinic acid) | Stable to heat, acids, and alkalis; converted to niacinamide by cells; synthesized from tryptophan | Part of coenzymes NAD and NADP needed for oxidation of glucose and synthesis of proteins, fats, and nucleic acids | Liver, lean meats, peanuts, legumes 20 mg | Hyperglycemia, vasodilation, gout | Pellagra, photosensitive dermatitis, diarrhea, mental disorders |
| Pantothenic acid | Destroyed by heat, acids, and bases | Part of coenzyme A needed for oxidation of carbohydrates and fats | Meats, whole-grain cereals, legumes, milk, fruits, vegetables 10 mg | None known | Rare, loss of appetite, mental depression, muscle spasms |
| Vitamin B6 | Group of three compounds; stable to heat and acids; destroyed by oxidation, bases, and ultraviolet light | Coenzyme needed for synthesis of proteins and various amino acids, for conversion of tryptophan to niacin, for production of antibodies, and for synthesis of nucleic acids | Liver, meats, bananas, avocados, beans, peanuts, whole-grain cereals, egg yolk 2 mg | Numbness | Rare, convulsions, vomiting, seborrhea lesions |
| Cyanoco-balamin (Vitamin B₁₂) | Complex, cobalt-containing compound; stable to heat; inactivated by light, strong acids, and strong bases; absorption regulated by intrinsic factor from gastric glands; stored in liver | Part of coenzyme needed for synthesis of nucleic acids and for metabolism of carbohydrates; plays role in synthesis of myelin | Liver, meats, milk, cheese, eggs 3–6 µg | None known | Pernicious anemia |
| Folacin (Folic acid) | Occurs in several forms; destroyed by oxidation in acid environment or by heat in alkaline environment; stored in liver where it is converted into folinic acid | Coenzyme needed for metabolism of certain amino acids and for synthesis of DNA; promotes production of normal red blood cells | Liver, leafy green vegetables, whole-grain cereals, legumes 0.4 mg | None known | Megaloblastic anemia |
| Biotin | Stable to heat, acids, and light; destroyed by oxidation and bases | Coenzyme needed for metabolism of amino acids and fatty acids, and for synthesis of nucleic acids | Liver, egg yolk, nuts, legumes, mushrooms 0.3 mg | None known | Rare, elevated blood cholesterol, nausea, fatigue, anorexia |
| Ascorbic acid (Vitamin C) | Closely related to monosaccharides; stable in acids but destroyed by oxidation, heat, light, and bases | Needed for production of collagen, conversion of folacin to folinic acid, and metabolism of certain amino acids; promotes absorption of iron and synthesis of hormones from cholesterol | Citrus fruits, tomatoes, potatoes, leafy green vegetables 60 mg | Exacerbates gout and kidney stone formation | Scurvy, lowered resistance to infection, wounds heal slowly |

*RDA = recommended dietary allowance.

| 1 | What factors destroy vitamin C? |

| 2 | What is the function of vitamin C? |

| 3 | Which foods are good sources of vitamin C? |

| 4 | What are symptoms of vitamin C deficiency? |

## MINERALS

Carbohydrates, lipids, proteins, and vitamins are all organic compounds. Dietary **minerals** are inorganic elements that are essential in human metabolism. These elements are usually extracted from the soil by plants, and humans obtain them from plant foods or from animals that have eaten plants.

### Characteristics of Minerals

Minerals are responsible for about 4% of body weight and are most concentrated in the bones and teeth. The minerals *calcium* and *phosphorus* are very abundant in these tissues.

Minerals are usually incorporated into organic molecules. For example, phosphorus is found in phospholipids, iron in hemoglobin, and iodine in thyroxine. However, some minerals are part of inorganic compounds, such as the calcium phosphate of bone. Other minerals are free ions, such as the sodium, chloride, and calcium ions in the blood.

Minerals are present in all body cells, where they comprise parts of the structural materials. They are also portions of enzyme molecules, contribute to the osmotic pressure of body fluids, and play vital roles in conduction of nerve impulses, contraction of muscle fibers, coagulation of blood, and maintenance of pH. The physiologically active form of minerals is the ionized form, such as $Ca^{+2}$.

Homeostatic mechanisms regulate the concentrations of minerals in body fluids. This ensures that excretion of minerals matches intake.

| 1 | How do minerals differ from other nutrients? |

| 2 | What are the major functions of minerals? |

| 3 | Which are the most abundant minerals in the body? |

### Major Minerals

Calcium and phosphorus account for nearly 75% by weight of the mineral elements in the body; thus, they are **major minerals** (macrominerals). Other major minerals, each of which accounts for 0.05% or more of body weight, include potassium, sulfur, sodium, chlorine, and magnesium. Descriptions of the major minerals follow:

1.  **Calcium.** Calcium (Ca) is widely distributed in cells and body fluids, even though 99% of the body's supply is in the inorganic salts of the bones and teeth. It is essential for nerve impulse conduction,

muscle fiber contraction, and blood coagulation. Calcium also decreases the permeability of cell membranes and activates certain enzymes.

The amount of calcium absorbed varies with a number of factors. For example, the proportion of calcium absorbed increases as the body's need for calcium increases. Vitamin D and high protein intake promote calcium absorption; increased motility of the digestive tract or an excess intake of fats decreases absorption. Consequently, the amount of dietary calcium needed to supply cells may vary. However, a daily intake of 800 mg is sufficient to cover adult requirements in spite of variations in absorption.

Only a few foods contain significant amounts of calcium. Milk and milk products and fish with bones, such as salmon or sardines, are the richest sources. Leafy green vegetables, such as mustard greens, turnip greens, and kale, are good sources, but because one must consume large amounts of these vegetables to obtain sufficient minerals, most people must regularly consume milk or milk products to maintain an adequate intake of calcium.

Calcium deficiency in children causes stunted growth, misshapen bones, and enlarged wrists and ankles. In adults, such a deficiency may remove calcium from the bones, thinning them and raising risk of fracture. Because calcium is required for normal closing of the sodium channels in nerve cell membranes, too little calcium (hypocalcemia) can cause tetany. Extra calcium demands in pregnancy can cause cramps.

2.  **Phosphorus.** Phosphorus (P) is responsible for about 1% of total body weight, and most of it is incorporated in the calcium phosphate of bones and teeth. The remainder serves as structural components and plays important roles in nearly all metabolic reactions. Phosphorus is a constituent of nucleic acids, many proteins, some enzymes, and some vitamins. It is also found in the phospholipids of cell membranes, in the energy-carrying molecule ATP, and in the phosphates of body fluids that regulate pH. (Review the molecular structure of ATP in fig. 4.7.)

The recommended daily adult intake of phosphorus is 800 mg, and since this mineral is abundant in protein foods, diets adequate in proteins are also adequate in phosphorus. Phosphorus-rich foods include meats, cheese, nuts, whole-grain cereals, milk, and legumes.

| 1 | What are the functions of calcium? |

| 2 | What are the functions of phosphorus? |

| 3 | Which foods are good sources of calcium and phosphorus? |

3. **Potassium.** Potassium (K) is widely distributed throughout the body and tends to be concentrated inside cells rather than in extracellular fluids. On the other hand, sodium, which has similar chemical properties, tends to be concentrated outside the cells. The ratio of potassium to sodium within a cell is 10:1, whereas the ratio outside the cell is 1:28.

Potassium helps maintain intracellular osmotic pressure and pH. It promotes reactions of carbohydrate and protein metabolism, and plays a vital role in establishing the membrane potential that occurs in nerve impulse conduction and muscle fiber contraction.

Nutritionists recommend a daily adult intake of 2.5 grams (2,500 mg) of potassium. Because this mineral is widely distributed in foods, a typical adult diet provides between 2 and 6 grams each day. Dietary potassium deficiency is rare, but it may occur for other reasons. For example, when a person has diarrhea, the intestinal contents may pass through the digestive tract so rapidly that potassium absorption is greatly reduced. Vomiting or using diuretic drugs may also deplete potassium. The consequences of such losses may include muscular weakness, cardiac abnormalities, and edema.

Foods rich in potassium are avocados, dried apricots, meats, milk, peanut butter, potatoes, and bananas. Citrus fruits, apples, carrots, and tomatoes provide lesser amounts.

**1** How is potassium distributed in the body?

**2** What is the function of potassium?

**3** Which foods are good sources of potassium?

4. **Sulfur.** Sulfur (S) is responsible for about 0.25% of body weight and is widely distributed through tissues. It is particularly abundant in skin, hair, and nails. Most sulfur is part of the amino acids *methionine* and *cysteine*. Other sulfur-containing compounds include thiamine, insulin, and biotin (fig. 18.14). In addition, sulfur is a constituent of mucopolysaccharides found in cartilage, tendons, and bones, and of sulfolipids that are in the liver, kidneys, salivary glands, and brain.

No daily requirement for sulfur has been established. It is thought, however, that a diet providing adequate amounts of protein will also meet the body's need for sulfur. Good food sources of this mineral include meats, milk, eggs, and legumes.

5. **Sodium.** About 0.15% of adult body weight is due to sodium (Na), which is widely distributed throughout the body. Only about 10% of this

**FIGURE 18.14**

Three examples of essential sulfur-containing nutrients.

mineral is inside the cells, and about 40% is within the extracellular fluids. The remainder is bound to the inorganic salts of bones.

Sodium is readily absorbed from foods by active transport. The blood concentration of this element is regulated by the kidneys under the influence of the adrenal cortical hormone *aldosterone*, which causes the kidneys to reabsorb sodium while expelling potassium.

Sodium makes a major contribution to the solute concentration of extracellular fluids and thus helps regulate water movement between cells and their surroundings. It is necessary for nerve impulse conduction and muscle fiber contraction, and plays a role in moving substances, such as chloride ions, through cell membranes (see chapter 21).

The usual human diet probably provides more than enough sodium to meet the body's needs. Sodium may be lost as a result of diarrhea, vomiting, kidney disorders, sweating, or using diuretics. Such losses may cause a variety of symptoms including nausea, muscular cramps, and convulsions.

The amount of sodium naturally present in foods varies greatly, and it is commonly added to foods in the form of table salt (sodium chloride). In some geographic regions, drinking water contains significant concentrations of sodium. Foods high in sodium include cured ham, sauerkraut, cheese, and graham crackers.

☐ In what compounds and tissues of the body is sulfur found?

☑ Which hormone regulates the blood concentration of sodium?

☒ What are the functions of sodium?

6. **Chlorine.** Chlorine (Cl) in the form of chloride ions is widely distributed throughout the body, although it is most highly concentrated in cerebrospinal fluid and in gastric juice. Together with sodium, chlorine helps to maintain the solute concentration of extracellular fluids, regulate pH, and maintain electrolyte balance. It is also essential for the formation of hydrochloric acid in gastric juice, and it functions in the transport of carbon dioxide by red blood cells.

Chlorine and sodium are usually obtained together in the form of table salt (sodium chloride), and as in the case of sodium, an ordinary diet usually provides considerably more chlorine than the body requires. Vomiting, diarrhea, kidney disorders, sweating, or using diuretics can deplete chlorine in the body.

7. **Magnesium.** Magnesium (Mg) is responsible for about 0.05% of body weight and is found in all cells. It is particularly abundant in bones in the form of phosphates and carbonates.

Magnesium is important in ATP-forming reactions in mitochondria, as well as in breaking down ATP to ADP. Therefore, it is important in providing energy for cellular processes.

Magnesium absorption in the intestinal tract adapts to dietary intake of the mineral. When the intake of magnesium is high, a smaller percentage is absorbed from the intestinal tract, and when the intake is low, a larger percentage is absorbed. Absorption increases as protein intake increases, and decreases as calcium and vitamin D intake increase. Bone tissue stores a reserve supply of magnesium, and excess is excreted in the urine.

The recommended daily allowance of magnesium is 300 mg for females and 350 mg for males. A typical diet usually provides only about 120 mg of magnesium for every 1,000 calories, barely meeting the body's needs. Good sources of magnesium include milk and dairy products (except butter), legumes, nuts, and leafy green vegetables. Table 18.8 summarizes the major minerals.

☐ Where are chloride ions most highly concentrated in the body?

☑ Where is magnesium stored?

☒ What factors influence the absorption of magnesium from the intestinal tract?

## Trace Elements

**Trace elements** (microminerals) are essential minerals found in minute amounts, each making up less than 0.005% of adult body weight. They include iron, manganese, copper, iodine, cobalt, zinc, fluorine, selenium, and chromium.

**Iron** (Fe) is most abundant in the blood, is stored in the liver, spleen, and bone marrow, and is found to some extent in all cells. Iron enables *hemoglobin* molecules in red blood cells to carry oxygen (fig. 18.15). Iron is also part of *myoglobin*, which stores oxygen in muscle cells. In addition, iron assists in vitamin A synthesis, is incorporated into a number of enzymes, and is included in the cytochrome molecules that participate in ATP-generating reactions.

An adult male requires from 0.7 to 1 mg of iron daily, and a female needs 1.2 to 2 mg. A typical diet supplies about 10 to 18 mg of iron each day, but only 2% to 10% of the iron is absorbed. For some people, this may not be enough iron. Eating foods rich in vitamin C along with iron-containing foods can increase absorption of this important mineral.

Pregnant women require extra iron to support the formation of a placenta, and the growth and development of a fetus. Iron is also required for the synthesis of hemoglobin in a fetus as well as in a pregnant woman, whose blood volume increases by a third.

Liver is the only really rich source of dietary iron, and since liver is not a very popular food, iron is one of the more difficult nutrients to obtain from natural sources in adequate amounts. Foods that contain some iron include lean meats; dried apricots, raisins, and prunes; enriched whole-grain cereals; legumes; and molasses.

**Manganese** (Mn) is most concentrated in the liver, kidneys, and pancreas. It is necessary for normal growth and development of skeletal structures and other connective tissues. Manganese is part of enzymes that are essential for the synthesis of fatty acids and cholesterol, for urea formation, and for the normal functions of the nervous system.

The daily requirement for manganese is 2.5–5 mg. The richest sources include nuts, legumes, and whole-grain cereals; leafy green vegetables and fruits are good sources.

| | | | | | |
|---|---|---|---|---|---|
| **Mineral** | **Distribution** | **Functions** | **Sources and RDA\* for Adults** | **Conditions Associated with** | |
| | | | | *Excesses* | *Deficiencies* |
| Calcium (Ca) | Mostly in the inorganic salts of bones and teeth | Structure of bones and teeth; essential for nerve impulse conduction, muscle fiber contraction, and blood coagulation; increases permeability of cell membranes; activates certain enzymes | Milk, milk products, leafy green vegetables 800 mg | Kidney stones | Stunted growth, misshapen bones, fragile bones |
| Phosphorus (P) | Mostly in the inorganic salts of bones and teeth | Structure of bones and teeth; component in nearly all metabolic reactions; constituent of nucleic acids, many proteins, some enzymes, and some vitamins; occurs in cell membrane, ATP, and phosphates of body fluids | Meats, cheese, nuts, whole-grain cereals, milk, legumes 800 mg | None known | Stunted growth |
| Potassium (K) | Widely distributed; tends to be concentrated inside cells | Helps maintain intracellular osmotic pressure and regulate pH; promotes metabolism; needed for nerve impulse conduction and muscle fiber contraction | Avocados, dried apricots, meats, nuts, potatoes, bananas 2,500 mg | None known | Muscular weakness, cardiac abnormalities, edema |
| Sulfur (S) | Widely distributed; abundant in skin, hair, and nails | Essential part of various amino acids, thiamine, insulin, biotin, and mucopolysaccharides | Meats, milk, eggs, legumes None established | None known | None known |
| Sodium (Na) | Widely distributed; large proportion occurs in extracellular fluids and bound to inorganic salts of bone | Helps maintain osmotic pressure of extracellular fluids and regulate water movement; needed for conduction of nerve impulses and contraction of muscle fibers; aids in regulation of pH and in transport of substances across cell membranes | Table salt, cured ham, sauerkraut, cheese, graham crackers 2,500 mg | Hypertension, edema | Nausea, muscle cramps, convulsions |
| Chlorine (Cl) | Closely associated with sodium; most highly concentrated in cerebrospinal fluid and gastric juice | Helps maintain osmotic pressure of extracellular fluids, regulate pH, and maintain electrolyte balance; essential in formation of hydrochloric acid; aids transport of carbon dioxide by red blood cells | Same as for sodium None established | Vomiting | Muscle cramps |
| Magnesium (Mg) | Abundant in bones | Needed in metabolic reactions in mitochondria associated with ATP production; helps breakdown of ATP to ADP | Milk, dairy products, legumes, nuts, leafy green vegetables 300–350 mg | Diarrhea | Neuromuscular disturbances |

\*RDA = recommended dietary allowance.

(a) Heme group

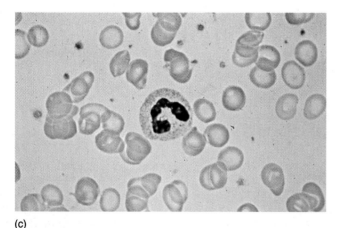

(b)

(c)

**FIGURE 18.15**

(a) A hemoglobin molecule contains four heme groups, each of which houses a single iron atom (Fe) that can combine with oxygen. Iron deficiency anemia can result from a diet poor in iron-containing foods. The blood cells in (b) are normal (400×), but many of those in (c) are small and pale (280×). They contain too little hemoglobin, because iron is lacking in the diet. Vegetarians must be especially careful to consume sufficient iron.

**1** What is the primary function of iron?

**2** Why does the usual diet provide only a narrow margin of safety in supplying iron?

**3** How is manganese utilized?

**4** Which foods are good sources of manganese?

**Copper** (Cu) is found in all body tissues, but is most highly concentrated in the liver, heart, and brain. It is essential for hemoglobin synthesis, bone development, melanin production, and formation of myelin within the nervous system.

A daily intake of 2 mg of copper is sufficient to supply cells. Because a typical adult diet has about 2–5 mg of this mineral, adults seldom develop copper deficiencies. Foods rich in copper include liver, oysters, crabmeat, nuts, whole-grain cereals, and legumes.

**Iodine** (I) is found in minute quantities in all tissues, but is highly concentrated within the thyroid gland. Its only known function is as an essential component of thyroid hormones. (The molecular structures of two of these hormones, thyroxine and triiodothyronine, are shown in figure 13.18.)

A daily intake of 1 microgram (0.001 mg) of iodine per kilogram of body weight is adequate for most adults. Since the iodine content of foods varies with the iodine content of soils in different geographic regions, many people use *iodized* table salt to season foods to prevent deficiencies.

**Cobalt** (Co) is widely distributed in the body, because it is an essential part of cyanocobalamin (vitamin $B_{12}$). It is also necessary for the synthesis of several important enzymes.

The amount of cobalt required in the daily diet is unknown. This mineral is found in a great variety of foods, and the quantity in the average diet is apparently sufficient. Good sources of cobalt include liver, lean meats, and milk.

TABLE 18.9  TRACE ELEMENTS

| Trace Element | Distribution | Functions | Sources and RDA* for Adults | Conditions Associated with | |
|---|---|---|---|---|---|
| | | | | *Excesses* | *Deficiencies* |
| Iron (Fe) | Primarily in blood; stored in liver, spleen, and bone marrow | Part of hemoglobin molecule; catalyzes formation of vitamin A; incorporated into a number of enzymes | Liver, lean meats, dried apricots, raisins, enriched whole-grain cereals, legumes, molasses 10–18 mg | Liver damage | Anemia |
| Manganese (Mn) | Most concentrated in liver, kidneys, and pancreas | Occurs in enzymes needed for synthesis of fatty acids and cholesterol, formation of urea, and normal functioning of the nervous system | Nuts, legumes, whole-grain cereals, leafy green vegetables, fruits 2.5–5 mg | None known | None known |
| Copper (Cu) | Most highly concentrated in liver, heart, and brain | Essential for synthesis of hemoglobin, development of bone, production of melanin, and formation of myelin | Liver, oysters, crabmeat, nuts, whole-grain cereals, legumes 2–3 mg | Rare | Rare |
| Iodine (I) | Concentrated in thyroid gland | Essential component for synthesis of thyroid hormones | Food content varies with soil content in different geographic regions; iodized table salt 0.15 mg | Decreased synthesis of thyroid hormones | Goiter |
| Cobalt (Co) | Widely distributed | Component of cyanocobalamin; needed for synthesis of several enzymes | Liver, lean meats, milk None established | Heart disease | Pernicious anemia |
| Zinc (Zn) | Most concentrated in liver, kidneys, and brain | Constituent of several enzymes involved in digestion, respiration, bone metabolism, liver metabolism; necessary for normal wound healing and maintaining integrity of the skin | Meats, cereals, legumes, nuts, vegetables 15 mg | Slurred speech, problems walking | Depressed immunity, loss of taste and smell, learning difficulties |
| Fluorine (F) | Primarily in bones and teeth | Component of tooth structure | Fluoridated water 1.5–4 mg | Mottled teeth | None known |
| Selenium (Se) | Concentrated in liver and kidneys | Occurs in enzymes | Lean meats, fish, cereals 0.05–2 mg | Vomiting, fatigue | None known |
| Chromium (Cr) | Widely distributed | Essential for use of carbohydrates | Liver, lean meats, wine 0.05–2 mg | None known | None known |

*RDA = recommended dietary allowance.

Zinc (Zn) is most concentrated in the liver, kidneys, and brain. It is part of many enzymes involved in digestion, respiration, and bone and liver metabolism. It is also necessary for normal wound healing and for maintaining the integrity of the skin.

The daily requirement for zinc is about 15 mg, and most diets provide 10–15 mg. Since only a portion of this amount may be absorbed, zinc deficiencies may occur. The richest sources of zinc are meats; cereals, legumes, nuts, and vegetables provide lesser amounts.

Fluorine (F), as part of the compound fluoroapatite, replaces hydroxyapatite in teeth, strengthening the enamel and preventing dental caries. Selenium (Se) is stored in the liver and kidneys. It is a constituent of certain enzymes and participates in heart function. This mineral is found in lean meats, whole-grain cereals, and onions. Chromium (Cr) is widely distributed throughout the body and regulates glucose utilization. It is found in liver, lean meats, yeast, and pork kidneys. Table 18.9 summarizes the characteristics of trace elements.

1 How is copper used?

2 What is the function of iodine?

3 Why might zinc deficiencies be common?

## HEALTHY EATING— THE FOOD PYRAMID AND READING LABELS

An adequate diet provides sufficient energy (calories), essential fatty acids, essential amino acids, vitamins, and minerals to support optimal growth and to maintain and repair body tissues. However, because individual nutrient requirements vary greatly with age, sex, growth rate, amount of physical activity, and level of stress, as well as with genetic and environmental factors, it is not possible to design a diet that is adequate for everyone. However, nutrients are so widely distributed in foods that satisfactory amounts and combinations can usually be obtained in spite of individual food preferences.

As the chapter so far shows, it is very difficult to keep track of the different nutrients in a diet and be certain that an adequate amount of each is consumed daily. Nutritionists have devised several ways to help consumers make healthy food choices.

Most familiar is the RDA guideline that has appeared on several charts in the chapter. *RDA* stands for United States Recommended Daily Allowance. An RDA is actually the upper limit of another measurement, called the Recommended Dietary Allowance, which lists optimal calorie intake for each sex at various ages, and the amounts of vitamins and minerals needed to avoid deficiency or excess conditions. The RDA values on food packages are set high, ensuring that most people who follow them receive sufficient amounts of each nutrient. Government panels meet every 5 years to evaluate the RDAs in light of new data.

Placing foods into groups is an easier way to follow a healthy diet. Figure 18.16 depicts the **food pyramid** system, which shows at a glance that complex carbohydrates should comprise the bulk of the diet. The four food group plan that the pyramid replaces, devised in the 1950s, put meat on equal footing with grains, dairy products, and fruits and vegetables in terms of number of servings. A food group plan from the 1940s had eight food groups, including separate groups for butter and margarine, and for eggs—foods now associated with the development of heart disease. In the 1920s, an entire food group was devoted to sweets!

When making individual food choices, it helps to read and understand food labels, which became much

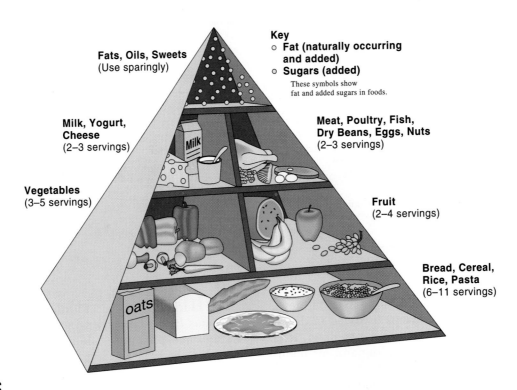

**Fats, Oils, Sweets** (Use sparingly)

**Key**
○ **Fat (naturally occurring and added)**
○ **Sugars (added)**

These symbols show fat and added sugars in foods.

**Milk, Yogurt, Cheese** (2–3 servings)

**Meat, Poultry, Fish, Dry Beans, Eggs, Nuts** (2–3 servings)

**Vegetables** (3–5 servings)

**Fruit** (2–4 servings)

**Bread, Cereal, Rice, Pasta** (6–11 servings)

**FIGURE 18.16**

The U.S. Department of Agriculture introduced the food pyramid in 1992. The chart is a guideline to healthy eating. In contrast to former food group plans, the pyramid gives an instant idea of which foods should make up the bulk of the diet—whole grains, fruits, and vegetables.

more informative in 1994. Disregard claims such as "light" and "low fat" and skip right to the calories of different ingredients; ideally, fat percentage should be below 30, and that of saturated fats below 10. Ingredients are listed in descending order by weight. Figure 18.17 shows where nutritional information can be found on a food label. Clinical Application 18.3 discusses some ways that understanding nutrition can help athletic performance.

## Malnutrition

**Malnutrition** is poor nutrition that results from a lack of essential nutrients or a failure to utilize them. It may result from *undernutrition* and produce the symptoms of deficiency diseases, or it may be due to *overnutrition* arising from excess nutrient intake.

The factors leading to malnutrition vary. For example, a deficiency condition may stem from lack of availability or poor quality of food. On the other hand, malnutrition may result from overeating or taking too many vitamin supplements. Malnutrition from diet alone is called *primary malnutrition*.

The Food and Drug Administration allows the following specific food and health claims:

- Dietary calcium decreases risk of osteoporosis (a bone-thinning condition).
- A low-fat diet lowers risk of some cancers.
- A diet low in saturated fat and cholesterol lowers risk of coronary heart disease.
- Fiber, fruits and vegetables, and whole grains reduce the risk of some cancers and coronary heart disease.
- Lowering sodium intake lowers blood pressure.
- Folic acid lowers the risk of neural tube defects.

Take any other health claims with a grain of salt!

*Secondary malnutrition* occurs when an individual's characteristics make a normally adequate diet insufficient. For example, a person who secretes low quantities of bile salts may develop a deficiency of fat-soluble vitamins because bile salts promote absorption of fats. Likewise,

### FIGURE 18.17

A key to healthy eating is to become familiar with the "Nutrition Facts" panel on food packages. Look for products with fewer than 30% of calories from fat.

severe and prolonged emotional stress may lead to secondary malnutrition, because stress can change hormonal concentrations, and such changes may result in amino acid breakdown or excretion of nutrients.

## Starvation

A healthy human can stay alive for 50 to 70 days without food. In prehistoric times, this margin allowed survival during seasonal famines. In some areas of Africa today, famine is not a seasonal event but a constant condition, and millions of people have starved to death. Starvation is also seen in hunger strikers, in inmates of concentration camps, and in sufferers of psychological eating disorders such as **anorexia nervosa** and **bulimia.**

Whatever the cause, the starving human body begins to digest itself. After only 1 day without eating, the body's reserves of sugar and starch are gone. Next, the body extracts energy from fat and then from muscle protein. By the third day, hunger ceases as the body uses energy from fat reserves. Gradually, metabolism slows to conserve energy, blood pressure drops, the pulse slows, and chills set in. Skin becomes dry and hair falls out as the proteins in these structures are broken down to release amino acids that are used for the more vital functioning of the brain, heart, and lungs. When the immune system's antibody proteins are dismantled for their amino acids, protection against infection declines. Mouth sores and anemia develop, the heart beats irregularly, and bone begins to degenerate. After several weeks without food, coordination is gradually lost. Near the end, the starving human is blind, deaf, and emaciated (fig. 18.18).

### Marasmus and Kwashiorkor

Lack of nutrients is called **marasmus,** and it causes people to resemble living skeletons (fig. 18.19). Children under the age of two with marasmus often die of measles or other infections, their immune systems too weakened to fight off normally mild viral illnesses.

Some starving children do not look skeletal but have protruding bellies. These youngsters suffer from a form of protein starvation called **kwashiorkor,** which in the language of Ghana means "the evil spirit which infects the first child when the second child is born." Kwashiorkor typically appears in a child who has recently been weaned from the breast, usually because of the birth of a sibling. The switch from protein-rich breast milk to the protein-poor gruel that is the staple of many developing nations is the source of this protein deficiency. The children's bellies swell with fluid, which is filtered from capillaries in greater than normal quantities that build up when protein is lacking. This is called **ascites.** Their skin may develop lesions. Infections overwhelm the body as the immune system becomes depleted of its protective antibodies.

**FIGURE 18.18**
A child dying of starvation.

### Anorexia Nervosa

**Anorexia nervosa** is self-imposed starvation. The condition is reported to affect 1 out of 250 adolescents, and 95% of them are female. The sufferer, typically a well-behaved adolescent girl from an affluent family, perceives herself to be overweight and eats barely enough to survive. She is terrified of gaining weight and usually loses 25% of her original body weight. In addition to eating only small amounts of low-calorie foods, she further loses weight by vomiting, by taking laxatives and diuretics, or by exercising intensely. Her eating behavior is often ritualized. She may meticulously arrange her meager meal on her plate or consume only a few foods. She develops low blood pressure, a slowed or irregular heartbeat, constipation, and constant chilliness. She stops menstruating as her body fat level plunges. Like any starving person, the hair becomes brittle and the skin dries out. She may develop soft, pale, fine body hair

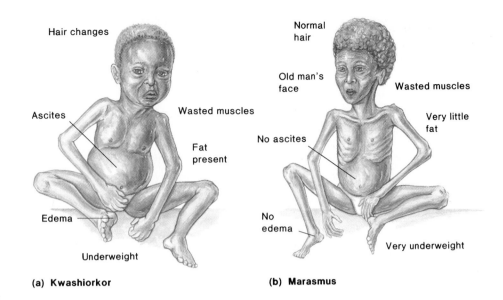

**FIGURE 18.19**

Two types of starvation in the young. (*a*) The child on the left suffers from kwashiorkor. Although he may have received adequate nourishment from breast milk early in life, he became malnourished when his diet switched to a watery, white extract from cassava that looks like milk but has very little protein. The lack of protein in the diet causes edema and the ascites that swells his belly. (*b*) The child on the right, suffering from marasmus, did not even have adequate nutrition as an infant.

called lanugo, normally seen only on a developing fetus, to preserve body heat.

When the anorectic reaches an obviously emaciated state, her parents usually have her hospitalized, where she is fed intravenously so that she does not starve to death or die suddenly of heart failure due to a mineral imbalance. She also receives psychotherapy and nutritional counseling. Despite these efforts, 15% to 21% of people with anorexia die.

Anorexia nervosa has no known physical cause. One hypothesis is that the person is rebelling against approaching womanhood. Indeed, her body is astonishingly childlike, with a flat chest and slim hips, and she has often ceased to menstruate. She typically has low self-esteem and believes that others, particularly her parents, are controlling her life. Her weight is something that she can control. Anorexia can be a one-time, short-term experience or a lifelong obsession.

## Bulimia

A person suffering from **bulimia** is often of normal weight. She eats whatever she wants, often in huge amounts, but she then rids her body of the thousands of extra calories by vomiting, taking laxatives, or exercising frantically. For an estimated one in five college students, the great majority of them female, "bingeing and purging" appears to be a way of coping with stress.

Sometimes a bulimic's dentist is the first to spot her problem by observing teeth decayed from frequent vomiting. The backs of her hands may bear telltale scratches from efforts to induce vomiting. Her throat is raw and her esophageal lining ulcerated from the stomach acid forced forward by vomiting. The binge and purge cycle is very hard to break, even with psychotherapy and nutritional counseling.

A person with bulimia tends to eat soft foods that can be consumed in large amounts quickly with minimal chewing. Reveals one athletic young woman, "For me a binge consists of a pound of cottage cheese, a head of lettuce, a steak, a loaf of Italian bread, a 10-ounce serving of broccoli, spinach or a head of cabbage, a cake, an 18-ounce pie, with a quart or half gallon of ice cream. When my disease is at its worst, I eat raw oatmeal with butter, laden with mounds of sugar, or a loaf of white bread with butter and syrup poured over it." She follows her "typical" 20,000-calorie binge with hours of bicycle riding, running, and swimming.

1. What is an adequate diet?
2. What factors influence individual needs for nutrients?
3. How can the food pyramid plan help consumers make wise food choices?
4. What is primary malnutrition? Secondary malnutrition?
5. What happens to the body during starvation?
6. How do marasmus and kwashiorkor differ?
7. How do anorexia nervosa and bulimia differ?

# CLINICAL APPLICATION

# Nutrition and the Athlete

Can a marathoner, cross-country skier, weight lifter, or competitive swimmer eat to win? A diet of 60% or more carbohydrate, 18% protein, and 22% fat should be adequate to support frequent, strenuous activity.

## MACRONUTRIENTS

As the source of immediate energy, carbohydrates are the athlete's best friend. Athletes should get the bulk of their carbohydrates from vegetables and grains in frequent meals, because the muscles can store only 1,800 calories worth of glycogen.

Many people erroneously believe that an athlete needs protein supplements. Excess dietary protein, however, can strain the kidneys in ridding the body of the excess nitrogen, dehydrating the athlete as more water is used in urine. The American Dietetic Association suggests that athletes eat 1 gram of protein per kilogram of weight per day, compared to 0.8 grams for nonathletes. Athletes should not rely solely on meat for protein, because these foods can be high in fat. Supple-

ments are only necessary for young athletes at the start of training, under a doctor's supervision. Too little protein in an athlete is linked to "sports anemia," in which hemoglobin levels decline and blood may appear in the urine.

The body stores 140,000 calories of fat, so it is clear why no one needs to replenish that constantly with fatty foods. Athletes should consume low-fat milk and meats.

## WATER

A sedentary person loses a quart of water a day as sweat; an athlete may lose 2 to 4 quarts of water an hour! To stay hydrated, athletes should drink 3 cups of cold water 2 hours before an event, then 2 more cups 15 minutes before the event, and small amounts every 15 minutes during the event. They should drink afterward too. An-

other way to determine water needs is to weigh in before and after training. For each pound lost, athletes should drink a pint of water. They should also avoid sugary fluids, which slow water's trip through the digestive system, and alcohol, which increases fluid loss.

## VITAMINS AND MINERALS

If an athlete eats an adequate, balanced diet, vitamin supplements are not needed. Supplements of sodium and potassium are usually not needed either, because the active body naturally conserves these nutrients. To be certain of enough sodium, athletes may want to salt their food; to get enough potassium, they can eat bananas, dates, apricots, oranges, or raisins.

A healthy pregame meal should be eaten 2 to 5 hours before the game, provide 500 to 1,500 calories, include 4 or 5 cups of fluid, and be high in carbohydrates, which taste good, provide energy, and are easy to digest. ■

## CLINICAL TERMS RELATED TO NUTRITION AND METABOLISM

**cachexia** (kah-kek′se-ah)   State of chronic malnutrition and physical wasting.

**casein** (ka′se-in)   Primary protein found in milk.

**celiac disease** (se′le-ak dĭ-zēz′)   Inability to digest or use fats and carbohydrates.

**emaciation** (e-ma″se-a′shun)   Extreme leanness due to tissue wasting.

**hyperalimentation** (hi″per-al″ĭ men-ta′shun)   Long-term intravenous nutrition.

**hypercalcemia** (hi″per-kal-se′me-ah)   Excess calcium in the blood.

**hypercalciuria** (hi″per-kal-se-u′re-ah)   Excess excretion of calcium in the urine.

**hyperglycemia** (hi″per-gli-se′me-ah)   Excess glucose in the blood.

**hyperkalemia** (hi″per-kah-le′me-ah)   Excess potassium in the blood.

**hypernatremia** (hi″per-nah-tre′me-ah)   Excess sodium in the blood.

**hypoalbuminemia** (hi″po-al-bu″mĭ-ne′me-ah)   Low level of albumin in the blood.

**hypoglycemia** (hi″po-gli-se′me-ah)   Low level of glucose in the blood.

**hypokalemia** (hi″po-kah-le′me-ah)  Low level of potassium in the blood.

**hyponatremia** (hi″po-nah-tre′me-ah)  Low level of sodium in the blood.

**isocaloric** (i″so-kah-lo′rik)  Containing equal amounts of heat energy.

**lipogenesis** (lip″o-jen′e-sis)  The formation of fat.

**nyctalopia** (nik″tah-lo′pe-ah)  Night blindness.

**polyphagia** (pol″e-fa′je-ah)  Overeating.

**proteinuria** (pro″te-i-nu′re-ah)  Protein in the urine.

**provitamin** (pro-vi′tah-min)  Biochemical precursor of a vitamin.

---

## CHAPTER SUMMARY

## Why We Eat  (page 704)

Nutrients include carbohydrates, lipids, proteins, vitamins, and minerals. The ways nutrients are used to support life processes constitute metabolism. Essential nutrients are needed for health and cannot be synthesized by body cells. Macronutrients include carbohydrates, lipids, and proteins. Micronutrients are vitamins and minerals. Water is also essential.

## Carbohydrates  (page 705)

Carbohydrates are organic compounds that are primarily used to supply cellular energy.

1. Carbohydrate sources
   a. Carbohydrates are ingested in a variety of forms.
   b. Starch, glycogen, disaccharides, and monosaccharides are carbohydrates.
   c. Cellulose is a polysaccharide that human enzymes cannot digest, but it is important in providing bulk that facilitates movement of intestinal contents.
2. Carbohydrate utilization
   a. Carbohydrates are absorbed as monosaccharides.
   b. The liver converts fructose and galactose to glucose.
   c. Oxidation releases energy from glucose.
   d. Excess glucose is stored as glycogen or converted to fat.
3. Carbohydrate requirements
   a. Most carbohydrates are used to supply energy, although some are used to produce sugars.
   b. Some cells depend on a continuous supply of glucose to survive.
   c. If glucose is scarce, amino acids may be converted to glucose.
   d. Humans survive with a wide range of carbohydrate intakes.
   e. Poor nutritional status is usually related to low intake of nutrients other than carbohydrates.

## Lipids  (page 706)

Lipids are organic compounds that supply energy and are used to build cell structures. They include fats, phospholipids, and cholesterol.

1. Lipid sources
   a. Triglycerides are obtained from foods of plant and animal origins.
   b. Cholesterol is mostly obtained in foods of animal origin.
2. Lipid utilization
   a. The liver and adipose tissue control triglyceride metabolism.
   b. The liver can alter the molecular structures of fatty acids.
   c. Linoleic acid is an essential fatty acid.
   d. The liver regulates the amount of cholesterol by synthesizing or excreting it.

3. Lipid requirements
   a. Humans survive with a wide range of lipid intakes.
   b. The amounts and types of lipids needed for health are unknown.
   c. Fat intake must be sufficient to carry fat-soluble vitamins.

## Proteins  (page 707)

Proteins are organic compounds that serve as structural materials, act as enzymes, and provide energy. Amino acids are incorporated into various structural and functional proteins, including enzymes. During starvation, tissue proteins may be used as energy sources; thus, the tissues waste away.

1. Protein sources
   a. Proteins are mainly obtained from meats, dairy products, cereals, and legumes.
   b. During digestion, proteins are broken down into amino acids.
   c. Eight amino acids are essential for adults, whereas ten are essential for growing children.
   d. All essential amino acids must be present at the same time in order for growth and repair of tissues to take place.
   e. Complete proteins contain adequate amounts of all the essential amino acids needed to maintain the tissues and promote growth.
   f. Incomplete proteins lack adequate amounts of one or more essential amino acids.
2. Nitrogen balance
   a. In healthy adults, the gain of protein equals the loss of protein, and a nitrogen balance exists.
   b. A starving person has a negative nitrogen balance; a growing child, a pregnant woman, or an athlete in training usually has a positive nitrogen balance.
3. Protein requirements
   a. Proteins and amino acids are needed to supply essential amino acids and nitrogen for the synthesis of nitrogen-containing molecules.
   b. The consequences of protein deficiencies are particularly severe among growing children.

## Energy Expenditures  (page 710)

Energy is of prime importance to survival and may be obtained from carbohydrates, fats, or proteins.

1. Energy values of foods
   a. The potential energy values of foods are expressed in calories.
   b. When energy losses due to incomplete absorption and incomplete oxidation are taken into account, 1 gram of carbohydrate or 1 gram of protein yields about 4 calories, whereas 1 gram of fat yields about 9 calories.

2. Energy requirements
   a. The amount of energy required varies from person to person.
   b. Factors that influence energy requirements include basal metabolic rate, muscular activity, body temperature, and nitrogen balance.
3. Energy balance
   a. Energy balance exists when caloric intake equals caloric output.
   b. If energy balance is positive, body weight increases; if energy balance is negative, body weight decreases.
4. Desirable weight
   a. The most common nutritional disorders involve caloric imbalances.
   b. Average weights of persons 25–30 years of age are desirable for older persons as well.
   c. Height–weight guidelines are based on longevity.
   d. A person who exceeds the desirable weight by 10%–20% is called overweight.
   e. A person whose body contains excess fatty tissue is obese.

# Vitamins  (page 713)

Vitamins are organic compounds (other than carbohydrates, lipids, and proteins) that are essential for normal metabolic processes and cannot be synthesized by body cells in adequate amounts.

1. Fat-soluble vitamins
   a. General characteristics
      (1) Fat-soluble vitamins are carried in lipids and are influenced by the same factors that affect lipid absorption.
      (2) They are fairly resistant to the effects of heat; thus, they are not destroyed by cooking or food processing.
   b. Vitamin A
      (1) Vitamin A occurs in several forms, is synthesized from carotenes, and is stored in the liver.
      (2) It functions in the production of pigments necessary for vision.
   c. Vitamin D
      (1) Vitamin D is a group of related steroids.
      (2) It is found in certain foods and is produced commercially; it can also be synthesized in the skin.
      (3) When needed, vitamin D is converted by the kidneys to an active form that functions as a hormone and promotes the intestine's absorption of calcium and phosphorus.
   d. Vitamin E
      (1) Vitamin E is a group of compounds that are antioxidants.
      (2) It is stored in muscles and adipose tissue.
      (3) Its precise functions are unknown, but it seems to prevent oxidation of vitamin A and polyunsaturated fatty acids, and to stabilize cell membranes.
   e. Vitamin K
      (1) Vitamin $K_1$ occurs in foods; vitamin $K_2$ is produced by certain bacteria that normally inhabit the intestinal tract.
      (2) It is stored to a limited degree in the liver.
      (3) It is used in the production of prothrombin that is needed for normal blood clotting.
2. Water-soluble vitamins
   a. General characteristics

   (1) Water-soluble vitamins include the B vitamins and vitamin C.
   (2) B vitamins make up a group called the vitamin B complex and oxidize carbohydrates, lipids, and proteins.
   b. Vitamin B complex
      (1) Thiamine
         (a) Thiamine functions as part of coenzymes that oxidize carbohydrates and synthesize essential sugars.
         (b) Small amounts are stored in the tissues; excess is excreted in the urine.
         (c) Quantities needed vary with caloric intake.
      (2) Riboflavin
         (a) Riboflavin functions as part of several enzymes and coenzymes that are essential to the oxidation of glucose and fatty acids
         (b) Its absorption is regulated by an active transport system; excess is excreted in the urine.
         (c) Quantities needed vary with caloric intake.
      (3) Niacin
         (a) Niacin functions as part of coenzymes needed for the oxidation of glucose and for the synthesis of proteins and fats.
         (b) It can be synthesized from tryptophan; daily requirement varies with the tryptophan intake.
      (4) Pantothenic acid
         (a) Pantothenic acid functions as part of coenzyme A; thus, it is essential for energy-releasing mechanisms.
         (b) Most diets provide sufficient amounts; deficiencies are rare.
      (5) Vitamin $B_6$
         (a) Vitamin $B_6$ is a group of compounds that function as coenzymes in metabolic pathways that synthesize proteins, certain amino acids, antibodies, and nucleic acids.
         (b) Its requirement varies with protein intake.
      (6) Cyanocobalamin
         (a) The cyanocobalamin molecule contains cobalt.
         (b) Its absorption is regulated by the secretion of intrinsic factor from the gastric glands.
         (c) It functions as part of coenzymes needed for the synthesis of nucleic acids and for the metabolism of carbohydrates and fats.
      (7) Folacin
         (a) Folacin is converted by the liver to physiologically active folinic acid.
         (b) It is a coenzyme needed for the metabolism of certain amino acids, the synthesis of DNA, and the normal production of red blood cells.
      (8) Biotin
         (a) Biotin is a coenzyme needed for the metabolism of amino acids and fatty acids, and for the synthesis of nucleic acids.
         (b) It is stored in metabolically active organs.
   c. Ascorbic acid (vitamin C)
      (1) Vitamin C is closely related chemically to monosaccharides.
      (2) It is needed for production of collagen, the metabolism of certain amino acids, and iron absorption.
      (3) It is not stored in large amounts; excess is excreted in the urine.

## Minerals  (page 722)

1. Characteristics of minerals
   a. Minerals are responsible for about 4% of body weight.
   b. About 75% by weight of the minerals are found in bones and teeth as calcium and phosphorus.
   c. Minerals are usually incorporated into organic molecules, although some occur in inorganic compounds or as free ions.
   d. They comprise structural materials, function in enzymes, and play vital roles in various metabolic processes.
   e. Mineral concentrations are generally regulated by homeostatic mechanisms.
   f. The physiologically active form of minerals is the ionized form.
2. Major minerals
   a. Calcium
      (1) Calcium is essential for the formation of bones and teeth, the conduction of nerve impulses, the contraction of muscle fibers, the coagulation of blood, and the activation of various enzymes.
      (2) Its absorption is affected by existing calcium concentration, vitamin D, protein intake, and motility of the digestive tract.
   b. Phosphorus
      (1) Phosphorus is incorporated for the most part into the salts of bones and teeth.
      (2) It plays roles in nearly all metabolic reactions as a constituent of nucleic acids, proteins, enzymes, and some vitamins.
      (3) It also occurs in the phospholipids of cell membranes, in ATP, and in phosphates of body fluids.
   c. Potassium
      (1) Potassium tends to be concentrated inside cells.
      (2) It functions in maintenance of osmotic pressure, regulation of pH, metabolism of carbohydrates and proteins, conduction of nerve impulses, and contraction of muscle fibers.
   d. Sulfur
      (1) Sulfur is incorporated for the most part into the molecular structures of certain amino acids.
      (2) It is also included in thiamine, insulin, biotin, and mucopolysaccharides.
   e. Sodium
      (1) Most sodium occurs in extracellular fluids or is bound to the inorganic salts of bone.
      (2) The blood concentration of sodium is regulated by the kidneys under the influence of aldosterone.
      (3) Sodium helps maintain solute concentration and regulates water balance.
      (4) It is essential for the conduction of nerve impulses, the contraction of muscle fibers, and the movement of substances through cell membranes.
   f. Chlorine
      (1) Chlorine is closely associated with sodium in the form of chloride ions.
      (2) It acts with sodium to help maintain osmotic pressure, regulate pH, and maintain electrolyte balance.
      (3) It is essential for the formation of hydrochloric acid and for the transport of carbon dioxide by red blood cells.
   g. Magnesium
      (1) Magnesium is particularly abundant in the bones as phosphates and carbonates.
      (2) It functions in the production of ATP and in the breakdown of ATP to ADP.
      (3) A reserve supply of magnesium is stored in the bones; excesses are excreted in the urine.
3. Trace elements
   a. Iron
      (1) Iron primarily occurs in hemoglobin of red blood cells and in myoglobin of muscles.
      (2) A reserve supply of iron is stored in the liver, spleen, and bone marrow.
      (3) It is needed to catalyze the formation of vitamin A; it is also incorporated into various enzymes and the cytochrome molecules.
   b. Manganese
      (1) Most manganese is concentrated in the liver, kidneys, and pancreas.
      (2) It is necessary for normal growth and development of skeletal structures and other connective tissues; it is essential for the synthesis of fatty acids, cholesterol, and urea.
   c. Copper
      (1) Most copper is concentrated in the liver, heart, and brain.
      (2) It is needed for synthesis of hemoglobin, development of bones, production of melanin, and formation of myelin.
   d. Iodine
      (1) Iodine is most highly concentrated in the thyroid gland.
      (2) It provides an essential component for the synthesis of thyroid hormones.
      (3) It is often added to foods in the form of iodized table salt.
   e. Cobalt
      (1) Cobalt is widely distributed throughout the body.
      (2) It is an essential part of cyanocobalamin and is probably needed for the synthesis of several important enzymes.
   f. Zinc
      (1) Zinc is most concentrated in the liver, kidneys, and brain.
      (2) It is a constituent of several enzymes involved with digestion, respiration, bone metabolism, and liver metabolism.
   g. Fluorine
      (1) The teeth concentrate fluorine.
      (2) It is incorporated into enamel and prevents dental caries.
   h. Selenium
      (1) The liver and kidneys store selenium.
      (2) It is a constituent of certain enzymes.
   i. Chromium
      (1) Chromium is widely distributed throughout the body.
      (2) It regulates glucose utilization.

## Healthy Eating—The Food Pyramid and Reading Labels  (page 728)

1. An adequate diet provides sufficient energy and essential nutrients to support optimal growth, as well as maintenance and repair, of tissues.

2. Individual needs vary so greatly that it is not possible to design a diet that is adequate for everyone.

3. Devices to help consumers make healthy food choices include Recommended Daily Allowances, Recommended Dietary Allowances, food group plans, the food pyramid, and food labels.

4. Malnutrition
   a. Poor nutrition is due to lack of foods or failure to make the best use of available foods.
   b. Primary malnutrition is due to poor diet.
   c. Secondary malnutrition is due to an individual characteristic that makes a normal diet inadequate.

5. Starvation
   a. A person can survive 50 to 70 days without food.
   b. A starving body digests itself, starting with carbohydrates, then proteins, then fats.
   c. Symptoms include low blood pressure, slow pulse, chills, dry skin, hair loss, and poor immunity. Finally, vital organs cease to function.
   d. Marasmus is lack of all nutrients.
   e. Kwashiorkor is protein starvation.
   f. Anorexia nervosa is a self-starvation eating disorder.
   g. Bulimia is an eating disorder characterized by bingeing and purging.

## CRITICAL THINKING QUESTIONS

1. For each of the following diets, indicate how the diet is nutritionally unsound (if it is) and why it would be easy or difficult to follow.
   a. The bikini diet consists of 500 calories per day, including 45% to 55% protein, 40% to 50% carbohydrate, and 4% to 6% fat.
   b. The Cambridge diet is a powder mixed with water and drunk three times a day. One day's intake equals 33 grams of protein, 40 grams of carbohydrate, and 3 grams of fat.
   c. For the first 10 days of the Beverly Hills diet, nothing but fruit is eaten. On day 10 you can eat a bagel and butter, and then it is back to fruit only until day 19, when you can eat steak or lobster. The cycle repeats, adding more meat. This diet is based on "conscious combining"—the idea that eating certain combinations of foods leads to weight loss.
   d. The Weight Loss Clinic diet consists of 800 calories per day, with 46.1% protein, 35.2% carbohydrate, and 18.7% fat.
   e. The macrobiotic diet includes 10% to 20% protein, 70% carbohydrate, and 10% fat, with a half hour of walking each day. Most familiar foods are forbidden, but you can eat many unusual foods—such as rice cakes, seaweed, barley stew, pumpkin soup, rice gruel, kasha and onions, millet balls, wheat berries, and parsnip chips.
   f. The No Aging diet maintains that eating foods rich in nucleic acids (RNA and DNA) can prolong life, since these are the genetic materials. Recommended foods include sardines, salmon, calves liver, lentils, and beets.

2. Why does the blood sugar concentration of a person whose diet is low in carbohydrates remain stable?

3. A young man takes several vitamin supplements each day, claiming that they give him energy. Is he correct? Why or why not?

4. A soccer coach advises his players to eat a hamburger and french fried potatoes about 2 hours before a game. Suggest a more sensible pregame meal.

5. Anorexia nervosa is a form of starvation. If it is a nutritional problem, then why should treatment include psychotherapy?

6. Why do starving children often die of infections that are usually mild in well-nourished children?

7. With the aid of nutrient tables available in reference books, calculate the carbohydrate, lipid, and protein content of your diet in grams for a 24-hour period. Also calculate the total calories represented by these foods. Assuming that this 24-hour sample is representative of your normal eating habits, what improvements could be made in its composition?

8. Examine the label information on the packages of a variety of dry breakfast cereals. Which types of cereals provide the best sources of vitamins and minerals? Which major nutrients are lacking in these cereals?

9. If a person decided to avoid eating meat and other animal products, such as milk, cheese, and eggs, what foods might be included in the diet to provide essential amino acids?

10. How might a diet be modified in order to limit the intake of cholesterol?

11. How do you think the nutritional requirements of a healthy 12-year-old boy, a 24-year-old pregnant woman, and a healthy 60-year-old man would differ?

## REVIEW EXERCISES

1. Define *essential nutrient*.
2. List some common sources of carbohydrates.
3. Summarize the importance of cellulose in the diet.
4. Explain what happens to excess glucose in the body.
5. Explain why a temporary drop in the blood glucose concentration may impair nervous system functioning.
6. List some of the factors that affect an individual's need for carbohydrates.
7. Define *triglyceride*.
8. List some common sources of lipids.
9. Describe the liver's role in fat metabolism.
10. Discuss the functions of cholesterol.
11. List some common sources of protein.
12. Distinguish between essential and nonessential amino acids.

13. Explain why all of the essential amino acids must be present before growth can occur.
14. Distinguish between complete and incomplete proteins.
15. Review the major functions of amino acids.
16. Define *nitrogen balance.*
17. Explain why a protein deficiency may be accompanied by edema.
18. Define *calorie.*
19. Explain how the caloric values of foods are determined.
20. Define *basal metabolic rate.*
21. List some of the factors that affect the BMR.
22. Define *energy balance.*
23. Explain what is meant by *desirable weight.*
24. Distinguish between overweight and obesity.
25. Discuss the general characteristics of fat-soluble vitamins.
26. List the fat-soluble vitamins, and describe the major functions of each vitamin.
27. List some good sources for each of the fat-soluble vitamins.
28. Explain what is meant by the *vitamin B complex.*
29. List the water-soluble vitamins, and describe the major functions of each vitamin.
30. List some good sources for each of the water-soluble vitamins.
31. Discuss the general characteristics of the mineral nutrients.
32. List the major minerals, and describe the major functions of each mineral.
33. List some good sources for each of the major minerals.
34. Distinguish between a major mineral and a trace element.
35. List the trace elements, and describe the major functions of each trace element.
36. List some good sources of each of the trace elements.
37. Define *adequate diet.*
38. Define *malnutrition.*
39. Distinguish between primary and secondary malnutrition.
40. Discuss bodily changes during starvation.
41. Distinguish among marasmus, kwashiorkor, anorexia nervosa, and bulimia.

**For More Information Related to This Chapter, Please Visit** *Hole's Human Anatomy and Physiology* **Home Page**
http://www.mhhe.com/biosci/abio/

# RESPIRATORY SYSTEM

**alveol-**, small cavity: *alveolus*—microscopic air sac within a lung.

**bronch-**, windpipe: *bronch*us—primary branch of the trachea.

**carcin-**, spreading sore: *carcin*oma—type of cancer.

**carin-**, keel-like: *carin*a—ridge of cartilage between the right and left bronchi.

**cric-**, ring: *cric*oid cartilage—ring-shaped mass of cartilage at the base of the larynx.

**epi-**, upon: *epi*glottis—flaplike structure that partially covers the opening into the larynx during swallowing.

**hem-**, blood: *hem*oglobin—pigment in red blood cells.

**inhal-**, to breathe in: *inhal*ation—to take air into the lungs.

**phren-**, mind, diaphragm: *phren*ic nerve—nerve associated with the cervical plexuses that stimulate the muscle fibers of the diaphragm to contract.

**tuber-**, swelling: *tuber*culosis—disease characterized by the formation of fibrous masses within the lungs.

## CHAPTER OBJECTIVES

AFTER YOU HAVE STUDIED THIS CHAPTER, YOU SHOULD BE ABLE TO:

1. List the general functions of the respiratory system.

2. Name and describe the locations of the organs of the respiratory system.

3. Describe the functions of each organ of the respiratory system.

4. Explain how inspiration and expiration are accomplished.

5. Name and define each of the respiratory air volumes and capacities.

6. Explain how the alveolar ventilation rate is calculated.

7. List several nonrespiratory air movements and explain how each occurs.

8. Locate the respiratory center and explain how it controls normal breathing.

9. Discuss how various factors affect the respiratory center.

10. Describe the structure and function of the respiratory membrane.

11. Explain how the blood transports oxygen and carbon dioxide.

12. Review the major events of cellular respiration.

13. Explain how cells use oxygen.

## KEY TERMS

**alveolus** (al-ve′o-lus)
**bronchial tree** (brong′ke-al trē)
**carbaminohemoglobin** (kar-bam″ĭ-no-he″mo-glo′bin)
**carbonic anhydrase** (kar-bon′ik an-hi′drās)
**cellular respiration** (sel′u-lar res″pĭ-ra′shun)
**citric acid cycle** (sit′rik as′id si′kl)
**expiration** (ek″spi-ra′shun)
**glottis** (glot′is)
**hemoglobin** (he″mo-glo′bin)

**hyperventilation** (hi″per-ven″tĭ-la′shun)
**inspiration** (in″spi-ra′shun)
**oxyhemoglobin** (ok″se-he′mo-glo′bin)
**partial pressure** (par′shil presh′ur)
**pleural cavity** (ploo′ral kav′ĭ-te)
**respiratory center** (re-spi′rah-to″re sen′ter)
**respiratory membrane** (re-spi′rah-to″re mem′brān)
**respiratory volume** (re-spi′rah-to″re vol′ūm)
**surface tension** (ser′fas ten′shun)
**surfactant** (ser-fak′tant)

A 78-year-old man arrived in the emergency department reporting chest pain, shortness of breath, and fatigue. He was confused and upset, with an irregular pulse, fast respirations, and abnormal heart rhythm. Physicians treated the man with oxygen, aspirin, and nitroglycerin, suspecting a heart condition. Then they noticed that his wife lay on a nearby stretcher, complaining of the same symptoms! The woman said that her symptoms had begun two hours after her husband fell ill. Her chest pain had begun in her left arm and intensified as she served him tea. She, too, was treated as if she had suffered heart damage. But the doctors were puzzled.

The couple was admitted to the hospital, even though their symptoms improved after spending two hours in the emergency department. They felt well when they returned to their apartment two days later, but in another two days, returned to the hospital, reporting the same symptoms. This time, medical workers suspected that something in their home environment was causing the symptoms. They were right.

Upon questioning, the couple mentioned that they had spent the winter in a small, unventilated apartment that was heated by two kerosene space heaters. The source of their medical problem became clear—carbon monoxide (CO) poisoning. This colorless, odorless gas was binding to their hemoglobin molecules, robbing tissues of their oxygen supply. But the man and woman were lucky to have developed symptoms suggestive of heart failure, because it sent them back to the hospital for help. They could have easily fallen unconscious, gone into a coma, and died. CO poisoning is known as "the great imitator," because its symptoms are easily attributed to other causes.

Detecting the colorless, ordorless gas carbon monoxide can be lifesaving, because this molecule prevents hemoglobin from adequately binding oxygen. Carbon monoxide is produced from the incomplete combustion of carbon-containing fuels. Common culprits include wood stoves, kerosene heaters, gasoline engines, blocked chimneys, malfunctioning fireplaces, faulty ice-resurfacing machines, gas-powered equipment used in poorly ventilated warehouses and garages, and even indoor tractor pulls.

The respiratory system consists of a group of passages that filter incoming air and transport it into the body, into the lungs, and to the many microscopic air sacs where gases are exchanged. The entire process of exchanging gases between the atmosphere and body cells is called **respiration.** It involves several events:

- Movement of air in and out of the lungs, commonly called breathing or *ventilation*.

- Exchange of gases between the air in the lungs and the blood, sometimes called *external respiration*.

- Transport of gases by the blood between the lungs and body cells.

- Exchange of gases between the blood and the body cells, sometimes called *internal respiration*.

- Oxygen utilization and production of carbon dioxide by body cells is part of the process of *cellular respiration*.

## WHY WE BREATHE

Respiration occurs on a macroscopic level—a function provided by an organ system. However, the reason that body cells need to exchange gases—that is, take up oxygen and rid themselves of carbon dioxide—is apparent at the cellular and molecular levels.

Respiration enables cells to harness the energy held in the chemical bonds of nutrient molecules. Recall from chapter 4 that energy is slowly liberated from food molecules by stripping off electrons and channeling them through a series of electron carriers. Without oxygen as a final electron acceptor, much energy remains locked in nutrient molecules. Recall also that in the reactions that remove electrons, carbons are cleaved, combined with oxygen, and released as carbon dioxide, a metabolic waste. Cellular respiration, then, explains why we must obtain oxygen and get rid of carbon dioxide.

## ORGANS OF THE RESPIRATORY SYSTEM

The organs of the respiratory system can be divided into two groups, or tracts. Those in the *upper respiratory tract* include the nose, nasal cavity, sinuses, and pharynx. Those in the *lower respiratory tract* include the larynx, trachea, bronchial tree, and lungs (fig. 19.1).

### Nose

The nose is covered with skin and is supported internally by muscle, bone, and cartilage. Its two *nostrils* (external nares) provide openings through which air can enter and leave the nasal cavity. Many internal hairs guard these openings, preventing entry of large particles carried in the air.

### Nasal Cavity

The **nasal cavity,** a hollow space behind the nose, is divided medially into right and left portions by the **nasal septum.** This cavity is separated from the cranial cavity by the cribriform plate of the ethmoid bone and from the mouth by the hard palate.

> The nasal septum is usually straight at birth, but may be bent during birth. It remains straight throughout early childhood, but then may bend toward one side or the other. Such a *deviated septum* may obstruct the nasal cavity, making breathing difficult.

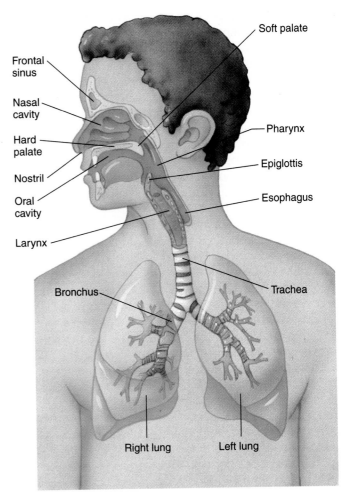

**FIGURE** 19.1
Organs of the respiratory system.

As figure 19.2 shows, **nasal conchae** (turbinate bones) curl out from the lateral walls of the nasal cavity on each side, dividing the cavity into passageways called the *superior, middle,* and *inferior meatuses* (see chapter 7). They also support the mucous membrane that lines the nasal cavity and help increase its surface area.

The upper posterior portion of the nasal cavity, below the cribriform plate, is slitlike, and its lining contains the olfactory receptors that provide the sense of smell. The remainder of the cavity conducts air to and from the nasopharynx.

The mucous membrane lining the nasal cavity contains pseudostratified ciliated epithelium that is rich in mucus-secreting goblet cells (see chapter 5). It also includes an extensive network of blood vessels, and normally appears pinkish. As air passes over the membrane, heat radiates from the blood and warms the air, adjusting its temperature to that of the body. In addition, evaporation of water from the mucous lining moistens the air. The sticky mucus the mucous membrane secretes entraps dust and other small particles entering with the air.

As the cilia of the epithelial cells move, a thin layer of mucus and any entrapped particles are pushed toward the pharynx (fig. 19.3). When the mucus reaches the pharynx, it is swallowed. In the stomach, gastric juice is likely to destroy microorganisms in the mucus, including disease-causing forms. Thus, the filtering that the mucous membrane provides not only prevents particles from reaching the lower air passages, but also helps prevent respiratory infections. Clinical Application 19.1 discusses how cigarette smoking impairs the respiratory system, beginning with the cleansing mucus and cilia.

**1** What is respiration?

**2** Which organs constitute the respiratory system?

**3** What is the function of the mucous membrane that lines the nasal cavity?

**4** What is the function of the cilia on the cells that line the nasal cavity?

**FIGURE 19.2**

Major features of the upper respiratory tract.

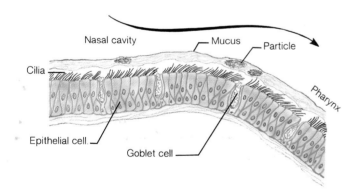

**FIGURE 19.3**

Cilia move mucus and trapped particles from the nasal cavity to the pharynx.

## Sinuses

Recall from chapter 7 that the *sinuses* (paranasal sinuses) are air-filled spaces located within the *maxillary*, *frontal*, *ethmoid*, and *sphenoid bones* of the skull (fig. 19.4). These spaces open into the nasal cavity and are lined with mucous membranes that are continuous with the lining of the nasal cavity. Consequently, mucus secretions drain from the sinuses into the nasal cavity. Mem-

branes that are inflamed and swollen because of nasal infections or allergic reactions (sinusitis) may block this drainage, increasing pressure within a sinus and causing headache.

The sinuses reduce the weight of the skull. They also serve as a resonant chamber that affects the quality of the voice.

It is possible to illuminate a person's frontal sinus in a darkened room by holding a small flashlight just beneath the eyebrow. Similarly, the maxillary sinuses can be illuminated by holding the flashlight in the mouth.

**1** Where are the sinuses located?

**2** What are the functions of the sinuses?

## Pharynx

The **pharynx** (throat) is located posterior to the oral cavity and between the nasal cavity and the larynx. It is a passageway for food traveling from the oral cavity to the esophagus and for air passing between the nasal cavity

# The Effects of Cigarette Smoking on the Respiratory System

Damage to the respiratory system from cigarette smoking is slow, progressive, and deadly. A healthy respiratory system is continuously cleansed. The mucus produced by the respiratory tubules traps dirt and disease-causing organisms, which cilia sweep toward the mouth, where they can be eliminated. Smoking greatly impairs this housekeeping. With the very first inhalation of smoke, the beating of the cilia slows. With time, the cilia become paralyzed and, eventually, disappear altogether. The loss of cilia leads to the development of smoker's cough. The cilia no longer effectively remove mucus, so the individual must cough it up. Coughing is usually worse in the morning because mucus has accumulated during sleep.

To make matters worse, excess mucus is produced and accumulates, clogging the air passageways. Pathogenic organisms that are normally removed now have easier access to the respiratory surfaces, and the resulting lung congestion favors their growth. This is why smokers are sick more often than nonsmokers. In addition, a lethal chain reaction begins. Smoker's cough leads to chronic bronchitis, caused by destroyed respiratory cilia. Mucus production increases and the lining of the bronchioles thickens, making breathing difficult. The bronchioles lose elasticity and are no longer able to absorb the pressure changes accompanying coughing. As a result, a cough can increase the air pressure within the alveoli (microscopic air sacs) enough to rupture the delicate alveolar walls; this condition is the hallmark of smoking-induced *emphysema*. The burst alveoli cause worsening of the cough, fatigue, wheezing, and impaired breathing. Emphysema is fifteen times more common among individuals who smoke a pack of cigarettes a day than among nonsmokers.

Simultaneous with the structural changes progressing to emphysema may be cellular changes leading to lung cancer. First, cells in the outer border of the bronchial lining begin to divide more rapidly than usual. Eventually, these cells displace the ciliated cells. Their nuclei begin to resemble those of cancerous cells—large and distorted with abnormal numbers of chromosomes. Up to this point, the damage can be repaired if smoking ceases. If smoking continues, these cells may eventually break through the basement membrane and begin reproducing within the lung tissue, forming a tumor with the potential of spreading throughout lung tissue (figs. 19A and 19B) and beyond, such as to the

---

and the larynx (see fig. 19.2). It also aids in producing the sounds of speech. The subdivisions of the pharynx—the nasopharynx, oropharynx, and laryngopharynx—are described in chapter 17.

## Larynx

The **larynx** is an enlargement in the airway superior to the trachea and inferior to the pharynx. It is a passageway for air moving in and out of the trachea and prevents foreign objects from entering the trachea. The larynx also houses the *vocal cords* (see reference plates 49 and 71).

The larynx is composed of a framework of muscles and cartilages bound by elastic tissue. The largest of the cartilages are the thyroid, cricoid, and epiglottic cartilages (fig. 19.5). These structures are single. The other laryngeal cartilages—the arytenoid, corniculate, and cuneiform cartilages—are paired.

The **thyroid cartilage** was named for the thyroid gland that covers its lower area. This cartilage is the shieldlike structure that protrudes in the front of the neck and is sometimes called the Adam's apple. The protrusion typically is more prominent in males than in females because of an effect of male sex hormones on the development of the larynx.

The **cricoid cartilage** lies inferior to the thyroid cartilage. It marks the lowermost portion of the larynx.

The **epiglottic cartilage,** the only one of the laryngeal cartilages that is elastic, not hyaline, cartilage, is attached to the upper border of the thyroid cartilage and supports a flaplike structure called the **epiglottis.** The epiglottis usually stands upright and allows air to enter the larynx. During swallowing, however, muscular contractions raise the larynx, and the base of the tongue presses the epiglottis downward. As a result, the epiglottis partially covers the opening into the larynx, brain or bones. Eighty percent of lung cancer cases are due to cigarette smoking. Only 13% of lung cancer patients live as long as 5 years after the initial diagnosis.

It pays to quit. Much of the damage to the respiratory system can be repaired. Cilia are restored and the thickening of alveolar walls due to emphysema can be reversed. But ruptured alveoli are gone forever. The nicotine in tobacco smoke causes a powerful dependency by binding to certain receptors on brain cells. ■

Normal lung tissue                    Cancerous lung tissue

—Tumor

**FIGURE 19A**

The lung on the left is healthy. A cancerous tumor has invaded the lung on the right, taking up nearly half of the lung space.

**FIGURE 19B**

Lung cancer may begin as a tiny tumor growing in an alveolus, a microscopic air sac (125×).

helping prevent foods and liquids from entering the air passages.

The pyramid-shaped **arytenoid cartilages** are located superior to and on either side of the cricoid cartilage. Attached to the tips of the arytenoid cartilages are the tiny, conelike **corniculate cartilages.** These cartilages are attachments for muscles that help regulate tension on the vocal cords during speech and aid in closing the larynx during swallowing.

The **cuneiform cartilages** are small, cylindrical structures in the mucous membrane between the epiglottic and the arytenoid cartilages. They stiffen the soft tissues in this region.

Inside the larynx, two pairs of horizontal folds composed of muscle tissue and connective tissue with a covering of mucous membrane extend inward from the lateral walls. The upper folds (vestibular folds) are called *false vocal cords* because they do not produce sounds.

Muscle fibers within these folds help close the larynx during swallowing.

The lower folds are the *true vocal cords*. They contain elastic fibers and are responsible for vocal sounds, which are created when air is forced between these folds, causing them to vibrate from side to side. This action generates sound waves, which can be formed into words by changing the shapes of the pharynx and oral cavity and by using the tongue and lips. Figure 19.6 shows both pairs of folds.

Changing tension on the vocal cords, by contracting or relaxing laryngeal muscles, controls *pitch* (musical tone) of the voice. Increasing the tension produces a higher pitch, and decreasing the tension creates a lower pitch.

The *intensity* (loudness) of a vocal sound depends upon the force of the air passing over the vocal cords. Stronger blasts of air result in greater vibration of the vocal cords and louder sound.

(a)

(b)

**FIGURE** 19.4

(a) Radiograph of a skull from the anterior and (b) from the lateral view, showing air-filled sinuses within the bones (arrows).

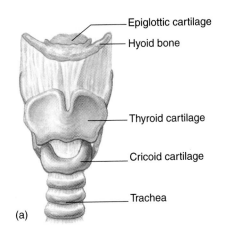

Epiglottic cartilage
Hyoid bone
Thyroid cartilage
Cricoid cartilage
Trachea

(a)

Hyoid bone
Epiglottic cartilage
Corniculate cartilage
Thyroid cartilage
Arytenoid cartilage
Cricoid cartilage
Trachea

(b)

**FIGURE** 19.5

(a) Anterior view and (b) posterior view of the larynx.

During normal breathing, the vocal cords remain relaxed, and the opening between them, called the **glottis,** is a triangular slit. However, when food or liquid is swallowed, muscles close the glottis within the false vocal folds. This action assists closing of the epiglottis, which prevents food or liquid from entering the trachea (fig. 19.7). The mucous membrane that lines the larynx continues to filter incoming air by entrapping particles and moving them toward the pharynx by ciliary action.

1   What part of the respiratory tract is shared with the alimentary canal?

2   Describe the structure of the larynx.

3   How do the vocal cords produce sounds?

4   What is the function of the glottis? Of the epiglottis?

## Trachea

The **trachea** (windpipe) is a flexible cylindrical tube about 2.5 centimeters in diameter and 12.5 centimeters in length. It extends downward anterior to the esophagus and into the thoracic cavity, where it splits into right and left bronchi (fig. 19.8 and reference plate 50).

(a)

(a)

(b)

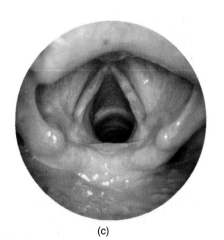

(c)

(b)

**FIGURE 19.6** 🏃

(a) Coronal section and (b) sagittal section of the larynx.

**FIGURE 19.7**

Vocal cords as viewed from above (a) with the glottis closed and (b) with the glottis open. (c) A photograph of the glottis and vocal folds.

The inner wall of the trachea is lined with a ciliated mucous membrane that contains many goblet cells. This membrane continues to filter the incoming air and to move entrapped particles upward into the pharynx where the mucus can be swallowed.

Within the tracheal wall are about twenty C-shaped pieces of hyaline cartilage, one above the other. The open ends of these incomplete rings are directed posteriorly, and the gaps between their ends are filled with smooth muscle and connective tissues (figs. 19.9 and 19.10). These cartilaginous rings prevent the trachea from collapsing and blocking the airway. At the same time, the soft tissues that complete the rings in the back allow the nearby esophagus to expand as food moves through it on the way to the stomach.

A blocked trachea can cause asphyxiation in minutes. If swollen tissues, excess secretions, or a foreign object obstruct the trachea, making a temporary, external opening in the tube so that air can bypass the obstruction is lifesaving. This procedure, shown in figure 19.11, is called a *tracheostomy*.

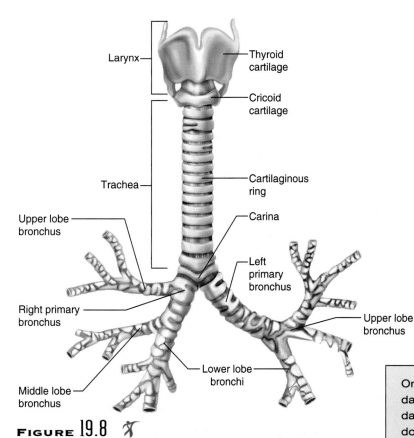

**FIGURE** 19.8

The trachea transports air between the larynx and the bronchi.

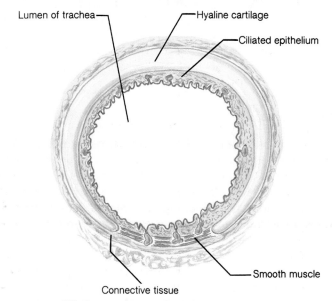

**FIGURE** 19.9

Cross section of the trachea. Note the C-shaped rings of hyaline cartilage in the wall.

**FIGURE** 19.10

Light micrograph of a section of the tracheal wall (63×).

On December 13, 1799, George Washington spent the day walking on his estate in a freezing rain. The next day, he had trouble breathing and swallowing. Several doctors were called in. One suggested a tracheostomy, in which a hole would be cut in the throat so that the president could breathe. He was voted down. The other physicians suggested bleeding the patient, plastering his throat with bran and honey, and placing blister beetles on his legs to produce blisters. No treatment was provided, and within a few hours, Washington's voice became muffled, breathing was more labored, and he was restless. For a short time he seemed euphoric, and then he died.

Today, Washington's problem is recognized as *epiglottitis*, an inflammation that swells the epiglottis to ten times its normal size. A tracheostomy might have saved his life.

## Bronchial Tree

The **bronchial tree** consists of branched airways leading from the trachea to the microscopic air sacs in the lungs. Its branches begin with the right and left **primary bronchi,** which arise from the trachea at the level of the fifth thoracic vertebrae. The openings of the primary bronchi are separated by a ridge of cartilage called the *carina* (see fig. 19.8). Each bronchus, accompanied by large blood vessels, enters its respective lung.

### Branches of the Bronchial Tree

A short distance from its origin, each primary bronchus divides into **secondary,** or **lobar, bronchi** (two on the left

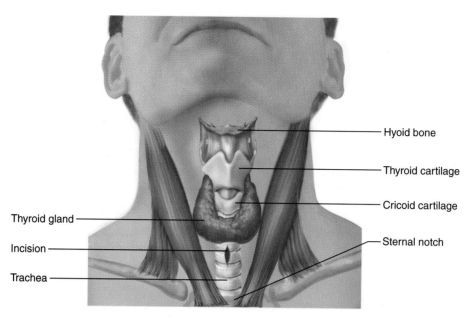

**FIGURE** 19.11

A tracheostomy may be performed to allow air to bypass an obstruction within the larynx.

and three on the right) that, in turn, branch repeatedly into finer and finer tubes (figs. 19.12 and 19.13). When stripped of their associated blood vessels and tissues, the airways appear as an upside down tree. The successive divisions of these branches from the trachea to the microscopic air sacs follow:

1. **Right** and **left primary bronchi.**

2. **Secondary,** or **lobar**, **bronchi.** Three branch from the right primary bronchus, and two branch from the left.

3. **Tertiary,** or **segmental**, **bronchi.** Each of these branches supplies a portion of the lung called a *bronchopulmonary segment.* Usually there are ten such segments in the right lung and eight in the left lung.

4. **Intralobular bronchioles.** These small branches of the segmental bronchi enter the basic units of the lung—the *lobules.*

5. **Terminal bronchioles.** These tubes branch from a bronchiole. Fifty to eighty terminal bronchioles occupy a lobule of the lung.

6. **Respiratory bronchioles.** Two or more respiratory bronchioles branch from each terminal bronchiole. Short and about 0.5 millimeter in diameter, these structures are called "respiratory" because a few air sacs bud from their sides, making them able to take part in gas exchange.

7. **Alveolar ducts.** Two to ten long, branching alveolar ducts extend from each respiratory bronchiole (fig. 19.14).

8. **Alveolar sacs.** Alveolar sacs are thin-walled, closely packed outpouchings of the alveolar ducts.

9. **Alveoli.** Alveoli are thin-walled, microscopic air sacs that open to an alveolar sac. Thus, air can diffuse freely from the alveolar ducts, through the alveolar sacs, and into the alveoli (fig. 19.15).

Dust particles, asbestos fibers, and other pollutants travel at speeds of 200 centimeters per second in the trachea, but slow to 1 centimeter per second when deep in the lungs. Gravity deposits such particles, particularly at branchpoints in the respiratory tree. It is a little like traffic backing up at an exit to a highway.

### Structure of the Respiratory Tubes

The structure of a bronchus is similar to that of the trachea, but the C-shaped cartilaginous rings are replaced with cartilaginous plates where the bronchus enters the lung. These plates are irregularly shaped, and completely surround the tube. However, as finer and finer branch tubes appear, the amount of cartilage decreases, and it finally disappears in the bronchioles, which have diameters of about 1 millimeter.

As the amount of cartilage decreases, a layer of smooth muscle that surrounds the tube just beneath the mucosa becomes more prominent. This muscular layer remains in the wall to the ends of the respiratory bronchioles, and only a few muscle fibers are in the walls of the alveolar ducts.

Elastic fibers are scattered among the smooth muscle cells and are abundant in the connective tissue that surrounds the respiratory tubes. These fibers play an

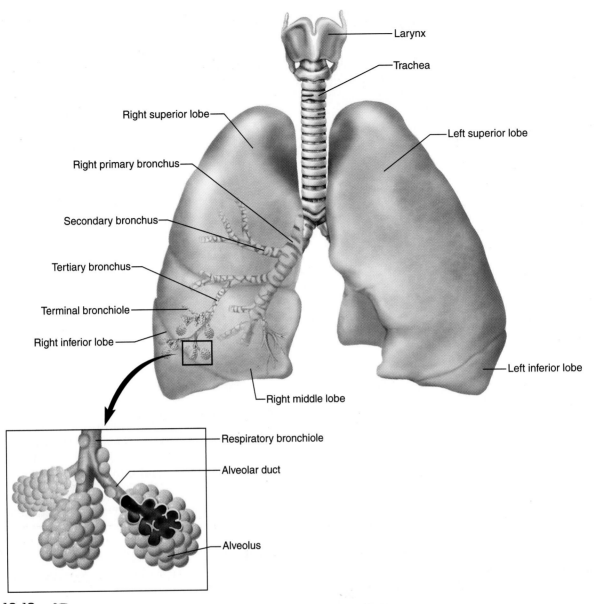

**FIGURE 19.12**

The bronchial tree consists of the passageways that connect the trachea and the alveoli. The alveolar duct and alveoli are enlarged to show their location.

important role in breathing, as is explained later in this chapter.

As the tubes become smaller in diameter, a change occurs in the type of cells that line them. The lining of the larger tubes consists of pseudostratified, ciliated columnar epithelium and mucus-secreting goblet cells. However, along the way the number of goblet cells and the height of the other epithelial cells decline, and cilia become scarcer. In the finer tubes, beginning with the respiratory bronchioles, the lining is cuboidal epithelium; and in the alveoli, it is simple squamous epithelium closely associated with a dense network of capillaries. The mucous lining gradually thins, until none appears in the alveoli.

> The trachea and bronchial tree can be examined directly using a flexible optical instrument called a *fiber-optic bronchoscope*. This procedure (bronchoscopy) is sometimes used in diagnosing tumors or other pulmonary diseases. It may also be used to locate and remove aspirated foreign bodies in the air passages.

## Functions of the Respiratory Tubes and Alveoli

The branches of the bronchial tree are air passages, which continue to filter the incoming air and distribute it to the alveoli in all parts of the lungs. The alveoli, in

**FIGURE 19.13**

A plastic cast of the bronchial tree.

turn, provide a large surface area of thin epithelial cells through which gas exchanges can occur (fig. 19.16). If the 300 million alveoli in the human lung were spread out, they would cover an area the size of half a tennis court—between 70 and 80 square meters.

During gas exchange, oxygen diffuses through the alveolar walls and enters the blood in nearby capillaries. Carbon dioxide diffuses from the blood through these walls and enters the alveoli (figs. 19.17 and 19.18).

In the inherited illness *cystic fibrosis*, airways become clogged with thick, sticky mucus, which attracts bacteria. As damaged white blood cells accumulate at the infection site, their DNA may leak out and further clog the area. A treatment that moderately eases breathing is deoxyribonuclease (DNase), an enzyme normally found in the body that degrades the accumulating extracellular DNA.

**1** What is the function of the cartilaginous rings in the tracheal wall?

**2** How do the right and left bronchi differ in structure?

**3** List the branches of the bronchial tree.

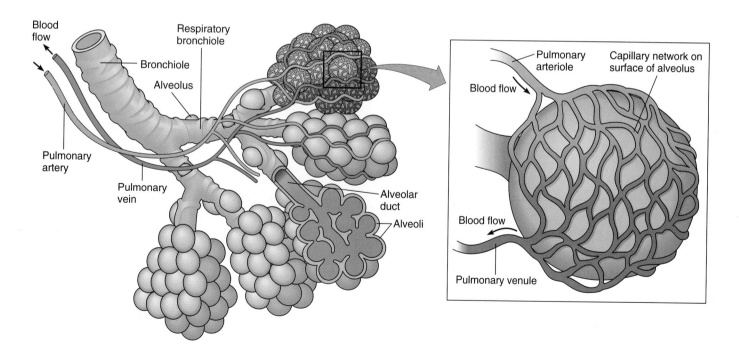

**FIGURE 19.14**

The respiratory tubes end in tiny alveoli, each of which is surrounded by a capillary network.

**4** Describe the changes in structure that occur in the respiratory tubes as their diameters decrease.

**5** How are gases exchanged in the alveoli?

## Lungs

The lungs are soft, spongy, cone-shaped organs located in the thoracic cavity. The right and left lungs are separated medially by the heart and the mediastinum, and they are enclosed by the diaphragm and the thoracic cage (see figs. 1.7, 19.19 and reference plates 56, 57, and 71).

Because the respiratory organs are vital, malfunction can be deadly. Temporarily maintaining respiratory function can be lifesaving, and can be accomplished in several ways.

In *artificial respiration*, a person blows into the mouth of a person who has stopped breathing. The oxygen in the rescuer's exhaled breath can keep the victim alive.

In *extracorporeal membrane oxygenation*, blood is pumped out of the body and across a gas-permeable membrane that adds oxygen and removes carbon dioxide, simulating lung function. Such a device can keep a person alive until he or she recovers from other problems but is too costly and cumbersome to maintain life indefinitely.

A lung assist device, called an *intravascular oxygenator*, is in clinical trials. It consists of hundreds of tiny porous hair-thin fibers surgically implanted in the inferior vena cava. Here, deoxygenated blood returning to the heart receives oxygen and is rid of carbon dioxide—but only at about 30% the capacity of a healthy respiratory system.

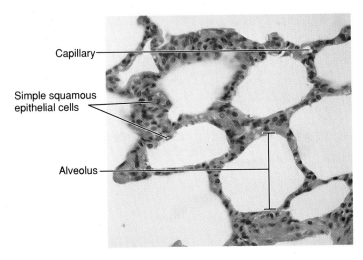

**FIGURE** 19.15

Light micrograph of alveoli (250×).

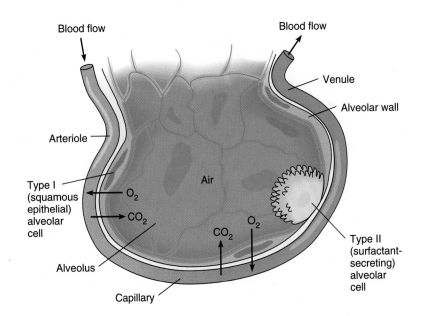

**FIGURE** 19.16

Oxygen diffuses from the air within the alveolus into the capillary, whereas carbon dioxide diffuses from the blood within the capillary into the alveolus.

Blood vessel    Capillary    Alveolus

**FIGURE 19.17**

Scanning electron micrograph of casts of alveoli and associated capillary networks. These casts were prepared by filling the alveoli and blood vessels with resin and later removing the soft tissues by digestion, leaving only the resin casts (420×).

*Tissues and Organs: A Text-Atlas of Scanning Electron Microscopy,* by Richard D. Kessel and Randy Kardon. © 1979 W. H. Freeman and Company.

Alveolus

Bronchiole

**FIGURE 19.18** 🏃

Scanning electron micrograph of lung alveoli and a bronchiole (70×).

Each lung occupies most of the thoracic space on its side and is suspended in the cavity by a bronchus and some large blood vessels. These tubular structures enter the lung on its medial surface through a region called the **hilus.** A layer of serous membrane, the *visceral pleura,* is firmly attached to the surface of each lung, and this membrane folds back at the hilus to become the *parietal pleura.* The parietal pleura, in turn, forms part of the mediastinum and lines the inner wall of the thoracic cavity.

There is no significant space between the visceral and parietal pleurae, but the potential space between them is called the *pleural cavity.* It contains a thin film of serous fluid that lubricates the adjacent pleural surfaces, reducing friction as they move against one another during breathing. This fluid also helps hold the pleural membranes together.

The right lung is larger than the left lung, and it is divided into three parts, called the superior, middle, and inferior lobes. The left lung consists of two parts, a superior and an inferior lobe.

A lobar bronchus of the bronchial tree supplies each lobe. A lobe also has connections to blood and lymphatic vessels, and is enclosed by connective tissues. Connective tissue further subdivides a lobe into **lobules,** each of which contains terminal bronchioles together with their alveolar ducts, alveolar sacs, alveoli, nerves, and associated blood and lymphatic vessels.

Table 19.1 summarizes the characteristics of the major parts of the respiratory system. Clinical Application 19.2 considers substances that irritate the lungs.

**1** Where are the lungs located?

**2** What is the function of the serous fluid within the pleural cavity?

**3** How does the structure of the right lung differ from that of the left lung?

**4** What kinds of structures make up a lung?

## BREATHING MECHANISM

Breathing, which is also called ventilation, is the movement of air from outside the body into the bronchial tree and alveoli, followed by a reversal of this air movement. The actions responsible for these air movements are termed **inspiration** (inhalation) and **expiration** (exhalation).

### Inspiration

Atmospheric pressure due to the weight of the air is the force that moves air into the lungs. At sea level, this pressure is sufficient to support a column of mercury about 760 millimeters high in a tube (fig. 19.20). Thus,

**Thyroid cartilage**

**Cricoid cartilage**

**Clavicle**

**Scapula**

**Superior lobe of right lung**

**Middle lobe of right lung**

**Inferior lobe of right lung**

**Rib cartilage**

**Trachea**

**Superior lobe of left lung**

**Inferior lobe of left lung**

**Sternum**

**FIGURE** 19.19

Locations of the lungs within the thoracic cavity.

| TABLE 19.1 | PART OF THE RESPIRATORY SYSTEM | |
|---|---|---|
| **Part** | **Description** | **Function** |
| Nose | Part of face centered above the mouth and inferior to the space between the eyes | Nostrils provide entrance to nasal cavity; internal hairs begin to filter incoming air |
| Nasal cavity | Hollow space behind nose | Conducts air to pharynx; mucous lining filters, warms, and moistens incoming air |
| Sinuses | Hollow spaces in various bones of the skull | Reduce weight of the skull; serve as resonant chambers |
| Pharynx | Chamber posterior to the oral cavity and between the nasal cavity and larynx | Passageway for air moving from nasal cavity to larynx and for food moving from oral cavity to esophagus |
| Larynx | Enlargement at the top of the trachea | Passageway for air; prevents foreign objects from entering trachea; houses vocal cords |
| Trachea | Flexible tube that connects larynx with bronchial tree | Passageway for air; mucous lining continues to filter air |
| Bronchial tree | Branched tubes that lead from the trachea to the alveoli | Conducts air to the alveoli; mucous lining continues to filter incoming air |
| Lungs | Soft, cone-shaped organs that occupy a large portion of the thoracic cavity | Contain the air passages, alveoli, blood vessels, connective tissues, lymphatic vessels, and nerves of the lower respiratory tract |

# Lung Irritants

The lungs are exquisitely sensitive to the presence of inhaled particles. Such exposures can cause a variety of symptoms, both acute and chronic, that range from a persistent cough to cancer.

## ASBESTOS

This naturally occurring mineral was once widely used in buildings and on various products because it resists burning and chemical damage. Absestos easily crumbles into fibers, which, when airborne, can enter human respiratory passages. Asbestos-related problems include:

- asbestosis (shortness of breath resulting from scars in lungs)
- lung cancer
- mesothelioma (a rare cancer of the pleural membrane)

Asbestos fibers that are longer than 5 micrometers (0.0002 inch) and thinner than 2 micrometers (0.00008 inch) can cause illness when inhaled. Table 19A indicates how risk of becoming ill rises with duration of exposure.

Although asbestos clearly causes respiratory illness, it does so only if it is disturbed, so that fibers break free and become airborne. Experts must determine whether it is safer to encapsulate asbestos in a building and leave it in place, or remove it. Often, removing asbestos is actually more dangerous because this releases fibers. Today, synthetic fiberglass or plastics stand in for asbestos.

## BERYLLIOSIS

Beryllium is an element used in fluorescent powders, metal alloys, and in the nuclear power industry. A small percentage of workers exposed to beryllium dust or vapor develop an immune response, which damages the lungs. Symptoms include cough, shortness of breath, fatigue, loss of appetite, and weight loss. Fevers and night sweats indicate the role of the immune system. Radiographs show granuloma scars in the lungs. Pulmonary function tests and simply listening to breath sounds with a stethoscope reveal impaired breathing.

Symptoms of berylliosis typically begin about a decade after the first exposure, but this response time can range from several months to as long as forty years. Many people who worked with the element in the Rocky Flats plant in Colorado have developed berylliosis, and they are being monitored at the National Jewish Medical and Research Center in Denver. Affected individuals include those who directly contacted the element frequently, as well as support staff such as secretaries who probably inhaled beryllium.

Berylliosis can be distinguished from other lung ailments with a blood test that detects antibodies to beryllium. Affected individuals and those who do not have symptoms but know that they were exposed to beryllium are advised to have periodic blood tests and chest radiographs to detect the condition early. The steroid drug prednisone is used to control symptoms.

## A DISORDER WITH MANY NAMES

Repeatedly inhaling dust of organic origin can cause a lung irritation called extrinsic allergic alveolitis. An acute form of this reaction impairs breathing and causes a fever a few hours after encountering dust. In the chronic form, lung changes occur gradually over several years. The condition is associated with several occupations, and has a variety of colorful names:

Bathtub refinisher's lung
Bird breeder disease
Cheese worker's lung
Enzyme detergent sensitivity
Farmer lung
Laboratory technician's lung
Maltworker lung
Maple bark stripper disease
Mushroom picker disease
Snuff taker's lung
Plastic worker's lung
Poultry raiser disease
Wheat weevil disease ∎

| TABLE 19A ASBESTOS-RELATED RESPIRATORY ILLNESS | | |
|---|---|---|
| Situation | Level of Exposure (fibers/cubic centimeter) | Cancer Cases per Million Exposed People |
| Asbestos workers with 20 years' exposure | 10 fibers/cc | 200,000 |
| Permissible upper limit in buildings today | 0.1 fibers/cc | 2,000 |
| Child in school with asbestos 6 hours/day | 0.0005 fibers/cc | 6 |
| Most modern buildings | 0.0002 fibers/cc | 4 |

normal air pressure equals 760 millimeters (mm) of mercury (Hg).

Air pressure is exerted on all surfaces in contact with the air, and because people breathe air, the inside surfaces of their lungs are also subjected to pressure. In other words, the pressures on the inside of the lungs and alveoli and on the outside of the thoracic wall are about the same (fig. 19.21).

If the pressure inside the lungs and alveoli (intra-alveolar pressure) decreases, outside air will then be pushed into the airways by atmospheric pressure. This is what happens during normal inspiration, and it involves the action of muscle fibers within the dome-shaped *diaphragm*.

The diaphragm is located just inferior to the lungs. It consists of an anterior group of skeletal muscle fibers (costal fibers), which originate from the ribs and sternum, and a posterior group (crural fibers), which originate from the vertebrae. Both groups of muscle fibers are inserted on a tendinous central portion of the diaphragm (reference plate 71).

The muscle fibers of the diaphragm are stimulated to contract by impulses carried by the *phrenic nerves*, which are associated with the cervical plexuses. When this occurs, the diaphragm moves downward, the thoracic cavity enlarges, and the intra-alveolar pressure falls about 2 mm Hg below that of atmospheric pressure. In response to this decreased pressure, air is forced into the airways by atmospheric pressure (fig. 19.22).

While the diaphragm is contracting and moving downward, the *external (inspiratory) intercostal muscles* and certain thoracic muscles may be stimulated to contract. This action raises the ribs and elevates the sternum, increasing the size of the thoracic cavity even more. As a result, the intra-alveolar pressure falls farther, and the greater atmospheric pressure forces more air into the airways.

Lung expansion in response to movements of the diaphragm and chest wall depends on movements of the pleural membranes as follows. The parietal pleura, on the inner wall of the thoracic cavity, and the visceral pleura, attached to the surface of the lungs, are separated only by a thin film of serous fluid. The water molecules in this fluid greatly attract one another, creating a force called **surface tension** that holds the moist surfaces of the pleural membranes tightly together, just as a wet

**FIGURE** 19.20

Atmospheric pressure at sea level is sufficient to support a column of mercury 760 millimeters high at sea level.

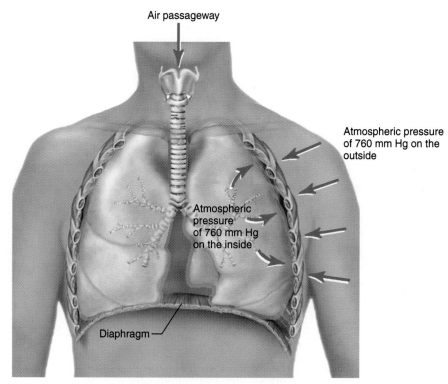

**FIGURE** 19.21

When the lungs are at rest, the pressure on the inside of the lungs is equal to the pressure on the outside of the thorax.

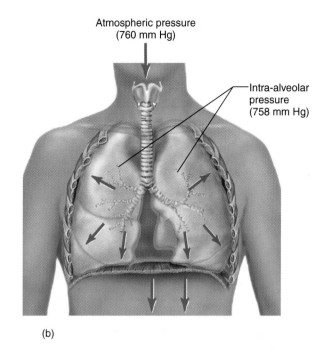

**FIGURE 19.22**

(*a*) Prior to inspiration, the intra-alveolar pressure is 760 mm Hg. (*b*) The intra-alveolar pressure decreases to about 758 mm Hg as the thoracic cavity enlarges, and atmospheric pressure forces air into the airways.

| TABLE 19.2 | MAJOR EVENTS IN INSPIRATION |
| --- | --- |

1. Nerve impulses travel on phrenic nerves to muscle fibers in the diaphragm, contracting them.
2. As the dome-shaped diaphragm moves downward, the thoracic cavity expands.
3. At the same time, the external intercostal muscles may contract, raising the ribs and expanding the thoracic cavity still more.
4. The intra-alveolar pressure decreases.
5. Atmospheric pressure, which is greater on the outside, forces air into the respiratory tract through the air passages.
6. The lungs fill with air.

coverslip sticks to a microscope slide. In addition, any separation of the pleural membranes will decrease pressure in the intrapleural space, also holding these membranes together. Consequently, when the intercostal muscles move the thoracic wall upward and outward, the parietal pleura moves too, and the visceral pleura follows it. This helps expand the lungs in all directions. Table 19.2 summarizes the steps in inspiration.

The surface tension between the moist membranes would be sufficient to collapse the alveoli, which have moist inner surfaces. However, certain alveolar cells (alveolar type II cells) synthesize a mixture of lipoproteins, called **surfactant** (see figure 19.16). Surfactant is continuously secreted into the alveolar air spaces. It re-

duces the surface tension and thus decreases the tendency of the alveoli to collapse when lung volume is low.

At birth, Benjamin McClatchey weighed only 2 pounds, 13 ounces, because he was born two months early. Like many of the 380,000 preemies born each year in the United States, Benjamin had *respiratory distress syndrome* (RDS). His lungs were too immature to produce sufficient surfactant, and as a result, they could not overcome the force of surface tension enough to inflate.

A decade ago, Benjamin might not have survived RDS. But with the help of a synthetic surfactant sprayed or dripped into his lungs through an endotracheal tube, and a ventilator machine designed to assist breathing in premature infants, he survived. Unlike conventional ventilators, which force air into the lungs at pressures that could damage delicate newborn lungs, the high-frequency ventilator used on preemies delivers the lifesaving oxygen in tiny, gentle puffs.

If a person needs to take a deeper than normal breath, the diaphragm and external intercostal muscles may contract to an even greater extent. Additional muscles, such as the pectoralis minors and sternocleidomastoids, can also be used to pull the thoracic cage further upward and outward, enlarging the thoracic cavity, and decreasing alveolar pressure even more (fig. 19.23).

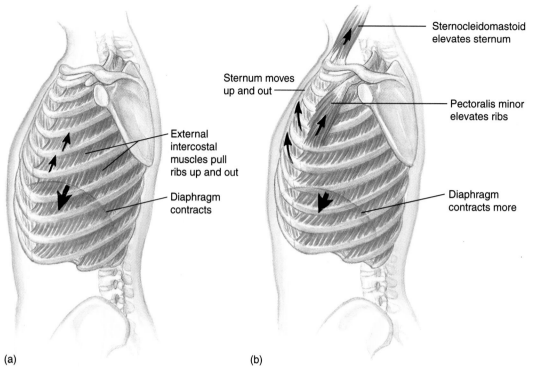

**FIGURE 19.23**

(a) Shape of the thorax at the end of normal inspiration. (b) Shape of the thorax at the end of maximal inspiration, aided by contraction of the sternocleidomastoid and pectoralis minor muscles.

Sternocleidomastoid elevates sternum

Sternum moves up and out

Pectoralis minor elevates ribs

External intercostal muscles pull ribs up and out

Diaphragm contracts

Diaphragm contracts more

The ease with which the lungs can be expanded as a result of pressure changes occurring during breathing is called *compliance* (distensibility). In a normal lung, compliance decreases as lung volume increases, because an inflated lung is more difficult to expand than a lung at rest. Conditions that obstruct air passages, destroy lung tissue, or impede lung expansion in other ways also decrease compliance.

## Expiration

The forces responsible for normal resting expiration come from *elastic recoil* of lung tissues and from surface tension. The lungs contain a considerable amount of elastic tissue, and lung expansion during inspiration stretches these tissues. When the diaphragm lowers, the abdominal organs inferior to it are compressed. As the diaphragm and the external intercostal muscles relax following inspiration, the elastic tissues cause the lungs to recoil, and they return to their original shapes. Similarly, elastic tissues within the abdominal organs cause them to spring back into their previous shapes, pushing the diaphragm upward. At the same time, surface tension that develops between the moist surfaces of the alveolar linings tends to shrink alveoli. Each of these factors increases the intra-alveolar pressure about 1 mm Hg above atmospheric pressure, so that the air inside the lungs is forced out through the respiratory passages. Thus, normal resting expiration is a passive process.

The recoil of the elastic fibers within the lung tissues reduces pressure in the pleural cavity. Consequently, the pressure between the pleural membranes (intrapleural pressure) is usually about 4 mm Hg less than atmospheric pressure.

Because surface tension adheres the visceral and parietal pleural membranes, no significant space normally separates them in the pleural cavity. However, if the thoracic wall is punctured, atmospheric air may enter the pleural cavity and create a substantial space between the membranes. This condition is called *pneumothorax*, and when it occurs, the lung on the affected side may collapse because of its elasticity.

Pneumothorax may be treated by covering the chest wound with an impermeable bandage, passing a tube (chest tube) through the thoracic wall into the pleural cavity, and applying suction to the tube. The suction reestablishes negative pressure within the cavity, and the collapsed lung expands.

If a person needs to exhale more air than normal, the posterior *internal (expiratory) intercostal muscles* can be contracted. These muscles pull the ribs and sternum downward and inward, increasing the pressure in the lungs. Also, the *abdominal wall muscles,* including the external and internal obliques, the transversus abdo-

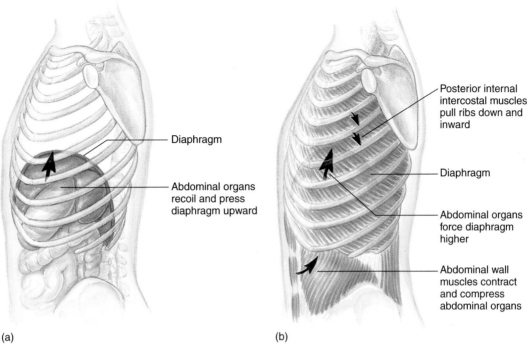

(a)

(b)

**FIGURE** 19.24 ✗

(a) Normal resting expiration is due to elastic recoil of the lung tissues and the abdominal organs. (b) Contraction of the abdominal wall muscles and posterior internal intercostal muscles aids maximal expiration.

| TABLE 19.3 | MAJOR EVENTS IN EXPIRATION |
|---|---|

1. The diaphragm and external respiratory muscles relax.
2. Elastic tissues of the lungs and thoracic cage, which were stretched during inspiration, suddenly recoil, and surface tension collapses alveolar walls.
3. Tissues recoiling around the lungs increase the intra-alveolar pressure.
4. Air is squeezed out of the lungs.

minis, and the rectus abdominis, can be used to squeeze the abdominal organs inward. Thus, the abdominal wall muscles can increase pressure in the abdominal cavity and force the diaphragm still higher against the lungs, squeezing additional air out of the lungs (fig. 19.24). Table 19.3 summarizes the steps in expiration.

**1** Describe the events in inspiration.

**2** How does surface tension aid in expanding the lungs during inspiration?

**3** What forces are responsible for normal expiration?

## Respiratory Volumes and Capacities

Different degrees of effort in breathing move different volumes of air in or out of the lungs. The measurement of such air volumes is called *spirometry*, and it describes four distinct *respiratory volumes*.

One inspiration plus the following expiration is called a **respiratory cycle.** The amount of air that enters or leaves during a respiratory cycle is termed the **tidal volume.** About 500 milliliters (mL) of air enter during a normal, resting inspiration. On the average, the same amount leaves during a normal, resting expiration. Thus, the tidal volume at rest is about 500 mL (fig. 19.25).

During forced maximal inspiration, a quantity of air in addition to the resting tidal volume enters the lungs. This additional volume is called the **inspiratory reserve volume** (complemental air) and it equals about 3,000 mL.

During a maximal forced expiration, about 1,100 mL of air in addition to the resting tidal volume can be expelled from the lungs. This quantity is called the **expiratory reserve volume** (supplemental air). However, even after the most forceful expiration, about 1,200 mL of air remains in the lungs. This is called the **residual volume.**

Residual air remains in the lungs at all times, and consequently, newly inhaled air always mixes with air already in the lungs. This prevents the oxygen and carbon dioxide concentrations in the lungs from fluctuating greatly with each breath.

Once the respiratory volumes are known, four *respiratory capacities* can be calculated by combining two or more of the volumes. Thus, if the inspiratory reserve volume (3,000 mL) is combined with the tidal volume (500 mL) and the expiratory reserve volume (1,100 mL), the total is termed the **vital capacity** (4,600 mL). This capacity is the maximum amount of air a person can exhale after taking the deepest breath possible.

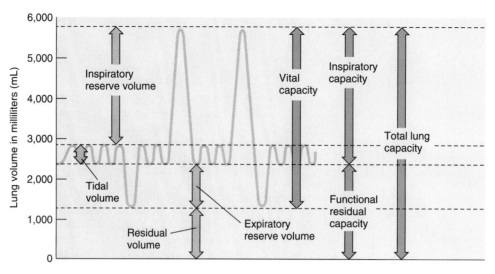

**FIGURE 19.25**

Respiratory volumes and capacities.

The tidal volume (500 mL) plus the inspiratory reserve volume (3,000 mL) gives the **inspiratory capacity** (3,500 mL), which is the maximum volume of air a person can inhale following a resting expiration. Similarly, the expiratory reserve volume (1,100 mL) plus the residual volume (1,200 mL) equals the **functional residual capacity** (2,300 mL), which is the volume of air that remains in the lungs following a resting expiration.

The vital capacity plus the residual volume equals the **total lung capacity** (about 5,800 mL). This total varies with age, sex, and body size.

Some of the air that enters the respiratory tract during breathing fails to reach the alveoli. This volume (about 150 mL) remains in the passageways of the trachea, bronchi, and bronchioles. Since gas exchanges do not occur through the walls of these passages, this air is said to occupy *anatomic dead space.*

Occasionally, air sacs in some regions of the lungs are nonfunctional due to poor blood flow in the adjacent capillaries. This creates *alveolar dead space.* The anatomic and alveolar dead space volumes combined equal *physiologic dead space.* In a normal lung the anatomic and physiologic dead spaces are essentially the same (about 150 mL).

A spirometer (fig. 19.26) is used to measure respiratory air volumes (except the residual volume). These measurements are used to evaluate the course of respiratory illnesses, such as emphysema, pneumonia, lung cancer, and bronchial asthma. Table 19.4 summarizes the respiratory air volumes and capacities.

**1** What is tidal volume?

**2** Distinguish between inspiratory and expiratory reserve volumes.

**FIGURE 19.26**

A spirometer can be used to measure respiratory air volumes.

**3** How is vital capacity measured?

**4** How is the total lung capacity calculated?

## Alveolar Ventilation

The amount of new atmospheric air that is moved into the respiratory passages each minute is called the *minute ventilation.* It equals the tidal volume multiplied by the breathing rate. Thus, if the tidal volume is 500 mL and the breathing rate is 12 breaths per minute, the minute ventilation is 500 mL × 12, or 6,000 mL per minute. However, much of the new air remains in the physiologic dead space.

## TABLE 19.4 — RESPIRATORY AIR VOLUMES AND CAPACITIES

| Name | Volume (Average) | Description |
|------|------------------|-------------|
| Tidal volume (TV) | 500 mL | Volume moved in or out of the lungs during a respiratory cycle |
| Inspiratory reserve volume (IRV) | 3,000 mL | Volume that can be inhaled during forced breathing in addition to tidal volume |
| Expiratory reserve volume (ERV) | 1,100 mL | Volume that can be exhaled during forced breathing in addition to tidal volume |
| Residual volume (RV) | 1,200 mL | Volume that remains in the lungs at all times |
| Inspiratory capacity (IC) | 3,500 mL | Maximum volume of air that can be inhaled following exhalation of tidal volume: IC = TV + IRV |
| Functional residual capacity (FRC) | 2,300 mL | Volume of air that remains in the lungs following exhalation of tidal volume: FRC = ERV + RV |
| Vital capacity (VC) | 4,600 mL | Maximum volume of air that can be exhaled after taking the deepest breath possible: VC = TV + IRV + ERV |
| Total lung capacity (TLC) | 5,800 mL | Total volume of air that the lungs can hold: TLC = VC + RV |

The volume of new air that does reach the alveoli and is available for gas exchange is calculated by subtracting the physiologic dead space (150 mL) from the tidal volume (500 mL). The resulting volume (350 mL) multiplied by the breathing rate (12 breaths per minute), is the *alveolar ventilation rate* (4,200 mL per minute). The alveolar ventilation rate is a major factor affecting the concentrations of oxygen and carbon dioxide in the alveoli.

In the microgravity environment of space, the gas exchange capacity of the alveoli increases by 28%. This is because blood flow in the pulmonary capillaries and ventilation of the alveoli are more uniform than on earth, in the presence of gravity. Eight astronauts aboard two space shuttle missions provided this information by performing various pulmonary function tests. They worked in Spacelab, a small, cylindrical pressurized laboratory that is taken into space aboard the shuttle.

## Nonrespiratory Air Movements

Air movements that occur in addition to breathing are called *nonrespiratory movements*. They are used to clear air passages, as in coughing and sneezing, or to express emotions, as in laughing and crying.

Nonrespiratory movements usually result from *reflexes*, although sometimes they are initiated voluntarily. A cough, for example, can be produced through conscious effort or may be triggered by a foreign object in an air passage.

*Coughing* involves taking a deep breath, closing the glottis, and forcing air upward from the lungs against the closure. Then the glottis is suddenly opened, and a blast of air is forced upward from the lower respiratory tract. Usually this rapid rush of air will remove the substance that triggered the reflex.

The most sensitive areas of the air passages are in the larynx, the carina, and in regions near the branches of the major bronchi. The distal portions of the bronchioles (respiratory bronchioles), alveolar ducts, and alveoli lack a nerve supply. Consequently, before any material in these parts can trigger a cough reflex, it must be moved into the larger passages of the respiratory tract.

A *sneeze* is much like a cough, but it clears the upper respiratory passages rather than the lower ones. This reflex is usually initiated by a mild irritation in the lining of the nasal cavity, and, in response, a blast of air is forced up through the glottis. This time, the air is directed into the nasal passages by depressing the uvula, thus closing the opening between the pharynx and the oral cavity.

*Laughing* involves taking a breath and releasing it in a series of short expirations. *Crying* consists of very similar movements, and sometimes it is necessary to note a person's facial expression in order to distinguish laughing from crying.

A *hiccup* is caused by sudden inspiration due to a spasmodic contraction of the diaphragm while the glottis is closed. Air striking the vocal folds causes the sound of the hiccup. We do not know the function, if any, of hiccups.

*Yawning* may aid respiration by providing an occasional deep breath. During normal, quiet breathing, not all of the alveoli are ventilated and some blood may pass through the lungs without becoming well oxygenated. This low blood oxygen concentration somehow triggers the yawn reflex, prompting a very deep breath that ventilates more alveoli.

Table 19.5 summarizes the characteristics of nonrespiratory air movements. Clinical Application 19.3 discusses respiratory problems that affect ventilation.

| TABLE 19.5 | NONRESPIRATORY AIR MOVEMENTS | | |
|---|---|---|---|
| **Air Movement** | **Mechanism** | | **Function** |
| Coughing | Deep breath is taken, glottis is closed, and air is forced against the closure; suddenly the glottis is opened and a blast of air passes upward | | Clears lower respiratory passages |
| Sneezing | Same as coughing, except air moving upward is directed into the nasal cavity by depressing the uvula | | Clears upper respiratory passages |
| Laughing | Deep breath is released in a series of short expirations | | Expresses happiness |
| Crying | Same as laughing | | Expresses sadness |
| Hiccuping | Diaphragm contracts spasmodically while glottis is closed | | No useful function known |
| Yawning | Deep breath is taken | | Ventilates a larger proportion of the alveoli and aids oxygenation of the blood |
| Speech | Air is forced through the larynx, causing vocal cords to vibrate; words are formed by lips, tongue, and soft palate | | Vocal communication |

1. How is the minute ventilation calculated? The alveolar ventilation rate?

2. Which nonrespiratory air movements help clear the air passages?

3. Which nonrespiratory air movements are used to express emotions?

4. What seems to be the function of a yawn?

## CONTROL OF BREATHING

Normal breathing is a rhythmic, involuntary act that continues when a person is unconscious.

### Respiratory Center

Groups of neurons in the brain stem comprise the **respiratory center,** which controls breathing. This center periodically initiates impulses that travel on cranial and spinal nerves to breathing muscles, causing inspiration and expiration. The respiratory center also adjusts the rate and depth of breathing to meet cellular needs for a supply of oxygen and removal of carbon dioxide, even during strenuous physical exercise.

The components of the respiratory center are widely scattered throughout the pons and medulla oblongata. However, two areas of the respiratory center are of special interest. They are the rhythmicity area of the medulla and the pneumotaxic area of the pons (fig. 19.27).

The **medullary rhythmicity area** includes two groups of neurons that extend throughout the length of the medulla oblongata. They are called the dorsal respiratory group and the ventral respiratory group.

The *dorsal respiratory group* is responsible for the basic rhythm of breathing. The neurons of this group emit bursts of impulses that signal the diaphragm and

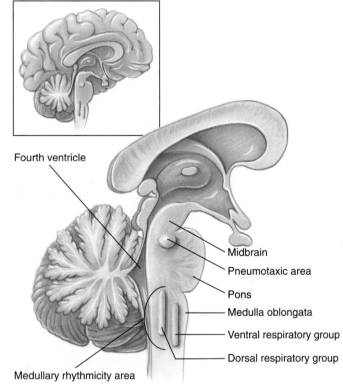

Fourth ventricle

Midbrain

Pneumotaxic area

Pons

Medulla oblongata

Ventral respiratory group

Dorsal respiratory group

Medullary rhythmicity area

**FIGURE 19.27**

The respiratory center is located in the pons and the medulla oblongata.

other inspiratory muscles to contract. The impulses of each burst begin weakly, strengthen for about two seconds, and cease abruptly. The breathing muscles that contract in response to the impulses cause the volume of air entering the lungs to increase steadily. The neurons remain inactive while expiration occurs passively, and then they emit another burst of inspiratory impulses so that the inspiration–expiration cycle repeats.

# Respiratory Disorders that Decrease Ventilation

Injuries to the respiratory center or to spinal nerve tracts that transmit motor impulses may paralyze breathing muscles. Paralysis may also be due to a disease, such as *poliomyelitis*, that affects parts of the central nervous system and injures motor neurons. The consequences of such paralysis depend on which muscles are affected. Sometimes, other muscles, by increasing their responses, can compensate for functional losses of a paralyzed muscle. Otherwise, mechanical ventilation is necessary.

*Bronchial asthma* is usually an allergic reaction to foreign antigens in the respiratory tract, such as plant pollen that enters with inhaled air. Cells of the larger airways secrete copious amounts of mucus, which traps allergens. Ciliated columnar epithelial cells move the mucus up and out of the bronchi, then up and out of the trachea. Thus, upper respiratory structures remain relatively clear. However, in the lower respiratory areas, mucus drainage plus edematous secretions accumulate because fewer cells are ciliated. The allergens and secretions irritate smooth muscles, stimulating bronchioconstrictions. Breathing becomes increasingly difficult, and it produces a characteristic wheezing sound as air moves through narrowed passages.

A person with asthma usually finds it harder to force air out of the lungs than to bring it in. This is because inspiration utilizes powerful breathing muscles, and, as they contract, the lungs expand, opening the air passages. Expiration, on the other hand, is a passive process due to elastic recoil of stretched tissues. Expiration also compresses the tissues and constricts the bronchioles, further impairing air movement through the narrowed air passages.

*Emphysema* is a progressive, degenerative disease that destroys many alveolar walls. As a result, clusters of small air sacs merge into larger chambers, which decreases the total surface area of the alveolar walls. At the same time, the alveolar walls lose their elasticity, and the capillary networks associated with the alveoli become less abundant (fig. 19C).

Because of the loss of tissue elasticity, a person with emphysema finds it increasingly difficult to force air out of the lungs. Abnormal muscular efforts are required to compensate for the lack of elastic recoil that normally contributes to expiration. Exposure to respiratory irritants, such as those in tobacco smoke and polluted air, is believed to cause emphysema. Rarely, emphysema is caused by an inherited enzyme deficiency. ■

(a)

(b)

**FIGURE 19C**

(a) Normal lung tissue (100×). (b) As emphysema develops, alveoli merge, forming larger chambers (100×).

The *ventral respiratory group* is quiescent during normal breathing. However, when more forceful breathing is necessary, the neurons in this group generate impulses that increase inspiratory movement. Other neurons in the group activate the muscles associated with forceful expiration as well.

The neurons in the **pneumotaxic area** continuously transmit impulses to the dorsal respiratory group and regulate the duration of inspiratory bursts originating from the dorsal group. In this way the pneumotaxic neurons control the rate of breathing. More specifically, when the pneumotaxic signals are strong, the inspiratory bursts are shorter, and the rate of breathing increases (tachypnea); when the pneumotaxic signals are weak, the inspiratory bursts are longer, and the rate of breathing decreases (bradypnea) (fig. 19.28).

**1** Where is the respiratory center located?

**2** Describe how the respiratory center maintains a normal breathing pattern.

**3** Explain how the breathing pattern may be changed.

## Factors Affecting Breathing

Factors in addition to the respiratory center influence breathing rate and depth. These include certain chemicals in body fluids, the degree to which lung tissues are stretched, and emotional state. For example, *chemosensitive areas* (**central chemoreceptors**) within the respiratory

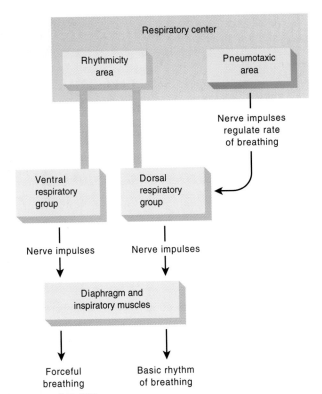

**FIGURE 19.28**

The medullary rhythmicity and pneumotaxic areas of the respiratory center control breathing.

center, located in the ventral portion of the medulla oblongata near the origins of the vagus nerves, are very sensitive to changes in the blood concentration of carbon dioxide and hydrogen ions. If carbon dioxide or hydrogen ion concentration rises, the central chemoreceptors signal the respiratory center, and breathing rate increases.

The similarity of the effects of carbon dioxide and hydrogen ions is a consequence of the fact that carbon dioxide combines with water in the cerebrospinal fluid to form carbonic acid ($H_2CO_3$):

$$CO_2 + H_2O \rightarrow H_2CO_3$$

The carbonic acid thus formed soon ionizes, releasing hydrogen ions ($H^+$) and bicarbonate ions ($HCO_3^-$):

$$H_2CO_3 \rightarrow H^+ + HCO_3^-$$

It is the presence of hydrogen ions rather than the carbon dioxide that influences the central chemoreceptors. In any event, breathing rate and tidal volume increase when a person inhales air rich in carbon dioxide or when body cells produce excess carbon dioxide or hydrogen ions. These changes increase alveolar ventilation. As a result, more carbon dioxide is exhaled, and the blood concentrations of carbon dioxide and hydrogen ions return toward normal.

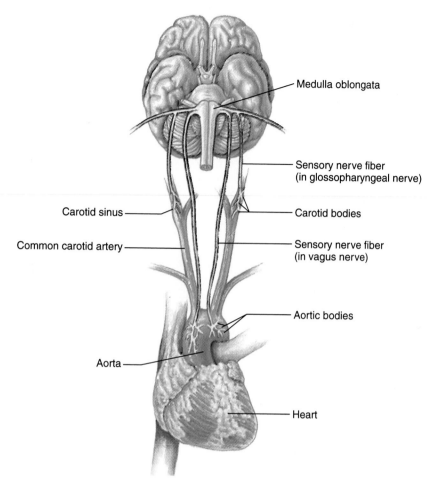

**FIGURE** 19.29

Decreased blood oxygen concentration stimulates peripheral chemoreceptors in the carotid and aortic bodies.

Adding carbon dioxide to air can stimulate the rate and depth of breathing. Ordinary air is about 0.04% carbon dioxide. If a patient inhales air containing 4% carbon dioxide, breathing rate usually doubles.

Low blood oxygen concentration seems to have little direct effect on the central chemoreceptors associated with the respiratory center. Instead, changes in the blood oxygen concentration are sensed by **peripheral chemoreceptors** in specialized structures called the *carotid bodies* and *aortic bodies*, which are located in the walls of the carotid sinuses and aortic arch (fig. 19.29). When decreased oxygen concentration stimulates these peripheral receptors, impulses are transmitted to the respiratory center, and the breathing rate and tidal volume increase, thus increasing alveolar ventilation. This mechanism is usually not triggered until the blood oxygen concentration reaches a very low level; thus, oxygen seems to play only a minor role in the control of normal respiration.

The peripheral chemoreceptors of the carotid and aortic bodies are also stimulated by changes in the blood concentrations of carbon dioxide and hydrogen ions. However, these substances have a much more powerful effect when they act on the central chemoreceptors of the respiratory center. The effects of carbon dioxide and hydrogen ions on the carotid and aortic bodies are relatively unimportant.

An exception to the normal pattern of chemical control may occur in patients who have chronic obstructive pulmonary diseases (COPD), such as asthma, bronchitis, and emphysema. Over a period of time, these patients seem to adapt to high concentrations of carbon dioxide, and for them, low oxygen concentrations may serve as a necessary respiratory stimulus. Paradoxically, when such a patient is placed on 100% oxygen, breathing may stop.

An *inflation reflex* (Hering-Breuer reflex) helps regulate the depth of breathing. This reflex occurs when stretch receptors in the visceral pleura, bronchioles, and alveoli are stimulated as lung tissues are stretched. The sensory impulses of the reflex travel via the vagus nerves to the pneumotaxic area of the respiratory center, and shorten the duration of inspiratory movements. This action prevents overinflation of the lungs during forceful breathing (fig. 19.30).

Emotional upset or strong sensory stimulation may alter the normal breathing pattern. Gasping and rapid breathing are familiar responses to fear, anger, shock, excitement, horror, surprise, sexual stimulation, or even the chill of stepping into a cold shower. Because control of the respiratory muscles is voluntary, we can alter breathing pattern consciously, or even stop it altogether for a short time. During childbirth, for example, women often concentrate on controlling their breathing, which distracts them from the pain.

If a person decides to stop breathing, the blood concentrations of carbon dioxide and hydrogen ions begin to rise, and the concentration of oxygen falls. These changes (primarily the increased $CO_2$) stimulate the respiratory center, and soon the need to inhale overpowers the desire to hold the breath—much to the relief of parents when young children threaten to hold their breaths until they turn blue! On the other hand, a person can increase the breath-holding time by breathing rapidly and deeply in advance. This action, termed **hyperventilation,** lowers the blood carbon dioxide concentration. Following hyperventilation, it takes longer than usual for the carbon dioxide concentration to reach the level needed to override the conscious effort of breath holding.

Table 19.6 discusses factors affecting breathing. Clinical Application 19.4 focuses on one influence on breathing—exercise.

> Sometimes a person who is emotionally upset may hyperventilate, become dizzy, and lose consciousness. This is due to a lowered carbon dioxide concentration followed by a rise in pH (respiratory alkalosis) and a localized vasoconstriction of cerebral arterioles, decreasing blood flow to nearby brain cells. Hampered oxygen supply to the brain causes fainting. Hyperventilation should never be used to help hold the breath while swimming, because the person may lose consciousness under water and drown.

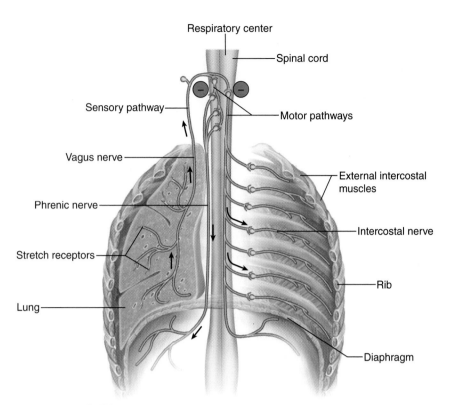

**FIGURE 19.30**

In the process of inspiration, motor impulses travel from the respiratory center to the diaphragm and external intercostal muscles, which contract and cause the lungs to expand. This expansion stimulates stretch receptors in the lungs to send inhibiting impulses to the respiratory center, thus preventing overinflation.

1 Describe the inflation reflex.

2 Which chemical factors affect breathing?

3 How does hyperventilation decrease respiratory rate?

## ALVEOLAR GAS EXCHANGES

The tubelike parts of the respiratory system move air in and out of the air passages. The alveoli are the sites of the vital process of gas exchange between the air and the blood.

### Alveoli

Alveoli are microscopic air sacs clustered at the distal ends of the finest respiratory tubes—the alveolar ducts. Each alveolus consists of a tiny space surrounded by a thin wall that separates it from adjacent alveoli. Minute openings, called **alveolar pores,** in the walls of some alveoli may permit air to pass from one alveolus to another (fig. 19.31). This arrangement provides alternate air pathways if the passages in some portions of the lung become obstructed.

Phagocytic cells called *alveolar macrophages* are in alveoli and in the pores connecting the air sacs. These macrophages phagocytize airborne agents, including bacteria, thereby cleaning the alveoli.

# CLINICAL APPLICATION

## Exercise and Breathing

Moderate to heavy physical exercise greatly increases the amount of oxygen skeletal muscles use. For example, a young man at rest will utilize about 250 milliliters of oxygen per minute, but he may require 3,600 milliliters per minute during maximal exercise. While oxygen utilization is increasing, the volume of carbon dioxide produced increases also. Since decreased blood oxygen and increased blood carbon dioxide concentration stimulate the respiratory center, it is not surprising that exercise is accompanied by an increased breathing rate. However, studies reveal that blood oxygen and carbon dioxide concentrations usually remain nearly unchanged during exercise—a reflection of the respiratory system's effectiveness in obtaining oxygen and releasing carbon dioxide to the outside.

What other factors might increase breathing rate during exercise? The cerebral cortex and the proprioceptors associated with muscles and joints seem to be involved (see chapter 12). Specifically, the cortex transmits stimulating impulses to the respiratory center whenever it signals the skeletal muscles to contract. At the same time, muscular movements stimulate the proprioceptors, triggering a *joint reflex.* In this reflex, sensory impulses are transmitted from the proprioceptors to the respiratory center, and breathing rate increases.

The increase in breathing rate during exercise requires an increase in blood flow to meet the requirements of skeletal muscles. Thus, physical exercise increases demand on both the respiratory and the circulatory systems. If either of these systems fails to keep up with cellular demands, the person will begin to feel out of breath. This sensation however, is usually due to the inability of the heart and circulatory system to move enough blood between the lungs and the cells, rather than to the inability of the respiratory system to provide enough air. ∎

## TABLE 19.6 FACTORS AFFECTING BREATHING

| Factors | Receptors Stimulated | Response | Effect |
|---|---|---|---|
| Stretch of tissues | Stretch receptors in visceral pleura, bronchioles, and alveoli | Inhibits inspiration | Prevents overinflation of lungs during forceful breathing |
| Low blood $P_{O_2}$ | Chemoreceptors in carotid and aortic bodies | Increases alveolar ventilation | Increases blood oxygen concentration |
| High blood $P_{CO_2}$ | Chemosensitive areas of the respiratory center | Increases alveolar ventilation | Decreases blood carbon dioxide concentration and CSF hydrogen ion concentration |
| High CSF hydrogen ion concentration | Chemosensitive areas of the respiratory center | Increases breathing rate and alveolar ventilation | Decreases blood carbon dioxide concentration and CSF hydrogen ion concentration |

## Respiratory Membrane

The wall of an alveolus consists of an inner lining of simple squamous epithelium (type I cells, see fig. 19.16) and a dense network of capillaries, which are also lined with simple squamous epithelial cells (fig. 19.32). Thin basement membranes separate the layers of these flattened cells, and in the spaces between them are elastic and collagenous fibers that help support the alveolar wall. At least two thicknesses of epithelial cells and basement membranes separate the air in an alveolus and the blood in a capillary. These layers make up the **respiratory membrane** (alveolar-capillary membrane), through which gas exchange occurs between the alveolar air and the blood (figs. 19.33 and 19.34).

## Diffusion through the Respiratory Membrane

Recall from chapter 3 that molecules diffuse from regions where they are in higher concentration toward regions where they are in lower concentration. For gases, it is

**FIGURE** 19.31

Alveolar pores (arrow) allow air to pass from one alveolus to another (300×).

**FIGURE** 19.32

The respiratory membrane consists of the walls of the alveolus and the capillary.

Labels: Cell of alveolar wall; Surfactant-secreting cell; Fluid with surfactant; Macrophage; Alveolus; Respiratory membrane; Capillary lumen; Cell of capillary wall

**FIGURE** 19.33

Electron micrograph of a capillary located between alveoli (7,000×). (*AS*, alveolar space; *RBC*, red blood cell; *BM*, basement membrane; *IS*, interstitial connective tissue; *EN*, epithelial nucleus.)

more useful to think of diffusion occurring from regions of higher pressure towards regions of lower pressure. The pressure of a gas determines the rate at which it diffuses from one region to another.

Measured by volume, ordinary air is about 78% nitrogen, 21% oxygen, and 0.04% carbon dioxide. Air has small amounts of other gases that have little or no physiological importance.

In a mixture of gases such as air, each gas accounts for a portion of the total pressure the mixture produces. The amount of pressure each gas contributes, called the **partial pressure**, is directly proportional to the concentration of the gas in the mixture. For example, because air is 21% oxygen, this gas accounts for 21% of the atmospheric pressure (21% of 760 mm Hg), or 160 mm Hg. Thus, the partial pressure of oxygen, symbolized $P_{O_2}$, in atmospheric air is 160 mm Hg. Similarly, the partial pressure of carbon dioxide ($P_{CO_2}$) in air is 0.3 mm Hg.

When a mixture of gases dissolves in blood, the resulting concentration of each gas is proportional to its partial pressure. Each gas diffuses between blood and its surroundings from areas of higher partial pressure to areas of lower partial pressure until the partial pressures in the two regions reach equilibrium. For example, the $P_{CO_2}$ of blood entering the pulmonary capillaries is 45 mm Hg, but the $P_{CO_2}$ in alveolar air is 40 mm Hg. Because of the difference in these partial pressures, carbon dioxide diffuses from blood, where its partial pressure is higher, across the respiratory membrane and into alveolar air. When blood leaves the lungs, its $P_{CO_2}$ is 40 mm Hg, which is the same as the $P_{CO_2}$ of alveolar air. Similarly, the $P_{O_2}$ of blood entering the pulmonary capillaries is 40 mm Hg, but that of alveolar air reaches 104 mm Hg as oxygen diffuses from alveolar air into blood. Thus, since equilibrium is reached, blood leaves the lungs with

**FIGURE** 19.34

Gas exchanges occur between the air of the alveolus and the blood of the capillary as a result of differences in partial pressures.

a $P_{O_2}$ of 104 mm Hg. (Because of the large volume of air always in the lungs, as long as breathing continues, alveolar $P_{O_2}$ stays relatively constant at 104 mm Hg.) Clinical Application 19.5 examines illnesses that result from impaired gas exchange.

**1** Describe the structure of the respiratory membrane.

**2** What is partial pressure?

**3** What causes oxygen and carbon dioxide to move across the respiratory membrane?

Exposure to high oxygen concentration (hyperoxia) for a prolonged time may damage lung tissues, particularly those that form the capillary walls. As a result of such damage, excess fluid may escape and flood the alveolar air spaces, interfering with gas exchange. This may lead to death. Similarly, hyperoxia can damage the retinal capillaries of premature infants causing *retrolental fibroplasia* (RLF), a condition that may lead to blindness.

## GAS TRANSPORT

The blood transports oxygen and carbon dioxide between the lungs and the body cells. As these gases enter the blood, they dissolve in the liquid portion, the plasma, or combine chemically with other atoms or molecules.

## Oxygen Transport

Almost all the oxygen (over 98%) is carried in the blood bound to the protein *hemoglobin* in red blood cells. The iron in hemoglobin provides the color of these blood cells. The remainder of the oxygen is dissolved in the blood plasma.

Hemoglobin consists of two types of components called *heme* and *globin* (see chapter 14). Globin is a pro-

tein that contains 574 amino acids, in four polypeptide chains. Each chain is associated with a heme group, and each heme group contains an atom of iron. Each iron atom can loosely combine with an oxygen molecule. As oxygen dissolves in blood, it rapidly combines with hemoglobin, forming a new compound called **oxyhemoglobin.** Each hemoglobin molecule can combine with a maximum of four oxygen atoms.

The $P_{O_2}$ determines the amount of oxygen that combines with hemoglobin. The greater the $P_{O_2}$, the more oxygen will combine with hemoglobin, until the hemoglobin molecules are saturated with oxygen (fig. 19.35).

Each year, about 100,000 mountain climbers and high-altitude exercisers experience varying degrees of altitude sickness. This is because at high altitudes, the proportion of oxygen in air remains the same (about 21%), but the $P_{O_2}$ decreases. Consequently, if a person breathes high-altitude air, oxygen diffuses less rapidly from the alveoli into the blood, and the hemoglobin is less likely to become saturated with oxygen. The body's efforts to get more oxygen—increased breathing and heart rate and enhanced red blood cell and hemoglobin production—cannot keep pace with the plummeting oxygen supply.

Symptoms of oxygen deficiency (hypoxia) may ensue. For example, at 8,000 feet a person may experience anxiety, restlessness, increased breathing rate, and rapid pulse. At 12,500 feet, where the $P_{O_2}$ is only 100 mm Hg rather than 160 mm Hg at sea level, symptoms include drowsiness, mental fatigue, headache, and nausea. Above 13,000 feet, the brain swells (cerebral edema), producing severe headache, vomiting, loss of coordination, and hallucinations. Finally, the person loses consciousness and may die at about 23,000 feet.

# Disorders Impairing Gas Exchange

Five-year-old Carly had what her parents at first thought was just a "bug" that was passing through the family. But after 12 hours of flulike symptoms, Carly's temperature shot up to 105°F, her chest began to hurt, and her breathing became rapid and shallow. Later that day, a chest radiograph confirmed what the doctor suspected—Carly had *pneumonia*. Apparently, the bacteria that had caused a mild upper respiratory infection in her parents and sisters had taken a detour in her body, infecting her lower respiratory structures instead.

Carly's bacterial pneumonia was successfully treated with antibiotics. Pneumonia can also be caused by a viral infection, or, as is often the case in people with AIDS, by *Pneumocystis carinii* infection. Whatever the cause, the events within the infected lung are similar: alveolar linings swell with edema and become abnormally permeable, allowing fluids and white blood cells to accumulate in the air sacs. As the alveoli fill, the surface area available for gas exchange diminishes. Breath-

ing becomes difficult. Untreated, pneumonia can kill.

*Tuberculosis* is a different type of lung infection, caused by the bacterium *Mycobacterium tuberculosis* (fig. 19D). In this condition, fibrous connective tissue develops around the sites of infection, forming structures called *tubercles*. By walling off the bacteria, the tubercles help inhibit their spread. Sometimes this protective mechanism fails, and the bacteria flourish throughout the lungs, and may even spread to other organs. In the later stages of in-

fection, other types of bacteria may cause secondary infections. As lung tissue is destroyed, the surface area for gas exchange decreases. In addition, the widespread fibrous tissue increases the thickness of the respiratory membrane, further restricting gas exchange. A variety of drugs can treat tuberculosis, but in recent years, strains resistant to drugs have arisen, and these can be swiftly deadly.

Another type of condition that impairs gas exchange is *atelectasis.* This is the collapse of a lung, or some part of it, together with the collapse of the blood vessels that supply the affected region. Obstruction of a respiratory tube, such as by an inhaled foreign object or excess mucus secretion, may cause atelectasis. The air in the alveoli beyond the obstruction is absorbed, and as the air pressure in the alveoli decreases, their elastic walls collapse, and they can no longer function. Fortunately, after a portion of a lung

---

The chemical bonds that form between oxygen and hemoglobin molecules are relatively unstable, and as the $P_{O_2}$ decreases, oxygen is released from oxyhemoglobin molecules (fig. 19.35). This happens in tissues, where cells have used oxygen in their respiratory processes. The free oxygen diffuses from the blood into nearby cells, as figure 19.36 shows.

Increasing blood concentration of carbon dioxide ($P_{CO_2}$), acidity, and temperature all increase the amount of oxygen that oxyhemoglobin releases (figs. 19.37, 19.38, and 19.39). These influences explain why more oxygen is released from the blood to the skeletal muscles

Oxyhemoglobin dissociation at 38°C

**FIGURE 19.35**

Hemoglobin is completely saturated at normal systemic arterial $P_{O_2}$, but readily releases oxygen at the $P_{O_2}$ of the body tissues.

collapses, the functional regions that remain are often able to carry on enough gas exchange to sustain the body cells.

*Adult respiratory distress syndrome* (ARDS) is a special form of atelectasis in which alveoli collapse. It has a variety of causes, all of which damage lung tissues. These include pneumonia and other infections, near drowning, aspiration of stomach acid into the respiratory system, or physical trauma to the lungs from an injury or surgical procedure. Anesthetics can suppress surfactant production, causing postsurgical difficulty breathing for 24 to 48 hours, or until surfactant production returns to normal. This damage disrupts the respiratory membrane that separates the air in the alveoli from the blood in the pulmonary capillaries, allowing protein-rich fluid to escape from the capillaries and flood the alveoli. They collapse in response, and surfactant is nonfunctional. Blood vessels and airways narrow, greatly elevating blood pressure in the lungs. Delivery of oxygen to tissues is seriously impaired. ARDS is fatal about 60% of the time. ∎

Healthy lungs　　　　　　　　　　　　　Tuberculosis

**FIGURE 19D**

Healthy lungs appear dark and clear on a radiograph. Lungs with tuberculosis have cloudy areas where fibrous tissue grows to wall off infected areas.

---

during periods of exercise. The increased muscular activity accompanied by increased oxygen use increases the $P_{CO_2}$, decreases the pH, and raises the local temperature. At the same time, less-active cells receive less oxygen.

## Carbon Monoxide

**Carbon monoxide** (CO) is a toxic gas produced in gasoline engines and some stoves as a result of incomplete combustion of fuels. It is also a component of tobacco smoke. Carbon monoxide is toxic because it combines with hemoglobin many times more effectively than does oxygen and therefore does not readily dissociate from hemoglobin. Thus, when a person breathes carbon monoxide, less hemoglobin is available for oxygen transport, and the body cells soon begin to suffer from oxygen deficiency. The effects of carbon monoxide on hemoglobin may be responsible for the lower average birth weights of infants born to women who smoked while pregnant.

Treatment for carbon monoxide poisoning is to administer oxygen in high concentration to replace some of the carbon monoxide bound to hemoglobin molecules. Carbon dioxide is usually administered simultaneously to stimulate the respiratory center which, in turn, increases breathing rate. Rapid breathing helps reduce the concentration of carbon monoxide in the alveoli.

1　How is oxygen transported from the lungs to body cells?

2　What factors affect the release of oxygen from oxyhemoglobin?

3　Why is carbon monoxide toxic?

## Carbon Dioxide Transport

Blood flowing through the capillaries of the body tissues gains carbon dioxide because the tissues have a relatively high $P_{CO_2}$. This carbon dioxide is transported to the lungs in one of three forms: as carbon dioxide dissolved in plasma, as part of a compound formed by bonding to hemoglobin, or as part of a bicarbonate ion (fig. 19.40).

The amount of carbon dioxide that dissolves in plasma is determined by its partial pressure. The higher the $P_{CO_2}$ of the tissues, the more carbon dioxide will go into solution. However, only about 7% of the carbon dioxide is transported in this form.

Unlike oxygen, which combines with the iron atoms of hemoglobin molecules, carbon dioxide bonds with the amino groups (—NH$_2$) of these molecules. Consequently, oxygen and carbon dioxide do not directly compete for binding sites—a hemoglobin molecule can transport both gases at the same time.

Carbon dioxide combining with hemoglobin forms a loosely bound compound called **carbaminohemoglobin.**

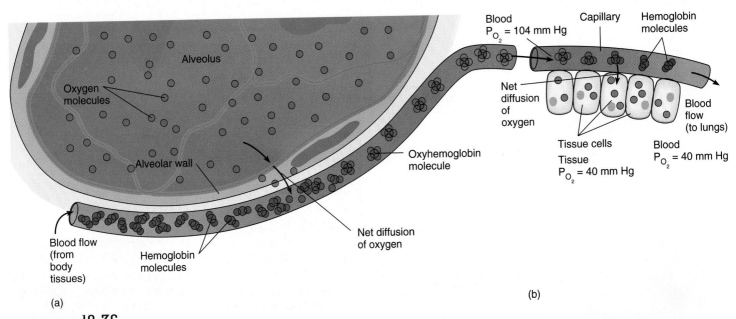

(a)

(b)

**FIGURE** 19.36

(a) Oxygen molecules, entering the blood from the alveolus, bond to hemoglobin, forming oxyhemoglobin. (b) In the regions of the body cells, oxyhemoglobin releases oxygen. Note that some oxygen is still bound to hemoglobin at the $P_{O_2}$ of systemic venous blood (see fig. 19.35).

Oxyhemoglobin dissociation at 38°C

**FIGURE** 19.37

The amount of oxygen released from oxyhemoglobin increases as the $P_{CO_2}$ increases.

Oxyhemoglobin dissociation at 38°C

**FIGURE** 19.38

The amount of oxygen released from oxyhemoglobin increases as the blood pH decreases.

This molecule readily decomposes in regions where the $P_{CO_2}$ is low, releasing its carbon dioxide. Although this method of transporting carbon dioxide is theoretically quite effective, carbaminohemoglobin forms relatively slowly. Only about 15%–25% of the total carbon dioxide is carried this way.

The most important carbon dioxide transport mechanism involves the formation of **bicarbonate ions** ($HCO_3^-$). Recall that carbon dioxide reacts with water to form carbonic acid ($H_2CO_3$). This reaction occurs slowly in the blood plasma, but much of the carbon dioxide diffuses into the red blood cells. These cells contain an en-zyme, **carbonic anhydrase**, which speeds the reaction between carbon dioxide and water.

The resulting carbonic acid dissociates almost immediately, releasing hydrogen ions ($H^+$) and bicarbonate ions ($HCO_3^-$):

$$H_2CO_3 \rightarrow H^+ + HCO_3^-$$

These new hydrogen ions might be expected to lower blood pH, but this reaction occurs in the systemic capillaries, where deoxyhemoglobin is generated. Deoxyhemoglobin is an excellent buffer because hydrogen ions combine readily with it. The bicarbonate ions diffuse out of the red blood cells and enter the blood plasma. At least 70% of the carbon dioxide transported in the blood is carried in this form.

As the bicarbonate ions leave the red blood cells and enter the plasma, *chloride ions*, which also have negative charges, are repelled electrically, and they move from the plasma into the red blood cells. This exchange in position of the two negatively charged ions, shown in figure 19.41, maintains the ionic balance between the red blood cells and the plasma. It is termed the **chloride shift.**

As blood passes through the capillaries of the lungs, the dissolved carbon dioxide diffuses into the alveoli, in response to the relatively low $P_{CO_2}$ of the alveolar air. As the plasma $P_{CO_2}$ drops, hydrogen ions and bicarbonate ions in the red blood cells recombine to form carbonic acid, and under the influence of carbonic anhydrase, the carbonic acid quickly yields new carbon dioxide and water:

$$H^+ + HCO_3^- \rightarrow H_2CO_3 \rightarrow CO_2 + H_2O$$

Oxyhemoglobin dissociation at various temperatures

**FIGURE** 19.39

The amount of oxygen released from oxyhemoglobin increases as the blood temperature increases.

**FIGURE** 19.40

Carbon dioxide produced by tissue cells is transported in the blood in a dissolved state, combined with hemoglobin or in the form of bicarbonate ions ($HCO_3^-$).

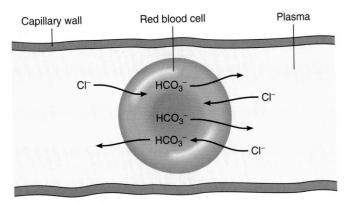

**FIGURE 19.41**

As bicarbonate ions ($HCO_3^-$) diffuse out of the red blood cell, chloride ions ($Cl^-$) from the plasma diffuse into the cell, thus maintaining the electrical balance between ions. This exchange of ions is called the chloride shift.

Carbaminohemoglobin also releases its carbon dioxide, and both of these events contribute to the $P_{CO_2}$ of the alveolar capillary blood. Carbon dioxide diffuses out of the blood until an equilibrium is established between the $P_{CO_2}$ of the blood and the $P_{CO_2}$ of the alveolar air. Figure 19.42 summarizes this process, and table 19.7 summarizes the transport of blood gases.

**1** Describe three ways carbon dioxide can be transported from body cells to the lungs.

**2** How can hemoglobin carry oxygen and carbon dioxide at the same time?

**3** How do bicarbonate ions help buffer the blood (maintain its pH)?

**4** What is the chloride shift?

**5** How is carbon dioxide released from the blood into the lungs?

**FIGURE 19.42**

In the lungs, carbon dioxide diffuses from the blood into the alveoli.

| | GASES TRANSPORTED IN BLOOD | |
|---|---|---|
| **Gas** | **Reaction** | **Substance Transported** |
| Oxygen | Combines with iron atoms of hemoglobin molecules | Oxyhemoglobin |
| Carbon dioxide | About 7% dissolves in plasma | Carbon dioxide |
| | About 23% combines with the amino groups of hemoglobin molecules | Carbaminohemoglobin |
| | About 70% reacts with water to form carbonic acid; the carbonic acid then dissociates to release hydrogen ions and bicarbonate ions | Bicarbonate ions |

**TABLE 19.7**

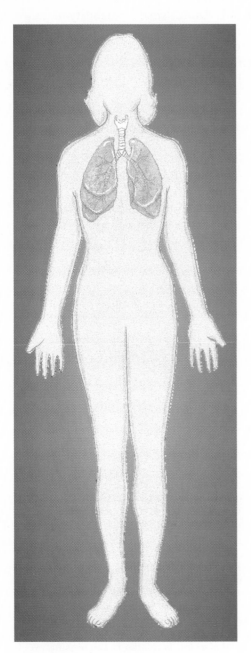

## RESPIRATORY SYSTEM

The respiratory system provides oxygen for the internal environment and excretes carbon dioxide.

## Integumentary System

Stimulation of skin receptors may alter respiratory rate.

## Skeletal System

Bones provide attachments for muscles involved in breathing.

## Muscular System

The respiratory system eliminates carbon dioxide produced by exercising muscles.

## Nervous System

The brain controls the respiratory system. The respiratory system helps control pH of the internal environment.

## Endocrine System

Hormonelike substances control the production of red blood cells that transport oxygen and carbon dioxide.

## Cardiovascular System

As the heart pumps blood through the lungs, the lungs oxygenate the blood and excrete carbon dioxide.

## Lymphatic System

Cells of the immune system patrol the lungs and defend against infection.

## Digestive System

The digestive system and respiratory system share openings to the outside.

## Urinary System

The kidneys and the respiratory system work together to maintain blood pH. The kidneys compensate for water lost through breathing.

## Reproductive System

Respiration increases during sexual activity. Fetal "respiration" begins before birth.

## CLINICAL TERMS RELATED TO THE RESPIRATORY SYSTEM

**anoxia** (ah-nok'se-ah)   Absence or a deficiency of oxygen within tissues.

**asphyxia** (as-fik'se-ah)   Deficiency of oxygen and excess carbon dioxide in the blood and tissues.

**atelectasis** (at″e-lek'tah-sis)   Collapse of a lung or some portion of it.

**bradypnea** (brad″e-ne'ah)   Abnormally slow breathing.

**bronchitis** (brong-ki'tis)   Inflammation of the bronchial lining.

**Cheyne-Stokes respiration**  (chān stōks res″pi-ra'shun)   Irregular breathing pattern of a series of shallow breaths that increase in depth and rate, followed by breaths that decrease in depth and rate.

**dyspnea** (disp'ne-ah)   Difficulty in breathing.

**eupnea** (up-ne'ah)   Normal breathing.

**hemothorax** (he″mo-tho'raks)   Blood in the pleural cavity.

**hypercapnia** (hi″per-kap'ne-ah)   Excess carbon dioxide in the blood.

**hyperoxia** (hi″per-ok'se-ah)   Excess oxygenation of the blood.

**hyperpnea** (hi″perp-ne'ah)   Increase in the depth and rate of breathing.

**hyperventilation** (hi″per-ven″ti-la'shun)   Prolonged, rapid, and deep breathing.

**hypoxemia** (hi″pok-se'me-ah)   Deficiency in the oxygenation of the blood.

**hypoxia** (hi-pok'se-ah)   Diminished availability of oxygen in the tissues.

**lobar pneumonia** (lo'ber nu-mo'ne-ah)   Pneumonia that affects an entire lobe of a lung.

**pleurisy** (ploo'rĭ-se)   Inflammation of the pleural membranes.

**pneumoconiosis** (nu″mo-ko″ne-o'sis)   Accumulation of particles from the environment in the lungs and the reaction of the tissues.

**pneumothorax** (nu″mo-tho'raks)   Entrance of air into the space between the pleural membranes, followed by collapse of the lung.

**rhinitis** (ri-ni'tis)   Inflammation of the nasal cavity lining.

**sinusitis** (si″nu-si'tis)   Inflammation of the sinus cavity lining.

**tachypnea** (tak″ip-ne'ah)   Rapid, shallow breathing.

**tracheotomy** (tra″ke-ot´o-me)   Incision in the trachea for exploration or the removal of a foreign object.

---

## CHAPTER SUMMARY

### Introduction  (page 739)

The respiratory system includes the passages that transport air to and from the lungs, and the air sacs in which gas exchanges occur. Respiration is the entire process by which gases are exchanged between the atmosphere and the body cells. Respiration is necessary because of cellular respiration. Cells require oxygen to extract maximal energy from nutrient molecules, and to rid themselves of carbon dioxide, a metabolic waste.

### Organs of the Respiratory System (page 740)

The respiratory system includes the nose, nasal cavity, sinuses, pharynx, larynx, trachea, bronchial tree, and lungs. The upper respiratory tract includes the nose, nasal cavity, sinuses, and pharynx; the lower respiratory tract includes the larynx, trachea, bronchial tree, and lungs.

1.  Nose
    a. Bone and cartilage support the nose.
    b. Nostrils provide entrances for air.
2.  Nasal cavity
    a. The nasal cavity is a space posterior to the nose.
    b. The nasal septum divides it medially.
    c. Nasal conchae divide the cavity into passageways and help increase the surface area of the mucous membrane.
    d. Mucous membrane filters, warms, and moistens incoming air.
    e. Particles trapped in the mucus are carried to the pharynx by ciliary action and are swallowed.

3.  Sinuses
    a. Sinuses are spaces in the bones of the skull that open into the nasal cavity.
    b. They are lined with mucous membrane that is continuous with the lining of the nasal cavity.
4.  Pharynx
    a. The pharynx is located posterior to the mouth and between the nasal cavity and the larynx.
    b. It provides a common passage for air and food.
    c. It aids in creating vocal sounds.
5.  Larynx
    a. The larynx is an enlargement at the top of the trachea.
    b. It is a passageway for air and helps prevent foreign objects from entering the trachea.
    c. It is composed of muscles and cartilages; some of these cartilages are single whereas others are paired.
    d. It contains the vocal cords, which produce sounds by vibrating as air passes over them.
       (1) The pitch of a sound is related to the tension on the cords.
       (2) The intensity of a sound is related to the force of the air passing over the cords.
    e. The glottis and epiglottis help prevent food and liquid from entering the trachea.
6.  Trachea
    a. The trachea extends into the thoracic cavity in front of the esophagus.
    b. It divides into the right and left bronchi.
    c. The mucous lining continues to filter incoming air.
    d. Incomplete cartilaginous rings support the wall.

7. Bronchial tree
   a. The bronchial tree consists of branched air passages that connect the trachea to the air sacs.
   b. The branches of the bronchial tree include primary bronchi, lobar bronchi, segmental bronchi, intralobular bronchioles, terminal bronchioles, respiratory bronchioles, alveolar ducts, alveolar sacs, and alveoli.
   c. Structure of the respiratory tubes
      (1) As tubes branch, the amount of cartilage in the walls decreases, and the muscular layer becomes more prominent.
      (2) Elastic fibers in the walls aid breathing.
      (3) The epithelial lining changes from pseudostratified and ciliated to cuboidal and simple squamous as the tubes become progressively smaller.
   d. Functions of the respiratory tubes include distribution of air and exchange of gases between the alveolar air and the blood.
8. Lungs
   a. The left and right lungs are separated by the mediastinum and are enclosed by the diaphragm and the thoracic cage.
   b. The visceral pleura is attached to the surface of the lungs; parietal pleura lines the thoracic cavity.
   c. The right lung has three lobes, and the left lung has two.
   d. Each lobe is composed of lobules that contain alveolar ducts, alveolar sacs, alveoli, nerves, blood vessels, lymphatic vessels, and connective tissues.

# Breathing Mechanism (page 751)

Inspiration and expiration movements are accompanied by changes in the size of the thoracic cavity.

1. Inspiration
   a. Atmospheric pressure forces air into the lungs.
   b. Inspiration occurs when the intra-alveolar pressure is reduced.
   c. The intra-alveolar pressure is reduced when the diaphragm moves downward, and the thoracic cage moves upward and outward.
   d. Surface tension holding the pleural membranes together aids lung expansion.
   e. Surfactant reduces surface tension within the alveoli.
2. Expiration
   a. The forces of expiration come from the elastic recoil of tissues and from surface tension within the alveoli.
   b. Expiration can be aided by thoracic and abdominal wall muscles that pull the thoracic cage downward and inward, and compress the abdominal organs inward and upward.
3. Respiratory volumes and capacities
   a. One inspiration followed by one expiration is called a respiratory cycle.
   b. The amount of air that moves in and out during a respiratory cycle is the tidal volume.
   c. Additional air that can be inhaled is the inspiratory reserve volume; additional air that can be exhaled is the expiratory reserve volume.
   d. Residual air remains in the lungs and is mixed with newly inhaled air.
   e. The inspiratory capacity is the maximum volume of air a person can inhale following exhalation of the tidal volume.

f. The functional residual capacity is the volume of air that remains in the lungs following the exhalation of the tidal volume.
   g. The vital capacity is the maximum amount of air a person can exhale after taking the deepest breath possible.
   h. The total lung capacity is equal to the vital capacity plus the residual air volume.
   i. Air in the anatomic and alveolar dead spaces is not available for gas exchange.
4. Alveolar ventilation
   a. Minute ventilation is calculated by multiplying the tidal volume by the breathing rate.
   b. Alveolar ventilation rate is calculated by subtracting the physiologic dead space from the tidal volume and multiplying the result by the breathing rate.
   c. The alveolar ventilation rate is a major factor affecting the exchange of gases between the alveolar air and the blood.
5. Nonrespiratory air movements
   a. Nonrespiratory air movements are air movements other than breathing.
   b. They include coughing, sneezing, laughing, crying, hiccuping, and yawning.

# Control of Breathing (page 760)

Normal breathing is rhythmic and involuntary, although the respiratory muscles can be controlled voluntarily.

1. Respiratory center
   a. The respiratory center is located in the brain stem, and includes parts of the medulla oblongata and pons.
   b. The medullary rhythmicity area includes two groups of neurons.
      (1) The dorsal respiratory group is responsible for the basic rhythm of breathing.
      (2) The ventral respiratory group increases inspiratory and expiratory movements during forceful breathing.
   c. The pneumotaxic area regulates the rate of breathing.
2. Factors affecting breathing
   a. Chemicals, lung tissue stretching, and emotional state affect breathing.
   b. Chemosensitive areas (central chemoreceptors) are associated with the respiratory center.
      (1) Carbon dioxide combines with water to form carbonic acid, which, in turn, releases hydrogen ions in the CSF.
      (2) Stimulation of these areas increases alveolar ventilation.
   c. Peripheral chemoreceptors are in the carotid bodies and aortic bodies of certain arteries.
      (1) These chemoreceptors sense low oxygen concentration.
      (2) When oxygen concentration is low, alveolar ventilation increases.
   d. Stretching the lung tissues triggers an inflation reflex.
      (1) This reflex reduces the duration of inspiratory movements.
      (2) This prevents overinflation of the lungs during forceful breathing.
   e. Hyperventilation decreases carbon dioxide concentration, but *this is very dangerous when associated with breath holding during underwater swimming.*

## Alveolar Gas Exchanges (page 764)

Gas exchanges between the air and the blood occur within the alveoli.

1. Alveoli
   a. The alveoli are tiny sacs clustered at the distal ends of the alveolar ducts.
   b. Some alveoli open into adjacent air sacs that provide alternate pathways for air when passages are obstructed.
2. Respiratory membrane
   a. The respiratory membrane consists of the alveolar and capillary walls.
   b. Gas exchanges take place through these walls.
3. Diffusion through the respiratory membrane
   a. The partial pressure of a gas is determined by the concentration of that gas in a mixture of gases or the concentration of gas dissolved in a liquid.
   b. Gases diffuse from regions of higher partial pressure toward regions of lower partial pressure.
   c. Oxygen diffuses from the alveolar air into the blood; carbon dioxide diffuses from the blood into the alveolar air.

## Gas Transport (page 767)

Blood transports gases between the lungs and the body cells.

1. Oxygen transport
   a. Oxygen is mainly transported in combination with hemoglobin molecules.
   b. The resulting oxyhemoglobin is relatively unstable and releases its oxygen in regions where the $P_{O_2}$ is low.
   c. More oxygen is released as the blood concentration of carbon dioxide increases, as the blood becomes more acidic, and as the blood temperature increases.
2. Carbon monoxide
   a. Carbon monoxide forms as a result of incomplete combustion of fuels.
   b. It combines with hemoglobin more readily than oxygen and forms a stable compound.
   c. Carbon monoxide is toxic because the hemoglobin with which it combines is no longer available for oxygen transport.
3. Carbon dioxide transport
   a. Carbon dioxide may be carried in solution, either as dissolved $CO_2$, $CO_2$ bound to hemoglobin, or as a bicarbonate ion.
   b. Most carbon dioxide is transported in the form of bicarbonate ions.
   c. Carbonic anhydrase, an enzyme, speeds the reaction between carbon dioxide and water to form carbonic acid.
   d. Carbonic acid dissociates to release hydrogen ions and bicarbonate ions.

## CRITICAL THINKING QUESTIONS

1. If the upper respiratory passages are bypassed with a tracheostomy, how might the air entering the trachea be different from air normally passing through this canal? What problems might this cause for the patient?
2. Certain respiratory disorders, such as emphysema, reduce the capacity of the lungs to recoil elastically. Which respiratory air volumes will this condition affect? Why?
3. What changes would you expect to occur in the relative concentrations of blood oxygen and carbon dioxide in a patient who breathes rapidly and deeply for a prolonged time? Why?
4. If a person has stopped breathing and is receiving pulmonary resuscitation, would it be better to administer pure oxygen or a mixture of oxygen and carbon dioxide? Why?
5. The air pressure within the passenger compartment of a commercial aircraft may be equivalent to an altitude of 8,000 feet. What problem might this create for a person with a serious respiratory disorder?
6. Patients experiencing asthma attacks are often advised to breathe through pursed (puckered) lips. How might this help reduce the symptoms of the asthma?

## REVIEW EXERCISES

1. Describe the general functions of the respiratory system.
2. Distinguish between the upper and lower respiratory tracts.
3. Explain how the nose and nasal cavity filter incoming air.
4. Name and describe the locations of the major sinuses, and explain how a sinus headache may occur.
5. Distinguish between the pharynx and the larynx.
6. Name and describe the locations and functions of the cartilages of the larynx.
7. Distinguish between the false vocal cords and the true vocal cords.
8. Compare the structure of the trachea with the structure of the branches of the bronchial tree.
9. List the successive branches of the bronchial tree, from the primary bronchi to the alveoli.
10. Describe how the structure of the respiratory tube changes as the branches become finer.
11. Explain the functions of the respiratory tubes.
12. Distinguish between visceral pleura and parietal pleura.
13. Name and describe the locations of the lobes of the lungs.
14. Explain how normal inspiration and forced inspiration are accomplished.
15. Define *surface tension*, and explain how it aids the breathing mechanism.
16. Define *surfactant*, and explain its function.
17. Define *compliance*.

18. Explain how normal expiration and forced expiration are accomplished.
19. Distinguish between vital capacity and total lung capacity.
20. Distinguish among anatomic, alveolar, and physiologic dead spaces.
21. Distinguish between minute respiratory volume and alveolar ventilation rate.
22. Compare the mechanisms of coughing and sneezing, and explain the function of each.
23. Explain the function of yawning.
24. Describe the location of the respiratory center, and name its major components.
25. Describe how the basic rhythm of breathing is controlled.
26. Explain the function of the pneumotaxic area of the respiratory center.
27. Explain why increasing blood concentrations of carbon dioxide and hydrogen ions have similar effects on the respiratory center.

28. Describe the function of the peripheral chemoreceptors in the carotid and aortic bodies of certain arteries.
29. Describe the inflation reflex.
30. Discuss the effects of emotions on breathing.
31. Define *hyperventilation*, and explain how it affects the respiratory center.
32. Define *respiratory membrane*, and explain its function.
33. Explain the relationship between the partial pressure of a gas and its rate of diffusion.
34. Summarize the gas exchanges that occur through the respiratory membrane.
35. Describe how oxygen is transported in blood.
36. List three factors that increase release of oxygen from the blood.
37. Explain why carbon monoxide is toxic.
38. List three ways that carbon dioxide is transported in blood.
39. Explain the function of carbonic anhydrase.
40. Define *chloride shift*.

 **For More Information Related to This Chapter, Please Visit** *Hole's Human Anatomy and Physiology* **Home Page**
http://www.mhhe.com/biosci/abio/

# URINARY SYSTEM

## UNDERSTANDING WORDS

**af-,** to: *af*ferent arteriole—arteriole that leads to a nephron.

**calyc-,** small cup: major *calyc*es—cuplike subdivisions of the renal pelvis.

**cort-,** covering: renal *cort*ex—shell of tissue surrounding the inner region of a kidney.

**cyst-,** bladder: *cyst*itis—inflammation of the bladder.

**detrus-,** to force away: *detrus*or muscle—muscle within the bladder wall that causes urine to be expelled.

**glom-,** little ball: *glom*erulus—cluster of capillaries within a renal corpuscle.

**juxta-,** near to: *juxta*medullary nephron—nephron located near the renal medulla.

**mict-,** to pass urine: *mict*urition—process of expelling urine from the bladder.

**nephr-,** pertaining to the kidney: *nephr*on—functional unit of a kidney.

**papill-,** nipple: renal *papill*ae—small elevations that project into a renal calyx.

**prox-,** nearest: *prox*imal tubule—coiled portion of the renal tubule leading from the glomerular capsule.

**ren-,** kidney: *ren*al cortex—outer region of a kidney.

**trigon-,** triangular shape: *trigon*e—triangular area on the internal floor of the bladder.

## CHAPTER OBJECTIVES

### AFTER YOU HAVE STUDIED THIS CHAPTER, YOU SHOULD BE ABLE TO:

1. Name the organs of the urinary system and list their general functions.

2. Describe the locations of the kidneys and the structure of a kidney.

3. List the functions of the kidneys.

4. Trace the pathway of blood through the major vessels within a kidney.

5. Describe a nephron and explain the functions of its major parts.

6. Explain how glomerular filtrate is produced and describe its composition.

7. Explain how various factors affect the rate of glomerular filtration and how this rate is regulated.

8. Discuss the role of tubular reabsorption in urine formation.

9. Explain why the osmotic concentration of the glomerular filtrate changes as it passes through a renal tubule.

10. Describe a countercurrent mechanism and explain how it helps concentrate urine.

11. Define tubular secretion and explain its role in urine formation.

12. Describe the structure of the ureters, urinary bladder, and urethra.

13. Discuss the process of micturition and explain how it is controlled.

## KEY TERMS

**afferent arteriole** (af′er-ent ar-te′re-ōl)

**autoregulation** (aw″to-reg″u-la′shun)

**countercurrent mechanism** (kown′ter-kur′ent mek′ah-nizm)

**detrusor muscle** (de-truz′or mus′l)

**efferent arteriole** (ef′er-ent ar-te′re-ōl)

**glomerular filtration** (glo-mer′u-lar fil-tra′shun)

**glomerulus** (glo-mer′u-lus)

**juxtaglomerular apparatus** (juks″tah-glo-mer′u-lar ap″ah-ra′tus)

**micturition** (mik″tu-rish′un)

**nephron** (nef′ron)

**peritubular capillary** (per″i-tū′bu-lar kap′ i-ler″e)

**renal corpuscle** (re′nal kor′pusl)

**renal cortex** (re′nal kor′teks)

**renal medulla** (re′nal mĕ-dul′ah)

**renal plasma threshold** (re′nal plaz′mah thresh′old)

**renal tubule** (re′nal tu′būl)

**retroperitoneal** (re″tro-per″ĭ-to-ne′al)

**tubular reabsorption** (tu′bu-lar re-ab-sorp′shun)

**tubular secretion** (tu′bu-lar se-kre′shun)

elicia had looked forward to summer camp all year, especially the overnight hikes. A three-day expedition in July was wonderful, but five days after returning to camp, Felicia developed severe abdominal cramps. So did seventeen other campers and two counselors, some of whom had bloody diarrhea, too. Several of the stricken campers were hospitalized, Felicia among them. Although the others improved in a few days and were released, Felicia suffered from a complication, called hemolytic uremic syndrome (HUS). Her urine had turned bloody, and she also had blood abnormalities—severe anemia and lack of platelets.

Camp personnel reported the outbreak to public health officials, who quickly recognized the signs of food poisoning and traced the illness to hamburgers cooked outdoors on the trip. The burgers were served rare, the red meat not hot enough to kill a strain of *Escherichia coli* bacteria that releases a poison called shigatoxin.

Most people who eat meat tainted with *E. coli* toxin become ill, but the damage usually is restricted to the digestive tract, producing cramps and diarrhea for several days. In about 6% of affected people, mostly children, HUS develops because the bloodstream transports the toxin to the kidneys. Here, the toxin destroys cells of the microscopic tubules that normally filter proteins and blood cells from forming urine. With the tubule linings compromised, proteins and blood cells, as well as damaged kidney cells, appear in the urine.

HUS is a leading cause of acute renal (kidney) failure, killing 3–5% of affected children. Felicia was in the lucky majority. Blood clotted around the sites of her damaged kidney cells, and over a few weeks, new cells formed. Three weeks after the bloody urine began, her urine was once again clear, and she was healthy.

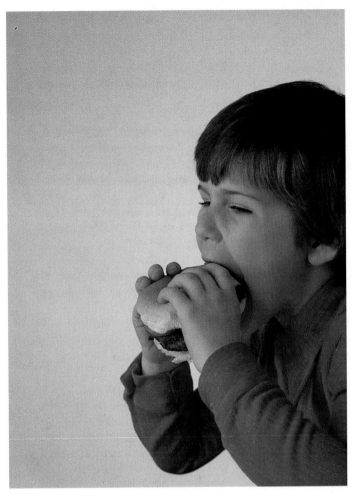

Cook that burger! Hemolytic uremic syndrome is a complication of infection by a strain of E. coli bacteria that produces shigatoxin. Destruction of the proximal portions of kidney tubules allows proteins and blood cells to enter urine.

---

The urinary system consists of a pair of glandular kidneys, which remove substances from the blood, form urine, and help regulate certain metabolic processes; a pair of tubular ureters, which transport urine from the kidneys; a saclike urinary bladder, which collects urine from the ureters and serves as a urine reservoir; and a tubular urethra, which conveys urine to the outside of the body. Figures 20.1 and 20.2 show these organs.

## KIDNEYS

A **kidney** is a reddish brown, bean-shaped organ with a smooth surface. It is about 12 centimeters long, 6 centimeters wide, and 3 centimeters thick in an adult, and it is enclosed in a tough, fibrous capsule (tunic fibrosa).

## Location of the Kidneys

The kidneys lie on either side of the vertebral column in a depression high on the posterior wall of the abdominal cavity. Although the positions of the kidneys may vary slightly with changes in posture and with breathing movements, their upper and lower borders are generally at the levels of the twelfth thoracic and third lumbar vertebrae, respectively. The left kidney is usually about 1.5 to 2 centimeters higher than the right one.

The kidneys are positioned *retroperitoneally,* which means they are behind the parietal peritoneum and against the deep muscles of the back. Connective tissue (renal fascia) and masses of adipose tissue (renal fat) surrounding the kidneys hold them in place (fig. 20.3 and reference plates 58, 59).

## Kidney Structure

The lateral surface of each kidney is convex, but its medial side is deeply concave. The resulting medial depression leads into a hollow chamber called the **renal sinus.** Through the entrance to this sinus, termed the *hilum,* pass blood vessels, nerves, lymphatic vessels, and the ureter (see fig. 20.1).

The superior end of the ureter expands to form a funnel-shaped sac called the **renal pelvis,** which is located inside the renal sinus. The pelvis is subdivided into two or three tubes, called *major calyces* (sing., *calyx*), and they, in turn, are subdivided into eight to fourteen *minor calyces* (fig. 20.4). A series of small projections called *renal papillae* project into each minor calyx.

The kidney includes two distinct regions: an inner medulla and an outer cortex. The **renal medulla** is composed of conical masses of tissue called *renal pyramids,* the bases of which are directed toward the convex surface of the kidney, and the apexes of which form the renal papillae. The tissue of the medulla appears striated because it consists of microscopic tubules that lead from the cortex to the renal papillae.

The **renal cortex,** which appears somewhat granular, forms a shell around the medulla. Its tissue dips into the medulla between the renal pyramids, forming *renal columns.* The granular appearance of the cortex is due to the random distribution of tiny tubules associated with the **nephrons,** which are the functional units of the kidney.

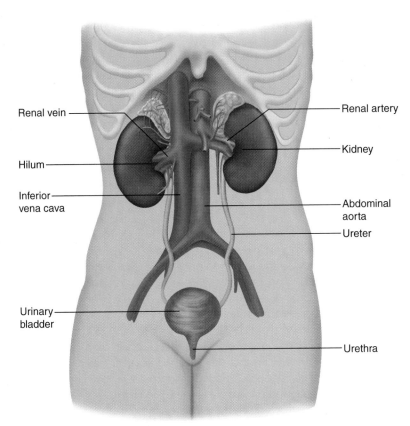

**FIGURE 20.1** ✗

The urinary system includes the kidneys, ureters, urinary bladder, and urethra. Notice the relationship of these structures to the major blood vessels.

> Two common inherited kidney abnormalities are *poly-cystic kidney disease,* which affects adults, and *Wilms tumor,* which affects young children. In polycystic kidney disease, cysts present in the kidneys since childhood or adolescence begin to produce symptoms in the thirties, including abdominal pain, bloody urine, and elevated blood pressure. The condition can lead to renal failure. In Wilms tumor, pockets of cells within a child's kidney remain as they were in the embryo—they are unspecialized and reproduce rapidly, forming a cancerous tumor. Loss of a tumor suppressor gene causes Wilms tumor.

## Functions of the Kidneys

The main function of the kidneys is to regulate the volume, composition, and pH of body fluids. In the process, the kidneys remove metabolic wastes from the blood and excrete them to the outside. These wastes include nitrogenous and sulfur-containing products of protein metabolism. The kidneys also help control the rate of red blood cell formation by secreting the hormone *erythro-*

**FIGURE 20.2**

Structures of the urinary system are visible in this false-color radiograph.

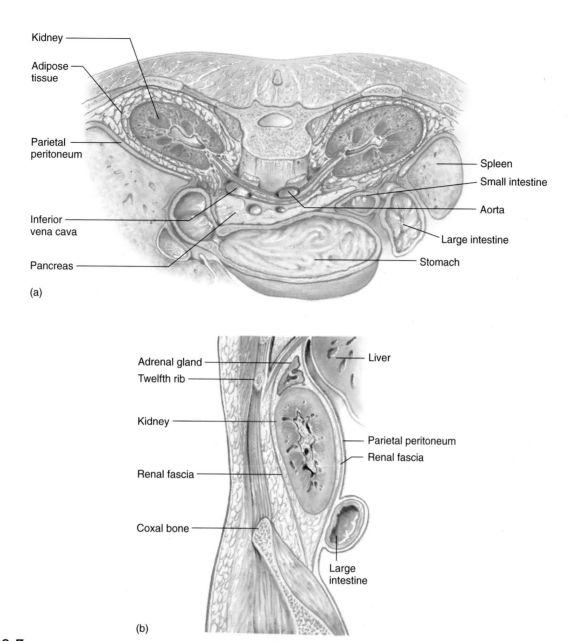

**FIGURE 20.3**

(a) Transverse sections of the kidneys, which are located behind the parietal peritoneum. Adipose and other connective tissues surround and support the kidneys. (b) Sagittal section through the posterior abdominal cavity showing the kidney.

*poietin* (see chapter 14), regulate blood pressure by secreting the enzyme *renin* (see chapter 15), and regulate absorption of calcium ions by activating *vitamin D* (see chapter 13).

Medical technology can take over the role of a kidney. In *hemodialysis,* a person's blood is rerouted across an artificial membrane that "cleanses" it, removing substances that would normally be excreted in the urine. A patient usually must use this artificial kidney three times a week, for several hours each time. Clinical Application 20.1 further discusses hemodialysis.

About two in every ten patients with renal failure can use a procedure that can be done at home called *continuous ambulatory peritoneal dialysis* instead of hemodialysis. The patient infuses a solution into the abdominal cavity through a permanently implanted tube. The solution stays in for 4 to 8 hours, while it takes up substances that would normally be excreted into urine. Then the patient drains the waste-laden solution out of the tube, replacing it with clean fluid. Infection is a risk associated with this procedure.

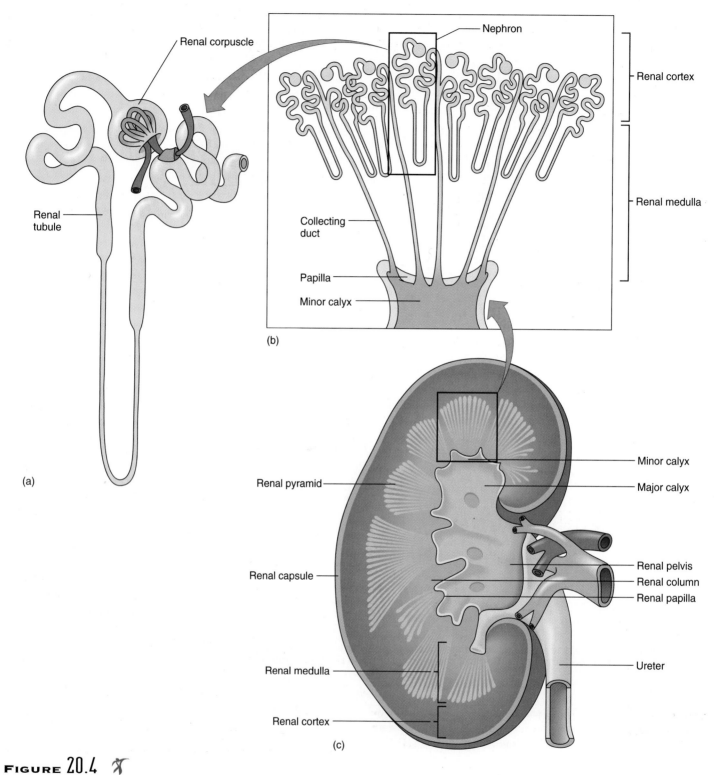

**FIGURE 20.4**

(a) A single nephron. (b) A renal pyramid containing nephrons. (c) Longitudinal section of a kidney.

1 Where are the kidneys located?

2 Describe the structure of a kidney.

3 Name the functional unit of the kidney.

4 What are the general functions of the kidneys?

## Renal Blood Vessels

The **renal arteries,** which arise from the abdominal aorta, supply blood to the kidneys (fig. 20.5). These arteries transport a large volume of blood. When a person is at rest, the renal arteries usually carry from 15% to 30% of

# CLINICAL APPLICATION

# Chronic Kidney Failure

Charles B., a 43-year-old muscular construction worker, had been feeling unusually tired for several weeks, with occasional dizziness and difficulty sleeping. More recently he had noticed a burning pain in his lower back, just below his rib cage, and his urine had darkened. In addition, his feet, ankles, and face seemed swollen. His wife suggested that he consult their family physician about these symptoms.

The physician found that Charles had elevated blood pressure (hypertension) and that the regions of his kidneys were sensitive to pressure. A urinalysis revealed excess protein (proteinuria) and blood (hematuria). Blood tests indicated elevated blood urea nitrogen (BUN), elevated serum creatinine, and decreased serum protein (hypoproteinemia) concentrations.

The physician told Charles that he probably had *chronic glomerulonephritis,* an inflammation of the capillaries within the glomeruli of the renal nephrons, and that this was an untreatable progressive degenerative disease. Microscopic examination of a small sample of kidney tissue (biopsy) later confirmed the diagnosis.

In spite of medical treatment and careful attention to his diet, Charles's condition deteriorated rapidly. When it appeared that most of his kidney function had been lost (end-stage renal disease, or ESRD), he was offered artificial kidney treatments (hemodialysis).

To prepare Charles for hemodialysis, a vascular surgeon created a fistula in his left forearm by surgically connecting an artery to a vein. The greater pressure of the blood in the artery that now flowed directly into the vein swelled the vein, making it more accessible.

During hemodialysis treatment, a hollow needle was inserted into the vein of the fistula near its arterial connection. This allowed the blood to flow, with the aid of a blood pump, through a tube leading to the blood compartment of a dialysis machine. Within this compartment, the blood passed over a selectively permeable membrane. On the opposite side of the membrane was a dialysate solution with a controlled composition. Negative pressure on the dialysate side of the membrane, created by a vacuum pump, increased the movement of fluid through the membrane. At the same time, waste and excess electrolytes diffused from the blood through the membrane and entered the dialysate solution. The blood was then returned through a tube to the vein of the fistula.

In order to maintain favorable blood concentrations of waste, electrolytes, and water, Charles had to undergo hemodialysis three times per week, with each treatment lasting three to four hours. During the treatments, he was given an anticoagulant to prevent blood clotting, an antibiotic drug to control infections, and an antihypertensive drug to reduce his blood pressure.

Charles was advised to carefully control his intake of water, sodium, potassium, proteins, and total calories between treatments. He was also asked to consider another option for the treatment of ESRD—a kidney transplant—which could free him from the time-consuming dependence on hemodialysis.

In this surgical procedure, a kidney from a living donor or a cadaver, whose tissues are antigenically similar (histocompatible) to those of the recipient, is placed in the depression on the medial surface of the right or left ilium (iliac fossa). The renal artery and vein of the donor kidney are connected to the recipient's iliac artery and vein, respectively, and the kidney's ureter is attached to the dome of the recipient's urinary bladder. The patient must then remain on immunosuppressant drugs to prevent rejection of the transplant. ∎

---

the total cardiac output into the kidneys although the kidneys account for only 1% of body weight.

A renal artery enters a kidney through the hilum and gives off several branches, called the *interlobar arteries,* which pass between the renal pyramids. At the junction between the medulla and the cortex, the interlobar arteries branch to form a series of incomplete arches, the *arcuate arteries* (arciform arteries), which, in turn, give rise to *interlobular arteries.* The final branches of the interlobular arteries, called **afferent arterioles,** lead to the nephrons, the functional units of the kidneys.

Venous blood is returned through a series of vessels that generally correspond to the arterial pathways. For example, the venous blood passes through interlobular, arcuate, interlobar, and renal veins. The **renal vein** then joins the inferior vena cava as it courses through the abdominal cavity. (Figs. 20.6 and 20.7 show branches of the renal arteries and veins.)

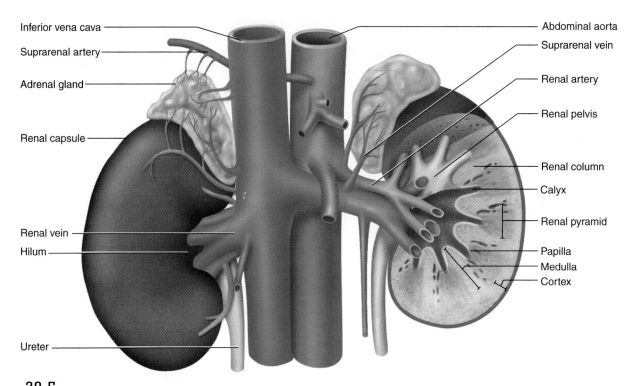

**FIGURE 20.5**

Blood vessels associated with the kidneys and adrenal glands. Note their relationship with the renal pelvis and ureters.

(a)

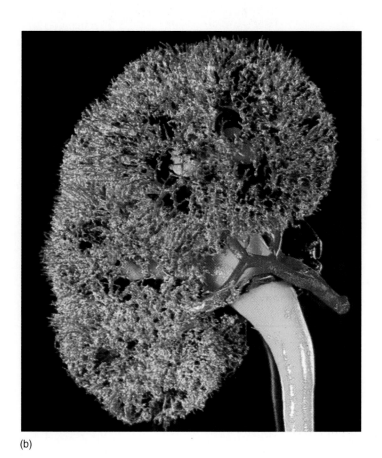

(b)

**FIGURE 20.6**

(a) Main branches of the renal artery and vein. (b) Corrosion cast of the renal arterial system. Not all blood vessels associated with the nephron are shown.

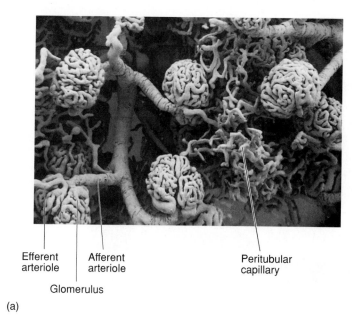

Efferent arteriole
Afferent arteriole
Glomerulus
Peritubular capillary

(a)

Renal tubules
Glomerulus
Glomerular capsule

(b)

**FIGURE 20.7**

(a) A scanning electron micrograph of a cast of the renal blood vessels associated with the glomeruli (200×). (b) Scanning electron micrograph of a glomerular capsule surrounding a glomerulus. (480×)

*Tissues and Organs: A Text-Atlas of Scanning Electron Microscopy,* by R. G.Kessel and R. H. Kardon. © 1979 W. H. Freeman and Company.

## Nephrons

### Structure of a Nephron

A kidney contains about one million nephrons. Each nephron consists of a **renal corpuscle** and a **renal tubule** (see fig. 20.4).

A renal corpuscle consists of a filtering unit composed of a tangled cluster of blood capillaries called a **glomerulus** and a surrounding thin-walled, saclike structure called a **glomerular** (Bowman's) **capsule.** Afferent arterioles give rise to these capillaries, which lead to **efferent arterioles** (fig. 20.7). Filtration of fluid from the glomerular capillaries is the first step in urine formation.

The glomerular capsule is an expansion at the end of a renal tubule that receives the fluid filtered at the glomerulus. It is composed of two layers of squamous epithelial cells: a visceral layer that closely covers the glomerulus, and an outer parietal layer that is continuous with the visceral layer and with the wall of the renal tubule.

The cells of the parietal layer are typical squamous epithelial cells; however, those of the visceral layer are highly modified epithelial cells called *podocytes.* Each podocyte has several primary processes extending from its cell body, and these processes, in turn, bear numerous secondary processes, or *pedicels.* The pedicels of each cell interdigitate with those of adjacent podocytes, and the clefts between them form a complicated system of *slit pores* (fig. 20.8).

The renal tubule leads away from the glomerular capsule and becomes highly coiled. This coiled portion of the tubule is named the *proximal convoluted tubule.*

Following the proximal convoluted tubule is a structure known as the **nephron loop** (loop of Henle). The proximal convoluted tubule dips toward the renal pelvis to become the *descending limb of the nephron loop.* The tubule then curves back toward its renal corpuscle and forms the *ascending limb of the nephron loop.*

The ascending limb returns to the region of the renal corpuscle, where it becomes highly coiled again and is called the *distal convoluted tubule.* This distal portion tends to be shorter and less convoluted than the proximal tubule.

Several distal convoluted tubules merge in the renal cortex to form a *collecting duct* (collecting tubule), which is technically not part of the nephron. The collecting duct passes into the renal medulla, increasing in diameter as it joins other collecting ducts. The resulting tube empties into a minor calyx through an opening in a renal papilla.

Figures 20.9 and 20.10 show the parts of a nephron. Clinical Application 20.2 examines glomerulonephritis, an inflammation of the glomeruli.

### Juxtaglomerular Apparatus

Near its origin, the distal convoluted tubule passes between the afferent and efferent arterioles and contacts them. At the point of contact, the epithelial cells of the distal tubule are quite narrow and densely packed. These cells comprise a structure called the *macula densa.*

Close by, in the wall of the afferent arteriole near its attachment to the glomerulus, are some large, vascular smooth muscle cells. They are called *juxtaglomerular cells,* and together with the cells of the macula

**FIGURE** 20.8

Scanning electron micrograph of a portion of a glomerulus (8,000×). Note the slit pores between the pedicels.

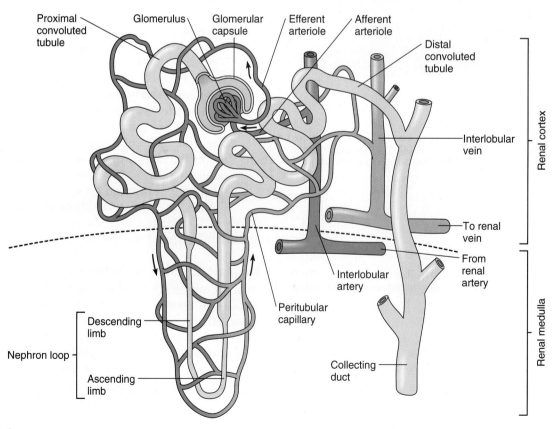

**FIGURE** 20.9

Structure of a nephron and the blood vessels associated with it.

# Glomerulonephritis

*Nephritis* refers to an inflammation of the kidney; *glomerulonephritis* is an inflammation of the glomeruli. This latter condition may be acute or chronic, and can lead to renal failure.

*Acute glomerulonephritis* (AGN) usually results from an abnormal immune reaction that develops one to three weeks following bacterial infection by beta-hemolytic streptococci. As a rule, the infection occurs in some other part of the body and does not affect the kidneys directly. Instead, bacterial antigens trigger an immune reaction. Antibodies against these antigens form insoluble immune complexes (see chapter 16) that travel in the bloodstream to the kidneys. The antigen-antibody complexes are deposited in and block the glomerular capillaries, which become further obstructed as the inflammatory response sends many white blood cells to the region. Those capillaries remaining open may become abnormally permeable, sending plasma proteins and red blood cells into the urine.

Most glomerulonephritis patients eventually regain normal kidney function; however, in severe cases, renal functions may fail completely, and without treatment, the person is likely to die within a week or so.

*Chronic glomerulonephritis* is a progressive disease in which increasing numbers of nephrons are slowly damaged until finally the kidneys are unable to function. This condition is usually associated with certain diseases other than streptococcal infections, and it also involves formation of antigen-antibody complexes that precipitate and accumulate in the glomeruli. The resulting inflammation is prolonged, and it is accompanied by fibrous tissue replacing glomerular membranes. As this happens, the functions of the nephrons are permanently lost, and eventually the kidneys fail. ■

(a)

Renal tubules — Glomerular capsule — Glomerulus

(b)

Collecting ducts

## FIGURE 20.10

(*a*) Light micrograph of a section of the human renal cortex (220×). (*b*) Light micrograph of the renal medulla (80× micrograph enlarged to 200×).

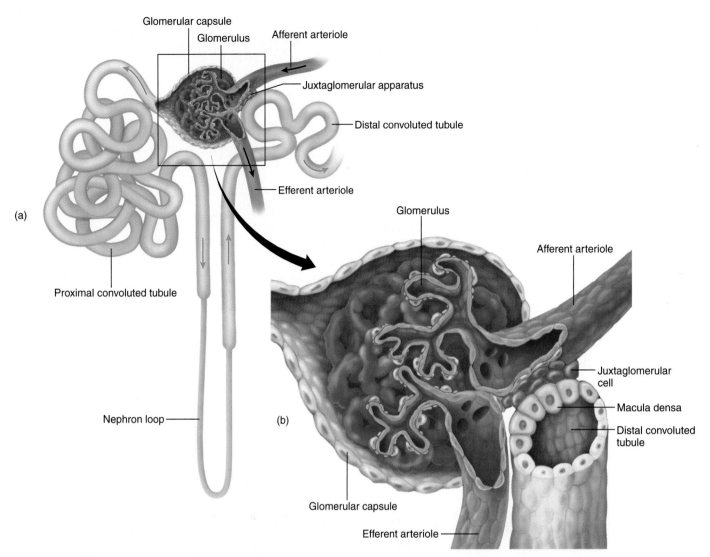

## Figure 20.11

(*a*) Location of the juxtaglomerular apparatus. (*b*) Enlargement of a section of the juxtaglomerular apparatus, which consists of the macula densa and the juxtaglomerular cells. Black arrows indicate direction of blood flow. Blue arrows indicate flow of glomerular filtrate and tubular fluid.

densa, they constitute the **juxtaglomerular apparatus** (complex). This structure is important in regulating the secretion of renin (see chapter 13) (fig. 20.11).

### Cortical and Juxtamedullary Nephrons

Most nephrons have corpuscles located in the renal cortex near the surface of the kidney. These *cortical nephrons* have relatively short nephron loops that usually do not reach the renal medulla.

Another group, called *juxtamedullary nephrons,* have corpuscles close to the renal medulla, and their nephron loops extend deep into the medulla. Although they represent only about 20% of the total, these nephrons are important in regulating water balance (fig. 20.12).

### Blood Supply of a Nephron

The cluster of capillaries that forms a glomerulus arises from an afferent arteriole whose diameter is greater than that of other arterioles. Blood passes through the capillaries of the glomerulus, then (minus any filtered fluid) enters an efferent arteriole (rather than a venule), whose diameter is smaller than that of the afferent arteriole. The greater resistance to blood flow of the efferent arteriole causes blood to back up into the glomerulus. This results in a relatively high pressure in the glomerular capillaries.

The efferent arteriole branches into a complex network of capillaries that surrounds the renal tubule called the **peritubular capillary system.** Blood in the system has passed through two arterioles and is under relatively low pressure (see fig. 20.9). Branches of this system that

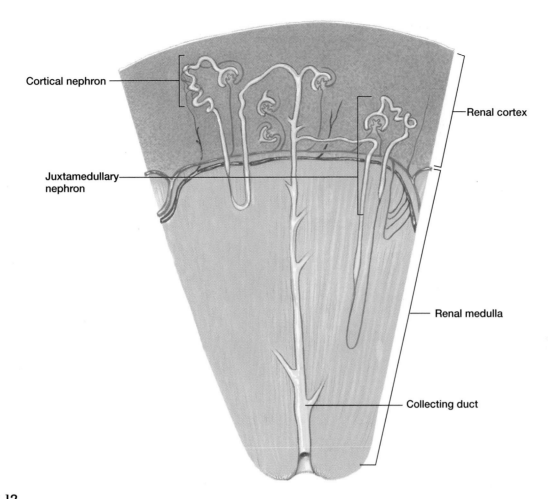

**FIGURE 20.12**

Cortical nephrons are close to the surface of a kidney; juxtamedullary nephrons are near the renal medulla.

primarily receive blood from the efferent arterioles of the juxtamedullary nephrons form capillary loops called *vasa recta.* These loops dip into the renal medulla and are closely associated with the loops of the juxtamedullary nephrons (fig. 20.13). After flowing through the vasa recta, blood returns to the renal cortex, where it joins blood from other branches of the peritubular capillary system and enters the venous system of the kidney. Figure 20.14 summarizes the pathway that blood follows as it passes through the blood vessels of the kidney and nephron.

1. Describe the system of vessels that supplies blood to the kidney.

2. Name the parts of a nephron.

3. Which structures comprise the juxtaglomerular apparatus?

4. Distinguish between a cortical nephron and a juxtamedullary nephron.

5. Describe the blood supply of a nephron.

## URINE FORMATION

The main function of the nephrons is to control the composition of body fluids and remove wastes from the blood. The product is **urine,** which is excreted from the body. It contains wastes and excess water and electrolytes.

Urine formation involves three processes. In *glomerular filtration,* certain chemicals move from the glomerular capillaries into the renal tubules. In *tubular reabsorption,* some of these chemicals move back into the blood plasma. In *tubular secretion,* certain other chemicals move from the peritubular capillaries into the renal tubules. In other words, the following relationship determines the amount of substance excreted in the urine:

$$\text{urinary excretion} = \text{glomerular filtration} \\ + \text{tubular secretion} \\ - \text{tubular reabsorption}$$

### Glomerular Filtration

Urine formation begins when water and other small dissolved molecules and ions are filtered out of the

**FIGURE 20.13**

The capillary loop of the vasa recta is closely associated with the nephron loop of a juxtamedullary nephron.

Renal artery

↓

Interlobar artery

↓

Arcuate artery

↓

Interlobular artery

↓

Afferent arteriole

↓

Glomerular capillary

↓

Efferent arteriole

↓

Vasa recta and peritubular capillary

↓

Interlobular vein

↓

Arcuate vein

↓

Interlobar vein

↓

Renal vein

**FIGURE 20.14**

Pathway of blood through the blood vessels of the kidney and nephron.

glomerular capillary plasma and into the glomerular capsules. The filtration of these materials through the capillary walls is much like the filtration that occurs at the arteriole ends of other capillaries throughout the body. The glomerular capillaries, however, are many times more permeable to small molecules than the capillaries in other tissues, due to the many tiny openings (fenestrae) in their walls (fig. 20.15).

The glomerular capsule receives the resulting **glomerular filtrate,** which has about the same composition as the filtrate that becomes tissue fluid elsewhere in the body. That is, glomerular filtrate is mostly water and the same solutes as in blood plasma, except for the larger protein molecules. More specifically, glomerular filtrate contains water, glucose, amino acids, urea, uric acid, creatine, creatinine, and sodium, chloride, potassium, calcium, bicarbonate, phosphate, and sulfate ions. Table 20.1 shows the relative concentrations of some of the substances in the blood plasma, glomerular filtrate, and urine.

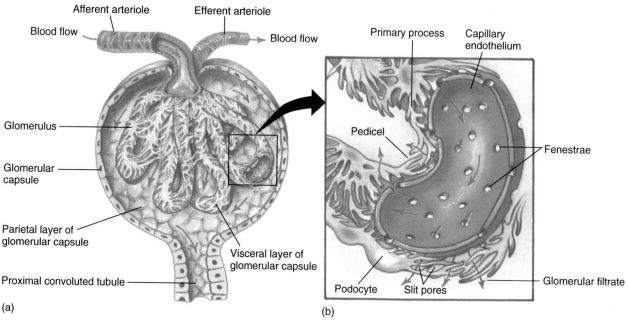

Afferent arteriole
Efferent arteriole
Blood flow
Blood flow
Glomerulus
Glomerular capsule
Parietal layer of glomerular capsule
Visceral layer of glomerular capsule
Proximal convoluted tubule
Primary process
Capillary endothelium
Pedicel
Fenestrae
Podocyte
Slit pores
Glomerular filtrate

(a)

(b)

**FIGURE 20.15**

(a) The first step in urine formation is filtration of substances through the glomerular membrane into the glomerular capsule.
(b) The glomerular filtrate passes through fenestrae of the capillary endothelium.

**TABLE 20.1 RELATIVE CONCENTRATIONS OF PLASMA, GLOMERULAR FILTRATE, AND URINE COMPONENTS**

| | Concentrations (mEq/L) | | |
|---|---|---|---|
| Substance | Plasma | Glomerular Filtrate | Urine |
| Sodium (Na$^+$) | 142 | 142 | 128 |
| Potassium (K$^+$) | 5 | 5 | 60 |
| Calcium (Ca$^{+2}$) | 4 | 4 | 5 |
| Magnesium (Mg$^{+2}$) | 3 | 3 | 15 |
| Chloride (Cl$^-$) | 103 | 103 | 134 |
| Bicarbonate (HCO$_3^-$) | 27 | 27 | 14 |
| Sulfate (SO$_4^{-2}$) | 1 | 1 | 33 |
| Phosphate (PO$_4^{-3}$) | 2 | 2 | 40 |
| | Concentrations (mg/100 mL) | | |
| Substance | Plasma | Glomerular Filtrate | Urine |
| Glucose | 100 | 100 | 0 |
| Urea | 26 | 26 | 1,820 |
| Uric acid | 4 | 4 | 53 |
| Creatinine | 1 | 1 | 196 |

## Filtration Pressure

As in the case of other capillaries, the main force responsible for moving substances by filtration through the glomerular capillary wall is the hydrostatic pressure of the blood inside. (Recall that glomerular capillary pressure is high compared to other capillaries.) Glomerular filtration is also influenced by the osmotic pressure of the blood plasma in the glomerulus and by the hydrostatic pressure inside the glomerular capsule.

Due to the plasma proteins, the osmotic pressure of the plasma is always higher than that of the glomerular filtrate (except in some kinds of kidney disease). This tends to draw water back into the glomerular capillaries, thus opposing filtration. Any increase in glomerular capsule hydrostatic pressure would also oppose filtration.

The net pressure that forces substances out of the glomerulus is called **filtration pressure,** and it is determined by the above factors. It can be calculated as follows:

**Filtration pressure =**

**force favoring filtration − forces opposing filtration**

(glomerular capillary hydrostatic pressure)

(capsular hydrostatic pressure and glomerular capillary osmotic pressure)

The concentrations of certain components of the blood plasma can be used to evaluate kidney functions. For example, if the kidneys are functioning inadequately, the plasma concentrations of urea (as indicated by a blood urea nitrogen test) and of creatinine may increase as much as tenfold above normal.

## Filtration Rate

The glomerular filtration rate (GFR) is directly proportional to the filtration pressure described in figure 20.16. Consequently, the factors that affect the glomerular hydrostatic pressure, glomerular plasma osmotic pressure, or hydrostatic pressure in the glomerular capsule will also affect the rate of filtration.

Normally, glomerular hydrostatic pressure is the most important factor determining filtration pressure and GFR. Since the glomerular capillary is located between two arterioles—the afferent and efferent arterioles—any change in the diameters of these vessels is likely to change glomerular hydrostatic pressure, causing a change in glomerular filtration rate. The afferent arteriole, through which the blood enters the glomerulus, may constrict in response to mild stimulation by sympathetic nerve impulses. If this occurs, filtration pressure decreases, and filtration rate drops. If, on the other hand, the efferent arteriole (through which the blood leaves the glomerulus) constricts, blood backs up into the glomerulus, filtration pressure increases, and filtration rate rises. Vasodilation of these vessels produces opposite effects.

If arterial blood pressure drops drastically, as may occur during *shock,* the glomerular hydrostatic pressure may fall below the level required for filtration, leading to acute renal failure. At the same time, the epithelial cells of the renal tubules may fail to receive sufficient nutrients to maintain their high rates of metabolism. As a result, cells may die (tubular necrosis), and renal functions may be lost permanently, resulting in chronic renal failure.

Recall that the osmotic pressure of the glomerular plasma also influences filtration pressure and the rate of filtration. In capillaries, the blood hydrostatic pressure, forcing water and dissolved substances outward, is opposed by the plasma osmotic pressure that attracts water inward. Actually, as filtration occurs through the capillary wall, proteins remaining in the plasma raise the osmotic pressure within the glomerular capillary. When this pressure reaches a certain high level, filtration ceases. Conversely, conditions that lower plasma osmotic pressure, such as a decrease in plasma protein concentration, increase filtration rate.

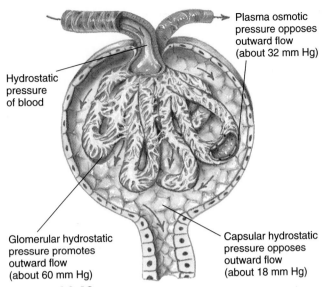

Plasma osmotic pressure opposes outward flow (about 32 mm Hg)

Hydrostatic pressure of blood

Glomerular hydrostatic pressure promotes outward flow (about 60 mm Hg)

Capsular hydrostatic pressure opposes outward flow (about 18 mm Hg)

**FIGURE 20.16**

The rate of glomerular filtration is affected by the hydrostatic and osmotic pressure of the plasma, and the hydrostatic pressure of the fluid in the glomerular capsule.

The hydrostatic pressure in the glomerular capsule is another factor that may affect filtration pressure and rate. This capsular pressure sometimes changes as a result of an obstruction, such as a stone in a ureter or an enlarged prostate gland pressing on the urethra. If this occurs, fluids back up into the renal tubules and raise the hydrostatic pressure in the glomerular capsules. Because any increase in capsular pressure opposes glomerular filtration, filtration rate may decrease significantly.

In an average adult, the glomerular filtration rate for the nephrons of both kidneys is about 125 milliliters per minute, or 180,000 milliliters (180 liters) in 24 hours. Assuming that the blood plasma volume is about 3 liters, the production of 180 liters of filtrate in 24 hours means that all of the plasma must be filtered through the glomeruli about 60 times each day (fig. 20.17). Since this 24-hour volume is nearly 45 gallons, it is obvious that not all of it is excreted as urine. Instead, most of the fluid that passes through the renal tubules is reabsorbed and reenters the plasma.

The volume of plasma the kidneys filter also depends on the *surface area* of the glomerular capillaries. This surface area is estimated to be about 2 square meters—approximately equal to the surface area of an adult's skin.

An injury to a kidney can be more dangerous than an injury to another organ. An injured kidney produces a protein called *transforming growth factor beta.* This protein causes scars to form. The scars further damage the kidney, impairing its function.

**FIGURE** 20.17

Relative amounts of (*a*) glomerular filtrate and (*b*) urine formed in 24 hours.

1 What processes occur in urine formation?

2 How is filtration pressure calculated?

3 What factors influence the rate of glomerular filtration?

## Regulation of Filtration Rate

In general, glomerular filtration rate remains relatively constant through a process called **autoregulation of GFR.** However, certain conditions override autoregulation. GFR may increase, for example, when body fluids are in excess, and decrease when the body must conserve fluid.

Recall from chapter 10 that sympathetic nervous system fibers synapse with the vascular smooth muscle of arterioles. Reflexes responding to changes in blood pressure and volume control the activity of these sympathetic fibers. If blood pressure and volume drop, vasoconstriction of the afferent arterioles results, decreasing filtration pressure and thus GFR. The result is an appropriate decrease in the rate of urine formation when the body must conserve water. If receptors detect excess

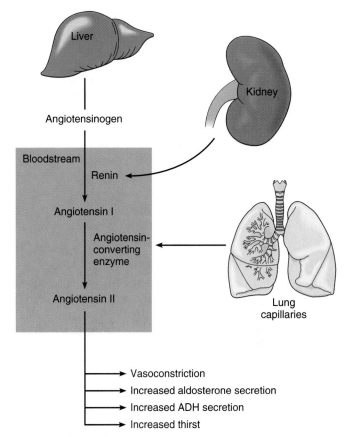

**FIGURE** 20.18

The formation of angiotensin II in the bloodstream involves several organs, and has multiple actions that conserve sodium and water.

body fluids, vasodilation of the afferent arteriole results, increasing filtration pressure and GFR. A second control of GFR is a hormone-like system called the **renin-angiotensin system,** whose primary function seems to be regulation of renal sodium excretion.

The juxtaglomerular cells of the afferent arterioles secrete an enzyme, **renin,** in response to stimulation from sympathetic nerves and pressure-sensitive cells called **renal baroreceptors** that are in the afferent arteriole. These factors stimulate renin secretion if blood pressure drops. The macula densa also controls renin secretion. Somehow the cells of the macula densa sense the concentration of sodium, potassium, and chloride ions in the distal renal tubule. Decreasing levels of these ions stimulate renin secretion.

Once in the bloodstream, *renin* reacts with the plasma protein *angiotensinogen* to form *angiotensin I.* An enzyme, *angiotensin-converting enzyme* (ACE), present on capillary endothelial cells (particularly in the lungs), rapidly converts angiotensin I to *angiotensin II.*

Angiotensin II has a number of renal effects that help maintain sodium balance, water balance, and blood pressure (fig. 20.18). As a vasoconstrictor, it affects both the afferent and efferent arterioles. Although afferent arteriolar constriction decreases GFR, efferent arteriolar constriction tends to minimize the decrease, thus contributing to autoregulation of GFR. Angiotensin II has a major effect on the kidneys through the adrenal cortical hormone aldosterone, which stimulates sodium reabsorption in the distal tubule. By stimulating aldosterone secretion, angiotensin II helps to reduce the amount of sodium excreted in the urine.

The hormone **atrial natriuretic peptide** (ANP) also affects sodium excretion. ANP secretion increases when the atria of the heart stretch due to increased blood volume. ANP stimulates sodium excretion through a number of mechanisms, including increasing GFR.

## Tubular Reabsorption

If the composition of the glomerular filtrate entering the renal tubule is compared with that of the urine leaving the tubule, it is obvious that the fluid changes as it passes through the tubule (see table 20.1). For example, glucose is present in the filtrate, but absent in the urine. In contrast, urea and uric acid are considerably more concentrated in urine than they are in the glomerular filtrate. Such changes in fluid composition are largely the result of **tubular reabsorption,** the process by which substances are transported out of the tubular fluid, through the epithelium of the renal tubule, and into the interstitial fluid. These substances then diffuse into the peritubular capillaries (fig. 20.19).

Peritubular capillary blood is under relatively low pressure because it has already passed through two arterioles. Also, the wall of the peritubular capillary is more permeable than that of other capillaries. Finally, the relatively high rate of glomerular filtration has increased the protein concentration of the peritubular capillary plasma. All of these factors enhance the rate of fluid reabsorption from the renal tubule.

Although tubular reabsorption occurs throughout the renal tubule, most of it occurs in the proximal convoluted portion. The epithelial cells in this portion have many *microvilli* that form a "brush border" on their free surfaces facing the tubular lumen. These tiny extensions greatly increase the surface area exposed to the glomerular filtrate and enhance reabsorption (see fig. 17.39).

Segments of the renal tubule are adapted to reabsorb specific substances, using particular modes of transport. Glucose reabsorption, for example, occurs through the walls of the proximal tubule by active transport. Water also is rapidly reabsorbed through the epithelium of the proximal tubule by osmosis; however, portions of the distal tubule and collecting duct are almost impermeable to water. This characteristic of the distal tubule is important in the regulation of urine concentration and volume, as we will soon see.

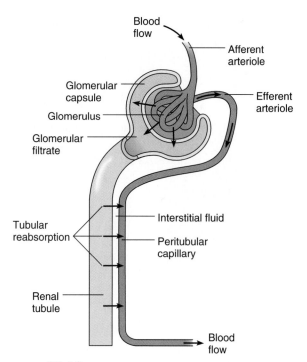

**FIGURE 20.19**

Reabsorption is the process by which substances are transported from the glomerular filtrate into the blood of the peritubular capillary.

As described in chapter 3, active transport depends on the presence of carrier proteins in a cell membrane. Exactly how these carrier molecules work is not known, but it appears to involve a binding site on the carrier and the ability of proteins to alter their shapes. The molecule to be transported binds to the carrier; the carrier alters its shape, releases the transported molecule on the other side of the cell membrane, and then returns to its original position and repeats the process. Such a mechanism has a *limited transport capacity;* that is, it can only transport a certain number of molecules in a given amount of time because the number of carriers is limited.

Usually all of the glucose in the glomerular filtrate is reabsorbed, because there are enough carrier molecules to transport it. When the plasma glucose concentration increases to a critical level, called the *renal plasma threshold,* more glucose molecules are in the filtrate than the active transport mechanism can handle. As a result, some glucose remains in the filtrate and is excreted in the urine. This explains why the elevated blood glucose of diabetes mellitus results in glucose in the urine.

Any increase in urine volume is called *diuresis.* Glucose in the filtrate will draw water into the renal tubule by osmosis, thus increasing urine volume. Such an increase is called an *osmotic diuresis.*

## CLINICAL APPLICATION

# The Nephrotic Syndrome

The *nephrotic syndrome* is a set of symptoms that often appears in patients with renal diseases. It causes considerable loss of plasma proteins into the urine (proteinuria), widespread edema, and increased susceptibility to infections.

Plasma proteins are lost into the urine because of increased permeability of the glomerular membranes, which accompanies renal disorders such as glomerulonephritis. As a consequence of a decreasing plasma protein concentration (hypoproteinemia), the plasma osmotic pressure falls, increasing filtration pressure in capillaries throughout the body. This may lead to widespread, severe edema as a large volume of fluid accumulates in the interstitial spaces within the tissues and in body spaces such as the abdominal cavity, pleural cavity, pericardial cavity, and joint cavities.

Also, as edema develops, blood volume decreases and blood pressure drops. These changes may activate the renin-angiotensin system, leading to the release of aldosterone from the adrenal cortex (see chapter 13), which, in turn, stimulates the kidneys to conserve sodium ions and water. This action reduces the urine output and may aggravate the edema.

The nephrotic syndrome sometimes appears in young children who have *lipoid nephrosis.* The cause of this condition is unknown, but it alters the epithelial cells of the glomeruli so that the glomerular membranes enlarge and distort, allowing proteins to leak through. ■

---

Glucose in the urine is called *glucosuria.* It may follow intravenous administration of glucose, epinephrine, or eating a candy bar, or it may occur in a person with diabetes mellitus. In the case of insulin-dependent (type I) diabetes, blood glucose concentration rises because of insufficient insulin secretion from the pancreas.

One in three people who have diabetes mellitus sustains kidney damage as part of their condition, as evidenced by protein in their urine. Following a low-protein diet can slow the loss of kidney function.

---

*Amino acids* also enter the glomerular filtrate and are reabsorbed in the proximal convoluted tubule. Three different active transport mechanisms are thought to reabsorb different groups of amino acids, whose members have similar structures. As a result, normally only a trace of amino acids remains in the urine.

Although the glomerular filtrate is nearly free of protein, a number of smaller protein molecules, such as albumin, somehow squeeze through the glomerular capillaries. These proteins are taken up by *pinocytosis* through the brush border of epithelial cells lining the proximal convoluted tubule. Once they are inside an epithelial cell, the proteins are degraded to amino acids and moved into the blood of the peritubular capillary.

The epithelium of the proximal convoluted tubule also reabsorbs creatine; lactic, citric, uric, and ascorbic (vitamin C) acids; and phosphate, sulfate, calcium, potassium, and sodium ions. Active transport mechanisms with limited transport capacities reabsorb all of these chemicals. Such substances begin to appear in the urine when their concentrations in the glomerular filtrate exceed their respective renal plasma thresholds. Clinical Application 20.3 discusses how the nephrotic syndrome causes plasma proteins to appear in the urine.

## Sodium and Water Reabsorption

Water reabsorption occurs passively by osmosis, primarily in the proximal convoluted tubule, and is closely associated with the active reabsorption of sodium ions. In the proximal tubule, if sodium reabsorption increases, water reabsorption increases; if sodium reabsorption decreases, water reabsorption decreases also.

Much of the sodium ion reabsorption occurs in the proximal segment of the renal tubule by active transport (sodium pump mechanism). When the positively charged sodium ions ($Na^+$) are moved through the tubular wall, negatively charged ions, including chloride ions ($Cl^-$), phosphate ions ($PO_4^{-3}$), and bicarbonate ions ($HCO_3^-$), accompany them. This movement of negatively charged ions is due to the electrochemical attraction between particles of opposite electrical charge. Although it depends on active transport of sodium, this movement of negatively charged ions is considered a passive process because it does not require a direct expenditure of cellular energy. Some of these ions (such as $HCO_3^-$ and $PO_4^{-3}$) are also reabsorbed by active transport.

As more and more sodium ions are reabsorbed into the peritubular capillary along with negatively charged ions, the concentration of solutes within the peritubular

Blood
flow

Glomerulus

Glomerular
capsule

Glomerular
filtrate

Proximal
tubule

Blood
flow

Na⁺ +
K⁺ +
Cl⁻ −
PO₄⁻³ −
Na⁺ +
HCO₃⁻ −

(1) Sodium and potassium ions
are reabsorbed by
active transport

(2) Negatively charged ions
are attracted to positively
charged ions
(passive transport)

Na⁺
Cl⁻
H₂O
H₂O
H₂O

(3) As the concentration of ions
(solute) increases in plasma,
osmotic pressure increases

(4) Water moves from renal
tubule to capillary by
osmosis

Peritubular capillary

Isotonic
tubular fluid

Blood
flow

**FIGURE 20.20**

Water reabsorption by osmosis occurs in response to the reabsorption of sodium and other solutes by active transport in the proximal tubule.

blood might be expected to increase. However, since water diffuses through cell membranes from regions of lesser solute concentration (hypotonic) toward regions of greater solute concentration (hypertonic), water moves by osmosis, following the ions from the renal tubule into the peritubular capillary.

The proximal tubule reabsorbs about 70% of the filtered sodium, other ions, and water. By the end of the proximal tubule, osmotic equilibrium is reached, and the remaining tubular fluid is isotonic (fig. 20.20).

Active transport continues to reabsorb sodium ions as the tubular fluid moves through the nephron loop, the distal convoluted tubule, and the collecting duct. Consequently, almost all of the sodium and water (97%–99%) that enters the renal tubules as part of the glomerular filtrate may be reabsorbed before the urine is excreted. However, aldosterone controls sodium reabsorption, and antidiuretic hormone controls water reabsorption. Under the influence of these hormones, reabsorption of these substances can change to keep conditions in the body fluids constant. The next chapter discusses this further.

Recall that the kidneys filter an extremely large volume of fluid (180 liters) each day. Thus, if 99% of the glomerular filtrate is reabsorbed, the remaining 1% excreted includes a relatively large amount of sodium and water. On the other hand, if sodium and water reabsorption decrease to 97% of the amount filtered, the amount excreted triples! Therefore, small changes in the tubular reabsorption of water and sodium result in large changes in urinary excretion.

**1** How is the peritubular capillary adapted for reabsorption?

**2** Which substances present in the glomerular filtrate are not normally present in urine?

**3** Which mechanisms reabsorb solutes from glomerular filtrate?

**4** Define *renal plasma threshold*.

**5** Describe the role of passive transport in urine formation.

## Regulation of Urine Concentration and Volume

In contrast to conditions in the proximal tubule, the tubular fluid reaching the distal convoluted tubule is hypotonic to its surroundings due to changes that occur through the loop segment of each nephron. The cells lining the distal tubule and the collecting duct that follows continue to reabsorb sodium ions (chloride ions follow passively) under the influence of aldosterone, which the adrenal cortex secretes (see chapter 13). In addition, the interstitial fluid surrounding the collecting ducts is hypertonic, particularly in the medulla. These might seem to be ideal conditions for water reabsorption as well. However, the cells lining the later portion of the distal convoluted tubule and the collecting duct are impermeable to water. Thus, water tends to accumulate inside the tubule and may be excreted as dilute urine.

As discussed in chapter 13, neurosecretory cells in the hypothalamus produce antidiuretic hormone (ADH). The posterior lobe of the pituitary gland releases ADH in response to decreasing concentration of water in the body fluids or to decreasing blood volume and blood pressure. When ADH reaches the kidney, it increases permeability of the epithelial cell linings of the distal convoluted tubule and the collecting duct; consequently, water rapidly moves out of these segments by osmosis, especially where the collecting duct passes through the extremely hypertonic medulla. Thus, the urine becomes more concentrated (hypertonic), and water is retained in the internal environment (fig. 20.21).

A **countercurrent mechanism** involving the nephron loops, particularly of the juxtamedullary nephrons, ensures that the medullary interstitial fluid becomes hypertonic. This mechanism is possible because

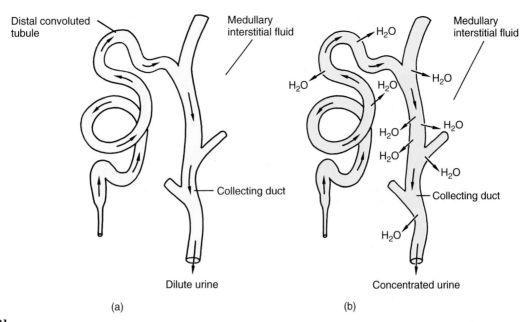

**FIGURE 20.21**

(a) The distal convoluted tubule and collecting duct are impermeable to water, so water may be excreted as dilute urine. (b) If ADH is present, however, these segments become permeable, and water is reabsorbed by osmosis into the hypertonic medullary interstitial fluid.

the descending and ascending limbs of the nephron loops lie parallel and very close to one another. The mechanism is named partly for the fact that fluid moving down the descending limb creates a current that is counter to that of the fluid moving up in the ascending limb.

The different parts of the nephron loop have important functional differences. For example, the epithelial lining in the thick upper portion of the ascending limb (thick segment) is relatively impermeable to water. However, the epithelium does actively reabsorb sodium and chloride ions (some potassium is actively reabsorbed as well). As these solutes accumulate in the interstitial fluid outside the ascending limb, it becomes hypertonic, while the tubular fluid inside becomes hypotonic because it is losing its solute.

In contrast, the epithelium of the descending limb (thin segment) is quite permeable to water, but relatively impermeable to solutes. Because this segment is surrounded by hypertonic fluid created by the ascending limb, water tends to leave the descending limb by osmosis. Thus, the contents of the descending limb become more and more concentrated, or hypertonic (fig. 20.22).

This very concentrated tubular fluid now moves into the ascending limb, and sodium chloride (NaCl) is again actively reabsorbed into the medullary interstitial fluid, raising the interstitial NaCl concentration even more. With the increased interstitial fluid solute concentration, even more water diffuses out of the descending limb, further increasing the salt concentration of the tubular fluid. Each time this circuit is completed, the concentration of NaCl increases, or multiplies. For this

reason, the mechanism is called a *countercurrent multiplier.* In humans, this mechanism creates a tubular fluid solute concentration near the tip of the loop that is more than four times the solute concentration of plasma (fig. 20.23a).

The solute concentration of the tubular fluid progressively decreases toward the renal cortex. Since the descending limb of the loop is permeable to water, the interstitial fluid at any level of the loop is essentially in equilibrium with the fluid in the tubule. Thus, the concentration gradient in the loop is also found in the interstitial fluid (fig. 20.23b).

The vasa recta is another countercurrent mechanism that maintains the NaCl concentration gradient in the renal medulla. Blood flows relatively slowly down the descending portion of the vasa recta, and NaCl enters it by diffusion. Then, as the blood moves back up toward the renal cortex, most of the NaCl diffuses from the blood and reenters the medullary interstitial fluid. Consequently, the bloodstream carries little NaCl away from the renal medulla, preserving the gradient (fig. 20.24).

To summarize, the countercurrent multiplier creates a large concentration gradient for water reabsorption in the interstitial fluid surrounding the distal convoluted tubules and the collecting ducts of the nephron. Although the epithelial lining of these structures is impermeable to water, in the presence of ADH it becomes permeable. The higher the blood levels of ADH, the more permeable it becomes. In this way, ADH stimulates water reabsorption and the production of concentrated urine, which contains soluble wastes and other

**FIGURE** 20.22 ▭

(a) Fluid in the ascending limb of the nephron loop becomes hypotonic as solute is reabsorbed. (b) Fluid in the descending limb becomes hypertonic as it loses water by osmosis.

**FIGURE** 20.23

(a) As NaCl completes the countercurrent circuit again and again, more water exits the descending limb, leaving more concentrated solute in the tubular fluid in the ascending limb. This, in turn, transports even more solute into the interstitial fluid. (b) As a result, an NaCl concentration gradient is established in the medullary interstitial fluid.

substances in a minimum of water, thus minimizing the loss of body water when dehydration is a threat. If the body fluids contain excess water, ADH secretion is decreased, and the epithelial linings of the distal segment and the collecting duct become less permeable to water. Thus less water is reabsorbed, and the urine will be more dilute. Table 20.2 summarizes the role of ADH in urine production.

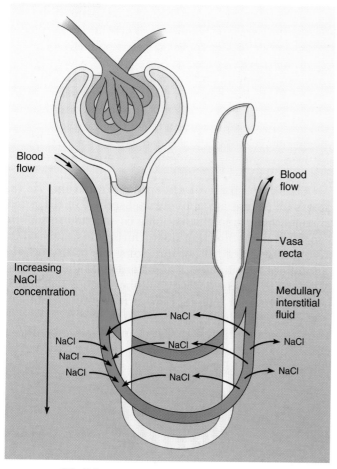

**FIGURE** 20.24

A countercurrent mechanism in the vasa recta helps maintain the NaCl concentration gradient in the medullary interstitial fluid (see fig. 20.13).

## Urea and Uric Acid Excretion

Because **urea** is a by-product of amino acid catabolism in the liver, its plasma concentration reflects the amount of protein in the diet. Urea enters the renal tubule by filtration. About 50% of it is reabsorbed (passively) by diffusion, but the remainder is excreted in the urine.

Urea moves within the kidney in other ways, including the countercurrent multiplier mechanism that concentrates the medullary interstitial fluid. As a result, urea contributes to the reabsorption of water from the collecting duct.

**Uric acid** is a product of the metabolism of certain organic bases (purines) in nucleic acids. Active transport completely reabsorbs the uric acid that is filtered. Since some uric acid (equal to approximately 10% of the amount filtered) is excreted in the urine, it is apparently secreted into the renal tubule.

In the inborn error of metabolism gout, uric acid crystals are deposited in certain joints, particularly of the great toe, causing severe pain. Treatments include taking drugs that increase the kidneys' excretion of uric acid and block an enzyme in the biosynthetic pathway for uric acid; limiting intake of foods that are sources of uric acid, including organ meats, anchovies, and sardines; maintaining a healthy weight; and drinking more fluids to dilute the urine, which enhances uric acid excretion.

Gout is an illness with a long history in medicine. Hippocrates mentioned it, and in 1793, English physician Alfred Baring Garrod isolated and implicated uric acid from the blood of a patient with gout, and noted that affected individuals often had relatives suffering from it too. At that time, gout was thought to be the result of being a lazy glutton!

**1** Describe a countercurrent mechanism.

**2** What role does the hypothalamus play in regulating urine concentration and volume?

**3** Explain how urea and uric acid are excreted.

| TABLE 20.2 | ROLE OF ADH IN REGULATING URINE CONCENTRATION AND VOLUME |
|---|---|

1. Concentration of water in the blood decreases.
2. Increase in the osmotic pressure of body fluids stimulates osmoreceptors in the hypothalamus.
3. Hypothalamus signals the posterior pituitary gland to release ADH.
4. Blood carries ADH to the kidneys.
5. ADH causes the distal convoluted tubules and collecting ducts to increase water reabsorption by osmosis.
6. Urine becomes more concentrated, and urine volume decreases.

## Tubular Secretion

In **tubular secretion,** certain substances move from the plasma of the peritubular capillary into the fluid of the renal tubule. As a result, the amount of a particular chemical excreted in the urine may exceed the amount filtered from the plasma in the glomerulus (fig. 20.25).

Active transport mechanisms similar to those that function in reabsorption secrete some substances. However, the *secretory mechanisms* transport substances in the opposite direction. For example, the epithelium of the proximal convoluted segment actively secretes certain organic compounds, including penicillin and histamine, into the tubular fluid.

Hydrogen ions are also actively secreted throughout the entire renal tubule. As a result, urine is usually acidic by the time it is excreted, although the urinary pH can vary considerably. The secretion of hydrogen ions is important in regulating the pH of body fluids, as chapter 21 explains.

Surgery is the primary treatment for cancer of the kidneys. However, in half of all cases, cancer returns, usually in the lungs. A promising new treatment for this metastatic kidney cancer is the immune system substance interleukin-2. It is administered intravenously in cycles in a hospital setting because of sometimes severe side effects. Interleukin-2 presumably stimulates the immune system to attack tumor cells. In about 15% of patients, this drug shrinks lung tumors.

Although most of the potassium ions in the glomerular filtrate are actively reabsorbed in the proximal convoluted tubule, some may be secreted in the distal segment and collecting duct. During this process, the active reabsorption of sodium ions out of the tubular fluid under the influence of aldosterone produces a negative electrical charge within the tube. Because positively charged potassium ions ($K^+$) are attracted to regions that are negatively charged, these ions move passively through the tubular epithelium and enter the tubular fluid. Potassium ions are also secreted by active processes (fig. 20.26).

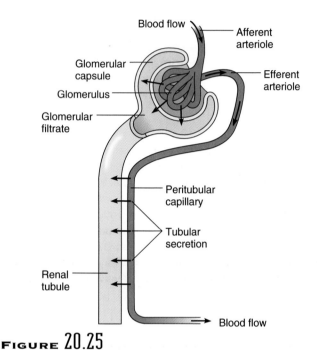

**FIGURE 20.25**

Secretory mechanisms move substances from the plasma of the peritubular capillary into the fluid of the renal tubule.

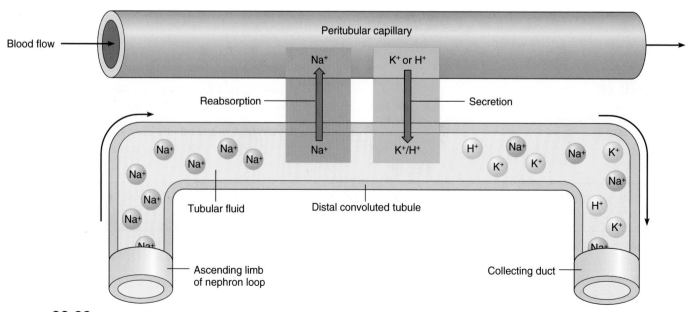

**FIGURE 20.26**

Passive secretion of potassium ions (or hydrogen ions) may occur in response to the active reabsorption of sodium ions.

To summarize, urine forms as a result of the following:

- Glomerular filtration of materials from blood plasma.

- Reabsorption of substances including glucose; water; urea; proteins; creatine; amino, lactic, citric, and uric acids; and phosphate, sulfate, calcium, potassium, and sodium ions.

- Secretion of substances including penicillin, histamine, phenobarbital, hydrogen ions, ammonia, and potassium ions (table 20.3).

**1** Define *tubular secretion.*

**2** Which substances are actively secreted? Passively secreted?

**3** How does the reabsorption of sodium affect the secretion of potassium?

## Urine Composition

Urine composition reflects the amounts of water and solutes that the kidneys must eliminate from the body or retain in the internal environment to maintain homeostasis. It varies considerably from time to time because of differences in dietary intake and physical activity. Urine is about 95% water, and usually also consists of urea from the catabolism of amino acids, uric acid from the catabolism of nucleic acids, and creatinine from metabolism of creatine. Urine may also contain a trace of amino acids, as well as electrolytes whose concentrations reflect the amounts included in the diet (see table 20.1). Appendix C lists the normal concentrations of urine components.

> Abnormal constituents of urine may not indicate illness. Glucose in the urine may result from a sugary meal or may occur toward the end of pregnancy; protein may appear in the urine following vigorous physical exercise; ketones appear in the urine during a prolonged fast or when a person follows a very low calorie or low carbohydrate diet.

The volume of urine produced usually varies between 0.6 and 2.5 liters per day. Such factors as fluid intake, environmental temperature, relative humidity of the surrounding air, and a person's emotional condition, respiratory rate, and body temperature influence the exact urine volume. An output of 50–60 milliliters of urine per hour is considered normal, and an output of less than 30 milliliters per hour may indicate kidney failure. Clinical Application 20.4 discusses renal clearance, which is a measure of kidney efficiency.

| TABLE 20.3 | FUNCTIONS OF NEPHRON COMPONENTS |
|---|---|
| **Part** | **Function** |
| *Renal Corpuscle* | |
| Glomerulus | Filtration of water and dissolved substances from the plasma |
| Glomerular capsule | Receives the glomerular filtrate |
| *Renal Tubule* | |
| Proximal convoluted tubule | Reabsorption of glucose; amino acids; creatine; lactic, citric, uric, and ascorbic acids; phosphate, sulfate, calcium, potassium, and sodium ions by active transport |
| | Reabsorption of proteins by pinocytosis |
| | Reabsorption of water by osmosis |
| | Reabsorption of chloride ions and other negatively charged ions by electrochemical attraction |
| | Active secretion of substances such as penicillin, histamine, creatinine, and hydrogen ions |
| Descending limb of nephron loop | Reabsorption of water by osmosis |
| Ascending limb of nephron loop | Reabsorption of sodium, potassium, and chloride ions by active transport |
| Distal convoluted tubule | Reabsorption of sodium ions by active transport |
| | Reabsorption of water by osmosis |
| | Active secretion of hydrogen ions |
| | Secretion of potassium ions both actively and by electrochemical attraction |
| *Collecting Duct* | Reabsorption of water by osmosis |
| (Note: Although the collecting duct is not anatomically part of the nephron, it is functionally very important.) | |

# Renal Clearance

The rate at which a particular chemical is removed from the plasma indicates kidney efficiency. This rate of removal is called *renal clearance.*

Tests of renal clearance can detect glomerular damage or judge the progress of renal disease. One such test, the *inulin clearance test,* uses *inulin* (not to be confused with insulin), a complex polysaccharide found in certain plant roots. In the test, a known amount of inulin is infused into the blood at a constant rate. The inulin passes freely through the glomerular membranes, so that its concentration in the glomerular filtrate equals that in the plasma. In the renal tubule, inulin is not reabsorbed to any significant degree, nor is it secreted. Consequently, the rate at which it appears in the urine can be used to calculate the rate of glomerular filtration.

Similarly, the kidneys remove creatinine from the blood. Creatinine is produced at a constant rate during muscle metabolism. Like inulin, creatinine is filtered, but neither reabsorbed nor secreted by the kidneys. Thus, the *creatinine clearance test,* which compares a patient's blood and urine creatinine concentrations, can also be used to calculate the GFR. A significant advantage is that the bloodstream normally has a constant level of creatinine. Therefore, a single measurement of plasma creatinine levels provides a rough index of kidney function. For example, significantly elevated plasma creatinine levels suggest that GFR is greatly reduced. Because

nearly all of the creatinine the kidneys filter normally appears in the urine, a change in the rate of creatinine excretion may reflect renal failure.

Another plasma clearance test uses *para-aminohippuric acid* (PAH), a substance that filters freely through the glomerular membranes. However, unlike inulin, any PAH remaining in the peritubular capillary plasma after filtration is secreted into the proximal convoluted tubules. Therefore, essentially all PAH passing through the kidneys appears in the urine. For this reason, the rate of PAH clearance can be used to calculate the rate of plasma flow through the kidneys. Then, if the hematocrit is known (see chapter 14), the rate of total blood flow through the kidneys can also be calculated. ■

---

Parents of infants may be startled when a physician hospitalizes their child for an illness that in an adult might be considered mild—a day or two of vomiting and diarrhea. Because the kidneys of infants and young children are unable to concentrate urine and conserve water as effectively as those of adults, they can lose water rapidly, which may lead to dehydration. A 20-pound infant can lose a pound in just a day of an acute viral illness, and this is a sufficiently significant proportion of body weight to warrant hospitalization, where intravenous fluids are given to restore water and electrolyte balance (see chapter 21).

**1** List the normal constituents of urine.

**2** What is the normal hourly output of urine? The minimal hourly output?

## ELIMINATION OF URINE

After forming along the nephrons, urine passes from the collecting ducts through openings in the renal papillae and enters the major and minor calyces of the kidney. From there it passes through the renal pelvis, into a ureter, and into the urinary bladder. The urethra delivers urine to the outside.

### Ureters

Each **ureter** is a tubular organ about 25 centimeters long, which begins as the funnel-shaped renal pelvis. It extends downward posterior to the parietal peritoneum and parallel to the vertebral column. Within the pelvic cavity, it courses forward and medially to join the urinary bladder from underneath.

The wall of a ureter is composed of three layers. The inner layer, or *mucous coat,* includes several thicknesses of transitional epithelial cells and is continuous

**FIGURE 20.27**

Cross section of a ureter (16×).

- Mucous coat
- Muscular coat
- Fibrous coat

**1** Describe the structure of a ureter.

**2** How is urine moved from the renal pelvis to the urinary bladder?

**3** What prevents urine from backing up from the urinary bladder into the ureters?

**4** How does an obstruction in a ureter affect urine production?

with the linings of the renal tubules and the urinary bladder. The middle layer, or *muscular coat,* largely consists of smooth muscle fibers arranged in circular and longitudinal bundles. The outer layer, or *fibrous coat,* is composed of connective tissue (fig. 20.27).

---

Because the linings of the ureters and the urinary bladder are continuous, bacteria may ascend from the bladder into the ureters, causing infection. An inflammation of the bladder, called *cystitis,* is more common in women than in men because the female urethral pathway is shorter. Inflammation of the ureter is called *ureteritis.*

---

Kidney stones, which are usually composed of calcium oxalate, calcium phosphate, uric acid, or magnesium phosphate, sometimes form in the renal pelvis. If such a stone passes into a ureter, it may produce severe pain, beginning in the region of the kidney and radiating into the abdomen, pelvis, and lower limbs. Nausea and vomiting may also occur.

About 60% of kidney stone patients pass their stones spontaneously. The others must have the stones removed. In the past, such removal required surgery or instruments that could be passed through the tubes of the urinary tract to capture or crush the stones. Today, shock waves applied from outside the body are used to fragment kidney stones. This procedure, called *extracorporeal shock-wave lithotripsy* (ESWL), focuses high-energy shock waves through water (either in a tub or in a water-filled sack placed against the patient). The shock waves break the stones into fragments small enough to be eliminated with the urine.

---

Muscular peristaltic waves, originating in the renal pelvis, help move the urine along the length of the ureter. The presence of urine in the renal pelvis initiates these waves, whose frequency keeps pace with the rate of urine formation. If this rate is high, a peristaltic wave may occur every few seconds; if the rate is low, a wave may occur every few minutes.

When such a peristaltic wave reaches the urinary bladder, it causes a jet of urine to spurt into the bladder. The opening through which the urine enters is covered by a flaplike fold of mucous membrane. This fold acts as a valve, allowing urine to enter the bladder from the ureter but preventing it from backing up from the bladder into the ureter.

If a ureter becomes obstructed, as when a small kidney stone (renal calculus) is present in its lumen, strong peristaltic waves are initiated in the proximal portion of the tube, which may help move the stone into the bladder. The presence of a stone usually also stimulates a sympathetic reflex (ureterorenal reflex) that constricts the renal arterioles and reduces urine production in the affected kidney.

## Urinary Bladder

The **urinary bladder** is a hollow, distensible, muscular organ. It is located within the pelvic cavity, posterior to the symphysis pubis, and inferior to the parietal peritoneum (fig. 20.28 and reference plate 52). In a male, the bladder lies posteriorly against the rectum, and in a female, it contacts the anterior walls of the uterus and vagina.

The pressure of surrounding organs alters the spherical shape of the bladder. When the bladder is empty, its inner wall forms many folds, but as it fills with urine, the wall becomes smoother. At the same time, the superior surface of the bladder expands upward into a dome.

When it is greatly distended, the bladder pushes above the pubic crest and into the region between the abdominal wall and the parietal peritoneum. The dome can reach the level of the umbilicus and press against the coils of the small intestine.

The internal floor of the bladder includes a triangular area called the *trigone,* which has an opening at each of its three angles (fig. 20.29). Posteriorly, at the base of the trigone, the openings are those of the ureters. Anteriorly, at the apex of the trigone, is a short, funnel-shaped extension called the *neck* of the bladder, which contains the opening into the urethra. The trigone generally remains in a fixed position, even though the rest of the bladder distends and contracts.

The wall of the urinary bladder consists of four layers. The inner layer, or *mucous coat,* includes several thicknesses of transitional epithelial cells, similar to those lining the ureters and the upper portion of the urethra (see chapter 5). The thickness of this tissue changes as the bladder expands and contracts. Thus, during distension, the tissue appears to be only two or three cells thick, but during contraction, it appears to be five or six cells thick (see fig. 5.9).

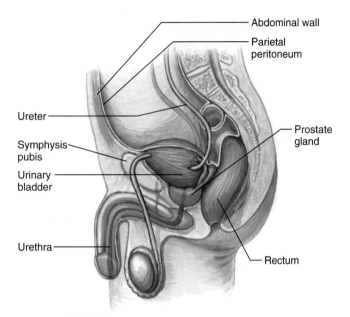

**FIGURE** 20.28

The urinary bladder is located within the pelvic cavity and behind the symphysis pubis. In a male, it lies against the rectum.

Former Vice President Hubert Humphrey died in 1978 of bladder cancer. He had been diagnosed in 1976, and an exam in 1973 had found a "borderline malignancy." In 1967, he had experienced his first symptom—blood in the urine. In 1994, researchers used the polymerase chain reaction to mass-produce a known cancer-causing gene variant (called p53) from preserved samples of the vice president's bladder tissue. The test revealed a mutation affecting only one DNA base, which caused the cancer.

Detecting a genetic change in cancer cells can be clinically useful in several ways. It can permit early diagnosis, perhaps even before symptoms appear, when treatment is more likely to be successful. Absence of the causative gene variant in cells bordering an excised tumor indicates that sufficient tissue has been removed to stop the cancer. Finally, a genetic test can detect a recurrence of cancer before symptoms arise.

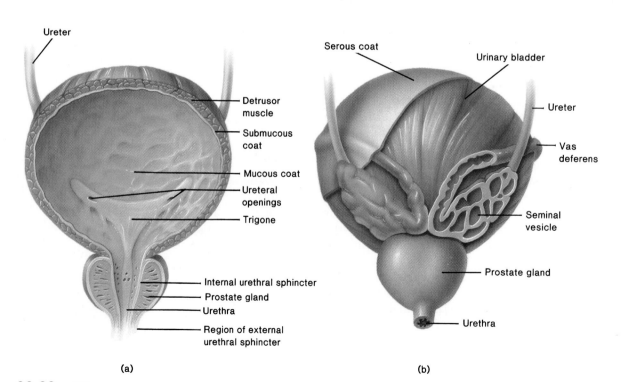

(a)

(b)

**FIGURE** 20.29

A male urinary bladder. (*a*) Coronal section. (*b*) Posterior view.

The second layer of the bladder wall is the *submucous coat*. It consists of connective tissue and contains many elastic fibers.

The third layer of the bladder wall, the *muscular coat*, is primarily composed of coarse bundles of smooth muscle fibers. These bundles are interlaced in all directions and at all depths, and together they comprise the **detrusor muscle.** The portion of the detrusor muscle that surrounds the neck of the bladder forms an *internal urethral sphincter.* Sustained contraction of this sphincter muscle prevents the bladder from emptying until the pressure within it increases to a certain level. The detrusor muscle has parasympathetic nerve fibers that function in the reflex that passes urine.

The outer layer of the wall, the *serous coat,* consists of the parietal peritoneum. It is found only on the upper surface of the bladder. Elsewhere, the outer coat is composed of connective tissue with numerous fibers (fig. 20.30).

**1** Describe the trigone of the urinary bladder.

**2** Describe the structure of the bladder wall.

**3** What kind of nerve fibers supply the detrusor muscle?

## Micturition

Urine leaves the urinary bladder by the **micturition** (urination) reflex. The detrusor muscle contracts, and contractions of muscles in the abdominal wall and pelvic floor may help, as well as fixation of the thoracic wall and diaphragm. In micturition, the *external urethral sphincter* also relaxes. This muscle, which is part of the urogenital diaphragm (see chapter 9), surrounds the urethra about 3 centimeters from the bladder and is composed of voluntary skeletal muscle tissue.

**FIGURE** 20.30

Light micrograph of the human urinary bladder wall (6×).

Muscular coat — Mucous coat — Submucous coat — Lumen

Distension of the bladder wall as it fills with urine stimulates the urge to urinate. The wall expands, stimulating stretch receptors, which triggers the micturition reflex.

The *micturition reflex center* is located in the sacral portion of the spinal cord. When sensory impulses from the stretch receptors signal the reflex center, parasympathetic motor impulses travel out to the detrusor muscle, which contracts rhythmically in response. A sensation of urgency accompanies this action.

Incontinence is the loss of control of micturition. Stress incontinence, caused by pressure on the bladder, is particularly common among women who have had children, especially if they have gained weight. An effective treatment is at least two months of doing Kegel exercises, in which a woman contracts the muscles that support the bladder, several times daily. Treatments for severe cases include a tampon-like cone inserted into the vagina to raise the pelvic floor; a small foam pad placed over the urethra to catch small amounts of urine; collagen injections around the urethra to tighten it; and surgery. Many people simply use absorbent pads that are widely available.

The urinary bladder may hold as much as 600 milliliters of urine. The desire to urinate usually appears when it contains about 150 milliliters. Then, as urine volume increases to 300 milliliters or more, the sensation of fullness becomes increasingly uncomfortable.

As the bladder fills with urine, and its internal pressure increases, contractions of its wall become more and more powerful. When these contractions become strong enough to force the internal urethral sphincter to open, another reflex signals the external urethral sphincter to relax, and the bladder may empty.

However, because the external urethral sphincter is composed of skeletal muscle, it is under conscious control. Thus, the sphincter muscle ordinarily remains contracted until a decision is made to urinate. Nerve centers in the brain stem and cerebral cortex that inhibit the micturition reflex aid this control. When a person decides to urinate, the external urethral sphincter relaxes, and the micturition reflex is no longer inhibited. Nerve centers within the pons and the hypothalamus heighten the micturition reflex. Consequently, the detrusor muscle contracts, and urine is excreted through the urethra. Within a few moments, the neurons of the micturition reflex tire, the detrusor muscle relaxes, and the bladder begins to fill with urine again. Table 20.4 outlines the micturition process, and Clinical Application 20.5 discusses inherited conditions that affect urine composition.

# Urinalysis: Clues to Health

Urine has long fascinated medical minds. As a folk remedy, urine has been used as a mouthwash, toothache treatment, and a cure for sore eyes. Hippocrates (460–377 B.C.) was the first to observe that the condition of the urine can reflect health, noting that frothy urine denoted kidney disease. During the Middle Ages, health practitioners consulted charts that matched certain urine colors to certain diseases. In the seventeenth century, British physicians diagnosed diabetes by having their medical students taste sugar in patients' urine. Today, urine composition is still used as a window on health, and also to check for illicit drug use.

Certain inherited disorders can alter urine quite noticeably. The name *maple syrup urine disease* vividly describes what this inborn error of metabolism does to the urine. This condition, which causes mental retardation, results from a block in the breakdown pathways for certain amino acids. In *alkaptonuria,* one of the first inborn errors to be described, urine turns black on standing. This condition also produces painful arthritis and blackened ear tips. People with *Wilson disease* have an inherited inability to excrete copper. If they are properly diagnosed and given the drug penicillamine, they excrete some of the built-up copper, and their urine takes on a coppery appearance.

Other genetic conditions alter urine without causing health problems. People with *beeturia* excrete dark pink urine after they eat beets. The problem for people with *urinary excretion of odoriferous component of asparagus* is obvious. Parents of newborns who have inherited *blue diaper syndrome* are in for a shock when they change their child's first diaper. Due to a defect in transport of the amino acid tryptophan in the small intestine, bacteria degrade the partially digested tryptophan, producing a compound that turns blue on contact with oxygen. ∎

| TABLE 20.4 | MAJOR EVENTS OF MICTURITION |
|---|---|

1. Urinary bladder distends as it fills with urine.
2. Stretch receptors in the bladder wall are stimulated and they signal the micturition center in the sacral spinal cord.
3. Parasympathetic nerve impulses travel to the detrusor muscle, which responds by contracting rhythmically.
4. The need to urinate is sensed as urgent.
5. Voluntary contraction of the external urethral sphincter and inhibition of the micturition reflex by impulses from the brain stem and the cerebral cortex prevent urination.
6. Following the decision to urinate, the external urethral sphincter is relaxed, and impulses from the pons and the hypothalamus facilitate the micturition reflex.
7. The detrusor muscle contracts, and urine is expelled through the urethra.
8. Neurons of the micturition reflex center fatigue, the detrusor muscle relaxes, and the bladder begins to fill with urine again.

Damage to the spinal cord above the sacral region may abolish voluntary control of urination. However, if the micturition reflex center and its sensory and motor fibers are uninjured, micturition may continue to occur reflexively. In this case, the bladder collects urine until its walls stretch enough to trigger a micturition reflex, and the detrusor muscle contracts in response. This condition is called an *automatic bladder.*

## Urethra

The **urethra** is a tube that conveys urine from the urinary bladder to the outside of the body. Its wall is lined with mucous membrane and contains a thick layer of longitudinal smooth muscle fibers. It also contains numerous mucous glands, called *urethral glands,* which secrete mucus into the urethral canal (fig. 20.31).

In a female, the urethra is about 4 centimeters long. It passes forward from the bladder, courses below the symphysis pubis, and empties between the labia minora. Its opening, the *external urethral orifice* (urinary meatus), is located anterior to the vaginal opening and about 2.5 centimeters posterior to the clitoris (fig. 20.32*a*).

In a male, the urethra, which functions both as a urinary canal and a passageway for cells and secretions from the reproductive organs, can be divided into three sections: the prostatic urethra, the membranous urethra, and the penile urethra (see fig. 20.32*b* and reference plate 60).

The **prostatic urethra** is about 2.5 centimeters long and passes from the urinary bladder through the *prostate gland,* which is located just below the bladder. Ducts from reproductive structures join the urethra in this region.

The **membranous urethra** is about 2 centimeters long. It begins just distal to the prostate gland, passes through the urogenital diaphragm, and is surrounded by the fibers of the external urethral sphincter muscle.

The **penile urethra** is about 15 centimeters long and passes through the corpus spongiosum of the penis, where erectile tissue surrounds it. This portion of the urethra terminates with the *external urethral* orifice at the tip of the penis.

**1** Describe micturition.

**2** How is it possible to consciously inhibit the micturition reflex?

**3** Describe the structure of the urethra.

**4** How does the urethra of a male differ from that of a female?

(a)

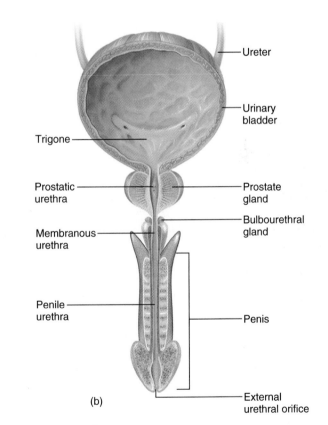

(b)

**FIGURE** 20.32

(*a*) Longitudinal section of the female urinary bladder and urethra. (*b*) Longitudinal section of the male urinary bladder and urethra.

**FIGURE** 20.31

Cross section of the urethra (10×).

**URINARY SYSTEM**

The urinary system controls the composition of the internal environment.

## Integumentary System

The urinary system compensates for water loss due to sweating. The kidneys and skin both play a role in vitamin D production.

## Skeletal System

The kidneys and bone tissue work together to control plasma calcium levels.

## Muscular System

Muscle tissue controls urine elimination from the bladder.

## Nervous System

The nervous system influences urine production and elimination.

## Endocrine System

The endocrine system influences urine production.

## Cardiovascular System

The urinary system controls blood volume. Blood volume and blood pressure play a role in determining water and solute excretion.

## Lymphatic System

The kidneys control extracellular fluid volume and composition (including lymph).

## Digestive System

The kidneys compensate for fluids lost by the digestive system.

## Respiratory System

The kidneys and the lungs work together to control the pH of the internal environment.

## Reproductive System

The urinary system in males shares common organs with the reproductive system. The kidneys compensate for fluids lost from the male and female reproductive systems.

## CLINICAL TERMS RELATED TO THE URINARY SYSTEM

**anuria** (ah-nu're-ah)   Absence of urine due to failure of kidney function or to an obstruction in a urinary pathway.

**bacteriuria** (bak-te″re-u're-ah)   Bacteria in the urine.

**cystectomy** (sis-tek'to-me)   Surgical removal of the urinary bladder.

**cystitis** (sis-ti'tis)   Inflammation of the urinary bladder.

**cystoscope** (sis'to-skop)   Instrument to visually examine the interior of the urinary bladder.

**cystotomy** (sis-tot'o-me)   Incision of the wall of the urinary bladder.

**diuresis** (di″u-re'sis)   Increased production of urine.

**diuretic** (di″u-ret'ik)   Substance that increases urine production.

**dysuria** (dis-u're-ah)   Painful or difficult urination.

**enuresis** (en″u-re'sis)   Uncontrolled urination.

**hematuria** (hem″ah-tu're-ah)   Blood in the urine.

**incontinence** (in-kon'ti-nens)   Inability to control urination and/or defecation reflexes.

**nephrectomy** (ne-frek'to-me)   Surgical removal of a kidney.

**nephrolithiasis** (nef″ro-li-thi'ah-sis)   Kidney stones.

**nephroptosis** (nef'rop-to'sis)   Movable or displaced kidney.

**oliguria** (ol″i-gu're-ah)   Scanty output of urine.

**polyuria** (pol″e-u're-ah)   Excess urine.

**pyelolithotomy** (pi″ĕ-lo-li-thot'o-me)   Removal of a stone from the renal pelvis.

**pyelonephritis** (pi″ĕ-lo-ne-fri'tis)   Inflammation of the renal pelvis.

**pyelotomy** (pi″ĕ-lot'o-me)   Incision into the renal pelvis.

**pyuria** (pi-u're-ah)   Pus (excess white blood cells) in the urine.

**uremia** (u-re'me-ah)   Condition in which substances ordinarily excreted in the urine accumulate in the blood.

**ureteritis** (u-re″ter-i'tis)   Inflammation of the ureter.

**urethritis** (u″re-thri'tis)   Inflammation of the urethra.

## CHAPTER SUMMARY

## Introduction (page 779)

The urinary system consists of the kidneys, ureters, urinary bladder, and urethra.

## Kidneys (page 779)

1. Location of the kidneys
   a. The kidneys are bean-shaped organs on either side of the vertebral column, high on the posterior wall of the abdominal cavity.
   b. They are positioned posterior to the parietal peritoneum, and held in place by adipose and connective tissue.
2. Kidney structure
   a. A kidney contains a hollow renal sinus.
   b. The ureter expands into the renal pelvis, which, in turn, is divided into major and minor calyces.
   c. Renal papillae project into the renal sinus.
   d. Kidney tissue is divided into a medulla and a cortex.
3. Functions of the kidneys
   a. The kidneys remove metabolic wastes from the blood and excrete them to the outside.
   b. They also help regulate red blood cell production, blood pressure, calcium ion absorption, and the volume, composition, and pH of the blood.
4. Renal blood vessels
   a. Arterial blood flows through the renal artery, interlobar arteries, arcuate arteries, interlobular arteries, afferent arterioles, glomerular capillaries, efferent arterioles, and peritubular capillaries.
   b. Venous blood returns through a series of vessels that correspond to those of the arterial pathways.
5. Nephrons
   a. Structure of a nephron
      (1) A nephron is the functional unit of the kidney.
      (2) It consists of a renal corpuscle and a renal tubule.
         (a) The corpuscle consists of a glomerulus and a glomerular capsule.
         (b) Portions of the renal tubule include the proximal convoluted tubule, the nephron loop (ascending and descending limbs), and the distal convoluted tubule.
      (3) The nephron joins a collecting duct, which empties into the minor calyx of the renal pelvis.
   b. Juxtaglomerular apparatus
      (1) The juxtaglomerular apparatus is located at the point of contact between the distal convoluted tubule and the afferent and efferent arterioles.
      (2) It consists of the macula densa and the juxtaglomerular cells.
   c. Cortical and juxtamedullary nephrons
      (1) Cortical nephrons are the most numerous and have corpuscles near the surface of the kidney.
      (2) Juxtamedullary nephrons have corpuscles near the medulla.
   d. Blood supply of a nephron
      (1) The glomerular capillary receives blood from the afferent arteriole and passes it to the efferent arteriole.
      (2) The efferent arteriole gives rise to the peritubular capillary system, which surrounds the renal tubule.
      (3) Capillary loops, called vasa recta, dip down into the medulla.

## Urine Formation (page 789)

Nephrons remove wastes from the blood and regulate water and electrolyte concentrations. Urine is the product of these functions. Kidney function includes filtration, reabsorption, and secretion of substances from renal tubules.

1. Glomerular filtration
   a. Urine formation begins when water and dissolved materials are filtered out of the glomerular capillary.
   b. The glomerular capillaries are much more permeable than the capillaries in other tissues.
2. Filtration pressure
   a. Filtration is mainly due to hydrostatic pressure inside the glomerular capillaries.
   b. The osmotic pressure of the blood plasma and hydrostatic pressure in the glomerular capsule also affect filtration.
   c. Filtration pressure is the net force acting to move material out of the glomerulus and into the glomerular capsule.
   d. The composition of the filtrate is similar to that of tissue fluid.
3. Filtration rate
   a. The rate of filtration varies with the filtration pressure.
   b. Filtration pressure changes with the diameters of the afferent and efferent arterioles.
   c. As the osmotic pressure in the glomerulus increases, filtration decreases.
   d. As the hydrostatic pressure in a glomerular capsule increases, the filtration rate decreases.
   e. The kidneys produce about 125 milliliters of glomerular fluid per minute, most of which is reabsorbed.
   f. The volume of filtrate varies with the surface area of the glomerular capillary.
4. Regulation of filtration rate
   a. Glomerular filtration rate (GFR) remains relatively constant, but may be increased or decreased when the need arises. Increased sympathetic nerve activity can decrease GFR.
   b. When tubular fluid NaCl concentration decreases, the macula densa causes the juxtaglomerular cells to release renin. This triggers a series of changes leading to vasoconstriction, which may affect GFR, and secretion of aldosterone, which stimulates tubular sodium reabsorption.
   c. Autoregulation is the ability of an organ or tissue to maintain a constant blood flow under certain conditions when the arterial blood pressure is changing.
5. Tubular reabsorption
   a. Substances are selectively reabsorbed from the glomerular filtrate.
   b. The peritubular capillary is adapted for reabsorption.
      (1) It carries low-pressure blood.
      (2) It is very permeable.
   c. Most reabsorption occurs in the proximal tubule, where the epithelial cells possess microvilli.
   d. Different modes of transport reabsorb various substances in particular segments of the renal tubule.
      (1) Glucose and amino acids are reabsorbed by active transport.
      (2) Water is reabsorbed by osmosis.
      (3) Proteins are reabsorbed by pinocytosis.
   e. Active transport mechanisms have limited transport capacities.
   f. If the concentration of a substance in the filtrate exceeds its renal plasma threshold, the excess is excreted in the urine.
   g. Substances that remain in the filtrate are concentrated as water is reabsorbed.

   h. Sodium ions are reabsorbed by active transport.
      (1) As positively charged sodium ions are transported out of the filtrate, negatively charged ions accompany them.
      (2) Water is passively reabsorbed by osmosis as sodium ions are actively reabsorbed.
6. Regulation of urine concentration and volume
   a. Most of the sodium ions are reabsorbed before the urine is excreted.
   b. Sodium ions are concentrated in the renal medulla by the countercurrent mechanism.
      (1) Chloride ions are actively reabsorbed in the ascending limb, and sodium ions follow them passively.
      (2) Tubular fluid in the ascending limb becomes hypotonic as it loses solutes.
      (3) Water leaves the descending limb by osmosis, and NaCl enters this limb by diffusion.
      (4) Tubular fluid in the descending limb becomes hypertonic as it loses water and gains NaCl.
      (5) As NaCl repeats this circuit, its concentration in the medulla increases.
   c. The vasa recta countercurrent mechanism helps maintain the NaCl concentration in the medulla.
   d. The distal tubule and collecting duct are impermeable to water, which therefore is excreted in urine.
   e. ADH from the posterior pituitary gland increases the permeability of the distal tubule and collecting duct, promoting water reabsorption.
7. Urea and uric acid excretion
   a. Urea is a by-product of amino acid metabolism.
      (1) It is passively reabsorbed by diffusion.
      (2) About 50% of the urea is excreted in urine.
      (3) A countercurrent mechanism involving urea helps in the reabsorption of water.
   b. Uric acid results from the metabolism of nucleic acids.
      (1) Most is reabsorbed by active transport.
      (2) Some is secreted into the renal tubule.
8. Tubular secretion
   a. Tubular secretion transports certain substances from the plasma to the tubular fluid.
   b. Some substances are actively secreted.
      (1) These include various organic compounds and hydrogen ions.
      (2) The proximal and distal segments of the renal tubule secrete hydrogen ions.
   c. Potassium ions are secreted both actively and passively in the distal tubule and collecting duct.
9. Urine composition
   a. Urine is about 95% water, and it usually contains urea, uric acid, and creatinine.
   b. It may contain a trace of amino acids and varying amounts of electrolytes, depending upon dietary intake.
   c. The volume of urine varies with the fluid intake and with certain environmental factors.

## Elimination of Urine (page 802)

1. Ureters
   a. The ureter is a tubular organ that extends from each kidney to the urinary bladder.
   b. Its wall has mucous, muscular, and fibrous layers.
   c. Peristaltic waves in the ureter force urine to the bladder.

d. Obstruction in the ureter stimulates strong peristaltic waves and a reflex that decreases urine production.
2. Urinary bladder
   a. The urinary bladder is a distensible organ that stores urine and forces it into the urethra.
   b. The openings for the ureters and urethra are located at the three angles of the trigone in the floor of the urinary bladder.
   c. Muscle fibers in the wall form the detrusor muscle.
   d. A portion of the detrusor muscle forms an internal urethral sphincter.
3. Micturition
   a. Micturition is the process of expelling urine.
   b. It involves contraction of the detrusor muscle and relaxation of the external urethral sphincter.
   c. Micturition reflex
      (1) Distension stimulates stretch receptors in the bladder wall.
      (2) The micturition reflex center in the sacral portion of the spinal cord sends parasympathetic motor impulses to the detrusor muscle.
      (3) As the bladder fills, its internal pressure increases, forcing the internal urethral sphincter open.
      (4) A second reflex relaxes the external urethral sphincter, unless its contraction is voluntarily controlled.
      (5) Nerve centers in the brain stem and cerebral cortex aid control of urination.
4. Urethra
   a. The urethra conveys urine from the bladder to the outside.
   b. In females, it empties between the labia minora.
   c. In males, it conveys products of reproductive organs as well as urine.
      (1) Three portions of the male urethra are prostatic, membranous, and penile.
      (2) The urethra empties at the tip of the penis.

## CRITICAL THINKING QUESTIONS

1. If an infant is born with narrowed renal arteries, what effect would this condition have on the volume of urine produced? Explain your answer.
2. Why are people with nephrotic syndrome, in which plasma proteins are lost into the urine, more susceptible to infections?
3. If a patient who has had major abdominal surgery receives intravenous fluids equal to the volume of blood lost during surgery, would you expect the volume of urine produced to be greater or less than normal? Why?
4. If a physician prescribed oral penicillin therapy for a patient with an infection of the urinary bladder, how would you describe for the patient the route by which the drug would reach the bladder?
5. If the blood pressure of a patient who is in shock as a result of a severe injury decreases greatly, how would you expect the volume of urine to change? Why?
6. Inflammation of the urinary bladder is more common in women than in men. How might this observation be related to the anatomy of the female and male urethras?

## REVIEW EXERCISES

1. Name the organs of the urinary system, and list their general functions.
2. Describe the external and internal structure of a kidney.
3. List the functions of the kidneys.
4. Name the vessels the blood passes through as it travels from the renal artery to the renal vein.
5. Distinguish between a renal corpuscle and a renal tubule.
6. Name the structures fluid passes through as it travels from the glomerulus to the collecting duct.
7. Describe the location and structure of the juxtaglomerular apparatus.
8. Distinguish between cortical and juxtamedullary nephrons.
9. Distinguish among filtration, reabsorption, and secretion as they relate to urine formation.
10. Define filtration pressure.
11. Compare the composition of the glomerular filtrate with that of the blood plasma.
12. Explain how the diameters of the afferent and efferent arterioles affect the rate of glomerular filtration.
13. Explain how changes in the osmotic pressure of the blood plasma may affect the rate of glomerular filtration.
14. Explain how the hydrostatic pressure of a glomerular capsule affects the rate of glomerular filtration.
15. Describe two mechanisms by which the body regulates the filtration rate.
16. Define autoregulation.
17. Discuss how tubular reabsorption is a selective process.
18. Explain how the peritubular capillary is adapted for reabsorption.
19. Explain how the epithelial cells of the proximal convoluted tubule are adapted for reabsorption.
20. Explain why active transport mechanisms have limited transport capacities.
21. Define renal plasma threshold, and explain its significance in tubular reabsorption.
22. Describe the renal defects associated with polycystic kidney disease, Wilms tumor, and alkaptonuria.
23. Explain how amino acids and proteins are reabsorbed.
24. Describe the effect of sodium reabsorption on the reabsorption of negatively charged ions.
25. Explain how sodium ion reabsorption affects water reabsorption.
26. Explain how hypotonic tubular fluid is produced in the ascending limb of the nephron loop.
27. Explain why fluid in the descending limb of the nephron loop is hypertonic.

28. Describe the function of ADH.
29. Explain how urine may become concentrated as it moves through the collecting duct.
30. Compare the processes by which urea and uric acid are reabsorbed.
31. Explain how the renal tubule is adapted to secrete hydrogen ions.
32. Explain how potassium ions may be secreted passively.
33. List the more common substances found in urine and their sources.
34. List some of the factors that affect the volume of urine produced each day.
35. Describe the structure and function of a ureter.
36. Explain how the muscular wall of the ureter aids in moving urine.
37. Discuss what happens if a ureter becomes obstructed.
38. Describe the structure and location of the urinary bladder.
39. Define *detrusor muscle.*
40. Distinguish between the internal and external urethral sphincters.
41. Describe the micturition reflex.
42. Explain how the micturition reflex can be voluntarily controlled.
43. Compare the urethra of a female with that of a male.

 **For More Information Related to This Chapter, Please Visit** *Hole's Human Anatomy and Physiology* **Home Page**
http://www.mhhe.com/biosci/abio/

# WATER, ELECTROLYTE, AND ACID-BASE BALANCE

## CHAPTER OBJECTIVES

*AFTER YOU HAVE STUDIED THIS CHAPTER, YOU SHOULD BE ABLE TO:*

1. Explain water and electrolyte balance, and discuss the importance of this balance.

2. Describe how the body fluids are distributed within compartments, how fluid composition differs between compartments, and how fluids move from one compartment to another.

3. List the routes by which water enters and leaves the body, and explain how water input and output are regulated.

4. Explain how electrolytes enter and leave the body, and how the input and output of electrolytes are regulated.

5. Explain acid-base balance.

6. Describe how hydrogen ion concentrations are expressed mathematically.

7. List the major sources of hydrogen ions in the body.

8. Distinguish between strong and weak acids and bases.

9. Explain how chemical buffer systems, the respiratory center, and the kidneys minimize changing pH values of the body fluids.

## KEY TERMS

**acid** (as'id)
**acidosis** (as"i-do'sis)
**alkalosis** (al"kah-lo'sis)
**base** (bās)
**buffer system** (buf'er sis'tem)
**electrolyte balance** (e-lek'tro-līt bal'ans)
**extracellular** (ek"strah-sel'u-lar)
**intracellular** (in"trah-sel'u-lar)
**osmoreceptor** (oz"mo-re-sep'tor)
**transcellular** (trans-sel'u-lar)
**water balance** (wot'er bal'ans)

## UNDERSTANDING WORDS

**de-,** separation from: *de*hydration—removal of water from the cells or body fluids.

**edem-,** swelling: *edem*a—swelling due to an abnormal accumulation of extracellular fluid.

**-emia,** a blood condition: hypoprotein*emia*—abnormally low concentration of blood plasma protein.

**extra-,** outside: *extra*cellular fluid—fluid outside of the body cells.

**im- (or in-),** not: *im*balance—condition in which factors are not in equilibrium.

**intra-,** within: *intra*cellular fluid—fluid within the body cells.

**neutr-,** neither one nor the other: *neutr*al—solution that is neither acidic nor basic.

**-osis,** a state of: acid*osis*—condition in which hydrogen ion concentration is abnormally high.

**-uria,** a urine condition: keto*uria*—presence of ketone bodies in the urine.

he summer of 1995 was a brutal time in Chicago. A fierce heat wave swept the city on July 12–16, with temperatures hovering between 93–104°F. The Cook County medical examiner's office worked round the clock during this time to handle 465 heat-related deaths. Several factors point to a heat-related death:

Body temperature at or soon after death of greater than 105°F

Evidence that heat caused the death

Decomposition of the corpse

No other identifiable cause of death

Evidence that the person was alive just before the heat wave

In the many blazing hot, un-air-conditioned apartments of Chicago, mostly elderly people fell victim. An untold number suffered severe dehydration, which may have been the direct cause of death. As body fluids became concentrated, with not enough water to form sweat to counter the rising and relentless heat, body temperatures soared. Wastes accumulating in extracellular fluids may have caused symptoms of cerebral disturbance, including mental confusion, delirium, and coma, preventing many people from seeking help.

Heat waves are deadly. It is important to keep cool.

Other communities also lost citizens to the heat wave of 1995. As a result, local governments now issue public service announcements informing people how to deal with heat and provide air-conditioned areas for people to stay. A piece of advice for all—when the mercury rises, drink! The human body cannot function for long with too little water.

---

The term *balance* suggests a state of equilibrium, and in the case of water and electrolytes, it means that the quantities entering the body equal the quantities leaving it. Maintaining such a balance requires mechanisms to ensure that lost water and electrolytes will be replaced and that any excesses will be excreted. As a result, the levels of water and electrolytes in the body remain relatively stable at all times.

It is important to remember that water balance and electrolyte balance are interdependent because electrolytes are dissolved in the water of body fluids. Consequently, anything that alters the concentrations of the electrolytes will necessarily alter the concentration of the water by adding solutes to it or by removing solutes from it. Likewise, anything that changes the concentration of the water will change the concentrations of the electrolytes by making them either more concentrated or more dilute.

## DISTRIBUTION OF BODY FLUIDS

Body fluids are not uniformly distributed. Instead, they occur in regions, or *compartments,* of different volumes that contain fluids of varying compositions. The movement of water and electrolytes between these compartments is regulated to stabilize their distribution and the composition of body fluids.

## Fluid Compartments

The body of an average adult female is about 52% water by weight, and that of an average male is about 63% water. The differences between the sexes is due to the fact that females generally have more adipose tissue, which has little water. Males have relatively more muscle tissue, which contains a great deal of water. Water in the body (about 40 liters), together with its dissolved electrolytes, is distributed into two major compartments: an intracellular fluid compartment and an extracellular fluid compartment (fig. 21.1).

The **intracellular fluid compartment** includes all the water and electrolytes that cell membranes enclose. In other words, intracellular fluid is the fluid within the cells, and, in an adult, it represents about 63% by volume of the total body water.

The **extracellular fluid compartment** includes all the fluid outside the cells—within the tissue spaces (interstitial fluid), the blood vessels (plasma), and the lymphatic vessels (lymph). Epithelial layers separate a specialized fraction of the extracellular fluid from other extracellular fluids. This *transcellular fluid* includes cerebrospinal fluid of the central nervous system, aqueous and vitreous humors of the eyes, synovial fluid of the joints, serous fluid within the body cavities, and fluid secretions of the exocrine glands. The fluids of the extracellular compartment constitute about 37% by volume of the total body water (fig. 21.2).

## Body Fluid Composition

*Extracellular fluids* generally have similar compositions, including high concentrations of sodium, chloride, and bicarbonate ions, and lesser concentrations of potassium,

calcium, magnesium, phosphate, and sulfate ions. The blood plasma fraction of extracellular fluid contains considerably more protein than do either interstitial fluid or lymph.

*Intracellular fluid* contains high concentrations of potassium, phosphate, and magnesium ions. It includes a greater concentration of sulfate ions, and lesser concentrations of sodium, chloride, and bicarbonate ions than does extracellular fluid. Intracellular fluid also has a greater concentration of protein than plasma. Figure 21.3 shows these relative concentrations.

**1** How are fluid balance and electrolyte balance interdependent?

**2** Describe the normal distribution of water within the body.

**3** Which electrolytes are in higher concentrations in extracellular fluids? In intracellular fluid?

**4** How does the concentration of protein vary in the various body fluids?

## Movement of Fluid between Compartments

Two major factors regulate the movement of water and electrolytes from one fluid compartment to another: hydrostatic pressure and osmotic pressure. For example, as explained in chapter 15, fluid leaves the plasma at the arteriolar ends of capillaries and enters the interstitial spaces because of the net outward force of *hydrostatic pressure* (blood pressure). Fluid returns to the plasma from the interstitial spaces at the venular ends of capillaries because of the net inward force of *osmotic pressure*. Likewise, as mentioned in chapter 16, fluid leaves the interstitial spaces and enters the lymph capillaries due to the hydrostatic pressure of the interstitial fluid. As a result of the circulation of lymph, interstitial fluid returns to the plasma.

Such pressures similarly control the movement of fluid between the intracellular and extracellular compartments. However, because hydrostatic pressure within

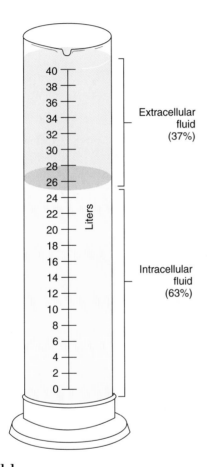

**FIGURE 21.1**

Of the 40 liters of water in the body of an average adult male, about two-thirds is intracellular and one-third is extracellular.

**FIGURE 21.2**

Cell membranes separate fluid in the intracellular compartment from fluid in the extracellular compartment. Approximately two-thirds of the water in the body is inside cells.

the cells and surrounding interstitial fluid is ordinarily equal and remains stable, any net fluid movement that occurs is likely to be the result of changes in osmotic pressure (fig. 21.4). Thus, although the composition of body fluids varies from one compartment to another, the total solute concentrations and the water concentrations are normally equal. A net gain or loss of water will therefore cause shifts affecting both the intracellular and extracellular fluids.

For example, because the sodium ion concentration in the extracellular fluids is relatively high, a decrease in this concentration will cause a net movement of water from the extracellular compartment into the intracellular compartment by osmosis. As a consequence, the cells will swell. Conversely, if the concentration of sodium ions in the interstitial fluid increases, the net movement of water will be outward from the intracellular compartment, and the cells will shrink as they lose water.

**1** Which factors control the movement of water and electrolytes from one fluid compartment to another?

**2** How does the sodium ion concentration within body fluids affect the net movement of water between the compartments?

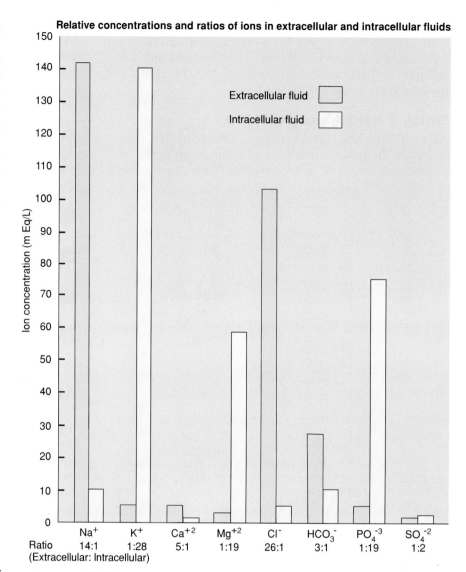

Relative concentrations and ratios of ions in extracellular and intracellular fluids

| | Extracellular fluid |
| | Intracellular fluid |

Ion concentration (m Eq/L)

| Ratio (Extracellular: Intracellular) | $Na^+$ 14:1 | $K^+$ 1:28 | $Ca^{+2}$ 5:1 | $Mg^{+2}$ 1:19 | $Cl^-$ 26:1 | $HCO_3^-$ 3:1 | $PO_4^{-3}$ 1:19 | $SO_4^{-2}$ 1:2 |

**FIGURE 21.3**

Extracellular fluid has relatively high concentrations of sodium, calcium, chloride, and bicarbonate ions. Intracellular fluid has relatively high concentrations of potassium, magnesium, phosphate, and sulfate ions.

## WATER BALANCE

Water balance exists when water intake equals water output. Homeostasis requires that both water intake and water output be controlled to maintain the constancy of the internal environment. Ultimately, maintenance of the internal environment depends on the kidneys' ability to vary water output, and thirst centers in the brain to vary water intake.

### Water Intake

The volume of water gained each day varies from individual to individual. An average adult living in a moderate environment takes in about 2,500 milliliters. Of this amount, probably 60% is obtained from drinking water or beverages, and another 30% comes from moist foods. The remaining 10% is a by-product of the oxidative metabolism of nutrients, which is called **water of metabolism** (fig. 21.5).

### Regulation of Water Intake

The primary regulator of water intake is thirst. The intense feeling of thirst derives from the osmotic pressure of extracellular fluids and a *thirst center* in the hypothalamus of the brain.

As the body loses water, the osmotic pressure of the extracellular fluids increases. Such a change is thought to stimulate *osmoreceptors* in the thirst center, and as a result, the hypothalamus causes the person to feel thirsty and to seek water. A thirsty person usually has a dry mouth, which is a consequence of the loss of extracellular water and resulting decreased flow of saliva.

The thirst mechanism is normally triggered whenever the total body water is decreased by as little as 1%.

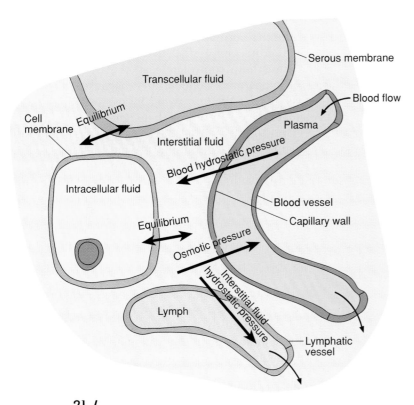

**Figure 21.4**
Net movements of fluids between compartments result from differences in hydrostatic and osmotic pressures.

**Average daily intake of water**

Water of metabolism (250 mL or 10%)

Water in moist food (750 mL or 30%)

Total intake (2,500 mL)

Water in beverages (1,500 mL or 60%)

**Figure 21.5**
Major sources of body water.

| TABLE 21.1 | REGULATION OF WATER INTAKE |
| --- | --- |

1. The body loses as little as 1% of its water.
2. An increase in the osmotic pressure of extracellular fluid due to water loss stimulates osmoreceptors in the thirst center.
3. Activity in the hypothalamus causes the person to feel thirsty and to seek water.
4. Drinking and the resulting distension of the stomach by water stimulate nerve impulses that inhibit the thirst center.
5. Water is absorbed through the walls of the stomach and small intestine.
6. The osmotic pressure of extracellular fluid returns to normal.

As a thirsty person drinks water, the act of drinking and the resulting distension of the stomach wall seem to trigger nerve impulses that inhibit the thirst mechanism. Thus, drinking stops long before the swallowed water is absorbed. This inhibition helps prevent the person from drinking more than is required to replace the quantity lost, avoiding development of an imbalance of the opposite type. Table 21.1 summarizes the steps in this mechanism.

**1** What is water balance?

**2** Where is the thirst center located?

**3** What stimulates fluid intake? What inhibits it?

## Water Output

Water normally enters the body only through the mouth, but it can be lost by a variety of routes. These include ob-vious losses in urine, feces, and sweat (sensible perspiration), as well as evaporation of water from the skin (insensible perspiration) and from the lungs during breathing.

If an average adult takes in 2,500 milliliters of water each day, then 2,500 milliliters must be eliminated to maintain water balance. Of this volume, perhaps 60% will be lost in urine, 6% in feces, and 6% in sweat. About 28% will be lost by evaporation from the skin and lungs (fig. 21.6). These percentages vary with such environmental factors as temperature and relative humidity, and with physical exercise.

If insufficient water is taken in, water output must be reduced to maintain balance. Water lost by sweating is a necessary part of the body's temperature control mechanism; water lost in feces accompanies the elimination of undigested food materials; and water lost by evaporation is largely unavoidable. Therefore, the primary means of regulating water output is control of urine production.

## Regulation of Water Output

The distal convoluted tubules and collecting ducts of the nephrons regulate the volume of water excreted in the urine. The epithelial linings of these segments of the renal tubule remain relatively impermeable to water unless antidiuretic hormone (ADH) is present.

Recall from chapter 13 that osmoreceptors in the hypothalamus help control release of ADH. If the blood plasma becomes more concentrated because of excessive water loss, these cells lose water by osmosis and shrink. This change triggers impulses that signal the posterior pituitary gland to release ADH. The ADH released into the bloodstream reaches the kidneys, where it increases the permeability of the distal tubule and collecting duct. Consequently, water reabsorption increases, and water is conserved. This action resists further osmotic change in the plasma. In fact, the *osmoreceptor-ADH mechanism* can reduce a normal urine production of 1,500 milliliters per day to about 500 milliliters per day when the body is dehydrating.

On the other hand, if a person drinks too much water, the plasma becomes less concentrated, and the osmoreceptors swell as they receive extra water by osmosis. In this instance, ADH release is inhibited, and the distal tubule and collecting duct remain impermeable to water. Consequently, less water is reabsorbed and more urine is produced. Table 21.2 summarizes the steps in this mechanism. Clinical Application 21.1 discusses disorders resulting from water imbalance.

**Average daily output of water**

Water lost in sweat (150 mL or 6%)

Water lost in feces (150 mL or 6%)

Water lost through skin and lungs (700 mL or 28%)

Total output (2,500 mL)

Water lost in urine (1,500 mL or 60%)

**FIGURE 21.6**

Routes by which the body loses water. Regulation of urinary water loss is most important in the control of water balance.

| TABLE 21.2 | EVENTS IN REGULATION OF WATER OUTPUT | |
| --- | --- | --- |
| **Dehydration** | | **Excess Water Intake** |
| **1.** Extracellular fluid becomes osmotically more concentrated. | | **1.** Extracellular fluid becomes osmotically less concentrated. |
| **2.** Osmoreceptors in the hypothalamus are stimulated by the increase in the osmotic pressure of body fluids. | | **2.** This change stimulates osmoreceptors in the hypothalamus. |
| **3.** The hypothalamus signals the posterior pituitary gland to release ADH into the blood. | | **3.** The posterior pituitary gland decreases ADH release. |
| **4.** Blood carries ADH to the kidneys. | | **4.** Renal tubules decrease water reabsorption. |
| **5.** ADH causes the distal convoluted tubules and collecting ducts to increase water reabsorption. | | **5.** Urine output increases, and excess water is excreted. |
| **6.** Urine output decreases, and further water loss is minimized. | | |

1. By what routes does the body lose water?

2. What is the primary regulator of water loss?

3. What types of water loss are unavoidable?

4. How does the hypothalamus regulate water balance?

## ELECTROLYTE BALANCE

An **electrolyte balance** exists when the quantities of electrolytes (molecules that release ions in water) the body gains equal those lost (fig. 21.7).

### Electrolyte Intake

The electrolytes of greatest importance to cellular functions release sodium, potassium, calcium, magnesium, chloride, sulfate, phosphate, bicarbonate, and hydrogen ions. These electrolytes are primarily obtained from foods, but they may also be found in drinking water and other beverages. In addition, some electrolytes are byproducts of metabolic reactions.

### Regulation of Electrolyte Intake

Ordinarily, a person obtains sufficient electrolytes by responding to hunger and thirst. However, a severe electrolyte deficiency may cause *salt craving,* which is a strong desire to eat salty foods.

### Electrolyte Output

The body loses some electrolytes by perspiring (sweat has about half the solute concentration of plasma). The quantities of electrolytes leaving vary with the amount of perspiration. More electrolytes are lost in sweat on warmer days and during strenuous exercise. Varying amounts of electrolytes are lost in the feces. The greatest electrolyte output occurs as a result of kidney function and urine production. The kidneys alter renal electrolyte losses to maintain the proper composition of body fluids.

1. Which electrolytes are most important to cellular functions?

2. Which mechanisms ordinarily regulate electrolyte intake?

3. By what routes does the body lose electrolytes?

### Regulation of Electrolyte Output

The concentrations of positively charged ions, such as sodium ($Na^+$), potassium ($K^+$), and calcium ($Ca^{+2}$), are particularly important. For example, certain concentrations of these ions are vital for nerve impulse conduction, muscle fiber contraction, and maintenance of cell membrane permeability.

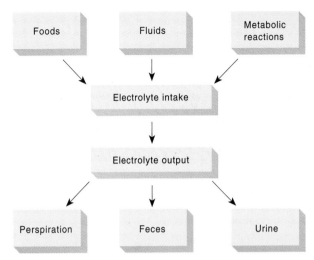

**FIGURE 21.7**

Electrolyte balance exists when the intake of electrolytes from all sources equals the output of electrolytes.

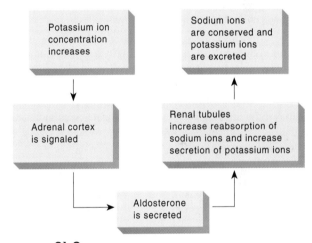

**FIGURE 21.8**

If the concentration of potassium ions increases, the kidneys conserve sodium ions and excrete potassium ions.

*Sodium ions* account for nearly 90% of the positively charged ions in extracellular fluids. The kidneys and the hormone aldosterone provide the primary mechanism regulating these ions. Aldosterone, which the adrenal cortex secretes, increases sodium ion reabsorption in the distal tubules and collecting ducts of the nephrons.

Aldosterone also regulates *potassium ions.* In fact, the most important stimulus for aldosterone secretion is a rising potassium ion concentration, which directly stimulates cells of the adrenal cortex. This hormone enhances the reabsorption of sodium ions and, at the same time, causes the renal tubules to excrete potassium ions (fig. 21.8).

Recall from chapter 13 that the calcium ion concentration dropping below normal directly stimulates the parathyroid glands to secrete parathyroid hormone. Parathyroid hormone increases activity in bone-resorbing cells (osteocytes and osteoclasts), which increases the

# Water Balance Disorders

Among the more common disorders involving an imbalance in the water of body fluids are dehydration, water intoxication, and edema.

## DEHYDRATION

In 1994, thousands of starving people died in the African nation of Rwanda. It wasn't lack of food that killed most of these people, but cholera, a bacterial infection that cripples the ability of intestinal lining cells to reabsorb water. The severe diarrhea that develops can kill in days, sometimes even hours. Dehydration is deadly.

*Dehydration* is a deficiency condition that occurs when output of water exceeds intake. It is a great problem for athletes, military personnel, and certain industrial workers. This condition may develop following excessive sweating, or as a result of prolonged water deprivation accompanied by continued water output. In either case, as water is lost, the extracellular fluid becomes increasingly more concentrated, and water tends to leave cells by osmosis (fig. 21A). Dehydration may also accompany illnesses in which prolonged vomiting or diarrhea depletes body fluids.

During dehydration, the skin and mucous membranes of the mouth feel dry, and body weight drops. Also, severe hyperthermia may develop as the body temperature-regulating mechanism becomes less effective due to a lack of water required for sweating. In severe cases, as waste products accumulate in the extracellular fluid, symptoms of cerebral disturbances, including mental confusion, delirium, and coma, may develop.

Because the kidneys of infants are less able to conserve water than are those of adults, infants are more likely to become dehydrated. Elderly people are also especially susceptible to developing water imbalances because the sensitivity of their thirst mechanisms decreases with age, and physical disabilities may make it difficult for them to obtain adequate fluids.

The treatment for dehydration is to replace the lost water and electrolytes. If only water is replaced, the extracellular fluid will become more dilute than normal. This may produce a condition called water intoxication.

## WATER INTOXICATION

Babies rushed to emergency rooms because they are having seizures sometimes are suffering from drinking too much water, a rare condition called *water intoxication*. This can occur when a baby under six months of age is given several bottles of water a day, or very dilute infant formula. The hungry infant gobbles down the water, and soon its tissues swell with the excess fluid. When the serum sodium level drops, the eyes begin to flutter and a seizure occurs. As extracellular fluid becomes hypotonic, water enters the cells rapidly by osmosis (fig. 21B). Coma resulting from swelling brain tissues may follow unless water intake is restricted and hypertonic salt solutions given to draw water back into the extracellular fluid. Usually, recovery is complete within a few days.

Water intoxication in infancy sometimes occurs when caregivers dilute formula to make it last longer. Another source of water intoxication in infants is bottled water products that are sold alongside infant formula. Their placement on grocery shelves leads some parents to believe that these products are adequate nutritional supplements. Used this way, they can be dangerous.

## EDEMA

*Edema* is an abnormal accumulation of extracellular fluid within the interstitial spaces. A variety of factors can cause it, including decrease in the plasma protein concentration (hy-

(1) Water loss from extracellular fluid compartment

Cell membrane

(2) Solute concentration increases in extracellular fluid compartment

Nucleus

(3) Water leaves intracellular fluid compartment by osmosis

**FIGURE 21A**

If excess extracellular fluids are lost, cells dehydrate by osmosis.

poproteinemia), obstructions in lymphatic vessels, increased venous pressure, and increased capillary permeability.

*Hypoproteinemia* may result from liver disease causing failure to synthesize plasma proteins: kidney disease (glomerulonephritis) that damages glomerular capillaries, allowing proteins to escape into the urine; or starvation, in which the intake of amino acids is insufficient to support synthesis of plasma proteins.

In each of these instances, the plasma protein concentration is decreased, which decreases plasma osmotic pressure, reducing the normal return of tissue fluid to the venular ends of capillaries. Tissue fluid consequently accumulates in the interstitial spaces.

As discussed in chapter 16, *lymphatic obstructions* may result from surgery or from parasitic infections of lymphatic vessels. Back pressure develops in the lymphatic vessels, which interferes with the normal movement of tissue fluid into them. At the same time, proteins that the lymphatic circulation ordinarily removes accumulate in the interstitial spaces, raising osmotic pressure of the interstitial fluid. This effect attracts still more fluid into the interstitial spaces.

If the outflow of blood from the liver into the inferior vena cava is blocked, the venous pressure within the liver and portal blood vessels increases greatly. This, in turn, raises pressure in liver sinusoids and intestinal capillaries. As a result, fluid with a high concentration of protein is exuded from the surfaces of the liver and intestine into the peritoneal cavity. This causes a rise in the osmotic pressure of the abdominal fluid, which, in turn, attracts more water into the peritoneal cavity by osmosis. This condition, called *ascites,* distends the abdomen. It is quite painful.

Edema may also result from increased capillary permeability accompanying *inflammation.* Recall that inflammation is a response to tissue damage and usually releases chemicals such as histamine from damaged cells. Histamine causes vasodilation and increased capillary permeability, so that excess fluid leaks out of the capillary and enters the interstitial spaces. Table 21A summarizes the factors that result in edema. ∎

(1) Excess water is added to extracellular fluid compartment

(2) Solute concentration of extracellular fluid compartment decreases

(3) Water moves into intracellular fluid compartment by osmosis

Cell membrane

Nucleus

**FIGURE 21B**

If excess water is added to the extracellular fluid compartment, cells gain water by osmosis.

| TABLE 21A | FACTORS ASSOCIATED WITH EDEMA | |
|---|---|---|
| **Factor** | **Cause** | **Effect** |
| Low plasma protein concentration | Liver disease and failure to synthesize proteins; kidney disease and loss of proteins in urine; lack of proteins in diet due to starvation | Plasma osmotic pressure decreases, less fluid enters venular ends of capillaries by osmosis |
| Obstruction of lymph vessels | Surgical removal of portions of lymphatic pathways; certain parasitic infections | Back pressure in lymph vessels interferes with movement of fluid from interstitial spaces into lymph capillaries |
| Increased venous pressure | Venous obstructions or faulty venous valves | Back pressure in veins increases capillary filtration and interferes with return of fluid from interstitial spaces into venular ends of capillaries |
| Inflammation | Tissue damage | Capillaries become abnormally permeable; fluid leaks from plasma into interstitial spaces |

concentrations of both calcium and phosphate ions in the extracellular fluids. Parathyroid hormone also indirectly stimulates calcium absorption from the intestine (see chapter 13). Concurrently, this hormone causes the kidneys to conserve calcium ions (through increased tubular reabsorption) and increases the urinary excretion of phosphate ions. The increased phosphate excretion offsets the increased plasma phosphate. Thus, the net effect of the hormone is to return the *calcium ion* concentration of the extracellular fluid to normal levels, but to maintain a normal *phosphate ion* concentration (fig. 21.9).

Abnormal increases in blood calcium (hypercalcemia) sometimes result from hyperparathyroidism, in which excess secretion of PTH increases bone resorption (see chapter 13). Hypercalcemia may also be caused by cancers, particularly those originating in the bone marrow, breasts, lungs, or prostate gland. Usually the increase in calcium occurs when bone tissue being invaded and destroyed by cancerous growths releases ions. In other cases, however, the blood calcium concentration increases when cancer cells produce biochemicals that have physiological effects similar to parathyroid hormone. This most often occurs in lung cancer. Symptoms of cancer-induced hypercalcemia include weakness and fatigue, impaired mental function, headache, nausea, increased urine volume (polyuria), and increased thirst (polydipsia).

Abnormal decreases in blood calcium (hypocalcemia) may result from a reduced availability of PTH following surgical removal of the parathyroid glands, or from vitamin D deficiency, which may result from decreased absorption following gastrointestinal surgery or excess excretion due to kidney disease. Hypocalcemia may be life-threatening because it may produce muscle spasms within the airways and cardiac arrhythmias. Administering calcium salts and high doses of vitamin D to promote calcium absorption can correct this condition.

**FIGURE 21.9**

If the concentration of calcium ions decreases, parathyroid hormone increases calcium ion concentration. Increased urinary phosphate excretion offsets bone resorption (which increases blood phosphate levels) to maintain a normal concentration of phosphate ions.

Generally, the regulatory mechanisms that control positively charged ions secondarily control the concentrations of negatively charged ions. For example, chloride ions ($Cl^-$), the most abundant negatively charged ions in the extracellular fluids, are passively reabsorbed from the renal tubules in response to the active reabsorption of sodium ions. That is, the negatively charged chloride ions are electrically attracted to the positively charged sodium ions and accompany them as they are reabsorbed.

Some negatively charged ions, such as phosphate ions ($PO_4^{-3}$) and sulfate ions ($SO_4^{-2}$), also are partially regulated by active transport mechanisms that have limited transport capacities. Thus, if the extracellular phosphate ion concentration is low, the phosphate ions in the renal tubules are conserved. On the other hand, if the renal plasma threshold is exceeded, the excess phosphate will be excreted in the urine. Clinical Application 21.2 discusses symptoms associated with sodium and potassium imbalances.

# Sodium and Potassium Imbalances

Extracellular fluids usually have high sodium ion concentrations, and intracellular fluid usually has high potassium ion concentrations. The renal regulation of sodium is closely related to that of potassium because active reabsorption of sodium (under the influence of aldosterone) is accompanied by secretion (and excretion) of potassium. Thus, it is not surprising that conditions that alter sodium ion balance also affect potassium ion balance.

Such disorders can be summarized as follows:

1. *Low sodium concentration* (hyponatremia). Possible causes of hyponatremia include prolonged sweating, vomiting, or diarrhea; renal disease in which sodium is inadequately reabsorbed; adrenal cortex disorders in which aldosterone secretion is insufficient to promote the reabsorption of sodium (Addison's disease); and drinking too much water.

   Possible effects of hyponatremia include the development of extracellular fluid that is hypotonic and promotes the movement of water into the cells by osmosis. This is accompanied by the symptoms of water intoxication previously described.

2. *High sodium concentration* (hypernatremia). Possible causes of hypernatremia include excessive water loss by evaporation and diffusion, as may occur during high fever, or increased water loss accompanying diabetes insipidus, in one form of which ADH secretion is insufficient to maintain water conservation by the renal tubules and collecting ducts. Possible effects of hypernatremia include disturbances of the central nervous system, such as confusion, stupor, and coma.

3. *Low potassium concentration* (hypokalemia). Possible causes of hypokalemia include excessive release of aldosterone by the adrenal cortex (Cushing's syndrome), which increases renal excretion of potassium; use of diuretic drugs that promote potassium excretion; kidney disease; and prolonged vomiting or diarrhea. Possible effects of hypokalemia include muscular weakness or paralysis, respiratory difficulty, and severe cardiac disturbances, such as atrial or ventricular arrhythmias.

4. *High potassium concentration* (hyperkalemia). Possible causes of hyperkalemia include renal disease, which decreases potassium excretion; use of drugs that promote renal conservation of potassium; insufficient secretion of aldosterone by the adrenal cortex (Addison's disease); or a shift of potassium from the intracellular fluid to the extracellular fluid, a change that accompanies an increase in plasma hydrogen ion concentration (acidosis). Possible effects of hyperkalemia include paralysis of the skeletal muscles and severe cardiac disturbances, such as cardiac arrest. ■

---

1. What is the action of aldosterone?

2. How are the concentrations of sodium and potassium ions controlled?

3. How is calcium regulated?

4. What mechanism regulates the concentrations of most negatively charged ions?

## ACID-BASE BALANCE

As discussed in chapter 2, electrolytes that ionize in water and release hydrogen ions are called acids. Substances that combine with hydrogen ions are called bases. Thus, acid-base balance primarily concerns regulation of the hydrogen ion concentration of body fluids. Regulation of hydrogen ions is very important, because slight changes in hydrogen ion concentrations can alter the rates of enzyme-controlled metabolic reactions, shift the distribution of other ions, or modify hormone actions. Recall from chapter 2 that the internal environment is normally maintained between pH 7.35 and 7.45.

## Sources of Hydrogen Ions

Most of the hydrogen ions in body fluids originate as by-products of metabolic processes, although the digestive tract may directly absorb small quantities. The

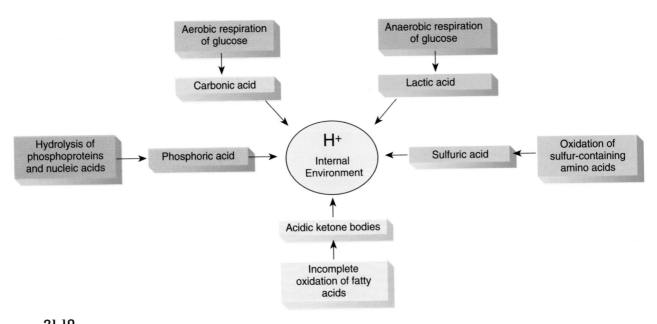

**FIGURE 21.10**
Some of the metabolic processes that provide hydrogen ions.

major metabolic sources of hydrogen ions include the following:

1. **Aerobic respiration of glucose.** This process produces carbon dioxide and water. Carbon dioxide diffuses out of the cells and reacts with water in the extracellular fluids to form *carbonic acid:*

$$CO_2 + H_2O \rightarrow H_2CO_3$$

The resulting carbonic acid then ionizes to release hydrogen ions and bicarbonate ions:

$$H_2CO_3 \rightarrow H^+ + HCO_3^-$$

2. **Anaerobic respiration of glucose.** Glucose metabolized anaerobically produces *lactic acid,* which adds hydrogen ions to body fluids.

3. **Incomplete oxidation of fatty acids.** The incomplete oxidation of fatty acids produces *acidic ketone bodies,* which increase hydrogen ion concentration.

4. **Oxidation of amino acids containing sulfur.** The oxidation of sulfur-containing amino acids yields *sulfuric acid* ($H_2SO_4$), which ionizes to release hydrogen ions.

5. **Breakdown (hydrolysis) of phosphoproteins and nucleic acids.** Phosphoproteins and nucleic acids contain phosphorus. Their oxidation produces *phosphoric acid* ($H_3PO_4$), which ionizes to release hydrogen ions.

The acids resulting from metabolism vary in strength. Thus, their effects on the hydrogen ion concentration of body fluids vary (fig. 21.10).

## Strengths of Acids and Bases

Acids that ionize more completely are strong acids, and those that ionize less completely are weak acids. For example, the hydrochloric acid (HCl) of gastric juice is a strong acid and releases a lot of $H^+$, but the carbonic acid ($H_2CO_3$) produced when carbon dioxide reacts with water is weak and releases less $H^+$.

Bases releases ions, such as hydroxyl ions ($OH^-$), which can combine with hydrogen ions, thereby lowering their concentration. Thus, sodium hydroxide (NaOH), which releases hydroxyl ions and sodium bicarbonate ($NaHCO_3$), which releases bicarbonate ions ($HCO_3^-$), are bases. Often, the negative ions themselves are called bases. For example, $HCO_3^-$ acting as a base combines with $H^+$ from the strong acid HCl to form the weak acid carbonic acid ($H_2CO_3$).

## Regulation of Hydrogen Ion Concentration

Either an acid shift or an alkaline (basic) shift in the body fluids could threaten the internal environment. However, normal metabolic reactions generally produce more acid than base. These reactions include cellular metabolism of glucose, fatty acids, and amino acids. Consequently, the maintenance of acid-base balance usually entails elimination of acid. This is accomplished in three ways:

1. Acid-base buffer systems
2. Respiratory excretion of carbon dioxide
3. Renal excretion of hydrogen ions

**1** Explain why the regulation of hydrogen ion concentration is so important.

**2** What are the major sources of hydrogen ions in the body?

## Acid-Base Buffer Systems

**Acid-base buffer systems** occur in all of the body fluids and involve chemicals that combine with acids or bases when they are in excess. More specifically, the chemical components of a buffer system can combine with strong acids to convert them into weak acids. Likewise, these buffers can combine with strong bases to convert them into weak bases. Such activity helps minimize pH changes in the body fluids. The three most important buffer systems in body fluids are the bicarbonate buffer system, the phosphate buffer system, and the protein buffer system.

1. **Bicarbonate buffer system.** The bicarbonate buffer system, which is present in both intracellular and extracellular fluids, involves the bicarbonate ion ($HCO_3^-$), acting as a weak base, and carbonic acid ($H_2CO_3$), acting as a weak acid. In the presence of excess hydrogen ions, bicarbonate ions will combine with hydrogen ions to form carbonic acid, thus minimizing any increase in the hydrogen ion concentration of the body fluids:

$$H^+ + HCO_3^- \rightarrow H_2CO_3$$

On the other hand, if conditions are basic or alkaline, carbonic acid will dissociate to release bicarbonate ion and hydrogen ion:

$$H_2CO_3 \rightarrow H^+ + HCO_3^-$$

It is important to remember that, even though this reaction releases bicarbonate ion, it is the increase of free hydrogen ions at equilibrium that is important in minimizing the shift toward a more alkaline pH.

2. **Phosphate buffer system.** The phosphate buffer system is also present in both intracellular and extracellular fluids. However it is particularly important in the control of hydrogen ion concentration in the intracellular fluid and in renal

tubular fluid and urine. This buffer system consists of two phosphate ions, dihydrogen phosphate ($H_2PO_4^-$) and monohydrogen phosphate ($HPO_4^{2-}$).

In the presence of excess hydrogen ions, monohydrogen phosphate ions act as a weak base, combining with hydrogen ions to form dihydrogen phosphate, thus minimizing any increase in the hydrogen ion concentration of the body fluids.

$$H^+ + HPO_4^{-2} \rightarrow H_2PO_4^-$$

On the other hand, if conditions are basic or alkaline, dihydrogen phosphate, acting as a weak acid, will dissociate to release hydrogen ion:

$$H_2PO_4^- \rightarrow H^+ + HPO_4^{-2}$$

3. **Protein buffer system.** The protein acid-base buffer system consists of the plasma proteins, such as albumins, and certain proteins within the cells, including the hemoglobin of red blood cells.

As described in chapter 2, proteins are chains of amino acids. Some of these amino acids have freely exposed groups of atoms, called carboxyl groups. If the $H^+$ concentration drops, a carboxyl group (—COOH) can become ionized, releasing a hydrogen ion, thus resisting the pH change:

$$—COOH \leftrightarrow —COO^- + H^+$$

Notice that this is a reversible reaction. In the presence of excess hydrogen ions, the —COO$^-$ portions of the protein molecules accept hydrogen ions and become —COOH groups again. This action decreases the number of free hydrogen ions in the body fluids and again minimizes the pH change.

Some of the amino acids within a protein molecule also contain freely exposed amino groups (—NH$_2$). If the $H^+$ concentration rises, these amino groups can accept hydrogen ions in another reversible reaction:

$$—NH_2 + H^+ \leftrightarrow —NH_3^+$$

In the presence of excess hydroxyl ions (OH$^-$), the —NH$_3^+$ groups within protein molecules give up hydrogen ions and become —NH$_2$ groups again. These hydrogen ions then combine with hydroxyl ions to form water molecules. Once again, pH change is minimized.

Thus protein molecules can function as acids by releasing hydrogen ions under alkaline conditions or as bases by accepting hydrogen ions under acid conditions. This special property allows protein molecules to operate as an acid-base buffer system.

Hemoglobin is an especially important example of a protein that buffers hydrogen ions. As explained in chapter 19, carbon dioxide, produced by cellular oxidation of glucose, diffuses through the capillary wall and enters the plasma and then the red blood cells. The red cells contain an enzyme, *carbonic anhydrase,* that speeds the reaction between carbon dioxide and water, producing carbonic acid:

$$CO_2 + H_2O \rightarrow H_2CO_3$$

The carbonic acid quickly dissociates, releasing hydrogen ions and bicarbonate ions:

$$H_2CO_3 \rightarrow H^+ + HCO_3^-$$

In the peripheral tissues, where $CO_2$ is generated, oxygen is used in the metabolism of glucose, hemoglobin has given up much of its oxygen and is in the form of deoxyhemoglobin. In this form, hemoglobin can bind hydrogen ions, thus acting as a buffer to minimize the pH change that would otherwise occur.

The above two reactions can be written as a single reversible reaction:

$$CO_2 + H_2O \leftrightarrow H_2CO_3 \leftrightarrow H^+ + HCO_3^-$$

Thus, in the peripheral tissues, where $CO_2$ levels are high, the reaction equilibrium shifts to the right, generating $H^+$, which is buffered by hemoglobin, and $HCO_3^-$, which becomes a plasma electrolyte. In the lungs, where oxygen levels are high, hemoglobin is no longer a good buffer and it releases its $H^+$. However the released $H^+$ combines with plasma $HCO_3$, causing the reaction equilibrium to shift to the left, generating carbonic acid, which quickly dissociates to form $CO_2$ and water. The water is added to the body fluids, and the $CO_2$ is exhaled. Because of this relationship to $CO_2$, carbonic acid is sometimes called a *volatile acid.*

Individual amino acids in body fluids can also function as acid-base buffers by accepting or releasing hydrogen ions, because every amino acid has an amino group (—$NH_2$) and a carboxyl group (—COOH).

To summarize, acid-base buffer systems take up hydrogen ions when body fluids are becoming more acidic, and give up hydrogen ions when the fluids are becoming more basic (alkaline). Buffer systems convert stronger acids into weaker acids or convert stronger bases into weaker bases, as table 21.3 summarizes.

In addition to minimizing pH fluctuations, acid-base buffer systems in body fluids buffer each other. Con-

| TABLE 21.3 | CHEMICAL ACID-BASE BUFFER SYSTEMS | |
|---|---|---|
| **Buffer System** | **Constituents** | **Actions** |
| Bicarbonate system | Bicarbonate ion ($HCO_3^-$) | Converts a strong acid into a weak acid |
| | Carbonic acid ($H_2CO_3$) | Converts a strong base into a weak base |
| Phosphate system | Monohydrogen phosphate ion ($HPO_4^{-2}$) | Converts a strong acid into a weak acid |
| | Dihydrogen phosphate ($H_2PO_4^-$) | Converts a strong base into a weak base |
| Protein system (and amino acids) | —$COO^-$ group of an amino acid or protein | Accepts hydrogen ions in the presence of excess acid |
| | —$NH_3^+$ group of an amino acid or protein | Releases hydrogen ions in the presence of excess base |

sequently, whenever the hydrogen ion concentration begins to change, the chemical balances within all of the buffer systems change too, resisting the drift in pH.

Neurons are particularly sensitive to changes in the pH of body fluids. For example, if the interstitial fluid becomes more alkaline than normal (alkalosis), neurons become more excitable, and seizures may result. Conversely, if conditions become more acidic (acidosis), neuron activity is depressed, and the level of consciousness may decrease.

1. What is the difference between a strong acid or base and a weak acid or base?

2. How does a chemical buffer system help regulate pH of body fluids?

3. List the major buffer systems of the body.

Chemical buffer systems only temporarily solve the problem of acid-base balance. Ultimately, the body must eliminate excess acid or base. This task becomes the job of the lungs (controlled by the respiratory center) and the kidneys.

## Respiratory Excretion of Carbon Dioxide

The **respiratory center** in the brain stem helps regulate hydrogen ion concentrations in the body fluids by controlling the rate and depth of breathing. Specifically, if

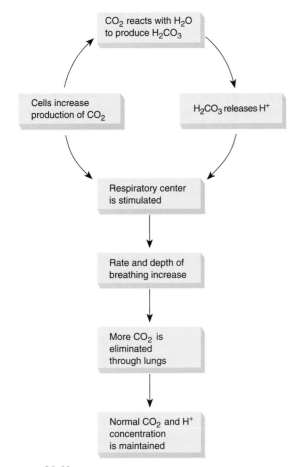

**FIGURE 21.11**

An increase in carbon dioxide elimination follows an increase in carbon dioxide production.

the body cells increase their production of carbon dioxide, as occurs during periods of physical exercise, carbonic acid production increases. As the carbonic acid dissociates, the concentration of hydrogen ions increases, and the pH of the internal environment drops (see chapter 19). Such an increasing concentration of carbon dioxide in the central nervous system and the subsequent increase in hydrogen ion concentration in the cerebrospinal fluid stimulate chemosensitive areas within the respiratory center.

In response, the respiratory center increases the depth and rate of breathing, so that the lungs excrete more carbon dioxide. As a result, hydrogen ion concentration in body fluids returns toward normal, because the released carbon dioxide is in equilibrium with carbonic acid and $CO_2$ (fig. 21.11):

$$CO_2 + H_2O \leftrightarrow H_2CO_3 \leftrightarrow H^+ + HCO_3^-$$

Conversely, if body cells are less active, concentrations of carbon dioxide and hydrogen ions in body fluids remain relatively low. As a result, breathing rate and depth fall. This increases the carbon dioxide level in the body fluids, returning pH to normal. If the pH drops below normal, the respiratory center is stimulated to increase the rate and depth of breathing once again.

Thus, the activity of the respiratory center changes in response to shifts in the pH of the body fluids, reducing these shifts to a minimum. Because most of the hydrogen ions in the body fluids originate from carbonic acid produced when carbon dioxide reacts with water, the respiratory regulation of hydrogen ion concentration is of considerable importance.

## Renal Excretion of Hydrogen Ions

Nephrons help regulate the hydrogen ion concentration of the body fluids by excreting hydrogen ions in the urine. Recall from chapter 20 that the epithelial cells that line the proximal and distal convoluted tubules and the collecting ducts secrete these ions into the tubular fluid. The kidneys also regulate the concentration of bicarbonate ions in body fluids. These mechanisms are important as well in balancing the quantities of sulfuric acid, phosphoric acid, and various organic acids that appear in body fluids as by-products of metabolic processes.

The metabolism of certain amino acids, for example, produces sulfuric and phosphoric acids. Consequently, a diet high in proteins may trigger excess acid formation. The kidneys compensate for such gains in acids by altering the rate of hydrogen ion secretion, thus resisting a shift in the pH of body fluids (fig. 21.12).

Once hydrogen ions are secreted, phosphates that were filtered into the fluid of the renal tubule buffer them. Ammonia aids in this buffering action.

Through deamination of certain amino acids, the cells of the renal tubules produce ammonia ($NH_3$), which diffuses readily through cell membranes and enters the urine. Because ammonia is a weak base, it can accept hydrogen ions to form *ammonium ions* ($NH_4^+$):

$$H^+ + NH_3 \rightarrow NH_4^+$$

Cell membranes are quite impermeable to ammonium ions. Consequently, these ions are trapped in the urine as they form, and they are excreted from the body with the urine. When increase in the hydrogen ion concentration of body fluids is prolonged, the renal tubules increase ammonia production. This mechanism helps to transport excess hydrogen ions to the outside and helps prevent the urine from becoming too acidic.

The various regulators of hydrogen ion concentration operate at different rates. Acid-base buffers, for example, function rapidly and can convert strong acids or bases into weak acids or bases almost immediately. For this reason, these chemical buffer systems are sometimes called the body's *first line of defense* against shifts in pH.

Physiological buffer systems, such as the respiratory and renal mechanisms, function more slowly, and constitute *secondary defenses*. The respiratory mechanism may require several minutes to begin resisting a

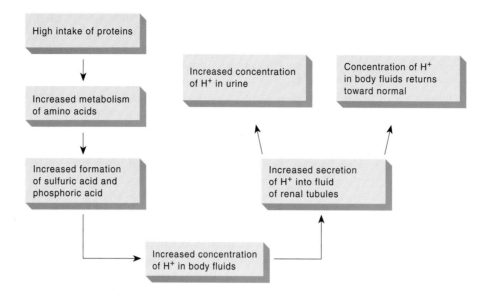

**FIGURE 21.12**

If the concentration of hydrogen ions in body fluids increases, the renal tubules increase their secretion of hydrogen ions into the urine.

**FIGURE 21.13**

Chemical buffers act rapidly, whereas physiological buffers may require several minutes to several days to begin resisting a change in pH.

change in pH, and the renal mechanism may require one to three days to regulate a changing hydrogen ion concentration (fig. 21.13). Clinical Application 21.3 examines the effects of acid-base imbalances.

**1** How does the respiratory system help regulate acid-base balance?

**2** How do the kidneys respond to excess hydrogen ions?

**3** How do the rates at which chemical and physiological buffer systems act differ?

## CLINICAL TERMS RELATED TO WATER, ELECTROLYTE AND ACID-BASE BALANCE

**acetonemia** (as″e-to-ne′me-ah)   Abnormal amounts of acetone in the blood.

**acetonuria** (as″e-to-nu′re-ah)   Abnormal amounts of acetone in the urine.

**albuminuria** (al-bu′m ĭ-nu′re-ah)   Albumin in the urine.

**anasarca** (an″ah-sar′kah)   Widespread accumulation of tissue fluid.

**antacid** (ant-as′id)   Substance that neutralizes an acid.

**anuria** (ah-nu′re-ah)   Absence of urine excretion.

**azotemia** (az″o-te′me-ah)   Accumulation of nitrogenous wastes in the blood.

**diuresis** (di″u-re′sis)   Increased production of urine.

**glucosuria** (glu″ko-su′re-ah)   Excess sugar in the urine.

**hyperglycemia** (hi″per-gli-se′me-ah)   Abnormally high concentration of blood sugar.

**hyperkalemia** (hi″per-kah-le′me-ah)   Excess potassium in the blood.

**hypernatremia** (hi″per-na″tre′me-ah)   Excess sodium in the blood.

**hyperuricemia** (hi″per-u″rĭ-se′me-ah)   Excess uric acid in the blood.

**hypoglycemia** (hi″po-gli-se′me-ah)   Abnormally low concentration of blood sugar.

**ketonuria** (ke″to-nu′re-ah)   Ketone bodies in the urine.

**ketosis** (ke″to′sis)   Acidosis due to excess ketone bodies in the body fluids.

**proteinuria** (pro″te-ĭ-nu′re-ah)   Protein in the urine.

**uremia** (u-re′me-ah)   Toxic condition resulting from nitrogenous wastes in the blood.

# CLINICAL APPLICATION

# Acid-Base Imbalances

Ordinarily, chemical and physiological buffer systems maintain the hydrogen ion concentration of body fluids within very narrow pH ranges. Abnormal conditions may disturb the acid-base balance. For example, the pH of arterial blood is normally 7.35–7.45. A value below 7.35 produces *acidosis*. A pH above 7.45 produces *alkalosis*. Such shifts in the pH of body fluids may be life threatening. In fact, a person usually cannot survive if the pH drops to 6.8 or rises to 8.0 for more than a few hours (fig. 21C).

Acidosis results from an accumulation of acids or a loss of bases, both of which cause abnormal increases in the hydrogen ion concentrations of body fluids. Conversely, alkalosis results from a loss of acids or an accumulation of bases accompanied by a decrease in hydrogen ion concentrations (fig. 21D).

The two major types of acidosis are *respiratory acidosis* and *metabolic acidosis*. Factors that increase carbon dioxide, also increasing the concentration of carbonic acid (the respiratory acid), cause respiratory acidosis. Metabolic acidosis is due to an abnormal accumulation of any other acids in the body fluids or to a loss of bases, including bicarbonate ions.

Similarly, the two major types of alkalosis are *respiratory alkalosis* and *metabolic alkalosis*. Excessive loss of carbon dioxide and consequent loss of carbonic acid cause respiratory alkalosis. Metabolic alkalosis is due to excessive loss of hydrogen ions or gain of bases.

Since in respiratory acidosis carbon dioxide accumulates, this can result from factors that hinder pulmonary ventilation (fig. 21E). These include the following:

1. Injury to the respiratory center of the brain stem, decreasing rate and depth of breathing.
2. Obstructions in air passages that interfere with air movement into the alveoli.

**FIGURE 21C**

If the pH of the arterial blood drops to 6.8 or rises to 8.0 for more than a few hours, the person usually cannot survive.

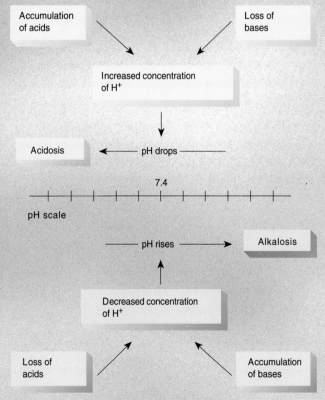

**FIGURE 21D**

Acidosis results from an accumulation of acids or a loss of bases. Alkalosis results from a loss of acids or an accumulation of bases.

*(Continued)*

# Acid-Base Imbalances (continued)

**3.** Diseases that decrease gas exchanges, such as pneumonia, or those that reduce surface area of the respiratory membrane, such as emphysema.

Any of these conditions can increase the level of carbonic acid and hydrogen ions in body fluids, lowering pH. Chemical buffers, such as hemoglobin, may resist this shift in pH. At the same time, increasing concentrations of carbon dioxide and hydrogen ions stimulate the respiratory center, increasing breathing rate and depth and thereby lowering carbon dioxide concentration. Also, the kidneys may begin to excrete more hydrogen ions.

Eventually, thanks to these chemical and physiological buffers, the pH of the body fluids may return to normal. When this happens, the acidosis is said to be *compensated.*

The symptoms of respiratory acidosis result from depression of central nervous system function, and include drowsiness, disorientation, and stupor. Evidence of respiratory insufficiency, such as labored breath-

ing and cyanosis, is usually also evident. In *uncompensated acidosis,* the person may become comatose and die.

Metabolic acidosis is due to either accumulation of nonrespiratory acids or loss of bases (fig. 21F). Factors that may lead to this condition include:

**1.** Kidney disease that reduces glomerular filtration and fails to excrete the acids produced in metabolism (uremic acidosis).
**2.** Prolonged vomiting that loses the alkaline contents of the upper intestine and stomach contents. (Losing only the stomach contents produces metabolic alkalosis.)
**3.** Prolonged diarrhea in which excess alkaline intestinal secretions are lost (especially in infants).
**4.** Diabetes mellitus, in which some fatty acids are converted into ketone bodies. These ketone bodies include *acetoacetic acid, beta-hydroxybutyric acid,* and *acetone.* Normally these molecules are

produced in relatively small quantities, and cells oxidize them as energy sources. However, if fats are being utilized at an abnormally high rate, as may occur in diabetes mellitus, ketone bodies may accumulate faster than they can be oxidized. At such times, these compounds may be excreted in the urine (ketonuria); in addition, acetone, which is volatile, may be excreted by the lungs and impart a fruity odor to the breath. More seriously, the accumulation of acetoacetic acid and beta-hydroxybutyric acid may lower pH (ketonemic acidosis).

---

**FIGURE 21E**

Some of the factors that lead to respiratory acidosis.

- Decreased rate and depth of breathing
- Obstruction of air passages
- Decreased gas exchange
  → Accumulation of $CO_2$
  → Respiratory acidosis

**FIGURE 21F**

Some of the factors that lead to metabolic acidosis.

- Kidney failure to excrete acids
- Excessive production of acidic ketones as in diabetes mellitus
  → Accumulation of nonrespiratory acids
  → Metabolic acidosis
- Excessive loss of bases
  ← Prolonged diarrhea with loss of alkaline intestinal secretions
  ← Prolonged vomiting with loss of intestinal secretions

---

These acids may also combine with bicarbonate ions in the urine. As a result, excess bicarbonate ions are excreted, interfering with the function of the bicarbonate acid-base buffer system.

In each case, the pH tends to shift toward lower values. However, the following factors resist this shift: chemical buffer systems, which accept excess hydrogen ions; the respiratory center, which increases the breathing rate and depth; and the kidneys, which excrete more hydrogen ions.

Respiratory alkalosis develops as a result of *hyperventilation,* described in chapter 19. Hyperventilation is accompanied by too great a loss of carbon dioxide and consequent decreases in carbonic acid and hydrogen ion concentrations (fig. 21G).

Hyperventilation may occur during periods of anxiety, although it may also accompany fever or poisoning from salicylates, such as aspirin. At high altitudes, hyperventilation may be a response to low oxygen pressure. Also, musicians, such as bass tuba players, who must provide a large volume of air when playing sustained passages, sometimes hyperventilate. In each case, rapid, deep breathing depletes carbon dioxide, and the pH of body fluids increases.

Chemical buffers, such as hemoglobin, that release hydrogen ions resist this pH change. Also, the lower concentrations of carbon dioxide and hydrogen ions stimulate the respiratory center to a lesser degree. This inhibits hyperventilation, thus reducing further carbon dioxide loss. At the same time, the kidneys decrease their secretion of hydrogen ions, and the urine becomes alkaline as bases are excreted.

The symptoms of respiratory alkalosis include lightheadedness, agitation, dizziness, and tingling sensations. In severe cases, impulses may be triggered spontaneously on peripheral nerves, and muscles may respond with tetanic contractions (see chapter 9).

Metabolic alkalosis results from a great loss of hydrogen ions or from a gain in bases, both of which are accompanied by a rise in the pH of the blood (alkalemia) (fig. 21H). This condition may occur following gastric drainage (lavage), prolonged vomiting in which only the stomach contents are lost, or the use of certain diuretic drugs. Because gastric juice is very acidic, its loss leaves the body fluids with a net increase of basic substances and a pH shift toward alkaline values. Metabolic alkalosis may also develop as a result of ingesting too much antacid, such as sodium bicarbonate, medication to relieve the symptoms of indigestion. The symptoms of metabolic alkalosis include a decrease in the breathing rate and depth, which, in turn, results in an increased concentration of carbon dioxide in the blood. ■

**FIGURE 21G**

Some of the factors that lead to respiratory alkalosis.

**FIGURE 21H**

Some of the factors that lead to metabolic alkalosis.

# CHAPTER SUMMARY

## Introduction (page 814)

The maintenance of water and electrolyte balance requires that the quantities of these substances entering the body equal the quantities leaving it. Altering the water balance necessarily affects the electrolyte balance.

## Distribution of Body Fluids (page 814)

1. Fluid compartments
   a. The intracellular fluid compartment includes the fluids and electrolytes cell membranes enclose.
   b. The extracellular fluid compartment includes all fluids and electrolytes outside cell membranes.
      (1) Interstitial fluid within tissue spaces
      (2) Plasma within blood
      (3) Lymph within lymphatic vessels
      (4) Transcellular fluid within body cavities
2. Body fluid composition
   a. Extracellular fluids
      (1) Extracellular fluids have high concentrations of sodium, chloride, calcium, and bicarbonate ions, with less potassium, calcium, magnesium, phosphate, and sulfate ions.
      (2) Plasma contains more protein than does either interstitial fluid or lymph.
   b. Intracellular fluid contains relatively high concentrations of potassium, magnesium, and phosphate ions; it also contains a greater concentration of sulfate ions and lesser concentrations of sodium, chloride, calcium, and bicarbonate ions than does extracellular fluid.
3. Movement of fluid between compartments
   a. Hydrostatic and osmotic pressure regulate fluid movements.
      (1) Fluid leaves plasma because of hydrostatic pressure and returns to plasma because of osmotic pressure.
      (2) Hydrostatic pressure drives fluid into lymph vessels.
      (3) Osmotic pressure regulates fluid movement in and out of cells.
   b. Sodium ion concentrations are especially important in fluid movement regulation.

## Water Balance (page 816)

1. Water intake
   a. The volume of water taken in varies from person to person.
   b. Most water comes from consuming liquid or moist foods.
   c. Oxidative metabolism produces some water.
2. Regulation of water intake
   a. The thirst mechanism is the primary regulator of water intake.
   b. Drinking and the resulting stomach distension inhibit the thirst mechanism.
3. Water output
   a. Water is lost in a variety of ways.
      (1) It is excreted in the urine, feces, and sweat.
      (2) Insensible loss occurs through evaporation from the skin and lungs.

   b. Urine production regulates water output.
4. Regulation of water output
   a. The distal convoluted tubules and collecting ducts of the nephrons regulate water output.
      (1) ADH from the hypothalamus and posterior pituitary gland stimulates water reabsorption in these segments.
      (2) The mechanism involving ADH can reduce normal output of 1,500 milliliters to 500 milliliters per day.
   b. If excess water is taken in, the ADH mechanism is inhibited.

## Electrolyte Balance (page 819)

1. Electrolyte intake
   a. The most important electrolytes in the body fluids are those that release ions of sodium, potassium, calcium, magnesium, chloride, sulfate, phosphate, and bicarbonate.
   b. These ions are obtained in foods and beverages or as by-products of metabolic processes.
2. Regulation of electrolyte intake
   a. Electrolytes are usually obtained in sufficient quantities in response to hunger and thirst mechanisms.
   b. In a severe electrolyte deficiency, a person may experience a salt craving.
3. Electrolyte output
   a. Electrolytes are lost through perspiration, feces, and urine.
   b. Quantities lost vary with temperature and physical exercise.
   c. The greatest electrolyte loss occurs as a result of kidney functions.
4. Regulation of electrolyte output
   a. Concentrations of sodium, potassium, and calcium ions in the body fluids are particularly important.
   b. The regulation of sodium ions involves the secretion of aldosterone from the adrenal glands.
   c. The regulation of potassium ions also involves aldosterone.
   d. Calcitonin from the thyroid gland and parathyroid hormone from the parathyroid glands regulate calcium ion concentration.
   e. The mechanisms that control positively charged ions secondarily regulate negatively charged ions.
      (1) Chloride ions are passively reabsorbed in renal tubules as sodium ions are actively reabsorbed.
      (2) Some negatively charged ions, such as phosphate ions, are reabsorbed partially by limited-capacity active transport mechanisms.

## Acid-Base Balance (page 823)

Acids are electrolytes that release hydrogen ions. Bases combine with hydrogen ions.

1. Sources of hydrogen ions
   a. Aerobic respiration of glucose
      (1) Aerobic respiration of glucose produces carbon dioxide, which reacts with water to form carbonic acid.
      (2) Carbonic acid dissociates to release hydrogen and bicarbonate ions.

b. Anaerobic respiration of glucose produces lactic acid.
c. Incomplete oxidation of fatty acids releases acidic ketone bodies.
d. Oxidation of sulfur-containing amino acids produces sulfuric acid.
e. Hydrolysis of phosphoproteins and nucleic acids gives rise to phosphoric acid.

2. Strengths of acids and bases
   a. Acids vary in the extent to which they ionize.
      (1) Strong acids, such as hydrochloric acid, ionize more completely.
      (2) Weak acids, such as carbonic acid, ionize less completely.
   b. Bases vary in strength also.
      (1) Strong bases, such as hydroxyl ions, combine readily with hydrogen ions.
      (2) Weak bases, such as bicarbonate ions, combine with hydrogen ions less readily.

3. Regulation of hydrogen ion concentration
   a. Acid-base buffer systems
      (1) Buffer systems are composed of sets of two or more chemicals.

   (2) They convert strong acids into weaker acids, or strong bases into weaker bases.
   (3) They include the bicarbonate buffer system, phosphate buffer system, and protein buffer system.
   (4) Buffer systems minimize pH changes.
   b. Respiratory excretion of carbon dioxide
      (1) The respiratory center is located in the brain stem.
      (2) It helps regulate pH by controlling the rate and depth of breathing.
      (3) Increasing carbon dioxide and hydrogen ion concentrations stimulate chemoreceptors associated with the respiratory center; breathing rate and depth increase, and carbon dioxide concentration decreases.
      (4) If the carbon dioxide and hydrogen ion concentrations are low, the respiratory center inhibits breathing.
   c. Renal excretion of hydrogen ions
      (1) Nephrons secrete hydrogen ions to regulate pH.
      (2) Phosphates buffer hydrogen ions in urine.
      (3) Ammonia produced by renal cells helps transport hydrogen ions to the outside of the body.
   d. Chemical buffers act rapidly; physiological buffers act more slowly.

## CRITICAL THINKING QUESTIONS

1. An elderly, semiconscious patient is tentatively diagnosed as having acidosis. What components of the arterial blood will be most valuable in determining if the acidosis is of respiratory origin?

2. Some time ago, several newborn infants died due to an error in which sodium chloride was substituted for sugar in their formula. What symptoms would this produce? Why do you think infants are more prone to the hazard of excess salt intake than adults?

3. Explain the threat to fluid and electrolyte balance in the following situation: A patient is being nutritionally maintained on concentrated solutions of hydrolyzed protein that are administered through a gastrostomy tube.

4. Describe what might happen to the plasma pH of a patient as a result of:
   a. prolonged diarrhea
   b. suction of the gastric contents
   c. hyperventilation
   d. hypoventilation

5. Radiation therapy may damage the mucosa of the stomach and intestines. What effect might this have on the patient's electrolyte balance?

6. If the right ventricle of a patient's heart is failing, increasing the venous pressure, what changes might occur in the patient's extracellular fluid compartments?

## REVIEW EXERCISES

1. Explain how water balance and electrolyte balance are interdependent.
2. Name the body fluid compartments, and describe their locations.
3. Explain how the fluids within these compartments differ in composition.
4. Describe how fluid movements between the compartments are controlled.
5. Prepare a list of sources of normal water gain and loss to illustrate how the input of water equals the output of water.
6. Define water of metabolism.
7. Explain how water intake is regulated.

8. Explain how the nephrons function in the regulation of water output.
9. List the most important electrolytes in the body fluids.
10. Explain how electrolyte intake is regulated.
11. List the routes by which electrolytes leave the body.
12. Explain how the adrenal cortex functions in the regulation of electrolyte output.
13. Describe the role of the parathyroid glands in regulating electrolyte balance.
14. Describe the role of the renal tubule in regulating electrolyte balance.
15. Distinguish between an acid and a base.

16. List five sources of hydrogen ions in the body fluids, and name an acid that originates from each source.

17. Distinguish between a strong acid and a weak acid, and name an example of each.

18. Distinguish between a strong base and a weak base, and name an example of each.

19. Explain how an acid-base buffer system functions.

20. Describe how the bicarbonate buffer system resists changes in pH.

21. Explain why a protein has acidic as well as basic properties.

22. Describe how a protein functions as a buffer system.

23. Describe the function of hemoglobin as a buffer of carbonic acid.

24. Explain how the respiratory center functions in the regulation of the acid-base balance.

25. Explain how the kidneys function in the regulation of the acid-base balance.

26. Describe the role of ammonia in the transport of hydrogen ions to the outside of the body.

27. Distinguish between a chemical buffer system and a physiological buffer system.

 **For More Information Related to This Chapter, Please Visit** *Hole's Human Anatomy and Physiology* **Home Page**
http://www.mhhe.com/biosci/abio/

## CHAPTER 22

# REPRODUCTIVE SYSTEMS

## CHAPTER OBJECTIVES

AFTER YOU HAVE STUDIED THIS CHAPTER, YOU SHOULD BE ABLE TO:

1. State the general functions of the male reproductive system.
2. Name the parts of the male reproductive system and describe the general functions of each part.
3. Outline the process of spermatogenesis.
4. Trace the path sperm cells follow from their site of formation to the outside.
5. Describe the structure of the penis and explain how its parts produce an erection.
6. Explain how hormones control the activities of the male reproductive organs and the development of male secondary sex characteristics.
7. State the general functions of the female reproductive system.
8. Name the parts of the female reproductive system and describe the general functions of each part.
9. Outline the process of oogenesis.
10. Describe how hormones control the activities of the female reproductive organs and the development of female secondary sex characteristics.
11. Describe the major events that occur during a menstrual cycle.
12. Describe the hormonal changes in the maternal body during pregnancy.
13. Describe the birth process and explain the role of hormones in this process.
14. List several methods of birth control and describe the relative effectiveness of each method.

## KEY TERMS

androgen (an′dro-jen)
cleavage (klēv′ij)
clitoris (klit′o-ris)
coitus (ko′ĭtus)
contraception (kon″trah-sep′shun)
ejaculation (e-jak″u-la′shun)
emission (e-mish′un)
epididymis (ep″ĭ-did′ĭ-mis)
estrogen (es′tro jen)
fertilization (fer″ti-li-za′shun)
fimbriae (fim′bre)
follicle (fol′ĭ-kl)
gonadotropin (go-nad″o-trōp′in)
gubernaculum (gu″ber-nak′u-lum)
implantation (im plan-ta′shun)

infundibulum (in″fun-dib′u-lum)
inguinal (ing′gwĭ-nal)
meiosis (mi-o′sis)
menopause (men′o-pawz)
menstrual cycle (men′stroo-al si′kl)
oogenesis (o″o-jen′ĕ-sis)
orgasm (or′gazm)
ovulation (o″vu-la′shun)
placenta (plah-sen′tah)
pregnancy (preg′nan-se)
progesterone (pro-jes′tĕ-rōn)
puberty (pu′ber-te)
semen (se′men)
spermatogenesis (sper″mah-to-jen′ĕ-sis)
testosterone (tes-tos′te-rōn)
vas deferens (vas def′er-enz)
zygote (zi′gōt)

## UNDERSTANDING WORDS

**andr-,** man: *andr*ogens—male sex hormones.
**contra-,** against, counter: *contra*ception—prevention of fertilization.
**crur-,** lower part: *crur*a—diverging parts at the base of the penis by which it is attached to the pelvic arch.
**ejacul-,** to shoot forth: *ejacul*ation—process by which semen is expelled from the male reproductive tract.
**fimb-,** fringe: *fimb*riae—irregular extensions on the margin of the infundibulum of the uterine tube.
**follic-,** small bag: *follic*le—ovarian structure that contains an egg cell.
**genesis,** origin: spermato*genesis*—process by which sperm cells are formed.
**gubern-,** to steer, to guide: *gubern*aculum—fibromuscular cord that guides the descent of a testis.
**labi-,** lip: *labi*a minora—flattened, longitudinal folds that extend along the margins of the vestibule.
**mamm-,** breast: *mamm*ary gland—female accessory gland that secretes milk.
**mast-,** breast: *mast*itis—inflammation of the mammary gland.
**mens-,** month: *mens*trual cycle—monthly female reproductive cycle.
**mons,** an eminence: *mons* pubis—rounded elevation of fatty tissue overlying the symphysis pubis in a female.
**oo-,** egg: *oo*genesis—formation of an egg.
**prim-,** first: *prim*ordial follicle—ovarian follicle composed of an egg surrounded by a single layer of cells.
**puber-,** adult: *puber*ty—time when a person becomes able to reproduce.
**zon-,** belt: *zon*a pellucida—transparent layer surrounding an oocyte.

he two cells approach each other, one 90,000 times the volume of the other. Although they look about as different as two cells can, they are alike in that each carries a half set of genetic instructions from each parent. The two cells meet and merge, launching a new life.

Each of us harbors many of these sex cells—sperm or eggs—that have the potential to participate in forming a new life. These very special cells are generated and nurtured in our reproductive systems, which are similarly organized in the male and female. Each system has paired structures where sperm or eggs are manufactured, a network of tubes to transport these cells, and hormones and exocrine secretions that oversee the entire process.

Our reproductive systems profoundly affect our lives, even when we are not parents. Many a teen has anguished over facial blemishes, turbulent and disturbing feelings, and a body seemingly out of control—all attributed to the hormonal upheaval that reflects a maturing reproductive system. We use our reproductive systems not only to make babies but also to intimately interact with another individual. But with sexual activity comes responsibility—to avoid pregnancy when one is not ready for parenthood, and to protect ourselves and our partners from sexually transmitted diseases.

## ORGANS OF THE MALE REPRODUCTIVE SYSTEM

The organs of the male reproductive system are specialized to produce and maintain the male sex cells, or *sperm cells;* to transport these cells, together with supporting fluids, to the female reproductive tract; and to secrete male sex hormones.

The *primary sex organs* (gonads) of this system are the two testes in which the sperm cells (spermatozoa) and the male sex hormones are formed. The other structures of the male reproductive system are termed *accessory sex organs* (secondary sex organs). They include the internal reproductive organs and the external reproductive organs (fig. 22.1).

## TESTES

The **testes** (sing., *testis*) are ovoid structures about 5 centimeters in length and 3 centimeters in diameter. Both testes, each suspended by a spermatic cord, are contained within the cavity of the saclike *scrotum* (see fig. 22.1 and reference plate 52).

### Descent of the Testes

In a male fetus, the testes originate from masses of tissue posterior to the parietal peritoneum, near the developing kidneys. Usually a month or two before birth, these organs descend to the lower abdominal cavity and pass through the abdominal wall into the scrotum.

The male sex hormone *testosterone,* which the developing testes secrete, stimulates the testes to descend. A fibromuscular cord called the **gubernaculum** aids movement of the testes. This cord is attached to each developing testis and extends into the inguinal region of the abdominal cavity. It passes through the abdominal wall and is fastened to the skin on the outside of the scrotum. The testis descends, guided by the gubernaculum, passing through the **inguinal canal** of the abdominal wall and entering the scrotum, where it remains anchored by the gubernaculum. Each testis carries a developing *vas deferens,* blood vessels, and nerves. These structures later form parts of the **spermatic cord** by which the testis is suspended in the scrotum (fig. 22.2).

If the testes fail to descend into the scrotum, they will not produce sperm cells because the temperature in the abdominal cavity is too high. If this condition, called *cryptorchidism,* is left untreated the cells that normally produce sperm cells degenerate, and the male is infertile.

During the descent of a testis, a pouch of peritoneum, called the *vaginal process,* moves through the inguinal canal and into the scrotum. In about one-quarter of males, this pouch remains open, providing a potential passageway through which a loop of intestine may be forced by great abdominal pressure, producing an *indirect inguinal hernia.* If the protruding intestinal loop is so tightly constricted within the inguinal canal that its blood supply stops, the condition is called a *strangulated hernia.* Without prompt treatment, the strangulated tissues may die.

1 What are the primary sex organs of the male reproductive system?

2 Describe the descent of the testes.

3 What is the function of the gubernaculum, both during and after the descent of the testes?

4 What happens if the testes fail to descend into the scrotum?

### Structure of the Testes

A tough, white, fibrous capsule called the *tunica albuginea* encloses the testis. Along its posterior border, the connective tissue thickens and extends into the organ, forming a mass called the *mediastinum testis.* From this structure, thin layers of connective tissue, called *septa,* pass into the testis and subdivide it into about 250 *lobules.*

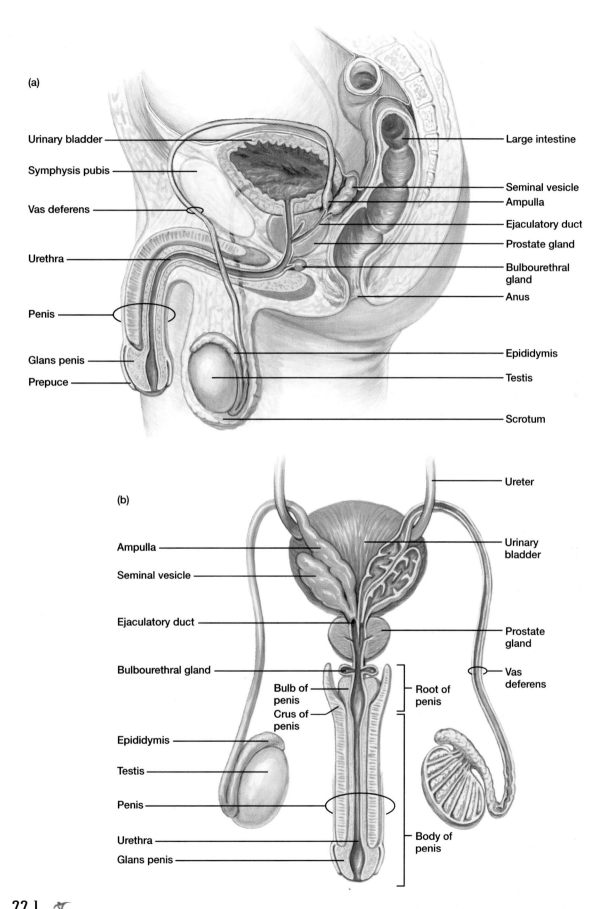

Urinary bladder

Symphysis pubis

Vas deferens

Urethra

Penis

Glans penis

Prepuce

Large intestine

Seminal vesicle

Ampulla

Ejaculatory duct

Prostate gland

Bulbourethral gland

Anus

Epididymis

Testis

Scrotum

Ampulla

Seminal vesicle

Ejaculatory duct

Bulbourethral gland

Bulb of penis

Crus of penis

Epididymis

Testis

Penis

Urethra

Glans penis

Ureter

Urinary bladder

Prostate gland

Vas deferens

Root of penis

Body of penis

**FIGURE** 22.1

(a) Sagittal view of male reproductive organs. (b) Posterior view.

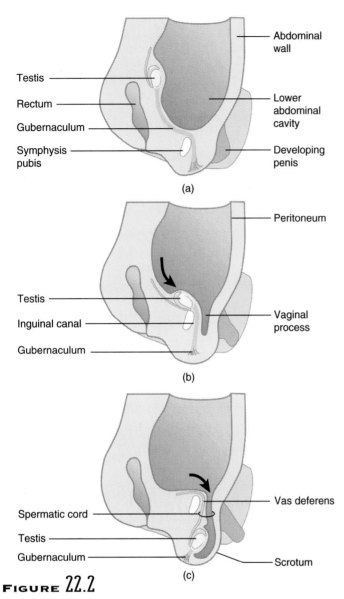

**Figure 22.2**

During fetal development, each testis descends through an inguinal canal and enters the scrotum (a–c).

A lobule contains one to four highly coiled, convoluted **seminiferous tubules,** each of which is approximately 70 centimeters long when uncoiled. These tubules course posteriorly and unite to form a complex network of channels called the *rete testis.* The rete testis is located within the mediastinum testis and gives rise to several ducts that join a tube called the *epididymis.* The epididymis, in turn, is coiled on the outer surface of the testis.

The seminiferous tubules are lined with a specialized stratified epithelium, which includes the spermatogenic cells that give rise to the sperm cells. Other specialized cells, called **interstitial cells** (cells of Leydig), lie in the spaces between the seminiferous tubules. Interstitial cells produce and secrete male sex hormones (figs. 22.3, 22.4, and 22.5*a*).

The epithelial cells of the seminiferous tubules can give rise to *testicular cancer,* a common cancer in young men. In most cases, the first sign is a painless testis enlargement or a scrotal mass that attaches to a testis.

If a biopsy (tissue sample) reveals cancer cells, surgery removes the affected testis (orchiectomy). Radiation and/or chemotherapy often prevent the cancer from recurring.

**1** Describe the structure of a testis.

**2** Where are the sperm cells produced within the testes?

**3** Which cells produce male sex hormones?

## Formation of Sperm Cells

The epithelium of the seminiferous tubules consists of supporting cells (*sustentacular cells,* or Sertoli's cells) and spermatogenic cells. The sustentacular cells are tall, columnar cells that extend the full thickness of the epithelium from its base to the lumen of the seminiferous tubule. Many thin processes project from these cells, filling the spaces between nearby spermatogenic cells. The sustentacular cells support, nourish, and regulate the *spermatogenic cells,* which give rise to sperm cells (spermatozoa).

In the male embryo, the spermatogenic cells are undifferentiated and are called *spermatogonia.* Each of these cells contains 46 chromosomes in its nucleus, the usual number for human cells (fig. 22.5).

Beginning during embryonic development, hormones stimulate the spermatogonia to become active. Some of the cells undergo mitosis, giving rise to new spermatogonia and providing a reserve supply of these undifferentiated cells. Other cells enlarge and become *primary spermatocytes.* Sperm production (spermatogenesis) (fig. 22.5) is arrested at this stage. At puberty, testosterone secretion rises, and the primary spermatocytes then reproduce by a special type of cell division called **meiosis,** which is described in detail in chapter 24.

During meiosis, the primary spermatocytes each reproduce to form two *secondary spermatocytes.* Each of these cells, in turn, reproduces to form two *spermatids,* which mature into sperm cells. Meiosis also reduces the number of chromosomes in each cell by one-half. Consequently, for each primary spermatocyte that undergoes meiosis, four sperm cells with 23 chromosomes in each of their nuclei are formed. Because the chromosome number is halved, when a sperm and egg join, the new individual has a complete set of 23 pairs of chromosomes.

The spermatogonia are located near the base of the germinal epithelium. As spermatogenesis occurs, cells in more advanced stages are pushed along the sides of sustentacular cells toward the lumen of the seminiferous tubule.

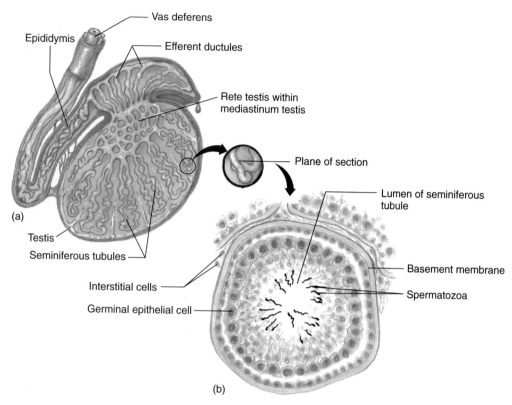

FIGURE 22.3

(*a*) Sagittal section of a testis. (*b*) Cross section of a seminiferous tubule.

FIGURE 22.4

Scanning electron micrograph of cross sections of human seminiferous tubules (150×).

*Tissues and Organs: A Text-Atlas of Scanning Electron Microscopy,* by R. G. Kessel and R. H. Kardon. © 1979 W. H. Freeman and Company.

Near the base of the epithelium, membranous processes from adjacent sustentacular cells fuse by specialized junctions (occluding junctions) into complexes that divide the tissue into two layers. The spermatogonia are on one side of this barrier, and the cells in more advanced stages are on the other side. This membranous complex helps maintain a favorable environment for development of sperm cells by preventing certain large molecules from moving from the interstitial fluid of the basal epithelium into the region of the differentiating cells.

Spermatogenesis occurs continually in a male. The resulting sperm cells collect in the lumen of each seminiferous tubule, then pass through the rete testis to the epididymis, where they accumulate and mature.

Sperm have fascinated biologists for centuries. Anton van Leeuwenhoek was the first to view human sperm under a microscope in 1678, concluding that they were parasites in semen. By 1685, he had modified his view, writing that sperm contain a preformed human being and are seeds requiring nurturing in a female to start a new life.

**Figure** 22.5

(a) Light micrograph of seminiferous tubules (200×).
(b) Spermatogonia give rise to primary spermatocytes by mitosis; the spermatocytes, in turn, give rise to sperm cells by meiosis.

(a)

Sperm cells

Spermatogonia

Seminiferous tubule

Interstitial cells

(b)

Sustentacular cell

Lumen of seminiferous tubule

Sperm cells (23 chromosomes)

Developmental sequence

Spermatid (23 chromosomes)

Secondary spermatocyte (23 chromosomes)

Wall of seminiferous tubule

Primary spermatocyte (46 chromosomes)

Spermatogonium (46 chromosomes)

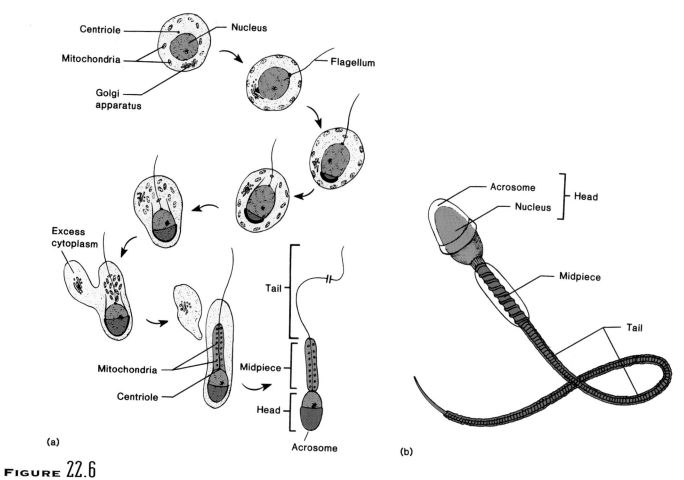

**FIGURE 22.6**

(a) The head of the sperm develops largely from the nucleus of the formative cell. (b) Parts of a mature sperm cell.

## Structure of a Sperm Cell

A mature sperm cell is a tiny, tadpole-shaped structure about 0.06 millimeters long. It consists of a flattened head, a cylindrical body, and an elongated tail.

The oval *head* of a sperm cell is primarily composed of a nucleus, and contains highly compacted chromatin consisting of twenty-three chromosomes. A small protrusion at its anterior end, called the *acrosome,* contains enzymes that aid the sperm cell in penetrating an egg cell (fig. 22.6).

The midpiece of a sperm contains a central, filamentous core and many mitochondria in a spiral. The *tail* (flagellum) consists of several microtubules enclosed in an extension of the cell membrane. The mitochondria provide ATP for the lashing movement of the tail that propels the sperm cell through fluid. The scanning electron micrograph in figure 22.7 shows a few mature sperm cells.

**FIGURE 22.7**

Scanning electron micrograph of human sperm cells (1,400×).

Many toxic chemicals that affect sperm hamper their ability to swim, so the cells cannot transmit the toxin to an egg. One notable exception is cocaine, which attaches to thousands of binding sites on human sperm cells, without apparently harming the cells or impeding their movements. Sperm can ferry cocaine to an egg, but scientists do not know what harm, if any, the drug causes. We do know that fetuses exposed to cocaine in the uterus may suffer a stroke, or, as infants, be unable to react normally to their surroundings.

**1** Explain the function of the sustentacular cells in the seminiferous tubules.

**2** Describe spermatogenesis.

**3** Describe the structure of a sperm cell.

## MALE INTERNAL ACCESSORY ORGANS

The internal accessory organs of the male reproductive system include the epididymides, vasa deferentia, ejaculatory ducts, and urethra, as well as the seminal vesicles, prostate gland, and bulbourethral glands.

### Epididymis

Each **epididymis** (pl., *epididymides*) is a tightly coiled, threadlike tube about 6 meters long (see figs. 22.1, 22.8, and reference plate 52). The epididymis is connected to ducts within a testis. It emerges from the top of the testis, descends along its posterior surface, and then courses upward to become the vas deferens.

The inner lining of the epididymis is composed of pseudostratified columnar cells that bear nonmotile cilia. These cells secrete glycogen and other substances that support stored sperm cells and promote their maturation.

When immature sperm cells reach the epididymis, they are nonmotile. However, as they travel through the epididymis as a result of rhythmic peristaltic contractions, they mature. Following this aging process, the sperm cells can move independently and fertilize egg cells (ova). However, they usually do not "swim" until after ejaculation.

### Vas Deferens

Each **vas deferens** (pl., *vasa deferentia*), also called ductus deferens, is a muscular tube about 45 centimeters long that is lined with pseudostratified columnar epithelium (fig. 22.9). It begins at the lower end of the epididymis and passes upward along the medial side of a testis to become part of the spermatic cord. Each vas deferens passes through the inguinal canal, enters the abdominal cavity outside the parietal peritoneum, and

Epithelial cells

Nonmotile cilia

Sperm cells

**FIGURE 22.8**

Cross section of a human epididymis (50× micrograph enlarged to 145×).

courses over the pelvic brim. From there, it extends backward and medially into the pelvic cavity, where it ends behind the urinary bladder.

Near its termination, the vas deferens dilates into a portion called the *ampulla.* Just outside the prostate gland, the tube becomes slender again and unites with the duct of a seminal vesicle. The fusion of these two ducts forms an **ejaculatory duct,** which passes through the prostate gland and empties into the urethra through a slitlike opening (see fig. 22.1).

### Seminal Vesicle

A **seminal vesicle** (see fig. 22.1) is a convoluted, saclike structure about 5 centimeters long that is attached to the vas deferens near the base of the urinary bladder. The glandular tissue lining the inner wall of the seminal vesicle secretes a slightly alkaline fluid. This fluid helps regulate the pH of the tubular contents as sperm cells travel to the outside. The secretion of the seminal vesicle also contains *fructose,* a monosaccharide that provides energy to the sperm cells, and *prostaglandins,* which stimulate muscular contractions within the female reproductive organs, aiding the movement of sperm cells toward the egg cell.

At emission, the contents of the seminal vesicles empty into the ejaculatory ducts. This greatly increases the volume of the fluid discharged from the vas deferens.

**1** Describe the structure of the epididymis.

**2** Trace the path of the vas deferens.

**3** What is the function of a seminal vesicle?

(a) Lumen
Epithelium
Smooth muscle
Sperm cell in lumen of vas deferens
Pseudostratified columnar epithelium
Smooth muscle layer

(b)

**FIGURE 22.9**

(a) Scanning electron micrograph of a cross section of the vas deferens (85×). (b) Light micrograph of the wall of the vas deferens (250× micrograph enlarged to 700×).

*Tissues and Organs: A Text-Atlas of Scanning Electron Microscopy*, by R. G. Kessel and R. H. Kardon. © 1979 W. H. Freeman and Company.

## Prostate Gland

The **prostate gland** (see figs. 22.1 and 22.10) is a chestnut-shaped structure about 4 centimeters across and 3 centimeters thick that surrounds the beginning of the urethra, just inferior to the urinary bladder. It is composed of many branched tubular glands enclosed in connective tissue. Septa of connective tissue and smooth muscle extend inward from the capsule, separating the tubular glands. The ducts of these glands open into the urethra.

The prostate gland secretes a thin, milky fluid. This alkaline secretion neutralizes the sperm cell-containing fluid, which is acidic from accumulation of metabolic wastes that stored sperm cells produce. Prostatic fluid also enhances the motility of sperm cells, which remain relatively nonmotile in the acidic contents of the epididymis. In addition, the prostatic fluid helps neutralize the acidic secretions of the vagina, thus helping to sustain sperm cells that enter the female reproductive tract.

The prostate gland releases its secretions into the urethra as a result of smooth muscle contractions in its capsular wall. As this release occurs the contents of the vas deferens and the seminal vesicles enter the urethra, which increases the volume of the fluid. Clinical Application 22.1 discusses the effects of prostate enlargement.

## Bulbourethral Glands

The **bulbourethral glands** (Cowper's glands) are two small structures, each about a centimeter in diameter. They are located inferior to the prostate gland lateral to

Secretory cells of the prostate gland

Smooth muscle

Urethra

**FIGURE 22.10**

Light micrograph of the prostate gland (10×).

the membranous urethra and are enclosed by fibers of the external urethral sphincter muscle (see fig. 22.1).

The bulbourethral glands are composed of many tubes whose epithelial linings secrete a mucuslike fluid. This fluid is released in response to sexual stimulation and lubricates the end of the penis in preparation for sexual intercourse (coitus). Females secrete most of the lubricating fluid for intercourse, however.

# Prostate Enlargement

The prostate gland is relatively small in boys, begins to grow in early adolescence, and reaches adult size several years later. Usually, the gland does not grow again until age 50, when in half of all men it enlarges enough to press on the urethra. This produces a feeling of pressure on the bladder because it cannot empty completely, and the man feels the urge to urinate frequently. Retained urine can lead to infection and inflammation, bladder stones, or kidney disease.

Medical researchers do not know what causes prostate enlargement. Risk factors include a fatty diet, having had a vasectomy, and possibly occupational exposure to batteries or the metal cadmium. The enlargement may be benign or cancerous. Because prostate cancer is nearly 100% treatable if detected early, men should have their prostates examined regularly.

Diagnostic tests include a rectal exam, as well as a blood test to detect prostate specific antigen (PSA), a cell surface protein normally found on prostate cells. Elevated PSA levels indicate an enlarged prostate, possibly from a benign or cancerous growth. Ultrasound may provide further information. Table 22A summarizes treatments for an enlarged prostate. ■

| TABLE 22A | MEDICAL TREATMENTS FOR AN ENLARGED PROSTATE GLAND |
|---|---|
| Surgical removal of prostate | |
| Radiation | |
| Drugs to block testosterone's growth-stimulating effect on the prostate | |
| Microwave energy delivered through a probe inserted into the urethra or rectum | |
| Balloon inserted into the urethra and inflated with liquid | |
| Liquid nitrogen delivered by a probe through the skin to freeze the tumor | |
| Device (stent) inserted between lobes of prostate to relieve pressure on the urethra | |

## Semen

The fluid the urethra conveys to the outside during ejaculation is called **semen.** It consists of sperm cells from the testes and secretions of the seminal vesicles, prostate gland, and bulbourethral glands. Semen is slightly alkaline (pH about 7.5), and its contents include prostaglandins and nutrients.

The volume of semen released at one time varies from 2 to 6 milliliters. The average number of sperm cells present in the fluid is about 120 million per milliliter.

Sperm cells remain nonmotile while they are in the ducts of the testis and epididymis, but become activated as they mix with the secretions of accessory glands. However, sperm cells cannot fertilize an egg cell until they enter the female reproductive tract. Development of this ability, called *capacitation,* entails changes that weaken the acrosomal membranes of the sperm cells. When sperm cells are placed with egg cells in a laboratory dish to achieve fertilization—a technique called in vitro fertilization, discussed in the next chapter—chemicals are added to simulate capacitation.

Although sperm cells can live for many weeks in the ducts of the male reproductive tract, they usually survive only a day or two after being expelled to the outside, even when they are maintained at body temperature. On the other hand, sperm cells can be stored and kept viable for years if they are frozen at a temperature below −100°C. Clinical Application 22.2 describes some causes of male infertility.

1 Where is the prostate gland located?

2 What is the function of the prostate gland's secretion?

3 What is the function of the bulbourethral glands?

4 What are the components of semen?

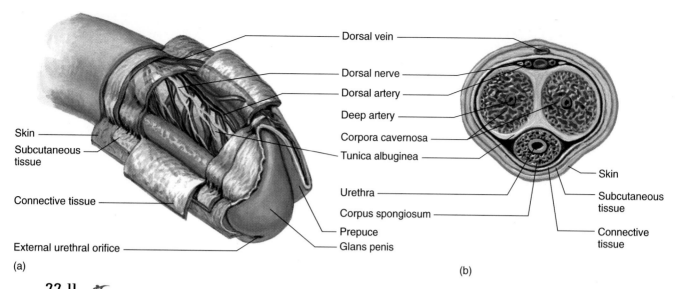

**FIGURE 22.11**

(a) Interior structure of the penis. (b) Cross section of the penis.

## MALE EXTERNAL REPRODUCTIVE ORGANS

The male external reproductive organs are the scrotum, which encloses the testes, and the penis. The urethra passes through the penis.

### Scrotum

The **scrotum** is a pouch of skin and subcutaneous tissue that hangs from the lower abdominal region posterior to the penis. The subcutaneous tissue of the scrotal wall lacks fat but contains a layer of smooth muscle fibers that constitute the *dartos muscle.* Exposure to cold stimulates these muscles to contract, the scrotal skin to wrinkle, and the testes to move closer to the pelvic cavity for heat absorption. Exposure to warmth stimulates the fibers to relax and the scrotum to hang loosely, and provides the testes with an environment 3°C (about 5°F) below body temperature, which is more conducive to sperm production and survival.

A medial septum divides the scrotum into two chambers, each of which encloses a testis. Each chamber also contains a serous membrane, which covers the front and sides of the testis and the epididymis, helping to ensure that the testis and epididymis move smoothly within the scrotum (see fig. 22.1).

### Penis

The **penis** is a cylindrical organ that conveys urine and semen through the urethra to the outside. It is also specialized to enlarge and stiffen by a process called **erection,** so that it can be inserted into the vagina during sexual intercourse.

The *body,* or shaft, of the penis is composed of three columns of erectile tissue, which include a pair of dorsally

located *corpora cavernosa* and a single, ventral *corpus spongiosum.* A tough capsule of white dense connective tissue called a *tunica albuginea* surrounds each column. Skin, a thin layer of subcutaneous tissue, and a layer of connective tissue enclose the penis (fig. 22.11).

The corpus spongiosum, through which the urethra extends, enlarges at its distal end to form a sensitive, cone-shaped **glans penis.** This glans covers the ends of the corpora cavernosa and bears the urethral opening—the *external urethral orifice.* The skin of the glans is very thin, hairless, and contains sensory receptors for sexual stimulation. A loose fold of skin called the *prepuce* (foreskin) begins just posterior to the glans and extends anteriorly to cover it as a sheath. The prepuce is sometimes removed by a surgical procedure called *circumcision.*

At the *root* of the penis, the columns of erectile tissue separate. The corpora cavernosa diverge laterally in the perineum and are firmly attached to the inferior surface of the pubic arch by connective tissue. These diverging parts form the *crura* (sing., crus) of the penis. The single corpus spongiosum is enlarged between the crura as the *bulb* of the penis, which is attached to membranes of the perineum.

1. Describe the structure of the penis.

2. What is circumcision?

3. How is the penis attached to the perineum?

### Erection, Orgasm, and Ejaculation

During sexual stimulation, parasympathetic nerve impulses from the sacral portion of the spinal cord release the vasodilator nitric oxide, causing the arteries leading into the penis to dilate. This causes increased blood flow into erectile tissues. At the same time, the increasing

# Male Infertility

Male infertility—the inability of sperm cells to fertilize an egg cell—has several causes. If during fetal development the testes do not descend into the scrotum, the higher temperature of the abdominal cavity or inguinal canal causes the developing sperm cells in the seminiferous tubules to degenerate. Certain diseases, such as mumps, may inflame the testes (orchitis), producing infertility by destroying cells in the seminiferous tubules.

Both the quality and quantity of sperm cells are essential factors in the ability of a man to father a child. If a sperm head is misshapen, if a sperm cannot swim, or if there are simply too few sperm cells, completing the arduous journey to the well-protected female cell may be impossible. Sometimes even a sperm cell that enters an egg is unsuccessful because it lacks the microtubules necessary to attract and merge the nuclei of the two cells.

Until recently, sperm analysis was rather subjective, based on a person's viewing the cells under a microscope. Now, computer-aided sperm analysis (CASA) is standardizing and expanding criteria for normalcy in human male seminal fluid and the sperm cells it contains.

To analyze sperm, a man abstains from intercourse for 2 to 3 days, then provides a sperm sample, which must be examined within the hour. The man must also provide information about his reproductive history and possible exposure

**FIGURE 22A**

Computer analysis improves the consistency and accuracy of describing sperm motility, morphology, and abundance.

pressure of the arterial blood entering the vascular spaces of the erectile tissue compresses the veins of the penis, reducing flow of venous blood away from the penis. Consequently, blood accumulates in the erectile tissues, and the penis swells and elongates, producing an **erection** (fig. 22.12).

The culmination of sexual stimulation is **orgasm**, a pleasurable feeling of physiological and psychological release. Orgasm in the male is accompanied by emission and ejaculation.

**Emission** is the movement of sperm cells from the testes and secretions from the prostate gland and seminal vesicles into the urethra, where they mix to form semen. Emission occurs in response to sympathetic nerve impulses from the spinal cord, which stimulate peristaltic contractions in smooth muscles within the walls of the testicular ducts, epididymides, vasa deferentia, and ejaculatory ducts. At the same time, other sympathetic impulses stimulate rhythmic contractions of the seminal vesicles and prostate gland.

to toxins. The sperm sample is placed on a slide under a microscope, and then technology intervenes. A video camera sends an image to a videocassette recorder, which projects a live or digitized image. The camera also sends the image to a computer, which traces sperm trajectories and displays them on a monitor or prints a hard copy. Figures 22A and 22B show a CASA of normal sperm, depicting how the swimming pattern alters as they travel.

Hundreds of CASA systems are now in use, mostly at fertility clinics. The devices are also helpful in using sperm as "biomarkers" of exposure to toxins. For example, in one study, the sperm of men who work in the dry cleaning industry and are exposed to the solvent perchloroethylene (believed to damage

sperm) were compared with sperm from men who work in the laundry industry and are exposed to many of the same chemicals except this one. CASA showed a difference in sperm motility that was directly related to level of exposure, as measured by exhalation of the chemical. This result

supported the reproductive evidence: Although the men in both groups had the same numbers of children, the dry cleaners' partners took much longer to conceive than did the launderers' wives. Table 22B lists the components of a sperm analysis. ■

| TABLE 22B SEMEN ANALYSIS | |
|---|---|
| **Characteristic** | **Normal Value** |
| Volume | 2–6 milliliters/ejaculate |
| Sperm density | 120 million cells/milliliter |
| Percent motile | > 40% |
| Motile sperm density | > 8 million/milliliter |
| Average velocity | > 20 micrometers/second |
| Motility | > 8 micrometers/second |
| Percent abnormal morphology | > 40% |
| White blood cells | > 5 million/milliliter |

(a)    (b)    (c)    (d)

**FIGURE 22B**

A computer tracks sperm movements. In semen, sperm swim in a straight line (a), but as they are activated by biochemicals normally found in the woman's body, their trajectories widen (b). The sperm in (c) are in the mucus of a woman's cervix, and the sperm in (d) are attempting to penetrate the structures surrounding an oocyte.

As the urethra fills with semen, sensory impulses are stimulated and pass into the sacral portion of the spinal cord. In response, motor impulses are transmitted from the cord to certain skeletal muscles at the base of the erectile columns of the penis, causing them to contract rhythmically. This increases the pressure within the erectile tissues and aids in forcing the semen through the urethra to the outside—a process called **ejaculation.**

The sequence of events during emission and ejaculation is coordinated so that the fluid from the bul-

bourethral glands is expelled first. This is followed by the release of fluid from the prostate gland, the passage of the sperm cells, and finally, the ejection of fluid from the seminal vesicles (fig. 22.13).

Immediately after ejaculation, sympathetic impulses constrict the arteries that supply the erectile tissue, reducing the inflow of blood. Smooth muscles within the walls of the vascular spaces partially contract again, and the veins of the penis carry the excess blood out of these spaces. The penis gradually returns to its

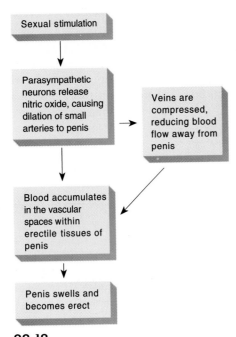

**FIGURE 22.12**

Mechanism of penile erection.

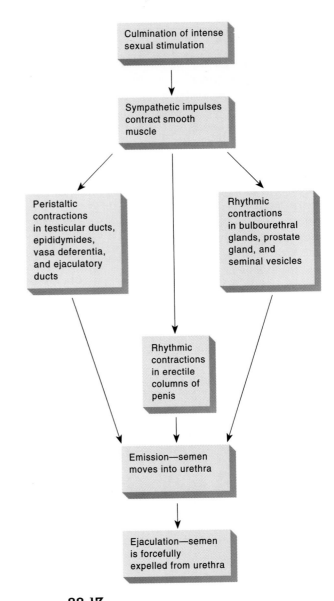

**FIGURE 22.13**

Mechanism of emission and ejaculation.

flaccid state, and usually another erection and ejaculation cannot be triggered for a period of 10 to 30 minutes or longer. Table 22.1 summarizes the functions of the male reproductive organs.

Spontaneous emissions and ejaculations commonly occur in adolescent males during sleep. Changes in hormonal concentrations that accompany adolescent development and sexual maturation cause these *nocturnal emissions.*

1 What controls blood flow into the erectile tissues of the penis?

2 Distinguish among orgasm, emission, and ejaculation.

3 Review the events associated with emission and ejaculation.

## HORMONAL CONTROL OF MALE REPRODUCTIVE FUNCTIONS

Hormones secreted by the *hypothalamus,* the *anterior pituitary gland,* and the *testes* control male reproductive functions. These hormones initiate and maintain sperm cell production and oversee the development and maintenance of male sex characteristics.

## Hypothalamic and Pituitary Hormones

Prior to 10 years of age, a boy's body is reproductively immature. During this period, the body is childlike, and the spermatogenic cells of the testes are undifferentiated. Then a series of changes leads to development of a reproductively functional adult. The hypothalamus controls many of these changes.

Recall from chapter 13 that the hypothalamus secretes gonadotropin-releasing hormone (GnRH), which enters the blood vessels leading to the anterior pituitary gland. In response, the anterior pituitary gland secretes the **gonadotropins** called *luteinizing hormone* (LH) and *follicle-stimulating hormone* (FSH). LH, which in males is called interstitial cell-stimulating

## TABLE 22.1 FUNCTIONS OF THE MALE REPRODUCTIVE ORGANS

| Organ | Function |
|---|---|
| Testis | |
|    Seminiferous tubules | Production of sperm cells |
|    Interstitial cells | Production and secretion of male sex hormones |
| Epididymis | Storage and maturation of sperm cells; conveys sperm cells to vas deferens |
| Vas deferens | Conveys sperm cells to ejaculatory duct |
| Seminal vesicle | Secretes an alkaline fluid containing nutrients and prostaglandins; fluid helps neutralize acidic semen |
| Prostate gland | Secretes an alkaline fluid that helps neutralize the acidic components of semen and enhances motility of sperm cells |
| Bulbourethral gland | Secretes fluid that lubricates end of the penis |
| Scrotum | Encloses and protects testes |
| Penis | Conveys urine and semen to outside of body; inserted into the vagina during sexual intercourse; the glans penis is richly supplied with sensory nerve endings associated with feelings of pleasure during sexual stimulation |

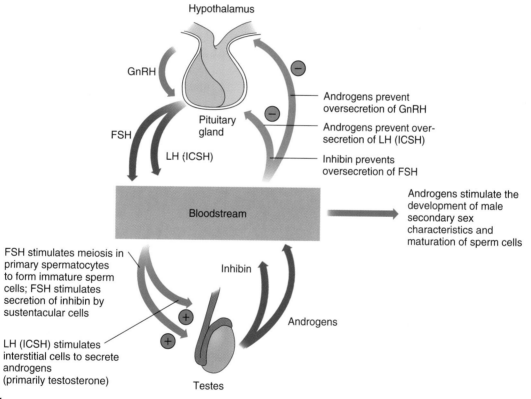

**FIGURE 22.14**

The hypothalamus controls maturation of sperm cells and development of male secondary sex characteristics. A negative feedback mechanism operating between the hypothalamus, the anterior lobe of the pituitary gland, and the testes controls the concentration of testosterone.

hormone (ICSH), promotes development of the interstitial cells (cells of Leydig) of the testes, and they, in turn, secrete male sex hormones. FSH stimulates the sustentacular cells of the seminiferous tubules to proliferate, grow, mature, and respond to the effects of the male sex hormone testosterone. Then, in the presence of FSH and testosterone, these cells stimulate the spermatogenic cells to undergo spermatogenesis, giving rise to sperm cells (fig. 22.14). The sustentacular cells also secrete a hormone called *inhibin,* which inhibits the anterior pituitary gland by negative feedback, and thus prevents oversecretion of FSH.

## Male Sex Hormones

Male sex hormones are termed **androgens.** The interstitial cells of the testes produce most of them, but small amounts are synthesized in the adrenal cortex (see chapter 13).

The hormone **testosterone** is the most abundant androgen. It is secreted and transported in the blood loosely attached to plasma proteins. Like other steroid hormones, testosterone combines with receptor molecules usually in the nuclei of its target cells (see chapter 13). However, in many target cells, such as those in the prostate gland, seminal vesicles, and male external accessory organs, testosterone is first converted to another androgen called **dihydrotestosterone,** which stimulates the cells of these organs. Androgen molecules that do not reach receptors in target cells are usually changed by the liver into forms that can be excreted in bile or urine.

Testosterone secretion begins during fetal development and continues for a few weeks following birth, then nearly ceases during childhood. Between the ages of 13 and 15, a young man's androgen production usually accelerates. This phase in development, when an individual becomes reproductively functional, is **puberty.** After puberty, testosterone secretion continues throughout the life of a male.

## Actions of Testosterone

Cells of the embryonic testes first produce testosterone after about eight weeks of development. This hormone stimulates the formation of the male reproductive organs, including the penis, scrotum, prostate gland, seminal vesicles, and ducts. Still later, testosterone causes the testes to descend into the scrotum.

During puberty, testosterone stimulates enlargement of the testes (the primary male sex characteristic) and accessory organs of the reproductive system, as well as development of male *secondary sex characteristics,* which are special features associated with the adult male body. Secondary sex characteristics in the male include:

1. Increased growth of body hair, particularly on the face, chest, axillary region, and pubic region, but sometimes accompanied by decreased growth of hair on the scalp.

2. Enlargement of the larynx and thickening of the vocal folds, accompanied by lowering of the pitch of the voice.

3. Thickening of the skin.

4. Increased muscular growth accompanied by broadening shoulders and narrowing of the waist.

5. Thickening and strengthening of the bones.

Testosterone also increases the rate of cellular metabolism and production of red blood cells (by stimulating erythropoietin, see chapter 19), so that the average number of red blood cells in a cubic millimeter of blood is usually greater in males than in females. Testosterone stimulates sexual activity by affecting certain portions of the brain.

## Regulation of Male Sex Hormones

The extent to which male secondary sex characteristics develop is directly related to the amount of testosterone that the interstitial cells secrete. A negative feedback system involving the hypothalamus regulates the quantity of testosterone (fig. 22.14).

As the concentration of testosterone in the blood increases, the hypothalamus becomes inhibited, and its stimulation of the anterior pituitary gland by GnRH decreases. As the pituitary's secretion of LH (ICSH) falls in response, the amount of testosterone released by the interstitial cells is reduced.

As the blood concentration of testosterone drops, the hypothalamus becomes less inhibited, and it once again stimulates the anterior pituitary gland to release LH. The increasing secretion of LH causes the interstitial cells to release more testosterone, and its blood concentration increases. Testosterone level decreases somewhat during and after the *male climacteric,* a decline in sexual function that occurs with aging. At any given age, the testosterone concentration in the male is regulated to remain relatively constant.

**1** Which hormone initiates the changes associated with male sexual maturity?

**2** Describe several male secondary sex characteristics.

**3** List the functions of testosterone.

**4** Explain the regulation of secretion of male sex hormones.

## ORGANS OF THE FEMALE REPRODUCTIVE SYSTEM

The organs of the female reproductive system are specialized to produce and maintain the female sex cells, or *egg cells;* to transport these cells to the site of fertilization; to provide a favorable environment for a developing offspring; to move the offspring to the outside; and to produce female sex hormones.

The *primary sex organs* (gonads) of this system are the ovaries, which produce the female sex cells and sex hormones. The other parts of the system comprise the internal and external *accessory organs.*

## OVARIES

The **ovaries** are solid, ovoid structures measuring about 3.5 centimeters in length, 2 centimeters in width, and 1 centimeter in thickness. An individual ovary is located in a shallow depression (ovarian fossa) on each side in the lateral wall of the pelvic cavity (fig. 22.15).

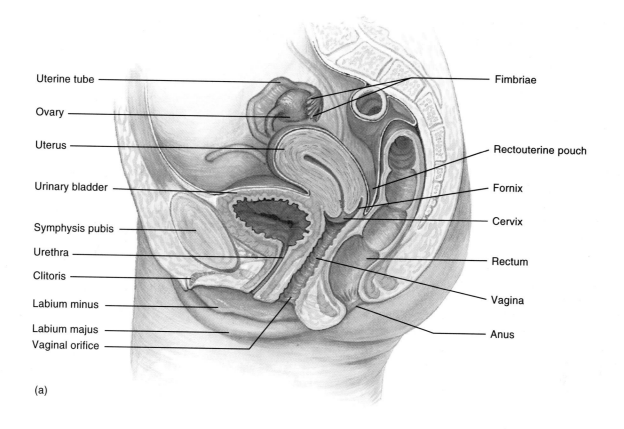

Uterine tube

Ovary

Uterus

Urinary bladder

Symphysis pubis

Urethra

Clitoris

Labium minus

Labium majus

Vaginal orifice

Fimbriae

Rectouterine pouch

Fornix

Cervix

Rectum

Vagina

Anus

(a)

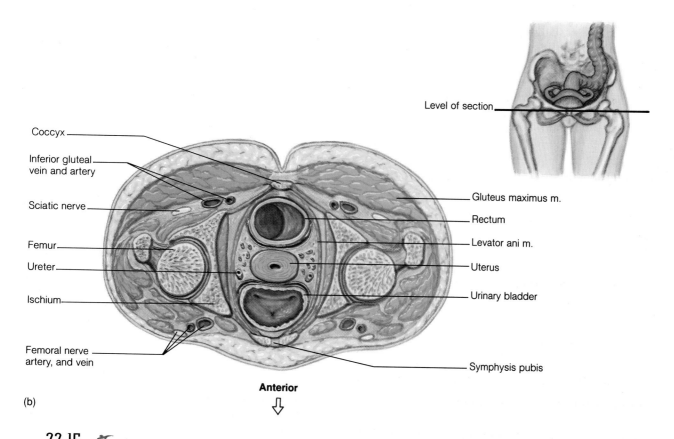

Level of section

Coccyx

Inferior gluteal vein and artery

Sciatic nerve

Femur

Ureter

Ischium

Femoral nerve artery, and vein

Gluteus maximus m.

Rectum

Levator ani m.

Uterus

Urinary bladder

Symphysis pubis

**Anterior**
⇩

(b)

**FIGURE** 22.15 ✗

(a) Sagittal view of the female reproductive organs. (b) Transverse section of the female pelvic cavity.

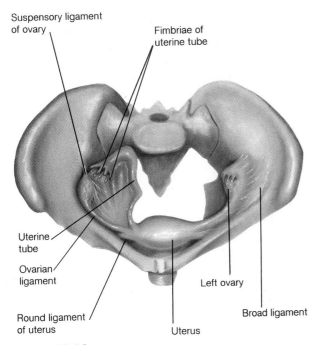

Suspensory ligament
of ovary

Fimbriae of
uterine tube

Uterine
tube

Ovarian
ligament

Round ligament
of uterus

Left ovary

Broad ligament

Uterus

**FIGURE** 22.16

The ovaries are located on each side against the lateral walls of the pelvic cavity. The right uterine tube is retracted to reveal the ovarian ligament.

## Ovary Attachments

Several ligaments help hold each ovary in position. The largest of these, formed by a fold of peritoneum, is called the *broad ligament.* It is also attached to the uterine tubes and the uterus.

A small fold of peritoneum, called the *suspensory ligament,* holds the ovary at its upper end. This ligament also contains the ovarian blood vessels and nerves. At its lower end, the ovary is attached to the uterus by a rounded, cordlike thickening of the broad ligament called the *ovarian ligament* (fig. 22.16).

## Ovary Descent

Like the testes in a male fetus, the ovaries in a female fetus originate from masses of tissue posterior to the parietal peritoneum, near the developing kidneys. During development, these structures descend to locations just inferior to the pelvic brim, where they remain attached to the lateral pelvic wall.

## Ovary Structure

The tissues of an ovary can be subdivided into two rather indistinct regions, an inner *medulla* and an outer *cortex.* The ovarian medulla is mostly composed of loose connective tissue, and contains many blood vessels, lymphatic vessels, and nerve fibers. The ovarian cortex consists of more compact tissue, and has a granular appearance due to tiny masses of cells called *ovarian follicles.*

A layer of cuboidal epithelium cells (germinal epithelium) covers the free surface of the ovary. Just beneath this epithelium is a layer of dense connective tissue called the *tunica albuginea.*

**1** What are the primary sex organs of the female?

**2** Describe the descent of the ovary.

**3** Describe the structure of an ovary.

## Primordial Follicles

During prenatal development (before birth), small groups of cells in the outer region of the ovarian cortex form several million **primordial follicles.** Each of these structures consists of a single, large cell called a *primary oocyte,* which is closely surrounded by a layer of flattened epithelial cells called *follicular cells.*

Early in development, the primary oocytes begin to undergo meiosis, but the process soon halts and does not continue until puberty. Once the primordial follicles appear, no new ones form. Instead, the number of oocytes in the ovary steadily declines, as many of the oocytes degenerate. Of the several million oocytes formed originally, only a million or so remain at the time of birth, and perhaps 400,000 are present at puberty. Of these, probably fewer than 400 or 500 will be released from the ovary during the reproductive life of a female. Probably fewer than ten will go on to form a new individual!

Scientists think that the reason for the increased incidence of chromosome defects in children of older mothers is that the eggs, having been present for several decades, had time to be extensively exposed to damaging agents, such as radiation, viruses, and toxins.

## Oogenesis

Beginning at puberty, some primary oocytes are stimulated to continue meiosis. As in the case of sperm cells, the resulting cells have one-half as many chromosomes (23) in their nuclei as their parent cells (see chapter 24). Egg formation is called **oogenesis.**

When a primary oocyte reproduces, the distribution of the cytoplasm is unequal. One of the resulting cells, called a *secondary oocyte,* is relatively large, and the other, called the *first polar body,* is very small.

The large secondary oocyte represents a future *egg cell* (ovum) that can be fertilized by uniting with a sperm cell. If this happens, the oocyte reproduces unequally to produce a tiny *second polar body* and a large fertilized egg cell, or **zygote,** that can give rise to an embryo (fig. 22.17). The polar bodies have no further function, and they soon degenerate.

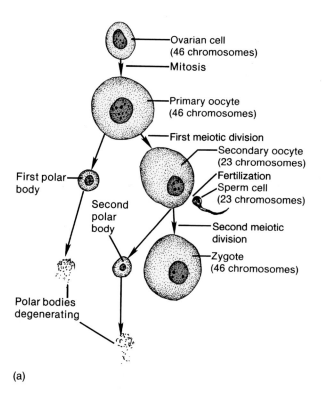

(a)

Ovarian cell
(46 chromosomes)

Mitosis

Primary oocyte
(46 chromosomes)

First meiotic division

Secondary oocyte
(23 chromosomes)

Fertilization
Sperm cell
(23 chromosomes)

First polar body

Second polar body

Second meiotic division

Zygote
(46 chromosomes)

Polar bodies degenerating

(b)

**FIGURE 22.17**

(a) During oogenesis, a single egg cell (secondary oocyte) results from the meiosis of a primary oocyte. If the egg cell is fertilized, it generates a second polar body and becomes a zygote. (b) Light micrograph of a secondary oocyte and a polar body (arrow) (700×).

Formation of polar bodies may appear wasteful, but it has an important biological function. It produces an egg cell that has the massive amounts of cytoplasm and abundant organelles required to carry a zygote through the first few cell reproductions.

**1** Describe the major events of oogenesis.

**2** How is polar body formation beneficial to an egg?

An experimental procedure called polar body biopsy allows couples to select an egg that does not carry a disease-causing gene that the woman carries. First, oocytes with attached first polar bodies are removed from the woman and cultured in a laboratory dish. Then the polar bodies are screened with a "DNA probe," which is a piece of genetic material that binds to a specific disease-causing gene and fluoresces or gives off radiation, which can be detected.

In polar body biopsy, bad news is really good news. Because of the laws of inheritance (discussed in chapter 24), if the defective gene is in a polar body, it is *not* in the egg cell that it is physically attached to. Researchers can then fertilize the egg with sperm in the laboratory and implant it in the woman who donated it, with some confidence that the disorder carried in the family will not pass to this particular future child.

## Follicle Maturation

At puberty, the anterior pituitary gland secretes increased amounts of FSH, and the ovaries enlarge in response. At the same time, some of the primordial follicles mature (fig. 22.18).

During maturation, the oocyte of a follicle enlarges, and the surrounding follicular cells proliferate by mitosis, giving rise to a stratified epithelium composed of *granulosa cells.* A layer of glycoprotein, called the *zona pellucida,* gradually separates the oocyte from the granulosa cells; at this stage, the structure is called a *primary follicle.*

Meanwhile, the ovarian cells outside the follicle organize into two cellular layers. The *inner vascular layer* (theca interna) is largely composed of steroid-secreting cells, plus some loose connective tissue and blood vessels. The *outer fibrous layer* (theca externa) is composed of tightly packed connective tissue cells.

The follicular cells continue to proliferate, and when there are six to twelve layers of cells, irregular, fluid-filled spaces appear among them. These spaces soon join to form a single cavity (antrum), and the oocyte is pressed to one side of the follicle. At this stage, the follicle is about 0.2 millimeters in diameter and is called a *secondary follicle.*

Maturation takes ten to fourteen days. The *mature follicle* (preovulatory, or Graafian, follicle) is about 10 millimeters or more across, and its fluid-filled cavity bulges outward on the surface of the ovary, like a blister. The oocyte within the mature follicle is a large, spherical cell, surrounded by a relatively thick zona pellucida, attached to a mantle of follicular cells (corona radiata). Processes from these follicular cells

**FIGURE 22.18**

Light micrograph of the surface of a mammalian ovary (50× micrograph enlarged to 200×).

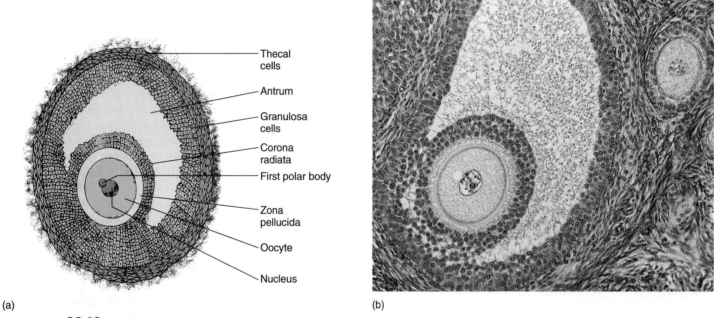

Thecal
cells

Antrum

Granulosa
cells

Corona
radiata

First polar body

Zona
pellucida

Oocyte

Nucleus

(a)                                                                                          (b)

**FIGURE 22.19**

(a) Structure of a mature (Graafian) follicle. (b) Light micrograph of a mature follicle (250×). The first polar body has just separated and has not been expelled from the cell.

extend through the zona pellucida and supply nutrients to the oocyte (fig. 22.19).

Although as many as twenty primary follicles may begin maturing at any one time, one follicle (dominant follicle) usually outgrows the others. Typically, only the dominant follicle fully develops, and the other follicles degenerate (fig. 22.20).

Certain drugs used to treat female infertility, such as Clomid (clomiphene), may cause a woman to "super-ovulate." More than one follicle grows, more than one oocyte is released, and the result is twins—or even triplets. In 1997, an Iowa couple had septuplets, after using a fertility drug.

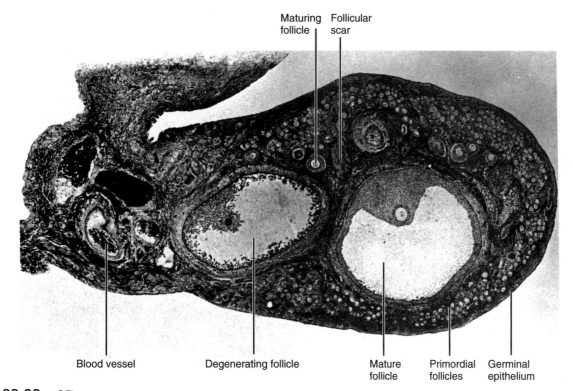

Maturing follicle · Follicular scar · Blood vessel · Degenerating follicle · Mature follicle · Primordial follicles · Germinal epithelium

**FIGURE 22.20**

Light micrograph of a mammalian (monkey) ovary (30×). If ovulation does not occur, the follicle degenerates.

Uterine tube · Oocyte · Ovary

**FIGURE 22.21**

Light micrograph of a follicle during ovulation (75×).

## Ovulation

As a follicle matures, its primary oocyte undergoes oogenesis, giving rise to a secondary oocyte and a first polar body. A process called **ovulation** releases these cells from the follicle.

Hormonal stimulation from the anterior pituitary gland triggers ovulation, causing the mature follicle to swell rapidly and its wall to weaken. Eventually the wall ruptures, and the follicular fluid, accompanied by the oocyte, oozes outward from the surface of the ovary. Figure 22.21 shows expulsion of a mammalian oocyte.

After ovulation, the oocyte and one or two layers of follicular cells surrounding it are usually propelled to the opening of a nearby uterine tube. If the oocyte is not fertilized within a relatively short time, it degenerates. Figure 22.22 illustrates maturation of a follicle and the release of an oocyte.

1 What changes occur in a follicle and its oocyte during maturation?

2 What causes ovulation?

3 What happens to an oocyte following ovulation?

## FEMALE INTERNAL ACCESSORY ORGANS

The internal accessory organs of the female reproductive system include a pair of uterine tubes, a uterus, and a vagina.

## Uterine Tubes

The **uterine tubes** (fallopian tubes, or oviducts) are suspended by portions of the broad ligament, and open near the ovaries. Each tube, which is about 10 centimeters long and 0.7 centimeters in diameter, passes medially to the uterus, penetrates its wall, and opens into the uterine cavity.

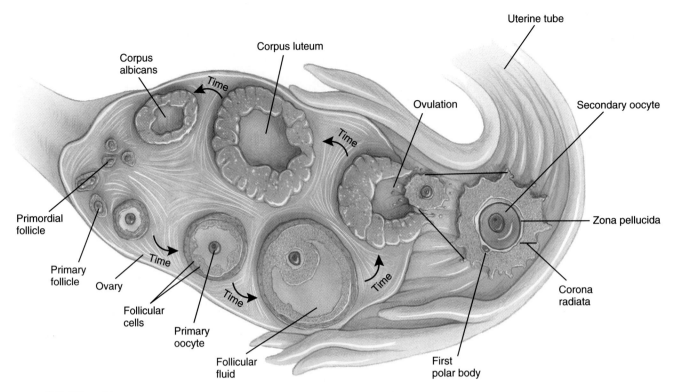

**FIGURE 22.22**

As a follicle matures, the egg cell enlarges and becomes surrounded by follicular cells and fluid. Eventually, the mature follicle ruptures, releasing the egg cell.

Near each ovary, a uterine tube expands to form a funnel-shaped *infundibulum,* which partially encircles the ovary medially. On its margin, the infundibulum bears a number of irregular, branched extensions called *fimbriae* (fig. 22.23). Although the infundibulum generally does not touch the ovary, one of the larger extensions (ovarian fimbria) connects directly to the ovary.

The wall of a uterine tube consists of an inner mucosal layer, a middle muscular layer, and an outer covering of peritoneum. The mucosal layer is drawn into many longitudinal folds and is lined with simple columnar epithelial cells, some of which are *ciliated* (fig. 22.24). The epithelium secretes mucus, and the cilia beat toward the uterus. These actions help draw the egg cell and expelled follicular fluid into the infundibulum following ovulation. Ciliary action and peristaltic contractions of the tube's muscular layer aid transport of the egg down the uterine tube.

## Uterus

The **uterus** receives the embryo that develops from an egg cell that has been fertilized in the uterine tube, and sustains its development. It is a hollow, muscular organ, shaped somewhat like an inverted pear.

The *broad ligament,* which also attaches to the ovaries and uterine tubes, extends from the lateral walls of the uterus to the pelvic walls and floor, creating a drape across the top of the pelvic cavity (fig. 22.23). A flattened band of tissue within the broad ligament, called the *round ligament,* connects the upper end of the uterus to the anterior pelvic wall (see figs. 22.16 and 22.23).

The size of the uterus changes greatly during pregnancy. In its nonpregnant, adult state, it is about 7 centimeters long, 5 centimeters wide (at its broadest point), and 2.5 centimeters in diameter. The uterus is located medially within the anterior portion of the pelvic cavity, superior to the vagina, and is usually bent forward over the urinary bladder.

The upper two-thirds, or *body,* of the uterus has a dome-shaped top, called the *fundus,* and is joined by the uterine tubes, which enter its wall at its broadest part. The lower one-third, or neck, of the uterus is called the **cervix.** This tubular part extends downward into the upper portion of the vagina. The cervix surrounds the opening called the *cervical orifice* (ostium uteri), through which the uterus opens to the vagina.

The uterine wall is thick and composed of three layers (fig. 22.25). The **endometrium** is the inner mucosal layer lining the uterine cavity. It is covered with columnar epithelium and contains abundant tubular glands. The **myometrium,** a very thick, muscular layer, largely consists of bundles of smooth muscle fibers in longitudinal, circular, and spiral patterns, and is interlaced with connective tissues. During the monthly female reproductive cycles and during pregnancy, the endometrium and myometrium extensively change. The **perimetrium** con-

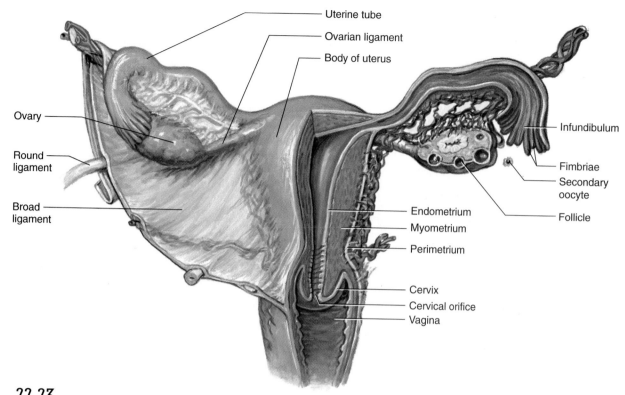

**FIGURE** 22.23

The funnel-shaped infundibulum of the uterine tube partially encircles the ovary.

(a)

(b)

**FIGURE** 22.24

(a) Light micrograph of a uterine tube (250× micrograph enlarged to 800×). (b) Scanning electron micrograph of ciliated cells that line the uterine tube (4,000×).

Lumen

Endometrium

Myometrium

Perimetrium

**FIGURE 22.25**

Light micrograph of the uterine wall (10.5×).

sists of an outer serosal layer, which covers the body of the uterus and part of the cervix.

## Vagina

The **vagina** is a fibromuscular tube, about 9 centimeters long, that extends from the uterus to the outside. It conveys uterine secretions, receives the erect penis during sexual intercourse, and provides the open channel for the offspring during birth.

The vagina extends upward and back into the pelvic cavity. It is posterior to the urinary bladder and urethra, anterior to the rectum, and attached to these structures by connective tissues. The upper one-fourth of the vagina is separated from the rectum by a pouch (rectouterine pouch). The tubular vagina also surrounds the end of the cervix, and the recesses between the vaginal wall and the cervix are termed *fornices* (sing., *fornix*).

The fornices are clinically important because they are relatively thin-walled and allow the physician to palpate the internal abdominal organs during a physical examination. Also, the posterior fornix, which is somewhat longer than the others, provides a surgical access to the peritoneal cavity through the vagina.

The *vaginal orifice* is partially closed by a thin membrane of connective tissue and stratified squamous epithelium called the **hymen.** A central opening of varying size allows uterine and vaginal secretions to pass to the outside.

The vaginal wall consists of three layers. The inner *mucosal layer* is stratified squamous epithelium and is drawn into many longitudinal and transverse ridges (vaginal rugae). This layer lacks mucous glands; the mucus found in the lumen of the vagina comes from the glands of the cervix and the vestibular glands at the mouth of the vagina.

Daughters of women who took the drug DES (diethylstilbestrol) while pregnant with them may develop a benign condition called *adenosis.* It arises when secretory columnar epithelium, resembling normal cells of the uterine lining, grow in the wrong place—in the vagina, up near the cervix. It is a little as if the lining of the mouth were to grow onto the face.

Adenosis may produce a slight vaginal discharge. It is detected with a procedure called the *Pap* (Papanicolaou) *smear test.* A doctor or nurse scrapes off a tiny sample of cervical tissue, smears the sample on a glass slide, stains it, and sends it to a laboratory, where cytotechnologists examine it for the presence of abnormal cells, or a computer with image-analysis software scans it.

If the Pap smear is abnormal, the doctor follows up with a direct observation with a special type of microscope called a *colposcope.* The physician paints the patient's cervix with acetic acid (vinegar), which stains a purplish blue in the presence of carbohydrate in the discharge. A laser can be used to painlessly remove the abnormally placed tissue.

The middle *muscular layer* of the vagina mainly consists of smooth muscle fibers in longitudinal and circular patterns. At the lower end of the vagina is a thin band of striated muscle. This band helps close the vaginal opening; however, a voluntary muscle (bulbospongiosus) is primarily responsible for closing this orifice.

The outer *fibrous layer* consists of dense fibrous connective tissue interlaced with elastic fibers. It attaches the vagina to surrounding organs.

1 How does an egg cell move into the infundibulum following ovulation?

2 How is an egg cell moved along a uterine tube?

3 Describe the structure of the uterus.

4 What is the function of the uterus?

5 Describe the structure of the vagina.

# FEMALE EXTERNAL REPRODUCTIVE ORGANS

The *external accessory organs* of the female reproductive system include the labia majora, the labia minora, the clitoris, and the vestibular glands. These structures that surround the openings of the urethra and vagina compose the **vulva** (fig. 22.26).

## Labia Majora

The **labia majora** (sing., *labium majus*) enclose and protect the other external reproductive organs. They correspond to the scrotum of the male, and are composed of rounded folds of adipose tissue and a thin layer of smooth muscle, covered by skin. On the outside, this skin includes hairs, sweat glands, and sebaceous glands, whereas on the inside, it is thinner and hairless.

The labia majora lie closely together and are separated longitudinally by a cleft (pudendal cleft), which includes the urethral and vaginal openings. At their anterior ends, the labia merge to form a medial, rounded elevation of adipose tissue called the *mons pubis,* which overlies the symphysis pubis. At their posterior ends, the labia taper and merge into the perineum near the anus.

## Labia Minora

The **labia minora** (sing., *labium minus*) are flattened longitudinal folds between the labia majora, and they extend along either side of the vestibule. They are composed of connective tissue richly supplied with blood vessels, causing a pinkish appearance. Stratified squamous epithelium covers this tissue. Posteriorly, the labia minora merge with the labia majora, whereas anteriorly, they converge to form a hoodlike covering around the clitoris.

## Clitoris

The **clitoris** is a small projection at the anterior end of the vulva between the labia minora. It is usually about 2 centimeters long and 0.5 centimeters in diameter, including a portion embedded in surrounding tissues. The clitoris corresponds to the penis and has a similar structure. It is composed of two columns of erectile tissue called *corpora cavernosa.* A septum separates these columns, which are covered with dense connective tissue.

At the root of the clitoris, the corpora cavernosa diverge to form *crura,* which, in turn, attach to the sides of the pubic arch. At its anterior end, a small mass of erectile tissue forms a **glans,** which is richly supplied with sensory nerve fibers.

## Vestibule

The labia minora enclose a space called the **vestibule.** The vagina opens into the posterior portion of the vestibule, and the urethra opens in the midline, just anterior to the vagina and about 2.5 centimeters posterior to the glans of the clitoris.

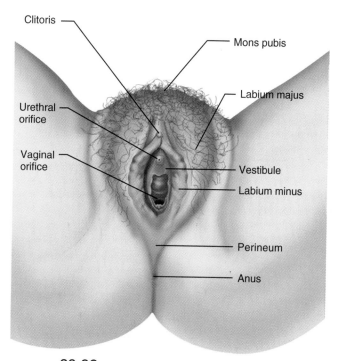

**FIGURE** 22.26
Female external reproductive organs.

A pair of **vestibular glands** (Bartholin's glands), corresponding to the bulbourethral glands, lie on either side of the vaginal opening. Their ducts open into the vestibule near the lateral margins of the vaginal orifice.

Beneath the mucosa of the vestibule on either side is a mass of vascular erectile tissue. These structures are called the *vestibular bulbs.* They are separated from each other by the vagina and the urethra, and they extend forward from the level of the vaginal opening to the clitoris.

**1** What is the male counterpart of the labia majora? Of the clitoris?

**2** What structures are within the vestibule?

## Erection, Lubrication, and Orgasm

Erectile tissues located in the clitoris and around the vaginal entrance respond to sexual stimulation. Following such stimulation, parasympathetic nerve impulses from the sacral portion of the spinal cord inhibit sympathetic control of the arteries associated with the erectile tissues, causing them to dilate. As a result, inflow of blood increases, and the erectile tissues swell. At the same time, the vagina begins to expand and elongate.

If sexual stimulation is sufficiently intense, parasympathetic impulses stimulate the vestibular glands to secrete mucus into the vestibule. This secretion moistens and lubricates the tissues surrounding the vestibule and the lower end of the vagina, facilitating insertion of the penis into the vagina. Mucus secretion continuing

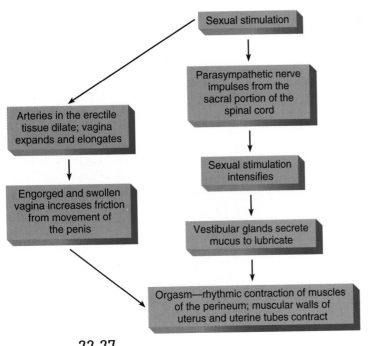

**FIGURE 22.27**

Mechanism of erection, lubrication, and orgasm in the human female.

| TABLE 22.2 | FUNCTIONS OF THE FEMALE REPRODUCTIVE ORGANS | |
|---|---|---|
| **Organ** | **Function** | |
| Ovary | Produces egg cells and female sex hormones | |
| Uterine tube | Conveys egg cell toward uterus; site of fertilization; conveys developing embryo to uterus | |
| Uterus | Protects and sustains embryo during pregnancy | |
| Vagina | Conveys uterine secretions to outside of body; receives erect penis during sexual intercourse; provides open channel for offspring during birth process | |
| Labia majora | Enclose and protect other external reproductive organs | |
| Labia minora | Form margins of vestibule; protect openings of vagina and urethra | |
| Clitoris | Glans is richly supplied with sensory nerve endings associated with feelings of pleasure during sexual stimulation | |
| Vestibule | Space between labia minora that includes vaginal and urethral openings | |
| Vestibular glands | Secrete fluid that moistens and lubricates vestibule | |

during sexual intercourse helps prevent irritation of tissues that might occur if the vagina remained dry.

The clitoris is abundantly supplied with sensory nerve fibers, which are especially sensitive to local stimulation. The culmination of such stimulation is orgasm, the pleasurable sense of physiological and psychological release.

Just prior to orgasm, the tissues of the outer third of the vagina engorge with blood and swell. This action increases the friction on the penis during intercourse. Orgasm initiates a series of reflexes involving the sacral and lumbar portions of the spinal cord. In response to these reflexes, the muscles of the perineum and the walls of the uterus and uterine tubes contract rhythmically. These contractions help transport sperm cells through the female reproductive tract toward the upper ends of the uterine tubes (fig. 22.27).

Following orgasm, the flow of blood into the erectile tissues slackens, and the muscles of the perineum and reproductive tract relax. Consequently, the organs return to a state similar to that prior to sexual stimulation. Table 22.2 summarizes the functions of the female reproductive organs.

**1** What events result from parasympathetic stimulation of the female reproductive organs?

**2** How does the vagina change just prior to and during female orgasm?

**3** How do the uterus and the uterine tubes respond to orgasm?

## HORMONAL CONTROL OF FEMALE REPRODUCTIVE FUNCTIONS

The *hypothalamus,* the *anterior pituitary gland,* and the *ovaries* secrete hormones that control development and maintenance of female secondary sex characteristics, maturation of female sex cells, and changes that occur during the monthly reproductive cycle.

### Female Sex Hormones

A girl's body is reproductively immature until about 10 years of age. Then, the hypothalamus begins to secrete increasing amounts of GnRH, which in turn stimulate the anterior pituitary gland to release the gonadotropins FSH and LH. These hormones play primary roles in controlling female sex cell maturation and in producing female sex hormones.

Several tissues, including the ovaries, the adrenal cortices, and the placenta (during pregnancy), secrete female sex hormones. These hormones belong to two major groups, called **estrogen** and **progesterone.**

The primary source of estrogen (in a nonpregnant female) is the ovaries, although some estrogen is also synthesized in adipose tissue from adrenal androgens. At puberty, under the influence of the anterior pituitary gland, the ovaries secrete increasing amounts of the hormone. Estrogen stimulates enlargement of accessory or-

860

UNIT SIX

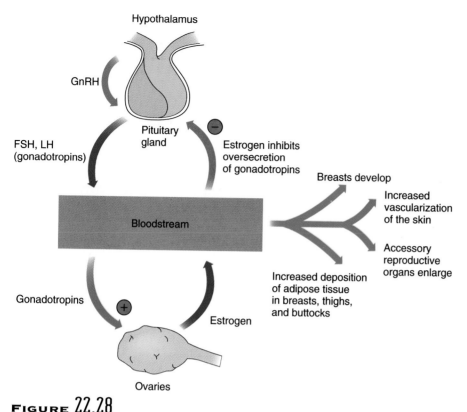

**FIGURE 22.28**

Control of female secondary sex development. Estrogen inhibits LH and FSH during most of the menstrual cycle.

**1** What stimulates sexual maturation in a female?

**2** Name the two major female sex hormones.

**3** What is the function of estrogen?

**4** What is the function of androgen in a female?

gans including the vagina, uterus, uterine tubes, and ovaries, as well as the external structures, and is also responsible for the development and maintenance of female *secondary sex characteristics.* These are listed in figure 22.28 and include the following:

1. Development of the breasts and the ductile system of the mammary glands within the breasts.

2. Increased deposition of adipose tissue in the subcutaneous layer generally and in the breasts, thighs, and buttocks particularly.

3. Increased vascularization of the skin.

The ovaries are also the primary source of progesterone (in a nonpregnant female). This hormone promotes changes that occur in the uterus during the female reproductive cycle, affects the mammary glands, and helps regulate secretion of gonadotropins from the anterior pituitary gland.

Certain other changes that occur in females at puberty are related to *androgen* (male sex hormone) concentrations. For example, increased growth of hair in the pubic and axillary regions is due to androgen secreted by the adrenal cortices. Conversely, development of the female skeletal configuration, which includes narrow shoulders and broad hips, seems to be a response to a low concentration of androgen.

## Female Reproductive Cycle

The female reproductive cycle, or **menstrual cycle,** is characterized by regular, recurring changes in the uterine lining, which culminate in menstrual bleeding (menses). Such cycles usually begin near the thirteenth year of life and continue into middle age, then cease.

Women athletes may have disturbed menstrual cycles, ranging from diminished menstrual flow (oligomenorrhea) to complete stoppage (amenorrhea). The more active an athlete, the more likely are menstrual problems. This effect results from a loss of adipose tissue and a consequent decline in estrogen, which adipose tissue synthesizes from adrenal androgens.

A female's first menstrual cycle (menarche) occurs after the ovaries and other organs of the female reproductive control system mature and respond to certain hormones. Then, the hypothalamic secretion of gonadotropin-releasing hormone (GnRH) stimulates the anterior pituitary gland to release threshold levels of FSH (follicle-stimulating hormone) and LH (luteinizing hormone). As its name implies, FSH stimulates maturation of an ovarian *follicle.* The granulosa cells of the follicle produce increasing amounts of estrogen and some

## TABLE 22.3 — HORMONAL CONTROL OF FEMALE SECONDARY SEX CHARACTERISTICS

1. The hypothalamus releases GnRH, which stimulates the anterior pituitary gland.
2. The anterior pituitary gland secretes FSH and LH.
3. FSH stimulates the maturation of a follicle.
4. Granulosa cells of the follicle produce and secrete estrogen; LH stimulates certain cells to secrete estrogen precursor molecules.
5. Estrogen is responsible for the development and maintenance of most of the female secondary sex characteristics.
6. Concentrations of androgen affect other secondary sex characteristics, including skeletal growth and growth of hair.
7. Progesterone, secreted by the ovaries, affects cyclical changes in the uterus and mammary glands.

progesterone. LH also stimulates certain ovarian cells (theca interna) to secrete precursor molecules (testosterone) used to produce estrogen.

In a young female, estrogen stimulates development of various secondary sex characteristics. Estrogen secreted during subsequent menstrual cycles continues development of these traits and maintains them. Table 22.3 summarizes the hormonal control of female secondary sex characteristics.

Increasing concentration of estrogen during the first week or so of a menstrual cycle changes the uterine lining, thickening the glandular endometrium (proliferative phase). Meanwhile, the developing follicle completes maturation, and by the fourteenth day of the cycle, the follicle appears on the surface of the ovary as a blisterlike bulge.

Within the follicle, the granulosa cells, which surround the oocyte and connect it to the inner wall, loosen. Follicular fluid accumulates rapidly.

While the follicle matures, estrogen that it secretes inhibits the release of LH from the anterior pituitary gland, but allows LH to be stored in the gland. Estrogen also makes the anterior pituitary cells more sensitive to the action of GnRH, which is released from the hypothalamus in rhythmic pulses about 90 minutes apart.

Near the fourteenth day of follicular development, the anterior pituitary cells finally respond to the pulses of GnRH and release stored LH. The resulting surge in LH concentration, which lasts for about 36 hours, weakens and ruptures the bulging follicular wall. At the same time, the oocyte and follicular fluid escape from the ovary (ovulation).

Following ovulation, the remnants of the follicle and the theca interna within the ovary rapidly change. The space containing the follicular fluid fills with blood, which soon clots, and under the influence of LH, the follicular and thecal cells expand to form a temporary glandular structure within the ovary, called a **corpus luteum** (see fig. 22.22).

Follicular cells secrete some progesterone during the first part of the menstrual cycle. However, corpus luteum cells secrete abundant progesterone and estrogen during the second half of the cycle. Consequently, as a corpus luteum is established, the blood concentration of progesterone increases sharply.

Progesterone causes the endometrium to become more vascular and glandular. It also stimulates the uterine glands to secrete more glycogen and lipids (secretory phase). As a result, the endometrial tissues fill with fluids containing nutrients and electrolytes, which provide a favorable environment for the development of an embryo.

Estrogen and progesterone inhibit the release of LH and FSH from the anterior pituitary gland. Consequently, no other follicles are stimulated to develop when the corpus luteum is active. However, if the oocyte released at ovulation is not fertilized by a sperm cell, the corpus luteum begins to degenerate (regress) about the twenty-fourth day of the cycle. Eventually, connective tissue replaces it. The remnant of such a corpus luteum is called a *corpus albicans* (see fig. 22.22).

When the corpus luteum ceases to function, concentrations of estrogen and progesterone decline rapidly, and in response, blood vessels in the endometrium constrict. This reduces the supply of oxygen and nutrients to the thickened endometrium, and these lining tissues (decidua) soon disintegrate and slough off. At the same time, blood escapes from damaged capillaries, creating a flow of blood and cellular debris, which passes through the vagina as the *menstrual flow* (menses). This flow usually begins about the twenty-eighth day of the cycle and continues for three to five days, while the estrogen concentration is relatively low.

The beginning of the menstrual flow marks the end of a menstrual cycle and the beginning of a new cycle. This cycle is diagrammed in figure 22.29 and summarized in table 22.4.

Because the blood concentrations of estrogen and progesterone are low at the beginning of the menstrual cycle, the hypothalamus and anterior pituitary gland are no longer inhibited. Consequently, the concentrations of FSH and LH soon increase, and a new follicle is stimulated to mature. As this follicle secretes estrogen, the uterine lining undergoes repair, and the endometrium begins to thicken again.

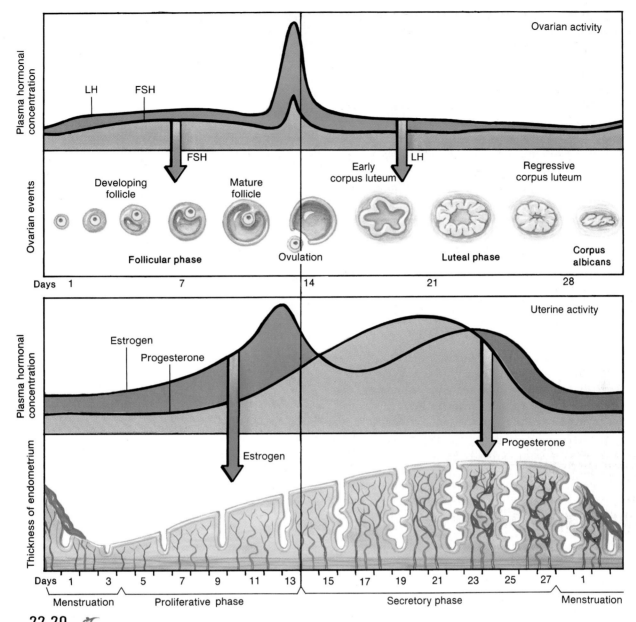

**FIGURE 22.29** ✗

Major events in the female ovarian and menstrual cycles.

A woman who becomes extremely irritable, fatigued, bloated, or easily upset at the same time each month is likely to have *premenstrual syndrome* (PMS). This condition may occur only during some months. The cause of PMS is unknown, but it is believed to involve hormonal imbalance.

## Menopause

After puberty, menstrual cycles continue at regular intervals into the late forties or early fifties, when they usually become increasingly irregular. Then, in a few months or years, the cycles cease altogether. This period in life is called **menopause** (female climacteric).

The cause of menopause is aging of the ovaries. After about 35 years of cycling, few primary follicles remain to respond to pituitary gonadotropins. Consequently, the follicles no longer mature, ovulation does not occur, and the blood concentration of estrogen plummets, although many women continue to synthesize some estrogen from adrenal androgens.

As a result of reduced estrogen concentration and lack of progesterone, the female secondary sex characteristics may change. The breasts, vagina, uterus, and uterine tubes may shrink, and the pubic and axillary hair may thin. The epithelial linings associated with urinary and reproductive organs may thin. There may be increased loss of bone matrix (osteoporosis) and thinning of the skin. Because the pituitary secretions of FSH and

**TABLE 22.4 MAJOR EVENTS IN A MENSTRUAL CYCLE**

1. The anterior pituitary gland secretes FSH and LH.
2. FSH stimulates maturation of a follicle.
3. Granulosa cells of the follicle produce and secrete estrogen.
   a. Estrogen maintains secondary sex traits.
   b. Estrogen causes the uterine lining to thicken.
4. The anterior pituitary gland releases a surge of LH, which stimulates ovulation.
5. Follicular and thecal cells become corpus luteum cells, which secrete estrogen and progesterone.
   a. Estrogen continues to stimulate uterine wall development.
   b. Progesterone stimulates the uterine lining to become more glandular and vascular.
   c. Estrogen and progesterone inhibit secretion of FSH and LH from the anterior pituitary gland.
6. If the egg cell is not fertilized, the corpus luteum degenerates and no longer secretes estrogen and progesterone.
7. As the concentrations of luteal hormones decline, blood vessels in the uterine lining constrict.
8. The uterine lining disintegrates and sloughs off, producing a menstrual flow.
9. The anterior pituitary gland, no longer inhibited, again secretes FSH and LH.
10. The menstrual cycle repeats.

LH are no longer inhibited, these hormones may be released continuously for some time.

About 50% of women reach menopause by age 50, and 85% reach it by age 52. Of these, perhaps 20% have no unusual symptoms—they simply stop menstruating. However, about 50% of women experience unpleasant vasomotor symptoms during menopause, including sensations of heat in the face, neck, and upper body called "hot flashes." Such a sensation may last for 30 seconds to 5 minutes and may be accompanied by chills and sweating. Women may also experience headache, backache, and fatigue during menopause. These vasomotor symptoms may result from changes in the rhythmic secretion of GnRH by the hypothalamus in response to declining concentrations of sex hormones. To reduce the unpleasant side effects of menopause, women are often treated with *estrogen replacement therapy* (ERT), which consists of estrogen delivered through a transdermal (skin) patch.

At age 53, Mary Shearing gave birth to twins, Amy Leigh and Kelly Ann. In the last stages of menopause, Mary already had three grown children from a previous marriage, and two grandchildren. She and her 32-year-old husband, Don, decided to try for pregnancy. Don's sperm was used to fertilize oocytes in a laboratory dish, and the fertilized ova were implanted in Mary's uterus. The Shearings' success showed that it is not the condition of the uterine lining in an older woman that makes conceiving, carrying, and delivering a healthy baby difficult, but the age of the oocyte. Women in their sixties have since given birth, using donated oocytes. However, most women who have the treatment that Shearing underwent are under 45 years of age, and have lost ovaries to disease. Often the donor is a sister.

1 Trace the events of the female menstrual cycle.

2 What effect does progesterone have on the endometrium?

3 What causes menstrual flow?

4 What are some changes that may occur at menopause?

## PREGNANCY

**Pregnancy** is the presence of a developing offspring in the uterus. It results from the union of the genetic packages of an egg cell and a sperm cell—an event called **fertilization.**

### Transport of Sex Cells

Ordinarily, before fertilization can occur, a secondary oocyte must be ovulated and enter a uterine tube. During sexual intercourse, the male deposits semen containing sperm cells in the vagina near the cervix. To reach the secondary oocyte, the sperm cells must then move upward through the uterus and uterine tube. Lashing of sperm tails and muscular contractions within the walls of the uterus and uterine tube, stimulated by prostaglandins in the semen, aid the sperm cells' journey. Also, under the influence of high estrogen concentrations during the first part of the menstrual cycle, the uterus and cervix contain a thin, watery secretion that promotes sperm transport and survival. Conversely, during the latter part of the cycle, when the progesterone concentration is high, the female reproductive tract secretes a viscous fluid that hampers sperm transport and survival (fig. 22.30).

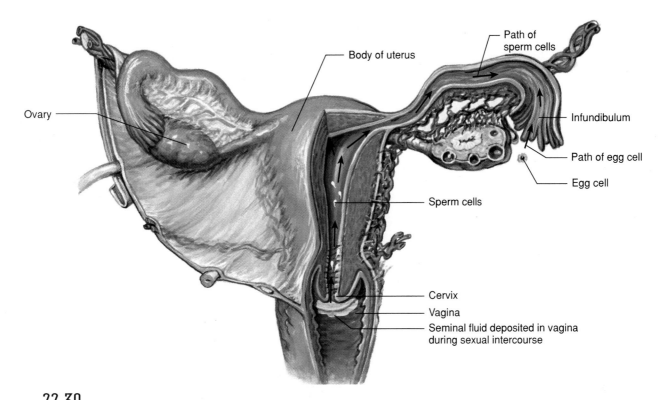

FIGURE 22.30
The paths of the egg and sperm cells through the female reproductive tract.

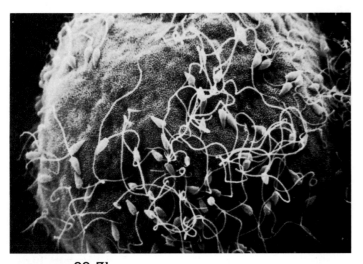

FIGURE 22.31
Scanning electron micrograph of sperm cells on the surface of an egg cell (1,200×).

Sperm movement is inefficient. Even though as many as 300 million to 500 million sperm cells may be deposited in the vagina by a single ejaculation, only a few hundred ever reach a secondary oocyte.

Sperm cells reach the upper portions of the uterine tube within an hour following sexual intercourse. Although many sperm cells may reach a secondary oocyte, only one sperm cell actually fertilizes it (fig. 22.31).

About one in a million births produces a severely deformed child who has inherited three sets of chromosomes. This may be the result of two sperm cells entering a single egg cell.

A secondary oocyte may survive for only 12 to 24 hours following ovulation, whereas sperm cells may live up to 72 hours within the female reproductive tract. Consequently, sexual intercourse probably must occur no later than 72 hours before ovulation or within 24 hours following ovulation if fertilization is to take place.

## Fertilization

When a sperm reaches a secondary oocyte, it invades the follicular cells that adhere to the oocyte's surface (corona radiata) and binds to the **zona pellucida** that surrounds the oocyte's cell membrane. The acrosome of a sperm cell attached to the zona pellucida releases an enzyme that helps the motile sperm penetrate the zona pellucida (fig. 22.32).

In "zona blasting," an experimental procedure to aid certain infertile men, an egg cell cultured in a laboratory dish is chemically stripped of its zona pellucida. The more vulnerable egg now presents less of a barrier to a sperm.

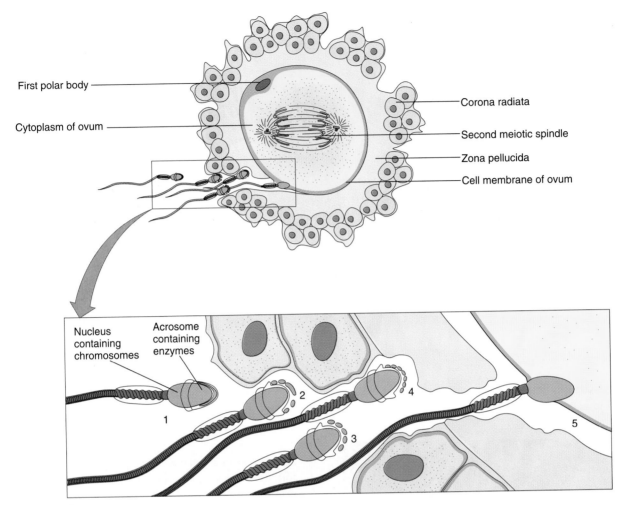

First polar body

Cytoplasm of ovum

Corona radiata

Second meiotic spindle

Zona pellucida

Cell membrane of ovum

Nucleus containing chromosomes

Acrosome containing enzymes

1
2
3
4
5

**FIGURE 22.32**

Steps in fertilization: (1) Sperm cell reaches corona radiata surrounding the egg cell. (2) Acrosome of sperm cell releases protein-digesting enzyme. (3 and 4) Sperm cell penetrates zona pellucida surrounding egg cell. (5) Sperm cell's plasma membrane fuses with egg cell membrane.

The sperm cell's plasma membrane fuses with that of the secondary oocyte, and sperm movement ceases. The sperm cell sheds its tail. At the same time, the oocyte cell membrane becomes unresponsive to other sperm cells. The union of the oocyte and sperm cell membranes also triggers some lysosomelike granules (cortical granules) just beneath the oocyte cell membrane to release enzymes that harden the zona pellucida. This reduces the chance that other sperm cells will penetrate, and it forms a protective layer around the newly formed fertilized egg cell.

Once the sperm enters the oocyte's cytoplasm, the nucleus within its head swells. The secondary oocyte then divides unequally to form a relatively large cell and a tiny second polar body, which is later expelled. Therefore, female meiosis completes only after the sperm enters the egg. Next, the nuclei of the male and female cells unite. Their nuclear membranes disassemble, and their chromosomes mingle, completing the process of fertilization, diagrammed in figure 22.32.

Because the sperm cell and the egg cell each provide twenty-three chromosomes, the product of fertilization is a cell with forty-six chromosomes—the usual number in a human cell. This cell, called a **zygote,** is the first cell of the future offspring.

**1** What factors aid the movements of the egg and sperm cells through the female reproductive tract?

**2** Where in the female reproductive system does fertilization normally take place?

**3** List the events of fertilization.

## Early Embryonic Development

About 30 hours after forming, the zygote undergoes *mitosis,* giving rise to two cells (blastomeres). These cells, in

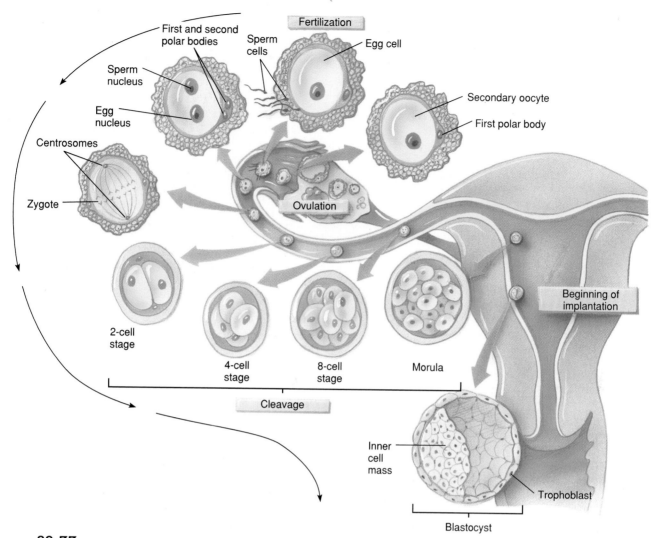

**FIGURE 22.33**

Stages in early human development.

turn, divide into four cells, which divide into eight cells, and so forth. These divisions take place rapidly with little time to grow. Thus, each subsequent division results in smaller and smaller cells. This phase of development is termed **cleavage.**

Meanwhile, the tiny mass of cells moves through the uterine tube to the uterine cavity, aided by the beating of cilia of the tubular epithelium and by weak peristaltic contractions of smooth muscles in the tubular wall. Secretions from the epithelial lining may provide the developing organism with nutrients.

The trip to the uterus takes about three days, and by then, the structure consists of a solid ball, called a **morula,** of about sixteen cells. The morula remains free within the uterine cavity for about three days. During this stage, the zona pellucida of the original egg cell degenerates, and the structure, which now consists of a

hollow ball of cells, called a **blastocyst,** drops into one of the tubules in the endometrium. By the end of the first week of development, it superficially *implants* in the endometrium (fig. 22.33).

About the time of implantation, certain cells within the blastocyst organize into a group, called the **inner cell mass,** that will give rise to the body. This marks the beginning of the *embryonic period* of development. The offspring is termed an **embryo** until the end of the eighth week, after which it is called a **fetus** up to the time of birth.

Eventually, the cells surrounding the embryo, with cells of the endometrium, form a complex vascular structure called the **placenta.** This organ attaches the embryo to the uterine wall and exchanges nutrients, gases, and wastes between the maternal blood and blood of the embryo. The placenta also secretes hormones. The placenta is described in more detail in chapter 23.

**FIGURE 22.34**

Mechanism that preserves the uterine lining during early pregnancy.

**FIGURE 22.35**

Relative concentrations of three hormones in the blood during pregnancy.

Occasionally, a blastocyst implants in tissues outside the uterus, such as those of a uterine tube, an ovary, the cervix, or an organ in the abdominal cavity. The result is called an *ectopic pregnancy.* If a fertilized egg implants within the uterine tube, it is called a *tubal pregnancy.* A tubal pregnancy is dangerous to a pregnant woman and the developing offspring because the tube usually ruptures as the embryo enlarges, and is accompanied by severe pain and heavy vaginal bleeding. Treatment is prompt surgical removal of the embryo and repair or removal of the damaged uterine tube.

**1** What is a zygote? A morula?

**2** How does an embryo implant in the uterine wall?

**3** How does a developing offspring obtain nutrients and oxygen?

## Hormonal Changes during Pregnancy

During a typical menstrual cycle, the corpus luteum degenerates about two weeks after ovulation. Consequently, estrogen and progesterone concentrations decline rapidly, the uterine lining is no longer maintained, and the endometrium sloughs off as menstrual flow. If this occurs following implantation, the embryo is lost (spontaneously aborted).

A hormone called hCG (**human chorionic gonadotropin**) normally helps prevent spontaneous abortion. A layer of embryonic cells called a **trophoblast** (see fig. 22.33), which secretes hCG and later helps form the placenta, surrounds the developing embryo (see chapter 23). This hormone has properties similar to those of LH, and it maintains the corpus luteum, which continues secreting estrogen and progesterone. Thus, the uterine wall continues to grow and develop. At the same time, release of FSH and LH from the anterior pituitary gland is inhibited, so that normal menstrual cycles cease (fig. 22.34).

Secretion of hCG continues at a high level for about two months, then declines to a low level by the end of four months. Although the corpus luteum persists throughout pregnancy, its function as a source of hormones becomes less important after the first three months (first trimester), when the placenta secretes sufficient estrogen and progesterone (fig. 22.35).

Detecting hCG in a woman's urine or blood is used to confirm pregnancy. The level of hCG in a pregnant woman's body fluids peaks at 50 to 60 days of gestation, then falls to a much lower level for the remainder of pregnancy. Later on, measuring hCG has other uses. If a woman miscarries but her blood still shows hCG, fetal tissue may remain in her uterus, and it must be removed. At the fifteenth week of pregnancy, most women have a blood test that measures levels of three substances produced by the fetus—alpha fetoprotein, estriol (a form of estrogen), and hCG. If alpha fetoprotein and estriol are low but hCG is high, the fetus may have an extra chromosome, with severe effects on health.

For the remainder of the pregnancy, *placental estrogen* and *placental progesterone* maintain the uterine wall. The placenta also secretes a hormone called **placental lactogen** that may stimulate breast development and prepare the mammary glands for milk secretion, with the aid of placental estrogen and progesterone. Placental progesterone and a polypeptide hormone called *relaxin* from the corpus luteum inhibit the smooth muscles in the myometrium, suppressing uterine contractions until the birth process begins.

The high concentration of placental estrogen during pregnancy enlarges the vagina and the external reproductive organs. Also, relaxin relaxes the ligaments holding the symphysis pubis and sacroiliac joints together. This action, which usually occurs during the last week of pregnancy, allows for greater movement at these joints, aiding passage of the fetus through the birth canal.

Other hormonal changes that occur during pregnancy include increased secretions of aldosterone from the adrenal cortex and of parathyroid hormone from the parathyroid glands. Aldosterone promotes renal reabsorption of sodium, leading to fluid retention. Parathyroid hormone helps to maintain a high concentration of maternal blood calcium, since fetal demand for calcium can cause hypocalcemia, which promotes cramps (see chapter 13). Table 22.5 summarizes the hormonal changes of pregnancy.

**1** What mechanism maintains the uterine wall during pregnancy?

**2** What is the source of hCG during the first few months of pregnancy?

**3** What is the source of the hormones that sustain the uterine wall during pregnancy?

**4** What other hormonal changes occur during pregnancy?

## Other Changes during Pregnancy

Other changes in a woman's body respond to the increased demands of a growing fetus. As the fetus increases in size, the uterus enlarges greatly, and instead of being confined to its normal location in the pelvic cavity, it extends upward and may eventually reach the level of the ribs. At the same time, the abdominal organs are displaced upward and compressed against the diaphragm. The enlarging uterus also presses on the urinary bladder.

A pregnant woman is well aware of the effects of her expanding uterus. She can no longer eat large meals, develops heartburn often as stomach contents are pushed up into the esophagus, and frequently has to urinate as her uterus presses on her bladder.

| TABLE 22.5 | HORMONAL CHANGES DURING PREGNANCY |
|---|---|

1. Following implantation, cells of the trophoblast begin to secrete hCG.
2. hCG maintains the corpus luteum, which continues secreting estrogen and progesterone.
3. As the placenta develops, it secretes large quantities of estrogen and progesterone.
4. Placental estrogen and progesterone
   a. stimulate the uterine lining to continue development.
   b. maintain the uterine lining.
   c. inhibit secretion of FSH and LH from the anterior pituitary gland.
   d. stimulate development of the mammary glands.
   e. inhibit uterine contractions (progesterone).
   f. enlarge the reproductive organs (estrogen).
5. Relaxin from the corpus luteum also inhibits uterine contractions and relaxes the pelvic ligaments.
6. The placenta secretes placental lactogen that stimulates breast development.
7. Aldosterone from the adrenal cortex promotes reabsorption of sodium.
8. Parathyroid hormone from the parathyroid glands helps maintain a high concentration of maternal blood calcium.

As the placenta grows and develops, it requires more blood, and as the fetus enlarges, it needs more oxygen and produces more waste that must be excreted. The pregnant woman's blood volume, cardiac output, breathing rate, and urine production all increase to handle fetal growth.

The pregnant woman must eat more to obtain adequate nutrition for the fetus. Her intake must supply sufficient vitamins, minerals, and proteins for herself and the fetus. The fetal tissues have a greater capacity to capture available nutrients than do the maternal tissues. Consequently, if the pregnant woman's diet is inadequate, her body will usually show symptoms of a deficiency condition before fetal growth is adversely affected.

## Birth

Pregnancy usually continues for forty weeks, or about nine calendar months, if it is measured from the beginning of the last menstrual cycle. Pregnancy terminates with the *birth process* (parturition).

Birth is a complex, little-understood process. Progesterone seems to play a major role in beginning parturition. During pregnancy, this hormone suppresses uterine contractions. As the placenta ages, the concentration of progesterone within the uterus declines, which may also stimulate synthesis of a prostaglandin that promotes uterine contractions.

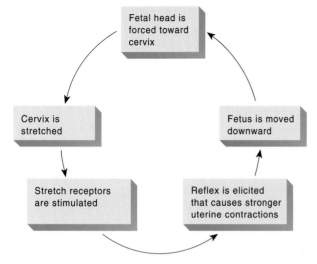

**FIGURE 22.36**

The birth process involves this positive feedback mechanism.

Diagram labels (clockwise): Fetal head is forced toward cervix → Fetus is moved downward → Reflex is elicited that causes stronger uterine contractions → Stretch receptors are stimulated → Cervix is stretched → Fetal head is forced toward cervix

| TABLE 22.6 | FACTORS CONTRIBUTING TO THE LABOR PROCESS |
|---|---|

1. As the time of birth approaches, secretion of progesterone declines, and its inhibiting effect on uterine contractions may lessen.
2. Decreasing progesterone concentration may stimulate synthesis of prostaglandins, which may initiate labor.
3. Stretching uterine tissues stimulates release of oxytocin from the posterior pituitary gland.
4. Oxytocin may stimulate uterine contractions and aid labor in its later stages.
5. As the fetal head stretches the cervix, a positive feedback mechanism results in stronger and stronger uterine contractions and a greater release of oxytocin.
6. Positive feedback stimulates abdominal wall muscles to contract with greater and greater force.
7. The fetus is forced through the birth canal to the outside.

Another stimulant beginning the birth process is stretching of the uterine and vaginal tissues late in pregnancy. This may initiate nerve impulses to the hypothalamus, which, in turn, signals the posterior pituitary gland to release the hormone **oxytocin.**

Oxytocin stimulates powerful uterine contractions. Combined with the greater excitability of the myometrium due to the decline in progesterone secretion, oxytocin aids labor in its later stages.

During *labor,* muscular contractions force the fetus through the birth canal. Rhythmic contractions that begin at the top of the uterus and travel down its length force the contents of the uterus toward the cervix.

Since the fetus is usually positioned head downward, labor contractions force the head against the cervix. This action stretches the cervix, which elicits a reflex that stimulates still stronger labor contractions. Thus, a *positive feedback system* operates in which uterine contractions produce more intense uterine contractions until a maximum effort is achieved (fig. 22.36). At the same time, dilation of the cervix reflexly stimulates an increased release of oxytocin from the posterior pituitary gland.

During childbirth, the tissues between the vulva and anus (perineum) are sometimes torn by the stretching that occurs as the infant passes through the birth canal. For this reason, an incision may be made along the midline of the perineum from the vestibule to within 1.5 centimeters of the anus before the birth is completed. This procedure, called an *episiotomy,* ensures that the perineal tissues are cut cleanly rather than torn, which aids healing.

As labor continues, positive feedback stimulates abdominal wall muscles to contract. These muscles also help move the fetus through the cervix and vagina to the outside.

Table 22.6 summarizes some of the factors promoting labor. Figure 22.37 illustrates the steps of the birth process.

Following birth of the fetus, the placenta, which remains inside the uterus, separates from the uterine wall and is expelled by uterine contractions through the birth canal. This expulsion, termed the *afterbirth,* is accompanied by bleeding, because vascular tissues are damaged in the process. However, the loss of blood is usually minimized by continued contraction of the uterus that compresses the bleeding vessels. The action of oxytocin stimulates this contraction.

For several weeks following childbirth, the uterus shrinks by a process called *involution.* Also, its endometrium sloughs off and is discharged through the vagina. The new mother passes a bloody and then yellowish discharge from the vagina for a few weeks. This is followed by the return of an epithelial lining characteristic of a nonpregnant female. Clinical Application 22.3 addresses some causes of infertility in the female.

**1** List some of the physiological changes that occur in a woman's body during pregnancy.

**2** Describe the role of progesterone in initiating labor.

**3** Explain how dilation of the cervix affects labor.

**4** Explain how bleeding is controlled naturally after the placenta is expelled.

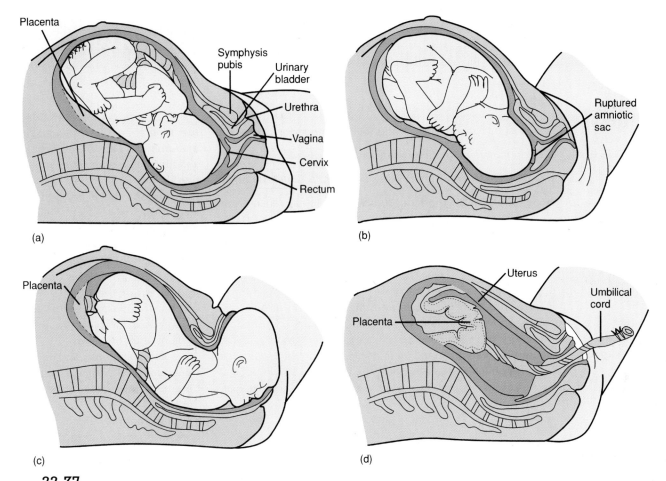

**FIGURE 22.37**

Stages in birth. (*a*) Fetal position before labor, (*b*) dilation of the cervix, (*c*) expulsion of the fetus, (*d*) expulsion of the placenta.

## MAMMARY GLANDS

The **mammary glands** are accessory organs of the female reproductive system that are specialized to secrete milk following pregnancy.

### Location of the Glands

The mammary glands are located in the subcutaneous tissue of the anterior thorax within the hemispherical elevations called *breasts.* The breasts overlie the *pectoralis major* muscles and extend from the second to the sixth ribs and from the sternum to the axillae.

A *nipple* is located near the tip of each breast at about the level of the fourth intercostal space. It is surrounded by a circular area of pigmented skin called the *areola* (fig. 22.38).

### Structure of the Glands

A mammary gland is composed of fifteen to twenty irregularly shaped lobes. Each lobe contains glands (alveolar glands) and a duct (lactiferous duct) that leads to the nipple and opens to the outside. Dense connective and adipose tissues separate the lobes. These tissues also support the glands and attach them to the fascia of the underlying pectoral muscles. Other connective tissue, which forms dense strands called *suspensory ligaments,* extends inward from the dermis of the breast to the fascia, helping support the breast's weight. Clinical Application 22.4 discusses a pressing problem concerning health—breast cancer.

### Development of the Breasts

The mammary glands of boys and girls are similar. As children reach *puberty,* the glands in males do not develop, whereas ovarian hormones stimulate development of the glands in females. As a result, the alveolar glands and ducts enlarge, and fat is deposited, so that the breasts become surrounded by adipose tissue, except for the region of the areola.

During pregnancy, placental estrogen and progesterone stimulate further development of the mammary glands. Estrogen causes the ductile systems to grow and branch, and deposits large quantities of fat around them. Progesterone stimulates the development of the alveolar glands at the ends of the ducts. Placental lactogen also promotes these changes.

Because of hormonal activity, the breasts may double in size during pregnancy. At the same time, glandular

# Female Infertility

For one out of six couples, trying for parenthood is a time of increasing concern, as pregnancy remains elusive. Physicians define infertility as the inability to conceive after a year of trying. A physical cause is found in 90% of cases, and 60% of the time, the abnormality lies in the female's reproductive system.

One of the more common causes of female infertility is hyposecretion of gonadotropic hormones from the anterior pituitary gland, followed by failure to ovulate (anovulation). This type of anovulatory cycle can sometimes be detected by testing the female's urine for *pregnanediol,* a product of progesterone metabolism. Since the concentration of progesterone normally rises following ovulation, no increase in pregnanediol in the urine during the latter part of the menstrual cycle suggests lack of ovulation.

Fertility specialists can treat absence of ovulation due to too little secretion of gonadotropic hormones by administering hCG (obtained from human placentas) or another ovulation-stimulating biochemical, human menopausal gonadotropin (hMG), which contains LH and FSH and is obtained from urine of women who are past menopause. However, either hCG or hMG may overstimulate the ovaries and cause many follicles to release egg cells simultaneously, resulting in multiple births.

Another cause of female infertility is *endometriosis,* in which tissue resembling the inner lining of the uterus (endometrium) is abnormally

present in the abdominal cavity. Small pieces of the endometrium may move up through the uterine tubes during menses and implant in the abdominal cavity. Here the tissue undergoes changes similar to those that take place in the uterine lining during the menstrual cycle. However, when the tissue begins to break down at the end of the cycle, it cannot be expelled to the outside. Instead, material remains in the abdominal cavity where it may irritate the lining

(peritoneum) and cause considerable abdominal pain. These breakdown products also stimulate formation of fibrous tissue (fibrosis). This material may encase the ovary and prevent ovulation, or obstruct the uterine tubes.

Some women become infertile as a result of infections, such as gonorrhea. Infections can inflame and obstruct the uterine tubes, or stimulate production of viscous mucus that can plug the cervix and prevent entry of sperm.

The first step in finding the right treatment for a particular patient is to determine the cause of the infertility. Table 22C describes diagnostic tests that a woman who is having difficulty conceiving may undergo. ■

| TABLE **22C** | TESTS TO ASSESS FEMALE INFERTILITY | |
|---|---|---|
| **Test** | **What It Checks** | |
| Hormone levels | If ovulation occurs | |
| Ultrasound | Placement and appearance of reproductive organs and structures | |
| Postcoital test | Cervix examined soon after unprotected intercourse to see if mucus is thin enough to allow sperm through | |
| Endometrial biopsy | Small piece of uterine lining sampled and viewed under microscope to see if it can support an embryo | |
| Hysterosalpingogram | Dye injected into uterine tube and followed with scanner shows if tube is clear or blocked | |
| Laparoscopy | Small, lit optical device inserted near navel to detect scar tissue blocking tubes, which could be missed in ultrasound | |
| Laparotomy | Scar tissue in tubes removed through incision made for laparoscopy | |

(Lewis, *Human Genetics* Table 20.1)

tissue replaces the adipose tissue of the breasts. Beginning about the fifth week of pregnancy, the anterior pituitary gland releases increasing amounts of *prolactin.* Prolactin is synthesized from early pregnancy throughout gestation, peaking at the time of birth. However, milk secretion does not begin until after birth. This is because during pregnancy, placental progesterone inhibits milk production, and placental lactogen blocks the action of prolactin. Consequently, even though the mammary glands can secrete milk, none is produced. The micrographs in figure 22.39 compare the mammary gland tissues of a nonpregnant woman with those of a lactating woman.

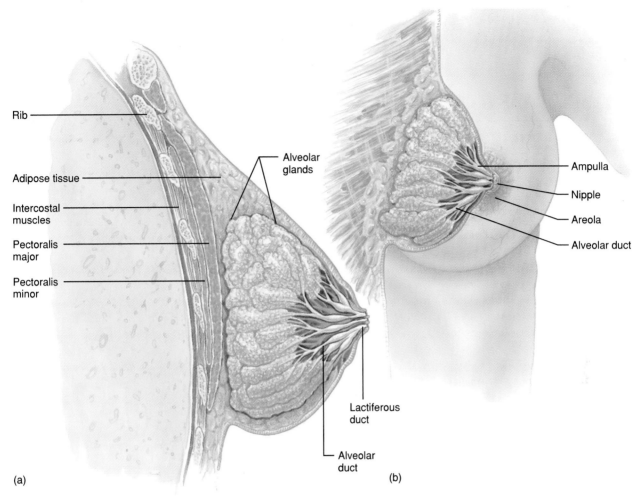

**FIGURE** 22.38

Structure of the breast. (*a*) Sagittal section. (*b*) Anterior view.

**FIGURE** 22.39

(*a*) Light micrograph of a mammary gland in a nonpregnant woman (60× micrograph enlarged to 160×). (*b*) Light micrograph of an active (lactating) mammary gland (60× micrograph enlarged to 160×).

# Breast Cancer Update

### FINDING THE LUMP

Few discoveries are as terrifying to a woman as finding a lump in her breast. The woman may first notice the irregularity as a small area of thickening, or, she may detect a dimple, a change in contour, or a nipple that is flatter than usual, points in an unusual direction, or produces a discharge.

The next step is having a physical exam. The doctor palpates the breast and performs a mammogram, an X-ray scan that pinpoints the location and approximate extent of abnormal tissue (fig. 22C). An ultrasound scan might be done to distinguish between a cyst (fluid-filled sac of glandular tissue) and a tumor (a solid mass). If an area might be cancerous, the next step is a biopsy, in which a very thin needle is used to sample the affected tissue.

Eighty percent of the time, a breast lump is a sign of fibrocystic breast disease, which is benign. Such a lump may be a cyst or a solid, fibrous mass of connective tissue (a fibroadenoma). Treatment includes taking vitamin E or synthetic androgens under a doctor's care, lowering caffeine intake, and examining unusual lumps with mammograms or biopsies. Primarily due to difficulty identifying new lumps, women with fibrocystic breasts are 1.6 times more likely to develop breast cancer than are other women.

### TREATMENT

If biopsied breast cells are cancerous, treatment is usually surgical. A *lumpectomy* removes a small tumor and some surrounding tissue; a *modified mastectomy* removes an entire breast; and a *radical mastectomy* removes the breast and surrounding lymph and muscle tissue. The type of follow-up treatment depends upon the extent of the tumor. If abnormal tissue is found in nearby lymph nodes, chemotherapy and possibly radiation therapy will follow surgery. Chemotherapy is also indicated if a bone scan or other test reveals that the cancer has spread. This happens in 7% of newly diagnosed cases of breast cancer.

The types of estrogen and progesterone receptors in the cancer cells help a doctor determine which drugs to use. Some drugs block estrogen or progesterone receptors, which blocks signals that stimulate the cells to reproduce. Tumor cells that lack hormone receptors are associated with a poor prognosis. Women with these tumors usually receive some type of chemotherapy

FIGURE 22C

Mammogram of a breast with a tumor (arrow).

even if the cancer has not spread to adjacent lymph nodes.

## CAUSES OF BREAST CANCER

A woman who has a close relative who has had breast cancer may be at a higher risk of developing the condition than a woman without a family history of breast cancer. Genetic tests can reveal the likelihood that a woman with a very strong family history of breast cancer will develop the inherited form of the illness. This was the case for a young woman whose mother and sister had died of breast cancer. Another sister had recently been diagnosed. Just before she was to have her breasts removed to avoid developing cancer, a genetic test showed that she had not inherited the responsible gene. The woman's cousin, who thought she could not inherit breast cancer because her father was related to the affected family, found that she had indeed inherited the gene. A mammogram revealed a tiny tumor, and surgery saved her life. Doctors must carefully screen women, considering family history and ethnic background, to determine which individuals might benefit from genetic testing.

Ten percent of all breast cancers are inherited, with the two major responsible genes—BRCA 1 and BRCA 2—discovered in 1993 and 1994. Researchers are still searching for the environmental triggers of the remaining 90% of cases. One candidate is prolonged exposure to estrogen. The estrogen link was first recognized in the 1970s, when researchers realized that young women who had had their

ovaries removed very rarely developed breast cancer. Did their lack of estrogen protect against breast cancer? It appears so.

Prolonged exposure to estrogen can occur in a variety of ways:

- Early menarche (first period) and late menopause (last period).
- Exposure to pesticide residues and other pollutants. These contaminants stimulate cell reproduction as estrogens do. Wildlife populations exposed to environmental estrogens have reproductive problems, such as infertility, undersized genitalia, and thin eggshells.
- Having no children, or having a first child after the age of 30.
- Not breast-feeding. Women who do not breast-feed have a higher breast cancer risk than women who do.

## BREAST CANCER STATISTICS

People frequently misunderstand the oft-quoted figure that a woman's lifetime risk of developing breast cancer is

1 in 8, believing that at any given time 1 in 8 women has breast cancer. This figure, however, refers to the lifetime risk for women who live to be 95 (table 22D).

Statistics indicate that breast cancer is on the rise. The lifetime risk of 1 in 8 was only 1 in 16 in the 1940s. It has increased by 1% a year since then, and by 4% a year since 1987. At least some of that rise is due to more wide-reaching and earlier diagnosis, a result of public health programs to educate women about breast self-exam and mammography. Thanks to these efforts, most women will recover from breast cancer. However, the illness is the second most common cancer in women after lung cancer.

## PREVENTION

Health agencies advise women to have mammograms yearly after age forty. A woman with a family history of breast cancer may benefit from having earlier mammograms. A mammogram may spot a tumor two years before a woman can feel it. ■

| TABLE 22D | BREAST CANCER RISK | | | |
|---|---|---|---|---|
| By Age | Odds | By Age | Odds |
| 25 | 1 in 19,608 | 60 | 1 in 24 |
| 30 | 1 in 2,525 | 65 | 1 in 17 |
| 35 | 1 in 622 | 70 | 1 in 14 |
| 40 | 1 in 217 | 75 | 1 in 11 |
| 45 | 1 in 93 | 80 | 1 in 10 |
| 50 | 1 in 50 | 85 | 1 in 9 |
| 55 | 1 in 33 | 95 or older | 1 in 8 |

## Milk Production and Secretion

Following childbirth and the expulsion of the placenta, the maternal blood concentrations of placental hormones decline rapidly. The action of prolactin is no longer inhibited.

Prolactin stimulates the mammary glands to secrete large quantities of milk. This hormonal effect does not occur until two or three days following birth, and in the meantime, the glands secrete a thin, watery fluid called *colostrum.* Although colostrum is rich in proteins, particularly protective antibodies, its concentrations of carbohydrates and fats are lower than those of milk.

Milk does not flow readily through the ductile system of the mammary gland, but must be actively ejected by contraction of specialized *myoepithelial cells* surrounding the alveolar glands. A reflex action controls this process and is elicited when the breast is suckled or the nipple or areola is otherwise mechanically stimulated (fig. 22.40). Then, impulses from sensory receptors within the breasts travel to the hypothalamus, which signals the posterior pituitary gland to release oxytocin. The oxytocin reaches the breasts by means of the blood and stimulates the myoepithelial cells to contract (in both breasts). As a result, milk is ejected into a suckling infant's mouth in about thirty seconds (fig. 22.41).

Sensory impulses triggered by mechanical stimulation of the nipples also signal the hypothalamus to continue secreting prolactin. Thus, prolactin is released as long as milk is removed from the breasts. However, if milk is not removed regularly, the hypothalamus inhibits the secretion of prolactin, and within about one week, the mammary glands lose their capacity to produce milk.

A woman who is breast-feeding feels her milk "let down," or flood her breasts, when her infant suckles. If the baby nurses on a very regular schedule, the mother may feel the letdown shortly before the baby is due to nurse. The connection between mind and hormonal control of lactation is so strong that if a nursing mother simply hears a baby cry, her milk may flow. If this occurs in public, she can keep from wetting her shirt by pressing her arms strongly against her chest.

To wean a nursing child, it is best to stop breast-feeding gradually, by eliminating one feeding per day each week, for example. If a woman stops nursing abruptly, her breasts will become painfully engorged for several days.

A woman who is breast-feeding usually does not ovulate for several months. This may be because prolactin suppresses release of gonadotropins from the anterior pituitary gland. When a woman discontinues breast-feeding, the anterior pituitary no longer secretes prolactin. Then, FSH is released and the menstrual cycle is activated. If a new mother does not wish to repeat her recent childbirth experience soon, she or her partner should be practicing contraception, because she will be fertile during the two weeks prior to the return of her menstrual period.

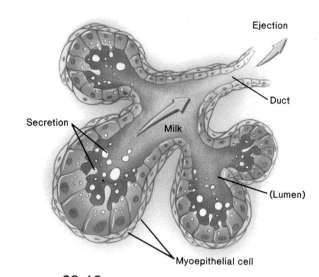

**FIGURE** 22.40

Myoepithelial cells eject milk from an alveolar gland.

**FIGURE** 22.41

Mechanism that ejects milk from the breasts.

## TABLE 22.7 HORMONAL CONTROL OF THE MAMMARY GLANDS

| Before Pregnancy (Beginning of Puberty) | Following Childbirth |
|---|---|
| Ovarian hormones secreted during menstrual cycles stimulate alveolar glands and ducts of mammary glands to develop. | 1. Placental hormonal concentrations decline, so that the action of prolactin is no longer inhibited. |
| **During Pregnancy** | 2. The breasts begin producing milk. |
| 1. Estrogen causes the ductile system to grow and branch. | 3. Mechanical stimulation of the breasts releases oxytocin from the posterior pituitary gland. |
| 2. Progesterone stimulates development of alveolar glands. | 4. Oxytocin stimulates ejection of milk from ducts. |
| 3. Placental lactogen promotes development of the breasts. | 5. As long as milk is removed, more prolactin is released; if milk is not removed, milk production ceases. |
| 4. Prolactin is secreted throughout pregnancy, but placental progesterone inhibits milk production. | |

## TABLE 22.8 AGENTS CONTRAINDICATED DURING BREAST-FEEDING

| Agent | Use | Effect on Lactation or Baby |
|---|---|---|
| Doxorubicin, methotrexate | Cancer chemotherapy | Immune suppression |
| Cyclosporine | Immune suppression in transplant patients | Immune suppression |
| Radioactive isotopes | Cancer diagnosis and therapy | Radioactivity in milk |
| Phenobarbitol | Anticonvulsant | Sedation, spasms on weaning |
| Oral contraceptives | Birth control | Decreased milk production |
| Caffeine (large amounts) | Food additive | Irritability, poor sleeping |
| Cocaine | Drug of abuse | Infant becomes intoxicated |
| Ethanol (alcohol) (large amounts) | Drug of abuse | Weak, drowsy; infant decreases in length but gains weight; decreased milk ejection reflex |
| Heroin | Drug of abuse | Tremors, restlessness, vomiting, poor feeding |
| Nicotine | Drug of abuse | Diarrhea, shock, increased heart rate; lowered milk production |
| Phencyclidine | Drug of abuse | Hallucinations |

Table 22.7 summarizes the hormonal control of milk production, and table 22.8 lists some agents that adversely affect lactation or harm the child. Clinical Application 22.5 explains the benefits of breast-feeding.

1 Describe the structure of a mammary gland.

2 How does pregnancy affect the mammary glands?

3 What stimulates the mammary glands to produce milk?

4 What causes milk to flow into the ductile system of a mammary gland?

5 What happens to milk production if milk is not regularly removed from the breast?

## BIRTH CONTROL

*Birth control* is the voluntary regulation of the number of offspring produced and the time they will be conceived. This control requires a method of **contraception** designed to avoid fertilization of an egg cell following sexual intercourse or to prevent implantation of a blastocyst. Table 22.9 describes several contraceptive approaches and devices and indicates their effectiveness.

### Coitus Interruptus

*Coitus interruptus* is withdrawing the penis from the vagina before ejaculation, preventing entry of sperm cells into the female reproductive tract. This method of

# Human Milk—The Perfect Food for Human Babies

The female human body manufactures milk that is a perfect food for a human newborn in several ways. Human milk is rich in the lipids required for rapid brain growth, and it is low in protein. Cow's milk is the reverse, with three times as much protein as human milk. Much of this protein is casein, which is fine to spur a calf's rapid muscle growth but forms hard-to-digest curds in a human baby's stomach. The protein in human milk has a balance of essential amino acids more suited to human growth and development than does the protein in cow's milk.

Human milk protects a newborn from many infections. For the first few days after giving birth, a new mother's breasts produce colostrum, which has less sugar and fat than mature milk but more protein, and is rich in antibodies. The antibodies protect the baby from such infections as *Salmonella* poisoning and polio. When the milk matures by a week to 10 days, it has antibodies, enzymes, and white blood cells from the mother that continue infection protection. A milk protein called lactoferrin binds iron, making it unavailable to microorganisms that might use it to thrive in the newborn's digestive tract. Another biochemical in human milk, bifidus factor, encourages the growth of the bacteria *Lactobacillus bifidus,* which manufactures acids in the baby's digestive system that kill harmful bacteria.

A breast-fed baby typically nurses until he or she is full, not until a certain number of ounces have been drunk, which may explain why breast-fed babies are less likely to be obese than bottle-fed infants. Babies nurtured on human milk are also less likely to develop allergies to cow's milk.

But breast-feeding is not the best choice for all women. It may be impossible to be present for each feeding or to provide milk. Also, many drugs a mother takes may enter breast milk and can affect the baby. A nursing mother must eat about 500 calories per day more than usual to meet the energy requirements of milk production—but she also loses weight faster than a mother who bottle-feeds, because the fat reserves set aside during pregnancy are used to manufacture milk. Another disadvantage of breast-feeding is that the father cannot do it.

An alternative to breast-feeding is infant formula, which is usually cow's milk plus fats, proteins, carbohydrates, vitamins, and minerals added to make it as much like breast milk as possible. Although infant formula is nutritionally sound, the foul-smelling and bulkier bowel movements of the bottle-fed baby compared to the odorless, loose, and less abundant feces of a breast-fed baby indicate that breast milk is a more digestible first food than infant formula. ∎

---

contraception often proves unsatisfactory and may result in pregnancy, since a male may find it difficult to withdraw just prior to ejaculation. Also, small quantities of semen containing sperm cells may reach the vagina before ejaculation occurs.

## Rhythm Method

The *rhythm method* (also called timed coitus or natural family planning) requires abstinence from sexual intercourse a few days before and a few days after ovulation. The rhythm method results in a relatively high rate of pregnancy because accurately identifying infertile times to have intercourse is difficult. Another disadvantage of the rhythm method is that it requires adherence to a particular pattern of behavior and restricts spontaneity in sexual activity.

**1** Why is coitus interruptus unreliable?

**2** Describe the idea behind the rhythm method of contraception.

**3** What factors make the rhythm method less reliable than some other methods of contraception?

The effectiveness of the rhythm method can sometimes be increased by measuring and recording the woman's body temperature when she awakes each morning for several months. Body temperature typically rises about 0.6°F immediately following ovulation. However, this technique does not work for all women. More helpful may be an "ovulation predictor kit" that detects the surge in LH preceding ovulation.

| TABLE 22.9 | BIRTH CONTROL METHODS |

| | Method | Mechanism | Advantages | Disadvantages | Pregnancies per Year per 100 Women* |
|---|---|---|---|---|---|
| | None | | | | 85 |
| **Barrier and Spermicidal** | Condom | Worn over penis or within vagina, keeps sperm out of vagina | Protection against sexually transmitted diseases | Disrupts spontaneity, can break, reduces sensation in male | 2–12 |
| | Condom and spermicide | Worn over penis or within vagina, keeps sperm out of vagina, and kills sperm that escape | Protection against sexually transmitted diseases | Disrupts spontaneity, reduces sensation in male | 2–5 |
| | Diaphragm and spermicide | Kills sperm and blocks uterus | Inexpensive | Disrupts spontaneity, messy, needs to be fitted by doctor | 6–18 |
| | Cervical cap and spermicide | Kills sperm and blocks uterus | Inexpensive, can be left in 24 hours | May slip out of place, messy, needs to be fitted by doctor | 6–18 |
| | Spermicidal foam or jelly | Kills sperm and blocks vagina | Inexpensive | Messy | 3–21 |
| | Spermicidal suppository | Kills sperm and blocks vagina | Easy to use and carry | Irritates 25% of users, male and female | 3–15 |
| **Hormonal** | Combination birth control pill | Prevents ovulation and implantation, thickens cervical mucus | Does not interrupt spontaneity, lowers risk of some cancers, decreases menstrual flow | Raises risk of cardiovascular disease in some women, causes weight gain and breast tenderness | 3 |
| | Medroxyprogesterone acetate (Depo-Provera) | Prevents ovulation, alters uterine lining | Easy to use | Menstrual changes, weight gain | 0.3 |
| | Progesterone implant | Prevents ovulation, thickens cervical mucus | Easy to use | Menstrual changes | 0.3 |
| **Behavioral** | Rhythm method | No intercourse during fertile times | No cost | Difficult to do, hard to predict timing | 20 |
| | Withdrawal (coitus interruptus) | Removal of penis from vagina before ejaculation | No cost | Difficult to do | 4–18 |
| **Surgical** | Vasectomy | Sperm cells never reach penis | Permanent, does not interrupt spontaneity | Requires minor surgery | 0.15 |
| | Tubal ligation | Egg cells never reach uterus | Permanent, does not interrupt spontaneity | Requires surgery, entails some risk of infection | 0.4 |
| **Other** | Intrauterine device | Prevents implantation | Does not interrupt spontaneity | Severe menstrual cramps, increases risk of infection | 3 |

*The lower figures apply when the contraceptive device is used correctly. The higher figures take into account human error.

## Mechanical Barriers

*Mechanical barriers* prevent sperm cells from entering the female reproductive tract during sexual intercourse. One such device used by males is a *condom*. It consists of a thin latex or natural membrane sheath that is placed over the erect penis before intercourse to prevent semen from entering the vagina upon ejaculation (fig. 22.42*a*).

A recently introduced *female condom* resembles a small plastic bag. A woman inserts it into her vagina prior to intercourse. The device blocks sperm from reaching the cervix.

A condom is inexpensive, and it may also help protect the user against contracting sexually transmitted diseases. However, men often feel that a condom decreases the sensitivity of the penis during intercourse. Also, its use interrupts the sex act.

Another mechanical barrier is the *diaphragm*. It is a cup-shaped structure with a flexible ring forming the

(a)

(b)

(c)

(d)

**FIGURE 22.42**

Devices and substances used for birth control include: (*a*) condom, (*b*) diaphragm, (*c*) spermicidal gel, (*d*) oral contraceptive, and (*e*) IUD.

(e)

rim. The diaphragm is inserted into the vagina so that it covers the cervix, preventing entry of sperm cells into the uterus (fig. 22.42*b*).

To be effective, a diaphragm must be fitted for size by a physician, be inserted properly, and be used in conjunction with a chemical spermicide that is applied to the surface adjacent to the cervix and to the rim of the diaphragm. The device must be left in position for several hours following sexual intercourse. A diaphragm can be inserted into the vagina up to 6 hours before sexual contact.

Similar to but smaller than the diaphragm is the *cervical cap,* which adheres to the cervix by suction. A woman inserts it with her fingers before intercourse. Cervical caps have been used for centuries in different cultures, and have been made of such varied substances as beeswax, lemon halves, paper, and opium poppy fibers.

### Chemical Barriers

*Chemical barrier* contraceptives include creams, foams, and jellies with spermicidal properties. Within the vagina, such chemicals create an environment that is unfavorable for sperm cells (fig. 22.42*c*).

Chemical barriers are fairly easy to use, but have a high failure rate when used alone. They are more effective when used with a condom or diaphragm.

### Oral Contraceptives

An *oral contraceptive,* or birth control pill, contains synthetic estrogenlike and progesteronelike substances.

When taken daily by a woman, these drugs disrupt the normal pattern of gonadotropin secretion and prevent the surge in LH release that triggers ovulation. They also interfere with buildup of the uterine lining that is necessary for implantation (fig. 22.42*d*).

Oral contraceptives, if used correctly, prevent pregnancy nearly 100% of the time. However, they may cause nausea, retention of body fluids, increased pigmentation of the skin, and breast tenderness. Also, some women, particularly those over 35 years of age who smoke, may develop intravascular blood clots, liver disorders, or high blood pressure when using certain types of oral contraceptives.

### Injectable Contraception

An intramuscular injection of Depo-Provera (medroxyprogesterone acetate) protects against pregnancy for three months by preventing maturation and release of a secondary oocyte. It also alters the uterine lining, making it less hospitable for a blastocyst. Because Depo-Provera is long-acting, it takes 10 to 18 months after the last injection for the effects to wear off.

Use of Depo-Provera requires a doctor's care, because of potential side effects and risks. The most common side effect is weight gain. Women with a history of breast cancer, depression, kidney disease, high blood pressure, migraine headaches, asthma, epilepsy, or diabetes, or strong family histories of these conditions, should probably not use this form of birth control.

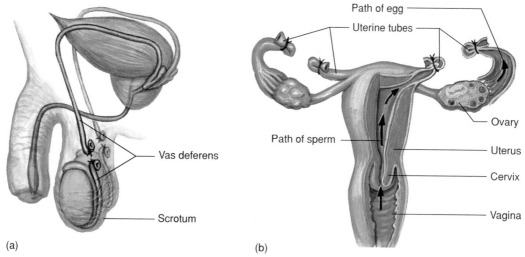

**FIGURE 22.43**

(*a*) Vasectomy removes a portion of each vas deferens. (*b*) Tubal ligation removes a portion of each uterine tube.

## Contraceptive Implants

A *contraceptive implant* is a set of small progesterone-containing capsules or rods, which are inserted surgically under the skin of a woman's arm or scapular region. The progesterone, which is released slowly from the implant, prevents ovulation in much the same way as do oral contraceptives. A contraceptive implant is effective for a period of up to 5 years, and its contraceptive action can be reversed by removing the device.

> A large dose of high-potency estrogens can prevent implantation of a blastocyst in the uterus. Such a "morning-after pill," taken shortly after unprotected intercourse, promotes powerful contractions of smooth muscle in a woman's reproductive tract. This may dislodge and expel a fertilized egg or blastocyst. However, if a blastocyst has already implanted, this treatment may injure it.

**1** Describe two methods of contraception that use mechanical barriers.

**2** What action can increase the effectiveness of chemical contraceptives?

**3** What substances are contained in oral contraceptives?

**4** How does an oral contraceptive prevent pregnancy? An injectable contraceptive? A contraceptive implant?

## Intrauterine Devices

An *intrauterine device,* or *IUD,* is a small solid object that a physician places within the uterine cavity. An IUD interferes with implantation, perhaps by inflaming the uterine tissues (fig. 22.42*e*).

Unfortunately, an IUD may be spontaneously expelled from the uterus or produce abdominal pain or excessive menstrual bleeding. It may also injure the uterus or produce other serious health problems, and should be checked at regular intervals by a physician. A few babies have been born with IUDs attached to them.

## Surgical Methods

*Surgical methods* of contraception sterilize the male or female. In the male, a physician removes a small section of each vas deferens near the epididymis, and ties the cut ends of the ducts. This is a *vasectomy,* and it is a simple operation that produces few side effects, although it may cause some pain for a week or two.

After a vasectomy, sperm cells cannot leave the epididymis, thus they are not included in the semen. However, sperm cells may already be present in portions of the ducts distal to the cuts. Consequently, the sperm count may not reach zero for several weeks.

The corresponding procedure in the female is called *tubal ligation.* The uterine tubes are cut and tied so that sperm cells cannot reach an egg cell.

Neither a vasectomy nor a tubal ligation produces any changes in hormonal concentrations or sexual drives. These procedures, shown in figure 22.43, provide the most reliable forms of contraception. Reversing them requires microsurgery.

**1** How does an IUD prevent pregnancy?

**2** Describe the surgical methods of contraception for a male and for a female.

| TABLE | 22.10 | SOME SEXUALLY TRANSMITTED DISEASES |

| Disease | Cause | Symptoms | Number of Reported Cases (U.S.) | Effects on Fetus | Treatment | Complications |
|---|---|---|---|---|---|---|
| Acquired immune deficiency syndrome | Human immuno-deficiency virus | Fever, weakness, infections, cancer | > 14 million (infected) | Exposure to AIDS virus and other infections | Drugs to treat or delay symptoms; no cure | Dementia |
| Chlamydia infection | Bacteria of genus *Chlamydia* | Painful urination and intercourse, mucus discharge from penis or vagina | 3–10 million | Prematurity, blindness, pneumonia | Antibiotics | Pelvic inflammatory disease, infertility, arthritis, ectopic pregnancy |
| Genital herpes | Herpes virus type II | Genital sores, fever | 20 million | Brain damage, stillbirth | Antiviral drug (acyclovir) | Increased risk of cervical cancer |
| Genital warts | Human papilloma virus | Warts on genitals | 1 million | None known | Chemical or surgical removal | Increased risk of cervical cancer |
| Gonorrhea | *Neisseria gonorrhoeae* bacteria | In women, usually none; in men, painful urination | 2 million | Blindness, stillbirth | Antibiotics | Arthritis, rash, infertility, pelvic inflammatory disease |
| Syphilis | *Treponema pallidum* bacteria | Initial chancre sore usually on genitals or mouth; rash 6 months later; several years with no symptoms as infection spreads; finally damage to heart, liver, nerves, brain | 90,000 | Miscarriage, prematurity, birth defects, stillbirth | Antibiotics | Dementia |

## SEXUALLY TRANSMITTED DISEASES

The twenty recognized **sexually transmitted diseases** (STDs) are often called "silent infections" because the early stages may not produce symptoms, especially in women (table 22.10). By the time symptoms appear, it is often too late to prevent complications or the spread of the infection to sexual partners. Because many STDs have similar symptoms, and some of the symptoms are also seen in diseases or allergies that are not sexually related, it is wise to consult a physician if one or a combination of these symptoms appears:

1. Burning sensation during urination

2. Pain in the lower abdomen

3. Fever or swollen glands in the neck

4. Discharge from the vagina or penis

5. Pain, itch, or inflammation in the genital or anal area

6. Pain during intercourse

7. Sores, blisters, bumps, or a rash anywhere on the body, particularly the mouth or genitals

8. Itchy, runny eyes

One possible complication of the STDs gonorrhea and chlamydia is **pelvic inflammatory disease,** in which bacteria enter the vagina and spread throughout the reproductive organs. The disease begins with intermittent cramps, followed by sudden fever, chills, weakness, and severe cramps. Hospitalization and intravenous antibiotics can stop the infection. The uterus and uterine tubes are often scarred, resulting in infertility and increased risk of ectopic pregnancy.

*Acquired immune deficiency syndrome* (AIDS) is a sexually transmitted disease. AIDS is a steady deterioration of the body's immune defenses, and is caused by a virus. The body becomes overrun by infection and often cancer, diseases that the immune system usually conquers. The AIDS virus (human immunodeficiency virus, or HIV) is passed from one person to another in body fluids such as semen, blood, and milk. It is most frequently transmitted during unprotected intercourse or by using a needle containing contaminated blood.

**1** Why are sexually transmitted diseases often called "silent infections"?

**2** Why are sexually transmitted diseases sometimes difficult to diagnose?

**3** What are some common symptoms of sexually transmitted diseases?

---

# CLINICAL TERMS RELATED TO THE REPRODUCTIVE SYSTEMS

**abortion** (ah-bor'shun)   Spontaneous or deliberate termination of pregnancy; a spontaneous abortion is commonly termed a miscarriage.

**amenorrhea** (a-men″o-re'ah)   Absence of menstrual flow, usually due to a disturbance in hormonal concentrations.

**Cesarean section** (sĕ-sa're-an sek'shun)   Delivery of a fetus through an abdominal incision.

**conization** (ko″nĭ-za'shun)   Surgical removal of a cone of tissue from the cervix for examination.

**curettage** (ku″rĕ-tahzh')   Surgical procedure in which the cervix is dilated and the endometrium of the uterus is scraped (commonly called D and C for dilation and curettage).

**dysmenorrhea** (dis″men-o-re'ah)   Painful menstruation.

**eclampsia** (ē-klamp'se-ah)   Condition characterized by convulsions and coma that sometimes accompanies toxemia of pregnancy.

**endometritis** (en″do-mē-tri'tis)   Inflammation of the uterine lining.

**epididymitis** (ep″ĭ-did″i-mi'tis)   Inflammation of the epididymis.

**gestation** (jes-ta'shun)   Entire period of pregnancy.

**hematometra** (hem″ah-to-me'trah)   Accumulation of menstrual blood within the uterine cavity.

**hydrocele** (hi'dro-seal)   Enlarged scrotum caused by accumulation of fluid along the spermatic cord.

**hyperemesis gravidarum** (hi″per-em'ĕ-sis grav'i-dar-um)   Vomiting associated with pregnancy; morning sickness.

**hypospadias** (hi″po-spay'dee-us)   Male developmental anomaly in which the urethra opens on the underside of the penis.

**hysterectomy** (his″te-rek'to-me)   Surgical removal of the uterus.

**mastitis** (mas″ti'tis)   Inflammation of a mammary gland.

**oophorectomy** (o″of-o-rek'to-me) or **ovariectomy** (o″va-re-ek'to-me)   Surgical removal of an ovary.

**oophoritis** (o″of-o-ri'tis)   Inflammation of an ovary.

**orchiectomy** (or″ke-ek'to-me)   Surgical removal of a testis.

**orchitis** (or-ki'tis)   Inflammation of a testis.

**prostatectomy** (pros″tah-tek'to-me)   Surgical removal of a portion or all of the prostate gland.

**prostatitis** (pros″tah-ti'tis)   Inflammation of the prostate gland.

**salpingectomy** (sal″pin-jek'to-me)   Surgical removal of a uterine tube.

**toxemia of pregnancy** (tok-se'me-ah)   Group of metabolic disturbances that may occur during pregnancy.

**vaginitis** (vaj″ĭ-ni'tis)   Inflammation of the vaginal lining.

**varicocele** (var'ĭ-ko-sēl″)   Distension of the veins within the spermatic cord.

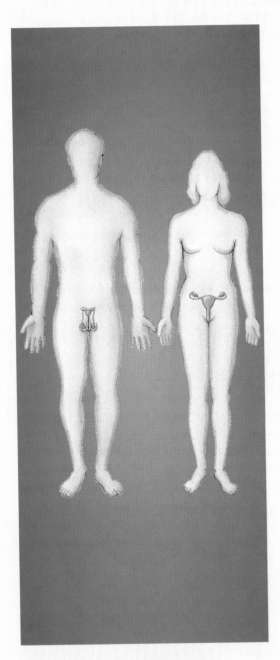

**REPRODUCTIVE SYSTEMS**

Gamete production, fertilization, fetal development, and childbirth are essential for survival of the species.

## Integumentary System

Skin sensory receptors play a role in sexual pleasure.

## Skeletal System

Bones can be a temporary source of calcium during lactation.

## Muscular System

Skeletal, cardiac, and smooth muscles all play a role in reproductive processes and sexual activity.

## Nervous System

The nervous system plays a major role in sexual activity and sexual pleasure.

## Endocrine System

Hormones control the production of ova in the female and sperm in the male.

## Cardiovascular System

Blood pressure is necessary for the normal function of erectile tissue in the male and female.

## Lymphatic System

Special mechanisms inhibit the female immune system from attacking sperm as foreign invaders.

## Digestive System

Proper nutrition is essential for the formation of normal gametes and for normal fetal development during pregnancy.

## Respiratory System

During pregnancy, the placenta provides oxygen to the fetus and removes carbon dioxide.

## Urinary System

Male urinary and reproductive systems share common structures. Kidneys compensate for fluid loss from the reproductive systems. Pregnancy may cause fluid retention.

## Introduction (page 836)

Various reproductive organs produce sex cells and sex hormones, sustain these cells, or transport them from place to place.

## Organs of the Male Reproductive System (page 836)

The male reproductive organs produce and maintain sperm cells, transport these cells, and produce male sex hormones. The primary male sex organs are the testes, which produce sperm cells and male sex hormones. Accessory organs include the internal and external reproductive organs.

## Testes (page 836)

1. Descent of the testes
   a. Testes originate posterior to the parietal peritoneum near the level of the developing kidneys.
   b. The gubernaculum guides the descent of the testes into the lower abdominal cavity and through the inguinal canal.
   c. Undescended testes fail to produce sperm cells because of the high abdominal temperature.
2. Structure of the testes
   a. The testes are composed of lobules separated by connective tissue and filled with the seminiferous tubules.
   b. The seminiferous tubules unite to form the rete testis that joins the epididymis.
   c. The seminiferous tubules are lined with epithelium, which produces sperm cells.
   d. The interstitial cells that produce male sex hormones occur between the seminiferous tubules.
3. Formation of sperm cells
   a. The epithelium lining the seminiferous tubules includes sustentacular cells and spermatogenic cells.
      (1) The sustentacular cells support and nourish the spermatogenic cells.
      (2) The spermatogenic cells give rise to spermatogonia.
   b. The process of spermatogenesis produces sperm cells from spermatogonia.
      (1) Meiosis reduces the number of chromosomes in sperm cells by one-half (forty-six to twenty-three).
      (2) Spermatogenesis produces four sperm cells from each primary spermatocyte.
   c. Membranous processes of adjacent sustentacular cells form a barrier within the epithelium.
      (1) The barrier separates early and advanced stages of spermatogenesis.
      (2) It helps provide a favorable environment for differentiating cells.
4. Structure of a sperm cell
   a. Sperm head contains a nucleus with twenty-three chromosomes.
   b. Sperm body contains many mitochondria.
   c. Sperm tail propels the cell.

## Male Internal Accessory Organs (page 842)

1. Epididymis
   a. The epididymis is a tightly coiled tube on the outside of the testis that leads into the vas deferens.
   b. It stores and nourishes immature sperm cells and promotes their maturation.
2. Vas deferens
   a. The vas deferens is a muscular tube that forms part of the spermatic cord.
   b. It passes through the inguinal canal, enters the abdominal cavity, courses medially into the pelvic cavity, and ends behind the urinary bladder.
   c. It fuses with the duct from the seminal vesicle to form the ejaculatory duct.
3. Seminal vesicle
   a. The seminal vesicle is a saclike structure attached to the vas deferens.
   b. It secretes an alkaline fluid that contains nutrients, such as fructose, and prostaglandins.
   c. This secretion is added to sperm cells during emission.
4. Prostate gland
   a. This gland surrounds the urethra just below the urinary bladder.
   b. It secretes a thin, milky fluid, which enhances the motility of sperm cells and neutralizes the fluid containing the sperm cells as well as acidic secretions of the vagina.
5. Bulbourethral glands
   a. These glands are two small structures inferior to the prostate gland.
   b. They secrete a fluid that lubricates the penis in preparation for sexual intercourse.
6. Semen
   a. Semen is composed of sperm cells and secretions of the seminal vesicles, prostate gland, and bulbourethral glands.
   b. This fluid is slightly alkaline and contains nutrients and prostaglandins.
   c. Semen activates sperm cells, but these sperm cells are unable to fertilize egg cells until they enter the female reproductive tract.

## Male External Reproductive Organs (page 845)

1. Scrotum
   a. The scrotum is a pouch of skin and subcutaneous tissue that encloses the testes.
   b. The dartos muscle in the scrotal wall causes the skin of the scrotum to be held close to the testes or to hang loosely, thus regulating the temperature for sperm production and survival.
2. Penis
   a. The penis conveys urine and semen.
   b. It is specialized to become erect for insertion into the vagina during sexual intercourse.
   c. Its body is composed of three columns of erectile tissue surrounded by connective tissue.
   d. The root of the penis is attached to the pelvic arch and membranes of the perineum.
3. Erection, orgasm, and ejaculation
   a. During erection, the vascular spaces within the erectile tissue become engorged with blood as arteries dilate and veins are compressed.
   b. Orgasm is the culmination of sexual stimulation and is accompanied by emission and ejaculation.

c. Semen moves along the reproductive tract as smooth muscle in the walls of the tubular structures contract, stimulated by a reflex.

d. Following ejaculation, the penis becomes flaccid.

## Hormonal Control of Male Reproductive Functions (page 848)

1. Hypothalamic and pituitary hormones
   The male body remains reproductively immature until the hypothalamus releases GnRH, which stimulates the anterior pituitary gland to release gonadotropins.
   a. FSH stimulates spermatogenesis.
   b. LH (ICSH) stimulates the interstitial cells to produce male sex hormones.
   c. Inhibin prevents oversecretion of FSH.

2. Male sex hormones
   a. Male sex hormones are called androgens.
   b. Testosterone is the most important androgen.
   c. Testosterone is converted into dihydrotestosterone in some organs.
   d. Androgens that fail to become fixed in tissues are metabolized in the liver and excreted.
   e. Androgen production increases rapidly at puberty.

3. Actions of testosterone
   a. Testosterone stimulates the development of the male reproductive organs and causes the testes to descend.
   b. It is responsible for the development and maintenance of male secondary sex characteristics.

4. Regulation of male sex hormones
   a. A negative feedback mechanism regulates testosterone concentration.
      (1) As the concentration of testosterone rises, the hypothalamus is inhibited, and the anterior pituitary secretion of gonadotropins is reduced.
      (2) As the concentration of testosterone falls, the hypothalamus signals the anterior pituitary to secrete gonadotropins.
   b. The concentration of testosterone remains relatively stable from day to day.

## Organs of the Female Reproductive System (page 850)

The primary female sex organs are the ovaries, which produce female sex cells and sex hormones. Accessory organs are internal and external.

## Ovaries (page 850)

1. Ovary attachments
   a. Several ligaments hold the ovaries in position.
   b. These ligaments include broad, suspensory, and ovarian ligaments.

2. Ovary descent
   a. The ovaries descend from posterior to the parietal peritoneum near the developing kidneys.
   b. They are attached to the pelvic wall just inferior to the pelvic brim.

3. Ovary structure
   a. The ovaries are subdivided into a medulla and a cortex.
   b. The medulla is composed of connective tissue, blood vessels, lymphatic vessels, and nerves.
   c. The cortex contains ovarian follicles and is covered by cuboidal epithelium.

4. Primordial follicles
   a. During development, groups of cells in the ovarian cortex form millions of primordial follicles.
   b. Each primordial follicle contains a primary oocyte and a layer of flattened epithelial cells.
   c. The primary oocyte begins to undergo meiosis, but the process is soon halted and is not continued until puberty.
   d. The number of oocytes steadily declines throughout the life of a female.

5. Oogenesis
   a. Beginning at puberty, some oocytes are stimulated to continue meiosis.
   b. When a primary oocyte undergoes oogenesis, it gives rise to a secondary oocyte in which the original chromosome number is reduced by one-half (from forty-six to twenty-three).
   c. A secondary oocyte may be fertilized to produce a zygote.

6. Follicle maturation
   a. At puberty, FSH initiates follicle maturation.
   b. During maturation, the oocyte enlarges, the follicular cells proliferate, and a fluid-filled cavity appears and produces a secondary follicle.
   c. Ovarian cells surrounding the follicle form two layers.
   d. Usually only one follicle reaches full development.

7. Ovulation
   a. Ovulation is the release of an oocyte from an ovary.
   b. The oocyte is released when its follicle ruptures.
   c. After ovulation, the oocyte is drawn into the opening of the uterine tube.

## Female Internal Accessory Organs (page 855)

1. Uterine tubes
   a. These tubes convey egg cells toward the uterus.
   b. The end of each uterine tube is expanded, and its margin bears irregular extensions.
   c. Ciliated cells that line the tube and peristaltic contractions in the wall of the tube move an egg cell into the tube's opening.

2. Uterus
   a. The uterus receives the embryo and sustains it during development.
   b. The vagina partially encloses the cervix.
   c. The uterine wall includes the endometrium, myometrium, and perimetrium.

3. Vagina
   a. The vagina connects the uterus to the vestibule.
   b. It receives the erect penis, conveys uterine secretions to the outside, and provides an open channel for the fetus during birth.
   c. The vaginal orifice is partially closed by a thin membrane, the hymen.
   d. Its wall consists of a mucosa, muscularis, and outer fibrous coat.

## Female External Reproductive Organs (page 859)

1. Labia majora
   a. The labia majora are rounded folds of adipose tissue and skin that enclose and protect the other external reproductive parts.
   b. The upper ends form a rounded elevation over the symphysis pubis.

2. Labia minora
   a. The labia minora are flattened, longitudinal folds between the labia majora.
   b. They are well supplied with blood vessels.
3. Clitoris
   a. The clitoris is a small projection at the anterior end of the vulva; it corresponds to the male penis.
   b. It is composed of two columns of erectile tissue.
   c. Its root is attached to the sides of the pubic arch.
4. Vestibule
   a. The vestibule is the space between the labia minora that encloses the vaginal and urethral openings.
   b. The vestibular glands secrete mucus into the vestibule during sexual stimulation.
5. Erection, lubrication, and orgasm
   a. During periods of sexual stimulation, the erectile tissues of the clitoris and vestibular bulbs become engorged with blood and swollen.
   b. The vestibular glands secrete mucus into the vestibule and vagina.
   c. During orgasm, the muscles of the perineum, uterine wall, and uterine tubes contract rhythmically.

## Hormonal Control of Female Reproductive Functions  (page 860)

Hormones from the hypothalamus, anterior pituitary gland, and ovaries play important roles in the control of sex cell maturation, and the development and maintenance of female secondary sex characteristics.

1. Female sex hormones
   a. A female body remains reproductively immature until about 10 years of age when gonadotropin secretion increases.
   b. The most important female sex hormones are estrogen and progesterone.
      (1) Estrogen is responsible for the development and maintenance of most female secondary sex characteristics.
      (2) Progesterone causes changes in the uterus.
2. Female reproductive cycle
   a. The menstrual cycle is characterized by regularly recurring changes in the uterine lining culminating in menstrual flow.
   b. A menstrual cycle is initiated by FSH, which stimulates maturation of a follicle.
   c. Granulosa cells of a maturing follicle secrete estrogen, which is responsible for maintaining the secondary sex traits and thickening the uterine lining.
   d. Ovulation is triggered when the anterior pituitary gland releases a relatively large amount of LH.
   e. Following ovulation, the follicular cells and thecal cells give rise to the corpus luteum.
      (1) The corpus luteum secretes estrogen and progesterone, which causes the uterine lining to become more vascular and glandular.
      (2) If an oocyte is not fertilized, the corpus luteum begins to degenerate.
      (3) As the concentrations of estrogen and progesterone decline, the uterine lining disintegrates, causing menstrual flow.
   f. During this cycle, estrogen and progesterone inhibit the release of LH and FSH; as the concentrations of these hormones decline, the anterior pituitary secretes FSH and LH again, stimulating a new menstrual cycle.

3. Menopause
   a. Eventually the ovaries cease responding to FSH, and cycling ceases.
   b. Menopause is characterized by a low estrogen concentration and a continuous secretion of FSH and LH.
   c. The female reproductive organs undergo varying degrees of regressive changes.

## Pregnancy  (page 864)

1. Transport of sex cells
   a. Ciliary action aids movement of the egg cell to the uterine tube.
   b. A sperm cell moves by its tail lashing and muscular contraction in the female reproductive tract.
2. Fertilization
   a. With the aid of an enzyme, a sperm cell penetrates the zona pellucida.
   b. When a sperm cell penetrates an egg cell membrane, changes in the egg cell membrane and the zona pellucida prevent entry of additional sperm.
   c. Fusion of the nuclei of a sperm and an egg cell complete fertilization.
   d. The product of fertilization is a zygote with forty-six chromosomes.
3. Early embryonic development
   a. Cells undergo mitosis, giving rise to smaller and smaller cells.
   b. The developing offspring (preembryo) moves down the uterine tube to the uterus, where it implants in the endometrium.
   c. The offspring is called an embryo from the second through the eighth week of development; thereafter it is called a fetus.
   d. Eventually some of the embryonic and maternal cells form a placenta.
4. Hormonal changes during pregnancy
   a. Embryonic cells produce hCG that maintains the corpus luteum.
   b. Placental tissue produces high concentrations of estrogen and progesterone.
      (1) Estrogen and progesterone maintain the uterine wall and inhibit secretion of FSH and LH.
      (2) Progesterone and relaxin inhibit contractions of uterine muscles.
      (3) Estrogen causes enlargement of the vagina.
      (4) Relaxin helps relax the ligaments of the pelvic joints.
   c. The placenta secretes placental lactogen that stimulates the development of the breasts and mammary glands.
   d. During pregnancy, increasing secretion of aldosterone promotes retention of sodium and body fluid, and increasing secretion of parathyroid hormone helps maintain a high concentration of maternal blood calcium.
5. Other changes during pregnancy
   a. The uterus enlarges greatly.
   b. The woman's blood volume, cardiac output, breathing rate, and urine production increase.
   c. The woman's dietary needs increase, but if intake is inadequate, fetal tissues have priority for use of available nutrients.
6. Birth
   a. Pregnancy usually lasts 40 weeks from the beginning of the last menstrual cycle.

b. During pregnancy, placental progesterone inhibits uterine contractions.
c. A variety of factors are involved with the birth process.
   (1) A decreasing concentration of progesterone and the release of prostaglandins may initiate the birth process.
   (2) The posterior pituitary gland releases oxytocin.
   (3) Uterine muscles are stimulated to contract, and labor begins.
   (4) A positive feedback mechanism causes stronger contractions and greater release of oxytocin.
d. Following the birth of the infant, placental tissues are expelled.

## Mammary Glands (page 871)

1. Location of the glands
   a. The mammary glands are located in the subcutaneous tissue of the anterior thorax within the breasts.
   b. The breasts extend between the second and sixth ribs, and from sternum to axillae.
2. Structure of the glands
   a. The mammary glands are composed of lobes that contain tubular glands.
   b. The lobes are separated by dense connective and adipose tissues.
   c. The mammary glands are connected to the nipple by ducts.
3. Development of the breasts
   a. Breasts of males remain nonfunctional.
   b. Estrogen stimulates breast development in females.
      (1) Alveolar glands and ducts enlarge.
      (2) Fat is deposited around and within the breasts.
   c. During pregnancy, the breasts change.
      (1) Estrogen causes the ductile system to grow.
      (2) Progesterone causes development of alveolar glands.
      (3) Prolactin is released during pregnancy, but progesterone inhibits milk production.
4. Milk production and secretion
   a. Following childbirth, the concentrations of placental hormones decline.
      (1) The action of prolactin is no longer blocked.
      (2) The mammary glands begin to secrete milk.
   b. Reflex response to mechanical stimulation of the nipple causes the posterior pituitary to release oxytocin, which causes milk to be ejected from the alveolar ducts.
   c. As long as milk is removed from glands, more milk is produced; if milk is not removed, production ceases.
   d. During the period of milk production, the menstrual cycle is partially inhibited.

## Birth Control (page 877)

Voluntary regulation of the number of children produced and the time they are conceived is called birth control. This usually involves some method of contraception.

1. Coitus interruptus
   a. Coitus interruptus is withdrawal of the penis from the vagina before ejaculation.

   b. Some semen may be expelled from the penis before ejaculation.
2. Rhythm method
   a. Abstinence from sexual intercourse a few days before and after ovulation is the rhythm method.
   b. It is almost impossible to accurately predict the time of ovulation.
3. Mechanical barriers
   a. Males and females can use condoms.
   b. Females use diaphragms.
4. Chemical barriers
   a. Spermicidal creams, foams, and jellies are chemical barriers to conception.
   b. These provide an unfavorable environment in the vagina for sperm survival.
5. Oral contraceptives
   a. Tablets that contain synthetic estrogenlike and progesteronelike substances are taken by the woman.
   b. They disrupt the normal pattern of gonadotropin secretion, and prevent ovulation and the normal buildup of the uterine lining.
   c. When used correctly, this method is almost 100% effective.
   d. Some women develop undesirable side effects.
6. Injectable contraceptives
   a. Intramuscular injection with medroxyprogesterone acetate every three months.
   b. High levels of hormone act similarly to oral contraceptives to prevent pregnancy.
   c. Very effective if administered promptly at the end of the three months.
   d. Women may experience side effects; in some women, use is contraindicated.
7. Contraceptive implants
   a. A contraceptive implant consists of a set of progesterone-containing capsules or rods that are inserted under the skin.
   b. Progesterone released from the implant prevents ovulation.
   c. The implant is effective for years, and its action can be reversed by having it removed.
8. Intrauterine devices
   a. An IUD is a solid object inserted in the uterine cavity.
   b. It is thought to prevent pregnancy by interfering with implantation.
   c. It may be expelled spontaneously or produce undesirable side effects.
9. Surgical methods
   a. These are sterilization procedures.
      (1) Vasectomy is performed in males.
      (2) Tubal ligation is performed in females.
   b. Surgical methods are the most reliable forms of contraception.

## Sexually Transmitted Diseases (page 882)

1. Sexually transmitted diseases are passed during sexual contact, and may go undetected for years.
2. The twenty such disorders share certain symptoms.

# CRITICAL THINKING QUESTIONS

1. What changes, if any, might occur in the secondary sex characteristics of an adult male following removal of one testis? Following removal of both testes? Following removal of the prostate gland?

2. How would you explain the fact that new mothers sometimes experience cramps in their lower abdomens when they begin to nurse their babies?

3. If a woman who is considering having a tubal ligation asks, "Will the operation cause me to go through my change of life early?" how would you answer?

4. What effect would it have on a woman's menstrual cycles if a single ovary were removed surgically? What effect would it have if both ovaries were removed?

5. Which methods of contraception are theoretically the most effective in preventing unwanted pregnancies? Which methods are theoretically the least effective?

6. As a male reaches adulthood, what will be the consequences if his testes have remained undescended since birth? Why?

7. Why does injecting a sperm cell into an egg cell not result in fertilization?

8. What types of contraceptives provide the greatest protection against sexually transmitted diseases?

# REVIEW EXERCISES

1. List the general functions of the male reproductive system.
2. Distinguish between the primary and accessory male reproductive organs.
3. Describe the descent of the testes.
4. Define *cryptorchidism*.
5. Describe the structure of a testis.
6. Explain the function of the sustentacular cells in the testis.
7. List the major steps in spermatogenesis.
8. Describe a sperm cell.
9. Describe the epididymis, and explain its function.
10. Trace the path of the vas deferens from the epididymis to the ejaculatory duct.
11. On a diagram, locate the seminal vesicles, and describe the composition of their secretion.
12. On a diagram, locate the prostate gland, and describe the composition of its secretion.
13. On a diagram, locate the bulbourethral glands, and explain the function of their secretion.
14. Describe the composition of semen.
15. Define *capacitation*.
16. Describe the structure of the scrotum.
17. Describe the structure of the penis.
18. Explain the mechanism that produces an erection of the penis.
19. Distinguish between emission and ejaculation.
20. Explain the mechanism of ejaculation.
21. Explain the role of GnRH in the control of male reproductive functions.
22. Distinguish between androgen and testosterone.
23. Define *puberty*.
24. Describe the actions of testosterone.
25. List several male secondary sex characteristics.

26. Explain the regulation of testosterone concentration.
27. List the general functions of the female reproductive system.
28. Distinguish between the primary and accessory female reproductive organs.
29. Describe how the ovaries are held in position.
30. Describe the descent of the ovaries.
31. Describe the structure of an ovary.
32. Define *primordial follicle*.
33. List the major steps in oogenesis.
34. Distinguish between a primary and a secondary follicle.
35. Describe a mature follicle.
36. Define *ovulation*.
37. On a diagram, locate the uterine tubes, and explain their function.
38. Describe the structure of the uterus.
39. Describe the structure of the vagina.
40. Distinguish between the labia majora and the labia minora.
41. On a diagram, locate the clitoris, and describe its structure.
42. Define *vestibule*.
43. Describe the process of erection in the female reproductive organs.
44. Define *orgasm*.
45. Explain the role of GnRH in regulating female reproductive functions.
46. List several female secondary sex characteristics.
47. Define *menstrual cycle*.
48. Explain how a menstrual cycle is initiated.
49. Summarize the major events in a menstrual cycle.
50. Define *menopause*.
51. Describe how male and female sex cells are transported within the female reproductive tract.

52. Describe the process of fertilization.
53. List the major functions of the placenta.
54. Explain the major hormonal changes that occur in the maternal body during pregnancy.
55. Describe the major nonhormonal changes that occur in the maternal body during pregnancy.
56. Describe the role of progesterone in initiating the birth process.

57. Discuss the events that occur during the birth process.
58. Describe the structure of a mammary gland.
59. Explain the roles of prolactin and oxytocin in milk production and secretion.
60. Define *contraception*.
61. List several methods of contraception, and explain how each prevents pregnancy.
62. List several sexually transmitted diseases.

---

**For More Information Related to This Chapter, Please Visit** *Hole's Human Anatomy and Physiology* **Home Page**

http://www.mhhe.com/biosci/abio/

## CHAPTER OBJECTIVES

**AFTER YOU HAVE STUDIED THIS CHAPTER, YOU SHOULD BE ABLE TO:**

1. Distinguish between growth and development.

2. Describe the major events of the period of cleavage.

3. Explain how the primary germ layers originate and list the structures each layer produces.

4. Describe the formation and function of the placenta.

5. Define *fetus* and describe the major events that occur during the fetal stage of development.

6. Trace the general path of blood through the fetal circulatory system.

7. Describe the major circulatory and physiological adjustments that occur in the newborn.

8. Name the stages of development between the neonatal period and death, and list the general characteristics of each stage.

## KEY TERMS

**allantois** (ah-lan′to-is)
**amnion** (am′ne-on)
**apoptosis** (ap″o-to′sis)
**chorion** (ko′re-on)
**cleavage** (klēv′ij)
**embryo** (em′bre-o)
**fetus** (fe′tus)
**germ layer** (jerm la′er)

**neonatal** (ne″o-na′tal)
**placenta** (plah-sen′tah)
**postnatal** (pōst-na′tal)
**prenatal** (pre-na′tal)
**senescence** (se-nes′ens)
**umbilical cord** (um-bil′ĭ-kal kord)
**zygote** (zi′gōt)

## UNDERSTANDING WORDS

**allant-,** sausage: *allant*ois— tubelike structure that extends from the yolk sac into the connecting stalk of an embryo.

**cleav-,** to divide: *cleav*age— period of development characterized by a division of the zygote into smaller and smaller cells.

**ect-,** outside: *ect*oderm— outermost germ layer of embryo.

**lacun-,** pool: *lacun*a—space between the chorionic villi that fills with maternal blood.

**lanug-,** down: *lanug*o—fine hair covering the fetus.

**mes-,** middle: *mes*oderm— middle germ layer of embryo.

**morul-,** mulberry: *morul*a— embryonic structure consisting of a solid ball of about sixteen cells, thus looking somewhat like a mulberry.

**nat-,** to be born: pre*nat*al— period of development before birth.

**ne-,** new, young: *ne*onatal period—period of development including the first four weeks after birth.

**post-,** after: *post*natal period— period of development after birth.

**pre-,** before: *pre*natal period— period of development before birth.

**sen-,** old: *sen*escence—process of growing old.

**troph-,** nuture: *troph*oblast— cellular layer that surrounds the inner cell mass and helps nourish it.

**umbil-,** navel: *umbil*ical cord— structure attached to the fetal navel (umbilicus) that connects the fetus to the placenta.

Reporters gathered in Arles, France, on March 1, 1995, to celebrate a very special birthday. Jeanne Calment, a woman with the distinction of being listed in *The Guinness Book of Records* as "oldest living human," was turning 120 years old. Although she was able to ride her bicycle daily up until her 100th birthday, after that Jeanne began to feel some physical effects of her many years. Her hearing diminished, and she lost her vision to cataracts. She hadn't been able to walk since breaking her hip in her 115th year. But this "oldest living human" still had a sharp wit. When reporters asked her what type of future she anticipated, she answered, "A very short one."

People over the age of 85 are the largest growing population group in many nations. Medical researchers are finding that many of these "oldest old" people are extraordinarily healthy. The reason—the most common disorders strike before this age, essentially "weeding out" much of the population. Heart disease and cancers tend to occur between the fifties and the eighties, and people who have developed Alzheimer's disease or other neurological disorders usually succumb by their eighties.

The lucky "oldest old," researchers suspect, probably inherit combinations of genes that maintain their health as they age and enable them to fight off free radicals, repair DNA adequately, prevent cholesterol buildup in arteries, and resist certain cancers. But the "oldest" old may have more in common than lucky genes. Studies show that they handle stress well (such as living longer than your loved ones); live a moderate lifestyle; exercise regularly; and are well educated. Jeanne Calment explained her longevity: "I took pleasure when I could. I acted clearly and morally and without regret. I'm very lucky." She died in 1997, at age 122.

Humans grow, develop, and age. Growth is an increase in size. It entails increase in cell numbers as a result of mitosis, followed by enlargement of the newly formed cells and of the body.

Development, which includes growth, is the continuous process by which an individual changes from one life phase to another (fig. 23.1). These life phases include a **prenatal period,** which begins with the fertilization of an egg cell and ends at birth, and a **postnatal period,** which begins at birth and ends with death.

(a)

(b)

**FIGURE 23.1**

The contrast between a human embryo at 28 days (a) and a six-month-old fetus (b) shows evidence of profound changes in development.

## PRENATAL PERIOD

The prenatal period of development usually lasts for thirty-eight weeks from conception and can be divided into a period of cleavage, an embryonic stage, and a fetal stage.

## Period of Cleavage

Conception occurs when the genetic packages of sperm and egg merge, forming a *zygote.* Thirty hours later, the zygote undergoes mitosis, giving rise to two new cells. These cells, in turn, reproduce to form four cells, which then reproduce to form eight cells, and so forth. These divisions take place rapidly with little time to grow. Thus, with each subsequent division, the resulting cells are smaller and smaller. This distribution of the zygote's cytoplasmic contents into progressively smaller cells is called **cleavage,** and the cells produced in this way are called *blastomeres* (fig. 23.2). Researchers can manipulate eight-celled cleavage embryos (Clinical Application 23.1).

The approximate time of conception (fertilization) can be calculated by adding 14 days to the date of the onset of the last menstrual period. The expected time of birth can be estimated by adding 266 days to the fertilization date. Most babies are born within 10 to 15 days of this calculated time.

Obstetricians estimate the date of conception by scanning the embryo with ultrasound and comparing the crown-to-rump length to known values that are the average for each day of gestation. This approach is inaccurate if an embryo is smaller or larger than usual due to a medical problem.

**FIGURE 23.2**

(*a*) During the period of cleavage, the cells reproduce rapidly and become smaller and smaller. (*b*) A light micrograph of a human egg cell surrounded by follicular cells and sperm cells (250×). (*c*) Two-cell stage of development (750×).

## CLINICAL APPLICATION

# Preimplantation Genetic Diagnosis

Chloe O'Brien is a very special, healthy little girl. When she was just a ball of eight cells, growing in a laboratory dish, she was tested to see if she would develop cystic fibrosis, the disorder that her parents carried and had passed to her brother. He had a severe case and had to be hospitalized frequently for respiratory problems. In cystic fibrosis, because of the absence of chloride channels in the cell membranes of cells lining the respiratory passages, thick mucus accumulates. This impairs breathing and creates an environment that increases the risk of infection.

technique is so technically difficult and costly, however, that such use is unlikely to occur. In addition, the U.S. government has banned the use of federal funds to perform the procedure, and a researcher who pioneered the technique was punished for defying the ban. However, several privately funded clinics do preimplantation genetic diagnosis for a variety of disorders. ■

One of Chloe-to-be's eight cells was removed and tested to see if it matched a DNA probe for the abnormal cystic fibrosis gene (fig. 23A). When that one cell was found to be free of the defective gene, doctors implanted the remaining seven-celled ball into Chloe's mother's uterus, where prenatal development continued. Chloe was born in March 1992. The procedure is called preimplantation genetic diagnosis because it is performed before the eight-celled prenatal human implants in the uterus. Several other healthy children have since been born following this prenatal procedure.

People fear that preimplantation genetic diagnosis might someday be abused, enabling parents to pick and choose trivial traits in their offspring, such as those affecting appearance rather than health. The

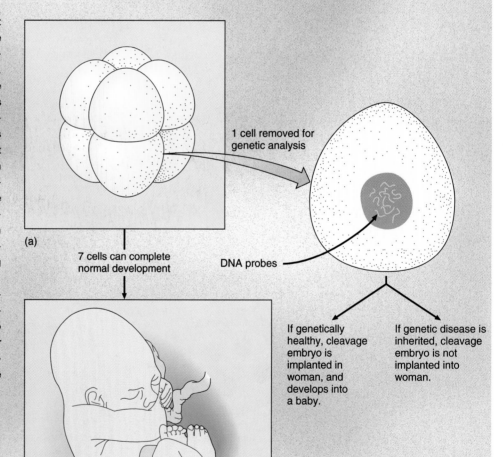

1 cell removed for genetic analysis

(a)

7 cells can complete normal development

DNA probes

If genetically healthy, cleavage embryo is implanted in woman, and develops into a baby.

If genetic disease is inherited, cleavage embryo is not implanted into woman.

## FIGURE 23A

Preimplantation genetic diagnosis probes disease-causing genes in an eight-celled cleavage embryo.

(b)

The mass of cells formed by cleavage is still enclosed in the zona pellucida of the egg cell. In about three days, it consists of a solid ball of about sixteen cells called a **morula.**

The morula enters the uterine cavity and remains unattached for about three days. During this time, the zona pellucida degenerates, and the morula develops a fluid-filled central cavity. Once this cavity appears, the morula becomes a hollow ball of cells called a *blastocyst* (fig. 23.3).

Within the blastocyst, cells in one region group to form an *inner cell mass* that eventually gives rise to the **embryo proper**—the body of the developing offspring. The cells that form the wall of the blastocyst make up the *trophoblast,* which forms structures that assist the embryo in its development.

(a) Morula (3 days)

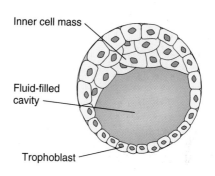

Inner cell mass

Fluid-filled cavity

Trophoblast

(b) Section of blastocyst (6 days)

Sometimes two ovarian follicles release egg cells simultaneously, and if both are fertilized, the resulting zygotes can develop into fraternal (dizygotic) twins. Such twins are no more alike genetically than any brothers or sisters. Twins may also develop from a single fertilized egg (monozygotic twins). This may happen if two inner cell masses form within a blastocyst and each produces an embryo. Twins of this type usually share a single placenta, and they are identical genetically. Thus, they are always the same sex and are very similar in appearance.

About the sixth day, the blastocyst begins to attach to the uterine lining (fig. 23.4), aided by its secretion of proteolytic enzymes that digest a portion of the endometrium. The blastocyst sinks slowly into the resulting depression, becoming completely buried in the uterine lining. At the same time, the uterine lining is stimulated to thicken below the implanting blastocyst, and cells of the trophoblast begin to produce tiny, finger-like processes (microvilli) that grow into the endometrium. This process of **implantation** begins near the end of the first week and is completed during the second week of development (fig. 23.5).

The trophoblast secretes the hormone hCG, which maintains the corpus luteum during the early stages of pregnancy and keeps the immune system from rejecting the blastocyst. This hormone also stimulates synthesis of other hormones from the developing placenta. Clinical Application 23.2 discusses an alternate way for a pregnancy to begin—in the laboratory.

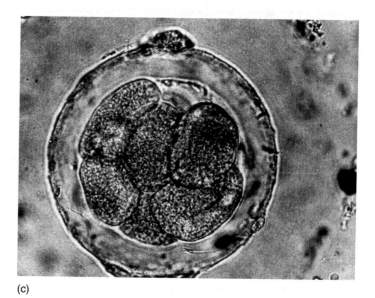

(c)

**FIGURE** 23.3

(*a*) A morula is a solid ball of cells. (*b*) A blastocyst has a fluid-filled cavity. (*c*) Light micrograph of a human morula (500×).

1. Distinguish between growth and development.

2. What changes occur during cleavage?

3. How does a blastocyst attach to the endometrium?

4. In what ways does the endometrium respond to the activities of the blastocyst?

(a)

(b)

(c)

**FIGURE** 23.4

(a) About the sixth day of development, the blastocyst contacts the uterine wall and (b) begins to implant. (c) Light micrograph of a blastocyst from a monkey in contact with the endometrium of the uterine wall (150×).

## Embryonic Stage

The **embryonic stage** extends from the beginning of the second week through the eighth week of development. During this time, the placenta forms, the main internal organs develop, and the major external body structures appear.

**FIGURE** 23.5

Light micrograph of a human cleavage embryo (arrow) implanting in the uterine wall (18×).

During the second week of prenatal development, the blastocyst completes implantation, and the inner cell mass changes. A space, called the *amniotic cavity,* forms between the inner cell mass and the portion of the trophoblast that "invades" the endometrium. The inner cell mass then flattens and is called the **embryonic disk.** By the start of the third week, layers form. A connecting stalk, which attaches the embryo to the developing placenta, appears.

The embryonic disk initially consists of two distinct layers: an outer *ectoderm* and an inner *endoderm.* A short time later, a third layer of cells, the *mesoderm,* forms between the ectoderm and endoderm. These three layers of cells are called the **primary germ layers** of the primordial embryo, or **gastrula.** All organs form from these germ layers (fig. 23.6).

Gastrulation is an important process in prenatal development because a cell's fate is determined by which layer it is in. *Ectodermal* cells give rise to the nervous system, portions of special sensory organs, the epidermis, hair, nails, glands of the skin, and linings of the mouth and anal canal. *Mesodermal cells* form all types of muscle tissue, bone tissue, bone marrow, blood, blood vessels, lymphatic vessels, connective tissues, internal reproductive organs, kidneys, and the epithelial linings of the body cavities. *Endodermal cells* produce the epithelial linings of the digestive tract, respiratory tract, urinary bladder, and urethra (fig. 23.7).

As the embryo implanted in the uterus, slender projections grew out from the trophoblast into the surrounding endometrium. These extensions, called **chorionic**

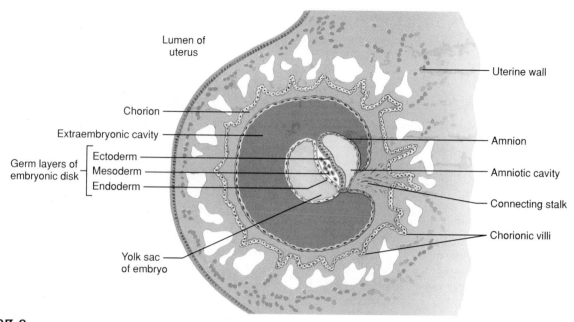

Lumen of
uterus

Uterine wall

Chorion

Extraembryonic cavity

Amnion

Germ layers of
embryonic disk
Ectoderm
Mesoderm
Endoderm

Amniotic cavity

Connecting stalk

Chorionic villi

Yolk sac
of embryo

**FIGURE 23.6**
Early in the embryonic stage of development, the three primary germ layers form.

**villi,** branch. By the end of the fourth week, they are well formed.

Also during the fourth week of development, the flat embryonic disk is transformed into a cylindrical structure (fig. 23.8). The head and jaws appear, the heart beats and forces blood through blood vessels, and tiny buds form, which will give rise to the upper and lower limbs (fig. 23.9).

During the fifth through the seventh weeks, as figure 23.10 shows, the head grows rapidly and becomes rounded and erect. The face, which is developing the eyes, nose, and mouth, appears more humanlike. The upper and lower limbs elongate, and fingers and toes form (fig. 23.11). By the end of the seventh week, all the main internal organs are established, and as these structures enlarge, the body takes on a humanlike appearance.

During the development of chorionic villi, the embryonic blood vessels appear within them, and these vessels are continuous with those passing through the connecting stalk to the body of the embryo. At the same time, irregular spaces called **lacunae** form around and between the villi. These spaces fill with maternal blood that escapes from eroded endometrial blood vessels.

A thin membrane separates embryonic blood within the capillary of a chorionic villus from maternal blood in a lacuna. This membrane, called the **placental membrane,** is composed of the epithelium of the villus and the epithelium of the capillary (fig. 23.12). Through this membrane, exchanges of substances take place between the maternal blood and the embryo's blood. Oxygen and nutrients diffuse from the maternal blood into

the embryo's blood, and carbon dioxide and other wastes diffuse from the embryo's blood into the maternal blood. Various substances also move through the placental membrane by active transport and pinocytosis.

A prenatal test called chorionic villus sampling (CVS) examines the chromosomes in chorionic villus cells to obtain information on fetal health. The idea behind the test is that villus cells are genetically identical to fetal cells because they derive from the same fertilized egg. Because the test carries a risk of causing miscarriage, it is usually offered to women who have previously had a child with a detectable chromosome abnormality. CVS is performed at the tenth week of gestation.

**1** Which major events occur during the embryonic stage of development?

**2** Which tissues and structures develop from ectoderm? From mesoderm? From endoderm?

**3** Describe the structure of a chorionic villus.

**4** What is the function of the placental membrane?

**5** How are substances exchanged between the embryo's blood and the maternal blood?

Until about the end of the eighth week, the chorionic villi cover the entire surface of the former trophoblast, which is now called the **chorion.** However, as

# Assisted Reproductive Technologies

Michele and Ray L'Esperance wanted children badly, but her uterine tubes had been removed due to scarring. A procedure called in vitro fertilization (IVF) solved the couple's problem.

First, Michele received human menopausal gonadotropin to stimulate development of ovarian follicles. When an ultrasound scan showed the follicles had grown to a certain diameter, she received human chorionic gonadotropin to induce ovulation. Then her physician used an optical instrument called a laparoscope to examine the interior of her abdomen and harvest the largest oocytes from an ovary. The oocytes were incubated at body temperature in a medium buffered at pH 7.4. When the oocytes had matured, they were mixed in a laboratory dish with Ray's sperm, which had been "washed" to remove various inhibitory factors. Secretions from Michele's reproductive tract were added to capacitate the sperm.

Next, fertilized eggs were selected and incubated in a special medium for about 60 hours. At this stage, five eight- to sixteen-cell cleavage embryos were transferred through Michele's cervix and into her uterus with the aid of a specially designed catheter. This many balls of cells were transferred to increase the odds that one or two would complete develop-

ment. Michele received progesterone to prepare her uterus to receive them.

IVF worked almost too well for the L'Esperances—they had quintuplets (fig. 23B)! In some cases where too many cleavage embryos implant

for any to survive, some are removed to allow the others to develop. Success rates for IVF vary from clinic to clinic, ranging from 0% to 40%, with the average about 14%. It costs thousands of dollars to become pregnant via IVF. Table 23A describes other assisted reproductive technologies, and table 23B lists some cases where assisted reproductive technology has raised some important issues. ■

**FIGURE 23B**

IVF worked too well for Michele and Ray L'Esperance. Five fertilized ova implanted in Michele's uterus are now Erica, Alexandria, Veronica, Danielle, and Raymond.

the embryo and the chorion surrounding it enlarge, only those villi that remain in contact with the endometrium endure. The others degenerate, and the portions of the chorion to which they were attached become smooth. Thus, the region of the chorion still in contact with the uterine wall is restricted to a disk-shaped area that becomes the **placenta.**

The embryonic portion of the placenta is composed of the chorion and its villi; the maternal portion is com-

posed of the area of the uterine wall (decidua basalis) to which the villi are attached. When it is fully formed, the placenta appears as a reddish-brown disk, about 20 centimeters long and 2.5 centimeters thick. It usually weighs about 0.5 kilogram. Figure 23.13 shows the structure of the placenta.

While the placenta is forming from the chorion, another membrane, called the **amnion,** develops around the embryo. This second membrane began to appear during

## TABLE 23A ASSISTED REPRODUCTIVE TECHNOLOGIES

| Technology | Procedure | Condition It Treats |
|---|---|---|
| Artificial insemination | Donated sperm or pooled specimens are placed near a woman's cervix. | Male infertility—lack of sperm or low sperm count |
| Surrogate mother | An oocyte fertilized in vitro is implanted in a woman other than the one who donated the oocyte. The surrogate, or "gestational mother," gives the newborn to the "genetic mother" and her partner, the sperm donor. | Female infertility—lack of a uterus |
| Gamete intrafallopian transfer (GIFT) | Oocytes are removed from a woman's ovary, then placed along with donated sperm into a uterine tube. | Female infertility—bypasses blocked uterine tube |
| Zygote intrafallopian transfer (ZIFT) | An oocyte fertilized in vitro is placed in a uterine tube. It travels to the uterus on its own. | Female infertility—bypasses blocked uterine tube |
| Embryo adoption | A woman is artificially inseminated with sperm from a man whose partner cannot ovulate healthy oocytes. If the woman conceives, the morula is flushed from her uterus and implanted in the uterus of the sperm donor's partner. | Female infertility—a woman has nonfunctional ovaries, but a healthy uterus |

## TABLE 23B ASSISTED REPRODUCTIVE ISSUES

1. A physician in California used his own sperm to artificially inseminate fifteen patients and told them that he had used sperm from anonymous donors.
2. A plane crash killed the wealthy parents of two cleavage embryos stored at −195°C (−320°F) in a hospital in Melbourne, Australia. Adult children of the couple were asked to share their estate with two eight-celled siblings-to-be.
3. Several couples in Chicago planning to marry discovered that they were half-siblings. Their mothers had been artificially inseminated with sperm from the same donor.
4. Two Rhode Island couples sued a fertility clinic for misplacing several cleavage embryos.
5. Several couples in California sued a fertility clinic for implanting their oocytes or cleavage embryos in other women without consent from the donors. One woman is requesting partial custody of the resulting children if her oocytes were taken and full custody if her cleavage embryos were used.
6. A man sued his ex-wife for possession of their frozen cleavage embryos as part of the divorce settlement.

the second week. Its margin is attached around the edge of the embryonic disk, and fluid called **amniotic fluid** fills the space between the amnion and the embryonic disk.

The developing placenta synthesizes progesterone from cholesterol in the maternal blood. Cells associated with the developing fetal adrenal glands use the placental progesterone to synthesize estrogens. The estrogens, in turn, promote changes in the maternal uterus and breasts, and influence the metabolism and development of various fetal organs.

As the embryo becomes more cylindrical, the margins of the amnion fold, enclosing the embryo in the amnion and amniotic fluid. The amnion envelops the tissues on the underside of the embryo, particularly the connecting stalk, by which it is attached to the chorion and the developing placenta. In this manner, the **umbilical cord** forms (fig. 23.14).

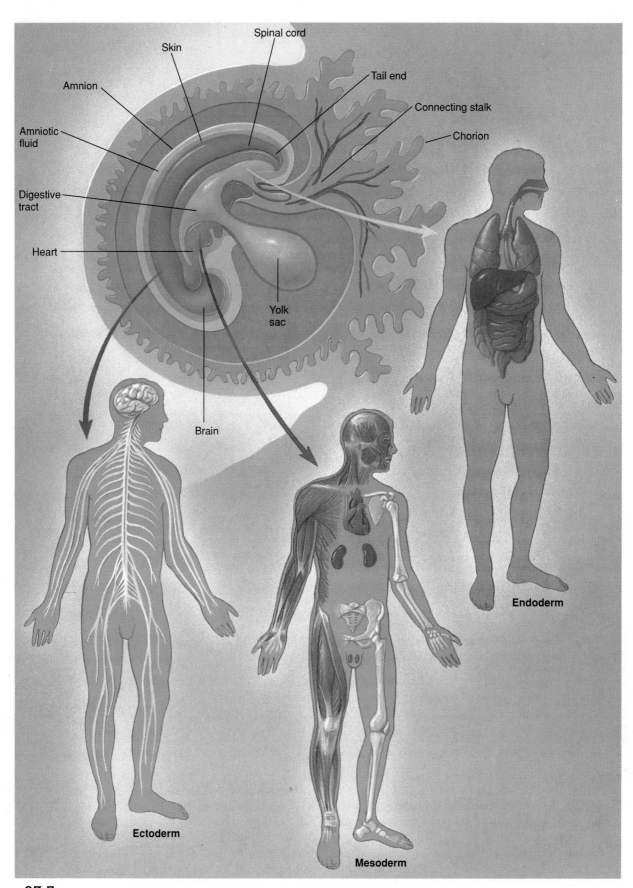

Skin

Spinal cord

Amnion

Tail end

Connecting stalk

Amniotic
fluid

Chorion

Digestive
tract

Heart

Yolk
sac

Brain

Ectoderm

Mesoderm

Endoderm

**FIGURE** 23.7

Each of the primary germ layers forms a particular set of organs.

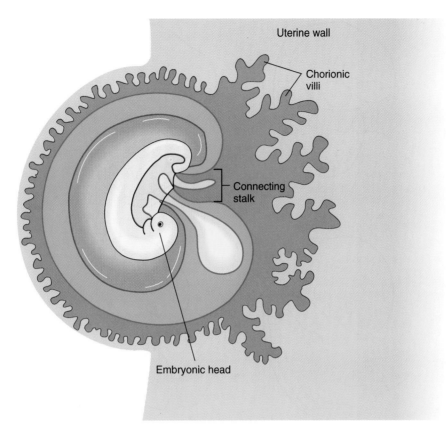

Uterine wall

Chorionic villi

Connecting stalk

Embryonic head

**FIGURE 23.8**

During the fourth week, the flat embryonic disk becomes a cylindrical structure.

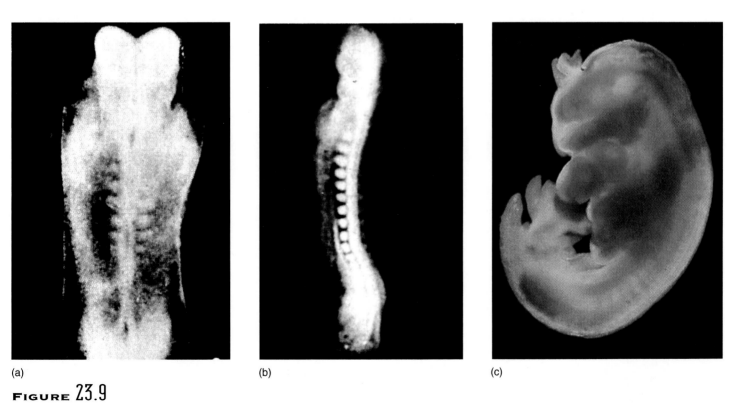

(a)    (b)    (c)

**FIGURE 23.9**

(*a*) A human embryo at three weeks, dorsal view; (*b*) at 3.5 weeks, lateral view; (*c*) at about four weeks, lateral view.

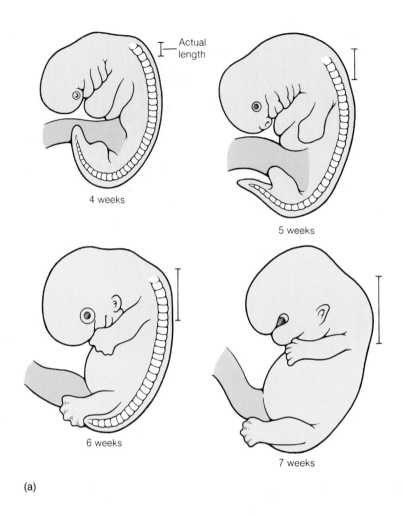

4 weeks

Actual length

5 weeks

6 weeks

7 weeks

(a)

(b)

## FIGURE 23.10

(a) In the fifth through the seventh weeks of gestation, the embryonic body and face develop a more humanlike appearance.
(b) A human embryo after about six weeks of development.

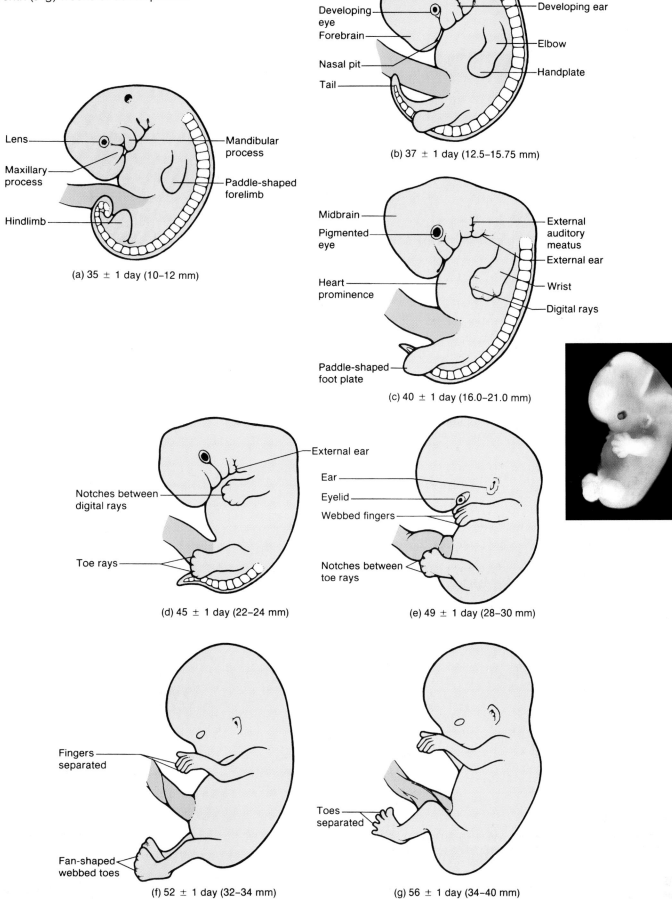

**FIGURE 23.11**

Changes occurring during the fifth (*a–c*), sixth (*d*), and seventh (*e–g*) weeks of development.

Lens

Maxillary process

Hindlimb

Mandibular process

Paddle-shaped forelimb

(a) 35 ± 1 day (10–12 mm)

Developing eye

Forebrain

Nasal pit

Tail

Developing ear

Elbow

Handplate

(b) 37 ± 1 day (12.5–15.75 mm)

Midbrain

Pigmented eye

Heart prominence

Paddle-shaped foot plate

External auditory meatus

External ear

Wrist

Digital rays

(c) 40 ± 1 day (16.0–21.0 mm)

Notches between digital rays

Toe rays

External ear

(d) 45 ± 1 day (22–24 mm)

Ear

Eyelid

Webbed fingers

Notches between toe rays

(e) 49 ± 1 day (28–30 mm)

Fingers separated

Fan-shaped webbed toes

(f) 52 ± 1 day (32–34 mm)

Toes separated

(g) 56 ± 1 day (34–40 mm)

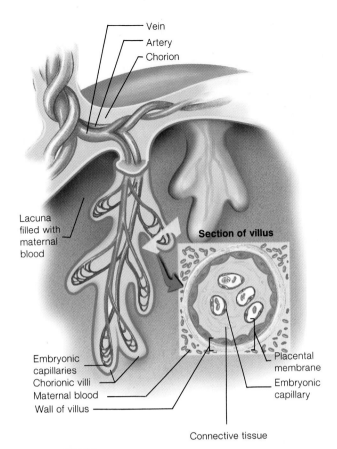

**FIGURE 23.12**

As is illustrated in the section of villus (lower part of figure), the placental membrane consists of the epithelial wall of an embryonic capillary and the epithelial wall of a chorionic villus.

**FIGURE 23.13**

The placenta consists of an embryonic portion and a maternal portion.

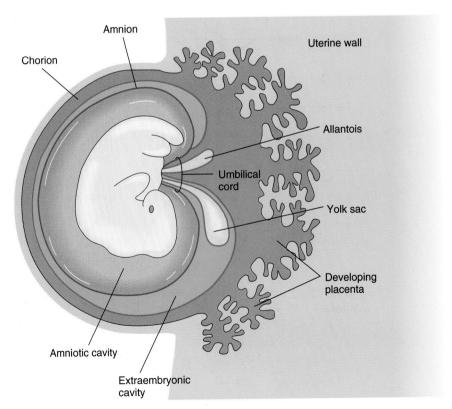

**FIGURE 23.14**

As the amnion develops, it surrounds the embryo, and the umbilical cord begins to form from structures in the connecting stalk.

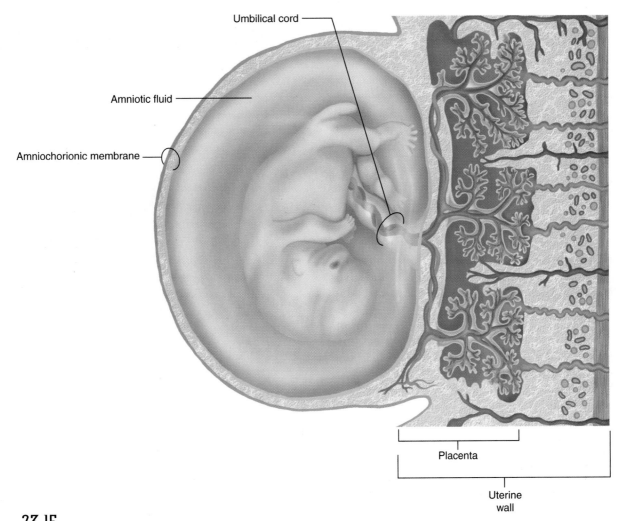

Umbilical cord

Amniotic fluid

Amniochorionic membrane

Placenta

Uterine wall

**FIGURE 23.15**

The developing placenta as it appears during the seventh week of development.

If a pregnant woman ingests an addictive substance, her newborn may suffer from withdrawal symptoms when amounts of the chemical it is accustomed to receiving suddenly plummet. Newborn addiction occurs with certain addictive drugs of abuse, such as heroin; with certain prescription drugs used to treat anxiety; and even with very large doses of vitamin C. Although vitamin C is not addictive, if a fetus is accustomed to megadoses, after birth the sudden drop in vitamin C level may bring on symptoms of vitamin C deficiency.

The fully developed umbilical cord is about 1 centimeter in diameter and about 55 centimeters in length. It begins at the umbilicus of the embryo and inserts into the center of the placenta. The cord contains three blood vessels—two *umbilical arteries* and one *umbilical vein*—

that transport blood between the embryo and the placenta (fig. 23.15).

The umbilical cord also suspends the embryo in the *amniotic cavity,* and the amniotic fluid provides a watery environment in which the embryo can grow freely without being compressed by surrounding tissues. The amniotic fluid also protects the embryo from being jarred by the movements of the woman's body.

In addition to the amnion and chorion, two other embryonic membranes form during development. They are the yolk sac and the allantois.

The **yolk sac** forms during the second week, and it is attached to the underside of the embryonic disk (see fig. 23.14). This structure forms blood cells in the early stages of development and gives rise to the cells that later become sex cells. The yolk sac also gives rise to the stem cells of the immune system. Portions of the yolk sac form the embryonic digestive tube as well. Part of the membrane derived from the yolk sac becomes incorporated into the umbilical cord, and the remainder lies in the cavity between the chorion and the amnion near the placenta.

> Researchers can remove the cells in the yolk sac that give rise to sex cells, culture them in the laboratory, and control the biochemicals in the dish in a way that stimulates the cells to specialize. Using these embryonic germ cells, researchers may be able to cultivate tissues and organs for transplant.

The **allantois** forms during the third week as a tube extending from the early yolk sac into the connecting stalk of the embryo. It, too, forms blood cells and gives rise to the umbilical arteries and vein (see figs. 23.14 and 23.15).

Eventually, the amniotic cavity becomes so enlarged that the membrane of the amnion contacts the thicker chorion around it. The two membranes fuse into an *amniochorionic membrane* (see fig. 23.15).

By the beginning of the eighth week, the embryo is usually 30 millimeters long and weighs less than 5 grams. Although its body is quite unfinished, it looks human (fig. 23.16).

The *embryonic stage* concludes at the end of the eighth week. It is the most critical period of development, because during it the embryo implants within the uterine wall, and all the essential external and internal body parts form. Disturbances to development during the embryonic stage can cause major malformations or malfunctions. This is why early prenatal care is very important.

Factors that cause congenital malformations by affecting an embryo during its period of rapid growth and development are called **teratogens.** Such agents include drugs, viruses, radiation, and even large amounts of otherwise healthful substances, such as fat-soluble vitamins. Each prenatal structure has a time in development, called its *critical period,* when it is sensitive to teratogens (fig. 23.17).

**FIGURE 23.16**

By the beginning of the eighth week of development, the embryonic body is recognizable as a human.

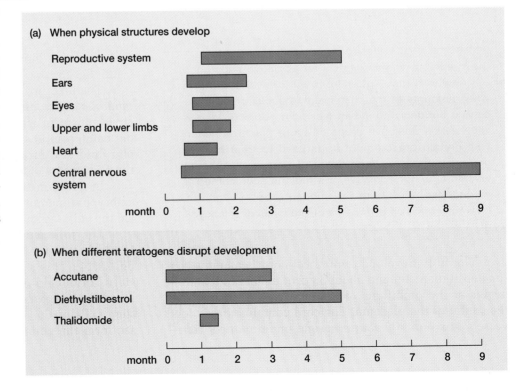

**FIGURE 23.17**

Structures in the developing embryo and fetus are sensitive to specific teratogens at different times in gestation.

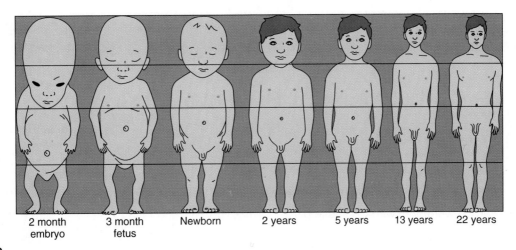

2 month embryo  3 month fetus  Newborn  2 years  5 years  13 years  22 years

**FIGURE 23.18**

During development, body proportions change considerably.

A critical period may extend over many months or be just a day or two. Neural tube defects, for example, are traced to day 28 in development, when a sheet of ectoderm called the neural tube normally folds into a tube, which then develops into the central nervous system. Occasionally something intervenes to disrupt this process, and an opening remains in the spine (spina bifida) or in the brain (anencephaly).

In contrast, the critical period for the brain begins when the anterior neural tube begins to swell into a brain, and continues throughout gestation. This is why so many teratogens affect the brain. Clinical Application 23.3 discusses some teratogens and their effects.

1. Describe the development of the amnion.

2. Which blood vessels are in the umbilical cord?

3. What is the function of amniotic fluid?

4. What types of cells and other structures are derived from the yolk sac?

5. How do teratogens cause birth defects?

## Fetal Stage

The **fetal stage** begins at the end of the eighth week of development and lasts until birth. During this period, growth is rapid, and body proportions change considerably. At the beginning of the fetal stage, the head is disproportionately large, and the lower limbs are relatively short (fig. 23.18). Gradually, proportions more closely resemble those of a child.

During the third month, body lengthening accelerates, but growth of the head slows. The upper limbs achieve the relative length they will maintain throughout development, and ossification centers appear in most of the bones. By the twelfth week, the external reproductive organs are distinguishable as male or female. Figure 23.19 illustrates how these external reproductive organs of the male and female differentiate from precursor structures.

In the fourth month, the body grows very rapidly and reaches a length of up to 20 centimeters and weighs about 170 grams. The lower limbs lengthen considerably, and the skeleton continues to ossify. The fetus has hair, eyebrows, lashes, nipples, and nails, and may even scratch itself.

Amniocentesis is a prenatal test performed after the fourteenth week of gestation. A physician inserts a needle into the amniotic sac and withdraws about 5 milliliters of fluid. Fetal fibroblasts in the fluid are cultured and their chromosomes checked. It takes about a week to grow the chromosomes. A faster technique uses DNA probes to highlight specific chromosomes.

Amniocentesis carries about a 0.5% chance of being followed by miscarriage, and so it is usually performed on women over the age of 35, when the risk of the fetus having a chromosomal anomaly equals the risk of the procedure. Women of any age who have already had a child with a detectable chromosomal abnormality are also candidates for amniocentesis.

An experimental prenatal test, fetal cell sorting, is safer than either chorionic villus sampling or amniocentesis. It detects rare fetal cells in the maternal bloodstream.

# Some Causes of Birth Defects

### THALIDOMIDE

The idea that the placenta protects the embryo and fetus from harmful substances was tragically disproven between 1957 and 1961, when 10,000 children in Europe were born with flippers in place of limbs. Doctors soon identified a mild tranquilizer, thalidomide, which all of the mothers of deformed infants had taken early in pregnancy, during the time of limb formation. Although some women in the United States did use thalidomide and had affected children, the United States was spared a thalidomide disaster because an astute government physician noted adverse effects of the drug on monkeys in experiments, and she halted use of the drug.

### RUBELLA

The virus that causes rubella (German measles) is a powerful teratogen. Australian physicians first noted its effects in 1941, and a rubella epidemic in the United States in the early 1960s caused 20,000 birth defects and 30,000 stillbirths. Exposure in the first trimester leads to cataracts, deafness, and heart defects, and later exposure causes learning disabilities, speech and hearing problems, and juvenile-onset diabetes. Widespread vaccination has slashed the incidence of congenital rubella syndrome, and today it occurs only where people are not vaccinated.

### ALCOHOL

A pregnant woman who has just one or two alcoholic drinks a day, or per- haps a large amount at a crucial time in prenatal development, risks *fetal alcohol syndrome* or the more prevalent *fetal alcohol effects* in her unborn child. Because the effects of small amounts of alcohol at different stages of pregnancy are not yet well understood, and because each woman metabolizes alcohol slightly differently, it is best to avoid drinking alcohol entirely when pregnant, or when trying to become pregnant.

A child with fetal alcohol syndrome has a characteristic small head, misshapen eyes, and a flat face and nose (fig. 23C). He or she grows slowly before and after birth. Intellect is impaired, ranging from minor learning disabilities to mental retardation.

Teens and young adults with fetal alcohol syndrome are short and have small heads. Many individuals remain at early grade school level. They often lack social and communication skills, such as understanding the consequences of actions, forming friendships, taking initiative, and interpreting social cues.

Problems in children of alcoholic mothers were noted by Aristotle more than 23 centuries ago. Fetal alcohol syndrome is the third most common cause of mental retardation in newborns. One to 3 in every 1,000 infants has the syndrome.

### CIGARETTES

Chemicals in cigarette smoke stress a fetus. Carbon monoxide crosses the placenta and plugs up the sites on the fetus's hemoglobin molecules that would normally bind oxygen. Other chemicals in smoke prevent nutrients from reaching the fetus. Studies comparing placentas of smokers and nonsmokers show that smoke-exposed placentas lack important growth factors. The result of all of these assaults is poor growth before and after birth. Cigarette smoking during pregnancy raises the risk of spontaneous abortion, stillbirth, prematurity, and low birth weight.

### NUTRIENTS

Certain nutrients in large amounts, particularly vitamins, act in the body as drugs. The acne medication *isotretinoin* (Accutane) is a derivative

---

In the fifth month, growth rate decreases somewhat. The lower limbs achieve their final relative proportions. The skeletal muscles become active, and the pregnant woman may feel fetal movements for the first time. Some hair grows on the fetal head, and fine, downy hair called lanugo covers the skin. A cheesy mixture of sebum from the sebaceous glands and dead epidermal cells (vernix caseosa) also coats the skin. The fetus, weighing about 450 grams and about 30 centimeters long, curls into the fetal position.

During the sixth month, the fetus gains a substantial amount of weight. The eyebrows and eyelashes appear.

of vitamin A that causes spontaneous abortions and defects of the heart, nervous system, and face. The tragic effects of this drug were noted exactly 9 months after dermatologists began prescribing it to young women in the early 1980s. Today, the drug package bears prominent warnings, and it is never prescribed to pregnant women. A vitamin A–based drug used to treat psoriasis, as well as excesses of vitamin A itself, also cause birth defects. This is because some forms of vitamin A are stored in body fat for up to 3 years after ingestion.

Malnutrition in a pregnant woman threatens the fetus. Obstetrical records of pregnant women before, during, and after World War II link inadequate nutrition early in pregnancy to an increase in the incidence of spontaneous abortion. The aborted fetuses had very little brain tissue. Poor nutrition later in pregnancy affects development of the placenta. The infant has a low birth weight and is at high risk for short stature, tooth decay, delayed sexual development, learning disabilities, and possibly mental retardation.

## OCCUPATIONAL HAZARDS

Some teratogens are encountered in the workplace. Increased rates of spontaneous abortion and birth defects have been noted among women who work with textile dyes, lead, certain photographic chemicals, semiconductor materials, mercury, and cadmium. We do not know much about the role of the male in environmentally caused birth defects. Men whose jobs expose them to sustained heat, such as smelter workers, glass manufacturers, and bakers, may produce sperm that can fertilize an oocyte and possibly lead to spontaneous abortion or a birth defect. A virus or a toxic chemical carried in semen may also cause a birth defect. ■

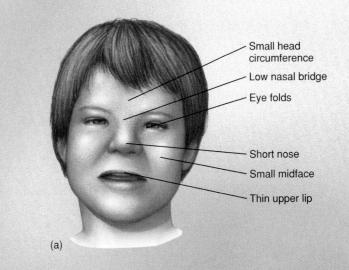

- Small head circumference
- Low nasal bridge
- Eye folds
- Short nose
- Small midface
- Thin upper lip

(a)

(b)

(c)

(d)

### FIGURE 23C

Fetal alcohol syndrome. Some children whose mothers drank alcohol during pregnancy have characteristic flat faces (a) that are strikingly similar in children of different races (b–d). Women who drink excessively while pregnant have a 30% to 45% chance of having a child who is affected to some degree by prenatal exposure to alcohol. Two mixed drinks per day seems to be the level at which damage occurs, but researchers are not yet certain of this.

The skin is quite wrinkled and translucent. Blood vessels in the skin give the fetus a reddish appearance.

In the seventh month, the skin becomes smoother as fat is deposited in the subcutaneous tissues. The eyelids, which fused during the third month, reopen. At the end of this month, a fetus is about 40 centimeters long.

In the final trimester, fetal brain cells rapidly form networks, as organs elaborate and grow. A layer of fat is laid down beneath the skin. The testes of males descend from regions near the developing kidneys, through the inguinal canal, and into the scrotum (see chapter 22). The digestive and respiratory systems mature last, which

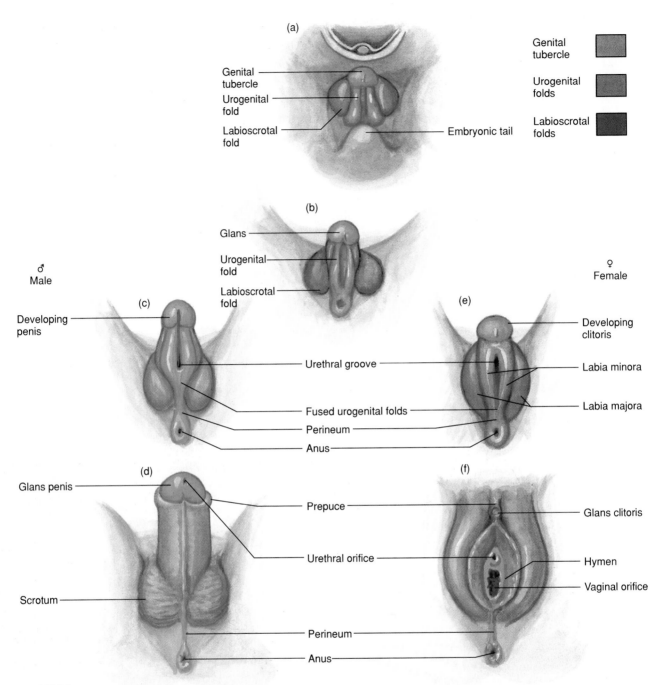

**FIGURE 23.19**

(a and b) The genital tubercle, urogenital fold, and labioscrotal folds that appear during the fourth week of development may differentiate into (c and d) male external reproductive organs or (e and f) female external reproductive organs.

is why infants born prematurely often have difficulty digesting milk and breathing.

Approximately 266 days after a single sperm burrowed its way into an oocyte, a baby is ready to be born. It is about 50 centimeters long and weighs 2.7 to 3.6 kilograms. The skin has lost its downy hair, but is still coated with sebum and dead epidermal cells. The scalp is usually covered with hair; the fingers and toes have well-developed nails; and the skull bones are largely ossified. As figure 23.20 shows, the fetus is usually positioned upside down with its head toward the cervix (*vertex position*).

Premature infants' survival chances increase directly with age and weight. Survival is more likely if the lungs are sufficiently developed with the thin respiratory membranes necessary for rapid exchange of oxygen and carbon dioxide, and if they produce enough surfactant to reduce alveolar surface tension (see chapter 19). A fetus of less than 24 weeks or weighing less than 600 grams at birth seldom survives, even with intensive medical care. Neonatology is the medical field that deals with premature and ill newborns.

**FIGURE 23.20**

A full-term fetus usually becomes positioned with its head near the cervix.

**TABLE 23.1 STAGES OF PRENATAL DEVELOPMENT**

| Stage | Time Period | Major Events |
|-------|-------------|--------------|
| Period of cleavage | First week | Cells undergo mitosis, blastocyst forms; inner cell mass appears; blastocyst becomes implanted in uterine wall<br>Size: 1/4 inch (0.63 centimeters), weight: 1/120 ounce (0.21 grams) |
| Embryonic stage | Second through eighth week | Inner cell mass becomes embryonic disk; primary germ layers form, embryo proper becomes cylindrical; main internal organs and external body structures appear; placenta and umbilical cord form, embryo proper is suspended in amniotic fluid<br>Size: 1 inch (2.5 centimeters), weight: 1/30 ounce (0.8 grams) |
| Fetal stage | Ninth through twelfth week | Ossification centers appear in bones, sex organs differentiate, nerves and muscles become coordinated so that the fetus can move its upper and lower limbs<br>Size: 4 inches (10 centimeters), weight: 1 ounce (28 grams) |
| | Thirteenth through sixteenth week | Body grows rapidly; ossification continues<br>Size: 8 inches (20 centimeters), weight: 6 ounces (170 grams) |
| | Seventeenth through twentieth week | Muscle movements are stronger and woman may be aware of slight flutterings; skin is covered with fine downy hair (lanugo) and coated with sebum mixed with dead epidermal cells (vernix caseosa)<br>Size: 12 inches (30.5 centimeters), weight: 1 pound (454 grams) |
| | Twenty-first through thirty-eighth week | Body gains weight, subcutaneous fat deposited; eyebrows and lashes appear; eyelids reopen; testes descend.<br>Size: 21 inches (53 centimeters), weight: 6 to 10 pounds (2.7 to 4.5 kilograms) |

The birth of a live, healthy baby is against the odds, considering human development from the beginning. Of every 100 secondary oocytes that are exposed to sperm, 84 are fertilized. Of these, 69 implant in the uterus, 42 survive 1 week or longer, 37 survive 6 weeks or longer, and only 31 are born alive. Of those that do not survive to birth, about half have chromosomal abnormalities that are too severe to maintain life. Table 23.1 summarizes the stages of prenatal development.

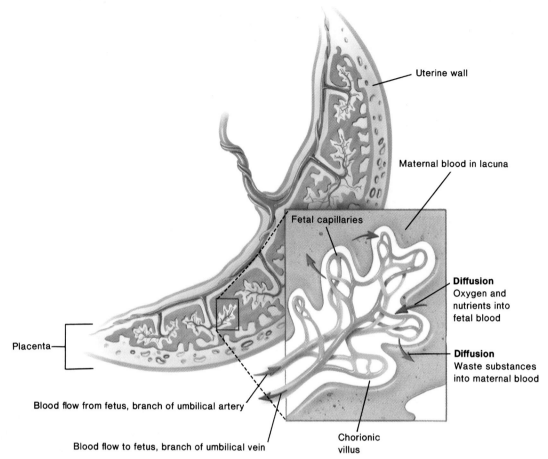

**FIGURE 23.21**

Oxygen and nutrients diffuse into the fetal blood from the maternal blood. Wastes diffuse into the maternal blood from the fetal blood.

Blake Schultz made medical history when he underwent major surgery 7 weeks before birth. Ultrasound had revealed that his stomach, spleen, and intestines protruded through a hole in his diaphragm, the muscle sheet that separates the abdomen from the chest. This defect would have suffocated him shortly after birth, were it not for pioneering surgery that exposed Blake's left side, gently tucked his organs in place, and patched the hole with a synthetic material.

Some prenatal medical problems can be treated by administering drugs to the pregnant woman or by altering her diet. An undersized fetus can receive a nutritional boost by putting the pregnant woman on a high-protein diet. It is also possible to treat prenatal medical problems directly: A tube inserted into the uterus can drain the dangerously swollen bladder of a fetus with a blocked urinary tract, providing relief until the problem can be surgically corrected at birth. A similar procedure can remove excess fluid from the brain of a fetus with hydrocephaly (a neural tube defect also called "water on the brain").

**1** What major changes occur during the fetal stage of development?

**2** When can the sex of a fetus be determined?

**3** How is a fetus usually positioned within the uterus at the end of pregnancy?

## Fetal Blood and Circulation

Throughout fetal development, the maternal blood supplies oxygen and nutrients, and carries away wastes. These substances diffuse between the maternal and fetal blood through the placental membrane, and the umbilical blood vessels carry them to and from the fetus (fig. 23.21). Consequently, the fetal blood and vascular system are adapted to intrauterine existence. For example, the concentration of oxygen-carrying hemoglobin in the fetal blood is about 50% greater than in the maternal blood. Also, fetal hemoglobin has a greater attraction for oxygen than an adult's hemoglobin. Thus, at the oxygen partial pressure of the placental capillaries, fetal hemoglobin can carry 20%–30% more oxygen than maternal hemoglobin.

In the fetal circulatory system, the *umbilical vein* transports blood rich in oxygen and nutrients from the placenta to the fetal body. This vein enters the body through the umbilical ring and travels along the anterior abdominal wall to the liver. About half the blood it carries passes into the liver, and the rest enters a vessel called the **ductus venosus,** which bypasses the liver.

The ductus venosus extends a short distance and joins the inferior vena cava. There, oxygenated blood from the placenta mixes with deoxygenated blood from the lower parts of the fetal body. This mixture continues through the vena cava to the right atrium.

In an adult heart, the blood from the right atrium enters the right ventricle and is pumped through the pulmonary trunk and pulmonary arteries to the lungs. In the fetus, however, the lungs are nonfunctional, and the blood largely bypasses them. As blood from the inferior vena cava enters the fetal right atrium, much of it is shunted directly into the left atrium through an opening in the atrial septum. This opening is called the **foramen ovale,** and the blood passes through it because the blood pressure in the right atrium is somewhat greater than that in the left atrium. Furthermore, a small valve (septum primum) located on the left side of the atrial septum overlies the foramen ovale and helps prevent blood from moving in the reverse direction.

The rest of the fetal blood entering the right atrium, including a large proportion of the deoxygenated blood entering from the superior vena cava, passes into the right ventricle and out through the pulmonary trunk. Only a small volume of blood enters the pulmonary circuit, because the lungs are collapsed, and their blood vessels have a high resistance to blood flow. However, enough blood reaches the lung tissues to sustain them.

Most of the blood in the pulmonary trunk bypasses the lungs by entering a fetal vessel called the **ductus arteriosus,** which connects the pulmonary trunk to the descending portion of the aortic arch. As a result of this connection, the blood with a relatively low oxygen concentration, which is returning to the heart through the superior vena cava, bypasses the lungs. At the same time, it is prevented from entering the portion of the aorta that branches to the heart and brain.

The more highly oxygenated blood that enters the left atrium through the foramen ovale mixes with a small amount of deoxygenated blood returning from the pulmonary veins. This mixture moves into the left ventricle and is pumped into the aorta. Some of it reaches the myocardium through the coronary arteries, and some reaches the brain tissues through the carotid arteries.

The blood the descending aorta carries is partially oxygenated and partially deoxygenated. Some of it is carried into the branches of the aorta that lead to the lower regions of the body. The rest passes into the *umbilical arteries,* which branch from the internal iliac ar-

| TABLE 23.2 | FETAL CIRCULATORY ADAPTATIONS | |
|---|---|
| **Adaptation** | **Function** |
| Fetal blood | Has greater oxygen-carrying capacity than adult blood |
| Umbilical vein | Carries oxygenated blood from the placenta to the fetus |
| Ductus venosus | Conducts about half the blood from the umbilical vein directly to the inferior vena cava, thus bypassing the liver |
| Foramen ovale | Conveys a large proportion of the blood entering the right atrium from the inferior vena cava, through the atrial septum, and into the left atrium, thus bypassing the lungs |
| Ductus arteriosus | Conducts some blood from the pulmonary trunk to the aorta, thus bypassing the lungs |
| Umbilical arteries | Carry the blood from the internal iliac arteries to the placenta |

teries and lead to the placenta. There the blood is reoxygenated (fig. 23.22).

The umbilical cord usually contains two arteries and one vein. A small percentage of newborns have only one umbilical artery. This condition is often associated with other cardiovascular or gastrointestinal disorders. Because of the possibility of these conditions, the vessels within the severed cord are routinely counted following a birth.

Table 23.2 summarizes the major features of fetal circulation. At the time of birth, important adjustments must occur in the circulatory system when the placenta ceases to function and the newborn begins to breathe. Clinical Application 23.4 describes the birth of conjoined twins.

**1** How does the pattern of fetal circulation differ from that of an adult?

**2** Which umbilical vessel carries oxygen-rich blood to the fetus?

**3** What is the function of the ductus venosus?

**4** How does fetal circulation allow blood to bypass the lungs?

**5** What characteristic of the fetal lungs tends to shunt blood away from them?

## POSTNATAL PERIOD

The postnatal period of development lasts from birth until death. It can be divided into the neonatal period, infancy, childhood, adolescence, adulthood, and senescence.

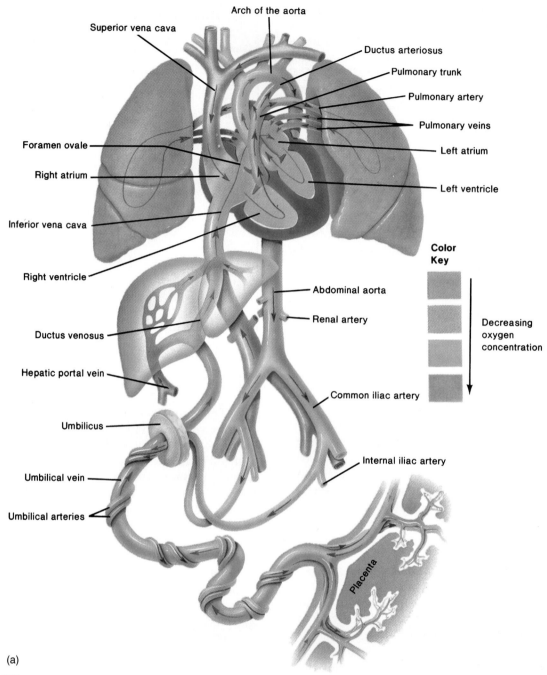

Arch of the aorta

Superior vena cava

Ductus arteriosus

Pulmonary trunk

Pulmonary artery

Pulmonary veins

Foramen ovale

Left atrium

Right atrium

Left ventricle

Inferior vena cava

Right ventricle

Abdominal aorta

Renal artery

Color Key

Decreasing oxygen concentration

Ductus venosus

Hepatic portal vein

Common iliac artery

Umbilicus

Umbilical vein

Internal iliac artery

Umbilical arteries

Placenta

(a)

**FIGURE** 23.22

The general pattern of fetal circulation is shown anatomically (*a*) and schematically (*b*).

## Neonatal Period

The **neonatal period,** which extends from birth to the end of the first four weeks, begins very abruptly at birth (fig. 23.23). At that moment, physiological adjustments must occur quickly, because the newborn must suddenly do for itself what the mother's body had been doing for it. Thus, the newborn (neonate) must respire, obtain and digest nutrients, excrete wastes, and regulate body temperature. However, a newborn's most immediate need is

to obtain oxygen and excrete carbon dioxide, so its first breath is critical.

The first breath must be particularly forceful, because the newborn's lungs are collapsed, and the airways are small, offering considerable resistance to air movement. Also, surface tension tends to hold the moist membranes of the lungs together. However, the lungs of a full-term fetus continuously secrete surfactant (see chapter 19), which reduces surface tension. After the first powerful breath begins to expand the lungs, breathing eases.

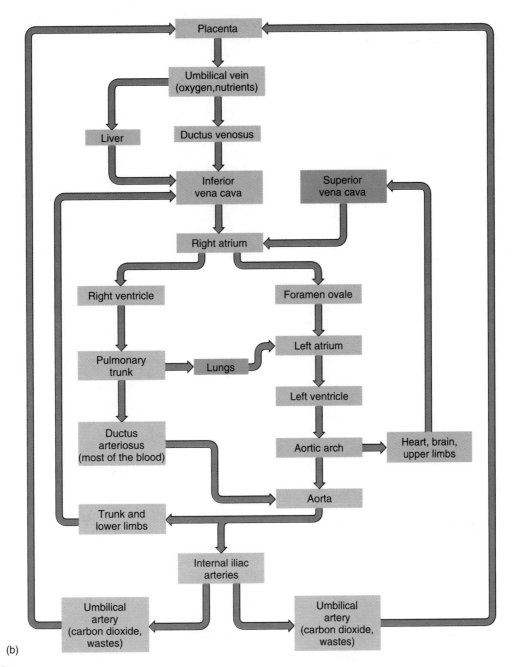

(b)

**FIGURE** 23.22   *Continued*

During birth, uterine contractions mechanically stimulate a newborn's lungs, which increases the quantity of surfactant secreted. This may explain why respiratory distress syndrome is more common in infants whose births do not include labor.

A newborn's first breath is stimulated by increasing concentration of carbon dioxide, decreasing pH, low oxygen concentration, drop in body temperature, and mechanical stimulation during and after birth. Also, in

response to the stress the fetus experiences during birth, blood concentrations of epinephrine and norepinephrine rise significantly (see chapter 13). These hormones promote normal breathing by increasing the secretion of surfactant and dilating the airways.

For energy, the fetus primarily depends on glucose and fatty acids in the pregnant woman's blood. The newborn, on the other hand, is suddenly without an external source of nutrients. The mother will not produce milk for two to three days, by which time the infant's gastrointestinal tract will be able to digest it. However, the mother's breasts secrete *colostrum*, a fluid rich in

# Joined for Life

Patty Hensel's pregnancy in 1990 was uneventful. An ultrasound scan revealed an apparently normal fetus, although at one medical exam, Mike Hensel thought he heard two heartbeats, but he dismissed it as an error. Mike's ears weren't deceiving him though—he *had* heard two distinct heartbeats.

A Caesarean section was necessary because the baby was positioned bottom-first. To everyone's amazement, the baby had two heads and two necks, yet it appeared to share the rest of the body, with two legs and two arms in the correct places, and a third arm between the heads. The ultrasound had probably imaged the twins from an angle that superimposed one head on the other. Patty, dopey from medication, recalls hearing the word "Siamese" and thinking she had given birth to cats. She had delivered conjoined, or Siamese, twins.

The baby was actually two individuals, named Abigail and Brittany. Each twin had her own neck, head, heart, stomach, and gallbladder. Remarkably, each also had her own nervous system. The twins shared a large liver, a single bloodstream, and all organs below the navel, including the reproductive tract. They had three lungs and three kidneys.

Abby and Britty were strong and healthy. Doctors suggested surgery to separate the twins. Aware that only one child would likely survive surgery, Mike and Patty chose to let their daughters be. In Ireland 2 years later, the parents of conjoined twins Eilish and Katie Holton faced the same agonizing choice, and they took the other option.

Eilish and Katie were similar to Abby and Britty but had four arms and a large, abnormal heart. In 1992, a team of 15 doctors separated them. Katie died of heart failure 4 days after the surgery. Eilish, now a healthy child who gets around quite well on an artificial leg, still looks over her shoulder for her missing twin.

In 1993, U.S. physicians attempted to separate Amy and Angela Lakeberg, infants who shared a liver and heart. Doctors determined that Angela had the better chance to survive, and so during surgery, they gave the heart to her. Amy died instantly. Angela lived until the following June, never able to breathe without the aid of a respirator.

Because Abby and Britty have separate hearts and nervous systems, they might fare better in surgery than the Holton or Lakeberg twins. But the girls are happy as they are, for now

(fig. 23D). They were particularly distressed to meet Eilish in 1995 to do a television program and learn the fate of Katie. Abby and Britty have very distinctive personalities and attend school, swim, ride bikes, and play, like any other kids.

Conjoined twins occur in 1 in 50,000 births, and about 40% are stillborn. Abby and Britty Hensel are rare among the rare, being joined in a manner seen only four times before. They are the result of incomplete twinning, which probably occurred during the first two weeks of gestation. Because the girls have duplicated tissue derived from ectoderm, mesoderm, and endoderm, the partial twinning event must have occurred before the three germ layers were established.

The term "Siamese twins" comes from Chang and Eng, who were born in Thailand, then called Siam, in 1811. They were joined by a ligament from the navel to the breastbone, which surgeons could easily correct today. Chang and Eng lived for 63 years, and each married. ∎

**FIGURE 23D**
Abby and Britty Hensel are conjoined twins, the result of incomplete twinning during the first two weeks of prenatal development.

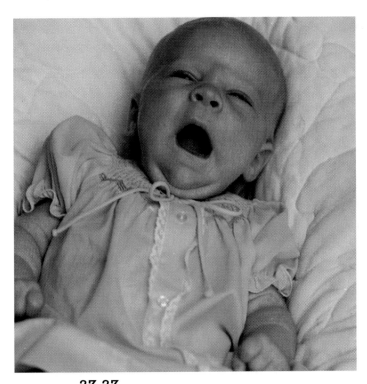

**FIGURE 23.23**
The neonatal period extends from birth to the end of the fourth week after birth.

nutrients and antibodies, until the mature milk comes in—an adaptation to the state of the newborn's digestive physiology. The newborn has a relatively high metabolic rate, and its liver, which is not fully mature, may be unable to supply enough glucose to support metabolism. Consequently, the newborn typically utilizes stored fat for energy.

**1** Define *neonatal period of development.*

**2** Why must the infant's first breath be particularly forceful?

**3** What factors stimulate the first breath?

**4** What does a newborn use for energy during the first few days after birth?

A newborn's kidneys are usually unable to produce concentrated urine, so they excrete a dilute fluid. For this reason, the newborn may become dehydrated and develop a water and electrolyte imbalance. Also, some of the newborn's homeostatic control mechanisms may not function adequately. For example, the temperature regulating system may be unable to maintain a constant body temperature. As a result, body temperature may respond to slight stimuli by fluctuating above or below the normal level during the first few days of life.

When the placenta ceases to function and breathing begins, changes also occur in the newborn's circu-latory system. Following birth, the umbilical vessels constrict. The arteries close first, and if the umbilical cord is not clamped or severed for a minute or so, blood continues to flow from the placenta to the newborn through the umbilical vein, adding to the newborn's blood volume.

The proximal portions of the umbilical arteries persist in the adult as the *superior vesical arteries* that supply blood to the urinary bladder. The more distal portions become solid cords (lateral umbilical ligaments). The umbilical vein becomes the cordlike *ligamentum teres* that extends from the umbilicus to the liver in an adult. The ductus venosus constricts shortly after birth and appears in the adult as a fibrous cord (ligamentum venosum) superficially embedded in the wall of the liver.

The foramen ovale closes as a result of blood pressure changes in the right and left atria as the fetal vessels constrict. As blood ceases to flow from the umbilical vein into the inferior vena cava, the blood pressure in the right atrium falls. Also, as the lungs expand with the first breathing movements, resistance to blood flow through the pulmonary circuit decreases, more blood enters the left atrium through the pulmonary veins, and blood pressure in the left atrium increases.

As the pressure in the left atrium rises and that in the right atrium falls, the valve (septum primum) on the left side of the atrial septum closes the foramen ovale. In most individuals, this valve gradually fuses with the tissues along the margin of the foramen. In an adult, the site of the previous opening is marked by a depression called the *fossa ovalis.*

The ductus arteriosus, like other fetal vessels, constricts after birth. After the ductus arteriosus closes, blood can no longer bypass the lungs by moving from the pulmonary trunk directly into the aorta. In an adult, the ductus arteriosus is represented by a cord called the *ligamentum arteriosum.*

In *patent ductus arteriosus* (PDA), the ductus arteriosus fails to close completely. This condition is common in newborns whose mothers were infected with rubella virus (German measles) during the first three months of pregnancy.

After birth, the metabolic rate and oxygen consumption in neonatal tissues increase, in large part to maintain body temperature. If the ductus arteriosus remains open, the neonate's blood oxygen concentration may be too low to adequately supply body tissues, including the myocardium. If PDA is not corrected surgically, the heart may fail, even though the myocardium is normal.

Changes in the newborn's circulatory system are gradual. Although constriction of the ductus arteriosus may be functionally complete within fifteen minutes, the permanent closure of the foramen ovale may take up to a year. These circulatory changes are illustrated in figure 23.24 and summarized in table 23.3.

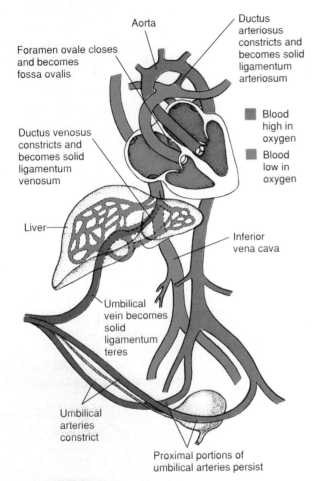

**FIGURE 23.24**

Major changes that occur in the newborn's circulatory system.

**1** How do the kidneys of a newborn differ from those of an adult?

**2** What portions of the umbilical arteries continue to function into adulthood?

**3** What is the fate of the foramen ovale? Of the ductus arteriosus?

**4** When is closure of the ductus arteriosus functionally complete?

## Infancy

The period of continual development extending from the end of the first four weeks to one year is called **infancy.** During this time, the infant grows rapidly and may triple its birth weight. Its teeth begin to erupt through the gums, and its muscular and nervous systems mature so that coordinated muscular activities become possible. The infant is soon able to follow objects visually; reach for and grasp objects; and sit, creep, and stand.

Infancy also brings the beginning of the ability to communicate. The infant learns to smile, laugh, and respond to some sounds. By the end of the first year, the infant may be able to say two or three words. Which parent's name is first spoken is sometimes a matter of great interest; often it is the name of a beloved pet.

Because infancy (as well as childhood) is a period of rapid growth, the infant has particular nutritional requirements. In addition to an energy source, the body requires proteins to provide the amino acids necessary to form new tissues; calcium and vitamin D to promote the development and ossification of skeletal structures; iron to support blood cell formation; and vitamin C for production of structural tissues such as cartilage and bone. By the time an infant is four months old, most of the circulating hemoglobin is the adult type.

## Childhood

**Childhood** begins at the end of the first year and ends at puberty. During this period, growth continues at a high

| | TABLE 23.3 | CIRCULATORY ADJUSTMENTS IN THE NEWBORN | |
|---|---|---|---|

| Structure | Adjustment | In the Adult |
|---|---|---|
| Umbilical vein | Constricts | Becomes ligamentum teres that extends from the umbilicus to the liver |
| Ductus venosus | Constricts | Becomes ligamentum venosum that is superficially embedded in the wall of the liver |
| Foramen ovale | Closes by valvelike septum primum as blood pressure in right atrium decreases and pressure in left atrium increases | Valve fuses along margin of foramen ovale and is marked by a depression called the fossa ovalis |
| Ductus arteriosus | Constricts | Becomes ligamentum arteriosum that extends from the pulmonary trunk to the aorta |
| Umbilical arteries | Distal portions constrict | Distal portions become lateral umbilical ligaments; proximal portions function as superior vesical arteries |

rate. The primary teeth appear, and then secondary teeth replace them. The child develops a high degree of voluntary muscular control and learns to walk, run, and climb. Bladder and bowel controls are established. The child learns to communicate effectively by speaking, and later, usually learns to read, write, and reason objectively. At the same time, the child is maturing emotionally.

**1** Define *infancy.*

**2** What developmental changes characterize infancy?

**3** Define *childhood.*

**4** What developmental changes characterize childhood?

## Adolescence

**Adolescence** is the period of development between puberty and adulthood. It is a time of anatomical and physiological changes that result in reproductively functional individuals. Most of these changes are hormonally controlled, and they include the appearance of secondary sex characteristics as well as growth spurts in the muscular and skeletal systems.

Females usually experience these changes somewhat earlier than males, so that early in adolescence, females may be taller and stronger than their male peers. On the other hand, females attain full growth at earlier ages, and in late adolescence, the average male is taller and stronger than the average female.

The periods of rapid growth in adolescence, which usually begin between the ages of 11 and 13 in females and between 13 and 15 in males, cause increased demands for certain nutrients. It is not uncommon for a teenager to consume a huge plate of food, go back for more—and still remain thin. In addition to energy sources, foods must provide ample amounts of proteins, vitamins, and minerals to support growth of new tissues. Adolescence also brings increasing levels of motor skills, intellectual ability, and emotional maturity.

Aging at the cellular, tissue, and organ level is a lifelong process. Human cells begin to die even before an individual is born. Fingers and toes, for example, are gradually carved from weblike extremities by cells in the webbing that die.

By the age of 2 years, a child's brain cells are as numerous as they will ever be. The brain has been growing over the past couple of years at the same rate as it did in the last 6 months in the uterus. At the age of 10, a person's hearing is the best that it will ever be. Shortly before sexual maturity, the *thymus gland* reaches its greatest size (about the size of a walnut) and then begins to shrink gradually until it is almost microscopic by about the seventh decade of life. The organ's declining activity may contribute to the increasing susceptibility to certain illnesses seen with advancing age.

## Adulthood

**Adulthood** (maturity) extends from adolescence to old age. By age 18, the human male is producing the highest level of the sex hormone testosterone that he will ever have in his lifetime, and sex drive is strong. In the 20s, muscle strength peaks in both sexes. Hair is its fullest, with each hair as thick as it will ever be. By the end of the third decade of life, obvious signs of aging may first appear as a loss in the elasticity of facial skin, producing small wrinkles around the mouth and eyes. Height is already starting to decrease, but not yet at a detectable level.

The age of 30 seems to be a developmental turning point. Hearing often becomes less acute. Heart muscle begins to thicken. The elasticity of the ligaments between the small bones in the back lessens, setting the stage for the slumping posture that becomes apparent in later years. Some researchers estimate that beginning roughly at age 30, the human body becomes functionally less efficient by about 0.8% every year.

During their 40s, many people weigh 10 to 20 pounds (4.5 to 9 kilograms) more than they did at the age of 20, thanks to a slowing of metabolism and decrease in activity level. They may be ⅛ inch (0.3 centimeter) shorter, too. Hair may be graying as melanin-producing biochemical pathways lose efficiency, and some hair may fall out. Vision may become farsighted or nearsighted. The immune system is less efficient, making the body more prone to infection and cancer. Skeletal muscles tend to lose strength as more and more connective tissue appears within the muscles; the circulatory system is strained as the lumens of arterioles and arteries narrow due to accumulations of fatty deposits; skin loosens and wrinkles as elastic fibers in the dermis break down.

The early 50s bring further declines in the functioning of the human body. Nail growth slows, taste buds die, and the skin continues to lose its elasticity. For most people, the ability to see close objects becomes impaired, but for the nearsighted, vision improves. Women stop menstruating, although this does not necessarily mean an end to or loss of interest in sex. Decreased activity of the pancreas may lead to diabetes. By the decade's end, muscle mass and weight begin to decrease. A male produces less semen but is still sexually active. His voice may become higher as his vocal cords degenerate. A man has half the strength in his upper limb muscles and half the lung function as he did at age 25. He is about ¾ inch (2 centimeters) shorter.

The 60-year-old may experience minor memory losses. A few million of the person's billions of brain cells have been lost over his or her lifetime, but for the most part, intellect remains quite sharp. By age 70, height decreases a full inch (2.5 centimeters). Sagging skin and loss of connective tissue, combined with continued growth of cartilage, make the nose, ears, and eyes more prominent. Figure 23.25 outlines some of the anatomical and physiological changes that accompany aging.

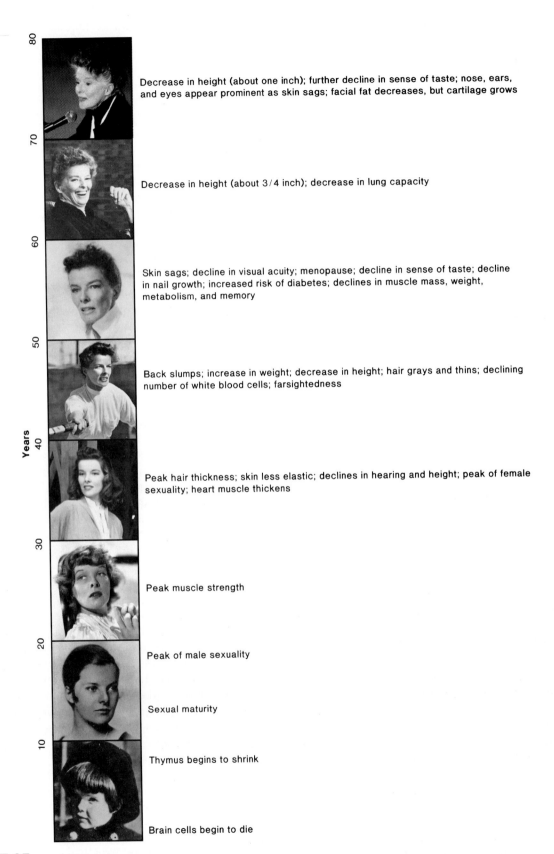

**FIGURE 23.25**

Although many biological changes ensue as we grow older, photographs of actress Katharine Hepburn at various stages of her life indicate that we can age with great grace and beauty.

## TABLE 23.4 STAGES IN POSTNATAL DEVELOPMENT

| Stage | Time Period | Major Events |
|---|---|---|
| Neonatal period | Birth to end of fourth week | Newborn begins to carry on respiration, obtain nutrients, digest nutrients, excrete wastes, regulate body temperature, and make circulatory adjustments |
| Infancy | End of fourth week to one year | Growth rate is high; teeth begin to erupt; muscular and nervous systems mature so that coordinated activities are possible; communication begins |
| Childhood | One year to puberty | Growth rate is high; deciduous teeth erupt and are replaced by permanent teeth; high degree of muscular control is achieved; bladder and bowel controls are established; intellectual abilities mature |
| Adolescence | Puberty to adulthood | Person becomes reproductively functional and emotionally more mature; growth spurts occur in skeletal and muscular systems; high levels of motor skills are developed; intellectual abilities increase |
| Adulthood | Adolescence to old age | Person remains relatively unchanged anatomically and physiologically; degenerative changes begin to occur |
| Senescence | Old age to death | Degenerative changes continue; body becomes less and less able to cope with the demands placed upon it; death usually results from mechanical disturbances in the cardiovascular system or from disease processes that affect vital organs |

---

Becoming a mother in the sixth decade of life is not normal, but it is possible with medical technology. After hormone treatments, Rosanna Della Corte, age 62, became pregnant, with a donated oocyte her husband Mauro's sperm fertilized in a laboratory dish. Her pregnancy proved that the uterine lining of an older woman can support a fetus to term if given appropriate hormones.

## Senescence

**Senescence** is the process of growing old. It is a continuation of the degenerative changes that begin during adulthood. As a result, the body becomes less and less able to cope with the demands placed on it by the individual and by the environment.

Senescence is a result of the normal wear-and-tear of body parts over many years. For example, the cartilage covering the ends of bones at joints may wear away, leaving the joints stiff and painful.

Other degenerative changes are caused by disease processes that interfere with vital functions, such as gas exchanges or blood circulation. Metabolic rate and distribution of body fluids may change. The rate of cell reproduction may decline, and the immune responses may weaken. As a result, the person becomes less able to repair damaged tissue and more susceptible to disease.

Decreasing efficiency of the central nervous system accompanies senescence. As a result, the person may lose some intellectual functions. Also, the physiological coordinating capacity of the nervous system may decrease, and homeostatic mechanisms may fail to operate effectively.

Sensory functions usually decline with age also. Perhaps one-third of older persons have significant hearing losses, difficulty focusing on objects at varying distances, and problems with depth perception. The senses of taste and smell diminish with age, and many older persons have decreased abilities to perceive environmental temperature accurately. They may also have impaired perception of touch and pressure as well as decreased sensitivity to painful stimuli.

Death usually results, not from these degenerative changes, but from mechanical disturbances in the cardiovascular system, failure of the immune system, or disease processes that affect vital organs. However, the loss of function that often precedes death in the elderly is associated with inactivity, poor nutrition, and chronic disease. Table 23.4 summarizes the major phases of postnatal life and their characteristics, while table 23.5 lists some aging-related changes.

1. How does the body change during adolescence?
2. Define *adulthood.*
3. What changes occur during adulthood?
4. What changes accompany senescence?

TABLE 23.5

## AGING-RELATED CHANGES

| Organ System | Aging-Related Changes |
|---|---|
| Integumentary system | Degenerative loss of collagenous and elastic fibers in dermis; decreased production of pigment in hair follicles; reduced activity of sweat and sebaceous glands |
| | Skin tends to become thinner, wrinkled, and dry; hair turns gray and then white |
| Skeletal system | Degenerative loss of bone matrix |
| | Bones become thinner, less dense, and more likely to fracture; stature may shorten due to compression of intervertebral disks and vertebrae |
| Muscular system | Loss of skeletal muscle fibers; degenerative changes in neuromuscular junctions |
| | Loss of muscular strength |
| Nervous system | Degenerative changes in neurons; loss of dendrites and synaptic connections; accumulation of lipofuscin in neurons; decreases in sensory sensitivities |
| | Decreasing efficiency in processing and recalling information; decreasing ability to communicate; diminished senses of smell and taste; loss of elasticity of lenses and consequent loss of ability to accommodate for close vision |
| Endocrine system | Reduced hormonal secretions |
| | Decreased metabolic rate; reduced ability to cope with stress; reduced ability to maintain homeostasis |
| Digestive system | Decreased motility in gastrointestinal tract; reduced secretion of digestive juices |
| | Reduced efficiency of digestion |
| Respiratory system | Degenerative loss of elastic fibers in lungs; reduced number of alveoli |
| | Reduced vital capacity; increase in dead air space; reduced ability to clear airways by coughing |
| Cardiovascular system | Degenerative changes in cardiac muscle; decrease in lumen diameters of arteries and arterioles |
| | Decreased cardiac output; increased resistance to blood flow; increased blood pressure |
| Lymphatic system | Decrease in efficiency of immune system |
| | Increased incidence of infections and neoplastic diseases; increased incidence of autoimmune diseases |
| Excretory system | Degenerative changes in kidneys; reduction in number of functional nephrons |
| | Reductions in filtration rate, tubular secretion, and reabsorption |
| Reproductive systems Female | Degenerative changes in ovaries; decrease in secretion of sex hormones |
| | Menopause; regression of secondary sex characteristics |
| Male | Reduced secretion of sex hormones; enlargement of prostate gland; decrease in sexual energy |

Recent public interest in physician-assisted suicide has focused attention on why severely ill people seek to end their lives. While the courts argue the legality of assisted suicide, the medical community is recognizing shortcomings in the treatment of the dying. Curricula for medical students and medical residents are being revamped to increase emphasis on providing palliative care for the terminally ill. Such care seeks to make a patient comfortable, even if the treatment does not cure the disease or extend life.

From 65 to 80% of all deaths in the United States take place in hospitals, often with painful and sometimes unwanted interventions to prolong life. One study found that about half of all conscious patients suffer severe pain prior to death. In Oregon, which has pioneered education on caring for the dying patient, a greater percentage of patients live out their last days at home, in nursing homes, or in hospices, which are facilities dedicated to providing comfort and support for the dying.

## AGING

The aging process is difficult to analyze because of the intricate interactions of the body's organ systems. Breakdown of one structure ultimately affects the functioning of others. The medical field of gerontology examines the biological changes of aging at the molecular, cellular, organismal, and population levels. Aging is both passive and active.

### Passive Aging

Aging as a passive process entails breakdown of structures and slowing of functions. At the molecular level, passive aging is seen in the degeneration of the elastin and collagen proteins of connective tissue, causing skin to sag and muscle tissue to lose its firmness.

During a long lifetime, biochemical abnormalities accumulate. Mistakes occur throughout life when DNA replicates in reproducing cells. Usually, repair enzymes correct this damage immediately. But over many years, exposure to chemicals, viruses, and radiation disrupt

DNA repair mechanisms, so that the error burden becomes too great to be fixed. The cell may die as a result of faulty genetic instructions.

Another sign of passive aging at the biochemical level is the breakdown of lipids. As aging membranes leak during lipid degeneration, a fatty, brown pigment called lipofuscin accumulates. Mitochondria also begin to break down in older cells, decreasing the supply of chemical energy to power the cell's functions.

The cellular degradation associated with aging may be set into action by highly reactive chemicals called **free radicals.** A molecule that is a free radical has an unpaired electron in its outermost valence shell. This causes the molecule to grab electrons from other molecules, destabilizing them, and a chain reaction of chemical instability begins that could kill the cell. Free radicals are a by-product of normal metabolism, and also form by exposure to radiation or toxic chemicals. Enzymes that usually inactivate free radicals diminish in number and activity in the later years. One such enzyme is *superoxide dismutase* (SOD).

## Active Aging

Aging also entails new activities or the appearance of new substances. Lipofuscin granules, for example, may be considered an active sign of aging, but they result from the passive breakdown of lipids. Another example of active aging is autoimmunity, in which the immune system turns against the body, attacking its cells as if they were invading organisms.

Active aging actually begins before birth, as certain cells die as part of the developmental program encoded in the genes. This process of programmed cell death, called **apoptosis,** occurs regularly in the embryo, degrading certain structures to pave the way for new ones. The number of neurons in the fetal brain, for example, is halved as those that make certain synaptic connections are spared from death. Apoptosis also occurs in the fetal thymus, where T cells that do not recognize "self" cell surfaces die, thereby building the immune system. In the adult, apoptosis removes brain, blood, and skin cells. Throughout life, apoptosis makes it possible for organs to maintain their characteristic shapes.

Mitosis and apoptosis are opposite, but complementary, processes. That is, as organs grow, the number of cells in some regions increases, but in others it decreases. Cell death, then, is not a phenomenon that is restricted to the aged. It is a normal part of life. Clinical Application 23.5 discusses genetic disorders that greatly accelerate aging.

## The Human Life Span

In the age-old quest for longer life, people have sampled everything from turtle soup to owl meat to human blood. A Russian-French microbiologist, Ilya Mechnikov, believed that a life span of 150 years could be achieved

| FIVE LEADING CAUSES OF DEATH WORLDWIDE: 1900, 1986, AND 1993 | | |
|---|---|---|
| **1900** | **1986** | **1993** |
| 1. Pneumonia and influenza | 1. Heart disease | 1. Heart and blood vessel disease |
| 2. Tuberculosis | 2. Cancer | 2. Diarrheal diseases |
| 3. Diarrheal diseases | 3. Stroke | 3. Cancer |
| 4. Heart disease | 4. Chronic respiratory disease | 4. Pneumonia |
| 5. Stroke | 5. Injuries | 5. Tuberculosis |

TABLE 23.6

Source: Date from Centers for Disease Control and Prevention, Atlanta, GA.

with the help of a steady diet of milk cultured with bacteria. He thought that the bacteria would live in the large intestine and somehow increase the human life span. (He died at 71.) Ironically, many people have died in pursuit of a literal "fountain of youth."

The human *life span*—the length of time that a human can theoretically live—is 120 years. Of course, most people succumb to disease or injury long before that point. *Life expectancy* is a more realistic projection of how long an individual will live, based on epidemiological information. In the United States, life expectancy for males is 79 years and for females 83 years.

Life expectancy approaches life span as technology conquers diseases. Technology also alters the most prevalent killers. Development of antibiotic drugs removed some infectious diseases such as pneumonia and tuberculosis from the top of the "leading causes of death" list, a position that heart disease filled. Cancer is currently overtaking heart disease as the most common cause of death in developed nations. Infections remain a major cause of death in less-developed countries.

**1** Why is it difficult to sort out the causes of aging?

**2** How is aging a passive process?

**3** How is aging an active process?

**4** Distinguish between life span and life expectancy.

Medical advances have greatly contributed to improved life expectancy. Antibiotics have tamed some once-lethal infections, drugs enable many people with cancer to survive, and such advances as beta blocker drugs and coronary bypass surgery have extended the lives of people with heart disease. However, a look at the five leading causes of death in the world at three points in recent history—1900, 1986, 1993—sounds a warning against complacency (table 23.6). The top two killers in

# CLINICAL APPLICATION

## Old Before Their Time

The progerias are inherited disorders that cause a person to literally live a lifetime in just a few years. The Luciano brothers appear much older than their years (fig. 23E). They are aging at an accelerated pace on the inside as well.

In *Hutchinson-Gilford syndrome,* which affects the Luciano brothers, a child appears normal at birth, but by the first birthday, growth is obviously retarded. Within just a few years, the child acquires a shocking appearance of wrinkles, baldness, and the prominent facial features characteristic of advanced age. Arteries clog with fatty deposits. The child usually dies of a heart attack or a stroke by the age of 12, although some patients live into their 20s. Only a few dozen cases of this syndrome have ever been reported.

An adult form of progeria called *Werner syndrome* becomes apparent before the twentieth birthday. The person usually dies before age 50 of disorders usually associated with advanced age, such as diabetes mellitus, atherosclerosis, and osteoporosis. Curiously, they usually do not develop hypertension or Alzheimer's disease.

The cells of progeria patients show profound aging-related changes. Normal cells in culture reproduce only fifty times before

dying. Cells from progeria patients die in culture after only ten to thirty reproductions, as if they were programmed to die prematurely. Certain structures

seen in normal cultured cells as they near the fifty-reproduction limit (glycogen particles, lipofuscin granules, many lysosomes and vacuoles, and a few ribosomes) appear early in the cells of people with progeria. Understanding the mechanisms that cause these diseased cells to race through the aging process may help us to understand the biology of normal aging. ■

**FIGURE 23E**

The Luciano brothers inherited progeria and appear far older than their years.

1900, pneumonia and tuberculosis, are absent from the 1986 list. However, they appear again as numbers 4 and 5 in the 1993 list, in drug-resistant forms! The appearance of new or renewed infectious diseases, such as AIDS and measles, also indicates that we cannot yet conquer all killers. Although we can alter our environment more than other species can, there are still forces of nature beyond our control.

## CLINICAL TERMS RELATED TO HUMAN GROWTH AND DEVELOPMENT

**abruptio placentae** (ab-rup′shē-o plah-cen′ta)
    Premature separation of the placenta from the uterine wall.

**amniocentesis** (am″ne-o-sen-te′sis)   Technique in which a sample of amniotic fluid is withdrawn from the amniotic cavity by inserting a hollow needle through the pregnant woman's abdominal wall.

**dizygotic twins** (di″zi-got′ik twinz)   Twins resulting from the fertilization of two ova by two sperm cells.

**hydatid mole** (hi′dah-tid mōl)   Type of uterine tumor that originates from placental tissue.

**hydramnios** (hi-dram′ne-os)   Excess amniotic fluid.

**intrauterine transfusion** (in″trah-u′ter-in trans-fu′zhun)   Transfusion administered by injecting blood into the fetal peritoneal cavity before birth.

**lochia** (lo′ke-ah)   Vaginal discharge following childbirth.

**meconium** (mě-ko′ne-um)   Anal discharge from the digestive tract of a full-term fetus or a newborn.

**monozygotic twins** (mon″o-zi-got′ik twinz)   Twins resulting from one sperm cell fertilizing one egg cell, which then splits.

**perinatology** (per″ĭ-na-tol′o-je)   Branch of medicine concerned with the fetus after twenty-five weeks of development and with the newborn for the first four weeks after birth.

**postpartum** (pōst-par′tum)   Occurring after birth.

**teratology** (ter″ah-tol′o-je)   Study of substances that cause abnormal development and congenital malformations.

**trimester** (tri-mes′ter)   Each third of the total period of pregnancy.

**ultrasonography** (ul″trah-son-og′rah-fe)   Technique used to visualize the size and position of fetal structures from patterns of deflected ultrasonic waves.

## CHAPTER SUMMARY

## Introduction  (page 892)

Growth refers to an increase in size; development is the process of changing from one phase of life to another.

## Prenatal Period  (page 892)

1. Period of cleavage
   a. Fertilization occurs in a uterine tube and results in a zygote.
   b. The zygote undergoes mitosis, and the newly formed cells divide mitotically too.
   c. Each subsequent division produces smaller and smaller cells.
   d. A solid ball of cells (morula) forms, and it becomes a hollow ball called a blastocyst.
   e. The inner cell mass that gives rise to the embryo proper forms within the blastocyst.
   f. The blastocyst implants in the uterine wall.
      (1) Enzymes digest the endometrium around the blastocyst.
      (2) Fingerlike processes from the blastocyst penetrate into the endometrium.
   g. The period of cleavage lasts through the first week of development.
   h. The trophoblast secretes hCG, which helps maintain the corpus luteum, helps protect the blastocyst against being rejected, and stimulates the developing placenta to secrete hormones.
2. Embryonic stage
   a. The embryonic stage extends from the second through the eighth weeks.
   b. It is characterized by the development of the placenta and the main internal and external body structures.
   c. The cells of the inner cell mass fold inward, forming a gastrula that has two and then three primary germ layers.
      (1) Ectoderm gives rise to the nervous system, portions of the skin, the lining of the mouth, and the lining of the anal canal.
      (2) Mesoderm gives rise to muscles, bones, blood vessels, lymphatic vessels, reproductive organs, kidneys, and linings of body cavities.

      (3) Endoderm gives rise to linings of the digestive tract, respiratory tract, urinary bladder, and urethra.
   d. The embryonic disk becomes cylindrical and is attached to the developing placenta by the connecting stalk.
   e. The embryo develops head, face, upper limbs, lower limbs, and mouth, and appears more humanlike.
   f. Chorionic villi develop and are surrounded by spaces filled with maternal blood.
   g. The placental membrane consists of the epithelium of the villi and the epithelium of the capillaries inside the villi.
      (1) Oxygen and nutrients diffuse from the maternal blood through the placental membrane and into the fetal blood.
      (2) Carbon dioxide and other wastes diffuse from the fetal blood through the placental membrane and into the maternal blood.
   h. The placenta develops in the disk-shaped area where the chorion contacts the uterine wall.
      (1) The embryonic portion consists of the chorion and its villi.
      (2) The maternal portion consists of the uterine wall and attached villi.
   i. A fluid-filled amnion develops around the embryo.
   j. The umbilical cord is formed as the amnion envelops the tissues attached to the underside of the embryo.
      (1) The umbilical cord includes two arteries and a vein.
      (2) It suspends the embryo in the amniotic cavity.
   k. The chorion and amnion fuse.
   l. The yolk sac forms on the underside of the embryonic disk.
      (1) It gives rise to blood cells and cells that later form sex cells.
      (2) It helps form the digestive tube.
   m. The allantois extends from the yolk sac into the connecting stalk.
      (1) It forms blood cells.
      (2) It gives rise to the umbilical vessels.
   n. By the beginning of the eighth week, the embryo is recognizable as a human.

3. Fetal stage
   a. This stage extends from the end of the eighth week and continues until birth.
   b. Existing structures grow and mature; only a few new parts appear.
   c. The body enlarges, upper and lower limbs reach final relative proportions, the skin is covered with sebum and dead epidermal cells, the skeleton continues to ossify, muscles contract, and fat is deposited in subcutaneous tissue.
   d. The fetus is full term at the end of the ninth month, which equals approximately 266 days.
      (1) It is about 50 centimeters long and weighs 6–8 pounds.
      (2) It is positioned with its head toward the cervix.
4. Fetal blood and circulation
   a. Umbilical vessels carry blood between the placenta and the fetus.
   b. Fetal blood carries a greater concentration of oxygen than does maternal blood.
   c. Blood enters the fetus through the umbilical vein and partially bypasses the liver by means of the ductus venosus.
   d. Blood enters the right atrium and partially bypasses the lungs by means of the foramen ovale.
   e. Blood entering the pulmonary trunk partially bypasses the lungs by means of the ductus arteriosus.
   f. Blood enters the umbilical arteries from the internal iliac arteries.

## Postnatal Period  (page 913)

1. Neonatal period
   a. This period extends from birth to the end of the fourth week.
   b. The newborn must begin to respire, obtain nutrients, excrete wastes, and regulate its body temperature.
   c. The first breath must be powerful in order to expand the lungs.
      (1) Surfactant reduces surface tension.
      (2) A variety of factors stimulate the first breath.
   d. The liver is immature and unable to supply sufficient glucose, so the newborn depends primarily on stored fat for energy.
   e. Immature kidneys cannot concentrate urine very well.
      (1) The newborn may become dehydrated.
      (2) Water and electrolyte imbalances may develop.
   f. Homeostatic mechanisms may function imperfectly, and body temperature may be unstable.
   g. The circulatory system changes when placental circulation ceases.
      (1) Umbilical vessels constrict.
      (2) The ductus venosus constricts.
      (3) The foramen ovale is closed by a valve as blood pressure in the right atrium falls and pressure in the left atrium rises.
      (4) The ductus arteriosus constricts.
2. Infancy
   a. Infancy extends from the end of the fourth week to one year of age.
   b. Infancy is a period of rapid growth.
      (1) The muscular and nervous systems mature, and coordinated activities become possible.
      (2) Communication begins.

   c. Rapid growth depends on an adequate intake of proteins, vitamins, and minerals in addition to energy sources.
3. Childhood
   a. Childhood extends from the end of the first year to puberty.
   b. It is characterized by rapid growth, development of muscular control, and establishment of bladder and bowel control.
4. Adolescence
   a. Adolescence extends from puberty to adulthood.
   b. It is characterized by physiological and anatomical changes that result in a reproductively functional individual.
   c. Females may be taller and stronger than males in early adolescence, but the situation reverses in late adolescence.
   d. Adolescents develop high levels of motor skills, their intellectual abilities increase, and they continue to mature emotionally.
5. Adulthood
   a. Adulthood extends from adolescence to old age.
   b. The adult remains relatively unchanged physiologically and anatomically for many years.
   c. After age thirty, degenerative changes usually begin to occur.
      (1) Skeletal muscles lose strength.
      (2) The circulatory system becomes less efficient.
      (3) The skin loses its elasticity.
      (4) The capacity to produce sex cells declines.
6. Senescence
   a. Senescence is the process of growing old.
   b. Degenerative changes continue, and the body becomes less able to cope with demands placed upon it.
   c. Changes occur because of prolonged use, effects of disease, and cellular alterations.
   d. An aging person usually experiences losses in intellectual functions, sensory functions, and physiological coordinating capacities.
   e. Death usually results from mechanical disturbances in the cardiovascular system or from disease processes that affect vital organs.

## Aging  (page 922)

1. Passive aging
   a. Passive aging entails breakdown of structures and slowing or failure of functions.
   b. Connective tissue breaks down.
   c. DNA errors accumulate.
   d. Lipid breakdown in aging membranes releases lipofuscin.
   e. Free radical damage escalates.
2. Active aging
   a. In autoimmunity, the immune system attacks the body.
   b. Apoptosis is a form of programmed cell death. It occurs throughout life, shaping organs.
3. The human life span
   a. The theoretical maximum life span is 120 years.
   b. Life expectancy, based on real populations, is 79 years for males and 83 years for females.
   c. Medical technology makes life expectancy more closely approach life span.

# CRITICAL THINKING QUESTIONS

1. How would you explain the observation that twins resulting from a single fertilized egg cell can exchange body parts by tissue or organ transplant procedures without experiencing rejection reactions?
2. One of the more common congenital cardiac disorders is a ventricular septum defect in which an opening remains between the right and left ventricles. What problem would such a defect create as blood moves through the heart?
3. What symptoms may appear in a newborn if its ductus arteriosus fails to close?
4. What technology would enable a fetus born in the fourth month to survive in a laboratory setting?

5. Why is it important for a middle-aged adult who has neglected physical activity for many years to have a physical examination before beginning an exercise program?
6. What precautions do you think should be taken to avoid conflicts between "gestational" (surrogate) mothers and genetic mothers?
7. If an aged relative came to live with you, what special provisions could you make in your household environment and routines that would demonstrate your understanding of the changes brought on by aging processes?

# REVIEW EXERCISES

1. Define *growth* and *development.*
2. Describe the process of cleavage.
3. Distinguish between a morula and a blastocyst.
4. Describe the formation of the inner cell mass, and explain its significance.
5. Describe the process of implantation.
6. List three functions of hCG.
7. Explain how the primary germ layers form.
8. List the major body parts derived from ectoderm.
9. List the major body parts derived from mesoderm.
10. List the major body parts derived from endoderm.
11. Describe the formation of the placenta, and explain its functions.
12. Define *placental membrane.*
13. Distinguish between the chorion and the amnion.
14. Explain the function of amniotic fluid.
15. Describe the formation of the umbilical cord.
16. Explain how the yolk sac and the allantois are related, and list the functions of each.
17. Explain why the embryonic period of development is so critical.
18. Define *fetus.*
19. List the major changes that occur during the fetal stage of development.

20. Distinguish among preimplantation genetic diagnosis, chorionic villus sampling, and amniocentesis.
21. Describe a full-term fetus.
22. Compare the properties of fetal hemoglobin with those of adult hemoglobin.
23. Explain how the fetal circulatory system is adapted for intrauterine life.
24. Trace the pathway of blood from the placenta to the fetus and back to the placenta.
25. Distinguish between a newborn and an infant.
26. Explain why a newborn's first breath must be particularly forceful.
27. List some of the factors that stimulate the first breath.
28. Explain why newborns tend to develop water and electrolyte imbalances.
29. Describe the circulatory changes that occur in the newborn.
30. Describe the characteristics of an infant.
31. Distinguish between a child and an adolescent.
32. Define *adulthood.*
33. List some of the degenerative changes that begin during adulthood.
34. Define *senescence.*
35. List some of the factors that promote senescence.
36. Cite evidence of passive aging and active aging.

 **For More Information Related to This Chapter, Please Visit** *Hole's Human Anatomy and Physiology* **Home Page**
http://www.mhhe.com/biosci/abio/

# 24 CHAPTER

# GENETICS

## UNDERSTANDING WORDS

**alb-**, white: *alb*inism—condition in which normal skin pigments are lacking.

**centr-**, point, center: *centr*omere—region at which paired chromatids are held together.

**hem-**, blood: *hem*ophilia—condition in which bleeding is prolonged due to a defect in the blood clotting mechanism.

**hetero-**, other, different: *hetero*zygous—condition in which the members of a gene pair are different.

**hom-**, same, common: *hom*ologous chromosomes—pair of chromosomes that contain similar genetic information.

**phen-**, show, be seen: *phen*otype—physical appearance that results from the way genes are expressed in an individual.

**syn-**, together: *syn*apsis—process by which homologous chromosomes become tightly intertwined.

**tri-**, three: *tri*somy—condition in which one kind of chromosome is triply represented.

## CHAPTER OBJECTIVES

AFTER YOU HAVE STUDIED THIS CHAPTER, YOU SHOULD BE ABLE TO:

1. Explain how gene discoveries are relevant to the study of anatomy and physiology and to health care.

2. Distinguish between genes and chromosomes.

3. Outline the process of meiosis and explain why meiotic products have different combinations of parental genes.

4. Define the two types of chromosomes.

5. Explain how genes can have many alleles (variants), but a person can have only two alleles of a particular gene.

6. Describe how alleles interact.

7. Distinguish between modes of inheritance by which traits pass from parents to offspring.

8. Explain how gene expression varies among individuals.

9. Describe how multiple genes and the environment interact to produce complex traits.

10. Describe how genes and chromosomes control the inheritance of gender and how traits are transmitted on the X chromosome.

11. Describe how gender affects gene expression.

12. Explain how deviations in chromosome number or arrangement can harm health.

13. Explain how conditions caused by extra or missing chromosomes reflect a meiotic error.

14. Describe the challenges that the ability to diagnose inherited disease presents.

15. Explain how gene therapy works.

## KEY TERMS

allele (ah-lēl′)
autosome (aw′to-sŏm)
dominant (dom′i-nant)
gene (jēn)
genetics (jĕ-net′iks)
genotype (je′no-tīp)
heredity (he-red′i-te)
heterozygous (het″er-o-zi′gus)

homozygous (ho″mo-zi′gus)
meiosis (mi-o′sis)
multiple allele (mul′ti-pl ah-lēl′)
mutation (mu-ta′shun)
nondisjunction (non″dis-jungk′shun)
phenotype (fe′no-tīp)
recessive (re-ses′iv)
sex chromosome (seks kro′mo-sōm)

The year is 2010, and the young pregnant woman has blood drawn for a routine prenatal exam. Among her cells are a few immature red blood cells (erythroblasts) from the fetus. These will be separated out and then scrutinized for clues to the health of the person-to-be. An initial peek at the chromosomes, using a fluorescent dye that homes in on the Y chromosome denoting maleness, shows that the fetus is male. At this news, the woman and her mate decide to name him Orwell. A closer look at the specific genes comprising his chromosomes reveals much more.

Happily, Orwell has not inherited any of the more common genetic disorders, nor has he inherited the gene that caused the early senility of Alzheimer's disease in a grandparent. Despite Orwell's apparently healthy background, some of his blood cells in the umbilical cord will be set aside and stored at birth, should he need a bone marrow transplant later in life. The information from the prenatal test also indicates that Orwell has inherited certain potentially unhealthy characteristics, but he can minimize their effects, and perhaps extend his life, by modifying his lifestyle in particular ways.

A panel of tests to type the genes whose protein products could predispose Orwell to heart disease—clotting factors, apoproteins, angiotensin-converting enzyme—make it clear that a lifetime of eating low-fat foods combined with regular exercise can add years to his life. The same measures may help prevent, or delay, development of colon cancer, for which he has inherited a pair of susceptibility genes. Finally, Orwell's parents know that their son should avoid drinking alcohol, smoking cigarettes, and abusing drugs, because a neurotransmitter receptor variant that strongly correlates to addictive behavior is also a part of his genetic makeup.

In 1990, when Cynthia Cutshall was eight years old, she became the first person to receive gene therapy. At the National Institutes of Health in Bethesda, Maryland, she received an infusion of her own white blood cells, bolstered with the gene specifying the enzyme adenosine deaminase, which her cells lacked due to an inborn error of metabolism. Correcting the genetic defect helped her body to develop an immune system, and as this photo taken 3 years later shows, she has done quite well.

## THE EMERGING ROLE OF GENETICS IN MEDICINE

Orwell's scenario takes place in the future, but each test described is possible today. The avalanche of genetic discoveries from the human genome project (see chapter 4) is providing a wealth of new diagnostic tests and therapies, for rare as well as common disorders. **Genetics,** the study of inheritance, will play an ever-increasing role in future health care.

New genetic information has explained many physiological processes at the cellular and molecular levels, as we have seen throughout this book. Figure 24.1 recaps some of these examples along a timeline, with the numbers referring to chapters where the disorder is discussed. For further descriptions of the disorders mentioned, see table 24.6 at the chapter's end.

Genes encode traits other than illness, including eye and hair color, singing ability, and hard-to-define personality traits. Clinical Application 24.1 highlights a few interesting human inherited traits.

**Heredity** is the transmission of genetic information from parents to offspring. This transfer occurs through DNA molecules in the nuclei of eggs and sperm that provide chemical instructions for human development. The environment, however, often influences how genes are expressed. The environment includes the chemical, physical, social, and biological factors surrounding an individual that influence his or her characteristics. For example, a person inheriting a gene conferring susceptibility to lung cancer may never develop the illness if he or she breathes clean air and never smokes.

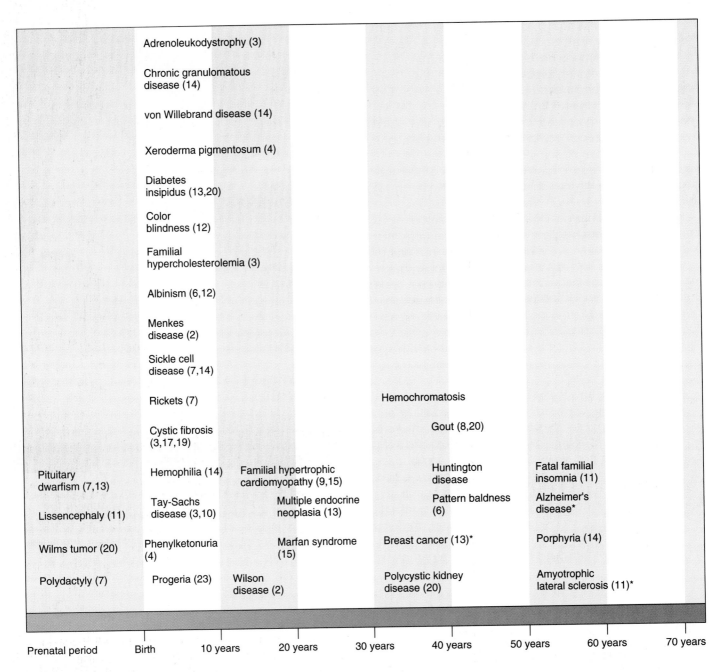

Adrenoleukodystrophy (3)

Chronic granulomatous disease (14)

von Willebrand disease (14)

Xeroderma pigmentosum (4)

Diabetes insipidus (13,20)

Color blindness (12)

Familial hypercholesterolemia (3)

Albinism (6,12)

Menkes disease (2)

Sickle cell disease (7,14)

Rickets (7)                    Hemochromatosis

Cystic fibrosis (3,17,19)                    Gout (8,20)

| | | | | |
|---|---|---|---|---|
| Pituitary dwarfism (7,13) | Hemophilia (14) | Familial hypertrophic cardiomyopathy (9,15) | Huntington disease | Fatal familial insomnia (11) |
| Lissencephaly (11) | Tay-Sachs disease (3,10) | Multiple endocrine neoplasia (13) | Pattern baldness (6) | Alzheimer's disease* |
| Wilms tumor (20) | Phenylketonuria (4) | Marfan syndrome (15) | Breast cancer (13)* | Porphyria (14) |
| Polydactyly (7) | Progeria (23) | Wilson disease (2) | Polycystic kidney disease (20) | Amyotrophic lateral sclerosis (11)* |

Prenatal period · Birth · 10 years · 20 years · 30 years · 40 years · 50 years · 60 years · 70 years

* These conditions may also result from nongenetic causes.

**FIGURE 24.1**
Genetic disorders described throughout this book are arranged along a timeline of a human lifetime, illustrating that many conditions affect children, but some may strike before birth or well into adulthood. Numbers in parentheses indicate chapters where each disorder is discussed.

## GENES AND CHROMOSOMES

A **gene** is the nucleotide sequence of a DNA molecule that tells a cell how to construct a specific protein molecule. A gene may also control (turn on or off) another gene, or encode an rRNA or tRNA molecule. Gene products—proteins—ultimately produce the trait associated with the gene, as figure 24.2 illustrates for the common inherited illness *cystic fibrosis* (CF). We will return to this condition throughout the chapter to illustrate various concepts.

All of the genes in a human cell—estimated to number about 70,000—constitute the human **genome.** These genes are distributed among 23 pairs of chromosomes. Recall from chapter 3 that *chromosomes* consist of proteins (histones) as well as a continuous, highly wound

(a)

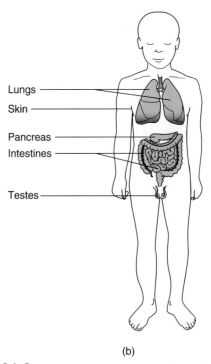

(b)

**FIGURE 24.2**

From gene to protein to person. (*a*) The gene encoding the CFTR protein, and causing cystic fibrosis when mutant, is on the seventh largest chromosome. CFTR protein folds into a channel that regulates the flow of chloride ions into and out of cells lining the respiratory tract, pancreas, intestines, and possibly elsewhere. (*b*) In cystic fibrosis, the CFTR protein is abnormal, usually missing an amino acid. Its shape is altered, which entraps the chloride ions inside cells. Water entering these cells leaves behind very thick mucus and other secretions in the places highlighted in the illustration. The sticky secretions cause the symptoms of the illness.

Source: Data from M. C. Iannuzi and F. S. Collins, "Reverse Genetics and Cystic Fibrosis" in *American Journal of Respiratory Cellular and Molecular Biology,* 2:309–316, 1990.

thread of DNA. Chromosomes appear as rod-shaped bodies in the nucleus when a cell reproduces (fig. 24.3). The number of chromosomes within cells varies with species. A human fertilized egg contains 46 chromosomes, 23 from the sperm and 23 from the egg.

## CHROMOSOMES AND SEX CELL PRODUCTION

Siblings, except for identical twins, are never exactly alike physically. The reason for this lies in **meiosis,** the special form of cell reproduction that produces sperm and eggs, which are also called *gametes.* An individual's gametes (sperm or eggs) differ from each other in that each contains a unique combination of the person's genes. That is, a man's sperm have variable combinations of genes, as do a woman's eggs.

## Meiosis

Meiosis is the part of *spermatogenesis* and *oogenesis* that halves the chromosome number (see chapter 22). This process includes two successive divisions, called the *first* and *second meiotic divisions.* The first meiotic division (meiosis I) separates homologous chromosome pairs. Homologous pairs are the same, gene for gene. At this stage, each homologous chromosome is replicated, so that it consists of two complete DNA strands called *chromatids.* The chromatids of a replicated chromosome are attached at special regions called the *centromeres.* Each of the cells that undergo the second meiotic division (*meiosis II*) have one representative of each homologous pair, and are thus haploid (having one copy of each chromosome). This second division separates chromatids, producing cells that are still haploid, but whose chromosomes are no longer replicated. After meiosis II, each of the chromatids is considered a chromosome.

# It's All in the Genes

Do you have uncombable hair, misshapen toes or teeth, a pigmented tongue tip, or an inability to smell skunk? Do you lack teeth, eyebrows, eyelashes, nasal bones, thumbnails, or fingerprints? Can you roll your tongue or wiggle your ears? If so, you may find your unusual trait described in a book called *Mendelian Inheritance in Man,* compiled by Johns Hopkins University geneticist Victor McKusick, which catalogs more than 5,000 known human genetic variants. Most of the entries include family histories, clinical descriptions, and how the trait is transmitted. Amidst the medical terminology can be found some fascinating inherited traits in humans, from top to toes.

Genes control whether hair is blond, brown, or black, whether or not it has red highlights, and whether it is straight, curly, or kinky. Widow's peaks, cowlicks, a whorl in the eyebrow, and white forelocks run in families, as do hairs with triangular cross sections. Some people have multicolored hairs like cats, and others have hair in odd places, such as on the elbows, nosetip, knuckles, palms of the hands, or soles of the feet. Teeth can be missing or extra, protuberant or fused, present at birth, or "shovel-shaped" or "snow-capped." One can have a grooved tongue, duckbill lips, flared ears, egg-shaped pupils, three rows of eyelashes, spotted nails, or "broad thumbs and great toes." Extra breasts have been observed in humans and guinea pigs, and one family's claim to fame is a double nail on the littlest toes.

Unusual genetic variants can affect metabolism, sometimes resulting in disease or other times producing harmless yet quite noticeable effects. Members of some families experience "urinary excretion of odoriferous component of asparagus" or "urinary excretion of beet pigment" after eating the vegetables in question. In "blue diaper syndrome," an infant's inherited inability to break down an amino acid results in urine that turns blue on contact with air.

One father and son were plagued with an inability to open their mouths completely. Some families suffer from "dysmelodia," the inability to carry a tune. Those who have inherited "odor blindness" are unable to smell either musk, skunk, cyanide, or freesia flowers. Motion sickness, migraine headaches, and stuttering may be inherited. Uncontrollable sneezing may be due to inherited hayfever or to Achoo syndrome (an acronym for "autosomal dominant compelling helioophthalmic outburst syndrome").

Perhaps the most bizarre inherited illness is the "jumping Frenchmen of Maine syndrome." This is an exaggerated startle reflex that was first noted among French-

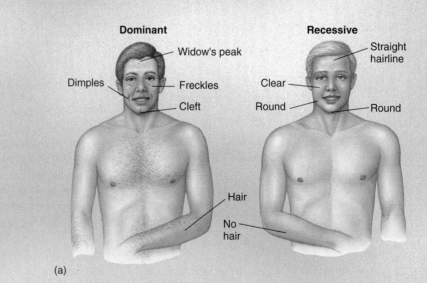

(a)

**FIGURE 24A**

Inheritance of some common traits: (*a*) Freckles, dimples, widow's peak, hairy elbows, and a cleft chin, (*b*) the ability to bend the thumb backward or forward, (*c*) widow's peak, and (*d*) attached or unattached earlobes.

Canadian lumbermen from the Moosehead Lake area of Maine, whose ancestors were from the Beauce region of Quebec. The disorder was first described at a meeting of the American Neurological Association in 1878 and was confirmed with the aid of videotape in 1980. McKusick describes the disorder as follows: "If given a short, sudden, quick command, the affected person would respond with the appropriate action, often echoing the words of command. . . . For example, if one of them was abruptly asked to strike another, he would do so without hesitation, even if it was his mother and he had an ax in his hand." This condition may be an extreme form of Tourette syndrome, which is common and causes people to utter animal noises or obscenities uncontrollably.

Figure 24A illustrates some more common genetic quirks. ■

(b)

(c)

(d)

**FIGURE 24.3**

(a) Chromosomes consist of a continuous DNA double helix and associated proteins, and are located in the cell's nucleus. (b) A transmission electron micrograph of a chromosome in its replicated form, just prior to cell division. Each half of the chromosome is a chromatid. Note the constriction, where the centromeres are attached (25,000×).

The steps of this complex process are clearer when considered in order. However, keep in mind that, like mitosis, meiosis is a continuous process. Considering it in steps simply makes it easier to follow. Refer to figure 24.4 throughout the following discussion.

1. **Prophase I.** Individual chromosomes appear as thin threads within the nucleus, then shorten and thicken. Nucleoli disappear, the nuclear membrane temporarily disassembles, and microtubules begin to build the spindle that will separate the

**FIGURE 24.4**

Stages in the first meiotic division.

**FIGURE 24.5**

(a) Pairing of homologous chromosomes, (b) chromatids crossing over, (c) results of crossing over.

chromosomes. The chromosomes have already been replicated.

As prophase continues, homologous chromosomes attract, lie side by side, and tightly intertwine. During this pairing, called *synapsis,* the chromatids of the homologous chromosomes contact one another at various points. Often, the chromatids break in one or more places and exchange parts, forming chromatids with new combinations of genetic information (fig. 24.5). Since one chromosome of a homologous pair is from a person's mother and the other is from the father, an exchange, or **crossover,** between

homologous chromosomes produces chromatids that contain genetic information from both parents.

2. **Metaphase I.** During the first metaphase, chromosome pairs line up about midway between the poles of the developing spindle, and they are held under great tension, like two groups of people playing tug-of-war. Each chromosome pair consists of two chromosomes, which equals four chromatids. Each chromosome attaches to spindle fibers from one pole. The chromosome alignment is random with respect to maternal and paternal origin of the chromosomes. That is, each of the 23 chromosomes contributed from the mother may be on the left or the right, and the same is true for the paternal chromosomes—it is similar to the number of ways that 23 pairs of children could line up, while maintaining the pairs. Chromosomes can line up with respect to each other in many, many ways.

3. **Anaphase I.** Homologous chromosome pairs separate, and each replicated member moves to one end of the spindle. Thus, each new cell receives only one replicated member of a homologous pair of chromosomes, reducing the chromosome number by one-half.

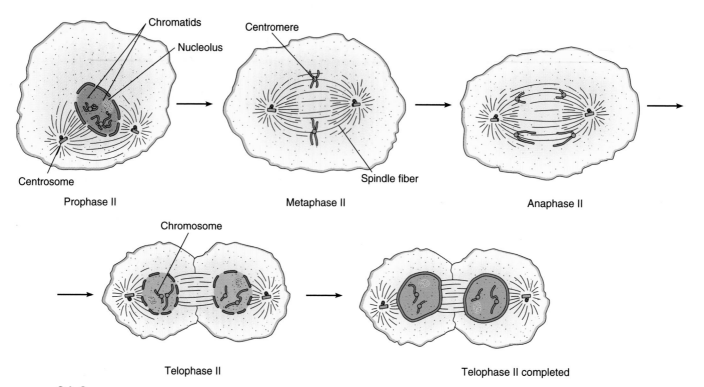

Chromatids
Nucleolus
Centromere

Centrosome

Prophase II

Metaphase II

Spindle fiber

Anaphase II

Chromosome

Telophase II

Telophase II completed

**FIGURE 24.6**

Stages in the second meiotic division.

4. **Telophase I.** The original cell divides in two. Nuclear membranes form around the chromosomes, nucleoli reappear, and the spindle fibers disassemble into their constituent microtubules. Then the second meiotic division begins.

Meiosis II (fig. 24.6) is very similar to a mitotic division. During *prophase II,* chromosomes condense so that they reappear, still replicated. They move into positions midway between the poles of the developing spindle. In *metaphase II,* the replicated chromosomes attach to spindle fibers. In *anaphase II,* centromeres separate, freeing the chromatids to move to opposite poles of the spindles. The former chromatids are now considered to be chromosomes. In *telophase II,* each of the two cells resulting from meiosis I divides to form two cells. Therefore, each cell undergoing meiosis has the potential to produce four gametes. In males, the gametes mature into four sperm cells. In females, three of the products of meiosis are "cast aside" as polar bodies and one cell becomes the egg.

## Meiosis Leads to Genetic Variability

Meiosis generates astounding genetic variety. Any one of a person's more than 8 million possible combinations of 23 chromosomes can combine with any one of the more than 8 million combinations of his or her mate, raising the potential variability to more than 70 trillion geneti-

cally unique individuals! Crossing over contributes even more genetic variability. Figure 24.7 illustrates in a simplified manner how maternal and paternal traits reassort during meiosis.

1 What are genes and chromosomes?

2 What is the role of meiosis in the transmission of traits?

3 Describe the major events that occur during meiosis.

4 How does meiosis provide genetic variability?

## Chromosomes and Genes Come in Pairs

From the moment of conception, a human cell is *diploid,* containing two copies of each of the 23 different chromosomes. Chromosome charts called **karyotypes** display the 23 chromosome pairs in size order (fig. 24.8). Pairs 1 through 22 are **autosomes,** which are chromosomes that do not determine sex. The other two chromosomes, the X and the Y, determine sex and are called **sex chromosomes.** They will be discussed later in the chapter. Computerized karyotype devices scan microscope slides of white blood cell preparations, and align the chromosomes into a size-order chart (fig. 24.9).

Since we have two copies of each chromosome, we also have two copies of each gene, each located at the same position on the chromosome. Sometimes the mem-

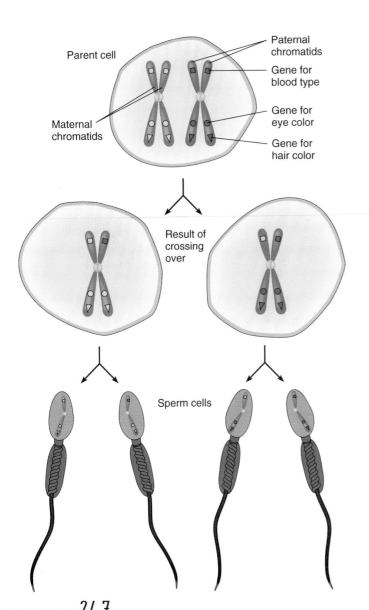

**FIGURE** 24.7

As a result of crossing over, the genetic information in sperm cells and egg cells varies from cell to cell.

**FIGURE** 24.8

A normal human karyotype has the 22 pairs of autosomes aligned in size order, plus the sex chromosomes. In this karyotype, fluorescently tagged pieces of DNA are used as "probes" to bind to specific chromosomes, imparting vibrant colors. This technique is called FISH, which stands for fluorescence in situ hybridization.

**FIGURE** 24.9

Inheritance of cystic fibrosis from carrier parents illustrates autosomal recessive inheritance. Sexes are affected with equal frequency, and symptoms usually begin in early childhood. Autosomal recessive conditions can skip generations.

bers of a gene pair are alike, their DNA sequences specifying the same amino acid sequence of the protein product. However, because a gene consists of hundreds of nucleotide building blocks, it may exist in several variant forms, called **alleles.** Thus, although homologous pairs of chromosomes are the same, gene for gene, they may have different alleles of those genes. An individual who has two identical alleles of a gene is said to be **homozygous** for that gene. A person with two different alleles for a gene is said to be **heterozygous** for it. A gene may have a huge number of alleles, but an individual person can only have two alleles for a particular gene. Clinical Application 24.2 discusses how localizing genes to particular chromosomes is providing new types of medical tests.

# Maps, Markers, and Medicine

Geneticists have "mapped" genes to particular chromosomes since early in the twentieth century. Until the 1980s, most maps were constructed indirectly. Working with fruit flies, researchers observed that two traits frequently occurred together, or became separated by crossing over. The closer two genes were located to each other along a chromosome, the less often crossing over occurred between them, simply because there was less space. Researchers took such crossover frequencies and converted them into map distances, constructing intricate "linkage" maps. To do this in humans required studying very large families, and so was not usually possible.

Today, researchers use a variety of biochemical techniques to first localize a gene to a particular part of a specific chromosome, and then to actually obtain the DNA sequence of the gene. Often such studies begin with people who have a particular inherited illness, and a chromosome that is broken at a particular point, leading researchers to the probable site of the disease-causing gene.

As genes associated with a particular chromosome accumulate, researchers build maps, called ideograms, to depict gene locations on the chromosome. Figure 24B shows an ideogram for the entire human genome. These "physical" maps tend to agree with the gene orders established with older linkage maps. Each October, *Science* magazine publishes ideograms for all human chromosomes, tracking progress of the human genome project. Each year, the ideograms become much more crowded.

Often discovery of a disease-causing gene in humans is preceded by detecting a genetic marker, which is a DNA sequence located very close to the gene of interest on its chromosome. Within a family, the presence of the marker indicates a high probability that the disease-causing gene has also been inherited, even if symptoms have not yet appeared. The first marker discovered was in 1983, for *Huntington disease*, which causes neurological degeneration in middle age. It took 10 years to "walk" down chromosome 4 to finally find the disease-causing gene. However, a marker for an inherited form of breast cancer was found in 1993, and the causative gene was discovered just a year later.

Knowing the sequences of genetic markers and disease-causing genes makes possible presymptomatic diagnoses, and sometimes even life-saving treatments. For example, researchers tested children who had lost a parent or sibling to a rare form of thyroid cancer and removed the thyroids of those who had inherited the gene and were therefore destined, with 100% certainty, to develop the cancer. This intervention saved lives.

The health practitioners of tomorrow will increasingly face situations where genetic tests indicate a future illness in a presently healthy person, or will help them to determine a precise diagnosis and thus better tailor treatment. People preparing to enter health professions must now take more courses in genetics, to enable them to integrate future genetic tests into medical practice. ■

1. Gaucher disease
2. familial colon cancer*
3. retinitis pigmentosa*
4. Huntington disease
5. familial polyposis of the colon
6a. spinocerebellar ataxia
6b. hemochromatosis

7. cystic fibrosis
8. multiple exostoses*
9. malignant melanoma
10. multiple endocrine neoplasia, type 2
11. sickle cell disease
12. PKU (phenylketonuria)

13. retinoblastoma
14. Alzheimer's disease*
15. Tay-Sachs disease
16. polycystic kidney disease
17. breast cancer*
18. amyloidosis
19a. familial hypercholesterolemia
19b. myotonic dystrophy

20. ADA immune deficiency
21. amyotrophic lateral sclerosis*
22. neurofibromatosis, type 2
Xa. muscular dystrophy
Xb. hemophilia

*Gene responsible for only some cases.

FIGURE 24B

Each year the human genome map becomes more crowded, as researchers map genes to chromosomes, then decipher the genes' roles in human anatomy and physiology, in sickness and in health. In this depiction of the human genome, the circles indicate the sites of specific genes on chromosomes.

The allele that causes most cases of cystic fibrosis was discovered in 1989, and researchers immediately began developing a test to detect it. However, other alleles were soon discovered. Today, hundreds of mutations (changes) in the cystic fibrosis gene are known. Different allele combinations produce different combinations and severities of symptoms, although most cases are quite serious. Inherited breast cancer is also caused by a gene that has many different alleles.

**1** What is the relationship between genes and chromosomes?

**2** What is an allele?

**3** What does it mean to be homozygous or heterozygous for a particular gene?

## MODES OF INHERITANCE

The particular combination of genes present in a person's cells constitutes the **genotype.** The appearance or health condition of the individual that develops as a result of the ways the genes are expressed is termed the **phenotype.** An allele is **wild-type** if its associated phenotype is either normal function or the most common expression in a particular population. An allele that is a change from wild-type, perhaps producing an uncommon phenotype, is **mutant.** Disease-causing alleles are mutant.

### Dominant and Recessive Inheritance

Often in heterozygotes, one allele determines the phenotype. Such an allele whose action masks that of another allele is termed **dominant.** The allele whose expression is masked is **recessive.** A gene that causes a disease can be recessive or dominant. It may also be *autosomal* (carried on a nonsex chromosome) or *sex-linked* (carried on the X chromosome). The Y chromosome carries only a few genes.

Whether a trait is dominant or recessive, autosomal or sex-linked is called its *mode of inheritance.* This designation has important consequences in predicting the chance that offspring will inherit an illness or trait. These rules emerge:

1. An autosomal condition is equally likely to affect both sexes. Sex-linked characteristics affect males much more often than females, a point we will return to later in the chapter.

2. A person most likely inherits a recessive condition from two parents who are each heterozygotes (carriers). The parents are usually healthy. For this reason, recessive conditions can "skip" generations.

3. A person who inherits a dominant condition has at least one affected parent. Therefore, dominant conditions do not skip generations. (An exception is if the dominant allele arises, as a new mutation, in the sperm or egg.)

Cystic fibrosis (CF) is an example of an autosomal recessive disorder. The wild-type allele for the CFTR gene, which is dominant over the disease-causing allele, specifies formation of protein chloride channels on cells lining the pancreas, respiratory tract, intestine, testes, and elsewhere (see fig. 24.2). Disease-causing recessive mutant alleles disrupt the structure and possibly the function of the chloride channels. An individual who inherits two such mutant alleles has cystic fibrosis. He or she is homozygous recessive. A person inheriting only one recessive mutant allele plus a dominant wild-type allele is a carrier, and transmits the disease-causing allele in half of the gametes. A person who has two wild-type alleles is homozygous dominant for the gene. Note that although three genotypes are possible, there are two phenotypes, because carriers and homozygous dominant individuals are healthy.

Using logic, understanding how chromosomes and genes are apportioned into gametes in meiosis, and knowing that the allele that causes cystic fibrosis is autosomal recessive, we can predict genotypes and phenotypes of the next generation. Figure 24.9 illustrates two people who are each heterozygous for a CF-causing allele. Half of the man's sperm contain the mutant allele, as do half of the woman's eggs. Since sperm and eggs combine at random, each child who these people conceive has a

- 1/4 chance of inheriting two wild-type alleles (homozygous dominant, healthy, and not a carrier)

- 1/2 chance of inheriting a mutant allele from either parent (heterozygous and a carrier, but healthy)

- 1/4 chance of inheriting a mutant allele from each parent (homozygous recessive, has CF)

In an autosomal recessive illness, an affected person's parents are carriers—they do not have the illness. In an autosomal dominant condition, however, an affected person typically has an affected parent. He or she need inherit only one copy of the mutant allele to have the associated phenotype; in contrast, expression of an autosomal recessive condition requires two copies of the mutant allele.

An example of an autosomal dominant condition is Huntington disease. Symptoms usually begin in the late 30s or early 40s, and include loss of coordination, uncontrollable dancelike movements, and personality changes. Figure 24.10 shows the inheritance pattern. Here, if one parent has the mutant allele, half of his or her gametes will have it. Assuming the other parent does not have a mutant allele, each child conceived has a 1 in 2 chance of inheriting the gene and, eventually, the condition.

Most of the 5,000 or so known human inherited disorders are autosomal recessive. These conditions tend to produce symptoms very early in life, sometimes before birth. Autosomal dominant conditions often have an adult onset.

Why would recessive alleles that cause illness remain in a population, if they so endanger health? One reason may be that carrying a genetic disease can protect against an infectious disease. The basis of this phenomenon lies in anatomy and physiology. For example, carriers of *sickle cell disease,* an inherited anemia, do not contract malaria. In sickle cell disease, a tiny genetic alteration causes the gene's product—the beta globin chain of hemoglobin—to aggregate, bending the red blood cell containing it into a sickle shape that can block blood flow when oxygen level is low. Carriers have only a few sickled cells, but these apparently are enough to somehow make red blood cells inhospitable to malaria parasites.

Similarly, carriers of cystic fibrosis resist diarrheal disorders, in which bacterial toxins open chloride channels in the small intestine. Carriers of CF have some abnormal chloride channels, which renders these toxins ineffective.

Carriers of *Tay-Sachs disease* may resist tuberculosis. This association was first noted in the Jewish ghettos of World War II, where some healthy relatives of children who died of Tay-Sachs disease did not contract tuberculosis, when many people around them did. Tay-Sachs disease causes excess myelin to accumulate on nerve cells. How this protects against tuberculosis is not known.

1  Distinguish between genotype and phenotype.

2  Distinguish between wild-type and mutant.

3  How do the modes of transmission of autosomal recessive and autosomal dominant inheritance differ?

## Different Dominance Relationships

Most genes exhibit complete dominance or recessiveness. Interesting exceptions are incomplete dominance and codominance. In **incomplete dominance,** the heterozygous phenotype is intermediate between that of either homozygote. For example, in **familial hypercholes-**

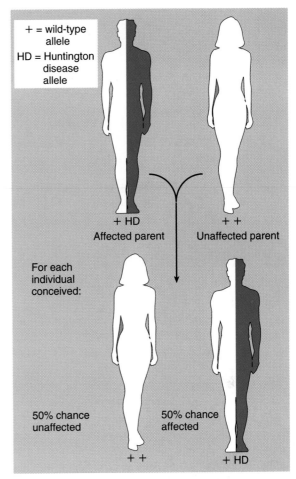

**FIGURE 24.10**

Inheritance of Huntington disease from a parent who will be affected in middle age illustrates autosomal dominant inheritance. Sexes are affected with equal frequency. Autosomal dominant conditions affect every generation until, by chance, none of the children in a generation inherits the mutant allele.

**terolemia** (FH), a person with two disease-causing alleles completely lacks LDL (low-density lipoprotein) receptors on liver cells that take up cholesterol from the bloodstream. A person with one disease-causing allele has half the normal number of cholesterol receptors. Someone with two wild-type alleles has the normal number of receptors. The associated phenotypes parallel the number of receptors—those with two mutant alleles die as children of heart attacks, individuals with one mutant allele die in young or middle adulthood, and people with two wild-type alleles do not develop this type of hereditary heart disease.

Incomplete dominance is seen in carriers of sickle cell disease at the cellular level. Some of their hemoglobin molecules are abnormal, and under low oxygen conditions, some of their red blood cells may sickle (fig. 24.11).

(a)

(b)

## FIGURE 24.11

(*a*) Falsely colored scanning electron micrograph of normal red blood cells (3,500×). (*b*) Red blood cells from a person with sickle cell disease (3,500×). A carrier (heterozygote) for sickle cell disease would have a few sickled red blood cells.

Sickle cell disease produces symptoms resulting from blocked circulation and impaired oxygen delivery to tissues: severe joint and abdominal pain, skin ulcers, chronic kidney disease, and frequent infections. Being a carrier of sickle cell disease very rarely, if ever, produces symptoms. However, in the early 1970s, when many people were tested for the disorder, people who were carriers were said to have "sickle cell trait." Because this implied that they were ill, the name led to discrimination. Today, most newborns are tested for sickle cell disease, because early treatment with penicillin can prevent severe infections and sometimes forestall symptoms. A bone marrow transplant can cure sickle cell disease, but it is a very risky procedure.

Different alleles that are both expressed in a heterozygote are **codominant.** For example, two of the three alleles of the I gene, which determines ABO blood type, are codominant. People of blood type A have a molecule called antigen *A* on the surfaces of their red blood cells. Blood type B corresponds to red blood cells with antigen *B*. A person with type AB has red blood cells with both the *A* and *B* antigens, and the red cells of a person with type O blood have neither antigen.

The I gene encodes the enzymes that place the *A* and *B* antigens on red blood cell surfaces. The three alleles are $I^A$, $I^B$, and $i$. People with type A blood may be either genotype $I^A I^A$ or $I^A i$; type B corresponds to $I^B I^B$ or $I^B i$; type AB to $I^A I^B$; and type O to $ii$.

## GENE EXPRESSION

### Penetrance and Expressivity

The same allele combination can produce different degrees of the phenotype in different individuals, even siblings, because of outside influences such as nutrition, toxic exposures, other illnesses, and the activities of other genes. Most disease-causing allele combinations are **completely penetrant,** which means that everyone who inherits it has some symptoms. A genotype is **incompletely penetrant** if some individuals do not express the phenotype. Polydactyly, having extra fingers or toes, is incompletely penetrant (see fig. 7.48). Some people who inherit the dominant allele have more than five digits on a hand or foot, yet others who are known to have the allele (because they have an affected parent and child) have the normal number of fingers and toes. The penetrance of a gene is described numerically. If 80 of 100 people who have inherited the dominant polydactyly allele have extra digits, the allele is 80% penetrant. Inherited breast cancer is also incompletely penetrant—that is, not all women who inherit a mutant form of a gene called BRCA1 develop the cancer, but they do have a much greater risk of doing so than a woman who inherits the wild-type allele.

A phenotype is **variably expressive** if the symptoms vary in intensity in different people. One person with polydactyly might have an extra digit on both hands and a foot; another might have two extra digits on both hands and both feet; a third person might have just one extra fingertip on a finger. *Penetrance* refers to the all-or-none expression of a genotype in an individual; *expressivity* refers to the severity of a phenotype.

## Pleiotropy

Some genetic disorders produce several symptoms. Family members who have different symptoms can appear to have different illnesses. This phenomenon is called **pleiotropy.**

Pleiotropy is seen in genetic diseases that affect a single protein found in different parts of the body. This is the case for *Marfan syndrome,* an autosomal dominant defect in an elastic connective tissue protein called fibrillin. The protein is abundant in the lens of the eye, in the aorta, and in the bones of the limbs, fingers, and ribs. This molecular-level abnormality explains the Marfan syndrome symptoms of lens dislocation, long limbs, spindly fingers, and a caved-in chest. The most serious symptom is a life-threatening weakening in the aorta wall, which sometimes causes the vessel to suddenly burst. However, if the weakening is found early, a synthetic graft can be used to patch that part of the vessel wall, saving the person's life.

| TABLE 24.1 | POLYGENIC MODEL OF SKIN COLOR INHERITANCE | |
|---|---|---|
| **Genotype** | | **Number of Dominant Pigment Alleles** |
| AABBCC | | 6 |
| AaBBCC, AABbCC, AABBCc | | 5 |
| aaBBCC, AAbbCC, AABBcc, AaBbCC, AaBBCc, AABbCc | | 4 |
| AaBbCc, aaBbCC, AAbbCc, AabbCC, AABbcc, aaBBCc, AaBBcc | | 3 |
| AaBbcc, AabbCc, aaBbCc, AAbbcc, aaBBcc, aabbCC | | 2 |
| Aabbcc, aaBbcc, aabbCc | | 1 |
| aabbcc | | 0 |

Studies that classify skin color by measuring the amount of light reflected from the skin surface suggest that three or four or more different genes, probably with several alleles each, produce pigment in the skin. The greater the number of pigment-specifying genes, the darker the skin. The inheritance of skin color is most likely even more complicated than this model of three genes.

The American Revolution has been indirectly blamed on a pleiotropic illness, *porphyria,* that affected British King George III's ability to rule. At age 50 he first experienced recurring cycles of symptoms, including abdominal pain, red urine, fever, and weakness, then insomnia, headache, visual disturbances, restlessness, delirium, convulsions, and stupor. The symptoms always appeared in this order. He was eventually dethroned because of his illness.

The king's crimson urine led twentieth-century physicians to diagnose porphyria. In this inborn error of metabolism, the organic portions of the hemoglobin molecule that surround iron atoms, called porphyrin rings, are excreted in the urine instead of being broken down in cells. Looking back at the royal family tree, researchers identified several relatives who experienced only some of the symptoms that plagued George III. One had red urine, another abdominal pain. The many guises of this pleiotropic illness had baffled physicians for decades.

## Genetic Heterogeneity

The same phenotype may result from the actions of different genes, a phenomenon called **genetic heterogeneity.** For example, nearly 200 forms of hereditary deafness are known, presumably due to impaired actions of many different genes. The eleven types of clotting disorders reflect the many factors and enzymes that control this process. Any of several genes may also cause cleft palate,

albinism, diabetes insipidus, colon cancer, and breast cancer.

**1** What is incomplete dominance?

**2** What is codominance?

**3** How do penetrance, expressivity, and pleiotropy affect gene expression?

## COMPLEX TRAITS

Most of the inherited disorders mentioned so far are *monogenic*—that is, they are determined by a single gene, and their expression is usually not greatly influenced by the environment. However, most characteristics and disorders reflect input from the environment as well as genes.

Traits determined by more than one gene are termed *polygenic.* Usually, several genes each contribute to differing degrees toward molding the overall phenotype, which may vary greatly among individuals. Such a trait, with many degrees of expression because of the input of several genes, is said to be continuously varying. Height and skin color are polygenic traits (fig. 24.12 and table 24.1).

Traits molded by one or more genes plus the environment are termed *multifactorial* or complex traits. Height and skin color are multifactorial as well as polygenic, because environmental factors influence them: good nutrition enables a person to reach the height dictated by genes, and sun exposure affects skin color. Most

(a)

(b)

**FIGURE 24.12**

Previous editions of this (and other) textbooks have used the photograph in (a) to illustrate the continuously varying nature of height. In the photo, taken around 1920, 175 cadets at the Connecticut Agricultural College lined up by height. In 1997, Professor Linda Strausbaugh asked her genetics students at the school, today the University of Connecticut at Storrs, to recreate the scene (b). They did, and confirmed the continuously varying nature of human height. But they also elegantly demonstrated how height has increased during the twentieth century. Improved nutrition has definitely played a role in expressing genetic potential for height. The tallest people in the old photograph (a) are 5'9" tall, whereas the tallest people in the more recent photograph (b) are 6'4"tall.

# Adoption and Twin Studies

Geneticists enlist two types of people to tease apart the inherited and environmental components of multifactorial traits—adopted individuals and twins.

A person who is adopted usually shares environmental influences with the adoptive parents, and shares genes with the biological parents. By comparing an adopted person's characteristics with those of both sets of parents, researchers can attempt to sort out inherited from environmental influences. Studies examining causes of death, for example, find that a person's risk of dying of infection before age 50 is more similar to the risk of the biological parents doing so than to that of

adoptive parents. Susceptibility to infection, then, is largely determined by genetics, most likely inherited variations in immune system function.

Identical twins arise from the same fertilized egg and are therefore genetically identical. Fraternal twins share half of their genes, and are no more alike genetically than are any two siblings. A trait has a large genetic component if it affects both members of identical twin pairs with greater frequency than it affects both members of fraternal twin pairs.

Even more informative are identical twins who were separated at birth and reared in different environments. Any distinctive traits they share are likely to be rooted in their genes. For example, a pair of identical male twins raised in different countries in different religions were astounded when they were reunited as adults to find that they both laugh when someone sneezes and flush the toilet before using it. A pair of girls reunited in late adolescence found they were both afraid of swimming. Twins who met for the first time in their 30s both paused for 30 seconds after being asked a question, each rotated a gold necklace she was wearing three times, and then answered. Coincidence, or genetics? ■

of the more common illnesses, including heart disease, diabetes, hypertension, and cancers, are multifactorial. Clinical Application 24.3 discusses how twins help geneticists understand complex traits.

**1** How does polygenic inheritance make possible many variations of a trait?

**2** How may the environment influence gene expression?

## MATTERS OF SEX

Human somatic (nonsex) cells include an X and a Y chromosome in males, and two X chromosomes in females. All eggs carry a single X chromosome, and sperm carry either an X or a Y chromosome. Sex is determined at conception: a Y-bearing sperm fertilizing an egg conceives a male, and an X-bearing sperm conceives a female (fig. 24.13).

### Sex Determination

Association of the Y chromosome with maleness has been known since 1959, but researchers did not identify the genes responsible for maleness until 1990. The **SRY gene**

(sex-determining region of the Y) encodes a type of protein called a *transcription factor,* which switches on other genes that direct development of male structures in the embryo, while suppressing formation of female structures. Figure 24.14 shows the X and Y chromosomes.

Some unusual individuals helped researchers identify the SRY gene—men who have two X chromosomes, and women who have one X and one Y chromosome, which is the reverse of normal. A close look at the composition of these unusual peoples' sex chromosomes revealed that the XX males actually had a small piece of a Y chromosome, and the XY females lacked a small part of the Y chromosome. The part of the Y chromosome that was present in the XX males was the same part that was missing in the XY females—and it contains the SRY gene.

### Sex Chromosomes and Sex-Linked Traits

The X and Y chromosomes carry genes, but they are inherited in different patterns than are autosomal genes because of the differing sex chromosome constitutions in males and females. The X chromosome has more than one thousand genes; the Y chromosome has very few genes. Traits transmitted on the X chromosome are said to be **sex-linked.**

Any gene on the X chromosome of a male is expressed in his phenotype, because he has no second allele to mask its expression. An allele on an X chromosome in a female may or may not be expressed depending upon whether it is dominant or recessive, and upon the nature of the allele on the second X chromosome. The human male is said to be **hemizygous** for sex-linked traits because he has half the number of genes on the X chromosome than the female has. *Red-green color blindness* and the most common form of the clotting disorder *hemophilia* are examples of recessive sex-linked traits.

It would appear that the sexes are unequal, as far as sex-linked traits go, because females have two gene-packed X chromosomes compared to the male's one. To remedy the inequity, the cells of female mammalian embryos undergo a process called *X inactivation,* which shuts off one X chromosome in each somatic cell. A gene called XIST on the X chromosome controls the shutdown. Which of a female's X chromosomes is silenced—the one she inherited from her mother or the one from her father—occurs randomly. Therefore, a female is a mosaic, with her father's X chromosome expressed in some cells, and her mother's in others. This can be vividly illustrated for heterozygous, sex-linked genes.

A woman who is a carrier (a heterozygote) for *Duchenne muscular dystrophy* has a wild-type allele for the dystrophin gene on one X chromosome, and a disease-causing allele on the other. Cells in which the X chromosome bearing the wild-type allele is inactivated do not produce the gene's protein product, dystrophin. However, cells in which the mutant allele is inactivated produce dystrophin. When a stain for dystrophin is applied to a sample of her muscle tissue, some cells turn blue, and others do not. This is the basis of a carrier detection test for this illness. If by chance many of such a woman's wild-type dystrophin alleles are turned off in her muscle cells, she may experience mild muscle weakness.

A male always inherits his Y chromosome from his father and his X chromosome from his mother. A female inherits one X chromosome from each parent. If a mother is heterozygous for a particular sex-linked gene, her son has a 50% chance of inheriting either allele from her. Sex-linked genes are therefore passed from mother to son. Because a male does not receive an X chromosome from his father (he inherits the Y chromosome from his father), a sex-linked trait is not passed from father to son.

Consider the sex-linked inheritance pattern of hemophilia A. It is passed from carrier mother to affected son with a risk of 50%, because he can inherit either her normal allele or the mutant one. A daughter has a 50% chance of inheriting the hemophilia allele and being a carrier like her mother, and a 50% chance of not carrying the allele.

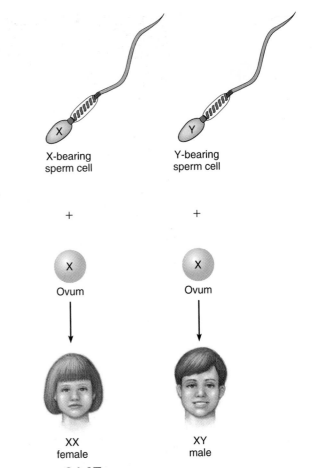

**FIGURE** 24.13

Sex determination. An egg contributes an X chromosome, and a sperm either an X or a Y. If a Y-bearing sperm cell fertilizes an egg, the zygote is male (XY). If an X-bearing sperm cell fertilizes an egg, the zygote is female (XX). Sex is actually determined by a gene on the Y chromosome.

A daughter can inherit a sex-linked recessive disorder or trait if her father is affected and her mother is a carrier. She inherits one affected X chromosome from each parent. Without a biochemical test, though, a woman would not know that she is a carrier of a sex-linked recessive trait unless she has an affected son.

A sex-linked recessive trait is more likely to be seen in females if it isn't serious enough to prevent an affected man from fathering children. This is the case for color blindness.

Dominant disease-causing alleles on the X chromosome are extremely rare. Generally in such cases, males are much more severely affected than females, who have a second X to offer a protective effect. In a condition called *incontinentia pigmenti,* for example, an affected girl has swirls of pigment on her skin where melanin in the epidermis leaks into the dermis. She may have abnormal teeth, sparse hair, visual problems, and seizures. However, males inheriting the dominant gene on their X chromosomes are so severely affected that they do not survive to be born.

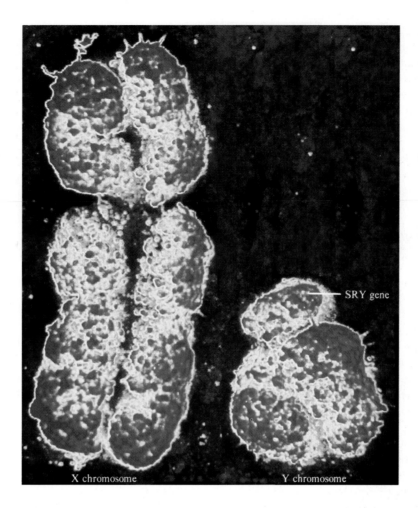

**FIGURE 24.14**

The X and Y chromosomes. The SRY gene, at the top of the short arm of the Y chromosome, starts the cascade of gene activity that directs development of a male (31,000×).

## GENDER EFFECTS ON PHENOTYPE

Sex-linked traits generally are more prevalent in males than females. Traits that are not sex-linked, however, can also be expressed differently in males and females, due to differences between the sexes.

### Sex-Limited Traits

A **sex-limited trait** affects a structure or function of the body that is present in only males or only females. Such a gene may be sex-linked or autosomal. Beard growth and breast size are sex-limited traits. A woman cannot grow a beard because she does not manufacture large amounts of the hormones required for facial hair growth. She can, however, pass to her sons the genes that specify heavy beard growth.

### Sex-Influenced Traits

In **sex-influenced inheritance,** an allele is dominant in one sex but recessive in the other. Again, such a gene may be sex-linked or autosomal. This difference in expression can be caused by hormonal differences between the sexes. For example, a gene for hair growth pattern has two alleles, one that produces hair all over the head and another that causes pattern baldness (fig. 24.15). The baldness allele is dominant in males but recessive in females, which is why more men than women are bald. A heterozygous male is bald, but a heterozygous female is not. A bald woman would have two mutant alleles.

In a little-understood phenomenon called *genomic imprinting,* the expression of a disorder differs depending upon which parent transmits the disease-causing gene or chromosome. The phenotype may differ in degree of severity, in age of onset, or even in the nature of the symptoms.

A fascinating example of genomic imprinting appears in *Angelman syndrome* and *Prader-Willi syndrome.* In these two disorders that result from abnormalities in the same region of chromosome 15, different sets of symptoms result depending upon the sex of the parent transmitting the gene. (Actually, the gene is

(a)

(b)

(c)

(d)

**FIGURE 24.15**

Pattern baldness is a sex-influenced trait, and was a genetic trademark of the illustrious Adams family. (a) John Adams (1735–1826) was the second president of the United States. He was the father of (b) John Quincy Adams (1767–1848), who was the sixth president. John Quincy was the father of (c) Charles Francis Adams (1807–1886), who was a diplomat and the father of (d) Henry Adams (1838–1918), who was a historian.

absent, or deleted.) Prader-Willi is apparently inherited from the mother. A child with Prader-Willi syndrome is obese, has small hands and feet, is mentally retarded, and does not develop signs of puberty. Angelman syndrome, inherited from the father, is sometimes called "happy puppet syndrome," which describes the phenotype. The child has an extended tongue, poor muscle coordination that gives a "floppy" appearance, and a large jaw, laughs uncontrollably, and has peculiar convulsions in which the upper limbs flap. Geneticists do not yet understand how deleting the same genes can have such different effects. Clinical Application 24.4 discusses a disorder that is more severe in males—fragile X syndrome.

**1** Which chromosomes and genes determine sex?

**2** What is sex-linked inheritance?

**3** Why do sex-linked recessive conditions appear most commonly in males?

**4** How can gender affect gene expression?

## CHROMOSOME DISORDERS

Deviations from the normal human chromosome number of 46 produce syndromes, because of the excess or deficit of genes. Rearrangement of chromosomes, such as an inversion of a section of a chromosome, or two nonhomologous chromosomes exchanging parts, may also cause symptoms. This may happen if the rearrangement disrupts a vital gene, or if it results in "unbalanced" gametes that contain too little or too much genetic material. Following is a closer look at specific types of chromosome aberrations.

### Polyploidy

The most drastic upset in chromosome number is an entire extra set, a condition called **polyploidy.** This results from formation of a diploid (rather than a normal haploid) gamete. For example, if a haploid sperm fertilizes a diploid egg, the fertilized egg is *triploid,* with three copies of each chromosome. Most human polyploids die as embryos or fetuses, but occasionally an infant survives

## TABLE 24.2 ANEUPLOIDS SURVIVING BIRTH

| Chromosome Constitution | Syndrome | Incidence | Phenotype |
|---|---|---|---|
| Trisomy 21 | Down | 1/770 | Mental retardation<br>Abnormal pattern of palm creases<br>Flat face<br>Sparse, straight hair<br>Short stature<br>High risk of:<br>   cardiac anomalies<br>   leukemia<br>   cataracts<br>   digestive blockages |
| Trisomy 13 | Patau | 1/15,000 | Mental and physical retardation<br>Skull and facial abnormalities<br>Defects in all organ systems<br>Cleft lip<br>Large, triangular nose<br>Extra digits |
| Trisomy 18 | Edward | 1/4,000–8,000 | Mental and physical retardation<br>Skull and facial abnormalities<br>Defects in all organ systems<br>Extreme muscle tone |
| Trisomy 22 | | Very few reported | Multiple defects<br>Large ears and nose<br>Narrow face<br>Small jaw<br>Excess muscle tone |
| Monosomy 21 | | Very few reported | Mental and physical retardation<br>Skull, jaw, and facial abnormalities |
| XO | Turner | 1/2,500–10,000 females | No sexual maturity<br>Short stature<br>Webbed neck<br>Wide-spaced nipples<br>Narrow aorta<br>Pigmented moles |
| XXY | Klinefelter | 1/500–2,000 males | Lack of secondary sexual characteristics<br>Breast swelling<br>No sperm |
| XXX | Triplo-X | 1/1,000–2,000 females | Tall and thin<br>Menstrual irregularity |
| XYY | Jacobs | 1/1,000 males | Tall |

for a few days, with defects in nearly all organs. However, many agriculturally important plants are polyploids. Some organs normally have a few polyploid cells, with no adverse effects on health. Liver cells, for example, may be tetraploid (4 chromosome sets) or even octaploid (8 chromosome sets).

## Aneuploidy

Cells missing a chromosome or having an extra one are **aneuploid.** A normal chromosome number is termed **eu-** ploid. Symptoms that result from aneuploidy depend upon which chromosome is missing or extra. Autosomal aneuploidy often results in mental retardation, possibly because so many genes affect brain function. Sex chromosome aneuploidy is less severe. Table 24.2 describes polyploid individuals who can survive to be born.

Extra genetic material is apparently less dangerous than missing material, and this is why most children born with the wrong number of chromosomes have an extra one, called a **trisomy** rather than a missing one,

# Fragile X Syndrome and Expanding Genes

An unusually breakable X chromosome is associated with a form of mental retardation that affects 1 in 1,500 individuals, most of them male. *Fragile X syndrome* has an interesting history.

In the 1940s, geneticists noticed that more males than females were mentally retarded. Might sex-linked inheritance explain this observation? In 1969, a clue emerged to the genetic basis of one form of sex-linked mental retardation, in a family with two affected brothers. Chromosomes from the brothers and their mother revealed an unusual X chromosome. One tip of it dangled, separated from the rest by a thin thread (fig. 24C). When grown under certain culture conditions, this part of the X chromosome was very prone to breaking—hence, the name fragile X.

Youngsters with fragile X syndrome look normal, but by adolescence, characteristic features appear. These include a very long, narrow face (fig. 24D), a long jaw, and protruding ears. The testicles are very large. Mental impairment varies greatly, and may include mental retardation, learning disabilities, poor speech, hyperactivity, shyness, social anxiety, and a short attention span.

In 1991 researchers discovered the molecular basis of fragile X, and in so doing stumbled upon a completely unknown way that genes can change: They can expand.

In unaffected individuals, the fragile X area contains a stretch of DNA consisting of five to fifty repeats of the sequence CGG. (See chapter 4 and table 4.1 for an explanation of the genetic code.) In people with full symptoms and fragile X chromosomes, this region is greatly expanded, harboring 230 to more than two thousand CGG repeats. This expansion attracts methyl groups ($CH_3$) to the fragile X gene, shutting it down and causing the symptoms. Most interesting is that mothers of affected boys often have a "premutation"—an expansion of sixty to 200 repeats. They usually do not have symptoms, but may be mentally slow. Occasionally a male has the premuta-

tion, which means that he can pass the expanding gene to a daughter, who will then have a 50% chance of passing it to a son, who will be affected.

Fragile X syndrome, as the first of the "triplet repeat" disorders, opened up a whole new area of genetic investigation. Expanding genes have since been implicated in Hunt-

X chromosome

Fragile site

**FIGURE 24C**

A constriction appears in the X chromosome of individuals who have symptoms of fragile X syndrome when the cells containing them are grown in a medium that lacks thymidine and folic acid.

---

called a **monosomy**. Trisomies and monosomies are named according to the implicated chromosome. In the past, the associated syndrome was named for the investigator who first described it. Down syndrome, for example, is also known as trisomy 21 (Clinical Application 24.5).

The meiotic error that results in aneuploidy is called **nondisjunction** (fig. 24.16). In normal meiosis, pairs of homologous chromosomes separate and each of the resulting gametes therefore contains only one member of each pair. In nondisjunction, a chromosome pair fails to separate, either at the first or at the second meiotic division, producing a sperm or egg that has two copies of a particular chromosome or none, rather than the normal one copy. When such a gamete fuses with its mate at fertilization, the zygote has either 47 or 45 chromosomes, instead of the normal 46.

Today, geneticists know that a man with an extra Y chromosome may be very tall, have acne, and also perhaps have speech and reading problems. They did not always understand this. A study published in 1965 identified twelve men of 197 at a Scottish high-security mental hospital who had chromosome abnormalities. Of the twelve, seven had the same anomaly, an extra Y chromosome. This led to the widely publicized idea that an extra Y chromosome causes aggressiveness and criminal behavior. For a short time, hospital nurseries in England, Canada, Denmark, and Boston screened newborn boys for the telltale extra chromosome, then sent psychologists to help parents cope with their "congenital criminals." It is more likely that such boys become aggressive because adults expect this behavior from children who are large for their age.

ington disease, the neuromuscular disorder myasthenia gravis, and more than a dozen other conditions, most of which affect brain function. On a whole-body level, expanding genes explain a common phenomenon called anticipation, in which symptoms of an inherited illness worsen with each generation. ■

**FIGURE 24D**

The characteristic facial structure and features of individuals who have fragile X syndrome become more pronounced with age.

---

1  Why do deviations from the normal chromosome number of 46 affect health?

2  Define *polyploidy*.

3  Define *aneuploidy*.

4  How do extra sets of chromosomes, or extra individual chromosomes, arise?

## MEDICAL GENETICS

This is an exciting time in the field of human genetics, with new discoveries reaching medical journals and headlines weekly. This tide of new genetic information has implications for the diagnosis, treatment, and better understanding of thousands of disorders.

### Diagnosis

Once a gene is localized to a specific part of a particular chromosome, researchers can develop tests to diagnose the associated condition. These tests are unlike other diagnostic tests in three ways. First, they can be performed on any cell type, because all of the cells of an individual contain all of the genes. Thus, a test to detect an allele that could cause cystic fibrosis, which predominantly affects the lungs, can be performed on cells scraped from the inside of the cheek, on white blood cells, or on cells shed from a fetus into amniotic fluid.

Second, the fact that all cells have all the genes means that a disease whose symptoms do not usually appear until middle age can be diagnosed much earlier—even in a cleavage embryo. A presymptomatic diagnosis is more a prediction than it is a demonstration of symptoms.

# CLINICAL APPLICATION

# Down Syndrome

The most common autosomal aneuploid is *trisomy 21* (fig. 24E). The characteristic slanted eyes and flat face of the Down syndrome patient prompted Sir John Langdon Haydon Down to coin the inaccurate term "mongolism" when he described the syndrome in 1866. As the medical superintendent of a facility for the profoundly mentally retarded, Down noted that about 10% of his patients resembled people of the Mongolian race. The resemblance is coincidental. Males and females of all races can have Down syndrome (fig. 24F).

A person with Down syndrome is short and has straight, sparse hair and a tongue protruding through thick lips. The hands have an abnormal pattern of creases, the joints are loose, and reflexes and muscle tone are poor. Developmental milestones (such as sitting, standing, and walking), are slow, and toilet training may take several years. Intelligence varies greatly, from profound mental retardation to following simple directions, reading and using a computer. People with Down syndrome have graduated from college, and have contributed to society as much as other people.

Down syndrome is associated with many physical problems (table 24A). Nearly 50% of affected people die before their first birthdays, often of heart or kidney defects, or a suppressed immune system that enables infections to rage out of control. Blockages in the digestive system are common and must be corrected surgically shortly after birth. A child with Down syndrome is 15 times more likely to develop leukemia (a cancer of the white blood cells) than a healthy child.

About 25% of people with Down syndrome, who live past age 40, develop the fibers and tangles of amyloid protein in their brains that

are also seen in the brains of people who have died of Alzheimer's disease. (The risk among the general population is 6%.) Both disorders seem to involve

accelerated aging of part of the brain and accumulation of the sticky amyloid protein.

By looking at people who have a third copy of only part of chromosome 21, researchers are identifying the genes whose malfunction causes the symptoms of Down syndrome. This tiny chromosome contains only 1.5% of the genome, or about 1,050 genes. The distal third of the long arm is consistently present in people with Down syndrome who have only part of the extra chromosome, a region that may house a few hundred genes. It appears that certain parts of the chromosome are more crucial

Sex Chromosomes

Abnormal karyotype and InSight analysis revealing trisomy 21.

## FIGURE 24E

The karyotype of an individual with trisomy 21 Down syndrome has an extra chromosome 21. Another way to detect the condition is to use a DNA probe, which is a piece of DNA normally found on chromosome 21, with a fluorescent molecule attached to it. Three flashes in the nucleus, rather than the normal two, indicate an extra chromosome 21.

than others in developing Down syndrome, because some people with large parts of chromosome 21 missing have normal intelligence.

The likelihood of giving birth to a child with Down syndrome increases dramatically with the age of the mother. For women under 30, the chances of conceiving such a child are 1 in 3,000. By age 48, the number jumps to 1 in 9 (fig. 24G). However, 80% of children with trisomy 21 are born to women under age 35. This is because younger women are more likely to become pregnant, and less likely to undergo amniocentesis. About 5% of cases can be traced to nondisjunction in the father.

The age factor in Down syndrome may be due to the fact that meiosis in the female is completed after conception. The older a woman is, the longer her oocytes have been arrested on the brink of completing meiosis. During this time, the oocytes may have been exposed to chromosome-damaging chemicals or radiation. Other trisomies are more likely to occur among the offspring of older women, too.

Many of the medical problems that people with Down syndrome suffer are treatable, so that life expectancy is now 55 years. In 1910, life expectancy was only to age nine! ■

| TABLE 24A | SIGNS AND SYMPTOMS USED TO DIAGNOSE DOWN SYNDROME IN NEWBORNS |
| --- | --- |

1. Poor muscle tone
2. Flat face
3. Eyes slant upward
4. Abnormally-shaped ears
5. Single deep crease across palms
6. Joints extremely flexible
7. Underdeveloped fifth fingers
8. Skin folds in inner corners or eyes (epicanthal folds)
9. Gap between first and second toes
10. Large tongue

**FIGURE 24F**
Wendy Weisz has Down syndrome. She enjoys studying art at Cuyahoga Community College.

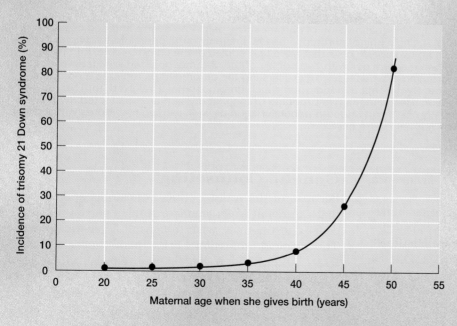

**FIGURE 24G**
Down syndrome risk increases with the age of the mother. The incidence of this form of Down syndrome in the general population is about 1 in every 770 births. Among women over the age of 35 years, however, the risk increases.

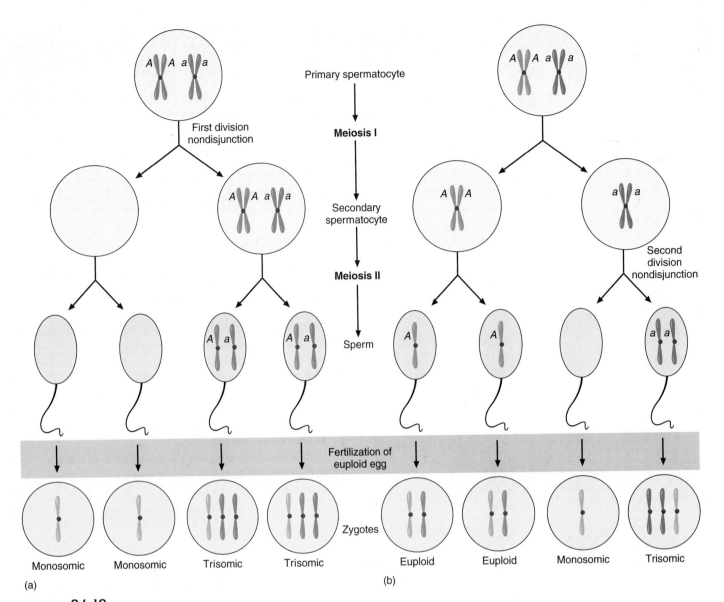

**FIGURE 24.16**

Extra or missing chromosomes constitute aneuploidy. Unequal division of chromosome pairs into sperm and egg cells can occur at either the first or the second meiotic division. (a) A single pair of chromosomes is unevenly partitioned into the two cells arising from the first division of meiosis in a male. The result: two sperm cells that have two copies of the chromosome, and two sperm cells that have no copies of that chromosome. When a sperm cell with two copies of the chromosome fertilizes a normal egg cell, the zygote produced is trisomic for that chromosome; when a sperm cell lacking the chromosome fertilizes a normal egg cell, the zygote is monosomic for that chromosome. Symptoms depend upon which chromosome is involved. (b) This nondisjunction occurs at the second meiotic division. Because the two products of the first division are unaffected, two of the mature sperm are normal and two are aneuploid. Egg cells can undergo nondisjunction as well, leading to zygotes with extra or missing chromosomes when they are fertilized by normal sperm cells.

It is also not absolute, because environmental factors can modify the expression of inherited traits.

A third distinction between diagnosis of an inherited illness compared to a noninherited illness is psychological. A genetic disease diagnosis—presymptomatically or after a person experiences symptoms—carries with it a burden of guilt not seen in acquired illnesses, because an affected person may have passed the gene causing his or her illness to children.

## Genetic Counseling

Because of the unique ethical questions and dilemmas that can result from genetic testing, a person undergoing genetic testing may wish to consult a **genetic counselor.** This medical professional can explain the facts, available tests, and options. A genetic counselor is trained in psychology, statistics, counseling techniques, genetics, molecular biology, and related areas. Working with an accurate diagnosis and a family history, the counselor

TABLE **24.3**

## GENETIC COUNSELING CASES

1. A mother has just found out that her 6-year-old son, whom she thought was merely clumsy, has Duchenne muscular dystrophy. She also has a 2-year-old son but is not certain she wants to know whether he has also inherited the disease, because she feels that knowing so would affect her ability to be a good parent. She also feels guilty because she knows she carries the gene that made her son ill.

2. A young man is distressed because his father has just been diagnosed with Huntington disease. His father's mother probably had the condition, although she was institutionalized and her diagnosis was uncertain. The man needs help in deciding whether to take a test that will tell him if he has inherited the dominant gene that causes the condition. He is only 23, and symptoms do not usually appear until the late thirties. He wants the information to help decide whether to have children.

3. A young couple meet with a genetic counselor because the woman's brother has a son with hemophilia. They are relieved to hear that their children's risk is not elevated, because the affected nephew received the hemophilia gene from his mother, who is not a blood relative.

4. A woman has a very mild case of an autosomal dominant disorder, neurofibromatosis I. Her only symptoms are a few brownish patches on her skin and two benign lumps under the skin on her upper left thigh. She is surprised to learn from the genetic counselor that if the condition appears in a child, it might not be so mild.

5. Amniocentesis indicates that a fetus has three X chromosomes. The parents-to-be are distraught, and can hardly believe the genetic counselor when she tells them that the child will probably be healthy.

6. A woman undergoes a genetic test that indicates that she has not inherited the gene that caused breast cancer in her mother and sisters. She becomes depressed because she feels guilty.

determines recurrence risks for certain conditions in specific relatives, and provides information on the illness so that families can make informed medical decisions. Table 24.3 describes some actual situations that a genetic counselor faces.

## Gene Therapy

Cynthia Cutshall and Ashanthi DeSilva made headlines in 1990 when they became the first two people to officially undergo **gene therapy.** Each girl received white blood cells bearing the gene for adenosine deaminase, which their bodies could not produce and which robbed them of an immune system.

Cutshall and DeSilva's treatment replaced genes, thus operating at the level of the genotype. However, physicians have treated some genetic disorders at the phenotype level for several years. A boy with hemophilia receives a clotting factor. A child with cystic fibrosis sprinkles powdered cow digestive enzymes onto applesauce, which she eats before each meal to replace the enzymes that her clogged pancreas cannot secrete. People with certain inborn errors of metabolism follow restrictive diets to counteract biochemical buildups (see Clinical Application 4.4). Even wearing eyeglasses is a way of altering the expression of one's inheritance.

Gene therapy is a more lasting treatment for genetic disease than correcting symptoms. There are two basic types of gene therapy. Effecting a change in all cells of an individual would require intervention at the fertilized egg stage. This **heritable gene therapy** is impractical in humans and is not being attempted in them. However, researchers use it to create *transgenic* experimental animals that serve as models of human disease. A mouse with muscular dystrophy, for example, has an abnormal human dystrophin gene in each of its cells.

In contrast to heritable gene therapy, **nonheritable gene therapy** targets only affected cells, such as young Cynthia and Ashanthi's impaired white blood cells. However, nonheritable gene therapy is not passed to subsequent generations, as is heritable gene therapy, because gametes are not altered. One form of nonheritable gene therapy, bone marrow transplants, have been performed for many years, but still have a high mortality rate. This procedure replaces blood stem cell populations in bone marrow, but leaves the patient vulnerable to infection for several months because the recipient's bone marrow must be destroyed first.

Laura Cay Boren has witnessed firsthand the evolution of therapies for her inherited adenosine deaminase (ADA) deficiency, which left her without an immune system since her birth in July 1982. In 1983 and again in 1984, she received bone marrow transplants from her father, which did not cure her underlying problem. Blood transfusions helped for awhile. But by the end of 1985, Laura was near death. Then, she was chosen to receive a new treatment—a form of the missing enzyme, altered so that it would remain in her bloodstream long enough to help her. It did. Within hours her ADA level increased 20-fold. By 3 months her blood no longer contained toxins, although immunity had not yet appeared. But by 6 months, she had a functioning immune system! Gene therapy will give Laura a longer-lasting treatment.

Both the successes and failures of bone marrow transplants have prompted researchers to alter genes in other types of cells (fig. 24.17). Researchers use several methods to introduce therapeutic genes into cells. Healing DNA is given linked to viruses, in fatty bubbles

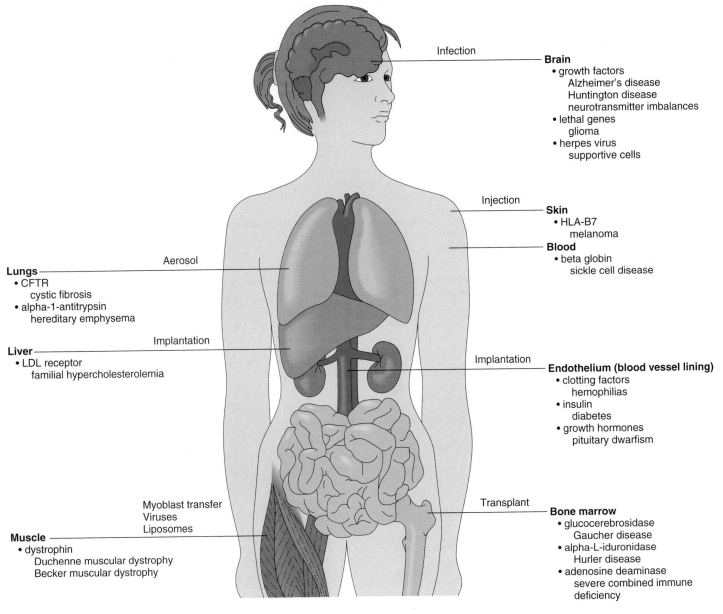

**Brain**
- growth factors
  Alzheimer's disease
  Huntington disease
  neurotransmitter imbalances
- lethal genes
  glioma
- herpes virus
  supportive cells

Infection

Injection

**Skin**
- HLA-B7
  melanoma

**Blood**
- beta globin
  sickle cell disease

Aerosol

**Lungs**
- CFTR
  cystic fibrosis
- alpha-1-antitrypsin
  hereditary emphysema

Implantation

**Liver**
- LDL receptor
  familial hypercholesterolemia

Implantation

**Endothelium (blood vessel lining)**
- clotting factors
  hemophilias
- insulin
  diabetes
- growth hormones
  pituitary dwarfism

Myoblast transfer
Viruses
Liposomes

Transplant

**Muscle**
- dystrophin
  Duchenne muscular dystrophy
  Becker muscular dystrophy

**Bone marrow**
- glucocerebrosidase
  Gaucher disease
- alpha-L-iduronidase
  Hurler disease
- adenosine deaminase
  severe combined immune
  deficiency

**FIGURE 24.17**

Sites of gene therapy and the methods used to introduce normal DNA.

called liposomes, "shot" with gold particles that blast into cells, and as "naked" preparations. A look at some of the ongoing work in gene therapy provides a nice ending to our survey of human genetics, as well as reviews organ systems discussed throughout the book. Most of these efforts are still experimental. A decade from now, gene therapy may be a routine part of medical practice.

**Skin**   Skin grafts genetically bolstered to secrete missing proteins, such as clotting factors or enzymes, may be a new route to drug delivery. So far, a gene encoding human growth hormone has been inserted into human skin cells growing in culture. When

grafted onto mice from a strain that does not reject foreign tissue, the grafts grow into normal appearing mouse epidermis—but they secrete human growth hormone. This is a first step to human gene therapy on skin.

**Endothelium**   Recall that endothelium forms the walls of capillaries and lines other blood vessels. This tissue can be engineered to secrete a substance directly into the bloodstream. In a person with diabetes, for example, altered endothelium might secrete insulin.

**Muscle**   Immature skeletal muscle cells, called myoblasts, bearing therapeutic genes inserted into their DNA, can be incorporated into the natural muscle in blood vessel walls, where they could secrete a needed protein into the bloodstream. Dogs with hemophilia, for example, are treated with myoblasts harboring functional clotting factor genes. Gene therapy experiments attempting to treat Duchenne muscular dystrophy use implanted myoblasts, and dystrophin genes ferried in on viruses or in liposomes. The challenge in effectively treating this condition is to restore dystrophin function in enough muscles to restore function.

**Liver** The liver is a very important focus of gene therapy because it controls many bodily functions, and because it can regenerate. A gene therapy that corrects just 5% of the 10 trillion cells of the liver could produce an effect. The first liver gene therapy targets heart disease. Normal liver cells have low-density lipoprotein (LDL) receptors on their surfaces, which bind cholesterol in the bloodstream and bring it into the cell. When liver cells lack LDL receptors, cholesterol accumulates on artery interiors. Liver cells genetically altered to have more LDL receptors can relieve the cholesterol buildup that causes cardiovascular disease. Such gene therapy could be lifesaving for children who have inherited familial hypercholesterolemia, in which liver cells lack LDL receptors. Existing treatment is a liver transplant or continually cleansing the blood of cholesterol. These individuals die of heart failure before age 20. Gene therapy removes about 15% of the liver, inserts the LDL receptor gene, and injects the corrected liver cells into a blood vessel in the patient's liver. Here, the cells attach to the vessel walls and produce LDL receptors that bind cholesterol.

**Lungs** The respiratory tract is a prime candidate for gene therapy because an aerosol can directly reach its lining cells, making it unnecessary to remove cells, alter them, and reimplant them. To treat cystic fibrosis, a modified virus that causes the common cold is genetically altered to not cause cold symptoms and to contain a functional CFTR gene. Once inhaled, lung-lining cells take up the gene and produce the protein missing or mutant in the inherited illness. Gene therapy is also being used to provide an enzyme whose absence causes a hereditary form of emphysema. The gene is introduced into fibroblasts or directly in an aerosol of tiny bubbles (liposomes). The challenge for both lung diseases is to alter enough cells to alleviate symptoms.

**Nerve Tissue** Gene therapy on neurons presents a problem, because these cells do not normally reproduce. Altering other cell types can circumvent this obstacle.

Consider a gene treatment for mice injured in the part of the brain that degenerates in Alzheimer's disease. When fibroblasts genetically engineered to contain a gene for nerve growth factor are implanted at the wound site, they secrete the growth factor, which neurons need in order to enlarge. The neighboring traumatized neurons grow and secrete their neurotransmitter, stimulated by the bath of nerve growth factor. This type of treatment may one day boost neurotransmitter levels in Huntington disease, Parkinson's disease, Alzheimer's disease, and clinical depression, which are caused by deficits of these important biochemicals.

Another route to nerve cell gene therapy is to send in a valuable gene attached to the herpes simplex virus, which remains in nerve cells after infection. Such a herpes gene carrier could alter a neuron's ability to secrete neurotransmitters. In experiments with rats, a growth factor gene introduced into damaged neurons in the central nervous system restored some movement following spinal cord injury.

**Gene Therapy against Cancer** Viruses may provide a treatment for a deadly type of brain tumor called a *glioma,* which affects the glial cells of the nervous system. Unlike neurons, glia can reproduce, and cancerous glia do so very rapidly, usually causing death within a year even with aggressive treatment. A new approach to treatment is to infect fibroblasts with a virus bearing a gene from a herpes virus that makes the cells sensitive to a drug called ganciclovir. The altered fibroblasts are implanted near the tumor. Then, the virus infects nearby rapidly reproducing glia—the cancer cells. Give the patient ganciclovir, and only the cells harboring the virus die—not healthy brain cells.

Another genetic approach to battling cancer is to enable tumor cells to produce immune system biochemicals, or to mark them so that the immune system more easily recognizes them. This approach is called a cancer vaccine. A treatment for the skin cancer *melanoma,* for example, alters tumor cells to display an antigen called HLA-B7, which stimulates the immune system to attack the cell (fig. 24.18). Other cancers may in the future be treated by altering cells in blood vessels that feed a tumor to secrete immune system biochemicals. Tables 24.4 and 24.5 highlight concerns about delivering gene therapy.

Not long ago, genetic diseases were thought to be extremely rare. The gene discoveries of the past decade, however, have revealed the contributions of genes to virtually every physiological process. Therefore, genes play

**FIGURE 24.18**

A gene encoding a cell surface protein that attracts the immune system's tumor-killing T cells is injected directly into a melanoma tumor. Such gene therapy against cancer is sometimes called a cancer vaccine. It doesn't prevent cancer from starting, but may stop it from recurring.

| TABLE **24.4** | **REQUIREMENTS FOR APPROVAL OF CLINICAL TRIALS FOR GENE THERAPY** |
| --- | --- |

1. Knowledge of defect and how it causes symptoms.
2. An animal model.
3. Success in human cells growing in vitro.
4. Either no alternate therapies, or a group of patients for whom existing therapies are not possible or have not worked.
5. Experiments must be safe.

| TABLE **24.5** | **GENE THERAPY CONCERNS** |
| --- | --- |

1. Which cells should be treated?
2. What proportion of the targeted cell population must be corrected to alleviate or halt progression of symptoms?
3. Is overexpression of the therapeutic gene dangerous?
4. If the engineered gene "escapes" and infiltrates other tissues, is there danger?
5. How long will the affected cells function?
6. Will the immune system attack the introduced cells?

a role in many, if not all, illnesses. Susceptibility to infection depends to an extent upon one's immune system, which is inherited. Cancers often reflect genetic alterations in somatic cells. Many genes encode proteins that affect the cardiovascular system. The enzymes that con-

trol biochemical pathways are the products of genes. Table 24.6 describes several of the more common inherited illnesses.

As the human genome project continues to offer genetic explanations for normal anatomy and physiology, as well as pathology, health practitioners will embrace this new information, this new molecular and cellular way of looking at the functioning of the human body. Soon, genetic tests may be a part of everyone's life, from conception to old age.

**1** What are the unique challenges associated with diagnosing genetic disease?

**2** What is genetic counseling?

**3** Why is heritable gene therapy in humans not practical?

**4** What are some of the ways that nonheritable gene therapy is being conducted?

**TABLE 24.6**

## SELECTED INHERITED DISORDERS

959

| Disorder | Cause and Symptoms | Mode of Inheritance* |
|---|---|---|
| Adrenoleukodystrophy | Accumulation of certain fatty acids causes seizures, nervous system degeneration, and death in childhood | xlr |
| Albinism | Lack of melanin pigment in skin and hair | ar |
| Alzheimer's disease | Progressive loss of memory and reasoning ability, beginning in late middle age | AD |
| Amyotrophic lateral sclerosis | Progressive muscle weakening, beginning in adulthood | AD |
| Breast cancer | Tumors in breast | AD |
| Chronic granulomatous disease | Phagocytes cannot destroy pathogenic bacteria, causing frequent infection | xlr |
| Color blindness | Inability to see red or green | xlr |
| Cystic fibrosis | Abnormal chloride channels in cell membranes clogs lungs, impairs pancreatic secretion, causes respiratory infection and male infertility | ar |
| Diabetes insipidus | Deficiency of antidiuretic hormone leads to copious urination | xlr |
| Duchenne muscular dystrophy | Lack of dystrophin protein collapses skeletal muscle cells, destroying muscles | xlr |
| Familial hypercholesterolemia | Lack of LDL receptors on liver cells leads to extreme cholesterol buildup | ar |
| Familial hypertrophic cardiomyopathy | Sudden death in young adulthood from heart failure | AD |
| Fatal familial insomnia | Degeneration of brain's sleep center causes neurological degeneration | AD |
| Gout | Enzyme deficiency causes buildup of uric acid in joints | AD |
| Hemochromatosis | Iron deposits in liver, heart, endocrine glands; may not produce symptoms | ar |
| Hemophilia | Absence of a clotting factor causes slow wound healing, internal bleeding | xlr |
| Huntington disease | Adult onset loss of coordination, uncontrollable movements, personality changes | AD |
| Lissencephaly | Lack of brain convolutions | ar |
| Marfan syndrome | Lack of fibrillin protein causes long limbs, lens dislocation, weakened aorta wall | AD |
| Menkes disease | Cells cannot utilize copper, causing twisted hair, weak arteries, scurvy, poor growth, brain degeneration | xlr |
| Multiple endocrine neoplasia | Cancers of endocrine glands | AD |
| Osteogenesis imperfecta | Collagen defect causes easily broken bones, possibly before birth | ar |
| Pattern baldness | Hair loss | AD (♂) ar (♀) |
| Phenylketonuria | Inborn error that causes phenylalanine to accumulate, causing mental retardation | ar |
| Pituitary dwarfism | Short stature due to lack of growth hormone | ar |
| Polycystic kidney disease | Adult-onset cysts in kidneys | AD |
| Polydactyly | Extra digits | AD |
| Porphyria | Abnormal porphyrin ring in hemoglobin leads to accumulation of hemoglobin breakdown products, producing red urine, abdominal pain, neurological impairment | ar |
| Progeria | Extreme premature aging | ar |
| Retinitis pigmentosa | Visual field constriction, night blindness, pigment clumps in eye | xlr |
| Rickets | Inability of cells to respond to vitamin D, weakening bones | ar |
| Tay-Sachs disease | Enzyme deficiency causes lipid to accumulate on nerve cells, impairing neurological development | ar |
| von Willebrand disease | Absent clotting factor prolongs bleeding | ar |
| Wilms tumor | Prenatal or childhood kidney cancer | AD |
| Wilson disease | Excess copper damages brain and liver, causing abdominal pain, headache, neurological symptoms | ar |
| Xeroderma pigmentosum | Absent DNA repair leads to freckling and high risk of skin cancer | ar |

*xlr = sex-linked recessive
ar = autosomal recessive
AD = autosomal dominant

# The Emerging Role of Genetics in Medicine (page 929)

Individual characteristics result from the interaction of heredity (transmission of genetic information from parents to offspring) and the environment (chemical, physical, social, and biological influences in an individual's surroundings). Genetics will play an increasingly important role in health care, as new discoveries explain physiological and pathological processes.

# Genes and Chromosomes (page 930)

1. A gene is a portion of a DNA molecule that contains information for producing one kind of protein molecule, or controls other genes.
2. All genes in a human cell constitute the genome.
3. The 70,000 or so human genes are distributed among 23 pairs of chromosomes, half of which are inherited from the mother and half from the father.

# Chromosomes and Sex Cell Production (page 931)

1. Meiosis
   a. Meiosis consists of two successive divisions, each progressing through prophase, metaphase, anaphase, and telophase.
   b. In the first meiotic division, homologous, replicated chromosomes separate and their number is halved. A replicated chromosome consists of two chromatids held together by a centromere.
   c. In the second meiotic division, the chromatids part, producing four haploid cells from each diploid cell undergoing meiosis.
   d. Meiosis yields sex cells. Each male cell undergoing meiosis produces four mature sperm cells; each female cell produces one oocyte and three polar bodies.
2. Meiosis leads to genetic variability
   a. Variation results from random alignment of maternally and paternally derived chromosomes in metaphase I.
   b. Crossing over mixes up contributions from the previous generation.
3. Chromosomes and genes come in pairs
   a. Chromosome charts are called karyotypes.
   b. Chromosomes 1 through 22, numbered in decreasing size order, are autosomes. They do not determine sex.
   c. The X and Y chromosomes are sex chromosomes. They determine sex.
   d. Because chromosomes are paired, the genes they carry are also paired.
   e. An alternate form of a gene is called an allele. An individual can have two different alleles for a particular gene. The gene itself can have many alleles, because a gene consists of many building blocks, any of which may be altered.
   f. An individual with a pair of identical alleles for a particular gene is homozygous; if the alleles are different, the individual is heterozygous.

# Modes of Inheritance (page 940)

The combination of genes present in an individual's cells constitutes a genotype; the appearance of the individual is its phenotype. A wild-type allele provides normal or the most common function. A mutant allele causes disease or an unusual trait; it represents a change from the wild-type condition.

1. Dominant and recessive inheritance
   a. In the heterozygous condition, an allele that is expressed when the other is not is dominant. The masked allele is recessive.
   b. Recessive and dominant genes may be autosomal or sex-linked.
   c. An autosomal recessive condition affects both sexes, and may skip generations. The homozygous dominant and heterozygous individuals have normal phenotypes. The homozygous recessive individual has the condition. The heterozygote is a carrier. An affected individual inherits one mutant allele from each parent.
   d. An autosomal dominant condition affects both sexes and does not skip generations. A person inherits it from one parent, who is affected.
2. Different dominance relationships
   a. In incomplete dominance, a heterozygote has a phenotype intermediate between those of both homozygotes.
   b. In codominance, each of the alleles in the heterozygote is expressed.

# Gene Expression (page 942)

1. Penetrance and expressivity
   a. A genotype is incompletely penetrant if not all individuals inheriting it express the phenotype.
   b. A genotype is variably expressive if it is expressed to different degrees in different individuals.
2. Pleiotropy
   a. A pleiotropic disorder has several symptoms, different subsets of which are expressed among individuals.
   b. Pleiotropy reflects a gene product that is part of more than one biochemical reaction, or is found in several organs or structures.
3. Genetic heterogeneity
   a. Genetic heterogeneity refers to a phenotype that can be caused by alterations in more than one gene.
   b. The same symptoms may result from alterations in genes whose products are enzymes in the same biochemical pathway.

## Complex Traits (page 943)

1. A trait caused by the action of a single gene is monogenic.
2. A trait caused by the action of more than one gene is polygenic.
3. A trait caused by the action of one or more genes and the environment is multifactorial or complex.
4. Height, skin color, and many common illnesses are complex traits.

## Matters of Sex (page 945)

A female has two X chromosomes; a male has one X and one Y chromosome. The X chromosome has many genes; the Y chromosome has very few genes.

1. Sex determination
   a. A male zygote forms when a Y-bearing sperm fertilizes an egg. A female zygote forms when an X-bearing sperm fertilizes an egg.
   b. A gene on the Y chromosome, called SRY, switches on genes in the embryo that promote development of male characteristics and suppresses genes that promote development of female characteristics.
   c. Researchers discovered the SRY gene by studying XY females lacking SRY, and XX males with a small bit of a Y chromosome.
2. Sex chromosomes and sex-linked traits
   a. Genes on the sex chromosomes are inherited differently than those on autosomes because the sexes differ in sex chromosome constitution.
   b. Traits transmitted on the X chromosome are termed sex-linked.
   c. Males are hemizygous for sex-linked traits; that is, they can have only one copy of a sex-linked gene, because they have only one X chromosome.
   d. Females can be heterozygous or homozygous for genes on the X chromosome, because they have two copies of it.
   e. A male inherits a sex-linked trait from a carrier mother. These traits are more common in males than in females.
   f. A female inherits a sex-linked mutant gene from her carrier mother, and/or from her father if the associated trait does not impair his ability to have children.
   g. Dominant sex-linked traits are rare. Affected males die before birth.

## Gender Effects on Phenotype (page 947)

Gender can affect gene expression.

1. Sex-limited traits
   a. Such a trait affects a structure or function seen in only one sex.
   b. Such a trait may be inherited according to any mode of inheritance.
2. Sex-influenced traits
   a. A sex-influenced trait is dominant in one sex and recessive in the other.
   b. In genomic imprinting, the severity, age of onset, or nature of symptoms varies according to which parent transmits the causative gene.

## Chromosome Disorders (page 948)

Extra, missing, or rearranged chromosomes or parts of them can cause syndromes, because they either cause an imbalance of genetic material or disrupt a vital gene.

1. Polyploidy
   a. Polyploidy is an extra chromosome set.
   b. Polyploidy results from fertilization of a diploid gamete.
   c. Human polyploids do not survive beyond a few days of birth.
2. Aneuploidy
   a. Cells with an extra or missing chromosome are aneuploid. Cells with the normal chromosome number are euploid.
   b. A cell with an extra chromosome is trisomic. A cell with a missing chromosome is monosomic. Individuals with trisomies are more likely to survive to be born than those with monosomies.
   c. Aneuploidy results from a meiotic error called nondisjunction, in which a chromosome pair does not separate, either in meiosis I or meiosis II, producing a gamete with a missing or extra chromosome. At fertilization, a monosomic or trisomic zygote results.

## Medical Genetics (page 951)

Frequent discoveries of new genes are providing new diagnostic tests and new treatments.

1. Diagnosis
   a. Diagnostic tests for inherited conditions present special challenges.
   b. Such tests can be performed on any cell type.
   c. Because diagnostic tests for genetic disease can be performed at any point in life, they can presymptomatically detect illness.
   d. Genetic disease often brings guilt feelings of responsibility for relatives' illness.
2. Genetic counseling
   a. A genetic counselor explains inheritance patterns, risks, and options to families with inherited disease.
   b. A genetic counselor must have an accurate diagnosis and a complete family history to provide useful information to patients.
3. Gene therapy
   a. Gene therapy corrects the genetic defect causing symptoms.
   b. Heritable gene therapy alters all genes in an individual, and therefore must be done on a fertilized egg. It is not being pursued in humans but is useful in research on other animals.
   c. Nonheritable gene therapy replaces or corrects defective genes in somatic cells, often those where symptoms occur.
   d. Gene therapy is being tested in skin, endothelium, muscle, liver, lungs, and to treat cancer.

## CRITICAL THINKING QUESTIONS

1. Can a person who has been successfully treated for sickle cell disease still transmit the disease-causing allele to offspring?

2. A young couple is devastated when their second child is born and has PKU. Their older child is healthy and no one else in the family has PKU. How is this possible?

3. A balding man undergoes a treatment that transfers some of the hair from the sides of his head, where it is still plentiful, to the top. Is he altering his phenotype or his genotype?

4. Bob and Joan know from a blood test that they are each heterozygous (carriers) for the autosomal recessive gene that causes sickle cell disease. If their first three children are healthy, what is the probability that their fourth child will have the disease?

5. A man with type AB blood has children with a woman who has type O blood. What are the chances that a child that they conceive has blood that is type A? B? AB? O?

6. A person has one X chromosome, and a Y chromosome missing the SRY gene. Describe this person's phenotype.

7. Traits that appear more frequently in one sex compared to the other may be caused by genes that are inherited in a sex-linked, sex-limited, or sex-influenced fashion. How might these possibilities be distinguished?

8. Cirrhosis of the liver, emphysema, and heart disease are all conditions that can be caused by a faulty gene, or by a dangerous lifestyle habit (drinking alcohol, smoking, following a poor diet). When gene therapies become available for these conditions, should people with gene-caused disease be given priority in receiving the treatments? If not, what other criteria should be used for deciding who should receive a limited medical resource?

## REVIEW EXERCISES

1. Identify two major factors that influence the development of individual characteristics.

2. Define *gene; chromosome; genome.*

3. Discuss the origin of the 46 chromosomes in a human zygote.

4. Define *homologous chromosomes.*

5. Outline the process of meiosis.

6. Explain the significance of synapsis during meiosis.

7. List two ways that meiosis provides genetic variability.

8. Distinguish between:
   - homozygote and heterozygote
   - autosome and sex chromosome
   - mutant and wild-type
   - homozygous dominant, homozygous recessive, and heterozygous
   - phenotype and genotype
   - incomplete dominance and codominance

9. Explain how a gene can have many alleles.

10. Describe the components of a complex trait.

11. Describe how the environment can influence gene expression.

12. Explain how genes and chromosomes determine gender.

13. Explain how the presence of the SRY gene causes embryonic reproductive organs to develop as male structures.

14. Define *sex-linked genes.*

15. Explain why recessive sex-linked genes are expressed more frequently in males than females.

16. Explain why a male cannot inherit a sex-linked trait from his father.

17. Explain why X-linked dominant traits are not seen in males.

18. Explain how an individual with an extra set of chromosomes arises.

19. Explain how nondisjunction leads to aneuploidy.

20. List three ways that diagnosis of genetic disease differs from diagnosis of other disorders.

21. Describe the services that a genetic counselor provides.

22. Describe why heritable gene therapy is impractical in humans.

23. Explain how nonheritable gene therapy is being attempted in various human tissues.

 **For More Information Related to This Chapter, Please Visit** *Hole's Human Anatomy and Physiology* **Home Page**
http://www.mhhe.com/biosci/abio/

# Human Cadavers

■ *The following set of illustrations includes medial sections, horizontal sections, and regional dissections of human cadavers. These photographs will help you visualize the spatial and proportional relationships between the major anatomic structures of actual specimens. The photographs can also serve as the basis for a review of the information you have gained from your study of the human organism.*

Cerebrum

Nasal cavity
Cerebellum
Tongue

Trachea

Aorta

Heart
Vertebral column
Spinal cord
Liver
Stomach

Small intestine

Greater omentum

Colon

Urinary bladder

Testis

**PLATE** Forty-Eight
Sagittal section of the head and trunk.

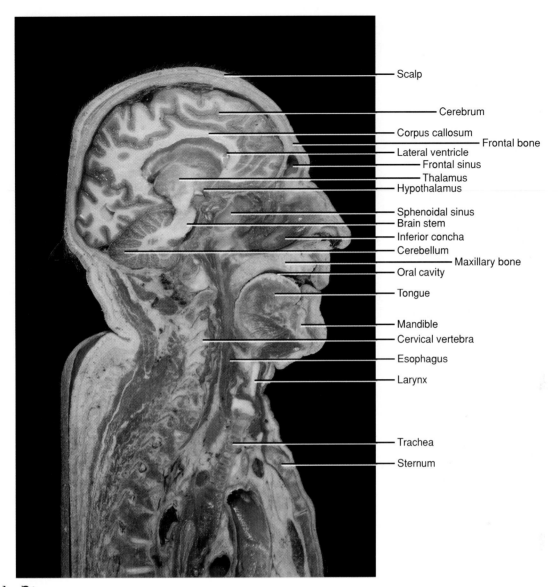

Scalp

Cerebrum

Corpus callosum
Frontal bone
Lateral ventricle
Frontal sinus
Thalamus
Hypothalamus

Sphenoidal sinus
Brain stem
Inferior concha
Cerebellum
Maxillary bone
Oral cavity

Tongue

Mandible
Cervical vertebra

Esophagus

Larynx

Trachea
Sternum

**PLATE** Forty-Nine

Sagittal section of the head and neck.

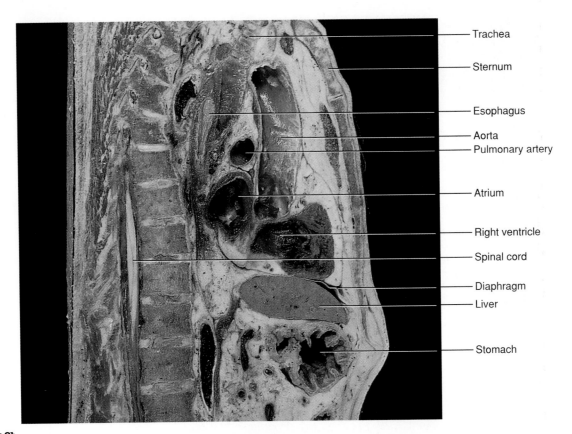

— Trachea

— Sternum

— Esophagus

— Aorta
— Pulmonary artery

— Atrium

— Right ventricle

— Spinal cord

— Diaphragm

— Liver

— Stomach

**PLATE** Fifty

Viscera of the thoracic cavity, sagittal section.

— Aorta

— Sternum

— Right ventricle
— Spinal cord

— Diaphragm
— Liver

— Stomach
— Abdominal aorta

— Transverse colon

— Cauda equina
— Small intestine
— Intervertebral disk

— Lumbar vertebral body

PLATE Fifty-One

Viscera of the abdominal cavity, sagittal section.

Small intestine
Intervertebral disk
Lumbar vertebral body

Cauda equina

Sacrum
Small intestine
Rectus abdominis

Sigmoid colon

Coccyx
Symphysis pubis
Urinary bladder
Rectum

Spermatic cord
Epididymis
Testis
Scrotum

PLATE Fifty-Two
Viscera of the pelvic cavity, sagittal section.

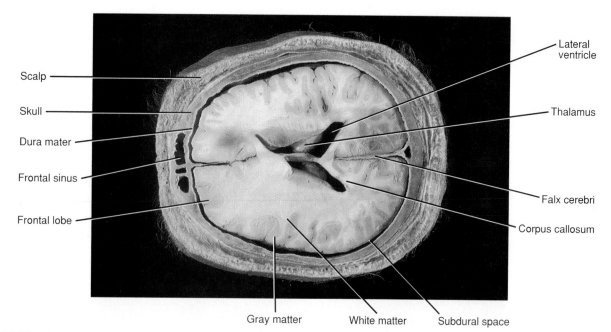

Scalp

Skull

Dura mater

Frontal sinus

Frontal lobe

Lateral ventricle

Thalamus

Falx cerebri

Corpus callosum

Gray matter    White matter    Subdural space

**PLATE Fifty-Three**

Transverse section of the head above the eyes, superior view.

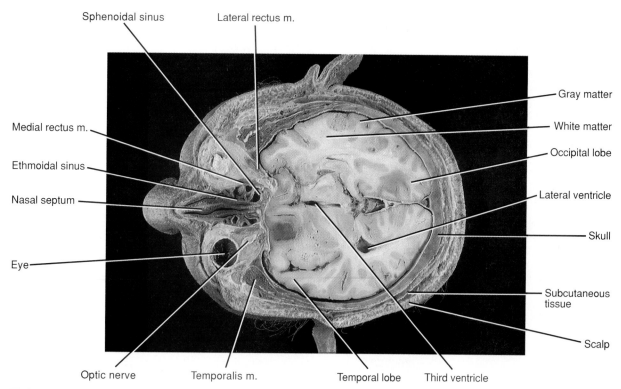

Sphenoidal sinus    Lateral rectus m.

Medial rectus m.

Ethmoidal sinus

Nasal septum

Eye

Gray matter

White matter

Occipital lobe

Lateral ventricle

Skull

Subcutaneous tissue

Scalp

Optic nerve    Temporalis m.    Temporal lobe    Third ventricle

**PLATE Fifty-Four**

Transverse section of the head at the level of the eyes, superior view.

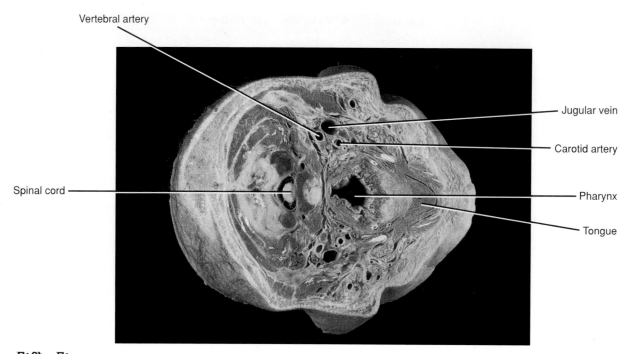

Vertebral artery

Jugular vein

Carotid artery

Spinal cord

Pharynx

Tongue

## PLATE Fifty-Five

Transverse section of the neck, inferior view.

Spinal cord

Lung

Aorta

Rib

Esophagus

Heart

Pericardial cavity

Pectoralis major m.

Sternum

## PLATE Fifty-Six

Transverse section of the thorax through the base of the heart, superior view.

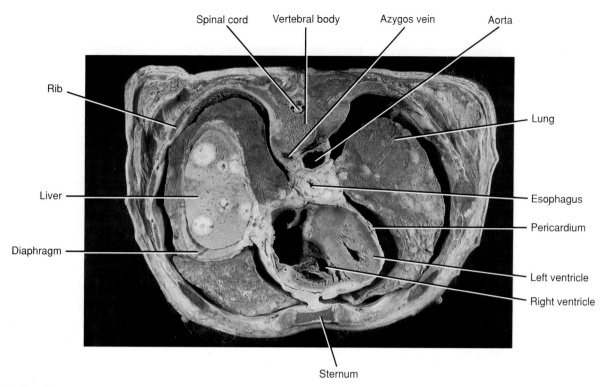

Spinal cord   Vertebral body   Azygos vein   Aorta

Rib

Lung

Liver

Esophagus

Pericardium

Diaphragm

Left ventricle

Right ventricle

Sternum

PLATE Fifty-Seven

Transverse section of the thorax through the heart, superior view.

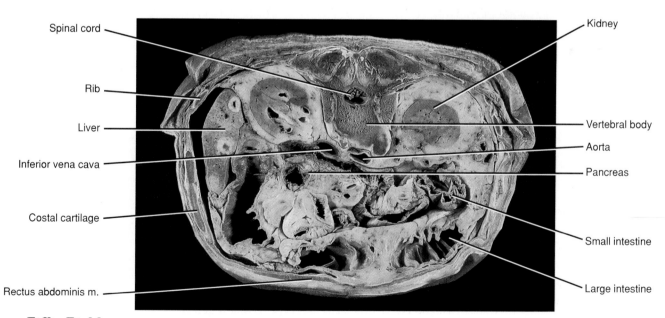

Spinal cord

Kidney

Rib

Liver

Vertebral body

Inferior vena cava

Aorta

Pancreas

Costal cartilage

Small intestine

Rectus abdominis m.

Large intestine

PLATE Fifty-Eight

Transverse section of the abdomen through the kidneys, superior view.

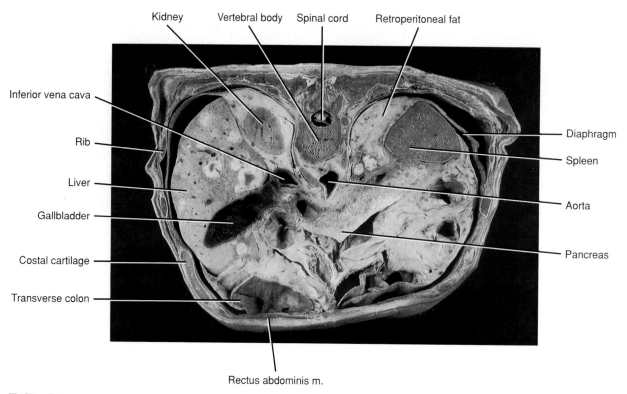

Kidney Vertebral body Spinal cord Retroperitoneal fat

Inferior vena cava

Rib

Liver

Gallbladder

Costal cartilage

Transverse colon

Diaphragm

Spleen

Aorta

Pancreas

Rectus abdominis m.

**PLATE Fifty-Nine**

Transverse section of the abdomen through the pancreas, superior view.

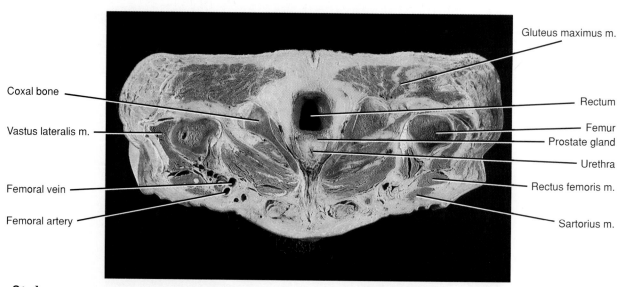

Coxal bone

Vastus lateralis m.

Femoral vein

Femoral artery

Gluteus maximus m.

Rectum

Femur
Prostate gland

Urethra

Rectus femoris m.

Sartorius m.

**PLATE Sixty**

Transverse section of the male pelvic cavity, superior view.

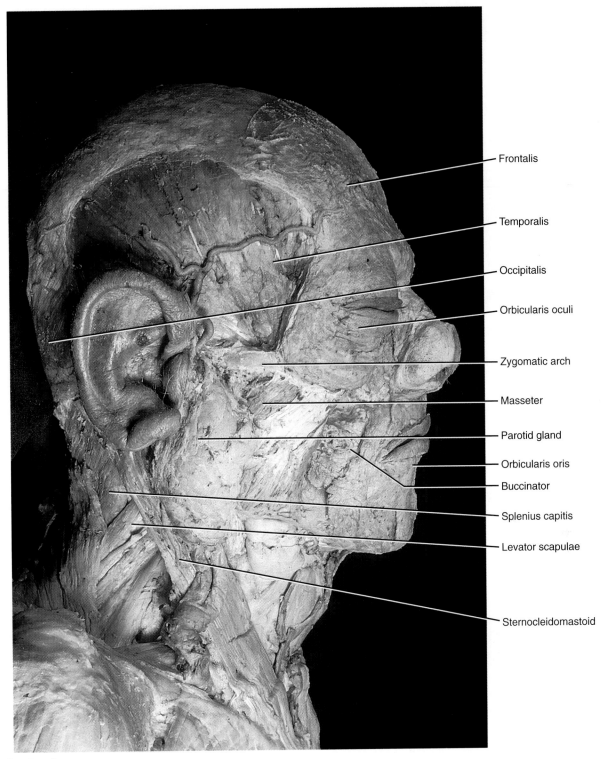

Frontalis

Temporalis

Occipitalis

Orbicularis oculi

Zygomatic arch

Masseter

Parotid gland

Orbicularis oris

Buccinator

Splenius capitis

Levator scapulae

Sternocleidomastoid

PLATE Sixty-One
Lateral view of the head.

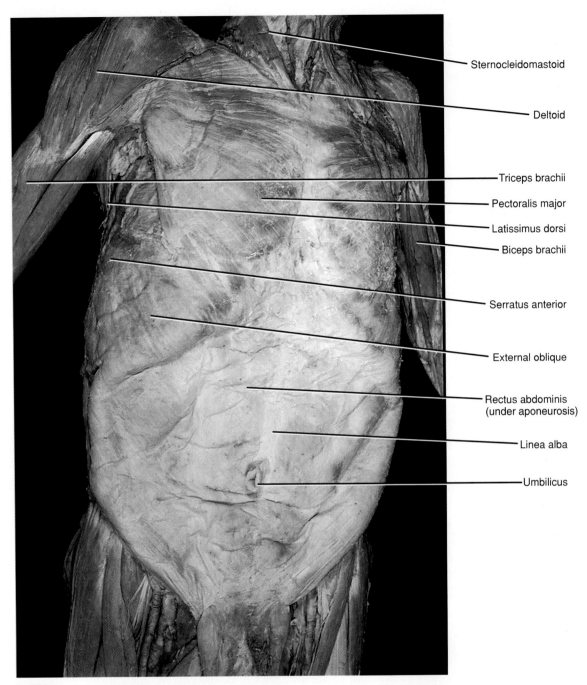

Sternocleidomastoid

Deltoid

Triceps brachii

Pectoralis major

Latissimus dorsi

Biceps brachii

Serratus anterior

External oblique

Rectus abdominis
(under aponeurosis)

Linea alba

Umbilicus

**PLATE** Sixty-Two

Anterior view of the trunk.

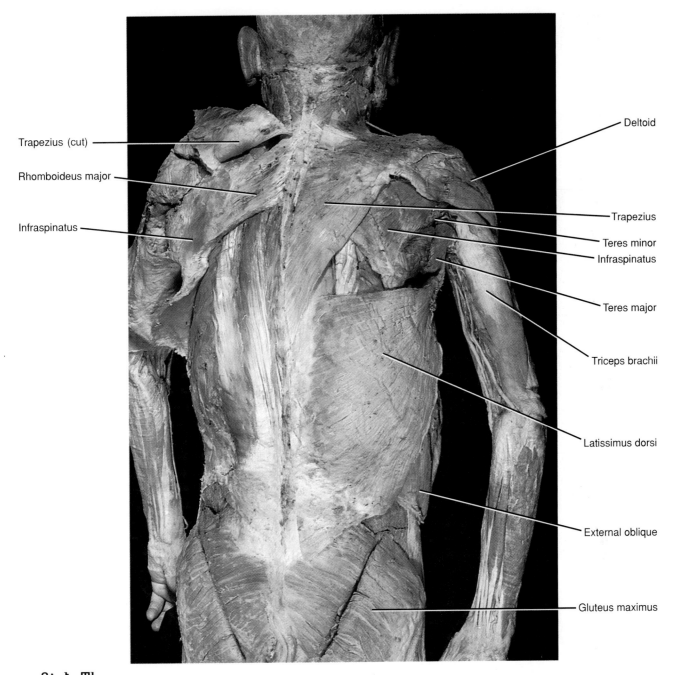

Trapezius (cut)

Rhomboideus major

Infraspinatus

Deltoid

Trapezius

Teres minor

Infraspinatus

Teres major

Triceps brachii

Latissimus dorsi

External oblique

Gluteus maximus

## PLATE Sixty-Three

Posterior view of the trunk, with deep thoracic muscles exposed on the left.

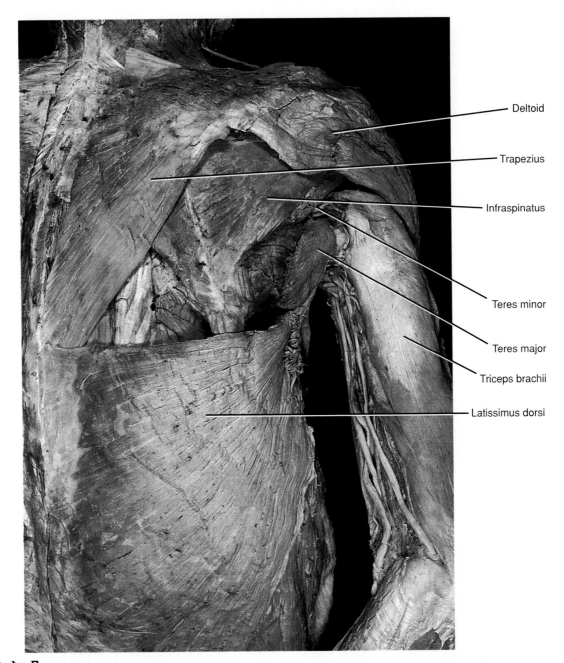

Deltoid

Trapezius

Infraspinatus

Teres minor

Teres major

Triceps brachii

Latissimus dorsi

**PLATE** Sixty-Four

Posterior view of the right thorax and arm.

Triceps brachii

Brachioradialis

Brachialis

Biceps brachii

Extensor carpi
radialis longus

Extensor carpi
radialis brevis

Extensor digitorum

Extensor carpi
ulnaris

Extensor
retinaculum

**PLATE Sixty-Five**
Posterior view of the right forearm and hand.

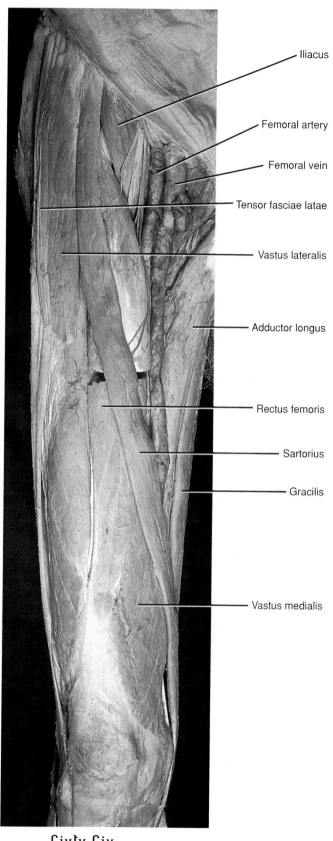

Iliacus

Femoral artery

Femoral vein

Tensor fasciae latae

Vastus lateralis

Adductor longus

Rectus femoris

Sartorius

Gracilis

Vastus medialis

**PLATE Sixty-Six**
Anterior view of the right thigh.

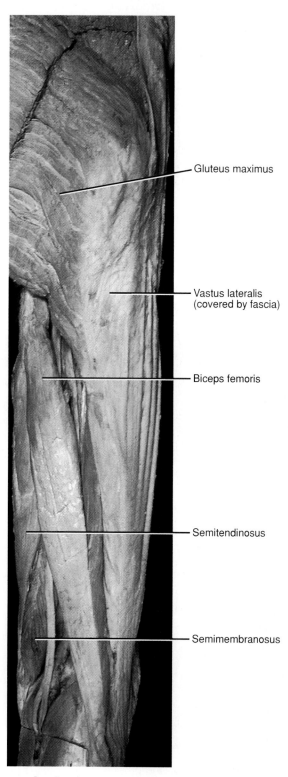

Gluteus maximus

Vastus lateralis
(covered by fascia)

Biceps femoris

Semitendinosus

Semimembranosus

**PLATE Sixty-Seven**
Posterior view of the right thigh.

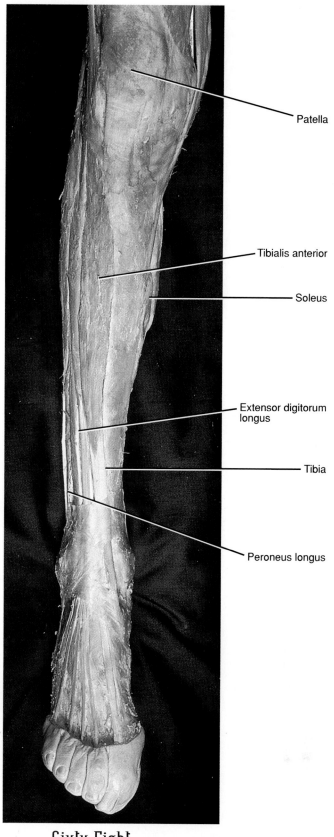

Patella

Tibialis anterior

Soleus

Extensor digitorum
longus

Tibia

Peroneus longus

**PLATE Sixty-Eight**
Anterior view of the right leg.

Patella

Patellar ligament

Tibialis anterior

Extensor digitorum
longus

Gastrocnemius

Peroneus longus

Soleus

Peroneus brevis

Calcaneal tendon

**PLATE** Sixty-Nine
Lateral view of the right leg.

Biceps femoris

Semimembranosus

Gastrocnemius

Soleus

Peroneus longus

Peroneus brevis

Calcaneal tendon

**PLATE** Seventy
Posterior view of the right leg.

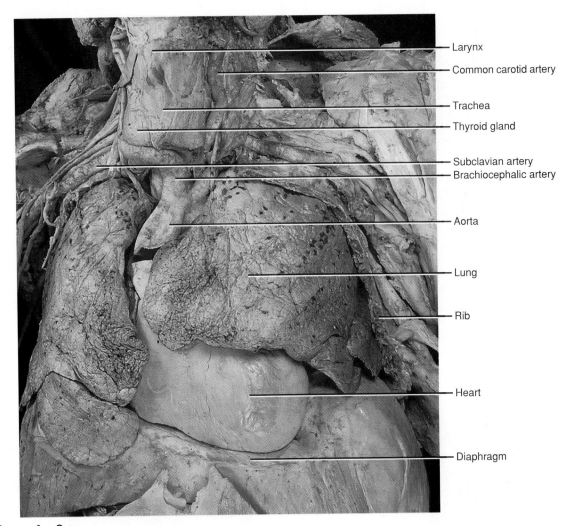

Larynx

Common carotid artery

Trachea

Thyroid gland

Subclavian artery

Brachiocephalic artery

Aorta

Lung

Rib

Heart

Diaphragm

**PLATE Seventy-One**

Thoracic viscera, anterior view. (Brachiocephalic vein has been removed to expose the aorta.)

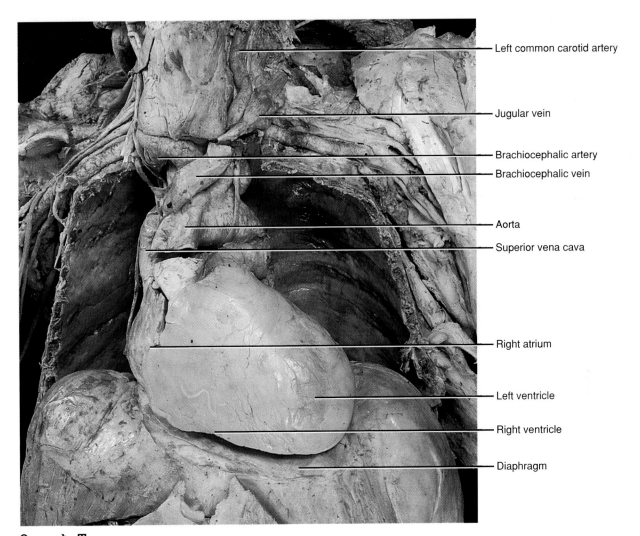

— Left common carotid artery

— Jugular vein

— Brachiocephalic artery
— Brachiocephalic vein

— Aorta
— Superior vena cava

— Right atrium

— Left ventricle

— Right ventricle

— Diaphragm

**PLATE** Seventy-Two
Thorax with the lungs removed, anterior view.

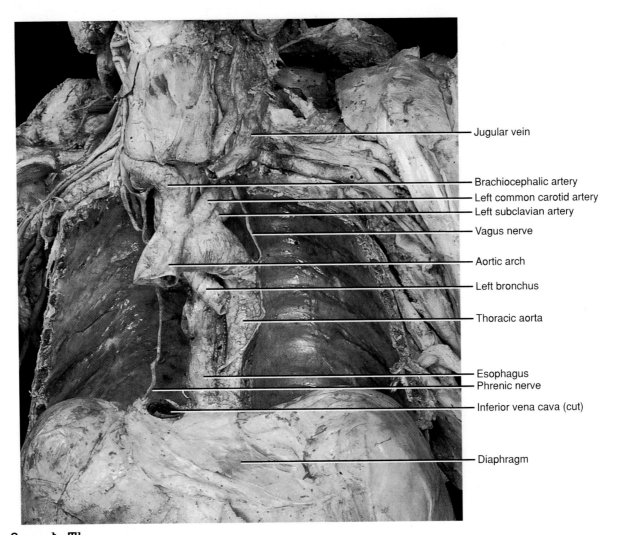

— Jugular vein

— Brachiocephalic artery
— Left common carotid artery
— Left subclavian artery

— Vagus nerve

— Aortic arch

— Left bronchus

— Thoracic aorta

— Esophagus
— Phrenic nerve

— Inferior vena cava (cut)

— Diaphragm

**PLATE Seventy-Three**
Thorax with the heart and lungs removed, anterior view.

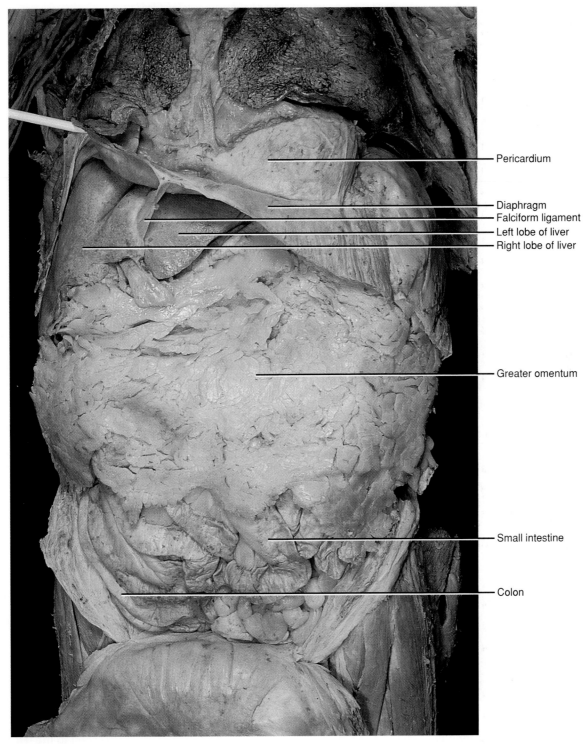

Pericardium

Diaphragm
Falciform ligament
Left lobe of liver
Right lobe of liver

Greater omentum

Small intestine

Colon

**PLATE** Seventy-Four
Abdominal viscera, anterior view.

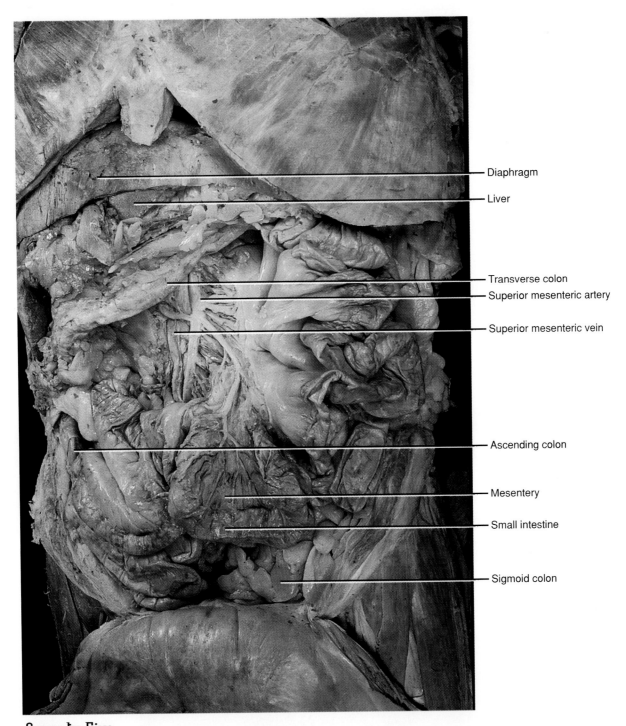

—Diaphragm

—Liver

—Transverse colon

—Superior mesenteric artery

—Superior mesenteric vein

—Ascending colon

—Mesentery

—Small intestine

—Sigmoid colon

PLATE Seventy-Five

Abdominal viscera with the greater omentum removed, anterior view. (Small intestine has been displaced to the left.)

Anterior

Posterior

Corpus callosum

Frontal lobe of cerebrum

Diencephalon/thalamus

Diencephalon/hypothalamus

Midbrain

Pons

Medulla oblongata

Occipital lobe of cerebrum

Cerebellum

Arbor vitae

**PLATE Seventy-Six**
Midsagittal section of the brain.

# APPENDIX A

## *Periodic Table of Elements*

**Representative Elements (*s* Series)**

**Representative Elements (*p* Series)**

**Key**

| 1 |
| Hydrogen |
| **H** |
| 1.0079 |

Atomic Number
Name
Symbol
Atomic Weight

**Transition Metals (*d* Series of Transition Elements)**

**Inner Transition Elements (*f* Series)**

*Lanthanides — 4f

**Actinides — 5f

## Units of Measurement and Their Equivalents

### Apothecaries' Weights and Their Metric Equivalents

**1 grain (gr) =**
0.05 scruple (s)
0.017 dram (dr)
0.002 ounce (oz)
0.0002 pound (lb)
0.065 gram (g)
65. milligrams (mg)

**1 scruple (s) =**
20. grains (gr)
0.33 dram (dr)
0.042 ounce (oz)
0.004 pound (lb)
1.3 grams (g)
1,300. milligrams (mg)

**1 dram (dr) =**
60. grains (gr)
3. scruples (s)
0.13 ounce (oz)
0.010 pound (lb)
3.9 grams (g)
3,900. milligrams (mg)

**1 ounce (oz) =**
480. grains (gr)
24. scruples (s)
0.08 pound (lb)
31.1 grams (g)
31,100. milligrams (mg)

**1 pound (lb) =**
5,760. grains (gr)
288. scruples (s)
96. drams (dr)
12. ounces (oz)
373. grams (g)
373,000. milligrams (mg)

### Apothecaries' Volumes and Their Metric Equivalents

**1 minim (min) =**
0.017 fluid dram (fl dr)
0.002 fluid ounce (fl oz)
0.0001 pint (pt)
0.06 milliliter (mL)
0.06 cubic centimeter (cc)

**1 fluid dram (fl dr) =**
60. minims (min)
0.13 fluid ounce (fl oz)
0.008 pint (pt)
3.70 milliliters (mL)
3.70 cubic centimeters (cc)

**1 fluid ounce (fl oz) =**
480. minims (min)
8. fluid drams (fl dr)
0.06 pint (pt)
29.6 milliliters (mL)
29.6 cubic centimeters (cc)

**1 pint (pt) =**
7,680. minims (min)
128. fluid drams (fl dr)
16. fluid ounces (fl oz)
473. milliliters (mL)
473. cubic centimeters (cc)

### Metric Weights and Their Apothecaries' Equivalents

**1 gram (g) =**
0.001 kilogram (kg)
1,000. milligrams (mg)
1,000,000. micrograms (μg)
1,000,000,000. nanograms (ng)
1,000,000,000,000. picograms (pg)
15.4 grains (gr)
0.032 ounce (oz)

**1 kilogram (kg) =**
1,000. grams (g)
1,000,000. milligrams (mg)
1,000,000,000. micrograms (μg)
32. ounces (oz)
2.7 pounds (lb)

**1 milligram (mg) =**
0.000001 kilogram (kg)
0.001 gram (g)
1,000. micrograms (μg)
0.0154 grains (gr)
0.000032 ounce (oz)

### Metric Volumes and Their Apothecaries' Equivalents

**1 liter (L) =**
1,000. milliliters (mL)
1,000. cubic centimeters (cc)
2.1 pints (pt)
270. fluid drams (fl dr)
34. fluid ounces (fl oz)

**1 milliliter (mL) =**
0.001 liter (L)
1. cubic centimeter (cc)
16.2 minims (min)
0.27 fluid dram (fl dr)
0.034 fluid ounce (fl oz)

### Approximate Equivalents of Household Measures

**1 teaspoon (tsp) =**
4. milliliters (mL)
4. cubic centimeters (cc)
1. fluid dram (fl dr)

**1 tablespoon (tbsp) =**
15. milliliters (mL)
15. cubic centimeters (cc)
0.5 fluid ounce (fl oz)
3.7 teaspoons (tsp)

**1 cup (c) =**
240. milliliters (mL)
240. cubic centimeters (cc)
8. fluid ounces (fl oz)
0.5 pint (pt)
16. tablespoons (tbsp)

**1 quart (qt) =**
960.0 milliliters (mL)
960. cubic centimeters (cc)
2. pints (pt)
4. cups (c)
32. fluid ounces (fl oz)

## Conversion of Units from One Form to Another

Refer to the preceding equivalency lists when converting one unit to another equivalent unit.

To convert a unit shown in bold type to one of the equivalent units listed immediately below it, multiply the first number (bold type unit) by the appropriate equivalent unit listed below it.

## Sample Problems

1. Convert 320 grains into scruples (1 gr = 0.05 s).

$$320 \text{ gr} \times \frac{0.05 \text{ s}}{1 \text{ gr}} = 16.0 \text{ s}$$

2. Convert 320 grains into drams (1 gr = 0.017 dr).

$$320 \text{ gr} \times \frac{0.017 \text{ dr}}{1 \text{ gr}} = 5.44 \text{ dr}$$

3. Convert 320 grains into grams (1 gr = 0.065 g).

$$320 \text{ gr} \times \frac{0.065 \text{ g}}{1 \text{ gr}} = 20.8 \text{ g}$$

## Body Temperatures in °Fahrenheit and °Celsius

| °F | °C | °F | °C |
|------|------|-------|------|
| 95.0 | 35.0 | 100.0 | 37.8 |
| 95.2 | 35.1 | 100.2 | 37.9 |
| 95.4 | 35.2 | 100.4 | 38.0 |
| 95.6 | 35.3 | 100.6 | 38.1 |
| 95.8 | 35.4 | 100.8 | 38.2 |
| 96.0 | 35.5 | 101.0 | 38.3 |
| 96.2 | 35.7 | 101.2 | 38.4 |
| 96.4 | 35.8 | 101.4 | 38.6 |
| 96.6 | 35.9 | 101.6 | 38.7 |
| 96.8 | 36.0 | 101.8 | 38.8 |
| 97.0 | 36.1 | 102.0 | 38.9 |
| 97.2 | 36.2 | 102.2 | 39.0 |
| 97.4 | 36.3 | 102.4 | 39.1 |
| 97.6 | 36.4 | 102.6 | 39.2 |
| 97.8 | 36.6 | 102.8 | 39.3 |
| 98.0 | 36.7 | 103.0 | 39.4 |
| 98.2 | 36.8 | 103.2 | 39.6 |
| 98.4 | 36.9 | 103.4 | 39.7 |
| 98.6 | 37.0 | 103.6 | 39.8 |
| 98.8 | 37.1 | 103.8 | 39.9 |
| 99.0 | 37.2 | 104.0 | 40.0 |
| 99.2 | 37.3 | 104.2 | 40.1 |
| 99.4 | 37.4 | 104.4 | 40.2 |
| 99.6 | 37.6 | 104.6 | 40.3 |
| 99.8 | 37.7 | 104.8 | 40.4 |
|      |      | 105.0 | 40.6 |

**To convert °F to °C**
Subtract 32 from °F and multiply by 5/9.
___ °F − 32 × 5/9 = ___ °C
**To convert °C to °F**
Multiply °C by 9/5 and add 32.
___ °C × 9/5 + 32 = ___ °F

## Laboratory Tests of Clinical Importance

### Common Tests Performed on Blood

| Test | Normal Values* (adult) | Clinical Significance |
|---|---|---|
| Albumin (serum) | 3.2–5.5 gm/100 mL | Values increase in multiple myeloma and decrease with proteinuria and as a result of severe burns. |
| Albumin-globulin ratio, or A/G ratio (serum) | 1.5:1 to 2.5:1 | Ratio of albumin to globulin is lowered in kidney diseases and malnutrition. |
| Ammonia | 12–55 $\mu$ mol/L | Values increase in severe liver disease, pneumonia, shock, and congestive heart failure. |
| Amylase (serum) | 4–25 units/mL | Values increase in acute pancreatitis, intestinal obstructions, and mumps. They decrease in chronic pancreatitis, cirrhosis of the liver, and toxemia of pregnancy. |
| Bilirubin, total (serum) | 0–1.0 mg/100 mL | Values increase in conditions causing red blood cell destruction or biliary obstruction. |
| Blood urea nitrogen, or BUN (plasma or serum) | 8–25 mg/100 mL | Values increase in various kidney disorders and decrease in liver failure and during pregnancy. |
| Calcium (serum) | 8.5–10.5 mg/100 mL | Values increase in hyperparathyroidism, hypervitaminosis D, and respiratory conditions that cause a rise in $CO_2$ concentration. They decrease in hypoparathyroidism, malnutrition, and severe diarrhea. |
| Carbon dioxide (serum) | 24–30 mEq/L | Values increase in respiratory diseases, intestinal obstruction, and vomiting. They decrease in acidosis, nephritis, and diarrhea. |
| Chloride (serum) | 100–106 mEq/L | Values increase in nephritis, Cushing's syndrome, dehydration, and hyperventilation. They decrease in metabolic acidosis, Addison's disease, diarrhea, and following severe burns. |
| Cholesterol, total (serum) | 120–220 mg/100 mL (below 200 mg/100 mL recommended by the American Heart Association) | Values increase in diabetes mellitus and hypothyroidism. They decrease in pernicious anemia, hyperthyroidism, and acute infections. |
| Cholesterol, high-density lipoprotein (HDL) | Women: 30–80 mg/100 mL Men: 30–70 mg/100 mL | Values increase in liver disease. Decreased values are associated with an increased risk of atherosclerosis. |
| Cholesterol, low-density lipoprotein (LDL) | 62–185 mg/100 mL | Increased values are associated with an increased risk of atherosclerosis. |
| Creatine (serum) | 0.2–0.8 mg/100 mL | Values increase in muscular dystrophy, nephritis, severe damage to muscle tissue, and during pregnancy. |
| Creatinine (serum) | 0.6–1.5 mg/100 mL | Values increase in various kidney diseases. |
| Ferritin (serum) | Men: 10–270 µg/100 mL Women: 5–280 /gu/g/100 mL | Values correlate with total body iron store. They decrease with iron deficiency. |
| Globulin (serum) | 2.3–3.5 gm/100 mL | Values increase as a result of chronic infections. |
| Glucose (plasma) | 70–110 mg/100 mL | Values increase in diabetes mellitus, liver diseases, nephritis, hyperthyroidism, and pregnancy. They decrease in hyperinsulinism, hypothyroidism, and Addison's disease. |

*These values may vary with hospital, physician, and type of equipment used to make measurements.

| Test | Normal Values* (adult) | Clinical Significance |
|------|------------------------|------------------------|
| Hematocrit (whole blood) | Men: 40–54 mL/100 mL<br>Women: 37–47 mL/100 mL<br>Children: 35–49 mL/100 mL<br>(varies with age) | Values increase in polycythemia due to dehydration or shock. They decrease in anemia and following severe hemorrhage. |
| Hemoglobin (whole blood) | Men: 14–18 gm/100 mL<br>Women: 12–16 gm/100 mL<br>Children: 11.2–16.5 gm/100 mL<br>(varies with age) | Values increase in polycythemia, obstructive pulmonary diseases, congestive heart failure, and at high altitudes. They decrease in anemia, pregnancy, and as a result of severe hemorrhage or excessive fluid intake. |
| Iron (serum) | 50–150 µg/100 mL | Values increase in various anemias and liver disease. They decrease in iron-deficiency anemia. |
| Iron-binding capacity (serum) | 250–410 µg/100 mL | Values increase in iron-deficiency anemia and pregnancy. They decrease in pernicious anemia, liver disease, and chronic infections. |
| Lactic acid (whole blood) | 0.6–1.8 mEq/L | Values increase with muscular activity and in congestive heart failure, severe hemorrhage, and shock. |
| Lactic dehydrogenase, or LDH (serum) | 45–90 U/L | Values increase in pernicious anemia, myocardial infarction, liver disease, acute leukemia, and widespread carcinoma. |
| Lipids, total (serum) | 450–850 mg/100 mL | Values increase in hypothyroidism, diabetes mellitus, and nephritis. They decrease in hyperthyroidism. |
| Magnesium | 1.3–2.1 mEq/L | Values increase in renal failure, hypothyroidism, and Addison's disease. They decrease in renal disease, liver disease, and pancreatitis. |
| Mean corpuscular hemoglobin (MCH) | 26–32 pg/RBC | Values increase in macrocytic anemia. They decrease in microcytic anemia. |
| Mean corpuscular volume (MCV) | 86–98 µ mm³/RBC | Values increase in liver disease and pernicious anemia. They decrease in iron-deficiency anemia. |
| Osmolality | 275–295 osmol/kg | Values increase in dehydration, hypercalcemia, and diabetes mellitus. They decrease in hyponatremia, Addison's disease, and water intoxication. |
| Oxygen saturation (whole blood) | Arterial: 96–100%<br>Venous: 60–85% | Values increase in polycythemia and decrease in anemia and obstructive pulmonary diseases. |
| pH (whole blood) | 7.35–7.45 | Values increase due to mild vomiting, Cushing's syndrome, and hyperventilation. They decrease as a result of hypoventilation, severe diarrhea, Addison's disease, and diabetic acidosis. |
| Phosphatase acid (serum) | Women: 0.01–0.56 Sigma U/mL<br>Men: 0.13–0.63 Sigma U/mL | Values increase in cancer of the prostate gland, hyperparathyroidism, certain liver diseases, myocardial infarction, and pulmonary embolism. |
| Phosphatase, alkaline (serum) | 13–39 U/L | Values increase in hyperparathyroidism (and in other conditions that promote resorption of bone), liver diseases, and pregnancy. |
| Phosphorus (serum) | 3.0–4.5 mg/100 mL | Values increase in kidney diseases, hypoparathyroidism, acromegaly, and hypervitaminosis D. They decrease in hyperparathyroidism. |
| Platelet count (whole blood) | 150,000–350,000/mm³ | Values increase in polycythemia and certain anemias. They decrease in acute leukemia and aplastic anemia. |
| Potassium (serum) | 3.5–5.0 mEq/L | Values increase in Addison's disease, hypoventilation, and conditions that cause severe cellular destruction. They decrease in diarrhea, vomiting, diabetic acidosis, and chronic kidney disease. |
| Protein, total (serum) | 6.0–8.4 gm/100 mL | Values increase in severe dehydration and shock. They decrease in severe malnutrition and hemorrhage. |

*These values may vary with hospital, physician, and type of equipment used to make measurements.

**A**

| Test | Normal Values* (adult) | Clinical Significance |
|------|------------------------|------------------------|
| Prothrombin time (serum) | 12–14 sec (one stage) | Values increase in certain hemorrhagic diseases, liver disease, vitamin K deficiency, and following the use of various drugs. |
| Red cell count (whole blood) | Men: 4,600,000–6,200,000/mm³<br>Women: 4,200,000–5,400,000/mm³<br>Children: 4,500,000–5,100,000/mm³<br>(varies with age) | Values increase as a result of severe dehydration or diarrhea, and decrease in anemia, leukemia, and following severe hemorrhage. |
| Red cell distribution width (RDW) | 8.5–11.5 microns | Variation in cell width changes with pernicious anemia. |
| Sedimentation rate, erythrocyte (whole blood) | Men: 1–13 mm/hr<br>Women: 1–20 mm/hr | Values increase in infectious diseases, menstruation, pregnancy, and as a result of severe tissue damage. |
| Serum glutamic pyruvic transaminase (SGPT) | Women: 4–17 U/L<br>Men: 6–24 U/L | Values increase in liver disease, pancreatitis, and acute myocardial infarction. |
| Sodium (serum) | 135–145 mEq/L | Values increase in nephritis and severe dehydration. They decrease in Addison's disease, myxedema, kidney disease, and diarrhea. |
| Thromboplastin time, partial (plasma) | 35–45 sec | Values increase in deficiencies of blood factors VIII, IX, and X. |
| Thyroid-stimulating hormone (TSH) | 0.5–5.0 µU/mL | Values increase in hypothyroidism and decrease in hyperthyroidism. |
| Thyroxine, or $T_4$ (serum) | 4–12 µg/100 mL | Values increase in hyperthyroidism and pregnancy. They decrease in hypothyroidism. |
| Transaminases, or SGOT (serum) | 7–27 units/mL | Values increase in myocardial infarction, liver disease, and diseases of skeletal muscles. |
| Triglycerides | 40–150 mg/100 mL | Values increase in liver disease, nephrotic syndrome, hypothyroidism, and pancreatitis. They decrease in malnutrition and hyperthyroidism. |
| Triiodothyronine, or T3 (serum) | 75–195 ng/100 mL | Values increase in hyperthyroidism and decrease in hypothyroidism. |
| Uric acid (serum) | Men: 2.5–8.0 mg/100 mL<br>Women: 1.5–6.0 mg/100 mL | Values increase in gout, leukemia, pneumonia, toxemia of pregnancy, and as a result of severe tissue damage. |
| White blood cell count, differential (whole blood) | Neutrophils 54–62%<br>Eosinophils 1–3%<br>Basophils <1%<br>Lymphocytes 25–33%<br>Monocytes 3–7% | Neutrophils increase in bacterial diseases; lymphocytes and monocytes increase in viral diseases; eosinophils increase in collagen diseases, allergies, and in the presence of intestinal parasites. |
| White blood cell count, total (whole blood) | 5,000–10,000/mm³ | Values increase in acute infections, acute leukemia, and following menstruation. They decrease in aplastic anemia and as a result of drug toxicity. |

## Common Tests Performed on Urine

| Test | Normal Values* (adult) | Clinical Significance |
|------|------------------------|------------------------|
| Acetone and acetoacetate | 0 | Values increase in diabetic acidosis. |
| Albumin, qualitative | 0 to trace | Values increase in kidney disease, hypertension, and heart failure. |
| Ammonia | 20–70 mEq/L | Values increase in diabetes mellitus and liver diseases. |
| Bacterial count | Under 10,000/mL | Values increase in urinary tract infection. |
| Bile and bilirubin | 0 | Values increase in melanoma and biliary tract obstruction. |
| Calcium | Under 300 mg/24 hr | Values increase in hyperparathyroidism and decrease in hypoparathyroidism. |
| Creatinine (24 hours) | 15–25 mg/kg body weight/day | Values increase in infections, and decrease in muscular atrophy, anemia, leukemia, and kidney diseases. |
| Creatinine clearance (24 hours) | 100–140 mL/min | Values increase in renal diseases. |

*These values may vary with hospital, physician, and type of equipment used to make measurements.

| Test | Normal Values* | Clinical Significance |
|---|---|---|
| Glucose | 0 | Values increase in diabetes mellitus and various pituitary gland disorders. |
| Hemoglobin | 0 | Blood may occur in urine as a result of extensive burns, crushing injuries, hemolytic anemia, or blood transfusion reactions. |
| 17-hydroxycorticosteroids | 3–8 mg/24 hr | Values increase in Cushing's syndrome and decrease in Addison's disease. |
| Osmolality | 850 osmol/kg | Values increase in hepatic cirrhosis, congestive heart failure, and Addison's disease. They decrease in hypokalemia, hypercalcemia, and diabetes insipidus. |
| pH | 4.6–8.0 | Values increase in urinary tract infections and chronic renal failure. They decrease in diabetes mellitus, emphysema, and starvation. |
| Phenylpyruvic acid | 0 | Values increase in phenylketonuria. |
| Specific gravity (SG) | 1.003–1.035 | Values increase in diabetes mellitus, nephrosis, and dehydration. They decrease in diabetes insipidus, glomerulonephritis, and severe renal injury. |
| Urea | 25–35 gm/24 hr | Values increase as a result of excessive protein breakdown. They decrease as a result of impaired renal function. |
| Urea clearance | Over 40 mL blood cleared of urea/min | Values increase in renal diseases. |
| Uric acid | 0.6–1.0 gm/24 hr as urate | Values increase in gout and decrease in various kidney diseases. |
| Urobilinogen | 0–4 mg/24 hr | Values increase in liver diseases and hemolytic anemia. They decrease in complete biliary obstruction and severe diarrhea. |

*These values may vary with hospital, physician, and type of equipment used to make measurements.

# GLOSSARY

Each word in this glossary is followed by a phonetic guide to pronunciation. In this guide, any unmarked vowel that ends a syllable or stands alone as a syllable has the long sound. Thus, the word *play* would be spelled *pla.* Any unmarked vowel that is followed by a consonant has the short sound. The word *tough,* for instance, would be spelled *tuf.* If a long vowel appears in the middle of a syllable (followed by a consonant), it is marked with the macron ( ̄ ), the sign for a long vowel. Thus, the word *plate* would be phonetically spelled plāt. Similarly, if a vowel stands alone or ends a syllable, but has the short sound, it is marked with a breve ( ̆ ).

## A

**abdomen** (ab-do′men)  Portion of the body between the diaphragm and the pelvis.  p. 22

**abdominal** (ab-dom′ĭ-nal)  Pertaining to the abdomen.  p. 20

**abdominal cavity** (ab-dom′ĭ-nal kav′ĭ-te)  The space between the diaphragm and the pelvic inlet that contains the abdominal viscera.  p. 13

**abdominopelvic cavity** (ab-dom″ĭ-no-pel′vik kav′ĭ-te)  The space between the diaphragm and the lower portion of the trunk of the body.  p. 11

**abduction** (ab-duk′shun)  Movement of a body part away from the midline.  p. 264

**absorption** (ab-sorp′shun)  The taking in of substances by cells or membranes.  p. 655

**accessory organs** (ak-ses′o-re or′ganz)  Organs that supplement the functions of other organs; accessory organs of the digestive and reproductive systems.  p. 457

**accommodation** (ah-kom″o-da′shun)  Adjustment of the lens for close vision.  p. 463

**acetone** (as′e-tōn)  One of the ketone bodies produced as a result of the oxidation of fats.  p. 991

**acetylcholine** (as″ĕ-til-ko′lēn)  A type of neurotransmitter, which is a biochemical secreted at the axon ends of many neurons. It transmits nerve impulses across synapses.  p. 289

**acetyl coenzyme A** (as′ĕ-til ko-en′zīm)  An intermediate compound produced during the oxidation of carbohydrates and fats.  p. 423

**acid** (as′id)  A substance that ionizes in water to release hydrogen ions.  p. 45

**acid-base buffer system** (as′id-bās buf′er sis′tem)  A pair of chemicals, one a weak acid, the other a weak base, that resists pH changes.  p. 825

**acidosis** (as″ĭ-do′sis)  A relative increase in the acidity of body fluids.  p. 45

**acromial** (ah-kro′me-al)  Pertaining to the shoulder.  p. 20

**ACTH**  Adrenocorticotropic hormone.  p. 489

**actin** (ak′tin)  A protein in a muscle fiber that, together with myosin, is responsible for contraction and relaxation.  p. 285

**action potential** (ak′shun po-ten′shal)  The sequence of electrical changes occurring when a nerve cell membrane is exposed to a stimulus that exceeds its threshold.  p. 355

**activation energy** (ak″tĭ-va′shun en′er-je)  Energy required to initiate a chemical reaction.  p. 634

**active site** (ak′tiv sīt)  Region of an enzyme molecule that temporarily combines with a substrate.  p. 104

**active transport** (ak′tiv trans′port)  Process that requires an expenditure of energy to move a substance across a cell membrane; usually moved against the concentration gradient.  p. 83

**acupuncture** (ak′u-pungk″chŭr)  Procedure in which needles are inserted into various tissues to control pain sensations.  p. 435

**adaptation** (ad″ap-ta′shun)  Adjustment to environmental conditions.  p. 5

**adduction** (ah-duk′shun)  Movement of a body part toward the midline.  p. 264

**adenoids** (ad′ĕ-noids)  The pharyngeal tonsils located in the nasopharynx.  p. 661

**adenosine diphosphate** (ah-den′o-sēn di-fos′fāt)  ADP; molecule produced when the terminal phosphate is lost from a molecule of adenosine triphosphate.  p. 108

**adenosine triphosphate** (ah-den′o-sēn tri-fos′fāt)  An organic molecule that stores energy and releases energy for use in cellular processes.  p. 72

**adenylate cyclase** (ah-den′ĭ-lāt si′klās)  An enzyme that is activated when certain hormones combine with receptors on cell membranes, causing ATP to release two phosphates and circularize, forming cyclic AMP.  p. 484

**ADH** Antidiuretic hormone.  p. 494

**adipose tissue** (ad′ĭ-pōs tish′u)  Fat-storing tissue.  p. 148

**adolescence** (ad″o-les′ens)  Period of life between puberty and adulthood.  p. 919

**ADP** Adenosine diphosphate.  p. 106

**adrenal cortex** (ah-dre′nal kor′teks)  The outer portion of the adrenal gland.  p. 502

**adrenal glands** (ah-dre′nal glandz)  Endocrine glands located on the superior portions of the kidneys.  p. 501

**adrenaline** (ah-dren′ah-lin)  Epinephrine. A hormone produced by the adrenal glands.  p. 502

**adrenal medulla** (ah-dre′nal me-dul′ah)  The inner portion of the adrenal gland.  p. 501

**adrenergic fiber** (ad″ren-er′jik fi′ber)  A nerve fiber that secretes norepinephrine at the axon terminal.  p. 423

**adrenocorticotropic hormone** (ah-dre″no-kor″te-ko-trōp′ik hor′mōn)  ACTH; hormone secreted by the anterior lobe of the pituitary gland that stimulates activity in the adrenal cortex.  p. 489

**aerobic respiration** (a″er-ōb′ik res″pĭ-ra′shun)  Phase of cellular respiration that requires oxygen.  p. 106

**afferent** (af′er-ent)  Conducting toward a center. For example, an afferent arteriole conveys blood to the glomerulus of a nephron within the kidney.  p. 781

**agglutination** (ah-gloo″ti-na′shun)  Clumping of blood cells in response to a reaction between an antibody and an antigen.  p. 542

**agranulocyte** (a-gran′u-lo-sīt)  A nongranular leukocyte.  p. 527

**albumin** (al-bu′min)  A plasma protein that helps regulate the osmotic concentration of blood.  p. 535

**aldosterone** (al-dos′ter-ōn)  A hormone, secreted by the adrenal cortex, that regulates sodium and potassium ion concentrations and water balance.  p. 503

**alimentary canal** (al″i-men′tar-e kah-nal′)  The tubular portion of the digestive tract that leads from the mouth to the anus.  p. 656

**alkaline** (al′kah-līn)  Pertaining to or having the properties of a base or alkali; basic.  p. 444

**alkaline tide** (al′kah-lin tīd)  An increase in the blood concentration of bicarbonate ions following a meal.  p. 673

**alkaloid** (al′kah-loid)  A group of organic substances that are usually bitter in taste and have toxic effects.  p. 444

**alkalosis** (al″kah-lo′sis)  A relative increase in the alkalinity of body fluids.  p. 45

**allantois** (ah-lan′to-is)  A structure that appears during embryonic development and functions in the formation of umbilical blood vessels.  p. 906

**allele** (ah-lēl′)  Different forms of a gene.  p. 937

**allergen** (al′er-jen)  A foreign substance capable of stimulating an allergic reaction.  p. 643

**all-or-none response** (al′or-nun′ re-spons′)  Phenomenon in which a muscle fiber completely contracts when it is exposed to a stimulus of threshold strength.  p. 294

**alpha receptor** (al′fah re-sep′tor)  Receptor on effector cell membrane that combines with epinephrine or norepinephrine.  p. 432

**alphatocopherol** (al″fah-to-kof′er-ol)  Vitamin E.  p. 716

**alveolar ducts** (al-ve′o-lar dukts′)  Fine tubes that carry air to the air sacs of the lungs.  p. 747

**alveolar pores** (al-ve′o-lar pōrz)  Minute openings in the walls of air sacs, which permit air to pass from one alveolus to another.  p. 764

**alveolar process** (al-ve′o-lar pros′es)  Projection on the border of the jaw in which the bony sockets of the teeth are located.  p. 661

**alveolus** (al-ve′o-lus)  An air sac of a lung; a saclike structure (pl., *alveoli*)  p. 747

**amacrine cell** (am′ah-krin sel)  A retinal neuron whose fibers pass laterally between other retinal cells.  p. 465

**amine** (am′in)  A type of nitrogen-containing organic compound, including the hormones secreted by the adrenal medulla.  p. 481

**amino acid** (ah-me′no as′id)  An organic compound of relatively small molecular size that contains an amino group ($-NH_2$) and a carboxyl group ($-COOH$); the structural unit of a protein molecule.  p. 51

**amniocentesis** (am″ne-o-sen-te′sis)  A procedure in which a sample of amniotic fluid is removed through the abdominal wall of a pregnant woman. Fetal cells in it are cultured and examined to check the chromosome complement.  p. 907

**amnion** (am′ne-on)  An extra embryonic membrane that encircles a developing fetus and contains amniotic fluid.  p. 898

**amniotic cavity** (am″ne-ot′ik kav′ĭ-te)  Fluid-filled space enclosed by the amnion.  p. 896

**amniotic fluid** (am″ne-ot′ik floo′id)  Fluid within the amniotic cavity that surrounds the developing fetus.  p. 899

**ampulla** (am-pul′ah)  An expansion at the end of each semicircular canal that contains a crista ampullaris.  p. 455

**amylase** (am′ĭ-lās)  An enzyme that hydrolyzes starch.  p. 665

**anabolism** (ah-nab′o-liz″em)  Metabolic process by which larger molecules are synthesized from smaller ones; anabolic metabolism.  p. 102

**anaerobic respiration** (an-a″er-ōb′ik res″pĭ-ra′shun)  Phase of cellular respiration that occurs in the absence of oxygen.  p. 106

**anal canal** (a′nal kah-nal′)  The most distal two or three inches of the large intestine that open to the outside as the anus.  p. 692

**anaphase** (an′ah-fāz)  Stage in mitosis when replicated chromosomes separate and move to opposite poles of the cell.  p. 91

**anastomosis** (ah-nas″to-mo′sis)  A union of nerve fibers or blood vessels to form a network.  p. 609

**anatomical position** (an″ah-tom′ĭ-kal po-zish′un)  A body posture with the body erect, the face forward, the arms at the sides with the palms facing forward, and the toes pointing straight ahead.  p. 18

**anatomy** (ah-nat′o-me)  Branch of science dealing with the form and structure of body parts.  p. 3

**androgen** (an′dro-jen)  A male sex hormone such as testosterone.  p. 850

**anemia** (ah-ne′me-ah)  A condition of red blood cell or hemoglobin deficiency.  p. 527

**aneuploid** (an′u-ploid)  A cell with one or more extra or missing chromosomes.  p. 949

**aneurysm** (an′u-rizm)  A saclike expansion of a blood vessel wall.  p. 146

**angiotensin** (an″je-o-ten′sin)  A vasoconstricting biochemical that is produced when blood flow to the kidneys is reduced, elevating blood pressure.  p. 504

**angiotensinogen** (an″je-o-ten-sin′o-jen)  A serum globulin the liver secretes that renin converts to angiotensin.  p. 504

**anion** (an′i-on)  An atom carrying a negative charge due to an extra electron.  p. 355

**anorexia nervosa** (an″o-rek′se-ah ner-vo′sa)  Disorder caused by the fear of becoming obese; includes loss of appetite and inability to maintain a normal minimum body weight.  p. 730

**anoxia** (an-ok′se-ah)  Abnormally low oxygen concentration of the tissues.  p. 774

**antagonist** (an-tag′o-nist)  A muscle that acts in opposition to a prime mover.  p. 300

**antebrachium** (an″te-bra′ke-um)  The forearm.  p. 316

**antecubital** (an″te-ku′bi-tal)  The region in front of the elbow joint.  p. 20

**anterior** (an-te′re-or)  Pertaining to the front.  p. 18

**anterior pituitary** (an-te′re-or pi-tu′i-tār″e)  The front lobe of the pituitary gland.  p. 488

**antibody** (an′ti-bod″e)  A protein that B cells of the immune system produce in response to the presence of a nonself antigen; it reacts with the antigen.  p. 530

**anticoagulant** (an″tĭ-ko-ag′u-lant)  A biochemical that inhibits blood clotting.  p. 538

**anticodon** (an″ti-ko′don)  Three contiguous nucleotides of a transfer RNA molecule that are complementary to a specific mRNA codon.  p. 122

**antidiuretic hormone** (an″tĭ-di″u-ret′ik hor′mōn) Hormone released from the posterior lobe of the pituitary gland that enhances the conservation of water by the kidneys; ADH. p. 489

**antigen** (an′tĭ-jen) A chemical that stimulates cells to produce antibodies. p. 542

**antigen binding site** (an′tĭ-jen bīn′ding sīt) Specialized ends of antibodies that bind specific antigens. p. 637

**antigen-presenting cell** (an′tĭ-jen-pre-sen′ting cel) The cell that displays the antigen to the cells of the immune system so they can defend the body against that particular antigen. p. 634

**antioxidant** (an′tĭ-ok′sĭ-dant) A substance that inhibits oxidation of another substance. p. 716

**antithrombin** (an″tĭ-throm′bin) A substance that inhibits the action of thrombin and thus inhibits blood clotting. p. 538

**anus** (a′nus) Inferior outlet of the digestive tube. p. 692

**aorta** (a-or′tah) Major systemic artery that receives blood from the left ventricle. p. 595

**aortic body** (a-or′tik bod′e) A structure associated with the wall of the aorta that contains a group of chemoreceptors. p. 595

**aortic sinus** (a-or′tik si′nus) Swelling in the wall of the aorta that contains baroreceptors. p. 595

**aortic valve** (a-or′tik valv) Flaplike structures in the wall of the aorta near its origin that prevent blood from returning to the left ventricle of the heart. p. 558

**apneustic area** (ap-nu′stik a′re-ah) A portion of the respiratory control center located in the pons. p. 402

**apocrine gland** (ap′o-krin gland) A type of sweat gland that responds during periods of emotional stress. p. 142

**aponeurosis** (ap″o-nu-ro′sis) A sheetlike tendon by which certain muscles are attached to other parts. p. 282

**apoptosis** (ap′o-to′-sis) Programmed cell death. p. 78

**appendicular** (ap″en-dik′u-lar) Pertaining to the arms or legs. p. 10

**appendix** (ah-pen′diks) A small, tubular appendage that extends outward from the cecum of the large intestine; vermiform appendix. p. 692

**aqueous humor** (a′kwe-us hu′mor) Watery fluid that fills the anterior and posterior chambers of the eye. p. 464

**arachnoid granulation** (ah-rak′noid gran″u-la′shun) Fingerlike structures that project from the subarachnoid space of the meninges into blood-filled dural sinuses and reabsorb cerebrospinal fluid. p. 397

**arachnoid mater** (ah-rak′noid ma′ter) Delicate, weblike middle layer of the meninges. p. 377

**arbor vitae** (ar′bor vi′ta) Treelike pattern of white matter in a section of cerebellum. p. 404

**areola** (ah-re′o-lah) Pigmented region surrounding the nipple of the mammary gland or breast. p. 29

**arrector pili muscle** (ah-rek′tor pil′i mus′l) Smooth muscle in the skin associated with a hair follicle. p. 162

**arrhythmia** (ah-rith′me-ah) Abnormal heart action characterized by a loss of rhythm. p. 574

**arterial pathway** (ar-te′re-al path′wa) Course blood follows as it flows from the heart to body cells. p. 583

**arteriole** (ar-te′re-ōl) A small branch of an artery that communicates with a capillary network. p. 573

**arteriosclerosis** (ar-te″re-o-sklĕ-ro′sis) Condition in which the walls of arteries thicken and lose their elasticity; hardening of the arteries. p. 573

**artery** (ar′ter-e) A vessel that transports blood away from the heart. p. 573

**arthritis** (ar-thri′tis) Joint inflammation. p. 262

**articular cartilage** (ar-tik′u-lar kar′tĭ-lij) Hyaline cartilage that covers the ends of bones in synovial joints. p. 185

**articulation** (ar-tik″u-la′shun) The joining of structures at a joint. p. 257

**ascending colon** (ah-send′ing ko′lon) Portion of the large intestine that passes upward on the right side of the abdomen from the cecum to the lower edge of the liver. p. 692

**ascending tracts** (ah-send′ing trakts) Groups of nerve fibers in the spinal cord that transmit sensory impulses upward to the brain. p. 378

**ascites** (ah-si′tez) Serous fluid accumulation in the abdominal cavity. p. 730

**ascorbic acid** (as-kor′bik as′id) One of the water-soluble vitamins; vitamin C. p. 720

**assimilation** (ah-sim″ĭ-la′shun) The action of chemically changing absorbed substances. p. 655

**association area** (ah-so″se-a′shun a′re-ah) Region of the cerebral cortex related to memory, reasoning, judgment, and emotional feelings. p. 407

**astigmatism** (ah-stig′mah-tizm) Visual defect due to errors in refraction caused by abnormal curvatures in the surface of the cornea or lens. p. 470

**astrocyte** (as′tro-sīt) A type of neuroglial cell that connects neurons to blood vessels. p. 351

**atherosclerosis** (ath″er-o-sklĕ-ro′sis) Condition in which fatty substances accumulate on the inner linings of arteries. p. 538

**atmospheric pressure** (at″mos-fer′ik presh′ur) Pressure exerted by the weight of the air; about 760 mm of mercury at sea level. p. 5

**atom** (at′om) Smallest particle of an element that has the properties of that element. p. 8

**atomic number** (ah-tom′ik num′ber) Number equal to the number of protons in an atom of an element. p. 38

**atomic weight** (ah-tom′ik wāt) Number approximately equal to the number of protons plus the number of neutrons in an atom of an element. p. 38

**ATP** Adenosine triphosphate, the biological energy molecule. p. 106

**ATPase** Enzyme that causes ATP molecules to release the energy stored in the terminal phosphate bonds. p. 288

**atrial natriuretic peptide** (a′trē-al na″trē-u-ret′ik pep′tīd) A family of polypeptide hormones that increase sodium excretion. p. 794

**atrioventricular bundle** (a″tre-o-ven-trik′u-lar bun′dl) Group of specialized fibers that conduct impulses from the atrioventricular node to the ventricular muscle of the heart; A-V bundle. p. 570

**atrioventricular node** (a″tre-o-ven-trik′u-lar nōd) Specialized mass of cardiac muscle fibers located in the interatrial septum of the heart; functions in transmitting cardiac impulses from the sinoatrial node to the ventricular walls; A-V node. p. 568

**atrioventricular orifice** (a″tre-o-ven-trik′u-lar or′i-fis) Opening between the atrium and the ventricle on one side of the heart. p. 557

**atrioventricular sulcus** (a″tre-o-ven-trik′u-lar sul′kus) A groove on the surface of the heart that marks the division between an atrium and a ventricle. p. 557

**atrioventricular valve** (a″tre-o-ven-trik′u-lar valv) Cardiac valve located between an atrium and a ventricle. p. 557

**atrium** (a′tre-um) A chamber of the heart that receives blood from veins (pl., *atria*). p. 556

**atrophy** (at′ro-fe) A wasting away or decrease in size of an organ or tissue. p. 297

**audiometer** (aw″de-om′ĕ-ter) An instrument used to measure the acuity of hearing. p. 448

**auditory** (aw′di-to″re) Pertaining to the ear or to the sense of hearing. p. 448

**auditory ossicle** (aw′di-to″re os′i-kl) A bone of the middle ear. p. 447

**auditory tube** (aw′di-to″re toob) The tube that connects the middle ear cavity to the pharynx; eustachian tube. p. 448

**auricle** (aw′ri-kl) An earlike structure; the portion of the heart that forms the wall of an atrium. p. 446

**autoantibody** (aw″to-an′tĭ-bod″e) An antibody produced against oneself. p. 646

**autocrine** (aw′to-krin) A hormone that acts on the same cell that secreted it. p. 480

**autoimmunity** (aw″to-ĭ-mu′ni-tē) An immune response against a person's own tissues; autoallergy. p. 642

**autonomic nervous system** (aw″to-nom′ik ner′vus sis′tem) Portion of the nervous system that controls the actions of the visceral organs and skin. p. 407

**autoregulation** (aw″to-reg″u-la′shun) Ability of an organ or tissue to maintain a constant blood flow in spite of changing arterial blood pressure. p. 793

**autosome** (aw′to-sōm) A chromosome other than a sex chromosome. p. 936

**A-V bundle** (bun′dl) A group of fibers that conduct cardiac impulses from the A-V node to the Purkinje fibers; bundle of His. p. 570

**A-V node** (nōd) Atrioventricular node. p. 570

**axial skeleton** (ak′se-al skel′ĕ-ton) Portion of the skeleton that supports and protects the organs of the head, neck, and trunk. p. 199

**axillary** (ak′sĭ-ler″e) Pertaining to the armpit. p. 20

**axon** (ak′son) A nerve fiber that conducts a nerve impulse away from a neuron cell body. p. 345

**axonal transport** (ak′so-nal trans′port) Transport of substances from the neuron cell body to an axon terminal. p. 346

## B

**baroreceptor** (bar″o-re-sep′tor) Sensory receptor in the blood vessel wall stimulated by changes in pressure (pressoreceptor). p. 433

**basal ganglion** (bas′al gang′gle-on) Mass of gray matter located deep within a cerebral hemisphere of the brain. p. 394

**basal metabolic rate** (ba′sal met″ah-bol′ic rāt) Rate at which metabolic reactions occur when the body is at rest. p. 710

**base** (bās) A substance that ionizes in water to release hydroxyl ions (OH⁻) or other ions that combine with hydrogen ions. p. 45

**basement membrane** (bās′ment mem′brān) A layer of nonliving material that anchors epithelial tissue to underlying connective tissue. p. 137

**basophil** (ba′so-fil) White blood cell containing cytoplasmic granules that stain with basophilic dye. p. 527

**beta oxidation** (ba′tah ok″sĭ-da′shun) Chemical process by which fatty acids are converted to molecules of acetyl coenzyme A. p. 113

**beta receptor** (ba′tah re-sep′tor) Receptor on an effector cell membrane that combines mainly with epinephrine and only slightly with norepinephrine. p. 424

**bicarbonate buffer system** (bi-kar′bo-nāt buf′er sis′tem) A mixture of carbonic acid and sodium bicarbonate that weakens a strong base and a strong acid, respectively; resists a change in pH. p. 825

**bicarbonate ion** (bi-kar′bon-āt i′on) HCO₃⁻. p. 771

**bicuspid tooth** (bi-kus′pid tooth) A premolar that is specialized for grinding hard particles of food. p. 661

**bicuspid valve** (bi-kus′pid valv) Heart valve located between the left atrium and the left ventricle; mitral valve. p. 558

**bile** (bīl) Fluid secreted by the liver and stored in the gallbladder. p. 681

**bilirubin** (bil″ĭ-roo′bin) A bile pigment produced from hemoglobin breakdown. p. 525

**biliverdin** (bil″ĭ-ver′din) A bile pigment produced from hemoglobin breakdown. p. 524

**biochemistry** (bi″o-kem′is-tre) Branch of science dealing with the chemistry of living organisms. p. 37

**biofeedback** (bi″o-fēd′bak) Procedure in which electronic equipment is used to help a person learn to consciously control certain visceral responses. p. 438

**biotin** (bi′o-tin) A water-soluble vitamin; a member of the vitamin B complex. p. 720

**bipolar neuron** (bi-po′lar nu′ron) A nerve cell whose cell body has only two processes, one an axon and the other a dendrite. p. 350

**blastocyst** (blas′to-sist) An early stage of prenatal development that consists of a hollow ball of cells. p. 867

**B lymphocyte** (lim′fo-sīt) Lymphocyte that reacts against foreign substances in the body by producing and secreting antibodies; B cell. p. 632

**BMR** Basal metabolic rate. p. 710

**bond** (bond) Connection between atoms in a compound. p. 39

**Bowman's capsule** (bo′manz kap′sŭl) Proximal portion of a renal tubule that encloses the glomerulus of a nephron; glomerular capsule. p. 785

**brachial** (bra′ke-al) Pertaining to the arm. p. 20

**bradycardia** (brad″e-kar′de-ah) An abnormally slow heart rate or pulse rate. p. 574

**brain stem** (brān stem) Portion of the brain that includes the midbrain, pons, and medulla oblongata. p. 387

**brain wave** (brān wāv) Recording of fluctuating electrical activity in the brain. p. 406

**Broca's area** (bro′kahz a′re-ah) Region of the frontal lobe that coordinates complex muscular actions of the mouth, tongue, and larynx, making speech possible. p. 391

**bronchial tree** (brong′ke-al trē) The bronchi and their branches that carry air from the trachea to the alveoli of the lungs. p. 746

**bronchiole** (brong′ke-ōl) A small branch of a bronchus within the lung. p. 747

**bronchus** (brong′kus) A branch of the trachea that leads to a lung (pl., *bronchi*). p. 746

**buccal** (buk′al) Pertaining to the mouth and the inner lining of the cheeks. p. 20

**buffer** (buf′er) A substance that can react with a strong acid or base to form a weaker acid or base, and thus resist a change in pH. p. 825

**bulbourethral glands** (bul″bo-u-re′thral glandz) Glands that secrete a viscous fluid into the male urethra at times of sexual excitement; Cowper's glands. p. 837

**bulimia** (bu-lim′e-ah) A disorder of binge eating followed by purging. p. 730

**bursa** (bur′sah) A saclike, fluid-filled structure, lined with synovial membrane, near a joint. p. 261

**bursitis** (bur-si′tis)   Inflammation of a bursa.   p. 276

# C

**calcification** (kal″sĭ-fĭ-ka′shun)   Deposition of calcium salts in a tissue.   p. 197

**calcitonin** (kal″sĭ-to′nin)   Hormone secreted by the thyroid gland that helps regulate the level of blood calcium.   p. 497

**calorie** (kal′o-re)   A unit used to measure heat energy and the energy contents of foods.   p. 710

**calorimeter** (kal″o-rim′ĕ-ter)   A device used to measure the heat energy content of foods; bomb calorimeter.   p. 710

**canaliculus** (kan″ah-lik′u-lus)   Microscopic canals that connect the lacunae of bone tissue.   p. 151

**cancellous bone** (kan′sĕ-lus bōn)   Bone tissue with a latticework structure; spongy bone.   p. 186

**capacitation** (kah-pas″i-ta′shun)   Activation of a sperm cell to fertilize an egg cell.   p. 844

**capillary** (kap′ĭ-ler″e)   A small blood vessel that connects an arteriole and a venule.   p. 573

**carbohydrate** (kar″bo-hi′drāt)   An organic compound that contains carbon, hydrogen, and oxygen, with a 2:1 ratio of hydrogen to oxygen atoms.   p. 47

**carbonic anhydrase** (kar-bon′ik an-hi′drās)   Enzyme that catalyzes the reaction between carbon dioxide and water to form carbonic acid.   p. 771

**carbon monoxide** (kar′bon mon-ok′sīd)   A toxic gas that combines readily with hemoglobin to form a relatively stable compound; CO.   p. 769

**carboxyhemoglobin** (kar″bok″se-he″mo-glo′bin)   Compound formed by the union of carbon dioxide and hemoglobin.   p. 767

**carboxypeptidase** (kar-bok″se-pep′ti-dās)   A protein-splitting enzyme in pancreatic juice.   p. 677

**cardiac center** (kar′de-ak sen′ter)   A group of neurons in the medulla oblongata that controls heart rate.   p. 402

**cardiac conduction system** (kar′de-ak kon-duk′shun sis′tem)   System of specialized cardiac muscle fibers that conducts cardiac impulses from the S-A node into the myocardium.   p. 568

**cardiac cycle** (kar′de-ak si′kl)   A series of myocardial contractions that constitutes a complete heartbeat.   p. 565

**cardiac muscle** (kar′de-ak mus′el)   Specialized type of muscle tissue found only in the heart.   p. 154

**cardiac output** (kar′de-ak owt′poot)   A measurement calculated by multiplying the stroke volume in milliliters by the heart rate in beats per minute.   p. 584

**cardiac sphincter** (kar′de-ak sfingk′ter)   A ring of muscle located at the junction of the esophagus and stomach that prevents food from re-entering the esophagus when the stomach contracts; gastroesophageal sphincter.   p. 669

**cardiac veins** (kar′de-ak vāns)   Blood vessels that return blood from the venules of the myocardium to the coronary sinus.   p. 563

**cardiovascular** (kar″de-o-vas′ku-lar)   Pertaining to the heart and blood vessels.   p. 16

**carina** (kah-ri′nah)   A cartilaginous ridge located between the openings of the right and left bronchi.   p. 747

**carotene** (kar′o-tēn)   A yellow, orange, or reddish pigment in plants and a precursor of vitamin A.   p. 714

**carotid bodies** (kah-rot′id bod′ēz)   Masses of chemoreceptors located in the wall of the internal carotid artery near the carotid sinus.   p. 600

**carpals** (kar′pals)   Bones of the wrist.   p. 200

**carpus** (kar′pus)   The wrist; the wrist bones as a group.   p. 226

**cartilage** (kar′tĭ-lij)   Type of connective tissue in which cells are located within lacunae and are separated by a semisolid matrix.   p. 150

**catabolism** (ka-tab′o-lizm)   Metabolic process that breaks down large molecules into smaller ones; catabolic metabolism.   p. 102

**catalase** (kat′ah-lās)   An enzyme that catalyzes the decomposition of hydrogen peroxide.   p. 104

**catalyst** (kat′ah-list)   A chemical that increases the rate of a chemical reaction, but is not permanently altered by the reaction.   p. 44

**cataract** (kat′ah-rakt)   Loss of transparency of the lens of the eye.   p. 464

**catecholamine** (kat″ĕ-kol′am-in)   A type of organic compound that includes epinephrine and norepinephrine.   p. 502

**cauda equina** (kaw′da ek-wīn′a)   A group of spinal nerves that extends below the distal end of the spinal cord.   p. 413

**cecum** (se′kum)   A pouchlike portion of the large intestine attached to the small intestine.   p. 692

**celiac** (se′le-ak)   Pertaining to the abdomen.   p. 20

**cell** (sel)   The structural and functional unit of an organism.   p. 5

**cell adhesion molecules** (sel ad-hee′zhon mol′ĕ-kūlz)   Proteins that guide cellular movement within the body.   p. 65

**cell body** (sel bod′e)   Portion of a nerve cell that includes a cytoplasmic mass and a nucleus, and from which the nerve fibers extend.   p. 345

**cell cycle** (sel sī-kl)   The life cycle of a cell consisting of G₁ (growth), S (DNA synthesis), G₂ (growth), and mitosis (division).   p. 88

**cell-mediated immunity** (sel me′de-ā-tĭd ĭ-mu′nĭ-te)   The body's attack on foreign antigens carried out by T-lymphocytes and their secreted products.   p. 633

**cell membrane** (sel mem′brān)   The selectively permeable outer boundary of a cell consisting of a phospholipid bilayer embedded with proteins.   p. 60

**cellular respiration** (sel′u-lar res″pĭ-ra′shun)   Process that releases energy from organic compounds in cells.   p. 106

**cellulose** (sel′u-lōs)   A polysaccharide very abundant in plant tissues that human enzymes cannot break down.   p. 705

**cementum** (se-men′tum)   Bonelike material that fastens the root of a tooth into its bony socket.   p. 663

**central canal** (sen′tral kah-nal′)   Tube within the spinal cord that is continuous with the ventricles of the brain and contains cerebrospinal fluid.   p. 378

**central nervous system** (sen′tral ner′vus sis′tem)   Portion of the nervous system that consists of the brain and spinal cord; CNS.   p. 344

**centriole** (sen′tre-ōl)   A cellular structure built of microtubules that organizes the mitotic spindle.   p. 73

**centromere** (sen′tro-mēr)   Portion of a chromosome to which spindle fibers attach during mitosis.   p. 90

**centrosome** (sen′tro-sōm)   Cellular organelle consisting of two centrioles.   p. 73

**cephalic** (sĕ-fal′ik)   Pertaining to the head.   p. 20

**cerebellar cortex** (ser″ĕ-bel′ar kor′teks)   The outer layer of the cerebellum.   p. 404

**cerebellar peduncles** (ser″ĕ-bel′ar pe-dung′kl)   A bundle of nerve fibers connecting the cerebellum and the brain stem.   p. 404

**cerebellum** (ser″ĕ-bel′um)  Portion of the brain that coordinates skeletal muscle movement.  p. 403

**cerebral aqueduct** (ser′ĕ-bral ak′wĕ-dukt″)  Tube that connects the third and fourth ventricles of the brain.  p. 395

**cerebral cortex** (ser′ĕ-bral kor′teks)  Outer layer of the cerebrum.  p. 390

**cerebral hemisphere** (ser′ĕ-bral hem′ĭ-sfēr)  One of the large, paired structures that together constitute the cerebrum of the brain.  p. 387

**cerebrospinal fluid** (ser′ĕ-bro-spi′nal floo′id)  Fluid occupying the ventricles of the brain, the subarachnoid space of the meninges, and the central canal of the spinal cord.  p. 377

**cerebrovascular accident** (ser″ĕ-bro-vas′ku-lar ak′si-dent)  Sudden interruption of blood flow to the brain; a stroke.  p. 395

**cerebrum** (ser′ĕ-brum)  Portion of the brain that occupies the upper part of the cranial cavity and provides higher mental functions.  p. 405

**cerumen** (sĕ-roo′men)  Waxlike substance produced by cells that line the external ear canal.  p. 446

**cervical** (ser′vĭ-kal)  Pertaining to the neck or to the cervix of the uterus.  p. 20

**cervix** (ser′viks)  Narrow, inferior end of the uterus that leads into the vagina.  p. 856

**chemoreceptor** (ke″mo-re-sep′tor)  A receptor that is stimulated by the binding of certain chemicals.  p. 432

**chemotaxis** (ke″mo-tak′sis)  Attraction of leukocytes to chemicals released from damaged cells.  p. 531

**chief cell** (chēf sel)  Cell of gastric gland that secretes various digestive enzymes, including pepsinogen.  p. 671

**chloride shift** (klo′rīd shift)  Movement of chloride ions from the blood plasma into red blood cells as bicarbonate ions diffuse out of the red blood cells into the plasma.  p. 771

**cholecystokinin** (ko″le-sis″to-ki′nin)  Hormone the small intestine secretes that stimulates release of pancreatic juice from the pancreas and bile from the gallbladder.  p. 673

**cholesterol** (ko-les′ter-ol)  A lipid produced by body cells used to synthesize steroid hormones and excreted into the bile.  p. 535

**cholinergic fiber** (ko″lin-er′jik fi′ber)  A nerve fiber that secretes acetylcholine at the axon terminal.  p. 423

**cholinesterase** (ko″lin-es′ter-ās)  An enzyme that catalyzes breakdown of acetylcholine.  p. 425

**chondrin** (kon′drin)  A protein in the intercellular substance of cartilage.  p. 150

**chondrocyte** (kon′dro-sīt)  A cartilage cell.  p. 150

**chorion** (ko′re-on)  Extra-embryonic membrane that forms the outermost covering around a fetus and contributes to formation of the placenta.  p. 897

**chorionic villi** (ko″re-on′ik vil′i)  Projections that extend from the outer surface of the chorion and help attach an embryo to the uterine wall.  p. 896

**choroid coat** (ko′roid kōt)  The vascular, pigmented middle layer of the wall of the eye.  p. 462

**choroid plexus** (ko′roid plek′sus)  Mass of specialized capillaries from which cerebrospinal fluid is secreted into a ventricle of the brain.  p. 395

**chromaffin cell** (kro-maf′in sel)  Cell of the adrenal medulla that produces, stores, and secretes epinephrine and norepinephrine.  p. 501

**chromatid** (kro′mah-tid)  One half of a replicated chromosome or a single unreplicated chromosome.  p. 934

**chromatin** (kro′mah-tin)  DNA and complexed protein that condenses to form chromosomes during mitosis.  p. 77

**chromosome** (kro′mo-sōm)  Rodlike structure that condenses from chromatin in a cell's nucleus during mitosis.  p. 930

**chylomicron** (kil″o-mi′kron)  A microscopic droplet of fat in the blood following fat digestion.  p. 536

**chyme** (kīm)  Semifluid mass of partially digested food that passes from the stomach to the small intestine.  p. 674

**chymotrypsin** (ki″mo-trip′sin)  A protein-splitting enzyme in pancreatic juice.  p. 677

**cilia** (sil′e-ah)  Microscopic, hairlike processes on the exposed surfaces of certain epithelial cells.  p. 73

**ciliary body** (sil′e-er″e bod′e)  Structure associated with the choroid layer of the eye that secretes aqueous humor and contains the ciliary muscle.  p. 462

**circadian rhythm** (ser″kah-de′an rithm)  A pattern of repeated behavior associated with the cycles of night and day.  p. 510

**circle of Willis** (sir′kl uv wil′is)  An arterial ring located on the ventral surface of the brain.  p. 599

**circular muscles** (ser′ku-lar mus′lz)  Muscles whose fibers are arranged in circular patterns, usually around an opening or in the wall of a tube; sphincter muscles.  p. 465

**circumduction** (ser″kum-duk′shun)  Movement of a body part, such as a limb, so that the end follows a circular path.  p. 265

**cisternae** (sis-ter′ne)  Enlarged portions of the sarcoplasmic reticulum near the actin and myosin filaments of a muscle fiber.  p. 69

**citric acid cycle** (sit′rik as′id si′kl)  A series of chemical reactions that oxidizes certain molecules, releasing energy; Krebs cycle.  p. 106

**cleavage** (klēv′ij)  The early successive divisions of the blastocyst cells into smaller and smaller cells.  p. 867

**clitoris** (kli′to-ris)  Small erectile organ located in the anterior portion of the vulva; corresponds to the penis.  p. 859

**clone** (klōn)  A group of cells that originate from a single cell and are therefore genetically identical.  p. 634

**CNS**  Central nervous system.  p. 344

**coagulation** (ko-ag″u-la′shun)  Blood clotting.  p. 537

**cocarboxylase** (ko″kar-bok′sĭ-lās)  A coenzyme synthesized from thiamine that oxidizes carbohydrates.  p. 717

**cochlea** (kok′le-ah)  Portion of the inner ear that contains the receptors of hearing.  p. 448

**codon** (ko′don)  A set of three nucleotides of a messenger RNA molecule corresponding to a particular amino acid.  p. 122

**coenzyme** (ko-en′zīm)  A nonprotein substance that is necessary for the activity of a particular enzyme.  p. 105

**coenzyme A** (ko-en′zīm)  Acetyl coenzyme A.  p. 719

**cofactor** (ko′fak-tor)  A nonprotein substance that must be combined with an enzyme for activity.  p. 105

**collagen** (kol′ah-jen)  Protein in the white fibers of connective tissues and in bone matrix.  p. 145

**collateral** (ko-lat′er-al)  A branch of a nerve fiber or blood vessel.  p. 421

**collecting duct** (ko-lek′ting dukt)  In the kidneys, a straight tubule that receives fluid from several nephrons.  p. 623

**colon** (kolon)  The large intestine.  p. 692

**colony stimulating factor** (ko′le-ne stim′yu-lay″ting fak′tor)  A protein that stimulates differentiation and maturation of white blood cells.  p. 522

**color blindness** (kul'er blīnd'nes) An inherited inability to distinguish certain colors. p. 432

**colostrum** (ko-los'trum) The first secretion of a woman's mammary glands after she gives birth. p. 876

**common bile duct** (kom'mon bīl dukt) Tube that transports bile from the cystic duct to the duodenum. p. 681

**compact bone** (kom'pakt bōn) Dense tissue in which cells are arranged in Haversian systems with no apparent spaces. p. 185

**compartment** (com-part'ment) A space occupied by a group of muscles, blood vessels, and nerves that is enclosed by fasciae. p. 283

**complement** (kom'plě-ment) A group of enzymes that are activated by the combination of antibody with antigen and enhance the reaction against foreign substances within the body. p. 637

**completely penetrant** (kom-plēt'le pen'e-trent) In genetics, indicates that the frequency of expression of a genotype is 100%. p. 942

**complete protein** (kom-plēt' pro'tēn) A protein that contains adequate amounts of the essential amino acids to maintain body tissues and to promote normal growth and development. p. 708

**compound** (kom'pownd) A substance composed of two or more chemically bonded elements. p. 37

**concussion** (kon-kush'un) Loss of consciousness due to a violent blow to the head. p. 395

**condom** (kon'dum) A latex sheath used to cover the penis during sexual intercourse; used as a contraceptive and to minimize the risk of transmitting infection. p. 879

**conduction** (kon-duk'shun) Movement of body heat into the molecules of cooler objects in contact with the body surface. p. 173

**condyle** (kon'dīl) A rounded process of a bone, usually at the articular end. p. 231

**cones** (kōns) Color receptors in the retina of the eye. p. 469

**conformation** (kon-for-ma'shun) The three-dimensional form of a protein, determined by its amino acid sequence and attractions and repulsions between amino acids. p. 51

**congenital** (kon-jen'ǐ-tal) A medical condition present at the time of birth not caused by a genetic defect. p. 955

**conjunctiva** (kon"junk-ti'vah) Membranous covering on the anterior surface of the eye. p. 458

**connective tissue** (kō-nek'tiv tish'u) One of the basic types of tissue that includes bone, cartilage, blood, loose and fibrous connective tissue. p. 144

**contraception** (kon"trah-sep'shun) A behavior or device that prevents fertilization. p. 877

**contralateral** (kon"trah-lat'er-al) Positioned on the opposite side of something else. p. 19

**convection** (kon-vek'shun) The transmission of heat from one substance to another through the circulation of heated air particles. p. 173

**convergence** (kon-ver'jens) Nerve impulses arriving at the same neuron. p. 365

**convolution** (kon"vo-lu'shun) An elevation on a structure's surface caused by infolding of it upon itself. p. 389

**cornea** (kor'ne-ah) Transparent anterior portion of the outer layer of the eye wall. p. 461

**coronary artery** (kor'o-na"re ar'ter-e) An artery that supplies blood to the wall of the heart. p. 562

**coronary sinus** (kor'o-na"re si'nus) A large vessel on the posterior surface of the heart into which the cardiac veins drain. p. 563

**corpus callosum** (kor'pus kah-lo'sum) A mass of white matter within the brain, composed of nerve fibers connecting the right and left cerebral hemispheres. p. 388

**corpus luteum** (kor'pus lu'te-um) Structure that forms from the tissues of a ruptured ovarian follicle and secretes female hormones. p. 862

**corpus striatum** (kor'pus stri-a'tum) Portion of the cerebrum that includes certain basal ganglia. p. 387

**cortex** (kor'teks) Outer layer of an organ such as the adrenal gland, cerebrum, or kidney. p. 501

**cortical nephron** (kor'tǐ-kl nef'ron) A nephron with its corpuscle located in the renal cortex. p. 788

**cortisol** (kor'ti-sol) A glucocorticoid secreted by the adrenal cortex. p. 504

**costal** (kos'tal) Pertaining to the ribs. p. 20

**countercurrent mechanism** (kown'ter-ker"ent me'kě-nǐ"zm) Process by which the kidneys concentrate urine. p. 796

**covalent bond** (ko'va-lent bond) Chemical bond formed by electron sharing between atoms. p. 41

**coxal** (kok'sel) Pertaining to the hip. p. 200

**cranial** (kra'ne-al) Pertaining to the cranium. p. 11

**cranial cavity** (kran'e-al kav'i-te) A hollow space in the cranium containing the brain. p. 11

**cranial nerve** (kra'ne-al nerv) Nerve that arises from the brain. p. 407

**creatine phosphate** (kre'ah-tin fos'fāt) A muscle biochemical that stores energy. p. 291

**crenation** (kre-na'shun) Shrinkage of a cell caused by contact with a hypertonic solution. p. 60

**crest** (krest) A ridgelike projection of a bone. p. 202

**cretinism** (kre'tǐ-nizm) A condition resulting from lack of thyroid secretion in an infant. p. 498

**cricoid cartilage** (kri'koid kar'tǐ-lij) A ringlike cartilage that forms the lower end of the larynx. p. 742

**crista ampullaris** (kris'tah am-pul-lah'ris) Sensory organ located within a semicircular canal that functions in the sense of dynamic equilibrium. p. 455

**crossing over** (kros'ing o'ver) The exchange of genetic material between homologous chromosomes during meiosis. p. 934

**crossmatching** (kros'mach"ing) A procedure used to determine whether donor and recipient blood samples will agglutinate. p. 542

**crural** (krur'al) Pertaining to the leg. p. 20

**cubital** (ku'bi-tal) Pertaining to the forearm. p. 20

**cuspid** (kus'pid) A canine tooth. p. 662

**cutaneous** (ku-ta'ne-us) Pertaining to the skin. p. 161

**cyanocobalamin** (si"ah-no-ko-bal'ah-min) Vitamin $B_{12}$. p. 719

**cyanosis** (si"ah-no'sis) Bluish skin coloration due to decreased blood oxygen concentration. p. 176

**cyclic AMP** (sik'lik AMP) A circularized derivative of ATP that responds to messages entering a cell and triggers the cell's response. p. 484

**cyclosporine** (si"klo-spor'in) A drug that suppresses the action of helper T cells, preventing rejection of transplanted tissue. p. 644

**cystic duct** (sis'tik dukt) Tube that connects the gallbladder to the common bile duct. p. 681

**cytochrome** (si'to-krōm) Protein within the inner mitochondrial membrane that is an electron carrier in aerobic respiration (electron transport chain). p. 112

**cytocrine secretion** (si'to-krin se-kre'shun) Transfer of granules of melanin from melanocytes into adjacent epithelial cells. p. 633

**cytokine** (si'to-kīn)  A type of protein secreted by a T lymphocyte that attacks viruses, virally infected cells, and cancer cells.  p. 633

**cytoplasm** (si'to-plazm)  The contents of a cell excluding the nucleus and cell membrane.  p. 60

**cytoskeleton** (si'to-skel″e-ton)  A system of protein tubules and filaments that reinforces a cell's three-dimensional form and provides scaffolding and transport tracts for organelles.  p. 68

### D

**deamination** (de-am″ĭ-na'shun)  Removing amino groups (NH₂) from amino acids.  p. 114

**deciduous teeth** (de-sid'u-us tēth)  Teeth that are shed and replaced by permanent teeth; primary teeth.  p. 662

**decomposition** (de-kom″po-zish'un)  The breakdown of molecules into simpler compounds.  p. 44

**defecation** (def″ĕ-ka'shun)  The discharge of feces from the rectum through the anus.  p. 695

**dehydration** (de″hi-dra'shun)  Excessive water loss.  p. 820

**dehydration synthesis** (de″hi-dra'shun sin'thĕ-sis)  Anabolic process that joins small molecules; synthesis.  p. 102

**dendrite** (den'drīt)  Nerve fiber that transmits impulses toward a neuron cell body.  p. 345

**densitometer** (den″si-tom'e-ter)  An instrument used to measure the density of bone tissue.  p. 198

**dental caries** (den'tal kar'ēz)  Decalcification and decay of teeth.  p. 664

**dentin** (den'tin)  Bonelike substance that forms the bulk of a tooth.  p. 663

**deoxyhemoglobin** (de-ok″sĭ-he″mo-glo'bin)  Hemoglobin to which oxygen is not bound.  p. 522

**deoxyribonucleic acid** (dē-ok'si-rī″bō-nu-klē″ik as'id)  The genetic material; a double-stranded polymer of nucleotides, each containing a phosphate group, a nitrogenous base (adenine, thymine, guanine, or cytosine), and the sugar deoxyribose.  p. 117

**depolarization** (de-po″lar-ĭ-za'shun)  The loss of an electrical charge on the surface of a membrane.  p. 355

**dermatome** (der'mah-tōm)  An area of the body supplied by sensory nerve fibers associated with a particular dorsal root of a spinal nerve.  p. 414

**dermis** (der'mis)  The thick layer of the skin beneath the epidermis.  p. 162

**descending colon** (de-send'ing ko'lon)  Portion of the large intestine that passes downward along the left side of the abdominal cavity to the brim of the pelvis.  p. 692

**descending tracts** (de-send'ing trakts)  Groups of nerve fibers that carry nerve impulses downward from the brain through the spinal cord.  p. 378

**desmosome** (des'mo-sōm)  A specialized junction between cells, which serves as a "spot weld."  p. 66

**detrusor muscle** (de-trūz'or mus'l)  Muscular wall of the urinary bladder.  p. 805

**dextrose** (dek'strōs)  Glucose.  p. 705

**diabetes insipidus** (di″ah-be'tēz in-sip'ĭ-dus)  Extremely copious urine produced due to a deficiency of antidiuretic hormone.  p. 495

**diabetes mellitus** (di″ah-be'tēz mel-li'tus)  High blood glucose level and glucose in the urine due to a deficiency of insulin.  p. 480

**dialysis** (di-al'ĭ-sis)  Separation of smaller molecules from larger ones in a liquid.  p. 42

**diapedesis** (di″ah-pĕ-de'sis)  Squeezing of leukocytes between the cells of blood vessel walls.  p. 531

**diaphragm** (di'ah-fram)  A sheetlike structure largely composed of skeletal muscle and connective tissue that separates the thoracic and abdominal cavities; also a caplike contraceptive device inserted in the vagina.  p. 11

**diaphysis** (di-af'ĭ-sis)  The shaft of a long bone.  p. 185

**diastole** (di-as'to-le)  Phase of the cardiac cycle when a heart chamber wall relaxes.  p. 566

**diastolic pressure** (di-a-stol'ik presh'ur)  Arterial blood pressure during the diastolic phase of the cardiac cycle.  p. 583

**diencephalon** (di″en-sef'ah-lon)  Portion of the brain in the region of the third ventricle that includes the thalamus and hypothalamus.  p. 398

**differentiation** (dif″er-en″she-a'shun)  Cell specialization.  p. 92

**diffusion** (dĭ-fu'zhun)  Random movement of molecules from a region of higher concentration toward one of lower concentration.  p. 78

**digestion** (di-jest'yun)  Breaking down of large nutrient molecules into smaller molecules that can be absorbed; hydrolysis.  p. 18

**digital** (di'ji-tal)  Pertaining to the finger.  p. 20

**dihydrotestosterone** (di-hi″dro-tes-tos'ter-ōn)  Hormone produced from testosterone that stimulates certain cells of the male reproductive system.  p. 850

**dipeptide** (di-pep'tīd)  A molecule composed of two joined amino acids.  p. 102

**disaccharide** (di-sak'ah-rīd)  A sugar produced by the union of two monosaccharides.  p. 47

**distal** (dis'tal)  Farther from the midline or origin; opposite of proximal.  p. 19

**diuretic** (di″u-ret'ik)  A substance that increases urine production.  p. 494

**divergence** (di-ver'jens)  A spreading apart.  p. 366

**DNA**  Deoxyribonucleic acid.  p. 53

**dominant gene** (dom'ĭ-nant jēn)  The member of a gene pair that is expressed while its allele is not expressed.  p. 940

**dorsal cavity** (dor'sal kav'i-te)  A hollow space in the posterior portion of the body containing the cranial cavity and vertebral canal.  p. 10

**dorsal root** (dor'sal rōot)  The sensory branch of a spinal nerve by which it joins the spinal cord.  p. 413

**dorsal root ganglion** (dor'sal rōot gang'gle-on)  Mass of sensory neuron cell bodies located in the dorsal root of a spinal nerve.  p. 413

**dorsiflexion** (dor″si-flek'shun)  Bending the foot upward.  p. 264

**dorsum** (dor'sum)  Pertaining to the back surface of a body part.  p. 20

**ductus arteriosus** (duk'tus ar-te″re-o'sus)  Blood vessel that connects the pulmonary artery and the aorta in a fetus.  p. 913

**ductus venosus** (duk'tus ven-o'sus)  Blood vessel that connects the umbilical vein and the inferior vena cava in a fetus.  p. 913

**duodenum** (du″o-de'num)  The first portion of the small intestine that leads from the stomach to the jejunum.  p. 685

**dural sinus** (du'ral si'nus)  Blood-filled channel formed by the splitting of the dura mater into two layers.  p. 376

**dura mater** (du'rah ma'ter)  Tough outer layer of the meninges.  p. 376

**dynamic equilibrium** (di-nam'ik e″kwĭ-lib're-um)  Maintenance of balance when the head and body are suddenly moved or rotated.  p. 455

**dystrophin** (dis'tre-fin)  A protein comprising only 0.002% of the total protein in skeletal muscle that supports the cell membrane. Its absence causes muscular dystrophy.  p. 288

**eccrine gland** (ek′rin gland)   Sweat gland that maintains body temperature.   p. 171

**ECG**   Electrocardiogram; EKG.   p. 571

**ectoderm** (ek′to-derm)   The outermost primary germ layer.   p. 900

**edema** (ĕ-de′mah)   Accumulation of fluid within the tissue spaces.   p. 535

**effector** (ĕ-fek′tor)   A muscle or gland that responds to stimulation.   p. 344

**efferent** (ef′er-ent)   Conducting away from the center. For example, an efferent arteriole conducts blood away from the glomerulus of a nephron.   p. 785

**ejaculation** (e-jak″u-la′shun)   Discharge of sperm-containing semen from the male urethra.   p. 847

**ejaculatory duct** (e-jak′u-lah-to″re dukt)   Tube, formed by the joining of the vas deferens and the tube from the seminal vesicle, that transports sperm to the urethra.   p. 842

**elastin** (e-las′tin)   Protein that comprises the yellow, elastic fibers of connective tissue.   p. 146

**electrocardiogram** (e-lek″tro-kar′de-o-gram″)   A recording of the electrical activity associated with the heartbeat; ECG or EKG.   p. 571

**electrolyte** (e-lek′tro-līt)   A substance that ionizes in water solution.   p. 44

**electrolyte balance** (e-lek′tro-līt bal′ans)   Condition when the quantities of electrolytes entering the body equal those leaving it.   p. 47

**electron** (e-lek′tron)   A small, negatively charged particle that revolves around the nucleus of an atom.   p. 37

**electron transport chain** (e-lek′tron trans′pohrt)   A series of oxidation-reduction reactions that takes high-energy electrons from the citric acid cycle and forms water and ATP.   p. 112

**electrovalent bond** (e-lek″tro-va′lent bond)   Chemical bond formed between two ions as a result of the transfer of electrons.   p. 41

**element** (el′ĕ-ment)   A basic chemical substance.   p. 37

**elevation** (el-e-vā-shun)   Upward movement of a part of the body.   p. 265

**embolus** (em′bo-lus)   A substance, such as a blood clot or bubble of gas, that is carried by the blood and obstructs a blood vessel.   p. 539

**embryo** (em′bre-o)   A prenatal stage of development after germ layers form but before the rudiments of all organs are present.   p. 867

**embryonic disk** (em″brē-on′ik disk)   A flattened area in the cleavage embryo from which the embryo arises.   p. 896

**emission** (e-mish′un)   The movement of sperm cells from the vas deferens into the ejaculatory duct and urethra.   p. 846

**emphysema** (em″fĭ-se′mah)   Abnormal enlargement of the air sacs of the lungs.   p. 761

**emulsification** (e-mul″sĭ-fĭ′ka′shun)   Breaking up of fat globules into smaller droplets by the action of bile salts.   p. 684

**enamel** (e-nam′el)   Hard covering on the exposed surface of a tooth.   p. 662

**endocardium** (en″do-kar′de-um)   Inner lining of the heart chambers.   p. 557

**endochondral bone** (en′do-kon′dral bōn)   Bone that begins as hyaline cartilage that is subsequently replaced by bone tissue.   p. 189

**endocrine gland** (en′do-krin gland)   A gland that secretes hormones directly into the blood or body fluids.   p. 142

**endocytosis** (en″do-si-to′sis)   Physiological process by which substances may move through a cell membrane and involves the formation of tiny, cytoplasmic vacuoles.   p. 83

**endoderm** (en′do-derm)   The innermost primary germ layer.   p. 900

**endolymph** (en′do-limf)   Fluid contained within the membranous labyrinth of the inner ear.   p. 448

**endometrium** (en″do-me′tre-um)   The inner lining of the uterus.   p. 856

**endomysium** (en″do-mis′e-um)   The sheath of connective tissue surrounding each skeletal muscle fiber.   p. 283

**endoneurium** (en″do-nu′re-um)   Layer of loose connective tissue that surrounds individual nerve fibers.   p. 407

**endoplasmic reticulum** (en-do-plaz′mic rĕ-tik′u-lum)   Organelle composed of a system of connected membranous tubules and vesicles along which protein synthesis occurs.   p. 68

**endorphin** (en-dor′fin)   A neuropeptide synthesized in the pituitary gland that suppresses pain.   p. 365

**endosteum** (en-dos′tē-um)   Tissue lining the medullary cavity within the bone.   p. 186

**endothelium** (en″do-the′le-um)   The layer of epithelial cells that forms the inner lining of blood vessels and heart chambers.   p. 956

**energy** (en′er-je)   An ability to cause something to move and thus to do work.   p. 106

**energy balance** (en′er-je bal′ans)   When the caloric intake of the body equals its caloric output.   p. 711

**enkephalin** (en-kef′ah-lin)   A neuropeptide that occurs in the brain and spinal cord; it inhibits pain impulses.   p. 365

**enterogastric reflex** (en-ter-o-gas′trik re′fleks)   Inhibition of gastric (stomach) peristalsis and secretions when food enters the small intestine.   p. 674

**enzyme** (en′zīm)   A protein that catalyzes a specific biochemical reaction.   p. 51

**eosinophil** (e″o-sin′o-fil)   White blood cell containing cytoplasmic granules that stain with acidic dye.   p. 527

**ependyma** (ĕ-pen′dĭ-mah)   A membrane composed of neuroglial cells that lines the ventricles of the brain.   p. 352

**epicardium** (ep″ĭ-kar′de-um)   The visceral portion of the pericardium on the surface of the heart.   p. 556

**epicondyle** (ep″ĭ-kon′dīl)   A projection of a bone located above a condyle.   p. 231

**epidermis** (ep″ĭ-der′mis)   Outer epithelial layer of the skin.   p. 162

**epididymis** (ep″ĭ-did′ĭ-mis)   Highly coiled tubule that leads from the seminiferous tubules of the testis to the vas deferens.   p. 842

**epidural space** (ep″ĭ-du′ral spās)   The space between the dural sheath of the spinal cord and the bone of the vertebral canal.   p. 376

**epigastric region** (ep″ĭ-gas′trik re′jun)   The upper middle portion of the abdomen.   p. 19

**epiglottis** (ep″ĭ-glot′is)   Flaplike cartilaginous structure located at the back of the tongue near the entrance to the trachea.   p. 742

**epimysium** (ep″ĭ-mis′e-um)   The outer sheath of connective tissue surrounding a skeletal muscle.   p. 282

**epinephrine** (ep″ĭ-nef′rin)   A hormone the adrenal medulla secretes during times of stress.   p. 421

**epineurium** (ep″ĭ-nu′re-um)   Outermost layer of connective tissue surrounding a nerve.   p. 407

**epiphyseal disk** (ep″ĭ-fiz′e-al disk)   Cartilaginous layer within the epiphysis of a long bone that grows.   p. 190

**epiphysis** (e-pif′ĭ-sis)   The end of a long bone.   p. 185

**epiploic appendage** (ep″i-plo′ik ah-pen′dij)   Small collections of fat in the serous layer of the large intestinal wall.   p. 694

**epithelium** (ep″ĭ-the′le-um) Tissue type that covers all free body surfaces. p. 136

**equilibrium** (e″kwĭ-lib′re-um) A state of balance between two opposing forces. p. 79

**erythroblast** (ĕ-rith′ro-blast) An immature red blood cell. p. 526

**erythroblastosis fetalis** (ĕ-rith″ro-blas-to′sis fe-tal′is) A life-threatening condition of massive agglutination of the blood in the fetus or neonate due to the mother's anti-Rh antibodies reacting with the baby's Rh-positive red blood cells. p. 547

**erythrocyte** (ĕ-rith′ro-sīt) A red blood cell. p. 522

**erythropoiesis** (ĕ-rith″ro-poi-e′sis) Red blood cell formation. p. 526

**erythropoietin** (ĕ-rith″ro-poi′ĕ-tin) (EPo) A kidney hormone that promotes red blood cell formation. p. 526

**esophageal hiatus** (ĕ-sof″ah-je′al hi-a′tus) Opening in the diaphragm through which the esophagus passes. p. 667

**esophagus** (ĕ-sof′ah-gus) Tubular portion of the digestive tract that leads from the pharynx to the stomach. p. 667

**essential amino acid** (ĕ-sen′shal ah-me′no as′id) Amino acid required for health that body cells cannot synthesize in adequate amounts. p. 114

**essential fatty acid** (ĕ-sen′shal fat′e as′id) Fatty acid required for health that body cells cannot synthesize in adequate amounts. p. 706

**essential nutrient** (ĕ-sen′shal nu′trē-ent) A nutrient necessary for growth, normal functioning, and maintaining life that the diet must supply because the body cannot synthesize it. p. 704

**estrogen** (es′tro-jen) Hormone that stimulates the development of female secondary sex characteristics. p. 860

**euploid** (u′ploid) Having a balanced set of chromosomes. p. 949

**eustachian tube** (yoo-sta′she-an tūb) Tube that connects the middle ear to the pharynx; auditory tube. p. 485

**evaporation** (e″vap′o-ra-shun) Changing a liquid into a gas. p. 173

**eversion** (e-ver′zhun) Outward turning movement of the sole of the foot. p. 265

**exchange reaction** (eks-chānj re-ak′shun) A chemical reaction in which parts of two kinds of molecules trade positions. p. 44

**excretion** (ek-skre′shun) Elimination of metabolic wastes. p. 827

**exocrine gland** (ek′so-krin gland) A gland that secretes its products into a duct or onto a body surface. p. 142

**expiration** (ek″spĭ-ra′shun) Expulsion of air from the lungs. p. 751

**extension** (ek-sten′shun) Movement increasing the angle between parts at a joint. p. 262

**extracellular** (ek″strah-sel′u-lar) Outside of cells; refers to the organism's internal environment, body fluids outside the individual cells. p. 814

**extrapyramidal tract** (ek″strah-pi-ram′i-dal trakt) Nerve tracts, other than the corticospinal tracts, that transmit impulses from the cerebral cortex into the spinal cord. p. 381

**extremity** (ek-strem′ĭ-te) A limb; an arm or leg. p. 179

# F

**facet** (fas′et) A small, flattened surface of a bone. p. 202

**facilitated diffusion** (fah-sil′ĭ-tāt′id dĭ-fu′zhun) Diffusion in which substances move through membranes from a region of higher concentration to a region of lower concentration by carrier molecules. p. 80

**facilitation** (fah-sil″ĭ-tā′shun) The hastening of any natural process, increasing efficiency of the natural process. p. 365

**fallopian tube** (fah-lo′pe-an tūb) Tube that transports an egg cell from the region of the ovary to the uterus; oviduct or uterine tube. p. 855

**fascia** (fash′e-ah) A sheet of fibrous connective tissue that encloses a muscle. p. 282

**fasciculus** (fah-sik′u-lus) A small bundle of muscle fibers. p. 378

**fat** (fat) Adipose tissue; or an organic substance whose molecules contain glycerol and fatty acids. p. 148

**fatty acid** (fat′e as′id) A building block of a fat molecule. p. 48

**fatty acid oxidase** (fat′e as′id ok′si-days″) An enzyme that catalyzes the removal of hydrogen or electrons from a fatty acid molecule. p. 113

**feces** (fe′sēz) Material expelled from the digestive tract during defecation. p. 695

**femoral** (fem′or-al) Pertaining to the thigh. p. 20

**ferritin** (fer′ĭ-tin) An iron-protein complex that stores iron in liver cells. p. 525

**fertilization** (fer″tĭ-lĭ-za′shun) The union of an egg cell and a sperm cell. p. 864

**fetoscopy** (fe-tos′ko-pe) The direct observation of a fetus using a fiber-optic device. p. 867

**fetus** (fe′tus) A human embryo after eight weeks of development. p. 867

**fever** (fe′ver) Elevation of body temperature above normal. p. 631

**fibril** (fi′bril) A tiny fiber or filament. p. 144

**fibrillation** (fi″brĭ-la′shun) Uncoordinated contraction of muscle fibers. p. 331

**fibrin** (fi′brin) Insoluble, fibrous protein formed from fibrinogen during blood coagulation. p. 538

**fibrinogen** (fi-brin′o-jen) Plasma protein converted into fibrin during blood coagulation. p. 535

**fibrinolysin** (fi″brĭ-nol′ĭ-sin) A protein-splitting enzyme that can digest fibrin in a blood clot. p. 535

**fibroblast** (fi′bro-blast) Cell that produces fibers and other intercellular materials in connective tissues. p. 144

**fibrocartilage** (fi″bro-kar′ti-lij) Strongest and most durable cartilage; made up of cartilage cells and many collagenous fibers. p. 150

**filtration** (fil-tra′shun) Movement of material through a membrane as a result of hydrostatic pressure. p. 81

**filtration pressure** (fil-tra′shun presh′ur) Equal to the hydrostatic pressure of the blood entering the glomerulus minus the pressure of the opposing forces (the hydrostatic pressure within the glomerular capsule and the plasma osmotic pressure of the blood in the glomerulus). p. 791

**fissure** (fish′ūr) A narrow cleft separating parts, such as the lobes of the cerebrum. p. 389

**flaccid paralysis** (flak′sid pah-ral′ĭ-sis) Total loss of muscle tone when nerve fibers are damaged. p. 382

**flagella** (flah-jel′ah) Relatively long, motile processes that extend from the surface of a cell. p. 73

**flexion** (flek′shun) Bending at a joint to decrease the angle between bones. p. 262

**follicle** (fol′ĭ-kl) A pouchlike depression or cavity. p. 168

**follicle-stimulating hormone** (fol′ĭ-kl stim′u-la″ting hor′mōn) A substance secreted by the anterior pituitary gland that stimulates development of an ovarian follicle in a female or production of sperm cells in a male; FSH. p. 489

**follicular cells** (fŏ-lik′u-lar selz) Ovarian cells that surround a developing egg cell and secrete female sex hormones. p. 642

**fontanel** (fon″tah-nel′)   Membranous region between certain cranial bones in the skull of a fetus or infant.   p. 213

**foramen** (fo-ra′men)   An opening, usually in a bone or membrane (pl., *foramina*).   p. 202

**foramen magnum** (fo-ra′men mag′num)   Opening in the occipital bone of the skull through which the spinal cord passes.   p. 205

**foramen ovale** (fo-ra′men o-val′e)   Opening in the interatrial septum of the fetal heart.   p. 913

**forebrain** (fōr′brān)   The anteriormost portion of the developing brain that gives rise to the cerebrum and basal ganglia.   p. 387

**formula** (fōr′mu-lah)   A group of symbols and numbers used to express the composition of a compound.   p. 42

**fossa** (fos′ah)   A depression in a bone or other part.   p. 202

**fovea** (fo′ve-ah)   A tiny pit or depression.   p. 202

**fovea centralis** (fo′ve-ah sen-tral′is)   Region of the retina, consisting of densely packed cones, that provides the greatest visual acuity.   p. 465

**fracture** (frak′chur)   A break in a bone.   p. 194

**free radical** (frē rad′eh-kel)   Highly reactive by-product of metabolism that can damage tissue.   p. 923

**frenulum** (fren′u-lum)   A fold of tissue that anchors and limits movement of a body part.   p. 660

**frontal** (frun′tal)   Pertaining to the forehead.   p. 20

**FSH**   Follicle-stimulating hormone.   p. 489

**functional syncytium** (funk′shun-al sin-sish′e-um)   Merging cells performing together as a unit; those of the heart are joined electrically. A syncytium lacks cell boundaries, appearing as a multinucleated structure.   p. 568

## G

**galactose** (gah-lak′tōs)   A monosaccharide component of the disaccharide lactose.   p. 705

**gallbladder** (gawl′blad-er)   Saclike organ associated with the liver that stores and concentrates bile.   p. 681

**gamete** (gam′ēt)   A sex cell; either an egg cell or a sperm cell.   p. 899

**ganglion** (gang′gle-on)   A mass of neuron cell bodies, usually outside the central nervous system (pl., *ganglia*).   p. 408

**gastric gland** (gas′trik gland)   Gland within the stomach wall that secretes gastric juice.   p. 671

**gastric juice** (gas′trik jōōs)   Secretion of the gastric glands within the stomach.   p. 671

**gastric lipase** (gas′trik lī′pās)   Fat-splitting enzyme in gastric juice.   p. 671

**gastrin** (gas′trin)   Hormone secreted by the stomach lining that stimulates secretion of gastric juice.   p. 672

**gastrula** (gas′troo-lah)   Embryonic stage following the blastula; cells differentiate into three layers: the endoderm, the mesoderm, and the ectoderm.   p. 896

**gene** (jēn)   Portion of a DNA molecule that encodes the information to synthesize a protein, a control sequence, or tRNA or rRNA. The unit of inheritance.   p. 117

**genetic code** (jĕ-net′ik kōd)   Information for synthesizing proteins that is encoded in the nucleotide sequence of DNA molecules.   p. 117

**genetic heterogeneity** (jĕ-net′ik het″er-o-je-ne′ĭ-te)   Different genotypes that have identical phenotypes.   p. 943

**genetics** (jĕ-net′iks)   The study of heredity.   p. 929

**genital** (jen′i-tal)   Pertaining to the genitalia (internal and external organs of reproduction).   p. 20

**genome** (jeh′nōm)   All of the DNA in a cell of an organism.   p. 930

**genotype** (je′no-tīp)   The combination of genes present within a zygote or within the cells of an individual.   p. 940

**germinal epithelium** (jer′mĭ-nal ep″ĭ-the′le-um)   Tissue within an ovary that gives rise to sex cells.   p. 852

**germ layers** (jerm la′ers)   Layers of cells within an embryo that form the body organs during development: ectoderm, mesoderm, and endoderm.   p. 896

**glans penis** (glanz pe′nis)   Enlarged mass of corpus spongiosum at the end of the penis; may be covered by the foreskin.   p. 845

**globin** (glo′bin)   The protein portion of a hemoglobin molecule.   p. 826

**globulin** (glob′u-lin)   A type of protein in blood plasma.   p. 535

**glomerulus** (glo-mer′u-lus)   A capillary tuft located within the glomerular capsule of a nephron.   p. 785

**glottis** (glot′is)   Slitlike opening between the true vocal folds or vocal cords.   p. 744

**glucagon** (gloo′kah-gon)   Hormone secreted by the pancreatic islets of Langerhans that releases glucose from glycogen.   p. 507

**glucocorticoid** (gloo″ko-kor′tĭ-koid)   Any one of a group of hormones secreted by the adrenal cortex that influences carbohydrate, fat, and protein metabolism.   p. 504

**gluconeogenesis** (gloo″ko-ne″o-jen′ĕ-sis)   Synthesis of glucose from noncarbohydrates such as amino acids.   p. 706

**glucose** (gloo′kōs)   A monosaccharide in the blood that is the primary source of cellular energy.   p. 705

**glucosuria** (gloo′ko-sur′e-ah)   Presence of glucose in urine.   p. 795

**gluteal** (gloo′te-al)   Pertaining to the buttocks.   p. 20

**glycerol** (glis′er-ol)   An organic compound that is a building block for fat molecules.   p. 48

**glycogen** (gli′ko-jen)   A polysaccharide that stores glucose in the liver and muscles.   p. 705

**glycolysis** (gli-kol′ĭ-sis)   The conversion of glucose to pyruvic acid during cellular respiration.   p. 106

**glycoprotein** (gli″ko-pro′te-in)   A compound composed of a carbohydrate combined with a protein.   p. 65

**goblet cell** (gob′let sel)   An epithelial cell that is specialized to secrete mucus.   p. 138

**goiter** (goi′ter)   An enlarged thyroid gland.   p. 492

**Golgi apparatus** (gol′jē ap″ah-ra′tus)   An organelle that prepares cellular products for secretion.   p. 69

**Golgi tendon organ** (gol′jē ten′dun or′gan)   Sensory receptors in tendons close to muscle attachments that are involved in reflexes that help maintain posture.   p. 438

**gomphosis** (gom-fo′sis)   Type of joint in which a cone-shaped process is fastened in a bony socket.   p. 258

**gonad** (go′nad)   A sex cell–producing organ; an ovary or testis.   p. 853

**gonadotropin** (go-nad″o-trōp′in)   A hormone that stimulates activity in the gonads.   p. 848

**granulocyte** (gran′u-lo-sīt)   A leukocyte that contains granules in its cytoplasm.   p. 527

**gray matter** (grā mat′er)   Region of the central nervous system that generally lacks myelin and thus appears gray.   p. 390

**gray ramus** (grā ra′mus)   A short nerve containing postganglionic axons returning to a spinal nerve.   p. 421

**groin** (groin)   Region of the body between the abdomen and thighs.   p. 853

**growth** (grōth)   Process by which a structure enlarges.   p. 96

**growth hormone** (grōth hōr′mōn)   A hormone released by the anterior lobe of the pituitary gland that promotes the growth of the organism; GH or somatotropin.   p. 193

**gubernaculum** (goo″ber′nak′u-lum)   A structure that guides another structure.   p. 836

## H

**hair cell** (hār sel)   Mechanoreceptor in the inner ear that lies between the basilar membrane and the tectorial membrane and triggers action potentials in fibers of the auditory nerve.   p. 450

**hair follicle** (hār fol′ĭ-kl)   Tubelike depression in the skin in which a hair develops.   p. 168

**half-life** (haf′līf)   The time it takes for one-half of the radioactivity of an isotope to be released.   p. 41

**hapten** (hap′ten)   A small molecule that, in combination with a larger one, forms an antigenic substance and can later stimulate an immune reaction by itself.   p. 632

**haustra** (haws′trah)   Pouches in the wall of the large intestine.   p. 694

**hematocrit** (he-mat′o-krit)   The volume percentage of red blood cells within a sample of whole blood.   p. 521

**hematoma** (he″mah-to′mah)   A mass of coagulated blood within tissues or a body cavity.   p. 376

**hematopoiesis** (hem″ah-to-poi-e′sis)   The production of blood and blood cells; hemopoiesis.   p. 197

**heme** (hēm)   The iron-containing portion of a hemoglobin molecule.   p. 525

**hemizygous** (hem″ĭ-zi′gus)   A gene carried on the Y chromosome in humans.   p. 946

**hemocytoblast** (he″mo-si′to-blast)   A cell that gives rise to blood cells.   p. 522

**hemoglobin** (he″mo-glo′bin)   Pigment of red blood cells responsible for the transport of oxygen.   p. 197

**hemolysis** (he-mol′ĭ-sis)   The rupture of red blood cells accompanied by the release of hemoglobin.   p. 81

**hemopoiesis** (he″mo-poi-e′sis)   The production of blood and blood cells; hematopoiesis.   p. 197

**hemorrhage** (hem′ō-rij)   Loss of blood from the circulatory system; bleeding.   p. 716

**hemostasis** (he″mo-sta′sis)   The stoppage of bleeding.   p. 537

**heparin** (hep′ah-rin)   A substance that interferes with the formation of a blood clot; an anticoagulant.   p. 542

**hepatic** (hĕ-pat′ik)   Pertaining to the liver.   p. 678

**hepatic lobule** (hĕ-pat′ik lob′ūl)   A functional unit of the liver.   p. 678

**hepatic sinusoid** (hĕ-pat′ik si′nŭ-soid)   Vascular channel within the liver.   p. 607

**heredity** (hĕ-red′ĭ-te)   The transmission of genetic information from parent to offspring.   p. 929

**heritable gene therapy** (her′ĭ-tah-bl jēn ther′ah-pe)   Manipulation of genes to cure a medical condition.   p. 955

**heterozygote** (het″er-o-zi′gōt)   An individual who possesses different alleles in a gene pair.   p. 937

**hilum** (hi′lum)   A depression where vessels, nerves, and other structures (bronchus, ureter, etc.) enter an organ.   p. 627

**hilus** (hi′lus)   Hilum.   p. 751

**hindbrain** (hīnd′brān)   Posteriormost portion of the developing brain that gives rise to the cerebellum, pons, and medulla oblongata.   p. 387

**hippocampus** (hip″o-kam′pus)   A part of the cerebral cortex where memories form.   p. 393

**histamine** (his′tah-min)   A substance released from stressed cells.   p. 531

**histology** (his-tol′o-je)   The study of the structure and function of tissues.

**homeostasis** (ho″me-o-sta′sis)   A state of equilibrium in which the internal environment of the body remains in the normal range.   p. 6

**homozygote** (ho″mo-zi′gōt)   An individual possessing identical genes in a gene pair.   p. 937

**hormone** (hor′mōn)   A substance secreted by an endocrine gland that is transmitted in the blood or body fluids.   p. 480

**humoral immunity** (hu′mor-al ĭ-mu′nĭ-te)   Destruction of cells bearing foreign (nonself) antigens by circulating antibodies.   p. 634

**hydrolysis** (hi-drol′ĭ-sis)   Splitting of a molecule into smaller portions by addition of a water molecule.   p. 102

**hydrostatic pressure** (hi′dro-stat′ik presh′ur)   Pressure exerted by fluids, such as blood pressure.   p. 5

**hydroxyapatite** (hi-drok″se-ap′ah-tīt)   A type of crystalline calcium phosphate found in bone matrix.   p. 197

**hydroxyl ion** (hi-drok′sil i′on)   OH⁻.   p. 45

**hymen** (hi′men)   A membranous fold of tissue that partially covers the vaginal opening.   p. 858

**hyperextension** (hi″per-ek-sten′shun)   Extreme extension; continuing extension beyond the anatomical position.   p. 264

**hyperglycemia** (hi″per-gli-se′me-ah)   Excess blood glucose.   p. 512

**hyperkalemia** (hi″per-kah-le′me-ah)   Elevated concentration of blood potassium.   p. 573

**hypernatremia** (hi″per-nah-tre′me-ah)   Elevated concentration of blood sodium.   p. 732

**hyperparathyroidism** (hi″per-par″ah-thi′roi-dizm)   Excess secretion of parathyroid hormone.   p. 501

**hyperplasia** (hi″per-pla′ze-ah)   Increased production and growth of new cells.   p. 96

**hyperpolarization** (hi″per-po″lar-i-za′shun)   An increase in the negativity of the resting potential of a cell membrane.   p. 355

**hypertension** (hi″per-ten′shun)   Elevated blood pressure.   p. 590

**hyperthyroidism** (hi″per-thi′roi-dizm)   Oversecretion of thyroid hormones.   p. 497

**hypertonic** (hi″per-ton′ik)   Condition in which a solution contains a greater concentration of dissolved particles than the solution with which it is compared.   p. 81

**hypertrophy** (hi-per′tro-fe)   Enlargement of an organ or tissue.

**hyperventilation** (hi″per-ven″tĭ-la′shun)   Breathing that is abnormally deep and prolonged.   p. 764

**hypervitaminosis** (hi″per-vi″tah-mĭ-no′sis)   Excessive intake of vitamins.   p. 715

**hypochondriac region** (hi″po-kon′dre-ak re′jun)   The portion of the abdomen on either side of the middle or epigastric region.   p. 19

**hypodermis** (hi″po-der′mis)   Mainly composed of fat, this loose layer is directly beneath the dermis; subcutaneous.   p. 162

**hypogastric region** (hi″po-gas′trik re′jun)   The lower middle portion of the abdomen.   p. 19

**hypoglycemia** (hi″po-gli-se′me-ah)   Abnormally low concentration of blood glucose.   p. 510

**hypokalemia** (hi″po-kah-le′me-ah)   A low concentration of blood potassium.   p. 573

**hyponatremia** (hi″po-nah-tre′me-ah)   A low concentration of blood sodium.   p. 733

**hypoparathyroidism** (hi″po-par″ah-thi′roi-dizm)   An undersecretion of parathyroid hormone.   p. 501

**hypophysis** (hi-pof′i-sis)   The pituitary gland.   p. 488

**hypoproteinemia** (hi″po-pro″te-ĭ-ne′me-ah)   A low concentration of blood proteins.   p. 821

**hypothalamus** (hi″po-thal′ah-mus)   A portion of the brain located below the thalamus and forming the floor of the third ventricle.   p. 399

**hypothyroidism** (hi″po-thi′roi-dizm)  A low secretion of thyroid hormones. p. 495

**hypotonic** (hi″po-ton′ik)  Condition in which a solution contains a lesser concentration of dissolved particles than the solution to which it is compared.  p. 81

**hypoxia** (hi-pok′se-ah)  A deficiency of oxygen in the tissues.  p. 767

## I

**idiotype** (id′e-o-tīp′)  The parts of an antibody's antigen binding site that are complementary in conformation to a particular antigen.  p. 637

**ileocecal valve** (il″e-o-se′kal valv)  Sphincter valve located at the distal end of the ileum where it joins the cecum.  p. 692

**ileum** (il′e-um)  Portion of the small intestine between the jejunum and the cecum.  p. 685

**iliac region** (il′e-ak re′jun)  Portion of the abdomen on either side of the lower middle, or hypogastric, region.  p. 19

**ilium** (il′e-um)  One of the bones of a coxal bone or hipbone.  p. 227

**immunity** (ĭ-mu′nĭ-te)  Resistance to the effects of specific disease-causing agents.  p. 632

**immunoglobulin** (im″u-no-glob′u-lin)  Globular plasma proteins that function as antibodies.  p. 634

**immunosuppressive drugs** (im″u-no-sŭ-pres′iv drugz)  Substances that suppress the immune response against transplanted tissue.  p. 644

**implantation** (im″plan-ta′shun)  The embedding of a cleavage embryo in the lining of the uterus.  p. 895

**impulse** (im′puls)  A wave of depolarization conducted along a nerve fiber or muscle fiber.  p. 358

**incisor** (in-si′zor)  One of the front teeth that is adapted for cutting food.  p. 662

**inclusion** (in-kloo′zhun)  A mass of lifeless chemical substance within the cytoplasm of a cell.  p. 77

**incomplete dominance** (in″kom-plēt′ do′meh-nents)  A heterozygote whose phenotype is intermediate between the phenotypes of the two homozygotes.  p. 941

**incompletely penetrant** (in″kom-plēt′le pen′e-trent)  When the frequency of genotype expression is less than 100%.  p. 942

**incomplete protein** (in″kom-plēt′ pro′tēn)  A protein that lacks adequate amounts of essential amino acids.  p. 708

**infection** (in-fek′shun)  The invasion and multiplication of microorganisms in body tissues. p. 630

**inferior** (in-fēr′e-or)  Situated below something else; pertaining to the lower surface of a part.  p. 18

**inflammation** (in″flah-ma′shun)  A tissue response to stress that is characterized by dilation of blood vessels and an accumulation of fluid in the affected region.  p. 177

**infrared ray** (in″frah-red′ ra)  A form of radiation energy, with wavelengths longer than visible light, by which heat moves from warmer surfaces to cooler surroundings.  p. 173

**infundibulum** (in″fun-dib′u-lum)  The stalk attaching the pituitary gland to the base of the brain.  p. 399

**ingestion** (in-jes′chun)  The taking of food or liquid into the body by way of the mouth.  p. 658

**inguinal** (ing′gwĭ-nal)  Pertaining to the groin region.  p. 20

**inguinal canal** (ing′gwĭ-nal kah-nal′)  Passage in the lower abdominal wall through which a testis descends into the scrotum.  p. 836

**inhibin** (in′hib′in)  A hormone secreted by cells of the testes and ovaries that inhibits the secretion of FSH from the anterior pituitary gland.  p. 849

**inorganic** (in″or-gan′ik)  Chemical substances that lack carbon and hydrogen.  p. 46

**insertion** (in-ser′shun)  The end of a muscle attached to a movable part.  p. 300

**inspiration** (in″spĭ-ra′shun)  Breathing in; inhalation.  p. 751

**insula** (in′su-lah)  A cerebral lobe located deep within the lateral sulcus.  p. 390

**insulin** (in′su-lin)  A hormone secreted by the pancreatic islets of Langerhans that controls carbohydrate metabolism.  p. 507

**integumentary** (in-teg-u-men′tar-e)  Pertaining to the skin and its accessory organs.  p. 14

**intercalated disk** (in-ter″kah-lāt′ed disk)  Membranous boundary between adjacent cardiac muscle cells.  p. 299

**intercellular** (in″ter-sel′u-lar)  Between cells.  p. 65

**intercellular fluid** (in″ter-sel′u-lar floo′id)  Tissue fluid located between cells other than blood cells.  p. 65

**interferon** (in″ter-fēr′on)  A class of immune system chemicals (cytokines)  that inhibit multiplication of viruses and growth of tumors.  p. 631

**interleukin** (in″ter-lu′kin)  A class of immune system chemicals (cytokines)  with varied effects on the body.  p. 634

**interneuron** (in″ter-nu′ron)  A neuron located between a sensory neuron and a motor neuron; intercalated; internuncial, or association neuron.  p. 350

**interphase** (in′ter-fāz)  Period between two cell divisions when a cell is carrying on its normal functions and prepares for division.  p. 88

**interstitial cell** (in″ter-stish′al sel)  A hormone-secreting cell located between the seminiferous tubules of the testis.  p. 838

**interstitial fluid** (in″ter-stish′al floo′id)  Same as intercellular fluid.  p. 65

**intervertebral disk** (in″ter-ver′tĕ-bral disk)  A layer of fibrocartilage located between the bodies of adjacent vertebrae.  p. 214

**intestinal gland** (in-tes′tĭ-nal gland)  Tubular gland located at the base of a villus within the intestinal wall.  p. 686

**intestinal juice** (in-tes′tĭ-nal jōōs)  The secretion of the intestinal glands.  p. 686

**intracellular** (in″trah-sel′u-lar)  Within cells.  p. 814

**intracellular fluid** (in″trah-sel′u-lar floo′id)  Fluid within cells.  p. 814

**intracellular junction** (in″trah-sel′u-lar jungk′shun)  A connection between the membranes of adjacent cells.  p. 814

**intramembranous bone** (in″trah-mem′brah-nus bōn)  Bone that forms from membranelike layers of primitive connective tissue.  p. 188

**intrathecal injection** (in″trah-the′kal in-jek′shun)  The infusion of a substance directly into the cerebrospinal fluid.  p. 378

**intrauterine device** (in″trah-u′ter-in de-vīs)  A solid object placed in the uterine cavity for purposes of contraception; IUD.  p. 881

**intrinsic factor** (in-trin′sik fak′tor)  A substance produced by the gastric glands that promotes absorption of vitamin $B_{12}$.  p. 672

**inversion** (in-ver′zhun)  Movement in which the sole of the foot is turned inward.  p. 265

**involuntary** (in-vol′un-tār″e)  Not consciously controlled; functions automatically.  p. 153

**iodopsin** (i″o-dop′sin)  A light-sensitive pigment within the cones of the retina.  p. 472

**ion** (i′on)  An atom or molecule with an electrical charge.  p. 41

**ionic bond** (i-on′ik bond)  A chemical bond formed by the transfer of electrons from one atom to another.  p. 41

**ionization** (i'on-i-za'shun)  Dissociation into ions.

**ipsilateral** (ip"si-lat'er-al)  On the same side.  p. 19

**iris** (i'ris)  Colored muscular portion of the eye that surrounds the pupil and regulates its size.  p. 464

**irritability** (ir"ĭ-tah-bil'ĭ-te)  The ability of an organism to react to changes in its environment.

**ischemia** (is-ke'me-ah)  A deficiency of blood in a body part.  p. 571

**isometric contraction** (i"so-met'rik kon-trak'shun)  Muscular contraction in which the muscle does not shorten.  p. 296

**isotonic contraction** (i"so-ton'ik kon-trak'shun)  Muscular contraction in which the muscle shortens.  p. 296

**isotonic solution** (i"so-ton'ik so-lu'shun)  A solution that has the same concentration of dissolved particles as the solution with which it is compared.  p. 81

**isotope** (i'so-tōp)  An atom that has the same number of protons as other atoms of an element but has a different number of neutrons in its nucleus.  p. 39

**IUD**  An intrauterine device.  p. 881

## J

**jejunum** (jĕ-joo'num)  Portion of the small intestine located between the duodenum and the ileum.  p. 685

**joint** (joint)  The union of two or more bones; an articulation.  p. 256

**joint capsule** (joint kap'sul)  An envelope, attached to the end of each bone at the joint, enclosing the cavity of a synovial joint.  p. 260

**juxtaglomerular apparatus** (juks"tah-glo-mer'u-lār ap"ah-ra'tus)  Structure located in the walls of arterioles near the glomerulus that regulates renal blood flow.  p. 788

**juxtamedullary nephron** (juks"tah-med'u-lār-e nef'ron)  A nephron with its corpuscle located near the renal medulla.  p. 788

## K

**karyotype** (kar'ē-o-tīp)  A set of chromosomes characteristic of a species arranged in homologous pairs. The human karyotype has 23 chromosome pairs.  p. 936

**keratin** (ker'ah-tin)  Protein present in the epidermis, hair, and nails.  p. 139

**keratinization** (ker"ah-tin"ĭ-za'shun)  The process by which cells form fibrils of keratin and harden.  p. 139

**ketogenesis** (ke"to-jen'ĕ-sis)  Formation of ketone bodies.  p. 113

**ketone body** (ke'tōn bod'e)  Type of compound produced during fat catabolism, including acetone, acetoacetic acid, and betahydroxybutyric acid.  p. 113

**ketosis** (ke"to'sis)  An abnormal elevation of ketone bodies in body fluids.  p. 828

**kilocalorie** (kil'o-kal"o-re)  One thousand calories.

**kilogram** (kil'o-gram)  A unit of weight equivalent to 1,000 grams.  p. 710

**kinase** (ki'nas)  An enzyme that converts an inactive or precursor form of another enzyme to an active form by adding a phosphate group.  p. 484

**Krebs cycle** (krebz sī'kl)  The citric acid cycle.  p. 106

**Kupffer cell** (koop'fer sel)  Large, fixed phagocyte in the liver that removes bacterial cells from the blood.  p. 680

**kwashiorkor** (kwash"e-or'kor)  Starvation resulting from a switch from breast milk to food deficient in nutrients.  p. 730

**kyphosis** (ki-fo'sis)  An abnormally increased convex curvature in the thoracic portion of the vertebral column.  p. 224

## L

**labor** (la'bor)  The process of childbirth.  p. 888

**labyrinth** (lab'ĭ-rinth)  The system of connecting tubes within the inner ear, including the cochlea, vestibule, and semicircular canals.  p. 448

**lacrimal gland** (lak'rĭ-mal gland)  Tear-secreting gland.  p. 458

**lactase** (lak'tās)  Enzyme that catalyzes breakdown of lactose into glucose and galactose.  p. 688

**lactate** (lak'tāt)  Lactic acid.  p. 990

**lactation** (lak-ta'shun)  Production of milk by the mammary glands.  p. 876

**lacteal** (lak'te-al)  A lymphatic capillary associated with a villus of the small intestine.  p. 686

**lactic acid** (lak'tik as'id)  An organic compound formed from pyruvic acid during anaerobic respiration.  p. 113

**lactose** (lak'tōs)  A disaccharide in milk; milk sugar.  p. 705

**lacuna** (lah-ku'nah)  A hollow cavity.  p. 897

**lamella** (lah-mel'ah)  A layer of matrix in bone tissue.  p. 151

**large intestine** (lahrj in-tes'tin)  The part of the gastrointestinal tract extending from the ileum to the anus; divided into the cecum, colon, rectum, and anal canal.  p. 692

**laryngopharynx** (lah-ring"go-far'ingks)  The lower portion of the pharynx near the opening to the larynx.  p. 666

**larynx** (lar'ingks)  Structure located between the pharynx and trachea that houses the vocal cords.  p. 742

**latent period** (la'tent pe're-od)  Time between the application of a stimulus and the beginning of a response in a muscle fiber.  p. 294

**lateral** (lat'er-al)  Pertaining to the side.  p. 19

**leukocyte** (lu'ko-sīt)  A white blood cell.  p. 527

**leukocytosis** (lu"ko-si-to'sis)  Too many white blood cells.  p. 532

**leukopenia** (lu"ko-pe'ne-ah)  Too few leukocytes in the blood.  p. 532

**lever** (lev'er)  A simple mechanical device consisting of a rod, fulcrum, weight, and a source of energy that is applied to some point on the rod.  p. 196

**ligament** (lig'ah-ment)  A cord or sheet of connective tissue binding two or more bones at a joint.  p. 260

**limbic system** (lim'bik sis'tem)  A group of connected structures within the brain that produces emotional feelings.  p. 401

**linea alba** (lin'e-ah al'bah)  A narrow band of tendinous connective tissue in the midline of the anterior abdominal wall.  p. 317

**lingual** (ling'gwal)  Pertaining to the tongue.  p. 599

**lipase** (lī'pās)  A fat-digesting enzyme.  p. 688

**lipid** (lip'id)  A fat, oil, or fatlike compound that usually has fatty acids in its molecular structure.  p. 47

**lipoprotein** (lip"o-pro'te-in)  A complex of lipid and protein.  p. 535

**liver** (liv'er)  A large, dark red organ in the upper part of the abdomen on the right side that detoxifies blood, stores glycogen and fat-soluble vitamins, and synthesizes proteins.  p. 678

**lobule** (lob'ul)  A small, well-defined portion of an organ.  p. 751

**lordosis** (lor-do'sis)  An abnormally increased concave curvature in the lumbar portion of the vertebral column.  p. 205

**lower limb** (loh'er lim)  Inferior appendage consisting of the thigh, leg, ankle, and foot.  p. 202

**lumbar** (lum'bar)  Pertaining to the region of the loins.  p. 19

**lumen** (lu'men)  Space within a tubular structure such as a blood vessel or intestine.  p. 686

**luteinizing hormone** (lu'te-in-īz"ing hor'mōn) A hormone secreted by the anterior pituitary gland that controls the formation of the corpus luteum in females and the secretion of testosterone in males; LH (ICSH in males). p. 489

**lymph** (limf) Fluid transported by the lymphatic vessels. p. 621

**lymph node** (limf nōd) A mass of lymphoid tissue located along the course of a lymphatic vessel. p. 627

**lymphocyte** (lim'fo-sīt) A type of white blood cell that provides immunity. p. 529

**lysosome** (li'so-sōm) Organelle that contains enzymes that degrade worn cell parts. p. 73

## M

**macrocyte** (mak'ro-sīt) A large red blood cell. p. 533

**macromineral** (mak'ro-min"er-al) An inorganic substance that is necessary for metabolism and is one of a group that accounts for 75% of the mineral elements within the body; major mineral. p. 722

**macrophage** (mak'ro-fāj) A large phagocytic cell. p. 144

**macroscopic** (mak"ro-skop'ik) Large enough to be seen with the unaided eye. p. 186

**macula** (mak'u-lah) A group of hair cells and supporting cells associated with an organ of static equilibrium. p. 453

**macula lutea** (mak'u-lah lu'te-ah) A yellowish depression in the retina of the eye that is associated with acute vision. p. 465

**malabsorption** (mal"ab-sorp'shun) Failure to absorb nutrients following digestion. p. 655

**malignant** (mah-lig'nant) The power to threaten life; cancerous. p. 96

**malnutrition** (mal"nu-trish'un) Physical symptoms resulting from lack of specific nutrients. p. 729

**maltase** (mawl'tās) An enzyme that catalyzes conversion of maltose into glucose. p. 688

**maltose** (mawl'tōs) A disaccharide composed of two glucose molecules. p. 689

**mammary** (mam'ar-e) Pertaining to the breast. p. 20

**mammillary body** (mam'ĭ-lar"e bod'e) Two small, rounded bodies posterior to the hypothalamus involved with reflexes associated with the sense of smell. p. 399

**marasmus** (mah-raz'mus) Starvation due to profound nutrient deficiency. p. 730

**marrow** (mar'o) Connective tissue that occupies the spaces within bones and includes stem cells. p. 186

**mast cell** (mast sel) A cell to which antibodies, formed in response to allergens, attach, bursting the cell and releasing allergy mediators, which cause symptoms. p. 144

**mastication** (mast"ĭ-ka'shun) Chewing movements. p. 303

**matrix** (ma'triks) The intercellular substance of connective tissue. p. 144

**matter** (mat'er) Anything that has weight and occupies space. p. 37

**meatus** (me-a'tus) A passageway or channel, or the external opening of such a passageway. p. 202

**mechanoreceptor** (mek"ah-no-re-sep'tor) A sensory receptor that is sensitive to mechanical stimulation such as changes in pressure or tension. p. 433

**medial** (me'de-al) Toward or near the midline. p. 19

**mediastinum** (me"de-ah-sti'num) Tissues and organs of the thoracic cavity that form a septum between the lungs. p. 13

**medulla** (mĕ-dul'ah) The inner portion of an organ. p. 380

**medulla oblongata** (mĕ-dul'ah ob"long-gah'tah) Portion of the brain stem located between the pons and the spinal cord. p. 402

**medullary cavity** (med'u-lār"e kav'ĭ-te) Cavity within the diaphysis of a long bone containing marrow. p. 186

**medullary rhythmicity area** (med'u-lār"e rith-mi'si-te ār'e-a) Area of the brain stem that controls the basic rhythm of inspiration and expiration. p. 760

**megakaryocyte** (meg"ah-kar'e-o-sīt) A large bone marrow cell that gives rise to blood platelets. p. 532

**meiosis** (mi-o'sis) A form of cell division that halves the genetic material, resulting in egg and sperm cells (gametes). p. 838

**Meissner's corpuscle** (mīs'nerz kor'pus-l) Sensory receptor close to the surface of the skin that is sensitive to light touch. p. 434

**melanin** (mel'ah-nin) Dark pigment found in skin and hair. p. 164

**melanocyte** (mel'ah-no-sīt) Melanin-producing cell. p. 164

**melatonin** (mel'ah-to'nin) A hormone secreted by the pineal gland. p. 510

**memory cell** (mem'o-re sel) B-lymphocyte or T-lymphocyte produced in a primary immune response that remains dormant and can be activated rapidly if the same antigen is encountered in the future. p. 530

**menarche** (mĕ-nar'ke) The first menstrual period. p. 861

**meninges** (mĕ-nin'jēz) Three membranes that cover the brain and spinal cord (sing., *meninx*). p. 376

**meniscus** (men-is'kus) Fibrocartilage that separates the articulating surfaces of bones in the knee (pl., *menisci*). p. 261

**menopause** (men'o-pawz) Termination of the menstrual cycle. p. 863

**menstrual cycle** (men'stroo-al si'kl) Recurring changes in the uterine lining of a woman of reproductive age. p. 861

**menstruation** (men"stroo-a'shun) Loss of blood and tissue from the uterine lining at the end of a female reproductive cycle. p. 861

**mental** (men'tal) Pertaining to the mind; pertaining to the chin body region. p. 20

**mesentery** (mes'en-ter"e) A fold of peritoneal membrane that attaches an abdominal organ to the abdominal wall. p. 685

**mesoderm** (mez'o-derm) The middle primary germ layer. p. 900

**messenger RNA** (mes'in-jer RNA) RNA that transmits information for a protein's amino acid sequence from the nucleus of a cell to the cytoplasm. p. 120

**metabolic rate** (met"ah-bol'ic rāt) The rate at which chemical changes occur within the body. p. 4

**metabolism** (mĕ-tab'o-lizm) All of the chemical reactions in cells that use or release energy. p. 4

**metacarpals** (met"ah-kar'pals) Bones of the hand between the wrist and finger bones. p. 200

**metaphase** (met'ah-fāz) Stage in mitosis when chromosomes align in the center of the cell p. 90

**metastasis** (mĕ-tas'tah-sis) The spread of disease from one body region to another; a characteristic of cancer. p. 137

**metatarsals** (met"ah-tar'sals) Bones of the foot between the ankle and toe bones. p. 200

**microfilament** (mi"kro-fil'ah-ment) A rod of the protein actin that provides structural support or movement in the cytoplasm. p. 75

**microglia** (mi-krog'le-ah) Neuroglial cells that support neurons and phagocytize. p. 351

**micromineral** (mi'kro-min"er-al) An essential mineral present in a minute amount within the body; trace element. p. 704

**microtubule** (mi"kro-tu'būl) A hollow rod of the protein tubulin in the cytoplasm. p. 75

**microvilli** (mi″kro-vil′i)  Cylindrical processes that extend from some epithelial cell membranes and increase the membrane surface area.  p. 62

**micturition** (mik″tu-rish′un)  Urination.  p. 805

**midbrain** (mid′brān)  A small region of the brain stem between the diencephalon and the pons.  p. 402

**mineralocorticoids** (min″er-al-o-kor′tĭ-koids)  Hormones the adrenal cortex secretes that influence the concentrations of electrolytes in body fluids.  p. 503

**mitochondrion** (mi″to-kon′dre-on)  Organelle housing enzymes that catalyze reactions of aerobic respiration (pl., *mitochondria*).  p. 72

**mitosis** (mi-to′sis)  Division of a somatic cell to form two genetically identical cells.  p. 88

**mitral valve** (mi′tral valv)  Heart valve located between the left atrium and the left ventricle; bicuspid valve.  p. 558

**mixed nerve** (mikst nerv)  Nerve that includes both sensory and motor nerve fibers.  p. 407

**molar** (mo′lar)  A rear tooth with a flattened surface adapted for grinding food.  p. 662

**molding** (mold′ing)  Changing in shape of the fetal skull during birth.  p. 213

**molecular formula** (mo-lek′u-lar for′mu-lah)  An abbreviation for the number of atoms of each element in a compound.  p. 42

**molecule** (mol′ĕ-kūl)  A particle composed of two or more joined atoms.  p. 8

**monoamine inhibitor** (mon″o-am′in in-hib′i-tor)  A substance that inhibits the action of the enzyme monoamine oxidase.  p. 364

**monocyte** (mon′o-sīt)  A type of white blood cell that functions as a phagocyte.  p. 529

**monosaccharide** (mon″o-sak′ah-rīd)  A single sugar, such as glucose or fructose.  p. 47

**monosomy** (mon′o-so″me)  A cell missing one chromosome.  p. 950

**morula** (mor′u-lah)  An early stage in prenatal development; a solid ball of cells.  p. 867

**motor area** (mo′tor a′re-ah)  A region of the brain from which impulses to muscles or glands originate.  p. 395

**motor end plate** (mo′tor end plāt)  Specialized portion of a muscle fiber membrane at a neuromuscular junction.  p. 286

**motor nerve** (mo′tor nerv)  A nerve that consists of motor nerve fibers.  p. 407

**motor neuron** (mo′tor nu′ron)  A neuron that transmits impulses from the central nervous system to an effector.  p. 286

**motor unit** (mo′tor unit)  A motor neuron and the muscle fibers associated with it.  p. 286

**mucosa** (mu-ko′sah)  The membrane that lines tubes and body cavities that open to the outside of the body; mucous membrane.  p. 656

**mucous cell** (mu′kus sel)  Glandular cell that secretes mucus.  p. 671

**mucous membrane** (mu′kus mem′brān)  Mucosa.  p. 161

**mucus** (mu′kus)  Fluid secretion of the mucous cells.  p. 665

**multiple alleles** (mul′tĭ-pl ah-lēls′)  Different forms of a gene.  p. 937

**multiple motor unit summation** (mul′tĭ-pl mo′tor u′nit sum-mā′shun)  A sustained muscle contraction of increasing strength in response to input from many motor units.  p. 295

**multipolar neuron** (mul′tĭ-po′lar nu′ron)  Nerve cell that has many processes arising from its cell body.  p. 350

**muscle impulse** (mus′el im′puls)  Impulse that travels along the sarcolemma to the transverse tubules.  p. 289

**muscle spindle** (mus′el spin′dul)  Modified skeletal muscle fiber that can respond to changes in muscle length.  p. 438

**muscle tone** (mus′el tōn)  The contraction of some fibers in skeletal muscle at any given time.  p. 296

**mutagen** (mu′tah-jen)  An agent that can cause mutations.  p. 129

**mutation** (mu-ta′shun)  A change in a gene.  p. 126

**myelin** (mi′ĕ-lin)  Fatty material that forms a sheathlike covering around some nerve fibers.  p. 348

**myocardium** (mi″o-kar′de-um)  Muscle tissue of the heart.  p. 557

**myofibril** (mi″o-fi′bril)  Contractile fibers within muscle cells.  p. 284

**myoglobin** (mi″o-glo′bin)  A pigmented compound in muscle tissue that stores oxygen.  p. 292

**myogram** (mi′o-gram)  A recording of a muscular contraction.  p. 294

**myometrium** (mi″o-me′tre-um)  The layer of smooth muscle tissue within the uterine wall.  p. 856

**myoneural junction** (mi″o-nu′ral jungk′shun)  Site of union between a motor neuron axon and a muscle fiber.  p. 286

**myopia** (mi-o′pe-ah)  Nearsightedness.  p. 471

**myosin** (mi′o-sin)  A protein that, together with actin, produces muscular contraction and relaxation.  p. 285

**myxedema** (mik″sĕ-de′mah)  A deficiency of thyroid hormones in an adult.  p. 193

# N

**nasal cavity** (na′zal kav′ĭ-te)  Space within the nose.  p. 740

**nasal concha** (na′zal kong′kah)  Shell-like bone extending outward from the wall of the nasal cavity; a turbinate bone.  p. 740

**nasal septum** (na′zal sep′tum)  A wall of bone and cartilage that separates the nasal cavity into two portions.  p. 740

**nasopharynx** (na″zo-far′ingks)  Portion of the pharynx associated with the nasal cavity.  p. 666

**natural killer cell** (nat′u-ral kil′er sel)  Lymphocyte that causes an infected or cancerous cell to burst.  p. 634

**negative feedback** (neg′ah-tiv fēd′bak)  A mechanism activated by an imbalance that corrects it.  p. 8

**neonatal** (ne″o-na′tal)  The first four weeks of life.  p. 914

**nephron** (nef′ron)  The functional unit of a kidney, consisting of a renal corpuscle and a renal tubule.  p. 780

**nerve** (nerv)  A bundle of nerve fibers in the PNS.  p. 344

**nerve fiber** (nerv fi′ber)  A thin process of a neuron (i.e., axon, dendrite).  p. 344

**nerve impulse** (nerv im′puls)  The electrochemical process of depolarization and repolarization along a nerve fiber.  p. 344

**nerve tract** (nerv trakt)  A long bundle of nerve fibers within the CNS having the same origin, function, and termination.  p. 391

**neurilemma** (nur″ĭ-lem′ah)  Sheath on the outside of some nerve fibers formed from Schwann cells.  p. 348

**neurofibrils** (nu′ro-fi′brils)  Fine cytoplasmic threads that extend from the cell body into the processes of neurons.  p. 345

**neuroglial cell** (nu-ro′gle-ahl sel)  Specialized cell of the nervous system that produces myelin, communicates between cells, and maintains the ionic environment, as well as provides other functions.  p. 154

**neuromodulator** (nu″ro-mod′u-lā-tor)  A substance that alters a neuron's response to a neurotransmitter.  p. 365

**neuromuscular junction** (nu″ro-mus′ku-lar jungk′shun)   Point of contact between a nerve and muscle cell.   p. 286

**neuron** (nu′ron)   A nerve cell that consists of a cell body and its processes.   p. 344

**neuropeptide** (nu″ro-pep′tīd)   A peptide in the brain that functions as a neurotransmitter or neuromodulator.   p. 365

**neurosecretory cell** (nu″ro-se-kre′to-re sel)   Cell in the hypothalamus that functions as a neuron at one end but like an endocrine cell at the other, by receiving messages and secreting the hormones ADH and oxytocin.   p. 489

**neurotransmitter** (nu″ro-trans-mit′er)   Chemical secreted by the end of an axon that stimulates a muscle fiber to contract or a neuron to fire an impulse.   p. 286

**neutral** (nu′tral)   Neither acid nor alkaline; pH 7.   p. 45

**neutron** (nu′tron)   An electrically neutral particle in an atomic nucleus.   p. 37

**neutrophil** (nu′tro-fil)   A type of phagocytic leukocyte.   p. 527

**niacin** (ni′ah-sin)   A vitamin of the B-complex group; nicotinic acid.   p. 718

**niacinamide** (ni″ah-sin′ah-mīd)   The physiologically active form of niacin.   p. 718

**nicotinic acid** (nik″o-tin′ik as′id)   Niacin.   p. 718

**Nissl bodies** (nis′l bod′ēz)   Membranous sacs within the cytoplasm of nerve cells that have ribosomes attached to their surfaces.   p. 345

**nitrogen balance** (ni′tro-jen bal′ans)   Condition in which the quantity of nitrogen ingested equals the quantity excreted.   p. 709

**node of Ranvier** (nōd of Ron′vee-ay)   A short region of exposed (unmyelinated) axon between Schwann cells on neurons of the peripheral nervous system.   p. 348

**nodule** (nod′ul)   A small mass of tissue detected by touch.   p. 628

**nondisjunction** (non″dis-jungk′shun)   Failure of a pair of chromosomes to separate during meiosis.   p. 950

**nonelectrolyte** (non″e-lek′tro-līt)   A substance that does not dissociate into ions when it is dissolved.   p. 44

**nonheritable gene therapy** (non-her′i-tah-bl jēn ther′ah-pe)   Manipulation of genes in somatic cells to correct the effects of a mutation.   p. 955

**nonprotein nitrogenous substance** (non-pro′tēn ni-troj′ĕ-nus sub′stans)   A substance, such as urea or uric acid, that contains nitrogen but is not a protein.   p. 536

**norepinephrine** (nor″ep-ĭ-nef′rin)   A neurotransmitter released from the axons of some nerve fibers.   p. 421

**nuclease** (nu′kle-ās)   An enzyme that catalyzes decomposition of nucleic acids.   p. 677

**nucleic acid** (nu-kle′ik as′id)   A substance composed of bonded nucleotides; RNA or DNA.   p. 52

**nucleolus** (nu-kle′o-lus)   A small structure within the nucleus of a cell that contains RNA and proteins (pl., *nucleoli*).   p. 77

**nucleoplasm** (nu′kle-o-plazm″)   The contents of the nucleus of a cell.   p. 77

**nucleosome** (nu′kle-o-sōm)   A beadlike part of a chromatin fiber consisting of DNA wrapped around protein octets.   p. 77

**nucleotide** (nu′kle-o-tīd″)   A building block of a nucleic acid molecule, consisting of a sugar, a nitrogenous base, and a phosphate group.   p. 52

**nucleus** (nu′kle-us)   A cellular organelle that is enclosed by a double-layered, porous membrane and contains DNA; the dense core of an atom that is composed of protons and neutrons (pl., *nuclei*).   p. 37

**nutrient** (nu′tre-ent)   A chemical substance that must be supplied to the body from the environment.   p. 704

**nutrition** (nu-trish′un)   The study of the sources, actions, and interactions of nutrients.   p. 704

# O

**obesity** (o-bēs′ĭ-te)   Excess adipose tissue; exceeding desirable weight by more than 20%.   p. 711

**occipital** (ok-sip′ĭ-tal)   Pertaining to the lower, back portion of the head.   p. 203

**olfactory** (ol-fak′to-re)   Pertaining to the sense of smell.   p. 408

**olfactory nerves** (ol-fak′to-re nervz)   The first pair of cranial nerves, which conduct impulses associated with the sense of smell.   p. 408

**oligodendrocyte** (ol′ĭ-go-den′dro-sīt)   A type of neuroglial cell that connects neurons to blood vessels and forms myelin.   p. 351

**oncogene** (ong′ko-jēn)   A gene that normally controls cell division but when overexpressed leads to cancer.   p. 97

**oocyte** (o′o-sīt)   An immature egg cell.   p. 852

**oogenesis** (o″o-jen′ĕ-sis)   Differentiation of an egg cell.   p. 852

**ophthalmic** (of-thal′mik)   Pertaining to the eye.   p. 599

**optic** (op′tik)   Pertaining to the eye.   p. 399

**optic chiasma** (op′tik ki-az′mah)   X-shaped structure on the underside of the brain formed by a partial crossing over of fibers in the optic nerves.   p. 399

**optic disk** (op′tik disk)   Region in the retina of the eye where nerve fibers leave to become part of the optic nerve.   p. 465

**oral** (o′ral)   Pertaining to the mouth.   p. 20

**orbital** (or′bĭ-tal)   Pertaining to the body region of the eyeball; a region, in the atom, containing the electrons.   p. 20

**organ** (or′gan)   A structure consisting of a group of tissues that performs a specialized function.   p. 8

**organ of Corti** (or′gan uv kor′te)   An organ in the cochlear duct that contains the receptors for hearing. It consists of hair cells and supporting cells.   p. 450

**organelle** (or″gah-nel′)   A part of a cell that performs a specialized function.   p. 8

**organic** (or-gan′ik)   Carbon-containing molecules.   p. 46

**organism** (or′gah-nizm)   An individual living thing.   p. 4

**orgasm** (or′gaz-em)   The culmination of sexual excitement.   p. 860

**orifice** (or′ĭ-fis)   An opening.   p. 846

**origin** (or′ĭ-jin)   End of a muscle that is attached to a relatively immovable part.   p. 300

**oropharynx** (o″ro-far′ingks)   Portion of the pharynx in the posterior part of the oral cavity.   p. 666

**osmoreceptor** (oz″mo-re-sep′tor)   Receptor that is sensitive to changes in the osmotic pressure of body fluids.   p. 494

**osmosis** (oz-mo′sis)   Diffusion of water through a selectively permeable membrane in response to a concentration gradient.   p. 80

**osmotic** (oz-mot′ik)   Pertaining to osmosis.   p. 81

**osmotic pressure** (oz-mot′ik presh′ur)   The amount of pressure needed to stop osmosis; the potential pressure of a solution caused by nondiffusible solute particles in the solution.   p. 81

**osseous tissue** (os′e-us tish′u)   Bone tissue.   p. 448

**ossification** (os″ĭ-fĭ-ka′shun)   The formation of bone tissue.   p. 188

**osteoblast** (os'te-o-blast") A bone-forming cell. p. 188

**osteoclast** (os'te-o-klast") A cell that erodes bone. p. 191

**osteocyte** (os'te-o-sīt) A mature bone cell. p. 189

**osteon** (os'te-on) A cylinder-shaped unit containing bone cells that surround an osteonic canal; Haversian system. p. 151

**osteonic canal** (os'te-o-nik ka-nal') A tiny channel in bone tissue that contains a blood vessel; Haversian canal. p. 151

**osteoporosis** (os"te-o-po-ro'sis) A condition in which bones break easily because calcium is removed from them faster than it is replaced. p. 198

**otic** (o'tik) Pertaining to the ear. p. 20

**otolith** (o'to-lith) A small particle of calcium carbonate associated with the receptors of equilibrium. p. 456

**otosclerosis** (o"to-sklĕ-ro'sis) Abnormal formation of spongy bone within the ear that may interfere with the transmission of sound vibrations to hearing receptors. p. 454

**oval window** (o'val win'do) Opening between the stapes and the inner ear. p. 447

**ovarian** (o-va're-an) Pertaining to the ovary. p. 852

**ovary** (o'var-e) The primary reproductive organ of a female; an egg cell–producing organ. p. 510

**oviduct** (o'vĭ-dukt) A tube that leads from the ovary to the uterus; uterine tube or fallopian tube. p. 850

**ovulation** (o"vu-la'shun) The release of an egg cell from a mature ovarian follicle. p. 855

**ovum** (o'vum) A mature egg cell. p. 842

**oxidase** (ok'sĭ-dās) An enzyme that promotes oxidation. p. 112

**oxidation** (ok"sĭ-da'shun) Process by which oxygen is combined with another chemical; the removal of hydrogen or the loss of electrons; the opposite of reduction. p. 112

**oxidative phosphorylation** (ok'sĭ-da-tiv fos"fo-ri-la'shun) The process of transferring electrons to form a high-energy phosphate bond by introducing a phosphate group to ADP and forming ATP. p. 106

**oxygen debt** (ok'sĭ-jen det) The amount of oxygen that must be supplied following physical exercise to convert accumulated lactic acid to glucose. p. 292

**oxyhemoglobin** (ok"sĭ-he"mo-glo'bin) Compound formed when oxygen combines with hemoglobin. p. 767

**oxytocin** (ok"sĭ-to'sin) Hormone released by the posterior lobe of the pituitary gland that contracts smooth muscles in the uterus and mammary glands. p. 489

# P

**pacemaker** (pās'māk-er) Mass of specialized cardiac muscle tissue that controls the rhythm of the heartbeat; the sinoatrial node. p. 568

**Pacinian corpuscle** (pah-sin'e-an kor'pusl) Nerve endings deep in the dermis providing perception of pressure. p. 434

**pain receptor** (pān re"sep'tor) Sensory nerve ending associated with the feeling of pain. p. 433

**palate** (pal'at) The roof of the mouth. p. 660

**palatine** (pal'ah-tīn) Pertaining to the palate. p. 211

**palmar** (pahl'mar) Pertaining to the palm of the hand. p. 20

**pancreas** (pan'kre-as) Glandular organ in the abdominal cavity that secretes hormones and digestive enzymes. p. 507

**pancreatic** (pan"kre-at'ik) Pertaining to the pancreas. p. 676

**pantothenic acid** (pan"to-the'nik as'id) A vitamin of the B-complex group. p. 718

**papilla** (pah-pil'ah) Tiny, nipplelike projection. p. 442

**papillary muscle** (pap'ĭ-ler"e mus'el) Muscle that extends inward from the ventricular walls of the heart and to which the chordae tendineae attach. p. 558

**paracrine** (par'ah-krin) A type of endocrine secretion in which the hormone affects nearby cells. p. 480

**paradoxical sleep** (par"ah-dok'se-kal slēp) Sleep in which some areas of the brain are active, producing dreams and rapid eye movements. p. 403

**paralysis** (pah-ral'i-sis) Loss of ability to control voluntary muscular movements, usually due to a disorder of the nervous system. p. 331

**parasympathetic division** (par"ah-sim"pah-thet'ik dĭ-vizh'un) Portion of the autonomic nervous system that arises from the brain and sacral region of the spinal cord. p. 419

**parathyroid glands** (par"ah-thi'roid glandz) Small endocrine glands embedded in the posterior portion of the thyroid gland. p. 499

**parathyroid hormone** (par"ah-thi'roid hor'mōn) Hormone secreted by the parathyroid glands that helps regulate the level of blood calcium and phosphate ions; PTH. p. 499

**paravertebral ganglia** (par"ah-ver'tĕ-bral gang'gle-ah) Sympathetic ganglia that form chains along the sides of the vertebral column. p. 419

**parietal** (pah-ri'ĕ-tal) Pertaining to the wall of an organ or cavity. p. 213

**parietal cell** (pah-ri'ĕ-tal sel) Cell of a gastric gland that secretes hydrochloric acid and intrinsic factor. p. 671

**parietal pleura** (pah-ri'ĕ-tal ploo'rah) Membrane that lines the inner wall of the thoracic cavity. p. 751

**parotid glands** (pah-rot'id glandz) Large salivary glands located on the sides of the face just in front and below the ears. p. 665

**partial pressure** (par'shal presh'ur) The pressure one gas produces in a mixture of gases. p. 766

**parturition** (par"tu-rish'un) Childbirth. p. 869

**pathogen** (path'o-jen) A disease-causing agent. p. 630

**pathology** (pah-thol'o-je) The study of disease. p. 630

**pectoral** (pek'tor-al) Pertaining to the chest. p. 20

**pectoral girdle** (pek'tor-al ger'dl) Portion of the skeleton that provides support and attachment for the arms. p. 222

**pedal** (ped'al) Pertaining to the foot. p. 20

**pedigree** (ped'ĭ-gre) A chart showing the relationships of relatives and which ones have a particular trait. p. 978

**pelvic** (pel'vik) Pertaining to the pelvis. p. 13

**pelvic girdle** (pel'vik ger'dl) Portion of the skeleton to which the legs are attached. p. 227

**pelvic inflammatory disease** (pel'vik in-flam'ah-tor'e dĭ-zēz') An ascending infection of the upper female genital tract. p. 882

**pelvis** (pel'vis) Bony ring formed by the sacrum and coxal bones. p. 200

**penis** (pe'nis) External reproductive organ of the male through which the urethra passes. p. 845

**pepsin** (pep'sin) Protein-splitting enzyme secreted by the gastric glands of the stomach. p. 672

**pepsinogen** (pep-sin'o-jen) Inactive form of pepsin. p. 672

**peptidase** (pep'tĭ-dās) An enzyme that catalyzes the breakdown of polypeptides. p. 688

**peptide** (pep′tīd) Compound composed of two or more bonded amino acids. p. 102

**peptide bond** (pep′tīd bond) Bond that forms between the carboxyl group of one amino acid and the amino group of another. p. 102

**perception** (per-sep′shun) Mental awareness of sensory stimulation. p. 433

**pericardial** (per″ĭ-kar′de-al) Pertaining to the pericardium. p. 13

**pericardium** (per″ĭ-kar′de-um) Serous membrane that surrounds the heart. p. 13

**perichondrium** (per″ĭ-kon′dre-um) Layer of fibrous connective tissue that encloses cartilaginous structures. p. 150

**perilymph** (per′ĭ-limf) Fluid contained in the space between the membranous and osseous labyrinths of the inner ear. p. 448

**perimetrium** (per-ĭ-me′tre-um) The outer serosal layer of the uterine wall. p. 856

**perimysium** (per″ĭ-mis′e-um) Sheath of connective tissue that encloses a bundle of striated muscle fibers. p. 282

**perineal** (per″ĭ-ne′al) Pertaining to the perineum. p. 20

**perineum** (per″ĭ-ne′um) Body region between the scrotum or urethral opening and the anus. p. 416

**perineurium** (per″ĭ-nu′re-um) Layer of connective tissue that encloses a bundle of nerve fibers within a nerve. p. 407

**periodontal ligament** (per″e-o-don′tal lig′ah-ment) Fibrous membrane that surrounds a tooth and attaches it to the bone of the jaw. p. 258

**periosteum** (per″e-os′te-um) Covering of fibrous connective tissue on the surface of a bone. p. 185

**peripheral** (pĕ-rif′er-al) Pertaining to parts located near the surface or toward the outside. p. 19

**peripheral nervous system** (pĕ-rif′er-al ner′vus sis′tem) The portions of the nervous system outside the central nervous system. p. 344

**peripheral protein** (pe-rif′er-al pro′tēn) A globular protein associated with the inner surface of the cell membrane. p. 68

**peripheral resistance** (pĕ-rif′er-al re-zis′tans) Resistance to blood flow due to friction between the blood and the walls of the blood vessels. p. 584

**peristalsis** (per″ĭ-stal′sis) Rhythmic waves of muscular contraction in the walls of certain tubular organs. p. 298

**peritoneal** (per″ĭ-to-ne′al) Pertaining to the peritoneum. p. 14

**peritoneal cavity** (per″ĭ-to-ne′al kav′ĭ-te) The potential space between the parietal and visceral peritoneal membranes. p. 15

**peritoneum** (per″ĭ-to-ne′um) A serous membrane that lines the abdominal cavity and encloses the abdominal viscera. p. 14

**peritubular capillary** (per″ĭ-tu′bu-lar kap′ĭ-ler″e) Capillary that surrounds a renal tubule and functions in reabsorption and secretion during urine formation. p. 788

**permeable** (per′me-ah-bl) Open to passage or penetration. p. 587

**peroxisome** (pĕ-roks′ĭ-sōm) Membranous cytoplasmic vesicle that contains enzymes that catalyze reactions that produce and decompose hydrogen peroxide. p. 73

**pH** A shorthand notation for the hydrogen ion concentration used to indicate the acid or alkaline condition of a solution; values range from 0 to 14. p. 45

**phagocytosis** (fag″o-si-to′sis) Process by which a cell engulfs and digests solid substances. p. 84

**phalanx** (fa′langks) A bone of a finger or toe (pl., *phalanges*). p. 200

**pharynx** (far′ingks) Portion of the digestive tube between the mouth and the esophagus. p. 666

**phenotype** (fe′no-tīp) The appearance of an individual due to the action of a particular set of genes. p. 940

**phosphate buffer system** (fos′fāt buf′er sis′tem) Consists of a mix of sodium monohydrogen phosphate and sodium dihydrogen phosphate to weaken a strong base and a strong acid, respectively; to resist changes in pH. p. 825

**phospholipid** (fos″fo-lip′id) A lipid that contains two fatty acid molecules and a phosphate group combined with a glycerol molecule. p. 49

**photoreceptor** (fo″to-re-sep′tor) A nerve ending that is sensitive to light energy. p. 433

**physiology** (fiz″e-ol′o-je) The branch of science that studies body functions. p. 3

**pia mater** (pi′ah ma′ter) Inner layer of meninges that encloses the brain and spinal cord. p. 377

**pineal gland** (pin′e-al gland) A small structure located in the central part of the brain. p. 399

**pinocytosis** (pin″o-si-to′sis) Process by which a cell engulfs droplets of fluid from its surroundings. p. 83

**pituitary gland** (pĭ-tu′ĭ-tār″e gland) Endocrine gland attached to the base of the brain that consists of anterior and posterior lobes; the hypophysis. p. 488

**placenta** (plah-sen′tah) Structure attaching a fetus to the uterine wall, providing a conduit to receive nutrients and excrete wastes. p. 510

**placental lactogen** (plah-sen′tahl lak′to-jen) Hormone secreted by the placenta to inhibit maternal insulin activity during pregnancy. p. 869

**plantar** (plan′tar) Pertaining to the sole of the foot. p. 20

**plantar flexion** (plan′tar flek′shun) Bending the foot downward. p. 264

**plasma** (plaz′mah) Fluid portion of circulating blood. p. 521

**plasma cell** (plaz′mah sel) Antibody-producing cell forms as a result of the proliferation of sensitized B cells. p. 523

**plasma protein** (plaz′mah pro′tēn) Proteins dissolved in blood plasma. p. 532

**platelet** (plāt′let) Cytoplasmic fragment formed in the bone marrow that functions in blood coagulation. p. 532

**pleiotropy** (plē′o-tro-pē) A gene that has several expressions (phenotypes). p. 943

**pleural** (ploo′ral) Pertaining to the pleura or membranes investing the lungs. p. 751

**pleural cavity** (ploo′ral kav′ĭ-te) Potential space between the pleural membranes. p. 751

**pleural membranes** (ploo′ral mem′brānz) Serous membranes that enclose the lungs. p. 13

**plexus** (plek′sus) A network of interlaced nerves or blood vessels. p. 415

**pneumotaxic area** (nu″mo-tax′ik ār′e-a) A portion of the respiratory control center in the pons of the brain. p. 762

**PNS** Peripheral nervous system p. 344

**polar body** (po′lar bod′e) Small, nonfunctional cell that is a product of meiosis in the female. p. 852

**polarization** (po″lar-ĭ-za′shun) Development of an electrical charge on the surface of a cell membrane due to an unequal distribution of ions on either side of the membrane. p. 354

**polydactyly** (pol″e-dak′ti-le) Extra fingers or toes. p. 226

**polymorphonuclear leukocyte** (pol″e-mor″fo-nu′kle-ar lu′ko-sīt) A white blood cell (leukocyte) with an irregularly lobed nucleus. p. 527

**polynucleotide** (pol″e-noo′-kle-o-tīd) A compound formed by the union of many nucleotides; a nucleic acid. p. 52

**polypeptide** (pol″e-pep′tīd) A compound formed by the union of many amino acid molecules. p. 102

**polyploidy** (pol′e-ploi″de) A condition in which a cell has one or more extra sets of chromosomes. p. 948

**polysaccharide** (pol″e-sak′ah-rīd) A carbohydrate composed of many joined monosaccharides. p. 47

**pons** (ponz) A portion of the brain stem above the medulla oblongata and below the midbrain. p. 402

**popliteal** (pop′lĭ-te′al) Pertaining to the region behind the knee. p. 20

**positive chemotaxis** (poz″ĭ-tiv ke″mo-tak′sis) Movement of a cell toward the greater concentration of a substance. p. 531

**positive feedback** (poz′ĭ-tiv fēd′bak) Process by which changes cause more changes of a similar type, producing unstable conditions. p. 538

**posterior** (pos-tēr′e-or) Toward the back; opposite of anterior. p. 18

**postganglionic fiber** (pōst″gang-gle-on′ik fi′ber) Autonomic nerve fiber located on the distal side of a ganglion. p. 419

**postnatal** (pōst-na′tal) After birth. p. 892

**postsynaptic neuron** (pōst″sĭ-nap′tik nu′ron) One of two adjacent neurons transmitting an impulse; cell situated after the synapse is crossed. p. 360

**postsynaptic potential** (pōst″sĭ-nap′tik po-ten′shal) Membrane polarization is increased (excitatory) or decreased (inhibitory) in the postsynaptic neuron with repeated stimulation over an excitatory or inhibitory pathway so that the neuron will either fire or have diminished responsiveness. p. 360

**precursor** (pre-ker′sor) Substance from which another substance forms. p. 396

**preganglionic fiber** (pre″gang-gle-on′ik fi′ber) Autonomic nerve fiber located on the proximal side of a ganglion. p. 419

**pregnancy** (preg′nan-se) The condition in which a female has a developing offspring in her uterus. p. 864

**prenatal** (pre-na′tal) Before birth. p. 892

**presbycusis** (pres″bĭ-ku′sis) Loss of hearing that accompanies old age. p. 454

**presbyopia** (pres″be-o′pe-ah) Loss of the eye's ability to accommodate due to declining elasticity in the lens; farsightedness of age. p. 478

**presynaptic neuron** (pre″sĭ-nap′tik nu′ron) One of two adjacent neurons transmitting an impulse; cell situated before the synapse is crossed. p. 360

**primary germ layers** (pri′ma-re jerm lā′erz) Three layers (endoderm, mesoderm, and ectoderm) of embryonic cells that develop into specific tissues and organs. p. 896

**primary immune response** (pri′ma-re ĭ-mūn′ re-spons′) The immune system's response to its first encounter with a foreign antigen. p. 640

**primary reproductive organs** (pri′ma-re re″pro-duk′tiv or′ganz) Sex cell–producing parts; testes in males and ovaries in females. p. 18

**prime mover** (prīm mōōv′er) Muscle that is mainly responsible for a particular body movement. p. 300

**profibrinolysin** (pro″fi-brĭ-no-li′sin) The inactive form of fibrinolysin. p. 551

**progeny** (proj′e-ne) Offspring. p. 929

**progesterone** (pro-jes′tĕ-rōn) A female hormone secreted by the corpus luteum of the ovary and by the placenta. p. 860

**projection** (pro-jek′shun) Process by which the brain causes a sensation to seem to come from the region of the body being stimulated. p. 433

**prolactin** (pro-lak′tin) Hormone secreted by the anterior pituitary gland that stimulates the production of milk from the mammary glands; PRL. p. 489

**pronation** (pro-na′shun) Movement of the palm downward or backward. p. 265

**prophase** (pro′fāz) Stage of mitosis when chromosomes become visible. p. 90

**proprioceptor** (pro″pre-o-sep′tor) A nerve ending that detects changes in tension of a muscle or tendon. p. 433

**prostacyclin** (pros″tah-si′klin) A substance released from endothelial cells that inhibits platelet adherence. p. 542

**prostaglandins** (pros″tah-glan′dins) A group of compounds that have powerful, hormonelike effects. p. 481

**prostate gland** (pros′tāt gland) Gland surrounding the male urethra below the urinary bladder that adds its secretion to semen during ejaculation. p. 843

**protein** (pro′tēn) Nitrogen-containing organic compound composed of joined amino acid molecules. p. 51

**protein buffer system** (pro′tēn buf′er sis′tem) The amino acids of a protein accept or donate hydrogen ions to keep the concentration of hydrogen ions in solution constant; resist changes in pH. p. 825

**prothrombin** (pro-throm′bin) Plasma protein that functions in the formation of blood clots. p. 538

**proton** (pro′ton) A positively charged particle in an atomic nucleus. p. 37

**protraction** (pro-trak′shun) A forward movement of a body part. p. 265

**proximal** (prok′sĭ-mal) Closer to the midline or origin; opposite of distal. p. 19

**pseudostratified** (soo″do-strat′ĭ-fīd) A single layer of epithelium appearing as more than one layer because the nuclei occupy different positions in the cells. p. 138

**puberty** (pu′ber-te) Stage of development in which the reproductive organs become functional. p. 850

**pulmonary** (pul′mo-ner″e) Pertaining to the lungs. p. 750

**pulmonary circuit** (pul′mo-ner″e ser′kit) System of blood vessels that carries blood between the heart and the lungs. p. 592

**pulmonary valve** (pul′mo-ner″e valv) Valve leading from the right ventricle to the pulmonary trunk (artery); pulmonary semilunar valve. p. 558

**pulse** (puls) The surge of blood felt through the walls of arteries due to the contraction of the ventricles of the heart. p. 583

**pupil** (pu′pil) Opening in the iris through which light enters the eye. p. 464

**Purkinje fibers** (pur-kin′je fi′berz) Specialized muscle fibers that conduct the cardiac impulse from the A-V bundle into the ventricular walls. p. 570

**pyloric sphincter muscle** (pi-lor′ik sfingk′ter mus′l) A ring of muscle located between the stomach and the duodenum; controls food entry into the duodenum. p. 670

**pyramidal cell** (pĭ-ram′ĭ-dal sel) A large, pyramid-shaped neuron in the cerebral cortex. p. 390

**pyridoxine** (pir′ĭ-dok′sēn) A vitamin of the B-complex group; vitamin $B_6$. p. 719

**pyrogen** (pi′ro-jen) A biochemical that increases body temperature. p. 631

**pyruvic acid** (pi-roo′vik as′id) An intermediate product of carbohydrate oxidation. p. 108

# R

**radiation** (ra″de-a′shun) A form of energy that includes visible light, ultraviolet light, and X rays; the means by which body heat is lost in the form of infrared rays. p. 173

**radioactive** (ra″de-o-ak′tiv) An atom that releases energy at a constant rate. p. 39

**rate-limiting enzyme** (rāt lim′ĭ-ting en′zīm) An enzyme, usually present in limited amount, that controls the rate of a metabolic pathway by regulating one of its steps. p. 115

**reabsorption** (re″ab-sorp′shun) Selective reuptake of substances into or across tissues. p. 887

**receptor** (re″sep′tor) A structure, usually protein, at the distal end of a sensory dendrite that can be stimulated. p. 344

**receptor-mediated endocytosis** (re″sep′tor-me′de-ā-tid en″do-si-to′sis) A selective uptake of molecules into a cell by binding to a specific receptor. p. 84

**recessive gene** (re-ses′iv jēn) An allele of a gene pair that is not expressed while the other member of the pair is expressed. p. 940

**recruitment** (re-krōōt′ment) Increase in number of motor units activated as intensity of stimulation increases. p. 297

**rectum** (rek′tum) The terminal end of the digestive tube between the sigmoid colon and the anus. p. 692

**red blood cell** (red blud sel) A disc-shaped cell, lacking a nucleus, that is packed with the oxygen-carrying molecule hemoglobin. p. 522

**red marrow** (red mar′o) Blood cell–forming tissue in spaces within bones. p. 197

**red muscle** (red mus′el) Slow-contracting postural muscles that contain abundant myoglobin. p. 293

**reduction** (re-duk′shun) A chemical reaction in which electrons are gained; to properly realign a fractured bone. p. 112

**referred pain** (re-ferd′ pān) Pain that feels as if it is originating from a part other than the site being stimulated. p. 435

**reflex** (re′fleks) A rapid, automatic response to a stimulus. p. 383

**reflex arc** (re′fleks ark) A nerve pathway, consisting of a sensory neuron, interneuron, and motor neuron, that forms the structural and functional bases for a reflex. p. 383

**refraction** (re-frak′shun) A bending of light as it passes from one medium into another medium with a different density. p. 468

**refractory period** (re-frak′to-re pe′re-od) Time period following stimulation during which a neuron or muscle fiber will not respond to a stimulus. p. 294

**relaxin** (re-lak′sin) Hormone from the corpus luteum that inhibits uterine contractions during pregnancy. p. 887

**releasing hormone** (re-le′sing hor′mōn) A substance secreted by the hypothalamus whose target cells are in the anterior pituitary gland. p. 399

**renal** (re′nal) Pertaining to the kidney. p. 780

**renal corpuscle** (re′nal kor′pusl) Part of a nephron that consists of a glomerulus and a glomerular capsule; Malpighian corpuscle. p. 785

**renal cortex** (re′nal kor′teks) The outer portion of a kidney. p. 780

**renal medulla** (re′nal mĕ-dul′ah) The inner portion of a kidney. p. 780

**renal pelvis** (re′nal pel′vis) The cavity in a kidney. p. 780

**renal tubule** (re′nal tu′būl) Portion of a nephron that extends from the renal corpuscle to the collecting duct. p. 785

**renin** (re′nin) Enzyme released from the kidneys that triggers a rise in blood pressure. p. 504

**replication** (rep″lĭ-ka′shun) Production of an exact copy of a DNA sequence. p. 126

**repolarization** (re-po″lar-ĭ-za′shun) Returning the cell membrane potential to resting potential. p. 357

**reproduction** (re″pro-duk′shun) Offspring formation. p. 18

**residual volume** (re-zid′u-al vol′ūm) The amount of air remaining in the lungs after the most forceful expiration. p. 757

**resorption** (re-sorp′shun) Decomposition of a structure as a result of physiological activity. p. 192

**respiration** (res″pĭ-ra′shun) Cellular process that releases energy from nutrients; breathing. p. 739

**respiratory center** (re-spi′rah-to″re sen′ter) Portion of the brain stem that controls the depth and rate of breathing. p. 402

**respiratory membrane** (re-spi′rah-to″re mem′brān) Membrane composed of a capillary wall and an alveolar wall through which gases are exchanged between the blood and the air. p. 765

**response** (re-spons′) The action resulting from a stimulus. p. 344

**resting potential** (res′ting po-ten′shal) The difference in electrical charge between the inside and outside of an undisturbed nerve cell membrane. p. 355

**reticular formation** (rĕ-tik′u-lar fōr-ma′shun) A complex network of nerve fibers within the brain stem that arouses the cerebrum. p. 403

**reticulocyte** (rĕ-tik′u-lo-sīt) An immature red blood cell that has a network of fibrils in its cytoplasm. p. 526

**reticuloendothelial tissue** (rĕ-tik″u-lo-en″do-the′le-al tish′u) Tissue composed of widely scattered phagocytic cells. p. 153

**retina** (ret′ĭ-nah) Inner layer of the eye wall that contains the visual receptors. p. 465

**retinal** (ret′ĭ-nal) A form of vitamin A; retinene. p. 470

**retinene** (ret′ĭ-nēn) Chemical precursor of rhodopsin. p. 470

**retraction** (rĕ-trak′shun) Movement of a part toward the back. p. 265

**retroperitoneal** (ret′ro-per″ĭ-to-ne′al) Located behind the peritoneum. p. 972

**rhodopsin** (ro-dop′sin) Light-sensitive biochemical in the rods of the retina; visual purple. p. 470

**rhythmicity area** (rith-mis′ĭ-te a′re-ah) A portion of the respiratory control center in the medulla. p. 298

**riboflavin** (ri″bo-fla′vin) A vitamin of the B-complex group; vitamin $B_2$. p. 718

**ribonucleic acid** (ri″bo-nu-kle′ik as′id) A nucleic acid that contains ribose sugar; RNA. p. 120

**ribose** (ri′bōs) A 5-carbon sugar in RNA. p. 705

**ribosomal RNA** (ri-bo-sōm′al) A type of RNA that forms part of the ribosome. p. 123

**ribosome** (ri′bo-sōm) An organelle composed of RNA and protein that is a structural support for protein synthesis. p. 69

**RNA** Ribonucleic acid. p. 53

**rod** (rod) A type of light receptor that provides colorless vision. p. 469

**rotation** (ro-ta′shun) Movement turning a body part on its longitudinal axis. p. 264

**rouleaux** (roo-lo′) Groups of red blood cells stuck together like stacks of coins. p. 522

**round window** (rownd win′do) A membrane-covered opening between the inner ear and the middle ear. p. 448

**rugae** (roo′je) Thick folds in the inner wall of the stomach that disappear when the stomach is distended. p. 670

## S

**saccule** (sak′ūl) A saclike cavity that makes up part of the membranous labyrinth of the inner ear. p. 453

**sacral** (sa′kral) Pertaining to the five fused (pelvic) vertebrae at the distal end of the spinal column. p. 200

**sagittal** (saj′i-tal) A plane or section that divides a structure into right and left portions. p. 19

**salivary gland** (sal′ĭ-ver-e gland) A gland, associated with the mouth, that secretes saliva. p. 665

**salt** (sawlt) A compound produced by a reaction between an acid and a base. p. 45

**saltatory conduction** (sal′tah-tor-e kon-duk′shun) Nerve impulse conduction that seems to jump from one node to the next. p. 359

**S-A node** (nōd) Sinoatrial node. p. 568

**sarcolemma** (sar″ko-lem′ah) The cell membrane of a muscle fiber. p. 289

**sarcomere** (sar′ko-mēr) The structural and functional unit of a myofibril. p. 285

**sarcoplasm** (sar′ko-plazm) The cytoplasm within a muscle fiber. p. 285

**sarcoplasmic reticulum** (sar″ko-plaz′mik rĕ-tik′u-lum) Membranous network of channels and tubules within a muscle fiber, corresponding to the endoplasmic reticulum of other cells. p. 285

**saturated fatty acid** (sat′u-rāt″ed fat′e as′id) Fatty acid molecule that lacks double bonds between the atoms of its carbon chain. p. 49

**Schwann cell** (shwahn sel) A type of neuroglial cell that surrounds a fiber of a peripheral nerve, forming the neurilemmal sheath and myelin. p. 348

**sclera** (skle′rah) White fibrous outer layer of the eyeball. p. 461

**scoliosis** (sko″le-o′sis) Abnormal lateral curvature of the vertebral column. p. 224

**scrotum** (skro′tum) A pouch of skin that encloses the testes. p. 845

**sebaceous gland** (sĕ-ba′shus gland) Gland of the skin that secretes sebum. p. 170

**sebum** (se′bum) Oily secretion of the sebaceous glands. p. 170

**secondary immune response** (sek′un-der″e i-mun′ re-spons′) The immune system's response to subsequent encounters with a foreign antigen. p. 641

**secretin** (se-kre′tin) Hormone secreted from the small intestine that stimulates the pancreas to release pancreatic juice. p. 677

**secretion** (se-kre′shun) Substance produced in and released from a gland cell. p. 142

**semen** (se′men) Fluid discharged from the male reproductive tract at ejaculation that contains sperm cells and secretions. p. 844

**semicircular canal** (sem″ĭ-ser′ku-lar kah-nal′) Tubular structure within the inner ear that contains the receptors providing the sense of dynamic equilibrium. p. 448

**seminal vesicle** (sem′ĭ-nal ves′ĭ-kel) One of a pair of pouches that adds fructose and prostaglandins to sperm as semen forms. p. 842

**seminiferous tubule** (sem′ĭ-nif′er-us tu′būl) Tubule within the testes where sperm cells form. p. 838

**semipermeable** (sem″ĭ-per′me-ah-bl) A membrane that allows some molecules through but not others. p. 64

**senescence** (sĕ-nes′ens) Aging. p. 921

**sensation** (sen-sa′shun) A feeling resulting from the brain's interpretation of sensory nerve impulses. p. 433

**sensory adaptation** (sen′so-re ad″ap-ta′shun) The phenomenon of a sensation becoming less noticeable once it has been recognized by constant repeated stimulation. p. 433

**sensory area** (sen′so-re a′re-ah) A portion of the cerebral cortex that receives and interprets sensory nerve impulses. p. 407

**sensory nerve** (sen′so-re nerv) A nerve composed of sensory nerve fibers. p. 407

**sensory neuron** (sen′so-re nu′ron) A neuron that transmits an impulse from a receptor to the central nervous system. p. 350

**sensory receptor** (sen′so-re re″sep′tor) A specialized dendrite that detects a particular stimulus and fires an action potential in response, which is transmitted to the central nervous system. p. 344

**serosa** (sēr-o′sah) Serous membrane composing the outer layer in the walls of the organs of digestion. p. 656

**serotonin** (se″ro-to′nin) A vasoconstrictor that blood platelets release when blood vessels break, controlling bleeding. p. 532

**serous cell** (se′rus sel) A glandular cell that secretes a watery fluid with a high enzyme content. p. 665

**serous fluid** (se′rus floo′id) The secretion of a serous membrane. p. 142

**serous membrane** (se′rus mem′brān) Membrane that lines a cavity without an opening to the outside of the body. p. 161

**serum** (se′rum) The fluid portion of coagulated blood. p. 539

**sesamoid bone** (ses′ah-moid bōn) A round bone in tendons adjacent to joints. p. 186

**sex chromosome** (seks kro′mo-sōm) A chromosome that carries genes responsible for the development of characteristics associated with maleness or femaleness; an X or Y chromosome. p. 936

**sex-influenced inheritance** (seks-in′floo-enst in-her′ĭ-tens) Transmission of a trait that is dominant in one sex but recessive in the other. p. 947

**sex-limited inheritance** (seks-lim′it-ed in-her′ĭ-tens) Transmission of a trait expressed in one sex only. p. 947

**sex-linked inheritance** (seks′-linkt in-her′ĭ-tens) Transmission of a trait controlled by a gene on a sex chromosome (X or Y). p. 945

**sexually transmitted disease** (sek′shoo-ah-le trans-mi′ted dĭ-zēz′) Infection transmitted from one individual to another by direct contact during sexual activity; STD. p. 882

**sigmoid colon** (sig′moid ko′lon) S-shaped portion of the large intestine between the descending colon and the rectum. p. 692

**simple sugar** (sim′pl shoog′ar) A monosaccharide. p. 705

**signal transduction** (sig′nahl trans-duk′shun) A series of biochemical reactions that allows cells to receive and respond to messages coming in through the cell membrane. p. 64

**sinoatrial node** (si′no-a′tre-al nōd) Specialized tissue in the wall of the right atrium that initiates cardiac cycles; the pacemaker; S-A node. p. 568

**sinus** (si′nus) A cavity or space in a bone or other body part. p. 741

**skeletal muscle** (skel′ĕ-tal mus′l) Type of muscle tissue found in muscles attached to bones. p. 153

**sliding filament theory** (slī'ding fil'eh-ment the'o-re) Muscles contract when the thin (actin) and thick (myosin) filaments move past each other, shortening the skeletal muscle cells. p. 288

**small intestine** (smawl in-tes'tin) Part of the digestive tract extending from the stomach to the cecum; consisting of the duodenum, jejunum, and ileum. p. 685

**smooth muscle** (smōōth mus'el) Type of muscle tissue found in the walls of hollow visceral organs; visceral muscle. p. 153

**sodium pump** (so'de-um pump) The active transport mechanism that concentrates sodium ions on the outside of a cell membrane. p. 63

**solute** (sol'ūt) The substance dissolved in a solution. p. 532

**solvent** (sol'vent) The liquid portion of a solution in which a solute is dissolved. p. 81

**somatic** (so-mat'ik) Pertaining to the body. p. 407

**somatic cell** (so-mat'ik sel) Any cell of the body other than the sex cells. p. 407

**somatomedin** (so"mah-to-me'din) A substance released from the liver in response to growth hormone that promotes the growth of cartilage. p. 492

**somatostatin** (so-mat'o-sta'tin) A hormone secreted by the islet of Langerhans cells that inhibits the release of growth hormone. p. 508

**somatotropin** (so"mah-to-tro'pin) Growth hormone. p. 490

**spastic paralysis** (spas'tik pah-ral'ĭ-sis) A form of paralysis characterized by an increase in muscular tone without atrophy of the muscles involved. p. 382

**special sense** (spesh'al sens) Sense that involves receptors associated with specialized sensory organs, such as the eyes and ears. p. 461

**species resistance** (spe'sēz re-zis'tans) A natural ability of one type of organism to resist infection by microorganisms that might cause disease in another type of organism. p. 630

**spermatic cord** (sper-mat'ik kord) A structure consisting of blood vessels, nerves, the vas deferens, and other vessels extending from the abdominal inguinal ring to the testis. p. 836

**spermatid** (sper'mah-tid) An intermediate stage in the formation of sperm cells. p. 838

**spermatocyte** (sper-mat'o-sīt) An early stage in the formation of sperm cells. p. 838

**spermatogenesis** (sper"mah-to-jen'ĕ-sis) The production of sperm cells. p. 838

**spermatogonium** (sper"mah-to-go'ne-um) Undifferentiated spermatogenic cell found in the germinal epithelium of a seminiferous tubule. p. 838

**spermatozoa** (sper"mah-to-zo'ah) Male reproductive cells; sperm cells. p. 838

**sphincter** (sfingk'ter) A circular muscle that closes an opening or the lumen of a tubular structure. p. 322

**sphygmomanometer** (sfig'mo-mah-nom'ĕ-ter) Instrument used for measuring blood pressure. p. 586

**spinal** (spi'nal) Pertaining to the spinal cord or to the vertebral canal. p. 377

**spinal cord** (spi'nal kord) Portion of the central nervous system extending downward from the brain stem through the vertebral canal. p. 377

**spinal nerve** (spi'nal nerv) Nerve that arises from the spinal cord. p. 377

**spleen** (splēn) A large, glandular organ located in the upper left region of the abdomen. p. 629

**spongy bone** (spunj'e bōn) Bone that consists of bars and plates separated by irregular spaces; cancellous bone. p. 185

**squamous** (skwa'mus) Flat or platelike. p. 205

**staircase effect** (stair'kays i-fekt') A gradual increase in contractile strength of a muscle in response to repeated stimuli of the same intensity. p. 295

**starch** (starch) A polysaccharide that is common in foods of plant origin. p. 47

**static equilibrium** (stat'ik e'kwĭ-lib're-um) The maintenance of balance when the head and body are motionless. p. 453

**stereocilia** (ste"re-o-sil'e-ah) The hairlike processes of the hair cells within the organ of Corti. p. 450

**stereoscopic vision** (ster"e-o-skop'ik vizh'un) Objects perceived as three-dimensional; depth perception. p. 473

**sternal** (ster'nal) Pertaining to the sternum. p. 20

**steroid** (ste'roid) A type of organic molecule including complex rings of carbon and hydrogen atoms. p. 49

**stimulus** (stim'u-lus) A change in the environmental conditions that is followed by a response by an organism or cell (pl., *stimuli*). p. 289

**stomach** (stum'ak) Digestive organ located between the esophagus and the small intestine. p. 670

**strabismus** (strah-biz'mus) Lack of visual coordination; crossed eyes. p. 460

**stratified** (strat'ĭ-fīd) Arranged in layers. p. 140

**stratum corneum** (stra'tum kor'ne-um) Outer horny layer of the epidermis. p. 163

**stratum germinativum** (stra'tum jer'mĭ-na"ti-vum) The deepest layer of the epidermis in which the cells divide; stratum basale. p. 163

**stress** (stres) Response to factors perceived as capable of threatening life. p. 511

**stressor** (stres'or) A factor capable of stimulating a stress response. p. 511

**stretch receptor** (strech re-sep'ter) Sensory nerve ending that responds to tension. p. 433

**stretch reflex** (strech re'fleks) Muscle contraction in response to stretching the muscle. p. 438

**stroke volume** (strōk vol'ūm) The amount of blood discharged from the ventricle with each heartbeat. p. 584

**structural formula** (struk'cher-al fōr'mu-lah) A representation of the way atoms bond to form a molecule, using the symbols for each element and lines to indicate chemical bonds. p. 44

**subarachnoid space** (sub"ah-rak'noid spās) The space within the meninges between the arachnoid mater and the pia mater. p. 377

**subcutaneous** (sub"ku-ta'ne-us) Beneath the skin. p. 162

**sublingual** (sub-ling'gwal) Beneath the tongue. p. 665

**submaxillary** (sub-mak'sĭ-ler"e) Below the maxilla. p. 208

**submucosa** (sub"mu-ko'sah) Layer of connective tissue underneath a mucous membrane. p. 656

**substrate** (sub'strāt) The substance upon which an enzyme acts. p. 104

**sucrase** (su'krās) Digestive enzyme that catalyzes the breakdown of sucrose. p. 688

**sucrose** (soo'krōs) A disaccharide; table sugar. p. 47

**sulcus** (sul'kus) A shallow groove, such as that between convolutions on the surface of the brain (pl., *sulci*). p. 389

**summation** (sum-ma'shun) Phenomenon in which the degree of change in membrane potential is directly proportional to the intensity of stimulation. p. 295

**superficial** (soo'per-fish'al) Near the surface. p. 19

**superior** (su-pe're-or) Pertaining to a structure that is higher than another structure. p. 18

**supination** (soo"pǐ-na'shun) Rotation of the forearm so that the palm faces upward when the arm is outstretched. p. 265

**surface tension** (ser'fas ten'shun) Force that holds moist membranes together when water molecules attract. p. 754

**surfactant** (ser-fak'tant) Substance the lungs produce that reduces the surface tension within the alveoli. p. 755

**suture** (soo'cher) An immovable joint, such as that between flat bones of the skull. p. 199

**sympathetic nervous system** (sim"pah-thet'ik ner'vus sis'tem) Portion of the autonomic nervous system that arises from the thoracic and lumbar regions of the spinal cord. p. 419

**symphysis** (sim'fǐ-sis) A slightly movable joint between bones separated by a pad of fibrocartilage. p. 259

**synapse** (sin'aps) The junction between the axon of one neuron and the dendrite or cell body of another neuron. p. 344

**synaptic cleft** (sǐ-nap'tik kleft) A narrow extracellular space between the presynaptic and postsynaptic neurons. p. 286

**synaptic knob** (sǐ-nap'tik nob) Tiny enlargement at the end of an axon that secretes a neurotransmitter. p. 346

**synaptic transmission** (sǐ-nap'tik trans-mish'un) Communication of an impulse from one neuron to the next. p. 360

**synchondrosis** (sin"kon-dro'sis) Type of joint in which bones are united by bands of hyaline cartilage. p. 258

**syncytium** (sin-sish'e-um) A mass of merging cells. p. 299

**syndesmosis** (sin"des-mo'sis) Type of joint in which the bones are united by relatively long fibers of connective tissue. p. 257

**syndrome** (sin'drōm) A group of symptoms that characterize a disease condition.

**synergist** (sin'er-jist) A muscle that assists the action of a prime mover. p. 300

**synovial fluid** (sǐ-no've-al floo'id) Fluid secreted by the synovial membrane. p. 261

**synovial joint** (sǐ-no've-al joint) A freely movable joint. p. 259

**synovial membrane** (sǐ-no've-al mem'brān) Membrane that forms the inner lining of the capsule of a freely movable joint. p. 161

**synthesis** (sin'thě-sis) The process by which substances unite to form more complex substances. p. 44

**system** (sis'tem) A group of organs that act together to carry on a specialized function. p. 14

**systemic circuit** (sis-tem'ik ser'kit) The vessels that conduct blood between the heart and all body tissues except the lungs. p. 592

**systole** (sis'to-le) Phase of the cardiac cycle during which a heart chamber wall contracts. p. 566

**systolic pressure** (sis-tol'ik presh'ur) Arterial blood pressure during the systolic phase of the cardiac cycle. p. 583

# T

**tachycardia** (tak"e-kar'de-ah) An abnormally rapid heartbeat. p. 574

**target tissue** (tar'get tish'u) Specific tissue on which a hormone exerts its effect. p. 480

**tarsus** (tar'sus) The ankle bones. p. 20

**taste bud** (tāst bud) Organ containing receptors associated with the sense of taste. p. 442

**telophase** (tel'o-fāz) Stage in mitosis when newly formed cells separate. p. 91

**tendon** (ten'don) A cordlike or bandlike mass of white fibrous connective tissue that connects a muscle to a bone. p. 282

**teratogen** (ter'ah-to-jen) A chemical or other environmental agent that causes a birth defect. p. 906

**testis** (tes'tis) Primary reproductive organ of a male; a sperm cell–producing organ (pl., *testes*). p. 510

**testosterone** (tes-tos'tě-rōn) Male sex hormone secreted by the interstitial cells of the testes. p. 850

**tetanus** (tet'ah-nus) A continuous, forceful muscular contraction (tetanic contraction) without relaxation. p. 295

**thalamus** (thal'ah-mus) A mass of gray matter located at the base of the cerebrum in the wall of the third ventricle. p. 390

**thermoreceptor** (ther"mo-re-sep'tor) A sensory receptor that is sensitive to changes in temperature; a heat receptor. p. 433

**thiamine** (thi'ah-min) Vitamin B₁. p. 717

**thoracic** (tho-ras'ik) Pertaining to the chest. p. 11

**threshold potential** (thresh'old po-ten'shal) Level of potential at which an action potential or nerve impulse is produced. p. 294

**threshold stimulus** (thresh'old stim'u-lus) The level of stimulation that must be exceeded to elicit a nerve impulse or a muscle contraction. p. 294

**thrombin** (throm'bin) Blood-clotting enzyme that catalyzes formation of fibrin from fibrinogen. p. 538

**thrombocyte** (throm'bo-sīt) A blood platelet. p. 532

**thrombocytopenia** (throm"bo-si"to-pe'ne-ah) A low number of platelets in the circulating blood. p. 542

**thrombus** (throm'bus) A blood clot in a blood vessel that remains at its site of formation. p. 539

**thymosin** (thi'mo-sin) A group of peptides secreted by the thymus gland that increases production of certain types of white blood cells. p. 510

**thymus** (thi'mus) A glandular organ located in the mediastinum behind the sternum and between the lungs. p. 510

**thyroglobulin** (thi"ro-glob'u-lin) Biochemical secreted by cells of the thyroid gland that stores thyroid hormones. p. 497

**thyroid gland** (thi'roid gland) Endocrine gland located just below the larynx and in front of the trachea that secretes thyroid hormones. p. 495

**thyrotropin** (thi"ro-trōp'in) Hormone secreted by the anterior pituitary gland that stimulates the thyroid gland to secrete hormones; TSH. p. 492

**thyroxine** (thi-rok'sin) A hormone secreted by the thyroid gland. p. 495

**tidal volume** (tīd'al vol'ūm) Amount of air that enters the lungs during a normal, quiet inspiration. p. 757

**tissue** (tish'u) A group of similar cells that performs a specialized function. p. 8

**T cell** (T sel) Lymphocytes that interact directly with antigen-bearing cells and particles and secrete cytokines, producing cell-mediated immunity. p. 632

**tonsils** (ton'silz) Collections of lymphatic tissue in the throat. p. 627

**trabecula** (trah-bek'u-lah) Branching bony plate that separates irregular spaces within spongy bone. p. 186

**trachea** (tra'ke-ah) Tubular organ that leads from the larynx to the bronchi. p. 744

**transcellular fluid** (trans"sel'u-lar floo'id) A portion of the extracellular fluid, including the fluid within special body cavities. p. 814

**transcription** (trans-krip′shun) Manufacturing RNA from DNA. p. 121

**transferrin** (trans-fer′rin) Blood plasma protein that transports iron. p. 524

**transfer RNA** (trans′fer RNA) Molecule of RNA that carries an amino acid to a ribosome in the process of protein synthesis. p. 122

**translation** (trans-la′shun) Assembly of an amino acid chain according to the sequence of base triplets in a molecule of mRNA. p. 122

**transverse** (trans-vers′) At right angles to the long axis of a part; crosswise. p. 19

**transverse colon** (trans-vers′ ko′lon) Portion of the large intestine that extends across the abdomen from right to left below the stomach. p. 692

**transverse tubule** (trans-vers′ tu′būl) Membranous channel that extends inward from a muscle fiber membrane and passes through the fiber. p. 285

**tricuspid valve** (tri-kus′pid valv) Heart valve located between the right atrium and the right ventricle. p. 558

**trigger zone** (trig′ger zōn) Site on the neuron where an action potential is generated. p. 350

**triglyceride** (tri-glis′er-īd) A lipid composed of three fatty acids combined with a glycerol molecule. p. 113

**triiodothyronine** (tri″i-o″do-thi′ro-nēn) A type of thyroid hormone. p. 495

**trisomy** (tri′so-me) Condition in which a cell contains three chromosomes of a particular type instead of two. p. 949

**trochanter** (tro-kan′ter) A broad process on a bone. p. 202

**trochlea** (trok′le-ah) A pulley-shaped structure. p. 226

**trophoblast** (trof′o-blast) The outer cells of a blastocyst that help form the placenta and other extra-embryonic membranes. p. 868

**tropic hormone** (trōp′ik hor′mōn) A hormone that has an endocrine gland as its target tissue. p. 488

**tropomyosin** (tro″po-mi′o-sin) Protein that blocks muscle contraction until calcium ions are present. p. 288

**troponin** (tro′po-nin) Protein that functions with tropomyosin to block muscle contraction until calcium ions are present. p. 288

**trypsin** (trip′sin) An enzyme in pancreatic juice that breaks down protein molecules. p. 677

**trypsinogen** (trip-sin′o-jen) Substance pancreatic cells secrete that is enzymatically cleaved to yield trypsin. p. 677

**tubercle** (tu′ber-kl) A small, rounded process on a bone. p. 202

**tuberosity** (tu″bĕ-ros′ĭ-te) An elevation or protuberance on a bone. p. 202

**tumor** (too′mor) A tissue mass formed when cells lose division control. p. 96

**twitch** (twich) A brief muscular contraction followed by relaxation. p. 294

**tympanic membrane** (tim-pan′ik mem′brān) A thin membrane that covers the auditory canal and separates the external ear from the middle ear; the eardrum. p. 446

## U

**umbilical cord** (um-bil′ĭ-kal kord) Cordlike structure that connects the fetus to the placenta. p. 899

**umbilical region** (um-bil′ĭ-kal re′jun) The central portion of the abdomen. p. 19

**umbilicus** (um-bil′ĭ-kus) Region to which the umbilical cord was attached; the navel. p. 29

**unipolar neuron** (un″ĭ-po′lar nu′ron) A neuron that has a single nerve fiber extending from its cell body. p. 350

**unsaturated fatty acid** (un-sat′u-rāt″ed fat′e as′id) Fatty acid molecule that has one or more double bonds between the atoms of its carbon chain. p. 49

**upper limb** (uh′per lim) The superior appendage consisting of the arm, forearm, wrist, and hand. p. 200

**urea** (u-re′ah) A nonprotein nitrogenous substance produced as a result of protein metabolism. p. 114

**ureter** (u-re′ter) A muscular tube that carries urine from the kidney to the urinary bladder. p. 802

**urethra** (u-re′thrah) Tube leading from the urinary bladder to the outside of the body. p. 806

**uric acid** (u′rik as′id) Product of nucleic acid metabolism in the body. p. 799

**urine** (u′rin) Wastes and excess water removed from the blood and excreted by the kidneys into the ureters to the urinary bladder and out of the body through the urethra. p. 789

**uterine** (u′ter-in) Pertaining to the uterus. p. 856

**uterine tube** (u′ter-in tūb) Tube that extends from the uterus on each side toward an ovary and transports sex cells; fallopian tube or oviduct. p. 855

**uterus** (u′ter-us) Hollow muscular organ within the female pelvis in which a fetus develops. p. 856

**utricle** (u′trĭ-kl) An enlarged portion of the membranous labyrinth of the inner ear. p. 453

**uvula** (u′vu-lah) A fleshy portion of the soft palate that hangs down above the root of the tongue. p. 660

## V

**vaccine** (vak′sēn) A substance that contains antigens used to stimulate an immune response. p. 642

**vacuole** (vak′u-ōl) A space or cavity within the cytoplasm of a cell. p. 76

**vagina** (vah-ji′nah) Tubular organ that leads from the uterus to the vestibule of the female reproductive tract. p. 858

**Valsalva's maneuver** (val-sal′vahz mah-noo′ver) Increasing the intrathoracic pressure by forcing air from the lungs against a closed glottis. p. 695

**varicose veins** (var′ĭ-kos vānz) Abnormally swollen and enlarged veins, especially in the legs. p. 582

**vasa recta** (va′sah rek′tah) A branch of the peritubular capillary that receives blood from the efferent arterioles of juxtamedullary nephrons. p. 789

**vascular** (vas′ku-lar) Pertaining to blood vessels. p. 573

**vas deferens** (vas def′er-ens) Tube that leads from the epididymis to the urethra of the male reproductive tract (pl., *vasa deferentia*). p. 842

**vasoconstriction** (vas″o-kon-strik′shun) A decrease in the diameter of a blood vessel. p. 575

**vasodilation** (vas″o-di-la′shun) An increase in the diameter of a blood vessel. p. 576

**vasomotor center** (vas″o-mo′tor sen′ter) Neurons in the brain stem that control the diameter of the arteries. p. 402

**vasopressin** (vas″o-pres′in) Antidiuretic hormone. p. 494

**vein** (vān) A vessel that carries blood toward the heart. p. 573

**vena cava** (vēn′ah kāv′ah) One of two large veins that convey deoxygenated blood to the right atrium of the heart. p. 558

**ventral** (ven′tral)   Pertaining to the front or anterior.   p. 10

**ventral root** (ven′tral rōōt)   Motor branch of a spinal nerve by which it attaches to the spinal cord.   p. 413

**ventricle** (ven′trĭ-kl)   A cavity, such as those of the brain that are filled with cerebrospinal fluid, or those of the heart that contain blood.   p. 387

**venule** (ven′ūl)   A vessel that carries blood from capillaries to a vein.   p. 573

**vermiform appendix** (ver′mĭ-form ah-pen′diks)   Appendix.   p. 692

**vertebral** (ver′te-bral)   Pertaining to the bones of the spinal column.   p. 11

**vesicle** (ves′ĭ-kal)   Membranous, cytoplasmic sac formed by an infolding of the cell membrane.   p. 75

**vestibule** (ves′tĭ-būl)   A space at the opening to a canal.   p. 448

**villus** (vil′us)   Tiny, fingerlike projection that extends outward from the inner lining of the small intestine (pl., *villi*).   p. 896

**visceral** (vis′er-al)   Pertaining to the contents of a body cavity.   p. 10

**visceral peritoneum** (vis′er-al per″ĭ-to-ne′um)   Membrane that covers the surfaces of organs within the abdominal cavity.   p. 15

**visceral pleura** (vis′er-al ploo′rah)   Membrane that covers the surfaces of the lungs.   p. 751

**viscosity** (vis-kos′ĭ-te)   The tendency for a fluid to resist flowing due to the internal friction of its molecules.p. 585

**vital capacity** (vi′tal kah-pas′i-te)   The maximum amount of air a person can exhale after taking the deepest breath possible.   p. 757

**vitamin** (vi′tah-min)   An organic compound other than a carbohydrate, lipid, or protein that is needed for normal metabolism but that the body cannot synthesize in adequate amounts.   p. 105

**vitreous body** (vit′re-us bod′e)   The collagenous fibers and fluid that occupy the posterior cavity of the eye.   p. 467

**vitreous humor** (vit′re-us hu′mor)   The substance that occupies the space between the lens and the retina of the eye.   p. 467

**vocal cords** (vo′kal kordz)   Folds of tissue within the larynx that produce sounds when they vibrate.   p. 742

**Volkmann's canal** (fōlk′mahnz kah-nal′)   A transverse channel that connects Haversian canals within compact bone.   p. 187

**volt** (vōlt)   Unit used to measure differences in electrical potential.   p. 365

**voluntary** (vol′un-tār″e)   Capable of being consciously controlled.

**vulva** (vul′vah)   The external reproductive parts of the female that surround the opening of the vagina.   p. 859

## W

**water balance** (wot′er bal′ans)   Condition in which the volume of water entering the body is equal to the volume leaving it.   p. 817

**water of metabolism** (wot′er uv mĕ-tab′o-lizm)   Water produced as a by-product of metabolic processes.   p. 816

**white blood cell** (whīt blud sel)   A cell that helps fight infection.   p. 527

**white muscle** (whīt mus′el)   Fast-contracting skeletal muscle.   p. 293

**wild type** (wīld tīp)   A phenotype or allele that is the most common for a certain gene in a population.   p. 940

## X

**X-linked trait** (eks-linkt′ trāt)   Trait determined by a gene on an X chromosome.   p. 945

**x-ray** (eks′ ray)   Used as a verb: to photograph using radiation.   p. 54

**X ray** (eks ray)   Used as a noun: a photograph produced by radiation. May also be used as an adjective: X-ray.   p. 54

## Y

**yellow marrow** (yel′o mar′o)   Fat storage tissue found in the cavities within certain bones.   p. 187

**yolk sac** (yōk sak)   An extraembryonic membrane connected to the embryo by a long, narrow tube.   p. 905

## Z

**zonular fiber** (zon′u-lar fi′ber)   A delicate fiber within the eye that extends from the ciliary process to the lens capsule.   p. 461

**zygote** (zi′gōt)   Cell produced by the fusion of an egg and sperm; a fertilized egg cell.   p. 852

**zymogen granule** (zi-mo′jen gran′ūl)   A cellular structure that stores inactive forms of protein-splitting enzymes in a pancreatic cell.   p. 677

# CREDITS

Medicine, Northeastern Ohio Universities College of Medicine and Research Associate, Carnegie Museum of Natural History.; **8.3a-b:** Courtesy of John W. Hole, Jr.; **8.13b, 8.15b:** © Paul Reimann; **8.17a-b** Courtesy of John W. Hole, Jr.; **8.18b, 8.20b:** © Paul Reimann.

## Chapter 9

**Opener:** © AP/Wide World Photos; **9.3:** © The McGraw-Hill Companies, Inc./Carol D. Jacobson, Ph.D., Department of Veterinary Anatomy, Iowa State University.; **9.5:** © H.E. Huxley; **9.8b:** © The McGraw-Hill Companies, Inc./Carol D. Jacobson, Ph.D., Department of Veterinary Anatomy, Iowa State University.; **9.18:** © Dr. Paul Heidger, University of Iowa College of Medicine; **p. 337(all):** © John C. Mese.

## Chapter 10

**Opener:** © Dr. Mark Mattson, University of Kentucky College of Medicine; **10.1:** © Ed Reschke; **10.4b:** © Biophoto Associates/Photo Researchers; **10.5:** © Dr. Dennis Emery, Department of Zoology and Genetics, Iowa State University; **10.8:** © R.G. Kessel and R.H. Kardon, *Tissues and Organs: A Text-Atlas of Scanning Electron Microscopy,* 1979.; **10.17c:** © Don Fawcett/Photo Researchers; **10A:** © Kobal Collection; **10B:** Courtesy, American Medical Association.

## Chapter 11

**Opener:** © John Allison/Peter Arnold; **p. 375:** From: Damasio H, Grabowski, T, Frank R, Galaburda AM, Damasio AR. 20 May 1994. "The return of Phineas Gage: Clues about the brain from the skull of a famous patient." *Science,* 264:1102-1105. Department of Neurology and Image Analysis Facility, University of Iowa. @1994 American Association for the Advancement of Science.; **11.4b:** © Per H. Kjeldsen; **11B:** © AP/Wide World Photos; **11C:** © M. Abbey/Photo Researchers; **11Ea-Eb:** © Dr. Kent M. Van De Graaff; **11.23:** © Martin Rotker/Photo Researchers; **11.25:** © R.G. Kessel and R.H. Kardon, *Tissues and Organs: A Text-Atlas of Scanning Electron Microscopy,* 1979.

## Chapter 12

**Opener:** © Richard Hutchings/Photo Researchers; **12.2a-b:** © Ed Reschke; **12.7:** © Dwight Kuhn; **12.9:** © Victor B. Eichler, Ph.D.; **12.16a:** © John D. Cunningham/Visuals Unlimited; **12.16b:** © Fred Hossler/Visuals Unlimited; **12.21:** © Dean E. Hillman; **12.29:** © SPL/Photo Researchers; **12.35:** © Per H. Kjeldsen; **12.36a:** © Carroll Weiss/Camera M.D. Studios; **12.40c:** © Frank S. Werblin.

## Chapter 13

**Opener (both):** Courtesy, Eli Lilly and Company Archives.; **13A:** © Gianni Giansanti/Sygma; **13.12:** © John Cunningham/Visuals Unlimited; **13B(all):** Reprinted by permission of the publisher from: Albert Mendeloff. "Acromegaly, diabetes, hypermetabolism, proteinuria, and heart failure." *American Journal of Medicine:* 20:1 (Jan. 1956), p. 135. © by Excerpta Medica Inc.; **13.17:** © Fred Hossler/Visuals Unlimited; **13.19-13.20:** From R.H. Kampmeier and T.M. Blake. *Physical Examination in Health and Disease* © 1952 F.A. Davis Company, Philadelphia; **13.21:** © Lester Bergman & Associates; **13.23:** © R. Calentine/Visuals Unlimited; **13.27, 13.33:** © Ed Reschke.

## Chapter 14

**Opener:** © Ken Eward/Science Source/Photo Researchers; **14.3b:** © J. & L. Weber/Peter Arnold; **14.4b:** © Bill Longcore/Photo Researchers; **14.8a:** © Fred Hossler/Visuals Unlimited; **14.8b:** © CNRI/Phototake; **14.9-14.13:** © Ed Reschke; **14Aa-b:** © John Forsythe/Visuals Unlimited; **14.18:** © SPL/Photo Researchers; **14.20a-b:** © Carroll Weiss/Camera M.D. Studios; **14B:** National Library of Medicine; **14C:** © Biopharm (USA) Limited 1994.

## Chapter 15

**Opener:** © D.W. Fawcett/Photo Researchers; **p. 554a-b:** From Leonard V. Crowley. *Introduction to Human Disease,* third edition, © Jones and Bartlett Publishers, Boston. Reprinted by permission.; **15.2:** © A. & F. Michler/Peter Arnold; **15.7-15.8:** © The McGraw-Hill Companies, Inc./University of Michigan Biomedical Communications; **15.14:** Courtesy, Eastman Kodak; **15.27:** © R.G. Kessel and R.H. Kardon, *Tissues and Organs: A Text-Atlas of Scanning Electron Microscopy,* 1979.; **15.29b-c:** © D.W. Fawcett; **15.30:** © T. Kuwabara, from Bloom, W. and Fawcett, D.W., *Texbook of Histology,* 10th edition., W.B. Saunders Co., 1975.; **15.32a:** © Ed Reschke; **15.32b:** © Victor B. Eichler, Ph.D.; **15Ea-d:** Courtesy, American Heart Association; **15.37a-b:** From Patricia Phelps and J. Luft. "Electron study of relaxation and constriction in frog arterioles." *American Journal of Anatomy,* 125:409, fig. 2-3, © 1969, Wiley-Liss, a division of John Wiley and Sons, Inc. Reprinted by permission.; **15H:** © NASA; **15.46b:** © Dr. Kent M. Van De Graaff; **15.48:** From J. Dankmeijer, H.G. Lambers, and J.M.F. Landsmeer, Practische Ontleedkunde, Bohn, Scheltema & Holkema, Publishers, Netherlands.; **15J:** © AP/Wide World Photos.

## Chapter 16

**Opener:** © Photo Works/Photo Researchers; **16.3:** © Ed Reschke; **p. 621:** Courtesy, The Hancock Family; **16.5:** Courtesy, Eastman Kodak; **16.9b:** © John Watney/Photo Researchers; **16.10:** © Dr. Kent M. Van De Graaff; **16.13:** © John D. Cunningham/Visuals Unlimited; **16.15:** © Biophoto Associates/Photo Researchers; **16.17:** Courtesy of Dr. Etienne de Harven and Miss Nina Lampen, from the Sloan-Kettering Institute for Cancer Research.; **16.19b:** © Manfred Kage/Peter Arnold; **16A:** © Courtesy of Schering-Plough Corporation/Photo by Phillip A. Harrington; **16.24b:** © Lennart Nilsson/Bonnierforlagen AB; **16B:** © Zeva Oelbaum/Peter Arnold.

## Chapter 17

**Opener:** © P. Motta/SPL/Photo Researchers; **17.8:** Courtesy of John W. Hole, Jr.; **17.9b:** © Biophoto Associates/Photo Researchers; **17A:** © SPL/Photo Researchers; **17.12a, 17.12c:** © Bruce Iverson; **17.12b:** © Biophoto Associates/Photo Researchers; **17.16:** © Ed Reschke; **17.18:** © Dr. Kent M. Van De Graaff; **17.20:** © Ed Reschke; **17.29c:** © Victor B. Eichler, Ph.D.; **17.31:** © Carroll Weiss/Camera M.D. Studios; **17.34:** Armed Forces Institute of Pathology; **17.38:** © Manfred Kage/Peter Arnold; **17.39b:** © Keith Porter; **17.46:** © James L. Shaffer; **17.48:** © Ed Reschke/Peter Arnold; **17.49:** © Ed Reschke; **17B:** © Susan Leavines/Photo Researchers; **17C:** From Leonard V. Crowley, *Introduction to Human Disease,* 3rd edition, fig. 23.22. © 1992 Jones and Bartlett Publishers, Boston. Reprinted by permission.

## Chapter 18

**Opener:** © Tony Freeman/Photo Edit; **p. 704:** © AP/Wide World Photos; **18A:** From B. T. Burton and W. R. Foster: "Health Implications of Obesity and NIH Consensus Conference." Copyright © The American Dietetic Association. Reprinted by permission from *Journal of the American Dietetic Association,* Vol. 85: 1985, 1117-21; **18.6a:** © Bob Daemmrich/Image Works; **18.6b:** © Bruce Curtis/Peter Arnold; **18.8:** © World Health Organization; **18.15b:** © Fred Hossler/Visuals Unlimited; **18.15c:** © CNRI/Phototake; **18.18:** © Reuters/Bettmann.

## Chapter 19

**Opener:** © Manfred Kage/Peter Arnold; **p. 739:** Courtesy, BRK Brand/First Alert; **19Aa-b:** Used with permission of the American Cancer Society; **19B:** © Moredun Animal Health/SPL/Photo Researchers; **19.4a-b:** Courtesy, Eastman Kodak; **19.7c:** © CNRI/Phototake; **19.10:** © Ed Reschke; **19.13:** © John Watney Photo Library; **19.15:** © Dwight Kuhn; **19.17:** R.G. Kessel and R.H. Kardon, *Tissues and Organs: A Text-Atlas of Scanning Electron Microscopy,* 1979.; **19.18:** Courtesy, American Lung Association; **19.26:** © Edward Lettau/Photo Researchers; **19Ca-b:** © Victor B. Eichler, Ph.D; **19.31:** Courtesy, American Lung Association; **19.33:** © From John F. Murray, *The Normal Lung,* second edition, fig. © 1986, W.B. Saunders Company.; **19D(1-2):** © CNRI/Phototake.

## Chapter 20

**Opener:** © Biophoto Associates/Photo Researchers; **20.2:** © CNRI/SPL/Photo Researchers; **p. 779:** © Edward Lettau/Photo Researchers; **20.6b:** © Lester Bergman & Associates; **20.7a:** © R.G. Kessel and R.H. Kardon, *Tissues and Organs: A Text-Atlas of Scanning Electron Microscopy,* 1979.; **20.7b:** Courtesy of R.B. Wilson, M.D./Eppley Institute for Research in Cancer, University of Nebraska Medical Center; **20.8:** © David M. Phillips/ Visuals Unlimited; **20.10a:** © Biophoto Associates/Photo Researchers; **20.10b:** © Manfred Kage/Peter Arnold; **20.27:** © Per H. Kjeldsen; **20.30:** © John D. Cunningham/Visuals Unlimited; **20.31:** © Ed Reschke.

## Chapter 21

**Opener:** © Bruce Ayres/Tony Stone Images; **p. 814:** © Corbis-Bettmann.

## Chapter 22

**Opener:** © Bob Schuchman/Phototake; **22.4:** © R.G. Kessel and R.H. Kardon, *Tissues and Organs: A Text-Atlas of Scanning Electron Microscopy,* 1979.; **22.5a:** © Biophoto Associates/Photo Researchers; **22.7:** © David M. Phillips/The Population Council/Photo Researchers; **22.8:** © Ed Reschke; **22.9a:** © R.G. Kessel and R.H. Kardon, *Tissues and Organs: A Text-Atlas of Scanning Electron Microscopy,* 1979.; **22.9b:** © Ed Reschke; **22.10:** © Manfred Kage/Peter Arnold; **22.17b:** © R.J. Blandau; **22.18:** © Ed Reschke; **22.19b:** © Ed Reschke/Peter Arnold; **22.20:** © H. Mizoguchi, from W. Bloom and D.W. Fawcett, *A Textbook of Histology,* tenth edition, W.B. Saunders Co., 1975 © D.W. Fawcett; **22.21:** © Dr. Landrum B. Shettles; **22.24a:** © Ed Reschke; **22.24b:** © Don Fawcett/Photo Researchers; **22.25:** © The McGraw-Hill Companies, Inc./Carol D. Jacobson, Ph.D., Department of Veterinary Anatomy, Iowa State University; **22.31:** From M. Tegner and D. Epel. 16 February 1973. "Sea Urchin Sperm." *Science,* 179:685-688. @ 1973 American Association for the Advancement of Science; **22.39a-b:** © Biophoto Associates/Photo Researchers; **22.39c:** Courtesy, Southern Illinois University School of Medicine; **22.42a-e:** © Bob Coyle.

## Chapter 23

**Opener:** © AP/Wide World Photos; **23.1a:** © Petit Format/Nestle/Photo Researchers; **23.1b:** © Richard Nowitz/Phototake; **23.2b:** © Alexander Tsiaras/Photo Researchers; **23.2c:** © Omikron/Photo Researchers; **23.3c:** © Petit Format/Nestle/Photo Researchers; **23.4c, 23.5:** © Courtesy of Ronan O'Rahilly, M.D., Carnegie Institute of Washington, Department of Embryology, Davis Division; **23B:** People Weekly © Taro Yamasaki, 1992; **23.9a-b:** © Dr. Landrum B. Shettles; **23.9c:** © Petit Format/Nestle/Photo Researchers; **23.10b:** © Donald Yaeger/Camera M.D. Studios; **23.11b:** © Dr. Landrum B. Shettles; **23.16:** © Donald Yaeger/Camera M.D. Studios; **23Cb:** From Anne Pytkowicz Streissguth. 18 July 1980. *Science,* 209:353-61. @1980 American Association for the Advancement of Science.; **23D:** © Steve Wernka/Impact Visuals; **23.23:** © Donna Jernigan; **23.25(1-5):** © UPI/Corbis-Bettmann; **23.25(6):** © AP/Wide World Photos; **23.25(7-8):** © Corbis-Bettmann; **23E:** © J.L. Bucao/Gamma Liaison.

## Chapter 24

**Opener:** © Jeff Greenberg/Photo Researchers; **p. 928:** © AP/Wide World Photos; **24Ac(1-2):** © Bob Coyle; **24Ad(1-2):** © Tom Ballard/EKM-Nepenthe; **24.3b:** © From E.J. Dupraw, *DNA and Chromosomes* © 1970, Holt, Rinehart & Winston, Inc.; **24.8:** Courtesy, David Ward and Dr. Michael Speicher; **24.11a-b:** © Bill Longcore/Photo Researchers; **24.12a:** Library of Congress; **24.12b:** © Peter Morenus/UConn at Storrs; **24.14:** © Biophoto Associates/Photo Researchers; **24.15a-d:** © Corbis-Bettmann; **24D(1-4):** From R. Simensen and R. Curtis Rogers. May 1989. "Fragile X Syndrome." *American Family Physician* 39(5):186. Published by the American Academy of Family Physicians; **24E(1-2):** Courtesy, Integrated Genetics; **24F:** Courtesy, Colleen Weisz; **24G:** From *Human Genetics: An Introduction to the Principles of Heredity,* 2/e by Singer. © Used with permission of W. H. Freeman and Company.

## Color Plates

Plates 8-36: Courtesy of John W. Hole, Jr.
Plates 37-47: © Dr. Sheril D. Burton.
Plates 48-75: © McGraw-Hill Companies, Inc./Karl Rubin.
Plate 76: © Martin Rotker/Photo Researchers.

## Clinical Application Boxes

© Comstock.

# INDEX

Note: Page numbers in *italics* refer to illustrations; page numbers in **boldface** refer to plates; page numbers followed by t refer to tables.

## A

A bands, of muscle fibers, 285, *285*
Abdomen, quadrants of, 22, *22*
Abdominal cavity, *12, 13*, **35**
   lymph nodes of, *627*, 628
Abdominal reflex, 386
Abdominal viscera, **31, 32, 33**. *See also*
    *specific organs*
Abdominal wall
   muscles of, 316–19, 317t, *318*, **974, 975**
   veins from, 606–7, *607*
Abdominopelvic cavity, 11, *12*, 13
   membrane of, 14
Abducens nerve (VI), 409, *409*, 412t
Abduction, 264, *265*, 267t
Abductors, of arm, *307, 309, 311*, 312
ABO blood group, 544–46, 544t, *545, 546*
Abortion, 883
Abruptio placentae, 924
Accelerator nerves, 572
Accelerator neuron, 350
Accessory nerve (XI), *409*, 411, 412t
   cranial branch of, 411, 412t
   spinal branch of, 411, 412t
Accommodation, of lens, 463–64, *464*
Acetabulum, *229*
Acetazolamide, in taste disorder, 445
Acetone, test for, 991
Acetonemia, 828
Acetonuria, 828
Acetyl coenzyme A, 109, 719
Acetylcholine, 423–25, *424*
   in muscle contraction, 289–90
   as neurotransmitter, 361
Acetylcholinesterase, 425
   in muscle contraction, 290
Achalasia, 697
Achilles tendon (calcaneal tendon), 327, *328,*
   *329*, **342, 979**
Achlorhydria, 697
Achondroplasia, 235
Acid-base balance, 823–28
   acid-base buffer systems and, 824–26, 826t
   carbon dioxide excretion and, 826–27
   disorders of, 829–31, *829, 830, 831*
   hydrogen ion sources and, 823–24
   regulation of, 824–28
Acid-base buffer systems, 825–26, 826t
Acid-base imbalances, 829–31, *829, 830, 831*
Acidosis, 45–46, 829–31, *829, 830*

Acids, 45, 45t, *46*, 46t
   metabolic production of, 823–24, *824*
   strength of, 824
Acinar cells, 676, *676*
Acne, 172, *172*, 172t
Acoustic meatus, *208, 210*, 213t, **251**
Acquired immune deficiency syndrome
    (AIDS), 648–49, 882t, 883
Acromegaly, 193, 235
Acromial process, *223, 224*, **339**
Acromioclavicular joint, 267t
Acrosome, 841, *841*
Actin, 285, 287–89
*Actinomyces,* in dental caries, *664*
Action potential, 355, 357–58, *358*
Active transport, 83, *83*, 87t
Addiction, 367–69
   in newborn, 905
Addison's disease, 506, 823
Adduction, 264, *265*, 267t
Adductor brevis muscle, **34**
Adductor longus muscle, **32, 33, 34, 35**, *301,*
   *321*, 322, 325t, **977**
Adductor magnus muscle, **35**, *301, 321*, 322,
   *324*, 325t
A-delta fibers, 436
Adenine, *117, 121*
Adenoids, 661, *661*
Adenosine deaminase deficiency, 955
Adenosine diphosphate (ADP),
    phosphorylation of, 106, 108, *108,*
    112, *112*
Adenosine monophosphate, cyclic (cAMP),
    484, *486*
Adenosine triphosphate (ATP), 106–8, *107,*
    *108*, 111–13, *112*
   in muscle contraction, 288, *289*, 291, *291*
Adenosine triphosphate (ATP) synthase, 112
Adenosis, vaginal, 858
Adenylate cyclase, 484, *486*
Adipose tissue, 148, *148*, 152t
Adolescence, 919, 921t
Adoption, 945
Adrenal cortex, 501, *502*
   hormones of, 483t, 503–6, *504, 505*, 506t
Adrenal glands, **34**, *481*, 501–6, *502*
Adrenal medulla, 501, *502*
   hormones of, 483t, 502–3, *502*, 503t
   preganglionic fibers to, 421
Adrenalectomy, 512
Adrenergic receptors, 424
Adrenocorticotropic hormone (ACTH), 483t,
    489, *491*, 492, 496t
Adrenogenital syndrome, 512
Adrenoleukodystrophy, 74–75, 959t
Adult respiratory distress syndrome, 769
Adulthood, 919, *920*, 921t
   skull of, **255**

Afferent arteriole, of kidney, 783, *784, 785*
Afferent fibers, 407–8
Afterbirth, 870
Ageusia, 445t
Agglutination, 542, 544, *546*
Aging, 919, *920*, 921t, 922–24, 922t
   active, 923
   passive, 922–23
Agranulocytes, 527
Alanine, 121t
Albinism, *130, 176*, 469, 959t
Albumin
   plasma, *534*, 535, 535t
   serum, 989
Albumin-globulin ratio, 989
Albuminuria, 828
Alcohol
   diuretic effects of, 494
   fetal exposure to, 908, *909*
Aldosterone, 483t, 503–4, *504, 505*, 506t, 723
   in electrolyte regulation, 819, *819*
Ali, Mohammed, 396
Alimentary canal, 656–58, *657*
   innervation of, 657
   layers of, 656, *659*, 659t
   length of, *658*
   movements of, 657, *660*
Alkaline phosphatase, serum, 990
Alkaline tide, in gastric secretion
    regulation, 673
Alkaloid poisons, 444
Alkalosis, 45, 829–31, *829, 831*
Alkaptonuria, *130*, 806
Allantois, *904, 905*, 906
Alleles, 937. *See also* Genes
   codominant, 942
   complete penetrance of, 942
   dominant, 940–41
   incomplete dominance of, 941–42
   incomplete penetrance of, 942
   mutant, 940
   recessive, 940–41
   wild-type, 940
Allergens, 643, *645*
Allergy, 643–44, *645*
Allograft, 644, 646t
Alopecia, androgenic, 170
Alopecia areata, 170
Alpha cells, of pancreas, 507
Alpha globulins, 535, 535t
Alpha waves, of brain, 406, *406*
Alphatocopherol, 716
Altitude sickness, 767
Altitude, tissue oxygen and, 526
Alveolar arch, *208, 210, 211*, **248**
Alveolar dead space, 758
Alveolar ducts, 747, *748, 749*
Alveolar gland, 142, *142*, 143t

Alveolar pores, 764, *766*
Alveolar process, 208, *211*
Alveolar sacs, 747
Alveolar ventilation, 758–59
Alveolar-capillary membrane, 765, *766, 767*
Alveoli, 747, *748, 749*
    gas exchange from, 764–67, *766*
    in space, 759
    oxygen diffusion from, *750*
    respiratory membrane of, 765, *766, 767*
    structure of, *750, 751*
Alzheimer's disease, 344, 363t, 959t
Amacrine cells, of retina, 465, *466*
Amblyopia, 461, 475
Ameboid motion, 531
Amenorrhea, 883
Amines, 481, 483t
Amino acids, 51
    deamination of, 114, *115*
    essential, 114, 708, 708t
    intestinal absorption of, 689, *690*, 691t
    as neurotransmitters, 361
    in nutrition, 707–9, *708*, 708t
    structure of, *51*
    urinary, 795
Ammonia, 989
    test for, 991
Ammonium, formation of, 827
Amniocentesis, 907, 925
Amnion, *897*, 898–99, *904*
Amniotic cavity, 896, *897*, 905
Amniotic fluid, 899
Amoeba, *6*
Amphetamine, 368
Ampulla, *449, 455*
Ampulla of Vater, 676, *676*
Amylase, 689t
    pancreatic, 677
    of serous cells, 665
    serum, 989
Amyloidosis, 565
Amyotrophic lateral sclerosis (Lou Gehrig's
    disease), 387, 959t
Anabolic steroids, 485
Anabolism, cellular, 102, *103*
Anaerobic threshold, 292
Anal canal, *657*, 692, *693*
Anal sphincters, *694*
Analgesia, 426
Anaphase, *88, 89*, 91, *91*, 92t
Anaphase I, 935, *935*
Anaphase II, 936, *936*
Anaphylactic shock, 644
Anasarca, 828
Anastomosis, 609
Anatomic dead space, 758
Anatomical terminology, 18–24
    for body regions, 22, *22*, 23–24, *23*
    for body sections, 19, 21–22
    for cylindrical structures, 22, *22*
    for relative position, 18–19
Anatomy, 3–4
Androgens, 850
    adrenal, 505, 506t
    bone effects of, 193
Anemia, 527, 529t
    hypochromic, *529*
    megaloblastic, 720
    pernicious, 720
Anencephaly, 388t, 720
Anesthesia, 426
Aneuploidy, 949–50, 949t
Aneurysm, 146, 582
Angelman syndrome, 947
Angina pectoris, 563, 614

Angiography, coronary, 562
Angiospasm, 609
Angiotensin I, 793, *793*
Angiotensin II, 504, *504*, 793–94, *793*
Angiotensin-converting enzyme
    inhibitors, 504
Angiotensinogen, 504, *504*
Anion, 355
Anisocytosis, 547
Ankle, 233, *234*, 235t, 267t
Ankle-jerk reflex, 386
Ankylosis, 275
Annular ligament, 270, *270*
Anopia, 475
Anorexia nervosa, 704, 730–31
Anosmia, 445t
Anoxia, 774
Antacids, 674, 828
Antagonists, of skeletal muscles, 300
Anterior cruciate ligament, *274*, 275
Anterior, definition of, 18
Anterior horn, of spinal cord, 378, *379*
Anterior median fissure, 378, *379*
Anti-aging collagen cream, 145
Antibodies, 530, 634–40, *635*, 637t
    actions of, 638, 640, 640t
    structure of, 637, *637*
Anticodon, 122
Antidepressants, tricyclic, 364t
Antidiuretic hormone (ADH), 483t, 489,
    494–95, 496t
    in renal regulation, 796, 797, *797*, 799, 799t
    in water regulation, 818
Antigen, 632, *635*
Antigen-presenting cells, 634
Antihemophilic factor (factor VIII), 541t
Antihemophilic plasma, 547
Antioxidants, 716
Antithrombin, 538–39, 542, 542t
Anuria, 828
Anus, 692, *694*
Aorta, 558, *559, 564, 565*, 595–97, *605*, **964,
    966, 967, 970, 971, 972, 980, 981**
    abdominal, **35**, 596, *597*, 598t, **967**
    aneurysm of, 146
    arch of, **32, 33, 34, 35**, 595, *596*, 598t
    ascending, 595, 598t
    descending, **34, 35**, 596, 598t
    thoracic, 596, 598t, *601*, **982**
Aortic aneurysm, 146
Aortic arch, **32, 33, 34, 35**, 595, *596*, 598t, **982**
Aortic bodies, 595
    in ventilation, 763, *763*, 765t
Aortic sinus, 595
Aortic sound, 568
Aortic valve, 558, *559, 560, 561*, 561t
Aphagia, 697
Aphasia, 426
Aplastic anemia, 529t
Apocrine gland, 142, *143*, 144t, 171, *171*,
    173, 173t
Aponeuroses, 282, *282*
    epicranial, 303
    of external oblique muscle, *308*
Apoprotein-B, 84
Apoproteins, 536
Apoptosis, 78, 923
Apothecaries' weights, 987
Appendicitis, 692
Apraxia, 426
Aquaporins, 818
Aqueduct of Sylvius (cerebral aqueduct), 395,
    *397, 398*, 402, *402*
Aqueous humor, *462, 464, 465*
Arachidonic acid, 707

Arachnoid granulations, 397, *398*
Arachnoid mater, *376, 377*
Arbor vitae, 404, **985**
Arcuate arteries, 783
Arcuate popliteal ligament, *274*, 275
Areola, **29**
Areolar tissue (loose connective tissue),
    147–48, *147*
Arginine, 121t
Arm
    arteries to, 600–601, *601*
    bones of, 200, 223–26, *225, 226, 227, 228*
    muscles of, *309, 310*, 311–12, *311*, 312t,
        **974, 975, 976**
    veins from, *60*, 606
Arrector pili muscle, *162, 163*, 168
Arrhythmias, 574–75
Arterial pulse, 583, *584*
Arteriography, 609
Arterioles, 573, 576–77, *576*, 581t
    renal
        afferent, 783, *784, 785*
        efferent, 785, *785*
    spasm of, 537
Arteriovenous shunt, 577, *577*
Artery (arteries), 573–77, *574*, 581t. *See also
    specific arteries*
    in atherosclerosis, *541*
    cross section of, *580*
    layers of, 573–75, *576*
    light microscopy of, *541*
    wall of, 573–75, *576*
Arthralgia, 275
Arthritis, 276–77, 276t
    joint aspiration in, 261
Arthrodesis, 275
Arthrogram, 275
Arthrology, 275
Arthropathy, 275
Arthroplasty, 275
Arthroscopy, 271
Arthrostomy, 275
Arthrotomy, 275
Articular cartilage, 186, *187*, 262
Articulating process, of vertebrae, 216, *216*
Artificial insemination, 899t
Artificial respiration, 899t
Arytenoid cartilages, 743, *744, 745*
Asbestos, 753, 753t
Ascending colon, 692, *693*
Ascites, 580, 821
Ascorbic acid (vitamin C), 193, 720, *720*, 721t
Asparagine, 121t
Aspartame, 705
Aspartic acid, 121t
Asphyxia, 774
Asplenia, 647
Assisted reproductive technologies, 898, 899t
Association neuron, 350, 351t
Asthma, 761
Astigmatism, 470
Astrocytes, 351, *352*, 353t
Asystole, 609
Ataxia, 426
Atelectasis, 768–69, 774
Atherosclerosis, 538, 582, *583*
    artery wall changes in, *541*
    dietary fat in, 49
Athlete, nutrition for, 732
Athlete's foot, 179
Atlantooccipital joint, 267t
Atlas (C1), 217, *217*
    dislocation of, 382, *382*
Atmospheric pressure, 5, 751, 754, *754*
Atomic number, 38

Atomic weight, 38
Atoms, 8, *9*, 9t, 37, 38t
  bonding of, 39–42
  structure of, 37–38
ATPase, in muscle contraction, 288
Atrial fibrillation, 574
Atrial flutter, 574, *575*
Atrial natriuretic peptide, 510, 558
  in urine production, 794
Atrial septal defect, 569
Atrial syncytium of, 568
Atrioventricular bundle (bundle of
  His), 570, *570*
Atrioventricular node, 568, *570, 573*
  as secondary pacemaker, 575
Atrioventricular orifice, 557
Atrioventricular sulcus, *556, 557*
Atrioventricular valve, 557
Atrium (atria), *556, 557, 560*, **966**
  left, **33**, *559, 564*
  right, **33**, *559, 564*, **981**
Audiometry, 475
Auditory cortex, 450, *453*
Auditory meatus, *204*, 205, **242, 244, 249,**
  446, *446*
Auditory ossicles, *446*, 447
Auditory tube, *446, 447*, 448
Auricle
  of ear, 446, *446*
  of heart, *556, 557, 564, 596*
Auricular artery, *598, 599*
Autoantibodies, 646
Autograft, 644, 646t
  for burns, 177
Autoimmunity, 643, 647t
Automatic bladder, 806
Autonomic nervous system, 418–26, *420*, 425t
  adrenergic fibers of, 423–24, *424*
  alimentary canal innervation by, 657
  cholinergic fibers of, 423–24, *424*
  drug effects on, 425
  in gastric secretion regulation, 672, *673*
  neurotransmitters of, 423–24, *424*
  parasympathetic division of, 419, 421, 423,
    *423, 424*
  regulation of, 425–26
  sympathetic division of, 419–21, *420, 422*
Autosomes, 936, *937*
Axilla, lymph nodes of, 627–28, *627*
Axillary artery, **32**, 600, *601, 605*
Axillary nerve, 416, *416, 417*
Axillary vein, **32**, 606, *606, 607, 611*
Axis, 217, *217*
Axon, 345–46, *345, 347*
  collaterals of, 346
Axon terminal, 346
Axonal hillock, *347*
Axonal transport, 346
Azotemia, 828
Azygos vein, 606, *607, 611*, **971**

# B

B lymphocytes, 530, 632, *633*, 634–42, *635,*
  *636, 637*, 637t, 639t, *641*
Babinski reflex, 386
Back pressure, 580
Bacteria, intestinal, 695
Ball-and-socket joint, 262, *263*, 264t
Bands, 527–28
Baroreceptors (pressoreceptors), 433, 572,
  *573*, 793
Barrett's esophagus, 667
Barrier birth control, 879–80, 879t, *880*
Bartholin's glands, 859, *859*

Basal body, *62, 75*
Basal ganglia, 394, *394*, 405t
Basal metabolic rate, 710
Bases, 45, 45t, *46*, 46t
  strength of, 824
Basement membrane, 137, *137, 138, 163*
Basilar artery, *598*, 599, *600*
Basilar membrane, 449–50
Basilic vein, 606, *606, 607*, 611
Basophils, 527, 528–29, *530, 534*, 534t
  origin of, *523*
Beeturia, 806
Behavior, reflex, 383–85
Benign joint hypermobility syndrome, 259
Beriberi, 717
Berylliosis, 753, 753t
Beryllium, 39t
Beta cells, 507
Beta endorphin, 365
Beta globulins, 535, 535t
Beta oxidation, 113, *114*
Beta waves, 406, *406*
Beta-carotene, *714*
Bicarbonate, 771, *772*
  cellular, 48t
  urinary, 791t
Bicarbonate buffer system, 825, 826t
Biceps brachii muscle, **30**, 300, *300*, 301, *301,*
  *310, 311*, 313, *314*, 316t, **339, 340, 341,**
  **974, 977**
  long head of, *310*
  short head of, *310*
Biceps femoris muscle, *301*, 322, *323, 324,*
  326t, *328, 329*, **341, 978, 979**
  tendon of, **341**
Biceps femoris, tendon of, **340**
Biceps-jerk reflex, 386
Bicuspid teeth, 662, *662*
Bile
  composition of, 681
  gallbladder release of, 683, *683*
  measurement of, 991
Bile canals, 680
Bile ducts, *679*, 681
Bile salts, 684
Bilirubin, 525, *525*
  measurement of, 989, 991
Biliverdin, 524, *525*
Biotin, 720, 721t, *723*
Birth, 869–70, *870*, 870t, *871*
Birth control, 877–81, 879t, *881*
Birthmark, 179
Bitter receptors, 444, *444*
Bladder, **31, 32, 33, 34**, 803–5, *804, 805,*
  **964, 968**
  automatic, 806
Bladder cancer, 804
Blastocyst, 895, *895*
Blastomere, 893, *893*
Blepharitis, 475
Blind spot, 465
Blindness, 474
Blood, 152, *152*, 152t, 521–32, *521*. *See also*
  *specific cells*
  circulation of, 592–94, *594, 595*
  composition of, 521–22, *521, 534*
  formation of, 197, *523*
  origin of, 522, *523–24*
  osmotic pressure of, 535
  oxygen-carrying capacity of, 524
  synthetic, 548
  from umbilical cord, 521
  viscosity of, 585
  volume of, 521–22, *521, 534*
Blood flow
  cardiac, 561–62, *562, 563*

  pulmonary, 592–94, *595*
  systemic, 594
  venous, 590–92
Blood gases, *534*, 535–36
Blood group(s), 542, 544–47
  ABO, 544–46, 544t, *545, 546*
  genes for, 942
  Rh, 546–47, *547*
  typing of, 544–46, 544t, *545, 546*, 548
Blood pressure, 582–92
  arterial, 582–83
  blood volume and, 584, *584*
  in capillary exchange, 579
  cardiac output and, 585–89, *589*
  elevation of, 590–91, 591t
  heart action and, 584, *584*
  homeostatic regulation of, 6, 8
  measurement of, 586–87, *587*
  peripheral resistance and, 584–89, *584,*
    *585, 589*
  regulation of, 585–89, *589*
Blood smear, *521*
Blood transfusion, 545–46, 546t, 548
  reaction to, 542, 544–47
  universal donor for, 545–46
  universal recipient for, 545
Blood typing, 544–46, 544t, *545, 546*, 548, 942
Blood urea nitrogen, 989
Blood volume, 581, *581*
  antidiuretic hormone secretion and, 495
  blood pressure and, 584, *584*
Blood-brain barrier, 68
Bloodhound, olfactory sense of, 442
Blue diaper syndrome, 806
Body, **29–35**
  abdominopelvic cavity of, 11, *12*, 13
  appendicular portion of, 10
  axial portion of, 10
  cavities of, 10–13, *12, 13*
  dorsal cavity of, 10, *12*
  organization of, 10–18, *12*
  pericardial cavity of, *12*
  thoracic cavity of, 11, *12*
  ventral cavity of, 10, *12*
Body mass index, *713*
Body temperature, 988
  elevation of, 175
  regulation of, 173–75, *174*
Boil, 179
Bomb calorimeter, 710, *710*
Bond
  atomic, 39–42
  covalent, 41, *42*
  hydrogen, 52, *52*
  ionic, 41, *42*
  peptide, 102, *103*
Bone(s), 151–52, *151*, 152t. *See also specific*
  *bones*
  articular cartilage of, 186, *187*
  atrophy of, 193
  in body movement, 196–97, *196, 197*
  calcium effects on, 197–99, *199*
  classification of, 186, *187*
  compact, 186, 187, *187, 188*
  deposition of, 192
  development of, 188–92, *189, 190*, 190t,
    *191*, 192t
  diaphysis of, 186, *187*
  endochondral, 189–90, *190*, 190t
  endosteum of, 187
  epiphysis of, 186, *187*
  exercise effect on, 193, *193*
  flat, 186, *187*
  fracture of, 194–95
  function of, 193, 196–99

growth hormone effects on, 193
growth of, 190–92, *191, 192, 193*
homeostasis of, 192
inorganic salt storage in, 197–99
intramembranous, 188–89, *189,* 190t
irregular, 186, *187*
long, 186, *187*
medullary cavity of, 187, *187*
microscopic structure of, 186–88, *187, 188*
number of, 199, 200t
ossification of, 188–90, *189, 190,* 190t, 192t
periosteum of, 186, *187*
processes of, 186
repair of, 194–95, *195*
resorption of, 192
sesamoid, 186
sex hormone effects on, 193
short, 186, *187*
spongy, 186–88, *188*
structure of, 186–88, *187, 188*
supportive function of, 193, 196
thyroid hormone effects on, 193
vitamin A effects on, 193
vitamin C effects on, 193
vitamin D effects on, 192–93
Bone marrow, 187, *187,* 197
of sternum, 222
Boron, 39t
*Borrelia burgdorferi* infection, 271
Botulism, 290
Bowman's capsule, 785, *785*
Brachial artery, **32,** 600, *601, 605*
Brachial plexus, **32,** 415–17, *416*
Brachial vein, 606, *606, 607, 611*
Brachialis muscle, *301, 311,* 313, *314,* 316t, **341, 977**
Brachiocephalic artery, **33, 35,** 596, *596, 598,* 598t, *599, 605,* **980, 981, 982**
Brachiocephalic vein, **32, 34,** 596, 604, *606, 607, 611,* **981**
Brachioradialis muscle, *301,* 313, *314, 315,* 316t, **339, 341, 977**
Bradycardia, 574, *575*
Bradypnea, 774
Brain, *345,* 387–406, 405t
arteries of, *600*
basal ganglia of, 394, *394*
cerebellum of, 403–7. *See also* Cerebellum
cerebrum of, 387–94. *See also* Cerebrum
coronal plane of, *21*
development of, 387, *387,* 388t
diencephalon of, 398–401. *See also* Diencephalon
dominant hemisphere of, 393
functions of, 390–94, *390, 391,* 392t
gray matter of, 348–49, *349*
injury to, 395
magnetic resonance imaging of, *11*
sagittal plane of, *21*
sagittal section of, *405*
structure of, 387–90, *388, 389, 390*
transverse plane of, *21*
veins from, 604, *606*
ventricles of, 394–95, 397–98. *See also* Cerebrospinal fluid; Ventricle(s).
white matter of, 348–49, *349*
Brain death, 406, *406*
Brain stem, 387, *387, 399,* 401–3, *405,* 405t, **965**
Brain waves, 406, *406*
Breast, **29**
milk secretion from, *71*
Breast cancer, 874–75, *874,* 875t, 959t
gene for, 938

Breast-feeding, 876–77, *876,* 877t
alternatives to, 878
benefits of, 878
Breath, first, 914–15
Breathing. *See* Ventilation
Broad ligament, 852, *852,* 856
Broca's area, 391
Bronchial artery, 598t
Bronchioles, 747, *748, 749, 751*
Bronchitis, 742, 774
Bronchoscopy, 748
Bronchus (bronchi), **34,** *740,* 746–50, 752t, **982**
branching of, 746–47, *748, 749*
functions of, 748–49
secondary (lobar), 746–47, *748*
structure of, 747–48
tertiary (segmental), 747, *748*
Brown-Squard's syndrome, 381
Brunner's glands, 688
Buccinator muscle, *302,* 303, 303t, **973**
Buffy coat, *521*
Bulbospongiosus muscle, 319–20, 319t, *320*
Bulbourethral gland, *837,* 843, 849t
Bulimia, 730, 731
Bulk elements, 37
Bundle branch block, 570
Bundle branch, left, *570*
Burns, 177, 179, *179*
Bursae, 261, *261,* 269, 275
Bursitis, 276

## C

C fibers, 436
Cachexia, 732
Cacosmia, 445
Calcaneal tendon (Achilles tendon), *301,* 327, *328, 329,* **342, 979**
Calcaneus, *234,* 235
Calcitonin, 483t, 497, 498t
Calcium
in bone formation, 197–99, *199*
cardiac cycle and, 572, 573
cellular, 48t
hormonal regulation of, *199*
intestinal absorption of, 500, *500*
in muscle contraction, 289–90
in nutrition, 722, 725t
parathyroid hormone regulation of, 822, *822*
serum, 989, 991
urinary, 791t
Calcium channels, 66–67
Calculus, renal, 803
Callus, cartilaginous, 194
Calluses, 164
Calmodulin, 487
in smooth muscle contraction, 298
Calories, 710, *710*
Calorimeter, bomb, 710, *710*
Calyces, renal, 780, *782*
Canal of Schlemm, *465*
Canaliculi, 151–52, *151,* 459, *459*
Cancer, 96–97, *96*
bladder, 804
breast, 874–75, *874,* 875t, 938, 959t
colon, 694, 697
endocrine paraneoplastic syndrome in, 500
gene therapy in, 957, *958*
immunotherapy in, 638–39
kidney, 800
lung, 742–43
pain in, 438
rectal, 697

skin, 166–67
spread of, 137
testicular, 838
vitamins and, 715
Capacitation, of sperm cells, 844
Capillary (capillaries), *577,* 577–79, 581t
blood flow of, *577,* 579
cross section of, *578*
exchange across, 579, *580*
lymphatic, 621, *622*
network of, 578–79, *578*
permeability of, 577
splenic, 629, *629*
Capitate, *228*
Capitulum, *226*
Carbohydrates
in cells, 47, *48, 49,* 53t
digestion of, 689, *689*
metabolism of, 108–13, *108, 109, 110, 116,* 681t
in nutrition, 705–6, *705,* 709t
requirements for, 706
sources of, 705
storage of, 113
utilization of, 705–6, *705*
Carbon, 39t
Carbon dioxide
blood transport of, 769–72, *771, 772,* 772t
cellular, 48t
respiratory excretion of, 826–27
serum, 989
three-dimensional model of, *56*
Carbon monoxide, 47, 769
Carbon monoxide poisoning, 739, 769
Carbonate, cellular, 48t
Carbonic acid, dissociation of, 826
Carbonic anhydrase, 771
in acid-base balance, 826
Carboxypeptidase, 677, 689t
Carbuncle, 179
Carcinoma, 140, 166. *See also* Cancer
Cardiac. *See also* Heart
Cardiac center, of medulla oblongata, 402
Cardiac conduction system, 568, 570, *570*
Cardiac cycle, 566–68, *567*
regulation of, 571–73
Cardiac muscle, 298–99, *299,* 299t. *See also* Heart
Cardiac output, 584, 586, *589*
blood pressure and, 585–89, *589*
Cardiac tamponade, 592, 609
Cardiac veins, 563, *564, 565*
Cardioaccelerator reflex, 587
Cardioinhibitor reflex, 587
Cardiomyopathy, hypertrophic, familial, 959t
Cardiovascular disease, 612–13
Cardiovascular system, 16–17, *17,* 19t, 553–615, *555. See also* Artery (arteries); Heart; Vein(s)
blood flow through, 561–62, *562,* 592–94, *594, 595, 609*
exercise and, 593
system innerconnections of, *615*
Caries, dental, 664
Carotenes, 714, *714*
Carotid artery, *596,* **970**
common, **30, 31, 33, 34, 35,** 596, *596, 598,* 598t, *599, 605,* **980, 981, 982**
external, *598,* 599, *599, 605*
internal, *598,* 599–600, *599, 605*
Carotid bodies, 600
in ventilation, 763, *763,* 765t
Carotid canal, *204,* 205, 213t, **245, 246**
Carotid foramen, **247**

Carotid sinus, *598*, 600
Carpal bones, 200, *201, 225*, 226, *228*, 229t, **341**
Carpometacarpal joints, 267t
Carpus, 226, *228*
Cartilage(s), 150, *150*
  articular, 186, *187*, 262
  arytenoid, 743, *744, 745*
  corniculate, 743, *744, 745*
  costal, *221*, **971, 972**
  cricoid, 742, *744, 745*
  cuneiform, 743, *744, 745*
  elastic, 150, *150*, 152t
  epiglottic, 742, *744*
  hyaline, 150, *150*, 152t
  thyroid, 742, *744, 745*
Cartilaginous joints, 258–59, *259, 260*, 264t
Casein, 732
Catabolism, cellular, 102–4
Catalase, 104
Catalysts, 44. *See also* Enzymes
Cataract, 464
Catecholamines, 502–3, *502*
Cation, 354
Cauda equina, 413, **967, 968**
Caudate nucleus, 394, *394*
Causalgia, 475
Cavities, body, 10–13, *12*
Cecum, **32, 33**, 692, *693*
Celiac artery, **34**, 596, *597*, 598t, *605*
Celiac disease, 732
Cell(s), *9*, 9t, 60–78. *See also* specific cells
  arachaeal, 60
  basal body of, 75
  carbohydrates in, 47, *48, 49*, 53t
  carbon dioxide in, 47, 48t
  centrioles of, *62*
  centrosome of, 73, *74*, 79t
  chemical constituents of, 46–56
  chromatin of, *62*, 77–78, *78*, 79t
  cilia of, *62*, 73, 75, *76*, 79t
  cytoplasmic organelles of, 60, *62*, 68–77
  cytoskeleton of, 68, *77*
  cytosol of, 68
  death of, 78
  differentiation of, 92–93, *94*
  endoplasmic reticulum of, 68–69, *70*, 79t
  eukaryotic, 60
  fixed, 144
  flagella of, 73, 75, *76*, 79t
  Golgi apparatus of, *62*, 69, *70*, 71–72, 79t
  inclusions of, 77
  inorganic salts in, 47
  inorganic substances in, 46–47, 48t
  lipids in, 47–49, *50*, 51t, 53t
  lysosomes of, *62*, 73, *73*, 79t
  membrane of, 60, *62*, 63–67, *64, 65*, 66t, 79t. *See also* Membrane, cell
  microfilaments of, 75–77, *76, 77*, 79t
  microtubules of, 75–77, *76, 77*, 79t
  microvilli of, *62*
  mitochondria of, 72, *72*, 79t
  mitochondrion of, *62*
  nuclear envelope of, *62*, 78, 79t
  nuclear pore of, *78*
  nucleic acids of, 52–53, *53*, 53t
  nucleolus of, *62*, 77, *78*, 79t
  nucleus of, 60, *62*, 77–78, *78*
  organic substances of, 47–53
  oxygen in, 47, 48t
  peroxisomes of, 73, 79t
  prokaryotic, 60
  proteins in, 51–52, *51, 52*, 53t
  reproduction of, 93, 96–97
  ribosomes of, *62*, 69, 79t

  rough endoplasmic reticulum of, *62*
  secretory vesicles of, *62, 70*, 71–72, *71*
  shape of, 60, *61*
  size of, 60, *61*
  smooth endoplasmic reticulum of, *62*
  vesicle trafficking within, 71, *71*
  vesicles of, 75, 79t
  wandering, 144
  water in, 46–47, 48t
Cell adhesion molecules, 65, 67, 69
Cell body, of neuron, 345, *345, 347, 352*
Cell cycle, 88–93, *88*
  anaphase of, *88, 89*, 91, *91*, 92t
  cytoplasmic division of, 92, *93*
  interphase of, 88, *88*
  metaphase of, *88, 89*, 90–91, *91*, 92t
  mitosis of, 88–92, *88, 89, 90, 91, 92*, 92t
  prophase of, *88, 89*, 90, *90*, 92t
  telophase of, *88, 89*, 91, *92*, 92t
Cell implant, 646
Cellulose, 705
Celsius temperature, 988
Cementum, 663
Central canal, of spinal cord, 378
Central nervous system, 344, *345. See also* Brain; Spinal cord
Central venous pressure, 592
Centrioles, *62, 73, 74*, 90, *90*
Centromere, 90, *90, 934*
Centrosome, 73, *74*, 79t
Cephalic vein, *60*, 606, *606, 607, 611*
Cerebellar cortex, 404
Cerebellum, 387, *387*, 403–7, *404, 405*, 405t, **964, 965, 985**
  functions of, 405, 405t
Cerebral aqueduct (aqueduct of Sylvius), 395, *397, 398*, 402, *402*
Cerebral artery, 599, 600, *600*
Cerebral cortex, 390, *390*
  association areas of, *390*, 392, 392t
  functions of, 390–94, *390, 391*, 392t
  motor areas of, 390–91, *390, 391*, 392t
  sensory areas of, *390, 391*, 392, 392t
Cerebral palsy, 395
Cerebral peduncles, *399*, 402
Cerebrospinal fluid, 394–95, 397–98
  composition of, 397
  pressure of, 400–401
  secretion of, 397
Cerebrovascular accident (stroke), 395, 591, 591t
Cerebrum, *405*, 405t, **964, 965, 985**
  convolutions of, 389
  functions of, 390–94, *390, 391*, 392t, 405t
  injury to, 395
  structure of, 387–90, *388, 389, 390*
Ceruminous glands, 173t
Cervical cap, 879t, 880
Cervical enlargement, 377, *378*
Cervical nerves, 412, *413*
Cervical plexus, 415, *416*
Cervical spine, *215*, 216–17, *217, 218*, 219t, **965**
Cervix, 856
Cesarean section, 883
Cheekbrush test, 658
Cheeks, 658, 663t
Chemical barriers, for contraception, 879t, 880, *880*
Chemical reactions, 44
Chemicals
  as infection barrier, 631, 632t
  inorganic, 46–47, 48t
  organic, 47–53, *48, 49, 50, 51*, 51t, *52, 53*, 53t

Chemoreceptors, 432
  central, 762, 765t
  peripheral, 763, *763*, 765t
Chemotaxis, in infection, 531, *531*
Cheyne-Stokes respiration, 774
Chicken pox, 164t
Chief cells, 671, *671*
Childhood, 918–19, 921t
  skull of, **255**
Chlamydia infection, 882t
Chloride
  cellular, 48t
  serum, 989
  urinary, 791t
Chloride channels, 66–67
Chlorine
  electrons of, 39, *42*
  in nutrition, 724, 725t
Chlorolabe, 472
Cholangiogram, 684
Cholecalciferol, 715–16
Cholecystectomy, 684
Cholecystitis, 684, 697
Cholecystokinin, 685t
Cholelithiasis, 684, 697
Cholestasis, 699
Cholesterol, 535, 536
  disorders of, 612–13
  receptor-mediated endocytosis of, 84–85, *86*
  serum, 989
  structure of, *50*
  synthesis of, 707, *707*
Cholinergic receptors, 424, *426*
Chondrocytes, 150, *150*
Chondrodysplasia, 147t
Chordae tendineae, 558, *559, 560*
Chorion, 897, *897*
Chorionic villi, 896–97, *897*
Chorionic villus sampling, 897
Choroid artery, *598*
Choroid coat, 462, *462, 463, 466*, 467t
Choroid plexus, 395, *397–98, 398*
  ependyma of, 352
Chromatid, *934*
Chromatin, *62*, 77–78, *78*, 79t
Chromium, in nutrition, 727, 727t
Chromosomes, 930–31, *934*, 936–37
  crossover of, 935, *935, 937*
  disorders of, 948–51, 949t, *951*
  in prophase, 90, *90*
  sex, 945, *946, 947*
Chronic fatigue syndrome, 646
Chronic granulomatous disease, 527, 959t
Chronic obstructive pulmonary disease, 763
Chylomicrons, 536, 536t, 689, *690*, 691
Chyme, 674, *675*
Chymotrypsin, 677, 689t
Cigarette smoking, 368–69, 742–43
  fetal effects of, 908
Cilia, *62*, 73, 75, *76*, 79t
  of pseudostratified columnar epithelium, 139, *139*
  of uterine tube, 856, *857*
Ciliary body, 462, *462, 463*, 465
Ciliary muscles, 462, *463*, 463–64, *464*
Ciliary process, 462, *463*
Circadian rhythms, 510
Circle of Willis, 599, *600*
Circular muscles, of iris, 465, *466*
Circumduction, 265
Circumflex artery, 562, *564, 601*
Cirrhosis, 699
Cisternae
  of Golgi apparatus, 69
  of sarcoplasmic reticulum, 285, *285*

Citric acid cycle, *110,* 111, *114, 116,* 292, *292*
Clavicle, **29, 30,** 200, *201,* 222, *223,* 229t, **339, 340**
Clavicular notch, *221*
Cleft palate, 209
Cleidocranial dysplasia, 223
Clinical laboratory tests, 989–91
Clitoris, *851,* 859, *859,* 860t
   erection of, 859–60, *860*
Cloning, 95
*Clostridium botulinum* poisoning, 290
Clot, 537–39, *538. See also* Coagulation
Clotting factors, 541t
Clubfoot, 235
Cluster-of-differentiation antigen, 634
Coagulation, 537–42, 538t, 541t
   extrinsic mechanism of, 538–39, *538, 540*
   inhibition of, 542, 542t
   intrinsic mechanism of, 539, *540*
Cobalt, in nutrition, 726, 727t
Cocaine, 364t, 368
   sperm binding of, 842
Coccygeal nerve, 412, *413*
Coccygeus muscle, 319, 319t, *320*
Coccyx, 200, *201,* 218–19, *219,* 219t, 229, **968**
Cochlea, *446,* 448, *449, 450*
   cross section of, *451*
Cochlear duct (scala media), 449, *449, 450, 451*
Cochlear implant, 454
Cochlear nerve, *449*
Codon, 121t, 122
Coenzyme A, 719
Coenzymes, 105
Cofactors, 105
Coitus interruptus, 877–78, 879t
Cold receptors, 434
Colitis, ulcerative, 697
Collagen
   disorders of, 146, *146,* 147t
   three-dimensional model of, *56*
Collagen cream, 145
Collagenous fibers, 145, *145,* 147t, *148,* 149, *149*
Collateral ganglia, 421, *421*
Collateral ligaments, *274,* 275
Colles fracture, 235
Colliculus, *399,* 402
Colloid, thyroid, 495
Colon, 692, *693,* **964, 983**
   ascending, **32, 33, 984**
   descending, **32, 33, 34**
   sigmoid, **33, 34, 968, 984**
   transverse, **32, 33, 967, 972, 984**
Colon cancer, 694, 697
Colony-stimulating factors, 522, 634t
Color blindness, 432, 946, 959t
Color vision, 473
Colostrum, 876, 915, 917
Colposcopy, 858
Coma, 403
   hepatic, 680
Commissure, gray, 378, *379*
Communicating artery, 600, *600*
Compact bone, 186, 187, *187, 188*
Compartment syndrome, 283
Complement, 637, 640
Compound, 37, 42, 44
Compound gland, 142, *142,* 143t
Computed tomography, 54
Concentration gradient, 79
Concussion, 395
Condom, 879–80, 879t, *880*
Conduction, in body temperature
   regulation, 173

Conductive deafness, 454
Condyle, 202t
   femoral, 231, *233*
   mandibular, 210, **248**
   occipital, *204,* 205, **246, 247, 249, 251**
   tibial, 232, *232, 233*
Condyloid joint, 262, *263,* 264t
Cones, *466,* 469
   iodopsins of, 472–73
Congestive heart failure, 609
Conization, 883
Conjunctiva, 458, *459, 463*
Conjunctivitis, 458, 475
Connective tissue, 136t, 144–52, 147t, 152t.
     *See also* Blood; Bone(s)
   adipose, 148, *148,* 152t
   cells of, 144–45, *145*
   collagenous fibers of, 145, *145,* 147t, *148,* 149, *149*
   dense, 145, 149, *149,* 152t
   elastic, *145,* 146, 147t, 149, *149,* 152t
   loose, 145, 147–48, *147,* 152t
   reticular, 146, 147t, 148, *148,* 152t
Connective tissue disease, 612
Constipation, 696–97
Constrictor muscles, of pharynx, 666–67, *667*
Contact dermatitis, 164
Continuous ambulatory peritoneal dialysis, 781
Contraception, 877–81, 879t, *881*
Contraceptive implant, 881
Contracture, 331
Contralateral, definition of, 19
Conus medullaris, 378, *378*
Convection, in body temperature
   regulation, 173
Convulsion, 331
Copper, in nutrition, 726, 727t
Cor pulmonale, 609
Coracobrachialis muscle, **30, 31, 32,** *310,* 311, *311,* 312t
Coracohumeral ligament, 268, *269*
Coracoid process, *223, 224*
Cordotomy, 426
Cornea, *459,* 461, *462, 463*
   transplantation of, 462
Corniculate cartilage, 743, *744, 745*
Corns, 164
Coronal plane, definition of, *21*
Coronal suture, *203, 204,* 205, **241, 242, 251**
Coronary angiography, 562, 614
Coronary artery (arteries), 562–63, *564, 565,* 598t, *605*
   angiography of, *565*
   left, *564*
   opening of, *561*
   right, *564*
Coronary artery disease, 614
Coronary bypass surgery, 614
Coronary ligament, 678, *679*
Coronary sinus, 558, *559,* 563, *564*
Coronary stent, 614
Coronary thrombosis, 591
Coronoid fossa, 224, *226*
Coronoid process, 210, *211,* 226, *227,* **248**
Corpora cavernosa, 845, *845*
Corpora quadrigemina, 402
Corpus albicans, *856, 862, 863*
Corpus callosum, 388t, *393,* **965, 969, 985**
   splenium of, *404*
Corpus luteum, *856, 862, 863*
Corpus spongiosum, 845, *845*
Cortical nephrons, 788, *789*
Corticospinal tracts (pyramidal tracts), 378, *379, 380,* 390

anterior, *379,* 381, 381t
   lateral, 381, 381t
Corticotropes, 490
Corticotropin-releasing hormone, 483t, *491,* 492
Cortisol, 483t, 504–5, *505,* 506t, 512
   in stress response, 512, *514*
   structure of, *482*
Costal cartilage, **971, 972**
Costal margin, **340**
Costocervical arteries, 599
Costocervical arteries, 599
Countercurrent mechanisms, of nephron, 796–99, *798, 799*
Covalent bond, 41, *42*
Coxal bones, 200, *201,* 218, 227–28, *230,* 235t, **972**
Cranial cavity, 11, *12, 13*
Cranial nerve(s), 407, 408–11
   I (olfactory), 408, 412t
   II (optic), 408, *409,* 412t
   III (oculomotor), 408–9, *409,* 412t
   IV (trochlear), 409, *409,* 412t
   V (trigeminal), 409, *409,* 412t
   VI (abducens), 409, *409,* 412t
   VII (facial), *409,* 410, *411,* 412t
   VIII (vestibulocochlear), *409,* 410, 412t, *450, 455*
   IX (glossopharyngeal), *409,* 410, 412t
   X (vagus), *409,* 410–11, *412,* 412t, 571–72, *573,* **982**
   XI (accessory), *409,* 411, 412t
   XII (hypoglossal), *409,* 411, 412t
Craniotomy, 426
Cranium, 201, *201,* 205–8. *See also* Skull
   floor of, *208*
Creatine phosphate, 536
   in muscle contraction, 291, *291*
Creatine, serum, 989
Creatinine
   serum, 989
   urinary, 791t
Creatinine clearance, 802, 991
Cremasteric reflex, 386
Crest, 202t
Cretinism, *498,* 498t
Cribriform plate, 206, *207, 208,* 210, 211
Cricoid cartilage, 742, *744, 745*
Crista ampullaris, 455, *457*
Crista galli, 206, *207, 208,* 210, 211, **249, 252, 253**
Cristae, of mitochondria, 72, *72*
Critical period, 906–7
Cross section, *22*
Crossed extensor reflex, 385, *386*
Cruciate ligament, *274,* 275
Crying, 760t
Cryptosporidiosis, 630
Crypts of Lieberkühn, 686
Cubital vein, 606, *606*
   median, *611*
Cuboid, *234*
Cuneiform bone, *234*
Cuneiform cartilage, 743, *745*
Cupula, 455
Curare, 364t
Curettage, 883
Cushing's syndrome, 506, 823
Cuspid teeth, 662, *662*
Cutaneous membrane, 161. *See also* Skin
Cyanide, 112
Cyanocobalamin (vitamin $B_{12}$), 719, *719,* 721t
   red blood cell effects of, 526–27, *528,* 528t
Cyanolabe, 472

Cyanosis, 176, 522
Cyst, 179
Cysteine, 121t
Cystic duct, **33**, 681
Cystic fibrosis, 67, 621, 677, 749, 959t
Cystic fibrosis transmembrane conductance
    regulator gene, 930, *931, 937, 940*
Cystitis, 803
Cytochromes, 112
Cytokines, 638–39
    T–cell secretion of, 633, 634t
Cytoplasmic division, of cell cycle, 92, *93*
Cytosine, *117, 121*
Cytoskeleton, 68, *77*
Cytosol, 68

## D

Dartos muscle, 845
Date of conception, 893
*De Humani Corporis, 3*
Dead space, 758
Deafness, 454
Deamination, 114, *115*
Decibels, 450
Decomposition, chemical, 44
Deep, definition of, 19
Defecation reflex, 695
Dehydration, 820, *820*
Dehydration synthesis, 102, *103*
Delta cells, of pancreas, 507
Delta waves, 406, *406*
Deltoid muscle, **29, 30**, 301, *301, 307, 308,
    309, 311*, 312, 312t, **339, 340, 974,
    975, 976**
    long head of, *310*
Deltoid tuberosity, 224, *226*
Dendrites, of neuron, 345, *345, 347*
Dense connective tissue, 149, *149*, 152t
Dental caries, 664
Dentate nucleus, 404
Dentin, 663
Deoxyhemoglobin, 522
Deoxyribonucleic acid (DNA), 53, 124t
    complementary base pairs of, 119
    mutations in, 126, *127*
        effects of, 129–30, 129t, *131*
    repair of, 129
    replication of, 126, *127*
    structure of, 117–20, *117, 118, 119*
    transcription of, 121, *122, 123,* 126t
Deoxyribose, *53,* 705
Depo-Provera, 880
Depression, 265, *266,* 363t
Dermatitis, contact, 164
Dermatomes, *414*
Dermis, *165,* 166–67
Descending colon, 692, *693*
Desmosomes, 65, 66t, *69*
Diabetes insipidus, 495, 818, 959t
Diabetes mellitus, 480, 509t
    juvenile, 647t
    maturity-onset, 508
Diacylglycerol, 487
Diapedesis, 531, *531*
Diaphragm, 11, *12, 31, 32, 33, 34, 35,* 879–80,
    879t, *880,* **966, 967, 971, 972, 980, 981,
    982, 983, 984**
    in inspiration, 754, *754, 755*
    pelvic, 319, 319t, *320*
    urogenital, 319–20, 319t, *320*
Diaphysis, of bone, 186, *187*
Diarrhea, 692, 696

Diastole
    atrial, 566, *567*
    ventricular, 566, *567*
Diastolic pressure, 583, 586, *587*
Diencephalon, 387, *387,* 388t, 398–401, *405,*
    405t, **985**
    functions of, 405t
Diet
    for athlete, 732
    in obesity treatment, 712
    vegetarian, 705t
Diethylstilbestrol, 858
Differential white blood cell count, 532
Diffusion, 87t
    across cell membrane, 78–80, *80*
    facilitated, 80, *80,* 87t
    net, 78–79
Digestion, 656, 689–71, *689, 690. See also*
        Metabolism; Nutrition
    of carbohydrates, 705–6, *705*
    of lipids, 706–7, *707*
    of proteins, 689, *690*
Digestive system, 17–18, *17, 19*t, 656–58, *657*
    esophagus of, 667, *668*
    gallbladder of, 681, 683, *683*
    intestine of, 685–97. *See also* Intestine
    liver of, 678–84. *See also* Liver
    mouth of, 658–64, *660, 661, 662,* 663t
    pancreas of, 676–77, *676, 677*
    pharynx of, 666–67, *667*
    salivary glands of, 665, *665, 666,* 666t
    stomach of, 670–75. *See also* Stomach
    structure of, 656–58, *657, 658*
    system interconnections of, 698
Digital artery, *601*
Dihydrotestosterone, 850
2,4-Dinitrophenol, 112
Dipeptide, 102
Diplopia, 461, 475
Disaccharides, *689,* 705
    cellular, 47, *49*
Disseminated intravascular clotting, 539
Distal, definition of, 19
Diuresis, 794, 828
Diuretics, 494, 818
Diverticulitis, 699
Diverticulosis, *696,* 697
Dopa, in catecholamine synthesis, 502
Dopamine, 394
    in Parkinson's disease, 396
Dorsal arch vein, *606*
Dorsal cavity, 10, *12*
Dorsal root, 413, *415*
Dorsal root ganglion, 413, *415*
Dorsalis pedis artery, *603,* 604, *605*
Dorsalis pedis vein, *610*
Dorsiflexion, 264, *265,* 267t
Down syndrome, 949t, 952–53
Drug addiction, 367–69
    in newborn, 905
Drug therapy, in obesity treatment, 712–13
Drugs
    breast-feeding and, 877t
    taste disorders with, 445
Duchenne muscular dystrophy, 946, 959t
Ductus arteriosus, 594, 913, 913t, 918t
Ductus venosus, 913, 913t, *914,* 918t
Dumping syndrome, 699
Duodenum, *33, 34, 657, 685, 685*
Dura mater, 376–77, *376, 377,* **969**
Dural sinuses, 376
Dwarfism, 193, 493, 959t
Dysentery, 699
Dysgeusia, 445t

Dyslexia, 393
Dysmenorrhea, 883
Dysosmia, 445t
Dyspepsia, 699
Dysphagia, 699
Dyspnea, 774
Dystrophic epidermolysis bullosa, 147t
Dystrophin, in muscle contraction, 288–89

## E

Ear. *See also* Hearing
    external, 446, *446*
    infection of, 205, 448
    inner, 448–50, *449, 450*
    middle, 446–48
Eccrine glands, 171, *171,* 173, 173t
Eclampsia, 883
Ectoderm, *900*
Ectodermal cells, 896, *897*
Ectopic beats, 574
Eczema, 179
Edema, 535, 580, 626, 820–21, *821,* 821t
    in nephrotic syndrome, 795
    nutritional, 709
    pulmonary, 594
Effectors, of nervous system, 344
Efferent arterioles, of kidney, 785, *785*
Efferent fibers, 407, 408
Ehlers-Danlos syndrome, 146, *146*
Ejaculation, 845–47, *848*
Ejaculatory duct, *837, 842*
Elastic cartilage, 150, *150,* 152t
Elastic fibers, *145,* 146, 147t, 149, *149*
Elastin, 146
Elbow joint, 267t, 269–71, *270*
Electrocardiogram, 571, *572*
Electrolyte balance, 47, 819, *819,* 822, *822*
Electrolytes, 44–45, 47
    intake of, 819, *819*
    intestinal absorption of, 691t
    output of, 819, *819*
    plasma, 537
    regulation of, 819, *819,* 822, *822*
Electromyography, 331
Electron microscope, 63, *63, 64*
Electron transport chain, 112
Electrons, 37–38, *38,* 38t
    shell distribution of, 39–41, *40, 42*
Elements, 37, 38t
    isotopes of, 38–39
    periodic table of, *986*
Elevation, 265, *266*
Emaciation, 732
Embolectomy, 609
Embolism, 538
Embryo, 866–67, *867, 892,* 896–907, 911t
    adoption of, 899t
    at eighth week, 906, *906*
    at fifth week, 897, *902, 903*
    at fourth week, 897, *901*
    genetic diagnosis in, 894
    at seventh week, *903*
    at sixth week, *902, 903*
    teratogen exposures of, 906, *906,* 908–9
Embryonic disk, 896
Emmetropia, 475
Emotion, diencephalon in, 400–401
Emphysema, 742, 761, *761*
Encephalitis, 426
Endarterectomy, 609
Endocardium, 557, *557,* 557t
Endochondral bone, 189–90, *190,* 190t
    growth of, 190–92, *191, 192, 193*

Endocrine paraneoplastic syndrome, 500
Endocrine system, 16, *16*, 19t, 480, *481*. See
    *also specific glands and hormones*
    system innerconnections of, 515
Endocytosis, 83–85, *84, 85,* 87t
    receptor-mediated, 84–85, *86,* 87t
Endoderm, *900*
Endodermal cells, 896, *897*
Endolymph, 448, *449*
Endometritis, 883
Endometrium, 856
Endomysium, 283, *283*
Endoneurium, 407, *407, 408*
Endoplasmic reticulum, *62,* 68–69, *70,* 79t
Endorphins, 365, *366*
    in pain inhibition, 437
Endoscopic papillotomy, 684
Endosteum, 187
Endothelium, 573–74, *576*
    as gene therapy target, 956
Energy, 710–12
    of food, 710, *710*
    requirements for, 710–11, 711t
Energy balance, 711
Enkephalins, 365
    in pain inhibition, 437
Enteritis, 699
Enterogastric reflex, 674, *675*
Enterokinase, 677, 689t
Enucleation, 475
Environment
    external, 5
    internal, 5–8, *7*
Enzymes, 51, 102, 104–5, *105*
    active sites of, 104
    denaturation of, 105
    names of, 105
    rate-limiting, 115, *115*
    regulation of, 115, *115*
Eosinophils, 527, 528, *530, 534,* 534t
    origin of, *523*
Ependyma, 352, 353t
Ependymal cells, 352, *352,* 353t
Epicardium, 556–57, *557,* 557t
Epicondyles, 202t
    femoral, 231, *233*
Epicranial aponeurosis, 303
Epicranius muscle, 303t
Epidermis, 163–66, *163, 165*. See also Skin
Epidermolysis bullosa, 162
Epididymis, **32,** *837,* 838, 842, *842,* 849t, **968**
Epididymitis, 883
Epidural space, 376
Epigastric artery, *602,* 604
Epigastric region, 22, *22*
Epiglottic cartilage, 742, *744*
Epiglottis, *661,* 742–43, *745,* 746
Epilepsy, 363t, 427
Epimysium, 282, *283*
Epinephrine, 483t, 502–3, *502,* 503t
Epineurium, 407, *407, 408*
Epiphyseal disk, 190, *190, 191*
    x-ray of, 192, *192*
Epiphysiolysis, 235
Epiphysis, of bone, 186, *187*
Episiotomy, 870
Epithelial tissues, 136–44, 136t
Epithelium
    columnar
        pseudostratified, 138–39, *139,* 141t
        simple, 137–38, *138,* 141t
        stratified, 140, *140,* 141t
    cuboidal
        simple, 137, *138,* 141t
        stratified, 139–40, *140,* 141t

glandular, 142–44, *142,* 143t
    squamous
        simple, 137, *137,* 141t
        stratified, 139, *139,* 141t
        transitional, 140, *141,* 141t
Equilibrium
    diffusion and, 79
    dynamic, 455–56, *457, 458*
    static, 453, 455, *456, 457*
Equilibrium sense, 451, 453, 455–56
Erection
    of clitoris, 859–60, *860*
    of penis, 845–47, *848*
Erector spinae muscles, 305t, 307
Ergocalciferol, 715–16
Erythema, 181
Erythroblastosis fetalis, 547, 630
Erythroblasts, 526
Erythrocytes. *See* Red blood cells
Erythrocytosis, 526
Erythrolabe, 472
Erythropoietin, 510, 526, *526*
*Escherichia coli,* 779
Esophageal artery, 598t
Esophageal sphincter, *669*
Esophagitis, 699
Esophagus, **34, 35,** *657, 661,* 667, *668,* **965,
    966, 970, 971, 982**
    Barrett's, 667
    cross section of, *668*
Estrogen, 492, 505, 860–61, *861*
    bone effects of, 193
    in female reproductive function, *863*
    in pregnancy, *868,* 869
Estrogen replacement therapy, 864
Ethmoid bone, *203, 204,* 205t, 206, *207, 211,*
    **241, 242, 243, 249, 251, 252**
    cribriform plate of, *210, 253*
    perpendicular plate of, *203, 210,* **241, 243**
Ethmoidal sinus, 206, *207, 208,* 208t, **249,
    252, 969**
Euploidy, 949, *954*
Eupnea, 774
Evaporation, in body temperature
    regulation, 173
Eversion, 265, *266*
Evertors, of leg, *328, 329,* 331
Exchange reaction, 44
Excitatory postsynaptic potential, 361
Exercise
    bone effects of, 193, *193*
    breathing and, 765
    cardiovascular system and, 593
    in obesity treatment, 712
Exocrine gland, 142–44, *142*
Exocytosis, 85, *86,* 87t
Exophthalmos, 475, 513
Expiration, 756–57, *757,* 757t
Expiratory reserve volume, 757, *758,* 759t
Extension, 262, 265, 267t, 300
Extensor carpi radialis brevis muscle, 315,
    *315,* 317t, **977**
Extensor carpi radialis longus muscle, *314,*
    315, *315,* 317t, **339, 977**
Extensor carpi ulnaris muscle, *315,* 315–16,
    317t, **977**
Extensor digitorum longus muscle, *301,* 326,
    *327,* 330t, **978, 979**
    tendons of, *342*
Extensor digitorum muscle, *301, 315, 316,*
    317t, **339, 977**
Extensor retinaculum, *315, 316, 327,* 331, **977**
Extensors
    of arm, *307, 309, 310,* 312
    of forearm muscles, *309, 310,* 313

of hand, 315–16, *315, 316*
    of leg, *321, 324, 325*
External auditory meatus, 446, *446*
External ear, 446, *446*
External oblique muscle, *301,* 302, *308,* 317,
    317t, *318,* 319, **340, 974, 975**
Extracorporeal membrane oxygenation, 750
Extracorporeal shock-wave lithotripsy, 803
Extrafollicular cells (C cells), of thyroid, 495
Eye, **969**
    anterior chamber of, *462, 463, 464, 465*
    color of, 465
    dark adapted, 472
    extrinsic muscles of, 460, *460,* 461t
    inner tunic of, 465–67, *466,* 467t
    light adapted, 472
    light refraction by, 468, *468*
    middle tunic of, 461–65, *462,* 467t
    orbit of, *203*
    outer tunic of, 461, *462,* 467t
    photoreceptors of, 469
    posterior chamber of, *462, 463,* 464,
        *465,* 467t
    structure of, 461–69, *462*
Eyelashes, *459*
Eyelid, 457–58, *459*

## F

Face, *201,* 208–12, 212t. See also Skull;
    *specific bones*
Facet, 202t
Facial artery, *598, 599, 599*
Facial expression, muscles of, 302–3, *302,* 303t
Facial nerve (VII), *409, 410, 411,* 412t
Facial vein
    anterior, *611*
    superior, *606*
Facilitation state, of neuron, 365
Fahrenheit temperature, 988
Falciform ligament, **31,** 678, *679,* **983**
Fallopian tubes, 855–56, *857*
Falx cerebelli, 403
Falx cerebri, 388t, **969**
Familial hemiplegic migraine, 346
Familial hypercholesterolemia, 941, 959t
Familial hypertrophic cardiomyopathy,
    612, 959t
Farsightedness, 470, *471*
Fascia, 282, *282, 283, 283*
Fascicle
    of muscle fiber, 282, *283, 284*
    of nerve fiber, *408*
Fasciculus cuneatus, 378, *379,* 380, *380,*
    381t, 392
Fasciculus gracilis, 378, *379,* 380, 381t
Fat (adipose tissue), 148, *149,* 152t
Fatal familial insomnia, 404t, 959t
Fatty acid oxidases, 113
Fatty acids, 48–49, *49, 50*
    essential, 706–7
    intestinal absorption of, 689, *690,* 691t
    saturated, 49, *49*
    three-dimensional model of, *56*
    unsaturated, 49, *49*
Feces, 695–96
Feedback
    in homeostasis, 7, 8
    negative, 488, *489,* 490
Femoral artery, **30, 31, 33,** *602, 603,* 603–4,
    *605,* **972, 977**
    deep, *603, 605*
    lateral, *603*
Femoral nerve, **30, 31, 32,** *416, 417, 418*

Femoral vein, **29, 30, 31, 33,** 608, *610, 611,* **972, 977**
Femoropatellar joint, 267t
Femur, **35,** 200, *201,* 231, *232, 233,* 235t, **972**
  greater trochanter of, **340**
Ferritin, 525, 681, 989
Fertilization, 865–66, *866,* 893
  in vitro, 898
Fetal alcohol syndrome, 908, 909
Fetal brain transplantation, in Parkinson's
  disease, 396
Fetal cell implant, 646
Fetus, 867, *892,* 907–12, *907,* 911t
  blood supply of, 912–13, *912*
  circulation of, *912,* 913, 913t, *914–15*
  at full-term, *911*
  skull of, **254**
  teratogen exposures of, 906, *906,* 908–9
  ultrasound examination of, *11*
Fever, in infection, 631, 632t
Fiberoptic bronchoscope, 748
Fibers
  of cardiac muscle, 568
  collagenous, 145, *145,* 147t, *148,* 149, *149*
  elastic, *145,* 146, 147t, 149, *149*
  reticular, 146, 147t
  of skeletal muscle, 184–85, *284*
  of ventricles, 570, *571*
Fibrillation, 331
Fibrillin, defect in, 612
Fibrin, 538, *538*
Fibrinogen (factor I), *534,* 535, 535t,
  538, 541t
Fibrin-stabilizing factor (factor XIII), 541t
Fibroblasts, 144, *145,* 147t
  in scar formation, 154
Fibrocartilage, 150, *151,* 152t
Fibrosis, 331
Fibrositis, 331
Fibrous joints, 257–58, *257, 258,* 264t
Fibula, 200, *201, 232, 233, 233,* 235t
Fibular collateral ligament, *274, 275*
Fifth disease, 164t
Fight-or-flight response, 512, *513*
Filtration pressure, in urine formation, 791
Filtration, through cell membrane, 81–82,
  *82,* 87t
Filum terminale, 378, *378*
Fimbriae, of infundibulum, 856, *857*
Fingers, 200t, *201,* 225, *226, 228,* 229t, **341**
Fissure, 202t
  cerebral, 389
  longitudinal, 389, *389*
  transverse, 389, *389*
Fissure of Rolando, 389, *389*
Fissure of Sylvius, 389, *389*
Flaccid paralysis, 382
Flagella, *62, 73, 75, 76,* 79t
  of sperm, 841, *841*
Flavine adenine dinucleotide (FAD), *110,* 112
Flavoproteins, 718
Flexion, 262, *265,* 267t, 300
Flexor carpi radialis muscle, 313, *314,* 317t
  tendon of, **341**
Flexor carpi ulnaris muscle, 313, *314,*
  *315,* 317t
  tendon of, **341**
Flexor digitorum longus muscle, 328,
  *329,* 330t
Flexor digitorum profundus muscle, 313, 317t
Flexor digitorum superficialis muscle, 313,
  *314, 316,* 317t
Flexor retinaculum, *329,* 331

Flexors
  of arm, *310, 311,* 311–12
  of foot, 326–28, *328, 329*
  of forearm muscles, *311,* 313, *314*
  of hand, 313, *314,* 315, *316*
  of leg, 322–23, *323, 324*
Floaters, vitreous, 467t
Fluid compartments, 814–15, *815*
  extracellular, 814–15, *815, 816*
  intracellular, 814, 815, *815, 816*
  movement between, 815–16, *817*
Fluids
  body
    composition of, 815, *816*
    distribution of, 814–16, *815, 816, 817*
  hypertonic, 81, *82*
  hypotonic, 81, *82*
  isotonic, 81, *82*
  serous, 161
  synovial, 161, 261
Fluorine
  atomic structure of, 39t
  in nutrition, 727, 727t
Folacin, 720, 721t
Folic acid, 720, 721t
  red blood cell effects of, 526–27, *528,* 528t
Folinic acid, 720
Follicles
  hair, *162, 168, 169*
  ovarian, 852
    degeneration of, 855, *855*
    maturation of, 853–54, *854, 856*
  thyroid, 495
Follicle-stimulating hormone, 483t, 489, *491,*
  492, 493, 496t
  in female reproductive function,
    861–62, *863*
  in male reproductive function, 848–49, *849*
Follicular cells, of thyroid gland, *496,* 497, *497*
Folliculitis, 168
Fontanels, 202t, 213
Food, 5, 6t
Food poisoning, 779
Food pyramid, 728–29, *728*
Foot, 233, *234*
  bones of, 233, *234*
  muscles of, 326–31, *327, 328, 329, 330,*
    **978,** 330t, **978, 979, 980**
  radiograph of, *234*
Foramen, 202t
  carotid, **247**
  incisive, *204,* 213t, **245, 247**
  infraorbital, *203, 204,* 213t, **243, 244**
  intervertebral, 216, *216,* 414
  jugular, *208,* 213t, **245, 246, 247, 252, 253**
  mandibular, *211,* 212, 213t, **248**
  mental, *203, 204, 211,* 212, 213t, **248**
  obturator, **35,** 228, *229, 230*
  olfactory, 206, *208*
  palatine, *204,* 213t, **245, 247**
  sacral, 218, *219*
  stylomastoid, **245, 247**
  supraorbital, *203,* 205, 213t, **241, 243, 244**
  transverse, 216
  vertebral, 216, *216*
Foramen lacerum, *204, 208,* 213t, **245, 246,
  247, 252, 253**
Foramen magnum, *204,* 205, *208,* 213t, **245,
  246, 247, 249, 252, 253**
Foramen ovale, *204, 208,* 213t, **245, 246, 247,
  250, 252, 253,** 913, 913t, 918t
  closure of, 917
Foramen rotundum, *208,* 213t, **250, 253**

Foramen spinosum, *204, 208,* 213t, **245, 246,
  247, 250, 252, 253**
Forearm
  cross section of, *316*
  muscles of, *311,* 312–13, *313, 315, 316,*
    316t, **975, 977**
Forebrain (prosencephalon), 387, *387,* 388t
Formula
  molecular, 42
  structural, 44
Fornices, 858
Fossa, 202t
Fossa ovalis, 917
Fovea, 202t
Fovea capitis, 231
Fovea centralis, *462,* 465, *467,* 469
Fracture, 194–95
  Colles, 235
  repair of, 194–95, *195*
  types of, *194*
Fragile X syndrome, 950, *951*
Free radicals, in aging, 923
Frenulum, 660
Frontal bone, *203, 204,* 205, 205t, *208, 210,
  211,* **241, 242, 243, 244, 249, 251, 252,
  253, 338, 965**
Frontal eye field, 391
Frontal lobe, 389, *389,* 392t, **969, 985**
  association areas of, 392
Frontal sinus, *13,* 205, *207, 208,* 208t, *210,*
  **251, 252, 253,** *741,* **965, 969**
Frontal suture, **249**
Frontalis muscle, *301, 302,* 303, **973**
Fructose, 705, *705*
Functional residual capacity, 758, *758,* 759t

## G

G protein, 484
GABA, 368
Gage, Phineas, 375
Galactose, 705, *705*
Gallbladder, **31, 32, 33,** *657,* 681, 683, **972**
  bile release by, 683, *683*
  disease of, 684
Gallstones, 681, 683, *683*
Gamete intrafallopian transfer, 899t
Gamma globulins, 535, 535t
Ganglia, 408
  collateral, 421, *421*
  paravertebral, 419–21, *420*
Ganglion cells, of retina, 465, *466*
Gap junction, 65, 66t, *69*
Gas, intestinal, 696
Gases
  alveolar exchange of, 764–67, *766, 767*
  blood, *534,* 535–36
  blood transport of, 767–72, *768, 770, 771,
    772,* 772t
  diffusion of, 766–67, *767*
  partial pressure of, 766–67, *767*
Gastrectomy, 699
Gastric artery, *597*
Gastric glands, 671, *671*
Gastric juice, 671, 672–73, 672t, *673,* 673t
Gastric pits, 671
Gastric secretions, 670–72
  regulation of, 672–73, *673*
Gastric veins, 607, *608*
Gastrin, 672, 673, *673,* 685t
Gastrocnemius muscle, *301,* 324, 326–27, *327,
  328, 329,* 330t, **341, 342, 979**
Gastrostomy, 699

Gastrula, 896
Gastrulation, 896
Gene therapy, 955–58, *956, 958,* 958t
Genes, 117, 930–31, *931,* 932–33. *See also*
    Alleles
    amplification of, 128
    dominant, 932–33
    expression of, 942–43
    heterogeneity of, 943
    mapping of, 938–39
    mutations in, 126, *127*
        effects of, 129–30, 129t, *131*
    recessive, 932–33
    ultraviolet radiation effect on, 129
General interpretative area, 392
General stress syndrome, 512, 514t
Genetic code, 119–20, 121t
Genetics. *See also* Chromosomes; Genes
    medical, 929, 930t, 938–39, 951–58, 959t
        for counseling, 954–55, 955t
        in diagnosis, 951, 954
        gene therapy in, 955–58, *956, 958,* 958t
Genicular artery, *603,* 604
Geniculate nucleus, 399
Genital herpes, 882t
Genital warts, 882t
Genome, 930, 939
Genomic imprinting, 947
Genotype, 940
Germ layers, 896, *897*
Gigantism, 493
    pituitary, 193
Gland(s)
    alveolar, 142, *142,* 143t
    apocrine, 142, *143,* 144t, 171, *171,* 173, 173t
    autocrine, 480
    ceruminous, 173t
    compound, 142, *142,* 143t
    eccrine, 171, *171,* 173, 173t
    endocrine, 480, *481*
    exocrine, 142–44, *142,* 480, *481*
    holocrine, 142, *143,* 144t
    mammary, 173t
    merocrine, 142, *143,* 144t
    paracrine, 480
    sebaceous, 142, *144, 162,* 170–71, *171,* 173t
    simple, 142, *142,* 143t
    sweat, 171, *171,* 173, 173t
    tubular, 142, *142,* 143t
Glans penis, *837, 845, 845*
Glaucoma, 464
Glenohumeral joint (shoulder joint), 267t,
    268–69, *268*
    dislocation of, 269
Glenohumeral ligament, 268, *269*
Glenoid fossa, *224*
Glenoidal labrum, 269, *269*
Gliadin, 708
Gliding joint, 262, *263,* 264t
Globus pallidus, 394, *394*
Globulins, *534,* 535, 535t, 989
Glomerular capsule, 785, *785,* 801t
Glomerular filtrate, 790, 791t, *793*
Glomerular filtration, in urine formation,
    789–90, *791,* 791t
Glomerular filtration rate
    measurement of, 802
    regulation of, 793–94, *793*
    in urine formation, 792, *792, 793*
Glomerulonephritis, 647t, 783, 787
Glomerulus, 785, *785,* 801t
Glossitis, 699
Glossopharyngeal nerve (IX), *409,* 410, 412t
Glottis, 744, *745*
Glucagon, 483t, 507, *508,* 509t

Gluconeogenesis, 706
Glucose, *689,* 705, *705*
    hepatic formation of, 292, *293*
    homeostatic regulation of, 8
    renal reabsorption of, 794
    serum, 989
    structure of, *48*
    three-dimensional model of, *56*
    urinary, 791t, 795, 992
Glucosuria, 791t, 795, 828
Glutamic acid, 121t
Glutamic pyruvic transaminase, 991
Glutamine, 121t
Gluteal artery, 602, *602*
Gluteal gait, 322
Gluteal nerve, 417, *418*
Gluteal vein, 608
Gluteus maximus muscle, *301, 320,* 322, *323,*
    *324,* 325t, **340, 972, 975, 978**
Gluteus medius muscle, **35,** *301,* 322, *323,*
    *324,* 325t, **340**
Gluteus minimus muscle, 322, 325t
Glycerol, 48, *50*
Glycine, *56,* 121t
Glycogen, 705, *705*
    cellular, 47
Glycolysis, 106, *107,* 109–11, *109*
    defects in, 111
Glycoprotein hormones, 481, 483t
Glycoproteins, of cell membrane, 65
Goblet cells, 138, *138*
Goiter, 492, 498t, *499*
Golgi apparatus, *62,* 69, *70,* 71–72, *71,* 79t
Golgi tendon organs, 438–39, *439,* 439t
Gomphosis, 258, *259,* 264t
Gonadal arteries, 597, *597,* 598t, *605*
Gonadal veins, 607, *611*
Gonadotropes, 490
Gonadotropin-releasing hormone, 483t,
    *491,* 493
Gonadotropins, 492
    in female reproductive function,
        861–62, *863*
    in male reproductive function, 848–49, *849*
Gonorrhea, 882t
Gout, 799, 959t
Graafian follicle, 853–54, *854*
Gracilis muscle, **32, 33, 34, 35,** *301, 321,* 322,
    *324,* 325t, *329,* **977**
Granulocytes, 527
Granulocytic leukemia, *533*
Grave's disease, 498t, 647t
Gravity
    absence of, 588
    in equilibrium sense, 453
Gray commissure, of spinal cord, 378, *379*
Gray matter, **969**
    of brain, 390, *390*
    of spinal cord, 378, *379*
Gray rami, 421, *421*
Growth
    of adolescents, 919, 921t
    of adults, 919, *920,* 921t
    of children, 918–19, 921t
    of infant, 918, 921t
    of neonate, 914–15, 917–18, *917*
    postnatal, 913–21, 921t
    prenatal, 893–913. *See also* Embryo; Fetus
Growth factors, 96
Growth hormone, 483t, 488, *491,* 492, 496t
    bone effects of, 193
    deficiency of, 493
    excess of, 493
Growth hormone-releasing hormone, 483t,
    *491,* 492

Guanine, *117, 121*
Guanosine monophosphate, cyclic (cGMP),
    471, 487
Gubernaculum, 836
Gyri, cerebral, 389

## H

H zone, of muscle fibers, 285
Hageman factor (factor XII), 539, 541t
Hair, 168, *169*
    loss of, 170
    transplantation of, 170
Hair cells
    of macula, 453, *457*
    of organ of Corti, 450, *451, 452*
Hair follicles, *162,* 168, *169*
Half-life, of radioactive isotope, 41
Hamate, *228*
Hamstring muscles, **340**
Hand, 4, 226, *228*
    muscles of, 313–16, *314, 315, 316,* 317t
    radiograph of, *228*
Hashimoto's disease, 498t
Head
    arteries of, 597–600, *598, 599*
    magnetic resonance imaging of, *11*
    muscles of, 305–7, 305t, *306, 307*
    veins from, 604, *606*
Hearing, 446–51, 453t. *See also* Ear
    auditory tube in, 448
    external ear in, 446, *446*
    inner ear in, 448–50, *449, 450*
    loss of, 454
    middle ear in, 446–48
    vestibulocochlear nerve in, 450
Hearing aid, 454
Heart, **4, 32,** 554–65, **964, 970, 980.** *See also*
    Cardiac muscle
    apex of, *564*
    of athlete, 593
    blood flow through, 561–62, *562, 563*
    blood supply to, 562–63, *564, 565, 566*
    chambers of, *556,* 557–58, *559*
    collateral circulation of, 563
    conduction system of, 568, 570, *570,*
        *571, 572*
    coverings of, 555–56, *556*
    cycle of, 566–68, *567,* 571–73, *573*
    fetal, 913
    fibers of, 154, *154*
    functional syncytium of, 568
    layers of, 557t
    location of, 554, *555*
    in Marfan syndrome, *613*
    muscle fibers of, 568
    neonatal, 917
    size of, 554, *555*
    skeleton of, 560, *561*
    valves of, 557–58, *559, 560*
    wall of, 556–57, *557*
Heart sounds, 568, *569*
Heart transplantation, 569
Heartbeat, 565. *See also* Cardiac cycle
    apical, 554
Heat, 5, 6t
    body, 173–74, *174*
Heat exhaustion, 174
Heat receptors, 434
Helicotrema, 449, *449, 450*
Helium, 39, 39t, *40*
Hemarthrosis, 275
Hematocrit, 521, 990
Hematoma, subdural, 376
Hematometra, 883

Hematopoiesis, 197
Heme, *525, 726*
Hemianopsia, 475
Hemiazygos veins, 607, *607*
Hemicellulose, 705
Hemochromatosis, 525, 959t
Hemocytoblasts, 522, *523, 524*
Hemodialysis, 781
Hemoglobin, 522, *525, 726*, 990
    in acid-base balance, 826
        fetal, 912
    in oxygen transport, 767, 768, *768*
    urinary, 992
Hemolytic anemia, 529t, 647t
Hemolytic uremic syndrome, 779
Hemophilia, 544, 946, 959t
Hemorrhagic disease of the newborn, 716
Hemorrhagic telangiectasia, 547
Hemorrhoids, 693
Hemostasis, 537–42. *See also* Coagulation
Hemothorax, 774
Heparin, 542, 542t
    mast cell release of, 145
Hepatic. *See also* Liver
Hepatic artery, *597, 679*
Hepatic ducts, 680
Hepatic flexure, *693*
Hepatic lobules, 678, *679, 680*
Hepatic portal vein, 607, *608, 609*
Hepatic sinusoids, 607, 678
Hepatic veins, 607, *611*, 678, *679*, 680
Hepatitis, 682
Hepatopancreatic ampulla, 676, *676*
Hepatopancreatic sphincter, 676, *676*
Hereditary osteoarthritis, 147t
Heredity, 929. *See also* Genes; Genetics
Hering-Breuer reflex, in ventilation, 764, *764*
Hernia
    hiatal, 667
    inguinal, 836
Herpes, 181, 882t
Heterozygous, definition of, 937
Hiatal hernia, 667
Hiccuping, 759, 760t
Hilum
    of lymph node, 627
    of renal sinus, 780
Hilus, 751
Hindbrain (rhombencephalon), 387, *387*, 388t
Hinge joint, 262, *263*, 264t
Hip joint, 267t, 271–73, *271, 272*
Hip pointer injury, 227
Hippocampus, in memory formation, 393, 394
Hirschsprung disease, 695
Hirsutism, 513
Hirudin, 543
*Hirudo medicinalis,* 543
Histamine
    in capillary exchange, 579
    in gastric acid regulation, 672
    in infection, 531, *531*
    mast cell release of, 145
Histidine, 121t
Holocrine gland, 142, *143*, 144t
Homeostasis, 5–8, *7*
    in body temperature regulation, 173–74, *174*
    in bone formation, 192
    feedback in, *7,* 8
Homograft, for burns, 177
Homozygous, definition of, 937
Horizontal cells, of retina, 465, *466*
Hormone(s), 16, 480–87. *See also specific*
       *hormones*
    actions of, 482–87
    amine, 481, 483t

chemistry of, 480–81
of female reproductive function, 860–64,
    *861, 862t, 863,* 864t
glycoprotein, 481, 483t
of male reproductive system, 848–50, *849*
negative feedback regulation of, 488, *489*
nonsteroid, 483–87, *486,* 487t
peptide, 481, 483t
pregnancy-related changes in, 868–69,
    *868,* 869t
protein, 481, 483t
receptors for, 483–84, *484*
secretion of, 487–88, *489*
in smooth muscle contraction, 298
steroid, 482–83, 483t, *484,* 484t
in stress response, 512
target cells for, 482
tropic, 488
Household measures, 987
Human chorionic gonadotropin, in pregnancy,
    868, *868*
Human genome project, 130–31
Human immunodeficiency virus (HIV)
      infection, 648–49, 649t
Human immunodeficiency virus (HIV),
      transcytosis of, 85, *87*
Humeral ligament, transverse, 268, *269*
Humerus, **32**, 200, *201*, 223–24, *225, 226,* 229t
    epicondyle of, **339, 340, 341**
Humor
    aqueous, *462, 464, 465*
    vitreous, *462, 465,* 467t
Huntington disease, 363t, 427, 959t
    gene for, 938, 940–41
Hutchinson-Gilford syndrome, 924
Hyaline cartilage, 150, *150,* 152t
Hybridoma, 638
Hydatid mole, 925
Hydramnios, 925
Hydrarthrosis, 275
Hydrocele, 883
Hydrocephalus, 401, *401*
    fetal, 911
Hydrochloric acid, 672, 672t
Hydrogen, 39, 39t, *40*
Hydrogen bonds, 52, *52*
Hydrogen ions
    in cells, 48t
    regulation of, 824–28
    renal excretion of, 827–28, *828*
    renal tubule, 800, *800*
    sources of, 823–24, *824*
Hydrolysis, 102–4
Hydrostatic pressure, 5
    in capillary exchange, 579, *580*
Hydroxyapatite, 197
Hydroxycholecalciferol, formation of, 500
17-Hydroxycorticosteroids, urinary, 992
Hymen, 858
Hyoid bone, 199, *201, 202, 661, 744, 745*
Hyperalgesia, 475
Hyperalimentation, 732
Hypercalcemia, 513, 573, 732, 822
Hypercalciuria, 732
Hypercapnia, 774
Hypercholesterolemia, 85, 613
    familial, 941, 959t
Hyperemesis gravidarum, 883
Hyperextension, 264, *265,* 267t
Hypergeusia, 445t
Hyperglycemia, 513, 732, 828
Hyperkalemia, 573, 732, 823, 828
Hyperkalemic periodic paralysis, 66–67
Hypernatremia, 732, 823, 828
Hyperopia, *471*

Hyperoxia, 774
Hyperparathyroidism, 501t
Hyperpnea, 774
Hyperreflexia, 382
Hypersensitivity reactions, 643–44
Hypersomia, 445t
Hypersomnia, 363t
Hypertension, 590–91, 591t
Hyperthermia, 175
Hyperthyroidism, 497, *498,* 498t
Hypertrophic pyloric stenosis, 670
Hyperuricemia, 828
Hyperventilation, 764, 774
    alkalosis with, 831
Hypervitaminosis A, 715
Hypervitaminosis D, 715
Hypoalbuminemia, 732
Hypocalcemia, 513, 573, 822
Hypochondriac regions, 22, *22*
Hypodermis, *162,* 168
Hypogastric region, 22, *22*
Hypogeusia, 445t
Hypoglossal canal, *210,* 213t
Hypoglossal nerve (XII), *409,* 411, 412t
Hypoglycemia, 510, 732, 828
Hypokalemia, 573, 733, 823
Hyponatremia, 733, 823
Hypoparathyroidism, 501t
Hypophyseal portal veins, 489
Hypophysectomy, 513
Hypophysis. *See* Pituitary gland
Hypopituitary dwarfism, 493
Hypoplastic left heart syndrome, 594
Hypoproteinemia, 821
Hyposmia, 445t
Hypospadias, 883
Hypothalamus, 399, *399, 481,* **965, 985**
    functions of, 400
    hormones of, 483t
Hypothermia, 174–75
Hypoxemia, 774
Hypoxia, 767, 774
    cyanosis with, 522
Hysterectomy, 883

## I

I bands, of muscle fibers, 285, *285*
Ileitis, 699
Ileocecal sphincter, 692, *693*
Ileum, **33**, 685, *685, 693*
Ileus, 699
Iliac artery, 598t
    circumflex, *602, 603,* 604
    common, **33**, 597, *597,* 602, *602, 603,* 605
        left, **34**
    external, 602, *602,* 603, *603,* 605
    internal, 602, *602, 603,* 605
Iliac crest, **35**, 227, *230,* **340**
Iliac regions, 22, *22*
Iliac spine, **29, 31, 35,** *230,* **340**
Iliac vein
    common, *610, 611*
    external, 608, *610, 611*
    internal, 608, 609, *610, 611*
Iliacus muscle, *321,* 322, 325t, **977**
Iliocostalis cervicis muscle, 305t, *306*
Iliocostalis lumborum muscle, 305t, *306*
Iliocostalis thoracis muscle, 305t, *306*
Iliofemoral ligament, 271, *272*
Iliolumbar artery, 602, *602*
Ilium, 227, *229, 230*
Immune response, 640–42, *641*
    primary, 640–41, *641*
    secondary, 641–42, *641*

Immunity, 630, 632–47
  active, 642, 643t
  antibody-mediated, 634–40, *635, 636,* 637t
  antigens in, 632
  cell-mediated, 633–34, 634t
  lymphocytes in, 632–34, *633*
  passive, 642, 643t
  polyclonal response in, 635
  types of, 642–43, 643t
Immunocompetence, 647
Immunodeficiency, 647
Immunoglobulin A (IgA), 637, 639t
Immunoglobulin D (IgD), 638, 639t
Immunoglobulin E (IgE), 638, 639t, 640
Immunoglobulin G (IgG), 637, 639t
Immunoglobulin M (IgM), 638, 639t
Immunoglobulins, 634–40
  structure of, 637, *637*
  types of, 637–38
Immunosuppressive drugs, in transplantation,
  644, 646
Immunotherapy, 638–39
Impetigo, 164t
Implantation, 895, *896*
In vitro fertilization, 898
Incisive foramen, 213t, **245, 247**
Incisor teeth, 662, *662*
Inclusions, cellular, 77
Incontinence, urinary, 805
Incontinentia pigmenti, 946
Incus, *446, 447, 447*
Infant, 918, 921t
  skull of, 213, *214*
Infarction, 538
Infection, 991
  chemical barriers to, 631, 632t
  defense against, 630
  fever with, 631, 632t
  inflammatory response to, 631, 631t
  mechanical barriers to, 631, 632t
  nonspecific defenses against, 630–32, 632t
  phagocytosis in, 632, 632t
  species defenses against, 630, 632t
  specific defenses against, 632–47. *See also*
    Immunity
  white blood cells in, 531, *531*
Inferior, definition of, 18
Inferior oblique muscle, 460, 461t
Inferior rectus muscle, 460, 461t
Inferior vena cava, **34, 35,** 558, *559, 564,* 609,
  *610, 611,* **971, 972, 982**
Infertility
  female, 854, 872, 872t
  male, 846–47, 847t, 865
Inflammation, 512, 531, *531,* 631, 631t, 632t
  cortisol in, 505
  edema in, 821, 821t
Inflation reflex, in ventilation, 764, *764*
Influenza vaccine, 642
Infraorbital foramen, *203, 204,* 213t,
  **243, 244**
Infraspinatus muscle, *301, 307, 309,* 312, 312t,
  **339, 975, 976**
Infundibulum, 399, *399,* 488
  of uterine tube, 856, *857*
Inguinal canal, **31,** 836
Inheritance, 940–42. *See also* Chromosomes;
  Genes; Genetics
  dominant, 940–41
  recessive, 940–41
Inhibin, 849
Inhibitory neuron, 350
Inhibitory postsynaptic potential, 361
Injection, subcutaneous, 168
Inner ear, 448–50, *449, 450*

Insomnia, 363t, 404t, 959t
Inspiration, 751, 754–56, *754, 755,* 755t, *756*
Inspiratory capacity, 758, *758,* 759t
Inspiratory reserve volume, 757, *758,* 759t
Insula, *389,* 390
Insulin, 480, 483t, 507–8, *508,* 509t
  in glucose homeostasis, 8
Integrins, *69*
Integumentary system, 14, *15,* 19t, 161–81.
  *See also* Skin
Intensity, of vocal sound, 743
Interatrial septum, 570
Intercalated disks, of cardiac muscle, 299, *299*
Intercarpal joints, 267t
Intercellular junctions, 65, 66t, *69*
Intercondylar eminence, of tibia, *233*
Intercostal artery, 598t, *605*
  anterior, 601, *601*
  external, *601*
  internal, *601*
  posterior, *601,* 602
Intercostal muscles
  external, **30, 31, 35**
  internal, **31, 35,** *308*
Intercostal nerve, *416,* 417
Intercostal vein, 607, *607*
Interferons, 631, 634t
Interleukins, 634t
Interleukin-2, in kidney cancer, 800
Interlobar artery, 783
Interlobular artery, 783, *784*
Interlobular vein, *784*
Internal oblique muscle, *308,* 317t, *318,* 319
Interneurons, 350, 351t
Internodal atrial muscle, 568
Interosseous ligament, 257, *257*
Interphalangeal joints, 267t
Interphase, of mitosis, 88, *88*
Interstitial cells, 838
Interstitial cell-stimulating hormone, in male
  reproductive function, 848–49, *849*
Intertarsal joint, 267t
Interventricular artery, 562–63, *564*
Interventricular septum, *559, 560,* 570
Interventricular sulci, *556,* 557
Intervertebral disk, **35,** 214, *260,* **967, 968**
Intervertebral foramen, 216, *216*
Intervertebral joint, 267t
Intestinal flora, 695
Intestinal gas, 696
Intestinal glands, 686
Intestine, *657, 659*
  large, *657,* 692–93, *693, 694,* **971**
    functions of, 694–95
    movements of, 695
    wall of, 693–94, *694*
  small, **31, 32,** 685–92, *686,* **964, 967, 968,**
    **971, 983, 984**
    absorption in, *670,* 689–71, *689*
    epithelial cells of, *687,* 688
    movements of, 691–92
    nutrient absorption by, *689,* 689–91,
     *690,* 691t
    peristaltic movements of, 691–92
    secretion of, 688
    segmentation movements of, 691
    structure of, 685–86, *685*
    wall of, 686–88, *687*
Intramembranous bone, 188–89, *189,* 190t
Intrapatellar bursa, 275
Intrauterine device, 879t, *880,* 881
Intrauterine transfusion, 925
Intravascular oxygenator, 750
Intrinsic factor, 527, 672, 672t, 719
Invertor, of leg, *329,* 331

Iodine
  deficiency of, 498t, *499*
  in nutrition, 726, 727t
Iodine-131, medical use of, 40–41
Iodine pump, of follicular cells, 497
Iodopsins, 472–73
Ion, 38t, 41
Ion channels, 66–67
  in nerve impulse propagation, 354–55, *355,*
   *356, 357*
Ionic bond, 41, *42*
Ionizing radiation, 43
Ipsilateral, definition of, 19
Iridectomy, 475
Iris, *462, 463, 464, 465*
  circular muscles of, 465, *466*
  radial muscles of, 464–65, *466*
Iritis, 475
Iron
  absorption of, 527
  liver storage of, 681
  in nutrition, 724, *726,* 727t
  red blood cell effects of, 527, 528t
  serum, 990
Iron deficiency anemia, 529t
Iron-binding capacity, 990
Ischemia, cardiac, 571, *572*
Ischial spine, *230*
Ischial tuberosity, *230,* **340**
Ischiocavernosus muscle, 319t, 320, *320*
Ischiofemoral ligament, *272, 273*
Ischium, 227, *229, 230*
Islets of Langerhans, 507, *507,* 509t
Isocaloric, 733
Isograft, 644, 646t
Isoleucine, 121t
Isotopes, 38–39
Isotretinoin, fetal effects of, 908–9

## J

Jacobs syndrome, 949t
Jaundice, 176, 681
Jejunum, **33,** 685, *685*
Joint(s), 256–77. *See also specific joints*
  ball-and-socket, 262, *263,* 264t
  cartilaginous, 258–59, *259, 260,* 264t
  classification of, 257–59
  condyloid, 262, *263,* 264t
  disorders of, 276–77, 276t
  fibrous, 257–58, *257, 258,* 264t
  gliding, 262, *263,* 264t
  hinge, 262, *263,* 264t
  movement of, 262, 264–65, *265, 266,* 267t
  pivot, 262, *263,* 264t
  replacement of, 273
  saddle, 262, *263,* 264t
  synovial, 259, 260–61, *260,* 264t, 267t
  types of, 262, *263,* 264t
Joint capsule, 260, *260*
  of knee, 273, *274*
Joint reflex, breathing rate and, 765
Jugular foramen, *204,* 205, *208, 210,* 213t, **245,**
  **246, 247, 252, 253**
Jugular vein, *606,* **970, 981, 982**
  external, **31,** *606, 607, 611*
  internal, **30, 31, 34,** *596,* 604, *606, 607, 611*
  superior, *606*
Juxtaglomerular apparatus, 785, 788, *788*
Juxtamedullary nephrons, 788, *789*

## K

Karyotypes, 936, *937*
Keloid, 181

Keratin, 139
Keratitis, 475
Keratoplasty, penetrating, 462
Ketone bodies, 113, 114, *114*
Ketonuria, 828
Ketosis, 828
Kidney(s), *481,* 779–89, **971, 972.** *See also*
　　Renal tubules
　baroreceptors of, 793
　cancer of, 800
　functions of, 780–81
　of infants, 802
　left, **34**
　location of, 779, *780, 781*
　longitudinal section of, *782*
　of newborn, 917
　right, **34**
　structure of, 780, *780*
　transverse sections of, *781*
Kidney stones, 803
Klinefelter syndrome, 949t
Knee joint, 267t, 273–75, *274*
　capsule of, *274*
　radiography of, *232*
Knee-jerk reflex (patellar tendon reflex),
　　383–85, *384*
Korotkoff's sound, 586
Kupffer cells, 607, 680
Kwashiorkor, 730, *731*

## L

Labia majus, *851, 859, 859,* 860t
Labia minus, *851, 859, 859,* 860t
Laboratory tests, 989–91
Labyrinth, 448–50, *449, 450*
　membranous, 448, *449, 450,* 453, *455*
　osseous, 448, *449*
Labyrinthectomy, 475
Labyrinthitis, 475
Lacrimal apparatus, 458–59, *459*
Lacrimal bones, *203, 204,* 209, 212t, **242,
　　243, 244**
Lacrimal ducts, 458–59, *459*
Lacrimal gland, 303, 458–59, *459*
Lacrimal sac, 459, *459*
Lactase, 688, 689t
Lactation, 876–77, *876,* 877t
Lacteal, of intestinal villi, 686
Lactic acid, 990
　formation of, 113
　hepatic glucose formation from, 292, *293*
　tolerance to, 293
Lactic dehydrogenase, serum, 990
Lactose, 705
Lactose intolerance, 688
Lacuna, of chondrocytes, 150, *150*
Lambdoidal suture, *204,* 205, *210,* **242**
Lamellae, of bone, 151, *151*
Laminectomy, 235, 427
Large intestine, *657,* 692–93, *693, 694,* **971**
　functions of, 694–95
　movements of, 695
　wall of, 693–94, *694*
Laryngopharynx, *661,* 666
Larynx, **30, 33,** *740, 741,* 742–44, *744, 746,*
　　752t, **965, 980**
Lateral, definition of, 19
Lateral horn, of spinal cord, 378, *379*
Lateral rectus muscle, 460, 461t
Latissimus dorsi muscle, **30, 31,** *301, 307, 310,*
　　312, 312t, **339, 340, 974, 975, 976**
Laughing, 759, 760t
Learning, positron emission tomography
　　during, 55

Leech, medicinal, 543
Left ventricular assist device, 569
Leg
　arteries to, 602–4, *603*
　bones of, 200, 231–35, *232, 233, 234,* 234t
　cross section of, *330*
　muscles of, *321, 322,* 322–24, *324, 325,*
　　326t, **977, 978**
　veins from, 608–9, *610*
Lens, *462, 463, 465*
　accommodation of, 463–64, *464*
　implantation of, 464
Lens capsule, 463
Lens fibers, 462, *463*
Leucine, 121t
Leukemia, 533
Leukocytes. *See* White blood cells
Leukocytosis, 532
Leukopenia, 532
Levator ani muscle, 319, 319t, *320*
Levator palpebrae superioris muscle, 458, 461t
Levator scapulae muscle, *307,* 308, *309,*
　　310t, **973**
Levers, in body movement, 196–97, *196, 197*
Lewy bodies, 396, *396*
Life
　characteristics of, 4, 5t
　maintenance of, 5–8, *6,* 6t, *7*
Life expectancy, 923–24
Life span, 923–24
Ligament, 260–61. *See also specific ligaments*
Ligamentum arteriosum, *564, 596,* 917
Ligamentum teres, 917
Light, refraction of, 468, *468*
Lignin, 705
Limbic system, 401
Linea, 202t
Linea alba, **30,** *308,* 317, *318,* **974**
Linea aspera, of femur, *233*
Linea terminalis, 228, *230*
Lingual artery, *598, 599, 599*
Linoleic acid, 706–7
Linolenic acid, 707
Lips, *658, 660, 661,* 663t
Lipase, 689t, *690*
　intestinal, 688
　pancreatic, 677
Lipids
　in cells, 47–49, *50,* 51t, 53t
　digestion of, 689, *690*
　metabolism of, 113–14, *114, 116,* 681t, 707
　in nutrition, 706–7, *707,* 709t
　plasma, 535–36
　requirements for, 707
　serum, 990
　sources of, 706
　synthetic, 707
　utilization of, 706–7, *707*
Lipogenesis, 733
Lipoid nephrosis, 795
Lipoprotein complexes, 535
Lipoproteins, 536, 536t
　high-density, 536, 536t
　low-density, 84, 536, 536t
　very low-density, 536, 536t
Lissencephaly, 389, 959t
Lithium, *38,* 39, 39t, *40*
Liver, **31, 32, 33,** *657, 678–84,* **964, 966,
　　967, 971, 972, 983, 984.** *See also*
　　Hepatic veins
　anatomy of, 678, *678*
　functions of, 680–81, 681t
　as gene therapy target, 957
　glucose formation in, 292, *293*
　glycogen stores of, 706
　inflammation of, 682

lipid metabolism of, 707
　structure of, 678–80, *679*
　transverse section of, 678, *678*
Lochia, 925
Longissimus capitis muscle, *306*
Longissimus cervicis muscle, 305t, *306*
Longissimus thoracis muscle, 305t, *306*
Longitudinal fissure, 389, *389*
Longitudinal section, *22*
Long-QT syndrome, 67
Long-term synaptic potentiation, in memory
　　formation, 393
Loose connective tissues, 152t
Louis, angle of, 221, *221*
Lower back pain, 438
Lower motor neuron syndrome, 382
Lucid, Shannon, 588
Lumbago, 235
Lumbar artery, 597, *597,* 598t, 602, *605*
Lumbar enlargement, 377, *378*
Lumbar nerves, 412, *413*
Lumbar puncture, 400–401
Lumbar regions, 22, *22*
Lumbar spine, **35,** *215,* 218, *218,* 219t, **967, 968**
Lumbar vein, 607, *611*
Lumbosacral plexus, *416,* 417, *418*
Lumen, of alimentary canal, 686
Lunate, *228*
Lung(s), **31, 32, 33,** *740,* 750–51, *752,* 752t,
　　**970, 971, 980.** *See also* Pulmonary value
　compliance of, 756
　elastic recoil of, 756
　as gene therapy target, 957
　irritants of, 753, 753t
　lobes of, 751, *752*
　surface tension of, 754–55
Lung cancer, 742–43
Luteinizing hormone, 483t, 489, *491,* 492, 496t
　in female reproductive function,
　　861–62, *863*
Luxation, 275
Lyme arthritis, 276t, 277
Lyme disease, 164t, 271
Lymph, 621, *622,* 625
　flow of, 624, *625,* 626
　formation of, 625, *625*
　function of, 625
　obstruction of, 626
Lymph node(s), 621, *622,* 626–28, 630t
　abdominal, *627,* 628
　axillary, 627–28, *627*
　cervical, 627, *627*
　cross section of, *626*
　functions of, 628
　inguinal, *627,* 628
　location of, 627–28, *627*
　pelvic, *627,* 628
　structure of, 626–27, *626*
　supratrochlear, *627,* 628
　thoracic, *627,* 628
Lymph nodules, 627
Lymph sinuses, 627
Lymphadenectomy, 647
Lymphadenitis, 627
Lymphadenopathy, 647
Lymphadenotomy, 647
Lymphangitis, 627
Lymphatic capillaries, 621, *622*
Lymphatic ducts, 621, 623–24, *623, 624*
Lymphatic system, 17, *17,* 19t, 621–24, *622,* 630t
　system interconnections with, 650
Lymphatic trunks, 621, 623–24, *623*
Lymphatic vessels, 621, *622, 623*
　afferent, 627
　efferent, 627

obstruction of, 821, 821t
Lymphocytes, *94,* 529–30, *530, 534,* 534t,
    632–40, *633*
    B, 530, 632, *633,* 634–42, *635, 636, 637,*
      637t, 639t, *641*
    origin of, *523*
    T, 530, 632, 633–34, *633,* 634t, *636*
Lymphocytopenia, 647
Lymphocytosis, 647
Lymphoid leukemia, 533
Lymphoma, 647
Lymphosarcoma, 647
Lysine, 121t
Lysosomes, *62,* 73, *73,* 79t
    abnormalities of, 75
Lysozyme, of tears, 459

# M

M line, of muscle fibers, *284,* 285
McCune-Albright syndrome, 485
Macromolecules, 8, *9,* 9t
Macronutrients, 704
Macrophages, 144, *145,* 147t, 530, *530*
    in immune response, *636*
    origin of, *523*
Macula
    in equilibrium, 453, 455, *456*
    of membranous labyrinth, 453, *455*
Macula densa, 785, *788*
Macula lutea, 465, *467,* 469
Magnesium
    atomic structure of, 39t
    cellular, 48t
    in nutrition, 724, 725t
    test for, 990
    urinary, 791t
Magnetic resonance imaging, 10–11
    of cardiac blood flow, 562
Major histocompatibility complex, 634
    in transplantation, 644
Malleolus
    lateral, 233, *233,* **342**
    medial, 232, *233,* **342**
Malleus, *446, 447,* 447
Malnutrition, 729–30
Maltase, 688, 689t
Maltose, *689*
Mammary glands, **29,** 173t, 871–77
    active, *873*
    development of, 871–72
    milk production by, 876–77, *876,* 877t
    structure of, 871, *873*
Mammatropes, 490
Mammillary bodies, 399
Mammography, 874, *874*
Mandible, 201, *203, 204,* 210–12, *211,*
    212t, **248, 251, 338,** *965*
Mandibular condyle, *204,* 210, **248**
Mandibular foramen, *211,* 212, 213t, **248**
Mandibular fossa, *204,* 205, **244, 245, 249**
Mandibular ramus, **248**
Manganese, in nutrition, 724, 727t
Mania, 363t
Manubrium, 221, *221,* **340**
Maple syrup urine disease, 806
Marasmus, 730, *731*
Marfan syndrome, 612, *613,* 943, 959t
Marginal artery, 563, *564*
Marrow, bone, 187, *187,* 197
Masseter muscle, *301, 302,* 303, 304t, **338,** *973*
Mast cells, 144–45, *145,* 147t, *645*
Mastication, muscles of, *302,* 303–4, 304t, **973**
Mastitis, 883
Mastoid process, *204,* 210, **242, 245, 249, 338**

Mastoiditis, 205
Matrix, 144
Matter, 37–46, 38t
    atomic structure of, 37–42, *38,* 39t, *40*
    chemical reactions of, 44–45, *45*
    molecular structure of, 42, *42,* 44, *44*
    pH of, 45–46
Maturity–onset diabetes mellitus of young, 508
Maxilla, *203, 204, 207,* 208, *210, 211,* 212t,
    **241, 242, 243, 244, 245, 338,** *965*
    palatine process of, *204, 211,* **247**
Maxillary artery, *598,* 599
Maxillary bones, 208, 212t
Maxillary sinus, 208, *208,* 208t, *209, 211,*
    **251, 252**
Mean arterial pressure, 587
Mean corpuscular hemoglobin, 990
Mean corpuscular volume, 990
Meatus, 202t
Mechanical barriers, against infection, 631
Mechanoreceptors, 433, 439t
Meconium, 925
Medial, definition of, 19
Medial rectus muscle, 460, 461t
Median nerve, 415, *416, 417*
Mediastinal artery, 598t
Mediastinum, *12,* 13
Mediastinum testis, 836
Medical genetics, 929, 930t, 938–39,
    951–58, 959t
    for counseling, 954–55, 955t
    in diagnosis, 951, 954
    gene therapy in, 955–58, *956, 958,* 958t
Medical sciences, 24
Medroxyprogesterone acetate (Depo-
    Provera), 880
Medulla oblongata, 380, *380,* 387, *387,* 402–3,
    *402,* 405t, **985**
    cardiac control center of, 572, 587
    functions of, 405t
    vasomotor center of, 587, 589, *589*
Medullary cavity, of bone, 187, *187*
Medullary rhythmicity area, in ventilation,
    760, 762, *762*
Megakaryocytes, 522, 532
Megaloblastic anemia, 720
Meiosis, 838, 932, 934–36, *935, 936*
    genetic variability and, 936, *937*
Meissner's corpuscles, 167, 434, *434,*
    *435,* 439t
Melanin, 164, *165*
    of iris, 465
Melanocytes, 164, *165*
Melanocyte-stimulating hormone, 489
Melanoma, cutaneous, 166–67, *167*
MELAS (*mitochondrial encephalomyopathy,
    lactic acidosis, and strokelike episodes*)
    syndrome, 74
Melatonin, 510, 511
Membrane, 161
    alveolar-capillary, 765, *766, 767*
    cell, *62,* 63–67, *64, 65,* 66t, 79t
      active transport across, 83, *83,* 87t
      desmosomes of, 65, 66t
      diffusion across, 78–80, *80,* 87t
      endocytosis by, 83–85, *84, 85, 86,* 87t
      exocytosis by, 85, *86,* 87t
      facilitate diffusion across, 87t
      facilitated diffusion across, 80, *80*
      filtration through, 81–82, *82,* 87t
      gap junctions of, 65, 66t
      intercellular junctions of, 65, 66t
      ion channels of, 66–67
      osmosis across, 80–81, *81,* 87t
      permeability of, 64, 79–80
      structure of, 64–65, *65*

      tight junctions of, 65, 66t
      transcytosis by, 85, *87,* 87t
    cutaneous, 161. *See also* Skin
    mucous, 161
    of nasal cavity, 740, *741*
    serous, 161
    synovial, 161, 261, *261*
Membrane potential, *155,* 354–60
    ion distribution in, 354–55
    resting, 355, *356*
Membranous labyrinth, 448, *449, 450,*
    453, *455*
Membranous urethra, 807
Memory, 393
Memory cells, in immune response,
    640–41, *641*
Menarche, 861
Ménière's disease, 475
Meningeal artery, *599*
Meninges, 376–77, *376,* 376t
Meningitis, 377
Menisci, 261, *261,* 275
Menkes disease, 959t
Menopause, 863–64
Menstrual cycle, 861–62, *864,* 864t
Mental foramen, *203, 204, 211,* 212, 213t, **248**
Merocrine gland, 142, *143,* 144t
Mesencephalon (midbrain), 387, *387,* 388t,
    402, *402*
Mesenteric artery, **34,** 598t, *602,* **984**
    inferior, **34,** 597, *597, 605*
    superior, 596, *597, 605*
Mesenteric vein, **34,** 607, *608,* **984**
Mesentery, **33,** 685, *685, 686,* **984**
Mesoderm, *900*
Mesodermal cells, 896, *897*
Metabolic pathway, 104, *105*
Metabolism, 4, 704
    of carbohydrates, 108–13, *108, 109, 110,*
      *116,* 681t
    cellular, 102–4, *103. See also* Respiration,
      cellular
      energy for, 106
      regulation of, 104–5, *105*
    inborn errors of, 129–30, *130*
    of lipids, 113–14, *114, 116,* 681t, 707
    of proteins, 114, *115, 116,* 681t
    vitamins in, 105
    water of, 816, *817*
Metacarpal bones, 200, *201, 225, 226, 228,*
    229t, **341**
Metacarpophalangeal joints, 267t
Metaphase, *88, 89,* 90–91, *91,* 92t
Metaphase I, *935*
Metaphase II, 936, *936*
Metarterioles, 577, *577*
Metatarsal bones, 200, *201,* 232, *234,* 235,
    235t, **342**
Metatarsophalangeal joint, 267t
Metencephalon, 387, *387,* 388t
Methionine, 121t, 723, *723*
Metric weights and measures, 987–88
Microcytosis, 547
Microfilaments, 75–77, *76, 77,* 79t
Microglia, 351, *352,* 353t
Micrometer, 60
Micronutrients, 704
Microtubules, *62,* 75–77, *76, 77,* 79t
    in prophase, 90, *90*
Microvilli, *62*
    intestinal, 686, *687*
    of simple columnar cells, 138, *138*
Micturition, 805, 806, 806t
Midbrain (mesencephalon), 387, *387,* 388t,
    402, *402,* 405t, **985**
    functions of, 405t

Middle ear, 13, *13*, 446–48
Migraine, 346, 438
Milk, human, *71*, 876–77, *876*, 877t
Minerals, in nutrition, 722–27, *723*, 725t, *726*, 727t
Minute ventilation, 758
Mitochondria, *62*, 72, *72*, 79t
abnormalities of, 74
maternal inheritance of, 73
Mitosis, 88–92, *88, 89, 90, 91, 92*, 92t
Mitral (bicuspid) valve, 558, *559, 561*, 561t
Mitral valve prolapse, 558
Mixed nerves, 407
Modiolus, *449*
Molars, 662, *662*
Mole, 181
Molecular formula, 42
Molecule, 8, *9*, 9t, 38t, 42, 44
Monoamine oxidase, 425
Monoamine oxidase inhibitors, 364t
Monoamines, as neurotransmitters, 361
Monoclonal antibodies, 638, *639*
Monocytes, 529–30, *530, 534*, 534t
Monosaccharides, *689*, 705, *705*
in cells, 47, *48, 49*
intestinal absorption of, 689, *689*, 691t
Monosomy, 949–50, *954*
Mons pubis, **29**
Mortality, 923–24, 923t
Morula, 867, *867*, 895, *895*
Motor end plate, 286, *286*
Motor nerves, 407
Motor neuron, *287*, 350, 351t
Motor unit, 286–87, *287*
recruitment of, 295
Mouth, *4*, 13, *13*, 657, 658–64, *660*, **965**. *See also specific components*
Movement, body, 196–97, *196, 197*
Mucin, 144
Mucosa, of alimentary canal, 656, *659*, 659t
Mucous cells
of gastric glands, 671, *671*
of salivary glands, 665
Mucous membrane, 161
of nasal cavity, 740, *741*
Mucus, 138, 144, 161, 665, 672, 672t
Mucus-secreting glands, 688
Multiple endocrine neoplasia, 480, 959t
Multiple sclerosis, 349
Multipolar neuron, 350, *350*, 351t
Murmur, 568
Muscarinic receptors, 424, *426*
Muscle(s), 136t, 153–54, 299t. *See also specific muscles*
cardiac, 154, *154*, 155t, 298–99, *299*, 299t. *See also* Heart
compartment for, 283
as gene therapy target, 956–57
involuntary, 153
skeletal, 153, *153*, 155t, 282–87, 299t. *See also* Skeletal muscle(s)
smooth, 153, *153*, 155t, 297–98, 299t
voluntary, 153
Muscle impulse, in muscle contraction, 289
Muscle spindle, 438, *439*, 439t
Muscle strain, 286
Muscular dystrophy, 331
Duchenne, 946, 959t
Muscular layer, of alimentary canal, 656, *659*, 659t
Muscular ridges, *560*
Muscular system, *15*, 16, 19t. *See also* Muscle(s); Skeletal muscle(s)
system innerconnections of, 332
Musculocutaneous nerve, **32**, 415, *416, 417*

Mutations, 126, *127*
effects of, 129–30, 129t, *131*
Myalgia, 331
Myasthenia gravis, 289, 331, 363t, 364, 647t
Myelencephalon, 387, *387*, 388t
Myelin, *347*, 348, *348*
Myelin sheath, 348, *348*
Myeloid leukemia, 533
Myocardial infarction, 563
Myocardium, 557, *557*, 557t
Myocytes, *94*
Myofascial syndrome, 438
Myofibrils, 75–76, *184*, 284
Myoglobulin, 292
Myogram, 294, 295, *295*
Myokymia, 331
Myology, 331
Myoma, 331
Myometrium, 856
Myopathy, 331
Myopia, *471*
Myosin, 285
defect in, 612
in muscle contraction, 287–89
Myositis, 331
Myositis ossificans, 324
Myotomy, 331
Myotonia, 331

# N

NADH, 112, *112*
Nails, 169, *169*
color of, 170
Narcolepsy, 404t
Nasal bones, 203, 204, 207, 209, 210, 212t, **241, 242, 243, 338**
Nasal cavity, 13, *13*, 740, *740, 741*, 752t, **964**
lateral wall of, *207*
Nasal conchae, 740, *741*
inferior, 203, 207, 210, 211, 212t, **243, 965**
middle, 203, 206, *207*, 211, **243, 249**
superior, 206, *207*
Nasal continuous positive airway pressure, 762
Nasal septum, 740, **969**
Nasolacrimal duct, 459
Nasopharynx, *661*, 666
Natural killer cells, 634
Nausea, 675
Navicular, *234*
Nearsightedness, 470, *471*
Neck
arteries of, 597–600, *598, 599*
veins of, 604, *606*
Necrosis, 78, 164
Negative feedback
in homeostasis, *7*, 8
in hormone regulation, 488, *489*, 490
*Neisseria gonorrhoeae* infection, 882t
Neon, 39t
Neonate, 914–15, 917–18, *917*, 921t
circulation of, 918, *918*, 918t
Nephron, 780, *782*, 785–89, 801t
blood supply of, 788–89, *790*
cortical, 788, *789*
countercurrent mechanisms of, 796–99, *798, 799*
juxtaglomerular apparatus of, 785, 788, *788*
juxtamedullary, 788, *789*
structure of, 785, *786, 787*
Nephron loop (loop of Henle), 785, *786*
Nephrotic syndrome, 795
Nerve(s), 344. *See also specific nerves*
classification of, 407–8

cranial, 407, 408–11. *See also* Cranial nerve(s)
spinal, 407, 412–17. *See also* Spinal nerves
structure of, 407, *407, 408*
Nerve cell. *See* Neuron(s)
Nerve fibers, 344, *367*
myelinated, 348–49, *348, 349*, 359, *359*
regeneration of, 353, *354*
unmyelinated, 348, 349, *349*
Nerve impulse, 16, 357–58, *358*, 358t
conduction of, 359–60, *359*
convergence of, 365, *367*
divergence of, 366, *367*
processing of, 365–66
Nervous system, 16, *16*, 19t, 344–70, 407t. *See also* Brain; Neuron(s); Peripheral nervous system; Spinal cord
autonomic, 407, 418–26. *See also* Autonomic nervous system
cells of. *See* Neuroglia; Neuron(s)
central. *See* Brain; Spinal cord
functions of, 344–45
peripheral, 344, *345*, 407–17. *See also specific nerves*
system innerconnections of, 370
Nervous tissue, 136t, 154, *155*, 155t
Neural tube defect, 388t, 720
Neuralgia, 427, 475
Neurilemma, 348, *348*
Neurilemmal sheath, 348, *348*
Neuritis, 427, 475
Neuroglia, 154, *155*, 344, 352, 353t
classification of, 351–52
implants of, 353
Neuroma, 354
Neuromuscular junction, 286, *286*, 287
Neuron(s), *94*, 154, *155*, 155t, 344, *345, 347*
accelerator, 350
action potentials of, 357–58, *358*
all-or-none response of, 359
association, 350, 351t
bipolar, 350, *350*, 351t
of retina, 465, *466*
classification of, 350, *350*, 351t
depolarization of, 355, *357*
facilitation state of, 365
as gene therapy target, 957
hyperpolarization of, 355
inhibitory, 350
local potential changes of, 355–56
membrane potential of, 354–55, *355*
memory formation by, 393
motor, 350, 351t
multipolar, 350, *350*, 351t
pH sensitivity of, 826
pools of, 365
postsynaptic, 360
presynaptic, 360, *362*
refractory period of, 359
repolarization of, *357*
resting potential of, 355, *356, 357*
retinal, 465, *466*
sensory, 350, 351t
structure of, 345–49
summated potentials of, 356
synapse of, 360–65, *361, 362, 363*, 364t
threshold potential of, 355–56
unipolar, 350, *350*, 351t
Neuronal pools, 365
Neuropeptides, 365
Neurosecretory cells, 489
Neurotransmitters, 286, 344, 361, *362, 363*, 364t, 365t
in muscle contraction, 289–90
reuptake of, 363

Neutrons, 37–38, *38*, 38t
Neutrophils, 527–28, *534*, 534t
  origin of, *523*
Neutrophilia, 547
Niacin (nicotinic acid), 718, *718*, 721t
Niacinamide, 718, *718*
Nicotine, 364t
Nicotine addiction, 368–69
Nicotinic receptors, 424, *426*
Niemann-Pick disease, 613
Night blindness, 715, 733
Nipple, **29**
Nissl bodies, of neuron, 345, *347*
Nitric oxide, 47
Nitrogen, 39t
Nitrogen balance, 709
Nociceptors, 433, 435, 439t
Nodes of Ranvier, *347*, 348, *348*, 359, *359*
Nondisjunction, 950, *954*
Nonprotein nitrogenous substances, in plasma, 536–37
Nonrespiratory air movements, 759, 760t
Norepinephrine, 423–25, *424*, 483t, 502–3, *502*, 503t
  in smooth muscle contraction, 298
  structure of, *482*
Normal range, *7, 8*
Nose, 740, *740*, 752t
Nuclear envelope, *62, 78*, 79t
Nuclear pore, 77, *78*
Nucleases, pancreatic, 677, 689t
Nuclei, brain stem, 401–3
Nucleic acids, 117–26
  of cells, 52–53, *53*, 53t
Nucleolus, *62*, 77, *78*, 79t, *347*
Nucleotides, 52, *53*
  structure of, *117*
Nucleus
  atomic, 37–38, *38*
  cell, 60, *62*, 77–78, *78*
Nucleus cuneatus, 402
Nucleus gracilis, 402
Nutrients, essential, 704
Nutrition, 704–33
  for athlete, 732
  carbohydrates in, 705–6, *705*, 709t
  food pyramid in, 728–29, *728*
  growth hormone release and, 492
  lipids in, 706–7, *707*, 709t
  minerals in, 722–27, *723*, 725t, *726*, 727t
  package information in, *729*
  proteins in, 707–9, *708*, 708t, 709t
  vitamins in, 713–21. *See also* Vitamins
Nyctalopia, 715, 733
Nystagmus, 469, 475

## O

Obesity, 113, 711–12, *711, 713*
  treatment for, 712–13
Oblique muscles
  external, **29, 30, 31**
  of eye, 460, *460*, 461t
  internal, **30, 31**
Oblique popliteal ligament, 273, *274*
Oblique section, *22*
Obstructive sleep apnea syndrome, 404t
Obturator artery, *602*
Obturator foramen, **35**, 228, *229, 230*
Obturator nerve, *416*, 417, *418*
Occipital artery, *598*, 599
Occipital bone, *199, 204*, 205, 205t, *208, 210*, **242, 245, 246, 247, 249, 251, 252, 338**

Occipital condyle, *204*, 205, **245, 246, 247, 249, 251**
Occipital lobe, *389*, 389–90, 392t, **969, 985**
  association areas of, 392
Occipitalis muscle, *301, 302*, 303, **973**
Oculomotor nerve (III), 408–9, *409*, 412t
Olecranon fossa, 224, *225, 226*
Olecranon process, 225, *225, 227*, **339, 340**
Olestra, 707
Olfactory bulb, 408, *409*, 442
Olfactory code, 442
Olfactory foramina, 206, *208*
Olfactory nerve (I), 408, 412t
Olfactory organs, 440–42, *441*
Olfactory receptor, 408, 440, *441*
Olfactory tract, 408, *409*, 442
Oligodendrocytes, 351, *352*, 353t
Olive, 402
Omentum, **31**, 686, *686*, **964, 983**
Oncogenes, 97, *97*
Oocytes, 852–53, *853*
  maturation of, 853–54, *854, 856*
  transport of, 864–65, *865*
Oogenesis, 852–53, *853*
Oophorectomy, 883
Oophoritis, 883
Ophthalmic artery, 599
Ophthalmic vein, *606*
Opiates, endogenous, 366
Opsin, 470, *471*
Optic canal, *203, 208*, 213t, **253**
Optic chiasma, 399, *399*, 474, *474*
Optic disk, *462*, 465, *467*
Optic nerve (II), 408, *409*, 412t, 461, *462, 467*, 473–75, *474*, **969**
Optic radiations, 474, *474*
Optic tract, 399, *399, 409*, 474, *474*
Oral cavity, *4*, 13, *13*, 657, 658–64, *660*, **965**. *See also specific components*
Oral contraceptives, 879t, 880, *880*
Orbicularis oculi muscle, *301, 302*, 303, 303t, 458, *459*, 461t, **973**
Orbicularis oris muscle, *301, 302*, 303, 303t
Orbit, 13, *13, 203*, **249**, 470
Orbital fissures, *203*, 213t, **243, 250, 253**
Orchiectomy, 883
Organ, 8, *9*, 9t, 161
Organ of Corti, 450, *452*
Organ systems, 8, *9*, 9t, 14–18, 19t, *20*
Organelles, 8, *9*, 9t
  cytoplasmic, 60, *62*, 68–77. *See also specific organelles*
Organism, 3–4
  organizational levels of, 8–9, *9*, 9t
  requirements of, 5, 6t
Organization
  body, 10–18
  levels of, 8–9, *9*, 9t
Orgasm
  female, 859–60
  male, 845–47
Oropharynx, *661*, 666
Orthopedics, 235
Orthostatic intolerance, 588
Osgood-Schlatter disease, 232
Osmolality
  test for, 990
  urinary, 992
Osmoreceptors, 494
  in water regulation, 816, 818
Osmosis, 80–81, *81*, 87t
Osmotic diuresis, 794
Osmotic pressure, 81
  of blood, 535
  in capillary exchange, 579, *580*

Osseous labyrinth, 448, *449*
Ossification, of bones, 188–90, *189, 190*, 190t, 192t
Ostealgia, 235
Ostectomy, 235
Osteitis, 235
Osteoarthritis, 276t, 277, *277*
  hereditary, 147t
Osteoblasts, 188–89
Osteochondritis, 235
Osteoclasts, 191, *191*
Osteocytes, 189, *189*
Osteogenesis, 235
Osteogenesis imperfecta, 147t, 237, 959t
Osteoma, 237
Osteomalacia, 192, 237, 716
Osteomyelitis, 237
Osteon, 151, 187, *187, 188*
Osteonecrosis, 237
Osteonic (Haversian) canals, 151, *151*, 187, *188*
Osteopathology, 237
Osteopenia, 237
Osteoporosis, 198, 237
Osteotomy, 237
Otitis media, 205, 448, 475
Otoliths, *456*
Otosclerosis, 454, 475
Oval window, *446*, 447, *447*
Ovarian follicles
  degeneration of, 855, *855*
  maturation of, 853–54, *854, 856*
Ovarian ligament, 852, *852*
Ovariectomy, 883
Ovaries, **33, 34**, *481*, 510, 850–55, *851, 852*, 860t
  descent of, 852
  follicles of, 852
  ligaments of, 852, *852*
  structure of, 852
Ovulation, 855, *855, 856*
Oxidation, 106, 112, *112*
  beta, 113, *114*
Oxygen, 5, 6t
  atomic structure of, 39t
  blood transport of, 767–69, *768, 770*, 772t
  in cells, 47, 48t
  tissue, 526
Oxygen debt, 292, *293*
Oxygen saturation, 990
Oxygenator, intravascular, 750
Oxyhemoglobin, 522, 767, 768, *768, 770*, 771
Oxytocin, 483t, 489, 494, *494*, 495, 496t
  in birth process, 870
  in smooth muscle contraction, 298
  structure of, *482*

## P

P wave, 571, *572*
Pacemaker, 568
  artificial, *575*
Pacinian corpuscles, 167, 434, *434, 435*, 439t
Packed cell volume, 521
Pain
  acute, 436
  cancer, 438
  chronic, 436, 438
  inhibition of, 437
  referred, 435–36, *436*
  visceral, 435–36, *436*
  withdrawal reflex in, *384*, 385
Pain fibers, 436–37, *437*
Pain receptors, 433, 439t
Pain sense, 435–37, *436*

Palate, *660,* 660–61, *661,* 663t
  cleft, 209
  hard, 208
Palatine bone, *204, 207,* 209, *210,* 212t, **245, 247**
Palatine foramen, *204,* 213t, **245, 247**
Palatine process, 208, *210, 211,* **245**
Palatine suture, *204,* **245, 247**
Palmaris longus muscle, 313, *314,* 317t
  tendon of, **341**
Palpitation, 610
Pancreas, **34,** *481,* 507–8, *507,* 509t, *657, 676–77,* **971, 972**
  acinar cells of, 676, *676*
  hormones of, 483t
  secretions from, 676–77, *677*
  structure of, 676, *676*
Pancreatic juice, 676–77, *677*
Pancreatitis, 677
Pancytopenia, 547
Pantothenic acid (vitamin B$_5$), 718–19, 721t
Pap smear test, 858
Papillae
  of dermis, *162,* 166
  of tongue, 442, *443,* 660, *661*
Papillary muscles, 558, *559, 560*
Papillotomy, endoscopic, 684
Para-aminohippuric acid clearance, 802
Paralysis, 331
  flaccid, 382
  sleep, 404t
  spastic, 382
Parasomnias, 404t
Parathyroid glands, *481,* 499–501, *499*
  disorders of, 501t
Parathyroid hormone, 483t, 499–500, *500, 501*
  in calcium homeostasis, 500, *500, 501*
  in electrolyte balance, 822, *822*
  structure of, *482*
Parathyroidectomy, 513
Paravertebral ganglia, 419–21, *420*
Paresis, 331
Parietal bone, *199, 203, 204,* 205, 205t, *208, 210,* **241, 242, 251, 252, 338**
Parietal cells, of gastric glands, 671, *671*
Parietal lobe, 389, *389,* 392t
  association areas of, 392
Parietal peritoneum, *15*
Parietal pleura, 13, *14,* 751
Parietal suture, *258*
Parkinson's disease, 363t, 396
Parotid duct, 665, *665*
Parotid gland, 665, *665, 666,* 666t, **973**
Partial thromboplastin time, 539
  plasma, 991
Parturition, 869–70, *870,* 870t, *871*
Patella, 200, *201,* 231, *232,* 235t, **341, 978, 979**
Patellar dislocation, 231
Patellar ligament, 273, *274,* **341, 979**
Patellar tendon reflex (knee-jerk reflex), 383–85, *384*
Patent ductus arteriosus, 917
Pathogen, 630
  chemical barriers to, 631
  immune response to, 632–47. *See also* Immunity
  inflammatory response to, 631
  mechanical barriers to, 631
  phagocytosis of, 632
  species resistance to, 630
Pattern baldness, 947, *948,* 959t
Pectin, 705
Pectineus muscle, **34,** *321,* 322, 325t
Pectoral girdle, 200, 222–23, *223*
  muscles of, *307,* 308–10, *308, 309,* **975, 976**

Pectoral nerves, 416
Pectoralis major muscle **29, 30,** 301, *301, 308, 310,* 311–12, 312t, **339, 340, 970, 974**
Pectoralis minor muscle, **30,** *308,* 309, 310t
Pedicels, of podocytes, 785, *786*
Pediculosis, 181
Peduncles, cerebellar, 404–5, *404*
Pellagra, 718
Pelvic brim, 228, *230*
Pelvic cavity, *12,* 13
Pelvic diaphragm, 319, 319t, *320*
Pelvic girdle, 200, 227–31, *229*
  radiograph of, *229*
Pelvic inflammatory disease, 882
Pelvic outlet, muscles of, 319–20, 319t, *320*
Pelvis, 200, 228, *230*
  arteries to, 602–4, *602*
  female, 231, 231t
  greater, 231
  lesser, 231
  lymph nodes of, *627,* 628
  male, 231, 231t
  veins from, 608–9, *610*
Penicillin, 105
Penis, **31, 32,** *837,* 849t
  erection of, 845–47, *848*
  structure of, 845, *845*
Pepsin, 672, 672t, 689t
Pepsinogen, 672, 672t
Peptidase, 688, 689t, *690*
Peptide bond, 102, *103*
Peptide hormones, 481, 483t
Peptide neurotransmitters, 361
Percutaneous transluminal coronary angioplasty, 614
Perforating canals, of bone, 187, *188*
Perforin, 634
Pericardial artery, 598t
Pericardial cavity, *12,* 13–14, **34,** 556, *556,* **970**
Pericardial membranes, 13
Pericardial sac, **31**
Pericardiectomy, 610
Pericarditis, 556
Pericardium, 555–56, *556,* **971, 983**
  fibrous, 14, *14*
  parietal, 13, *14*
  visceral, 13, *14*
Perichondrium, 150
Perilymph, of inner ear, 448, *449*
Perimetrium, 856, 858
Perimysium, 282, *283*
Perinatology, 925
Perineal artery, *602*
Perineurium, 407, *407, 408*
Periodic table, *986*
Periodontal ligament, 258, *259,* 663
Periosteum, 186, *187*
Peripheral nerve, 407–8, *407, 408*
Peripheral nervous system, 344, *345,* 407–17.
  *See also specific nerves*
  autonomic, 407, 418–26. *See also* Autonomic nervous system
  cranial nerves of, 407, 408–11. *See also* Cranial nerve(s)
  somatic, 407
  spinal nerves of, 407, 412–17. *See also* Spinal nerves
Peripheral resistance, blood pressure and, 584–85, *584, 585,* 586–89, *589*
Peristalsis, 298
  of alimentary canal, 657, *660*
  of small intestine, 691–92
  of ureter, 803
Peritoneal cavity, *15*
Peritoneum, *15,* 656

Peritonitis, 692
Pernicious anemia, 527, 529t, 647t, 720
Peroneal artery, *603, 604, 605*
Peroneal nerves, 417
Peroneal retinacula, *328, 329,* 331
Peroneal vein, *610, 611*
Peroneus brevis muscle, 327, *328, 329,* **979**
Peroneus longus muscle, *301,* 327, *328, 329,* 330t, 331, **341, 978, 979**
Peroneus tertius muscle, 326, *328,* 330t
Peroxisomes, 73, 79t
  abnormalities of, 74–75
Perpendicular plate, 206, *207, 211,* **249**
Persistent vegetative state, 375, 403
Peyer's patches, 627
pH, 45–46, *46,* 46t. *See also* Acid-base balance
  test for, 990
  urinary, 992
Phagocytosis, 84, *84, 85,* 87t, 632, 632t
Phalanges
  of fingers, 200t, *201,* 225, 226, *228,* 229t, **341**
  of toes, 200t, *201, 232, 234,* 235, 235t, **342**
Pharyngitis, 699
Pharynx, *657,* 666–67, *667,* 740, 741–42, *741,* 752t, **970**
  muscles of, 666–67, *667*
Phenotype, 940
  gender effects on, 947–48, *948*
  genetic heterogeneity and, 943
  variable expressiveness of, 942–43
Phenylalanine, 121t
Phenylketonuria, *130,* 131, 959t
Phenylpyruvic acid, 992
Pheochromocytoma, 513
Phlebitis, 582, 610
Phlebotomy, 610
Phosphate
  cellular, 48t
  urinary, 791t
Phosphate buffer system, 825, 826t
Phosphaturic factor, 500
Phosphodiesterase, 485
Phospholipids, 49, *50,* 51t, 535
  of cell membrane, 64–65, *64, 65*
Phosphorus
  in nutrition, 722, 725t
  serum, 990
Phosphorylation, of adenosine diphosphate, 106, 108, *108,* 112, *112*
Photoreceptors, 433
Phrenic artery, 596, *597,* 598t, 602
Phrenic nerve, 415, *416,* **982**
Phrenic vein, 607
Physical stress, 511
Physician-assisted suicide, 922
Physiologic dead space, 758
Physiology, 3–4
Pia mater, *376,* 377
Pica, 726
Pineal gland, 399, *481,* 510
Pinkeye, 458
Pinocytosis, 83–84, *84,* 87t
Pisiform, *228*
Pitch, of voice, 743
Pituitary dwarfism, 959t
Pituitary gland, *481,* 488–95, *488*
  anterior lobe (adenohypophysis) of, 488–89, *488,* 490–93, *491, 492,* 496t
  hormones of, 483t
  intermediate lobe (pars intermedia) of, 489
  posterior lobe (neurohypophysis) of, 399, 488, *488,* 494–95, *495,* 496t
  regulation of, 489, *490*
Pivot joint, 262, *263,* 264t

Placenta, 510, 867, 898, *904, 905*
Placental membrane, 897, *904*
Plantar artery, *603*
Plantar vein, *610*
Plaque, in atherosclerosis, 582, *583*
Plasma, 521, *521, 522,* 532, *534,* 535–37
 dried, 547
 electrolytes of, 537
 lipids of, 535–36, 536t
 nonprotein nitrogenous substances of, 536–37
Plasma cell, origin of, *523*
Plasma proteins, 532, 535
Plasma thromboplastin antecedent (factor XI), 541t
Plasma thromboplastin component (factor IX), 541t
Plasminogen (profibrinolysin), 538
Platelet count, 532, 990
Platelet plug, 537, *537,* 538t
Platelet transfusion, 542
Platelet-derived growth factor, 538
Platelets, 152, *152,* 532, 534t
 in blood smear, *521*
 origin of, *523*
Platysma muscle, *302,* 303, 303t
Pleiotropy, 943
Pleura, 751
Pleural cavity, *12, 13, 14,* **34,** 751
Pleural membranes, 13
Pleurisy, 774
Plexuses
 brachial, 415–17, *416*
 cervical, 415, *416*
 lumbosacral, *416,* 417, *418*
Plicae circulares, 686, 688, *688*
Pneumoconiosis, 774
Pneumonia, 768, 774
Pneumotaxic area, in ventilation, 762, *762*
Pneumothorax, 756, 774
Podocytes, 785, *786*
Poikilocytosis, *549*
Polar body, 852–53, *854*
Polar body biopsy, 853
Poliomyelitis, 287, 761
Polycystic kidney disease, 780, 959t
Polydactyly, 226, *228,* 959t
Polygenic traits, 943–45, 943t, *944*
Polymerase chain reaction, 128
Polymorphonuclear leukocytes, 527–28
Polynucleotide, 52, *53*
Polypeptide, 102
Polyphagia, 513, 733
Polyploidy, 948–49
Polysaccharides, in cells, 47, *49*
Pons, 387, *387, 399,* 402, *404,* 405t, **985**
 functions of, 405t
Popliteal artery, *603,* 604, *605*
Popliteal ligament, 273
Popliteal vein, 608, *610, 611*
Pore, 171, *171*
Porphyria, 529, 943, 959t
Portal vein
 hepatic, 607, *608, 609*
 hypophyseal, 489, *490*
Positive feedback, in homeostasis, 8
Positron emission tomography, 54–55
Posterior cruciate ligament, *274,* 275
Posterior, definition of, 19
Posterior horn, of spinal cord, 378, *379*
Posterior median sulcus, 378, *379*
Postganglionic fibers, sympathetic, 419, *420, 421, 421*
Postnatal development, 913–21, 921t
 of adolescents, 919

of adults, 919, *920*
of children, 918–19
of infant, 918
of neonate, 914–15, 917–18, *917. See also* Neonate
Postnatal period, 892
Postpartum, definition of, 925
Postsynaptic neuron, 360
Potassium
 cardiac cycle and, 572, 573
 cellular, 48t
 imbalances of, 823
 in nutrition, 723, 725t
 renal tubule, 800, *800*
 serum, 990
 urinary, 791t
Potassium channels, 66–67
Potential difference, 355
P-Q interval, 571, *572*
P-R interval, 571, *572*
Prader-Willi syndrome, 947–48
Precapillary sphincters, 579
Preganglionic fibers
 parasympathetic, 423, *423*
 sympathetic, 419, *420,* 421, *421*
Pregnancy, 864–70
 ectopic, 868
 embryo in, 866–67, *867*
 fertilization in, 865–66, *866*
 hormonal changes of, 868–69, *868,* 869t
 tubal, 868
Preimplantation genetic diagnosis, 894
Premature beat, 574
Prematurity, 910
Premenstrual syndrome, 863
Prenatal period, 892, 893–913. *See also* Embryo; Fetus
 cleavage in, 893, *893, 895*
Prepatellar bursa, 275
Prepuce, *837,* 845, *845*
Presbycusis, 454
Pressure, 5, 6t
Pressure sense, 434, *434, 435*
Pressure ulcers, 164
Presynaptic neuron, 360
Prime mover, 300
Proaccelerin (factor V), 541t
Process, of bone, 186, 202t
Products, chemical, 44
Profunda femoral artery, *603,* 604
Progeria, 959t
Progesterone, 860–61
 in female reproductive function, *863*
 in pregnancy, *868,* 869
Projection, 433
Prolactin, 483t, 489, *491, 492,* 496t
Prolactin release-inhibiting hormone, 483t, *491,* 492
Prolactin-releasing hormone, 483t, *491,* 492
Proline, 121t
Pronation, 265, *266*
Pronator quadratus muscle, 313, *314,* 316t
Pronator teres muscle, 313, *314,* 316t
Prophase, *88, 89,* 90, *90,* 92t
Prophase I, 934–35, *935*
Prophase II, 936, *936*
Proprioceptors, 433
Prosencephalon (forebrain), 387, *387,* 388t
Prostacyclin, 542, 542t
Prostaglandin E$_2$, *482*
Prostaglandins, 481, 487
Prostate gland, *837,* 843, *843,* 849t, **972**
 enlargement of, 844, 844t
Prostatectomy, 883
Prostatitis, 883

Proteins
 of cell membrane, 64–65, *64, 65*
 in cells, 51–52, *51, 52,* 53t
 chaperone, 123
 complete, 708
 conformation of, 51–52, *52*
 denaturation of, 52
 digestion of, 689, *690*
 dynamic equilibrium of, 709
 hydrogen bonds of, 52, *52*
 incomplete, 708
 integral, 64–65
 metabolism of, 114, *115, 116,* 681t
 in nutrition, 707–9, *708,* 708t, 709t
 partially complete, 708
 peripheral, 65
 pinocytosis of, 795
 plasma, *534,* 535, 535t
 primary structure of, 51, *52*
 requirements for, 709
 secondary structure of, 51, *52*
 serum, 990
 sources of, 708
 synthesis of, 122–25
 tertiary structure of, 51, *52*
Protein buffer system, 825–26, 826t
Protein hormones, 481, 483t
Protein kinases, 484
Proteinuria, 733, 828
 in nephrotic syndrome, 795
Prothrombin (factor II), 538, 541t
Prothrombin activator, 538
Prothrombin conversion accelerator (factor VII), 541t
Prothrombin time, 539, 991
Protons, 37–38, *38,* 38t
Protraction, 265, *266*
Provitamin, 733
Proximal, definition of, 19
Prozac, 364t
Pruritus, 181
Psoas muscle, *321,* 322, 325t
Psoriasis, 163
Psychological stress, 511–12
PTC-tasters, 444
Pterygium, 475
Pterygoid muscles, *302,* 304, 304t
Pubic arch, 228, *229, 230*
Pubis, 228, *229*
Pubofemoral ligament, 271, *272*
Pudendal artery, 602, *602, 603,* 604
Pudendal nerve, 417, *418*
Pudendal vein, 608
Pulled hamstrings, 323
Pulmonary. *See also* Lung(s)
Pulmonary arteries, **33,** 558, *559, 564, 596,* **966**
Pulmonary circuit, of blood flow, 592–94, *595*
Pulmonary edema, 594
Pulmonary embolism, 582
Pulmonary trunk, **32, 33,** 558, *559, 564, 596*
Pulmonary valve, 558, *559, 560, 561,* 561t
Pulmonary veins, **33,** 558, *559, 564,* 594, *596*
Pulmonic sound, 568
Pulse, 583, *584*
Pulse pressure, 586
Pupil, *462, 464*
 constriction of, 464
 dilation of, 464–65
Purkinje fibers, 557, 570, *570*
Purpura, *549*
Pus, 532
Pustule, 181
Putamen, 394, *394*
Pyloric sphincter, 670, *670*

Pylorospasm, 699
Pyorrhea, 699
Pyramids, renal, 780, *782*
Pyramidal cells, of cerebral cortex, 390
Pyridostigmine bromide, 289
Pyridoxal, 719, *719*
Pyridoxamine, 719, *719*
Pyridoxine, 719, *719*
Pyrogen, endogenous, 631

## Q

Q wave, 571, *572*
QRS complex, 571, *572*
Quadrants, abdominal, 22, *22*
Quadriceps muscle, 324
Quinlan, Karen Ann, 375

## R

R wave, 571, *572*
Radial artery, 601, *601, 605*
Radial collateral ligament, 270, *270*
Radial muscles, of iris, 464–65, *466*
Radial nerve, 416, *416, 417*
Radial notch, 227
Radial pulse, 583, *584*
Radial tuberosity, 225, *227*
Radial vein, 606, *606, 611*
Radiation
    atomic, 39
    in body temperature regulation, 173
    ionizing, 43
Radioactive isotopes, 39
    medical applications of, 40–41
Radiogastrogram, 657
Radioulnar joint, 267t
Radius, 200, *201,* 225, *225, 227,* 229t
    styloid process of, **340, 341**
Ramus, of bone, 202t
Reactants, chemical, 44
Receptors
    adrenergic, 424
    of cell membrane, 64
    cholinergic, 424, *426*
    muscarinic, 424, *426*
    nicotinic, 424, *426*
    olfactory, 440
    pain, 435
    retinal, 465, *466*
    sensory, 344, 432–33, 433t, 439t
    stretch, 438–39
    taste, 442–44, *444*
Receptor-mediated endocytosis, 84–85, *86,* 87t
Recruitment, of motor units, 295
Rectal artery, *602,* 603
Rectal cancer, 697
Rectal vein, 608
Rectouterine pouch, *851*
Rectum, **33, 35,** *657,* 692, *693, 694,* **968, 972**
Rectus abdominis muscle, **29, 30,** *301,* 308, 317t, *318,* 319, **340, 968, 971, 972, 974**
Rectus femoris muscle, **30, 32, 33, 34,** *301, 321, 323,* 326t, **341, 972, 977**
Rectus muscles, **969**
    of eye, *460,* 460, 461t
Recurrent artery, *601*
Red blood cells, 152, *152,* 522, *522, 524,* 534t
    antigens of, 542, 544–47, 544t, *545, 546*
    in blood smear, *521*
    destruction of, 524–25, 525t
    dietary effects on, 526–27, *528*
    folic acid effects on, 526–27, *528,* 528t
    iron effects on, 527, 528t

life cycle of, *528*
life span of, 526
light microscopy of, *529*
microscopic appearance of, 63, *63*
normal, *529, 533*
origin of, *523*
packed, 547
production of, 526, 527t
shape of, 524, *524*
vitamin B$_{12}$ effects on, 526–27, *528,* 528t
Red blood cell count, 524, 991
Red blood cells, *94*
Red cell distribution width, 991
Red nucleus, 402
Red pulp, of spleen, 629, *629*
Red-green color blindness, 946
Reduction, *110,* 112
Reeve, Christopher, 382
Referred pain, 435–36, *436*
Reflex arc, 383–86, *383,* 386t
Reflex center, 383
    cardiac, 572, 587
Reflexes, 383–85
    abdominal, 386
    ankle-jerk, 386
    Babinski, 386
    biceps-jerk, 386
    cremasteric, 386
    crossed extensor, 385, *386*
    defecation, 695
    enterogastric reflex, 674, *675*
    Hering-Breuer, 764, *764*
    inflation, 764, *764*
    joint, breathing rate and, 765
    knee-jerk (patellar tendon), 383–85, *384*
    micturition, 805, 806t
    plantar, 386
    spinal, 383, *383*
    stretch, 438
    swallowing, *668*
    triceps-jerk, 386
    tympanic, 447–48
    withdrawal, *384,* 385
Refractory period, of nerve cells, 359
REM-sleep behavior disorder, 404t
Renal. *See also* Kidney(s)
Renal artery, 597, *597,* 598t, *605,* 782–84, *784*
Renal clearance, 802
Renal corpuscle, *782, 785*
Renal cortex, 780, *782, 786, 787*
Renal failure, 781, 783
Renal medulla, 780, *782, 786, 787*
Renal pelvis, 780, *782*
Renal pyramids, 780, *782*
Renal sinus, 780
Renal tubules, *782, 785, 785, 786,* 801t
    glucose reabsorption by, 794
    peritubular capillary system of, *786,* 788–89, *790*
    reabsorption by, 794–96, *794, 796*
    secretions into, 800–801, *800*
    sodium reabsorption by, 795–96
    water reabsorption by, 795–96, *796*
Renal vein, 607, *611, 784*
Renin, 504, *504,* 793–94, *793*
Renin-angiotensin system, 503–4, *504,* 793–94, *793*
Reproductive system, 18, *18,* 19t
    female, 850–60, *851. See also specific components*
        development of, *910*
        hormonal regulation of, 860–64, *861,* 862t, *863,* 864t
    male, 836–48, *837. See also specific components*

development of, *910*
    hormonal regulation of, 848–50, *849*
    system innerconnections with, 884
Reserpine, 364t
Residual volume, 757, *758,* 759t
Respiration. *See also* Ventilation
    cellular, *17,* 106–16, *108*
        aerobic, 106, *292*
        anaerobic (glycolysis), 106, *107,* 111, 292, *292*
        carbohydrate pathways of, 108–13, *108, 109, 110, 116*
        lipid pathways of, 113–14, *114, 116*
        protein pathways of, 114, *115, 116*
        regulation of, 115, *115*
        in skeletal muscle, 292, *292*
Respiratory capacities, 757–58, *758,* 759t
Respiratory center, 402, 760, *760,* 762
Respiratory cycle, 757, *758*
Respiratory distress syndrome, 755
Respiratory membrane
    alveolar, 765, *766, 767*
    diffusion through, 765–67, *767*
Respiratory system, *17,* 18, 19t, 738–74. *See also* Ventilation
    alveolar gas exchange of, 764–67, *766, 767*
    bronchial tree of, 746–50, *749, 750, 751,* 752t
    bronchus of, *740*
    cigarette smoking effects on, 742–43
    gas transport of, 767–72, *768, 769, 770, 771, 772,* 772t
    larynx of, *740, 741,* 742–44, *744, 745,* 752t
    lungs of, *740,* 750–51, *752,* 752t
    nasal cavity of, *740, 740,* 752t
    nose of, *740, 740,* 752t
    pharynx of, *740,* 741–42, *741,* 752t
    sinuses of, 741, *744,* 752t
    system innerconnections of, 773
    trachea of, *740,* 744–45, *746, 747,* 752t
    upper, *740, 741*
Respiratory volumes, 757–58, *758,* 759t
Resting potential, 355, *356*
Restless-leg syndrome, 404t
Rete testis, 838
Reticular connective tissue, 148, *148,* 152t
Reticular fibers, 146, 147t
Reticular formation, *402,* 403
Reticulocytes, 526
Reticulospinal tracts (extrapyramidal tracts), 391
    anterior, *379,* 381, 381t
    lateral, *379,* 381, 381t
    medial, *379,* 381, 381t
Retina, *462, 463,* 465, 467t
    cells of, 465, *466,* 469
    detachment of, 470
    image formation on, 468, *469*
Retinal (retinene), 470, *714,* 714–15
Retinitis pigmentosa, 475, 959t
Retinoblastoma, 475
Retinol, *714,* 714–15
Retraction, 265, *266*
Retroperitoneal fat, **972**
Rh blood group, 546–47, *547*
Rheumatic fever, 647t
Rheumatic heart disease, 554, *554*
Rheumatoid arthritis, 276t, 277, 647t
Rhinitis, 774
Rhodopsin, 470–72, *473,* 714
Rhombencephalon (hindbrain), 387, *387,* 388t
Rhomboideus muscle, *301,* 307, 308, 310t, **975**
Rhythm method, 878, 879t
Rib, **35, 970, 971, 972, 980**
Riboflavin (vitamin B$_2$), 718, 721t

Ribonucleic acid (RNA), 53, 120–22, 124t
  messenger (mRNA), 120–21, *123, 124*
  ribosomal (rRNA), 123
  transfer (tRNA), 122–25, *125*
  translation of, 122, *124,* 126t
Ribose, *53,* 705
Ribosomes, *62,* 69, 79t
Ribs, 200, *201,* 220, *221*
  structure of, *222*
Rickets, 192, 716, *716,* 959t
Rigor mortis, 293
Rinne test, 454
Rods, *466,* 469
  rhodopsin of, 470–72, *473*
Rogaine, 170
Rolando, fissure of, 389, *389*
Romanovs, genetic studies of, 120
Rosacea, 164t
Roseola infantum, 164t
Rotation, 264, *266,* 267t
Rotator cuff, 268
Rotators, of arm, 307, *309, 311,* 312, 313
Rough endoplasmic reticulum, *62,* 69, *70,* 79t
Round ligament, **33,** 856
Round window, *446, 447,* 448, *449, 450*
Rubella, fetal, 908, 917
Rubrospinal tracts (extrapyramidal tracts),
  *379,* 381, 381t, 391
Rule of nines, for burns, 179, *179*

Saccharin, 705
Saccule, *449,* 453, *455, 457*
Sacral artery, 598t
  lateral, *602*
  middle, 597, *597, 602*
Sacral canal, 218, *229*
Sacral foramen, **35,** 218, *219*
Sacral hiatus, 218, *219,* 229
Sacral nerves, 412, *413*
Sacral plexus, *418*
Sacral promontory, 218, *229,* 230
Sacroiliac joint, 227, *229,* 267t
Sacrum, **35,** 200, *201,* 218, *219,* 219t, *229,*
  **340, 968**
Saddle joint, 262, *263,* 264t
Sagittal, definition of, 19
Sagittal plane, definition of, *21*
Sagittal suture, 205, **242**
Salivary glands, *657,* 665, *665, 666*
Salpingectomy, 883
Salt receptors, 444, *444*
Saltatory conduction, of nerve impulse,
  359, *359*
Salts, 45, 45t
  in body fluids, 47
Saphenous nerve, *418*
Saphenous vein, **29, 30, 33,** 608, *610, 611*
Sarcomere, *284,* 285, *285*
Sarcoplasmic reticulum, 285, *285*
Sartorius muscle, **29, 30, 31, 33, 34,** *301, 321,*
  *323, 323, 324,* 326t, *329,* **341, 972, 977**
Scabies, 181
Scala tympani, 448, *449, 450*
Scala vestibule, *449*
Scala vestibuli, 448, *450*
Scalp, **965, 969**
Scanning electron microscope, 63, *63*
Scaphoid, *228*
Scapula, 200, *201,* 222, *223, 224,* 229t
Scapular nerves, 416
Scarlet fever, 164t
Schizophrenia, 363t

Schwann cell, 348, *348*
  nucleus of, *347*
Sciatic nerve, *416,* 417, *418*
Sciatic notch, *230*
Sclera, 461, *462, 463, 465, 466,* 467t
Scleroderma, 647t
Scrotum, **32,** *837,* 845, 849t, **968**
Scurvy, 720
Seasonal affective disorder, 510
Sebaceous gland, 142, *144, 162,* 170–71,
  *171,* 173t
Seborrhea, 181
Sebum, 170, *171*
Second messengers, 483–87, *486*
Secretin, 677, *677,* 685t
Secretory vesicles, *62, 70,* 71–72, *71*
Selectins, 67, *69*
Selenium, in nutrition, 727, 727t
Sella turcica, 206, *206, 207, 208, 210,* **250,**
  **252, 253**
Semen, 844
  analysis of, 846–47, 847t
Semicircular canal, *446,* 448, *449,* 455
Semimembranosus muscle, *301,* 322–23, *324,*
  326t, *329,* **978, 979**
Seminal vesicle, *837,* 842, 849t
Seminiferous tubule, 838, *840*
  cross section of, *839*
Semispinalis capitis muscle, 305t, *306,* 307
Semitendinosus muscle, *301,* 322, *324,* 326t,
  *329,* **978**
  tendon of, **340**
Senescence, 921, 921t, 922t
Sensation, 433
  pain, 435–37, *436*
  pressure, 434, *434, 435*
  temperature, 434–35
  touch, 434, *434, 435*
Sensorineural deafness, 454
Sensory adaptation, 433–34
Sensory impulse, 433
Sensory nerve, 407
Sensory nerve fibers, 434, *434*
Sensory neuron, 350, 351t
Sensory receptors, 344, 432–33, 433t, 439t
Septicemia, *549*
Septum, nasal, 740, **969**
Serine, 121t
Serosa, of alimentary canal, 656, *659,* 659t
Serotonin
  in pain inhibition, 437
  platelet release of, 532
Serous cells, of salivary glands, 665
Serous fluid, 142, 144, 161
Serous membrane, 161
Serratus anterior muscle, **29, 30,** *301,* 308,
  *308, 310,* 310t, **339, 340, 974**
Serum, 538
Sesamoid bones, 186
Set point, in homeostasis, 6, *7*
Severe combined immune deficiency, 642
Sex chromosomes, 936, *937,* 945, *946, 947*
Sex-influenced traits, 947–48
Sex-limited traits, 947
Sex-linked traits, 945–46, *947*
Sexually transmitted diseases, 882–83, 882t
Shin splints, 331
Shingles, 164t
Shock, renal failure in, 792
Shoulder
  arteries to, 600–601, *601*
  muscles of, *307,* 308–9, *309*
  veins from, *60,* 606
Shunt, arteriovenous, 577, *577*
Siamese twins, 916

Sickle cell disease, 522, 529t, 941, 942, *942*
  bone pain in, *188*
Sight, 457–74. See also Eye
  accessory organs of, 457–60, *459*
Sigmoid colon, *693,* **968, 984**
Signal transduction, 64
Simple gland, 142, *142,* 143t
Sinoatrial node, 568, *570, 573*
Sinus rhythm, 575, 610
Sinuses, 201, 202t, 205, 741, *744,* 752t. *See
  also specific sinuses*
  illumination of, 741
Sinusitis, 741, 774
Skeletal muscle(s), 282–87, 299t, 301–31, *301.
  See also specific muscles*
  all-or-none response of, 294
  antagonistic, 300
  atrophy of, 297
  connective tissue coverings of, 282–83, *282*
  contraction of, 287–94, *289, 290, 291,* 291t
    observation of, 294–97, *295*
    types of, 296–97, *296*
  energy sources for, 291–94, *291*
  fast-contracting (fast-twitch), 293–94
  fatigue of, 293
  fibers of, 184–85, *284*
  heat production by, 294
  hypertrophy of, 297
  insertion of, 300, *300*
  interaction of, 300
  isokinetic contraction of, 296–97
  isometric contraction of, 296, *296*
  isotonic contraction of, 296, *296*
  latent period of, 294, *295*
  of mastication, *302,* 303–4, 304t, **973**
  motor units of, 286–87, *287,* 295
  in movement, 300, *300*
  neuromuscular junction of, 286, *286*
  origin of, 300, *300*
  oxygen debt of, 292, *293*
  oxygen supply for, 292, *292*
  as prime mover, 300
  red, 293
  refractory period of, 294
  sliding filament theory of, 288, *289, 290*
  slow-contracting (slow-twitch), 293
  staircase effect of, 295, *295*
  structure of, 282–87, *283, 284, 285,*
    *286, 287*
  summation of, 295, *295*
  sustained contractions of, 295–96
  synergistic, 300
  tetanic contraction of, 295, *295,* 296
  threshold stimulus of, 294
  tone of, 296
  twitch of, 294, *295*
  white, 293–94
Skeletal system, 14–16, *15,* 19t, *236. See also*
  Bone(s); Joint(s)
  appendicular, 200
  axial, 199–200, *200*
  divisions of, 199–200, 200t. *See also
    specific bones*
  female, **35**
  male, **31**
  system innerconnections of, 236
Skin, 161–81, *180*
  accessory organs of, *162,* 168–73, *169,*
    *171,* 173t
  aging of, 166
  apocrine glands of, 171, *171,* 173, 173t
  burns to, 177, 179, *179*
  cancer of, 166–67
  carcinoma of, 166
  color of, 175–76, *176*

cuts of, 177, *178*
dermis of, 162, *162, 165,* 166–67
disorders of, 179–80
eccrine glands of, 171, *171,* 173, 173t
epidermis of, 162, *162,* 163–66, *163, 165*
free nerve endings of, 434, *434*
glands of, *162,* 170–73, *171*
hair follicles of, *162,* 168, *169*
healing of, 177–79, *178*
melanoma of, 166–67, *167*
sebaceous glands of, *162,* 170–71, *171,* 173t
sensory receptors of, 434–35, *434*
subcutaneous layer of, 162, *162,* 168
sweat glands of, 171, *171,* 173, 173t
system interconnections of, 180
in temperature regulation, 173–75, *174*
tissues of, 161–68, *162, 163*
Skin graft, 956
for burns, 177
Skull, 199, *201,* 201–14, **969.** *See also* Face
anterior view of, *203*
anterolateral view of, **242**
of child, **255**
coronal section of, *211*
cranium of, 201, 205–8
of elderly person, **255**
ethmoidal region of, **251**
fetal, **254**
floor of, **252**
frontal region of, **253**
frontal view of, **241**
infantile, 213, *214*
inferior view of, *204,* **245**
lateral view of, *204*
mandibular region of, **248**
maxillary region of, **247**
nasal region of, **243**
occipital region of, **246**
orbital region of, **243**
posteriolateral view of, **242**
radiographs of, *212*
reference plates for, **241–55**
sagittal section of, *210,* **251**
sphenoidal region of, **245, 252, 253**
temporal region of, **244**
zygomatic region of, **244**
Sleep, 403
disorders of, 404t
non-REM, 403
REM, 403
Sleep apnea, 404t, 762
Sleep deprivation, 403
Sleep paralysis, 404t
Sliding filament theory, of muscle contraction, 288, *289, 290*
Small intestine, 685–92, *686,* **964, 967, 968, 971, 983, 984**
absorption in, *670,* 689–91, *689*
epithelial cells of, *687,* 688
movements of, 691–92
nutrient absorption by, *689,* 689–91, *690,* 691t
peristaltic movements of, 691–92
secretion of, 688
segmentation movements of, 691
structure of, 685–86, *685*
wall of, 686–88, *687*
Smell, 440–42, *441*
disorders of, 445, 445t
Smooth endoplasmic reticulum, *62,* 69, *70,* 79t
Smooth muscle, 297–98, 299t
Sneezing, 759, 760t
Sodium
atomic structure of, 39t
cellular, 48t

electrons of, 39–40, *42*
imbalances of, 823
in nutrition, 723, 725t
renal reabsorption of, 795–96
serum, 991
urinary, 791t
Sodium channels, 66–67
Sodium chloride, renal regulation of, 796–99, *798, 799*
Soleus muscle, *301, 327,* 328, *328, 329,* 330t, **342, 978, 979**
Solutions
hypertonic, 81, *82*
hypotonic, 81, *82*
isotonic, 81, *82*
Somatic afferent fibers, 407–8
Somatic efferent fibers, 407
Somatomedin, 492
Somatostatin, 483t, *491,* 492, 508, 509t, 685t
Somatotropes, 490
Sour receptors, 444, *444*
Spain, 276
Spastic paralysis, 382
Speaking, 760t
Species resistance, to infection, 630, 632t
Specific gravity, of urine, 992
Sperm cells
cocaine binding to, 842
emission of, 846, *848*
flagella of, *76*
formation of, 838–39, *840*
structure of, 841, *841*
transport of, 864–65, *865*
Spermatic arteries, 597
Spermatic cord, **30, 31,** 836, **968**
Spermatids, 838, *840*
Spermatocytes, 838
Spermatogenesis, 838–39
Spermatogenic cells, 838
Spermatogonia, 838, *840*
Spermatozoa, 838–39
Spermicide, 879–80, 879t, *880*
Sphenoid bone, *203, 204,* 205t, 206, *206, 208,* **241, 242, 244, 245, 247, 250, 251, 252, 253**
Sphenoidal sinus, *13,* 206, *207, 208,* 208t, *210,* **250, 251, 252, 965, 969**
Sphenomandibularis muscle, 304
Spherocytosis, *549*
Sphincter
anal, 692, *694*
esophageal, *669*
ileocecal, 692, *693*
precapillary, 579
pyloric, 670, *670*
urethral, 319t, 320, 805
Sphincter of Oddi, 676, *676*
Sphincter urethrae, 319t, 320
Spina bifida, 388t, 720
Spinal artery, *600*
Spinal cord, *345,* 377–83, **964, 966, 967, 970, 971, 972**
ascending tracts of, 378–81, *379,* 381t
cross section of, 378, *379*
descending tracts of, *379,* 381, 381t
ependyma of, 352
functions of, 378–81, *379, 380,* 381t
injury to, 382
structure of, 377–78
Spinal nerves, 377, 407, 412–17, *413*
anterior branch of, 415, *415*
meningeal branch of, 414
posterior branch of, 415, *415*
visceral branch of, 415, *415*
Spinal shock, 382

Spinalis capitis muscle, 305t, *306*
Spinalis cervicis muscle, 305t
Spinalis thoracis muscle, 305t, *306*
Spindle fibers
in anaphase, 91, *91*
in metaphase, 90–91, *91*
Spine. *See* Vertebrae; Vertebral column
Spinocerebellar tract
anterior, *379,* 381, 381t
posterior, *379,* 381t
Spinothalamic tract, 378, 380, *380*
anterior, *379,* 380, 381t
lateral, *379,* 380
posterior, 381t
Spinous process, of vertebrae, 216, *216*
Spiral ganglion, *449*
Spirometer, 758, *758*
Spirometry, 757–58, *758*
Spleen, **32, 33, 34,** 629, *629,* 630t, **972**
Splenectomy, 647
Splenic artery, *597*
Splenic flexure, *693*
Splenic vein, 607, *608*
Splenitis, 647
Splenius capitis muscle, 305, 305t, *306, 307,* **973**
Splenius cervicis muscle, *306*
Splenomegaly, 647
Splenotomy, 647
Spondylolysis, 216
Spongy bone, 186–88, *188*
Squamosal suture, *203, 204,* 205, *210, 242, 244*
Stanozolol, 485
Stapedius, 447, *447*
Stapes, *446,* 447, *447, 450*
Starling's law, 586
Starvation, 730–31, *730, 731*
Stereopsis, 473, *473*
Sternal angle, 221, *221*
Sternal notch, *221*
Sternal puncture, 222
Sternoclavicular joint, 267t
Sternocleidomastoid muscle, **29, 30,** *301,* 302, *302,* 305, 305t, *308,* **338, 973, 974**
Sternocostal joint, 267t
Sternum, **31,** 200, *201,* 220–22, *221,* **340, 965, 966, 967, 970, 971**
Steroids, 49, *50,* 51t, 480–81, 482–83, 483t, *484,* 484t
structure of, *482*
Stickler syndrome, 147t
Stomach, **31, 32, 33,** *657, 669,* 670–75, **964, 966, 967**
emptying movements of, 675
mixing movements of, 674, *675*
peristalsis of, 657
radiograph of, *670*
secretions of, 670–73, 672t, *673,* 673t
structure of, 670, *670, 671*
Stomachache, 674
Stomachbeat, 657
Stomatitis, 699
Strabismus, 460
Stratum basale, 163, *163,* 164t
Stratum corneum, 163, *163,* 164t
Stratum granulosum, 163, *163,* 164t
Stratum lucidum, 163, *163,* 164t
Stratum spinosum, 163, *163,* 164t
Streptokinase, 538
Stress, 511–12, *513, 514,* 514t
responses to, 512
Stress incontinence, 805
Stressor, 511
Stretch, in smooth muscle contraction, 298

Stretch receptors, 433, 438–39
  of venae cavae, 572
Stretch reflex, 438
Stroke, 591, 591t
Stroke volume, 584
Structural formula, 44
Stuart-Prower factor (factor X), 541t
Styloid process, *204, 205, 210,* 226, *227,* **341**
Stylomastoid foramen, *204,* 213t, **245, 247**
Subacromial bursa, 269
Subarachnoid space, 377
Subclavian artery, **32, 33, 34, 35,** 596, *596,*
  598, 598t, *599,* 605, **980, 982**
Subclavian vein, **31, 32, 34,** 60, 606, *606,*
  *607,* 611
Subcoracoid bursa, 269
Subcutaneous injection, 168
Subdeltoid bursa, 269
Subdural hematoma, 376
Subdural space, **969**
Sublingual gland, 665, *665, 666,* 666t
Subluxation, 275
Submandibular gland, 665, *665, 666,* 666t
Submucosa, of alimentary canal, 656,
  *659,* 659t
Submucosal plexus, of alimentary canal, 657
Subscapular bursa, 269
Subscapularis muscle, **31,** *311,* 312, 312t
Substance P, 365
Substrate, for enzymes, 104, *105*
Suckling, 495
Sucrase, 688, 689t
Sudden infant death syndrome, 363t
Sudoriferous glands, 171, *171*
Suicide, physician-assisted, 922
Sulcus, cerebral, 389, *389*
Sulfate
  cellular, 48t
  urinary, 791t
Sulfur, in nutrition, 723, 725t
Superficial, definition of, 19
Superficial transversus perinei muscle, 319,
  319t, *320*
Superior, definition of, 18
Superior oblique muscle, 460, 461t
Superior rectus muscle, 460, 461t
Superior vena cava, **33, 34,** 558, *559, 564,* 596,
  604, *606, 607,* 611, **981**
Supination, 265, *266*
Supinator muscle, 313, *314,* 316t
Supraorbital foramen, *203,* 205, 213t, **241,**
  **243, 244**
Supraorbital notch, *203,* **249, 338**
Suprapatellar bursa, 275
Suprarenal artery, 597, *597,* 598t, *605, 784*
Suprarenal vein, 607, *784*
Suprascapular nerve, 417
Supraspinatus muscle, *307, 309,* 312, 312t
Surfactant, *750,* 755
Surgery, in obesity treatment, 713
Suspensory ligament, 462, *462, 463,* 852, *852*
Sustentacular (Sertoli's) cells, 838, *840*
Sutural ligament, 257, *258*
Suture, 199, *199,* 202t, 257, *258,* 264t
  coronal, 205, *210,* **241, 242, 251**
  frontal, **249**
  lambdoidal, 205, *210,* **242**
  metopic, 213
  palatine, **245, 247**
  parietal, *258*
  sagittal, 205, **242**
  squamosal, 205, *210,* **242, 244**
Swallowing, 667, *668*
Sweat gland, 171, *171,* 173, 173t
Sweet receptors, 444, *444*

Sylvius
  aqueduct of (cerebral aqueduct), 395, *397,*
    *398,* 402, *402*
  fissure of, 389, *389*
Symphysis pubis, **34, 35,** 228, *229, 230,* 259,
  *260,* 264t, 267t, **968**
Synapse, 344, 360–65, *361*
Synaptic cleft, 286, *286,* 346, *362*
Synaptic knob, 346, *347,* 360, *362, 363*
Synaptic potential, 360–61, *361*
Synaptic transmission, 360
Synaptic vesicles, *362*
Synchondrosis, 258, *259,* 264t
Syncytium
  of cardiac muscle, 299
  functional, 568
Syndesmosis, 257, *257,* 264t
Synergists, of skeletal muscles, 300
Synesthesia, 440
Synovectomy, 275
Synovial fluid, 161, 261
Synovial joints, 259, 260–61, *260,* 264t, 267t.
  *See also specific joints*
  aspiration of, 261
  movements of, 262, 264–65, *265, 266*
  types of, 262, *263*
Synovial membrane, 161, 261, *261*
  of elbow, 271
Synthesis, chemical, 44
Syphilis, 882t
Systemic circuit, of blood flow, 594
Systemic lupus erythematosus, 647t
Systole
  atrial, 566, *567*
  ventricular, 566, *567*
Systolic pressure, 582–83, 586, *587*

## T

T lymphocytes, 530, 632, 633–34, *633,* 634t
  cytotoxic, 634
  helper, 634, *636*
  memory, 634
T wave, 571, *572*
Tachycardia, 574, *575*
Tachypnea, 774
Talocrural joint, 267t
Talus, 233, *234*
Tardive dyskinesia, 363t
Target cells, for hormones, 16, 480, 482
Tarsal bones, 200, *201, 232,* 233, *234,*
  235t, **342**
Tarsometatarsal joint, 267t
Taste, 442–45, *443, 444,* 445t
  disorders of, 445, 445t
Taste buds, 442, *443, 444, 444,* 660, *661*
Taste cells (gustatory cells), 442
Taste pore, 442
Taste receptors, 442–43
  patterns of, 444, *444*
Tay-Sachs disease, 75, 941, 959t
Tears, 458–59
Tectorial membrane, 450, *451*
Teeth, *657,* 661–63, 663t
  caries of, 664
  primary (deciduous), 662, *662,* 663t
  secondary (permanent), 662, *662,* 663t
  structure of, 662, *663*
Telencephalon, 387, *387,* 388t
Telophase, *88, 89,* 91, *92,* 92t
Telophase I, *935,* 936
Telophase II, 936, *936*
Temperature, 5
  body, 988
    cardiac cycle and, 572

  elevation of, 175
  homeostatic regulation of, 6, *7*
  regulation of, 173–75, *174*
Temperature sense, 434–35
Temporal artery, superficial, *598,* 599, *605*
Temporal bone, *199, 203, 204,* 205, 205t, *208,*
  *210,* **241, 242, 244, 245, 249, 338**
  zygomatic process of, *204*
Temporal lobe, 389, *389,* 392t, **969**
  association areas of, 392
Temporal vein, *611*
Temporalis muscle, *301, 302,* 303, 304t, **338,**
  **969, 973**
Temporomandibular joint, 267t
Tendinitis, 282
Tendon, 282, *282, 283*
Tenia coli, *693*
Tenosynovitis, 282
Tensor fasciae latae muscle, **30, 33, 34,** *301,*
  *321, 322, 323,* 325t, **977**
Tensor tympani, 447, *447*
Teratogens, 906, *906,* 908–9
Teratology, 925
Teres major muscle, **31,** *301, 307, 309, 310,*
  312, 312t, **339, 976**
Teres minor muscle, *301, 307, 309,* 312, 312t,
  **975, 976**
Testes, **32,** *481,* 510, 836–38, *837,* 849t,
  **964, 968**
  descent of, 836, *838*
  sagittal section of, *839*
  structure of, 836, 838
Testicular cancer, 838
Testosterone, 836, 850
  synthetic, 485
Tetany, 360
Thalamus, 398–99, *399, 404,* **965, 969, 985**
Thalassemia, 529t, *549*
Thalidomide, 908
Thermoreceptors, 433, 439t
Theta waves, 406, *406*
Thiamine (vitamin $B_1$), 717, 721t, *723*
Thigh
  cross section of, *325*
  muscles of, 320–22, *321, 322, 323, 324,*
    *325,* 325t, **977, 978**
Thirst center, 816
Thoracic artery, 601, *601*
Thoracic cage, 220–22, *221, 222*
Thoracic cavity, 11, *12,* **35**
  membrane of, 13, *14*
Thoracic duct, 623, *623*
Thoracic nerves, 412, *413*
Thoracic spine, *215,* 217, *218,* 219t
Thoracic wall
  arteries to, 601–2, *601*
  veins from, 606–7, *607*
Thoracodorsal nerve, 417
Thorax, 200
  lymph nodes of, *627,* 628
Threonine, 121t
Thrombin (factor IIa), 538
Thrombocytopenia, 542
Thrombophlebitis, 582, 610
Thrombopoietin, 522
Thrombus, 538
Thymectomy, 513, 647
Thymine, *117*
Thymitis, 647
Thymus gland, *481,* 510, 628–29, *628, 629,*
  630t, 919
Thyrocervical arteries, 599
Thyrocervical axis, *598*
Thyroglobulin, 497
Thyroid artery, *598,* 599, *599*

Thyroid cartilage, **31**, 742, *744, 745*
Thyroid gland, **30, 31**, *481*, 495–98, *496, 497*, **980**
   disorders of, *498*, 498t
   hormones of, 483t
   structure of, 495
Thyroid hormones, 193, 495, 497, *497*, 498t, 991
Thyroid scan, 40–41
Thyroidectomy, 513
Thyroiditis, 513
Thyroid-stimulating hormone (thyrotropin), 483t, 489, *491*, 492, *492*, 496t
   test for, 991
Thyrotropes, 490
Thyrotropin-releasing hormone, 483t, *491*, 492
Thyroxine ($T_4$), 483t, 495, 497, *497*, 498t
   test for, 991
Tibia, 200, *201*, 232, *232, 233*, 235t, **341, 342, 978**
Tibial artery, *603*, 604, *605*
Tibial collateral ligament, *274, 275*
Tibial nerve, *418*
Tibial tuberosity, *233*, **341**
Tibial vein, *610, 611*
Tibialis anterior muscle, *301*, 326, *327*, 330t, **341, 342, 978, 979**
   tendon of, **342**
Tibialis posterior muscle, *329*, 330t, 331
Tibiofemoral joint, 267t
Tibiofibular joint, 267t
Tidal volume, 759t
Tight junction, 65, 66t, *69*
Tinnitus, 475
Tissue, 8, *9*, 9t, 136. *See also specific tissues*
Tissue fluid, 625, *625*
Tissue plasminogen activator, 538
Tissue rejection reaction, 644, 646
Tissue thromboplastin (factor III), 538, 541t
Toes, 200t, *201*, 232, *234*, 235, 235t, **342**
Tongue, *657*, 658, 660, *660, 661*, 663t, **964, 965, 970**
   papillae of, 442, *443*
   taste buds of, 442, *443*, 444, *444*
Tonometry, 475
Tonsillectomy, 660
Tonsillitis, 660
Tonsils, 627, 660, *660*, 661, *661*
Tooth. *See* Teeth
Torticollis, 331
Total lung capacity, 758, *758*, 759t
Touch sense, 434, *434, 435*
Trabeculae, of spongy bone, 187, *188*
Trace elements, 37, 38t, 724, 726
Trachea, **31, 33, 34**, *661*, 740, *741*, 744–45, *746*, 752t, **964, 965, 966, 980**
   cross section of, 745, *746*
   C-shaped rings of, 745, *746*
Tracheostomy, 745, *747*
Tracheotomy, 774
Trachoma, 475
Traits
   polygenic, 943–45, 943t, *944*
   sex-influenced, 947–48
   sex-limited, 947
   sex-linked, 945–46, *947*
Transaminases, 991
Transcytosis, 85, *87*, 87t
Transferrin, 524–25
Transforming growth factor beta, 792
Transfusion
   blood, 542, 545–47, 546t, 548
   intrauterine, 925
Transient ischemic attack, 395, 591
Transmission electron microscope, 63, *63, 64*

Transplantation, 644, 646, 646t
Transverse colon, 692, *693*
Transverse fissure, 389, *389*
Transverse foramen, 216
Transverse humeral ligament, 268, *269*
Transverse plane, definition of, *21*, 22
Transverse process, of vertebrae, 216, *216*
Transverse tubules, 285, *285*
Transversus abdominis muscle, **30**, *308*, 317t, *318*, 319
Transversus perinei muscle, 319, 319t, *320*
Trapezium, *228*
Trapezius muscle, **29**, *301, 307*, 308, *308*, 310t, *311*, **338, 339, 340, 975, 976**
Trapezoid, *228*
*Treponema pallidum* infection, 882t
Triangle of auscultation, 309
Triceps brachii muscle, *301, 310*, 313, *315*, 316t, **339, 340, 974, 975, 976, 977**
   lateral head of, *309*
   long head of, *309*
Triceps-jerk reflex, 386
Tricuspid valve, 558, *559, 560, 561*, 561t
Trigeminal nerve (V), 409, *409, 410*
   mandibular division of, 409, *410*, 412t
   maxillary division of, 409, *410*, 412t
   ophthalmic division of, 409, *410*, 412t
Trigeminal neuralgia (tic douloureux), 409
Triglycerides (fats), 49, *50*, 51t
   plasma, 535
   serum, 991
Trigone, of urinary bladder, 803–4, *804*
Triiodothyronine ($T_3$), 483t, 495, 497, *497*, 498t
   serum, 991
Triplo-X syndrome, 949t
Triquetrum, *228*
Trisomy, 949–50, *954*
Trisomy 13, 949t
Trisomy 18, 949t
Trisomy 21 (Down syndrome), 949t, 952–53
Trisomy 22, 949t
Trochanter, 202t, *233*
Trochlea, *226*
Trochlear nerve (IV), 409, *409*, 412t
Trochlear notch, 225, *227*
Trophoblast, 868, 895
Tropomyosin, 288, *288*
Troponin, 288, *288*
Trypsin, 677, 689t
Trypsinogen, 677
Tryptophan, 121t, 364t, 718
Tubal ligation, 879t, 881
Tubercle, 202t
Tuberculosis, 768, *769*
Tuberosity, 202t
Tubular gland, 142, *142*, 143t
Tumor, 96–97, *96, 97*. *See also* Cancer
Tumor necrosis factor, 634t, 639
Tumor suppressor genes, 97, *97*
Tunica albuginea, 836, 845, *845*
Tunica externa, 575, *576*
Tunica media, 574–75, *576*
Turbinate bones, 740, *741*
Turner syndrome, 949t
Twins, 895, 925
   Siamese, 916
   studies of, 945
Tympanic cavity, 446, *446, 447*
Tympanic membrane, 446, *446, 447*
Tympanic reflex, 447–48
Tympanoplasty, 475
Tyrosine, 121t, 502
Tyrosinemia, *130*

## U

Ulcer, 181
   pressure, 164
Ulcerative colitis, 647t, 697
Ulna, 200, *201, 225*, 225–26, *227*, 229t
   head of, **341**
Ulnar artery, 600–601, *601, 605*
Ulnar collateral ligament, 270, *270*
Ulnar nerve, 415, *416, 417*
Ulnar notch, *227*
Ulnar vein, 606, *606, 611*
Ultrasonography, 10, 925
Ultratrace elements, 37
Ultraviolet radiation, in mutation, 129
Umbilical arteries, 913, 913t, *914*, 918t
Umbilical cord, 899, *904, 905*
   blood from, 521
Umbilical region, 22, *22*
Umbilical vein, 913t, *914*, 918t
Umbilicus, **29, 340, 974**
Unipolar neuron, 350, *350*, 351t
Upper motor neuron syndrome, 382
Uracil, *121*
Urea, 791t, 799, 992
Urea clearance, 992
Uremia, 828
Ureters, **33, 34**, 802–3, *803*
Urethra, **35**, 806–7, *807*, **972**
   cross section of, *807*
   female, *851*
   longitudinal section of, *807*
   male, *837*
   membranous, 807
   penile, 807
   prostatic, 807
Urethral sphincter, 319t, 320, 805
Uric acid, 799
   serum, 991
   urinary, 791t, 992
Urinalysis, 806
Urinary system, *17*, 18, 19t, 778–809
   kidneys of, 779–89. *See also* Kidney(s)
   system innerconnections of, 808
Urine
   amino acids in, 795
   composition of, 801, 806
   concentration of, 796–99, *797, 798, 799*, 799t
   elimination of, 802–7, 806t
      bladder in, 803–5, *804, 805*
      micturition reflex in, 805, 806t
      ureters in, 802–3, *803*
      urethra in, 806–7, *807*
   formation of, 789–802
      filtration pressure in, 791
      filtration rate in, 792–94, *792, 793*
      glomerular filtration in, 789–90, *791*, 791t
      regulation of, 796–99, *797, 798, 799*
      tubular reabsorption in, 794–96, *794*
      tubular secretion in, 800–801, *800*
   glucose in, 795
   tests on, 992
   volume of, 801
Urobilinogen, urinary, 992
Urogenital diaphragm, 319–20, 319t, *320*
Urokinase, 538
Urticaria, 181
Uterine artery, 603
Uterine tube, **33**, 855–56, *857*, 860t
Uterine vein, 608
Uterus, **33, 34**, *851*, 856–58, *857, 858*, 860t
   involution of, 870
Utricle, *449*, 453, *455, 457*

Uveitis, 475
Uvula, 660, *660, 661*

## V

Vaccine, 642, 643t
Vagina, **35**, *851*, 858, 860t
Vaginal adenosis, 858
Vaginal vein, 608
Vaginitis, 883
Vagotomy, 427
Vagus nerve (X), *409*, 410–11, *412*, 412t, 571–72, *573*, **982**
Valine, 121t
Valium, 364t
Valves, venous, 580–81, *581*
Valvotomy, 610
Varicocele, 883
Varicose veins, 582–83
Vas deferens, **32**, 836, *837*, 842, *843*, 849t
Vasa recta, 789, *790*
    countercurrent mechanism of, 797, *799*
Vasectomy, 879t, 881, *881*
Vasoconstriction, 575–76, 589, *589*
Vasodilation, 576, 589, *589*
Vasomotor center, of medulla oblongata, 402
Vasomotor fibers, 575
Vasopressin, 494
Vasospasm, 537, 538t
Vastus intermedius muscle, **34**, *321*, 326t
Vastus lateralis muscle, **32, 33, 34**, *301*, *321*, *323, 324*, 326t, *328*, **341, 972, 977, 978**
Vastus medialis muscle, **32, 33**, *301*, *321*, 326t, **341, 977**
Vegetarian diet, 705t
Vein(s), 580–81, *580, 581*, 581t, 604, 606–9, *606, 607, 608, 609, 610, 611*. *See also specific veins*
    as blood reservoir, 581
    cross section of, *580*
    valves of, 580–81, *581*
    wall of, *576*
Vena cava
    inferior, **34, 35**, 558, *559, 564*, 609, *610, 611*, **971, 972, 982**
        stretch receptors of, 572
    superior, **33, 34**, 558, *559, 564*, 596, 604, *606, 607, 611*, **981**
        compression of, 604
Venipuncture, 606
Venoconstriction, 591–92
Venography, 610
Ventilation (breathing), 739, 751, 754–59
    alveolar, 758–59
    artificial, 750
    disease-related decrease in, 761, *761*
    exercise and, 765
    expiration of, 756–57, *757*, 757t
    inspiration of, 751, 754–56, *754, 755*, 755t, *756*
    measurement of, 757–58, *758*, 759t
    medullary rhythmicity area in, 760, 762, *762*
    nonrespiratory movements with, 759, 760t
    pneumotaxic area in, 762, *762*
    regulation of, 760, *760*, 762, 762–64, *763, 764*, 765t
Ventral cavity, 10, *12*
Ventral root, 413–14
Ventricle(s), 394–95, *397*, 397–98
    cardiac, *556*, 557
        left, **33**, *559, 564*, **981**

muscle fibers of, 570, *571*
    right, **33**, *559, 564*, **966, 967, 981**
    cerebral
        ependyma of, 352
        fluid accumulation in, 401, *401*
        fourth, 395, *397, 398*
        lateral, 394–95, *397*, **965, 969**
        left, **971**
        right, **971**
        third, 395, *397, 398*, **969**
Ventricular fibrillation, 574, *574*
Venules, 580, 581
    spasm of, 537
Vermiform appendix, **32, 33**, 692, *693*
Vermis, 403
Vertebra prominens, 216–17
Vertebrae, 214, 215–16, *215, 216*
    body of, 215–16, *216*
    cervical, *215*, 216–17, *217, 218*, 219t, **965**
    lumbar, *215*, 218, *218*, 219t, **967, 968**
        fifth, **35**
    spinous processes of, **339**
    thoracic, *215*, 217, *218*, 219t
Vertebral artery, 597–99, *598, 599, 600*, *605*, **970**
Vertebral body, **971, 972**
Vertebral canal, 11, *12*, 214
Vertebral column, 199–200, *201*, 214–19, *215*, **339, 964**. *See also* Vertebrae
    curvatures of, 214–15, *215*
    disorders of, 220
    muscles of, 305–7, 305t, *306, 307*
Vertebral foramen, 216, *216*
Vertebral vein, *606*
Vertebrocostal joint, 267t
Vertigo, 475
Vesical artery, 602, *602*, 917
Vesical vein, 608
Vesicles, 75, 79t
Vestibular bulbs, 859
Vestibular glands, 859, *859*, 860t
Vestibular membrane (Reissner's membrane), 449
Vestibule, 448, *449*, 859, *859*, 860t
Vestibulocochlear nerve (VIII), *409*, 410, 412t, 450, *455*
    cochlear branch of, 409, 412t
    vestibular branch of, 409, 412t
Villi
    chorionic, 896–97, *897*
    intestinal, 686, *687*
Viral hepatitis, 682
Viscera, 10
Visceral afferent fibers, 408
Visceral efferent fibers, 407, 408
Visceral pain, 435–36, *436*
Visceral peritoneum, *15*
Visceral pleura, 13, *14*, 751
Viscosity, of blood, 585
Vision. *See also* Eye
    stereoscopic, 473, *473*
    visual nerve pathways in, 473–75, *474*
    visual pigments in, 470–73, *473*
Visual cortex, 474, *474*
Visual pigments, 470–73
Vital capacity, 757, *758*, 759t
Vital signs, 4
Vitamins, 713–21, 714t
    fat-soluble, 713, 714–16
    liver storage of, 681
    in metabolism, 105
    water-soluble, 713, 717–21, 721t

Vitamin A, 714–15, 717t
    bone effects of, 193
    cancer and, 715
Vitamin B$_1$ (thiamine), 717, 721t, *723*
Vitamin B$_2$ (riboflavin), 718, 721t
Vitamin B$_5$ (pantothenic acid), 718–19
Vitamin B$_6$, 719, *719*, 721t
Vitamin B$_{12}$ (cyanocobalamin), 719, *719*, 721t
    red blood cell effects of, 526–27, *528*, 528t
Vitamin C (ascorbic acid), 720, *720*, 721t
    bone effects of, 193
Vitamin D, 500, 715–16, 717t
    in bone development, 192–93
Vitamin E, 716, 717t
    cancer and, 715
Vitamin K, 716, 717t
Vitreous body, 467t
Vitreous humor, *462, 465*, 467t
Vocal cords, 742
    false, 743, *745*
    true, 743, *745*
Volar arch artery, *601*
Volkmann's canals, 187, *188*
Vomer bone, *203, 204, 210, 211*, 212t, **241, 243, 245, 247**
Vomiting, 675
Von Willebrand disease, 544, 959t

## W

Wart, 181
Water, 5, 6t
    cellular, 46–47, 48t
    intestinal absorption of, 691t
    renal reabsorption of, 795–96, *796*
    three-dimensional model of, *56*
Water balance, 816–18, *817*, 817t, *818*, 818t
    disorders of, 494, 820–21
    water intake and, 816–17, *817*
    water output and, 817–18, *818*, 818t
Water intoxication, 820
Water of metabolism, 816, *817*
Weber test, 454
Weight, body, 711–12, *711*
White blood cell count, 532, 532t, 991
White blood cells, 152, *152*, 527–30, 534t
    in blood smear, *521*
    in infection, 531, *531*
    origin of, *523*
White matter, **969**
    of brain, 390, *390*
    of spinal cord, 378, *379*
White pulp, of spleen, 629, *629*
Whole blood, 547
Wilms tumor, 780, 959t
Wilson disease, 806, 959t
Withdrawal reflex, *384*, 385
Wounds, healing of, 177–79, *178*
Wrist, 267t

## X

X chromosome, 945, *946, 947*
X inactivation, 946
Xenograft, 644, 646t
Xeroderma pigmentosum, 166, *166*, 959t
Xiphoid process, *221, 222*, **340**
XIST gene, 946
X-ray fluoroscopy, 614

# Y

Y chromosome, 945, *946, 947*
Yawning, 759, 760t
Yolk sac, *904,* 905–6

# Z

Z lines, of muscle fibers, 285, *285*
Zein, 708
Zinc, in nutrition, 727, 727t
Zona pellucida, 865

Zygomatic arch, *204,* **242, 244, 338, 973**
Zygomatic bone, *204,* 209, *211,* 212t, **241, 242, 243, 244, 245**
  temporal process of, *204*
Zygomatic process, 206, **244, 249**
Zygomaticus muscle, *301, 302,* 303, 303t
Zygote, 893
Zygote intrafallopian transfer, 899t
Zymogen granules, 677

# UNDERSTANDING WORDS

*a-* negative prefix: amorphous
*ab-* away from: abduction
*acet-* vinegar: acetabulum
*ad-* to: adduction
*adip-* fat: adipose tissue
*aer-* air: aerobic respiration
*af-* to: afferent arteriole
*agglutin-* to glue together: agglutination
*alb-* white: albino
*aliment-* food: alimentary canal
*allant-* sausage: allantois
*alve-* trough, channel, cavity: alveolus
*an-*-1 see a-
*an-*-2 see ana-
*ana-* up, positive: anabolic
*andr-* man: androgens
*angi-* vessel: angiogram
*annul-* ring: annular ligament
*ante-* before: antebrachium
*append-* to attach, hang, or fix to: appendicular
*-ar* pertaining to: lobar bronchus
*arbor-* tree: arbor vitae
*arc-* bow: arciform artery
*areol-* open space: areola
*arthr-* joint: arthrology
*astr-* star: astrocyte
*athero-* porridge: atherosclerosis
*atri-* entrance: atrium
*aud-* hearing: auditory canal
*aur-* ear: auricle
*auto-* self: autoimmune
*ax-* axis: axial skeleton
*axill-* armpit: axillary artery
*bas-* base: basilar artery
*bi-* two: bicuspid valve
*bicep-* two heads: biceps brachii
*bil-* bile: bilirubin
*bio-* life: biology
*-blast* bud, child, a growing thing in early stages: osteoblast
*blephar-* eyelid: blepharitis
*brachi-* arm: brachial region
*brady-* slow: bradycardia
*bronch-* windpipe: bronchus
*burs-* bag, purse: bursa
*calat-* something inserted: intercalated disk
*calc-* heel: calcaneus
*calor-* heat: calorie
*calyc-* cup: major calyx
*canal-* channel: canaliculus
*carcin-* crab, cancer: carcinoma
*cardi-* heart: pericardium
*cari-* decay: dental caries
*carin-* hull, keel: carina
*carot-* carrot: carotene
*carpus-* wrist: carpals

*cata-* down: catabolic
*caud-* tail: cauda equina
*cav-* hollow: corpora cavernosa
*cec-* blind: cecum
*centr-* point, center: centromere
*cephal-* head: cephalic region
*cerebr-* brain: cerebrum
*cervic-* neck: cervical vertebrae
*chiasm-* cross-shaped: optic chiasma
*chondr-* cartilage: chondrocyte
*chori-* protective fetal membrane: chorion
*chrom-* color: chromatin
*chym-* juice: chyme
*-clas* break: osteoclast
*clav-* bar: clavicle
*cleav-* to divide: cleavage
*co-* with, together: coenzyme
*cochlea* snail: cochlea
*col-* lower intestine: colon
*condyl-* knuckle: condyle
*contra-* against, counter: contraception
*corac-* raven: coracoid process
*corn-* horn: cornea
*corpor-* body: corpus callosum
*cortic-* bark, rind: cortex
*crani-* skull: cranial
*cribrum-* sieve: cribriform plate
*cric-* ring-shaped: cricoid cartilage
*-crin* distinguish, separate off: endocrine
*crista* crest: crista galli
*crur-* shin, leg: crura
*cusp-* point, apex: tricuspid valve
*cut-* skin: subcutaneous
*cyst-* bladder: cystitis
*cyt-* cell: cytoplasm
*dart-* like skin: dartos muscle
*de-* to remove: deamination
*decidu-* falling off: deciduous
*deglut-* to get rid of: deglutination
*dendr-* tree: dendrite
*derm-* skin: dermis
*detrus-* to force away: detrusor muscle
*di-* two: diencephalon
*diastol-* dilation: diastole
*digit-* finger: digital artery
*diuret-* to pass urine: diuretic
*dors-* back: dorsal
*ect-* outside: ectoderm
*ectop-* out of place: ectopic beat
*ede-* swelling: edema
*ejacul-* to shoot forth: ejaculation
*embol-* stopper: embolus
*-emia* in the blood: hypoproteinemia
*encephal-* the brain: encephalitis
*end-* inside: endoplasmic reticulum

*ependym-* tunic: ependyma
*epi-* upon, after, in addition: epithelial tissue
*erg-* work, deed: synergist
*erythr-* red: erythrocyte
*ethmo-* sieve: ethmoid bone
*exo-* outside: exocrine gland
*extra-* outside of, beyond: extracellular
*falx-* a sickle: falx cerebelli
*fasci-* band: fasciculus
*femur-* thigh: femur
*fimb-* fringe: fimbriae
*flacc-* flabby: flaccid
*flagell-* whip: flagellum
*follic-* small bag: hair follicle
*fore-* before, the front part of: forebrain
*foss-* ditch, trench: fossa ovalis
*fovea-* pit: fovea capitis
*frenul-* bridle, restraint: frenulum
*funi-* small cord or fiber: funiculus
*gangli-* swelling: ganglion
*gastr-* stomach: gastric gland
*-gen* become, be produced: allergen
*genesis* origin: spermatogenesis
*gladi-* sword: gladiolus
*glen-* joint socket: glenoid cavity
*-glia* glue: neuroglia
*glom-* little ball: glomerulus
*glut-* buttocks: gluteus maximus
*glyc-* sweet: glycogen
*gracil-* slender: gracilis
*-gram* letter, drawing: myogram
*gubern-* to steer, to guide: gubernaculum
*gust-* to taste: gustatory receptor
*gyn-* woman: gynecology
*hem-* blood: hemoglobin
*hemi-* half: hemiplegia
*hepar-* liver: heparin
*hepat-* liver: hepatic duct
*hetero-* other, different: heterozygous
*hiat-* opening: esophageal hiatus
*hist-* web, tissue: histology
*hol-* entire, whole: holocrine gland
*hom-* same, common: homeostasis
*horm-* impetus, impulse: hormone
*humor-* moisture, fluid: humoral
*hy-* U-shaped: hyoid bone
*hyal-* glass: hyaline cartilage
*hyper-* above, beyond: hypertonic
*hypo-* under, below: hypotonic
*im-* negative prefix: imbalance
*immun-* free, exempt: immunity
*in-* see im-
*inflamm-* to set on fire: inflammation
*infra-* beneath: infraorbital nerve
*inhal-* to breathe in: inhalation